# Periodic Table (Taken from IUPAC, 1995 – Updated through June 1999)

http://www.chem.qmw.ac.uk/iupac2/atwt.html

| 1 | 2 | 3 | 4 | 5 | 6 | 7 | 8 | 9 | 10 | 11 | 12 | 13 | 14 | 15 | 16 | 17 | 18 |
|---|---|---|---|---|---|---|---|---|---|---|---|---|---|---|---|---|---|
| $s^1$ | $s^2$ | $s^2d^1$ | $s^2d^2$ | $s^2d^3$ | $s^2d^4$ | $s^2d^5$ | $s^2d^6$ | $s^2d^7$ | $s^2d^8$ | $s^2d^9$ | $s^2d^{10}$ | $s^2p^1$ | $s^2p^2$ | $s^2p^3$ | $s^2p^4$ | $s^2p^5$ | $s^2$ or $s^2p^6$ |
| 1<br>**H**<br>Hydrogen<br>gm 1.00794 | | | | | | | | | | | | | | | | | 2<br>**He**<br>Helium<br>g 4.002602 |
| 3<br>**Li**<br>Lithium<br>gm 6.941 | 4<br>**Be**<br>Beryllium<br>9.012182 | | | | | | | | | | | 5<br>**B**<br>Boron<br>gm 10.811 | 6<br>**C**<br>Carbon<br>g 12.0107 | 7<br>**N**<br>Nitrogen<br>g 14.00674 | 8<br>**O**<br>Oxygen<br>g 15.9994 | 9<br>**F**<br>Fluorine<br>18.9984032 | 10<br>**Ne**<br>Neon<br>gm 20.1797 |
| 11<br>**Na**<br>Sodium<br>22.989770 | 12<br>**Mg**<br>Magnesium<br>24.3050 | | | | | | | | | | | 13<br>**Al**<br>Aluminum<br>26.981538 | 14<br>**Si**<br>Silicon<br>28.0855 | 15<br>**P**<br>Phosphorus<br>30.973761 | 16<br>**S**<br>Sulfur<br>g 32.066 | 17<br>**Cl**<br>Chlorine<br>m 35.4527 | 18<br>**Ar**<br>Argon<br>g 39.948 |
| 19<br>**K**<br>Potassium<br>g 39.0983 | 20<br>**Ca**<br>Calcium<br>g 40.078 | 21<br>**Sc**<br>Scandium<br>44.955910 | 22<br>**Ti**<br>Titanium<br>47.867 | 23<br>**V**<br>Vanadium<br>50.9415 | 24<br>**Cr**<br>Chromium<br>51.9961 | 25<br>**Mn**<br>Manganese<br>54.938049 | 26<br>**Fe**<br>Iron<br>55.845 | 27<br>**Co**<br>Cobalt<br>58.933200 | 28<br>**Ni**<br>Nickel<br>58.6934 | 29<br>**Cu**<br>Copper<br>63.546 | 30<br>**Zn**<br>Zinc<br>65.39 | 31<br>**Ga**<br>Gallium<br>69.723 | 32<br>**Ge**<br>Germanium<br>72.61 | 33<br>**As**<br>Arsenic<br>74.92160 | 34<br>**Se**<br>Selenium<br>78.96 | 35<br>**Br**<br>Bromine<br>79.904 | 36<br>**Kr**<br>Krypton<br>gm 83.80 |
| 37<br>**Rb**<br>Rubidium<br>g 85.4678 | 38<br>**Sr**<br>Strontium<br>g 87.62 | 39<br>**Y**<br>Yttrium<br>88.90585 | 40<br>**Zr**<br>Zirconium<br>g 91.224 | 41<br>**Nb**<br>Niobium<br>92.90638 | 42<br>**Mo**<br>Molybdenum<br>95.94 | 43<br>**Tc**<br>Technetium<br>[98] | 44<br>**Ru**<br>Ruthenium<br>g 101.07 | 45<br>**Rh**<br>Rhodium<br>102.90550 | 46<br>**Pd**<br>Palladium<br>g 106.42 | 47<br>**Ag**<br>Silver<br>g 107.8682 | 48<br>**Cd**<br>Cadmium<br>g 112.411 | 49<br>**In**<br>Indium<br>114.818 | 50<br>**Sn**<br>Tin<br>g 118.710 | 51<br>**Sb**<br>Antimony<br>g 121.760 | 52<br>**Te**<br>Tellurium<br>g 127.60 | 53<br>**I**<br>Iodine<br>126.90447 | 54<br>**Xe**<br>Xenon<br>gm 131.29 |
| 55<br>**Cs**<br>Cesium<br>132.90545 | 56<br>**Ba**<br>Barium<br>137.327 | 57-71<br>* | 72<br>**Hf**<br>Hafnium<br>178.49 | 73<br>**Ta**<br>Tantalum<br>180.9479 | 74<br>**W**<br>Tungsten<br>183.84 | 75<br>**Re**<br>Rhenium<br>186.207 | 76<br>**Os**<br>Osmium<br>g 190.23 | 77<br>**Ir**<br>Iridium<br>192.217 | 78<br>**Pt**<br>Platinum<br>195.078 | 79<br>**Au**<br>Gold<br>196.96655 | 80<br>**Hg**<br>Mercury<br>200.59 | 81<br>**Tl**<br>Thallium<br>204.3833 | 82<br>**Pb**<br>Lead<br>g 207.2 | 83<br>**Bi**<br>Bismuth<br>208.98038 | 84<br>**Po**<br>Polonium<br>[209] | 85<br>**At**<br>Astatine<br>[210] | 86<br>**Rn**<br>Radon<br>[222] |
| 87<br>**Fr**<br>Francium<br>[223] | 88<br>**Ra**<br>Radium<br>[226] | 89-103<br>† | 104<br>**Rf**<br>Rutherfordium<br>[261] | 105<br>**Db**<br>Dubnium<br>[262] | 106<br>**Sg**<br>Seaborgium<br>[263] | 107<br>**Bh**<br>Bohrium<br>[262] | 108<br>**Hs**<br>Hassium<br>[265] | 109<br>**Mt**<br>Meitnerium<br>[266] | 110<br>**Uun**<br>Not named<br>[269] | 111<br>**Uuu**<br>Not named<br>[272] | 112<br>**Uub**<br>Not named<br>[277] | | 114?<br>**Uuq**<br>Not named<br>[289] | | 116<br>**Uuh**<br>Not named<br>[289] | | |

| | | | | | | | | | | | | | | | |
|---|---|---|---|---|---|---|---|---|---|---|---|---|---|---|---|
| * | 57<br>**La**<br>Lanthanum<br>g 138.9055 | 58<br>**Ce**<br>Cerium<br>g 140.116 | 59<br>**Pr**<br>Praeseodymium<br>140.90765 | 60<br>**Nd**<br>Neodymium<br>g 144.24 | 61<br>**Pm**<br>Promethium<br>[145] | 62<br>**Sm**<br>Samarium<br>g 150.36 | 63<br>**Eu**<br>Europium<br>g 151.964 | 64<br>**Gd**<br>Gadolinium<br>g 157.25 | 65<br>**Tb**<br>Terbium<br>158.92534 | 66<br>**Dy**<br>Dysprosium<br>g 162.50 | 67<br>**Ho**<br>Holmium<br>164.93032 | 68<br>**Er**<br>Erbium<br>g 167.26 | 69<br>**Tm**<br>Thulium<br>168.93421 | 70<br>**Yb**<br>Ytterbium<br>173.04 | 7<br>**L**<br>Lutet<br>174.9 |
| † | 89<br>**Ac**<br>Actinium<br>[227] | 90<br>**Th**<br>Thorium<br>g 232.0381 | 91<br>**Pa**<br>Protactinium<br>231.03588 | 92<br>**U**<br>Uranium<br>gm 238.0289 | 93<br>**Np**<br>Neptunium<br>[237] | 94<br>**Pu**<br>Plutonium<br>[244] | 95<br>**Am**<br>Americium<br>[243] | 96<br>**Cm**<br>Curium<br>[247] | 97<br>**Bk**<br>Berkelium<br>[247] | 98<br>**Cf**<br>Californium<br>[251] | 99<br>**Es**<br>Einsteinium<br>[252] | 100<br>**Fm**<br>Fermium<br>[257] | 101<br>**Md**<br>Mendelevium<br>[258] | 102<br>**No**<br>Nobelium<br>[259] | 10<br>**Lr**<br>Lawren<br>[262 |

The typical uncertainty in the listed atomic weights is 1-3 in the least significant digit. See the IUPAC table for the actual, experimental uncertainties.

g Geologically exceptional specimens are known in which the atomic weight is outside the IUPAC specified uncertainty range.

m Modified atomic weight outside the IUPAC specified uncertainty range may be found in commercial samples of this element that have undergone isotope separation or extraction.

[ ] Brackets denote a radioactive element that lacks a characteristic natural mixture of isotopes. The value stated is the nucleon number of the element's most stable nuclide. Th, Pa, and U have no stable isotopes, but their isotopes are so long lived that we can treat their atomic weights as meaningful.

C
Jim

# General Chemistry for Engineers

James O. Glanville

Virginia Polytechnic Institute
and State University

PRENTICE HALL, Upper Saddle River, NJ 07458

Acquisitions Editor: *John Challice*
Editorial Assistant: *Gillian Buonanno*
Special Projects Manager: *Barbara A. Murray*
Formatter: *Allyson Graesser*
Cover Design: *Joseph Sengotta*
Manufacturing Buyer: *Trudy Pisciotti*



Printed in the United States of America

10 9 8 7 6 5 4 3 2

ISBN 0-13-032514-7

Prentice-Hall International (UK) Limited, *London*
Prentice-Hall of Australia Pty. Limited, *Sydney*
Prentice-Hall Canada, Inc., *Toronto*
Prentice-Hall Hispanoamericana, S.A., *Mexico*
Prentice-Hall of India Private Limited, *New Delhi*
Pearson Education Asia Pte. Ltd., *Singapore*
Prentice-Hall of Japan, Inc., *Tokyo*
Editora Prentice-Hall do Brazil, Ltda, *Rio de Janeiro*

*For Deena*

# Table of Contents

# Preface

A decent respect for the opinions of humankind demands that an author give some accounting of himself. Why this book? The answer is straightforward: 1. Because engineering colleagues asked for a one-semester replacement course for the two-semester course sequence that their students had traditionally taken, and 2. Because no suitable book existed. This book began with me sitting down with my engineering colleagues, asking what are the essential, bedrock materials, and then writing a curriculum to match it. The book was born as a series of web-based chapters in the Fall of 1998, grew to a self-published volume in the Fall of 1999, and was heavily revised to become this preliminary edition for the Fall of 2000.

ABET describes itself this way: "The Accreditation Board for Engineering and Technology (ABET) is primarily responsible for monitoring, evaluating, and certifying the quality of engineering, engineering technology, and engineering-related education in colleges and universities in the United States." Here's how ABET defines engineering: "Engineering is the profession in which a knowledge of the mathematical and natural sciences gained by study, experience, and practice is applied with judgment to develop ways to utilize, economically, the materials and forces of nature for the benefit of mankind." Engineers are the people who make things with stuff, and chemists make the stuff.

The theme of the book is simple: All the stuff that engineers use and apply is composed of chemical entities (atoms, molecules, and ions). The bulk properties of matter derive from those entities and the forces among them. To understand chemistry is to understand stuff. The book is organized into five parts, broadly going from the more simple to the more complex.

Part 1. Chemical Fundamentals: Atoms (including nuclides and isotopes), molecules and ions, moles and stoichiometry, and units and measurements.
Part 2. The Nature of Chemical Bonding: Gases and intermolecular forces. Electrons in atoms and chemical bonding.
Part 3. Chemical Thermodynamics and Chemical Equilibrium. Characterization of the properties of chemical entities and the nature of a chemical reaction process.
Part 4. Properties of Matter: Properties of liquids and liquid mixtures, solids and the basics of materials science. Relationship among the states of aggregation of matter.
Part 5. Applied Chemistry. Descriptive chemistry and the periodic table. The nature of redox chemistry and electrochemistry. Rate processes and the aging of materials.

Throughout the book I have stressed problem solving and engineering approximation. The students in my classroom want to know why the material I am presenting is relevant to their lives and to their professional careers. So I try to answer those questions as we go along, often and in various ways. Experimental uncertainty is not something to be discussed in Chapter 1 and then dropped. It is an ongoing topic and one with very distinguished roots in engineering, as I point out in Appendix B.

Likewise, throughout the book I have tried to use punctiliously what IUPAC calls the algebra-of-quantities. Experimental uncertainty and the algebra-of-quantities are given extensive treatment in the Solutions Manual, which I wrote as an integral part of writing the text, and not as an afterthought. (To order this for yourselves or your students, please contact your Prentice-Hall representative and ask about ISBN 0-13-032514-7).

Designing, preparing, and interpreting graphs remains an important part of science and engineering. Most of the graphs in the text were prepared by the author, and graphing exercises are included to encourage the student to do likewise.

Engineering students are computer literate and internet savvy. I have learned a lot from them. To take advantage of their skill I introduced internet searching as a key component of the book. Throughout the book you'll find entries such as [❁kws +topic + "another topic"]. The initials "kws" stand for "key word search." The symbol ❁kws suggests you do an internet search with your favorite search engine. The internet is a gigantic, dynamic, rapidly accessible encyclopedia. Many times in the course of preparing material I have been able to find needed information quickly and efficiently by using the internet. I am convinced that learning how to search the internet effectively is an important 21st-century skill that, once learned, becomes invaluable.

Obviously, there can be no complete agreement on either the content or the order of presentation for a course so new. However, I have been gratified by the response of my reviewers, several of whom told me that I have independently arrived at the sequence they themselves concluded was appropriate for a one-semester general chemistry course for engineers. But authors notwithstanding, the best teaching takes place when the instructor does it her or his way and is deeply involved in developing and organizing the course. A wise editor once wrote: "Kinetics before or after equilibrium? This point always comes up. It can be argued both ways. Remember the folks who will swap the two around and see if you can't engineer the discussion to make such a swap less unruly." In writing and revising material I've tried to keep that advice firmly in mind.

## Acknowledgments

In preparing this revised edition I have been greatly helped by two old friends, Luther Brice and David West. My heartfelt thanks to them both. In their different ways they have worked very hard to keep me free from error. I, and not they, am responsible for all the mistakes that remain.

At Prentice-Hall I begin by thanking John Challice, who has placed in me a level of confidence that I can only aspire to deserve. Thanks in turn to Paul Corey and Tim Bozik, whose support was no less palpable for being largely silent. On a day-to-day basis my thanks to Gillian Buonanno, who steered me through the shoals. My thanks also to Don Beville for much wise and business-like advice. Thanks to the entire Prentice-Hall chemistry team. Thanks to Prentice-Hall production staff, and particularly my page formatter, Allyson Graesser, for fast, professional turnaround.

I also thank my many professional reviewers, many of whom still remain unknown to me. Their kind words of encouragement sustained me in moments of self-doubt; their stern words of criticism challenged me to try ever harder. My thanks to my colleagues at Virginia Tech, both faculty and staff, in the Chemistry Department and in the College of Engineering.

Friends, supporters, and teachers: Kershi Cambata, Monty Crewe, Julia Coyne, Mike Curry, Bob Eiffert, Dick Fichter, Sam Grim, Ken King, John Schug, Eric Templeton (and family), Jimmy Viers, Bob Willis, and Karen Willis. My committee of engineers: Greg Adel, Bill Conger, Theo Dillaha, David Dillard, and Bev Watford. People who helped: Patricia Amateis, Bill Bebout, Harold Bell, Tom Bell, Anne Campbell, Jeff Clark, Jim Coulter, Daniel Crawford, Paul Deck, Jerry Diffell, Cindy Dillard, John Dillard, Harry Dorn, William Ducker, Jeannine Eddleton, Paul Field, Rich Gandour, Tom Glass, Jack Good, Hayden Griffin, Jim Hall, Brian Hanson, Travis Heath, Wanda Hensley, Milos Hudlicky, Vicki Hutchison, Geno Iannaccone, Darryl Iler, Larry Jackson, Kezia Johnson, Mike Johnson, Michael Jordan, Ed Lener, Owen Lofthus, Judi Lynch, Herve Marand, Angela McCracken, Harold McNair, Joe Merola, Angie Miller, John Murray, Susan Nichols, Gwen Nickerson, Carl Nocera, Neil Patterson, Tim Pickering, The Pioneers, Patti Prevo, Eric Remy, Judy Riffle, Rob Russell, George Sanzone, Linda Sheppard, Carla Slobodnick, Mark Stone, Jim Tanko, Larry Taylor, Jim Wightman, Tom Ward, Sarah Wheeler, and Jim Wolfe.

## Feedback

As a teacher, nothing gratifies me more than interactions with students and colleagues. This preliminary edition has been made available precisely so that you, students and instructors, can view it, ponder it, use it, and form an opinion about it. Tell me how I can make it work better for you. Please know that all feedback is welcome.

Since the advent of the web it's become a tradition for me to maintain a "boo file" for my courses. That's a place when I post all the mistakes and fixes, along with the names of the people who point them out. Send in the boos.

Jim Glanville
jglanvil@vt.edu
Blacksburg
July 2000

# Chapter 1

# Atoms, Elements, and Measurements

## Section 1.1 Chemistry and Its Evolution as a Science

### Introductory Remarks

We live in a time of remarkable scientific and technological achievement. Many technological developments of the past 120 years—such as telephones, automobiles, radio, television, computers, nuclear energy, the internet, space travel, etc.—were made possible by applying scientific principles to solve practical problems. Chemistry is an important science and its applications and products fill our daily lives. The role of the products of chemical industry is profound: we use manufactured fibers in our clothing and carpets; we take chemically synthesized drugs for our health; we construct vehicle components from engineering plastics; we use fuels produced by catalytic chemical processes; we use "vinyl" pipe to distribute water around our homes; and on and on. Most of the materials of our daily world are the products of chemical industry.

The profession of engineering requires the engineer to understand the properties of materials: How much load will the beam support? How can the information density of a silicon wafer be increased? What type of paint and what thickness of paint are needed to provide years of corrosion protection for a buried steel tank? The answers to these questions, and countless others, require a good working knowledge of chemistry and sound scientific principles. There is even a branch of engineering devoted to applying chemistry: **chemical engineering**.

This chapter describes: chemistry as a science; how chemistry developed; the reasons why your study of chemistry is fundamental to many of your other studies; and why chemistry is so important for twenty-first-century technology. It also describes the origin, discovery, and distribution of the chemical elements.

The **periodic table** summarizes how the elements are related to one another. There are many forms of the periodic table, some of which are shown in Appendix H, etc. Detailed information about the properties of atoms, elements, and isotopes is in the table in Appendix D. **Isotopes** are atoms of a given element with different masses.

The chapter concludes with a discussion of measurements. Having reliable measurements is fundamental to the progress of science in general, and to the progress of chemistry in particular. In Section 1.5 we discuss experimental uncertainty: the use of significant figures and standard deviation to address the questions "What do we know, and how well do we know it?" We also deal with units. Every time we make a measurement there is a third question to be answered: "What units did we measure it in?" Thus, any proper measurement has three components: 1. its value, 2. its uncertainty and 3. its unit. Units are the subject of Appendix A. Experimental uncertainty is the subject of Appendix B.

### Definitions of Chemistry

"Chemistry is the science of atoms and molecules." A pithy definition.

Next, a 200-year old definition of chemistry from Lecture 1 of Joseph Black's General Chemistry course taught at Edinburgh, Scotland from 1766-1799: "Chemistry is the science or study of...the effects produced by heat and by mixture, in all bodies, or mixtures of bodies, natural or artificial,...with a view to the improvement of arts, and the knowledge of nature." [*Joseph Black MD's Lectures on the Elements of Chemistry*. Prepared by John Robison with a dedication to James Watt. Published in Edinburgh, Scotland, 1803.]

Joseph Black (1728-1799) Chemist, physician, and teacher of James Watt.

Figure 1.1

The engineer James Watt (1736-1819) developed the steam engine. Black and Watt enjoyed many years of friendship. Of Black, Watt wrote: “To him, I owe in great measure, my being what I am; he taught me to reason and experiment in natural philosophy, and was always my true friend and adviser.” Of Watt, Black wrote: “I found him to be a young man possessing most uncommon talents for mechanical knowledge and practice...which often surprised and delighted me in our frequent conversations together.” You can see that chemistry and engineering have long had a productive and supportive relationship. [✽kws “James Watt” +“Joseph Black”]. (✽kws is explained on page xv.)

James Watt (1736-1819) Instrument maker, entrepreneur, and developer of the steam engine.

Figure 1.2

Here’s a definition of chemistry for engineers: Chemistry is the molecular-level science that underlies and unifies all aspects of modern technology. More specifically, chemistry is the science that describes how atoms are bonded together, either to form the molecules of gases or liquids, or to make various solid materials. The most characteristic aspect of chemistry is the study of chemical reactions, or the way changes in external conditions cause chemical changes via the rearrangements of atoms. Chemistry combines chemical theory, chemical synthesis (making chemical compounds), and the characterization of matter. The major goals of chemistry are: (1) to understand the principles that govern the nature and properties of materials, (2) to know the chemical structure of all materials, (3) to prepare and manufacture specific chemical substances and mixtures, and (4) to use and apply chemical materials for practical purposes.

## A Description of Chemistry

Chemistry is the science that deals with the composition, properties, and behavior of materials, and the ways materials can be transformed. **Matter** is the stuff that makes up our universe; it is anything that has mass and takes up space. **Space** is what is occupied by matter with mass. **Mass** is a property of matter occupying space. A **material** is any form of matter. **Properties** are the attributes of a material: its mass, color, strength, density, chemical reactivity, etc. Different kinds of material have different properties, so properties distinguish one material from another. **Composition** describes what kinds of matter a material contains. A material with a particular composition is a **substance**.

Chemistry has both pure and applied aspects. **Pure chemistry** is research performed for the purpose of better understanding nature. **Applied chemistry** is the use of chemical principles and methods for practical purposes. The raw materials of chemistry are the resources of our planet: metals, minerals, coal, petroleum, sea water, the gases of the atmosphere, and materials obtained from plants and animals. Pure chemistry provides an understanding of these materials. Applied chemistry uses them to produce the **synthetic**, or human-made, materials that make our advanced civilization possible.

Figure 1.3

Engineers are usually specially interested in **materials chemistry**. Engineers make things with stuff. So engineers must understand the properties of the materials they use. Chemistry is the science of atoms and molecules and it’s chemical bonds that connect atoms to make molecules. The properties of *all* materials derive directly from the type and extent of the chemical bonds within them; that’s an important practical reason why engineers need to understand chemistry. The properties of materials can be modified in many ways. On the left, it’s demonstrated that a banana sufficiently chilled makes a workable hammer.

Chemistry is an **experimental science;** it progresses by means of the interpretation of experiments and observations. Chemistry is also a quantitative science. A **quantitative experiment** measures numbers: Joseph Black (1728-1799) was one of the first chemists to make careful quantitative measurements. James Watt was a student of Black's and some wag once wrote that Black's most famous discovery was not carbon dioxide (which indeed Black did discover) but James Watt. Another of the 18th century chemists to emphasize the importance of experiments was Antoine Lavoisier (1743-1794), who wrote in his famous book *Elements of Chemistry*, published in 1789, "We must trust in nothing but facts. These are presented to us by nature and cannot deceive. We ought in every instance to submit our reasoning to the test of experiment." The simplest kind of experiment is to make a qualitative observation. For example, you might notice that gold is a yellow metal that does not corrode, or that liquid water turns to ice when it is cooled below its freezing point. More sophisticated experiments involve measurements. To measure something is to quantify it. Again, to quote Lavoisier: "As the usefulness and accuracy of chemistry depend entirely upon the determination of the weights of the ingredients and products, too much precision cannot be employed in this part of the subject, and for this purpose we must be provided with good instruments."

Figure 1.4 Lavoisier studying human respiration. His wife, seated at the table on the right, records the experiment.

As a science, chemistry was a late bloomer. Because of the landmark advance represented by Lavoisier's book, historians of chemistry mark 1789 as the beginning of the *chemical revolution*. Modern chemists employ many sophisticated and ingenious instruments that enable them to achieve levels of precision far beyond what Lavoisier might have imagined. You can find information about modern chemical instruments at the author's website. The importance of obtaining accurate, quantitative results remains as vital today as in Lavoisier's time, because it is only through experiment and measurement that chemists come to understand the laws of chemistry and advance their science.

A **chemical law** (or any scientific law) is a summary statement about the way nature behaves. For example, during the period of the chemical revolution it was realized (within the limits of experimental uncertainty) that a given chemical substance studied in many different laboratories, and by many different chemists, has a definite composition. This realization became the statement of a chemical law (**law of constant composition**). The interpretation of scientific laws leads to scientific theories. **Chemical theories** are intellectual formulations of the principles that underlie the facts and laws of chemistry.

## Historical Prelude to Chemistry

Although chemistry was a slow-developing science, the origins of practical chemistry are probably as old as the discovery and use of fire—and certainly lie beyond the horizon of recorded history. Chemical processes used by ancient civilizations include pottery glazing, dyeing, making detergent agents, and making bronze from mixed ores of tin and copper. By 400 BC many Greek philosophers had proposed theories of matter. They introduced the idea of elements as basic kinds of matter from which the entire universe is constructed. Plato concluded that there were four elements—air, earth, fire, and water. Demokritos of Abdera imagined that matter was created by particles of these four elements moving in a void. An important aspect of Greek philosophy was the belief that elements could be transformed into one another. The Pythagorean school of philosophers

Figure 1.5 An Egyptian metal worker.

introduced the use of letters to signify the elements. The ancient world still influences modern chemistry. For example, the Greek word for element is *stoicheion*, and the modern word stoichiometry (it refers to mass and number relationships in chemical reactions) serves to remind us of that distant influence on our modern science. [❁kws "Greek chemistry"]

During the middle ages, the knowledge that would eventually form the basis for the science of chemistry was found principally in three areas: alchemy, medicine, and technology.

## The Alchemical Roots of Chemistry

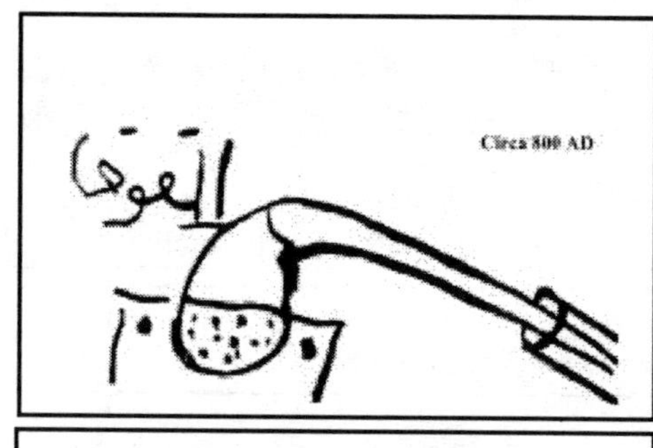


Figure 1.6. Distillation shown in an Arabic manuscript circa 800 AD.

Alchemy was a curious mixture of practical science, occultism, the search for the philosopher's stone (a magical substance that supposedly would convert base metal to pure gold), and, all too frequently, outright fraud. Alchemy arose among the Greek scholars of Alexandria soon after the death of Christ. It was taken over by Arab scholars and transmitted to the West following the passage of Arabic science to Italy and Spain. The powerful and complex ideas of alchemy influenced human minds for well over a thousand years. The alchemists made valuable practical contributions to chemistry, such as their use of distillation to prepare brandy and whiskey, but they left a residue of distrust, suspicion, and error that hampered the emergence of modern chemistry. Figure 1.6 pictures a retort that dates from about the year 800, a time when Arabic chemistry was flowering (retorts allow liquids to be distilled into a receiving vessel). [❁kws "Muslim chemistry" or " Moslem chemistry"]

## The Medical Roots of Chemistry

Medical use of chemical substances had been practiced in many societies throughout history, but in the hands of the Swiss physician Theophrastus Bombast von Hohenheim (1490-1541), who was generally known by his scholarly name of Paracelsus, the use of chemical substances was developed into an actual system of medicine. Paracelsus was a hot-tempered, controversial man, prominent in the history of both medicine and chemistry. Some of Paracelsus' writings were mystical and fantastic, but his chemical theory of medicine was the origin of modern chemotherapy. After Paracelsus, until its eventual separation as an independent science in the nineteenth century, chemistry remained closely associated with medicine and pharmacy. Even today, in England, one buys one's prescriptions at the "chemist's shop."

Figure 1.7 Paracelsus (1490-1541) Founder of chemical pharmacology. He advanced medicine by quarreling about it and eventually died in a bar brawl.

## The Technological Roots of Chemistry

By the sixteenth century, technological applications of chemistry could be found in metallurgy, dyeing, metal assaying, perfume making, glass making, gunpowder manufacture, and the making of salts and acids. Much of this technology was sound: the need to test an ore for its precious metal content (assaying), for example, stimulated the development of methods of chemical analysis that could be used today with good results. **Chemical analysis** is the laboratory procedure to determine what is in substance and in what proportion (**qualitative** and **quantitative analysis**). Unfortunately, chemical knowledge learned in the pursuit of these practical aims did not become organized in a comprehensive way, so no general theory of chemistry arose from the efforts of many skillful, practical people.

As noted, it was the work of Lavoisier that set in motion the chemical revolution. Lavoisier acquired a taste for chemistry in his youth after being taught in an apothecary's shop. He trained first as a lawyer, but subsequently abandoned law for science. In 1775 he became administrator of gunpowder at the French arsenal, where he set up a laboratory to improve the quality of the powder. This laboratory became a center of research, and a meeting place for scientific leaders—Benjamin Franklin and Thomas Jefferson were among its visitors. Lavoisier stressed the value of precise measurement to chemistry and was a leader in quantitative chemistry. He discarded the outdated alchemical ideas, brought organization to the facts of chemistry, presented a table of elements for the first time in a book, and emphasized the importance of experiments. Lavoisier set chemistry on the correct path. Unfortunately, he lived in turbulent times, and as a much-despised tax official, he was one of many who fell victim to the guillotine during the French revolution.

## Dalton's Atomic Theory

In 1806, John Dalton (1766-1844) took a decisive step by postulating the atomic theory of matter. Dalton studied the gases of the air and the relative masses of elements that combined chemically with one another. He also made use of new quantitative data produced by chemists under the influence of Lavoisier. Dalton's atomic theory states:

- Every element is made up of indivisible particles called atoms,
- all atoms of a given element have identical chemical properties, and
- atoms of different elements are different.

In other words, elements are composed of atoms of a single chemical type, and an atom is the smallest possible sample of an element. Pure substances, other than elements, contain at least two different kinds of elements in chemical combination and are chemical compounds. The term compound derives directly from John Dalton who spoke of two atoms joining to make a compound atom. The changing of one chemical compound into another, or the combining of elements into a compound is a chemical reaction. Chemical reactions are the essence of chemistry. Alternatively, a chemical reaction is a process that leaves atoms unchanged but transforms one or more substances into other substances by changing how the atoms are combined. Dalton was the first person to understand that chemical reactions are "atom scrambling" processes.

Dalton knew that pairs of elements made compounds, but he did not know how many atoms of each element went into making the compound. Lacking such information, he often assumed that atoms joined 1:1. Sometimes he was right and sometimes he was wrong. Here's a table that shows a few simple chemical compounds.

Table 1.1 **Some Simple Chemical Compounds**

| Name | Formula | Number of elements | Number of atoms |
|---|---|---|---|
| Hydrogen chloride | HCl | 2 | 2 |
| Water | $H_2O$ | 2 | 3 |
| Carbon monoxide | CO | 2 | 2 |
| Carbon dioxide | $CO_2$ | 2 | 3 |
| Propane | $C_3H_8$ | 2 | 11 |
| Methanol | $CH_3OH$ | 3 | 6 |
| Urea | $CO(NH_2)_2$ | 4 | 8 |

## The Scope of Applied Chemistry

**Applied chemistry** is the basis for the large and important chemical industry. The products of the chemical industry (either alone or as ingredients in a variety of other products) are widely used in modern, technological economies, and no study of chemistry is complete unless it describes some of the ways that chemistry is applied. Appendix F summarizes a good deal of information about the global chemical industry. From that appendix you will get a feel for the wide-ranging technological importance of chemistry and learn that your annual consumption of chemical materials is about 3000 pounds. Many everyday store-bought products have a chemical ingredient list on their label. Reading such labels is an excellent way to get a sense of the pervasive influence of chemistry in our daily life.

Table 1.2 below gives you a sense of the scale of the chemical industry. Sixteen major areas of industrial chemistry and its applications are ranked from top to bottom by their economic value. Petroleum refining and processing is not only the largest chemical industry, it's also the largest industry of any kind.

Table 1.2 **Overview of the Chemical Industry**

| Application/Industry | Purposes/Products |
|---|---|
| Petroleum refining and processing | Chemical modification of crude oil to produce fuels, lubricants and chemical raw materials. |
| Inorganic chemical industries | Manufacture of acids, alkalis, salts, and other inorganic materials. |
| Agrichemicals | Manufacture of fertilizers, herbicides, and pesticides. |
| Plastics industries | Production of synthetic materials such as plastics (resins), fibers, films, etc. |
| Electrochemical industries | Production of aluminum, magnesium, and other metals. Manufacture of batteries. Electroplating. Corrosion control processes. |
| Mining and minerals processing | Separation and enrichment of minerals and ores. Production of metals. |
| Pharmaceuticals | Manufacture of drugs. |
| Crop product industries | Manufacture of alcohol, turpentine, vegetable oils and other chemicals from plants. |
| Food industry | Manufacture of emulsifiers, stabilizers, preservatives, coloring agents, etc. |
| Paint and coatings industries | Manufacture of paints, varnishes, lacquers and other coating materials. Production of glues and adhesives. |
| Rubber | Manufacture of synthetic rubber. Chemical modification of natural rubber. Compounding rubber with other chemicals to make finished rubber products. |
| Electronic industries. Computers. | Manufacture of semiconductors and computer chips. Printed circuit manufacture. Production of optical fibers. |
| Cleaning agents | Manufacture of soaps and detergents. Blending and mixing chemical products to produce cleaning agents. |
| Personal care products | Production of perfumes, cosmetics, aerosols and related products. |
| Photographic industries | Manufacture and processing of photographic films. |
| Chemical analysis and testing | Applications of chemical analysis and chemical testing occur in every area of chemistry and many other fields of science. |

Figure 1.8 A chemical plant visible from the New Jersey turnpike about 20 miles south of New York City.

## Section 1.2 The Chemical Elements

Everything in our world is composed of atoms. Every substance we see, touch, or handle, including the substance of our own brains and bodies, is an assembly of vast numbers of atoms. There are 115 known elements. Of these, about 90 are **metals**; about 17 are **nonmetals** and about 8 have both metallic and nonmetallic properties and are **metalloids**. Under ordinary conditions of temperature and pressure at the earth's surface, all of the metals except one are solids: mercury is a liquid. Of the nonmetals, about half are gases and half are solids, and bromine is a liquid. Thus, elements occur in all three **states of aggregation** of matter: solid, liquid, and gas.

As we know them on Earth, the elements vary widely in abundance and occurrence. The percentages by mass of the ten most abundant elements in the earth's crust is shown in Table 1.3 on the right.

Table 1.3
**Occurrence of Elements Earth's Crust—Mass percent**

| Rank | Element | Mass Percent |
|---|---|---|
| 1 | Oxygen | 46.1 |
| 2 | Silicon | 28.2 |
| 3 | Aluminum | 8.23 |
| 4 | Iron | 5.63 |
| 5 | Calcium | 4.15 |
| 6 | Sodium | 2.36 |
| 7 | Magnesium | 2.33 |
| 8 | Potassium | 2.09 |
| 9 | Titanium | 0.565 |
| 10 | Hydrogen | 0.140 |
| | Total | 99.8% |

The one- or two-letter abbreviations for the elements are their **chemical symbols**. Most of the chemical symbols for the elements are taken from the letters of their English names. The names of eight elements: Fe (ferrum), Cu (cuprum), Ag (argentum), Sn (stannum), Sb (stibium), Au (aurum), Hg (hydrargyrum = watery silver), and Pb (plumbum), are taken directly from the Latin words. Na (natrium), K (kalium), were transmitted through Latin from originally Arab roots. The symbol for Tungsten, W, is derived from the German word wolfram.

### Discovery of the Elements

Although the concept of an element as an irreducible form of substance goes back to the Greek philosophers, it was not until 1661 that Robert Boyle (1627-91) rejected the Greek concept of four elements, proposed that different elements could not be transformed into one another, and advanced the modern concept of an **element**: a substance incapable of being chemically split into simpler substances.

Boyle listed 12 elements in his 1661 book *The Sceptical Chemist*. By 1789 Lavoisier had drawn up a table of 33 elements, correctly identifying about 25 modern elements. During the nineteenth century, many chemists added elements to the list and searched for ways to classify them and explain their existence. Around 1870, the Russian chemist Dmitri Mendeleev (1834-1907) promulgated the idea that if the elements are arranged in order of their increasing relative atomic weights (atomic weight will be defined in Section 1.4), there is a repetitive pattern (a periodic pattern) in their chemical and physical properties. Mendeleev arranged the 64 then-known elements into the first **periodic table**.

The story of the discovery of the chemical elements is a long and fascinating one, and much of the progress of chemical science can be followed through it. We don't know the dates of discovery of the first 13 elements. But since the discovery of phosphorus in 1669 history has recorded their discovery dates and the names of the men and women who discovered them. The history of the discovery of the chemical elements is summarized in Table 1.4 on the following page.

| Table 1.4 **The Chronology of the Discovery of the Chemical Elements** | | | | |
|---|---|---|---|---|
| Period | Description | Elements found (date) | # | Total |
| Antiquity | Elements known to the ancients | C S Fe Cu Ag Sn Au Hg Pb | 9 | 9 |
| The Period of Alchemistic Chemistry | Elements known in the Middle Ages | Zn As Sb Bi | 4 | 13 |
| The Beginnings of Modern Chemistry | Elements discovered 1669 to 1774 | P (1669) Co (1735) Pt (1748) Ni (1751) H (1766) F (1771) N (1772) Ba (1774) O (1774) Cl (1774) Mn (1774) | 11 | 24 |
| The Period of the Chemical Revolution | Elements discovered 1776 to 1805 | Mo (1778) W (1781) Te (1782) Zr (1789) U (1789) Sr (1790) Y (1794) Ti (1795) Cr (1797) Be (1798) Nb (1801) Ta (1802) Pd (1803) Ce (1803) Os (1804) Rh (1804) Ir (1804) | 17 | 41 |
| The Foundation of Electrochemistry | Elements isolated using electric current 1806 to 1808 | Na (1807) K (1807) Mg (1808) Ca (1808) | 4 | 45 |
| The Foundation Period of Chemical Analysis | Elements discovered in minerals 1808 to 1860 | B (1808) I (1811) Cd (1817) Se (1817) Li (1817) Si (1823) Al (1825) Br (1826) Th (1828) V (1830) La (1839) Er (1843) Tb (1843) Ru (1844) | 14 | 59 |
| The Beginning of Spectroscopy | Elements discovered from their flame colors 1860 to 1865 | Cs (1861) Rb (1861) Tl (1861) In (1863) He (1868) | 5 | 64 |
| The Period of Systematic Searching | Elements discovered in efforts to complete the periodic table 1875 to 1890 | Ga (1875) Yb (1878) Ho (1879) Tm (1879) Sc (1879) Sm (1879) Pr (1885) Nd (1885) Dy (1886) Gd (1886) Ge (1886) | 11 | 75 |
| The Noble Gases and the First Radioactive Elements | 1894 to 1900 | Ar (1894) Ne (1898) Kr (1898) Xe (1898) Po (1898) Ra (1898) Ac (1899) Rn (1899) | 8 | 83 |
| The Hard to Find Elements and Some Synthetic Elements | 1900 to 1940 | Eu (1901) Lu (1907) Pa (1918) Hf (1923) Re (1925) Tc (1937) Fr (1939) Np (1940) Pu (1940) At (1940) | 10 | 93 |
| The Period of Synthetic Elements | 1944 to 2000 | Cm (1944) Pm (1945) Am (1945) Bk (1950) Cf (1950) Ei (1952) Fm (1952) Mv (1955) Lw (1961) No (1964) Rf (1964) Db (1970)Sg (1974) Bh (1976) Mt (1982) Hs (1984) Uun (1994) Uuu (1994) Uub (1996) Uuq (1999) Uuh (1999), Uuo (1999) | 22 | 115 |

## Section 1.3 The Periodic Table

The periodic table is one of the great triumphs of science because it organizes our knowledge of the elements in a fundamental way. Mendeleev's periodic table of 1871 summarized **Mendeleev's periodic law**: that the chemical properties of the elements are periodic properties of their atomic weights. A 1909 version of that table is posted at the author's website. Mendeleev left unfilled positions or "gaps," in his table that he confidently *and correctly* predicted would eventually be filled by yet to be discovered elements. He made specific predictions about the elements that would fill three of the gaps; he predicted their densities, atomic weights, and chemical behavior. When isolated, it was found that each of these element's properties was close to Mendeleev's prediction. Gallium was isolated in France in 1877, scandium in Sweden in 1879, and germanium in Germany also in 1879. [❁kws +"periodic table" +Mendeleev]

As chemists later realized, the true periodic relationship is between chemical properties and atomic number. Actually, Mendeleev himself had foreshadowed this when he exercised superb chemical judgment to place some elements into his table in *reversed order* of their atomic weights (AW). For example, he placed tellurium (element 52, AW = 127.60) ahead of iodine (element 53, AW = 126.904) because tellurium's properties resembled those of sulfur and selenium while iodine's resembled those of chlorine and bromine. There are two other such atomic weight reversals in the periodic table: cobalt (element 27, AW = 58.93) and nickel (element 28, AW = 58.70), and argon (element 18, AW = 39.948) and potassium (element 19, AW = 39.0983). The **modern periodic law** states: the properties of the elements are periodic functions of their atomic numbers. Much of chemistry (and much of what we will study) is concerned with understanding and using the periodic table. **Periodicity**, the study of relationships among the elements, is a central organizing principle of chemistry, and a powerful scientific generalization.

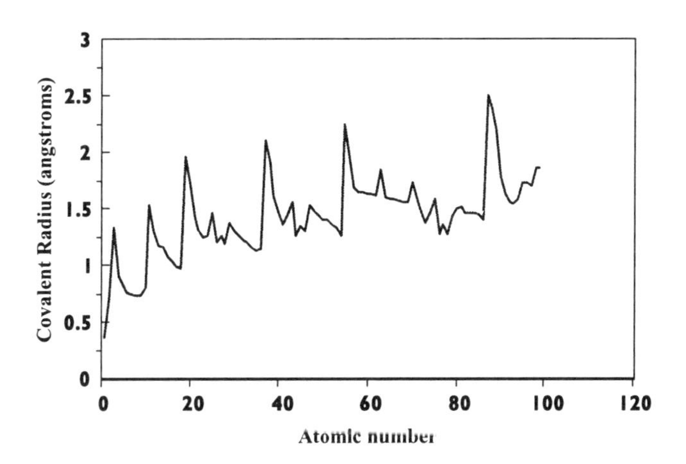


Figure 1.9 The covalent radii of the elements plotted as a function of the atomic numbers of the elements. Covalent radius is a measure of the size of the atoms of an element.

Anything periodic recurs. You can see from the recurring peaks and valleys in Figure 1.9 why it's called the *periodic* table. Chemists use considerable terminology when they speak of the periodic table, as described below.

The elements in each column of the periodic table behave alike, and are known as a family or a **group.** Elements on the left and right sides of the table (groups 1 and 2 and 13-18) are **main group** elements. For reasons that will become clear during Chapter 5, groups 1 and 2 are the s-block elements and groups 13-18 are the p-block elements. Elements in the middle region of the periodic table are **transition elements**. Many of the groups are given names. The elements of Group 1 (excluding hydrogen) are the **alkali metals**. Group 2 elements are the **alkaline earth metals**. Group 17 elements are the **halogens**. Group 18 elements are the **noble gases.**

The elements from scandium (Sc) to zinc (Zn) are known as the **first transition series**; elements yttrium (Y) through cadmium (Cd) constitute the **second transition series**; and elements lanthanum (La) through mercury (Hg) are the **third transition series**. These elements are collectively the **d-block elements** or the d-block transition metals. Each group of d-block transition metals consists of three elements arranged vertically and labeled as shown on the periodic table. For example, the d-block group containing Cr, Mo, and W is designated group 6.

The two rows of elements at the bottom of the conventional table (or the elements in unnumbered groups of the very long periodic table) are the inner transition elements, or the **f-block elements**. The elements cerium (Ce) through lutetium (Lu) are the **lanthanides**; they would follow lanthanum (La) if they were placed in the table in order of increasing atomic number. The long form of the periodic table has the lanthanide and actinide elements correctly placed in the body of the table (Appendix H). The elements thorium (Th) through lawrencium (Lr) follow actinium (Ac) and are the **actinides**. Elements beyond uranium (element 92) are the **trans-uranic** or **synthetic elements**. The synthetic elements are created by humans.

## Radioactive Elements

Why is the number of chemical elements shown in Table 1.4 limited to 115? Why can't there be as many elements as we'd like? The answer is **radioactivity**—the spontaneous decomposition or decay of an atom's nucleus. Beyond element 83 (bismuth) every isotope of element is radioactive, and to be radioactive is to be unstable. Radioactive isotopes spontaneously and uncontrollably decay, forming other isotopes. Thus no element beyond bismuth in the periodic table is permanently stable. However, three such isotopes are very long lived: thorium-232, uranium-235, and uranium-238. These three isotopes are sources of the radioactive elements Po (1898), Ra (1898), Ac (1899), Rn (1899), Pa (1918), Fr (1939), and At (1940), listed with their discovery dates.

In 1898, Marie Sklodowska Curie (1867-1934) discovered traces of the radioactive elements polonium (named after her native Poland) and radium (because it makes rays) in uranium ores. Today 115 element are known, but adding new elements to the periodic table is becoming increasingly difficult. The further we go, the more radioactive, and the less stable, the elements become.

## The Origin of the Chemical Elements

As recently as 40 years ago it would not have been possible to give an account of how the chemical elements formed in the universe. Today, it is possible to sketch out a history of the universe that explains the origin of the elements. Much of the experimental evidence comes from the analysis of starlight, which tells what elements are in the stars. In the laboratory, nuclear physicists have reproduced the highly energetic processes which build elements. From these and many other lines of research, scientists who study the physics and chemistry of the universe (astrophysicists and cosmologists) have developed a convincing account of the origin of the universe and the subsequent formation of the chemical elements.

According to the "standard model," the universe began with a "big bang." During its first three minutes of existence, the temperature of the universe was so hot that the elements were not stable. The beginning of element formation (the start of chemistry) occurred at the end of this time. After about an hour, the chemical composition of the universe became temporarily fixed. Hydrogen and helium were the only elements formed during this period, in a ratio of 75% hydrogen to 25% helium by weight. This time was the period of **cosmological synthesis** and hydrogen and helium are termed the **cosmological elements**.

As the universe subsequently expanded and cooled, stars formed from the cosmological hydrogen and helium and in stars were formed all the elements in the periodic table beyond H, He, (and a little Li). The subject of how the elements form in stars is **stellar nucleosynthesis**. All stars, including our sun, consume their hydrogen and helium (via nuclear fusion reactions) forming heavier elements and releasing energy. Stars that are ten or more times heavier than the sun derive energy from creating all the elements as far as iron (element 26) in the periodic table. But, beyond iron, nuclear fusion processes fail to yield energy. So stars sufficiently massive to produce iron eventually perish in spectacular supernova explosions, scattering a debris of elements back into space. The elements beyond iron primarily form during supernova explosions and this process was the origin, for example, of the uranium today present in the earth's crust. Isotopes of some lighter elements were formed when cosmic rays bombarded atoms in dust in the dark regions among the stars. The present observed abundances of the elements in the universe are well explained by our knowledge of how the universe works. [❁kws +nucleosynthesis +"origin of the elements"]

| Table 1.5 **Elements in the Sun (Number percent)** | | |
|---|---|---|
| Rank | Element | % of atoms |
| 1 | Hydrogen | 91.0 |
| 2 | Helium | 8.9 |
| 3 | Oxygen | 0.078 |
| 4 | Carbon | 0.033 |
| 5 | Neon | 0.0112 |
| 6 | Nitrogen | 0.0102 |
| 7 | Magnesium | 0.00350 |
| 8 | Silicon | 0.00326 |
| 9 | Iron | 0.00294 |
| 10 | Sulfur | 0.00168 |
| | Total | 100.04% |

Table 1.5 shows the composition of the sun. The composition of the universe as a whole is similar to that of the sun. Our sun and solar system formed from dust and the recycled debris of earlier stars. The material that formed the earth was particularly rich in iron and nickel, and those elements compose our planet's core. Compounds containing the other elements formed a crust over this core, a crust that in time cooled and solidified to form rocks and minerals. Since the formation of the earth, geological activity has racked its crust. Often, chemical processes within the earth's crust (**geochemical processes**) caused an element to accumulate in a concentrated deposit. These deposits became the mineral ores in which many elements were eventually to be discovered. Using nuclear science and geology, it is today possible to give a good accounting of why the various elements in the earth's crust have the abundances that they do.

Table 1.3 shows the major elements in the crust. The Geological Survey is the US federal agency responsible for being knowledgeable about mineral resources. [❁kws "Geological Survey" +"minerals information"].

## Section 1.4 Atoms, Isotopes, and Substances, and Their Properties

### Atoms and Isotopes

Our periodic table shows the atomic numbers and atomic weights of the chemical elements. Every atom of any given element has the same **atomic number**—the number of protons in its nucleus. An element's atomic number is also its number of electrons if the atom is an electrically neutral atom. An **electrically neutral** atom has no net electrical charge; an entity with a charge is an **ion**. Every element has a characteristic atomic number and a characteristic atomic weight. On the periodic table the atomic numbers of the elements begin with 1 at hydrogen and continue one at a time up to element 112 (beyond that are 114, 116, and 118). Atomic numbers are integers (whole numbers); any property of nature that occurs as discrete quantities or a steady whole number progression, without fractional values, is said to be **quantized**. Thus, atomic numbers are quantized. Excepting ordinary hydrogen, all atomic nuclei contain neutrons in addition to protons.

The sum of the number of protons and the number of neutrons in an atom's nucleus is the atom's **nucleon number**. Most elements are composed of atoms with varying numbers of neutrons: so the element's atoms have a range of slightly different masses. Atoms of the identical element but with different numbers of neutrons and different masses are **isotopes**. A sample of any element (meaning the same number of protons) composed entirely of atoms with the same number of neutrons is **an isotope** of that element. In other words, two atoms (or samples) with the *same* number of protons but *different* numbers of neutrons are isotopes of one another. Chemically, the isotopes of an element are virtually identical. The proton number (atomic number) alone controls most of the chemical properties of an atom; neutrons change its mass, but little else.

Collectively, all the isotopes of all the elements are **nuclides**. The nucleon number of an isotope is often written as a leading superscript: for example, $^{11}B$ is called "boron-eleven;" its atoms contain 11 nucleons (5 protons and 6 neutrons). In nature, most elements exist as a mixture of isotopes. Carbon, for example, exists naturally in as three isotopes; of every 10,000 carbon atoms 9,890 are carbon-12 or $^{12}C$ while 100 are carbon-13 or $^{13}C$. Carbon-12 has a natural abundance of 98.90%, and carbon-13 has a natural abundance of 1.0%. There is also a trace of $^{14}C$. A table of nuclides and isotopes is in Appendix D.

**Example 1.1**: State the number of protons and neutrons of each of the following nuclides: $^{12}C$, $^{13}C$, $^{82}Rb$, and $^{238}U$.

Strategy: On the periodic table look up each atom's atomic number; it equals the protons in its nucleus. To get the neutron count subtract the atomic number from the nucleon number shown in the symbols above.

$^{12}C$ has an atomic number of 6. So it contains 6 protons and 12 - 6 = 6 neutrons.

$^{13}C$ has an atomic number of 6. So it contains 6 protons and 13 - 6 = 7 neutrons.

$^{82}Rb$ has an atomic number of 37. So it contains 37 protons and 82 - 37 = 45 neutrons.

$^{238}U$ has an atomic number of 92. So it contains 92 protons and 238 - 92 = 146 neutrons.

## The Unified Atomic Mass Scale

Because atoms are extremely small it is inconvenient to state their masses in everyday units such as grams, kilograms, or pounds. For example, a single atom of $^{12}C$ weighs approximately $1.99 \times 10^{-23}$ g. So, for convenience, we choose to base the atomic mass scale on an extremely small unit. That unit is called the **unified atomic mass unit** and is given the symbol "u." It's called unified because for many years there were two slightly different mass scales in use. Eventually, cooperative action by scientists led to a single, unified scale—different from any of the earlier ones—that all agreed on.

The mass of a single atom of the isotope $^{12}C$ is the basis for the standard of the **atomic mass scale** or **u scale**. For chemistry, the **International Union of Pure and Applied Chemistry (IUPAC)** is the body that decides such matters as selecting a mass standard. By definition:

$$\text{The mass of one atom of } ^{12}\text{C} = 12 \text{ u (exactly)} \quad \text{or} \quad ^{12}\text{C} = 12.0000000000000\ldots \text{ u}$$

where u stands for **unified atomic mass unit**. The unified atomic mass unit is also sometimes called the **dalton** (especially in biochemistry).

$$\text{The definition of the u is thus, } 1 \text{ u} \; = \; \tfrac{1}{12} \text{ the mass of one } ^{12}\text{C atom (exactly)}$$

On the atomic mass scale $^{13}C$ atoms weigh 13.003354826 u. Note that this value, like the mass of all nuclides, is very close to a whole number. The **nucleon number** of an isotope is the integer near its mass on the atomic mass scale. Thus, an atom of nitrogen that weighs 15.00011 u has a mass-number of 15 and is $^{15}N$; an atom of mercury that weighs 201.9706 u has a mass-number of 202 and is $^{202}Hg$.

## Atomic Weight

Because the isotopes of an element are chemically alike but have different masses, the **atomic weight** (AW) of an element is the *average* of the weights of its isotopes taken in proportion to their **natural abundance** (that's the so-called **weighted average**). If the masses of the isotopes that compose an element and their natural abundances are known (Appendix D), then it is possible to calculate the atomic weight of the element. Such calculations are shown in Examples 1.2 and 1.3.

**A Note about Significant Figures.** Immediately we begin to manipulate isotope masses to calculate atomic weights, issues arise as to how accurately it is possible to calculate them and how to use significant figures to specify their uncertainty. Experimental uncertainty and significant figures are discussed in detail in Appendix B. Incidentally, Lavoisier was one of the first scientists to stress the importance of making measurements with many significant figures.

The **atomic mass** of a sample is strictly the average mass of the atoms in that particular sample. But that definition is unnecessarily restrictive for our purposes, and I will use the terms atomic weight and atomic mass synonymously. I will favor "atomic weight" if it is important to emphasize we are talking about a natural sample of an element, and I will use either **isotopic mass** or **nuclidic mass** to refer to the mass of the atoms of a specific isotope or nuclide. A **nuclide** is any particular isotope of any particular element.

The mathematical formula to calculate an atomic weight from isotopic mass and abundance data is

$$\text{Atomic weight} = \Sigma \mathbf{m}_i \mathbf{f}_i$$

where i refers to the ith stable isotope of the element, $m_i$ is the mass of the ith isotope, and $f_i$ is the fractional abundance (the percent abundance ÷ 100) of the ith isotope.

**Example 1.2**: Copper exists naturally in the form of two isotopes. $^{63}$Cu has an isotopic mass 62.9295989 and is 69.17% naturally abundant. $^{65}$Cu has an isotopic mass of 64.92779679 and is 30.83% naturally abundant. Calculate the atomic weight of copper.

Strategy: For i = 1 and i = 2 plug into the formula Atomic weight = $\Sigma m_i f_i$ .

$$\text{Atomic weight copper} = m_1 f_1 + m_2 f_2$$

$$= 62.9295989 \times 0.6917 + 64.92779679 \times 0.3083$$

$$= 43.528 + 20.017 \quad \text{(using five significant figures—temporarily carrying one extra)}$$

$$= 63.54_5 \text{ u} \quad \text{(Here reported to “4½” significant figures)}$$

The current value for copper’s atomic weight is 63.546 ± 0.003 u. You see that the rules of significant figures cause the 9 significant figure uncertainty of the isotopic masses to get washed out (reduced to only four or four and a half significant figures) during the atomic weight calculation. Unfortunately that’s inevitable because isotopic abundance measurements are much less reliable (more uncertain) than isotopic mass measurements.

Example 1.2 shows that even though the atomic weight of copper is 63.54 u, no individual atom of copper actually weighs very close to this amount. 69.17% of the atoms actually weigh about 0.5 u less whereas 30.17% weigh about 1.5 u more.

**Rule**: If you have a problem that requires you to use atomic weights, then use atomic weight values with one more significant figure than the data specified in the problem.

**Informal Example**: I’ve got a couple of hundred dogs. When I shipped them I discovered their average weight was 25 lbs. But because I have only St. Bernards and Pekes, none of them actually weighs anywhere near 25 lb.

The atomic weight of an element is the average mass of its atoms in its naturally occurring isotope mixture. Some elements have no stable isotopes and therefore do not occur in nature in significant quantity; so by the foregoing definition they have no atomic weight. Examples include element 43, technetium, element 61, promethium, element 85, astatine, and all the synthetic elements. So do we put them on the periodic table? Yes, but for these elements either the nucleon number of the most stable isotope is placed in parentheses and reported, or, if the element has a long-lived isotope (**pseudostable isotope**) then the actual mass of that isotope is placed on the table. For example, in the case of plutonium, element 94, the value (244) is the mass number of plutonium's most stable isotope $^{244}Pu$. On the other hand, protactinium (element 91) has the mass of its unstable, but long-lived, isotope $^{231}Pa$ reported to eight significant figures. The footnotes to the periodic table refer to the experimental uncertainty of atomic weight measurements. Be sure to read the definitions of g, m and [ ] beneath our periodic table. Note, however, that trace amounts of Tc, Pm, At, Np, and Pu do actually exist naturally; an account of this situation is given in Appendix D.

How variable is the natural mixture of isotopes and how reliable are the atomic weights of samples of an element obtained from different sources? The answers to these questions are that natural isotopic variations are common and for many elements those variations limit the certainty with which atomic weight can be measured. You saw this situation in Example 1.2 for copper isotopes. The atomic weight of lead is a particularly instructive example. Example 1.3 shows the calculation of lead's atomic weight with the proper number of significant figures.

**Example 1.3**: Use the mass of lead's isotopes and their natural abundance to calculate the atomic weight of lead and its number of significant figures.

Strategy: Plug into the formula Atomic weight = $\Sigma m_i f_i$ for i = 1, 2, 3, and 4.

| Isotope | Mass (u) | Abundance |
|---|---|---|
| $^{204}_{82}Pb$ | 203.973020 ± 0.000005 | 1.4% ± 0.1% |
| $^{206}_{82}Pb$ | 205.974440 ± 0.000004 | 24.1% ± 0.1% |
| $^{207}_{82}Pb$ | 206.975872 ± 0.000004 | 22.1% ± 0.1% |
| $^{208}_{82}Pb$ | 207.976627 ± 0.000004 | 52.4% ± 0.1% |

Using the formula: Atomic weight = $m_1f_1 + m_2f_2 + m_3f_3 + m_4f_4$

$$203.973020 \times .014 + 205.974440 \times 0.241 + 206.975872 \times .221. + 207.976627 \times .524$$

$$2.9\ u + 49.6\ u + 45.7\ u + 109.0\ u$$

AW = 207.2 u (4 significant figures)

Example 1.3 shows that the atomic weight of lead is known with low accuracy—only four significant figures. It turns out that lead samples show significant variation in lead's atomic mass depending on the source of the ore from which the lead sample was obtained. The reason for this variability is that lead's three most abundant isotopes $^{206}Pb$, $^{207}Pb$, and $^{208}Pb$ are all continuously formed in nature as the stable end products of natural breakdown of pseudostable elements (**natural decay chains**). The footnote "g" to the periodic table reminds us that samples of elements from some geologic sources might have atomic weights significantly different from the correct value of the element's atomic weight.

## Statistics and Experimental Uncertainty

When you report the results of trials or experiments your readers or listeners (or management) will want to know "How good are your data?" They will be interested in your values (in your numbers) but they will *also* want to know how reliable are those values (and numbers). The best way to make that report is to use the statistical quantity called **standard deviation**.

In Example 1.3 you'll see the measured property or physical quantity 207.976627 ± 0.000004 u written for the mass of the isotope $^{208}$Pb. The ± 0.000004 is the standard deviation of the physical quantity of mass, 207.976627, which is measured in the unit of u.

Measurement theory is just as much a part of engineering science as it is of chemistry. So the information presented in Appendix B about standard deviation and measurement theory is important to you as you begin a professional career. But rather than try to convince you of that myself, I'd rather have you read the comments of a working engineer who applies statistics to engineering problems. His statement is at the end of Appendix B.

The atomic weight of uranium is an interesting case. With uranium, human activities change its atomic mass. Natural uranium contains 99.2745% $^{238}$U and 0.7200% $^{235}$U. The lighter $^{235}$U isotope is useful for military and civilian nuclear power, and so large quantities of uranium are processed to extract it. The uranium, thus depleted of its U-235 content, serves perfectly well for ordinary chemical purposes, though now with a slightly increased atomic weight. The footnote "m" to the periodic table reminds us that humans change atomic weights. Incidentally, because it is readily available to military organizations, and because it is very dense, depleted uranium (uranium from which the $^{235}$U isotope has been removed for nuclear purposes) is used to make armor-piercing projectiles.

## Substances and Properties

Chemistry deals with the composition and properties of substances and with the changes of one substance into another. A **substance** is matter that has a specific chemical composition. All material things are made of matter. In ordinary life, matter is any gas, liquid, or solid, or any mixture of these. At a sufficiently high temperature, matter exists as ionized atoms in the form of a plasma. Figure 1.10 shows a scheme for the classification of matter.

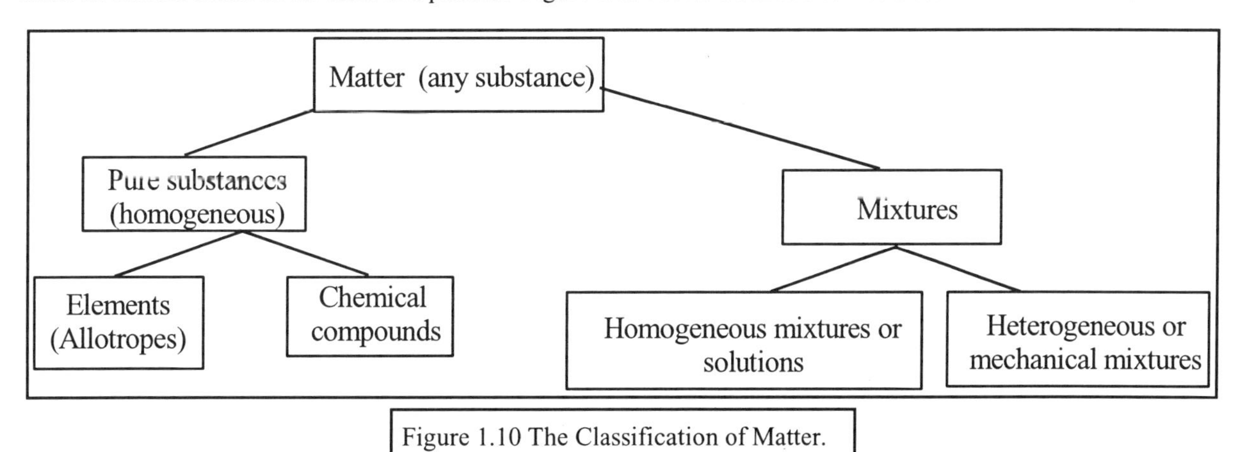


Figure 1.10 The Classification of Matter.

Substances are divided into two broad classes: pure and mixtures. Pure substances and solutions are homogeneous. **Homogenous substances** are those within which no physical boundaries can be detected. Any well-mixed gas or liquid is homogeneous, as are many solid substances. **Heterogeneous substances** have internal physical boundaries. For example, soda water is a heterogeneous substance with boundaries between the gas bubbles and the liquid; wood, with its obvious grain structure, is a heterogeneous substance. Heterogeneous substances are **mixtures** or **mechanical mixtures** if they're made by stirring more than one substance together. Paint, for example, is a mechanical mixture of pigment particles in an oil or water base. At the boundary between homogeneous or heterogeneous the distinction blurs; this is the realm of the **colloids**. Milk is an everyday colloidal substance—it's neither a proper solution nor a proper mixture.

Homogeneous substances are divided into two groups: solutions and pure substances. **Solutions** are homogeneous mixtures of variable composition. Wine and whiskey are both solutions of alcohol in water (with minor amounts of flavoring agents), but there is a good deal more alcohol in a gallon of whiskey than in a gallon of wine. **Pure substances** also divide into two groups: elements and chemical compounds. The molecules of a pure substance are all *chemically* alike.

Finally, in this scheme of classification, note that some elements occur or exist in more than one form. Diamond and graphite are both composed entirely of carbon atoms; oxygen gas and ozone gas are composed entirely of oxygen atoms. Collectively, the two or more forms of a chemical element are called **allotropes**. Thus, both diamond and graphite are allotropes of the element carbon. About 20 years ago, a new family of carbon allotropes was discovered. Humorously named "buckyballs" these allotropes consist of cages of carbon atoms such as $C_{60}$. (In structure, buckyballs resemble geodesic domes, for which the futurist Buckminster Fuller was known, and that's how they got their name.)

Any quality of a substance is a **property** of that substance. A metal's melting point is a property of the metal. Chlorine is a greenish-yellow gas, bromine is a brown liquid, iodine is a purple solid. Its color, its state of aggregation (gas, liquid, or solid), and its boiling and melting points are **physical properties** of a substance. Metals react (or don't react) with water in a characteristic way. Titanium, gold, silver, tantalum, etc., don't react with water. Sodium, rubidium, and barium do react and react violently, forming new substances. How a metal reacts with water is said to be a **chemical property** of the metal. **Intensive properties**, such as density, color, or boiling point, do not depend on the amount of substance of a sample. **Extensive properties**, such as mass, volume, and the number of atoms in a sample, increase when the amount of substance is increased

Changes that occur to a substance without its chemical composition changing are called **physical changes**. The melting of a solid, the boiling of a liquid, and the formation of a solution by mixing alcohol and water are examples of physical changes. As Dalton realized, a process that changes the chemical composition of a substance is a **chemical change** or chemical reaction. During a chemical change, some substances are consumed (the **reactants**) and others are formed (the **products**). The burning of a candle or the corrosion of iron are examples of chemical changes. When the candle burns, the substance wax and the substance oxygen are converted to the substances carbon dioxide and water. When the substance iron corrodes it combines with the substances oxygen and water to form the substance rust.

## Section 1.5 Units and Properties for Chemistry

Because of the importance of measuring things, chemists use a lot of equipment and instruments. Mostly, chemists use **SI** (*Le Système International d'Unités*) for the values of their measurements. SI is a coherent system founded on seven **base units**. A description of the SI system of units and how you can manipulate SI quantities is described in the Appendix A. Some important properties chemists measure are: amount of substance (by count and mass), volume, density, temperature and energy. Pressure is an important property, but we'll wait until we begin our study of gases in Chapter 4 before discussing pressure and its units. IUPAC bases its units on the SI units and publishes them in the "Green Book." It's called that for the obvious reason: it's printed with a green cover. The Green Book stresses the handling of units via the method called the **algebra-of-quantities**. The method is excellent and is described in Appendix A, which includes six worked examples. Essentially, the algebra-of-quantities method treats every measured property or physical quantity as the product of a numeric value multiplied by a unit: physical quantity = value × unit. The author strongly recommends the method and will make much use of it; in use it's similar to the method of dimensional analysis or the factor-label method. [❁kws IUPAC].

### Time

Everyone is familiar with measuring time, but few people are aware that time's base unit, the second, is defined as the time it takes for 9,192,631,770 vibrations of certain types of radiation from a $^{133}Cs$ atom. (Appendix A). Sometimes, time referred to this standard is said to be measured by an "atomic clock." Time is one SI base unit that ordinary people can get quite close to. You can calibrate the clock of your computer against an atomic clock by connecting to a standard time server maintained in the US by NIST (the US National Institute of Standards and Technology). If you wish, you can download a program for this purpose. Try searching for [❁kws +"nistime"]. Time is often important in chemistry, for example: when we measure rates of radioactive decay, rates of chemical reactions, or when we are interested in studying how long materials will hold up in use.

## Amount of Substance

Chemists weigh in grams and count in moles.

In the laboratory, chemists typically manipulate masses of material in the range of about 0.0001 g to 500 g. Quantities of compounds in this range contain huge numbers of molecules and to "count" them chemists use the mole. The **mole** is the SI base unit of amount of substance. The formal definitions of the mole and the other base units of the SI system are in Appendix A. The **amount of substance** of a sample is its quantity of matter. An amount of substance can be specified *either* by mass or by a count of its entities. An entity is whatever unit we choose to count. In chemistry, an **entity** is usually an atom, molecule, or ion.

You are already perfectly familiar with the concept of amounts of substances being specified either by mass or by count. If I say "I need a specified amount of nail substance" it sounds odd, but all I mean is "I need some nails." How to get them? If they are big nails I might count them and buy exactly the number I need. But if I need a bunch of tiny nails I buy them by weight. The idea is illustrated in everyday terms by Example 1.4.

Because atoms are small the chemist's counting unit is big. One mole is $6.0221367 \pm 0.000000036 \times 10^{23}$ entities $mol^{-1}$ (approximately!), where *you* have to specify what the entity is. The SI definition states that **one mole** is the number of carbon atoms in exactly 12.0000000... g of $^{12}C$. The number of $^{12}C$ atoms in such a sample is **Avogadro's number**. Avogadro's number is measured by experiment, and historically represented the number of hydrogen atoms in one gram of hydrogen. When used as a unit in mathematical expressions, the mole is usually abbreviated to mol.

**Example 1.4**: You need many 1.00 g nails. How to buy them?

Strategy: Be aware of how nails are sold and that there are 454 g $lb^{-1}$.

You can *either* buy a pound of nails *or* 454 nails (454 g is equivalent to one pound). The amount of "nail substance" is the same either way.

Next time you're in a supermarket observe that some things you buy by weight (mass) and others by count. By mass: meat, potatoes, carrots, etc. By count: eggs (dozen), beer (six-pack), soda (24 in a big box), etc. You can add many more examples to these two lists.

A mole is a counting measure like a dozen. A dozen is twelve of the specified objects; a score is 20 of them; a gross is 144 of them; and a mole is $6.022 \times 10^{23}$ of them. There can be moles of atoms, moles of ions, moles of molecules, or moles of any combination of these entities. Using moles to refer to anything much larger than molecules is rarely useful because one mole is such a large number. For example, a mole of sand grains is a much greater quantity of sand than can be found on all the earth's beaches. We'll return to moles in Chapter 3.

## Mass

Mass is an inherent property of an amount of substance. Often the word weight is used when mass is meant. It happens all the time in everyday life—nobody ever says "I massed the tomatoes." Weight and mass are not the same—though on the earth's surface they usually have the same value (but different units). To weigh things chemists use **balances**, so-named because traditional versions balance standard weights against the unknown weight that is being measured. A modern, single-pan, electronic balance is shown in Figure 1.11.

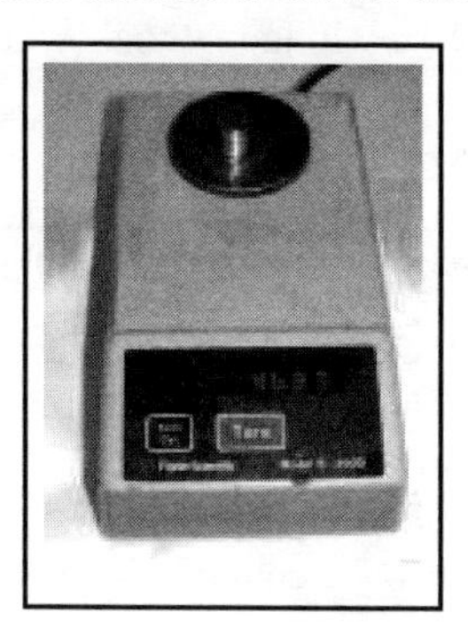

Figure 1.11

The SI base unit of mass is the **kilogram**. The international prototype kilogram (the original model), the SI standard of mass kept in Paris, is made of a platinum-iridium alloy. The common unit of mass in the United States is the pound (lb). The conversion factor from kilograms to pounds is approximately 0.4536 kg $lb^{-1}$. The reciprocal of the conversion factor from kilograms to pounds is the conversion factor from pounds to kilograms: $(\frac{1}{0.4536})$ lb $kg^{-1}$ or 2.205 lb $kg^{-1}$.

**Example 1.5**: What is the mass of a 95 lb girl in kg?

Strategy: Multiply the mass in pounds by the appropriate conversion factor. Pay attention to the use of proper significant figures.

$$\text{mass (kg)} = 95 \text{ lb} \times 0.4536 \text{ kg lb}^{-1} = \underline{43 \text{ kg}}$$

You already know that the atomic mass unit is exactly $\frac{1}{12}$th the mass of a $^{12}C$ atom. The u is an *alternative and independent* non-SI mass unit. As chemists make increasingly better measurements the conversion factor from u to kilograms is adjusted to keep the two scales of units identical. That conversion factor is Avogadro's number.

## Volume

The volume of a body is the space that it occupies in three dimensions. Volume has units of $length^3$. The SI unit of volume is the cubic meter ($m^3$). The cubic centimeter ($cm^3$) is in widespread use, and the volume 1000 $cm^3$ is given the special name of liter (L). The milliliter (1 mL = 1 $cm^3$) and microliter (μL) are used also. You will encounter many volume measuring devices in your laboratory course. In the United States the everyday measure of volume is the US gallon that contains 128 fluid ounces (fl oz). The conversion factor to SI is 3.785412 liters $gallon^{-1}$ I'll usually round this to 3.785 L $gallon^{-1}$. Application of the algebra-of-quantities method to a volume conversion is shown in the following example.

**Example 1.6**: Calculate the volume in liters and milliliters of 0.50 gallons (gal).

Strategy: Use conversion factors. Include and manipulate all units to check for correctness. Pay attention to the use of proper significant figures.

$$\text{Volume in L} = 0.50 \text{ gal} \times 3.785 \text{ L gal}^{-1}$$

$$= \underline{1.9 \text{ L}} \quad \text{(only 2 significant figures)}$$

The conversion factor from L to mL is 1000 mL $L^{-1}$, which is used to convert 1.9 L to mL.

$$\text{Volume in mL} = 1.9 \text{ L} \times 1000 \text{ mL L}^{-1}$$

$$= \underline{1900 \text{ mL}} \quad \text{(still only 2 significant figures)}$$

The volumes of various regularly shaped containers are derived from geometry. Recall that the volume of a cube is simply the cube of its edge length. The volume of a box with 90° angles is its length × breath × width. The volume of a cylinder of radius r is $\pi r^2$ × its length, and the volume of a sphere with radius r is $\frac{4}{3}\pi r^3$.

## Density

Density is the ratio of the mass (m) of a quantity of material to its volume, $d = \frac{m}{V}$. Density's SI unit is kilograms per cubic meter ($kg\ m^{-3}$). Chemists mostly use $g\ L^{-1}$ and $g\ mL^{-1}$ as density units. The density of regularly shaped objects can be found by measuring their dimensions, weighing them, and plugging into the density formula. The density of a liquid or an irregularly shaped solid can be measured using a graduated cylinder for its volume and a balance for its mass.

Here are two examples of density calculations:

**Example 1.7**: A cube of lead measures 1.533 cm along each edge. Weighing the cube gives its mass as 40.67 g. Calculate the density of lead in the units $g\ mL^{-1}$. Pay attention to the use of proper significant figures.

Strategy: First, find the volume of the cube. Second, calculate the density dividing that mass by the volume.

$$\text{volume of a cube} = (\text{edge length})^3$$

$$\text{volume} = (1.533\ \text{cm})^3$$

$$= 3.603\ \text{cm}^3 \text{ or mL}$$

$$\text{Density} = \text{mass} \div \text{volume} \quad = \quad 40.67\ \text{g} \div 3.603\ \text{mL}$$

$$\text{density} = \underline{11.29\ \text{g mL}^{-1}}$$

**Example 1.8**: An empty graduated cylinder weighs 60.52 g. Some gasoline is added to the cylinder, and its volume measured to be 33.27 mL. Together, the graduated cylinder and the gasoline weigh 83.14 g. Calculate the density of this sample of gasoline. As usual, pay attention to the significant figures.

Strategy: Use the difference of mass between the full and empty cylinder to calculate the mass of gasoline.

mass of gasoline = mass of cylinder plus gasoline - mass of the empty cylinder

$$\text{mass of gasoline} = 83.14\ \text{g} - 60.52\ \text{g} = 22.62\ \text{g}$$

$$\text{density of gasoline} = \frac{\text{mass}}{\text{volume}} = \frac{22.62\ \text{g}}{33.27\ \text{mL}}$$

$$= \underline{0.6799\ \text{g mL}^{-1}}$$

## Temperature

Temperature scales measure the level of a body's thermal energy. Hot objects transfer energy in the form of heat to cold bodies. Two objects are at the same temperature if no heat flows between them when they are placed in contact: they are in **thermal equilibrium**. The **zeroth law of thermodynamics** states that if A is in thermal equilibrium with B and B, in turn, is in thermal equilibrium with C, then A and C are also in thermal equilibrium. The commonest device to measure temperature is a thermometer. Two temperature scales are in common use in chemistry: Celsius (or centigrade) and Kelvin.

The Celsius scale is defined using water: the freezing point is 0°C and the boiling point is 100°C (both measured at 1.00 atmosphere pressure). The Kelvin scale of temperature is named for William Thomson, later Lord Kelvin (1824-1907). Its unit is the **kelvin**, the SI base unit of temperature. The freezing point of water at 1.00 atmosphere is defined to be 273.15 K. So the Celsius and Kelvin scales differ by 273.15 degrees. The Celsius degree is identical in size to the kelvin. Pilots of US domestic airlines began using "centigrade" temperatures in 1998; the centigrade scale is the same as the Celsius scale.

The Celsius scale is in everyday use throughout most of the world outside the United States. However, in the United States, the Fahrenheit scale is the scale of everyday use. The Fahrenheit scale was originally established so that the average temperature of a human body was about 100°F. Its precise definition is that the freezing and boiling points of water are 32°F and 212°F, respectively.

The relation between the Celsius and Fahrenheit scales is illustrated in Figure 1.12. If the two thermometers are identical, the fluid (alcohol or mercury, etc.) will move up and down in them identically. To equate $t_c$ °C on the Celsius scale and $t_f$ °F on the Fahrenheit scale, recognize that the measured temperature, expressed as a fraction of the difference between the freezing and boiling points of water, must be the same on either scale.

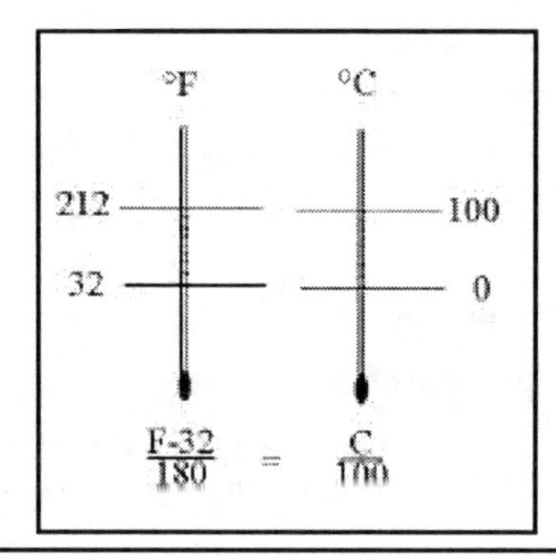


Figure 1.12 The relationship between the Fahrenheit and Celsius temperature scales.

On the Celsius scale, the fraction is $(t_c\,°C - 0\,°C) \div (100\,°C - 0\,°C)$
and on the Fahrenheit scale it is $(t_f\,°F - 32\,°F) \div (212\,°F - 32\,°F)$

Because this fraction must be the same on either scale, it becomes the conversion formula to get from one temperature unit to the other:

$$\frac{t_f - 32}{180} = \frac{t_c}{100}$$

**Example 1.9**: Meat is roasted at 350°F. What is this temperature on the Celsius and Kelvin scales?

Strategy: Plug into the Celsius to Fahrenheit conversion formula to get the temperature in Celsius. Then convert to K by adding 273.15.

$$t_c \div 100 = (350 - 32) \div 180$$

$$\underline{t_c = 180\ °C} \quad \text{(2 significant figures)}$$

To convert to K add 273.15 (exactly)

$$K = 273.15 + 180\ K = \underline{450\ K} \quad \text{(still 2 significant figures)}$$

## Energy

The SI unit of **energy, work,** or **quantity of heat** is a derived unit given the special name of the joule. It's named after James Prescott Joule (1818-1889), who was the first person to establish that heat and work are different forms of energy. Energy is a property of great importance for both chemists and engineers.

In terms of the SI base units, 1 joule is 1 kg $m^2\,s^{-2}$. Historically, chemists used calories to measure energy. The conversion factor is exactly 4.184000.... joules per calorie (J $cal^{-1}$). Calories are still widely used in discussing food energy, and they remain in use is some branches of chemistry—such as organic chemistry. **Organic chemistry** is the chemistry of carbon and its compounds. I assume you already know about **kinetic energy**, the energy of motion, and **potential energy**, the energy of position. Chemical energy is exceedingly important; we run our society on it—engineers (such as James Watt) build power plants and engines. So we'll study chemical energy a lot in our course. **Chemical energy** is energy stored in substances until liberated by a chemical reaction. Burning a fuel releases chemical energy as heat.

The formula for kinetic energy (KE) is one-half the mass of a moving object times its speed squared ($KE = ½ mv^2$). The kinetic energy of a 2-kilogram mass traveling at 1 meter per second is one joule.

$$KE = ½\ mv^2$$

$$= ½ \times 2\ kg \times (1\ m\ s^{-1})^2$$

$$= 1\ kg\ m^2\ s^{-2} = 1\ J$$

I use this formula as a quick way to remember the SI base units for the joule. Speed is $m\ s^{-1}$ and mass is kilograms. I do the algebra-of-quantities in my head and immediately get the correct base units of the joule.

**Example 1.10:** Calculate the kinetic energy in kilojoules of a 1.00 ton automobile traveling at 50.0 miles per hour.

Strategy: To solve this problem convert tons to kilograms, miles to meters, and hours to seconds. That puts everything into SI base units. Plug everything into $KE = ½ mv^2$ to get joules. Finally, convert from J to kJ.

$$KE = \frac{1}{2} \times \frac{1.00\ \text{ton}}{1} \times \frac{2000\ \text{lb}}{1\ \text{ton}} \times \frac{0.454\ \text{kg}}{1\ \text{lb}} \times \left(\frac{50.0\ \text{mile}}{1\ \text{hour}} \times \frac{5280\ \text{feet}}{1\ \text{mile}} \times \frac{12\ \text{inches}}{1\ \text{foot}} \times \frac{2.54\ \text{cm}}{1\ \text{inch}} \times \frac{1\ \text{m}}{100\ \text{cm}} \times \frac{1\ \text{hour}}{3600\ \text{s}}\right)^2$$

$$= 2.27 \times 10^5\ \text{kg m}^2\ \text{s}^{-2}\ \text{or joules}$$

$$= 2.27 \times 10^5\ \text{joules} \times 0.001\ \text{kJ J}^{-1} = \underline{227\ \text{kJ}}$$

We'll see later that 227 kJ is just about the right amount of energy to break one mole of chemical bonds. It's useful to convert all sorts of energy units to kJ per mole for convenient comparisons.

Example 1.11 shows the use of an engineering energy unit, the Btu or **British thermal unit**:

**Example 1.11**: The heat of combustion of acetylene is 50.0 kJ $g^{-1}$. Express this value in the engineering unit of Btu per pound. IUPAC specifies that 1 Btu is equivalent to 1055.06 J (exactly).

Strategy: Use conversion factors. Round 1055 J $Btu^{-1}$ to four significant figures—one more than the three specified in 50.0. Rounding to four figures allows me to minimize the entry of useless digits while maintaining the best possible accuracy available in the problem. Round the final answer to three significant figures.

$$\text{Heat of combustion} = 50.0\ \text{kJ g}^{-1} \times 1000\ \text{J kJ}^{-1} \times 454\ \text{g lb}^{-1} \times 1055\ \text{Btu J}^{-1}$$

$$= \underline{21{,}500\ \text{Btu lb}^{-1}}$$

The energy contents of fuels such as oil and coal are frequently stated by combustion engineers in Btu $lb^{-1}$.

**Petroleum Refining—Chemical Energy in the Everyday World**

As noted earlier, petroleum refining is an enormous business. Shown in Figure 1.13 is the Motiva Enterprises, LLC/USA refinery in Convent, Louisiana. It was formerly operated by Texaco and Star. It has a nominal capacity of 225,000 barrels per day. A US liquid barrel is 31.5 gallons or 119.2 L. So, assuming a liquid density of 0.8 g $mL^{-1}$, this refinery processes about 35,000 tons of crude oil per day.

Motiva was formed in 1998 as a joint venture. The plant is now owned by Shell (35%), Texaco (32.5%), and Saudi Refining, a subsidiary of Saudi Aramco (32.5%).

In the foreground, immediately beyond Louisiana Highway 70, is a field of sugar cane. Large chemical plants surrounded by cane fields are a common feature of the lower Mississippi region of Louisiana that stretches from north of Baton Rouge to New Orleans 60 miles away to the southeast.

The final example of Chapter 1 shows the relation between power and work in traditional engineering units and the SI metric system.

Figure 1.13 The Motiva Refinery, Convent, Louisiana.

**Example 1.12**: According to a 50-year old textbook of engineering thermodynamics there are 550 ft-$lb_f$ $sec^{-1}$ $hp^{-1}$. Explain the notation and the units of this conversion factor. Calculate the number of watts equivalent to one horsepower and compare your answer with the modern IUPAC conversion factor 1 hp = 745.7 watts (exactly).

Strategy: First, recognize that the notation ft-$lb_f$ means "foot-pounds force" which in terms of the equations, force = mass × acceleration and work = force × distance, is the work done by a one pound mass accelerated a distance of one foot by standard gravity. (The term foot-pounds force arose at the beginning of the 18th century as mining engineers developed methods to lift water from mines.)

Second, the IUPAC value of standard acceleration of free fall is 9.806 65 m $s^{-2}$ (exactly). The problem is, thus, a units conversion problem that can be solved by the algebra-of-quantities. I'll use engineering units up to the final conversion.

The stated conversion factor shows that 1 hp = 550 ft-$lb_f$ $sec^{-1}$, so you need to add the appropriate conversion factors on the right to obtain the units of watts or kg $m^2$ $s^{-2}$. Necessary conversion factors are 1 foot = 0.3048 m (exactly) or 1 m = 3.2808399... ft, and 1 kg = 0.453 592 37 lb (exactly).

Gravitational acceleration = 9.806 65 m $s^{-2}$ ÷ 0.3048 m $ft^{-1}$

= 32.17404856... ft $s^{-2}$

So, 1 ft-$lb_f$ $sec^{-1}$ = 1 lb × 1 foot × 32.17404856... ft $s^{-2}$

or, 1 ft-$lb_f$ $sec^{-1}$ = 32.17404856... lb $ft^2$ $s^{-2}$ (exactly)

Note that the quantity immediately above has the same value as gravitational acceleration but *different* units.

Now in the equation 1 hp = 550 ft-$lb_f$ $sec^{-1}$ make a substitution from the line above. Doing this gives:

1 hp = 550 × 32.1740856 lb $ft^2$ $s^{-2}$

1 hp = 17695.72671... lb $ft^2$ $s^{-2}$ (that's exact)

This relation can now be converted using the ordinary conversion factors to go from English to SI units.

1 hp = 17695.72671 lb $ft^2$ $s^{-2}$ × (0.3048 m $ft^{-1})^2$ × 0.453 592 37 kg $lb^{-1}$

So, 1 hp = 745.6998716... kg $m^2$ $s^{-2}$ or watts

The calculated value agrees with the IUPAC value to one part in 7.5 million. Many units have changed slightly over the past 50 years. In this case, the change was caused by a tiny redefinition of horse power.

## Essential Knowledge—Chapter 1

**Glossary words:** isotopes, matter, space, mass, material, properties, composition, substance, pure chemistry, applied chemistry, synthetic, materials chemistry, experimental science, quantitative experiment, quantitative and qualitative chemical analysis, chemical industry, metals, nonmetals, metalloids, states of aggregation, chemical symbols, periodic table, periodicity, periodic law, group, main group, transition elements, alkali metals, alkaline earth metals, halogens, noble gases, first transition series, d-block elements, f-block elements, lanthanides, actinides, synthetic elements, radioactivity, atomic number, nucleon number, isotopes, nuclides, unified atomic mass unit, natural decay chains, significant figures, standard deviation, pseudostable isotope, homogenous substances, heterogeneous substances, mixtures, colloids, solutions, pure substances, elements and chemical compounds, allotropes, property, physical properties, intensive and extensive properties, physical changes, chemical change, reactants, products, SI metric units, base units, algebra-of-quantities, conversion factor, mole, amount of substance, entity, Avogadro's number, petroleum refining.

**Key Concepts:** Dalton's atomic theory, scope of applied chemistry, periodic table, radioactivity, unified atomic mass scale, atomic weight and isotopic mass, substances and properties, mole concept, use of significant figures, units conversion problems.

**Key Equations:**

Atomic weight = $\Sigma \mathbf{m}_i \mathbf{f}_i$

physical quantity = value × unit

V(cylinder) $= \pi r^2 \times l$

V(sphere) $= \frac{4}{3}\pi r^3$

Density $= \frac{m}{v}$

$$\frac{t_f - 32}{180} = \frac{t_c}{100}$$

KE = $\frac{1}{2} mv^2$

## Questions and Problems

### Chemistry as a Science

1.1. What were the four elements of Greek science?

1.2. What are the four traditional branches of chemistry. Hint: Begin in the glossary with organic chemistry.

1.3. Use internet resources to locate an on-line biography of the engineer James Watt.

1.4. Discuss what is meant by the term "Chemical law."

1.5. Chemistry is an experimental science. What does this mean?

1.6. Carefully describe what is meant by the term "reliable measurement." Explain the factors that contributed to the reliability of measurements that led in turn to the chemical revolution.

1.7. Write a definition of each of the following terms used in the chapter: a. matter, b. mass, c. properties, d. composition, e. substance, f. pure chemistry, g. applied chemistry, h. raw materials.

1.8. a. List three important sources of chemical raw materials. b. Distinguish between a raw material and a synthetic material. c. Distinguish between a natural and synthetic element.

1.9. During the middle ages, what activities would eventually form the basis for the science of chemistry? State three technological applications of chemistry during the sixteenth century.

1.10. State the main contributions to the development of chemical science of Robert Boyle, John Dalton, and Dmitri Mendeleev.

1.11. Name any engineering discipline that relies heavily on chemistry.

### Industrial Chemistry

1.12. The scope of applied chemistry is very wide—as shown in Table 1.2. Describe five ways in which your personal life has this week been touched by some aspect of the chemical industry.

1.13. Use Table F-1 in Appendix F and write the chemical formulas of each of the following important industrial chemical compounds: a. sulfuric acid, b. ethylene, c. ammonia, d. nitric acid, e. sodium carbonate, and f. sodium hydroxide.

1.14. A hydrocarbon is a compound that contains only two elements: hydrogen and carbon. The simplest way to write the formula of a hydrocarbon is $C_xH_y$, where x is the number of carbon atoms in one of its molecules and y is the number of hydrogen atoms. Use Table F-1 in Appendix F for reference and write the $C_xH_y$ formulas for the following industrially important hydrocarbons: a. ethylene, b. propylene, c. benzene, d. ethylbenzene, e. 1,3-butadiene, and f. cumene.

1.15. Name six chemical compounds that are either used as fertilizers or used in the manufacture of fertilizers. You'll see from Appendix Table F-1 that many fertilizers are among the most important industrial chemicals.

1.16. Synthetic organic polymers play a large role in our daily lives. We'll meet many examples in due course. In Appendix Table F-1 you can identify polymers because their name begins with the prefix poly-, as in polyethylene or polypropylene. Approximately how many pounds of polymers are consumed by the average American citizen each year?

1.17. Name the five companies with the largest chemical sales in 1997. Name five large scale industrial chemical compounds. You will find answers in Appendix F.

### The Chemical Elements

1.18. Identify the names and chemical symbols of the following elements: a. an artificial element named after one of the United States, b. an element named after the sun, c. the most abundant element in the earth's crust, d. the two elements that mainly comprise the earth's core.

1.19. a. What are the three most abundant elements in the earth's crust? b. What are the two most abundant elements in the universe?

1.20. Define carefully each of the following terms: a. element, b. state of aggregation, c. metalloid, d. group, e. periodicity, f. alkali metal, g. lanthanide, h. actinide, i. d-block element, and j. chemical symbol.

1.21. In two or three sentences describe the origin of the chemical elements. Explain what is meant by the term cosmological synthesis.

1.22. Francium is a highly unstable (radioactive) element yet trace amounts of it are always present in the earth's crust. What is the explanation for this situation?

## The Periodic Table

1.23. The atomic numbers of the elements increase steadily through the periodic table whereas the atomic weights do not. Give an example of a pair of elements which show this weight reversal. State Mendeleev's original periodic law and the later modern revision of the law.

1.24. Use information from Appendix D to calculate the atomic weight of tungsten from the masses of its isotopes and their abundances. State how your answer compares with the IUPAC tabulated value of tungsten's atomic weight shown on our periodic table.

## Atoms and Isotopes

1.25. Distinguish between an isotope and a nuclide.

1.26. Use Appendix D to select an element with six or more isotopes. Use data from the same source to demonstrate that the mass of an isotope is always close to a whole number.

1.27. State the number of neutrons, protons, and electrons in one electrically neutral atom of each of the following: $^{31}P$, $^{107}Ag$, and $^{144}Sm$.

## Units and Properties

1.28. Write a short definition or description of each of the following terms: a. heterogeneous substance, b. mechanical mixture, c. colloid, d. solution, e. allotrope.

1.29. State two physical properties and one chemical property of the elements oxygen, iodine, gold, and mercury.

1.30. When using a computer it is common to speak of the properties of an object. What, in this context, are properties?

1.31. a. State the mass of an atom of $^{15}N$ in atomic mass units and in grams. b. State the mass of 10 atoms of $^{172}Yb$ in u and g. c. State the mass of a single (average) water molecule in u and g. d. Why does part c. of this question tell you it is an average molecule?

1.32. a. A mineral has a density of 3.2 g $cm^{-3}$. Calculate the mass of a perfect sphere of this mineral that has a radius of 4.81 inches. Express your answer in pounds and kilograms. b. A metal sphere has a radius of 2.17 inches and a density of 11.3 g $cm^{-3}$. What is the volume of the sphere in milliliters?

1.33. The density of the platinum-iridium alloy used to make the international prototype kilogram is 21.55 g $mL^{-1}$. The actual standard is a cylinder 4.00 cm high. Calculate its diameter.

1.34. The density of plutonium is 17.1 g $mL^{-1}$. What is the radius in centimeters of a 10.0 kg sphere of this plutonium?

1.35. A liquid with a density of 1.22 g $mL^{-1}$ is used to fill a cylinder with a length of 2.44 inches and a diameter of 1.08 inches. What is the mass of liquid used in grams?

1.36. A cubic centimeter of a sample of iron contains $8.48 \times 10^{22}$ atoms. Calculate the volume occupied by a single iron atom: a. in $cm^3$, b. in $m^3$, and c. in $pm^3$.

1.37. The density of solid lithium is 0.53 g $cm^{-3}$ and its atomic weight is 6.941 g $mol^{-1}$. The atomic radius of lithium is reported to be 2.08 Å units (1 Å = $10^{-10}$ m). Calculate the volume of a mole of lithium and then use Avogadro's number to calculate the volume of a single lithium atom in a sample of solid lithium. Take the cube root of the latter value and compare it to the reported atomic radius. Discuss your result.

1.38. The property called concentration gradient has the units: mol $m^{-4}$. Discuss the unit and in particular suggest an interpretation for length raised to the negative fourth power.

1.39. State the zeroth law of thermodynamics. This law is unprovable but not in doubt. Discuss.

1.40. a. How many hydrogen atoms are in a mole of hydrogen atoms? b. How many hydrogen molecules are in a mole of hydrogen molecules? c. How many hydrogen atoms are in a mole of propane?

## Units Conversion Problems

1.41. Avogadro's number is described in the chapter as a conversion factor. Explain why.

1.42. The speed of light is $3.00 \times 10^{10}$ cm $s^{-1}$. What is it in miles per hour?

1.43. If an automobile gets a gas mileage of 31 mpg (miles $gallon^{-1}$) what is its gas mileage in km $L^{-1}$?

1.44. The speed of light is defined by SI as 299792458 m $s^{-1}$ (exactly) in vacuum. You live 1430 miles from a NIST time server. Assuming that there is no net lag, how long in milliseconds does it take for a signal to travel from the NIST time server to your computer?

1.45. a. How many $cm^3$ are equivalent to a $m^3$? b. To convert a millimole to a megamole you would multiply by what conversion factor? c. How many micrograms are equivalent to one kilogram?

1.46. a. Convert 32.7 mph to m $s^{-1}$. b. Convert $3.5 \times 10^{-3}$ pounds to milligrams. c. Convert -35.2°F to Celsius and Kelvin. d. Convert 15 gallons to liters. e. Convert $8.1 \times 10^7$ J to Btu.

1.47. There are exactly 2.54 cm $inch^{-1}$, exactly 12 inch $foot^{-1}$, and exactly 5280 feet $mi^{-1}$. Derive the conversion factor for miles to kilometers. How many significant figures does that conversion factor have?

1.48. There is one temperature that is exactly the same on the Celsius and Fahrenheit scales. What is it?

1.49. There is exactly one temperature that has equal absolute value but opposite signs on the Celsius and Fahrenheit scales. What is it?

1.50. Calculate the temperature for which the Fahrenheit value is exactly three times the Celsius value.

1.51. The density of liquid helium is 7.62 lb $ft^{-3}$. Convert this value to the density unit g $cm^{-3}$. Calculate a single conversion factor to make the units change.

1.52. Professor Carla Slebodnick's favorite unit of length is the angstrom (Å). 1 Å = $10^{-8}$ cm. She is a crystallographer and crystallographers have traditionally favored angstrom units. a. How many Å are equivalent to 1 mile. b. How many $pm^3$ = 1 $Å^3$?

1.53. Surface tension is the force that makes liquid particles contract into little spheres. It causes water to form droplets. The surface tension of acetone is 23.7 dyne $cm^{-1}$. Convert this old style unit to its modern SI equivalent.

1.54. The planet Uranus has a density of 1.2 g $cm^{-3}$ and a diameter of 26,000 miles. What is the mass of Uranus in grams?

1.55. The solar system is composed of 91% hydrogen atoms and 8.9% helium atoms (not leaving much room for everything else). Use atomic weights to calculate the percentage by mass of everything else. (Assume everything else is carbon.) Strictly, considering significant figures, one cannot do this problem. Discuss.

1.56. The SI unit of power is the watt. Power is the rate of production or consumption of energy: 1 watt = 1 J $s^{-1}$ (joule per second). Calculate the total energy output of a 75 watt light bulb that operates for 12 hours. State the answer in joules, kilojoules, and calories.

1.57. If 1 horsepower is equivalent to 42.418 Btu $min^{-1}$, how many watts are produced by a motor operating at 300 hp? Approximately how many 100 watt bulbs can be simultaneously lit by this motor?

## Significant Figures

1.58. a. What is number style? Hint: use the glossary. b. Does changing a number's style change its number of significant figures?

1.59. Which of the following are exact numbers: a. the number of hours in a day, b. the number of days in a year, c. the number of centimeters in an inch, d. the melting point of chromium, e. the atomic number of any element, f. the density of mercury, g. the atomic weight of tin.

1.60. State the number of significant figures in each of the following numbers: a. $1.27 \times 10^3$, b. $9.004 \times 10^{-6}$, c. $4.00023 \times 10^{-12}$.

1.61. State the number of significant figures in each of the following numbers: a. 12.6, b. 1380, c. 0.41944, d. 0.0047, e. 0.001009, f. 0.01050, g. $4.300 \times 10^{-5}$.

1.62. For each of the following numbers state the least significant digit: a. 817, b. 34.560, c. 34000, d. $9.2402 \times 10^{-9}$, e. 0.00201, f. 900, g. 800.

1.63. Perform the following operations and round the answer to the proper number of significant figures: a. $0.345 \times 9.761$, b. $5 \div 1.0078$, c. $(56.00 + 4.567) \times 8.9$, d. $10.67 \times 19.4529 \times 0.130$.

1.64. Round the value 23.98571 to 6, 5, 4, and 3 significant figures. Round the value 0.037619385 to 7, 6, 5, 4, and 3 significant figures.

1.65. Round the value $1.2912 \times 10^{17}$ to 4, 3, and 2 significant figures.

## Measurement and Experimental Uncertainty

1.66. Calculate the difference and the percentage difference between the following pairs of numbers: a. 123 and 125, b. $1.03 \times 10^9$ and $9.88 \times 10^8$, c. the atomic weight of copper and 63.5, d. the atomic number of magnesium and 12. e. the mass of a proton and the mass of a neutron. Hint: percentage difference is defined in the glossary.

1.67. Use five significant figure values of the masses of the isotopes (Appendix D) to calculate the percentage mass difference between: a. $^1H$ and $^2H$, and b. $^{207}Pb$ and $^{208}Pb$. c. Comment on the relative sizes of the two percentage differences.

1.68. An automobile has a digital speedometer. The driver discovers that switching from 65 mph (mi $hr^{-1}$) in US customary units to the metric display mode gives a corresponding speed of 104 kph (km $hr^{-1}$). Calculate the percentage difference between the US and metric modes.

1.69. Calculate the mean and standard deviation of the following set of replicated experimental values: 3.213, 3.225, 3.189, 3.227, and 3.192. Normally, standard deviations are reported to one significant figure. Convert your standard deviation to one significant figure and discuss how many significant figures should be stated when you report the mean. Do this calculation using a calculator.

1.70. Repeat the previous problem using the following data set: 18.23, 17.91, 16.82, 17.85, 15.99.

1.71. a. What percentage of a normally distributed data set lies ±1 standard deviation (sigma) from the mean value of the data set? b. What percentage lies ±2 sigma from the mean? c. What percentage lies ±3 sigma from the mean?

1.72. 127 students took an exam consisting of 50 multiple choice questions. The average (mean) number of correct answers was 33/50. The standard deviation of the mean was ±5 questions. Assuming that the student's scores are normally distributed, a. approximately how many students scored 43 points or above? b. approximately how many students scored 23 or below?

1.73. The charge on an electron is one of the fundamental constants of nature. It is $1.60217733 \pm 0.000000049 \times 10^{-19}$ C. Is the chance greater or less than 5 in 100 that the correct value of the electron's charge is $1.60217748 \times 10^{-19}$ C.

1.74. Standards organizations often report the experimental uncertainties of fundamental constants to two significant figures. The rest of us use one significant figure to report experimental uncertainties. Discuss.

## Graphical Problem

1.75. A metal candle holder weighed 48.3 g. A candle weighing 32.2 g was put in the candle holder and the candle in its holder put on a single pan balance. The initial mass of the candle and holder was 80.5 grams. The mass was noted approximately every 30 minutes as shown in the following table:

| Data point | Time (minutes) | Total mass (g) |
|---|---|---|
| 1 | 30.2 | 76.9 |
| 2 | 60.5 | 73.3 |
| 3 | 90.3 | 70.2 |
| 4 | 121.1 | 69.5 |
| 5 | 150.8 | 62.8 |
| 6 | 178.2 | 59.3 |

Make a graph of the data. Discard one data point. a. Which one and why? b. Calculate the rate of loss of mass in the unit g $min^{-1}$. c. For how many minutes after reading data point six do you predict the candle will continue to burn? d. Why does a candle lose mass when it burns?

## Problem Solving

1.76. Electroplating involves creating a thin layer of metal covering to protect an object that might otherwise corrode or degrade in its environment. A gold electroplate of 2.47 micrometer thickness is uniformly applied to a base metal object that measures 12.0 inches × 3.50 inches × 0.125 inches. Calculate the mass of the gold plate. [❁kws +gold +density]

1.77. Convert the temperature 72°C to °F. Has the number of significant figures increased because of this conversion? Discuss.

1.78. Convert the temperature 72°C to kelvins. Has the number of significant figures increased because of this conversion? Discuss.

1.79. The projected world production of scandium for the year 2000 is 37 kg. Scandium has a density of 2.99 g $cm^{-3}$. Will 37 kg of scandium fit inside a briefcase measuring 3 inches × 12 inches × 24 inches?

1.80. At 25°C, a container filled with water (density 0.998 g $mL^{-1}$) weighs 88.4 grams. The same container weighs 179.8 g when filled with mercury (density 13.54 g $mL^{-1}$). Calculate the mass of the container.

1.81. A substance is a mixture that contains A, B, and C. The percentage by mass of A in the sample is the sum of twice the percentage by mass of B plus the percentage by mass of C in the substance. The percentage by mass of A in the sample is four times the percentage by mass of C in the substance. Calculate the percentage composition of the substance.

1.82. A production chemist begins the manufacture of a cleaning liquid by pumping 2300 gallons of deionized water into a 3000 gallon tank. Using the big pump, 22.5 minutes is required to deliver the needed 2300 gallons. Using the small pump the 2300 gallons is delivered in 57.3 minutes. If the chemist uses both pumps simultaneously how long does it take to deliver the requisite 2300 gallons?

1.83. Comment on the phrases "22.5 minutes is" and "2300 gallons is" used in the previous problem. Are they grammatically correct?

1.84. At 20°C water's density is 0.9983 g $mL^{-1}$; ethyl alcohol's density is 0.7893 g $mL^{-1}$. Calculate the total volume of 50.00 g of water and 50.00 g of ethyl alcohol at 20°C. The two liquids are mixed to make a solution of ethyl alcohol and water. The density of the mixture is 0.9139 g $mL^{-1}$; calculate its volume. Discuss the "*law of conservation of volume.*"

1.85. The specific heat of liquid benzene is 0.41 cal $g^{-1}$ $°C^{-1}$. This value is interpreted to mean that it takes 0.41 calories to raise the temperature of 1 gram of benzene by 1 degree Celsius. How many Btu does it take to heat 240 pounds of benzene from 15 °F to 35 °F? The needed relationship is heat (q) = m × specific heat × ΔT.

1.86. To heat water requires approximately 1.0 cal $g^{-1}$ $°C^{-1}$. How long will it take the energy from a 325 watt electrical heater to boil 378 g of water in a thermos bottle? Assume that no heat is lost from the system to its environment and that the water's initial temperature is 23°C.

1.87. An example calculation in Chapter 1 showed that a one-ton automobile traveling at 50 mph has a kinetic energy of 227 kJ. The combustion energy content of gasoline is about 40 kJ $mL^{-1}$. Assuming an efficiency of 3% (i.e. only 3% of the fuel's energy is converted to kinetic energy of motion with the other 97% being lost to the radiator, friction, etc.), how many times could a motorist accelerate from zero to 50 miles an hour on one gallon of gas? Work this problem using two significant figures and round the final answer to one significant figure.

1.88. The density of the planet Mercury is 5.43 g $cm^{-3}$. Mercury is believed to have an iron core with a radius that is 75% of the planet's radius; assume that it does. The density of iron is 7.87 g $cm^{-3}$. Calculate the density of the rock that composes that portion of planet Mercury that is not iron.

1.89. In Chapter 1 it is stated "a mole of sand grains is a much greater quantity of sand than can be found on all the earth's beaches." Use reasonable estimates to evaluate this contention.

# Chapter 2

# Nuclides, Molecules, and Ions

This chapter is divided into two parts. The first part of the chapter deals with **nuclear reactions**—the reactions of single atoms—and with the stability of nuclides. In Chapter 1 we saw that some elements are radioactive. Radioactivity, the spontaneous decomposition or **decay** of an atom's nucleus, involves nuclear reactions, and any radioactive decay process can be represented by a **nuclear equation**. If the reactants and the products of a nuclear reaction are properly accounted for, then we say the nuclear equation is balanced. The stability of nuclides depends on the numbers of protons and neutrons they contain, and we will relate nuclide stability to the isotopes of the elements listed in Appendix D. Radioactive nuclides are called **radionuclides**. The less stable is a radionuclide the faster is its rate of decay. We'll examine decay rates and see how to do rate calculations using half-life. The **half-life** of a radionuclide is the time period during which half of the nuclide will decay. Our discussion of nuclides will conclude with a brief consideration of nuclear energy and the question of how many elements exist.

In the second part of the chapter we start joining atoms together to make molecules and introduce the topic of chemical bonding. To conclude the chapter we'll make ions by adding electrons to (or taking them away from) atoms and molecules. Adding an electron to an atom or molecule creates a **negative ion**; removing an electron creates a **positive ion**.

## Section 2.1 Nuclear Reactions and Radioactivity

### Radioactivity

In 1895, Wilhelm Roentgen (1845-1923) accidentally discovered that electrical **gas discharge tubes** (similar to modern fluorescent lights) emit penetrating x-rays (x = the unknown ray) that would make a phosphor glow. A **phosphor** is a material that emits light when struck by x-rays or a beam of electrons. If you're looking right now at a computer monitor, then what you see is the phosphors coated inside the picture tube glowing under the continuous impact of an electron beam. It didn't take Roentgen long to figure out that these new x-rays could create images of bones inside the human body. X-ray medical imaging is widely used today; it is shown schematically in the diagram on the right. Almost everyone has "had an x-ray."

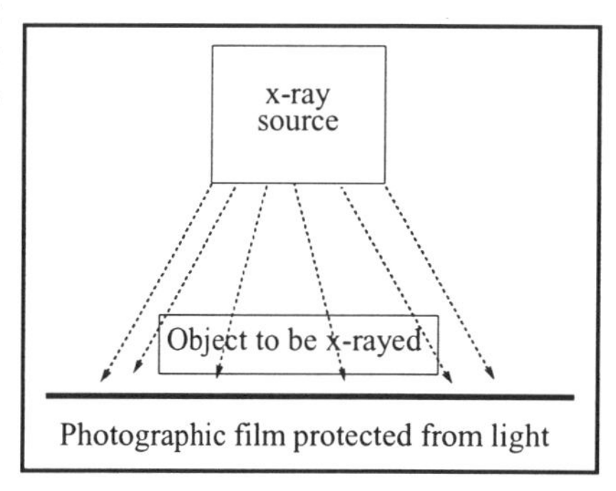


Figure 2.1 Making an"x-ray".

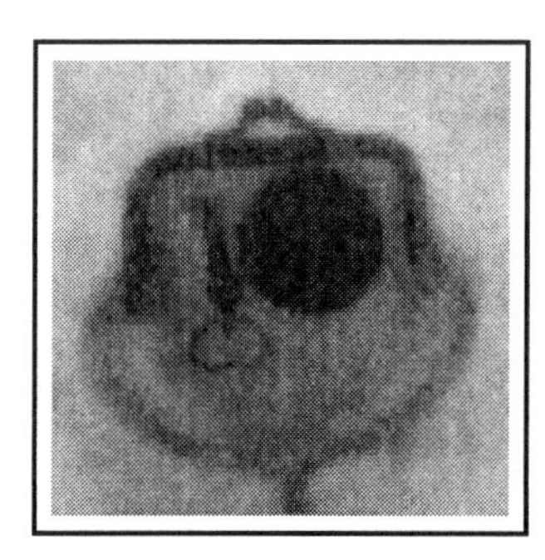

Figure 2.2 shows a "radiograph" of a coin purse taken by Marie Sklodowska Curie around 1903. A tiny sample of radium irradiated the purse, behind which was placed photographic film protected from light by black paper. The image was produced when the film was developed. The image shown is a negative. X-rays and radioactivity pass more easily through low density material than through high, creating a shadow picture. Engineers today routinely use radiography. For example, critical welds on a 10-story-high oil rig can be examined in precisely the manner described and illustrated. $^{192}$In and $^{60}$Co are two radionuclides used by engineers for weld radiography.

Figure 2.2 An early radiograph [❁kws +engineering +radiography +weld].

X-rays are electromagnetic radiation similar to, but much more energetic than, visible light. One of the earliest scientific consequences of x-rays was that they led to the discovery of radioactivity. Radioactivity was discovered in 1896 by the French physicist Henri Becquerel (1852-1908). He had been trying to produce x-rays by exposing various substances to sunlight. To test for x-rays he used photographic film covered with black paper. One of the effects of the newly discovered x-rays was that they penetrated black paper to fog (expose) the apparently protected photographic film beneath. When uranium compounds in sunlight caused fogging Becquerel thought he had achieved his objective. But he soon discovered that uranium compounds stored in a dark drawer with black paper covered film also caused fogging; sunlight was an unnecessary factor. He deduced that uranium compounds spontaneously emit rays and thereby discovered uranium's radioactivity. Shortly there followed the discovery of comparable radioactivity in thorium by Marie Curie (1867-1934) in Paris and G. C. Schmidt in Germany. Radioactivity was crucial in the early 20th century to the development of chemistry and remains today a phenomenon with many important and practical applications.

In the late 19th and early 20th centuries, many ingenious experiments had to be performed before subatomic particles were characterized. Their key properties are shown in Table 2.1. For most of chemistry we need to take account only of three subatomic particles: the electron, proton, and neutron.

Table 2.1 **Properties of Subatomic Particles**

| Particle (symbol) | Rest Mass in u ( ) uncertainty | Unit Charge |
|---|---|---|
| Electron ($e^-$) or $\beta^-$ particle | $5.485779903(13) \times 10^{-4}$ | -1 |
| Positron ($e^+$) or $\beta^+$ particle | $5.485779903(13) \times 10^{-4}$ | +1 |
| Proton (p or $H^+$) | 1.007276470(12) | +1 |
| Neutron (n) | 1.008664904(14) | 0 |
| Deuteron (d) | 2.013553214(24) | +1 |
| Alpha particle, ($\alpha$) or $He^{2+}$ | 4.001506170(50) | +2 |
| Photon ($\gamma$) | 0 | 0 |

## Alpha, Beta, and Gamma Radiation

In 1898 the young New Zealander Ernest Rutherford (1871-1937) was studying the formation of ions (electrically charged particles) in the air by x-rays. Because radioactivity also ionizes the air, Rutherford began to study it as well. In the summer of 1898 Rutherford moved to Montreal, where he met Frederick Soddy. Together, Rutherford and Soddy established that radioactivity involves three characteristic types of emissions: alpha, beta, and gamma rays ($\alpha$, $\beta$, $\gamma$). They also proved that radioactivity involves one element changing into another and so fulfilled the alchemist's 1,500-year-old dream of the transmutation of elements. Rutherford, a physicist, was awarded the 1908 Nobel prize for *chemistry* for this work. Soddy got the same prize in 1921—the year that Einstein won the physics prize.

The radiodecay of a parent nuclide produces a **daughter nuclide** that can be represented by a balanced **nuclear equation**. The general equation is:

Parent nuclide → Daughter nuclide (with rays and particles depending on the exact process)

Whether or not there are electrons around the nucleus doesn't really matter in nuclear processes. Radioactivity is a nuclear reaction, not a chemical one. Because nuclear reactions involve a limited number of reactants and products, learning how to balance nuclear reaction equations is a not too difficult and is a good introduction to the considerably more complicated problem of how to balance chemical reaction equations.

To balance a nuclear equation you use two rules: On both sides of a nuclear equation, 1. The sums of the nucleon numbers are equal, and 2. The sums of the charges are equal.

## Alpha Decay

An alpha particle is the nucleus of a $^{4}_{2}He$ atom; it has a charge of 2+ and a nucleon number of 4. As a consequence of **alpha decay**, the parent's atomic number is reduced by two and its nucleon number is reduced by four. For example, in his later experiments that led to the nuclear theory of the atom, Rutherford used radium-226 as the alpha particle source:

$$\underset{\text{Parent}}{^{226}_{88}Ra} \rightarrow \underset{\text{alpha}}{^{4}_{2}He} + \underset{\text{Daughter}}{^{222}_{86}Rn}$$

According to the balancing rules, the sum of the nucleon numbers (226 = 4 + 222) and also the sum of the charges (88 = 2 + 86) are equal on both sides of the equation.

## Beta Decay

A **beta particle** is simply an electron created during a radioactive decay. We often call it a beta particle because that was the name given to it when it was first discovered. When a beta particle, symbolized by $\beta^-$ or $^{0}_{-1}e$, is emitted, a daughter nuclide is formed with the same nucleon number as the parent but with its atomic number increased by one. For example:

$$^{14}_{6}C \rightarrow {}^{0}_{-1}e + {}^{14}_{7}N$$

The sums of the nucleon numbers (14 = 14 + 0) and also the sums of the charges (6 = -1 + 7) on both sides of the equation are equal. The daughter nuclide $^{14}N$ has one more proton and one less neutron than the parent $^{14}C$. The *net* effect of beta decay of a nuclide is the loss of one neutron and the gain of one proton. It can be represented by the nuclear equation:

$$^{1}_{0}n \rightarrow {}^{1}_{+1}p + {}^{0}_{-1}e$$

The electron does not exist in the radioactive nucleus, rather it is created at the instant it is emitted. Once emitted, however, the beta particle is simply an electron—though usually an electron with a lot of kinetic energy. Beta particle emission creates a daughter nuclide with a decreased neutron to proton (n/p) ratio. That ratio is important in studying nuclear stability—as we will discuss shortly.

Proton = Hydrogen Ion

In the equation above, the symbol $^{1}_{+1}p$ represents a proton. A proton is exactly the same thing as a hydrogen atom that has lost an electron by the process of ionization. The equation below shows a hydrogen atom undergoing ionization to produce a hydrogen ion.

$$H + \text{energy} \rightarrow H^+ + e^-$$

So when chemists talk about a "proton" or a "hydrogen ion" they are talking about the same thing.

The Thorium Natural Decay Series.

Sequences of α and β decays from U and Th isotopes are the source of natural radioactive elements. Here's the series that begins at $^{232}Th$ and ends at $^{208}Pb$. Each step is either an α or β decay.

$$^{232}Th \xrightarrow{\alpha} {}^{228}Ra \xrightarrow{\beta} {}^{228}Ac \xrightarrow{\beta} {}^{228}Th \xrightarrow{\alpha} {}^{224}Ra \xrightarrow{\alpha} {}^{220}Rn \xrightarrow{\alpha} {}^{216}Po \xrightarrow{\alpha} {}^{212}Pb \xrightarrow{\beta} {}^{212}Bi \xrightarrow{\beta} {}^{212}Po \xrightarrow{\alpha} {}^{208}Pb$$

## Historic Experiments to Characterize Subatomic Particles

The British physicist, J. J. Thomson (1856-1940) is credited with the discovery of the long expected electron. His method was to study the effects of electric and magnetic fields on electron beams. Electron beams are known as **cathode rays**, because when a large voltage is applied to a vacuum tube, electrons flow from the negative end (the cathode) to the positive end (the anode).

Charged particles are deflected if they pass through electric or magnetic fields oriented perpendicular to the particle's direction of motion. If a particle's mass and charge are known, the deflection of its path can be calculated for any field strength. When Thomson began his studies, cathode rays were known to consist of negatively charged particles, but their mass and charge was unknown. Thomson showed that, when a charged particle passes through crossed electric and magnetic fields, the radius of curvature of its path depends only on the ratio of its charge to its mass. Therefore, his experiments permitted him to determine that ratio (e/m); it is $1.76 \times 10^{11}$ C $kg^{-1}$. Amusingly, J. J. Thomson won the Nobel prize for proving the electron is a particle; some years later his son G. P. Thomson won the Nobel prize for proving the electron is a wave. More about particles and waves in Chapter 5.

Shown on the right is the rear of the picture tube of a computer monitor. Observe the copper coils that make a yoke around the rear of the tube. Fields from the coils control the path of the electron beam. The beam forms the picture when it selectively strikes the phosphors inside the front of the tube. In effect, every tv or video monitor tube recreates the Thomson experiment

In 1909, the oil drop experiments of Robert Millikan (1868-1953) revealed the electron's charge. Millikan created small oil droplets (using a perfume atomizer), and determined their masses by studying the rate at which they fell in air. A radioactive substance placed in the apparatus ionized the air, and some of the oil drops became charged. By studying the effects of electric fields on the motions of the charged oil droplets, Millikan measured the charge of many different drops. All the measured charges were multiples of a single value, namely $1.592 \times 10^{-19}$ C, that he calculated must be the charge of a single electron. The modern value of the electron's charge is $1.602\ 177\ 33 \times 10^{-19}$ C, close to Millikan's first value. The mass of the electron could be calculated from Millikan's value of the electron's charge and Thomson's value of its charge/mass ratio.

The discovery of the atomic nucleus occurred around 1910-11 at Cambridge in England. The key experiments were done by Ernest Rutherford and his collaborators. They fired a beam of alpha particles at thin metal foils (the gold foil experiment). Most of the alpha particles passed directly through the foil; however, some were dramatically scattered: a few were even back-scattered (deflected back toward their source). Rutherford realized such scattering must be caused by a massive, highly charged region within the atoms and by 1911 had devised a model in which all of the positive charge, and most of the mass, of the atom is in an exceedingly small nucleus. Confirmation came when his alpha particle scattering calculations reproduced the experimental results.

After Rutherford's discovery of the positively charged atomic nucleus it did not take long for scientists to suspect that the nuclei of atoms also contained neutral particles. In 1932, James Chadwick (1891-1974) examined the recent work of some French and German scientists and concluded that their incorrectly interpreted experimental results were actually producing free neutrons. One such nuclear reaction occurred when $^{11}B$ was bombarded with alpha particles. In this manner, Chadwick discovered the neutron.

$$^{4}_{2}He + ^{11}_{5}B \rightarrow ^{14}_{7}N + ^{1}_{0}n$$

Free neutrons have a half life of about 12 minutes, so they don't stay around too long before decaying into protons and electrons. But 12 minutes is sufficiently long that they can play the key role of sustaining a nuclear chain reaction in reactors and weapons.

| | Table 2.2 **A Very Short History of Radioactivity** |
|---|---|
| 1895 | The phenomenon of radioactivity accidentally discovered by Henri Becquerel in Paris. |
| 1900-1905 | $\alpha$, $\beta$, and $\gamma$ radiation identified by Ernest Rutherford and Frederick Soddy in Montreal. |
| 1903 | Transmutation of elements confirmed. William Ramsey and Frederick Soddy. |
| 1898-1913 | Radioactive elements found and placed in the periodic table. Marie Curie, et al. |
| 1921 | Marie Curie in America to raise funds for the Radium Institute. |
| 1934 | Ida Noddack proposes that neutron bombardment of nuclei causes "fragmentation." Thus is conceived the idea of nuclear fission. Her proposal does not impress the physicists. |
| 1935 | Synthetic elements made but not recognized (in the fishpond of the University of Rome). Enrico Fermi, et al. |
| 1938 | In Berlin, Germany, Otto Hahn reluctantly concludes that what he first thought was radium is in fact barium. He thereby obtains the first chemical evidence for nuclear fission. |
| 1939 | Lise Meitner, Hahn's now exiled collaborator, recognizes the nature of fission and gives it its name. A Jew, Meitner had recently left Berlin for neutral Sweden to escape Nazi racial laws. |
| 1940 | Controlled, induced nuclear fission is achieved under Stagg Field, University of Chicago. Enrico Fermi et al. "The Italian navigator has landed." |
| 1941-1945 | Nuclear weapons using fission. The Manhattan project. Robert Oppenheimer et al. Fission products become a source of radionuclides. Much new chemistry. |
| 1950 | Nuclear fusion weapon (H-Bomb). Edwin Teller and Stanislaw Ulam. |
| 1945-1963 | Period of unconfined atmospheric nuclear weapons testing. |
| 1955-1960 | Nuclear power developed for electric generating plants and naval propulsion. |
| 1980's | Nuclear accidents at Three Mile Island and Chernobyl raise nuclear safety questions. The US backs off building nuclear power plants. |
| 1980-1999 | Radioisotopes increasingly used in medicine and biochemistry. Many applications are found in medical treatments and as tracers. Nuclear waste becomes an issue. |
| 1997 | Element 109 is named in honor of Lise Meitner. |

## Gamma Decay

In many radioactive decay processes, gamma rays ($\gamma$) are concurrently emitted. Like x-rays, **gamma rays** are electromagnetic radiation with neither rest mass nor charge; they are even more energetic than x-rays. Gamma ray emission occurs when a parent nuclide with a large excess of energy decays to give a daughter nuclide, simultaneously liberating the excess energy. $\gamma$ emission changes neither the nucleon number nor the atomic number of the nuclide. For example:

$$^{107}_{49}In \rightarrow {}^{107}_{49}In + {}^{0}_{0}\gamma$$

Gamma emission frequently accompanies other decay processes, such as alpha emission:

$$^{226}_{88}Ra \rightarrow {}^{4}_{2}He + {}^{222}_{86}Rn + {}^{0}_{0}\gamma$$

During the 20th century, the phenomenon of radioactivity had a profound impact on the course of human history as summarized in Table 2.2. The development of nuclear weapons and nuclear power plants flowed inevitably from the initial discovery of radioactivity. For chemists, it is necessary to understand radioactivity and its consequences in order to understand the periodic table.

During almost all decay processes the ratio of neutrons to protons (**n/p ratio**) changes from the parent to the daughter nuclide. In the following section we'll see how the n/p ratio is related to the type of decay.

## Positron Emission and Electron Capture

Two closely related types of radiodecay are positron emission ($^{0}_{+1}e$ or $\beta^+$) and electron capture (EC). Both of these decay processes yield the identical daughter nuclide from a specified parent. A **positron** is an antiparticle with identical mass to an electron but a positive charge. **Positron emission** occurs when a positron is formed by a radioactive decay event. An example of a nuclide that decays by positron emission is $^{11}C$:

$$^{11}_{6}C \rightarrow {}^{0}_{+1}e + {}^{11}_{5}B$$

Here's the equation for positron emission decay of the proton-rich, highly radioactive isotope neodymium-136 for which 6% of the decays are by positron emission and 94% by electron capture:

$$^{136}_{60}Nd \rightarrow ^{136}_{59}Pr + ^{0}_{+1}e$$

Because a positively charged particle is emitted, the charge of the resulting daughter nucleus is reduced by one. The nucleon number, however, is unchanged by $\beta^+$ decay. The *net* effect of the change of nucleons involved in positron decay is

$$^{1}_{1}p \rightarrow ^{0}_{+1}e + ^{1}_{0}n$$

Positron emission increases the number of neutrons and decreases the number of protons. The daughter nuclide has a increased neutron to proton (n/p) ratio. In modern medicine the technique known as PET (positron emission tomography) is an important imaging method that relies on the gamma rays formed by particle - antiparticle annihilation event between a positron and an ordinary electron:

$$^{0}_{+1}e + ^{0}_{-1}e \rightarrow 2\ ^{0}_{0}\gamma$$

**Electron capture (EC)** is a process in which a nucleus decays by capturing one of the atom's electrons. About one decay out of every fifteen of neodymium-136 occurs via the EC mode:

$$^{136}_{60}Nd + ^{0}_{-1}e \rightarrow ^{136}_{59}Pr$$

In this case the *net* change in nucleon composition is represented by the nuclear reaction equation

$$^{1}_{+1}p + ^{0}_{-1}e \rightarrow ^{1}_{0}n$$

Electron capture, like positron emission, results in an increase in the n/p ratio of the daughter nuclide. There are many nuclides that decay by a combination of positron emission ($\beta^+$) and EC. Rhodium-99, for another example, decays with a half-life of 15 days: 2.6% via positron emission and 97.4% via EC.

## Fission

**Spontaneous fission** is a rare decay mode in which a heavy nucleus breaks up into two roughly equal fragments and several free neutrons. It occurs only with the actinide elements. **Induced fission** occurs when neutron bombardment of nuclei splits them. Of course, it is the deliberately induced fission of nuclides that is the basis for nuclear energy. $^{238}U$ is a nuclide that dominantly decays by alpha emission. However, it undergoes a small amount of spontaneous fission. Because $^{238}U$ breaks up in many different ways, an enormous number of possible fission reactions can be written. Here's one spontaneous fission equation:

$$^{238}_{92}U \rightarrow ^{130}_{50}Sn + ^{104}_{42}Mo + 4\ ^{1}_{0}n$$

On the right are two **fission products** (radioactive waste) along with four free neutrons. The two fission products are short-lived radioactive beta emitters. Their decay yields other, more stable, radionuclides. Collectively, fission products tend to be long-lived and dangerous. Their safe disposal is a matter for national concern. Nuclear power plants and naval power reactors operate because of induced nuclear fission reactions.

An example of one of many possible induced nuclear fission is the reaction that occurs when a slow-moving neutron strikes a $^{235}U$ atom.

$$^{235}_{92}U + ^{1}_{0}n \rightarrow ^{131}_{56}Ba + ^{101}_{36}Kr + 4\ ^{1}_{0}n$$

Because more neutrons are produced than are consumed in the above reaction, there exists the possibility of a reaction chain. In a nuclear power reactor the chain is carefully controlled. In a nuclear weapon the chain grows uncontrollably. Nuclear decay processes are summarized in Table 2.3.

| Table 2.3 **Summary of Nuclear Decay Processes** | | |
|---|---|---|
| Type | The Process | Example |
| Alpha emission, $\alpha$ | A nuclide emits an alpha particle, $^{4}_{2}He$ | $^{204}_{85}At \rightarrow ^{200}_{83}Bi + ^{4}_{2}He$ |
| Beta emission, $\beta^-$ | A nuclide emits a beta particle, $\beta^-$ or $e^-$ | $^{115}_{46}Pd \rightarrow ^{115}_{47}Ag + ^{0}_{-1}e$ |
| Gamma decay, $\gamma$ | A nuclide emits a gamma particle, $\gamma$ | $^{107}_{49}In \rightarrow ^{107}_{49}In + ^{0}_{0}\gamma$ |
| Positron emission, $\beta^+$ | A nuclide emits a positron, $\beta^+$ or $e^+$ | $^{35}_{18}Ar \rightarrow ^{35}_{17}Cl + ^{0}_{+1}e$ |
| Electron capture (EC) | A nuclide's nucleus captures an electron, $e^-$ | $^{72}_{36}Kr + ^{0}_{-1}e \rightarrow ^{72}_{35}Br$ |
| Spontaneous fission (SF) | Spontaneous nuclide break up | $^{255}_{104}Rf \rightarrow ^{112}_{50}Sn + ^{137}_{54}I + 6\ ^{1}_{0}n$ |
| Induced fission | Neutron-caused nuclide break up | $^{235}_{92}U + ^{1}_{0}n \rightarrow ^{104}_{44}Ru + ^{129}_{48}Cd + 3\ ^{1}_{0}n$ |

## The Stability of Nuclides

There are three factors that control whether or not a given nuclide will be stable:

First, its n/p ratio. That is, its number of neutrons divided by its number of protons. Stable nuclides occur with specific values of this ratio.

Second, whether its neutron count and proton count are odd or even. Nuclides with even numbers of protons and neutrons tend to be more stable than the others.

Third, its total number of nucleons. All nuclides with more than 209 nucleons are unstable.

A good way to represent nuclear stability is to make a proton count - neutron count map called a **nuclide chart** as shown in Figures 2.3a and 2.3b. Summary data for the n/p ratios of stable nuclides and their odd-even proton and neutron counts are shown in Table 2.3.

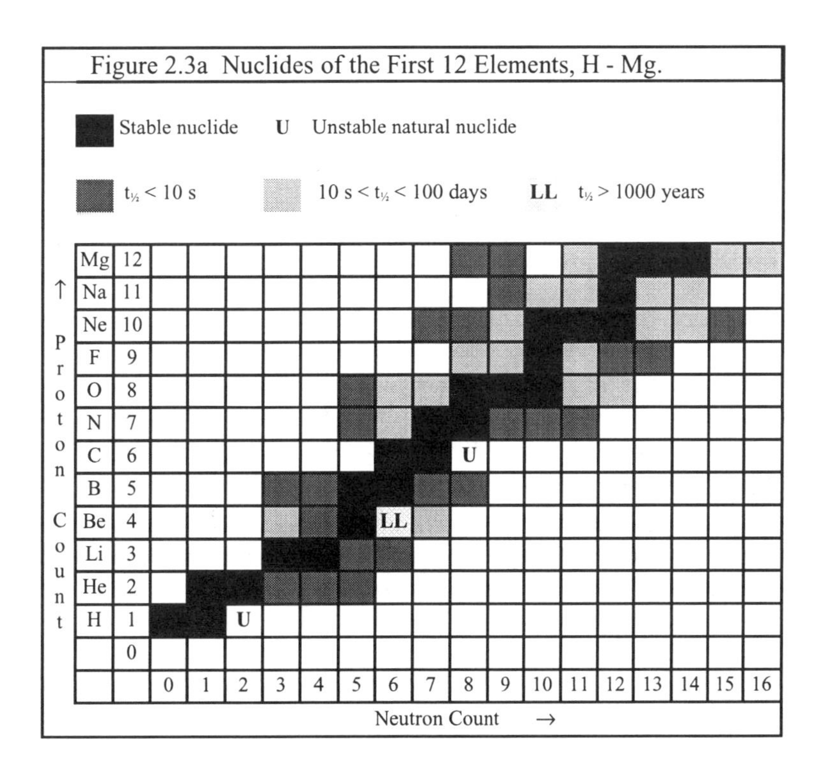


Figure 2.3a Nuclides of the First 12 Elements, H - Mg.

Figure 2.3a shows the nuclide chart of the first twelve elements. Each isotope of the first twelve elements is coded to show whether or not it is stable, and if it is radioactive to indicate its approximate half life. For example, Beryllium (element 4) has one stable isotope, $^{9}Be$, and four radioactive isotopes (**radioisotopes**): $^{7}Be$, $^{8}Be$, $^{10}Be$, and $^{11}Be$. $^{10}Be$ is rather stable with a half life of $2.5 \times 10^{6}$ years. $^{7}Be$, $^{8}Be$, and $^{11}Be$ have half lives respectively of 53.4 days, $2 \times 10^{-16}$ s, and 13.6 s. When comparing two radionuclides we say that the one with the longer half life is more stable. Nuclides with half-lives in the range of millions of years and longer are **pseudostable**.

Radionuclides decay by the various modes we've discussed yielding daughter products that are "closer" to being stable. The band of black squares on a nuclide chart is the **belt of stability**. Radionuclides decay in the manner that tends to bring their daughters toward, and eventually into, the belt of stability. [❁kws "Chart of the Nuclides"].

Figure 2.3b shows the chart of the nuclides extended across the periodic table. On both nuclide charts, the isotopes of any given element form a horizontal set of squares. Note that we eventually "run out" of stable nuclides (the black squares end). There comes a point (beyond bismuth, $^{209}Bi$, the heaviest stable nuclide known) where all nuclides are unstable.

The general shape of the belt of stability is strongly related to the n/p ratio of the nuclides. The ratio begins at 1.0, has reached 1.25 halfway through the list of elements, and has reached 1.6 at the end of the list. Nature requires heavy nuclides to have *relatively* more neutrons to make them stable. You can see how the n/p varies if you note how the belt moves further and further from the 1:1 diagonal line ratio. Radionuclides that lie above and to the left of the belt are said to be **proton rich**. Radionuclides that lie below and to the right of the belt are said to be **neutron rich**.

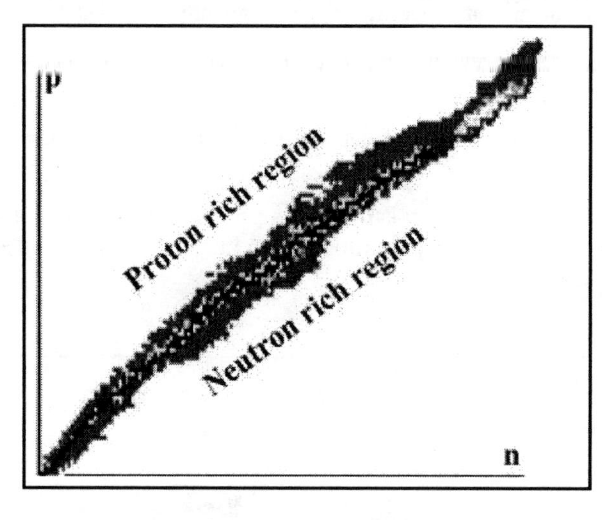


Figure 2.3b Chart of the Nuclides

When a radionuclide decays it "chooses" a decay mode that "improves" its n/p ratio. After decay the daughter nuclide has "moved" toward the belt of stability. Proton-rich nuclides will lose positive charge—which they can do by positron emission or electron capture. Neutron rich nuclides will gain positive charge—which they usually do by emitting negatively charged beta particles. Radionuclide decay produces a diagonal movement toward the belt of stability: $\beta^{-}$ emission moves a parent nuclide one step up and one step left—diagonally upwards to the left. Both $\beta^{+}$ emission and EC move a parent nuclide one step down and one step right—diagonally downwards to the right.

For reasons involving the fundamental physics of the nucleus, nuclei with an even number of protons combined with an even number of neutrons are most likely to be stable. The obvious evidence for this fact is the distribution of the stable nuclides shown in Table 2.4 (see also Appendix D).

| Table 2.4 **All Stable Nuclides** | | | |
|---|---|---|---|
| Proton Count | Neutron Count | Number of known nuclides of this type | Examples |
| even | even | 149 | $^{12}_{6}C$ $^{16}_{8}O$ $^{144}_{62}Sm$ |
| even | odd | 59 | $^{13}_{6}C$ $^{47}_{22}Ti$ $^{111}_{48}Cd$ |
| odd | even | 52 | $^{19}_{9}F$ $^{23}_{11}Na$ $^{203}_{81}Tl$ |
| odd | odd | 4 | $^{2}_{1}H$ $^{6}_{3}Li$ $^{189}_{73}Ta$ |

There are 149 + 59 + 52 + 4 = 264 total stable nuclides known. That fact means (given that there are 83 elements with stable nuclides, that the elements average about 3 stable isotopes per element. However, some elements (such as P and F, etc.) exist naturally as just one stable nuclide (they are **mononuclidic**), while some (such as tin) have as many as 10 stable isotopes.

**n/p Ratio and Nuclear Stability Summarized**

* For light elements 1.0 - 1.1 ratio
* For elements in the middle of the periodic table 1.2-1.4 ratio
* For heavy elements 1.4-1.6 ratio
* heavier nuclei require relatively more neutrons for stability
* unstable neutron-rich radionuclides decay by $\beta^-$ emission
* unstable proton rich nuclides decay by EC or positron ($\beta^+$) emission
* radionuclides formed as fission products typically are neutron rich

## Trans-Uranic Elements

A **trans-uranic element** is an element beyond uranium in the periodic table; none is stable, all of their isotopes are radioactive. Trans-uranic elements tend to become less and less stable (shorter and shorter half lives) the further we go beyond uranium. Neptunium, plutonium, and americium can be made by the neutron irradiation or bombardment of uranium isotopes. The elements beyond americium are made either by ion or neutron bombardment experiments. **Ion bombardment reactions** are ones in which highly accelerated, light nuclei ions strike atomic targets and form heavy nuclides. Table 2.5 shows the ion bombardment reactions used to form trans-uranic elements. For example, the formation reaction of nobelium's most stable isotope involves the bombardment of a curium isotope by a highly accelerated carbon-12 nucleus. The process can be represented by the ion bombardment equation:

$$^{246}_{96}Cm \quad + \quad ^{12}_{6}C \quad \rightarrow \quad ^{254}_{102}No \quad + \quad 4\ ^{1}_{0}n$$

In Darmstadt, Germany in 1994, the still unnamed element 111 was produced by the $^{64}Ni$ ion bombardment of $^{209}_{83}Bi$ according to the following bombardment equation:

$$^{209}_{83}Bi \quad + \quad ^{64}_{38}Ni \quad \rightarrow \quad ^{271}_{111}Uun \quad + \quad 2\ ^{1}_{0}n$$

The formation of element 118 (Uuo) in 1999 occurred via the krypton ion bombardment of a lead target. The nuclear equation for the formation reaction may be written as:

$$^{86}_{36}Kr + ^{208}_{82}Pb \rightarrow ^{293}_{118}Uuo \quad + \quad ^{1}_{0}n$$

The formation of element 116 (Uuh) occurs when a Uuo-118 nuclide undergoes alpha particle decay. This reaction was first observed in 1999.

$$^{293}_{118}Uuo \quad \rightarrow \quad ^{289}_{116}Uuh + ^{4}_{2}He$$

Table 2.5 **Trans-Uranic Elements**

| Atomic Number | Name | Symbol | First isotope | Date | Discovery method | Most Stable isotope | Half Life |
|---|---|---|---|---|---|---|---|
| 93 | Neptunium | Np | Np-239 | 1940 | Neutron irradiation of U | Np-237 | $2.14 \times 10^6$ yr |
| 94 | Plutonium | Pu | Pu-238 | 1941 | Neutron irradiation of U | Pu-244 | $80.0 \times 10^6$ yr |
| 95 | Americium | Am | Am-241 | 1945 | Neutron irradiation of Pu | Am-243 | 7370 yr |
| 96 | Curium | Cm | Cm-242 | 1944 | He ion bombardment of Pu | Cm-247 | $16 \times 10^6$ yr |
| 97 | Berkelium | Bk | Bk-243 | 1949 | He ion bombardment of Am | Bk-247 | 1380 yr |
| 98 | Californium | Cf | Cf-245 | 1950 | He ion bombardment of Cm | Cf-251 | 898 yr |
| 99 | Einsteinium | Es | Es-253 | 1952 | Formed in H-bomb test | Es-252 | 1.3 yr |
| 100 | Fermium | Fm | Fm-255 | 1952 | Formed in H-bomb test | Fm-257 | 101 days |
| 101 | Mendelevium | Md | Mv-256 | 1955 | He ion bombardment of Ei | Mv-258 | 55 days |
| 102 | Nobelium | No | No-254 | 1958 | C ion bombardment of Cm | No-259 | 58 min |
| 103 | Lawrencium | Lr | Lr-258 | 1961 | B ion bombardment of Cf | Lr-262 | 216 min |
| 104 | Rutherfordium | Rf | Rf-260 | 1964 | Ne ion bombardment of Pu | Rf-261 | 1.1 min |
| 105 | Dubnium | Db | Db-260 | 1967 | Ne ion bombardment of Am | Db-262 | 32 s |
| 106 | Seaborgium | Sg | Sg-263 | 1974 | O ion bombardment of Cf | Sg-263 | 0.9 s |
| 107 | Bohrium | Bh | Bh-258 | 1976 | Cr ion bombardment of Bi | Bh-262 | 102 ms |
| 108 | Hassium | Hs | Hs-267 | 1984 | Fe ion bombardment of Pb | Hs-265 | 1.8 ms |
| 109 | Meitnerium | Mt | Mt-266 | 1982 | Fe ion bombardment of Bi | Mt-266 | 3.4 ms |
| 110 | Unnamed | Uun | 110-269 | 1994 | Ni ion bombardment of Pb | 110-269 | < 1 ms |
| 111 | Unnamed | Uuu | 111-272 | 1994 | Ni ion bombardment of Bi | 111-272 | < 1 ms |
| 112 | Unnamed | Uub | 112-277 | 1996 | Zn ion bombardment of Pb | 112-277 | < 1 ms |
| 113 | Unknown | | | | | | |
| 114 | Unnamed | UUq | 114-289 | 1999 | Ca ion bombardment of Pu | 112-289 | ≈30 s |
| 115 | Unknown | | | | | | |
| 116 | Unnamed | Uuh | 116-289 | 1999 | α Decay of 118-293 | | ≈600 μs |
| 117 | Unknown | | | | | | |
| 118 | Unnamed | Uuo | 118-293 | 1999 | $^{86}$Kr bombardment of $^{208}$Pb | 118-293 | ≈120 μs |

Ion bombardment is one method of producing heavy nuclides. A second method is **successive neutron capture** that involves a sequence of neutron captures punctuated by $\beta^-$ emissions. Successive neutron capture occurs in the fuel of any operating nuclear reactor. It also occurs during supernova events in stars, and created the first man-made samples of einsteinium (element 99) and fermium (element 100) in the first US hydrogen bomb test over Eniwetok atoll in the Pacific Ocean in 1952. [❁kws "Eniwetok atoll"].

Successive neutron capture produces trans-uranic elements:

$^{239}_{94}Pu + ^{1}_{0}n \rightarrow ^{240}_{94}Pu$ followed by $^{240}_{94}Pu \rightarrow ^{240}_{95}Am + \beta^-$

$^{240}_{95}Am + ^{1}_{0}n \rightarrow ^{241}_{95}Am$ followed by $^{241}_{95}Am \rightarrow ^{241}_{96}Cm + \beta^-$

$^{241}_{96}Cm + ^{1}_{0}n \rightarrow ^{242}_{96}Cm$ followed by $^{242}_{96}Cm \rightarrow ^{242}_{97}Bk + \beta^-$

$^{242}_{97}Bk + ^{1}_{0}n \rightarrow ^{243}_{97}Bk$ followed by $^{243}_{97}Bk \rightarrow ^{243}_{98}Cf + \beta^-$

...and so on to Es and Fm

The reactions in the box above were those that took place in the explosion shown in Figure 2.5.

Figure 2.4 A commercial nuclear power reactor.

Figure 2.4 shows a British "Magnox," carbon dioxide cooled, $^{235}U$ type nuclear reactor from the 1970's. On the right is the 1952 Mike shot—the first hydrogen bomb or fusion test weapon (US Department of Energy photograph). You can't see them in the picture, but the world's first samples of the synthetic elements einsteinium and fermium are in that cloud. It was the US Air Force flying specially equipped planes that brought back some of those first-ever atoms of Es and Fm for study. The nuclear test was conducted under the overall control of the Panda committee; there was some sentiment among the scientists who worked on the test project to name one of these new elements *pandemonium*, :-).

Figure 2.5 The first hydrogen bomb.

## Section 2.2 Rate Processes: Radioactive Decay

A rate process occurs when any property of a substance or material changes with time. Radioactive decay and chemical changes that occur over time are rate processes. Controlling the rates of chemical reactions is a crucial aspect of chemistry. Engineers are also deeply interested in rate processes: how an engineering material ages is an example of a rate process of engineering interest. Food products, photographic film, and flashlight batteries are all date stamped. These and many other materials are subject to the rate process of degradation or aging. For example, the shelf life or use life of a material is often a key factor in determining its suitability for a particular use. The Society of Plastics Engineers publishes lots of information about the aging of plastics. [❅kws "Society of Plastics Engineers"].

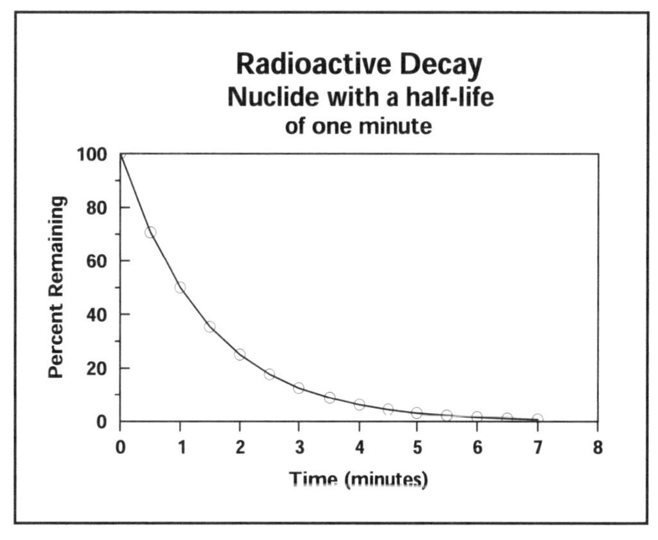


Figure 2.6 The decay of a nuclide with a one-minute half-life.

Radioactive decay is a simple rate process that can be completely explained mathematically; thus it's a good example to start with. The rates of chemical reactions are the subject of Chapter 15.

Radioactive decay is characterized by a half life. The **half-life** ($t_{½}$) of a radionuclide is the length of time it takes for one half of the sample to decay—for 100% of an original sample to decay to 50%. Sometimes the process is called exponential decay (as shown in Figure 2.6) because of the form of the mathematical law that governs this rate process:

$$c = c_0e^{-kt}$$

The variable c is the percentage of an original sample that remains after time t has elapsed, $c_0$ is 100%, the original "concentration" of the sample, and k is the specific rate constant of the process. Each radionuclide has its own k value. The specific rate constant and the half life are related by the expression $k = (\ln 2)/t_{½}$, where ln 2 = 0.693. Once you have a value of k, then solving the exponential decay equation is a straightforward plug 'n' chug. The procedure is illustrated in Examples 2.1 and 2.2.

**Example 2.1**: A radionuclide has a half life of 23.4 days. Calculate its specific rate of decay constant in the units of $s^{-1}$.

Strategy: Convert days to seconds looking ahead to the specified unit. Then substitute into the expression $k = (\ln 2)/t_{1/2}$.

$$k = (\ln 2) \div (23.4 \text{ days} \times 24 \text{ hour day}^{-1} \times 3600 \text{ s hour}^{-1})$$

$$= 0.693 \div 2.02 \times 10^6 \text{ s (Three significant figures is correct. Why?)}$$

$$\underline{k = 3.43 \times 10^{-7} \text{ s}^{-1}}$$

**Example 2.2**: A Radioactive Decay Calculation. $^{222}Rn$ (half-life = 3.82 days) is the most stable isotope of the noble gas radon. The presence of this nuclide in homes located near uranium-bearing minerals is a potential health hazard and a source of environmental concern. Suppose that a sample of air is collected from a home, and is monitored for a period of time. Calculate the percentage of the original $^{222}Rn$ that remains in the sample after 10 days.

Strategy: First, you can estimate the answer by using the half life. 50% will remain after one half life, 25% after two half lives, and so on. The following table shows the percentage remaining as a function of elapsed time:

| Number of half lives | Time elapsed | Percentage remaining |
|---|---|---|
| 0 | 0 | 100% |
| 1 | 3.82 days | 50% |
| 2 | 7.64 days | 25% |
| 3 | 11.46 days | 12.5% |

So even without doing a calculation you can see that after 10 days less than 25% but more than 12.5% of the original $^{222}Rn$ will remain.

More Strategy: To solve the equation, first calculate k, then plug 'n' chug.

$$k = 0.693/t_{1/2} = 0.693/3.82 \text{ days} = 0.181 \text{ days}^{-1}$$

Substituting into $c = c_0e^{-kt}$ $c = 100\% \times e^{-(0.181 \text{ day}^{-1} \times 10.0 \text{ days})}$ $\underline{x = 16.2\%}$

In connection with Example 2.2, note that radon is an ongoing human health concern. A colorless, odorless noble gas radon is not detected by the human senses. In some places radon leaks from the earth, having been formed in uranium minerals underground. If the radon leaks into a closed space, such as a basement, dangerous levels of contamination can occur. Such radon problems in the Reading Prong region of eastern Pennsylvania surfaced around 1990 when a nuclear power plant engineer set off the radiation alarm on his way *into* the plant, having been exposed to dangerous levels of radiation in his own basement.

## Radiocarbon Dating

The ages of once living objects can be measured using either their $^{14}C/^{12}C$ ratio or the actual amount of $^{14}C$ in the sample. The method is known as **radiocarbon dating** and is based on the fact that the proportion $^{14}C$ in living organisms is controlled by the current rate of $^{14}C$ formation. After an organism dies, the amount of $^{14}C$ in its remains decays in accord with $^{14}C$'s half life.

Carbon-14 is continuously produced in our atmosphere by the bombardment of $^{14}N$ by high energy neutrons arriving as part of the cosmic radiation from outer space. The amount of $^{14}C$ on planet earth is in **secular equilibrium,** a steady state situation in which the rate of formation of a substance balances its rate of disappearance. Protons are also produced in this nuclear reaction

$$^{14}_{7}N + {}^{1}_{0}n \rightarrow {}^{14}_{6}C + {}^{1}_{1}p$$

The $^{14}C$ is incorporated into carbon dioxide as $^{14}CO_2$ that is taken in by plants during photosynthesis. Thus, a small fraction of the carbon in plants is $^{14}C$. Carbon-14 decays by beta emission with a half life of 5,730 years.

$$^{14}_{6}C \rightarrow \, ^{14}_{7}N + \, ^{0}_{-1}e \qquad t_{½} = 5{,}730 \text{ years}$$

The $^{14}C$ is constantly decaying, but is taken into the plant by photosynthesis so long as the plant is alive. A steady state equilibrium is established in which the $^{14}C/^{12}C$ ratio is about $1 \times 10^{-12}$, the same as that found in the atmosphere. Living plants have a $^{14}C$ beta activity of about $13.6 \pm 0.1$ disintegrations per minute (dpm) per gram of carbon, irrespective of the type of plant. Going up the food chain, herbivores incorporate the $^{14}C$ into their bodies, as in turn do carnivores. So living organisms have a carbon beta activity of $13.6 \pm 0.1$ dpm $g^{-1}$.

When the plant or animal dies, its $^{14}C$ activity slowly drops off, with the characteristic half life of 5,730 years. If we assume that the $^{14}C$ activity in the past was the same as it is today, it is possible to determine how long ago the organism that provided the sample died by measuring the modern value of its $^{14}C$ decay rate in dpm per gram of carbon.

**Example 2.3**: A small piece of parchment from a cave near the Dead Sea was found to have a $^{14}C$ specific activity of 10.2 dpm $g^{-1}$. Determine the time elapsed since the death of the animal whose skin was used make the parchment. Assume that the atmospheric $^{14}C$ specific activity at the time the animal died was the same as it is today, 13.6 dpm $g^{-1}$ of C.

Strategy: Apply the basic equation for radioactive decay

$$c = c_0 e^{-kt} \qquad \text{or} \qquad \ln(c/c_o) = -kt$$

Radiocarbon dating rests on the assumption that the rate of production of $^{14}C$ has remained constant. Thus, we'll assume that the $c/c_o$ ratio is the current $^{14}C$ activity (10.2 dpm $g^{-1}$) in the sample divided by the $^{14}C$ activity in a living animal (13.6 dpm $g^{-1}$). Note that the units cancel to give us the concentration ratio $c/c_o$. We find k from the half life of $^{14}C$ using the relationship $k = (\ln 2) \div t_{½}$

$$\text{Thus, } k = 0.693/5730 \qquad \text{so, } k = 1.21 \times 10^{-4} \text{ year}^{-1}$$

Rearranging $\ln(c/c_o) = -kt$, for time we get: $t = (\ln c/c_o) \div -k$

And, plugging the values, $t = \ln(10.2/13.6) \div (-1.21 \times 10^{-4})$

Time elapsed = 2,380 y

The accuracy of radiocarbon dating rests on the assumption that the original activity of the $^{14}C$ is known, and on the errors inherent in the techniques used to measure the modern activity. Comparisons have been made between tree ages determined by counting their annual rings and ages obtained by the radiocarbon method. The two methods show about a 10% discrepancy. Radiocarbon dating is now more than 50 years old. Recently, mass spectroscopic measurements (to be discussed in the next section) have proved to be able to measure the needed $^{14}C/^{12}C$ or $^{13}C/^{12}C$ ratios for reliable dating back as far as 50,000 years ago.

## Section 2.3 Mass-Energy Relationships

Modern chemistry more-or-less began with the law of conservation of mass. However, mass is *not* conserved during nuclear reactions, when some is converted to energy according to the well-known equation $E=mc^2$.

### Energy Changes of Nuclear Reactions

The greatest social consequence of nuclear reactions has been the enormous energy they can supply. A controlled, induced fission reaction is a **nuclear reactor**. An uncontrolled, rapid chain nuclear reaction is a **nuclear bomb**. Why there is such a thing as nuclear energy can be understood from a study of the mass of nuclides. Have you ever wondered why *both* nuclear fission and nuclear fusion produce energy? Doesn't seem to make sense, does it? In this section we'll see why it does.

The first nuclear energy change we'll study is purely hypothetical—it's for a reaction we cannot actually do, the formation of one neutral carbon-12 atom from its constituent subatomic particles (six each of protons, neutrons, and electrons) according to the nuclear reaction equation:

$$6\ {}^{1}_{1}p + 6\ {}^{1}_{0}n + 6\ {}^{0}_{-1}e \rightarrow {}^{12}_{6}C$$

The total mass is obtained by summing up the masses of these components (six significant figure accuracy is sufficient):

| | | | |
|---|---|---|---|
| electrons | $6 \times 0.00054858$ u | = | 0.00329 u |
| protons | $6 \times 1.00728$ u | = | 6.04368 u |
| neutrons | $6 \times 1.00867$ u | = | 6.05202 u |
| | Total | | 12.09899 u |

The sum of the masses is 12.09899. *But,* and it's a big but, the mass of a $^{12}C$ nuclide is exactly 12.00000... Recall $^{12}C$ is the nuclide we choose for our atomic mass standard: 1 u = one twelfth the mass of a single atom of $^{12}C$.

So where is the missing mass? Why does the nuclide weigh less than the sum of the masses of its component parts? Answer: the missing mass got converted to energy. To what energy? To the energy needed to hold the $^{12}C$ nucleus together, or to break it apart. The missing mass is called the mass defect. The **mass defect** of a nuclide is the sum of the masses of its neutrons, protons, and electrons minus its experimental mass. Mass defects are always positive. The mass defect of $^{12}C$ is 12.09899 - 12.00000, or 0.09899 u. To calculate the energy involved we use $E = mc^2$, as shown in Example 2.4. The formation of a $^{12}C$ atom can be visualized according to the **energy level diagram** shown in Figure 2.7.

Figure 2.7 Hypothetical Formation of a Carbon-12 Atom. An energy level diagram.

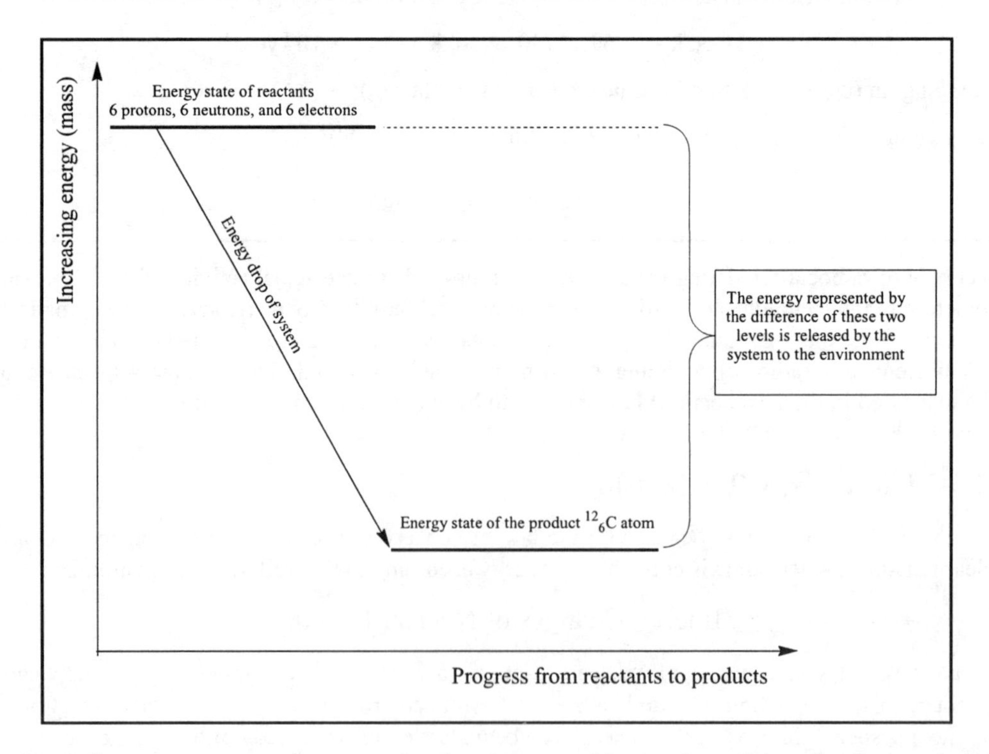


Energy level diagrams are very important in chemistry, so it's worth studying the features of this simple example. First, note that we can draw a perfectly correct diagram *even though* the reaction is hypothetical. The mass difference between the products and reactants is definite, so the energy "gap" between the upper and lower states is also definite. Second, note that the horizontal axis is arbitrary, we are really dealing with a one-dimensional graph, but spreading out the energy values as horizontal lines allows us to see the energy difference better. Third, note that the energy drop of the system (reactants becoming products) equals the energy released to the environment. We'll meet all of these facets of energy level diagrams again later, when we discuss electronic energy states and thermodynamic energy states.

**Example 2.4**: Calculate the binding energy of $^{12}C$ in kJ per mole of atoms and in kJ per mole of nucleons.

Strategy: We have already calculated the $^{12}C$'s mass defect is 0.09899 u atom$^{-1}$. So we plug this mass into $E = mc^2$ and because there are 12 nucleons in the atom we divide by 12. The only minor problem is keeping the units straight. If we use mass in kilograms and speed in m s$^{-1}$ then the energy will be in joules.

$$E = mc^2$$

$$E = \left(\frac{0.09899\ \text{u atom}^{-1}}{6.022 \times 10^{23}\ \text{u g}^{-1}}\right) \times \left(\frac{1\ \text{kg}}{1000\ \text{g}}\right) \times (2.997 \times 10^8)^2 \text{m}^2\text{s}^{-2}$$

$$\text{So,}\quad E = 1.476 \times 10^{-11}\ \text{kg m}^2\ \text{s}^{-2}\ \text{(joules) atom}^{-1}$$

Now I'll convert the J to kJ and put the result on a molar basis by multiplying by Avogadro's number. That gives us the chemist's customary unit of kJ mol$^{-1}$:

$$E = 1.476 \times 10^{-11}\ \text{J} \times \left(\frac{1\ \text{kJ}}{1000\ \text{J}}\right) \times 6.022 \times 10^{23}\ \text{mol}^{-1}$$

Total binding energy of a mole of $^{12}C$ atoms $= 8.888 \times 10^9$ kJ mol$^{-1}$

Dividing by 12 gives carbon-12's binding energy per mole of nucleons as: $\underline{7.407 \times 10^8\ \text{kJ mol}^{-1}}$

Multiplied by $10^8$ (for convenience) the value is shown on the left hand axis of Figure 2.8.

We can do the same type of calculation for any nuclide and obtain both the nuclide's mass defect and the energy involved in the nuclide formation from its constituent subatomic particles. The energy equivalent to the mass defect is the **binding energy** (BE) of the nucleus—the energy input needed to unbind it and convert it to its subatomic constituents. To compare all nuclides it is customary to calculate their **binding energy per nucleon**. That is, BE ÷ (n+p), where (n+p) is the total number of nucleons. These values when they are graphed (as shown in Figure 2.8) provide an insight into the origin of nuclear energy.

## The Curve of Binding Energy

A chart of the binding energy nucleon$^{-1}$ across the periodic table is shown in Figure 2.8. The values of binding energy nucleon$^{-1}$ have been multiplied by $10^{-8}$; to get the actual values multiply by $10^8$. Check for yourself that carbon's value calculated in Example 2.4 is in the correct place on the diagram. The most strongly bound nucleus is $^{56}Fe$, so $^{56}Fe$ must be the most stable nuclide. Other nuclides are less stable, and will tend to "move" closer to iron if there is a mechanism for them to do that. Nuclides of lower mass than iron can in principle fuse to make an iron nucleus with a release of energy to their environment. Likewise, heavy nuclides can split into two fragments with a release of energy to their environment. So we now have the explanation for both fission and fusion energy. Light elements (below iron) fuse to make the more stable heavier elements (up to iron) while heavy elements (well above iron) undergo fission to make more stable lighter elements. The fusion reactions

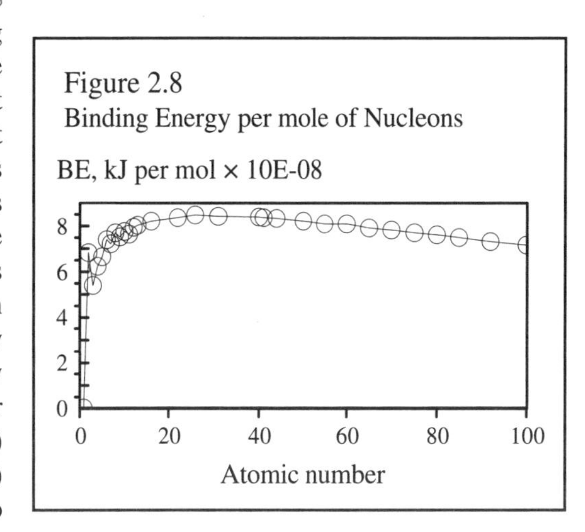

Figure 2.8
Binding Energy per mole of Nucleons

of stars or hydrogen bombs are an example of the former process. The nuclear fission of $^{235}U$ is an example of the latter process.

Now we've studied electrons, protons, neutrons, atoms and nuclides. From the point of view of a chemist, these are the simplest particles. Atoms join to make molecules and polyatomic ions. So next we'll take up molecules and their electrically charged relatives, the ions.

## Mass Spectrometry

### Mass Spectrometry and Isotopes

A **mass spectrum** is a graph or display (plotted according to their mass and abundance) of the various ions formed when a substance is energized, for example by being struck by the impacts of a beam of electrons. **Mass spectrometry** is an experiment used to generate a mass spectrum. Because the key difference among the isotopes of an element is their masses, mass spectrometry was used historically to sort out the isotopes. The data in Appendix D derives from highly accurate mass spectrometric experiments. Today, the method remains a powerful technique for chemical analysis. An instrument is shown in Figure 2.21.

In 1912-14 Francis Aston (1877-1945), working as an assistant to J. J. Thomson, was studying the behavior of positive ions in electrical and magnetic fields and learning how to separate them and then focus them onto a collector. By around 1913, Aston realized that his techniques could separate elements into their isotopes. Figure 2.9 shows a schematic diagram of the signals resolved from a sample of natural germanium. Each block represents a $Ge^+$ ion with a specific mass; together, the blocks form the mass spectrum of natural germanium. The heights of the blocks correspond to the relative abundance of the five natural isotopes of germanium. It is by means of such measurements, using high quality instruments, that accurate isotopic masses are established.

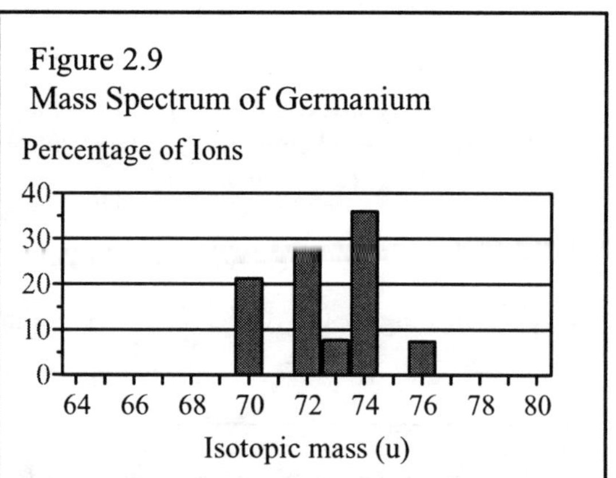

Figure 2.9
Mass Spectrum of Germanium

### Mass Spectrometry and Fragmentation

Up to this point in the chapter we've dealt with atoms. Now we are ready to move on to entities composed of multiple atoms: molecules and polyatomic ions.

When a molecular substance is impacted with a beam of electrons (or otherwise ionized) it breaks down into a set of ionized fragments. If those ions are passed down the vacuum tube of a mass spectrometer, a mass spectrum is obtained that shows a characteristic fragmentation pattern for the molecular substance under the conditions of the experiment. A **fragmentation pattern** is the set of ions that form as a molecule breaks up into ions in a mass spectrometer. Studying fragmentation patterns is an important technique chemists use to identify organic compounds. For example, allied with separation techniques, it allows minute traces of prohibited steroids to be detected in an athlete's urine.

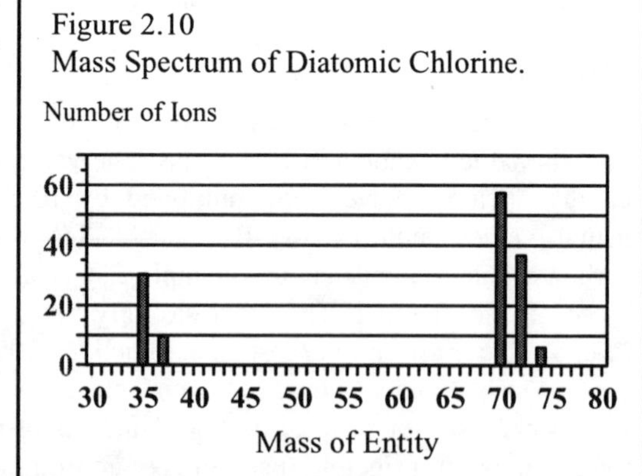

Figure 2.10
Mass Spectrum of Diatomic Chlorine.

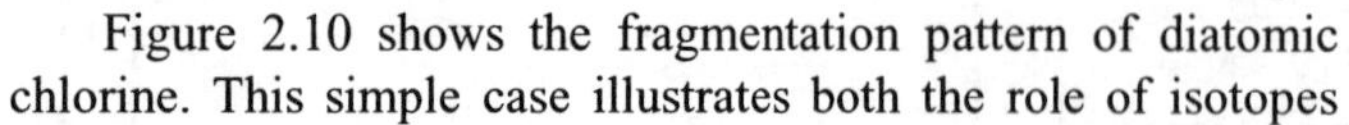

Figure 2.10 shows the fragmentation pattern of diatomic chlorine. This simple case illustrates both the role of isotopes and the role of fragmentation in the creation of a mass spectrum. At the right of Figure 2.10 are molecular ions with the formula $Cl_2^+$; at the left are monatomic ions, $Cl^+$. The interpretation of this mass spectrum is described in Example 2.5. The relative proportions of diatomic and monatomic ions depends on the intensity with which the chlorine molecules are energized. The greater the energy applied to the sample the greater the fraction of chlorine molecules that will break up into monatomic $Cl^+$ ions. We can write equations to represent the formation of the ions as follows:

Formation of a molecular ion: $e^- + Cl_2 \rightarrow Cl_2^+ + 2\,e^-$

Fragmentation to give a monatomic ion: $Cl_2^+ \rightarrow Cl^+ + Cl$

**Example 2.5**: As shown in Appendix D, natural chlorine consists of two isotopes: $^{35}Cl$ with an isotopic mass of 35.0 u and an abundance of 75.8%, and $^{37}Cl$ with an isotopic mass of 37.0 u and an abundance of 24.2%. In a sample of ordinary diatomic chlorine, $Cl_2$, calculate the percentage of the molecules that a. contain two $^{35}Cl$ atoms, b. two $^{37}Cl$ atoms, and c. one each of $^{35}Cl$ and $^{37}Cl$ atoms. Assume that for every 100 diatomic ions that pass through the spectrometer, there are 20 molecules that form two monatomic ions; calculate the relative abundance of those ions.

Strategy: Use the percentage abundances divided by 100 as values of the probability that each of the two atoms in the diatomic molecule will be of the type requested.

a. For $^{35}Cl$–$^{35}Cl$ the probability of the first being $^{35}Cl$ is 0.758, the probability of the second being $^{35}Cl$ is also 0.758. The product of the probabilities is 0.575, or 57.5%.

b. For $^{37}Cl$–$^{37}Cl$ the probability of the first being $^{37}Cl$ is 0.242, the probability of the second being $^{37}Cl$ is also 0.242. The product of the probabilities is 0.059, or 5.9%.

c. For $^{35}Cl$–$^{37}Cl$ the probability of the first being $^{35}Cl$ is 0.758, the probability of the second being $^{37}Cl$ is 0.242. The product of the probabilities is 0.183, or 18.3%. But that value must be doubled because there is also the molecule $^{37}Cl$–$^{35}Cl$. The total probability for a molecule with one atom of each type is 36.7%

Check: Total percentage = 57.3% + 5.9% + 36.7% = 99.9% (okay, rounding error)

The results are summarized in the table below. Compare them with the mass spectrum shown in Figure 2.10.

| Diatomic Entities in the Mass Spectrum | | | |
|---|---|---|---|
| Entity | Type | Mass (u) | Peak height |
| $^{37}Cl$–$^{37}Cl$ | Molecule | 74.0 | 5.9 |
| $^{35}Cl$–$^{37}Cl$ and $^{37}Cl$–$^{35}Cl$ | Molecule | 72.0 | 36.7 |
| $^{35}Cl$–$^{35}Cl$ | Molecule | 70.0 | 57.4 |

d. The relative abundances in the table above sum to 100% so the relative abundances of the monatomic ions will sum to 40% (20 molecules have decomposed to give 40 monatomic ions for every 100 above). They will be in relative proportion to the isotopic abundances of chlorine. $^{37}Cl$ will be 40 × 0.758 abundant and $^{35}Cl$ will be 40 × 0.242 abundant.

| Monatomic Entities in the Mass Spectrum | | | |
|---|---|---|---|
| Entity | Type | Mass (u) | Peak height |
| $^{37}Cl$ | Atom | 37.0 | 9.7 |
| $^{35}Cl$ | Atom | 35.0 | 30.3 |

## Section 2.4 Molecules

When two atoms link together X + X → X-X they have formed both a **chemical bond** and a **diatomic molecule**. The elements that form diatomic molecules at ordinary temperature and pressure are: $H_2$, $N_2$, $O_2$, $F_2$, $Cl_2$, $Br_2$ and $I_2$. **Molecules** are stable chemical units composed of atoms held together by chemical bonds. Molecules are known that contain from two atoms to millions of atoms.

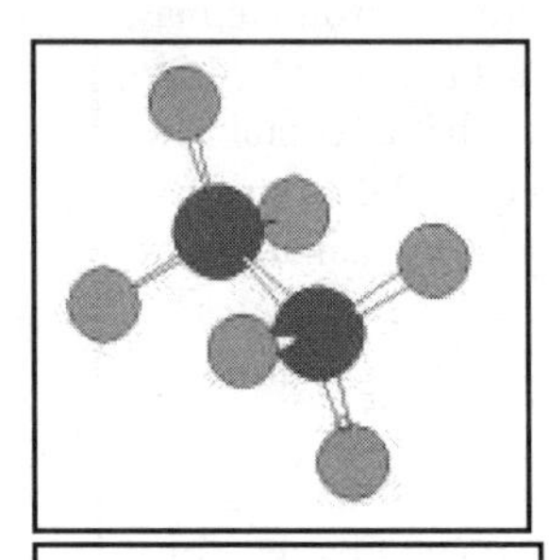

Figure 2.11 Ethane ball-and-stick representation.

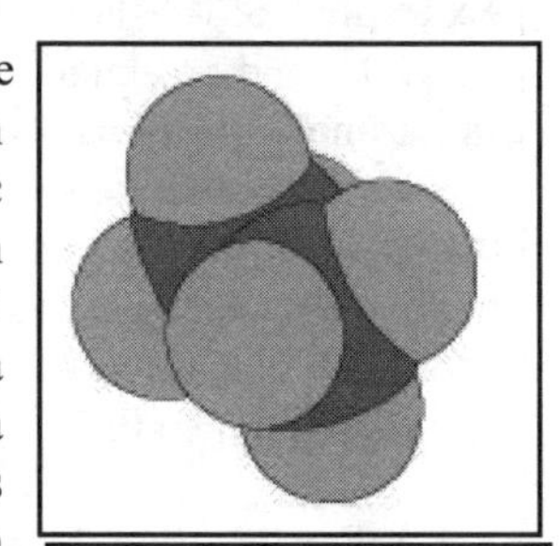

Figure 2.12 Ethane space filling representation.

High in the atmosphere, the element oxygen exists as the **triatomic** molecule $O_3$, ozone. Probably the best known molecular substance is water, $H_2O$, composed of **triatomic** molecules that contain two atoms of hydrogen and one atom of oxygen. (Water could be written $H_2O_1$ but the subscript 1 on oxygen is conventionally omitted.) Figure 2.11 shows a molecule of the fuel gas ethane. It has the chemical formula $C_2H_6$. The carbon atoms are black and the hydrogen atoms gray; this is a **ball-and-stick** view. Ethane is shown in Figure 2.12 in a **space-filling** view. Molecules are much too small to be seen, but chemists like to visualize them.

Using various methods and formats to display molecules is **molecular visualization**. In the 1990s increasing computer power made on-screen molecular display commonplace. For example, do the search [❁kws +chime +chemical +download] and download and install a plug in browser so you can explore for yourself the role of molecular visualization on the web.

Figure 2.13 Photograph of a ball-and-stick model of $CCl_4$.

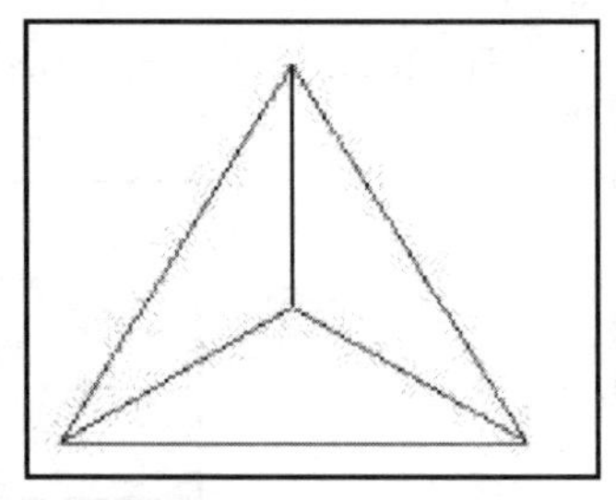

Figure 2.14 Top view of a tetrahedron.

It was around the middle of the 19th century that chemists began to realize that molecules have "architecture" or structure: that they have bonds of definite lengths pointing in directions that make definite angles. The use of painted wooden ball-and-stick models came into use at that time. Figure 2.13 is a model of the tetrachloromethane (or carbon tetrachloride) molecule, $CCl_4$. Carbon is black; the chlorine atoms are gray. All the bond angles in carbon tetrachloride are 109° 28' and all the bond lengths are 177 pm. All the atoms bound to a central atom are **ligands**, from the same root word found in ligature or ligation and meaning binding. The $CCl_4$ molecule is as symmetrical as four ligand atoms connected to a single central atom can ever get. So carbon tetrachloride is said to show **ideal geometry**. If you imagine lines connecting the nuclei of the chlorine atoms, or the centers of the gray balls you can visualize a tetrahedron. A regular tetrahedron is a three-sided pyramid. Figure 2.14 shows a sketch of a tetrahedron viewed from overhead. Molecules that adopt this general structure are said to be **tetrahedral.**

Sulfur tetrafluoride, $SF_4$, Figure 2.15, is another example of a molecule in which one central atom is surrounded by four ligand atoms. $SF_4$ is much less symmetrical than $CCl_4$. The angle F1-S-F4 is 179°, the angle F2-S-F3 is 103°, and all the other bond angles are close to 90°. The F1-S and the F4-S bond lengths are both 164 pm; the F2-S and F3-S bond lengths are both 154 pm. Chemists refer to $SF_4$ as a "teeter-totter" molecule. To visualize the teeter-totter imagine that the atoms F2 and F3 are placed on a horizontal surface and see that F1 and F2 become the seats of the teeter-totter. The wedge shaped bonds show perspective. Atom F3 comes toward you out of the plane of the paper; atom F2 is goes back behind the plane of the paper. [❁kws +chime +"sulfur tetrafluoride"].

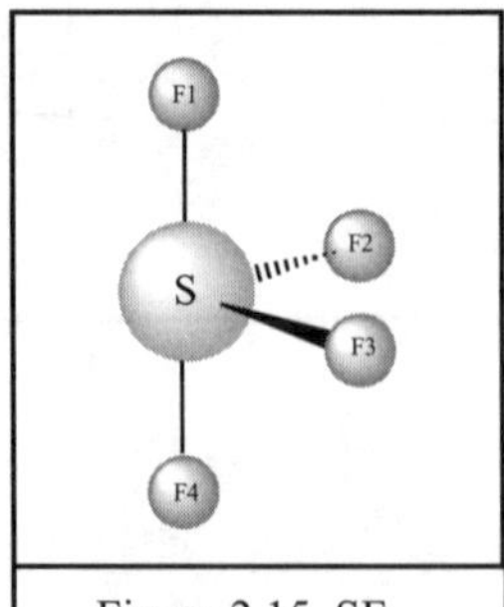


Figure 2.15 $SF_4$.

There are thousands of millions of different molecules. A **molecular substance** is one composed of molecules that are alike—when we speak of a molecular substance isotopic differences are ignored. Common examples of molecular substances are nitromethane, $CH_3NO_2$, and table sugar $C_{12}H_{22}O_{11}$. Each molecule of nitromethane contains four elements (C, H, N, and O) and 7 atoms (1 atom of C, 3 of H, 1 of N, and 2 of O). Each molecule of table sugar is composed of three elements and 45 atoms. Chemical formulas that show the actual number of atoms of each type in a molecule are **molecular formulas**. Formulas that show the positions of the atoms are **structural formulas**.

A **chemical compound** is a pure substance composed of two or more elements with a definite composition and a definite set of physical and chemical properties; it is composed of molecules that are chemically alike. Note, however, that every molecule of a naturally occurring pure compound is not necessarily identical to all the other molecules. Chemically alike does *not* mean identical. Most elements are mixtures of isotopes, so molecules that contain atoms of these elements show isotopic variations. For example, natural hydrogen contains $^1H$ and $^2H$ (called D for deuterium), so any sample of ordinary water contains the molecules $H_2O$, HDO, and $D_2O$ (if we included the isotopes of oxygen there would be even more types of water molecules).

**Example 2.6**: Hydrogen and chlorine form the diatomic molecule HCl, called hydrogen chloride as a gas and hydrochloric acid when dissolved in water. Using the stable nuclides of these elements calculate the mass of every possible HCl molecule.

Strategy: Look up the stable nuclides and their masses in Appendix D. They are $^1H$, $^2H$, $^{35}Cl$, and $^{37}Cl$, and they weigh respectively 1.008, 2.014, 34.969, and 36.967 u. Make all possible combinations. Adding gives:

$$^1H^{35}Cl = 35.977 \text{ u} \qquad ^1H^{37}Cl = 37.975 \text{ u}$$

$$^2H^{35}Cl = 36.983 \text{ u} \qquad ^2H^{37}Cl = 38.981 \text{ u}$$

In addition to water, hydrogen and oxygen also form hydrogen peroxide, $H_2O_2$. Every molecule of hydrogen peroxide contains two hydrogen and two oxygen atoms. The chemical and physical properties of water and hydrogen peroxide are significantly different. Some pairs of elements are capable of forming a wide range of molecules. For example, there are many different compounds with the formula $C_nH_m$ where n and m are whole numbers. Any compound with the formula $C_nH_m$ is a **hydrocarbon**.

Chemical formulas with the smallest possible whole number ratio of atoms are **empirical formulas**. The empirical formula CH tells us that the numbers of carbon and hydrogen atoms in the compound are equal so the molecular formula is $(CH)_n$, where n is an integer. No stable molecular substance with the formula CH exists on earth; the molecular substance represented by $(CH)_n$ with n = 2 would be $C_2H_2$, acetylene; the molecular substance with n = 6 is $C_6H_6$, benzene; the molecular substance with n = 8 is $C_8H_8$, cyclooctatetrene; the molecular substance with n = 1000s is polyethylene.

If the subscripts of a molecular formula have common factors, factoring gives the substance's empirical formula. The sugar glucose has the molecular formula $C_6H_{12}O_6$. Factoring the 6 gives an empirical formula of $CH_2O$. The formula $CH_2O$ looks like carbon combined with water, which is historically why sugars wound up in the carbohydrate food group. The compound with the *molecular* formula $CH_2O$ is formaldehyde.

Some substances exist as very large molecules called polymers. A **polymer** is by definition a substance composed of very large molecules. The importance of polymers for engineers is made clear in Appendix F, and in particular in Table F-1, where you will learn that synthetic polymers provide many engineering materials. For polymers, such as plastics and resins, only empirical formulas, not molecular formulas, can be written. Usually, it is not possible to measure the exact number of atoms present in the molecules of polymers, and anyway, most polymers are composed of a range of closely related but different molecules. The common plastic wrapping material, polyethylene, is represented by its empirical formula $CH_2$, or $(CH_2)_n$, where the value of n could be 10,000 - 50,000.

When writing on paper or a chalkboard chemists use two dimensional representations of molecular structure. The easiest method is just to use keyboard characters. So the atomic connections in a water molecule are represented by the structural formula: H–O–H. The structure is okay, but doesn't accurately represent water's H–O–H angle, which is 103.5°.

The atomic connections in a hydrogen peroxide molecule, $H_2O_2$, are represented by the structural formula: H–O–O–H. Dashes (–) show chemical bonds between the atoms.

Many computer programs are now available to draw two- and pseudo three-dimensional chemical structures. Better yet, you can do it easily for yourself using any of the browser plug-ins such as the Chime program mentioned earlier. There is no single "correct" way to represent a molecule's structure on paper or a video screen. There are simply lots of ways to do it; some ways better emphasize one aspect of a molecule's structure, some another. True three dimensional representations can be achieved by physical models. Students of organic chemistry are well advised to buy models, but engineering students taking a general chemistry course probably do not need them. Figure 2.16 shows various representation of a molecule of the over-the-counter drug ibuprofen, $C_{13}H_{18}O_2$. Different parts of the molecule are given names as shown.

Figure 2.16 Some representations of ibuprofen, $C_{13}H_{18}O_2$.

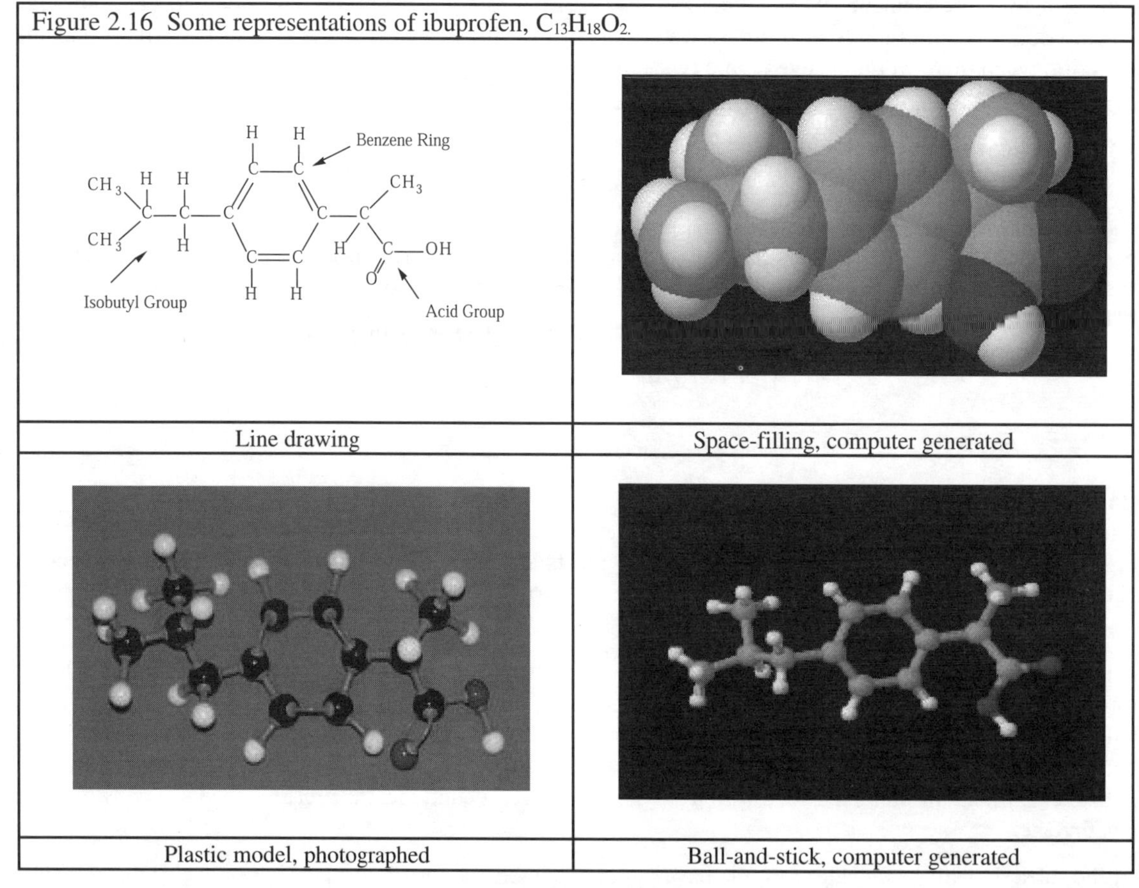


Line drawing

Space-filling, computer generated

Plastic model, photographed

Ball-and-stick, computer generated

In Figure 2.16 you can see that $-CH_3$ (called a methyl group) in the upper right of the molecule is a tetrahedral group made up of one carbon and three hydrogen atoms and is attached to a second carbon atom. At the lower right is the -COOH group consisting of a carbon atom doubly bonded to one oxygen atom and singly bonded to a second oxygen atom with a hydrogen atom also attached to the second oxygen atom. Collectively, the way all the atoms are joined is called ibuprofen's chemical structure. **Chemical structure** is a broad term used to describe the arrangement of the atoms in molecules and ions.

## Molecular and Molar Weight (Molar Mass)

As shown in Example 2.7, the mass (weight) of any molecule is the sum of the atomic weights of the atoms that compose it.

$$\text{Molecular weight (MW)} = \sum n_i AW_i$$

where $n_i$ is the number of ith atoms in the molecule and $AW_i$ is the atomic weight of the ith atom.

**Example 2.7**: Calculate the molecular weight of iboprufen, $C_{13}H_{18}O_2$.

Strategy: Use the atomic weights and the formula to substitute into the equation

$$\text{Molecular weight (MW)} = \sum n_i AW_i$$

$$\text{Molecular weight} = n_1 \times AW_1 + \quad n_2 \times AW_2 + \quad n_3 \times AW_3$$

$$MW = 13 \times AW\,\text{carbon} + 18 \times AW\ \text{hydrogen} + 2 \times AW\ \text{oxygen}$$

$$MW = 13 \times 12.01 \quad + 18 \times 1.008 \quad + 2 \times 16.00$$

$$= 156.13\ u \quad + 18.14\ u \quad + \ 32.00\ u$$

$$= \underline{206.27}\ u$$

The molecular weight of ibuprofen is 206.27 u. Ibuprofen is sold in over-the-counter medications such as Advil® and Motrin®. In its pure form, it is a colorless, crystalline solid with a melting point of 76°C.

Because of the existence of isotopes of C, H, and O, ibuprofen molecules can vary by a few atomic mass units. So strictly, molecular weight is an *average* weight. Chemists can avoid this technical problem by using the alternative unit of **molar weight**. A mole of a molecular compound contains $6.022 \times 10^{23}$ molecules and when the molecular weight is expressed in grams the masses of the individual molecules "average out." So we have two units:

$$\text{Molecular weight (MW)} = 206.27\ u\ (\text{average molecule})^{-1}, \text{ or just u molecule}^{-1}$$

$$\text{Molar weight (MW)} = 206.27\ g\ mol^{-1}$$

You can tell from the units whether molar or molecular is intended. How can a single value have two different units? The conversion factors between each pair of units must be identical. Thus,

$$1\ g = 6.022 \times 10^{23}\ u \qquad \text{and} \qquad 1\ \text{mole} = 6.022 \times 10^{23}\ \text{molecules}$$

The **molar mass** of a sample is strictly the average mass of the molecules (or specified entities) in that particular sample. But that definition is unnecessarily restrictive for our purposes here and we will use the terms molar weight and molar mass synonymously. The author will most favor "molar weight" to emphasize that we are talking about a substance composed of natural elements, which we generally are.

## Single, Delocalized, Double, and Triple Bonds

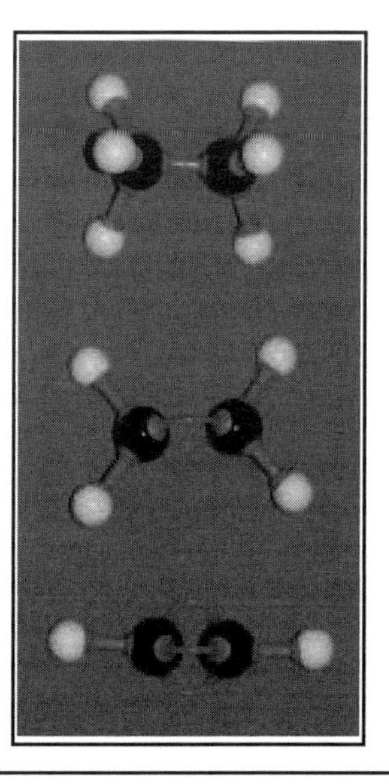

Figure 2.17 Single, double, and triple bonds.

The experimental facts are that carbon atoms can bond to other carbon atoms in four different ways. Ethane, $C_2H_6$, at the top has a carbon-carbon **single bond** (C-C). Ethene (ethylene), $C_2H_4$, in the middle has a carbon-carbon **double bond** (C=C). Ethyne (acetylene), $C_2H_2$, at the bottom has a carbon-carbon **triple bond** (C≡C). The fourth way they bond is intermediate between a single and double bond. Its called a **delocalized** carbon-carbon bond and its characteristic of the benzene ring of atoms. You can see a benzene ring in the middle of the ibuprofen structure shown in Figure 2.16.

The order of length of the bonds is single > delocalized > double > triple. The order of the strength of the bonds is reversed triple > double > delocalized > single. These properties are summarized in the Table 2.6. The bond energy is given in kJ $mol^{-1}$. It's the amount of energy needed to break one mole of the bonds. Bond energy measures the strength of bonds. If you remember the kinetic energy example from Chapter 1, you can figure out that it takes the kinetic energy of a 1-ton car traveling at 97 mph to break a mole of C-C bonds. But hitting ethane with a speeding vehicle is not a useful way to break bonds—it's easier just to set fire to the ethane.

Bond length and bond energy values are known for millions of molecules. Bond lengths and angles are measured by the technique of x-ray crystallography as described in Appendix G. The bond in the table shown as ≃ is a delocalized carbon-carbon bond.

Table 2.6 **Types and Properties of Some Carbon-Carbon Bonds**

| Bond type | Structure | Bond length pm | Bond energy kJ $mol^{-1}$ |
|---|---|---|---|
| Single | C–C | 154.1 | 368 |
| Double | C=C | 133.7 | 720 |
| Triple | C≡C | 120.4 | 962 |
| Delocalized | C≃C | 139.5 | about 550 |

Benzene

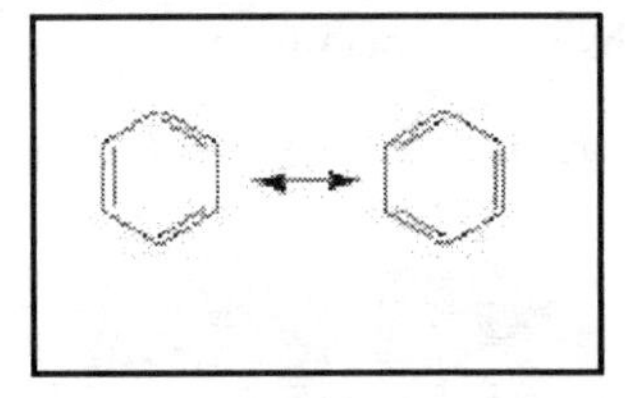

A benzene molecule, shown above, is a hexagon of six carbon atoms each joined to one hydrogen atom. In this conventional diagram the hydrogen atoms are omitted and one carbon atom is understood to be at each point of the hexagon. Experimentally, all the carbon-carbon bonds are equal at 139 pm. So benzene is at odds with the usual experimental situation that double bonds are shorter then single bonds. Actually, benzene's carbon-carbon bonds are intermediate between single and double—they are sort of 1½ bonds. Benzene is best represented by both structures *simultaneously*. In other words, the double bonds in benzene are not strictly local, they "smear out" around the benzene ring—they are delocalized.

## Covalent Bonds

Chemists are extremely interested the nature of chemical bonding. Bonding is a key aspect of chemistry that we'll take up in more detail later. For now, we'll regard the bonds between carbon atoms as covalent. A **covalent bond** is one imagined as two atoms sharing electrons. A C–C single bond involves the sharing of two electrons (one pair); a delocalized C≃C bond involves the sharing of three electrons (1½ electron pairs); a C=C double bond involves sharing four electrons (two pairs); a C≡C triple bond involves sharing six electrons (three pairs). **Molecular compounds** are composed of molecules held together by covalent bonds and are therefore also called **covalent compounds**.

## Valency

Is there some simple way to figure out how atoms join to make molecules? No, nature is pretty complicated, and to understand molecules a chemist has to be cognizant of the formidable science of quantum mechanics. However, the sometimes maligned historical idea of valency is helpful in understanding many compounds of carbon (**organic compounds**). The term valency arose around 1860, meaning the chemical combining power of an element. For carbon and many elements to which it bonds, the valency is derived from the formula of the simplest compound of that element with hydrogen. An atom's **valency** is the maximum number of hydrogen atoms it combines with.

Figure 2.18. The Valency of Some Atoms

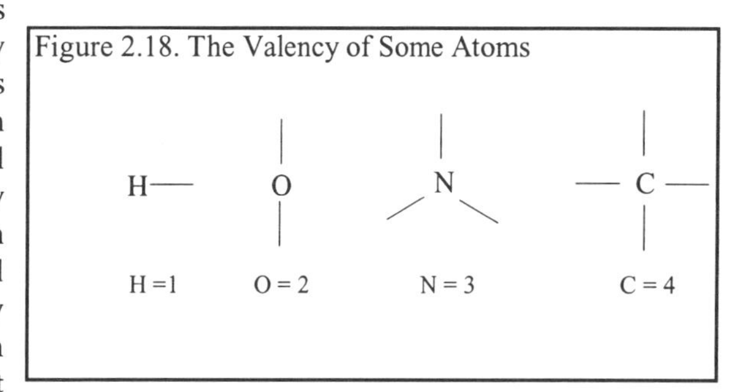


The Origin of the Word Valency.

Amusingly, valency, and its derivative covalent, have their formal origin in an *engineering* journal. "The molecule...is therefore a body in which all the attractions or valencies are satisfied, leaving the combined atoms to act as a whole from one center." Engineering Mechanics, November 1869, cited as the primary source by the *Oxford English Dictionary*. Valency derives from the Latin *valentia*, meaning vigor or capacity. Hence the English sense meaning chemical combining capacity.

Ball and stick models show valency quite well. The valency of a wooden ball atom is the number of holes it has for sticks. Balls for hydrogen have one hole, balls for oxygen have two holes, balls for nitrogen have three holes, and balls for carbon have four holes. If you're making a model with balls and sticks you know that the valencies are okay when you've filled all the holes with sticks. Figure 2.18 shows the valencies of hydrogen = 1, oxygen = 2, nitrogen = 3, and carbon = 4. Mendeleev positioned elements in the periodic table in large measure based on their valencies. If you examine the Mendeleev-type periodic table in Appendix H you'll see the column headings: Group I RH, Group II $RH_2$, Group III $RH_3$, and Group IV $RH_4$. A modern interpretation of these designations is that a hydrogen atom joins just one other hydrogen atom, oxygen joins 2, nitrogen joins 3, and carbon 4. The corresponding molecules are: $H_2$, $OH_2$, $NH_3$, and $CH_4$.

You can now understand that when carbon forms a double bond, each carbon atom uses two of its valencies to make that bond, leaving each carbon atom able to join with two hydrogen atoms, as in ethylene ($C_2H_4$):

```
  H H
  | |
H-C=C-H
```

You know that carbon dioxide is $CO_2$. The valency concept explains $CO_2$ neatly in that the carbon uses two of its four valencies with each oxygen, forming a triatomic molecule containing two double bonds: O=C=O. If you look at the ibuprofen structure described earlier you'll see that the simple valency rule is followed by every atom in that molecule.

**Example 2. 8**: Use the valencies C = 4, H = 1, and O = 2 to make sketches of the following: a. A $CH_2O$ (formaldehyde) molecule. b. Two different molecules (isomers) with the formula $C_3H_6$.

Strategy: Assemble the atoms and join them in such a way that all the atoms' valencies are "satisfied."

a. In formaldehyde the carbon and oxygen must be joined by a double bond with the third and fourth valencies of carbon being satisfied by hydrogen atoms.

b. $C_3H_6$ exists either as propylene with a C=C double bond, or as cyclopropane with a ring of three carbon atoms.

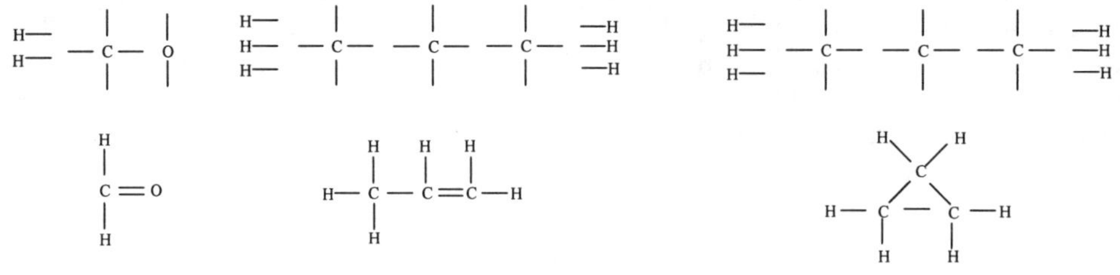

Formaldehyde

Propylene

Cyclopropane

The families of organic compounds and functional groups are shown in Table 2.7. This table summarizes the general formulas of 14 classes of organic chemical compounds

Table 2.7 **Families of Organic Compounds and Functional Groups**

| Class | Specific Example | IUPAC Name | Common Name | General Formula | Functional Group |
|---|---|---|---|---|---|
| Alkane | $CH_3CH_3$ | Ethane | Ethane | RH | C-H and C-C |
| Alkene | $H_2C{=}CH_2$ | Ethene | Ethylene | $RCH{=}CH_2$<br>$RCH{=}CHR$<br>$RCH{=}CHR_2$<br>$R_2C{=}CR_2$ | -C=C- |
| Alkyne | $HC{\equiv}CH$ | Ethyne | Acetylene | $RC{\equiv}CH$<br>$RC{\equiv}CR$ | $-C{\equiv}C-$ |
| Arene | | Benzene | Benzene | ArH | Aromatic ring |
| Haloalkane | $CH_3CH_2Cl$ | Chloroethane | Ethyl chloride | RX | -C-X |
| Alcohol | $CH_3CH_2OH$ | Ethanol | Ethyl alcohol | ROH | -C-OH |
| Ether | $H_3C\text{-}O\text{-}CH_3$ | Methoxy methane | Dimethyl ether | R-O-R | -C-O-C- |
| Amine | $CH_3NH_2$ | Methanamine | Methyl amine | $RNH_2$ | -C-N-H<br>|<br>H |
| Aldehyde | $CH_3CHO$ | Ethanal | Acetaldehyde | RCHO | H<br>|<br>-C=O |
| Ketone | O<br>||<br>$H_3C\text{-}C\text{-}CH_3$ | Propanone | Acetone | R-CO-R' | |<br>-C=O<br>| |
| Carboxylic acid | $CH_3COOH$ | Ethanoic acid | Acetic acid | RCOOH | OH<br>|<br>-C=O |
| Ester | $CH_3COOCH_3$ | Methyl ethanoate | Methyl acetate | RCOOR' | OR<br>|<br>-C=O |
| Amide | $CH_3CONH_2$ | Acetamide | Acetamide | RCONHR' | O<br>||<br>R-C-N-R' |
| Nitro-compound | $CH_3NO_2$ | Nitromethane | Nitromethane | $R\text{-}NO_2$ | $-C\text{-}NO_2$ |

In Table 2.7 "R" stands for the "rest of the molecule" and X stands for a halogen. The actual origin of "R" is probably the word radical, a word more commonly used nowadays to describe a reactive molecular fragment containing an unpaired electron. R is often referred to as the **R group**. Organic molecules are classified by their functional groups. A **functional group** is a group of atoms bonded in a particular way in a molecule and conferring specific properties (functions) on that molecule. Thus, the functional group -COOH is characteristic of carboxylic acids; all carboxylic acids have the functional property of neutralizing the strong base sodium hydroxide. RCOOH represents the entire family of carboxylic acids that begins $CH_3COOH$, $C_2H_5COOH$, $C_3H_7COOH$, $C_4H_9COOH$, etc., where R is successively $-CH_3$ the **methyl group**, $-C_2H_5$ the **ethyl group**, $-C_3H_7$ the **propyl group**, and $-C_4H_9$ the **butyl group**. $C_3H_7OH$ combines the alcohol functional group with a propyl group and is propyl alcohol. Its IUPAC name is propanol. You know it in one of its forms as isopropanol or rubbing alcohol. The designation R' means either a different group than R or possibly the same group. Collectively, methyl, ethyl, propyl, butyl, etc., are called the family of **alkyl groups.**

# Isomers

Be aware that the families in Table 2.7 only begin to skim the surface of what nature can do. Consider these factors: (1) There are many other functional groups, (2) the atoms in a given molecule can usually be rearranged with a different architecture to yield **structural isomers**, and (3) molecules with more than one functional group are common. Each of these factors contributes to the existence of millions of organic compounds.

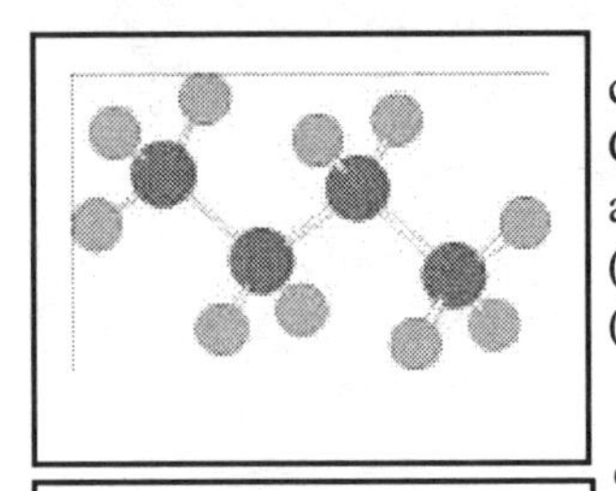

Figure 2.19a
Butane.

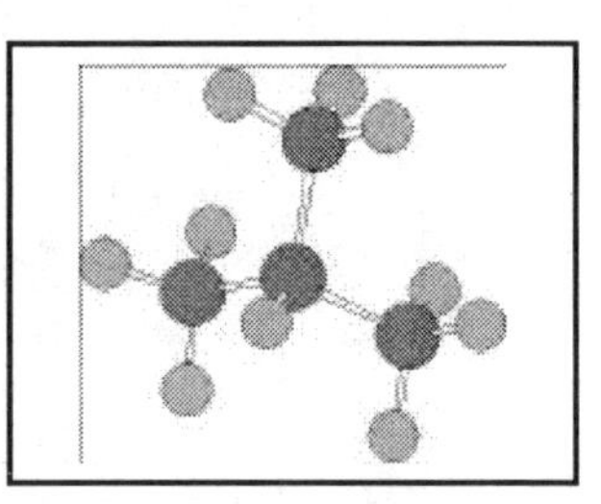

Figure 2.19b
Isobutane.

**Alkanes**, the first family in the Table of organic compounds, have the collective chemical formula $C_nH_{2n+2}$, where n is any integer 1, 2, 3... When n = 4 the alkane is called butane. It exists as two isomers, butane (Figure 2.19a) and isobutane or 2-methyl propane (Figure 2.19b).

The number of isomers in any of the families of organic compounds containing a particular functional group goes up dramatically as the number of carbon atoms in the molecules increase. Because of this fact, organic chemistry is a vast subject. For example, the number of possible alkanes for some given carbon counts ($C_n$) is shown in Table 2.8.

Table 2.8 **Total Numbers of Alkane Structural Isomers for n Carbon Atoms**

| Name | n | Alkane | Isomer count |
|---|---|---|---|
| Methane | 1 | $CH_4$ | 1 |
| Ethane | 2 | $C_2H_6$ | 1 |
| Propane | 3 | $C_3H_8$ | 1 |
| Butane | 4 | $C_4H_{10}$ | 2 |
| Pentane | 5 | $C_5H_{12}$ | 3 |
| Hexane | 6 | $C_6H_{14}$ | 5 |
| Heptane | 7 | $C_7H_{16}$ | 9 |
| Octane | 8 | $C_8H_{18}$ | 18 |
| Nonane | 9 | $C_9H_{20}$ | 35 |
| Decane | 10 | $C_{10}H_{22}$ | 75 |
| Undecane | 11 | $C_{11}H_{24}$ | 159 |
| Dodecane | 12 | $C_{12}H_{26}$ | 355 |
| Tridecane | 13 | $C_{13}H_{28}$ | 802 |
| Tetradecane | 14 | $C_{14}H_{30}$ | 1,858 |
| Pentadecane | 15 | $C_{15}H_{32}$ | 4,347 |
| Eicosane | 20 | $C_{20}H_{42}$ | 366,319 |
| Pentacosane | 25 | $C_{25}H_{52}$ | 36,797,588 |
| Triacontane | 30 | $C_{30}H_{62}$ | 4,111,846,763 |
| Heptacontane | 40 | $C_{40}H_{82}$ | 62,491,178,805,831 |

Organic chemistry is a large and complex subject mainly because a small number of atoms can combine with many different structures to produce a diverse range of isomers.

A economical method used by organic chemists is to represent molecules by just their bonds. These bond-only representations follow simple rules:

1. A carbon atom is present at any intersection of two or more bonds or at the end of any unjoined bond.

2. Every carbon atom in the molecule is attached to sufficient hydrogen atoms to satisfy the carbon atom's valency of four.

The use of a bond-only representation is shown in Example 2.9.

**Example 2.9**: Draw the bond-only representations of butane, isobutane, and two isomers of butene.

Strategy: Use the stated rules. The structures are shown below in both bond-only and atom-visible modes.

Butane Isobutane 1-Butene 2-Butene

Compare the butane and isobutane structures with Figures 2.19a and b. Note that the two butene isomers differ in that the double bond is in a different position and the hydrogen atoms are differently distributed.

Bond-only alkane representations are convenient for showing alkane isomers as illustrated in Example 2.10.

**Example 2.10**: Draw the bond-only representations of the isomers of hexane, $C_7H_{16}$.

Strategy: Use the stated rules. Hexane has a total of five isomers, as shown below.

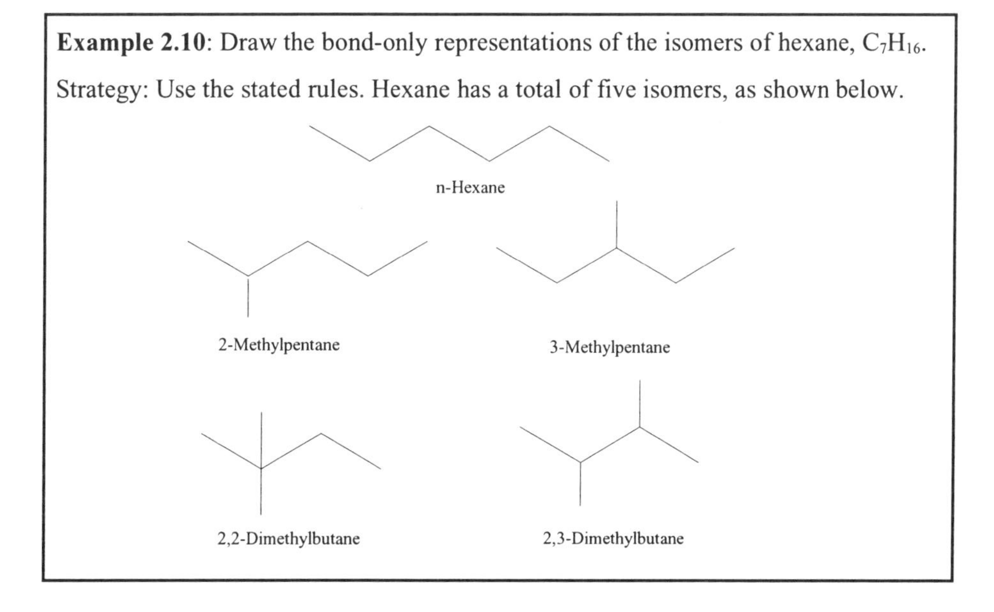


## Naming Molecular Chemical Compounds

Chemists have to be able to refer to specific chemical compounds by name. There are around 20 million named compounds, so having reliable rules to name them is important. There are two separate classes to be named: inorganic compounds and organic compounds. Historically, we've been stuck with many names from the past, so the rules are sometimes messy. Two organizations involved in making the rules are International Union of Pure and Applied Chemistry (IUPAC) and the Chemical Abstracts Service (CAS) of the American Chemical Society. Many common compounds that have been known for a long time have traditional (so-called trivial) names. Examples are water ($H_2O$), ammonia ($NH_3$), and methane ($CH_4$). However, most compounds are named using systematic methods. Because naming chemical compounds is more a problem for chemists than for engineers, we will only brush on the topic. As you can see in Example 2.10, naming organic compounds is reasonably sensible and systematic.

### Binary Compounds of Nonmetals

**Binary compounds** contain exactly two different elements.

Most pairs of nonmetals form molecular compounds. The simplest way to name these compounds is to use Greek numerical prefixes to denote the subscripts in their formulas: mono- 1, di- 2, tri- 3, tetra- 4, penta- 5, hexa- 6, etc. The prefix mono- is generally omitted; however, it may be used for emphasis if desired. By convention, the second listed element in the compound is given its -ide ending. Thus, the halogen elements that we met in Chapter 1 change names to the halide form. So chlorine becomes chloride, bromine becomes bromide, etc. Examples are shown in Table 2.9.

Table 2.9 **Names of Binary Nonmetallic Compounds**

| | |
|---|---|
| HCl | hydrogen chloride |
| IBr | iodine monobromide |
| $N_2O$ | dinitrogen monoxide |
| $NO_2$ | nitrogen dioxide |
| $SO_3$ | sulfur trioxide |
| $S_2Cl_4$ | disulfur tetrachloride |
| $N_2O_4$ | dinitrogen tetroxide |
| $N_2S_5$ | dinitrogen pentasulfide |
| $Cl_2O_7$ | dichlorine heptoxide |

**Example 2.11**: Name the following binary compounds using the Greek prefix method: a. $PBr_3$ b. $CCl_4$ c. $XeF_6$

Strategy: In each case the second element is a halogen. The stem name is of the halide form. The names are therefore

a. Phosphorus tribromide
b. Carbon tetrachloride
c. Xenon hexafluoride

## Section 2.5 Monatomic Ions

An **ion** is an entity with an electric charge. Atoms, being composed of electrons and protons, remain electrically neutral only so long as their number of electrons is unchanged. An atom that gains electrons becomes a negatively charged ion, or an **anion**. An atom that loses electrons becomes a positively charged ion, or a **cation.** From single atoms arise **monatomic** ions. The charges of ions are measured in units of e, the charge of a single electron or proton. Ions of opposite electrical charge attract one another and ions with the same electrical charge repel one another. **Coulomb's law** states that the force is proportional to the charge of the ions and the inverse square of the distance between the ions. Such forces are called coulombic forces.

Ions have already been mentioned: studying how radioactive substances caused ion formation in the air led Rutherford and Soddy to discover $\alpha$, $\beta$, and $\gamma$ radiation. Also, a beam of ions is separated in a mass spectrometer. This section describes chemical compounds that contain ions.

Whether or not an atom of an element has a tendency to form an ion by gaining or losing electrons correlates strongly with the element's position in the periodic table. More specifically, it correlates strongly with an element's group number.

The elements of Group 18—The Noble Gases—are indifferent to forming ions in chemical combination. The elements of Group 17—The Halogens—readily form ions with a single negative charge. (A **halide** is a halogen element in a compound and more specifically the form of anion with a negative charge.) The elements of Group 16—readily form ions with a doubly negative charge. The elements of Group 1—The Alkali Metals—readily form ions with a single positive charge. The elements of Group 2—The Alkaline Earths—readily form ions with a doubly positive charge. Main group elements tend to gain or lose electrons to become *noble gas-like*. Here are example equations:

| | |
|---|---|
| $Xe \rightarrow Xe$ | Noble gases of Group 18 are chemically unreactive |
| $Cl + e^- \rightarrow Cl^-$ | Chlorine, a Group 17 halogen element, gains one electron |
| $O + 2e^- \rightarrow O^{2-}$ | Oxygen, a group 16 element, gains two electrons |
| $Na \rightarrow Na^+ + e^-$ | Sodium, a Group 1 alkali metal, loses one electron |
| $Ca \rightarrow Ca^{2+} + 2e^-$ | Calcium, a Group 2 alkaline earth metal, loses two electrons |

Solid sodium chloride $Na^+Cl^-$ (table salt) is depicted in Figure 2.20. The darker gray spheres represent sodium cations and the lighter gray spheres represent chloride ions ($Cl^-$ is called chloride). When anions and cations have been formed, they come together to form compounds because of the mutual electrical attraction of the positive ions and negative ions. What's shown is a tiny fragment, greatly magnified. On the same scale a cube-shaped 1 mm salt crystal would be about 400 miles on each edge.

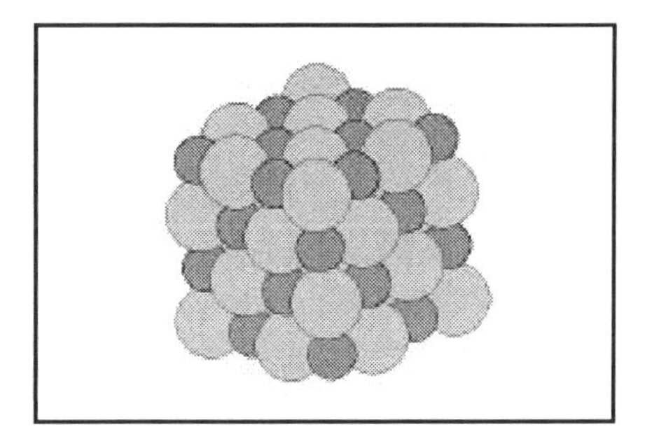

Figure 2.20 Sodium chloride represented in a hard-sphere space filling manner.

Many common substances are composed of anions and cations. Such substances are called **ionic compounds**. There is a perfect balance of the *total* charge of all the anions and cations in every ionic compound. In ionic compounds nothing comparable to a molecule exists, there is simply an alternating array of ions repeating and extending throughout space. The whole array is held together by a balance of **Coulombic forces**. In $CaCl_2$ ($Ca^{2+}$ + 2 $Cl^-$), which like ordinary salt is often spread on icy roads, calcium exists as $Ca^{2+}$ ions in an array along with *twice* the number of Cl ions.

> When ions are formed, the nucleus of the atom from which they are formed remains *unchanged*. In the case of monatomic cations, this limits the maximum possible positive charge to the nuclear charge.

A lithium atom, for example, can form three different ions:

| | | | | | |
|---|---|---|---|---|---|
| $Li$ | $\rightarrow$ | $Li^+$ | + | $e^-$ | First ionization step |
| 3 protons in nucleus<br>3 electrons | | 3 protons in nucleus<br>2 electrons | | first electron lost | |

| | | | | | |
|---|---|---|---|---|---|
| $Li^+$ | $\rightarrow$ | $Li^{2+}$ | + | $e^-$ | Second ionization step |
| 3 protons in nucleus<br>2 electrons | | 3 protons in nucleus<br>1 electron | | second electron lost | |

| | | | | | |
|---|---|---|---|---|---|
| $Li^{2+}$ | $\rightarrow$ | $Li^{3+}$ | + | $e^-$ | Third ionization step |
| 3 protons in nucleus<br>1 electron | | 3 protons in nucleus<br>0 electrons | | third electron lost | |

Thus, the *maximum* possible charge for a lithium ion is 3+. However like all alkali metals, lithium in its compounds exists only as a 1+ charged ion. In the everyday world there is not sufficient energy to cause lithium to lose more that one electron.

In forming compounds the main group metals of Groups 1 and 2 tend to lose electrons until they have the same number of electrons as the preceding noble gas in the periodic table. The ion formed is then said to be **isoelectronic** with the noble gas having the same number of electrons. $Na^+$ has ten electrons; an atom of the noble gas neon has also has 10 electrons. Sodium ions are isoelectronic with neon. Dipositive calcium ions, $Ca^{2+}$, form from calcium atoms (element 20) by the loss of two electrons. A calcium ion with 18 (20 - 2) electrons is isoelectronic with argon (element 18).

Likewise, nonmetals tend to gain electrons to form negatively charged ions that are isoelectronic with the nearest noble gas. A chlorine atom (element 17) gains one electron to become isoelectronic with argon (element 18). Oxygen (element 8) in many compounds exists as the $O^{2-}$, or oxide, ion that is isoelectronic with Ne (element 10). Knowing how to write the common ions of the main group elements allows you readily to write down formulas for ionic compounds they form. The formulas for ionic compounds are conventionally written with the cation on the left and the anion on the right. All ionic compounds are solids at room temperature. Furthermore, the total positive charges on the cations are balanced by the total negative charges of the anions, making ionic compounds electrically neutral. The table below shows how anions combine with cations:

| | | | |
|---|---|---|---|
| $Ca^{2+}$ | $Cl^-$ | leads to | $CaCl_2$ calcium chloride |
| $Al^{3+}$ | $O_2^-$ | leads to | $Al_2O_3$ aluminum oxide |
| $Mg^{2+}$ | $N_3^-$ | leads to | $Mg_3N_2$ magnesium nitride |
| $Ca^{2+}$ | $O_2^-$ | leads to | CaO calcium oxide |

**Example 2.12**: Write the formulas for the ionic compounds formed between:
a. Potassium and sulfur. b. Calcium and nitrogen. c. Aluminum and oxygen.

Strategy: Use the periodic table to establish the proper charge on the ions by adding or removing electrons to become isoelectronic with the nearby noble gas:

Potassium (group 1 metal) forms $K^+$ cations and sulfur (group 16 nonmetal) forms $S^{2-}$ anions. Together these ions make the compound $K_2S$, called potassium sulfide.

Calcium (group 2 metal) forms $Ca^{2+}$ cations and nitrogen (group 15 nonmetal) forms $N^{3-}$anions. Together these ions make the compound $Ca_3N_2$, called calcium nitride.

Aluminum (group 13 metal) forms $Al^{3+}$ cations and oxygen (group 16 nonmetal) forms $O^{2-}$ anions. Together these ions make the compound $Al_2O_3$, called aluminum oxide.

Binary ionic compounds combine cations with anions. As shown in Example 2.12, the cation is named first, followed by the anion. Cations are named for the element from which they are derived; anions are named using the stem name of the nonmetal with -ide replacing the ending. The stem name is a shortened form of the element's name.

Figure 2.21 A Modern Mass Spectrometer.

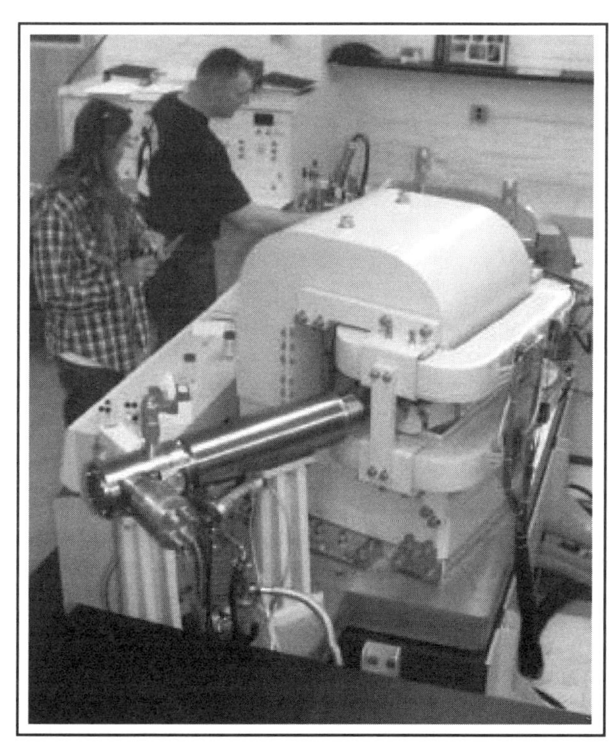

Mass spectrometers are routine tools in modern, well-instrumented chemistry laboratories. The photograph pictures a high resolution mass spectrometer.

This particular instrument spent several years in service for the Bristol-Myers Squibb pharmaceutical company. The operator is adjusting the inlet port where samples are introduced into the instrument. In the foreground you see the large magnet that separates the ions according to their different masses.

The ion detector is located at the left hand end of the shiny, metallic tube in the lower left-hand corner if the photograph.

Considerable skill and sophistication is needed to operate an instrument of this complexity.

**Ion Propulsion Engines**

Ion engines are the latest thing in space propulsion systems. The stable, noble gas xenon is the favored propellant. A xenon ion propulsion system (XIPS) begins by impinging high energy electrons on atoms of xenon creating xenon ions

$$e^- + Xe \rightarrow Xe^+ + 2e^-$$

An electric field accelerates the Xe ions. The fast-moving ions are neutralized as they leave the engine, so the actual propulsion is the momentum of a jet of xenon atoms. Xenon is used because it is the heaviest, non-radioactive noble gas. Xenon increases attainable levels of thrust compared with lighter gases and, because it is an inert gas, its storage and transportation costs are modest.

At the end of 1998, five on-orbit Hughes Space and Communications Company satellites were using XIPS for station keeping. Many more are being constructed. In normal operation, a typical ion thruster will operate for approximately 5 hours a day, using 500 watts from the satellite's 8-kilowatt solar array.

NASA's Deep Space 1 launched October 24, 1998 from the Cape Canaveral Air Station, is the first spacecraft to use ion thrusters as a primary propulsion system. The ion engine on board the 8½ foot, 1,000-pound spacecraft is fed by 81.5 kilograms of xenon propellant carried on the spacecraft. This quantity of xenon can provide up to 20 months of continuous thrust (more than enough for the mission). The engine will eventually increase the craft's speed beyond launch velocity by 4.5 kilometers per second, the equivalent of 10,000 miles per hour. Deep Space 1 reactivated its ion propulsion system on March 15, 1999 and the gentle, steady ion thrust has been operating since then. Xenon ion propulsion offers the possibility that humans may eventually venture beyond the solar system.

By May of 2000, Deep Space 1 had logged over 4000 flawless hours of thrust performance. However, navigational problems and software have degraded its mission.

[❁kws +"Deep Space 1" +"ion propulsion"]

## Section 2.6 Polyatomic Ions

**Polyatomic ions** are charged entities that contain two or more atoms. Simply diatomic ions are formed when x-rays or radioactive emissions pass through air. In these experiments, cations are produced by the removal of one or more electrons from gaseous molecules in the air. The following equations represent the formation of diatomic ions:

$$N_2 + \text{energy} \rightarrow N_2^+ + e^-$$
$$O_2 + \text{energy} \rightarrow O_2^{2+} + 2e^-$$

However, of much more interest to chemists are the common polyatomic ions which occur widely in minerals and simple laboratory compounds. For example, carbonate ions, $CO_3^{2-}$, composed of one carbon atom, three oxygen atoms, and two electrons, exist in minerals such as limestone and marble, in wood ashes, and in soda water.

### Naming Polyatomic Ions

Some important polyatomic anions are listed in Table 2.10. It's not particularly elegant learning, but you should memorize the formulas, names and charges of these ions. Naming ionic compounds containing polyatomic ions follows the same pattern as that for simple ions. The name of the cation is written first, followed by the name of the anion; either or both may be a polyatomic ion.

| | | |
|---|---|---|
| $Na^+ (CO_3)^{2-}$ | $Na_2CO_3$ | Sodium carbonate |
| $NH_4^+ (PO_4)^{3-}$ | $(NH_4)_3PO_4$ | Ammonium phosphate |
| $Al^{3+} (SO_4)^{2-}$ | $Al_2(SO_4)_3$ | Aluminum sulfate |

Parentheses are put around the formula unit for polyatomic ions when more than one is used in the formula. In this way the subscripts that show the number of each type of atom in the polyatomic ion are distinguished from the subscripts that show the number of polyatomic ions in the formula.

Table 2.10 **Common Inorganic Polyatomic Ions**

| Polyatomic anions with a charge of -1. | | Polyatomic anions with a charge of -2. | | Polyatomic anions with a charge of -3. | |
|---|---|---|---|---|---|
| $OH^-$ | hydroxide | $CO_3^{2-}$ | carbonate | $PO_4^{3-}$ | phosphate |
| $NO_2^-$ | nitrite | $SO_4^{2-}$ | sulfate | | |
| $NO_3^-$ | nitrate | $SO_3^{2-}$ | sulfite | | |
| $HCO_3^-$ | hydrogen carbonate or bicarbonate | $S_2O_3^{2-}$ | thiosulfite | Polyatomic cation with a charge of +1 | |
| $C_2H_3O_2^-$ | acetate | $C_2O_4^{2-}$ | oxalate | $NH_4^+$ | Ammonium |
| $ClO^-$ | hypochlorite | $Cr_2O_7^{2-}$ | dichromate | | |
| $ClO_3^-$ | chlorate | $HPO_4^{2-}$ | hydrogen phosphate | | |
| $ClO_4^-$ | perchlorate | | | | |
| $CN^-$ | cyanide | | | | |
| $MnO_4^-$ | permanganate | | | | |
| $H_2PO_4^-$ | dihydrogen phosphate | | | | |

**Example 2.13**: Write the formulas for: a. cesium sulfate, b. barium hydroxide, c. aluminum nitrate.

Strategy: Use the periodic table and Table 2.10 to obtain the electrical charges. Balance the positive and negative charges using subscripts.

Cesium (a group 1 metal) forms ions with a 1+ charge. Sulfate anion is $SO_4^{2-}$. Together they form $Cs_2SO_4$, cesium sulfate.

Barium (a group 2 metal) forms ions with a 2+ charge. Hydroxide anion is $OH^-$. Together they form $Ba(OH)_2$, barium hydroxide.

Aluminum (a group 13 metal) forms ions with a 3+ charge. Nitrate anion is $NO_3^-$. Together they form $Al(NO_3)_3$, aluminum nitrate.

## Hydrates

Many common solid chemical compounds and minerals occur with water molecules more or less firmly combined with them. Compounds of this type are called hydrates and the water they contain is called water of hydration. In writing the formulas of hydrates, the water of hydration is appended to the chemical formula following a midline period. Common examples of hydrates are copper(II) sulfate pentahydrate, $CuSO_4{\cdot}5H_2O$ and calcium sulfate dihydrate, $CaSO_4{\cdot}2H_2O$ (the mineral gypsum).

## Essential Knowledge—Chapter 2

**Glossary words:** nuclear reactions, radioactive decay, nuclear equations, radionuclides, half-life, properties of subatomic particles $\alpha$, $\beta$, and $\gamma$ decay, daughter nuclides, natural decay series, positron emission, electron capture, spontaneous fission, induced fission, fission products, stability of nuclides, nuclide chart, radioisotopes, belt of stability, nuclide n/p ratio, ion bombardment reactions, trans-uranic elements, successive neutron capture, rates processes, radiocarbon dating, $^{14}C/^{12}C$ and $^{13}C/^{12}C$ ratios, energy changes of nuclear reactions, nuclear reactor, mass defect, energy level diagrams, binding energy per nucleon, mass spectrometry, fragmentation pattern, molecular formulas, ball-and-stick and space-filling representations, molecular visualization, central atoms, ligands, ideal geometry, molecular substance, molecular formulas, structural formulas, hydrocarbons, empirical formulas, polymers, chemical structure, molecular and molar weight (molar mass), valency, single bonds, double bonds, triple bonds, delocalized bonds, covalent bonds, molecular compounds or covalent compounds, organic compounds, families of organic compounds and functional groups, R group. alkyl groups, structural isomers, alkanes, monatomic ions, ionic compounds, polyatomic ions.

**Key Concepts:** Natural radioactivity, transmutation, radioelements and the periodic table. Making synthetic elements. The range of known nuclides and isotopes. Nuclear reactions and their energy changes. Mass spectrometry and its role in modern chemistry. Nature and structure of molecular compounds. Nature and structure of ionic compounds.

**Key Equations:**

Concentration of a decaying radionuclide at a specified time: $c = c_o e^{-kt}$

Radioactive rate constant related to half-life: $k = (\ln 2)/t_{½}$

Energy from mass calculation: $E = mc^2$

Molecular weight (MW) = $\Sigma\, n_i AW_i$

## Questions and Problems

### Nuclear Reactions and Radioactivity

2.1. Natural boron contains 19.600% boron-10, which has an isotopic mass of 10.013 u; the other natural isotope is boron-11. Use the atomic weight of boron to determine the isotopic mass of boron-11 to five significant figures.

2.2. Why is the mass of an isotope (expressed in u) approximately equal to its nucleon number?

2.3. State the mass of an electron to 10 significant figures. State the current uncertainty in the value of the electron's mass.

2.4. Actinium is continuously formed in uranium minerals by the decay sequence $^{235}U$ (-α) → $^{231}Th$ (−β) → $^{231}Pa$ (-α) → $^{227}Ac$. Write a series of balanced nuclear equations to show the formation of $^{227}Ac$.

2.5. Complete the following table:

| Nuclide | Mass (u) to five significant figures | Mass (u) to three significant figures |
|---|---|---|
| $^{103}Rh$ | | |
| $^{108}Cd$ | | |
| $^{119}Sn$ | | |
| $^{123}Te$ | | |
| $^{127}I$ | | |
| $^{173}Yb$ | | |
| $^{204}Hg$ | | |
| $^{209}Bi$ | | |

2.6. The naturally occurring radionuclide $^{232}Th$ decays in multiple steps, eventually forming the stable nuclide $^{208}Pb$ as shown in Section 2.1. Write a balanced nuclear equation for each of these steps.

2.7. Write a balanced nuclear equation to show the positron emission and the electron capture decay of neodymium-129.

2.8. Write a balanced nuclear equation to represent the spontaneous fission of a $^{238}U$ nuclide forming a $^{137}I$ nuclide, three neutrons, and another nuclide whose formula you are to deduce.

2.9. The last known element in the periodic table was made in 1999. It is element 118 with a nucleon number of 293: $^{293}_{118}Uuo$. This nuclide decays by successive alpha particle emissions. How many alpha particles are emitted before a daughter product with a name is reached?

2.10. a. Rutherford and Soddy discovered that one of the three components of radioactivity is unaffected by passing through a electrical field. Was it α, β, or γ? b. Calculate the ratio of the mass of an α particle to the mass of an electron.

2.11. Make a graph of the masses of the natural ruthenium isotopes plotted as a function of their neutron number. Estimate the masses of $^{97}Ru$ and $^{103}Ru$.

2.12. Look in Appendix D and examine the masses of the nuclides $^{92}Nb$ and $^{146}Sm$. Comment on their accuracy. Can you improve on these values using internet resources?

### The Stability of Nuclides

2.13. Describe three important factors that control the stability of a nuclide.

2.14. Explain why neutron-rich nuclides tend to decay via $\beta^-$ emission.

2.15. Explain why radionuclides exist naturally.

2.16. Starting with $^{239}_{94}Pu$ write a series of nuclear equations alternating neutron capture and $\beta^-$ emission. Continue until an isotope of Es has been formed. What isotope is it?

## Rates of Radioactive Decay

2.17. The half-life of $^{224}Ra$ is 3.66 days. What percentage of a sample of $^{224}Ra$ remains after 3.5 days?

2.18. A radionuclide has a half life of 45 days. How long will it take for the radioactivity of this nuclide to decrease to 10.0% of its original value?

2.19. The nuclide $^{243}Es$ has a half life of 20.47 days and decays by alpha particle emission. If an airborne sampler gathered $8 \times 10^{10}$ atoms of $^{243}Es$ how many atoms would remain in the sample 2 weeks later?

2.20. A sample of charcoal from an ancient campfire in Australia had a $^{14}C$ specific activity of 5.2 dpm $g^{-1}$ (disintegrations $min^{-1}$ $g^{-1}$). When did the plant that was the source of the charcoal die?

2.21. Write balanced nuclear equations for each of the following processes: a. the $\beta$ decay of $^{194}Ir$ with a half life of 19.2 hours, b. the decay of $^{129}Pr$ which occurs by both electron capture and positron emission with a half life of 24 seconds, and c. the $\alpha$ particle decay of $^{237}Np$ which occurs with a half life of $2.14 \times 10^6$ years.

2.22. Use the half life data in the previous problem to calculate the time needed for $^{194}Ir$ and $^{237}Np$ to reach the same percentage of decay as $^{129}Pr$ does in 3.45 seconds.

2.23. The rate of decay of a radionuclide is often stated in disintegrations per second (dps). The rate law is dc/dt = -kc, where k is the specific rate constant of the decay and c is the number of atoms in the sample. Calculate the number of dps in a sample that contains $2.2 \times 10^7$ atoms of a radionuclide with a half life of 22.6 days.

## Subatomic Particles

2.24. Briefly describe the key contributions to our knowledge of subatomic particles made by Ernest Rutherford, J. J. Thomson, Robert Millikan, James Chadwick, and Francis Aston.

2.25. Write an equation to show the formation of a tripositive oxygen molecule from a neutral oxygen molecule, such as might happen when radioactive rays pass through air.

2.26. Early version cathode ray tubes were used by J. J. Thomson to measure the e/m ratio of an electron. Describe three applications of cathode ray tubes in your everyday life.

2.27. A TV tube is a cathode ray tube. Electrons produced by electron guns in the tube are focused by electromagnets onto a fluorescent screen that emits light and produces the picture. b. If the vacuum tube develops a small leak, it will soon stop working properly. Why is this? b. How does a color TV produce color?

2.28. To determine the mass of an atom using a mass spectrometer, the atom has first to be converted to an ion. Why is this necessary and how is this process accomplished in a mass spectrometer?

## The Curve of Binding Energy and Mass-Energy Relationships

2.29. Calculate the binding energy per nucleon for the nuclides $^{4}He$ and $^{238}U$. Express your answer in the units of kilojoules per mole and compare your answers with data on the Curve of Binding energy.

2.30. Calculate the energy in kilojoules per mole of uranium nuclei when the induced fission reaction: $^{235}_{92}U + ^{1}_{0}n \rightarrow$ $^{131}_{56}Ba + ^{105}_{36}Kr + 4\ ^{1}_{0}n$ occurs. To do this use mass data from Appendix D to calculate $\Delta$mass and then substitute that into $E = mc^2$.

2.31. Consider the nuclear fusion equation: $4\ ^{1}_{1}H \rightarrow ^{4}_{2}He + 2\beta^+$. Use information Appendix D to calculate the energy liberated in kilojoules per mole of helium formed. How many times is this energy greater than the energy of a mole of bonds with a chemical bond energy of 350 kJ $mol^{-1}$?

## Mass Spectrometry

2.32. What is a mass spectrum?

2.33. An amusing, antique way to determine the isotopic composition of an element was to plot the peaks of its mass spectrum on graph paper and then cut the peaks out with scissors and weigh them. From the following data, estimate the atomic weight of magnesium. You may assume that the nuclidic mass of these nuclides is the same as their nucleon number.

2.34. Make an enlarged copy of the figure in the chapter that shows a block diagram of the isotopes of germanium. Use an ordinary ruler to calculate the area of each of the five peaks and express each as a percentage of the total area of all the peaks. Make a side-by-side table of your data and the isotopic abundance for germanium from Appendix D.

## Molecules

2.35. Hydrogen has two stable isotopes $^{1}H$ and $^{2}H$. Oxygen has three stable isotopes $^{16}O$. $^{17}O$, and $^{18}O$. How many "different" water molecules exist, that is molecules with different isotopic contents?

2.36. a. What is meant by the term "molecular visualization"? b. Describe the Netscape plug-in CHIME.

2.37. Use the concept of valency to explain why the molecule $CH_3$ does not exist whereas $C_2H_6$ does.

2.38. Use the concept of valency to explain why the molecule $C_2H_5$ does not exist but that the molecule $C_2H_4$ and two different isomers of the molecule $C_4H_{10}$ do exist. Also, sketch the two-dimensional structures of butane and 2-methylpropane. These molecules are pictured earlier.

2.39. Draw the structural formulas of each of the following: ethanol, 1-propanol, and 2-propanol. Draw the structures of an isomer of ethanol, and an isomer of 1-propanol that is not 2-propanol.

2.40. The molecule $C_2H_4O$ is both ethylene oxide and vinyl alcohol. Sketch the structural formulas of this pair of isomers.

2.41. The compound $C_6H_4Cl_2$ is called dichlorobenzene. It consists of a benzene ring the six carbon atoms of which are joined to four hydrogen atoms and two chlorine atoms. Show that the valency of each of the atoms is satisfactory in the molecule. Sketch three isomers that involve changing the relative positions of the two chorine atom substituents.

2.42. Draw the structural formulas of each of the following: a. a ketone that contains four carbon atoms, b. the simplest aldehyde that contains a benzene ring, c. formic acid (the simplest carboxylic acid containing only one carbon atom), d. oxalic acid, which consists of two carboxylic acid groups joined together, e. the pair of isomers 1-propylamine and 2-propylamine.

2.43. Each of the following diagrams is a conventional chemists' representation of a saturated hydrocarbon with the general chemical formula $C_nH_m$. Write the specific chemical formula for each compound.

2.44. Each of the following diagrams is a conventional chemists' representation of an unsaturated, cyclic hydrocarbon with the general chemical formula $C_nH_m$. Write the specific chemical formula for each compound.

2.45. Calculate the molecular weight and the molar weight of quinine, $C_{20}H_{24}O_2N_2$.

2.46. Compare and contrast the lengths and strengths of carbon-carbon single, double, and triple bonds. What is a delocalized carbon-carbon bond? What is a covalent bond?

2.47. Each of the following compounds is found in food. State how many elements are present and how many total atoms each molecule of the compound contains. a. citric acid, $C_6H_8O_7$, b. caffeine, $C_8H_{10}O_2N_4$ and, c. quinine, $C_{20}H_{24}O_2N_2$. Which has the smallest molar mass? Which the greatest?

2.48. For each of the following industrially important molecular compounds (Appendix F) how many elements are present and how many total atoms are in each molecule of the compound: ammonia, propylene, ethylene dichloride, toluene, phenol, and vinyl acetate.

2.49. The ratios of the numbers of atoms in a compound's empirical and molecular formulas are always the same. Explain.

2.50. The molecule $S_8$ contains exactly eight chemical bonds and all the sulfur atoms are joined. Draw a structural formula to represent the molecule.

2.51. Name the following molecular compounds: a. $P_2O_5$, b. $BrF_3$, c. $N_2O$, d. $NF_3$.

2.52. Name the following molecular compounds: a. $Cl_2O_7$, b. $OF_2$, c. $N_2O_3$, d. $CBr_4$

## Ions

2.53. Write balanced equations to show the formation of: a. a tripositive ion from an aluminum atom by loss of three electrons, b. a dipositive ion from a barium atom by loss of two electrons, c. a sulfide ion from a sulfur atom by the gain of two electrons.

2.54. The SI base unit of electrical charge is the coulomb. In chemistry, the charges of ions are rarely stated in coulombs. a. In what units are they stated? b. Why is the SI unit for charge rarely used by chemists to specify the charges on ions?

2.55. State the formula of the ion of maximum positive charge that can be produced by removal of electrons from an atom of oxygen. Write an equation to represent the formation of such an ion. Do the same for a sulfur atom and a krypton atom. Are such reactions more likely to occur in the sun than on Earth?

2.56. State the formulas of a. potassium carbonate b. magnesium nitrate c. ammonium phosphate d. diammonium hydrogen phosphate e. ammonium dihydrogen phosphate.

2.57. Name the ionic compounds: a. $Al_2(SO_4)_3$, b. $KMnO_4$, c. $K_2Cr_2O_7$, d. $CaC_2O_4$, e. $NH_4C_2H_3O$, f. RbI,g. $LiClO_4$

2.58. Name the following binary ionic compounds: a. $Cs_2Se$, b. $MgI_2$ ,c. $Li_3N$, d. $Al_2S_3$.

2.59. Write the chemical formulas for barium iodide dihydrate and strontium chloride hexahydrate. Calculate the molar weights of these hydrates.

## Problem Solving

2.60. During the period 1945-1963 nuclear weapons were the object of unconfined testing. In this context, what was "fallout?" Home insurance policies frequently exclude liability for damage from ionizing radiation. Under what circumstances might one's home be exposed to large amounts of ionizing radiation?

2.61. Do an internet search and explain the phrase "Tickling the tail of the dragon."

2.62. In his famous experiment, Rutherford aimed a beam of alpha particles at a gold foil. The radius of a gold atom is about 144 pm. The radius of the nucleus of any nuclide (r) can be estimated (in centimeters) from the empirical equation $r = 1.2 \times 10^{-15} \times$ (its nucleon number)$^{1/3}$. Use geometric arguments based on areas to estimate the chance that a single alpha particle passing through a single gold atom will hit the atom's nucleus. Incidentally, does this empirical equation make logical sense to you?

2.63. Use information from the previous problem to estimate the density of a gold nucleus. Assume the nucleus is a sphere. Express your answer both in g cm$^{-3}$ and tons per cubic inch. The formula for the volume of a sphere is $V = 4/3\ \pi r^3$.

2.64. The depiction of the sodium chloride structure shown in this chapter is a cube with an edge length of about 0.75 inches. The true size of the edge length of the fragment shown is about 500 pm. Use the algebra-of-quantities (Appendix A) to convert pm to miles and to test the assertion in the chapter that a 1 mm salt cube would be 400 miles wide on the same scale.

2.65. A polymer molecule is composed of 22,000 methylene ($-CH_2-$) units linked in a chain. Calculate the molar mass of this molecules. What did you do about the "end groups?"

2.66. Make a sketch that joins: methylene group ($-CH_2-$), ester group, methylene group, ester group ... for a total of 12 carbon atoms. Explain why what you have sketched is a polyester.

2.67. The methane molecule is tetrahedral. A regular tetrahedron can be visualized inside a cube. If you imagine the carbon atom in the very center of the cube and hydrogen atoms at four of the eight corners of the cube, you should be able to "see" the tetrahedron. Based on that view, calculate the H-C-H angle. In purely geometric terms this question asks you to calculate the angle subtended at the center of a cube by its face diagonal.

2.68. The carbon-hydrogen bond length in methane is about 110 pm. Calculate the distance between any pair of hydrogen atoms in the $CH_4$ molecule.

2.69. Calculate the F2-F3 distance in the $SF_4$ molecule pictured in Figure 2.15. Estimate the maximum F-F distance in this molecule.

2.70. When atmospheric testing of nuclear devices was carried out in the 1950s and 1960s by the US and the Soviet Union, one radioactive isotope, strontium-90, was especially feared. Describe the makeup of neutral strontium-90 in terms of protons, neutrons and electrons. Knowing that calcium ions are an important component of bones, why is there a very good reason to be particularly concerned about strontium-90 in nuclear fall-out?

2.71. The equation $E = mc^2$ demonstrates that mass and energy are equivalent. Calculate the conversion factor that changes u to kJ.

2.72. Written in the international style, the mass of a proton is 1.007 276 470 u and the mass of a neutron is 1.008 664 904 u. An alpha particle is composed of two protons and two neutrons and has a mass of 4.001 506 170 u. An alpha ($\alpha$) particle is a $^4He$ nucleus. Calculate the mass taken together of two protons and two neutrons. Calculate the difference in mass between an $\alpha$-particle and two protons and two neutrons together, reporting your answer to seven significant figures. Also calculate the percentage difference between the two values.

2.73. The missing mass (see Section 2.3) of a helium nucleus is 0.030377 u. Use $E = mc^2$ to calculate the energy equivalent to this missing mass.

2.74. The radioisotope $^{244}Pu$, with a half-life of $8.3 \times 10^7$ years, was first identified in the debris of the 1952 Mike nuclear test. Several years later a team of chemists at the Los Alamos National Laboratory was able to extract 20 million atoms of $^{244}Pu$ from a sample of commercial thorium ore. The concentration of the $^{244}Pu$ in the ore was estimated to be 1 part in $10^{18}$. Deduce the mass of the ore sample.

2.75. Use internet resources to learn about the ionosphere. What is it and why does it get its name?

2.76. Find internet sites where you can use the chemical Chime program to manipulate protein molecules.

2.77. Natural carbon consists of 98.90% of the stable nuclide $^{12}C$ and 1.10% of the stable nuclide $^{13}C$. It also contains a small amount of the radionuclide $^{14}C$. A carbon sample taken from the biosphere (a living plant or animal) has a beta emission rate of 13.6 disintegrations $s^{-1}$ $g^{-1}$ of carbon according to the nuclear equation $^{14}_{6}C \rightarrow {}^{14}_{7}N + {}^{0}_{-1}\beta$. Calculate the ratio of the number of atoms of $^{13}C/^{14}C$ in a sample of carbon from the biosphere. The half life of $^{14}C$ is 5730 years.

2.78. The $\frac{^{13}C}{^{14}C}$ isotope ratio in biosphere carbon (modern living material) is $9.33 \times 10^9$. Accurate measurement of this ratio in archeological materials requires the technique of accelerator mass spectrometry (AMS). [❁kws "accelerator mass spectrometry"]. Only small samples (sometimes as little as 50 μg of carbon) are required for AMS carbon isotope ratio measurements. However, at low sample sizes careful sample preparation and pretreatment are important factors. A small sample of white pine charcoal from the lowest level of human habitation at the Cactus Hill site near Emporia, in Sussex County, Virginia, gave an AMS $\frac{^{13}C}{^{14}C}$ isotope ratio of $7.12 \times 10^{10}$. The half life of $^{14}C$ is 5730 years. How long ago was this habitation? (Ignore the corrections. Correcting radiocarbon dating is a complicated business because, among other considerations, nuclear explosions have added to the earth's inventory of $^{14}C$ while extensive fossil fuel burning has diluted it.)

# Chapter 3

# Chemical Reactions and Stoichiometry

## Introduction

**Stoichiometry** is the study of mass and number relationships in chemistry. It's the boot camp of our chemistry course. Stoichiometry's principles are: conservation of mass, conservation of atoms, and balancing chemical equations. With those principles, and the ability to use units or ratio and proportion, you should be able to work almost any stoichiometry problem. Chemical stoichiometry is not rocket science—though it is the science of rocket fuels. During Chapter 3 you will see lots of stoichiometry problems; there are 41 example problems in this chapter. Many students find stoichiometry to be review material—especially if they've had a strong high school course. Your studies in this chapter should give you lots of practice at problem solving. Whether or not an engineer makes much daily use of chemistry, every engineer makes regular use of problem-solving skills. The key to becoming a good problem solver is simple: practice, practice, practice.

## Section 3.1 Chemical Reactions

Recall that a change in the composition of a substance is both a chemical change and a chemical reaction. During a chemical reaction, **reactants** get consumed and **products** get formed. We say "reactants yield products" and write it:

$$\text{Reactants} \rightarrow \text{Products}$$

A chemical reaction is something that takes place "out there" in nature. It's a real process. To represent chemical reactions, chemists write chemical equations. A **chemical equation** is a symbolic written representation of a chemical reaction. There are an enormous number of possible chemical reactions. Fortunately, they can be organized into a few types, some of which are introduced here.

### Decomposition and Addition Reactions

A simple type of chemical reaction is a **decomposition reaction,** in which a chemical entity breaks down or decomposes into two or more smaller fragments. Here's the equation that represents the simplest conceivable decomposition reaction:

$H_2 \rightarrow 2\ H$ decomposition of a hydrogen molecule into two hydrogen atoms

A molecule of diatomic hydrogen (the simplest element) breaks down to form two hydrogen atoms. At room temperature this reaction requires the input of energy, specifically, at 25°C it needs 436 kJ $mol^{-1}$, which is the hydrogen molecule's molar bond energy. **Bond energy** is the energy needed to break a mole of bonds of a specified type. If you reverse the above equation you get the simplest conceivable addition reaction. An **addition** reaction occurs when two or more smaller units combine to make a single, larger molecule.

$2\ H \rightarrow H_2$ addition of two hydrogen atoms to form a hydrogen molecule

At room temperature the addition reaction releases energy, specifically at 25°C it produces 436 kJ $mol^{-1}$, which is also the hydrogen molecule's molar bond energy.

Which reaction happens faster, decomposition or addition? Both reactions occur continually in any hydrogen sample, but which one is faster depends on the temperature. At 25°C, $H_2$ is formed much faster than 2 H, the sample exists mostly as $H_2$, and we say addition is the favored reaction; at 2000°C, 2 H is formed much faster than $H_2$, the sample exists mostly as 2 H, and we say decomposition is the favored reaction. Chemists call this a reversible reaction. A **reversible reaction** is one in which the amounts of products and reactant change if conditions, such as the temperature, are altered. Here's how we write a reversible reaction using a "both ways" arrow:

$H_2 \rightleftarrows 2\ H$ both addition and decomposition are represented

In the equation $H_2 \rightarrow 2\ H$ the "2" is the coefficient of the hydrogen atom. **Coefficients** are the numbers used to balance equations in accord with the principle of conservation of atoms. A **balanced chemical equation** is one in which the number of atoms of each element in the reactants equals the number of atoms of each element in the products. The coefficients in a *balanced* chemical equation are called **stoichiometric coefficients**.

**The Role of Scientific Inquiry in the Progress of Humankind**

THE

AUTHOR'S

# PREFACE.

AN hundred and fifty years are ſcarce elapſed ſince the clouds of prejudice, which had long overſpread the world, began to clear up, and men were convinced, by cultivating the Sciences and attending to Nature, that no fanciful hypotheſes would ever lead them to the true cauſes of thoſe various phenomena that inceſſantly and every where meet the obſerver's eye; but that the narrow limits of the human underſtanding confine the courſe of our reſearches to one ſingle path; namely, that of Experiment, or the Uſe of our Senſes. Yet, in this ſhort period, Natural Philoſophy hath riſen to a high pitch of improvement, and may with truth be ſaid to have made much greater advances towards perfection, ſince the experimental method was introduced, than in the many ages before.

This is true with regard to every branch of Natural Philoſophy; but more particularly with regard to Chymiſtry. Though this Science cannot be ſaid to have ever exiſted without experiments, yet it laboured under the ſame diſadvantages with the reſt; becauſe thoſe who ſtudied it made all their experiments with a view to confirm their own Hypotheſes, and in conſequence of principles which had no foundation but in their imaginations.

In Chapter 1 the birth of chemistry is dated to the rise of quantitative experiments at the end of the 18th-century. In this chapter we study quantitative chemistry in detail. The preface above comes from P. J. Macquer (1718-1784), *Elements of the Theory of Chemistry*, published in translation in London, 25 March 1758. Macquer speaks loudly and clearly to us across 250 years.

Table 3.1 shows some decomposition, addition, and reversible reactions. Check for yourself that these equations are balanced.

Table 3.1 **Some Reversible Reactions**

| Decomposition reaction | Addition reaction | As reversible reaction |
|---|---|---|
| $2\ HgO(s) \rightarrow 2\ Hg(l) + O_2(g)$ | $2\ Hg(l) + O_2(g) \rightarrow 2\ HgO(s)$ | $2\ Hg(l) + O_2(g) \rightleftarrows 2\ HgO(s)$ |
| $N_2O_4(g) \rightarrow 2\ NO_2(g)$ | $2\ NO_2(g) \rightarrow N_2O_4(g)$ | $2\ NO_2(g) \rightleftarrows N_2O_4(g)$ |
| $CaCO_3(s) \rightarrow CaO(s) + CO_2(g)$ | $CaO(s) + CO_2(g) \rightarrow CaCO_3(s)$ | $CaO(s) + CO_2(g) \rightleftarrows CaCO_3(s)$ |
| $C_2H_6(g) \rightarrow H_2(g) + C_2H_4(g)$ | $H_2(g) + C_2H_4(g) \rightarrow C_2H_6(g)$ | $H_2(g) + C_2H_4(g) \rightleftarrows C_2H_6(g)$ |
| $2\ ICl_3(s) \rightarrow I_2(s) + 3\ Cl_2(g)$ | $I_2(s) + 3\ Cl_2(g) \rightarrow 2\ ICl_3(s)$ | $I_2(s) + 3\ Cl_2(g) \rightleftarrows 2\ ICl_3(s)$ |
| $NH_4Cl(s) \rightarrow NH_3(g) + HCl(g)$ | $NH_3(g) + HCl(g) \rightarrow NH_4Cl(s)$ | $NH_3(g) + HCl(g) \rightleftarrows NH_4Cl(s)$ |

The symbols (s) = solid, (l) = liquid, and (g) = gas specify the states of aggregation of each reactant and product. Here's the equation for production of $CO_2$ when lime is made by heating limestone rock in a kiln.

$$CaCO_3(s) \xrightarrow{\Delta} CaO(s) + CO_2\uparrow \text{ (carbon dioxide is a gas)}$$

The symbol $\Delta$ shows that heat is needed to cause the reaction to happen. The up arrow ($\uparrow$) is an alternative way of telling that the product carbon dioxide is a gas.

## Combustion Reactions

Arguably, the single most important type of chemical reaction is burning fuel to meet society's demand for energy. The world's largest corporations are the petrochemical companies, and their main products are fuels. You see the company names on the gas stations of America and they are listed in Appendix F. Fuels combust and liberate heat. It's this heat that is the object of fuel burning. A **combustion reaction** is the reaction (burning) of a substance (a fuel) with oxygen. Gaseous and liquid fuels are typically mixtures of hydrocarbons. The equation for the combustion reaction of natural gas ($CH_4$, methane) that many people use for home heating is:

$$CH_4(g) + 2\ O_2(g) \rightarrow CO_2(g) + 2\ H_2O(g)$$

You know from your own experience that some ignition source, such as a spark or match, is needed to light a gas flame. A fuel represents a potential source of chemical energy that can be tapped if a small amount of energy is supplied to activate the combustion reaction.

Burning a hydrocarbon fuel using plenty of oxygen (or air) yields carbon dioxide and water as the principal products. The general chemical equation for the combustion of a hydrocarbon is:

$$C_nH_m + \frac{(4n + m)}{4} O_2(g) \rightarrow n\ CO_2(g) + \frac{m}{2} H_2O(g)$$

where n and m are integers.

Here's the equation for the combustion of octane, $C_8H_{18}$, using an excess of oxygen. In this example n = 8 and m = 18. Burning octane makes the same products as burning methane (carbon dioxide and water)—but in a different proportion, 16:18 rather than 1:2.

$$2\ C_8H_{18}(l) + 25\ O_2(g) \rightarrow 16\ CO_2(g) + 18\ H_2O(g)$$

Mixtures of a particular isomer of heptane, $C_7H_{16}$ (Figure 3.1), and a particular isomer of octane called 2,2,4-trimethylpentane, $C_8H_{18}$, (Figure 3.2), are used to make comparison standards for gasoline. Hence the term "octane rating." There is probably some octane in most gasolines, but if you consider isomers (Section 2.4) it will come as no surprise to learn that gasoline contains thousands of different molecules.

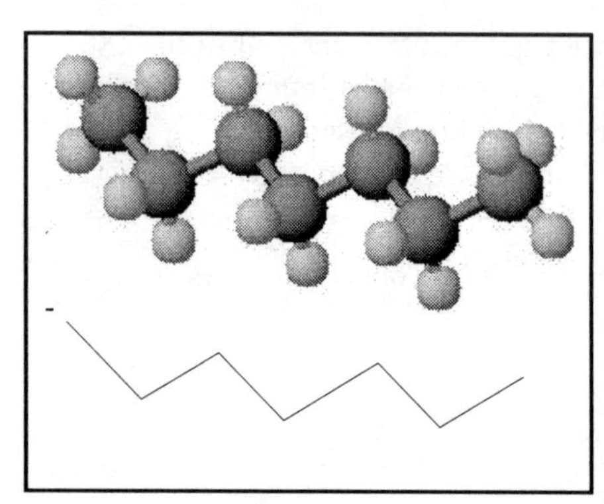

Figure 3.1 n-Heptane

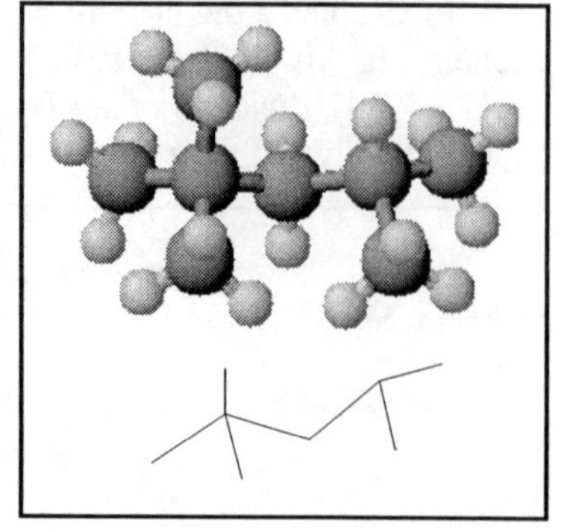

Figure 3.2 isooctane

## Octane Rating:

In 1926, the two hydrocarbons heptane (boiling point 98.4°C. Figure 3.1) and 2,2,4-trimethylpentane or "isooctane" (boiling point 99.3°C, Figure 3.2) were chosen by Graham Edgar as standards to rate automobile fuel. It is important that the two test compounds have similar volatility—that's why they have almost the same boiling points. Pure isooctane was a better fuel than most gasoline then available, while heptane was worse. Hence, if Edgar properly mixed his two hydrocarbons he could reproduce any actual fuel. Thus an 80-octane fuel would be one that was an equivalent fuel to an 80:20 isooctane:heptane mixture. Incidentally, Edgar synthesized isooctane specifically for this purpose.

The primary benefit of a high octane fuel is not related to the fuel's energy content. Rather, it derives from the fuel's ability to resist "pre-ignition." Pre-ignition manifests itself as "knocking," a sort of engine rattle that you might have heard if you've put low-grade fuel in an engine with a high compression ratio. Knocking can destroy engines in short order. The gasoline additive tetraethyl lead, $Pb(C_2H_5)_4$, was once used in the US as an octane rating booster. Nowadays lead-free gasoline is the US norm and non-lead compounds such as methylcyclopentadienyl manganese tricarbonyl $CH_3C_5H_4Mn(CO)_3$ are used as fuel octane boosters. [❁kws +"motor fuel" +"octane rating" +testing]

During hydrocarbon combustion, a **lean** fuel condition exists when there is relative excess of oxygen compared to the fuel. Engineers say a **complete combustion** has occurred when the only reaction products are water and carbon dioxide. When the fuel is in relative excess compared to the oxygen then the reaction is said to take place under **rich** fuel conditions. The terms lean and rich are in the province of combustion engineering. Chemists say instead that one reactant or the other in a combustion reaction is in relative **excess** while the other reactant is simultaneously **limiting**. When the supply of oxygen is limited (so the reaction is fuel rich) then reaction produces significant amounts of carbon monoxide, CO, and engineers say the combustion of fuel is **incomplete**. (The precise definition of what chemists mean by a complete reaction involves stoichiometry, a subject that we'll take up later in this chapter.) Here's the equation for the incomplete combustion of octane to make carbon monoxide as the only carbon-containing product.

$$2\ C_8H_{18}(l) + 17\ O_2(g) \rightarrow 16\ CO(g) + 18\ H_2O(g)$$

For efficiency, the complete combustion of any hydrocarbon fuel is always an objective. In the real world hydrocarbon combustion typically produces a mixture of carbon dioxide, carbon monoxide, and water. There may even be traces of unburned fuel and polluting gases such as $NO_2$ formed from air at high combustion temperatures. The relative proportions of carbon dioxide and carbon monoxide are controlled by the reaction conditions: the relative amounts of oxygen and hydrocarbon, how well mixed they were before combustion, the efficiency of the ignition source, the presence or absence of a catalyst, etc. Whole engineering organizations, such as the Gas Institute, are devoted to the study of combustion reactions. [❁kws "The Gas Institute"]

### Rich and Lean

The flame in Figure 3.3 is burning with plenty of oxygen (an excess of oxygen); it's fuel lean. The flame in Figure 3.4 is burning without sufficient oxygen (the oxygen is limiting); it's fuel rich. A screw mechanism near the burner's gas inlet regulates the amount of air that gets mixed with the gas allowing you to adjust the burner to burn either fuel rich or fuel lean.

Figure 3.3 A lean flame.

Figure 3.4 A rich flame.

When a hydrocarbon is burned with a severely limited amount of oxygen, it produces carbon. Candles don't use oxygen efficiently. Perhaps you've seen a candle burning with a yellow flame and making black wisps of soot. The yellow color is caused by light from hot, glowing particles of carbon. If you hold a ceramic tile in a candle flame you can make a visible deposit of soot, as shown in Figure 3.5.

Carbon black (a kind of soot) is manufactured in this way on an industrial scale. Carbon black is used in printing inks, rubber tires, etc. Equations to represent soot-forming reactions are:

Figure 3.5 A very rich flame.

$CH_4(g) + O_2 \rightarrow C(s) + 2\ H_2O(g)$ Making carbon black from natural gas.

$C_{20}H_{42}(g) + 10\frac{1}{2}\ O_2 \rightarrow C(s) + 21\ H_2O(g)$ Making candle soot.

Engineers sometimes want to boost the performance of fuels to get the maximum possible energy to power a motor. Chemical methods to boost engine horsepower are described in the following box.

## Chemical Methods to Boost Engine Horsepower

Model-plane hobbyists, funny-car drivers, NASCAR racers, hot-rodders, and others use chemistry to increase engine power beyond what's available from conventional gasoline and air combustion. There are various ways power can be increased. Turbocharging is a mechanical method for power boosting.

One chemical method is to replace or supplement air with nitrous oxide ($N_2O$). Ordinary air contains approximately 76.7% by weight of nitrogen and 23.3% by weight of oxygen. (Incidentally, nitrous oxide, or "laughing gas" is an anesthetic gas once widely used during dental surgery.) However, above 550°C in an operating engine nitrous oxide decomposes to produce nitrogen and oxygen: $2\ N_2O \rightarrow 2\ N_2 + O_2$. The mixture $2\ N_2 + O_2$ contains 33% by volume of oxygen, so this mixture is oxygen-enriched compared to air (air is only about 21% by volume oxygen). Furthermore, because $N_2O$ is triatomic, it's about 1.5 times as dense as air. So more oxidant gets into the cylinder for each intake stroke. The combination of higher oxygen percentage and greater oxidant density means that as much as a 50% power increase can be achieved if the proper ratio of fuel to oxidant is used.

$N_2O$ injection was first used by the British and German air forces in World War II. During the 1950s stock car drivers transferred the technology and won a number of races with it before $N_2O$ was banned by NASCAR. Today, commercial suppliers offer "nitrous kits" for many current model cars. But the method is not without its drawbacks. Using an $N_2O$-boosted-fuel explosion increases cylinder pressure and strains the engine. Many motors have literally been blown by improper boosting. Also, a pound of $N_2O$ gives only about a second of boost. Because it's not practical to carry tanks containing much more than 20-50 pounds of $N_2O$, boosting can only be for a brief time.

A second chemical method to boost engine power is to use an exotic fuel. Nitromethane, $CH_3NO_2$, is one such fuel. Nitromethane is actually the chemically simplest nitro explosive compound. It is capable of producing power by detonation even in the absence of air. The product mixture is complex, but detonation can be approximated by the equation: $2\ CH_3NO_2 \rightarrow N_2 + 3\ H_2O + CO + C$. In practice, nitromethane is usually mixed with methanol, $CH_3OH$, and that mixture is combusted with oxygen from air. Model aircraft and drag-racing vehicles often use nitromethane as part of their fuel blend. As with nitrous oxide boosting, successful use of exotic fuels requires close attention to the fuel-oxidant ratios. If the stoichiometry gets out of line, detonation occurs, engine temperatures soar, and motors blow.

[❁kws +"nitrous oxide" or "nitromethane" +fuel +chemical]

## Acid-Base Reactions

Acids and bases have been studied for thousands of years. Throughout most of history they were esoteric substances whose powerful properties were regarded with awe. Today we take them for granted and remain largely ignorant of their important roles in the economy.

### Acids

The simplest definition of an **acid** is any compound which produces hydrogen ion ($H^+$) or protons when dissolved in water. In everyday life we know that acids are sharp-tasting (vinegar, lemon juice) or corrosive (sulfuric or battery acid, $H_2SO_4$) substances, that can be dangerous if misused. We will study acids briefly now and return to them later. The names of some of the most common acids are given in the Tables 3.2 and 3.3. The formulas of the common inorganic or "mineral" acids are usually written with hydrogen as the first element in the chemical formula. As you know from the table of families of organic compounds in Chapter 2, this rule does not apply to organic acids. **Binary** acids contain hydrogen and one other element. The most important binary acids are the **hydrohalic acids**; they contain hydrogen combined with a halide anion. The halide anions are $F^-$, $Cl^-$, $Br^-$, and $I^-$; $X^-$ means any of the halide ions, where halide derives from the group name of halogens. In the gaseous state of aggregation, all the H-X compounds are molecular, i.e., the H and X atoms are joined by a covalent bond. But they all readily dissolve in water to form ions. $H\text{-}X \rightarrow H^+(aq) + X^-(aq)$. In water they are named as acids. The notation **(aq)** means aqueous solution, something dissolved in water

Table 3.2 **Hydrohalic acids**

| Chemical formula HX | Name of binary gaseous compound | Name of compound when dissolved in water |
|---|---|---|
| HF | Hydrogen fluoride | Hydrofluoric acid |
| HCl | Hydrogen chloride | Hydrochloric acid |
| HBr | Hydrogen bromide | Hydrobromic acid |
| HI | Hydrogen iodide | Hydroiodic acid |

For example, hydrogen chloride (HCl) exists as a gas under ordinary conditions. When the gas contacts water it immediately dissolves, yielding ions via an ionization reaction. An **ionization reaction** is one in which a substance dissolves in water to form ions. The equation is:

$$HCl(g) \xrightarrow{H_2O} H^+(aq) + Cl^-(aq)$$ an ionization reaction in water,

where the formula of water written as an annotation above the arrow shows that an aqueous solution is being formed. Hydrochloric acid is a strong acid. A **strong acid**, such as HCl, completely ionizes in water and abundantly displays the characteristic properties of an acid.

Many common acids are composed of hydrogen ion and an oxoanion. An oxoanion is a negatively charged polyatomic ion that contains oxygen. The acid contains as many hydrogen ions as are needed to neutralize the charge of the oxoanion. Examples are shown in Table 3.3:

Table 3.3 **Oxyanions and their Parent Acids**

| Oxoanion | Related acid | Acid Strength |
|---|---|---|
| $NO_3^-$ (nitrate) | $HNO_3$ (nitric acid) | strong |
| $SO_4^{2-}$ (sulfate) | $H_2SO_4$ (sulfuric acid) | strong |
| $ClO_4^-$ (perchlorate) | $HClO_4$ (perchloric acid) | strong |
| $NO_2^-$ (nitrite) | $HNO_2$ (nitrous acid) | weak |
| $SO_3^{2-}$ (sulfite) | $H_2SO_3$ (sulfurous acid) | weak |

Note that prefixes in the name of the anion (such as the *per-* in *per*chlorate) are retained in the name of the acid. If the name of the oxoanion has an *-ate* ending, the corresponding acid is given an *-ic* acid ending. If the name of the oxoanion has an *-ite* ending, the corresponding acid is given an *-ous* ending. Table 2.10 shows oxoanions and their names.

There are six common strong acids: hydrochloric, hydrobromic, hydriodic, nitric, sulfuric, and perchloric acids. It is characteristic of strong acids that most or all of their molecules have become ionized when they are dissolved in water. In such a case, chemists say "the reaction lies on the right," meaning that the acid is mainly in the form of ions, which are on the right-hand side of the chemical equation. In contrast, a weak acid such as hydrocyanic (HCN) ionizes only slightly, so the acid is mainly in the form of unionized reactant and the reaction "lies on the left." Here are the contrasting ionization equations for strong and weak acids:

| | | |
|---|---|---|
| Strong acid | $HBr \rightarrow H^+ + Br^-$ | the reaction lies entirely on the right |
| Strong acid | $H_2SO_4 \rightarrow H^+ + HSO_4^-$ | the reaction lies entirely on the right |
| Weak acid | $HNO_2 \leftrightarrows H^+ + NO_2^-$ | the reaction lies substantially on the left |
| Weak acid | $HCN \leftrightarrows H^+ + CN^-$ | the reaction lies substantially on the left |

Note the use of $\rightarrow$ for a complete reaction and $\leftrightarrows$ for an incomplete reaction.

**Example 3.1**: Write ionization equations to represent dissolving the strong acids nitric and sulfuric in water.

Strategy: Look up the formulas of the acids and also how many hydrogen ions they form per molecule when dissolved in water.

Nitric acid ($HNO_3$) yields one hydrogen ion and one nitrate ion for every molecule that ionizes:

$$HNO_3(l) \rightarrow H^+(aq) + NO_3^-(aq)$$

In two steps, sulfuric acid ($H_2SO_4$) yields two hydrogen ions and one sulfate ion for every molecule of sulfuric acid that ionizes:

$$H_2SO_4(l) \rightarrow H^+(aq) + HSO_4^-(aq)$$

$$HSO_4^-(aq) \leftrightarrows H^+(aq) + SO_4^{2-}(aq)$$

The first step lies entirely on the right. The second step lies substantially but not completely to the right.

A second class of acids are substances that, although they don't contain hydrogen ions, are able to form them by reacting with water. Metal ions are often acids of this type. For example, aluminum ion, $Al^{3+}$, is such an acid. The equation for the reaction of aluminum ion reacting as an acid in water is

$$Al^{3+} + H_2O \rightarrow AlOH^{2+} + H^+$$

## Bases

The simplest definition of a **base** is any compound that produces hydroxide ($OH^-$) ions when it's dissolved in water. The hydroxides of the group 1 metals, MOH (M = metal = Li, Na, K, Rb, and Cs) are bases. Bases are found in many household cleaning products; great-great-grandma used lye to make soap as many people still do; spray-on oven cleaners typically contain lye. Most of the group 1 and group 2 metal hydroxides are **strong bases** that completely ionize in water. The formation of solutions of bases can be represented by ionization equations:

$$\text{In water, } NaOH(s) \rightarrow Na^+(aq) + OH^-(aq)$$

The lower group 2 metal hydroxides $M(OH)_2$ (M = Ca, Sr, and Ba) are also strong bases.

$$\text{In water, } Ca(OH)_2(s) \rightarrow Ca^{2+}(aq) + 2\ OH^-(aq)$$

A second class of bases are substances that don't contain hydroxide ions but are able to form them by reacting with water. Ammonia, $NH_3$, is an example of a common household substance that behaves in this way. The equation for the reaction of ammonia with water is

$$NH_3(g) + H_2O(l) \rightarrow NH_4^+(aq) + OH^-(aq)$$

A solution of ammonia in water may also be represented as $NH_3(aq)$ or $NH_4OH(aq)$. When ammonia gas is added to water, only a small fraction of the ammonia molecules react to yield hydroxide ion. For this reason, ammonia is a **weak** base.

### Neutralization Reactions

In a **neutralization reaction** an acid reacts with a base. A **neutral solution** is neither acidic nor basic; later we'll see that a neutral solution has a pH of 7. The general reaction between an acid and a base in aqueous solution can be written:

$$\text{acid} + \text{base} \rightarrow \text{salt} + \text{water}$$

The equation for the specific reaction between hydrochloric acid and sodium hydroxide solutions is

$$HCl(aq) + NaOH(aq) \rightarrow NaCl(aq) + H_2O(l)$$

**Example 3.2**: Write balanced equations to illustrate the neutralization of a. sulfuric acid with potassium hydroxide, and b. hydroiodic acid with barium hydroxide.

Strategy: Look up the formulas and adjust the numbers of hydrogen and hydroxide ions to be equal.

a. Sulfuric acid ($H_2SO_4$) liberates two hydrogen ions per molecule and potassium hydroxide (KOH) liberates one hydroxide ion. They react, therefore, in a 1:2 ratio

$$2\ H^+(aq) + 2\ OH^-(aq) \rightarrow 2\ H_2O(l)$$

$$H_2SO_4(aq) + 2\ KOH(aq) \rightarrow K_2SO_4(aq) + 2\ H_2O(l)$$

b. Hydriodic acid (HI) liberates one hydrogen ion per molecule and barium hydroxide ($Ba(OH)_2$) liberates two hydroxide ions. They react, therefore, in a 2:1 ratio

$$2\ H^+(aq) + 2\ OH^-(aq) \rightarrow 2\ H_2O(l)$$

$$2\ HI(aq) + Ba(OH)_2(aq) \rightarrow BaI_2(aq) + 2\ H_2O(l)$$

**Exact neutralization** occurs when neither an excess of the acid nor an excess of the base remains at the end of the reaction; exact neutralization requires that chemically equivalent (or stoichiometrically equivalent) amounts of an acid and a base be mixed. The *net* neutralization reaction of any acid-base pair can be summarized by the equation for the addition reaction between hydrogen ion and hydroxide ion:

$$H^+(aq) + OH^-(aq) \rightarrow H_2O(l) \quad \text{net equation for neutralization}$$

## Reaction of an Active Metal with Water

Metals in group 1 and in the lower part of group 2 (Ca, Sr, and Ba) of the periodic table react vigorously with water; they are **active metals**. The products of the reaction of an active metal and water are diatomic hydrogen ($H_2$) and the hydroxide of the metal. Sodium is an active metal, and the equation for its reaction with water is

$$2\ Na(s) + 2\ H_2O(l) \rightarrow 2\ NaOH(aq) + H_2(g) \quad \text{sodium reacts with water}$$

The other alkali metals (Li, K, Rb, and Cs) react with water with the same stoichiometry as sodium. Thus, the reactions of all the alkali metals with water can be summarized by the single general equation:

$$2\ M(s) + 2\ H_2O(l) \rightarrow 2\ MOH(aq) + H_2(g) \quad \text{any alkali metal M reacts with water}$$

The lower metals of group 2 (Ca, Sr, and Ba) undergo a reaction with water that can be represented by the general equation:

$$M(s) + 2\ H_2O(l) \rightarrow M(OH)_2(aq) + H_2(g) \quad \text{an alkaline earth metal reacts with water}$$

Chemists favor the use of general equations where possible because they summarize more information than specific equations.

## Precipitation Reactions

A **precipitation reaction** is one that produces an insoluble precipitate. A **precipitate** is a solid that forms when certain pairs of aqueous solutions are mixed. For example, when a silver(I) nitrate solution is mixed with a hydrochloric acid solution:

$$AgNO_3(aq) + HCl(aq) \rightarrow HNO_3(aq) + AgCl\downarrow$$

A down arrow (↓) shows the formation of a precipitate of silver(I) chloride.

Silver(I) chloride is insoluble in water and forms as a precipitate when a solution is prepared that contains both $Ag^+$ ions and $Cl^-$. Writing $AgCl\downarrow$ is equivalent to writing AgCl(s). The net equation for the precipitation reaction is $Ag^+ + Cl^- \rightarrow AgCl(s)$. The hydrogen ions and nitrate ions do not participate in this reaction. Ions present in a reaction mixture but not participating are **spectator ions**.

Manganese(II) sulfide is insoluble in water and precipitates when a solution is made that contains both $Mn^{2+}$ ion and $S^{2-}$ ion. Three equivalent ways of writing that precipitation are shown below:

$$MnCl_2(aq) + Na_2S(aq) \rightarrow MnS\downarrow + 2NaCl(aq)$$

Written with the reactants in their ionized states the equation becomes:

$$Mn^{2+}(aq) + 2\ Cl^-(aq) + 2\ Na^+(aq) + S^{2-}(aq) \rightarrow MnS\downarrow + 2\ Na^+(aq) + 2\ Cl^-(aq)$$

Omitting the spectator ions ($Na^+$ and $Cl^-$) the same reaction can be written more economically as:

$$Mn^{2+}(aq) + S^{2-}(aq) \rightarrow MnS\downarrow$$

**Example 3.3**: When solutions of potassium phosphate and calcium chloride are mixed, calcium phosphate precipitates. Write a balanced equation for this precipitation reaction.

Strategy: Look up the formulas and balance using ionic charges.

$$2\ K_3PO_4(aq) + 3\ CaCl_2(aq) \rightarrow Ca_3(PO_4)_2(s) + 6\ KCl(aq)$$

## Oxidation-Reduction (Redox) Reactions

Oxidation-reduction reactions form a large and important reaction class. Many chemical reactions are redox reactions. *All* combustion and explosive reactions are redox reactions. The chemistry of the transition metals is largely a study in redox chemistry. Most biochemical reactions that occur in your body are redox reactions.

An atom, molecule, or ion that loses electrons is said to have been **oxidized**. An atom, molecule, or ion that gains electrons is said to have been **reduced**. A **redox reaction** occurs when one reactant is oxidized while a second reactant is simultaneously reduced. Reduction + Oxidation = redox. It is useful to catalog and rank chemical compounds in terms of their ability to undergo either oxidation or reduction. For these and other reasons we'll eventually study many more redox reactions. Here are some examples:

$Rb \rightarrow Rb^+ + e^-$ A rubidium atom is oxidized to a rubidium ion

$Fe^{2+} \rightarrow Fe^{3+} + e^-$ An iron(II) ion is oxidized to an iron(III) ion

$Cl_2 + 2e^- \rightarrow 2\ Cl^-$ A chlorine molecule is reduced to two chloride ions

$Sn^{4+} + 2e^- \rightarrow Sn^{2+}$ A tin(IV) ion is reduced to a tin(II) ion

Observe the use of the Roman numerals II, III, etc., as a conventional way of designating the charge on a positive ion. Note that if a chemical entity is getting oxidized the electrons are on the right of the equation; if the entity is getting reduced the electrons are on the left of the equation.

The general form of any redox reaction is shown below:

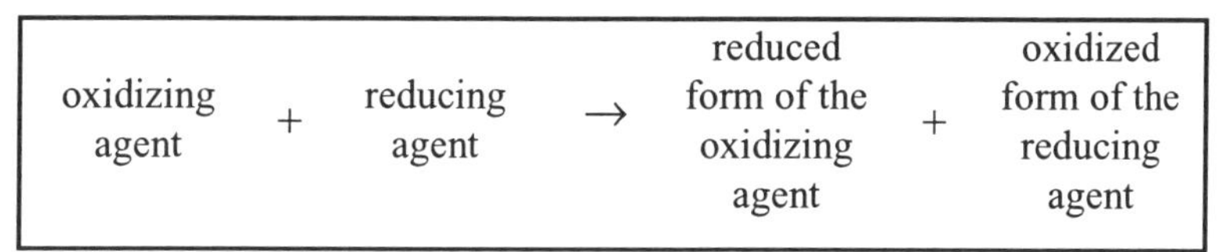


or, even simpler:

| oxidant | + | reductant | → | reduced oxidant | + | oxidized reductant |
|---|---|---|---|---|---|---|

Oxygen is the eponymous oxidizing agent. Fuels are the most common reductants. Here are some more redox equations:

$$Cu^0 + 2\,Fe^{3+} \rightarrow Cu^{2+} + 2\,Fe^{2+}$$

An atom of copper metal ($Cu^0$) is oxidized to a copper(II) ion while two ions of iron(III) are simultaneously reduced to iron(II) ions. This reaction is used in the electronic industry when an etching solution (iron(III) or ferric chloride, $FeCl_3(aq)$) is sprayed on the boards to dissolve copper metal in the manufacture of "printed" circuit boards. Home electronic enthusiasts can buy bottles of ferric chloride at their favorite electronics supply store.

$$2\,Mg + O_2 \rightarrow 2\,MgO$$

Magnesium oxide is an ionic compound: $Mg^{2+}\,O^{2-}$. Thus, in the reaction, two atoms of magnesium metal become oxidized to two dipositive magnesium ions while simultaneously one molecule of oxygen is reduced to two oxide ions. The equation represents the combustion of magnesium; the reaction is used in old-style flash bulbs and some types of military flares.

## Section 3.2 Balancing Chemical Equations

### Balancing Equations by Inspection

In a chemical equation there must be exactly the same number of atoms of each type in the reactants and products. Many simple equations can be balanced merely by looking at them; doing this is **balancing by inspection**. The equation for the reaction of hydrogen and oxygen to make water is:

$$H_2(g) + O_2(g) \rightarrow H_2O(l)$$

That's an unbalanced equation: there are more oxygen atoms on the left than on the right. To balance an unbalanced equation coefficients must be added. Equations *cannot* be balanced by changing any subscripts of reactants or products, because doing that arbitrarily changes the reactants and products. Adding a coefficient of 2 in front of the $H_2$ and 2 in front of the $H_2O$ balances the preceding equation as:

$$2\,H_2(g) + O_2(g) \rightarrow 2\,H_2O(l)$$

Many simple equations can be balanced by inspection, especially if you first balance the molecule with the largest number of atoms and leave balancing any elements until last. Balancing by inspection is shown in Example 3.4.

**Example 3.4**: The combustion reaction of pentene, $C_5H_{10}$, with excess oxygen, $O_2$, yields carbon dioxide, $CO_2$, and water. Write its equation. The unbalanced equation is

$$C_5H_{10}(g) + O_2(g) \rightarrow CO_2(g) + H_2O(l)$$

Strategy: Inspect the equation.

Pentene is the reactant with the largest number of atoms—so do it first. To balance the 5 carbons and 10 hydrogens on the opposite side needs the coefficient 5 in front of each product:

$$C_5H_{10}(g) + O_2(g) \rightarrow 5\ CO_2(g) + 5\ H_2O(g)$$

To balance the oxygen atoms total the O on the right. There are 10 O from the 5 $CO_2$ and 5 O from the 5 $H_2O$, a total of 15 O. The needed 15 O atoms require a coefficient of $\frac{15}{2}$ in front of the $O_2$ on the left.

$$C_5H_{10}(g) + \tfrac{15}{2}\ O_2(g) \rightarrow 5\ CO_2(g) + 5\ H_2O(g)$$

You could equally well have used the general equation for balancing hydrocarbon combustion that was stated earlier:

$$C_nH_m + \frac{(4n + m)}{4}\ O_2(g) \rightarrow n\ CO_2(g) + \tfrac{m}{2}\ H_2O(g)$$

That's done in Example 3.5.

The use of fractional coefficients is acceptable for balancing chemical equations; for example, IUPAC provides many fractional examples. However, equations are sometimes written without fractions. You can clear the equation of fractions by multiplying each coefficient by a factor divisible by the denominators of all the fractional coefficients. In the example above, multiply by 2. Doing this gives another balanced form of the equation:

$$2\ C_5H_{10}(g) + 15\ O_2(g) \rightarrow 10\ CO_2(g) + 10\ H_2O(g)$$

**Example 3.5**: Balance the equation that represents the complete combustion of the hydrocarbon hexane, $C_6H_{14} + O_2 \rightarrow CO_2 + H_2O$, using the general formula for a hydrocarbon combustion reaction.

Strategy: Use the general equation

$$C_nH_m(s,\ l,\ \text{or}\ g) + \frac{(4n + m)}{4}\ O_2(g) \rightarrow n\ CO_2(g) + \tfrac{m}{2}\ H_2O(g)$$

So obviously, n = 6 and m = 14. Substituting and doing the algebra gives:

$$C_6H_{14}(l) + \tfrac{19}{2}\ O_2(g) \rightarrow 6\ CO_2(g) + 7\ H_2O(g)$$

It can be doubled, if you desire, to remove the fractional coefficient and give:

$$2\ C_6H_{14}(l) + 19\ O_2(g) \rightarrow 12\ CO_2(g) + 14\ H_2O(g)$$

If you multiply all the coefficients of any balanced equation by any constant value you get another balanced equation. Learning to balance equations by inspection is largely a trial and error procedure. Using the guidelines suggested earlier will help, but it's often necessary to cycle through a reaction several times to get the atoms of all types balanced. Practice. Actually, there isn't any chemistry in balancing chemical equations; that's a "dirty little secret" of chemistry. Balancing chemical equations is actually an algebra problem as described in the following section.

## The Algebraic Method for Balancing Equations

Chemists use various systematic methods to balance chemical equations. In Chapter 13 we'll meet the oxidation number method. The most general way to balance equations is simply to use algebraic equations to express the conservation of atoms. The resulting equations are then solved to obtain the values of the coefficients. This technique is the **algebraic method**. As a simple example, reconsider the combustion of pentane, an equation that we balanced earlier by inspection in Example 3.4. Let a, b, c, and d, be the unknown coefficients. Then

$$a\ C_5H_{10}(g) + b\ O_2(g) \rightarrow c\ CO_2(g) + d\ H_2O(g)$$

Balancing the carbon atoms requires that $5\ a = c$

Balancing the hydrogen atoms requires that $10\ a = 2\ d$

And balancing the oxygen atoms requires that $2\ b = 2\ c + d$

So we have three equations and four unknowns. But we saw earlier that balanced equations have fixed ratios of coefficients, not fixed values. Thus we are free to assign arbitrarily any one of the coefficients, and then use the algebraic equations to find the other coefficients.

For example choose a = 1. Then the first two equations give c = 5 and d = 10/2 = 5. The third equation then yields $b = (2 \times 5 + 5)/2 = \frac{15}{2}$. Using these coefficients we therefore have the balanced equation

$$C_5H_{10}(g) + \frac{15}{2}\ O_2(g) \rightarrow 5\ CO_2(g) + 5\ H_2O(g)$$

Example 3.6 shows a second case of using the algebraic method to balance an equation.

**Example 3.6**: Concentrated nitric acid ($HNO_3$) reacts violently with hydrazine ($N_2H_4$). The reaction was used in early liquid-fueled rockets. The products of the reaction are diatomic nitrogen and steam. Write the equation for the reaction and balance it using the algebraic method.

Strategy: Apply the algebraic method. From the data given write the unbalanced equation as

$$HNO_3(l) + N_2H_4(l) \rightarrow N_2(g) + H_2O(g)$$

applying unknown coefficients a, b, c, and d, gives

$$a\ HNO_3(l) + b\ N_2H_4(l) \rightarrow c\ N_2(g) + d\ H_2O(g)$$

Balancing the hydrogen atoms requires $a + 4b = 2d$

Balancing the nitrogen atoms requires $a + 2b = 2c$

Balancing the oxygen atoms requires $3a = d$

Because we have free choice for one single coefficient we'll choose a = 1. Then, the third equation immediately gives d = 3. Next, the values of a and d are substituted into the first equation to give 1 + 4b = 6, or 4b = 5 and b = 1.25. Finally, substituting the values of a and b into the second equation gives $2c = 1 + 2 \times 1.25$, or 2c = 3.5 and c = 1.75. This gives the balanced equation:

$$1\ HNO_3(l) + 1.25\ N_2H_4(l) \rightarrow 1.75\ N_2(g) + 3\ H_2O(g)$$

Multiplying by 4 rounds the fractional parts to whole numbers to give the final form of the balanced equation:

$$4\ HNO_3(l) + 5\ N_2H_4(l) \rightarrow 7\ N_2(g) + 12\ H_2O(g)$$

One might say that the algebraic method for balancing chemical equations is a mathematical generalization of the inspection method.

## Section 3.3 Stoichiometry

**Stoichiometry** is the study of mass and number relationships in chemistry. It deals with the relationships among elements in chemical compounds and the relationships among elements and compounds as they participate in chemical reactions. In this section we discuss relationships among elements in chemical compounds. In Section 3.4 we will discuss the stoichiometry of chemical reactions.

> The word stoichiometry was coined in 1792 by the chemist F. A. Richter from Greek words that he thought meant roughly "equivalence measurement." Richter had been studying the reactions of acids and bases and had correctly concluded that a definite amount of a given base would exactly neutralize a definite amount of a given acid.

Historically, it was studies of stoichiometry that led John Dalton to the atomic theory of matter. We'll reverse the historical development, study the known principles of stoichiometry first, and see how they summarize the laws of chemical combination. Stoichiometry is best learned by working lots of example problems. Chemical reactions conserve mass—that's the **law of conservation of mass**. (Recall, the law's *not* true for nuclear reactions.) Antoine Lavoisier suggested conservation of mass in his 1789 book, stating that the mass of sugar consumed in wine-making would be equal to the sum of the masses of the alcohol and carbon dioxide produced.

$$C_6H_{12}O_6 \rightarrow 2\,C_2H_6O + 2\,CO_2$$

By the way, it was Lavoisier who named $C_2H_6O$ alcohol—deriving its name from an Arabic word. Since Dalton's work we've known that in any chemical reaction all the atoms are conserved. Because chemical reactions merely scramble atoms without changing them, the law of conservation of mass in chemistry is inevitable.

Because one mole of carbon dioxide contains 12.0 g of carbon and 32.0 g of oxygen, its weight percentage is $\frac{12.0}{12.0+32.0} \times 100 = 27.3\%$ carbon and 100% - 27.3% oxygen. Because carbon dioxide is the same compound in *any* laboratory or in *anyone's* breath it is *always* 27.3% carbon and 72.7% oxygen. This fact was enshrined in chemical history as the **law of constant or definite composition**: a given chemical compound has a definite composition.

Early chemists realized that some pairs of elements formed compounds with two or more *different* compositions. Nowadays that's obvious. There are such examples as $H_2O$ and $H_2O_2$; $CH_4$ and $C_2H_4$; $N_2O$, NO, and $NO_2$ (three of those). Here's the calculation for water and hydrogen peroxide:

$H_2O$ contains $(\frac{2 \times 1.00}{2 \times 1.0 + 16.0}) \times 100 = 11.1\%$ hydrogen and thus 100 - 11.1 or 88.9% oxygen,

$H_2O_2$ contains $(\frac{2 \times 1.00}{2 \times 1.0 + 2 \times 16.0}) \times 100 = 5.88\%$ hydrogen and 94.12% oxygen.

In $H_2O$ the ratio of %O÷%H is 88.9/11.1 = 8.01. In $H_2O_2$ the ratio of %O÷%H is 94.12/5.88 = 16.0. Within experimental uncertainty, the ratio of these ratios, 16.0/8.01, is a whole number, 2.0. For a specified amount of hydrogen, there are relatively twice as many oxygen atoms in $H_2O_2$ as there are in $H_2O$. This experimental situation became known as the **law of multiple proportions**: when pairs of elements form two or more distinct compounds the mass ratios between one compound and the other are simple whole numbers. Today we see that these simple whole numbers are merely the ratios of the number of atoms between the two different compounds. The law is illustrated in Example 3.7.

**Example 3.7**: Show that the hydrocarbons benzene, $C_6H_6$, and ethylene, $C_2H_4$, illustrate the law of multiple proportions.

Strategy: (1) Calculate the percentage by mass of hydrogen in each compound and the percentage by mass of carbon in each compound, (2) calculate the ratio %C÷%H for each compound, and (3) make a ratio of the two ratios.

In $C_6H_6$: $\%H = \left(\frac{6 \times 1.00}{6 \times 1.00 + 6 \times 12.0}\right) \times 100$ = 7.69% H and 92.31% carbon

In $C_2H_4$: $\%H = \left(\frac{4 \times 1.00}{4 \times 1.00 + 2 \times 12.0}\right) \times 100$ = 14.3% H and 85.7% carbon

For $C_6H_6$ the ratio %C÷%H is 92.31/7.69 = 12.0

For $C_2H_4$ the ratio %C÷%H is 85.7/14.3 = 5.99

So the two ratios are in a 12.0/5.99 or 2:1 ratio.

This simple, whole-number, ratio confirms the law of multiple proportions.

John Dalton was probably the first human to clearly perceive the law of multiple proportions. It led him to the atomic theory. Looking back it all seems obvious, though it surely wasn't. On the basis of experimental weights of products and reactants, Dalton made the incredible leap to postulate the actual existence of atoms.

### The Law of Multiple Proportions

"If two elements form more than one compound, the weight ratios of the elements in the compounds are small whole-number multiples of one another."

John Dalton, circa 1805. He's pictured on the right.

During the early nineteenth century, most scientists viewed atoms as merely constructs of the human mind, but to Dalton atoms were real entities. He said, "We might as well attempt to introduce a new planet into the solar system, or to annihilate one already in existence, as to create or destroy a particle of hydrogen."

Dalton was right about atoms but wrong about how hydrogen and oxygen combine. All his life he thought water was HO and that oxygen's atomic weight was 8, or one-half of its modern value. As will be described in Chapter 4, over 50 years were to pass after the birth of the atomic theory before chemists understood the truth that water is $H_2O$.

It was Joseph Louis Proust (1754-1826), building on the work of Lavoisier, who deserves most of the credit for establishing the laws of chemical composition. Proust was a thoroughly reliable quantitative investigator. During the period of the French revolution he fled from the turbulence to Spain, but his Spanish laboratory, where he did splendid work, was eventually destroyed by invading Napoleonic troops in 1808.

Recall that in a balanced equation the coefficients are called stoichiometric coefficients. The first scientist to obtain direct experimental evidence for stoichiometric coefficients in chemical equations was Joseph Gay-Lussac (1778-1850), who studied reactions between gases. He found that when hydrogen gas and chlorine gas react to form hydrogen chloride gas, one volume of hydrogen gas reacts with one volume of chlorine gas to form two volumes of gaseous hydrogen chloride, a 1:1:2 volume ratio (the gas volumes were always measured at the same temperature and pressure). In another experiment, two volumes of hydrogen combined with one volume of oxygen to make two volumes of water vapor, a 2:1→2 volume ratio. His study of these and other reactions led Gay-Lussac to formulate his law of combining volumes:

> "...measured under the same conditions of pressure and temperature, the volumes of gases that are consumed or produced in a chemical reaction occur in ratios of small whole numbers."

These small whole numbers turn out to be the coefficients of chemical equations.

In 1811 the Italian scientist Amadeo Avogadro (1776-1856) realized that Gay-Lussac's observations were consistent with John Dalton's atomic theory. Avogadro reasoned that the volume relationships observed by Gay-Lussac must be reflections of changes that occur at the molecular level. The simplest way to tie Gay-Lussac's findings to the atomic theory was to assume that the relationship between reactant and product gas volumes was identical to that between the number of reactant and product molecules. Avogadro summarized this in a short statement that has become known as Avogadro's law:

> "When they are measured at the same conditions of pressure and temperature, equal volumes of gases contain equal numbers of molecules."

Having made his hypothesis, Avogadro found himself obliged to introduce another new idea: the diatomic molecule. In 1811, the molecules of the gases hydrogen, chlorine, and oxygen were thought to be monatomic, but Avogadro realized that Gay-Lussac's results that *two* volumes of hydrogen combine with only *one* volume of oxygen to give *two* volumes of water could be explained only by assuming that the reactants were diatomic (composed of two atoms). So Avogadro understood by 1811 that water was $H_2O$.

**Informal Example**: Consider the following "reaction."

1 pair of gloves + 1 person → 1 gloved person

It's obviously correct. But why? It's correct because persons have two hands. It obviously doesn't work for horses, dogs, or three-legged stools, and a one-handed person leaves the equation unbalanced with an excess glove. Such thinking was the essence of Avogadro's reasoning that eluded some of the best minds of the 19th century.

However, the course of human events moves neither smoothly nor sensibly. So, despite the fact that Avogadro's reasoning was sound, the chemical community ignored his ideas for almost 50 years. Dalton died believing that water was HO, not $H_2O$. During those fifty years, the only other scientist to accept Avogadro's hypothesis was André Ampere (1775-1836)—after whom the SI unit of electric current, the amp, is named. Unfortunately, Ampere could not convince anyone else that Avogadro was correct. The man who finally changed the collective mind of chemists was the Sicilian, Stanislao Cannizzaro (1826-1910). Cannizzaro discussed Avogadro's hypothesis in his lectures at the University of Genoa. Cannizzaro's ideas were well publicized at an 1860 conference in Karlsruhe, Germany, attended by many leading chemists. At last, chemists were ready for Avogadro's ideas. They said a collective "how stupid we were to overlook the obvious," and within two years the existence of diatomic molecules and a new, correct scale of atomic weight was accepted by almost everyone.

## Stoichiometry and Molecular Weights

As we saw in Chapter 1, in the SI metric system **amount of substance** can be expressed either in mass units or in mole units. Molar weight is a conversion factor from one to the other—so this section is mostly about mole and mass interconversions.

We've seen that chemists use two, *independent*, mass standards: the international kilogram and the unified atomic mass unit (u, amu, dalton). The conversion factor from grams to u has the value (but not the units) of Avogadro's number. So you use the same numerical value but change the units to express either the molecular weight of an average molecule or the molar weight of a compound. One mole of a monatomic substance contains Avogadro's number of atoms and has a mass equal to the element's atomic weight in grams. One mole of a molecular substance contains Avogadro's number of molecules and has a mass equal to the compound's molar weight in grams. The four mass/mole conversions are summarized Table 3.4.

Table 3.4 **Conversions Factors for Mass and Moles**

| To convert from | To | Do this |
|---|---|---|
| g | u | Multiply by $6.022 \times 10^{23}$ u $g^{-1}$ |
| u | g | Divide by $6.022 \times 10^{23}$ u $g^{-1}$ |
| moles | molecules | Multiply by $6.022 \times 10^{23}$ molecules $mol^{-1}$ |
| molecules | moles | Divide by $6.022 \times 10^{23}$ molecules $mol^{-1}$ |

Recall that naturally occurring carbon is a mixture of $^{12}C$ and $^{13}C$ isotopes. It has an atomic weight of 12.01 u, so 1 mole of natural C atoms weighs 12.01 g. Here are some examples of molecular weight and molar weight specified with identical values but alternate units:

| Substance | Molecular weight | Molar weight |
|---|---|---|
| Xenon, Xe | 131.29 u (average atom)$^{-1}$ | 131.29 g $mol^{-1}$ |
| Propane, $C_3H_8$ | 44 u (average molecule)$^{-1}$ | 44 g $mol^{-1}$ |
| Phosphoric acid, $H_3PO_4$ | 98 u (average molecule)$^{-1}$ | 98 g $mol^{-1}$ |

Because using it is cumbersome, the word "average" is almost always omitted. Except for Example 3.12 this book will rarely use it again.

Having a property that has the same numerical value when expressed in two sets of units is not common. Historically, the root of this duplication of units goes back to 1803 when John Dalton set up the atomic mass scale based on hydrogen, the simplest element, being assigned a mass of one. At about the same time (1792), the metric system was set up with the kilogram as its base unit. About 70 years later, following the Karlsruhe conference and the acceptance of Avogadro's hypothesis and the mole, it became possible to see that $6.02 \times 10^{23}$ was the conversion factor between gram units and atomic mass units. The ratio of measurements on the two mass scales is today the basis for the *experimental* measurement of Avogadro's number.

**Example 3.8**: Starting with the fact that one atom of $^{12}C$ weighs 12 u, prove that 1 mol of $^{12}C$ weighs 12 g.

Strategy: Note that the values of *both* the conversion factor from atoms to moles *and* the conversion factor from atomic mass units to grams are Avogadro's number. Use conversion factors to change and u to g (step 1) and then atoms to moles (step 2)

$$^{12}C \text{ weighs } 12 \text{ u atom}^{-1}$$

$$\text{Step 1: } 1 \text{ atom weighs } 12 \text{ u atom}^{-1} \div 6.022 \times 10^{23} \text{ u g}^{-1} = 1.993 \times 10^{-23} \text{ g atom}^{-1}$$

$$\text{Step 2: } 1.993 \times 10^{-23} \text{ g atom}^{-1} \times 6.022 \times 10^{23} \text{ atoms mol}^{-1} = 12 \text{ g mol}^{-1}$$

So 1 mol of $^{12}C$ is indeed equivalent to 12 g of $^{12}C$.

The atomic weight of an element, with units of g $mol^{-1}$, serves as a conversion factor to convert amounts of substance of atoms and molecules between moles and mass. We'll soon see lots of examples of its use. For atoms the equation to remember is $n = \frac{m}{AW}$. For molecules the equation is $n = \frac{m}{MW}$ where m is the mass of the sample in grams, AW is the molar weight of the atoms (atomic weight), and MW is the molar weight of the molecules.

It is easy to construct two alternative units with the same numerical value for a property. Here's an informal example:

**Informal Example**: If eggs weigh 2.1 ounces each, what is the weight in pounds of a package of 16 eggs?

Strategy: Use the algebra of quantities and do a units conversion.

$$= 2.1\frac{\text{oz}}{\text{egg}} \times 16\,\frac{\text{eggs}}{\text{package}} \times \frac{1}{16}\frac{\text{lb}}{\text{oz}}$$

$$= 2.1 \text{ lb package}^{-1}$$

Thus "egg weight" $= 2.1$ oz egg$^{-1}$ *or* $= 2.1$ lb package$^{-1}$

Because the number of eggs in the package (16 egg package$^{-1}$) is the same as the value of the conversion factor between pounds and ounces (16 oz lb$^{-1}$), the identical numerical value has two alternative units.

Note: In computations that involve atomic weights, enough significant figures should be used so that the atomic weights are not the limiting source of uncertainty in your answer.

Here are five examples (Examples 3.9 through 3.13) of amount of substance calculations:

**Example 3.9**: How many moles and how many atoms are in: a. 30.0 g of iron; and, b. 1.000 cm$^3$ of liquid mercury, which has a density of 13.60 g mL$^{-1}$?

Strategy: The molar amount of substance in a sample of an element is found using $n = \frac{m}{AW}$, where n is moles, m is the mass of the sample in grams, and AW is the atomic weight of the element. The number of atoms is then obtained by multiplying the number of moles by Avogadro's number. The molar weight of iron is 55.85 g mol$^{-1}$ and the molar weight of mercury is 200.59 g mol$^{-1}$. Thus:

a. $30.0 \text{ g Fe} \times \frac{1 \text{ mol Fe atoms}}{55.85 \text{ g Fe}} = 0.537 \text{ mol Fe atoms}$

and, $0.537 \text{ mol Fe atoms} \times 6.022 \times 10^{23} \text{ atoms mol}^{-1} = \underline{3.23 \times 10^{23} \text{ Fe atoms}}$

b. $13.60 \text{ g Hg} \times \frac{1 \text{ mol Hg atoms}}{200.59 \text{ g Hg}} = 0.06780 \text{ mol Hg atoms}$

and $0.06780 \text{ mol Hg atoms} \times 6.022 \times 10^{23} \text{ atoms mol}^{-1} = \underline{4.083 \times 10^{22} \text{ Hg atoms}}$

**Example 3.10**: Calculate the mass in g of a single atom of $^{19}$F.

Strategy: The molar mass of $^{19}$F is 19.00 g mol$^{-1}$ and there are $6.022 \times 10^{23}$ atoms of fluorine in a mole (atoms mol$^{-1}$). Dividing the former quantity by the latter leaves the desired units of g atom$^{-1}$.

$$\text{Mass} = \frac{19.00 \text{ g mol}^{-1}}{6.022 \times 10^{23} \text{ atoms mol}^{-1}} = \underline{3.155 \times 10^{-23} \text{ g atom}^{-1}}$$

**Example 3.11**: Urea, $CO(NH_2)_2$, which is excreted in the urine of humans and other mammals, is a useful fertilizer. a. How many moles is 12.0 g of urea? b. What is the mass of 2.23 moles of urea?

Strategy: a. Calculate the molar weight using $MW = \sum_i n_i AW_i$. Then use the relationship $n = \frac{m}{MW}$,

$$MW = 12.01 \text{ g mol}^{-1} + 16.00 \text{ g mol}^{-1} + 2 \times 14.01 \text{ g mol}^{-1} + 4 \times 1.01 \text{ g mol}^{-1}$$

$$= 60.07 \text{ g mol}^{-1}$$

Substitute into the equation $n = \frac{m}{MW}$,

$$\frac{12.0 \text{ g } CO(NH_2)_2}{60.07 \text{ g mol}^{-1}} = \underline{0.200 \text{ mol}}\ CO(NH_2)_2 \text{ (three significant figures)}$$

Strategy: b. Rearranging $n = \frac{m}{MW}$ to $m = n \times MW$ allows for direct substitution to get the desired answer:

$$\text{mass} = n \times MW$$

$$2.23 \text{ mol} \times 60.07 \text{ g mol}^{-1}\ CO(NH_2)_2 = \underline{134 \text{ g } CO(NH_2)_2}$$

**Example 3.12**: Glucose is a sugar with the molecular formula $C_6H_{12}O_6$. a. Calculate the mass of one (average) molecule of glucose in u. b. Calculate the mass of one (average) molecule of glucose in grams. c. calculate the mass of 1.000 mol of glucose in grams. d. Calculate the mass of 1.000 mol of glucose in u.

Strategy: a. Calculate the molar weight using $MW = \sum_i n_i AW_i$ and express the result in u molecule$^{-1}$

$$6 \times 12.01 + 12 \times 1.01 + 6 \times 16.00 = \underline{180.18 \text{ u molecule}^{-1}}$$

Strategy: b. Use the conversion factor $6.022 \times 10^{23}$ u g$^{-1}$ to convert from u to grams.

$$1 \text{ molecule} \times 180.18 \text{ u molecule}^{-1} \div 6.022 \times 10^{23} \text{ u g}^{-1} = \underline{2.992 \times 10^{-22} \text{ g}}$$

c. The mass of 1.000 mole of glucose is given directly by changing from molecular weight to molar weight by changing the units. The molar weight is $\underline{180.18 \text{ g mol}^{-1}}$.

d. Again, usc the conversion factor $6.022 \times 10^{23}$ u g$^{-1}$

$$180.18 \text{ g mol}^{-1} \times 6.022 \times 10^{23} \text{ u g}^{-1} = \underline{1.085 \times 10^{26} \text{ u mol}^{-1}}$$

Avogadro's number and the definition of the mole can also be used to calculate the numbers of atoms of individual elements present in a specified quantity of substance.

**Example 3.13**: Glycerol ($C_3H_8O_3$) is a sweet, syrupy liquid. slow to evaporate, and used, among other purposes, as an ingredient in inks. In 124 g of glycerol, a. How many molecules are present? and b. How many atoms of each element are present?

Strategy: a. Convert from mass to number of moles using $n = \frac{m}{MW}$, and then to number of molecules using Avogadro's number as the conversion factor $6.022 \times 10^{23}$ atoms mol$^{-1}$. b. Each glycerol molecule contains 3 carbon atoms, 8 hydrogen atoms, and 3 oxygen atoms. Multiply the number of molecules by these numbers of atoms molecule$^{-1}$ to get the number of atoms of each element.

$$MW = (3 \times 12.01 + 8 \times 1.01 + 3 \times 16.00)\ \text{g mol}^{-1} \quad = 92.11\ \text{g mol}^{-1}$$

$$n = \ 124\ \text{g} \div 92.11\ \text{g mol}^{-1} = 1.35\ \text{mol } C_3H_8O_3$$

$$1.35\ \text{mol } C_3H_8O_3 \times 6.022 \times 10^{23}\ \text{molecules mol}^{-1}$$

$$= \underline{8.13 \times 10^{23}\ \text{molecules } C_3H_8O_3}$$

Total atoms of each element = number of molecules × atoms molecule$^{-1}$

Carbon: $8.13 \times 10^{23}$ molecules × 3 C atoms molecule$^{-1}$ = $\underline{2.44 \times 10^{24}\ \text{C atoms}}$

Hydrogen: $8.13 \times 10^{23}$ molecules × 8 H atoms molecule = $\underline{6.50 \times 10^{24}\ \text{H atoms}}$

Oxygen: $8.13 \times 10^{23}$ molecules × 3 O atoms molecule$^{-1}$ = $\underline{2.44 \times 10^{24}\ \text{O atoms}}$

## Elemental Analysis of Compounds

The **elemental analysis** of a compound is a specification by weight percentage of the elements in the compound. **Weight percentage** is the percentage contribution that individual elements make to a compound's mass.

**Example 3.14**: Calculate the weight percent of each element in calcium phosphate, $Ca_3(PO_4)_2$, the major ingredient of "phosphate rock," the chief ore of phosphorus and the largest single volume industrial substance—your annual use rate is about 400 pounds.

Strategy: First figure the molar weight and then express the individual weights of Ca, P, and O as percentages of that, using percentage = (part ÷ whole) × 100.

$$MW = 3 \times 40.08 + 2 \times 30.97 + 8 \times 16.00 \ = \ 310.18\ \text{g mol}^{-1}$$

In one mole of $Ca_3(PO_4)_2$, the elemental masses are

| | | | | |
|---|---|---|---|---|
| mass of calcium | = | 3 mol × 40.08 g mol$^{-1}$ Ca | = | 120.24 g Ca |
| mass of phosphorus | = | 2 mol × 30.97 g mol$^{-1}$ P | = | 61.94 g P |
| mass of oxygen | = | 8 mol × 16.00 g mol$^{-1}$ O | = | 128.00 g O |

The weight percentages are: Ca = 100 × (120.24 ÷ 310.18) = $\underline{38.76\%}$

P = 100 × ( 61.94 ÷ 310.18) = $\underline{19.97\%}$

O = 100 × (128.00 ÷ 310.18) = $\underline{41.27\%}$

Sum = $\underline{100.00\%}$

The fact that the values sum to 100% shows that the math was right.

In the following example we show how weight percentage can be applied to the calculation of the amount of iron potentially available from an iron ore.

**Example 3.15**: The mineral hematite has the formula $Fe_2O_3$. Determine the weight percent composition of hematite and calculate how much iron could be obtained from 1000. kg (a "metric ton") of the ore. Assume that the ore is 54.5% $Fe_2O_3$.

Strategy: Calculate the percentage composition of Fe in $Fe_2O_3$ and take that percentage of the actual mass of $Fe_2O_3$ in the ore.

$$\text{Wt. \% (Fe in } Fe_2O_3) = (2 \times 55.85 \div 159.70) \times 100 = 69.95\ \%$$

If 1000. kg of the 54.5% ore were processed and the iron completely recovered, the amount obtained would be 69.95 % of 545 kg, or 381 kg.

High grade iron ores from the Lake Superior region of the US and from the country of Ukraine contain 50-60% iron.

## Empirical Formula

We met empirical formula in Chapter 2. Now we'll show how the empirical formulas of compounds are obtained from experiments. Recall, the **empirical formula** of a chemical compound is the simplest ratio of its atoms obtained by factoring its molecular formula.

Getting empirical and molecular formulas from compositions expressed in weight percentages reverses the procedure for getting mass percentage composition from molecular formulas. It is a common practical problem, and was for many years (arguably from 1770 to 1970) the primary method by which chemists deduced the composition of substances from experimental data. We'll do some examples because the calculations well illustrate the principles of stoichiometry and are an excellent problem-solving exercise.

**Example 3.16**: Chemical analysis of a hydrocarbon shows its weight percentage is 92.3% carbon and 7.7% hydrogen. What's the empirical formula of the hydrocarbon?

Strategy: 1. Assume a 100 g sample. The percentages are thus automatically converted to grams. 2. Use the formula $n = \frac{m}{MW}$ to covert the gram masses to moles. 3. Scale the mole ratios to obtain whole number values. 4. Adjust the ratios to integers.

Carbon $92.3\text{ g} \div 12.01\text{ g mol}^{-1} = 7.69$ mol C atoms

Hydrogen $7.7\text{ g} \div 1.01\text{ g mol}^{-1} = 7.6$ mol H atoms

The immediate formula of the hydrocarbon is $C_{7.69}H_{7.6}$. It becomes an empirical formula of CH, rounding because the analysis has experimental uncertainty.

Elemental analyses are experimental results and are therefore uncertain. Uncertainty arises both from sample contamination (impurities) and manipulation errors (equipment and human limitations) in the analysis. As a result, the calculated relative numbers of atoms will rarely be exactly whole number ratios; rounding will usually be required and the rounding procedure involves judgment. Because 7.69 and 7.6 are close to one another, we conclude that the ratio of carbon atoms to hydrogen atoms in the example above is one-to-one and the empirical formula of the hydrocarbon is CH.

The molecular formula for our hydrocarbon example above can be written $(CH)_n$, with n being any integer. Measuring the molecular weight of the substance enables its true molecular formula to be deduced. We can determine n by dividing the molecular weight by the weight of the empirical formula as shown in the following examples:

**Example 3.17**: A hydrocarbon has a molecular weight of around 80 g mol$^{-1}$ and an empirical formula CH, so its molecular formula must be $(CH)_n$. It's a liquid. What is this compound?

Strategy: Calculate the empirical molar weight and divide that into the molecular weight.

The empirical molar weight of CH, is 12.01 + 1.01 = 13.02 g mol$^{-1}$.

$$n = 80 \div 13.02 = 6 \text{ (one significant figure)}$$

so the liquid is $(CH)_6$, or $C_6H_6$.

A stable cyclic, molecule with that molecular formula is benzene. H-C≡C-CH=CH-CH=$CH_2$ is an isomer of benzene, and has the same empirical formula, but it's very unstable.

**Example 3.18**: The common organic solvent acetone, sometimes used as a nail-polish remover, has a molecular weight of 58.09 g mol$^{-1}$. Its elemental analysis is: 62.0 % carbon; 27.6 % oxygen; and 10.4 % hydrogen by weight. What is its molecular formula.

Strategy: 1. Assume a 100 g sample. The percentages are thus automatically converted to grams. 2. Use the formula $n = \frac{m}{MW}$ to covert the gram masses to moles. 3. Scale the mole ratios to obtain whole number values. 4. Adjust the ratios to integers.

| | | | |
|---|---|---|---|
| Carbon | 62.0 g ÷ 12.01 g mol$^{-1}$ | = | 5.16 mol C atoms |
| Hydrogen | 10.4 g ÷ 1.01 g mol$^{-1}$ | = | 10.3 mol H atoms |
| Oxygen | 27.6 g ÷ 16.00 g mol$^{-1}$ | = | 1.73 mol O atoms |

To scale, divide by the smallest of the three to obtain ratios:

C: 5.16 mol ÷ 1.73 = 2.98
H: 10.3 mol ÷ 1.73 = 5.95
O: 1.73 mol ÷ 1.73 = 1.00

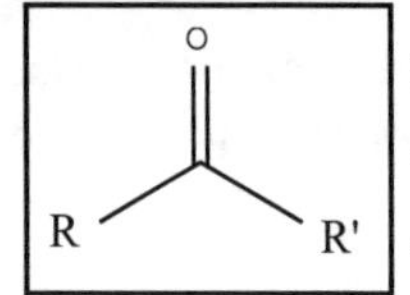


Rounding the first two numbers gives the empirical formula $C_3H_6O$,

The empirical formula weight = $3 \times 12.01 + 6 \times 1.01 + 1 \times 16.00 = 58.09$ g mol$^{-1}$

The result agrees with the given MW, so $\underline{C_3H_6O}$ is molecular formula of acetone.

Acetone is dimethyl ketone $CH_3COCH_3$. If you use $-CH_3$ to replace both R and R' in the general formula for a ketone (RCOR') shown in Table 2.7, you get the correct structure of acetone.

Benzene (Example 3.17) was discovered by Michael Faraday in 1825 when he was consulted to explain a liquid that was contaminating the tanks of illuminating gas sold by the Portable Gas Company in the early days of artificial lighting. The company was making gas by strongly heating or "cracking" whale oil, and this annoying liquid was interfering with their business. Faraday identified it as benzene. The original sample of benzene that Faraday analyzed is kept in the basement museum in the Royal Institution in London.

In many laboratories, the elemental analysis of hydrocarbons is a routine procedure. A carefully weighed sample is completely burned, using excess oxygen, to produce water vapor and carbon dioxide. The products are separately trapped and weighed. Stoichiometric calculation then gives the elemental analysis of the hydrocarbon as shown in the following example:

**Example 3.19**: A 4.9960 mg hydrocarbon sample was burned. The recovered products were 17.1675 mg of $CO_2$ and 2.8110 mg of $H_2O$. If the MW of the compound is known to be approximately 130 g $mol^{-1}$, what is the hydrocarbon's molecular formula?

Strategy: 1. Assume that all of the carbon from the hydrocarbon is converted to carbon dioxide, and that all of the hydrogen is converted to water. 2. Calculate the mass fraction of carbon and the mass fraction of hydrogen in the sample. 3. Use these mass fractions to compute the percentages of hydrogen and carbon in the sample and proceed as with a percentage composition problem. The molar weight of carbon dioxide is 44.01 g $mol^{-1}$; the molar weight of water is 18.01 g $mol^{-1}$.

In one mole (44.01 g) of carbon dioxide there is 12.01 g of carbon = (12.01/44.0) = 0.2729 mass fraction C.

In one mole (18.01 g) of water there is (2 × 1.008 g of hydrogen) ÷ 18.01 = 0.1119 mass fraction H.

$$\text{Percentage of carbon in the sample} = \frac{\text{mass of carbon in the } CO_2}{\text{mass of the sample}} \times 100\%$$

$$= \frac{17.1675 \text{ mg} \times .2729}{4.9960 \text{ mg}} \times 100 = 93.78\%$$

$$\text{Percentage of hydrogen in the sample} = \frac{\text{mass of hydrogen in the } H_2O}{\text{mass of the sample}} \times 100\%$$

$$= \frac{2.8110 \text{ mg} \times 0.1119}{4.9960 \text{ mg}} \times 100 = 6.296\%$$

Now, follow the procedure to use weight percentage data to next get the moles:

Carbon 93.78 g ÷ 12.01 g $mol^{-1}$ = 7.808 mol C atoms

Hydrogen 6.296 g ÷ 1.008 g $mol^{-1}$ = 6.246 mol H atoms

and then scale the mole ratios by dividing by the smallest number

C: 7.808 mol ÷ 6.246 = 1.250

H: 6.246 mol ÷ 6.246 = 1.000

So the formula is $C_{1.250}H_1$. It contains 1.250 carbons per hydrogen. Multiply by 4 to make the subscripts integers, $C_5H_4$. Figure the empirical weight as 64 g $mol^{-1}$, so the empirical formula must be doubled to give the MW of 128 g $mol^{-1}$, which is near the stated approximate value of 130 g $mol^{-1}$.

The compound is $\underline{C_{10}H_8}$

One handbook of chemistry lists two chemical compound with the formula $C_{10}H_8$, naphthalene and azulene.

## Section 3.4 Stoichiometry and Chemical Reactions

In the previous section we focused on the stoichiometry of individual compounds. In this section the same ideas are extended to chemical reactions. Either moles or grams can be used to specify the amounts of substance involved in chemical reactions. Moles measure numbers of atoms and molecules, so they are related to equation coefficients and express number relationships in chemical reactions. The reaction of carbon and oxygen to form carbon dioxide is described by the equation

$$C + O_2 \rightarrow CO_2,$$ which is read as

1 C atom + 1 $O_2$ molecule → 1 $CO_2$ molecule

The stoichiometric coefficients are only relative numbers, so they can be scaled by any desired factor. For example, multiplying each coefficient in the previous equation by twelve, we have

$$12\,C + 12\,O_2 \rightarrow 12\,CO_2,$$ which is read as

1 dozen C atoms + 1 dozen $O_2$ molecules → 1 dozen $CO_2$ molecules

If each of the original coefficients is multiplied by Avogadro's number, the equation becomes:

$$6.022 \times 10^{23}\ C + 6.022 \times 10^{23}\ O_2 \rightarrow 6.022 \times 10^{23}\ CO_2,$$ which is read as

1 mol C atoms + 1 mol $O_2$ molecules → 1 mol $CO_2$ molecules

This latter equation is the same as the original equation with the coefficients interpreted as moles directly, instead of as individual atoms or molecules. Because the coefficients in any chemical equation are just a set of proportional numbers, they can be read with equal correctness as molecules, dozens, or moles (or any other "counting" unit for that matter).

For any balanced chemical equation, if the amount of any *one* reactant or product is specified, then stoichiometry fixes the amounts of every other reactant and product.

A balanced chemical equation summarizes the stoichiometric relationships among the amounts of reactants and products. **Reaction stoichiometry** is the study of the mass and number relationships among reactants and products in a reaction. Consider the following balanced equation that we discussed earlier in connection with the algebraic method of balancing equations (Example 3.6). All the relationships are laid out in the framework below. If you understand this one then you understand them all!

**Reaction Stoichiometry Framework**

| | | | | | | | | |
|---|---|---|---|---|---|---|---|---|
| 1 | Names of reactants and products | Hydrazine | | Nitric acid | | Water | | Nitrogen |
| 2 | Balanced chemical equation | 5 $N_2H_4$ | + | 4 $HNO_3$ | → | 12 $H_2O$ | + | 7 $N_2$ |
| 3 | Moles of reactants and products | 5 mol | | 4 mol | | 12 mol | | 7 mol |
| 4 | Molar masses of reactants and products | 32.0 g mol$^{-1}$ | | 63.0 g mol$^{-1}$ | | 18.0 g mol$^{-1}$ | | 28.0 g mol$^{-1}$ |
| 5 | Stoichiometric masses | 160. g | + | 252 g | | 216 g | + | 196 g |
| 6 | Total mass | reactants = 412 g | | | | products = 412 g | | |

In the framework the sum of the masses of the reactants 160 g + 252 g = 412 g equals the sum of the masses of the products 216 g + 196 g = 412 grams. That they are equal does not prove the law of conservation of mass—that proof can only come in the laboratory. However, their being equal does prove that the math was done correctly (☺). One way to do stoichiometry calculations is simply to use the above framework and ratio and proportion. That's it. Here are a couple of examples:

**Example 3.20**: In the hydrazine-nitric acid reaction you use 15.0 moles of hydrazine. According to stoichiometry, how many moles of everything else must there be?

Strategy: Use the ratio of moles among all reactants and products (line 3 in the framework).

From the balanced equation line 3: 5 mol + 4 mol → 12 mol + 7 mol

So given 15.0 mol of $N_2H_4$ you write 15.0 mol x mol → y mol + z mol

Thus $x = 4 \times (15.0 \div 5) = \underline{12.0\text{ moles}}$ of nitric acid

and $y = 12 \times (15.0 \div 5) = \underline{36.0}$ moles of water

and $z = 7 \times (15.0 \times 5) = \underline{21.0}$ moles of nitrogen molecules

The next example applies the same concept but using masses instead of moles:

**Example 3.21**: In the hydrazine-nitric acid reaction you use 724 g of hydrazine. How many grams of everything else must there be if the reaction is stoichiometric?

Strategy: Use the ratios of the stoichiometric masses in the reaction from line 5 of the reaction stoichiometry framework.

| | | | | | | | |
|---|---|---|---|---|---|---|---|
| From line 5. | 160.0 g | + | 252.0 g | $\rightarrow$ | 216.0 g | + | 196.0 g |
| So given 724 g you write | 724 g | + | x g | $\rightarrow$ | y g | + | z g |

Thus $x = 724 \times (252.0 \div 160.0)$ = 1140 g of nitric acid

and $y = 724 \times (216.0 \div 160.0)$ = 977 g of water

and $z = 724 \times (196.0 \div 160.0)$ = 887 g of nitrogen

You can check the arithmetic using the conservation of mass principle. That shows I'm okay within the rounding errors:

Mass reactants = 724 + 1140 g = 1870 g (watch the significant digits)
Mass of products = 977 + 887 g = 1864 g

By now it will be obvious to you that you can mix moles and mass and you can do every problem in either mass or moles. If you were asked what mass of water is produced by the stoichiometric reaction of 20.0 moles of hydrazine with an excess of nitric acid you'd see that's a (20.0 ÷ 5) molar ratio, or 4.00 times the stoichiometric reaction. So the mass of water produced will be 4.00 × 216 g = 864 g.

## Section 3.5 Excess and Limiting Reactants and Reaction Yield.

### Excess and Limiting Reactants

When reactants are supplied in the precise proportions required by the balanced equation they are said to be present in **stoichiometric amounts**. But it is impossible to guarantee stoichiometric amounts because real-world samples contain so many molecules that we cannot count them sufficiently accurately. The **limiting reactant** is the one that will be used up. Some of the **excess reactant** remains after the reaction is complete.

Excess amounts of one or more reactants are often deliberately provided. For example, when a fuel is burned using oxygen from the air, excess air helps to burn the fuel completely. We will see later an excess reactant often helps to make more of the desired products and make them more rapidly. (Looking ahead, those topics are chemical equilibrium and chemical kinetics.) Here's an example of an excess reactant calculation with an engineering flavor:

**Example 3.22**: A chemical engineer has to design a plant to convert ethylene, $C_2H_4$, to ethylene oxide, $C_2H_4O$. The available supply of ethylene is 75 tons per hour. What will be the needed oxygen supply assuming that a 10% oxygen excess will be required to successfully perform the reaction?

Strategy: 1. Write a balanced chemical equation for the reaction.

2. Use stoichiometric relationships to figure the exact amount of oxygen need.

3. Provide for a 10% excess

The balanced equation is $C_2H_4 + \frac{1}{2} O_2 \rightarrow C_2H_4O$

Stoichiometric masses 28 tons + 16 tons → 44 tons

Note that we can just put the stoichiometric mass directly in tons. Obviously, mass stoichiometry can be in any desired mass unit. Thus for stoichiometric equivalence, 28 tons of ethylene needs 16 tons of oxygen, so 75 tons needs $75 \times (16 \div 28) = 43$ tons.

To provide for 10% excess I'll multiply by 1.1 and round up to be conservative.

Mass of oxygen needed = $43 \times 1.1$ = 48 tons per hour.

Note: Engineers tend to round conservatively.

**Non-stoichiometric** amounts of reactants exist if they are not provided in exactly the proportion required by the balanced equation. When the reactants are provided in non-stoichiometric amounts, one reactant is the limiting reactant and the other is in excess. The limiting reactant limits the amount of product that can be formed. It is the reactant that is entirely consumed as the reaction progresses. After it is depleted, no further reaction can occur. The concept of limiting reactant is illustrated in Example 3.23.

**Example 3.23**: When heated, zinc dust and powdered sulfur react to yield zinc sulfide (ZnS) according to the equation Zn + S → ZnS. In an experiment, 52.7 g of zinc dust is mixed with 45.2 g of sulfur powder; the mixture is heated and the elements react. Assume that the reaction goes to its stoichiometric conclusion. How much of what reactant remains and how much product is formed?

Strategy: 1. Calculate the moles of each reactant using $n = \frac{m}{AW}$ 2. Establish which one is in excess and which one is limiting from the moles. 3. Do a stoichiometric calculation based on the limiting reactant. 4. Check by doing a mass balance.

moles zinc $= 52.7 \text{ g} \div 65.39 \text{ g mol}^{-1} = 0.806$ mol
moles sulfur $= 45.2 \text{ g} \div 32.07 \text{ g mol}^{-1} = 1.41$ mol

So, sulfur is in excess and zinc limits. At the end of the reaction, all the zinc will be gone along with 0.806 mol of sulfur: 0.806 mol of sulfur was needed for each 0.806 mol of Zn. Thus the excess reactant is sulfur, and 1.41 - 0.806 mols = 0.60 mols remain. In grams that's $0.60 \text{ mol} \times 32.07 \text{ g mol}^{-1}$ = 19 g.

Stoichiometry requires that 0.806 mols of ZnS product are formed. MW ZnS = $97.46 \text{ g mol}^{-1}$. So the mass of product is $0.806 \text{ mol} \times 97.46 \text{ g mol}^{-1}$ = 78.6 g.

Finally, do a mass balance calculation to check:

| original mass of zinc | + | original mass of sulfur | = | mass of product | + | mass of excess sulfur |
|---|---|---|---|---|---|---|
| 52.7 g | + | 45.2 g | | 78.6 g | + | 19 g |
| Total reactants 97.9 g | | | | Total products 98 g | | |

The result shows satisfactory agreement considering the experimental uncertainty.

**Example 3.24**: Compound A reacts with compound B according to the balanced chemical equation A + 3 B $\rightarrow$ 2 P. In a reaction, 19.2 moles of A and 47.2 moles of B are allowed to react to the maximum possible extent. a. Which reactant is limiting? b. Which reactant is in excess? c. What amount of product is formed? d. What amount of the excess reactant remains at the end of the reaction?

Strategy: Observe from the coefficients of the balanced equation that stoichiometry requires three times as many moles of B as of A. So triple the given moles of A and compare that answer with the given number of moles of B. Select the excess reactant. Do the necessary arithmetic using the limiting reactant as the basis for the calculation.

a. and b. 19.2 moles of A would require 3 × 19.2 or 57.6 moles of B.

If 57.6 moles of B is needed and only 47.2 has been given, then there is insufficient B provided.

Conclusions: 1. B limits and A is present in excess. 2. Base the subsequent calculations on the amount of B.

c. According to stoichiometry, 3 moles of B react to give 2 moles of P.

So 47.2 moles of B will give (47.2 ÷ 3) × 2 moles of product, P. 31.5 moles of P will be formed.

d. The amount of A that will react is one third of the number of moles of the limiting reactant:

Moles of A that react = 47.2 ÷ 3 = 15.7 moles

Excess amount of reactant A = Moles provided - moles consumed = 19.2 - 15.7 = 3.5 moles of excess A.

**Example 3.25**: A 37.3 g sample of methane ($CH_4$, MW = 16.0 g mol$^{-1}$) is combusted with 287.3 g of oxygen ($O_2$, MW = 32.0 g mol$^{-1}$). a. Use reaction stoichiometry to demonstrate that oxygen is the excess reactant. b. Calculate the mass of oxygen that remains if the combustion of the methane is complete to form water and carbon dioxide.

Strategy: Write a balanced equation for the reaction. Use that equation to calculate the mass of oxygen need to combust the given mass of methane. Calculate the mass of excess oxygen by subtraction.

a. The balanced combustion equation is: $CH_4 + 2\ O_2 \rightarrow CO_2 + 2\ H_2O$

Thus, 1 mole of methane, or 16.0 g, requires 2 moles or 2 × 32.0 = 64.0 g of oxygen .

That's $\frac{64.0 \text{ g of } O_2}{16.0 \text{ g } CH_4}$ or 4.00 g of $O_2$ per g of $CH_4$.

Mass of oxygen consumed = 37.3 g of methane × $\frac{4.00 \text{ g of } O_2}{1.00 \text{ g } CH_4}$ = 149 g of $O_2$ consumed.

b. Mass of excess oxygen = Mass given - mass consumed

= 287.3 g - 149 g = 138 g

The concept of limiting reactant applied to a problem involving the reaction of two solutions is illustrated in Example 3.26.

**Example 3.26**: In aqueous solution, silver nitrate reacts with hydrochloric acid to produce insoluble silver chloride. The equation is: $AgNO_3(aq) + HCl(aq) \rightarrow AgCl(s) + HNO_3(aq)$. Mixing two solutions, one containing 1.000 gram of $AgNO_3$ and the other containing 1.000 g of HCl, causes immediate precipitation of AgCl(s). How many grams of the excess reactant will remain in solution?

Strategy: 1. Calculate the moles of each reactant using $n = \frac{m}{AW}$ 2. Establish which reactant is in excess and which is limiting from the moles. 3. Do a stoichiometric calculation based on the limiting reactant. Note that the mole ratio $AgNO_3(aq)$:HCl(aq) is 1:1.

$$\text{moles of } AgNO_3 = 1.00 \text{ g} \div 169.87 \text{ g mol}^{-1} = 0.00589 \text{ mol}$$

$$\text{moles of HCl} = 1.00 \text{ g} \div 36.46 \text{ g mol}^{-1} = 0.0274 \text{ mol}$$

So HCl is the excess reactant and silver nitrate is the limiting reactant. The reaction stoichiometry thus shows that the maximum amount of silver chloride that can precipitate is 0.00589 moles, equal to the complete precipitation of the reactant silver nitrate. Silver chloride has a molar weight of 143.32 g $mol^{-1}$. So the mass of the precipitate is:

$$0.00589 \text{ mol} \times 143.32 \text{ g mol}^{-1} = \underline{0.844 \text{ g}}$$

$$\text{The excess HCl is } (0.0274 - 0.00589 \text{ mol}) \times 36.46 \text{ g mol}^{-1} = \underline{0.785 \text{ g}}$$

When the stoichiometric coefficients of two reactants are the same, as in the previous examples, the limiting reagent is the reactant present in the smallest mole quantity. It becomes a little more challenging when the stoichiometric coefficients are different.

The Example 3.27 illustrates the same idea for a reaction in which the reactants have a 3:1 mole ratio as compared to the 1:1 ratio illustrated in Example 3.26.

**Example 3.27**: To a solution containing 2.44 millimoles of phosphoric acid is added 0.503 g of solid potassium hydroxide (KOH). Complete neutralization of *one* of the reactants occurs. Calculate the mass of the excess reactant.

Strategy: 1. Write the balanced equation to represent neutralization. 2. Calculate the number of moles of each reactant and decide which reactant is limiting and which is in excess. 3. Use reaction stoichiometry to deduce the moles of excess reactant. 4. Use molar weight to convert that to grams.

1. The balanced neutralization equation is $3\ KOH + H_3PO_4 \rightarrow K_3PO_4 + 3\ H_2O$

2. Moles of KOH $= 0.503 \text{ g} \div 56.1 \text{ g mol}^{-1} = 8.97 \times 10^{-3}$ mol

Moles of $H_3PO_4 = 2.44 \times 10^{-3}$ mol (stated)

The number of moles of KOH must be at least three times the number of moles of phosphoric acid. Multiplying the number of moles of phosphoric acid by three gives $(3 \times 2.44) \times 10^{-3}$, or $7.32 \times 10^{-3}$ mol. That value is less than the $8.97 \times 10^{-3}$ mol of potassium hydroxide. Thus, potassium hydroxide is in excess. The amount of the excess is $8.97 \times 10^{-3}$ mol minus $7.32 \times 10^{-3}$ mol.

3. Moles of excess reactant $= 8.97 \times 10^{-3}$ mol KOH $- 7.32 \times 10^{-3}$ mol $H_3PO_4$

$= 1.65 \times 10^{-3}$ mol potassium hydroxide

4. Converting to mass: mass of excess reactant $= 1.65 \times 10^{-3} \text{ mol} \times 56.1 \text{ g mol}^{-1}$

= <u>92.6 milligrams</u>

## Reaction Yield

The maximum amount of product that can be obtained from a reaction depends on the available amount of the limiting reactant. The **theoretical yield** of a reaction product is the maximum amount of it that can be obtained presuming that all of the limiting reactant is consumed. Here's a straightforward yield calculation:

**Example 3.28**: A copper ore contains 2.3% copper. What is the theoretical yield of copper from 1.00 kg of this ore? What is the percentage yield if 19 g was actually obtained?

Strategy: 1. Calculate the maximum amount of copper in the ore. 2 Express the actual yield as a percentage of the maximum possible yield.

$1.00 \text{ kg} \times \frac{2.3}{100} \times 1000 \text{ g kg}^{-1} = 23 \text{ g}$. The theoretical yield is $\underline{23 \text{ g}}$

The percentage yield is $\frac{19}{23} \times 100 = \underline{83\%}$

The percentage yield is calculated using the relationship:

$$\text{Percentage yield} = \frac{\text{mass of product obtained}}{\text{mass of theoretical yield of product}} \times 100\%$$

Theoretical yield can be calculated even if a balanced chemical equation is not known for a reaction. All that is necessary is to specify the amount of the limiting reactant as shown in Example 3.28.

**Example 3.29**: Every year, a large amount of sulfur is converted to sulfuric acid by US chemical companies. You personally consume about 380 pounds of sulfuric acid each year. What is the theoretical yield of sulfuric acid that can be obtained from 1.00 metric ton (= 1000 kg) of sulfur and a sufficient amount of water and oxygen so that sulfur is the limiting reactant. Molar weight $H_2SO_4 = 98.0 \text{ g mol}^{-1}$.

Strategy: 1. Calculate the moles of sulfur. 2. Based on S, the partial equation is $S \rightarrow H_2SO_4$, so a reaction stoichiometry of 1:1 is used to calculate the moles of sulfuric acid produced. 3. Convert moles to mass.

1. Moles of sulfur $= 1.00 \text{ ton} \times 1000 \text{ kg ton}^{-1} \times 1000 \text{ g kg}^{-1} \div 32.07 \text{ g mol}^{-1}$

$= 3.12 \times 10^4$ mols

2. The reaction stoichiometry is 1:1 so $3.12 \times 10^4$ mols of sulfuric acid are produced.

3. $3.12 \times 10^4 \text{ moles} \times 98.0 \text{ g mol}^{-1} \times 0.001 \text{ g kg}^{-1} \times 0.001 \text{ kg ton}^{-1} = \underline{3.06 \text{ tons.}}$

**Example 3.30**: Ammonium dichromate is an orange, crystalline material. When heated, a small pile of ammonium dichromate decomposes, with a characteristic "volcano" effect, according to the following balanced chemical equation:

$$(NH_4)_2Cr_2O_7(s) \rightarrow Cr_2O_3(s) + N_2(g) + 4\,H_2O(g)$$

If 32.4 g of ammonium dichromate yields 15.3 g of $Cr_2O_3(s)$, what is the percentage yield of the reaction? The molar weights are ammonium dichromate = $292.0 \text{ g mol}^{-1}$ and chromium oxide = $152.0 \text{ g mol}^{-1}$.

Strategy: Use the reaction stoichiometry to calculate the theoretical yield and then make the actual yield a percentage of that. If x is the theoretical yield, then the theoretical yield can be directly obtained from the ratio:

$$\frac{292.0}{32.4} = \frac{152.0}{x}$$

and x = $\underline{16.9 \text{ g}}$.

The percentage yield is $(15.3 \div 16.9) \times 100\%$ or $\underline{90.5\%}$

For commercial purposes, reactions with high yields are desirable. Maximizing the yield of products such as pharmaceuticals, pesticides, etc., not only improves profits, it also means less waste has to be treated. In many organic reactions, there is only one carbon-containing reactant (which, by design, is made the limiting reactant) and only one carbon-containing product. Here's an example of a yield calculation based on an organic compound being the limiting reactant.

**Example 3.31**: A sample of benzene ($C_6H_6$, MW = 78.11 g mol$^{-1}$) weighing 51.3 g was heated with excess bromine ($Br_2$) and iron(III) bromide as a catalyst. After the reaction had taken place, the reaction mixture was subjected to the required recovery and separation steps (the "work-up"), and 79.2 g of the only carbon containing product, bromobenzene ($C_6H_5Br$, MW = 157.02 g mol$^{-1}$), was recovered. Calculate the percentage yield of the reaction.

$$C_6H_6 + Br_2 \text{ (excess)} \xrightarrow{FeBr_3 \text{ catalyst}} C_6H_5Br + HBr$$

Strategy: Each mole of benzene that reacts produces one mole of bromobenzene. Use the mass stoichiometric ratio to calculate the theoretical yield and then express the actual yield as a percentage of that.

The theoretical yield can be directly obtained from the ratio:

$$\frac{78.11}{51.3} = \frac{157.02}{x} \quad \text{and } x = \underline{103\ g}.$$

The percentage yield is thus $(79.2 \div 103) \times 100 = \underline{76.9\%}$

## Section 3.6 Solution Stoichiometry and Molarity

A liquid, homogeneous mixture containing two or more substances is a **liquid solution**. An **aqueous solution** is water containing one or more dissolved substances. The pure liquid that does the dissolving is the **solvent**. Substances that dissolve are **solutes**. Reactions among dissolved substances are common. Indeed, because of their convenience, most chemical reactions are carried out in liquid solutions.

**Concentration** refers to how much solute is in a solution. A large amount of solute in a solvent makes a **concentrated** or **strong** solution. A small amount solute in a solvent makes a **dilute** or **weak** solution. There are many different quantitative measures of concentration. The most important chemical unit of concentration is molarity. **Molarity** (M) is the concentration of a solute in the units of moles of solute per liter of solution.

$$M = \frac{n}{V_L} = \frac{\text{moles of solute}}{\text{volume of solution in liters}}$$

**Example 3.32**: Calculate the molarity of an aqueous solution that contains 24.7 g of the strong base sodium hydroxide NaOH in 873 mL of solution.

Strategy: 1. Calculate the number of moles of NaOH, 2. Convert the volume to liters, 3. Plug into the equation $M = n \div V_L$.

$$\text{Molar weight NaOH} = 23.0 + 16.0 + 1.01 = 40.0\ \text{g mol}^{-1}$$

$$\text{moles} = \text{mass} \div \text{molar weight} = (24.7\ \text{g} \div 40.0\ \text{g mol}^{-1}) = 0.618\ \text{mols}$$

$$V_L = 873\ \text{mL} \times 0.001\ \text{L mL}^{-1} = \underline{0.873\ \text{L}}$$

$$\text{So } M = n \div V_L = (0.618\ \text{mol} \div 0.873\ \text{L}) = \underline{0.708\ M}$$

**Example 3.33**: A solution of nitric acid ($HNO_3$) is 3.24 M. How many moles of the acid are in a 14.8 mL sample of the solution? How many grams?

Strategy: Use the definition of molarity to find the moles. Then use the molar weight of nitric acid to convert that to grams.

$$\text{Number of moles, } n = M \times V_L. \text{ So, } n = 3.24 \text{ mol L}^{-1} \times 14.8 \text{ mL} \times 10^{3} \text{ L mL}^{-1}$$

$$= \underline{0.0480 \text{ mol}}$$

$$\text{The molar weight of nitric acid is } 1.01 + 14.0 + 3 \times 16.0 = 63.0 \text{ g mol}^{-1}$$

$$\text{Mass of nitric acid} = 0.0480 \text{ mol} \times 63.0 \text{ g mol}^{-1} = \underline{3.02 \text{ g}}$$

To make a solution of specified molarity, a chemist uses a volumetric flask. A volumetric flask has a narrow neck on which is etched a circle indicating the liquid level that corresponds to the calibrated volume. The procedure is to weigh the required amount of solute and put it in the flask, add sufficient water to dissolve the solute, shake thoroughly to dissolve, and then carefully top off until the solution level is at the calibration mark. The solution is ready for use after a final shaking. Volumetric glassware has not changed much in over a century. A modern flask is shown in Figure 3.6. The flask in Figure 3.7 was pictured in Francis Sutton's 1871 book titled *Volumetric Analysis*.

Figure 3.6 A modern one-liter volumetric flask.

Figure 3.7 Sutton's 1871 volumetric flask.

**Example 3.34**: A chemist needs 435 mL of a 0.231 M solution of potassium chloride (KCl, molar weight 74.6 g $mol^{-1}$). How should this solution be made?

Strategy: It should be done by first selecting a volumetric flask: 500.0 mL flasks are common, and big enough for this job. Next, calculate the needed mass of potassium chloride solute, weigh it out, put in the flask and fill to the line using the procedure described in the text.

The needed mass of KCl in 500.0 mL (0.5000 L) of a 0.231 M solution is

$$\text{molarity} \times \text{volume} \times \text{molar weight of solute}$$

$$0.231 \text{ mol L}^{-1} \times 0.5000 \text{ L} \times 74.6 \text{ g mol}^{-1} = \underline{8.62 \text{ g}}$$

(That's the amount to weigh out.)

Note: Only 435 mL of the solution is needed, so the chemist has an extra 65 mL.

## Dilution Calculations

In handling solutions, it is convenient to make and store them in concentrated form and later dilute portions of them to the desired concentration for use. Suppose we have a concentrated solution of molarity $M_1$, and we wish to dilute this to the molarity $M_2$. A simple procedure is to take a specified volume of the concentrated solution and add water to make up the desired volume of working solution

The mathematical relationship you use for doing dilution calculations is; $M_1V_1 = M_2V_2$, where $M_1$ is the original molarity, $M_2$ the new molarity, and $V_1$ and $V_2$ the original and new volumes.

If you think about it, you can see why this equation works. It simply states moles of solute = moles of solute. Which is obviously true if what's going on is only dilution.

A point of notation: The notation of square brackets [ ] is used to designate molarity. Thus $[H^+]$ means the molarity of hydrogen ion, $[CO_2]$ means the molarity of carbon dioxide, etc.

For combining dilute aqueous solutions it is okay to assume the volumes are additive:

$$V_1 + V_2 = V_{total}$$

But that's *not* true in general for mixing liquids. Go back and review Problem 1.59 that deals with volume change when mixing alcohol and water.

**Example 3.35**: A 5.00 mL sample of a 6.00 M solution is diluted to a final volume of 100.00 mL. What is the resulting molarity?

Strategy: Plug into the dilution relationship, $M_1V_1 = M_2V_2$

$$M_1 = 6.00 \text{ M};\ V_1 = 5.00 \text{ mL};\ V_2 = 100.0 \text{ mL};\ M_2 = x$$

$$6.00 \text{ mol L}^{-1} \times 5.00 \text{ mL} = 100.0 \text{ mL} \times x$$

Resulting molarity = <u>0.300 M</u>

In this type of problem it is not necessary to convert all the volumes to L. As long as $V_1$ and $V_2$ are in the same units those units cancel.

When ionic substances dissolve in water they break up to produce ions via ionization reactions, as described earlier. Here's an example of a solution concentration calculation involving two ionic substances.

**Example 3.36**: A mixture is made from 37.6 L of 0.221 M sodium chloride (NaCl) and 23.3 L of 0.318 M potassium chloride (KCl). These ionic substances are 100% ionized in aqueous solution. Calculate the concentrations of each of the ions in the mixture. Assume the volumes are additive. Here are the ionization equations:

$$NaCl(s) + H_2O \xrightarrow{H_2O} Na^+(aq) + Cl^-(aq) \qquad KCl(s) + H_2O \xrightarrow{H_2O} K^+(aq) + Cl^-(aq)$$

In the solution of NaCl or KCl the molarity of *each* of its constituent ions is the stated molarity.

Strategy: First calculate the total volume ($V_2$ = 37.6 + 23.3 L = 60.9 L). Next, *separately* dilute each given solution to that total volume using $M_1V_1 = M_2V_2$. For sodium ion and potassium ion that's the answer. Get the concentration of chloride ion by adding together the chloride ion from sodium chloride and the chloride ion from potassium chloride.

$$\text{For sodium and potassium ion, } M_2 = (M_1V_1) \div V_2,$$

$$\text{Molarity of sodium ion} = (0.221 \text{ mol L}^{-1} \times 37.6 \text{ L}) \div 60.9 \text{ L} = 0.136 \text{ M}$$

$$\text{Molarity of potassium ion} = (0.318 \text{ mol L}^{-1} \times 23.3 \text{ L}) \div 60.9 \text{ L} = 0.122 \text{ M}$$

To get the chloride ion concentration you could repeat the above procedure and add. But it's easier if you see that $[Cl^-] = [Na^+] + [K^+]$.

So the chloride ion concentration in the final solution is 0.136 M + 0.122 M = 0.258 M

## Stoichiometry of Reactions in Solution

A brief recapitulation: Many chemical reactions take place in solution. There are also many reactions in which a solution reacts with a solid. For now, we're only dealing with water as the solvent; it makes aqueous solutions. Most ordinary acids and bases used by chemists are in the form of aqueous solutions. Many chemical reactions involve the mixing of two solutions. The reactants are separately dissolved and the reaction takes place when the solutions are mixed. Water often does not participate in reactions involving aqueous solutions. If water doesn't participate it's said to be an **inert solvent**.

Aqueous solutions of acids are able to dissolve many solids. For example, geologists use hydrochloric acid solution to test rocks to see if they contain calcium carbonate. As we saw earlier, the balanced chemical equation for the reaction is

$$CaCO_3(s) + 2\ HCl(aq) \rightarrow CaCl_2(aq) + CO_2\uparrow + H_2O$$

The "fizzing" of the rock caused by the evolution of carbon dioxide ( $\uparrow$ ) is the basis of this test. The weight of rock that will be dissolved can be calculated from the stoichiometry of the above equation, as shown in the following example:

**Example 3.37**: A volume of 3.45 mL of 8.95 M aqueous hydrochloric acid solution is poured onto a pile of solid $CaCO_3$. What mass of solid dissolves? What mass of carbon dioxide is evolved? Assume the reaction occurs according to the equation above and that the yield is 100%.

Strategy: 1. Recognize that because the pile of $CaCO_3$ is large, the hydrochloric acid is the limiting reactant. So we base the calculation on the number of moles of HCl. 2. Calculate the moles of HCl from $n = M \times V_L$. 3. Use the stoichiometry to get the needed masses.

$$\text{Moles of HCl} = 8.95 \text{ mol L}^{-1} \times 3.45 \text{ mL} \times 10^{-3} \text{ L mL}^{-1} = 0.0309 \text{ mol}$$

The reaction stoichiometry shows that for every two moles of hydrochloric acid reacted, one mole of calcium carbonate is dissolved and one mole of carbon dioxide is evolved. The molar masses of $CaCO_3$ and $CO_2$ are 100.1 g mol$^{-1}$ and 44.01 g mol$^{-1}$, respectively. Thus one-half of the number of moles of HCl gives the number of moles of $CaCO_3$ and $CO_2$:

$$\text{Moles } CaCO_3 \text{ needed and } CO_2 \text{ produced} = 0.0309 \div 2 = 0.0154 \text{ mols}$$

$$\text{Mass of } CaCO_3 \text{ dissolved} = 0.0154 \text{ mol} \times 100.1 \text{ g mol}^{-1} = \underline{1.54 \text{ g } CaCO_3}$$

$$\text{Mass of } CO_2 \text{ gas evolved} = 0.0154 \text{ mol} \times 44.01 \text{ g mol}^{-1} = \underline{0.678 \text{ g } CO_2}$$

Many metals dissolve in acid solution, forming a salt of the metal and the simultaneous liberation of hydrogen gas. Iron reacts with sulfuric acid according to the following equation:

$$Fe(s) + H_2SO_4(aq) \rightarrow FeSO_4(aq) + H_2\uparrow$$

The following example is based on the stoichiometry of this equation.

**Example 3.38**: A sample of iron powder weighing 0.456 g is added to 985 mL of 1.20 M sulfuric acid solution. What is the molarity of the $FeSO_4(aq)$ now present? Assume the limiting reactant is completely consumed. The mol weight of iron is 55.85 g mol$^{-1}$.

Strategy: Calculate the number of moles of iron. Because of the 1:1 reaction stoichiometry this is the number of moles of $FeSO_4$. 2. Calculate molarity from $M = n \div V_L$.

$$n = m \div AW = 0.456 \text{ g} \div 55.85 \text{ g mol}^{-1} = 0.00816 \text{ mol Fe}$$

Thus, 0.00816 moles of $FeSO_4$ are produced, and the molarity of the solution is

$$\text{molarity} = n \div V_L = 0.00816 \text{ mol } FeSO_4 \div (985 \text{ mL} \div 1000 \text{ mL L}^{-1})$$

$$= \underline{0.00829 \text{ M } FeSO_4}$$

There are many practical examples of the application of the principles of stoichiometry to reactions involving the mixing of two solutions. Many students have used simple tests of this type to measure water quality in fish tanks or swimming pools. Solution stoichiometry is the basis of **volumetric analysis**, a procedure in which a measured volume of a reactant solution provides a quantitative analysis of a target substance (the **analyte**).

Mixing solutions of a strong acid and a strong base yields a neutral solution (pH = 7) if stoichiometric amounts of acid and base are used. It's easy in the lab to test when an acid and base are neutralized by using an **indicator dye** that changes color when sufficient acid has been added to neutralize a base, or vice versa. The **endpoint** of an acid-base reaction occurs when the indicator dye changes color. If the proper dye has been selected, the endpoint is also the point of stoichiometric equivalence between the acid and the base.

Figure 3.8 Many commercial dyes change color if sprayed with oven cleaner (a strong base), or with battery acid (a strong acid).

Many dyes are used in everyday life. Some of these are inadvertent acid-base indicators. Onto the sheets of colored paper in Figure 3.8 has been sprayed a little oven cleaner. Oven cleaner contains sodium hydroxide (to dissolve baked fat), and is strongly basic. You can see that the dyes have changed where the spray hit the paper. So these dyes are acid-base indicators.

For the solution reaction between an acid and a base the fundamental relationship that applies when they have exactly neutralized one another is:

$$M_aV_a = M_bV_b$$

where $M_a$ is the effective molarity of the *hydrogen ion* provided by the acid and $M_b$ is the effective molarity of the *hydroxide ion* provided by the base. (Effective in this context means if all the acid or base has been reacted.) For an acid, $M_a$ is the labeled molarity of the acid multiplied by the number of hydrogen ions that the acid yields when it ionizes.

Recall that nitric acid ($HNO_3$) yields one hydrogen ion and one nitrate ion for every molecule that ionizes. We could also say it reacts *as though* it produces one hydrogen ion.

$$HNO_3(aq) \rightarrow H^+(aq) \quad + \quad NO_3^-(aq)$$

Sulfuric acid ($H_2SO_4$) yields two hydrogen ions and one sulfate ion for every molecule of sulfuric acid that ionizes. We could also say it reacts *as though* it produces two hydrogen ions.

$$H_2SO_4(aq) \rightarrow \quad 2\,H^+(aq) \quad + \quad SO_4^{2-}(aq)$$

Thus, the effective molar hydrogen ion concentration, $[H^+]$, of a 1.44 M nitric acid solution is 1.44 M. Whereas the effective molar hydrogen ion concentration, $[H^+]$, of a 1.44 M sulfuric acid solution is 2.88 M.

As we'll see later, most acids (including sulfuric) are significantly less than 100% ionized. So their solutions are *less* acidic, contain less $[H^+]$, than we would expect just from reading the molarity on the label of their bottles. However, that does not present any difficulty when we do stoichiometric acid-base calculations. Whether the acid is ionized or not, it reacts with the same amount of hydroxide ions, as shown by the following pair of equations:

$2\,H^+ + 2\,OH^- \rightarrow 2\,H_2O$ ...two hydrogen ions from an ionized acid molecule neutralize two hydroxide ions

$H_2SO_4 + 2\,OH^- \rightarrow 2\,H_2O + SO_4^{2-}$ ...a molecule of sulfuric acid itself neutralizes two hydroxide ions

There's a parallel situation with bases and hydroxide ions. For an base, $M_b$ is the labeled molarity of the base multiplied by the number of hydroxide ions that the base yields when it ionizes. A 0.992 M sodium hydroxide solution has an $M_b$ of 0.992 M; a 1.33 M solution of $Ba(OH)_2$ has $M_b = 2.66$ M. The relationship between label acid or base molarity and the corresponding effective hydrogen ion and hydroxide ions molarities is summarized in Table 3.5 below:

Table 3.5 **Label and Effective Molarities of Acids and Bases**

| Acid | Label Molarity $M_a$ | Effective Hydrogen ion molarity | Base | Label Molarity $M_b$ | Effective Hydroxide ion molarity |
|---|---|---|---|---|---|
| HCl | 1.0 | 1.0 | NaOH | 1.00 | 1.00 |
| $HNO_3$ | 1.0 | 1.0 | KOH | 1.00 | 1.00 |
| $H_2SO_4$ | 1.0 | 2.0 | $Ca(OH)_2$ | 1.00 | 2.00 |
| $H_2C_2O_4$ | 1.0 | 2.0 | $Ba(OH)_2$ | 1.00 | 2.00 |
| $H_3PO_4$ | 1.0 | 3.0 | $Al(OH)_3$ | 1.00 | 3.00 |

**Example 3.39**: What volume of 0.378 M nitric acid is needed to exactly neutralize 30.6 mL of 0.519 M lithium hydroxide solution?

Strategy: 1. Recognize that for nitric acid, $M_a = 0.378$, and that for LiOH $M_b = 0.519$ M. Plug these values and the volume of the LiOH base into $M_aV_a = M_bV_b$

$$0.378 \text{ mol L}^{-1} \times V_a = 0.519 \text{ mol L}^{-1} \times 30.6 \text{ mL}$$

$$V_a = \underline{42.0 \text{ mL}}$$

**Example 3.40**: Calculate the molarity of a sulfuric acid solution knowing that 25.00 mL of this solution requires 32.77 mL of a 0.213 M sodium hydroxide solution to reach the endpoint.

Strategy: Apply $M_aV_a = M_bV_b$ to calculate $M_a$ and then take one-half of that value because the ionization equation below shows that the effective acid concentration of sulfuric acid is twice its label concentration, or, put the other way around, its label concentration will be one-half its effective hydrogen ion concentration.

$$H_2SO_4(aq) \rightarrow 2\ H^+(aq) + SO_4^{2-}(aq)$$

$$M_a \times 25.00 \text{ mL} = 0.213 \text{ mol L}^{-1} \times 32.77 \text{ mL}$$

$$M_a = 0.280 \text{ M}$$

$$M(H_2SO_4) = \tfrac{1}{2} \times 0.280 \text{ M} = \underline{0.140 \text{ M}}$$

Here's a final example of solution stoichiometry.

**Example 3.41**: An aqueous solution of 0.117 M hydrochloric acid with a volume of 56.8 mL is mixed with 45.7 mL of 0.213 M potassium hydroxide solution. Calculate the molarity of the excess reactant (acid or base) in the resulting mixture. Also calculate the molarity of the salt (KCl) in the resulting mixture. The equation is: $HCl(aq) + KOH(aq) \rightarrow H_2O(l) + KCl(aq)$

Strategy: 1. Decide which reactant is limiting by calculating the number of moles of each using $n = M \times V_L$. 2. Do a stoichiometric calculation based on the limiting reactant. 3. Convert moles to molarity. Assume that the volumes are additive.

Moles of hydrochloric acid = $0.0568\ L \times 0.117\ mol\ L^{-1}$ = 0.00665 mol HCl
Moles of potassium hydroxide = $0.0457\ L \times 0.213\ mol\ L^{-1}$ = 0.00973 mol KOH

Because the stoichiometry of the reaction is 1:1, the KOH is the excess reactant:

Thus, moles of salt (KCl) formed = 0.00665

Moles of KOH in excess = 0.00973 - 0.00665 mol = 0.00308 mol

Volume of solution after reaction = 0.0568 L + 0.0457 L = 0.1025 L

Molarity of KOH in final solution = 0.00308 mol ÷ 0.1025 L = <u>0.0301 M KOH</u>

Molarity of KCl in final solution = 0.00665 mol ÷ 0.1025 L = <u>0.0649 M KCl</u>

## Essential Knowledge—Chapter 3

**Glossary words**: Stoichiometry, reactants, products, chemical equation, decomposition reaction, bond energy, addition reaction, reversible reaction, balanced chemical equation, stoichiometric coefficients, lean and rich, complete combustion, limiting and excess reactant, incomplete reaction, strong acid, binary acid, ionization reaction, weak acid, strong base, weak base, neutralization reaction, neutral solution, exact neutralization, active metal, precipitation reaction, spectator ions, oxidized, reduced, redox reaction, balancing by inspection, algebraic method for balancing equations, amount of substance, elemental analysis, weight percentage, empirical formula, reaction stoichiometry, stoichiometric amount, limiting reactant, excess reactant, complete reaction, non-stoichiometric amount, actual yield, theoretical yield, percentage yield, aqueous solution, solvent, solute, concentration, molarity, inert solvent, indicator dye, endpoint.

**Key Concepts**: Types of chemical reactions. Law of conservation of mass. Law of constant or definite composition. Law of multiple proportions. Acid and bases and their nature. Stoichiometric relationships in compounds and reactions. Balancing chemical equations by inspection and the algebraic method. Elemental analysis of compounds and empirical formula. Stoichiometry and excess and limiting reactants. Reaction yield. Solution stoichiometry and molarity.

### Key Equations:

$$M = \frac{n}{V_L} = \frac{\text{moles of solute}}{\text{volume of solution in liters}}$$

$$M_aV_a = M_bV_b$$

## Questions and Problems

3.1. On a passenger plane there are 135 women with an average mass of 112 pounds and 125 men with an average mass of 174 pounds. What is the average mass of a passenger? Which sex is in mass excess? Which sex is in number excess.

3.2. The term **Amount of Substance** has a very special meaning in SI. Consider your answers to the previous question and state two methods for specifying the amount of substance in any sample of matter.

3.3. Define or describe in 25 words or less each of the following: reactant, product, reversible reaction, balanced equation, combustion reaction, fuel.

### Chemical Reactions and Equations

3.4. Write balanced chemical equations to represent the following processes: the addition reaction of one hydrogen molecule to an ethylene ($C_2H_4$) molecule to yield ethane, the addition reaction between three hydrogen molecules and one molecule of benzene to yield cyclohexane, the decomposition reaction of heptane to yield pentane and ethylene, the decomposition reaction of calcium carbonate to yield calcium oxide and carbon dioxide, the decomposition of ammonium chloride to yield the gases ammonia and hydrogen chloride.

3.5. Write a balanced chemical equation to represent the complete combustion of cyclohexane, $C_6H_{12}$.

3.6. Write a balanced chemical equation to represent the combustion of propane where the only carbon containing product is carbon monoxide.

3.7. Write equations to show the ionization in water of perchloric acid, nitric acid, and nitrous acid. The first two of these are strong acids, the third is weak.

3.8. Write balanced chemical equations for the following reactions: the neutralization of sulfuric acid by barium hydroxide, the neutralization of strontium hydroxide by hydriodic acid. Name the salts produced by the neutralizations.

3.9. Write a balanced chemical equation to represent the reaction of strontium metal with water.

3.10. Define or describe in 25 words or less each of the following: precipitation reaction, spectator ion, redox, oxidizing agent, oxidant, reducing agent, reductant.

### Balancing Chemical Equations

3.11. Write a balanced chemical equation to represent the following reactions: the combustion of sodium metal yielding sodium oxide as the only product, the combustion of magnesium to produce magnesium oxide as the only product.

3.12. Add coefficients to balance the following unbalanced chemical equations:

a. $C_2H_4 + O_2 \rightarrow C_2H_4O$
b. $NaOH + CO_2 \rightarrow Na_2CO_3 + H_2O$
c. $KClO_3 + CH_2O \rightarrow KCl + CO_2 + H_2O$

3.13. Try to balance the following equations by the method of inspection. If you are unable to do that, use the algebraic method:

### Stoichiometry and Molecular Weight

3.14. State Gay-Lussac's law of combining volumes and Avogadro's law.

3.15. State the following laws: law of constant composition, law of definite composition, law of conservation of mass, and law of multiple proportions.

3.16. When all of the volumes are measured at the same temperature and pressure, three liters of the diatomic gas ($F_2$) and one liter of a second diatomic gas ($Cl_2$) react to form two liters of a product gas ($Cl_mF_n$). What must be the values of m and n?

3.17. The gas phase reaction between phosphorus trichloride and chlorine is represented by the equation: $PCl_3(g) + Cl_2(g) \rightarrow PCl_5(g)$. What volume of chlorine can react with 1.55 L of phosphorus trichloride, and what volume of phosphorus pentachloride is formed when all of the gases are measured at the same temperature and pressure?

3.18. The combustion of ethane is represented by the equation: $2\ C_2H_6(g) + 7\ O_2(g) \rightarrow 4\ CO_2(g) + 6\ H_2O(g)$. What volume of oxygen gas is required to burn 5.0 L of ethane gas, and what total volume of gaseous products will be formed? All of the volumes are measured at the same temperature and pressure.

3.19. Tin forms both tin(II) oxide and tin(IV) oxide. Calculate the ratio of the mass of tin to the mass of oxygen for each oxide. Show that the results illustrate the law of multiple proportions.

## Elemental Analysis of Compounds and Empirical Formula

3.20. In his 1992 book, *To Engineer is Human—The Role of Failure in Successful Design*, the civil engineer Henry Petroski discusses cracking and failure in steel bus frames and kitchen knives. He gives the following recipe for type 304 austenitic stainless steel: 1,400 lb iron, 360 lb chromium, 160 lb nickel, 40 lb manganese, 20 lb silicon, 16 lb carbon, 2 lb phosphorus, and 2 lb sulfur. Calculate the mass percentage and the mole fraction of each of the elements in this complex mixture.

3.21. What is the percentage by weight of hydrogen and carbon in hexene ($C_6H_{12}$)?

3.22. Astronomers study "molecules" far away in space. The word molecules is here put in quotation marks because terrestrial chemists wouldn't necessarily call them that—they'd probably call them radicals. Most deep space molecules would be totally unstable on earth because they would immediately react with oxygen or something else in the earth's environment. Calculate the percentage composition of the radical $HC_4N$, known by astronomers to exist in deep space.

3.23. Ancient Egyptian art relies for much of its beauty on blue paint. The blue pigment that the Egyptians used in the paint was a synthetic form of the rare mineral cuprorivaite made by hearing a mix of ingredients to about 900°C. Cuprorivaite has the chemical formula $CaCuSi_4O_{10}$. a. Assuming that the Egyptians started with a mixture of sand ($SiO_2$), lime ($CaO$), and copper(II) oxide ($CuO$), write a balanced chemical equation to represent the manufacture of synthetic cuprorivaite. b. Calculate the percentage composition of the mineral.

3.24. Calculate the elemental analysis of nitromethane ($CH_3NO_2$) and glycerol ($C_3H_8O_3$).

3.25. Calculate the percentage composition of the compound methylcyclopentadienyl manganese tricarbonyl which has the chemical formula $C_9H_7MnO_3$. Do an internet search to discover the toxicity this anti-knock fuel additive compound.

3.26. State the molecular formulas of the four DNA bases: adenine, guanine, cytosine, and thymine. Calculate each of their molar weights and their percentage compositions.

3.27. What are the empirical formulas of pentene ($C_5H_{10}$), potassium peroxydisulfate, ($K_2S_2O_8$), and trinitrobenzene ($C_6H_3N_3O_6$)?

3.28. A compound contains 52.4% by weight of potassium, and 47.6% by weight of chlorine. What is its empirical formula?

3.29. A compound contains 13.20% by weight of magnesium, and 86.80% by weight of bromine. What is its empirical formula?

3.30. The analysis of a compound containing C, H, and O yields the following percentage composition by weight: carbon 71.4%, hydrogen 9.6%, and oxygen 19.0%. What is the empirical formula of the compound?

3.31. The simplest organic compound that contains the nitrile or cyano ($-C\equiv N$) functional group has the percentage composition C, 58.51%; H, 7.37%; and N, 34.12%. Its molar weight is 41 g $mol^{-1}$. Deduce its chemical formula and draw its structure.

3.32. The analysis of a compound containing C, H, and O shows that it contains 54.4% carbon and 9.2% hydrogen; the balance is oxygen. The molecular weight of the compound is determined to be about 90 g $mol^{-1}$. What are the empirical and molecular formulas of the compound?

3.33. A binary compound contains one mole of nitrogen atoms and three moles of metal atoms. Analysis shows that it contains 6.85% nitrogen by weight. Calculate the atomic weight of the metal and suggest what metal it is.

3.34. A 5.00 g sample of $X_2O_3$ decomposes completely and produces 3.07 g of $O_2(g)$. What is the atomic weight of X?

3.35. The drug Prozac® has the elemental analysis C 66.01%, H 5.86%, F 18.43%, N 4.53%, and O 5.17%. What is its empirical formula?

3.36. The drug ibuprofen (Advil® is a popular brand) has the analysis: carbon 75.69%, hydrogen 8.80%, and oxygen 15.51%. Its molecular weight is 206.27 g $mol^{-1}$. What is the molecular formula of ibuprofen?

3.37. Tylenol® has the analysis: carbon 63.56%, hydrogen 6.00%, nitrogen 9.27%, and oxygen 21.17%. Its molecular weight is 151.16 g $mol^{-1}$. What is the molecular formula of Tylenol®?

3.38. A 5.00 mg sample of a compound composed of C,H, and O was burned with an excess of oxygen. Carbon dioxide (9.55 mg) and water (5.87 mg) were trapped by chemical absorbents. What is the empirical formula of the compound? (Hint: the percentage of oxygen in the compound can be found by difference after the percentages of C and H have been obtained.)

3.39. A compound has a mole weight of 61.0 g $mol^{-1}$ and an elemental analysis of C 19.67%, H 4.95%, N 22.95%, and O 52.42%. What is the name of the compound?

## Reaction Stoichiometry

3.40. How many atoms of oxygen are needed to completely combust 57.5 moles of butane ($C_4H_{10}$)?

3.41. What is the maximum mass of sodium sulfate ($Na_2SO_4$) that can be made from 0.455 g of sulfur and sufficient amounts of the other elements?

3.42. What is the maximum number of molecules of oxygen ($O_2$) that can be obtained from the complete decomposition of 34.67 g of sulfuric acid ($H_2SO_4$)?

3.43. Water reacts with calcium to yield calcium hydroxide and hydrogen gas. How many moles of hydrogen will be produced if 489 milligrams of calcium is dropped into water?

3.44. What mass of water is produced by the complete combustion of 35.4 g of benzene ($C_6H_6$)?

3.45. How many moles of water are produced by the complete combustion of 48.9 millimoles of acetylene ($C_2H_2$)?

3.46. Hydrogen and oxygen react violently, when ignited, to yield water. A mixture composed of 1.78 g of hydrogen and 43.5 g of oxygen is reacted—assume 100% yield of water. Which gas is in excess? How many moles of water are produced? How many grams of water are produced? How many moles of the excess gas are left unreacted at the end of the reaction? What was the limiting reagent?

3.47. Given 22.7 moles of Ag(s) and 31.3 mol of $Cl_2(g)$, a. calculate the amount of AgCl(s) that could be produced assuming that the reactants combine to the maximum possible extent, and b. calculate the amount of excess reactant that remains at the end of the reaction. The equation for the reaction is $2\ Ag(s) + Cl_2(g) \rightarrow 2\ AgCl(s)$

3.48. Given a 227 mg sample of phosphorus (P, MW = 31.0 g $mol^{-1}$) and a 917 mg sample of fluorine ($F_2$, MW = 38.0 g $mol^{-1}$) calculate the maximum mass of phosphorus pentafluoride ($PF_5$, MW = 126 g $mol^{-1}$) that could be prepared.

## Yield

3.49. A 54.6 g sample of magnesium was burned in air making magnesium oxide (MgO). The product weighed 73.4 g. Calculate the theoretical yield of the reaction and the percentage yield of the reaction.

3.50. What mass of silver can be recovered from a metric ton of ore that contains 7.8 ppm of silver if the process recovery yield is 89%? (ppm = parts per million, i.e., grams of silver in a million grams of ore.)

3.51. From 1.00 g of benzene a chemist makes 0.98 g of bromobenzene. Calculate the theoretical yield and the percentage yield of the reaction.

3.52. A chemist uses pure gold and other reactants to make gold(III) chloride. From 238 mg of gold 359 mg of the chloride is obtained. What percentage of the gold was converted to gold(III) chloride?

3.53. State two reasons why an excess of one of the reactants is often used in commercial chemical reactions.

## Solution Stoichiometry

3.54. Two moles of potassium hydroxide are needed to neutralize one mole of sulfuric acid. What mass of potassium hydroxide (KOH) is needed to neutralize 2340 g of sulfuric acid?

3.55. What is the molarity of a potassium hydroxide solution which contains 0.565 g $L^{-1}$ of KOH?

3.56. A solution of nitric acid is 0.945 molar. How many grams of nitric acid are in 573 mL of the solution?

3.57. To 45.8 mL of 0.657 molar potassium chromate solution is added 200.5 mL of water. What is the molarity of the resulting solution?

3.58. If 34.7 mL of a 0.500 M solution is mixed with 56.1 mL of a 0.800 M solution of the same solute, what is the molarity of the solute in the mixture?

3.59. 18.7 mL of a 2.67 M solution of lithium bromide is mixed with 31.3 mL of a 1.96 M solution of sodium bromide. Calculate the molarity of the three ions in the resulting solution.

3.60. What volume of 0.00458 molar sulfuric acid can be prepared from 45.8 mL of 0.820 molar sulfuric acid and distilled water?

3.61. What volume of 2.00 molar potassium hydroxide solution can be prepared from 23.6 g of 90.0% pure potassium hydroxide. (In solid KOH the other 10.0% is water.)

3.62. 2.00 moles of hydrochloric acid (HCl) dissolve 1.00 mole of zinc. How many grams of zinc will dissolve in 34.9 mL of 0.568 molar HCl solution if the reaction is 100% efficient?

3.63. Given 1.00 M solutions of nitric, hydrochloric, and sulfuric acids, which has the greatest hydrogen ion concentration in g $L^{-1}$?

## Problem Solving

3.64. How is the octane rating of gasoline established? Under what circumstances is a high-octane gas more desirable that a low-octane gas? Under what circumstances is a low-octane gasoline more desirable than a high-octane gasoline?

3.65. A solution can be infinitely dilute, but not infinitely concentrated. Explain.

3.66. In a hydrocarbon, primary (1°) carbon atoms are connected to only one carbon atom; secondary (2°) carbon atoms are connected to exactly two carbon atoms; and tertiary (3°) carbon atoms are connected to exactly three carbon atoms. How many carbon atoms of each type are there in molecules of heptane and isooctane?

3.67. Sketch an isomer of octane that has none of its carbon atoms connected to more than two ligand carbon atoms.

3.68. Write a balanced chemical equation to represent the reaction between nitrous oxide and heptane to form nitrogen, carbon dioxide, and water.

3.69. A compound contains sodium, chlorine, and 45.09% oxygen. Heated to 300°C it decomposes to give pure oxygen and a new compound that contains 39.34% Na and 60.66% Cl. What is the empirical formula of the original compound?

3.70. Compounds A and B each contain only carbon, hydrogen, and oxygen. The number of carbon atoms in a B molecule is twice the number in an A molecule. Each molecule contains two oxygen atoms. The mole weight of B is 28.0 g $mol^{-1}$ greater than the mole weight of A. Calculate the empirical formulas of the two compounds.

3.71. The compound $NaMO_3$ contains sodium, oxygen, and the metal M. The percentage by mass of M in the compound is 2.216 times the percentage by mass of sodium in the compound. What metal is M?

3.72. When heated, 1.000 g of the compound $KClO_x$ decomposes to produce 0.538 g of KCl and oxygen. What is the empirical formula of the compound?

3.73. Suppose that acid A and base B react in a 1:1 molar ratio. Acid A is a 0.200 M solution and base B is a 0.300 M solution. A mixture is made from 500. mL of each solution and the resulting solution has a volume of 1.0 L. What is the molarity of unreacted B in the mixture?

3.74. $H_2SO_4$(aq) and NaOH(aq) react in a 1:2 mole ratio: $H_2SO_4(aq) + 2\ NaOH(aq) \rightarrow Na_2SO_4(aq) + 2\ H_2O$. If 250. mL of a 0.200 M $H_2SO_4$(aq) solution is mixed with 250. mL of a 0.500 M NaOH(aq), what are the concentrations of the resulting chemicals present in the mixture?

3.75. The 1998 US hydrogen peroxide production capacity consisted of eight plants with a nominal capacity of 1205 million pounds per year of 100% $H_2O_2$. If these plants operated at 85% of capacity, how many tons of 100% hydrogen peroxide were produced in the US in 1998?

3.76. In 1998 the production of phosphoric acid ($H_3PO_4$, MW = 98.0 g $mol^{-1}$) was reported to be 14.4 million tons calculated as $P_2O_5$ (MW = 142 g $mol^{-1}$). Calculate the actual production of phosphoric acid in tons.

3.77. NASA's long term plans for sending astronauts to Mars involves an "ISRU" strategy. ISRU is in-situ resource utilization and means that to the extent possible, chemical resources already on Mars will be utilized. Mars' atmosphere contains about 95.3% carbon dioxide. According to one NASA plan, hydrogen will be send to Mars from Earth before astronaut crews arrive. Fuel for the Mars Ascent Vehicle (MAV) will consist of methane-oxygen combustion: $CH_4 + 2\ O_2 \rightarrow CO_2 + 2\ H_2O$. The fuel and oxygen will be made by the following reactions $CO_2 + 4\ H_2 \rightarrow CH_4 + 2\ H_2O$ (The Sabatier reaction) and $2\ H_2O \rightarrow 2\ H_2 + O_2$ (the electrolysis of water). Energy to run these latter reactions will come perhaps from $^{238}Pu$ power sources or solar cells. Making use of the above reactions, calculate the minimum mass of hydrogen that will have to be sent from Earth to Mars in order to produce 500 kg of methane and the necessary oxygen to combust it.

3.78. Discuss experimental uncertainty in the context of calculating chemical formulas from experimental percentage mass data.

3.79. The formula of the compound in Example 3.19 turned out to be $C_{10}H_8$ and you were told that this the compounds naphthalene and azulene have this formula. Both of these molecules are bicyclic. Naphthalene has the same number of atoms in each ring. Azulene has two more atoms in one ring than the other. Draw the structures of these two molecules using this information and the valencies of four for carbon and one for hydrogen.

# Chapter 4

# Gases

## Section 4.1 Historical Background and Uses of Gases

We are all intimately familiar with gases from "the wind in our face." The word gas was coined by the Flemish chemist J. B. van Helmont (1577-1644) using the Greek word "chaos." There can be few things better named than a gas—we'll see just how chaotic a gas is towards the end of this chapter. The **gaseous state of aggregation** is the least dense form in which ordinary matter can exist. Gases completely fill their container and have no surface other than its inside walls. Because gases flow they are fluids. Gases are the most fluid form in which matter exists at ordinary temperatures and pressures. Of course, all gases can be condensed by chilling them, so in everyday language when we speak about a substance being a gas we imply ordinary temperature and pressure, or more particularly **room temperature**.

Gases obey simple laws that relate the properties: pressure, volume, temperature, and amount of substance. These measurements are easy to make, so knowledge about gas behavior developed early in the history of science. Because they are simple, gases are the easiest state of matter to understand. Thus, studies of gases have often spearheaded progress in the history of physics and chemistry. Boyle's Law, that the pressure of a gas is inversely related to its volume, $P \propto \frac{1}{V}$, was known by around the middle of the 17th century. Every high school chemistry student is familiar with the ideal gas law, $PV = nRT$. For the engineer, the behavior of expanding gases in engines is not only one of the roots of engineering science, it's the starting point for an understanding of thermodynamics. In this chapter we'll examine the behavior of gases and see what it tells us about the nature of matter.

Under ordinary conditions, eleven elements are gases: He, Ne, Ar, Kr, Xe, and Rn are monatomic; the others, $H_2$, $N_2$, $O_2$, $F_2$, and $Cl_2$ are diatomic. How a gas behaves (its physical properties) depends on the size of its molecules and forces among them. Because there are larger forces of attraction among larger molecules, substances that are gases at ordinary temperatures and pressures rarely have more than 15 atoms in their molecules—most have fewer than 10. **Intramolecular forces** hold individual molecules together; the intramolecular forces among the atoms in gas molecules *are* the covalent bonds. Between any pair of molecules in a gas there exists an intermolecular force. An **intermolecular force*** is the attraction or repulsion of one molecule for another. Roughly speaking, the *intra*molecular forces within a single molecule are a hundred times greater than the *inter*molecular force between two separate molecules. It will be the study of the intermolecular forces among gas molecules that gives us our first real insight into how such forces control the bulk properties of matter. By its very nature, a gas *is* a gas because of its intermolecular forces are weak.

Some key dates and events in the study of gases are outlined in the following box. The method of trapping gases underwater for study (shown in Figure 4.7) was developed around 1750. Today it seems remote and quaint. However, at the time it was state-of-the-art technology and enabled chemists to establish that different gases are different chemical compounds.

*It gets cumbersome to have to say interatomic forces when speaking about the monatomic noble gases and intermolecular forces about everything else. So I'll regard the atoms of a noble gas as monatomic "molecules."

## Milestones of Gases in Science and Technology

1. In 1662, the first mathematical relationship between two natural properties was discovered when Robert Boyle proved experimentally that the volume of a gas is inversely proportional to the pressure applied to the gas.

2. In the period 1760-1780, Joseph Black, Henry Cavendish, and Joseph Priestley studied gases trapped in containers by water seals—in pneumatic troughs. During this period the fact that there actually are different gases was established and Boyle's idea that air is an element was dispelled. The work on gases culminated in 1782 when Antoine Lavoisier explained that water is formed by the chemical combination of two different gases: hydrogen and oxygen. Thus was finally established the idea that different chemical compounds exist.

3. In 1769, James Watt, who worked with students of pneumatic chemistry, applied for a patent on a steam engine with an external condenser. Watt's patent built on the earlier "atmospheric" engines and revealed the great potential of expanding gases (steam) as a source of mechanical energy.

4. In 1808, Joseph Gay-Lussac showed that many gas reactions involve simple, whole-number volume ratios of reactants. In this way, the idea of the coefficient of a chemical reaction entered chemistry.

5. In 1824, the French military engineer Sadi Carnot published a paper on the efficiency of steam engines. At the time, the significance of his work went unnoticed. It lays the foundations of thermodynamics.

6. In 1860, Stanislao Cannizzarro finally convinced the European community of scientists that hydrogen and oxygen are diatomic. This realization solved the problem of atomic weights once and for all.

7. 1738-1910, Development of the kinetic molecular theory of gases. In 1738 Daniel Bernoulli derived Boyle's law by applying Newtonian mechanics to molecules. His work was ignored for over a century. In the period 1850-1900 James Prescott Joule, Rudolph Clausius, James Clerk Maxwell, Ludwig Boltzman, and others developed Bernoulli's idea into a fully fledged theory of molecular behavior. By 1910 the proponents of kinetic-molecular theory had finally convinced the worst skeptics of the reality of molecules (some who were never convinced simply died). Kinetic-molecular theory became the basis for heat theory and a model of physics that persisted long into the 20th century. Speaking loosely, the kinetic-molecular theory of gases was the first theory of everything.

8. In 1873, Johannes Diderik van der Waals published a Ph. D thesis in which he described a modified ideal gas equation. The interpretation of this equation led directly to experimental estimates of intermolecular forces. van der Waals became a leader in the "molecules are real" movement that eventually triumphed. All modern science recognizes the "van der Waals" forces among atoms.

9. In the twentieth century, the use of gas expansion for power generation, and of petroleum gases for fuels and as a source of synthetic materials, became the basis for of advanced industrial economies. Most electricity is generated in steam expansion generating plants, and most transportation relies on piston or gas turbine engines. Ethylene became the largest volume industrial gas. Petroleum-derived ethylene was the primary original raw material for the plastics that dominated consumer economies after 1950.

## Gases in the Everyday World

There are only about 250 substances that are gases at ordinary temperatures and pressures. Most have atomic or molar weights less than 100 g $mol^{-1}$. Table 4.1, abstracted from Appendix F, shows some of the most important industrial gases and their classification.

Oxygen, argon, and nitrogen are obtained by distilling liquid air. In their cold, liquid states, both of these are familiar items of commerce: liquid nitrogen for quick-freeze refrigeration, and liquid oxygen (LOX) as an oxidizing agent for rocket engines. Gaseous oxygen is used in large quantities during the manufacture of steel. Argon, being the cheapest noble gas, is used as an inert blanketing atmosphere during welding of aluminum and other active metals. Totaling the values in Table 4.1 shows that the average American consumes about 600 pounds of various gases each year.

| Table 4.1 **Industrially Important Gases** (estimated annual US consumption per capita) | | | |
|---|---|---|---|
| Rank | Material | lbs | Classification |
| 3 | Ethylene | 200 | Organic chemical |
| 5 | Ammonia | 145 | Inorganic chemical |
| 7 | Propylene | 110 | Organic chemical |
| 10 | Chlorine ($Cl_2$) | 100 | Inorganic chemical |
| 23 | Ethylene oxide | 33 | Organic chemical |
| 27 | Nitrogen ($N_2$) | 22 | Inorganic chemical |
| 29 | Oxygen ($O_2$) | 17 | Inorganic chemical |
| 39 | Hydrogen | 7 | Inorganic chemical |

Small amounts of three other gases (neon, krypton, and xenon) are also obtained from the air. The optical properties of these gases make them useful in various specialized lamps, of which the neon tube is the best known. It is convenient to store gases in liquid form because their volume is much reduced and they do not have to be pressurized. A liquid nitrogen storage facility is shown below in Figure 4.1. Observe the frost on the pipe that leads from the tank to the heat exchanger.

Figure 4.1 Gases such as nitrogen (shown here), oxygen, and methane are often stored in liquid form. Doing this saves space and, in the case of liquid nitrogen, provides for a convenient form of refrigeration.

Helium is the most unreactive of all the chemical elements. It is used as the lifting gas in blimps In its ultra-cold liquid state helium serves in a number of unique low temperature applications. Figure 4.2 shows a stainless steel liquid helium tank (in front) replenishing coolant for the superconducting magnets of the nmr spectrometer (behind). The proper functioning of superconducting magnets requires the low temperature that liquid helium can provide.

Figure 4.2 Topping up the superconducting magnets of an nmr spectrometer with liquid helium. The vessel in the foreground contains liquid helium, which is shipped all around the US from its primary source in Amarillo, Texas.

Helium is extracted commercially from natural gas that comes from particular gas fields. (Amarillo, Texas, bills itself as "The Helium Capital of the World.") In the earth's crust, decay of natural radioactive atoms produces alpha particles that lose energy, gain two electrons, and remained trapped as helium atoms; when the gas field is tapped for fuel, the helium comes along for the ride.

Hydrogen is also a lifting gas but, unlike helium, is highly flammable. The fiery fate of the airship Hindenberg serves as a reminder of the explosive potential of gaseous hydrogen. Liquefied hydrogen was the fuel for the second stage of the Saturn rocket that boosted men to the moon. Hydrogen was the primary fuel of the gigantic explosion that destroyed the space shuttle Challenger in January 1986. Fluorine, the most reactive of the chemical elements, is prepared almost solely for the purpose of making useful compounds such as the fluorinated hydrocarbon gases, pharmaceuticals, and plastics such as Teflon®. Chlorine is also a reactive element that has many uses. Its disinfecting power is put to use in water treatment, and its bleaching ability is used in laundering, paper-making and related processes. For use around the home, Clorox® is made by adding chlorine gas to a dilute solution of sodium hydroxide.

Low molar weight hydrocarbons ($C_xH_y$ with x usually less than 5) constitute an extremely important group of gases. They are used in large quantities as fuels (Figure 4.3) and a source of synthetic organic (carbon containing) compounds. Methane (natural gas, $CH_4$), ethane ($C_2H_6$), propane ($C_3H_8$), and butane ($C_4H_{10}$), are colorless, odorless, highly flammable compounds that burn cleanly. Ethylene ($H_2C{=}CH_2$) is the third largest volume industrial gas, and petroleum-derived ethylene is the primary original raw material for plastics. Acetylene gas ($C_2H_2$) burns with a particularly hot flame and finds use as the fuel in welding and cutting torches. The general chemical formulas of hydrocarbons such as alkanes, alkenes, and alkynes are shown in Table 2.7.

Figure 4.3 The hydrocarbon fuel gas propane liquefies when it is placed under pressure, a convenient phenomenon because liquefaction substantially reduces the volume of the container need to store the propane. Propane fuel tanks, such as those seen above in a rack outside a store, are widely available.

Carbon dioxide can be chilled to produce solid carbon dioxide, known as "dry-ice." The use of carbon dioxide to produce an artificially carbonated beverage (soda-water) was initiated by Joseph Priestley (1733-1804) in 1770 in imitation of the natural soda waters such as Evian® and Perrier®.

Figure 4.4 The carbon dioxide in a $CO_2$ fire extinguisher exists in liquid form. However, as we'll see later, when the extinguisher is used, the carbon dioxide actually vents from the tank in both the gaseous and solid states (not the liquid state!). Nature often surprises us.

In carbon dioxide fire extinguishers the carbon dioxide exists as a liquid. Yet carbon dioxide is a gas at room temperature and pressure, so how come it's liquid in a fire extinguisher tank? Well, because its molecules attract one another; as carbon dioxide is increasingly pressurized there comes a point when the molecules are squeezed sufficiently close that the forces cause them to "stick together" and aggregate to the liquid state of aggregation. Gases that can be "pressure condensed" have useful applications. Example of such applications are butane lighters and products in aerosol cans—items in which liquids spontaneously vaporize to produce gases when their container pressure is released by depressing a valve. Pressure condensation is also the basis for air conditioning, as we'll see at the end of this chapter.

Air is a mixture of many gases; the composition of dry air at sea level is shown in Table 4.2. Together, $N_2$, $O_2$, Ar, and $CO_2$ account for 99.997% of the total volume of dry air. Many other gases occur in small amounts in the atmosphere. Examples are the oxides of nitrogen: nitrous oxide ($N_2O$); nitric oxide (NO); and nitrogen dioxide ($NO_2$). The latter two oxides are pollutants that occur in photochemical smog; they are collectively termed $NO_x$. Sulfur forms a dioxide ($SO_2$) and a trioxide ($SO_3$); the combustion of coal that contains sulfur leads to the release of these polluting sulfur oxides to the air. In combination with hydrogen, sulfur forms the foul-smelling hydrogen sulfide ($H_2S$). At high altitudes oxygen exists in the form of ozone ($O_3$).

| Gas | Percent by volume or by moles |
|---|---|
| Nitrogen, $N_2$ | 78.084 |
| Oxygen, $O_2$ | 20.946 |
| Argon, Ar | 0.934 |
| Carbon dioxide, $CO_2$ | 0.033 |
| Neon | 0.00182 |
| Helium, He | 0.000524 |
| Methane, $CH_4$ | 0.0002 |
| Krypton, Kr | 0.000114 |
| Nitrous oxide, $N_2O$ | 0.00005 |
| Hydrogen, $H_2$ | 0.00005 |
| Xenon, Xe | 0.0000087 |

Table 4.2 **The Composition of Dry Air**

## Section 4.2 Gas Pressure

There are a number of important physical properties of gases. In Chapter 1 you learned about volume and temperature. In Chapters 2 and 3 you learned about the mole and the concept of amount of substance. Now, as a prelude to studying the gas laws, it's time to discuss pressure.

**Pressure** is defined as force per unit area. Its base unit in the SI metric system is the Pascal (Pa) with units N $m^{-2}$ or kg $s^{-2}$ $m^{-1}$. Chemists tend to favor the pressure unit of atmospheres. **Standard pressure** of 1.000 atmosphere is defined by IUPAC as 101,325 Pa (exactly). US engineers are more likely to use psi (pounds-force per square inch). The mere fact that gases exert pressure has long been recognized as evidence that molecules exist.

**The Units and Value of Atmospheric Pressure**

The base units of pressure can be derived from the fundamental definition Pressure = force ÷ area. In the SI metric system that becomes

$$\text{units of pressure} = \frac{\text{kg m s}^{-2}}{\text{m}^2} = \text{kg m}^{-1}\ \text{s}^{-2} = \text{pascal (Pa)}$$

Chemists often use the traditional unit of atmospheres, where 1 atmosphere (atm) = 101,325 Pa or 1.01325 bar, or in the international number style 101 325 Pa.

For engineers, standard atmospheric pressure is either 14.70 psia or 0 psig. **psia** is pounds per square inch absolute; **psig** is pounds per square inch gauge. The distinction becomes clear if you think about an ordinary tire-pressure gauge that reads zero pressure despite the fact that it is obviously experiencing 1 atmosphere's pressure from the surrounding air. The conversion equation is:

$$1 \text{ psig} + 14.70 = 1 \text{ psia}$$

It's air molecules bouncing off objects that creates atmospheric pressure. Devices that measure atmospheric pressure are **barometers**. In 1643 Evangelista Torricelli (1608-1647), a pupil of Galileo's, invented the mercury barometer and used it to show that atmospheric pressure near sea-level supports a mercury column approximately 76 centimeters in height.

Shown diagrammatically in Figure 4.5, a barometer consists of a glass tube, sealed at one end, containing liquid mercury, above which is a vacuum. The vacuum is made by holding the tube open end up, filling it with mercury to expel air, and inverting the filled tube into a reservoir. When the mercury level stabilizes, the pressure exerted by the mercury column must be equal to that which the atmosphere exerts on the surface of the reservoir. Note that mercury forms a meniscus (a curved surface); later we'll talk about why. The height of the mercury column measures external pressure. An actual barometer is pictured in Figure 4.6. The pressure unit torr (symbol Torr), is named for Torricelli. An equivalent and more descriptive symbol for a torr is mmHg or millimeter of mercury; 1 Torr = 1 mmHg.

Because it is convenient to do so, pressure is often specified in terms of the length of a mercury column. Example 4.1 shows the relationship between pressure measured by a mercury column and the SI unit pascal. Standard atmospheric pressure is 760.000 Torr (or mmHg); that's six significant figure uncertainty, as shown in example 4.1.

Incidentally, the quantity $10^5$ Pa is a **bar**. **Standard atmosphere pressure** is equivalent to 1.01325 bar; so with only 1% discrepancy we can say 1 atmosphere ≈ 1 bar. You may have heard weather reports where millibars are used for pressure. In the United States, weather reports quote barometric pressure in terms of the equivalent height of a mercury column measured in inches. Near sea level, depending on the state of the weather, mercury barometer pressures are typically 29-30 inches.

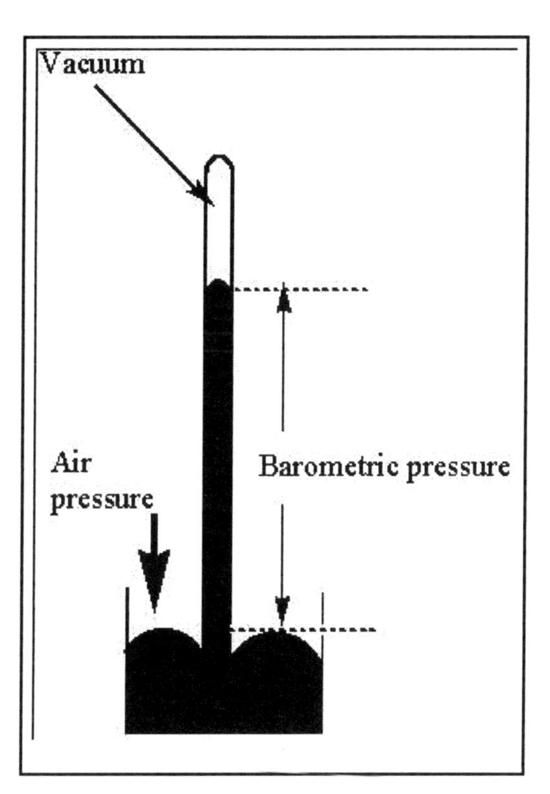


Figure 4.5 A schematic diagram of a barometer of the type first used by Evangelista Torricelli around 1643. The pressure of the air working against the liquid mercury surface just balances the force created by the attraction of gravity for the mercury in the column above the mercury pool. In Figure 4.6 you mainly see the protective brass housing; but the mercury reservoir at the bottom is clearly visible.

Figure 4.6 A traditional mercury barometer. The design is essentially unchanged in 350 years.

At low altitudes air is more dense than it is at high altitudes; atmospheric pressure decreases with increasing altitude. This fact was first demonstrated in 1647 by Blaise Pascal (1623-1662), who had his brother-in-law François Périer carry a barometer to the top of a mountain and make the necessary measurements. Today we know that air pressure fluctuates with changing weather and that the changes are used in weather forecasting. High atmospheric pressure usually means good weather.

Figure 4.7 A Pneumatic Trough. Around 1770 the development of the pneumatic trough revolutionized the experimental handling of gases and led almost immediately to the birth of modern chemistry. This illustration is from a chemistry book published in 1871.

Pierre Macquer's 1758 chemistry book devoted only one and a half pages (out of 800) to air—which means all gases—because, being pre-1782, the concept that *different* gases even exist is entirely absent from the book.

**Example 4.1**: At 25°C mercury has a density of 13.5951 g $cm^{-3}$. Standard gravitational acceleration is 9.80665 m $s^{-2}$ (exactly). A column of mercury exactly 76 cm high exerts very close to standard pressure. Use this information to calculate the standard pressure of the atmosphere in SI units.

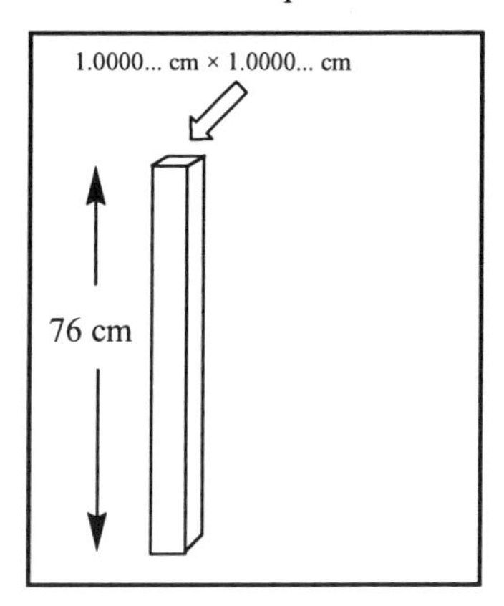


Strategy: Imagine a column of mercury standing on an exactly 1 $cm^2$ area of mercury as shown in the diagram above. Pressure has the units of force ÷ area. In this situation force is mass × the gravitational attraction of the earth, F = mg. The gas pressure equals the force exerted by the mercury. Get mass from mercury's density and volume. Plug into F = mg and do a conversion to obtain the required units.

The mass of the mercury is obtained from its density and volume: m = d V,

the volume of the mercury column is $1 \times 1 \times 76$ $cm^3$ (exactly).

Thus, mass of mercury = 13.5951g $cm^{-3}$ × 76 $cm^3$

= 1033.23 g or 1.03323 kg (6 significant figures)

The force that the column exerts is F = m g

$$\text{Force} = 1.03323 \text{ kg} \times 9.80665 \text{ m s}^{-2} = 10.1325 \text{ kg m s}^{-2} \text{(newtons)}$$

$$\text{Pressure is force} \div \text{area} = 10.1325 \text{ kg m s}^{-2} \div (1 \text{ cm}^2 \times 10^{-4} \text{ m}^2 \text{ cm}^{-2})$$

$$\text{Thus, 1 atmosphere pressure} = \underline{1.01325 \times 10^5 \text{ kg m}^{-1} \text{ s}^{-2}} \text{ or pascals (Pa).}$$

Recall that one atmosphere is defined by IUPAC to be exactly 101325 Pa. As shown by this example, at 25°C, this corresponds, with six significant figure accuracy, to a 760.000. mm height of mercury. The uncertainty between 760.000 and 760 (exactly) arises from our limited knowledge of mercury's density.

The huge force that the pressure of the atmosphere exerts was demonstrated in a famous experiment at Magdeburg, Germany in 1654 by Otto von Guericke (1602-1686) as shown in Figure 4.8. He assembled two snugly fitting brass hemispheres into a sphere about 20 inches in diameter and pumped out most of the air. With almost zero pressure on the inside, the hemispheres were held together by the atmospheric pressure on the outside. Two eight-horse teams could not pull the hemispheres apart. The same atmospheric pressure can be demonstrated in your kitchen by upturning an empty but steam-filled soda can into cold water (Figures 4.9a and 4.9b).

Figure 4.8 Von Guericke's famous 1654 Magdeburg experiment to show the force of atmospheric pressure. Two eight-horse teams were unable to pull apart two hemispheres joined by a vacuum into a sphere of about 20 inches.

Figure 4.9a In a convincing "kitchen" experiment a few milliliters of water are placed in an empty soda can and boiled on the stove. When plenty of steam is boiling from the can it is upturned in cold water…

Figure 4.9b ...with a sharp whacking sound, the can collapses as the steam inside it rapidly condenses and the pressure of the air crushes the can.

The magnitude of atmospheric pressure explains the once puzzling fact, long-known to mining engineers, that water cannot be lifted by a suction pump for a height much more than thirty feet. A suction pump works by removing the air from the surface of the liquid in a tube so that the atmospheric pressure on other parts of the liquid can *push* the liquid up the tube. Eventually, the weight of the liquid column that has been raised equals the available force of the atmosphere, and further suction has no effect. Deep wells require positive pressure pumping, either by means of an immersion pump or high pressure air supplied from the surface.

The height of a column of liquid that can be "raised" by the atmosphere is inversely proportional to the density of the liquid as illustrated in Example 4.2. The mathematical relationship is height (h) $\alpha \frac{1}{\text{density (d)}}$ and for two liquids (1 and 2) we may write:

$$\frac{h_2}{h_1} = \frac{d_1}{d_2}$$

Or, as we'll apply it in Examples 4.2 and 4.4

$$\text{Height of liquid 2} = \text{Height of liquid 1} \times \frac{\text{density of liquid 1}}{\text{density of liquid 2}}$$

**Example 4.2**: Calculate the height of a column of water in feet that can be "raised" by standard atmospheric pressure. Mercury has a density of 13.60 g $cm^{-3}$ at 25°C. Water has a density of 0.997 g $cm^{-3}$ at 25°C.

Strategy: Because mercury is much denser than water, the water column will be much longer. Multiply the length of the mercury column by the ratio of the densities of mercury and water. Do the proper units conversion to get feet.

$$\text{Height of water} = \text{Height of mercury} \times \frac{\text{density of mercury}}{\text{density of water}}$$

$$\text{Height of water} = 76.0 \text{ cm} \times \frac{13.60 \text{ g cm}^{-3}}{0.997 \text{ g cm}^{-3}} \div 30.48 \text{ cm foot}^{-1}$$

$$= \underline{34.0 \text{ feet}}$$

Therefore a suction pump cannot theoretically raise water from a depth greater than 34 feet. The practical working limit for a suction pump is 25-30 feet.

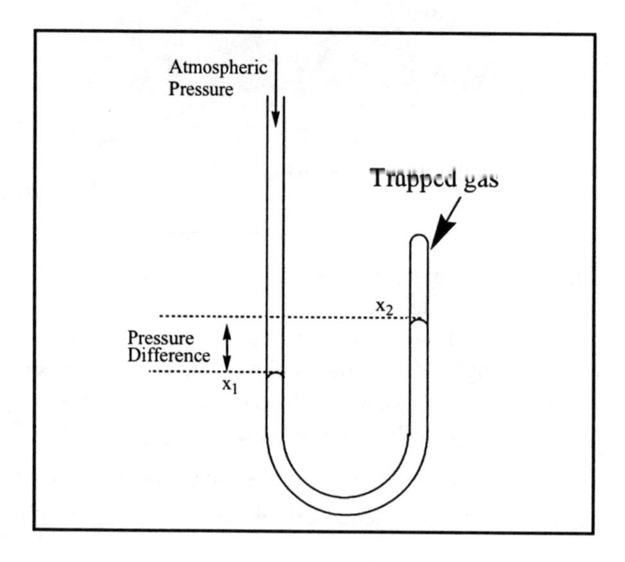


Figure 4.10 A J-tube manometer.

Torricelli's barometer measures the absolute pressure of the atmosphere; the relative pressure of two gases is measured with a manometer. A **manometer** is a device that measures pressure differences by comparing liquid levels. A simple mercury J-tube manometer is shown in Figure 4.10. The pressure difference between the air and the trapped gas is equivalent to the pressure generated by the level difference between the two columns of mercury in the J-tube. Thus, the pressure of the trapped gas equals the pressure of the atmosphere plus or minus the pressure attributable to the difference of mercury levels,

$$P_{gas} = P_{atm} + (x_1 - x_2)$$

where $x_1$ is the level on the left and $x_2$ is the level on the right. If $x_2$ is greater than $x_1$, as it is here, then $P_{gas}$ is < $P_{atm}$. The reverse is true if $x_1$ is greater than $x_2$. Example 4.3 shows a level difference calculation of pressure.

**Example 4.3**: Gas at 25°C is trapped in the closed side of a mercury J-tube manometer. Atmospheric pressure is 717 Torr. Calculate the pressure of the gas if the liquid level on the open side is 15.2 mm lower than the level in the closed side.

Strategy: Recognize that 15.2 mm of mercury is 15.2 Torr. Then substitute into the equation: $P_{gas} = P_{atm} + (x_1 - x_2)$.

Given $P_{atm} = 717$, $P_{gas} = x$, $x_1 = -15.2$, $x_2 = 0.00$.

$x = 717$ Torr + (-15.2 - 0) Torr = 702 Torr

In principle, any liquid can be used in barometers and manometers. Mercury is frequently used, but because mercury's density is high, pressure changes cause only small changes in its column heights. Dibutyl phthalate (DBP, density = 1.046 g $cm^{-3}$), a liquid which is slow to evaporate and chemically inert, is often used to measure small pressure differences. Because the density of dibutyl phthalate is much less than mercury, gas pressure changes cause much larger changes in its column heights than in a mercury manometer, as shown in Example 4.4.

**Example 4.4**: A gas at 25°C is trapped in the closed end of dibutyl phthalate (DBP) J-tube manometer. Atmospheric pressure is 722.4 Torr. The DBP level in the closed side of the manometer is 3.5 mm lower than that in the open side. Calculate a. the pressure difference between the closed and open sides, and b. the pressure of the trapped gas.

Strategy: Use the ratio of the density of mercury and DBP to convert the level difference to Torr. Then add that value to the atmospheric pressure because the level of DBP in the closed arm is below that in the open arm.

Height difference $x_1 - x_2 = 3.5$ mm of DBP

Converting to mmHg with the density ratio: $= 3.5 \text{ mmDBP} \times \frac{1.046 \text{ g cm}^{-3}}{13.6 \text{ g cm}^{-3}} \frac{\text{mmHg}}{\text{mmDBP}}$

= 0.27 mmHg or Torr

The actual pressure of the gas is 722.4 Torr + 0.27 Torr = 722.7 Torr

## Section 4.3 The Gas Laws

### Boyle's Law

In 1662, replying to criticism of his recently published book, *New Experiments Physico-Mechanical touching the spring of the air and its effects*, Robert Boyle (1627-1691) wrote:

> "As long as the temperature remains constant, the volume
> of a body of gas varies inversely with the pressure upon it."

This statement is Boyle's law. Mathematically it can be expressed in a number of ways: $P \propto \frac{1}{V}$ or $P = \frac{\text{constant}}{V}$ or PV = constant. Boyle was describing the results of some of his experiments, which are reproduced in the table below. He studied air trapped in a mercury J-tube manometer, at a constant temperature. He measured the air's volume at several different pressures by changing the amount of mercury in the open end side of the manometer. Some of Boyle's original observations are shown in Table 4.3 (the author has converted Boyle's fractions to decimals and edited the significant figures). You will probably replicate Boyle's experiment and calculations if you study gases in the laboratory.

Table 4.3 **Boyle's Law—Some of His Original 1660 Data**
**Pressure and Volume of Trapped Air**
**Fixed amount of gas, temperature held constant**

| Experiment | Pressure of the trapped air in inches of mercury, $P_{in}$ | *Length of air in column, inches | Product of length and $P_{in}$ (arbitrary unit of inchs$^2$) |
|---|---|---|---|
| 1 | 11.5 | 30.5 | 351 |
| 2 | 9.5 | 37 | 350 |
| 3 | 7.5 | 47.1 | 350 |
| 4 | 5.75 | 61.3 | 352 |
| 5 | 3.0 | 117.6 | 350 |

*Boyle measured the height (length) of the "cylinder" containing the trapped air sample. The height could have been converted into a volume by multiplying by the cross-sectional area of the manometer tube. However, it was not necessary and Boyle didn't do it. The cross-sectional area of his J-tube was constant; so his values in column 4 (on the right) prove that PV = constant (within experimental error).

Figure 4.11 shows a picture of Boyle himself and a graph of his law for one mole of an ideal gas at 273 K.

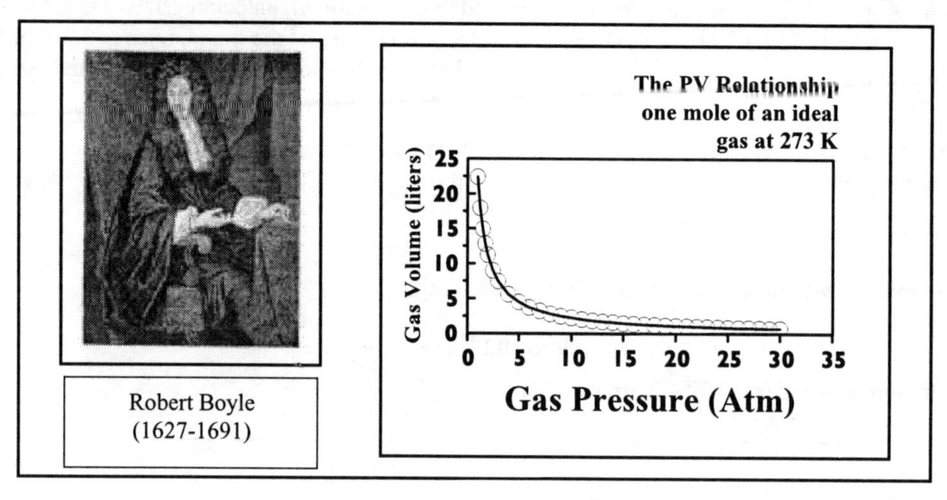


Figure 4.11 Robert Boyle and a graphical representation of his famous law.

Because the PV product is constant, if a gas sample has volume $V_1$ when the pressure is $P_1$ and volume $V_2$ when the pressure is $P_2$, then

$$P_1V_1 = P_2V_2$$

**Example 4.5**: A sample of 15.2 liters of nitrogen gas at 25°C has its pressure increased from 745 Torr to 895 Torr. Calculate the new volume of the gas.

Strategy: First note that the volume will go down because the pressure has gone up. Then plug into $P_1V_1 = P_2V_2$

$$P_1 = 745 \text{ Torr}, P_2 = 895 \text{ Torr}, V_1 = 15.3 \text{ L}, V_2 = x$$

$$15.2 \text{ L} \times 745 \text{ Torr} = x \times 895 \text{ Torr and } x = \underline{12.7 \text{ L}}$$

Careful experiments show that Boyle's law is not perfect. All gases obey Boyle's law at sufficiently low pressures, but at moderate and high pressures they deviate. As we'll see toward the end of this chapter, deviations from Boyle's law tend to increase as the pressure increases and provide clues to understanding intermolecular forces.

## Charles's Law

In the late 18th century Jacques Charles (1746-1823) studied the dependence of gas volume on temperature. Charles had considerable interest in gases and was the first person to suggest using hydrogen in balloons. In December 1783 he participated in the second ascent ever of a hydrogen-filled balloon, and later flew solo to a height of almost two miles. Charles did not publish his findings on volume-temperature relationships of gases, but his experiments were repeated by Joseph Gay-Lussac, who in 1802 did publish the results. The data established Charles's law:

"For a fixed amount of gas maintained at constant pressure, the volume increases linearly with temperature."

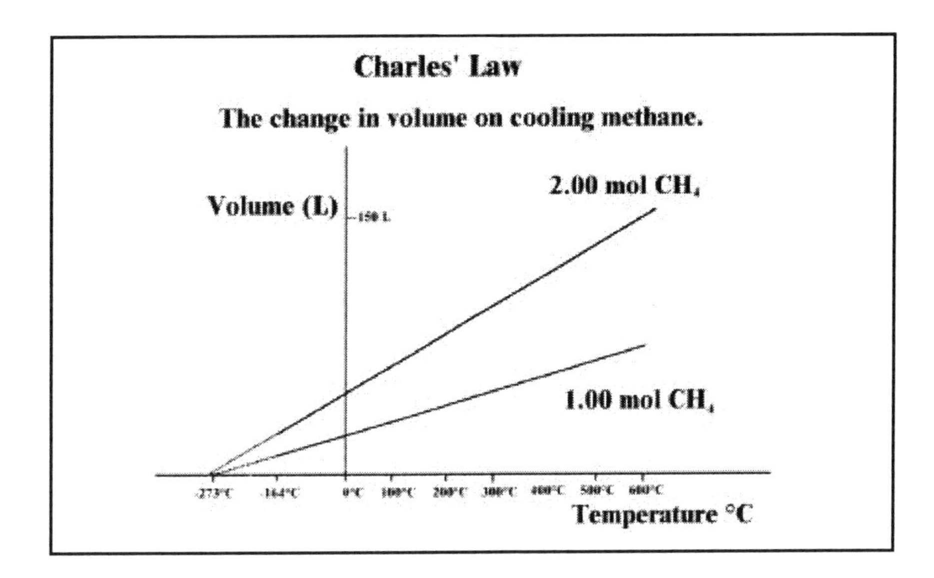


Figure 4.12 Charles' law suggests that gases will shrink to zero volume if sufficiently cooled. It doesn't actually happen that way because gases condense to liquids somewhere along the way.

In Figure 4.12 are plotted volume-temperature data at 1.00 atm pressure for 1.00 mol and 2.00 mol of methane, $CH_4$. The boiling point of methane is -164°C, so at this temperature methane liquefies and its volume drops to nearly zero as that temperature is reached. However, it is a simple matter to extend the data by extrapolation. Charles's law states that the volume of a fixed amount of gas at a constant pressure is directly proportional to its kelvin temperature.

Observe that at -273°C any amount of methane would have shrunk to zero volume—provided it stayed a gas. Exactly -273.15°C is now by definition the absolute zero of temperature and 273.15 K = 0°C. Absolute zero is the beginning point of the kelvin temperature scale. Both Joseph Gay-Lussac and John Dalton independently made the extrapolation around 1800.

Equivalent mathematical statements of Charles's law are $V = k\,T$ (fixed mass of gas at constant P), $V \propto t$, and $\frac{V}{T}$ = a constant, where k is the Charles's law constant for a specific mass of gas at a specific pressure. If the properties of a fixed amount of gas at constant pressure are changed from ($V_1$, $T_1$) to ($V_2$, $T_2$), then

$$\frac{V_1}{T_1} = \frac{V_2}{T_2}$$

You *must* use kelvin temperatures in these Charles's law expressions.

**Example 4.6**: A trapped gas is cooled from 95°C to 0°C, while its pressure is kept constant. a. What happens to its volume? b. If its initial volume is 2.50 L, what will be its final volume?

Strategy: In part a. the volume must decrease because the temperature is lowered. Part b. is done by plugging into the equation $\frac{V_1}{T_1} = \frac{V_2}{T_2}$ *after* converting the temperatures to kelvins.

$$T_1 = 273.15 + 95 = 368\text{ K},\quad T_2 = 273.15 + 0 = 273\text{ K},\quad V_1 = 2.50\text{ L}.$$

$$\frac{2.50}{368} = \frac{V_2}{273}$$

$$V_2 = \underline{1.86\text{ L}}$$

## Avogadro's Law

We met Avogadro's law in its original formulation in Section 3.3 in connection with the development of the concepts of reaction stoichiometry and stoichiometric coefficients. Recast in modern parlance and applied to a particular sample of a gas, Avogadro's law states that at any specified temperature and pressure the volume of the gas is directly proportional to the number of moles of the gas, the proportionality constant being *the same* for all gases.

$$V \propto n \text{ or } V = k\,n \quad \text{(at a specified temperature and pressure)}$$

Notice that Avogadro's law reveals that the Boyle's and Charles' law constants of gases are independent of the type of gas. The constant k in the above equation is the gas's molar volume, which at 0°C and 1 atmosphere pressure is approximately 22.4 L $mol^{-1}$.

## Section 4.4 The Ideal Gas Law

Boyle's law and Charles's law describe the dependence of gas volume on pressure and temperature. Avogadro's law describes the dependence of gas volume on the number of moles of the gas. Rewriting these three laws together using k, k' and k" to represent the three different constants of proportionality gives:

$V \propto 1/P$ is Boyle's Law, or $V = k/P$ (k depends on T and the amount of gas)
$V \propto T$ is Charles's Law, or $V = k'T$ (k' depends on P and the amount of gas, T in kelvins)
$V \propto n$ is Avogadro's Law, or $V = k''n$ (k" depends on T and P and is *independent of the nature of the gas*)

The dependence of V on 1/P, T, and n, can be represented in a single equation with a single constant, named the **ideal gas equation**, that is a combination of Boyle's, Charles's, and Avogadro's laws:

$$V = \frac{nRT}{P} \quad \text{or } PV = nRT \qquad \text{(ideal gas equation)}$$

where R is an important constant in chemistry called the **gas constant**. The usual way we write the ideal gas equation is: PV = nRT and a gas that obeys it is an **ideal gas**.

By definition, an **ideal gas** is a gas that obeys the ideal gas law.

The value of the constant R can be determined by measuring the properties of an ideal gas, where:

$$R = \frac{PV}{nT}$$

R has the units of energy $K^{-1}$ $mol^{-1}$. The most accurate measurements of R come from studying gases at low pressures. The 1998 IUPAC value is R = 0.08205783 L atm $K^{-1}$ $mol^{-1}$. In SI units R equals 8.314510 J $K^{-1}$ $mol^{-1}$. The correct value of R has seven significant figures.

The ideal gas equation is a powerful generalization. It applies with reasonable accuracy to a wide variety of gases under broadly ranging conditions of temperature and pressure. The ideal gas equation predicts the behavior of real gases to a sufficiently good approximation for many purposes. The use of the ideal gas equation is shown in Examples 4.7 and 4.8.

**Example 4.7**: Calculate the volume of 1.00 mole of an ideal gas at a pressure of 1.00 atm and 298.15 K.

Strategy: Plug the given values into the ideal gas equation and solve it for one mole of gas:

$$PV = nRT$$

$$P = 1.00 \text{ atm}, \quad R = 0.0821 \text{ L atm K}^{-1} \text{ mol}^{-1},$$
$$n = 1.00 \text{ mol}, \quad T = 298.15 \text{ K}, \quad \text{and } V = x$$

$$1.00 \text{ atm} \times x = 0.0821 \text{ L atm K}^{-1} \text{ mol}^{-1} \times 298.15 \text{ K}$$

$$\text{Volume} = \underline{22.5 \text{ L}} \quad \text{(three significant figures)}$$

Many common gases have a volume close to 22.5 L per mole at 1.00 atm and 298.15 K (25.00°C), which shows that their behavior is close to that of an ideal gas.

The correct IUPAC value of R, 0.08205783 L atm $K^{-1}$ $mol^{-1}$, has seven significant figures. Because the pressure in this problem was specified to only three significant figures, it was satisfactory to use a rounded value for R.

**Example 4.8**: A tank of helium is at 1950 psia. The volume of the tank is 56.0 liters and its temperature is 25°C. Calculate the pressure in the tank in atmospheres and then calculate the number of moles of helium in the tank and the mass of helium in the tank.

Strategy: use the conversion factor 14.70 psi $atm^{-1}$ to convert to the requested unit. Plug that and the other values into PV = nRT and solve for n.

$$\text{Pressure} = 1950 \text{ psi} \div 14.70 \text{ psi atm}^{-1} = 133 \text{ atm}$$

Plugging into PV = nRT

With P in atm and V in L, R = 0.0821 L atm $K^{-1}$ $mol^{-1}$. If x is the number of moles:

$$133 \text{ atm} \times 56.0 \text{ L} = x \times 0.0821 \text{ L atm K}^{-1} \text{ mol}^{-1} \times (273 + 25) \text{ K}$$

$$x = \underline{304 \text{ mol}}$$

The mass of helium is 304 mol × 4.00 g $mol^{-1}$ = 1220 g, or about 2.7 pounds. So storing gases in heavy, pressurized tanks is not a particularly efficient procedure.

For changes in the state of a trapped or fixed amount of an ideal gas you can use the combined gas law. If $P_1$, $V_1$, and $T_1$ are the initial state properties and $P_2$, $V_2$, and $T_2$ are the final state properties, then we can write the ideal gas equation twice and eliminate n and R:

$$P_1V_1 = nRT_1$$
$$P_2V_2 = nRT_2$$

$$\text{So, eliminating n and R} \quad \frac{P_1V_1}{T_1} = \frac{P_2V_2}{T_2}$$

The equation above combines the effects of Boyle's law and Charles's law and so is the **combined gas law**. Its use is illustrated in Example 4.9.

**Example 4.9**: Consider a fixed amount of gas at a pressure of 1.50 atm, a temperature of 25°C and a volume of 17.5 L. If the pressure is increased to 3.00 atm and the temperature is increased to 125°C, what will be the final volume of the gas?

Strategy: Convert the temperatures from °C to kelvins and plug into the combined form of the gas laws: $\frac{P_1V_1}{T_1} = \frac{P_2V_2}{T_2}$.

$$T_1 = 273 + 25\ K$$
$$T_2 = 273 + 125\ K$$
$$V_1 = 17.5\ L$$
$$P_1 = 1.50\ atm$$
$$P_2 = 3.00\ atm$$
$$V_2 = V_2.$$
$$\frac{P_1V_1}{T_1} = \frac{P_2V_2}{T_2}$$
$$\frac{1.50 \times 17.5}{298} = \frac{3.00 \times V_2}{398} \qquad \underline{V_2 = 11.7\ L}$$

As always, the number of significant figures in the answer is controlled by the number of significant figures in the data given.

## Gas Density

The density of an ideal gas at any specified pressure and temperature depends *only* on the molar weight of the gas. To obtain an equation for calculating the density of an ideal gas, you combine the ideal gas equation with the definition of the number of moles of a substance: $n = \frac{m}{MW}$. Substituting for n in the ideal gas equation, PV = nRT gives the equation:

$$PV = \frac{m}{MW}RT$$

which can be rearranged to make $\frac{m}{V}$ the subject

$$\frac{m}{V} = \frac{P(MW)}{RT}$$

But $\frac{m}{V}$ is density, d. So for an ideal gas:

$$d = \frac{P(MW)}{RT}$$

The use of the above equation for a specific gas at a specified P and T is illustrated in Example 4.10.

**Example 4.10**: Assuming it is an ideal gas, calculate the density of propane ($C_3H_8$) at 83.0°C and a pressure of 721 Torr.

Strategy: Figure the molar weight of propane to be 44.1 g $mol^{-1}$. Convert the pressure to atmospheres, the temperature to kelvins, and plug into the ideal gas density equation.

$$d = \frac{P(MW)}{RT}$$

$$P = 721\ Torr \div 760.0\ Torr\ atm^{-1} = 0.949\ atm$$
$$T = 273.2 + 83.0 = 356.2\ K$$
$$MW = 44.1\ g\ mol^{-1}$$
$$R = 0.0821\ L\ atm\ K^{-1}\ mol^{-1}$$

$$d = \frac{0.949\ atm \times 44.1\ g\ mol^{-1}}{0.0821\ L\ atm\ K^{-1}\ mol^{-1} \times 356.2\ K}$$
$$= \underline{1.43\ g\ L^{-1}}$$

## Molar Weight from Gas Measurements

You can use the ideal gas equation to calculate the molar weight of a gas if the temperature, pressure, and density of the gas are known. In the development of an equation for gas density in the preceding section we saw:

$$\frac{m}{V} = \frac{P(MW)}{RT}$$

Which rearranges to: $MW = \frac{mRT}{PV}$, or $MW = \frac{dRT}{P}$

To determine the molar weight of a gas experimentally, the needed properties are measured and their values substituted into the above equation, as shown in Example 4.11.

**Example 4.11**: Phosgene gas was employed as a weapon during World War I. A sample of phosgene gas weighing 0.152 g has a volume of 48.3 mL at 100.°C and 740. Torr. Calculate the molar weight of phosgene.

Strategy: Substitute values in the correct units into the right hand side of $MW = \frac{mRT}{PV}$

$m = 0.152$ g
$R = 0.0821$ L atm $K^{-1}$ $mol^{-1}$
$T = 100. + 273 = 373$ K
$P = 740.$ Torr $\div$ 760 Torr $atm^{-1}$ $= 0.974$ atm
$V = 48.3$ mL $\div$ 1000 mL $L^{-1}$ $= 0.0483$ L

$$MW = \frac{0.152 \text{ g} \times 0.0821 \text{ L atm K}^{-1} \text{ mol}^{-1} \times 373 \text{ K}}{0.974 \text{ atm} \times 0.0483 \text{ L}}$$

MW of phosgene = 98.9 g $mol^{-1}$

The formula of phosgene is $COCl_2$; its molar weight is $12.01 + 16.00 + 2 \times 35.45 = 98.91$ g $mol^{-1}$. This value agrees well with the above calculation. Based on the data given, phosgene behaves as an ideal gas under the specified conditions.

Air, or any gas mixture, can be treated as a pure gas with a mean molar weight equal to the weighted mean of the molar weights of the gases in the mixture. In an assembly of entities the **weighted mean** of a property is the mean or average of that property among *all* the entities. The **mean molar weight**, $MW_{mean}$, of a mixture is its total mass in grams divided by the total number of moles present. Dry air is a mixture in which 78% of the molecules (or moles) are $N_2$ and 21% are $O_2$. Mean molar weight is independent of the sample size; assume an exactly one mole air sample. Its mean molar weight is then:

$$MW_{mean} = 0.79 \text{ mol } N_2 \times 28.0 \text{ g mol}^{-1} N_2 + 0.21 \text{ mol } O_2 \times 32.0 \text{ g mol}^{-1} O_2 = 29.0 \text{ g mol}^{-1}.$$

Example 4.12 shows the use of the mean molar weight of air to calculate the lifting capacity of a lighter-than-air balloon.

**Example 4.12**: Consider a spherical balloon with a radius of 50.0 feet. Assuming standard pressure and 25°C, calculate a. the mass of hydrogen gas that can be contained within the balloon, b. the mass of air that can be contained within the balloon, and c. the lifting capacity of the balloon when it is filled with hydrogen.

Strategy: The volume of a sphere is $V = \frac{4}{3}\pi r^3$. Calculate V and then substitute values into the equation that comes from combining the ideal gas equation with the definition of the mole:

$$\text{mass of ideal gas (m)} = \frac{PV(MW)}{RT},$$

to calculate the masses of air and hydrogen that can be contained in the balloon. Finally use Archimedes' principle that states that the mass that can be lifted is equal to the difference between the mass of the displaced fluid (air) and the mass of the fluid contained in the balloon (hydrogen)

$$\text{radius of balloon, } r = 50.0 \text{ ft} \times 12 \text{ inches ft}^{-1} \times 2.54 \text{ cm inch}^{-1} = 1520 \text{ cm}$$

$$V = \frac{4}{3}\pi r^3$$

$$\text{Therefore, its volume is } \frac{4}{3}\pi\,(1520)^3 \text{ cm}^3 \div 1000 \text{ cm}^3 \text{ L}^{-1}$$

$$\text{Volume} = 1.47 \times 10^7 \text{ L}$$

$$\text{Mass hydrogen } (H_2) = \frac{PV(MW)}{RT} = \frac{1.00 \text{ atm} \times 1.47 \times 10^7 \text{ L} \times 2.02 \text{ g mol}^{-1}}{0.0821 \text{ L atm K}^{-1} \text{ mol}^{-1} \times 298 \text{ K}} = \underline{1.22 \times 10^6 \text{ g } H_2}.$$

Repeating for air, using the mean molar weight of air as calculated previously

$$\text{Mass air} = \frac{PV(MW)}{RT} = \frac{1.00 \text{ atm} \times 1.47 \times 10^7 \text{ L} \times 29.0 \text{ g mol}^{-1}}{0.0821 \text{ L atm K}^{-1} \text{ mol}^{-1} \times 298 \text{ K}} = \underline{1.74 \times 10^7 \text{ g air}}.$$

The difference between the masses of air and hydrogen contained in the balloon is its lifting capacity:

$$\text{Lifting capacity} = 1.74 \times 10^7 - 1.22 \times 10^6 \text{ g} = 1.62 \times 10^7 \text{ g}$$

$$1.62 \times 10^7 \text{ g} \times 0.001 \text{ kg g}^{-1} = \underline{16{,}200 \text{ kg}}$$

In English units, 16,200 kg converts to 35,700 pounds or nearly 18 US tons. The lifting capacity of this large, hydrogen-filled balloon is impressive.

## Dalton's Law of Partial Pressures

In ideal gas mixtures, the component gases behave independently of each other. Thus, in a mixture of two ideal gases A and B, the individual pressures, $P_A$ and $P_B$, are the **partial pressures** of the component gases. The partial pressure of the first gas is the pressure it alone would exert even if the second gas were absent. Likewise, the partial pressure of the second gas is the pressure it alone would exert even if the first gas were absent. The total pressure of the gas mixture, $P_{tot}$, is the sum of the individual gas partial pressures. Thus,

$$P_{tot} = P_A + P_B$$

The equation above is John Dalton's law of partial pressures (the same John Dalton who first proposed atomic theory). The law is exact only for mixtures of ideal gases, but it can often be used with little error for real gases.

Using the ideal gas equation independently for each of the two gases in the mixture we can write:

$$P_{tot} = P_A + P_B = \frac{n_A RT}{V} + \frac{n_B RT}{V} = (n_A + n_B) \times \frac{RT}{V}$$

The ratio $n_A/(n_A + n_B)$ is the mole fraction of A and is given the symbol, $X_A$. The ratio $n_B/(n_A + n_B)$ is the mole fraction of B and is given the symbol, $X_B$. The **mole fraction** of the ith component of a mixture (A, B, C...) is, in general, given by the equation:

$$X_i = \frac{n_i}{n_A + n_B + n_C + \ldots}$$

In a mixture of ideal gases, the partial pressure of each gas equals its mole fraction multiplied by the total pressure $P_i = P_{tot} X_i$ where the summation of all the mole fractions is, of course, one, $\Sigma X_i = 1$. In other words, Dalton's law of partial pressure for a many gas system is $P_i = P_{tot} X_i$.

Dalton's law of partial pressure is important in meteorology and studies of the air. Ignoring its minor components, dry air is 0.78 mole fraction $N_2$, 0.21 mole fraction $O_2$, and 0.01 mole fraction argon. Observe that the sum of the mole fractions equals 0.78 + 0.21 + 0.01 = 1.00. To a good approximation Dalton's law for dry air is

$$P\text{ (dry air)} = P_{N_2} + P_{O_2} + P_{Ar}$$

Example 4.13 shows the application of Dalton's law of partial pressures to the composition of a diver's gas.

**Example 4.13**: A breathing mixture of oxygen (2.0 mole%) and helium (98.0 mole%) is sometimes used by divers. Calculate the partial pressure of oxygen in this mixture at a depth of 300. feet in a freshwater lake. Assume ideal gas behavior and a barometric pressure of 1.00 atm.

Strategy: When descending into water, the pressure rises by one atmosphere for every 34.0 feet of water (as we calculated Example 4.2 ). The total gas pressure is a combination of the depth pressure and the surface pressure. Add the pressures and plug the result into Dalton's law of partial pressures.

$$P_{tot} = \frac{300.\text{ feet water}}{34.0\text{ foot water atm}^{-1}} + 1.00\text{ atm}$$

$$P_{tot} = 9.82\text{ atm}$$

The mole fraction of oxygen in the mixture is 2.0% ÷ 100% = 0.020, therefore, the partial pressure of oxygen is

$$P_{O_2} = P_{tot} X_{O_2} = 9.82\text{ atm} \times 0.020 = \underline{0.20\text{ atm}}$$

Note that the partial pressure of oxygen in the gas mixture at this depth is about the same as it is in the atmosphere on the earth's surface. The human body functions best when the partial pressure of oxygen is close to its customary value.

Divers use helium-oxygen mixtures because such mixtures are nitrogen free. Surprisingly, because we normally think of it as an inert gas lacking physiological properties, nitrogen has narcotic effects when at high pressures it dissolves to a considerable extent in the human bloodstream. The condition is called nitrogen narcosis and can cause a diver to behave bizarrely and possibly self-destructively.

[❁kws +"nitrogen narcosis"]

Dalton discovered the law of partial pressures in 1801 during his studies of the dew point of moist air. The **dew point** is the temperature to which an air sample must be cooled to cause liquid water to condense (form a dew). For wet air (air containing water vapor), Dalton's law becomes

$$P_{atm} = P_{N_2} + P_{O_2} + P_{Ar} + P_{H_2O} \quad \text{or,} \quad P_{atm} = P_{air} + P_{H_2O}$$

The above equation is useful to "correct" the properties of gases collected over water and is a fundamental relation of meteorology. Its use is shown in Example 4.14.

A **vapor** is simply a gas. However, the word vapor is often used in a special sense to refer to gas created by the evaporation of a liquid. Inside any closed container of liquid, there exists a continual dynamic equilibrium among molecules in the gas and liquid states of aggregation. The rate at which molecules vaporize is equal to the rate at which they condense, so the pressure of the vapor is a constant value at any specified temperature. Dynamic vapor equilibrium is shown schematically in Figure 4.13.

The study of vapor pressure has many applications, as we'll see in the following section. Studies of vapors are also useful if we want to learn about the intermolecular forces in liquids.

We'll take up the topic of measuring intermolecular forces via vapor pressure measurements in Chapter 10. For now, we'll just note that liquids with strong intermolecular forces are far more difficult to vaporize than liquids with weak intermolecular forces.

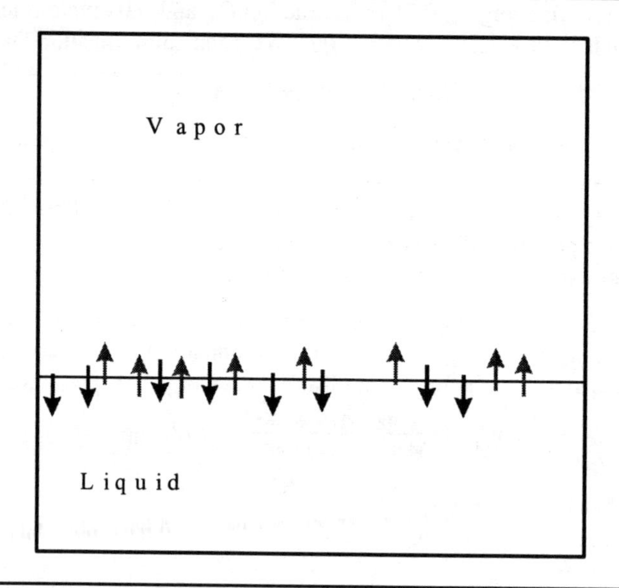


Figure 4.13 The dynamic equilibrium that exists between a liquid and the vapor saturated head space above it. Evaporation, the transport of liquid phase molecules into the gas phase (shown by the up arrows) occurs at precisely the same rate as condensation, the transport of gas phase molecules into the liquid phase (shown by the down arrows). The important topic of chemical equilibrium is the subject of Chapter 9.

**Example 4.14:** On a 20°C day, atmospheric pressure was 747.3 Torr and the pressure due to water vapor was 17.5 Torr. Assume ideal gas behavior and the composition of dry air, and calculate the partial pressures of oxygen and nitrogen.

Strategy: Calculate the pressure of dry air using Dalton's law of partial pressures to subtract out the pressure due to water vapor. Assign the pressure of the dry air to oxygen and nitrogen in proportion to their known mole fractions as stated in Table 4.2.

$$P_{\text{dry air}} = P_{\text{tot}} - P_{H_2O}$$

$$P_{\text{dry air}} = 747.3 \text{ Torr} - 17.5 \text{ Torr}$$

$$= 729.8 \text{ Torr}$$

Dry air contains 78 mole% nitrogen and 21 mole% oxygen. Thus,

$$P_{N_2} = .78 \times 729.8 \text{ Torr} = \underline{580 \text{ Torr}}$$

$$\text{and } P_{O_2} = .21 \times 729.8 \text{ Torr} = \underline{150 \text{ Torr}}$$

Check: Total pressure = 580 + 150 +17.5 = 750 Torr. That's not 747.3, but it's a perfectly satisfactory result considering the significant figures.

## Section 4.5 Vapor Pressure, Relative Humidity, and Dew Point

Figure 4.14 shows the vapor pressure of water in Torr between 0°C and 100°C. The data used to make the diagram are in Table 4.4. Figure 4.14 is an important diagram for you to understand. We'll begin our discussion of the diagram here and will return to it later. The horizontal axis of Figure 4.14 shows the temperature, and the vertical axis shows water's vapor pressure. The **vapor pressure of water** is the pressure exerted by water vapor that is in equilibrium with liquid water at a specified temperature. Observe that at 100°C the vapor pressure of water reaches 760.0 Torr. You can now understand why water's boiling point is 100°C at standard atmospheric pressure of 760 Torr. At 100°C water's vapor achieves sufficient pressure to "push back" the blanket of the atmosphere.

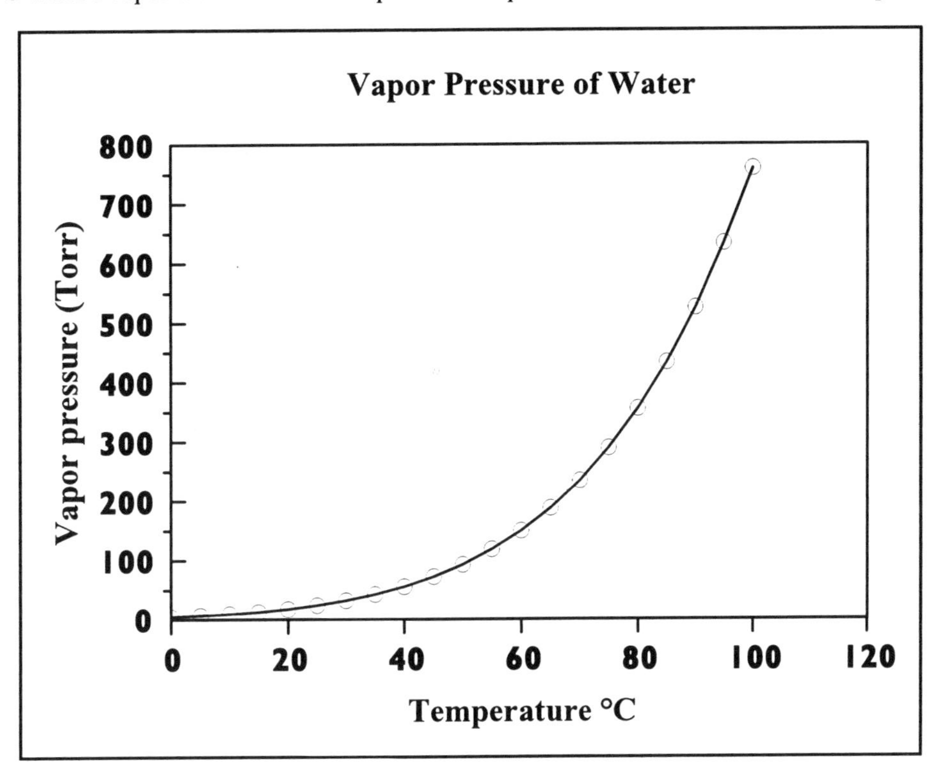


Figure 4.14 The vapor pressure of water plotted as a function of temperature. At standard pressure water' s boiling point is 100°C. That's the temperature at which water's vapor pressure reached standard pressure of 760 Torr.

| Temperature °C | Pressure Torr (or mmHg) |
|---|---|
| 0 | 0 |
| 5 | 6.5 |
| 10 | 9.2 |
| 15 | 12.8 |
| 20 | 17.5 |
| 25 | 23.8 |
| 30 | 31.8 |
| 35 | 42.4 |
| 40 | 55.3 |
| 45 | 71.9 |
| 50 | 92.5 |
| 55 | 118.0 |
| 60 | 149.4 |
| 65 | 187.5 |
| 70 | 233.7 |
| 75 | 289.1 |
| 80 | 355.1 |
| 85 | 433.6 |
| 90 | 525.8 |
| 95 | 633.9 |
| 100 | 760.0 |

Table 4.4 The vapor pressure of water. The graphical representation of these data is shown in Figure 4.14.

**Empirical Equations**

Given experimental data, such as the vapor pressure curve in Figure 4.14, every engineer knows that there is an equation that must "fit" the data. Scientists call equations obtained by fitting experimental data **empirical equations**. The data of Figure 4.14 are fitted by the empirical equation $\ln P = 20.5 - \frac{5170}{T}$, where P is the vapor pressure in Torr and T is the temperature in kelvins. We will interpret one of the constants in this empirical equation in Chapter 10.

Empirical equations are important. For example, we'll meet the van der Waals equation at the end of this chapter and the Balmer-Rydberg equation in the following chapter.

We all have observed that water evaporates. To **evaporate** is to transform spontaneously from a liquid to a vapor while simultaneously absorbing environmental heat. When weather reporters speak of "wind chill" they acknowledge that water evaporating from human skin lowers the temperature below what a nearby thermometer registers. Simply waving your wet hands in the air allows you to experience that cooling effect. Evaporation *stops* when the air has become saturated with water vapor. At a personal level you experience saturation when your perspiration fails to evaporate in a sauna or tropical swamp. Air is **saturated** (with water vapor) when the partial pressure of water vapor in the air equals the equilibrium vapor pressure of water (as shown in Figure 4.14 and Table 4.3). A closed bottle of water, or any other liquid, exists in a state of dynamic liquid-gas equilibrium represented by the equilibrium equation: L $\leftrightarrows$ G. **Relative humidity** is the existing water vapor pressure in the air divided by the saturated vapor pressure and expressed as percentage.

$$\text{Relative humidity} = \frac{\text{existing vapor pressure of water in the air}}{\text{saturated vapor pressure of water in the air}} \times 100\%$$

Here's a relative humidity example:

**Example 4.15:** The air temperature is 15.0°C. A measurement of water's vapor pressure that day shows it to be 9.6 Torr. What is the relative humidity?

Strategy: Look up the saturated vapor pressure of water in Table 4.4. Express the measured vapor pressure as a percentage of that value.

At 15.0°C water's vapor pressure is 12.8 Torr.

$$\text{Relative humidity} = \frac{9.6}{12.8} \times 100\ \% = \underline{75\%}$$

A simple way to measure the existing vapor pressure of the air is to use a wet-bulb and dry-bulb thermometer. A dry-bulb thermometer is an ordinary mercury thermometer. A wet-bulb thermometer is a second, identical thermometer, with its bulb covered by a wet cloth sleeve. The two thermometers are mounted on a board and slung like a noisemaker at an athletic event. The apparatus is called a **sling psychrometer** because it's a psychrometer you sling. Evaporation cools the wet-bulb thermometer, and if the humidity is not 100% the two bulbs record different temperatures. The two temperatures are noted and the humidity is looked up in tables designed for that purpose. If you think about it, you'll see that the lower the humidity, the faster moisture evaporates from the cloth sleeve, the colder the sleeve gets, and the greater the temperature difference between the dry and wet thermometers.

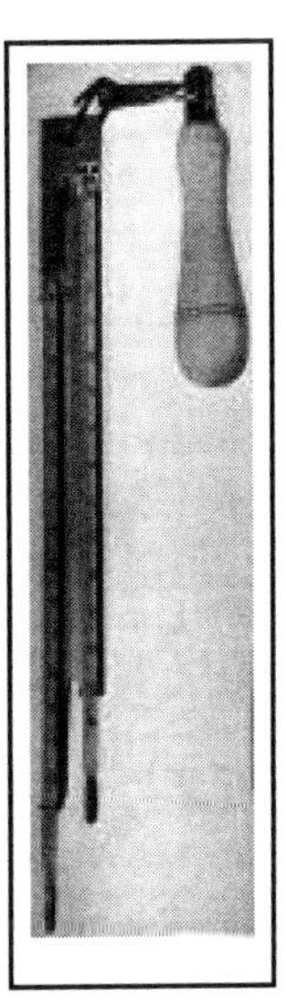

Figure 4.15 A sling psychrometer. The wet bulb is on the left. The dry bulb is on the right. Slung like a noisemaker, the wet bulb registers a lower temperature than the dry one because of evaporative cooling. Specially prepared tables allow the relative humidity to be obtained from the two temperatures.

**Example 4.16**: On a summer day the temperature is 30°C and the humidity is 35%. Estimate the dew point.

Strategy: First calculate the vapor pressure in the air using vapor pressure data and the humidity. Then go back and examine the vapor pressure data to see at what temperature this amount of vapor will saturate the air.

$$\text{Vapor pressure} = 31.8\ \text{Torr} \times \frac{35}{100} = 11\ \text{Torr}$$

From the table of water's vapor pressure we see that 11 Torr corresponds to about 11°C

The dew point is about 11°C

One way that working engineers encounter vapor pressure is if they have to deal with the potentially flammable vapors that come from industrial solvents. The danger of potentially flammable vapors is rated by a practical measurement called flash points, described in the following box.

## Flash Point and Vapor Pressure

The **flash point** of a flammable liquid is the temperature at which it is sufficiently volatile for its vapor to ignite by a spark or flame. The adjective "flash" refers to the flash of light and flame that accompanies ignition. Flash points are important to engineers responsible for the safe use of industrial solvents throughout manufacturing industries. The flash point of a liquid depends on two factors: 1. Its inherent flammability, and 2. Its vapor pressure.

Inherent flammability is related the fuel value of the liquid. For example, organic compounds containing carbon and only hydrogen as the second element (hydrocarbons) are invariably flammable. Organic compounds that contain no hydrogen are usually not flammable; for example: $CO_2$ is a fire extinguisher and $CF_4$ won't burn. Organic compounds containing some hydrogen may burn with difficulty. An illustrative series of compounds is: $CH_4$, highly flammable > $CH_2F_2$, somewhat flammable > $CF_4$, nonflammable.

Vapor pressure strongly affects flash point because ignition occurs between oxygen in the air and vapor from the liquid. A flammable liquid that has a low vapor pressure is hard to ignite—as you'll discover if you try to light liquid kerosene with a match.

There are many test methods for flash points: Tag closed tester; Cleveland open cup, Pensky-Martens closed tester; Setaflash closed tester; etc. Specialists in flash-point measurements debate the merits and demerits of the various methods. Some typical liquid boiling points and flash points are shown in the table below. Note that flammable liquids' flash points are always well below their boiling points.

Table 4.5 **The Flash Points of Some Common Liquids**

| Substance | BP °C | Flash point °C | Comment |
|---|---|---|---|
| Ether | 35 | -45 | Highly flammable. High vapor pressure. |
| Pentane | 36 | -40 | Highly flammable. High vapor pressure. |
| Gasoline | 40-200 | ≈ -40 | Highly flammable and somewhat variable. |
| Acetone | -38 | -18 | Readily flammable. |
| Toluene | 111 | 4 | Flammable. |
| Ethyl alcohol | 78 | 13 | Flammable. |
| Kerosene | 150-300 | 40-70 | Flammable but difficult to ignite at ordinary temperatures. |
| Ethylene glycol | 198 | 111 | Low vapor pressure allows for significant vapor formation only at high temperature. Very difficult to ignite. |
| Carbon dioxide | sublimes | none | Nonflammable substance |
| Carbon tetrachloride | 76.5 | none | Nonflammable substance |

## Section 4.6 The Work of Expanding Gases

In this section we will see how the expansion of a gas produces work and begin a discussion of thermodynamics that will extend over several chapters. Because of their simplicity, gases are almost always used to begin such discussions. **Thermodynamics** (thermos + dynamos) is literally the science of heat + work. The word appears to have been coined around 1854 by Macquorn Rankine (1820-1872), who is known for a particular type of power cycle and as the promulgator of the Rankine temperature scale used for many years by engineers. Rankine was appointed Professor of Engineering at the University of Glasgow in 1855. In thermodynamics, many ordinary-sounding words are given precise, technical definitions. Here are some such words:

The **thermodynamic state** of a gas is a specification of the properties needed to characterize it uniquely—the usual four properties are P, V, T, and amount of gas (moles or mass), but various combinations are possible. The defining properties are **state functions** or **state properties** and depend only on the thermodynamic state of a system. The gas sample itself is a **thermodynamic system**: any precisely defined experimental setup we choose to study. Changes in state properties occur independently of the path taken to go from state 1 to state 2. For example, when a gas is compressed from $P_1$ to $P_2$ the change in pressure $\Delta P$ is identical regardless of how and how fast the pressure was changed. A force that moves an object performs work: work = force × distance. Almost all the work done by machines in modern society involves the expansion of hot gases. The traditional diagram to illustrate the work of expansion of a gas is shown in Figure 4.16. When a gas expands against a piston with a cross-sectional area of A, against $P_{ext}$, the force on the piston, then $F = P_{ext} \times A$.

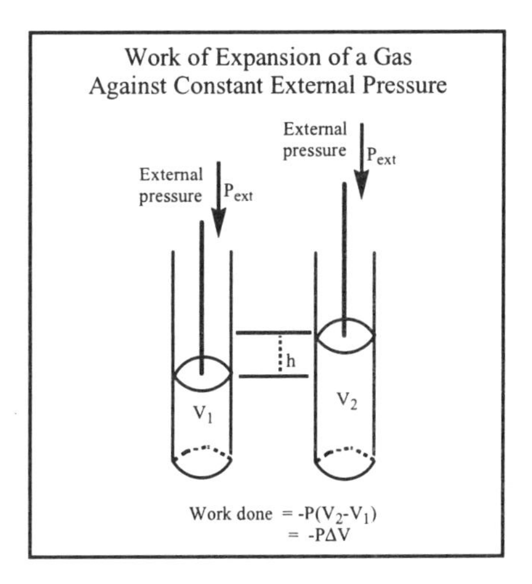


Figure 4.16 Hot gases expanding inside pistons are a prime source of motive power in modern society. Highway vehicles, boats, and many airplanes have piston engines.

When the piston moves a distance h, as is illustrated in Figure 4.16, the work done is equal to

$$\text{work} = \text{force} \times \text{distance} = P_{ext} \times A \times h$$

But A × h (area × length = volume) corresponds to a change in the volume of the gas, $\Delta V$. So:

$$\text{work} = -P_{ext}\,\Delta V$$

If the pressure is in atmospheres and the volume is in liters then the work has the units of liter atmospheres (L atm). Using the algebra of quantities it is straightforward to show that L atm is a proper energy (work) unit. The minus sign is written by convention; we'll explain it in Chapter 7.

Pressure = force ÷ area and F = mass × acceleration so:

$$-P\Delta V = \frac{(\text{mass} \times \text{acceleration})}{\text{area}} \times \text{length}^3$$

$$\text{or, expressed in base SI units: } -P\Delta V = \frac{(\text{kg m s}^{-2})}{\text{m}^2} \times \text{m}^3$$

$$= \text{kg m}^2\,\text{s}^{-2} \text{ or joules.}$$

So $-P\Delta V$ has the units of energy, as we said. We will return to thermodynamic work in Chapter 7. As we saw in Chapter 1, power has the units of energy per time: $1\ W = 1\ J\ s^{-1}$. Table 4.6 shows the output in watts of various types of "machine."

Table 4.6 **Comparative Power Outputs**

| |
|---|
| Human 25 - 150 W |
| Horse 300 - 600 W (1 hp ≈ 746 W) |
| Big windmill 5,000 - 8,000 W |
| Big waterwheel 2,000 - 10,000 W |
| Otto gas engine (1881) 15,000 W |
| Early one-cylinder gasoline engine for car 2500 W (Karl Benz, 1885) |
| Early Watt steam engine 10,000 - 40,000 W |
| First steam turboelectric generator (Parsons, 1888) 75,000 W |
| Modern race car 200,000 - 500,000 W |
| Jumbo jet 5,000,000 W (5 MW) |
| Big, modern, electric power plant 800,000,000 W (800 MW) |

Example 4.17 shows a somewhat oversimplified example of how the power of an engine can be calculated from the work done by an expanding gas.

**Example 4.17**: An imaginary one cylinder engine is expanding by 3.0 L 1,000 times per minute against an external pressure of 3.0 atm. Calculate the power generated in watts and horsepower.

Strategy: Use the equation $w = -P_{ext}\,\Delta V$ and calculate the work of a single stroke in joules. Use 1 watt = 1 joule $second^{-1}$ to obtain power. Convert that value to horsepower as in Example 1.12; 1 hp = 745.7 watts (exactly). The needed conversion factor is $745.7\ w\ hp^{-1}$.

The work of a single stroke is $w = -3.0\ L \times 3.0\ atm \times 101.3\ J\ L^{-1}\ atm^{-1}$

$= -910\ J\ stroke^{-1}$

Power = work ÷ time (and we now, again by convention, drop the negative sign)

Power $= 910\ J\ stroke^{-1} \times 1{,}000\ strokes \div 60\ s$

Power = $\underline{15{,}000\ J\ s^{-1}\ \text{or watts}}$

In horsepower it's $15{,}000\ \text{watts} \div 745.7\ w\ hp^{-1} = \underline{20\ hp}$

## Section 4.7 The Kinetic-Molecular Theory of Gases

Kinetic-molecular theory relates the bulk properties of gases to the random motions of their individual molecules. The complete theory developed over a very long period of time, 1738-1910.

One aspect of the kinetic-molecular theory is that it allows us to understand the limiting nature of the gas laws. Avogadro's law, Boyle's law, Charles's law, and the ideal gas equation are **limiting laws** because they are obeyed accurately by real gases only in the limit of extremely low pressure. The existence of limiting gas laws tells a great deal about the nature of gases and about molecules in general. Furthermore, gas pressure, Brownian motion, diffusion, and many other gas phenomena can be interpreted and made understandable by kinetic-molecular theory.

Pressure, something that we normally think of as being uniform, is actually the collective result of billions and billions of individual molecular collisions. The origin of the pressure of a gas inside a container is schematically represented in Figure 4.17. In 1827, the botanist Robert Brown (1773-1858) observed the random (Brownian) motion of pollen grains suspended in water. The pollen grain suffers repeated collisions with molecules of the fluid and is bumped about randomly as illustrated in Figure 4.18; any small particle suspended in a gas or liquid will perpetually move along a random zigzag path. Brown was probably the first person to observe direct evidence that matter is composed of molecules in motion.

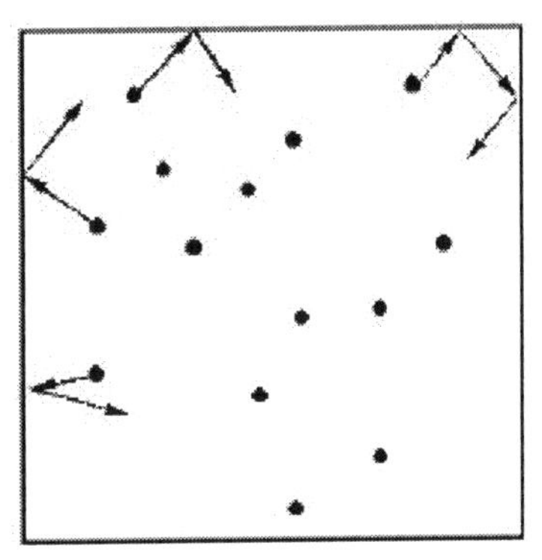

Figure 4.17 The kinetic behavior of a gas is the origin of gas pressure. Its molecules are in perpetual motion, changing direction when they strike one another or the walls of their container.

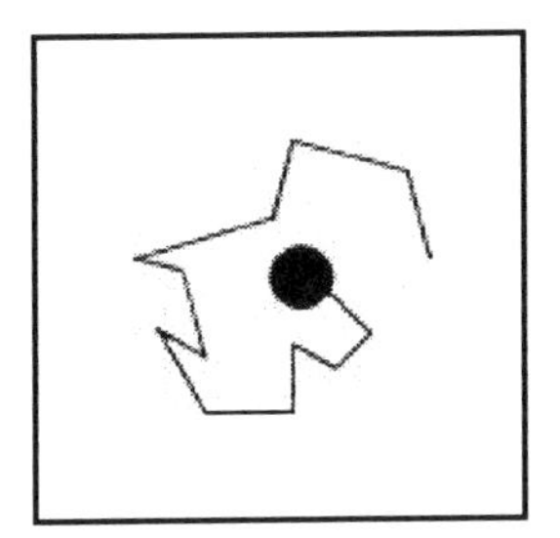

Figure 4.18 Brownian motion. Small particles of dust or pollen exhibit constant random motion as a consequence of collisions with the fluid that surrounds them. Each change of direction shows when a preponderance of molecular collisions occurred on one side of the particle.

When sufficiently cooled, all gases condense. Sometimes a gas can be condensed simply by increasing its pressure without lowering its temperature at all. Pressure-caused condensation was described earlier in Section 4.1 in connection with a carbon dioxide fire extinguisher. From this evidence alone you know that molecules *must* sometimes attract another sufficiently strongly to “stick” together.

At ordinary pressures and temperatures, the distances among the molecules in a gas are perhaps 10 times greater than the molecule’s size. As the pressure of a gas is lowered, the gas expands and the distances among its molecules become larger. Finally, after sufficient expansion, the molecules are so far apart that the long range forces become negligible. This line of reasoning allows us to understand why gases obey limiting laws. Furthermore, because different gases have different condensation temperatures (boiling points), the size of the intermolecular forces of attraction must be different for different molecules.

**Kinetic-Molecular Theory** (KMT) is a theory of gas behavior based on the assumption that, at a sufficiently low pressure, every gas behaves as though it is composed of widely separated, non-interacting (except during collisions) molecules in continuous and random motion. The properties of an ideal gas can be calculated from the kinetic-molecular theory of gases. Recall that the kinetic-molecular theory of gases began in 1738 with Bernoulli's derivation of Boyle's law based on applying Newtonian mechanics to gas particles. Bernoulli's method can be understood if we begin with four assumptions. These assumptions are the postulates of KMT.

## Postulates of the Kinetic-Molecular Theory of Gases

1. Gases are composed of molecules in constant, random motion.
2. The molecules undergo random, elastic collisions with each other and with the walls of their container. Elastic means without a loss of energy. They exhibit perpetual motion.
3. There are no forces among the particles or between the particles and the walls of their container (except during a collision).
4. The particles have no size. They are point-like.

From these postulates the gas laws can be derived. A couple of rhetorical questions here will help you visualize how KMT works. What happens to pressure when the temperature of a gas is increased? Answer: Heating a gas makes its molecules move faster as input heat energy becomes kinetic energy of the molecules. Faster molecules means that there are more frequent and more violent collisions against the walls, and so pressure rises with increasing temperature. What happens when more gas is injected into a fixed volume tank at a fixed temperature? Answer: The energy of an average collision remains the same, but pressure increases because more molecules will be colliding with the walls.

## The Speeds of Gas Molecules

In a fluid, each molecule undergoes random collisions with other molecules. The assumption of elastic collisions means that when two molecules collide no kinetic energy is lost. However, the total kinetic energy will be redistributed during a collision; the two molecules have different speeds, and different *individual* energies, after the collision than before. The continual random exchange of energy in collisions leads to the most important general result of kinetic-molecular theory: gas molecules develop a broad distribution of speeds. Some molecules move rapidly, some move slowly, and some move at virtually every intermediate speed.

Around 1870, James Clerk Maxwell (1831-1879) and Ludwig Boltzmann (1844-1906) derived the mathematical function that describes the distribution of molecular speeds. Their result is displayed in Figure 4.19 for nitrogen gas at 300 K and 1,000 K. The diagram displays the **Maxwell-Boltzmann distribution**, a graph of the relative numbers of molecules that travel at different speeds. The area under the curve between any two speeds, $v_1$ and $v_2$, is equal to the fraction of the molecules that are moving at speeds between $v_1$ and $v_2$. There is a maximum in the distribution, and the speed at which the maximum occurs is the **most probable speed**. More molecules travel at the most probable speed than at any other speed. The value of the most probable speed increases as the temperature is raised. Note also that the distribution of molecular speeds broadens at higher temperatures.

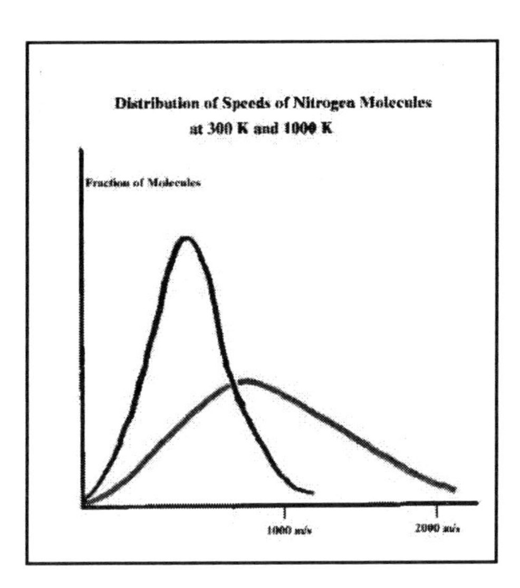


Figure 4.19 Distribution of molecular speeds for nitrogen gas at 300 K (left peak) and 1,000 K (right peak). The fraction of the molecules (vertical axis) is plotted against their speed in meter per second (horizontal axis).

Conceptually, the mathematics of the Maxwell-Boltzmann distribution function is similar to that of the curve of normal error described in the appendix that deals with experimental uncertainty (Appendix B). Kinetic-Molecular Theory provides a number of different measures of molecular speeds. One measure useful in understanding an ideal gas is the mean square speed. **Mean square speed**, $v^2_{mean}$, is the mean of the squares of the speeds of the molecules. Its square root is the root mean square (rms) speed, $v_{rms}$, of the gas molecules. The relationship between the two is $v_{rms} = \sqrt{v^2_{mean}}$ .

> **Informal Example:** Calculate the mean, the mean square, and the root mean square of the integers 2, 4, 6, and 8.
>
> $$\text{Mean} = (2 + 4 + 6 + 8) \div 4 = 20 \div 4 = 5$$
>
> $$\text{Mean square} = (2^2 + 4^2 + 6^2 + 8^2) \div 4 = 120 \div 4 = 30$$
>
> $$\text{Root mean square} = \sqrt{30} = 5.47\ldots$$
>
> The root mean square is somewhat larger than the mean.

Kinetic-molecular theory provides an equation that allows you to calculate the rms speed of gas molecules: $v_{rms} = \sqrt{\frac{3RT}{MW}}$ , where $v_{rms}$ is the root mean square speed, R is the gas constant, T the kelvin temperature, and MW the molar weight of the gas. It's a testament to the power of using units in scientific calculations to see how proper speed units arise out of a square root term that involves quantities such as R and MW, as shown in Example 4.18.

**Example 4.18**: Calculate the root-mean-square speed of nitrogen molecules at 300. K.

Strategy: Using quantities expressed in the proper units, substitute into the relationship: $v_{rms} = \sqrt{\frac{3RT}{MW}}$

For this situation the appropriate value of R is 8.314 J $K^{-1}$ $mol^{-1}$; the units of the joule are kg $m^2$ $s^{-2}$. To be consistent use the molar weight in kg $mol^{-1}$.

$$R = 8.314 \text{ kg m}^2 \text{ s}^{-2} \text{ K}^{-1} \text{ mol}^{-1}$$
$$MW = 28.0 \text{ g mol}^{-1} \times 0.001 \text{ kg g}^{-1} = 0.0280 \text{ kg mol}^{-1}$$
$$T = 300. \text{ K}$$

$$v_{rms} = \sqrt{\frac{3RT}{MW}} = \sqrt{\frac{3 \times 8.314 \text{ kg m}^2 \text{ s}^{-2} \text{ K}^{-1} \text{ mol}^{-1} \times 300 \text{ K}}{0.0280 \text{ kg mol}^{-1}}}$$

Root-mean-square speed = 517 m $s^{-1}$ (about 1160 miles per hour)

The maximum of the left hand peak in Figure 4.20 is called the **most probable speed** and is near 500 m $s^{-1}$. Note that the root-mean-square-speed is somewhat larger because the distribution curve is not symmetrical: the number of molecules with speeds more than the most probable speed is somewhat greater than the number of molecules with speeds less than the most probable speed.

## The Kinetic Energy of an Ideal Gas

The kinetic energy of a mole of gas molecules is

$$KE_{mean} = ½ \times MW\ v^2_{mean}$$

where the units of MW must be kg $mol^{-1}$ if $v^2_{mean}$ is in meters per second and the kinetic energy is in joules. It's because kinetic energy has a $v^2$ term that we use the mean square velocity and not the mean velocity.

Combining the equations $KE_{mean} = ½ \times MW\ v^2_{mean}$ and $v^2_{mean} = \frac{3RT}{MW}$ gives the relationship

$$KE = \tfrac{3}{2}RT$$

The above equation shows that the mean kinetic energy of gases is *independent* of their molar weight, and depends only on their temperature. The molar kinetic energy of an ideal gas is directly proportional to its temperature. It follows that at any specified temperature, the molecules in a gas with a low molar weight move faster than those of a gas with a higher molar weight. As a gas's temperature is increased its molecules gain kinetic energy. A molar kinetic energy calculation is illustrated in Example 4.19.

**Example 4.19**: Calculate the kinetic energy of a mole of any ideal gas at 300. K. For comparison, calculate at what velocity a 1.00 kg mass would have the same amount of kinetic energy.

Strategy: Substitute with appropriate units into $KE = \frac{3}{2}RT$. Then plug that value into $KE = ½ mv^2$ and solve for v.

$$KE = \tfrac{3}{2}RT \qquad \tfrac{3}{2} \times 8.314 \text{ kg m}^2 \text{ s}^{-2} \text{ mol}^{-1} \text{ K}^{-1} \times 1 \text{ mol} \times 300. \text{ K}$$

$$= 3740 \text{ kg m}^2 \text{ s}^{-2} \text{ or joules}$$

$$KE = ½ mv^2 \quad \text{so } v = \sqrt{\frac{2 \times KE}{m}}.$$

$$v = \sqrt{(2 \times 3740 \text{ kg m}^2 \text{ s}^{-2} \div 1.00 \text{ kg})} = 86.5 \text{ m s}^{-1} \quad (\approx 200 \text{ mph})$$

The molar volume of a gas at 300. K is about 25 L or 6-7 gallons. If you imagine a 6-gallon pail and think about a kilogram mass traveling at 200 mph inside it, you'll be pleased that the molecules of the gas move in all possible directions rather than all in the same direction. [❁kws +"armor piercing" +projectile]

## Thomas Graham's Law of Effusion and Diffusion

**Effusion** occurs when a gas escapes through a pinhole leak into a vacuum. Gaseous **diffusion** occurs when a volatile substance placed in a gas spreads throughout the gas or when gases are mixed. Effusion and gas diffusion are closely related processes; they depend on the speeds of gas molecules. The phenomena were studied by Thomas Graham (1805-1869) in the middle of the 19th century. He found that rates of effusion/diffusion of a particular gas vary inversely with the square root of the molar weight (MW) of the gas at any specified temperature and pressure. Graham's law is thus:

$$\text{Rate of effusion/diffusion} \propto \sqrt{\frac{1}{\text{MW}}}$$

If a comparison is made between the rates ($r_1$ and $r_2$) of effusion/diffusion of two gases at the same pressure and temperature, the relative rates are given by

$$\frac{r_1}{r_2} = \sqrt{\frac{\text{MW}_2}{\text{MW}_1}}$$

The existence of Graham's law is experimental support for the kinetic-molecular theory of gases. The rate of effusion/diffusion of a gas is obviously related to the speed of its molecules. The faster gas molecules move, the more rapidly they effuse or diffuse. Graham's law is illustrated in Example 4.20.

**Example 4.20**: Calculate the relative rates of effusion of methane ($CH_4$) and butane ($C_4H_{10}$) at a specified temperature and pressure.

Strategy: Calculate the molar weights and plug into Graham's law $\frac{r_1}{r_2} = \sqrt{\frac{\text{MW}_2}{\text{MW}_1}}$.

MW methane = 16.0 g mol$^{-1}$
MW butane = 58.1 g mol$^{-1}$
Let gas 1 be methane and gas 2 be butane

Then, $\frac{r_1}{r_2} = \sqrt{\frac{\text{MW}_2}{\text{MW}_1}} = \sqrt{\frac{58.1 \text{ g mol}^{-1}}{16.0 \text{ g mol}^{-1}}}$ So methane effuses 1.91 times faster than butane.

Techniques of gas separations can be developed using the principle that different gases have different rates of effusion. However, its usually much easier and cheaper to use chemical separations, procedures based on different physical properties such as boiling point differences, or one of the family of procedures known collectively a s chromatographic methods [❁kws "chromatographic separation"]. There is, however, one historic separation of great importance based on relative rates of effusion, that's the separation of $^{235}U$ from $^{238}U$.

Uranium occurs with isotopic abundances of 99.27% $^{238}U$, 0.72% $^{235}U$, and less than 0.01% $^{234}U$. It's the $^{235}U$ isotope that's needed for atomic weapons and power plants. Isotopes are incapable of separation via conventional chemical and physical methods—but they can be separated by exploiting their tiny mass difference. During the early days of World War II, a process for the separation of uranium isotopes (in the form of $UF_6$) by gas effusion was developed as one of a series of crash programs that were part of the US government's Manhattan Project.

Uranium hexafluoride is a toxic, reactive substance with a boiling point of 56.2°C. Successful separation of large quantities of the hexafluorides of the two major uranium isotopes required the construction of a gigantic plant at Oak Ridge, Tennessee. The plant contained miles of pipe and thousands of pumps. The successful solution of the problems involved was a triumph of chemical engineering. In Example 4.21 Graham's law is used to show why the separation of $^{235}UF_6$ from $^{238}UF_6$ by their small mass difference is a difficult and costly process:

**Example 4.21**: Calculate the relative rates of effusion of $^{235}UF_6$ and $^{238}UF_6$.

Strategy: Calculate the molar weight of each gas and substitute into the equation for relative rates of effusion:

$$\frac{r_1}{r_2} = \sqrt{\frac{MW_2}{MW_1}}$$

Molar weight $^{235}UF_6$ = 349 g mol$^{-1}$ Molar weight $^{238}UF_6$ = 352 g mol$^{-1}$

$$\frac{r_1}{r_2} = \sqrt{\frac{MW_2}{MW_1}} = 1.0043$$

So the hexafluoride of the lighter isotope effuses only 1.0043 times faster than that of the heavier uranium isotope.

Because the relative rate of effusion is only slightly greater for the $^{235}U$ hexafluoride, a gas effusion separation plant must have many stages, or cascades, to produce useful levels of isotopic enrichment. During wartime at Oak Ridge it required some 4,000 stages to yield $^{235}UF_6$ with greater than 99% purity.

Amusingly, the Oak Ridge plant was always called the gas *diffusion* plant, never the gas effusion plant. Engineers, being practical people, don't always respect the hairsplitting distinctions made by chemistry professors.

## Rationalization of Graham's Law

Recall that the equation $KE = \frac{3}{2}RT$ demonstrates that the mean kinetic energy of gases is independent of molar weight and depends only on the temperature. At any specified temperature, the molecules in a gas with a low molar weight move faster than those of a gas with a higher molar weight.

So let's consider two gases (1 and 2). They have molar weights $MW_1$ and $MW_2$ and they have mean square speeds of $v^2_{av,1}$ and $v^2_{av,2}$. Our premise is that at a fixed temperature the molecules of the two gases have the same average kinetic energy. That's a consequence of the many collisions. So, using the equation for KE, we can write:

$$\tfrac{1}{2} MW_1 v^2_{av,1} = \tfrac{1}{2} MW_2 v^2_{av,2}$$

That equation rearranges to: $$\frac{v_{av,1}}{v_{av,2}} = \sqrt{\frac{MW_2}{MW_1}}$$

Guess whose law that is? Graham's of course.

## Section 4.8 Real Gases and the van der Waals Equation

A **real gas** is one that *does not* follow the ideal gas equation. To see how good an approximation is the ideal gas equation for real gases, we'll look at some experimental results for nitrogen at two temperatures across a range of pressures. Making a graph of the ratio PV/nRT is a clear-cut way of showing deviations from ideal behavior. The ratio PV/nRT, that must be 1 for an ideal gas under any conditions, is termed the **compressibility factor** and given the symbol z.

$$z = \frac{PV}{nRT}$$

The compressibility factor is a sort of "ideality index." The closer z is to 1, the more like an ideal gas the actual gas is behaving. The further z is from 1, the less like an ideal gas the actual gas is behaving. Experiments on many different gases show that ideal gas behavior is approached at low pressures and high temperatures, but that deviations from ideal gas behavior become large at low temperatures and high pressures. Figure 4.20 shows compressibility factor data as a function of pressure for nitrogen. Figure 4.21 shows compressibility factor data for as a function of temperature for carbon dioxide.

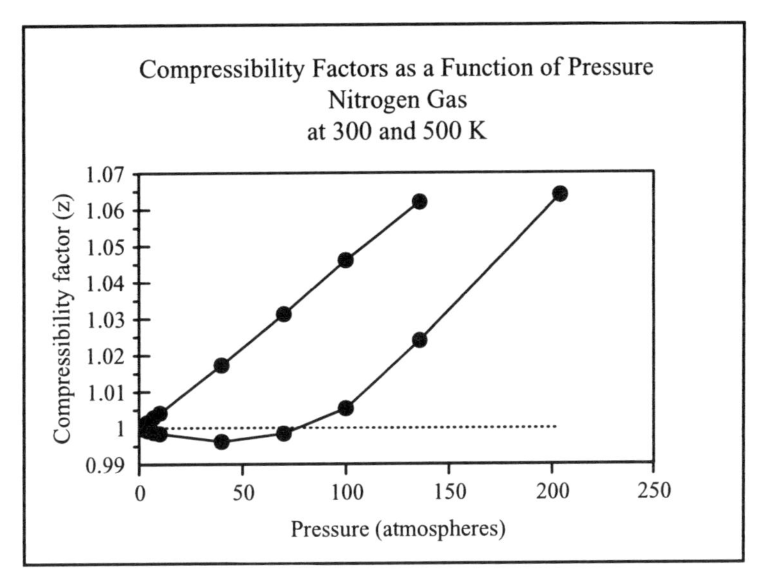


Figure 4.20 The compressibility factor (vertical axis) of nitrogen at various pressures (horizontal axis) and two specific temperatures. Left hand line at 300 K, right hand line at 500 K. If nitrogen were an ideal gas it would follow the horizontal dotted line, z = 1. Data from the Handbook of Tables for Applied Engineering Science.

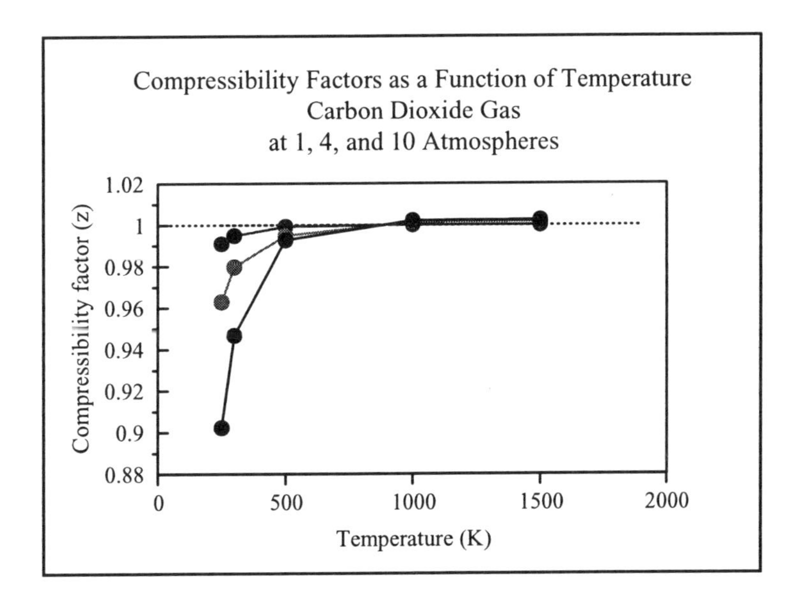


Figure 4.21 The compressibility factor (vertical axis) of carbon dioxide at various temperatures (horizontal axis) and three specific pressures. Upper line at 1 atm, middle line at 4 atm and lower line at 10 atm. If carbon dioxide were an ideal gas it would follow the horizontal dotted line, z = 1. Data from the Handbook of Tables for Applied Engineering Science.

Based on studies of the compressibility factors of gases and the small sample of data shown in Figures 4.20 and 4.21 a number of points can be made.

1. Ideal gases would produce graphs with a single horizontal line. The horizontal dashed line (-----) represents ideal behavior, because along that line $\frac{PV}{nRT} = 1$.
2. As its pressure is reduced, nitrogen approaches ideal gas behavior. At the limit of zero pressure (on the left hand side of Figure 4.22) it *is* an ideal gas. At high pressures nitrogen's deviations from ideality are greatest.
3. Nitrogen's deviations can be *both* positive and negative. At 300 K, and pressures below about 80 atm, nitrogen has z values less than 1 (negative deviation). At 300 K, and pressures above about 80 atm, nitrogen has z values greater than 1 (positive deviation). Because of the existence of both positive and negative deviations we conclude that there must be at least two physical factors at work controlling the properties of nitrogen, and other real gases. At 300 K and 80 atm nitrogen behaves accidentally as an ideal gas, the two physical factors offset. On Figure 4.22 that's the point where the curved line crosses the dashed line.
4. It often (but not always) happens that heating a gas causes it to behave in a more ideal fashion. Figure 4.22 shows clearly that carbon dioxide's behavior (shown at 1, 4, and 10 atm) comes closer to ideality at high temperatures.

Experimental data for real gases, such as that shown in Figures 4.21 and 4.22, can be fitted with equations that are modified versions of the ideal gas equation. Such fits have helped us to understand how gases "work." One of the first modified versions of the ideal gas equation was proposed by Rudolf Clausius (1822-1888). Clausius said that the molecules themselves are not compressible and proposed that a term (nb) be subtracted from the volume of the gas. In the nb term, n is the number of moles of gas and b the molar volume of the gas molecules themselves. In other words, the volume available to a gas is the volume of its container *minus* the volume of the molecules themselves. The Clausius equation for real gases is

$$P(V-nb) = nRT$$

Every gas has its own "b" constant related to the actual molar volume of the molecules themselves—not the space they occupy.

A second correction term may be inserted to account for intermolecular forces of attraction among real gas molecules. To do this, P in the ideal gas equation is replaced by the measured pressure *plus* a correction term. The correction term is additive because intermolecular attractive forces have the same effect as a small pressure increase. The correction term that van der Waals chose was the square of the number of moles of gas per unit volume: $\frac{n^2}{V^2}$. The inverse dependence of this term on volume squared makes sense if you recall that gases expanded to large volumes approach ideal behavior. As a gas expands $V^2$ becomes increasingly large and the correction term $\frac{n^2}{V^2}$ approaches zero. Thus, the second correction term fits in with the idea that as distances between the molecules become large the attractive forces weaken. Combined with the Clausius correction, this second modification yields the best-known of the non-ideal gas equations, the **van der Waals equation**:

$$\left(P + \frac{n^2a}{V^2}\right)(V - nb) = nRT$$

The symbols n, P, V, and T have their usual meanings; "a" and "b" are constants for any particular gas and are its **van der Waals constants**. Values of the van der Waals "a" and "b" constants for several gases are shown in Table 4.7. The van der Waals "a" constant is a rough measure of attractive intermolecular forces. The larger the value of "a" the greater the intermolecular forces. The van der Waals "b" constant approximates the volume of the molecules themselves.

Table 4.7 **Values of van der Waals a and b constants for Selected Gases**

| Gas | "a" $L^2$ atm $mol^{-2}$ | "b" L $mol^{-1}$ |
|---|---|---|
| Helium | 0.034 | 0.0237 |
| Neon | 0.211 | 0.0171 |
| Argon | 1.340 | 0.0322 |
| Oxygen | 1.360 | 0.0318 |
| Krypton | 2.318 | 0.0400 |
| Carbon dioxide | 3.592 | 0.0427 |
| Methane | 2.253 | 0.0429 |
| Ethane | 5.489 | 0.0638 |
| Propane | 8.664 | 0.0846 |
| Butane | 14.470 | 0.1226 |

From Table 4.7 you'll see that bigger molecules have stronger forces of intermolecular attraction, as measured by their van der Waals "a" constant. They also have a larger volume, as measured by the van der Waals' "b" constant. The use of the van der Waals equation is shown in Example 4.21.

The existence of van der Waals forces and pressure-condensable gases is crucial for air conditioning. When the pressure above a pressure-condensed liquid is released the liquid spontaneously boils and chills. That phenomenon is well illustrated by the activities of some creative marines (as you can read in the box titled Instant Propane Refrigeration at the end of the chapter). The chapter concludes with a diagrammatic explanation of how an air conditioner works, how it relies on the principles we have studied in the chapter, and how we must always be prepared for unexpected environmental consequences from our use of applied technology.

Conclusion: Historically, studies of gases provided the means for advancement in many areas of science and engineering. Gases have important roles in modern economies and the gases of the air create our weather. The simplicity of the ideal gas law is remarkable. But even more remarkable is how detailed studies of gas data lead to fundamental ideas about intermolecular forces. Today, we speak routinely of van der Waals forces and of the van der Waals radii of molecules.

**Example 4.22**: Calculate the pressure exerted by 2.00 moles of oxygen gas at 27.0°C in a 24.0 liter container. Use a. the ideal gas equation, and b. the van der Waals equation. Repeat the calculations with the volume changed to 0.240 L. Discuss the values obtained.

Strategy: This problem is a simple plug in. Using the proper units for the properties substitute values into PV = nRT and $\left(P + \frac{n^2a}{V^2}\right)(V - nb) = nRT$

a. Ideal gas equation: P = (nRT)/V

Oxygen in 24.0 L container: $P = (2.00 \text{ mol} \times 0.0821 \text{ L atm K}^{-1} \text{ mol}^{-1} \times 300.2 \text{ K}) \div 24.0 \text{ L}$

$$P = \underline{2.05 \text{ atm}}$$

Oxygen in .240 L container: $P = (2.00 \text{ mol} \times 0.0821 \text{ L atm K}^{-1} \text{ mol}^{-1} \times 300.2 \text{ K}) \div 0.240 \text{ L}$

$$\underline{P = 205 \text{ atm}}$$

b. van der Waals equation $P = \frac{nRT}{V - nb} - \frac{n^2a}{V^2}$ . From Table 4.7 the van der Waals constants for oxygen are a = 1.36 atm $L^2$ $mol^{-2}$ and b = 0.0318 L $mol^{-1}$.

Oxygen in 24.0 L container

$$P = \frac{2.00 \text{ mol} \times 0.0821 \text{ L atm mol}^{-1} \text{ K}^{-1} \times 300.2 \text{ K}}{24.0 \text{ L} - 2.00 \times 0.0318 \text{ L mol}^{-1}} - \frac{2.00^2 \text{ mol}^2 \times 1.36 \text{ atm L}^2 \text{ mol}^{-2}}{2.40^2 \text{ L}^2}$$

$$\underline{P = 2.04 \text{ atm}}$$

Oxygen in 0.240 L container

$$P = \frac{2.00 \text{ mol} \times 0.0821 \text{ L atm mol}^{-1} \text{ K}^{-1} \times 300.2 \text{ K}}{0.240 \text{ L} - 2.00 \times 0.0318 \text{ L mol}^{-1}} - \frac{2.00^2 \text{ mol}^2 \times 1.36 \text{ atm L}^2 \text{ mol}^{-2}}{0.240^2 \text{ L}^2}$$

$$\underline{P = 184 \text{ atm}}$$

Discussion: At the larger volume, the two equations give a similar result (2.04 versus 2.05 atm). With the volume reduced 100-fold, the oxygen behaves much less like an ideal gas, and the pressure computed from the van der Waals equation is significantly lower than that computed from the ideal gas equation (184 atm versus 205 atm).

**Instant Propane Refrigeration**

Propane is a convenient clean burning fuel with the chemical formula $C_3H_8$. Many outdoor cooks favor propane heating. The combustion reaction of propane is:

$$C_3H_8 + 5\ O_2 \rightarrow 3\ CO_2 + 4\ H_2O$$

Because of propane's well-known use for heating, it comes as a bit of a surprise to learn that propane can also be used for cooling. Propane has a critical temperature of 96.8°C and a critical pressure of 42 atmospheres. Those values mean that propane will be a liquid if it is stored below 96.8°C at a sufficient pressure. The lower the temperature, the lower the pressure needed to keep the gas in the liquid state of aggregation. So the propane in propane tanks is ordinarily liquid. You can't see inside a steel propane tank, but you can see the liquid state of the closely related fuel gas butane, $C_4H_{10}$, in a Bic® lighter. Butane has a critical temperature of 152°C and a critical pressure of 37.5 atmospheres—so butane's even easier to liquefy than propane.

Both propane and butane can be "pressure condensed." When compressed at a temperature below their respective critical temperatures they condense or liquefy. When the pressure on a propane tank is released, simply by opening the valve, the propane boils off. The propane molecules are held in the liquid state by intermolecular forces, so boiling is an endothermic process (it takes energy to overcome the forces) and heat is absorbed by the propane from its environment. In other words, releasing the pressure on a tank of propane gas has a chilling effect. If you've ever seen a carbon dioxide ($T_c = 31°C$, $P_c$ 72.9 atm) fire extinguisher used, you've seen the same phenomenon at work.

The Yuma, Arizona, air station is the home to several thousand marines who perform extend periods of desert duty in the Mohave Desert and often spend months away from the amenities of civilization. So when a cold beer sounds good, and conventional refrigeration is not available, ingenious marines turn to propane cooling.

The procedure is to take a metal ammunition box and put in four six-packs of beer. Then a propane tank is inverted and its delivery hose aimed at the bottom of the box. Three or four 20-second squirts of liquid propane are delivered, and *voilà*: instant cold beer.

Actually, every refrigerator or air conditioner operates on the same principle—by boiling a pressure condensable gas. Thanks Professor van der Waals, those intermolecular forces you discovered are pretty useful.

**HFC's, Air Conditioning, and the Environment**

Most air conditioning is made possible because the intermolecular forces among the molecules of refrigerant gases allow then to be condensed by pressurizing them. Modern refrigerating gases are hydrofluorocarbons (HFCs). The key step in the refrigerating cycle occurs when a liquid vaporizes under reduced pressure, does internal work overcoming its attractive intermolecular forces, and thereby cools. Here's a schematic diagram that shows how an air conditioning unit works:

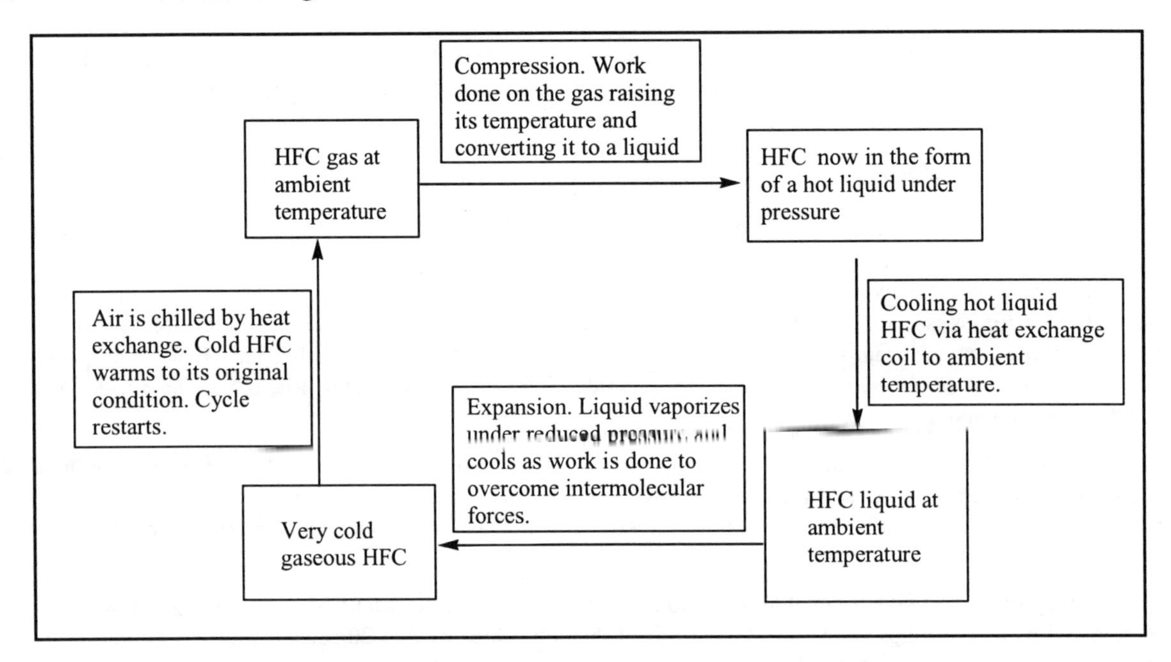


Fluorocarbon refrigerant gases were introduced around 1935. For many years the most widely used was dichlorodifluoromethane, $CF_2Cl_2$, a CFC or chlorofluorocarbon. This compound seemed like a miracle fifty years ago when it began to replace dangerous, toxic refrigerant gases such as ammonia and sulfur dioxide. Because $CF_2Cl_2$ seems to be such a safe compound (early demonstrations of its lack of toxicity involved people inhaling it and extinguishing flames by breathing out on to them) it was hard to foresee that it would present a serious threat to humans.

However, $CF_2Cl_2$ has two serious environmental consequences: First, it is itself a powerful "greenhouse" gas that helps trap solar heat on earth and may lead to global warming and, second, at high altitudes it sets up chemical processes that deplete the earth's ozone layer. In 1987 an international agreement called the Montreal Protocol was signed by major industrial countries. The protocol banned the use of $CF_2Cl_2$ beginning in 1996.

Today, HFC blends are used that operate reasonably well in existing ac units and have reduced potential to damage the ozone layer. One such blend is a mixture of $C_2HF_5$, $CH_3CF_3$, and $C_2H_2F_4$. However, the use of this blend is only a part way solution and research continues, with a target date of 2010, to find a permanent solution.

[❁kws "Montreal Protocol"] or [❁kws +"ozone layer" +chlorofluorocarbons].

## Essential Knowledge—Chapter 4

**Glossary words:** Intramolecular force, intermolecular force, fuel gas, pressure, standard atmosphere pressure, psia, psig, Torr, pascal, barometer, combined gas law, mean molar weight, partial pressures, mole fraction, dew point, humidity, vapor, vapor pressure, dynamic equilibrium, evaporate, saturated, empirical equations, sling psychrometer, flash point, expansion work, thermodynamics, thermodynamic state, state functions or state properties, thermodynamic system, kinetic-molecular theory, limiting laws, Brownian motion, pressure condensation, Maxwell-Boltzmann distribution, most probable speed, mean square speed, root mean square speed, effusion, diffusion, compressibility factor, empirical equations for gas behavior, van der Waals equation, van der Waals constants, air conditioning.

**Key Concepts**: Role of gases in the everyday world. Nature of gas pressure. Use of ideal gas equation. Meaning of partial pressure. Water vapor pressure and its role in weather. How expanding gases do work. Significance of the kinetic-molecular theory of gases. Real gases, the van der Waals Equation, and the discovery of intermolecular forces.

**Key Equations**:

Relative height rise of two barometric liquids: $\frac{h_2}{h_1} = \frac{d_1}{d_2}$

Pressure measured by a mercury barometer: $P_{gas} = P_{atm} + (x_1 - x_2)$,

Ideal gas laws: $P_1V_1 = P_2V_2$, $\frac{V_1}{T_1} = \frac{V_2}{T_2}$, $PV = nRT$, $\frac{P_1V_1}{T_1} = \frac{P_2V_2}{T_2}$

Properties of ideal gases: $d = \frac{P(MW)}{RT}$, $MW = \frac{dRT}{P}$

Dalton's law of partial pressure: $P_{tot} = P_A + P_B$,

Definition of mole fraction: $X_i = \frac{n_i}{n_A + n_B + n_C + \ldots}$

Weather: $P(\text{dry air}) = P_{N_2} + P_{O_2} + P_{Ar}$, Relative humidity = $\frac{\text{existing vapor pressure of water in the air}}{\text{saturated vapor pressure of water in the air}} \times 100\%$

Basic physics: Pressure = force ÷ area, Force = mass × acceleration,

Work of expanding gas: $w = -P_{ext}\,\Delta V$,

Kinetic-molecular theory: $v_{rms} = \sqrt{v^2_{mean}}$, $v_{rms} = \sqrt{\frac{3RT}{MW}}$, $KE_{mean} - \frac{1}{2} \times MW \times v^2_{mean}$, $KE = \frac{3}{2}RT$.

Graham's law: $\frac{r_1}{r_2} = \sqrt{\frac{MW_2}{MW_1}}$

Definition of compressibility factor: $z = \frac{PV}{nRT}$

Clausius equation: $P(V-nb) = nRT$,

van der Waals equation: $\left(P + \frac{n^2a}{V^2}\right)(V - nb) = nRT$

## Questions and Problems

### Gases in the Everyday World

4.1. Describe or define in 25 words or less each of the following: a. A pneumatic trough, b. LOX, c. a noble gas, d. $NO_x$, and e. liquid oxygen.

4.2. In 1662 Robert Boyle published the results of his studies of pressure-volume relationships in a sample of air trapped in a "crooked tube fit for the purpose" (a J-tube). The atmospheric pressure during his first series of experiments was "29 and two-sixteenths inches of quicksilver" (mercury). Express this value in Torr, atmospheres, psia, and kilopascals.

4.3. Describe the uses of each of the following: a. liquid helium, b. ethylene, c. propane, d. butane, and e. carbon dioxide.

4.4. Describe the commercial uses of chlorine gas.

4.5. Liquid nitrogen is valuable as a refrigerant. In contrast, liquid methane is never used as a refrigerant. So how come many commercial gas companies maintain a huge liquid methane tanks?

### Gas Pressure

4.6. A tank of gas has an internal pressure of 2100 psi. Calculate the pressure of the gas in atmospheres and kPa.

4.7. The gauge reads 27 psi when an ordinary measurement is made of automobile tire pressure. a. What is the air pressure inside the tire? b. Convert this pressure to atm and mmHg.

4.8. Approximately how many atmospheres of pressure are exerted on the hull of a submarine when it is about 100 ft below the surface of the ocean? (Hint: it's one atmosphere plus the pressure due to a 100 foot column of water).

4.9. A J-tube manometer had a height of Hg in the closed side connected to the gas sample of 22.0 cm. The side of the J-tube open to the atmosphere had a height of Hg of 52.5 cm. The atmospheric pressure was 756 mmHg. Calculate the pressure of the gas sample in Torr.

4.10. DBP is dibutyl phthalate, a liquid sometimes used in manometers, as described in the Chapter. A DBP-filled J-tube has a level of DBP in the open side that is 12.5 cm lower than the level on the closed side. The barometer reads 739 mmHg. What is the pressure of the trapped gas in Torr and atmospheres?

### The Gas Laws

4.11. Describe or define in twenty-five words or less each of the following: a. Boyle's law, b. Charles's law, c. Avogadro's law.

4.12. Given below are pressure and volume data for 16.00 g of $O_2$ at 25°C. Show from the data that $O_2$ obeys Boyle's law under these conditions.

| Experiment | 1 | 2 | 3 | 4 | 5 | 6 |
|---|---|---|---|---|---|---|
| P(atm) | 0.25 | 0.50 | 0.75 | 1.00 | 1.25 | 1.50 |
| V(L) | 48.92 | 24.45 | 16.30 | 12.22 | 9.77 | 8.14 |

4.13. A sample of gas is placed in a cylinder with a tight-fitting piston. The volume of gas is 3.75 L when the pressure is 1.00 atm. How much pressure must be exerted on the piston in order to reduce the gas volume to 1.75 L? The temperature of the gas is maintained at a constant value.

4.14. A fixed mass of gas is held at constant pressure. The absolute temperature of the gas is doubled. What happens to the volume of the gas?

4.15. A fixed mass of a gas is held at a constant temperature. The pressure exerted on the gas is tripled. What happens to its volume?

4.16. The Celsius temperature of a fixed mass of gas is increased from 100°C to 180°C at constant pressure. What happens to the volume of the gas?

4.17. At a temperature of 23.5°C, a gas has a volume of 2.55 L. If heated to 100.°C, what will its volume be?

4.18. The pressure exerted upon a fixed mass of an ideal gas is quadrupled, while simultaneously the absolute temperature of the gas is halved. What happens to the volume of the gas?

## The Ideal Gas Law

4.19. At standard temperature and pressure, the molar volume of an ideal gas is 22.4 L. Calculate the molar volume of the gas, a. at 25.0 °C and 720. Torr, and b. at 100°C and 735 Torr.

4.20. What is the density of gaseous nitrogen, $N_2$, in g $L^{-1}$ at 25°C and 0.970 atm?

4.21. A sample of an unknown gaseous substance has a density of 1.227 g $L^{-1}$ at 25°C and 0.938 atm. What is the molecular weight of the substance?

4.22. The chemical analysis of a hydrocarbon gas shows it contains 85.6% carbon and 14.4% hydrogen by mass. A sample of the gas weighing 2.75 g was found to have a volume of 1.34 liters at 52.7°C and 743 Torr. What is the molecular formula of the gas?

4.23. How many grams of propane, $C_3H_8$, are in a 25.0 L tank when the pressure gauge reads 224 lb $in^{-2}$, the atmospheric pressure is 14.6 lb $in^{-2}$, and the temperature is 18.4°C?

4.24. How many liters of $O_2$ gas can be obtained at 50.°C and 727 mm pressure by heating 52.0 grams of HgO? The decomposition reaction is: $2\ HgO(s) \rightarrow 2\ Hg(l) + O_2(g)$

4.25. A cylinder used to store gas must be constructed to withstand pressure generated by the gas. If a cylinder with a volume of 1,020 cubic meters contains 14,000 pounds of methane, $CH_4$, what pressure in psi must the container withstand when the temperature is 34.0°C?

4.26. Germane ($GeH_4$) is a toxic, flammable, colorless gas with a characteristic pungent odor. It is used chiefly for the production of high purity germanium semiconductors. The experimental density of germane at 0°C and 1.00 atmosphere pressure is 3.43 g $L^{-1}$. Calculate the density of germane using the ideal gas equation at these conditions of T and P. Under these conditions, can germane be considered to be an ideal gas?

## Partial Pressure

4.27 a. State Dalton's law of partial pressures. b. Define the property "mole fraction."

4.28. Deep-sea divers sometimes breathe a mixture of helium, nitrogen, and oxygen. The helium is used in the breathing mixture to help prevent the "bends," a nasty condition in which gas bubbles form in the diver's blood and which occurs if a diver returns to the surface too quickly after deep submersion. If the composition of the mixture used was 80.0 mole % He, 16.0 mole % $N_2$, and 4.0 mole % $O_2$, what would be the partial pressures of He, $N_2$, and $O_2$ when the diver is 200 feet below the surface, where the pressure is 6.90 atm?

4.29. A 5.0 L cylinder contains 0.625 g $CO_2$, 0.986 g of $O_2$ and 1.224 g of $N_2$ at 22°C. What are the partial pressures of $CO_2$, $O_2$, and $N_2$ in the cylinder? What is the total pressure of gases in the cylinder?

## Vapor Pressure

4.30. At 25°C a sample of air is saturated with water vapor. The sample is heated to 35°C (out of contact with liquid water, as in the case when air is being brought into a building from the outside and is heated via a heat exchanger). What is the percentage relative humidity in the sample after it has been heated?

4.31. At 99.7°F air contains water vapor with a pressure of 21 Torr. Calculate the relative humidity and the dew point of this air.

4.32. A partly filled, closed, water bottle is a dynamic equilibrium system. Explain.

4.33. Why is wind-chill most severe on cold days when there is wind and the air has a low dew point?

## Work of Expanding Gases

4.34. How much work is required to raise a 25 kg mass by a height of 25 m?

4.35. List three ways that useful energy is obtained from the expansion of hot gases in our society. State three energy sources that do not involve the work of hot gases expanding.

4.36. A gas expands from a volume of 2.34 liters to a volume of 5.23 liters against an external pressure of 0.88 atm. Calculate the work in joules.

4.37. Describe the manner in which a gas turbine does work.

4.38. 1.00 mole of an ideal gas at a pressure of 3.50 atm at 25°C is in a cylinder with a movable piston. The pressure on the piston is kept constant. The gas is heated until its volume triples. How much work does the gas do in L atm and kJ?

4.39. How much work does a gas perform if it is allowed to double its volume by expanding into a vacuum?

4.40. Carefully define: a. thermodynamic system, b. a state property, and c. path.

## Kinetic Theory of Gases

4.41. Describe or define in 25 words or less each of the following: a. Brownian motion, b. c. root-mean-square speed, d. the Maxwell-Boltzmann distribution.

4.42. Calculate the root-mean square speeds of a. helium, b. argon, and, c. xenon atoms at 25.0°C.

4.43. Introduced into clinical practice in 1956, the anesthetic halothane, 1,1,1- trifluoro-2-bromo-2-chloroethane ($CF_3CHClBr$), has proved to be extremely valuable. Calculate the rms speed of a halothane molecule at 37°C.

4.44. Sevoflurane® is a modern anesthetic. Chemically it is fluoromethyl 2,2,2-trifluoro-1-(trifluoromethyl) ethyl ether, $FH_2C\text{-}O\text{-}CH(CF_3)_2$, or $C_4H_3OF_7$. Calculate the rms speed of a $C_4H_3OF_7$ molecule at 37°C.

4.45. Calculate the kinetic energy of a mole of gas at a. 25.0°C and b. 300.°C.

4.46. Explain why at any particular temperature the molar kinetic energies of *different* gases are the same.

4.47. Show that kinetic energy divided by volume has the same units as pressure. Suggestion: Use SI base units and the method of algebra-of-quantities.

## Graham's Law of Effusion/Diffusion

4.48. Calculate the relative rates of effusion of helium and xenon at the same temperature and pressure.

4.49. What is the molecular weight of a gas that effuses (at the same temperature and pressure) at about one-ninth the rate at which helium effuses? Is there such a gas?

4.50. Heavy hydrogen (deuterium, $D_2$) can be separated from ordinary, light hydrogen by effusion. An atom of deuterium has a mass of 2.014 u. Calculate the relative rates of effusion of $H_2$ and $D_2$, and the average root-mean-square speeds of their molecules at 25°C.

4.51. An unknown gas effuses 1.10 times faster than CO at the same temperature and pressure. a. Is the molar weight of the unknown gas larger or smaller than the molar weight of CO? b. Calculate the molar weight of the unknown gas.

## Real Gases

4.52. Under what conditions do real gases deviate most from ideal behavior? Discuss.

4.53. A sample of $O_2(g)$ is at 100°C and 10 atm pressure. Would the gas act more or less ideally at the following sets of conditions: a. 300°C and 10 atm, b. 100°C and 20 atm, and c. 300°C and 1 atm?

4.54. The compressibility factor of $CF_4(g)$ (tetrafluoromethane gas) at 225 K and 10 atm is 0.8957 (no units). a. Calculate the density of $CF_4(g)$ from the given information. b. Calculated the density of $CF_4(g)$ at the same temperature and pressure assuming it acts as an ideal gas. c. Calculate the percentage difference between the two values.

4.55. The normal boiling point of $N_2$ is -196°C; the normal boiling of water is 100°C. Which gas would act most nearly ideal at 110°C and 1 atm pressure?

4.56. In the van der Waals equation, why is a corrective term added to the measured pressure and subtracted from the measured volume?

4.57. a. Use the van der Waals equation to calculate the pressure exerted by 12.25 g of ethane ($C_2H_6$) in a 0.500 L container at 0°C. b. Compare this pressure to that calculated from the ideal gas law. c. Calculate the percentage difference between the two values.

4.58. Use the van der Waals equation to calculate the pressure exerted by 0.200 moles of a. He and b. $CO_2$ in a 100. mL container at 0°C. Compare the calculated pressures to that calculated for 0.200 moles of an ideal gas and comment on the relative differences.

## Problem Solving

4.59. An air bubble has 10 mL volume 200 feet under water. What volume will it have when it reaches the surface? What does your answer imply for sailors who escape from a crippled submarine?

4.60. Hydrogen gas ($H_2$) can be produced by reacting magnesium metal with a solution of hydrochloric acid according to the following balanced equation: $Mg(s) + 2HCl(aq) \rightarrow MgCl_2(aq) + H_2(g)$. The hydrogen gas is collected by displacing water from a container so that it is saturated with water vapor. A volume of 194 mL of gas was collected at 23°C and 756 mm pressure. What mass of hydrogen was collected?

4.61. The following data table shows some properties of the noble gas elements:

| Gas | Atomic number | BP °C | Liquid Density at the BP, g $mL^{-1}$ | Gas Density at 1 bar and 273 K, g $L^{-1}$ |
|---|---|---|---|---|
| Helium | 2 | -268.90 | 0.12 | 0.18 |
| Neon | 10 | -246.07 | 1.21 | 0.89 |
| Argon | 18 | -185.88 | 1.40 | 1.78 |
| Krypton | 36 | -153.60 | 2.41 | 3.74 |
| Xenon | 54 | -108.06 | 3.05 | 5.84 |
| Radon | 86 | -62.00 | 3.40 | 9.91 |

Make graphs for each of the three listed properties as a function of the atomic weight of each gas.

4.62. Estimate the mass of air in a cubic mile of the atmosphere measured at a pressure of 1.00 atmosphere and a temperature of 298 K.

4.63. Gases released from aerosol cans containing pressure condensed liquids feel cool. Explain why.

4.64. A metal spray can is equipped with a pressure-relief valve. When the valve is briefly squeezed a pssst of gas emerges. Do you have sufficient information to decide if the pressure inside the can has gone up, gone down, or remained constant?

4.65. Isotopes of oxygen can be separated by gaseous effusion of carbon monoxide containing the isotopes. a. Calculate the relative rates of effusion of $^{12}C^{16}O$, $^{12}C^{17}O$ and $^{12}C^{18}O$. b. What is the advantage of using CO rather than $CO_2$ for these separations.

4.66. A quantity of a liquid was placed in a 100.0 mL flask. The flask was then covered by a piece of aluminum foil with a small pinhole in the top, and heated in boiling water at a temperature of 98°C until all the liquid disappeared and the vapor filled the flask. At this point the pressure of the vapor was equal to atmospheric pressure, which was 714 mmHg. The flask was then cooled and weighed. Subtracting the mass of the empty flask (which had been previously determined), gave the mass of the vapor, which was 0.474 g. What is the molecular weight of the unknown liquid? (The method described in this example is the Dumas method for measuring the mole weights of volatile liquids. Long obsolete, in its day it was a very important method.)

4.67. A tv picture tube with a volume of 2.50 L has a pressure of $1.0 \times 10^{-4}$ atm at 23°C. Assume that nitrogen is the only gas in the tube. How many $N_2$ molecules are present? What is the total volume divided by the number of nitrogen molecules in $pm^3$? If you imagine two nitrogen molecules occupying adjacent cubes, how far apart on average are the molecules?

4.68. Pipeline gas is often measured in the units of standard cubic feet (SCF), which in this context means cubic feet of gas measured at 0°C and 1.00 atmospheres pressure. Assuming that a typical pipeline gas is composed of 95.0 mole % methane and 5.0 mole % ethane, calculate the mass in pounds of exactly 1,000 SCF of the gas.

4.69. A mixture of helium and neon is in a 500.0 mL vessel at 25°C and 713 Torr. If the mixture contains 0.045 g of helium what is the mass of neon in the vessel?

4.70. Accurate molecular weights for gases can be measured by making use of the fact that real gases approach ideal behavior at low pressure. A 1.20 g sample of an unknown gas at 99°C gives the following data:

| P(atm) | V (mL) |
|---|---|
| 1.00 | 115.26 |
| 0.75 | 155.89 |
| 0.50 | 237.78 |
| 0.25 | 482.12 |

Use the ideal gas law to calculate the apparent molecular weight at each of the pressures listed. Estimate the gas's molecular weight under hypothetical conditions of zero pressure by making a plot of the apparent molecular weights, versus the pressure. Draw a straight line through the four points and extrapolate the line to zero pressure. The intercept when P = 0 is the limiting molecular weight.

4.71. For a spacecraft or a molecule to leave the Earth, it must reach the escape velocity, which is 11200 m $s^{-1}$. Determine the rms speed of $O_2$ in the upper atmospheric region called the mesosphere, where the temperature is about -80°C. Compare this value to the escape velocity. Would you predict that a very large fraction of $O_2$ molecules would escape under these temperature conditions? How about helium atoms?

4.72. At what temperature do helium atoms have the same average kinetic energy as do argon atoms at 250°C?

4.73. In 1978 the Pioneer Venus orbiter measured the chemical composition and physical conditions of the planet's atmosphere. The Venusian atmosphere contains 96.0 mole % carbon dioxide and 3.5 mole % nitrogen with traces of other gases. The pressure was measured to be 90 atm, and the temperature to be 475°C. Calculate the density of the atmosphere of Venus.

4.74. Use internet or library resources to find out what job is performed by "Maxwell's demon."

4.75. Use the data in Table 4.7 and plot the van der Waals "a" and "b" constants for the four gaseous hydrocarbons, methane, ethane, propane, and butane, using the number of carbon atoms (1,2,3, and 4 respectively) in each of these compounds as the x axis. Comment on the graphs you obtain.

4.76. Rates of diffusion are independent of pressure. Rates of effusion are dependent on pressure. Discuss.

# Chapter 5

# Atomic Spectra and Quantum Mechanics

## Introduction

Looking at the opening section of this chapter you might ask: why the sudden shift of focus from studying gases to studying light? Well, it's time to ask the question "How do atoms work?" It turns out that the best way to "look inside an atom" and examine how they work is to study the way they absorb or emit light. An atom that **absorbs** light gains energy; an atom that **emits** light loses energy. The **energy state** or **energy level** of an atom increases if it absorbs light and decreases if it emits light. A change in the energy of an atom changes the behavior of the electrons in the atom. Because an atom's chemical properties derive from the behavior and properties of its electrons, it follows logically that the study of electrons in atoms is fundamental to chemistry. Equally, the chemical properties of polyatomic ions and molecules derive from the behavior and properties of *their* electrons. However, in this chapter we're going to stick to atoms and monatomic ions.

Light, broken up into its component parts by a prism or grating, yields a spectrum (plural, spectra). Light from atoms generates **atomic spectra**. During the nineteenth century many atomic spectra had been painstakingly analyzed and an abundance of data accumulated. The interpretation of these data had to await the arrival of twentieth century physics and a theory of "how atoms work."

Hydrogen, the simplest atom, consists of just one proton and one electron. Niels Bohr developed the first good theory of electrons in an atom around 1912. In this chapter, like Bohr, we'll begin with hydrogen. Remarkably, it took less than 20 years for Bohr's original conception to be transformed into the bedrock theory of nature called **quantum mechanics**. Quantum mechanics is pretty mathematical, but hydrogen is a sufficiently simple atom to be amenable to a full quantum mechanical treatment. Much of what we'll cover in this chapter will be a non-mathematical account of the quantum mechanical treatment of the hydrogen atom. Once we've got hydrogen nailed down we'll do what chemists have been doing for over 70 years: we'll take things we learn about a hydrogen atom and extend them to the entire periodic table—to all the other elements. If you look at the conventional periodic table you'll see notations along the top such as $s^1$, $s^2$, $s^2d^3$, $s^2p^6$, etc. These are notations derived from hydrogen's quantum mechanics. In this chapter we'll explain where those notations come from.

The grandest goal of the chapter is to show why the periodic table is the way it is: to explain why it has the form that it does, and to explain **periodicity** or periodic relationships. I think the fact that the periodic table actually exists is the best single piece of evidence for the correctness of quantum mechanics—though the totality of the evidence for quantum mechanics is overwhelming.

A note of caution: quantum mechanics is an odd subject. In this chapter you may meet some quite peculiar ideas for the first time. Don't worry—you won't be alone. Albert Einstein himself engaged in debates with Niels Bohr about quantum mechanics. Einstein said, "I shall never believe that God plays dice with the world," speaking about the probabilistic nature of quantum mechanics. But Bohr won the debates. [❁kws +Bohr + Einstein +"Solvay conference" or +"Solvay Institute" +1927]

## Section 5. 1 Electromagnetic Radiation and Spectra

### Electromagnetic Radiation

Light is one type of **electromagnetic radiation**: the propagation of energy through space by oscillating waves of electric and magnetic fields. The entire **electromagnetic spectrum** stretches from radio waves to gamma radiation. Figures 5.1a and 5.1b represent two different electromagnetic waves frozen at a single instant in time:

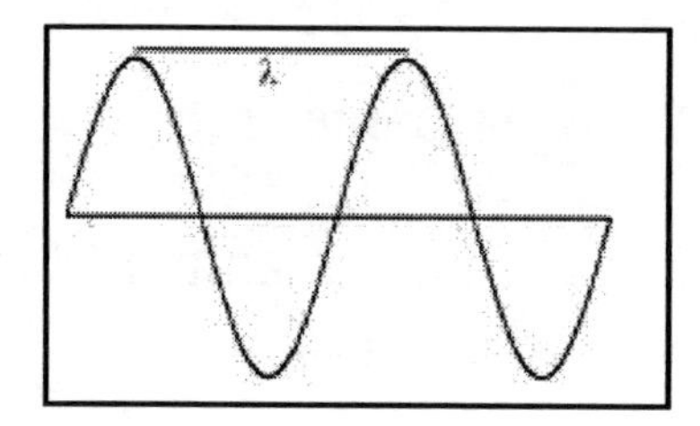


Figure 5.1a Electromagnetic radiation with longer wavelength (λ) and lower frequency (ν).

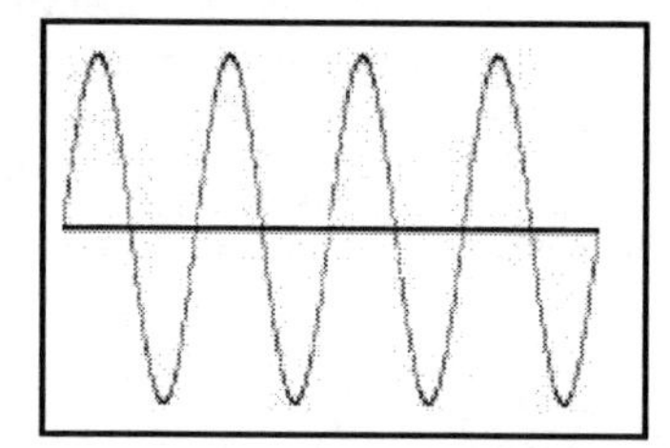

Figure 5.1b Electromagnetic radiation with shorter wavelength (λ) and higher frequency (ν).

The electric and magnetic fields of the electromagnetic wave are perpendicular to one another, and both fields are perpendicular to the wave's direction of propagation. A wave is characterized by a frequency, ν, a wavelength, λ, and an amplitude (or height). **Frequency** (ν) is the number of times per second that the field oscillates as the wave passes any point. **Wavelength** (λ) is the distance between any two successive peaks. All electromagnetic radiation travels through a vacuum at the speed of light, c (= $2.998 \times 10^8$ m $s^{-1}$). Frequency has units of $s^{-1}$, in the SI system the unit $s^{-1}$ is called the hertz (Hz). The frequency of the wave times its wavelength equals the speed of light:

$$\nu\lambda = c$$

Radio waves are one type of electromagnetic radiation. You probably know the broadcast frequencies of several radio stations. Their wavelengths can be calculated from their frequencies as shown in Example 5.1.

**Example 5.1**: Calculate the wavelength of an FM (frequency modulated) radio station broadcasting at 99.5 MHz. 1 MHz = $10^6$ $s^{-1}$.

Strategy: Using the proper units plug into the equation $\nu\lambda = c$. The speed of light = $2.99792458 \times 10^8$ m $s^{-1}$ (exactly); for this problem round the value of the speed of light to four significant figures (that's one more significant figure than the question demands).

$$\text{Frequency, } \nu = 99.5 \times 10^6 \text{ s}^{-1}$$

$$\text{Speed of light, } c = 2.998 \times 10^8 \text{ m s}^{-1}$$

Substituting into $\nu\lambda = c$ gives: $99.5 \times 10^6 \text{ s}^{-1} \times \lambda = 2.998 \times 10^8 \text{ m s}^{-1}$

$$\lambda = \underline{3.01 \text{ m}}$$

Radio broadcast stations typically operate at wavelengths between one and one thousand meters.

Electromagnetic radiation occurs at all frequencies. Radiation with high frequency has a short wavelength and radiation with low frequency has a long wavelength. Figure 5.2 shows the names given to radiation in different regions of the electromagnetic spectrum. Radio waves have low frequencies; x-rays and gamma rays have high frequencies. The human eye is sensitive to electromagnetic radiation in the narrow wavelength range between about $4 \times 10^{-7}$ m to $7 \times 10^{-7}$ m (400 to 700 nm)—the range of visible light. Visible light is labeled "Vis" in Figure 5.2. **Ultraviolet radiation** (UV) lies on the left beyond the violet end of the visible spectrum (<400 nm). **Infrared** (IR) radiation lies beyond the red end (>700 nm) on the right.

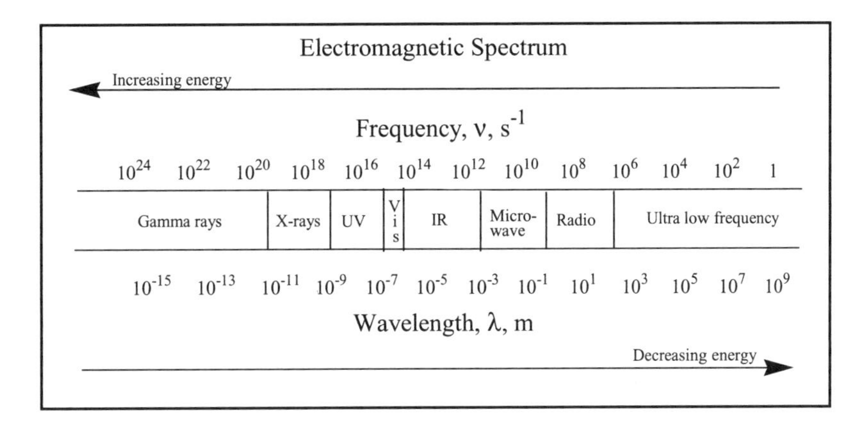


Figure 5.2 The Electromagnetic Spectrum. Electromagnetic radiation occurs naturally across an exceedingly wide range of frequencies, wavelength, and energies.

Figure 5.3 shows the familiar spectrum of white light. It is a **continuous spectrum** and composed of the rainbow of colors: ROYGBIV, or VIBGYOR as presented between 400 nm (blue) and 700 nm (red).

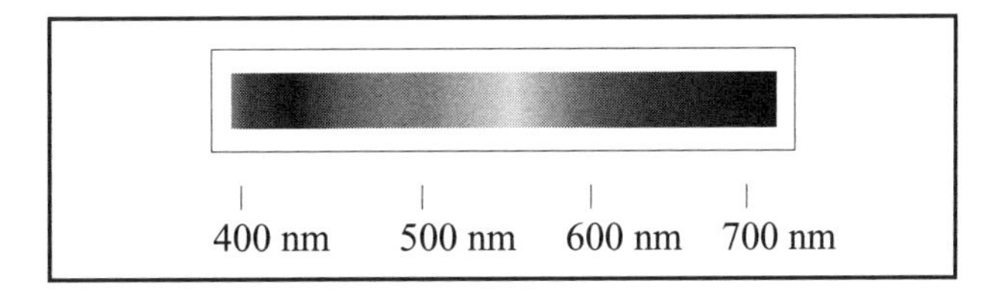


Figure 5.3 The Visible Spectrum. Human eyes are sensitive to an incredibly narrow but crucial strip of the electromagnetic spectrum. Compared with Figure 5.2 the horizontal scale here is expanded by a factor of about $10^{30}$.

White light is broken up into a continuous spectrum of colors by a prism or diffraction grating. **A diffraction grating** is a material (such as a piece of glass, or a plastic film) covered with many closely ruled lines that cause white light to interfere with itself. The result is to form a typical rainbow spectrum. If you own a CD you own a diffraction grating, as demonstrated in Figure 5.4.

Figure 5.4 Ordinary white light is diffracted by a CD, creating a rainbow of colors as described in the text.

The wavelike character of electromagnetic radiation is evidenced by diffraction. **Diffraction** happens when a beam of waves is multiply scattered by objects in its path and the separate scattered waves interfere. **Interference** is a characteristic property of electromagnetic radiation and may be either constructive or destructive. For substantial interference to occur, the size of the disturbing objects must approximate the wavelength of the radiation. The spectrum from the compact disk arises because its pits approximate in size the wavelength of visible light (about 550 nm for the midpoint of the visible spectrum). Figure 5.5 shows how x-ray diffraction from a crystal is caused by in-phase reinforcement and out-of-phase destruction of the x-rays.

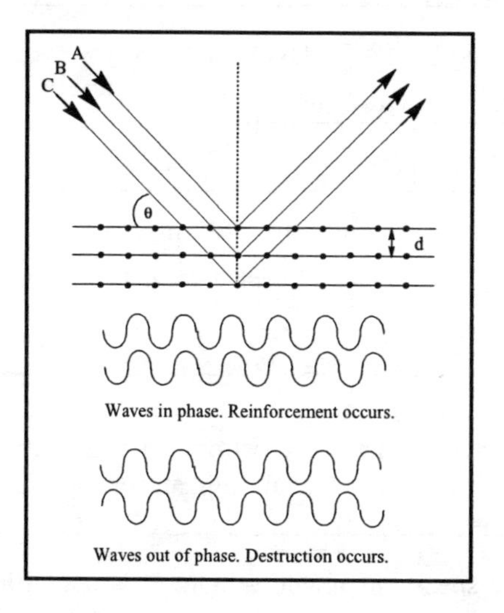


Figure 5.5 Interference between in-phase waves is reinforcing. Interference between out-of-phase waves is destructive. Interference created by an object with lines or rows of atoms is called diffraction and the overall result of a diffraction experiment is called a diffraction pattern.

In aggregate, the way electromagnetic radiation is scattered by an object in its path is called the **diffraction pattern** of the radiation by the object. Later we'll see how chemists use the diffraction pattern of x-rays by crystals to study the structure of molecules. (See also Appendix G.) Visible light containing a narrow range of wavelengths can be selected from a continuous spectrum using a narrow slit that passes only light of a single color. Radiation that contains only a single wavelength is **monochromatic**. Interference or diffraction of monochromatic light produces light and dark regions.

Incidentally, have you ever wondered why it is possible to hear around corners but not to see around them? Sound waves have long wavelengths and are diffracted by objects such as buildings and trees; light waves have wavelengths far too small for the waves to be diffracted by such objects. Hence, sound travels around corners but light does not. (Of course, sound waves are not electromagnetic radiation, but because they are waves they show diffraction patterns.)

## Atomic Spectra

**Spectroscopy** is the study of the absorption and emission of electromagnetic radiation by matter; spectroscopy produces spectra. A **spectroscope** separates electromagnetic radiation by its frequency and wavelength. Spectroscopy has yielded much information about atomic (and molecular structure) and forms the basis for many methods of chemical analysis. For now, we are concerned only with the different types of spectra of atoms. Figure 5.6 shows a simple, hand-held grating spectroscope. Its main features are: (1) a slit where light enters. (2) A diffraction grating to separate the light according to its color or frequency, and (3) A scale marked 4, 5, 6, 7, corresponding to wavelengths of 400 nm, 500 nm, 600 nm, and 700 nm.

The observer looks in through the grating and sees lines of color. Each line is an image of the slit formed by light of a specific wavelength; these images appear as separated lines because light of different wavelengths gets diffracted at slightly different angles. Each line is a different wavelength component of the light under study.

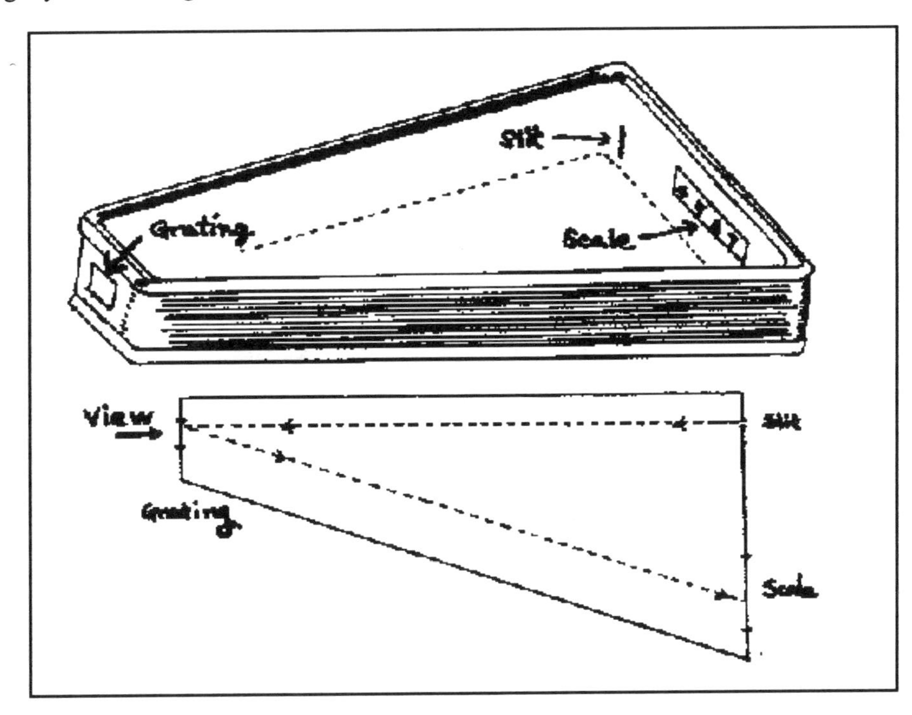


Figure 5.6 It requires only a simple, inexpensive spectroscope to reveal the characteristic colored lines of an atomic spectrum.

Types of Atomic Spectra

In 1752, Thomas Melvill (1726-1753) used a prism to study the colored light generated by metal compounds in flames. He saw brightly colored lines against a dark background; these are **bright line spectra**. Figure 5.7 shows a continuous spectrum (upper) and a bright line spectrum (lower). If one line is dominant, it colors the whole flame. Strontium, for example, generates a brilliant red flame. The ingredient strontium nitrate is used to produce the vivid red color of traffic flares and fireworks. It's the bright yellow-colored sodium spectrum you see when boiling, salted water splashes onto a gas flame. You also see it in yellow "sodium" street lamps.

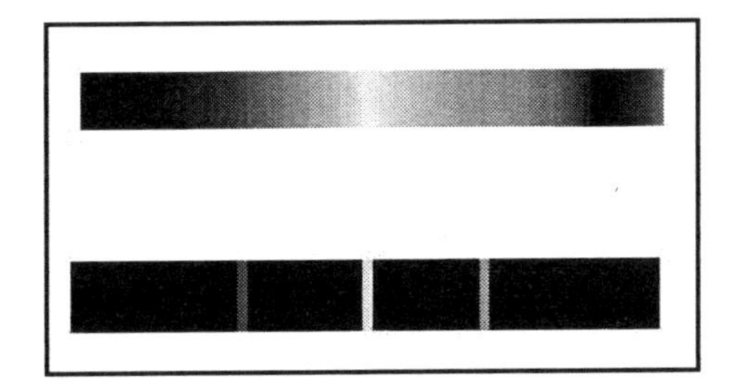

Figure 5.7 Compared with the upper continuous spectrum, the lower bright line spectrum shows only narrow bands (or lines) against a dominantly black backdrop. The fact that atomic spectra show such lines proves that electrons exist in atoms in quantized energy states.

Atoms excited in flames emit characteristic bright colors observable as spectroscopic lines. To **excite** an atom is to supply energy to it; heating in a flame or using an electrical discharge (the method used in a neon sign) are two common ways to excite atoms. The bright lines produced by excited atoms are called **emission spectra**. Trace amounts of dissolved metals in water can be analyzed by injecting the water into a flame and studying the bright line spectra produced by the metals. To get good results, the flame itself should be colorless. For this purpose, Robert Bunsen (1811-1899) invented the Bunsen burner in 1856. Together, Bunsen and Robert Kirchoff (1824-1887) developed an improved spectroscope and used it to discover the elements rubidium and cesium. These and other elements so discovered are listed in the section *Elements discovered from their flame colors 1860 to 1865* in Table 1.4.

An everyday source of bright line spectra are "neon" signs. The atoms of the gas in a neon tube or sign are continually being excited by an electric discharge (current). The excited atoms in the tubes emit their characteristic bright line spectra that the mind integrates and perceives as a single color.

Figure 5.8 Neon signs, such as this one that advertises a Cajun restaurant, are a widespread application of colored light generated by an atomic spectrum. Different filling gasses combined with tube coatings allow a wide range of available colors.

As a photographic positive is to a photographic negative, so a bright line spectrum is to a dark line spectrum. A **dark line spectrum** is a continuous spectrum of visible light interrupted by a pattern of narrow, dark lines. Such a spectrum is shown in Figure 5.9.

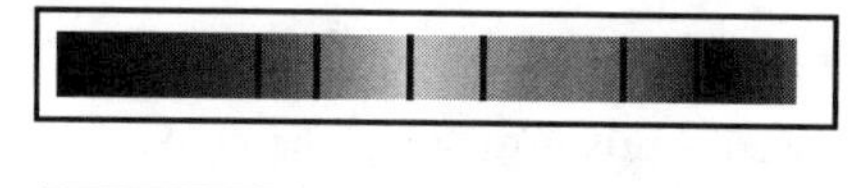

Figure 5.9 A dark line spectrum.

In 1814, Joseph von Fraunhofer (1787-1826) cataloged 574 dark lines in the sun's spectrum. The dark lines are caused by the selective absorption of light of specific wavelengths by the "cold" gas in the sun's outer regions. In 1868 studies of the solar dark lines led to the discovery of helium in the sun before it was known on earth—the only case of an element discovered extraterrestrially. Unexcited atoms of an element absorb at some of the *same* wavelengths that excited atoms of the element emit. The formation of the dark lines in the solar spectrum is shown in Figure 5.10.

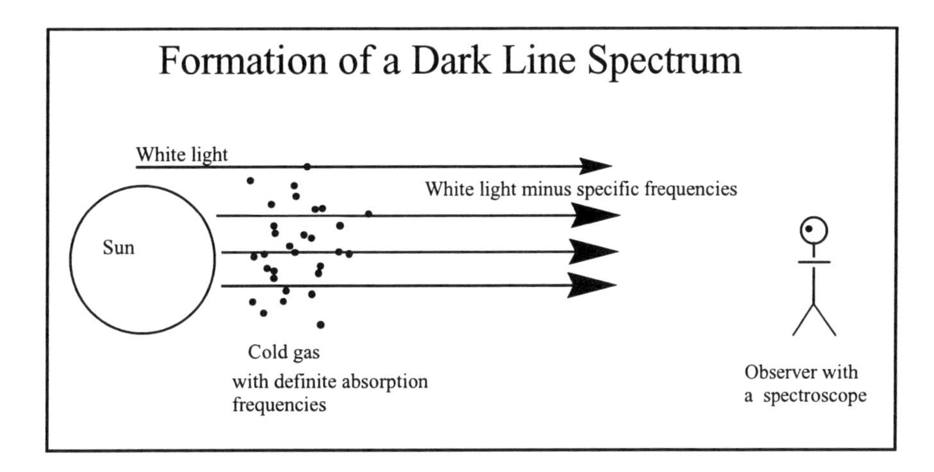


Figure 5.10 The creation of a dark line spectrum. Cold, unexcited atoms between the source of a continuous spectrum and an observer absorb quanta of characteristic wavelengths. In consequence the observer sees the missing quanta as "dark lines." The pattern of dark lines identifies the chemical nature of the cold gas.

Today, the analytical chemistry technique **atomic absorption spectroscopy**, based on the principle illustrated in Figure 5.10, is useful for measuring such things as extremely low level mercury pollution. Pushed to its limit, the technique can find one mercury atom in a sample containing perhaps $10^{13}$ total atoms.

Why then are there bright line and dark line spectra? The spectra arise because of energy levels in atoms (and molecules and ions). Light interacts with those levels in two different ways: via absorption or via emission. Absorption and emission are shown in Figure 5.11.

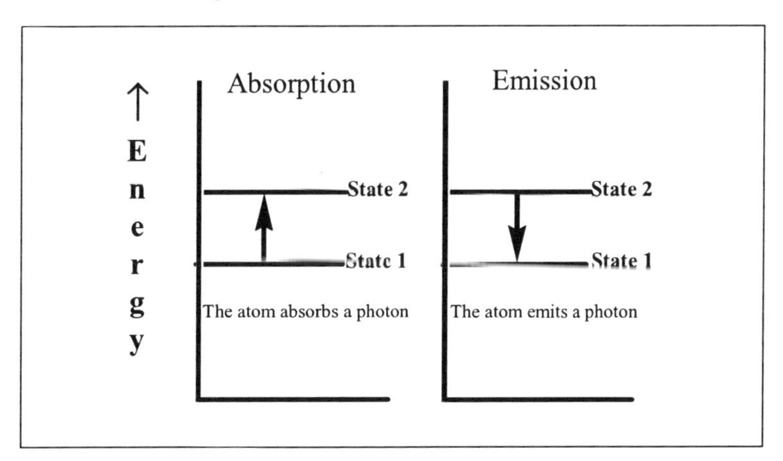


Figure 5.11 Absorption and emission should be seen as fundamentally identical processes that differ only in being the gain versus loss of an identical photon. Both processes are called a transition.

A hydrogen atom with its electron in state 1 in Figure 5.11 is in the ground state. The **ground state** of a hydrogen atom (or its electron) is the one in which the atom has the lowest possible energy. A hydrogen atom with its electron in state 2 is in an excited state. The **excited state** of a hydrogen atom (or its electron) is any state in which the atom has more than the lowest possible energy. There are many, many excited states of a hydrogen atom (in principle an infinite number), all with higher energy than the ground state. Absorption is the process of an atom gaining light energy $\Delta E$, corresponding to $E_1 \rightarrow E_2$. Emission is the process of an atom losing light energy $\Delta E$, corresponding to $E_2 \rightarrow E_1$. The arrows show an electron changing from one energy level to another. When an electron changes levels the atom is said to have undergone a **transition**. Bright line spectra arise from emission of energy: they are **emission spectra**. Dark line spectra arise from absorption of energy: they are **absorption spectra**.

Every atom, molecule, and ion has its own characteristic emission and absorption spectrum. The lines in a spectrum can be used as a sort of "fingerprint" to detect elements in the sample. Studies of absorption and emission spectra not only provide a powerful method of chemical analysis on earth but they also enable astronomers to measure the chemical composition of the stars. By 1872 many atomic spectra were well cataloged.

Studies of **molecular spectroscopy**, the absorption by molecules in various regions of the electromagnetic spectrum, are very important. The experimental techniques are extremely powerful and sensitive, and the large databases of information that now exist make identification of compounds almost routine.

Figure 5.12 shows the (somewhat enhanced) frontispiece of *Fourteen Weeks in Chemistry*, by J. Dorman Steele, published in 1872. Spectrum 1 is a continuous spectrum; spectra 2 and 5 are dark line spectra; and spectra 3, 4, and 6 are bright line spectra. Despite the fact that these are essentially modern spectra, there is no periodic table in Steele's book. Indeed, when this book was published in 1872 Bohr's atomic model and the explanation for the existence of atomic spectra lay 40 years in the future. After a brief diversion, our next step is to examine in detail the bright lines in the hydrogen spectrum.

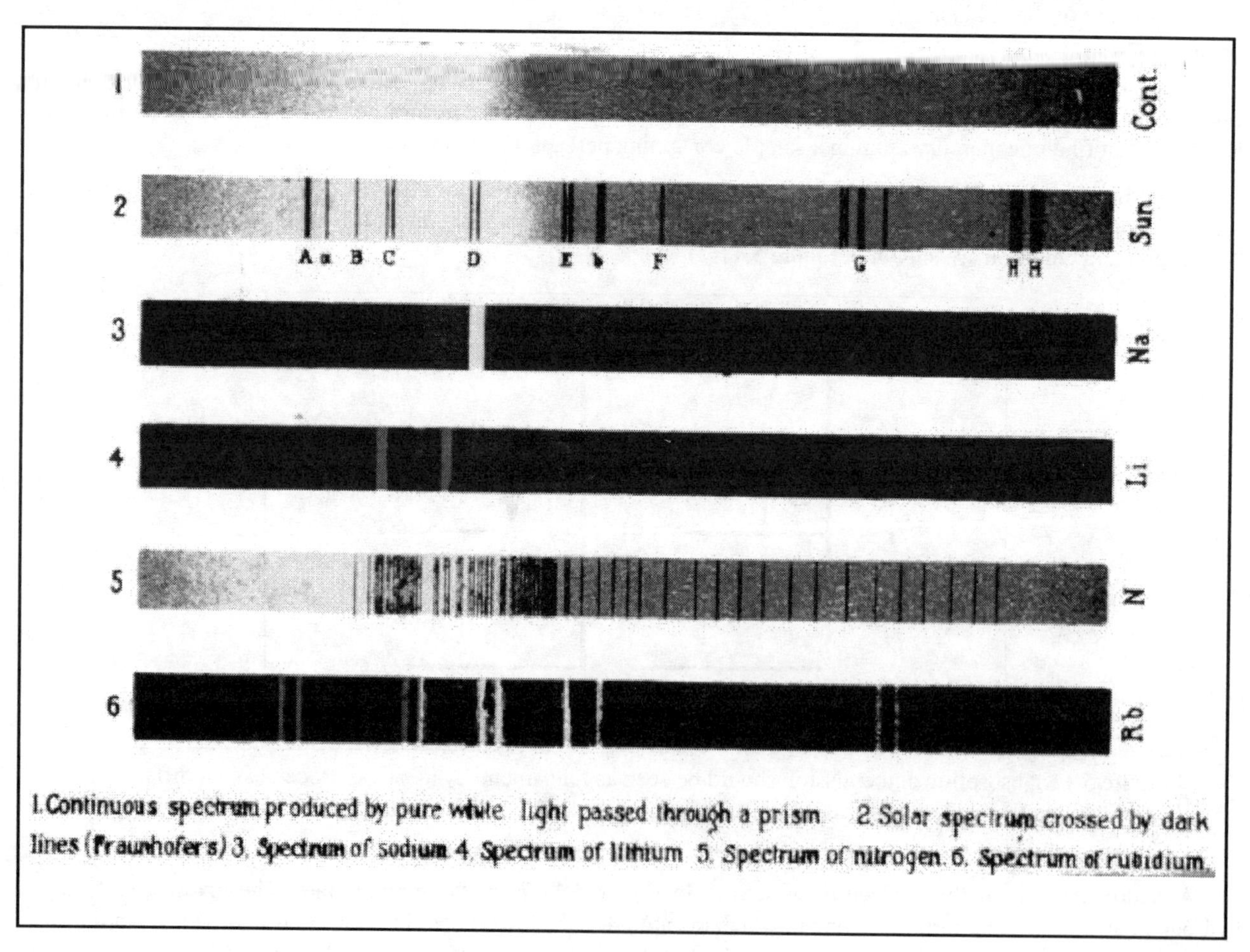


Figure 5.12 Atomic Spectra of various types from 1872.

**Atomic Spectroscopy in the 21st Century—The Atomic Line List**

As noted in the text, line spectra have long been part of astronomy because of studies of starlight. Careful analysis of the spectra of distant stars, and measurements of the wavelengths of their spectral lines, allow astronomers to detect the elements in the stars and the state of excitation of those elements.

As noted, helium was first discovered through studies of the dark lines in sunlight. In Appendix D it's noted that technetium, an element all but unknown on earth, can be shown by atomic spectroscopy to exist in the sun and other stars.

The Atomic Line List [❁kws "the atomic line list"] is a web based compilation of over 900,000 atomic transitions cataloged by wavelength with $\lambda$ ranging from $10^{-10}$ m to $10^{-3}$ m. The wavelengths in the List are entirely calculated via quantum mechanics and are mostly based on atomic energy level data taken from the US National Institute of Standards and Technology (NIST) Atomic Spectroscopic Database.

To use the List an astronomer enters information into a selection criteria data form that includes such items as the estimated experimental uncertainty of the measurement. In this way, unidentified lines in an experimental stellar line spectrum can be assigned to a specific transition of a specific element in a specific state of ionization.

## Section 5.2 The Hydrogen Spectrum and Energy Quantization

### The Hydrogen Spectrum

Because atoms have many energy levels they produce a wealth of lines in their spectra. Sometimes thousands of lines can be detected from a single, glowing, gas-filled tube. By 1890, after much trial and error, a general formula had been discovered to reproduce the experimental frequencies of the lines in the hydrogen spectrum:

$$\nu = R_H\left(\frac{1}{n_1^2} - \frac{1}{n_2^2}\right)$$

where $R_H$ is called the Rydberg constant and is equal to $3.290 \times 10^{15}$ s$^{-1}$, and $n_1$ and $n_2$ are positive integers (1,2,3,4…), with $n_1 < n_2$. This formula was the work of many scientists, but perhaps most credit for its discovery goes to J. J. Balmer (1825-1898) and J. R. Rydberg (1854-1919). We'll call it the **Balmer-Rydberg equation**. The Balmer-Rydberg equation is an **empirical equation**—one developed to fit experimental data. Its use is shown in the Example 5.2.

**Example 5.2**: Calculate the frequency and the wavelength (in nanometers) of the lines in the hydrogen spectrum when $n_1 = 1$, and $n_2 = 4$. In what region of the electromagnetic spectrum is this transition found?

Strategy: Use the Balmer-Rydberg formula to get the frequency in s$^{-1}$. Then use $\nu\lambda = c$ to calculate the wavelength. The Rydberg constant is $3.29 \times 10^{15}$ s$^{-1}$.

$$\nu = R_H\left(\frac{1}{n_1^2} - \frac{1}{n_2^2}\right)$$

$$\nu = 3.29 \times 10^{15}\left(\frac{1}{1^2} - \frac{1}{4^2}\right) \text{ s}^{-1}$$

$$\text{Frequency} = 3.084 \times 10^{15} \text{ s}^{-1}$$

$$\lambda = \frac{c}{\nu} = (2.998 \times 10^8 \text{ m s}^{-1} \times 10^9 \text{ nm m}^{-1}) \div (3.084 \times 10^{15} \text{ s}^{-1}) = \underline{97.21 \text{ nanometers}}$$

If you look at the electromagnetic spectrum you'll see that this transition is in the ultraviolet region.

If $n_1$ is fixed and $n_2$ ranges over positive integer values, the set of frequencies obtained from the Balmer-Rydberg formula for all possible values of $n_2$ is called a **series** of spectral lines. Those series with $n_1$ taking values of 1-5, and various values of $n_2$, are named after scientists, as listed in Table 5.1 below:

Table 5.1 **Named Series of Lines in the Hydrogen Spectrum**

| $n_1$ | $n_2$ | Named after | Spectrum region |
|---|---|---|---|
| 1 | 2,3,4,5... | Lyman | Ultraviolet |
| 2 | 3,4,5,6... | Balmer | Visible |
| 3 | 4,5,6,7... | Paschen | Infrared |
| 4 | 5,6,7,8... | Brackett | Infrared |
| 5 | 6,7,8,9... | Pfund | Infrared |

In addition to hydrogen itself, with minor modification the Balmer-Ryberg equation can be applied to any ion that has just one electron. So that means any atom that has a positive charge of one less than its atomic number. For example, ordinary carbon forms the one-electron $C^{5+}$ ion in stars via the reaction: $C \rightarrow C^{5+} + 5\ e^-$.

We next see how these series of lines in the hydrogen spectrum were explained.

## Energy Quantization

The quantum-of-energy concept that arose at the dawn of the 20th century was a truly revolutionary idea that changed all branches of science. Quantization is a common concept. Anything that can be expressed in terms of measurable increments is said to be **quantized;** the smallest increment is a quantum (plural, quanta). A **quantum** of energy is a single unit or particle of energy. It still surprises us today that energy really is quantized. It seems reasonable to us that atoms are quantized—we know that matter comes as tiny lumps—but energy? When we accelerate smoothly away in our car, the tiny energy jumps that accompany the pick up of speed are utterly impossible to notice, but they are there!

Energy quantization was first announced in 1900 by Max Planck (1858-1947) to explain the spectrum of electromagnetic radiation coming from a black surface. Classical (pre-quantum) physics predicted that the energy spectrum of a black surface should show increasing energy at increasing frequency—a prediction that was catastrophically wrong. When Planck took the approach that energy exchange with the surface is quantized, rather than continuous, he was able to develop an equation that exactly described the experimental situation. This equation introduced energy increments ($h\nu$), that he named quanta (singular, quantum). He didn't much like his own explanation, but it matched the facts. The constant, h, is Planck's constant. It is a fundamental constant of nature and equals $6.626 \times 10^{-34}$ J s (or J $Hz^{-1}$). [❁kws"catastrophe" +"ultra violet" or +ultraviolet or + ultra-violet]

In 1905, Albert Einstein (1879-1955) realized that electromagnetic radiation itself is quantized and that energy quantization explains the photoelectric effect. The **photoelectric effect** occurs when light falling on a suitable surface causes electrons to be ejected. Practical uses of the photoelectric effect include light-meters for cameras, night vision goggles, and charge capture devices (CCDs), such as those in a modern digital camera. It takes light of sufficient energy to eject electrons from a surface; often, what can be done by a single blue photon can't be done by an infinite supply of red photons. A **photon** (*photo* + electr*on*) is a quantum of electromagnetic energy that has both wave and particle characteristics. Einstein realized that the energy of a photon is proportional to its frequency and stated the fundamental equation that relates energy and frequency:

$$E = h\nu$$

The equation $E = h\nu$ provides the theoretical basis for spectroscopy. The energy change of an atom or molecule is directly related to the energy of the photon that is absorbed *or* emitted. We can show that relationship by the equations $E_{photon} = |\Delta E_{atom}|$ or $E_{photon} = |\Delta E_{molecule}|$. The use of absolute value bars allows the same equation to stand for *both* absorption processes and emission processes. $\Delta E_{atom}$ is negative for an emission process and positive for an absorption process.

Light had long been known to be a wave phenomenon. However, Einstein's work on the photoelectric effect established that in some experiments light behaves like a beam of particles. Anything that exhibits both wave and particle aspects is said to show **wave-particle duality**. In 1905 it came as a surprise that photons have both wave-like and particle-like qualities. Remarkably, only 20 years later came the reverse concept: that particles have waves associated with them called **matter-waves**. We'll return to matter-waves shortly. The use of the equation $E = h\nu$ is shown in the following example:

**Example 5.3**: A mercury lamp has an ultraviolet line in its spectrum at 254 nm. Calculate the energy of one quantum of light with this wavelength, and one mole of photons with this wavelength.

Strategy: Use $\lambda\nu = c$ to convert wavelength to frequency. Then plug into $E = h\nu$. Use Avogadro's number to convert to moles in the second step.

$$\nu = \frac{c}{\lambda} = 2.998 \times 10^{8}\ \text{m s}^{-1} \div (254\ \text{nm} \div 10^{9}\ \text{nm m}^{-1})$$

$$\text{Frequency} = 1.18 \times 10^{15}\ \text{s}^{-1}$$

$$\text{Plugging into } E = h\nu,\ E = 6.626 \times 10^{-34}\ \text{J s} \times 1.18 \times 10^{15}\ \text{s}^{-1}$$

$$\text{Thus, for a single photon, } E = \underline{7.82 \times 10^{-19}\ \text{J}}$$

For one mole, we multiply that answer by Avogadro's number

$$\text{For a mole of photons, } E = 7.82 \times 10^{-19}\ \text{J} \times 6.022 \times 10^{23}\ \text{mol}^{-1}$$

$$= 4.71 \times 10^{5}\ \text{J mol}^{-1}$$

$$= \underline{471\ \text{kJ mol}^{-1}}$$

If you review bond energies in Table 2.6 you'll see that 417 kJ $mol^{-1}$ is greater than the bond energy of a C-C bond (368 kJ $mol^{-1}$). Hence, ultraviolet light from the sun's rays contains sufficient energy to break such bonds and thus degrade organic materials such as rubber, cotton, and paper. Likewise, UV light is dangerous to people because it breaks chemical bonds in their biological molecules—it causes molecular injury. Photon energies across the entire electromagnetic spectrum are shown in Table 5.2.

Table 5.2 **The Energy and Frequency of Quanta**

| Radiation | $\lambda$(m) | $\nu(s^{-1})$ | $E_{photon}$ (J) | $E_{mol}$ (kJ $mol^{-1}$) |
|---|---|---|---|---|
| radio-frequency | 10 | $3.0 \times 10^{7}$ | $2.0 \times 10^{-26}$ | $1.2 \times 10^{-5}$ |
| microwave | 0.01 | $3.0 \times 10^{10}$ | $2.0 \times 10^{-23}$ | $1.2 \times 10^{-2}$ |
| infrared | $10^{-5}$ | $3.0 \times 10^{13}$ | $2.0 \times 10^{-20}$ | 12 |
| red light | $7 \times 10^{-7}$ | $4.3 \times 10^{14}$ | $2.8 \times 10^{-19}$ | 190 |
| blue light | $4 \times 10^{-7}$ | $7.5 \times 10^{14}$ | $5.0 \times 10^{-19}$ | 300 |
| ultraviolet | $10^{-7}$ | $3.0 \times 10^{15}$ | $2.0 \times 10^{-18}$ | 1,200 |
| x-ray | $10^{-10}$ | $3.0 \times 10^{18}$ | $2.0 \times 10^{-15}$ | $1.2 \times 10^{6}$ |
| gamma | $10^{-22}$ | $3.0 \times 10^{30}$ | $2.0 \times 10^{-3}$ | $1.2 \times 10^{21}$ |

Table 5.2 shows that visible and UV light energies match the range of chemical bond energies (100-1100 kJ $mol^{-1}$). At 190 kJ $mol^{-1}$ red light is comparable to a weak single bond, at 300 kJ $mol^{-1}$ blue light is comparable to a fairly strong single bond, and at 1200 kJ $mol^{-1}$ ultraviolet light is capable of breaking even the strongest triple bond. Thus, while white light is essentially harmless to human skin, the higher frequency ultraviolet waves in sunlight are damaging. The enormous energy of x-ray (or gamma) photons or rays is extremely damaging. A single x-ray photon has sufficient energy to break thousands of bonds, and a single gamma photon sufficient energy to break millions. To avoid radiation damage, people must limit their exposure to x-rays and gamma rays. Microwave photons lack sufficient energy to break chemical bonds, but they do cause molecules to rotate and vibrate faster. Energy going to these increased motions quickly manifests itself as heat—that's the reason that a microwave oven works.

As another example of the use of the fundamental equations that govern light energy, and to illustrate the units involved, let's calculate the approximate number of photons generated by an ordinary light bulb. That's done in Example 5.4.

**Example 5.4**: An ordinary 100 watt light bulb consumes 100 J $s^{-1}$ of electrical energy. Of this, only 2.5 watts, or 2.5 J $s^{-1}$, appear as visible light—the rest is "wasted" as heat. Assuming that the average wavelength of visible light is 570 nm, calculate how long the bulb must burn to generate a mole of photons.

Strategy: Use $\nu\lambda = c$ together with $E = h\nu$ to calculate the amount of energy in one mole of photons with $\lambda = 570$ nm. Then calculate how long it takes the bulb to deliver this much light energy.

$$\nu = \frac{c}{\lambda}, \quad \text{so } \nu = c \div \lambda, \quad = 2.998 \times 10^8 \text{ m s}^{-1} \div (570 \text{ nm} \div 10^9 \text{ nm m}^{-1}) = 5.3 \times 10^{14} \text{ s}^{-1}$$

$$\text{Energy per mole} = h\nu \times \text{Avogadro's number}$$

$$= 6.626 \times 10^{-34} \text{ J s} \times 5.3 \times 10^{14} \text{ s}^{-1} \times 6.022 \times 10^{23} \text{ mol}^{-1}$$

$$= 2.1 \times 10^5 \text{ J mol}^{-1}$$

$$\text{Time required} = \frac{\text{energy needed for one mole}}{\text{rate of energy delivery}} = (2.1 \times 10^5 \text{ J mol}^{-1}) \div (2.5 \text{ J s}^{-1} \times 3600 \text{ s hr}^{-1})$$

$$= \underline{23 \text{ hours}}$$

It takes almost a day for a 100 watt light bulb to produce a mole of photons.

## Section 5.3 Electron Theories of Atomic Behavior

### Bohr's Atomic Model

The first electron theory of how atoms work was developed by Niels Bohr (1885-1962) in 1912, shortly after he arrived in Manchester, England to study with Ernest Rutherford. If there ever was a case of the right man being in the right place at the right time, it was Bohr in Manchester in 1912. Bohr had a fresh Ph.D. in physics and had studied the mathematics of water jets for his thesis. He was well prepared for what lay ahead.

There were three key foundation stones on which Bohr would build his theory: (1) the atomic nucleus, which had been discovered just one year earlier in Rutherford's own laboratory; (2) the idea of the quantum which had been applied to light particles only seven years earlier by Albert Einstein; (3) the vast body of accumulated data about the spectra of atoms, especially the extensive and detailed data for the spectrum of hydrogen.

Bohr was the first person to recognize two key facts about the hydrogen atom which you've already been told: (1) that it can exist only with specific amounts of energy (i.e. its energy is quantized), and (2) hydrogen's spectrum forms when hydrogen atoms change from one energy state to another by absorbing or emitting a single photon during the transition.

Bohr developed a mathematical treatment that applied the quantum idea to Rutherford's nuclear atom. With this treatment (the so-called **Bohr model** of the atom), he was able to explain the Balmer-Rydberg formula for the lines in the hydrogen atom spectrum and to calculate a value for the Rydberg constant. Bohr's **atomic model** was received with acclaim and universal acceptance.

To show that an atom can only exist in definite, quantized energy states we use an **energy level diagram**, as shown in Figure 5.13. Recall, we described the general features of an energy level diagram in connection with Figure 2.8. There are no diagrams in Bohr's seminal 1913 paper, but this diagram is implied. The energy states or levels are numbered starting at the bottom as $n_1$, $n_2$, $n_3$, $n_4$, $n_5$, $n_6$.... Observe that at the bottom the spacing of the energy levels is the greatest, and that the spacing quickly shrinks and finally reaches a continuum. In a **continuum** there are an enormous number of energy states squashed together. That's why we don't notice quantum jerks when we push on the accelerator of our car. The energy gap $\Delta E$, corresponds to $E_2 - E_1$, the energy difference between the two energy levels. In the transition from level $n_2 \rightarrow n_1$ or from energy level $E_2 \rightarrow E_1$ the atom loses energy and the environment gains energy; it's an emission. Absorption is the reverse process.

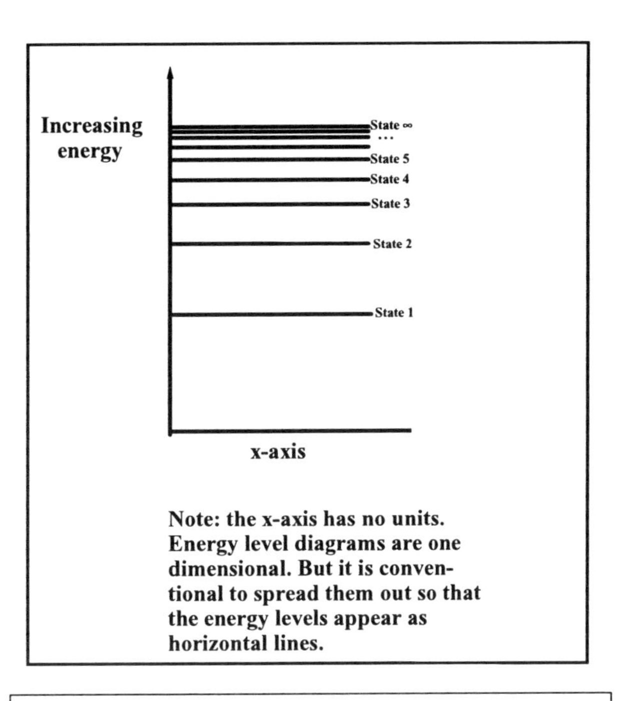


Figure 5.13 Bohr's energy level diagram for hydrogen.

In Figure 5.14 you see how hydrogen's energy level diagram can be used to interpret the empirical Balmer-Rydberg equation. If you study the diagram you'll see that $n_1$ in the equation represents the lower energy level and that $n_2$ represents the higher energy level. The values $n_1$ and $n_2$ are the **quantum numbers** of hydrogen—though in 1890 they were not recognized as such because there was as yet no quantum theory.

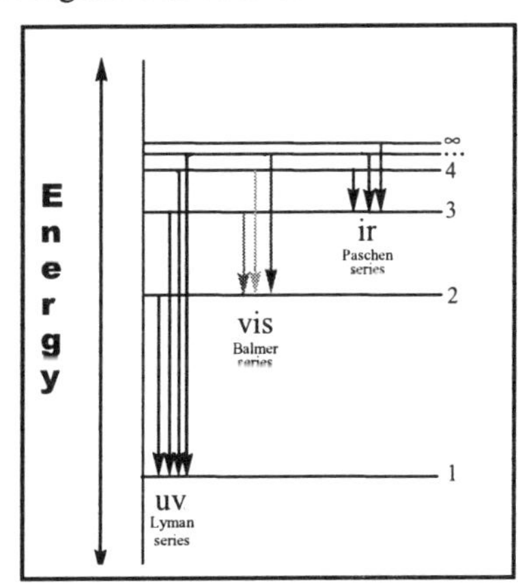


Figure 5.14. The Bohr explanation for the empirical Balmer-Rydberg equation, $\nu = R_H\left(\frac{1}{n_1^2} - \frac{1}{n_2^2}\right)$. Bohr's quantum numbers provided a theoretical explanation for the empirical quantities $n_1$ and $n_2$ in this equation by interpreting them as quantum numbers.

Bohr assumed that the electron in a hydrogen atom behaves as a particle and travels in one of many definite circular orbits, each with a specific energy. The formula for the energies of the orbits is

$$E_n = -R_H \frac{h}{n^2}$$

where $E_n$ is the energy of the nth energy state ($n = 1, 2, 3, 4 \ldots \infty$), h is Planck's constant, and $R_H$ is the Rydberg constant. The quantity n is the **principal quantum number**. Observe that $E_\infty = 0$ and because all the other energy states are lower than this they have negative values.

Immediate proof of this radically new view of atoms came when Bohr calculated the value of the Rydberg constant from the mass and charge of the electron and Planck's constant. Depending on whether the energy is regarded either from the point of view of the atom or the environment it will be positive or negative.

The energy change during a transition is $\Delta E = E_2 - E_1$. Obviously, the largest possible value of $\Delta E$ is for the transition $E_1 \rightarrow E_\infty$. That transition corresponds to the ionization reaction $H \rightarrow H^+ + e^-$. So $\Delta E$ in this case is ionization energy of hydrogen. Using $\Delta E = E_2 - E_1$ in conjunction with the equation $\Delta E = h\nu$ leads to the theoretical derivation of the Balmer-Rydberg equation:

If, $$E_2 = -R_H \times \frac{h}{n_2^2} \quad \text{and} \quad E_1 = -R_H \times \frac{h}{n_1^2}$$

then, $$h\nu = \Delta E = E_2 - E_1 = -R_H \times h \times \left(\frac{1}{n_2^2} - \frac{1}{n_1^2}\right)$$

and, $$\nu = R_H\left(\frac{1}{n_1^2} - \frac{1}{n_2^2}\right)$$

Not only did Bohr derive the empirical Balmer-Rydberg equation, he also calculated $R_H$, the Rydberg constant, and got close to the experimental value of $3.290 \times 10^{13}\ s^{-1}$ with a calculation of $3.1 \times 10^{13}\ s^{-1}$. Later calculations, using better values of the electron's charge and mass and a better value of Planck's constant, gave four significant figure agreement with $3.290 \times 10^{13}\ s^{-1}$. Writing at the time, Bohr said "The agreement between the theoretical and observed values is inside the uncertainty due to experimental errors in the constants entering into the expression for the theoretical value." He was quite correct. (The general topic of experimental uncertainty is the subject of Appendix B.). Let's use the above equation in Example 5.5 to calculate the ionization energy of hydrogen.

**Example 5.5**: Calculate the ionization energy of hydrogen in kJ $mol^{-1}$.

Strategy. Substitute into the Balmer-Rydberg equation the quantum numbers 1 and $\infty$. Convert that answer to energy in the desired units.

The needed equation is $$\Delta E = R_H \times h \times \left(\frac{1}{n_2^2} - \frac{1}{n_1^2}\right) \text{ per photon}$$

Convert to a molar basis by multiplying by Avogadro's number.

$$\Delta E = 3.290 \times 10^{15}\ s^{-1} \times 6.626 \times 10^{-34}\ J\ s \times 10^{-3}\ kJ\ J^{-1} \times \left(\frac{1}{1^2} - \frac{1}{\infty^2}\right) \times 6.022 \times 10^{23}\ mol^{-1}$$

$$= \underline{1313\ kJ\ mol^{-1}}$$

The value 1313 kJ $mol^{-1}$ is *also* the experimental value of hydrogen's ionization energy. By 1913 Bohr had already calculated a value of 1300 kJ $mol^{-1}$ from his atomic theory, although he stated it as 13 eV, where eV is the older energy unit called the electron-volt.

In summary, because hydrogen atoms have definite energy levels, the atoms can absorb and emit light of specific frequencies. The designations of the energy levels are the principal quantum numbers of hydrogen.

The Bohr atomic model was an important step: it brought science truly into the twentieth century. But succeeding events moved swiftly and Bohr's model was soon swallowed up and integrated into the broader theory of quantum mechanics. The Bohr model is successful only for hydrogen and, even there, it cannot explain why some lines in the hydrogen spectrum are bright and others faint, or why additional lines appear when the atoms are studied in a magnetic or electric field. It was the need to understand and explain such details that motivated the development of quantum mechanics.

## Matter Waves

A **matter wave** is observed when a particle shows wave-like behavior such as interference or diffraction effects. Matter waves are an example of wave-particle duality. Quantum mechanics treats electrons inside atoms as matter waves.

Because its most familiar properties are particle-like, the electron was at first thought to be just a particle. But in 1924, Louis Victor de Broglie (1892-1960) realized that electrons, and indeed all particles, must have wave properties. He stated a mathematical relationship between the particle's wavelength and its mass and velocity. For a particle of mass m, traveling at velocity v, its wavelength λ is $\lambda = \frac{h}{mv}$. That's the de Broglie relationship.

To correctly derive the de Broglie relationship is beyond our scope, but we can get a little insight into the situation by combining two familiar equations for a photon: E = hν and E = mc$^2$. So, obviously mc$^2$ = hν, and, rearranging, $\frac{c}{\upsilon} = \frac{h}{mc}$ and, because $\frac{c}{\upsilon} = \lambda$, we get $\lambda = \frac{h}{mc}$. From here, de Broglie now postulated that for a particle that moves at less than the speed of light, c can be replaced with v (the speed of the particle). Doing this yields his relationship: $\lambda = \frac{h}{mv}$.

Thinking about hydrogen's electron as a matter wave helps us to rationalize why the energy states of the hydrogen atom are quantized. Loosely speaking, an electron *inside an atom* behaves as a wave that must "meet up with itself" around the nucleus.

**Historical Development of Matter Waves**

By a strange irony of history, experimental evidence for electron waves had been obtained even before de Broglie derived the relationship that bears his name. In 1921, the American scientist Clinton Davisson (1881-1958), working at Bell Telephone Laboratories (now Lucent Technologies), began a series of experiments to study the scattering of a beam of electrons from a nickel surface. By 1925 Davisson had been joined in his work by Lester Germer, and in April of that year they found that an electron beam showed strong scattering at some angles and weak scattering at others. They had observed a diffraction pattern created by the interference of matter waves. They had no immediate explanation for their observations. The diffraction pattern was puzzling enough, but even more inexplicable was the fact that the pattern shifted as they changed the velocity of the electron beam. In 1928, at a scientific meeting in London, Davisson heard about de Broglie's hypothesis. A quick calculation established that the shifting diffraction pattern precisely followed the de Broglie prediction that the electron's wavelength depends on its velocity.

De Broglie was awarded the 1929 Nobel Prize for physics. Davisson received the 1939 Prize, shared with G. P. Thompson, who also made an early demonstration of electron diffraction. Amusingly, as mentioned earlier whereas G. P. Thomson won the Nobel Prize for proving the electron is a wave, his father, J. J. Thomson, earlier won the same Prize for proving the electron is a particle.

Matter waves don't have much effect in the everyday world. Things in our everyday lives have large masses, so their wavelengths are far too small to be detected. Matter waves for electrons in atoms have wavelengths comparable to the size of atoms, as shown in Example 5.6. Today, electron matter waves are routinely put to use for electron microscopes, for creating images of microelectronic devices, etc.

**Example 5.6**: Calculate the de Broglie wavelength of an electron traveling at 10$^7$ m s$^{-1}$. This velocity was typical of those used by Davisson and Germer in their electron diffraction experiments described in the "historical development" box. The mass of the electron is 9.11 × 10$^{-31}$ kg.

Strategy: Substitute into the de Broglie relationship. $\lambda = \frac{h}{mv}$

$$\lambda = \frac{6.626 \times 10^{-34}\ \text{J s} \times 10^{12}\ \text{m}^{-1}\ \text{pm}}{9.11 \times 10^{-31}\ \text{kg} \times 10^{7}\ \text{m s}^{-1}} = \underline{72.7\ \text{pm}}$$

72.7 pm is comparable to the size of a hydrogen atom. So the answer is reasonable.

The ideas of de Broglie firmly established that a wave-based theory would be needed to account for atomic spectra and atomic behavior. That theory is wave or quantum mechanics.

> What's that? You don't think quantum mechanics has anything to do with engineering? Try [❁kws +engineering +"quantum mechanics"]. Doing that will get you to many course descriptions for upper level courses in materials, electrical and electronic, chemical, and optical engineering. Quantum mechanics provides the theory for modern semiconductor, opto-electronic, and photonic devices.

Below is a summary of the development of quantum mechanics and its applications to chemistry.

**A Brief Historical Context—Quantum Mechanics and Chemistry**

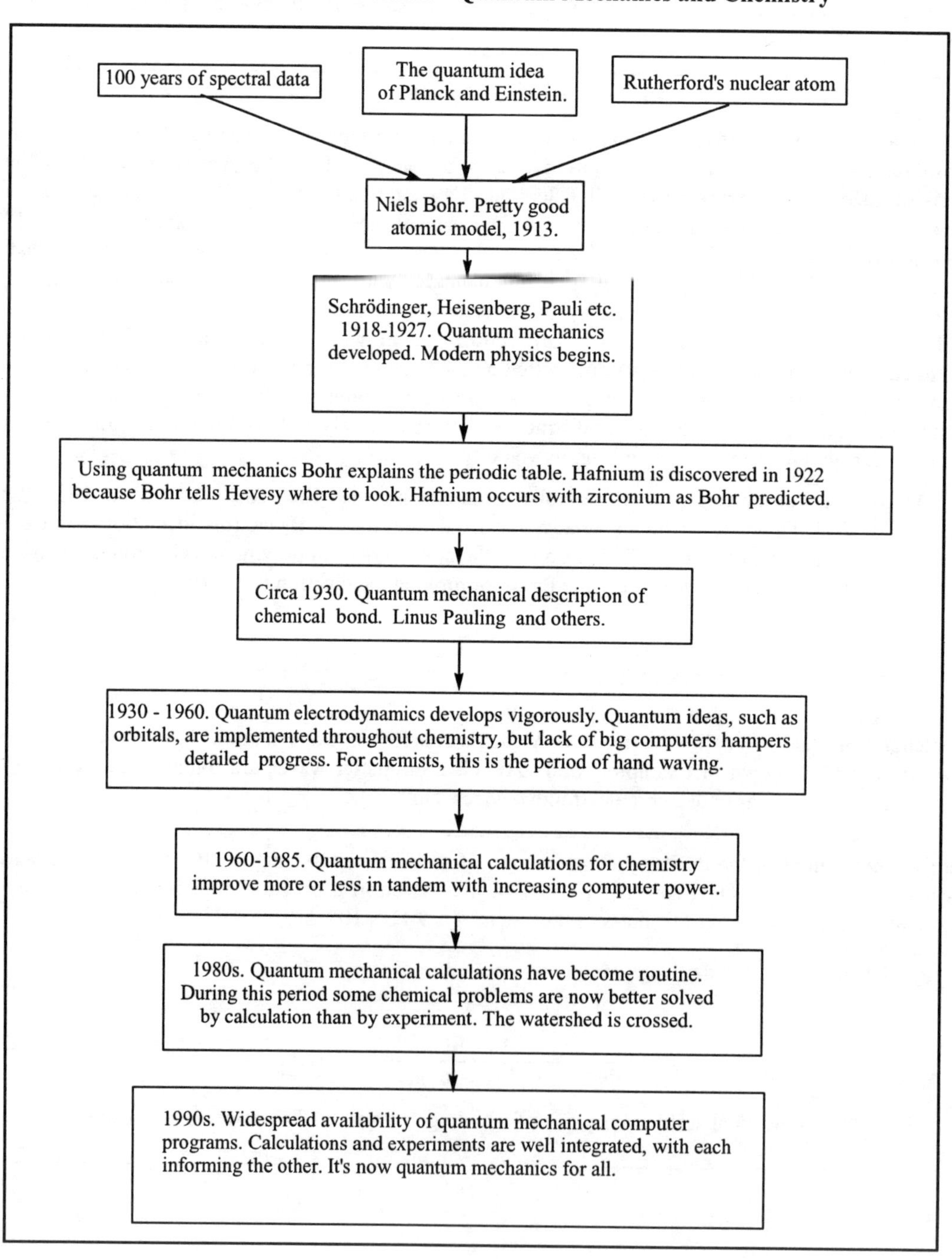

## Quantum Mechanics

Once Bohr's model was in place, and matter waves were accepted, the path forward opened. From 1924-1926 the theory of quantum mechanics arose out of the Bohr model by a combination of theoretical development and ever more detailed analysis of atomic spectra. Like all scientific theories, quantum mechanics was developed to interpret experiments. Formally, quantum mechanics was introduced by Erwin Schrödinger (1887-1961) in 1926 when he described the wave equation.

By roughly 1930, quantum mechanics was already able to give an essentially complete account of electrons in atoms. The theory remains today essentially as it was first formulated. In the beginning, the main objective of quantum mechanics was to explain the spectra of the elements, but its applications turned out to be far, far wider than that. **Quantum mechanics** is a mathematically-oriented, wide-ranging physical theory of nature; its correctness is not in doubt. Any measurable property of a hydrogen atom or hydrogen molecule can be calculated to within the limit of experimental uncertainty.

For chemists, quantum mechanics explains the behavior of electrons in atoms, molecules, and ions. So quantum mechanics is absolutely fundamental to chemistry. For many years, only the hydrogen atom could be correctly treated by quantum mechanics—and that is the reason why much of modern chemical terminology derives from hydrogen. Quantum mechanics enables chemists to "picture" the structure of the hydrogen atom and, by extension, all other atoms. Anyone who wishes to understand chemistry deeply must be familiar with its quantum mechanical underpinnings.

In the past 20 years, rapidly growing computer power has made quantum mechanical calculations accessible to many chemists. Today we live in a time of "quantum mechanics for all." In Chapter 6 we'll discuss how quantum mechanics explains chemical bonding.

The behavior of the electron in a hydrogen atom is described by a three-dimensional electron standing wave. This mathematical representation is Schrödinger's **wave equation**. For a hydrogen atom the equation is:

$$\frac{\partial^2\psi}{\partial x^2} + \frac{\partial^2\psi}{\partial y^2} + \frac{\partial^2\psi}{\partial z^2} + \frac{8\pi^2 m}{h}\left(E + \frac{e^2}{r}\right)\psi = 0$$

The solutions to Schrödinger's equation for the hydrogen atom are mathematical functions that can be interpreted as standing (or stationary) waves. A **standing wave** is one not going anywhere; one confined to a definite region in space. A plucked guitar string and its family of overtones are standing waves. You might have seen two dimensional standing waves on the surface of water. Such waves often form if a water glass is exposed to vibrations, for example, if the glass is placed on the hood of a car with its motor running.

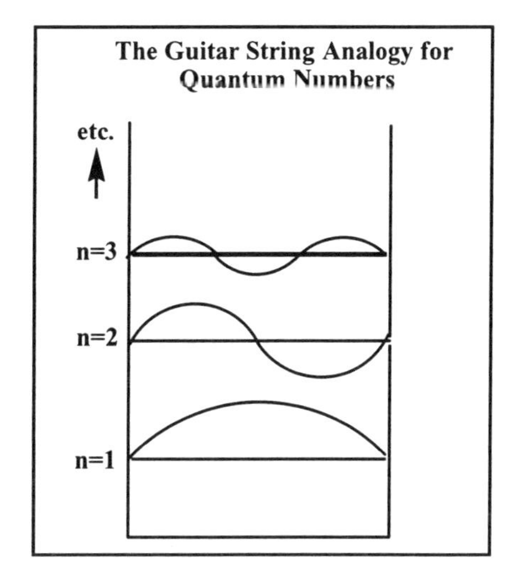


Figure 5.15 The Guitar String Analogy for Quantum Numbers

A guitar string is routinely taken as an analogy to an electron matter wave in an atom. If you study Figure 5.15, you'll see that the fundamental vibration (n=1) and the two overtones (n=2, and n=3) all contain an integral number of half wavelengths. Indeed, n is just the number of half waves along the string. So "n" is a quantum number.

Standing waves can only occur in strings if the string is clamped or fixed at each end: they become possible once the system is **bound**. Because electrons in atoms are standing waves, they are said to be in **bound states**.

A three-dimensional standing electron wave requires three quantum numbers to describe it, so Schrödinger's wave equation generates three quantum numbers for each energy state of hydrogen. These quantum numbers are designated as n, $\ell$, and m. We've already met the principal quantum number, n, in the Bohr model. For the electron in a hydrogen atom, the solutions to the wave equation are families of **wave functions**. Each wave function is denoted by the symbol, $\psi$ or psi, and has a particular allowed energy value.

$$\psi_{(n,\ell,m)} = \text{a hydrogen wave function for the quantum numbers } (n,\ell,m)$$

For hydrogen (with its one electron), any wave function is an orbital. An **orbital** is a one-electron wave function. The name orbital reminds you that it is related to an orbit, but it's something rather different. Shown below are four of hydrogen's wave functions or orbitals. The subscripts on the left (1,0,0), (2,0,0) (2,1,0) and (2,1,1) are quantum numbers. The subscripts 1s, 2s, $2p_x$, and $2p_y$ are the same quantum numbers written slightly differently; $\ell = 0$ is written as "s" and $\ell = 1$ is written as "p."

$$\psi_{(1,0,0)} = \frac{1}{\sqrt{\pi}}\left(\frac{Z}{a_o}\right)^{\frac{3}{2}} e^{-\left(\frac{Zr}{a_o}\right)} \quad \text{for } \psi_{1s}$$

$$\psi_{(2,0,0)} = \frac{1}{4\sqrt{2\pi}}\left(\frac{Z}{a_o}\right)^{\frac{3}{2}}\left(2 - \frac{Zr}{a_o}\right)e^{-\left(\frac{Zr}{2a_o}\right)} \quad \text{for } \psi_{2s}$$

where Z is +1, the charge on the hydrogen nucleus, $a_0$ is a constant of length roughly the size of the hydrogen atom, and r is the electron's distance from the nucleus. The equations above are for the 1s and 2s orbitals. Every atom has an infinite set of s-orbitals. Those orbitals may or may not be occupied by electrons. An empty orbital contains no electrons and a filled orbital contains two electrons.

Below are the $2p_x$ and $2p_y$ p-orbital wave functions or orbitals. The first depends on the angle $\theta$; the second depends on both the angles $\theta$ and $\phi$. Unlike the s-orbitals above, the p-orbitals below depend on angle, and therefore lack spherical symmetry. The distinction will become clear with the diagram that follows.

$$\psi_{(2,1,0)} = \frac{1}{4\sqrt{2\pi}}\left(\frac{Z}{a_o}\right)^{\frac{3}{2}} \frac{Zr}{a_o} e^{-\left(\frac{Zr}{2a_o}\right)} \cos\theta \quad \text{for } \psi_{2px}$$

$$\psi_{(2,1,1)} = \frac{1}{4\sqrt{2\pi}}\left(\frac{Z}{a_o}\right)^{\frac{3}{2}} \frac{Zr}{a_o} e^{-\left(\frac{Zr}{2a_o}\right)} \sin\theta\cos\phi \quad \text{for } \psi_{2py}$$

A wave function gives the amplitude of the standing electron wave at all points in space around the nucleus. But what does that sentence mean? Well, the probability of finding the electron at any point in space is proportional to the square of its wave function at that point. The wave function is interpreted as a probability density, $|\psi|^2$. This correct interpretation was first stated by Max Born (1882-1970) in 1926. The greater the value of $|\psi|^2$ at a specified point in space, the greater the chance of finding hydrogen's electron there. In other words, the square of the wave function at any point in space is a **probability density**, or a probability per unit volume. For a hydrogen atom, a graph of the probability density is also a graph of **electron density**. An **electron density map** of a hydrogen atom in a given quantum state is, thus, a probability "picture" of the atom in that state.

The probability interpretation of quantum mechanics shows why it is not correct to imagine an electron in an atom as a particle traveling in a fixed trajectory or orbit. The situation is one aspect of the **Heisenberg Uncertainty Principle** that places natural limits on the amount of knowledge that we can have about any particle. Its simplest and best-known statement is:

> "The exact position and exact momentum of a particle cannot be known simultaneously." (The momentum of a particle is the product of its mass and its velocity.)

In other words, you can't look at an electron in an atom *even if you want to*. The only way to try and find it is to fire photons at the electron—and they move it. The following informal example gives a slightly commonsense interpretation of the uncertainty principle.

> **Informal Example**: I've lost my black cat in a dark cellar and there's no source of light. To find the cat I throw rocks. Eventually there's a loud squeal. All I can say now is "there it was." That's my idea of Heisenberg's cat. Schrödinger's cat is a lot more popular. [☸kws +"Schrödinger's cat"]

Several months before Schrödinger described the wave equation, the German theoretical physicist Werner Heisenberg (1901-1976), along with Max Born and Pascual Jordan, developed an alternative quantum theory called matrix mechanics. At first it appeared that matrix mechanics and quantum mechanics were quite different; but it quickly turned out that they are equivalent. At the time, this equivalence gave strong support for the remarkable, new interpretation of nature that was quantum mechanics. In conception, quantum mechanics has remained unchanged since 1926—although today we know much more about its details and applications.

To summarize Section 5.3 so far: solving the Schrödinger equation for hydrogen yields hydrogen's orbitals. Chemists use orbitals to understand and interpret many chemical properties, particularly chemical bonding. So we are primarily studying quantum mechanics to understand chemical bonding.

### Electron Spin

There is one aspect of atomic spectra that cannot be explained by quantum mechanics, and that is the frequent appearance of closely spaced doubled lines or doublets. The explanation for this situation was offered in 1925 by G. E. Uhlenbeck and S. Goudschmit, who proposed that electrons have a property called "spin" which can be assigned one of two values: either +½ or -½. Every bound electron exists in one of those two spin states, which creates a situation in which there are two slightly different energy states in any given orbital and provides an explanation for the observed doublet spectral lines. Because of electron spin, every orbital can have two, but no more than two, electrons.

## Section 5.4 Quantum Mechanical Results for the Hydrogen Atom

To apply quantum mechanics to chemistry means solving Schrödinger's wave equation for atoms, ions, and molecules—a task that typically requires extensive computation. However, the Schrödinger equation can be solved exactly for a single electron bound to any nucleus; the simplest one-electron example is the hydrogen atom. Calculations of exact solutions for one-electron cases are important in themselves, but, even more importantly, their results provide a starting point for the approximations needed to deal with atoms and ions with more than one electron and with molecules. Figure 5.16 shows two hydrogen orbitals represented as orbital pictures with their wave functions written below them. An **orbital picture** is a stylized representation of a hydrogen wave function.

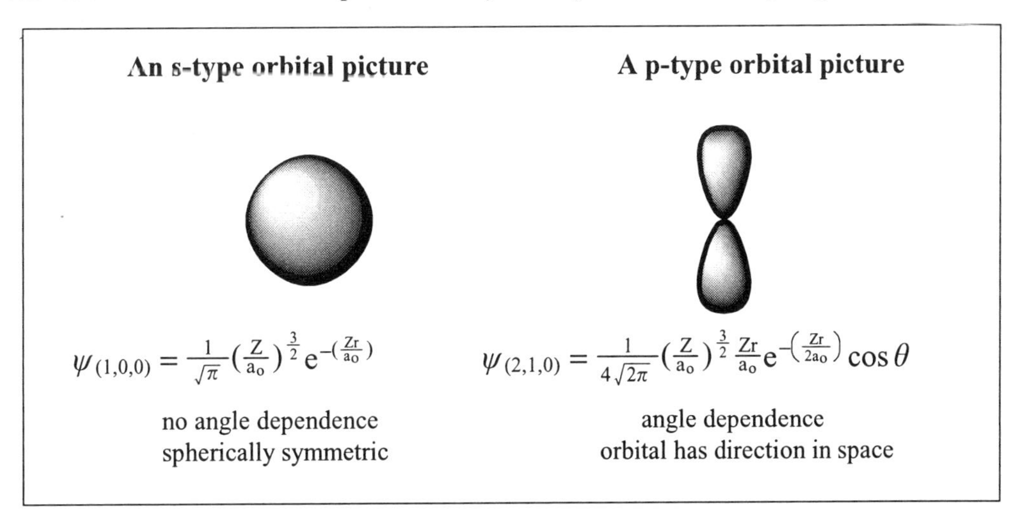


Figure 5.16 s-Type and p-Type Orbital Pictures and their Related Wave Functions. It is convenient to call the orbital pictures just orbitals, even if it's not strictly correct.

By now perhaps you can see where we're heading. Hydrogen has an infinite set of orbitals that we can calculate quite explicitly. So we'll generate that set of orbitals and then, in one fell swoop, apply them to the atoms of all other elements in the periodic table.

Two key properties of orbitals important for chemistry are their *shape* and their *energy*:

**Orbital shape** is a quantum mechanical idea difficult to define but easy to see: it is the general appearance of an orbital's electron density map. Orbital shape is a loosely defined term that refers to the way a particular wave function is represented in space. The different shapes of s- and p-orbitals are obvious in Figure 5.16.

**Orbital energy** in a hydrogen atom is a measure of the energy of the electron in that particular orbital or energy state; specifically, it is energy released when an ionized electron falls into that particular orbital, or as we saw earlier $E_n = -R_H \frac{h}{n^2}$. Orbital energies in a hydrogen atom increase as shown in the Bohr energy level diagram (Figure 5.13) and in Table 5.4. If you examine Figure 5.13 you'll see the higher the energy of an orbital the less energy is required to remove an electron from it. Orbital energies can't be quite so precisely defined for entities other than hydrogen, but the general concept remains the same.

## Rules for Hydrogen's Quantum Numbers

The element with the most electrons has maybe 118 of them. With each orbital being able to accommodate at most two electrons we're going to need at least ½ × 118 orbitals to complete the periodic table. So we need some way to generate them. Fortunately, we don't have to go back and resolve the wave equation—though that's where the rules really come from. Instead we'll just state the rules deduced for hydrogen and apply them to all the other elements. The actual process of applying them is simple arithmetic.

As we've seen, because an atom or ion is three-dimensional, all the wave functions of its possible states involve three coordinates of position (x, y, and z) and three quantum numbers. Recall, the quantum numbers are designated n, $\ell$, and m or (n,$\ell$,m); the set (2,1,1) is an example. The **principal quantum number** is designated n. The **secondary quantum number** is designated $\ell$ ("ell"). The third or **magnetic quantum number** is designated m. The property called electron spin has the effect of providing a fourth quantum number, s, designated as the **spin quantum number**; s can have two values and therefore "allows" two electrons to be in the same orbital. All the possible combinations of n, $\ell$, and m can be obtained by using the following three rules for quantum numbers:

Rule 1: $n = 1, 2, 3, 4\ldots$
Rule 2: $\ell = 0, 1, 2, 3, \ldots (n - 1)$
Rule 3: $m = -\ell, -\ell+1, -\ell+2, \ldots, 0, +1, +2, \ldots, +\ell$

To which can be added a fourth simple rule for the for the spin quantum number:

Rule 4: $s = \pm½$

Rule 1 states that the allowed energies of a hydrogen atom depend only on n, the principal quantum number; n = 1, 2, 3, 4.... There are $n^2$ states of identical energy in the hydrogen atom for each value of n. These states are called **electronic states**. Two different electronic states with the same energy are said to be **degenerate**, and a given set of degenerate energy levels in an atom is called a **shell**. The lowest or ground state level, with n = 1, is the K-shell. It contains just the 1s orbital. The first excited level, with n = 2, is the L-shell that contains four degenerate states: 2s, $2p_x$, $2p_y$, and $2p_z$. The second excited level, with n = 3, is the M-shell with nine degenerate states: 3s, $3p_x$, $3p_y$, $3p_z$, and five d-orbitals. The third excited level, with n = 4, is the N-shell with sixteen degenerate states: 4s, $4p_x$, $4p_y$, $4p_z$, five d-orbitals and seven f-orbitals.

Rule 2 for quantum numbers defines how the possible values of $\ell$ depend on n. Every different value of $\ell$ gives a wave function of a particular "shape" that can be visualized when the wave function is plotted. States with $\ell = 0$ are s-orbitals, those with $\ell = 1$ are p-orbitals, those with $\ell = 2$ are d-orbitals, and those with $\ell = 3$ are f-orbitals. Orbitals for $\ell > 3$ are labelled alphabetically. A family of wave functions with specific n and $\ell$ values is a **subshell**. For example, the three wave functions in the 2p subshell all have n = 2 and $\ell = 1$; the five wave functions in the 3d subshell all have n = 3 and $\ell = 2$, etc. Each orbital can hold a maximum of 2 electrons. The maximum electron occupancy is the greatest number electrons that can be in any given subshell and is, of course, double the number of orbitals in the same subshell. Properties of subshells are summarized in table 5.3.

Table 5.3 **Properties of Subshells**

| | | | | | | | |
|---|---|---|---|---|---|---|---|
| Subshell letter designation | s | p | d | f | g | h | ... |
| $\ell$ value | 0 | 1 | 2 | 3 | 4 | 5 | ... |
| Number of orbitals of this type for a given $\ell$ value | 1 | 3 | 5 | 7 | 9 | 11 | ... |
| Maximum electron occupancy | 2 | 6 | 10 | 14 | 18 | 22 | ... |

Rule 3 for quantum numbers defines how the possible values of m depend on $\ell$: For a particular value of $\ell$, the quantum number m may have $(2\ell + 1)$ different integral values ranging between $-\ell$ and $+\ell$. The magnetic quantum number describes the orientation of the wave function in space. The different orientations may be visualized when wave functions are plotted. Table 5.4 shows some of the various combinations of n, $\ell$, and m that arise for the hydrogen atom. In addition to the three quantum numbers shown in the table, the spin quantum number can adopt one of two values (+½ or -½) and so allows any orbital to hold a maximum of two electrons.

Table 5.4
**Quantum Numbers (n,$\ell$,m) and State Designations for the First 55 Orbitals of the Hydrogen Atom**

| Energy ↑ | (orbital energy increases from the bottom up) | n | $\ell$ | Orbital type | m | Orbital set | Total number of orbitals |
|---|---|---|---|---|---|---|---|
| | Fourth level (O-shell) | 5 | 4 | 5g | -4,-3, -2, -1, 0, +1, +2, +3,+4 | (nine 5g states) | 25 |
| | | 5 | 3 | 5f | -3, -2, -1, 0, +1, +2, +3 | (seven 5f states) | |
| | | 5 | 2 | 5d | -2, -1, 0, +1, +2 | (five 5d states) | |
| | | 5 | 1 | 5p | -1, 0, +1 | (three 5p states) | |
| | | 5 | 0 | 5s | 0 | (one 5s state) | |
| | Third level (N-shell) | 4 | 3 | 4f | -3, -2, -1, 0, +1, +2, +3 | (seven 4f states) | 16 |
| | | 4 | 2 | 4d | -2, -1, 0, +1, +2 | (five 4d states) | |
| | | 4 | 1 | 4p | -1, 0, +1 | (three 4p states) | |
| | | 4 | 0 | 4s | 0 | (one 4s state) | |
| | Second level (M-shell) | 3 | 2 | 3d | -2, -1, 0, +1, +2 | (five 3d states) | 9 |
| | | 3 | 1 | 3p | -1, 0, +1 | (three 3p states) | |
| | | 3 | 0 | 3s | 0 | (one 3s state) | |
| | First level (L-shell) | 2 | 1 | 2p | -1, 0, +1 | (three 2p states) | 4 |
| | | 2 | 0 | 2s | 0 | (one 2s state) | |
| | Ground state (K-shell) | 1 | 0 | 1s | 0 | (one 1s state) | 1 |

The letter designations s, p, d, and, f are historical holdovers from the days when quantum mechanics was solving the problems of lines in atomic spectra. The letters are the initials of various characteristics of the series of spectral lines. For example, s = sharp, p = principal, and d = diffuse. With f onward the designation follows alphabetically, g, h, i…∞.

A shorthand method to indicate the actual electron occupancy of a subshell is to write the number of electrons as a superscript. Thus, $2p^3$ means that there are three electrons in the 2p orbital subshell and $4d^8$ means there are eight electrons in the 4d orbital subshell.

**Example 5.7**: Explain what is meant by the designations 2s and 4d. Also explain what $3d^7$ means.

Strategy: Study Table 5.4 and the accompanying text.

2s means an orbital with a principal quantum number of 2 and a secondary quantum number of 0. 4d means an orbital with a principal quantum number of 4 and a secondary quantum number of 2.

In $3d^7$, the 3 refers to the principal quantum number being three, the d refers to the secondary quantum number $\ell$ being 2, and the superscript 7 refers to seven electrons (out of a possible total of 10) in this subshell.

At long last, you can see where the designations at the top of our periodic table come from. They are hydrogen's orbital designations. The superscripts are the number of electrons in the designated orbitals. Why do we put these designations on the periodic table? You probably remember that an atom's, ion's, or molecule's chemical properties are caused by the behavior of its electrons, so the designations at the top summarize the behavior of the elements in the group below.

Applying the rules for quantum numbers is quite straightforward, as you will see from Examples 5.8 and 5.9.

**Example 5.8**: What subshells are present in the n = 6 principal state?

Strategy: Apply the rules of quantum numbers to write the list.

For a given n, $\ell$ = n-1, n-2, n-3,... 0, thus possible values of $\ell$ are 5, 4, 3, 2, 1, 0. These are respectively designated h, g, f, d, p, s

There are six subshells in the n = 6 principal shell, designated 6h, 6g, 6f, 6d, 6p, and 6s.

**Example 5.9**: State the values of n, $\ell$, and m for all of the orbitals in the 4f subshell?

Strategy: Apply the rules of quantum numbers. The 4 in 4f designates the principal quantum number n = 4. The f in 4f designates the secondary quantum number $\ell$ = 3.

For a subshell with a secondary quantum number 3, the values of m range from -3 to +3 in steps of one. Thus, m = +3, +2, +1, 0, -1, -2, -3. Hence there are seven subsubshells or atomic orbitals in the 4f subshell. The quantum number designations of the seven orbitals are shown in the table below:

| n | $\ell$ | m |
|---|---|---|
| 4 | 3 | +3 |
| 4 | 3 | +2 |
| 4 | 3 | +1 |
| 4 | 3 | 0 |
| 4 | 3 | -1 |
| 4 | 3 | -2 |
| 4 | 3 | -3 |

## The Shapes of Orbitals

Because of Heisenberg's uncertainty principle, the probability distribution of the electron in a hydrogen atom *cannot* be determined experimentally. However, quantum mechanics does give an extremely detailed picture by mapping the calculated values of $|\psi|^2$. As we've seen, this map is, loosely speaking, the shape of an orbital.

s-Orbitals have spherical symmetry. They are round. As its principal quantum number increases an electron moves on average farther and farther from the nucleus, so the probability distribution grows larger, as you see in Figure 5.17. Shading is used to give the impression of the three-dimensional electron density distribution that comes by plotting the $\psi_{1s}$, $\psi_{2s}$, and $\psi_{3s}$ wave functions.

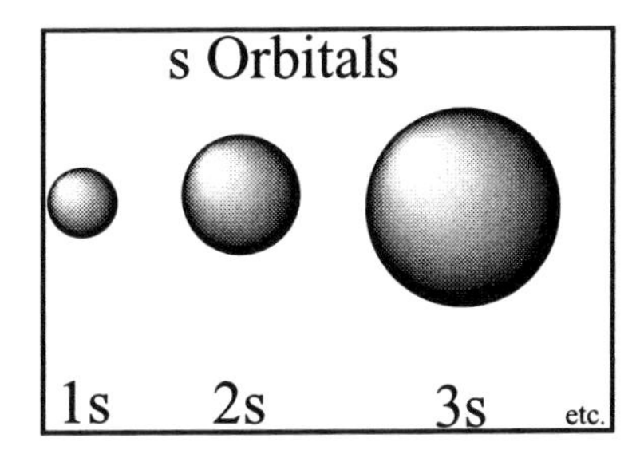


Figure 5.17 s-Type Orbital Pictures. The purpose of the shading is to give the viewer the sense that the orbitals are three-dimensional in quality.

Each family of three p-orbitals is designated 2p, 3p, 4p, 5p, 6p, ...etc. The three 2p orbitals are identical in shape but "point" in different directions. Because they point along the Cartesian axes they are often designated as $2p_x$, $2p_y$, and $2p_z$. We can imagine that the subscripts x, y, and z correspond to the magnetic quantum numbers -1, 0, and +1 also use to describe the set of p orbitals. However, we can associate any letter x, y, or z, with any magnetic quantum number -1, 0, or +1.

There are three p-orbitals in a given p-subshell. The three 2p orbitals of the 2p subshell are shown in outline representation in Figure 5.18. Figure 5.18—which does not use shading—is an alternative orbital representation to the scheme used in the Figure 5.17. Neither is "correct." There is no single best way to make pictorial representations of the squares of wave functions, though many methods have their partisans.

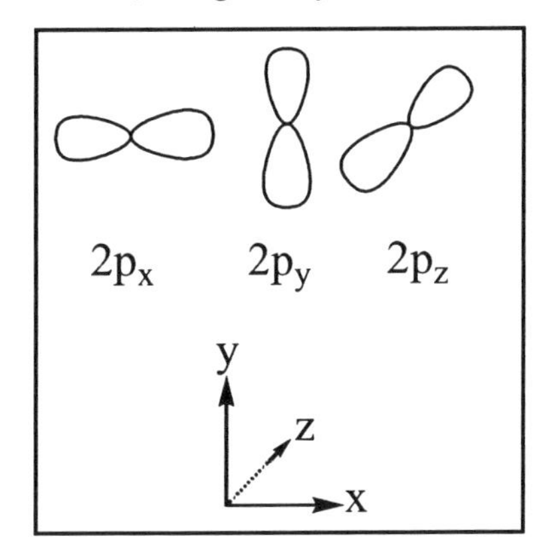


Figure 5.18 p-Type Orbitals in Outline or Contour Line Representation

The edges of the shaded regions of Figure 5.17 and the lines in Figure 5.18 represent where electron density plots cut off at some arbitrary contour (cutting off to include 95% of the total electron density within the contour is a common choice). In quantum mechanics, **contours** are three dimensional surfaces that connect points of equal electron density. For example, in Figure 5.18 an electron designated $2p_x$ will have 95% probability of being within the contour shown. Observe that each p orbital has two lobes. A **lobe** (often shaped like an ear lobe) is a designated sub region of an atomic orbital.

All orbitals of a given subshell type grow in size as the principal quantum number increases. Figure 5.19 shows the three orbitals in the 2p set and the three orbitals in the 3p set (all presented in a shaded representation). Understand that all six of these orbitals are *on the same atom*. Indeed, the six orbitals shown here and *all* the other hydrogen orbitals are superimposed on the same atom. You might ask how is it that the electrons doesn't get mixed up about what orbital each is "supposed" to be in? The answer is that the quantum numbers keep them straight, though the Uncertainty Principle guarantees that we humans can never have any detailed knowledge of where the electron is at any moment.

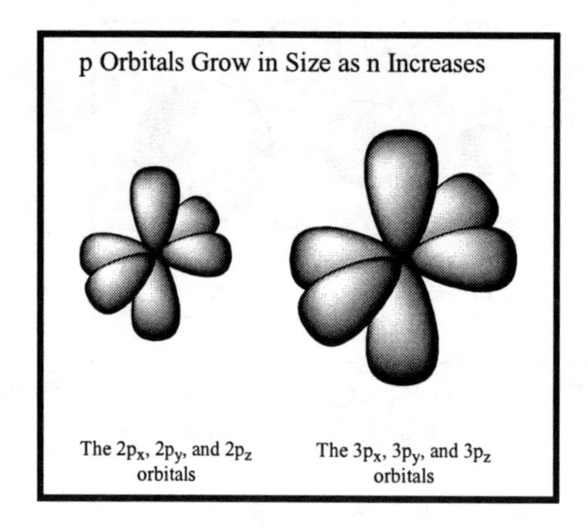


Figure 5.19 2p-Type Orbitals Compared to 3p-Type Orbitals

d Orbitals

Any d-subshell, with $\ell = 2$, is composed of 5 orbitals. Shaded three-dimensional electron density impressions of the 3d subshell of orbitals are shown in Figure 5.20. The d-orbitals have more lobes than do the p-orbitals and, as can be seen from Figure 5.20, are somewhat more complicated in shape. Observe that $d_{xy}$, $d_{xz}$, and $d_{yz}$ orbitals have the same shape and size, but differ in their alignment with the x, y and z, axes. The $d_{x^2-y^2}$orbital and the $d_{z^2}$ orbital have their lobes directed respectively along the x and y, and the z axes. As with s and p orbitals, the smallest d orbital set is the one with the lowest allowed principal quantum number. Thus the size relationship among the d subshells is 3d < 4d < 5d < 6d....

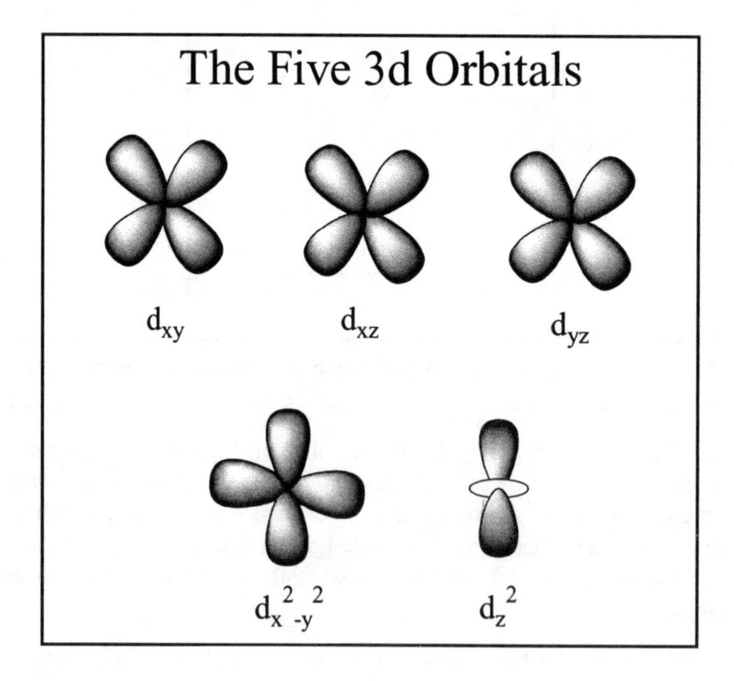


Figure 5.20 The set of five orbitals that constitute a d subshell

> **Informal Question**: How does an electron in an atom know what it's "supposed" to be doing? Well, it looks at its shirt. What? Every electron in an atom has a label, such as (3,2,-1,+½) on its shirt. Those are its four quantum numbers; they tell the electron what energy it has and where it's supposed to travel. If it checks them on its shirt it knows how to behave.
>
> Editorial comment: That's a pretty silly answer.

Orbitals from f and higher subshells have yet more complex shapes than the d-orbitals. Approximate representations of the seven 4f orbitals are shown below in Figure 5.21.

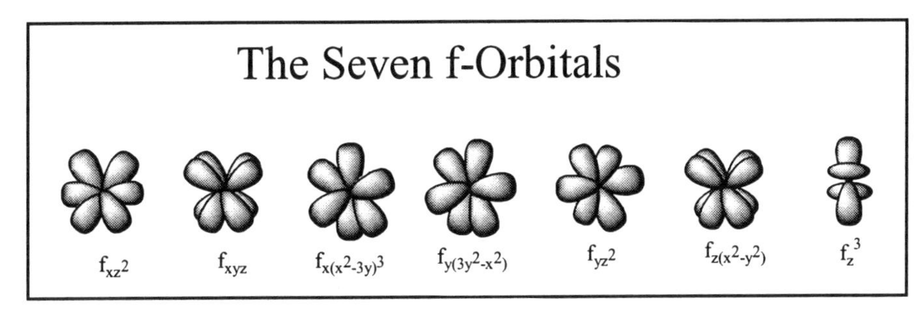


Figure 5.21 The set of seven orbitals that constitute an f subshell.

Chemists use many different shading schemes, coloring schemes, and drawing methods to represent orbitals. In every case, the objective is to help them visualize the electron density distribution in atoms, molecules, and ions. Sometimes the electron density distribution is called an **electron cloud**. An example of the use of shading to visualize electron clouds is shown in Figure 5.22.

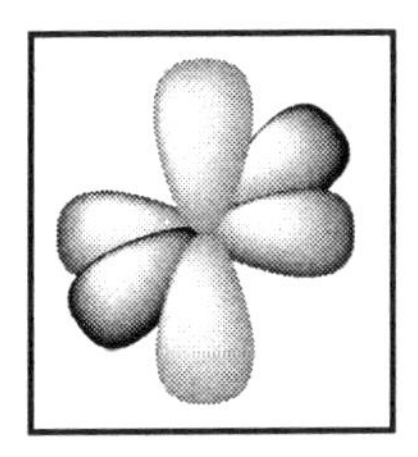

Figure 5.22 The three 2p orbitals shown by different shading.

## Section 5.5 Hydrogen's Energy Level Diagram

Every horizontal line in Figure 5.23 represents a single hydrogen orbital. Each can hold a maximum of two electrons. All the sets of orbitals with the same principal quantum number are **degenerate**—the subshells are at the same horizontal level. Each orbital can hold two electrons, but the electrons must have different spin quantum numbers (s). To designate an electron state in a hydrogen atom requires a total of four quantum numbers (n, $\ell$, m, and s). Recall, the spin quantum number is arbitrarily designated +½ or -½.

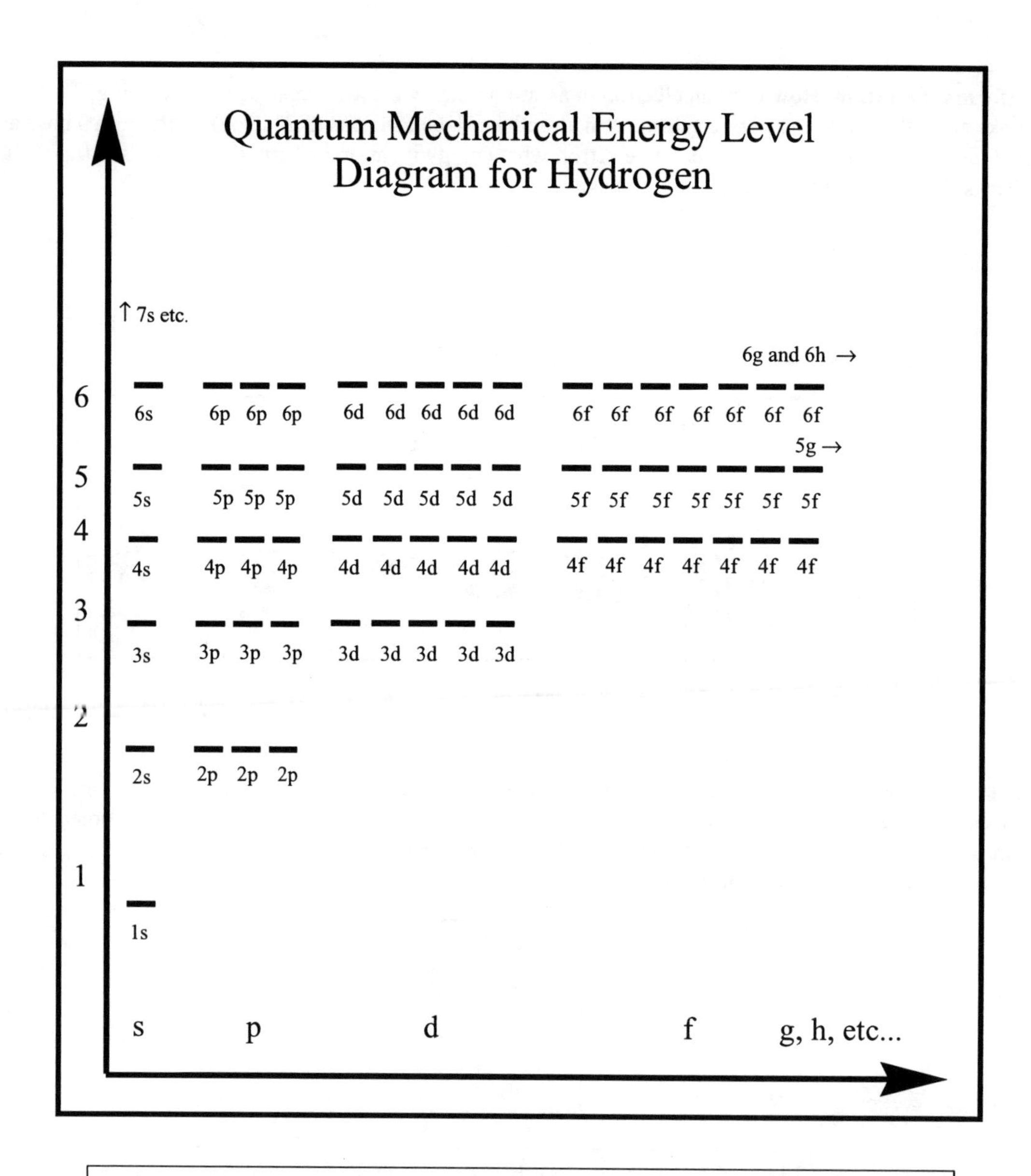


Figure 5.23 The quantum mechanical energy level diagram for the hydrogen atom.

Figure 5.24 shows *both* the shapes and energies of the hydrogen orbitals as they are calculated from quantum mechanics. It is similar to Figure 5.23, but includes tiny pictures of hydrogen's orbital shapes. Not shown are the 7s and higher s-levels above, nor the 5g, 6g, 6h levels and levels to the right of them.

Understand that *all* these orbitals exist in every atom; an electron can potentially be in any of them. It's a philosophical point whether an orbital exists in an absolute sense. John Glenn was the first American to orbit the Earth. Does his orbit exist?

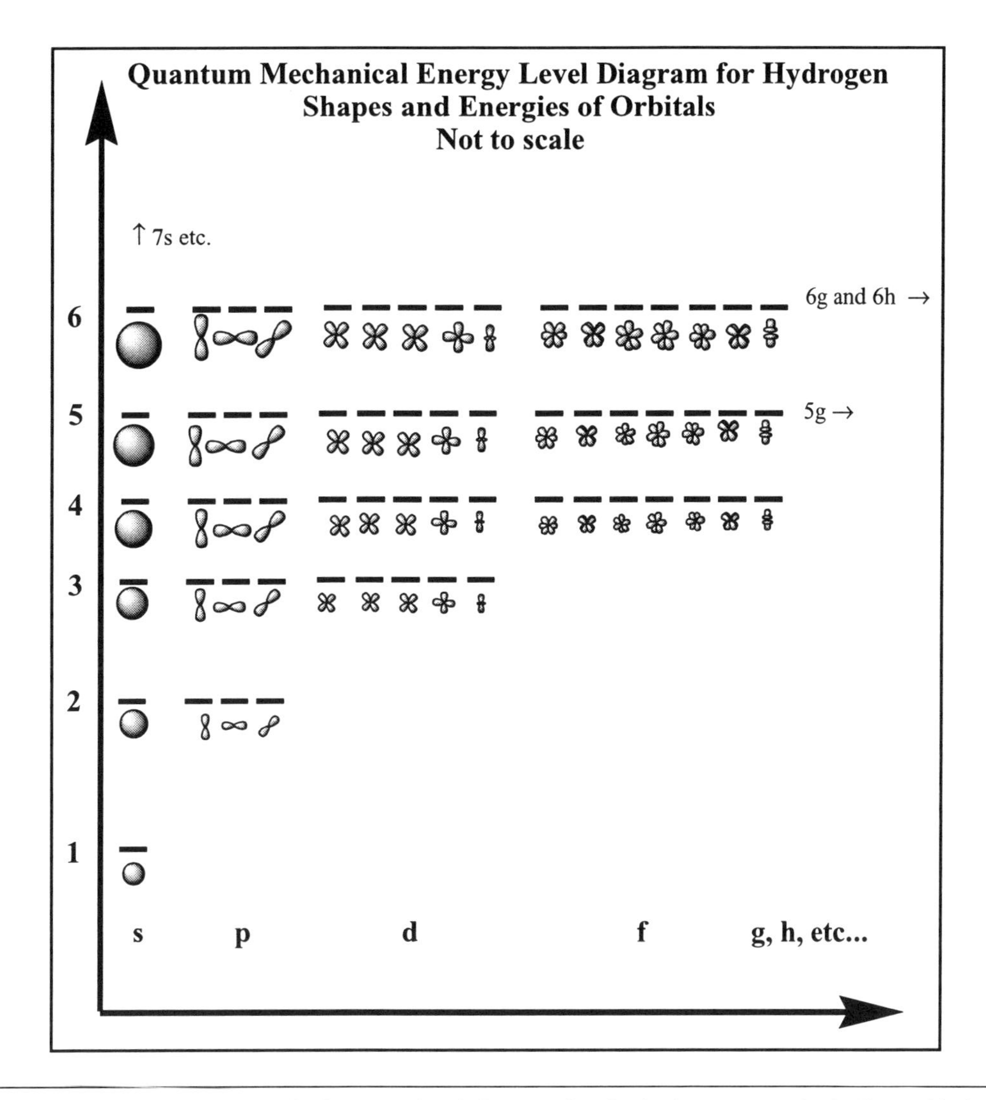


Figure 5.24 The quantum mechanical energy level diagram for the hydrogen atom including orbital shapes. A total of 62 levels are actually shown. It turns out that 62 levels are enough to accommodate the electrons of all known elements. When this diagram is generalized and the orbitals slightly rearranged (in Figure 5.27) it will be ready to explain the periodic table.

So far everything quantum mechanical has been for hydrogen. But quantum mechanics and the Schrödinger equation apply to all elements and throughout chemistry. What we're going to say is that the quantum mechanics of hydrogen is a such good starting point that we'll use it talk about all atoms, ions, and molecules.

Now we make a leap and simply assert that *with minor modifications* Figure 5.24 can be used for all the elements no matter *how many* electrons their atoms have. Making this leap is not too much of a gamble. Bohr did it around 1920, and these days we have lots of actual quantum mechanical calculations that support the leap. Conceptually, the process is simple: starting from the ground state, the 1s orbital in Figure 5.24, add successive electrons with into the lowest available energy level until you've added all of them for the needed atom. Doing that gives the electron configuration of an atom. The disposition or arrangement of electrons in an atom is called the atom's **electron structure** or electron configuration. An **electron configuration** is a shorthand notation that designates an atom's electrons by placing them into hydrogen-like atomic orbitals. Example 5.10 shows a simple case of electron configuration as a prelude to taking up the entire periodic table in Section 5.6.

**Example 5.10**: The electron configuration of a lithium atom is $1s^22s^1$. Clearly explain every facet of this notation.

The 1s refers to the lowest energy state of a hydrogen atom. The 1 means principal quantum number =1 and the s means the secondary quantum number is 0. So n = 1, and $\ell = 0$. Its similar with 2s, where n = 2 and $\ell = 0$.

The superscript 2 means that there are two electrons in the 1s orbital. The superscript 1 means that there is one electron in the 2s orbital.

The total number of electrons is 2 + 1, or 3. Three is lithium's atomic number and hence its number of electrons.

## Section 5.6 Electron Configurations and the Periodic Table

Figure 5.24 summarizes both the energy states and shapes for 62 hydrogen orbitals. When we start adding electrons into the levels of this diagram we begin to generate the periodic table. Doing that is sometimes called the **aufbau procedure**. Aufbau is a German word that means something like "building up." It's a procedure to build up the periodic table based on quantum mechanical principles. Figure 5.25 shows the electron configuration of lithium:

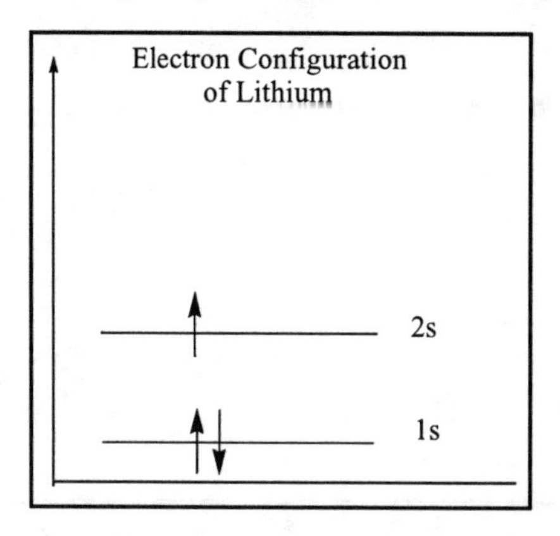


Figure 5.25 The electron configuration of a lithium atom generated by adding three electrons into hydrogen's orbital energy level diagram. Each electron is represented by an arrow with the direction of the arrow being related to the sign of the electron's spin quantum number, s. Two electrons in the same orbital must have opposite spins.

As usual for an energy level diagram, Figure 5.25 is a one-dimensional representation which shows energy increasing up the vertical axis, while the horizontal axis has no definite units but serves to make the energy levels visible as lines of arbitrary length.

Arrows are the conventional way to represent electrons in an energy level diagram. The up arrows ↑ and the down arrow ↓ correspond to spin quantum numbers of $\pm\frac{1}{2}$. We can say that an up arrow represents an electron with a spin of $+\frac{1}{2}$, but it's arbitrary. Any electron in an atom must have a unique set of four quantum numbers (n, $\ell$, m, s); that's called the **Pauli exclusion principle**. The sets of four quantum numbers for the three electrons in lithium are $(1,0,0,+\frac{1}{2})$, $(1,0,0,-\frac{1}{2})$, and $(2,0,0,+\frac{1}{2})$.

Table 5.5 shows the details of the building-up and the electron configurations of the first 18 elements that the process produces. Appendix E shows electron configurations for all the known elements.

| Table 5.5 **Electron Configurations of the first Eighteen Elements** | | | | | | | | | | |
|---|---|---|---|---|---|---|---|---|---|---|
| Element | Hydrogen Orbital Occupancy | | | | | | | | | Electron Configuration |
| | 1s | 2s | $2p_x$ | $2p_y$ | $2p_z$ | 3s | $3p_x$ | $3p_y$ | $3p_z$ | |
| Argon | ↑↓ | ↑↓ | ↑↓ | ↑↓ | ↑↓ | ↑↓ | ↑↓ | ↑↓ | ↑↓ | [Argon core] |
| Chlorine | ↑↓ | ↑↓ | ↑↓ | ↑↓ | ↑↓ | ↑↓ | ↑↓ | ↑↓ | ↑ | [Neon core]$3s^2 3p^5$ |
| Sulfur | ↑↓ | ↑↓ | ↑↓ | ↑↓ | ↑↓ | ↑↓ | ↑↓ | ↑ | ↑ | [Neon core]$3s^2 3p^4$ |
| Phosphorus | ↑↓ | ↑↓ | ↑↓ | ↑↓ | ↑↓ | ↑↓ | ↑ | ↑ | ↑ | [Neon core]$3s^2 3p^3$ |
| Silicon | ↑↓ | ↑↓ | ↑↓ | ↑↓ | ↑↓ | ↑↓ | ↑ | ↑ | | [Neon core]$3s^2 3p^2$ |
| Aluminum | ↑↓ | ↑↓ | ↑↓ | ↑↓ | ↑↓ | ↑↓ | ↑ | | | [Neon core]$3s^2 3p^1$ |
| Magnesium | ↑↓ | ↑↓ | ↑↓ | ↑↓ | ↑↓ | ↑↓ | | | | [Neon core]$3s^2$ |
| Sodium | ↑↓ | ↑↓ | ↑↓ | ↑↓ | ↑↓ | ↑ | | | | [Neon core]$3s^1$ |
| Neon | ↑↓ | ↑↓ | ↑↓ | ↑↓ | ↑↓ | | | | | $1s^2 2s^2 2p^6$ |
| Fluorine | ↑↓ | ↑↓ | ↑↓ | ↑↓ | ↑ | | | | | $1s^2 2s^2 2p^5$ |
| Oxygen | ↑↓ | ↑↓ | ↑↓ | ↑ | ↑ | | | | | $1s^2 2s^2 2p^4$ |
| Nitrogen | ↑↓ | ↑↓ | ↑ | ↑ | ↑ | | | | | $1s^2 2s^2 2p^3$ |
| Carbon | ↑↓ | ↑↓ | ↑ | ↑ | | | | | | $1s^2 2s^2 2p^2$ |
| Boron | ↑↓ | ↑↓ | ↑ | | | | | | | $1s^2 2s^2 2p^1$ |
| Beryllium | ↑↓ | ↑↓ | | | | | | | | $1s^2 2s^2$ |
| Lithium | ↑↓ | ↑ | | | | | | | | $1s^2 2s^1$ |
| Helium | ↑↓ | | | | | | | | | $1s^2$ |
| Hydrogen | ↑ | | | | | | | | | $1s^1$ |

The configurations generated and shown in Table 5.5 turn out to be the experimental electron configurations! When the details of the atomic spectra of these eighteen elements are studied, the spectra can be interpreted by energy levels that occur in exactly the same sequence as they do in Table 5.5. In other words, the hydrogen orbitals provide a perfect template for the first 18 elements. Note that we use the notation [noble gas **core**] as a shorthand way of designating portions of the electron configuration that are repetitive. Thus, [Neon core] = $1s^2 2s^2 2p^6$. **Core** electrons are not chemically reactive. Electrons not in an atom's core are **valence** electrons. Notice also throughout Table 5.5 that successive electrons that enter degenerate energy levels do so with maximum separation and parallel electron spins. Those two facts are known as **Hund's rules** and derive originally from experimental spectroscopy.

Figure 5.26 shows argon's ground state electron configuration deduced by adding electrons into the energy levels of hydrogen. Spectroscopic studies of argon confirm that its experimental electron configuration is indeed $1s^2 2s^2 2p^6 3s^2 3p^6$ as shown in the diagram. The last added electron in argon goes into the $3p_z$ orbital. The four quantum numbers of that electron are (3, 1, +1, -½).

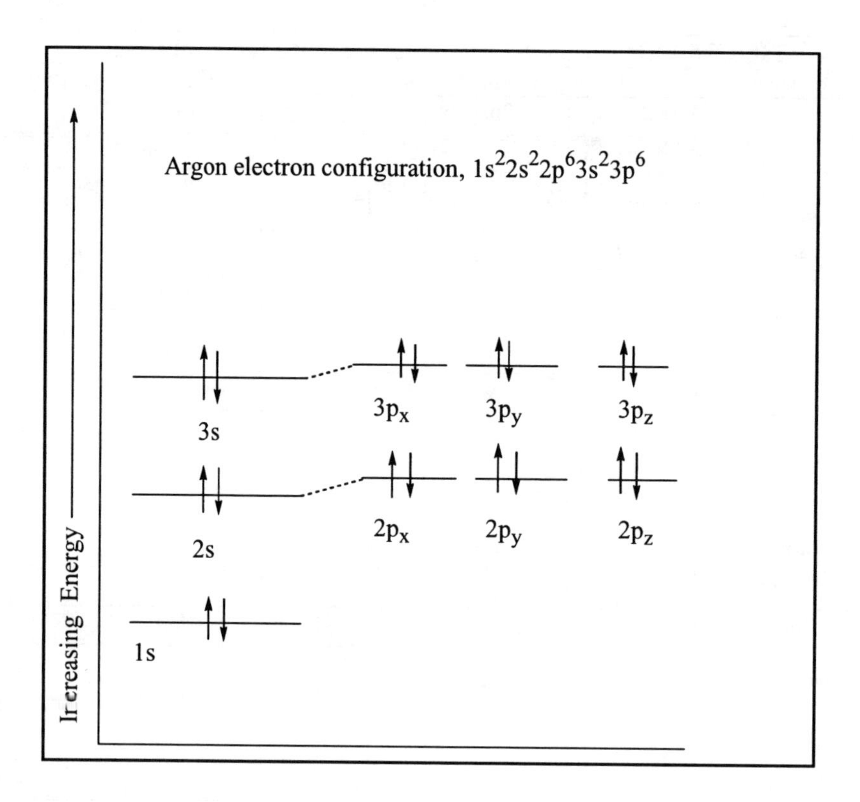


Figure 5.26 The electron configuration of an argon atom generated by adding 18 electrons into hydrogen's orbital energy level diagram. Here, for the first time, we encounter a hydrogen orbital energy splitting. Note that the 2s and 2p orbitals and the 3s and 3p orbitals are not degenerate (not at the same energy level) in argon as they are in the single-electron hydrogen atom. The splitting or degeneracy is shown by the dotted lines.

The orbital energy level diagram for hydrogen enables us to rationalize the properties of the spectra, and hence the electron configurations, of the elements up to argon. It also explains why the noble gas elements are chemically inert. Each has a closed shell electron configuration: He, $1s^2$; Ne, $1s^2 2s^2 2p^6$; Ar, $1s^2 2s^2 2p^6, 3s^2, 3p^6$. But after argon the situation gets a little more complicated. In a hydrogen atom, after the 3p comes the 3d orbitals. But element 19, potassium, has chemical properties extremely similar to the element directly above it in the periodic table: namely sodium, element 11. So the chemical evidence is that potassium's "last" electron is like sodium's. And sodium's last electron is an "s" not a "d," i.e., it's in the 4s orbital, not one of the 3d orbitals.

We now encounter a key point: when we extend the quantum mechanics of the hydrogen atom to the entire periodic table the relative energies of the orbitals shift. Orbitals sets that are degenerate in hydrogen are not degenerate in any other elements. Up to argon the relative energies match hydrogen's sequence, but beyond argon the sequencing is different. For this reason we replace hydrogen's atomic energy level diagram with a generalized energy diagram that works rather well for the entire periodic table; it's shown in Figure 5.27.

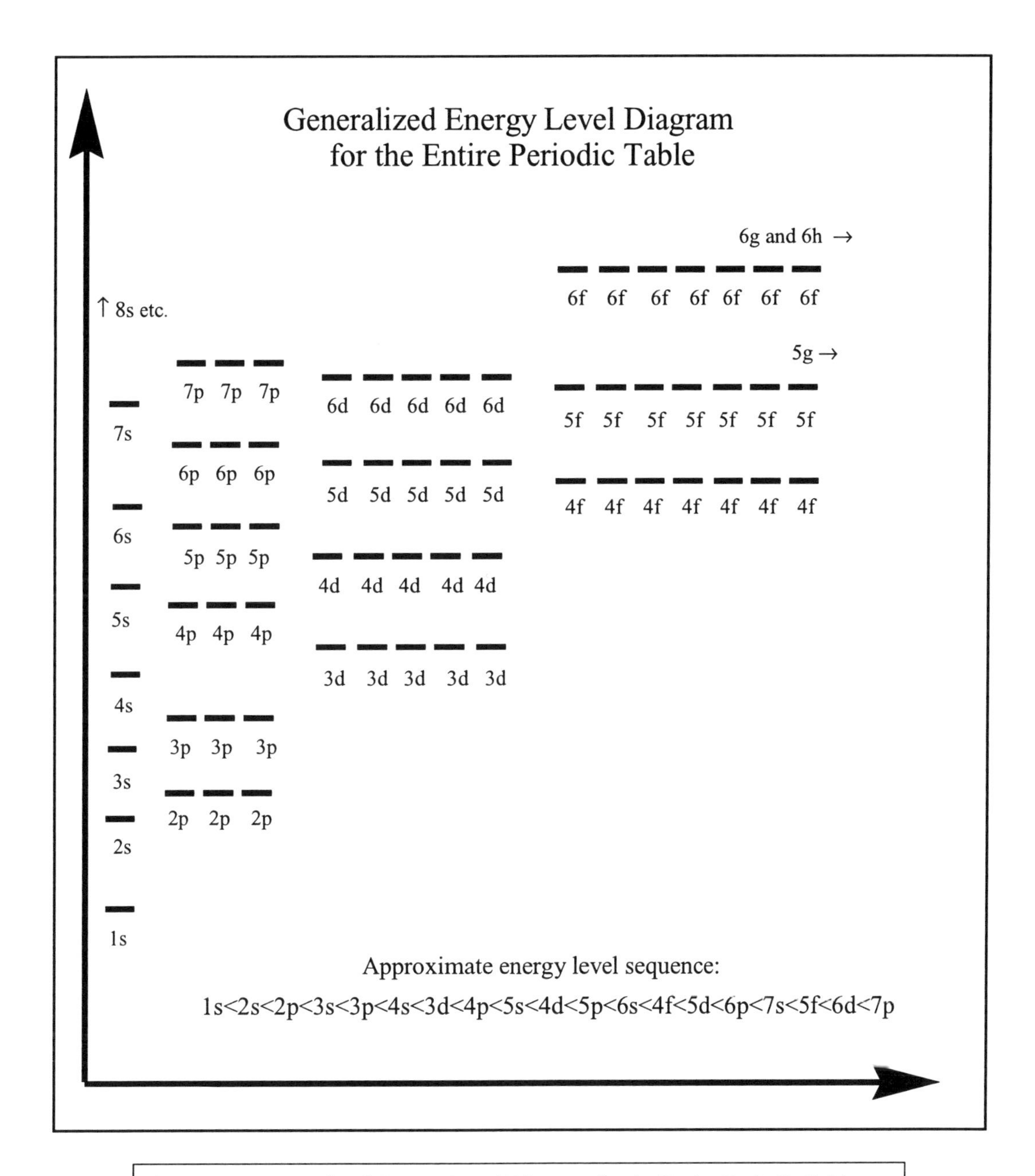


Figure 5.27 Generalized Energy Level Diagram for the Entire Periodic Table.

If you imagine the energy level diagram in Figure 5.27 being filled up from the bottom with successive electrons then you can understand how the periodic table arises. That is shown schematically on a long form of the periodic table in Figure 5.28, where the differently shaded regions are perfectly matched with the quantum numbers at the top.

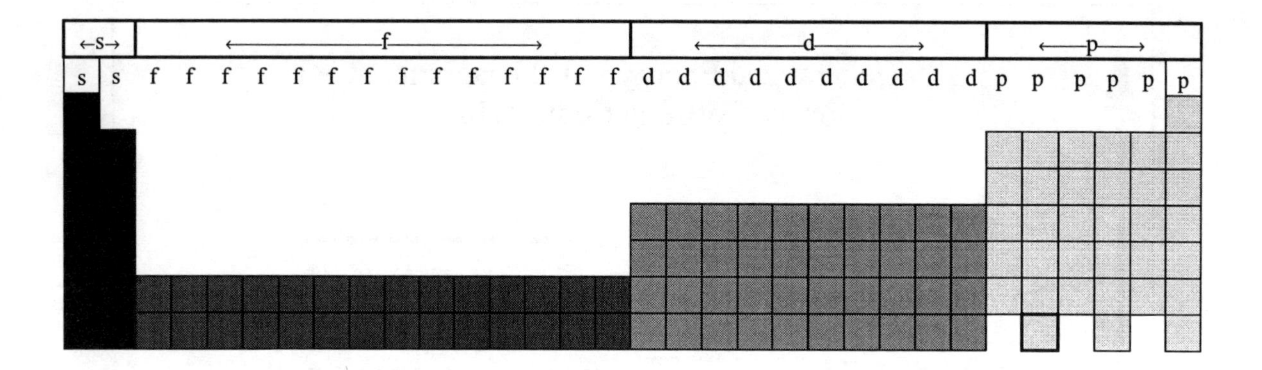


Figure 5.28 The very long form of the periodic table, showing the relation between the regions on the table and the secondary quantum number. A larger version of this table including orbital designations can be found in Appendix H.

Figure 5.29 is a collapsed form of Table 5.28 and is the more common space saving version, with the lanthanide and actinide elements written separately below the main table. Figure 5.29 shows the formal orbital designation of each table position.

| 1s | | | | | | | | | | | | | | | | | 1s |
|---|---|---|---|---|---|---|---|---|---|---|---|---|---|---|---|---|---|
| 2s | 2s | | | | | | | | | | | 2p | 2p | 2p | 2p | 2p | 2p |
| 3s | 3s | | | | | | | | | | | 3p | 3p | 3p | 3p | 3p | 3p |
| 4s | 4s | 3d | 3d | 3d | 3d | 3d | 3d | 3d | 3d | 3d | 3d | 4p | 4p | 4p | 4p | 4p | 4p |
| 5s | 5s | 4d | 4d | 4d | 4d | 4d | 4d | 4d | 4d | 4d | 4d | 5p | 5p | 5p | 5p | 5p | 5p |
| 6s | 6s | 5d | 5d | 5d | 5d | 5d | 5d | 5d | 5d | 5d | 5d | 6p | 6p | 6p | 6p | 6p | 6p |
| 7s | 7s | 6d | 6d | 6d | 6d | 6d | 6d | 6d | 6d | 6d | 6d | | 7p | | 7p | | 7p |
| | | | | | | | | | | | | | | | | | |
| | | | 4f | 4f | 4f | 4f | 4f | 4f | 4f | 4f | 4f | 4f | 4f | 4f | 4f | 4f | |
| | | | 5f | 5f | 5f | 5f | 5f | 5f | 5f | 5f | 5f | 5f | 5f | 5f | 5f | 5f | |

Figure 5.29 Subshell secondary quantum number designations shown on the conventional form of the periodic table. More specifically, the designations show the quantum numbers of the electron that is the last to fill the energy level diagram during the building-up procedure.

If you examine the periodic table in Figure 5.28 four questions arise: why are there *two* columns on the left? Why are those followed by *fourteen*? Why are those followed by *ten*? And why are those followed by *six*? The answer is quantum numbers and how they generate energy levels. Quantum mechanics shows that s-levels can hold two electrons, three p-levels can hold six electrons, five d-levels can hold ten electrons, and seven f-levels can hold fourteen electrons. So the rules of quantum numbers produce the form of the periodic table. The periodic table is a wonderful and experimental fact of nature; it remains the jewel in the crown of chemistry. But as remarked at the beginning of this chapter, the fact that the jewel exists is, in the author's opinion, the best single piece of experimental evidence that quantum mechanics is correct.

Electron configurations can be read directly from Figure 5.28. With Figure 5.28 you just look for the element you need and fill in all the electrons moving steadily through the table until you reach the element you want. Often you will have available a periodic table such as shown in Figure 5.29, in that case your task is a little more complicated because you'll have to watch exactly where the sequence of levels begins and ends.

A hundred years ago it was common to encounter a wide range of types of periodic tables. There have even been books written about the many types. Figure 5.30 shows a new type of periodic table devised by the author to bring together the best aspects of both chemical relationships and the theoretical insight of quantum mechanics.

**Example 5.11**: Write the electron configurations of sulfur, vanadium, and tungsten.

Strategy: Write the electron configurations following the energy levels shown on Figure 5.30.

S = $1s^2 2s^2 2p^6 3s^2 3p^4$, it can also be written as [Ne core]$3s^2 3p^4$

V = $1s^2 2s^2 2p^6 3s^2 3p^6 4s^2 3d^3$, it can also be written as [Ar core]$4s^2 3d^3$

W = $1s^2 2s^2 2p^6 3s^2 3p^6 4s^2 3d^{10} 4p^6 5s^2 4d^{10} 5p^6 6s^2 4f^{14} 5d^3$ or [Xe core]$6s^2 4f^{14} 5d^3$

Check: Add the total number of electrons in the above configurations to see that they match the atomic numbers of $_{16}S$, $_{23}V$, and $_{73}W$.

S = 2 + 2 + 6 + 2 + 4 = 16

V = 2 + 2 + 6 + 2 + 6+ 2 + 3 = 23

W = 2 + 2 + 6 + 2 + 6 + 2 + 10 + 6 + 2 + 10 + 6 + 2 + 14 +3 = 73

All the totals are okay.

| 8 | | | | | | | 119 ? | 120 ? | | | | | | | 8s |
|---|---|---|---|---|---|---|---|---|---|---|---|---|---|---|---|
| | | | | | 113 ? | 114 nn | 115 ? | 116 nn | 117 ? | 118 nn | | | | | 7p |
| | | | 103 Lr | 104 Rf | 105 Db | 106 Sg | 107 Bh | 108 Hs | 109 Mt | 110 nn | 111 nn | 112 nn | | | 6d |
| | 89 Ac | 90 Th | 91 Pa | 92 U | 93 Np | 94 Pu | 95 Am | 96 Cm | 97 Bk | 98 Cf | 99 Es | 100 Fm | 101 Md | 102 No | 5f |
| 7 | | | | | | | 87 Fr | 88 Ra | | | | | | | 7s |
| | | | | | 81 Tl | 82 Pb | 83 Bi | 84 Po | 85 At | 86 Rn | | | | | 6p |
| | | | 71 Lu | 72 Hf | 73 Ta | 74 W | 75 Re | 76 Os | 77 Ir | 78 Pt | 79 Au | 80 Hg | | | 5d |
| | 57 La | 58 Ce | 59 Pr | 60 Nd | 61 Pm | 62 Sm | 75 Re | 64 Gd | 65 Tb | 66 Dy | 67 Ho | 68 Er | 69 Tm | 70 Yb | 4f |
| 6 | | | | | | | 55 Cs | 56 Ba | | | | | | | 6s |
| | | | | | 49 In | 50 Sn | 51 Sb | 52 Te | 53 I | 54 Xe | | | | | 5p |
| | | | 39 Y | 40 Zr | 41 Nb | 42 Mo | 43 Tc | 44 Ru | 45 Rh | 46 Pd | 47 Ag | 48 Cd | | | 4d |
| 5 | | | | | | | 37 Rb | 38 Sr | | | | | | | 5s |
| | | | | | 31 Ga | 32 Ge | 33 As | 34 Se | 35 Br | 36 Kr | | | | | 4p |
| | | | 21 Sc | 22 Ti | 23 V | 24 Cr | 25 Mn | 26 Fe | 27 Co | 28 Ni | 29 Cu | 30 Zn | | | 3d |
| 4 | | | | | | | 19 K | 20 Ca | | | | | | | 4s |
| | | | | | 13 Al | 14 Si | 15 P | 16 S | 17 Cl | 18 Ar | | | | | 3p |
| 3 | | | | | | | 11 Na | 12 Mg | | | | | | | 3s |
| | | | | | 5 B | 6 C | 7 N | 8 O | 9 F | 10 Ne | | | | | 2p |
| 2 | | | | | | | 3 Li | 4 Be | | | | | | | 2s |
| 1 | | | | | | | 1 H | 2 He | | | | | | | 1s |

Figure 5.30 An inverted tree type periodic table. This type of periodic table shows the strong link between classical chemistry and its later quantum mechanical interpretation via energy level diagrams. nn = not named.

## The Discovery of Hafnium

Probably the first successful application of the aufbau principle was Niels Bohr's prediction in 1922 that the then missing element hafnium would be found in the same geologic sources (ores) as zirconium. Two chemists working in Bohr's Copenhagen laboratory, Georg von Hevesy and Dirk Coster, followed up on Bohr's prediction and quickly found hafnium where Bohr had predicted it would be found. They named the element hafnium from the Latin name for Copenhagen (Hafnia). Bohr didn't get an element named after him (Bh, element 107) until 1977. The element hafnium is a much better testament to Bohr's role in helping quantum mechanics explain the periodic table.

## Experimental Electron Configurations

The experimental configurations of the elements have all been worked out from a detailed study of their atomic spectra. But if you take a look at Appendix H and the periodic table that shows experimental electron configurations, you'll discover that the simple building up procedure described in this chapter correctly predicts 91/103 ground state electron configurations of the elements. In Figure 5.31 (Appendix H) the experimental electron configurations of the elements are shown below each element's symbol. Copper is one of the ten elements that shows a minor discrepancy. From the table, it would have the configuration [Argon core]$4s^2 3d^9$. Silver, immediately below copper in the periodic table, has an $s^2 d^9$ configuration. However, experimentally, copper's d-level fills at the expense of the s-level and it has the configuration [Argon core]$4s^1 3d^{10}$.

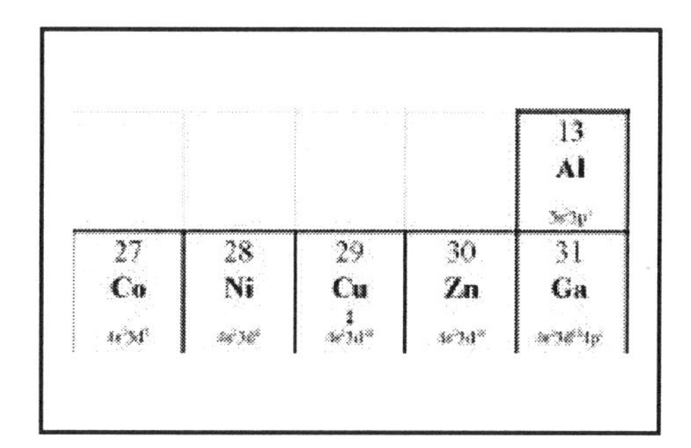


Figure 5.31 A piece of the electron configuration version of the periodic table. The full table is shown in Appendix H.

The most straightforward way to explain the minor differences between experimental electron configurations and the aufbau predicted configurations is to recognize that it is quite a stretch to assume that hydrogen's energy levels work for everything. They don't. As we progress from one element to the next, moving across the periodic table along the horizontal axis, s levels stay relatively fixed in energy, p levels drop slightly, d levels drop rather substantially, and f levels drop rather drastically. Thus, the energy levels shift in *relative* energy depending on where we are in the periodic table. It is these shifts that explain the experimental electron configurations of the elements. Shown in Figure 5.32 are the elements with a slight discrepancy between their experimental configurations and those read from Figure 5.28. They are not a big deal, but you might like to see them.

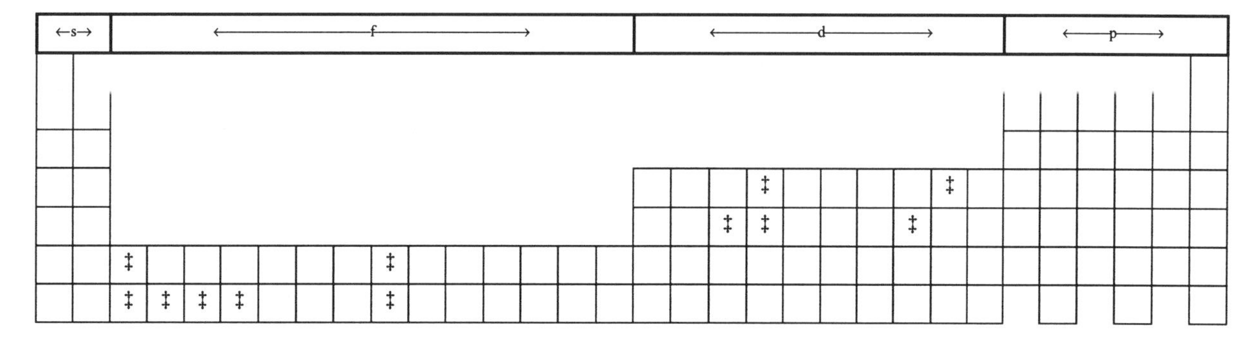


Figure 5.32 Only a handful of elements show minor discrepancies between their experimental electron configurations and those obtained from a generalized energy level diagram. In consequence, it is possible to "read" the correct electron configurations of most elements directly from a periodic table.

Of course the charge on the atomic nucleus grows as we progress through the periodic table. One effect of the increased nuclear charge is to "shrink" the sizes of the orbitals, and make all the electron clouds smaller. So, even though the relative energy states resemble hydrogen's, the actual energy states are quite different from hydrogen. Mostly the energy states don't change their order, which is why there are few exceptions in Figure 5.32.

Now that chemists have formidable computer power it is possible to calculate the electron configurations of any of the elements. If enough computer time is devoted to any one element, it is possible to show that theory and experiment match. However, doing such calculations is rather a sterile exercise, except for new elements where the experimental configurations are not available. For example, in the Spring of 1999 a detailed calculation of the electron configuration of element 114 predicted it to be chemically like lead. That's a good result, because the natural place for element 114 on the periodic table is directly below lead.

## Electron Configurations of Ions

The approximate electron configurations of ions are written either by subtracting electrons in the case of positive ions or adding electrons in the case of negative ions.

**Example 5.12**: Write the electron configuration of $S^{2-}$ and $Ti^{4+}$.

Strategy: Look up the atomic number of the elements and add or subact electrons; S = 16, Ti = 22. Write the electron configurations following the energy levels shown on the periodic table earlier.

$S^{2-}$ has 16 + 2 = 18 total electrons. Its configuration is [Ne core]$3s^2 3p^6$

$Ti^{4+}$ has 22 - 4 = 18 total electrons. Its configuration is also [Ne core]$3s^2 3p^6$

In chemical compounds, what ions are stable and why? The answer to this question can be evaluated by counting the electrons in ions. The rule is: monatomic ions with an identical number of electrons to any of the noble gas atoms are stable. They are exceptionally stable if positively charged and not greater than +3 charge, or, if negatively charged and not greater than -2 charge. Two ions that you know are particularly stable are sodium ion, $Na^+$, and chloride ion $Cl^-$. How do you know they are stable? The sea is full of them. Sodium ion's configuration is [Neon core]$3s^1$ and when it ionizes it becomes just the neon core. A sodium ion and a neon atom are **isoelectronic**—they have the same number and arrangement of electrons. Chlorine has the configuration [Neon core]$3s^2 3p^5$. Adding an electron gives the chloride ion with the configuration [Neon core]$3s^2 3p^6$ or just [Argon core], so $Cl^-$ is isoelectronic with an argon atom.

Rules: Positive ions isoelectronic with a noble gas atom and with a charge of 1+, 2+, or 3+ are common, widely dispersed in nature, and stable. Negative ions isoelectronic with a noble gas atom and with a charge of 1- or 2- are common, widely dispersed in nature, and stable.

The noble gases have great stability. Another name for them is the inert gases; the noble gases are chemically far less reactive than all the other elements in the periodic table. Many of the reactions and properties of non-noble gas elements can be interpreted as those elements "trying" to change their own electron configurations to noble gas atom electron configurations by gaining, losing, or sharing electrons.

That idea, of atoms gaining and losing electrons for stability, brings us to the threshold of chemical bonding, which we will take up in Chapter 6, but this time not with valency and wooden sticks but the proper way, with quantum mechanics.

## Essential Knowledge—Chapter 5

**Glossary words**: Energy state, energy level, atomic spectra, quantum mechanics, periodicity, electromagnetic radiation, electromagnetic spectrum, frequency, wavelength, ultraviolet radiation, infrared radiation, diffraction grating, interference, diffraction pattern, monochromatic light, spectroscopy, spectroscope, bright line spectra, dark line spectra, continuous spectrum, atomic absorption spectroscopy, ground state, excited state, transition, emission spectra, absorption spectra, molecular spectroscopy, energy quantization, Balmer-Rydberg equation, quantum of energy, photoelectric effect, photon, wave-particle duality, matter-waves, Bohr's atomic model, energy level diagram, continuum, quantum mechanics, standing wave, bound states, wave functions, orbital, probability density, electron density map, Heisenberg Uncertainty Principle, electron spin, orbital picture, orbital shape, orbital energy, principal quantum number, secondary quantum number, magnetic quantum number, spin quantum number, electronic states, degenerate, shell, electron configuration, aufbau procedure, valence electrons, core electrons, Hund's rules, isoelectronic.

**Key Concepts**: Electromagnetic radiation and its interaction with atoms. Formation of atomic spectra. Energy quantization and energy states in atoms. Bohr's atomic model. Wave-particle duality and the basis for quantum mechanics. Wave functions as the solution to the Schrödinger equation for the hydrogen atom. Quantum numbers and the shapes and energy states of orbitals. Origin of hydrogen's energy level diagrams. Hydrogen's energy level diagram as a template for the periodic table. Electron configurations, hydrogen's energy level diagram, and the periodic table.

**Key Equations**:

Basic physics: $c = \lambda\nu$, $E = h\nu$. Balmer-Rydberg Equation: $\nu = R_H\left(\frac{1}{n_1^2} - \frac{1}{n_2^2}\right)$

Bohr energy level formula for hydrogen: $E_n = -R_H \frac{h}{n^2}$

De Broglie relationship: $\lambda = \frac{h}{mv}$.

Rule for principal quantum number: $n = 1, 2, 3, 4\ldots$

Rule for secondary quantum number: $\ell = 0, 1, 2, 3, \ldots (n - 1)$

Rule for magnetic quantum number: $m = -\ell, -\ell+1, -\ell+2, \ldots, 0, +1, +2, \ldots, +\ell$

Rule for spin quantum number: $s = \pm\frac{1}{2}$

### Other Equations

Schrödinger equation: $\frac{\partial^2\psi}{\partial x^2} + \frac{\partial^2\psi}{\partial y^2} + \frac{\partial^2\psi}{\partial z^2} + \frac{8\pi^2 m}{h}\left(E + \frac{e^2}{r}\right)\psi = 0$

Wave function for hydrogen's 1s orbital: $\psi_{(1,0,0)} = \frac{1}{\sqrt{\pi}}\left(\frac{Z}{a_o}\right)^{\frac{3}{2}} e^{-\left(\frac{Zr}{a_o}\right)}$

Wave function for hydrogen's 2s orbital: $\psi_{(2,0,0)} = \frac{1}{4\sqrt{2\pi}}\left(\frac{Z}{a_o}\right)^{\frac{3}{2}}\left(2 - \frac{Zr}{a_o}\right)e^{-\left(\frac{Zr}{2a_o}\right)}$

Wave function for hydrogen's $2p_x$ orbital: $\psi_{(2,1,0)} = \frac{1}{4\sqrt{2\pi}}\left(\frac{Z}{a_o}\right)^{\frac{3}{2}} \frac{Zr}{a_o} e^{-\left(\frac{Zr}{2a_o}\right)} \cos\theta$

Wave function for hydrogen's $2p_y$ orbital: $\psi_{(2,1,1)} = \frac{1}{4\sqrt{2\pi}}\left(\frac{Z}{a_o}\right)^{\frac{3}{2}} \frac{Zr}{a_o} e^{-\left(\frac{Zr}{2a_o}\right)} \sin\theta\cos\phi$

## Questions and Problems

### Electromagnetic Radiation

5.1. Consider the following list of types of electromagnetic radiation: radio waves, gamma rays, red light, blue light, ultraviolet light. a. List them in order of decreasing wavelength, and b. list them in order of decreasing frequency.

5.2. The distance between the earth and the sun is about 93 million miles. How long does it take for visible light to travel this distance? How long for radio waves?

5.3. Recalling SI definitions of base units (Appendix A), the second is the duration of 9,192,631,770 periods of the radiation corresponding to the transition between the two hyperfine levels of the ground state of the $^{133}Cs$ atom. The speed of light is 299,792,458 m $s^{-1}$. Calculate the wavelength of the transition.

5.4. Describe the region of the electromagnetic spectrum between 400 nm and 700 nm.

5.5. Radiation with a frequency of 1000 megahertz is in the microwave region of the electromagnetic spectrum. a. What is its wavelength in centimeters? b. Explain why microwave radiation heats foods?

5.6. a. An AM (amplitude modulated) radio station broadcasts on an assigned frequency of 1240 kilohertz. At what wavelength in meters is the AM station broadcasting? b. An FM (frequency modulated) radio station broadcasts at a frequency of 103.5 MHz. At what wavelength in meters is the FM station broadcasting?

5.7. What is the frequency of an x-ray with a wavelength of 54.6 pm?

5.8. Under what circumstances do water droplets create a visible, continuous spectrum?

5.9. a. What is diffraction? b. What is interference of electromagnetic radiation? c. Why do chemists use x-ray diffraction to study crystals? d. What is monochromatic radiation?

### Atomic Spectra

5.10. The Balmer-Rydberg formula is an empirical formula. Explain what the adjective empirical means in this context.

5.11. a. Bright line spectra are produced by excited atoms. What is an excited atom? b. State two methods that are commonly used to excite atoms. c. What problem stimulated the development of the Bunsen burner?

5.12. Describe the way in which the bright line spectrum and the dark line spectrum of a gas are related to each other. What are Fraunhofer lines?

5.13. a. Use the Balmer-Rydberg formula to calculate the frequency of the emitted light when an electron falls from $n = 6$ to $n = 4$ in a hydrogen atom. b. To what region of the electromagnetic spectrum does the emitted photon belong? c. What is the wavelength of this radiation?

5.14. The element helium was discovered in the sun before it was known on earth. How was this accomplished?

### Hydrogen Spectrum and Energy Quantization

5.15. In 1885, Johann Balmer, a Swiss high-school mathematics teacher, derived the following relationship for the wavelengths of the lines in the visible spectrum of hydrogen: $\lambda = 3.646 \times 10^{-7} [n^2/(n^2-4)]$ where lambda is in meters and $n = 3, 4, 5$ etc. Show that this equation is equivalent to the Balmer-Rydberg formula.

5.16. What experimental information was Max Planck trying to explain when he introduced the idea of the quantum into science in 1900?

5.17. Describe the photoelectric effect. Intense red light containing many, many photons cannot eject a single electron from the surface of a metal; however, a single quantum of ultraviolet light can. Explain.

5.18. Calculate the energy of a single photon that has a frequency of $3.2 \times 10^{14}$ $s^{-1}$.

5.19. Calculate the total energy of a mole of photons of red light of wavelength 680 nm. Expresss your answer in the units kJ $mol^{-1}$. Could a single photon of this wavelength break a chemical bond with a bond energy of 255 kJ $mol^{-1}$?

5.20. What is the energy of the photon emitted when an electron falls from energy level n = 4 to n = 3 in a hydrogen atom?

5.21. Approximately how many moles of photons of visible light are produced by a 50 watt light bulb in a week? Assume that 3% of the energy is converted to visible light.

## The Bohr Model of Atoms

5.22. In what way did Bohr make use of the quantum idea to account for the spectrum of the hydrogen atom?

5.23. Bohr's atomic model of 1913 assumed that electrons move in circular orbits around the nucleus. Within just a few years it was realized that this view was wrong. How does the Heisenberg Uncertainty Principle provide a better view?

5.24. If a hydrogen atom is in its first excited level (n=2), how much energy is required to ionize the atom? Calculate the frequency and wavelength of a photon with this much energy.

## Quantum Mechanics

5.25. a. What is the difference between a classical orbit and a quantum mechanical orbital? b. What is meant by the expression "wave-particle duality?" c. How would you make a graph of a rain cloud?

5.26. Define or describe in 25 words or less: a. an orbital, b. a wave function, c. a quantum number, d. electron probability density.

5.27. Calculate the de Broglie wavelength for a hydrogen molecule whose speed corresponds to hydrogen gas's root mean square (rms) speed at 25°C.

5.28. At what speed must a neutron be moving in order that its de Broglie wavelength be equal to 100 pm?

5.29. We think of electrons as particles when they are ejected and free from an atom, but as waves when they are in bound states in an atom. Discuss.

5.30. How is the probability density of an electron in a hydrogen atom interpreted? What is the relationship between the electron's probability density and the wave function?

## Quantum Numbers and Orbitals

5.31. The Pauli Exclusion Principle states that any electron in an atom is uniquely identified by a set of our quantum numbers. Make a table that lists the four quantum numbers and tells their "jobs." I.e., tell what information each of the four quantum numbers provides about the orbital occupied by the electron.

5.32. Explain the meaning of degenerate states. Which of the following orbitals or orbital sets correspond to degenerate states of the hydrogen atom: 3s, $2p_y$, $3p_z$, 5s, $3d_{yz}$, 4d?

5.33. State the rule by which the allowed values of the secondary quantum number, $\ell$, can be derived from a specified value of the principal quantum number, n.

5.34. State the rule by which the allowed values of the magnetic quantum number, m, can be derived from a specified value of the secondary quantum number, $\ell$.

5.35. What are the possible values of the $\ell$ quantum number for a. n = 3, b. n = 4, c. n = 5, d. n = 6?

5.36. What are the possible values of the m quantum number for a. $\ell = 3$, b. $\ell = 4$, c. $\ell = 6$, d. n = 4 (all subshells)?

5.37. What is a. the minimum n value for $\ell = 3$? b. the maximum n value for $\ell = 3$? c. the minimum n value for m = 3? d. the $\ell$ value for a subshell which contains exactly 9 orbitals?

5.38. What is the meaning of the term "ground state of an atom?"

5.39. Which symbol in an orbital designation (3s, $3p_x$, $3d_{xy}$, etc.) indicates the value of the principal quantum number?

5.40. Which symbol in an orbital designation (3s, $3p_x$, $3d_{xy}$, etc.) indicates the value of the secondary quantum number? What information about the orbital is given by the secondary quantum number?

5.41. Which symbol in an orbital designation (3s, $3p_x$, $3d_{xy}$, etc.) indicates the value of the magnetic quantum number? What information about the orbital is given by the magnetic quantum number?

5.42. State the number of orbitals that are possible with the following quantum numbers: a. n = 5; b. n = 5 and $\ell = 4$; c. m = 4; d. n = 5, $\ell = 4$, and m = 3.

5.43. State the number of orbitals which are possible for each of the following orbital designations: a. 4f, b. 6d, c. 5p, d. $3d_{xy}$.

5.44. Each of the following orbital designations is incorrect. Explain in what way each violates the rules of quantum numbers. a. 2d (m = 0), b. 6s (m = 1), c. 5p (m = 5), d. 3f (m = 3).

5.45. Which of the following sub-shell designations are incorrect: 4s, 2f, 3p, 4d, 1p, 5s?

5.46. Describe how the features of the plot of the square of the wave function versus distance from the nucleus for the 1s orbital of the $He^+$ ion will differ compared with the same plot for the hydrogen atom.

5.47. In what ways are the 1s, 2s, and 3s orbitals of the hydrogen atom similar to one another? In what ways are they different?

5.48. Sketch the shapes of the $2p_x$, $2p_y$, and $2p_z$ orbitals of the hydrogen atom. Sketch the shapes of the five 3d orbitals in a hydrogen atom.

## Electron Configurations

5.49. Carefully describe the difference between hydrogen's energy level diagram and the generalized energy level diagram for all the other elements.

5.50. Give an account of why the periodic table takes the shape it does. More specifically, discuss the periodic tables in Figure 5.27 and 5.28.

5.51. Write the electron configurations of the following elements and ions using the positions of the elements in the periodic table: Na, $K^+$, $Al^{3+}$, $O^-$, V, $V^{5+}$, Tc, Hf, Gd, and U.

5.52. Discrepancies between aufbau predicted electron configurations and experimental electron configurations are sometimes explained on the basis of "filled" subshells. Discuss the discrepancy using as your example the difference between the aufbau-predicted and experimental configurations of copper.

5.53. Name two positive and two negative ions that are isoelectronic with xenon. Name the dipositive metal ion that is isoelectronic with a chloride ion.

## Problem Solving

5.54. a. Discuss why sodium ions and chloride ions have great chemical stability. b. Why is a sodium ion difficult to reduce? c. Why is a chloride ion difficult to oxidize?

5.55. As pictured earlier, a CD recording is a common example of an object that behaves like a diffraction grating; name some other everyday examples of diffraction. Make a diagram you could use to explain interference to a fifth-grader.

5.56. Calculate the frequencies and wavelengths for the first three emission lines of the hydrogen atom spectrum that terminate at n = 7. In which region of the electromagnetic spectrum do they occur?

5.57. Infrared radiation is absorbed by black objects and its energy converted to heat. How much heat is produced when 12.8 moles of infrared photons with an average wavelength of 2200 nm is absorbed? If this energy were used to heat 100 kg of water, what would the be temperature rise?

5.58. The minimum energy required to remove an electron from cesium metal is 206 kJ $mol^{-1}$. Calculate the maximum possible kinetic energy for an electron which is ejected when ultraviolet light with a wavelength of 254 nanometers strikes the cesium surface.

5.59. A classic World War II vintage searchlight burns by arcing 150 amps of electric current at 80 volts between a pair of graphite electrodes. If the searchlight is 10% efficient, how long does it need to produce a mole of photons (assume their wavelength averages 550 nm).

5.60. Use your favorite search engine to learn about the following people who have some relevance to the history of electron theories of atoms: Eugene Wigner, Sam Goudsmit, Johannes Stark, P. Zeeman, and Wolfgang Pauli. Which of these, if any, were engineers?

5.61. Predict the electron configuration of a $Ca^{+}$ ion using the periodic table and calcium's ground state electron configuration.

# Chapter 6

# Chemical Bonding

## Introduction

Nature is remarkable. There are 115 elements, but it needs only 10 of them to account for 99.8% of the mass of the earth's crust. There are millions and millions of different compounds—both natural and synthetic—that contain atoms of just these ten elements. Carbon, with its chain-making ability, is a particularly prolific former of compounds. Together, carbon and hydrogen form the largest group of compounds, the hydrocarbons. We saw in Chapter 2 that the hydrocarbon $C_{40}H_{82}$ potentially exists as 62,491,178,805,831 different isomers. In a distant second place are minerals that contain silicon and oxygen, of which the silicates are a major class. Silicon-oxygen compounds form 90% of the earth's crust. Both hydrocarbons and silicates are classes of substances with great diversity. Molecular diversity occurs because atoms can bond in myriad ways. To understand the nature of chemical bonding is one of the central objectives of chemical science. **Chemical bonds** are said to exist whenever atoms join to make a molecule or polyatomic ion. Chemical bonds do not literally exist, you can't have *just* a chemical bond; they are rather a manifestation of the tendency of atoms to join together.

Pairs of elements tend to combine with specific stoichiometric ratios: 1:1, 1:2, 1:3, 2:3, etc. So the idea developed that elements have "combining capacities." Recall from Chapter 2 that an atom's **valency** is the maximum number of hydrogen atoms it combines with. The elements' combining capacities, or valencies, were key experimental facts in developing the periodic table. If you look at the 1908 periodic table in Appendix H you'll see that in 1871 Mendeleev placed valencies as RH, $RH_2$, RO, $RO_2$ etc., at the top of the groups of the periodic table. It became clear about the same time that atoms combine to form molecules with definite sizes, shapes, and geometric arrangements of their atoms. Around 1874, Joseph Achille le Bel (1847-1930) and Jacobus Henricus van't Hoff (1852-1911) independently concluded that molecules containing four atoms bound to a central carbon atom must have tetrahedral geometry—a crucial step in understanding organic chemistry.

By 1861 Alexander Crum Brown (1838-1922) had started to use lines connecting letters to represent bonds between pairs of atoms. In 1865, August Wilhelm von Hoffman (1818-1892) built models using sticks and croquet balls to illustrate his ideas of chemical bonding. His model of the "methide of methyl," that today we call ethane, is represented in Figure 6.1. Ball and stick models remained a primary way to visualize chemical bonding for 150 years. However, the past decade of rapidly advancing computer animation has dramatically improved our ability to view bonding and structure using on-screen chemical visualization programs. The marvelous images of chemical structures that you today can view and manipulate at home or on the internet derive directly from Hoffman's models. [❁kws +chime +download].

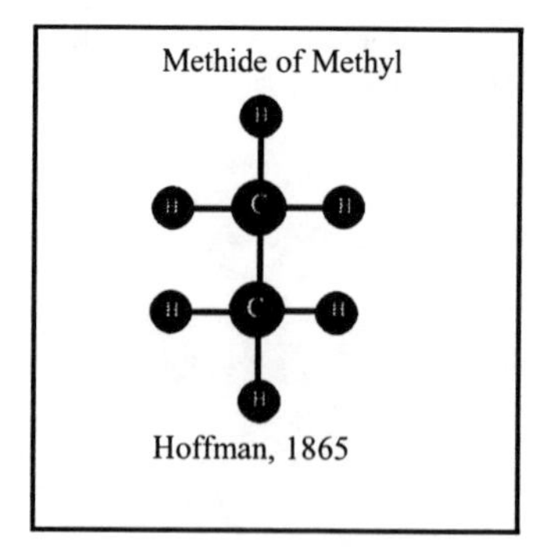


Figure 6.1 Some of the earliest models of chemical structure were made with croquet balls and wooden sticks. In the development of scientific thought, the whole notion of chemical structure came very late. Today we well understand that all of biology and all of materials science are aspects of chemical structure.

Although the concept of valency is both simple and helpful, it's obvious that wooden sticks and balls with holes can't possibly be the basis of chemical bonding. Bonding must depend on the inherent properties of atoms, and more particularly on their electrons. By 1905 some chemists were beginning to suspect that electrons are involved in forming chemical bonds. But it was not until the arrival of Bohr's model of the quantized atom, and subsequently quantum mechanics, that chemists could truly begin to understand chemical bonding. The American chemist Linus Pauling (1901-1994), a native Oregonian who started his career as a crystallographer at the California Institute of Chemistry, traveled to Europe in 1925 and worked with Arnold Sommerfeld for two years, putting him in close contact with the European physicists who were developing quantum mechanics.

Back in the States, Pauling published seminal papers between 1931 and 1933. Then, in 1939, he published the first edition of his classic book *The Nature of the Chemical Bond.* In the preface to that book Pauling wrote "...results of direct chemical interest have been obtained by the accurate solution of the Schrödinger equation." After the book was published, many chemists realized that to understand chemical bonding it would be necessary to apply the Schrödinger equation to molecules and polyatomic ions.

Modern quantum chemistry applies the Schrödinger equation by using it to calculate approximate wave functions of molecules and ions. As computers become ever more powerful so the approximations get ever better, and more and more chemical problems become accessible to accurate calculations. Today, the deepest basis for understanding chemical bonding comes from sophisticated computer programs that perform elaborate quantum mechanical calculations. However, despite the mathematical sophistication involved, most human chemists continue to interpret the results of such calculations in terms of electron energy states and orbital shapes, as they were described in Chapter 5.

In Chapter 1 we saw that there are two major types of chemical compounds: covalent and ionic. In this chapter we'll see that covalent compounds have covalent bonds, whereas ionic compounds have ionic bonds, and also that there are many compounds that are neither purely covalent not purely ionic. We'll see further that the concepts of ionic and covalent bonding can be cast in quantum mechanical terms.

## Section 6.1 Properties of Chemical Bonds

Chemistry, it's worth repeating, is an experimental science. So we'll begin this chapter with a survey of the facts about chemical bonds. The questions to be addressed are: How long are chemical bonds? How strong are chemical bonds? What is the polarity of chemical bonds or are they ionic or covalent? And, when we have at least three atoms, what are the angles formed by chemical bonds?.

Recall, the simplest conceivable chemical bond is H-H in the $H_2$ molecule. Figure 6.2 shows the bond-forming reaction that yields a hydrogen molecule from two atoms of hydrogen. Energy is plotted on the vertical axis; the distance between the two atoms is plotted on the horizontal axis. As the atoms approach, a bond forms and energy is released. The distance d is called the **equilibrium bond length**. Though often omitted, the qualification "equilibrium" is strictly needed because hydrogen molecules vibrate, and their bond length is actually a time-averaged distance. The reaction $H + H \rightarrow H_2$ is exothermic at 298 K. When it happens, the bond energy of 436 kJ $mol^{-1}$ is released. Conversely, bond breaking is an endothermic process. To break a mole of H-H bonds requires an input of 436 kJ. So 436 kJ $mol^{-1}$ is hydrogen's **bond energy** or **bond strength**.

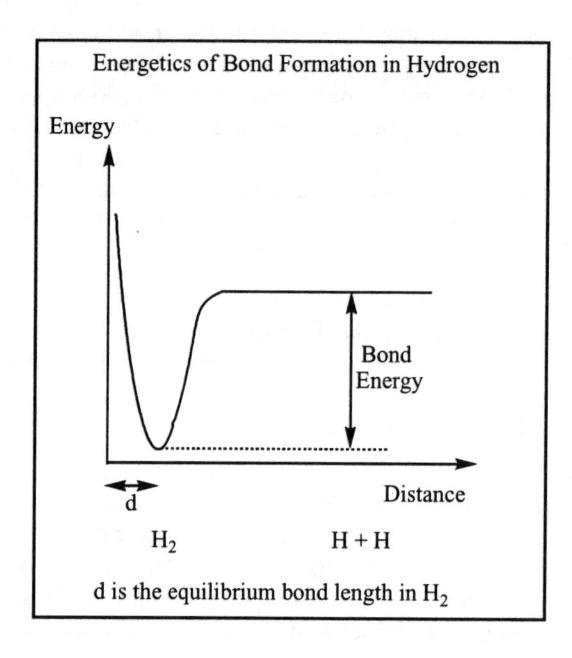


Figure 6.2 Bond formation in a hydrogen molecule. As two hydrogen atoms approach one another they release energy to the environment and join in a chemical bond. Averaged over time, there is a specific distance at which maximum energy has been released; that's the bond length of a hydrogen molecule. If the atoms approach closer than the bond length the energy of the system rises rapidly, as shown by the steeply rising region of the graph on the left..

Hydrogen's electrons must be responsible for the chemical bonding in $H_2$ because of Coulombic attraction. As shown in Figure 6.3 by diagonal lines, the electron of each hydrogen atom is attracted both to its own nucleus and to the nucleus of the other hydrogen atom. The dotted lines show the repulsions of the two electrons for one another and of the two protons for one another.

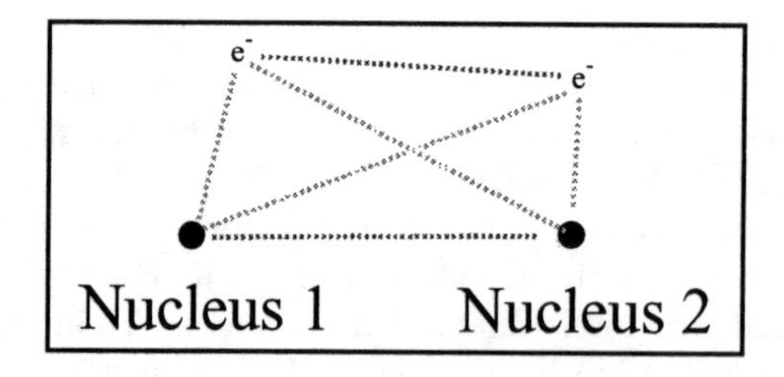


Figure 6.3 In purely Coulombic terms, a chemical bond between two hydrogen atoms arises from the attraction of both atom's positively charged nuclei for the negatively charged electrons of the other atom.

Note from Figure 6.2 that if two hydrogen atoms approach closely, the repulsive force grows large. Observe on the graph that energy rises steeply at short distances. The practical consequence of this short range energy rise is that people do not fall through their seats or through the floor. Such Coulombic repulsions make liquid and solid matter almost incompressible. Automobile engineers take advantage of this fact when they use a liquid in a hydraulic system to transfer pressure from a foot on a brake pedal to a car's brake pads.

## Bond Strengths

The strength of a chemical bond is the energy required to break it. Bond energies are measured experimentally by calorimetric methods, so they are a topic of chemical thermodynamics, which we'll take up in the next chapter. Bond strengths depend on the electron structures of the combining atoms. According to both historic and quantum mechanical bonding theories, the more electrons in the region between the bonded atoms the stronger their bond.

The bond energy in a diatomic molecule has a single value. In contrast, the energy of a specified bond in a molecule with three or more atoms varies from one molecule to another. So chemists make tables that have a mixture of *actual* and *average* bond energies. Also, as you've seen with carbon-carbon bonds, some pairs of atoms can form single, double, and even triple bonds, all with different energies. Typical bond energies are shown in Table 6.1.

Table 6.1 **Bond Energies**

| Single Bonds | Energy kJ mol$^{-1}$ | Double Bonds | Energy kJ mol$^{-1}$ | Triple Bonds | Energy kJ mol$^{-1}$ |
|---|---|---|---|---|---|
| H-H (diatomic) | 437 | | | | |
| H-O (average) | 460 | | (diatomic) means the energy is for a specific diatomic molecule as shown. | | |
| H-N (average) | 390 | | | | |
| H-C (average) | 415 | | | | |
| H-S (average) | 370 | | | | |
| H-P (average) | 326 | | (average) means the energy shown is taken as the average value of that particular bond energy across a group of different molecules containing that bond. | | |
| H-F (diatomic) | 568 | | | | |
| H-Cl (diatomic) | 432 | | | | |
| H-Br (diatomic) | 366 | | | | |
| H-I (diatomic) | 298 | | | | |
| Cl-Cl (diatomic) | 243 | | | | |
| Br-Br (diatomic) | 193 | | | | |
| I-I (diatomic) | 151 | | | | |
| C-N (average) | 275 | C=N (average) | 615 | C≡N (average) | 890 |
| C-C (average) | 368 | C=C (average) | 720 | C≡C (average) | 962 |
| C-O (average) | 351 | C=O (average) | 780 | C≡O (diatomic) | 1,077 |
| C-S (average) | 255 | C=S (average) | 480 | | |

Bonds with energies of 100 kJ mol$^{-1}$ or less are weak. Compounds containing such bonds rarely survive in the Earth's environment; however, such compounds may be quite stable in the vast, cold gas clouds found in many remote regions of the universe. Bonds around 400 kJ mol$^{-1}$ can be interpreted as single bonds; bonds around 600 kJ mol$^{-1}$ can be interpreted as double bonds; and, bonds around 900-1100 kJ mol$^{-1}$ can be interpreted as triple bonds.

Table 6.2 shows the bond energies of homonuclear diatomic molecules of the first three rows of the main group elements. A **homonuclear diatomic molecule** (A-A or $A_2$) is one containing two identical atoms.

Table 6.2
**Experimental Bond Energies of Main Group Diatomic Molecules**

| $Li_2$ | $Be_2$ | $B_2$ | $C_2$ | $N_2$* | $O_2$* | $F_2$* | $Ne_2$ |
|---|---|---|---|---|---|---|---|
| 106.5 | 59 | 297 | 607 | 945 | 498 | 158.3 | 3.9 |
| $Na_2$ | $Mg_2$ | $Al_2$ | $Si_2$ | $P_2$ | $S_2$ | $Cl_2$* | $Ar_2$ |
| 77 | 8.6 | 186 | 326.8 | 489.5 | 425.2 | 243 | 4.7 |
| $K_2$ | $Ca_2$ | $Ga_2$ | $Ge_2$ | $As_2$ | $Se_2$ | $Br_2$* | $Kr_2$ |
| 57.3 | 15.0 | 138 | 273.6 | 382.0 | 332.6 | 194 | 5.2 |

Molecules shown with a * in Table 6.2 are ordinary, stable diatomic molecules. The small values shown for the diatomic noble gases are the consequence of the weak van der Waals forces of intermolecular attraction described in

Chapter 4; there is no real bond between noble gas atoms. The other molecules in Table 6.2 are studied in gas discharge tubes, outer space, etc.

Some trends can be discerned among the homonuclear bond energies in Table 6.2, but it's obvious that the theory that explains the bond energy values will not be a simple one. That theory is of course chemical quantum mechanics. After we've developed some quantum mechanical theory for molecules we'll come back to interpret the data in the table above.

## Bond Lengths

Like bond strengths, bond lengths are capable of direct experimental measurement. There are many methods of measuring bond lengths, but by far the most important is **x-ray crystallography**. The technique of x-ray crystallography is described in Appendix G.

Because electron distribution in an atom is a probability function that tails away, we can't state the exact size of an isolated atom, or ion; you saw in Chapter 5 that orbital pictures have arbitrary contour lines. But measurable interatomic distances *do* exist once atoms or ions either bond to other atoms in molecules or form an array in an ionic solid. For example, the **atomic radius** of hydrogen is taken as one-half the H-H distance in a hydrogen molecule. The **covalent radius** of an atom is derived from a comprehensive study of its bond distances in a range of molecules. Likewise, the **ionic radius** of an ion of specified charge is derived from a comprehensive study of its interionic distances in a range of solid compounds. The shortest known bond length, 74.6 pm, is in the hydrogen molecule, H-H. One of the longest known bond lengths, about 530 pm, occurs between pairs of cesium atoms in cesium metal. Other bond lengths fall between these extremes. Atomic and ionic radii correlate with electron configurations and show strong periodicity.

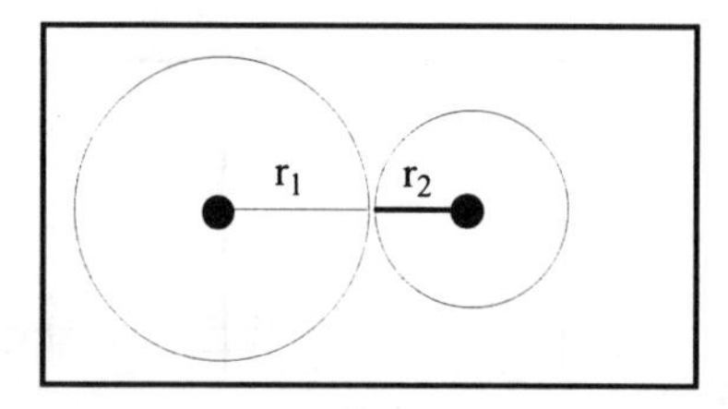


Figure 6.4 A simple view of atomic radius derives from imagining the two atoms in a diatomic molecule as hard spheres that touch. In this context, bond length is the sum of the radii of the hard spheres.

If we imagine that atoms are spherical, the length of the bond between them is the sum of the radii of the two atoms, as illustrated in Figure 6.4. For covalent chemical compounds we use tables of average covalent radii. The covalent radii in picometers of many elements shown in periodic table format in Table 6.3.

Table 6.3 **Atomic Radii (Single Covalent Bond Radii) of the Elements (pm)**

| | | | | | | | | | | | | | | | | | |
|---|---|---|---|---|---|---|---|---|---|---|---|---|---|---|---|---|---|
| 37 | | | | | | | | | | | | | | | | | |
| 134 | 90 | | | | | | | | | | | 82 | 77 | 75 | 73 | 64 | |
| 154 | 130 | | | | | | | | | | | 118 | 117 | 108 | 103 | 99 | |
| 196 | 174 | 144 | 132 | 125 | 127 | 146 | 120 | 126 | 120 | 138 | 131 | 126 | 122 | 120 | 116 | 114 | 115 |
| 211 | 192 | 162 | 148 | 137 | 145 | 156 | 126 | 135 | 131 | 153 | 148 | 144 | 140 | 140 | 137 | 133 | 126 |
| 225 | 198 | 164 | 149 | 138 | 146 | 159 | 128 | 137 | 128 | 143 | 151 | 152 | 147 | 146 | | | |
| | | | | | | | | | | | | | | | | | |

From Table 6.3 you can see that atoms get larger going down any periodic group. They also generally get smaller going from left to right—but it's not a smooth trend. Across any row of the table, the largest atom is the alkali metal in the first group—large atoms tend to lose electrons relatively easily. The smallest atom is the halogen in the penultimate group—small atoms tend to gain electrons relatively easily. With a table of covalent radii you can add them to make a good estimate of actual single bond lengths as shown in Example 6.1.

**Example 6.1**: Estimate the length of an F-Cl bond from the data provided. Estimate the lengths of the bonds in ammonia ($NH_3$) and methane ($CH_4$). The needed atomic radii are H = 37 pm, C = 77 pm, N = 75 pm. Finally, estimate the length of the C-C bond in ethane ($C_2H_6$).

Strategy: Use the relationship bond length = $r_1 + r_2$. Look up the atomic radii of F , Cl, C, H, and N.

Bond length (F-Cl) = 64 pm + 99 pm = 163 pm

Bond length (C-H) = 77 pm + 37 pm = 114 pm

Bond length (N-H) = 75 pm + 37pm = 112 pm

Bond length (C-C ) = 77 pm + 77 pm = 154 pm

Note that the C-C single bond length is in good agreement with the value in Table 2.6.

The sizes of ions are *very* different in size from the atoms they come from. Positive ions are much smaller, and negative ions are much larger, than the atoms they come from. Table 6.4 compares ionic and covalent radii.

Table 6.4 **Comparison of Covalent and Ionic Radii in Picometers (pm)**

| Group 1 Elements | | | | | Group 2 Elements | | | | | Group 17 Elements | | | |
|---|---|---|---|---|---|---|---|---|---|---|---|---|---|
| Li | 134 | $Li^+$ | 68 | | Be | 90 | $Be^{2+}$ | 35 | | F | 64 | $F^-$ | 133 |
| Na | 154 | $Na^+$ | 97 | | Mg | 130 | $Mg^{2+}$ | 82 | | Cl | 99 | $Cl^-$ | 181 |
| K | 196 | $K^+$ | 133 | | Ca | 174 | $Ca^{2+}$ | 99 | | Br | 114 | $Br^-$ | 196 |
| Rb | 211 | $Rb^+$ | 147 | | Sr | 192 | $Sr^{2+}$ | 112 | | I | 133 | $I^-$ | 220 |
| Cs | 225 | $Cs^+$ | 167 | | Ba | 198 | $Ba^{2+}$ | 134 | | | | | |

Comparing the size of an atom and the ions that are isoelectronic with it provides a clear-cut example of the effect of nuclear charge on the size of atoms and ions. For atoms and ions with a given electron configuration, those with greater nuclear charge will be smaller. Table 6.5 shows the sizes of argon and the ions that are isoelectronic with it.

Table 6.5 **Radii of Various Entities with 18 Electrons**

| | $S^{2-}$ | $Cl^-$ | Ar | $K^+$ | $Ca^{2+}$ | $Sc^{3+}$ | $Ti^{4+}$ |
|---|---|---|---|---|---|---|---|
| Nuclear Charge | 16 | 17 | 18 | 19 | 20 | 21 | 22 |
| Electron count | 18 | 18 | 18 | 18 | 18 | 18 | 18 |
| Ionic radius (pm) | 184 | 181 | 150? | 133 | 99 | 73 | 68 |

**Editorial comment:** It's an ironic fact that the key to a lot of chemistry is that the noble gases don't have any chemistry (well, hardly any!). So the number one rule of chemical behavior it that all the other elements spend their time trying to get to be like (isoelectronic with) the noble gas nearest to them in the periodic table.

## Bond Polarity

**Polarity** means separation of electric charge. A purely ionic bond, such as $Na^+Cl^-$, has the maximum possible polarity for a Na-Cl bond because one electron has been completely transferred from a sodium atom to a chlorine atom forming a $Na^+$ ion and a $Cl^-$ ion. A covalent bond between the two *identical* atoms in a homonuclear diatomic molecule has zero separation of charge: any homonuclear diatomic molecule is **non polar** and its bond is said to be a **pure covalent bond**. All homonuclear diatomic molecules, $H_2$, $N_2$, $O_2$, $Cl_2$, etc, are non-polar and have pure covalent bonds.

A **polar covalent bond** is intermediate between a pure covalent bond and an ionic bond. A polar covalent diatomic molecule (AB) can be represented as $A^{\delta+}$-$B^{\delta-}$ where $\delta^+$ signifies a partial positive charge and $\delta^-$ signifies a partial negative charge.

**Example 6.2**: Show an example of a polar covalent bond.

Strategy: Select the diatomic molecule HCl and represent it in the $A^{\delta+}$–$B^{\delta-}$ format.

The polar covalent bond in HCl can be written as $H^{\delta+}$–$Cl^{\delta-}$

To perform an experimental study of bond polarity it is convenient to study diatomic molecules. A diatomic molecule contains just one chemical bond. Bond polarities in compounds composed of diatomic molecules are obtained experimentally from the compound's dipole moment (μ). A **moment** is a twisting force (as every engineer knows) and **dipole** refers to two poles, or separated electrical charges. A dipole moment is shown schematically in Figure 6.5, where the labels $\delta^+$ and $\delta^-$ refer to a partial charges +q and -q. A polar diatomic molecule (one with a dipole moment) put in an electric field will tend to align itself with the field. This behavior is measured and used to calculate a molecule's dipole moment. For the dipole system in Figure 6.5 the bond length is R. One atom has the partial charge +q, the other -q. The **dipole moment** of a diatomic molecule is the product of the charge separation in the molecule and the distance between its atoms (the molecule's bond length):

$$\mu = q\,R$$

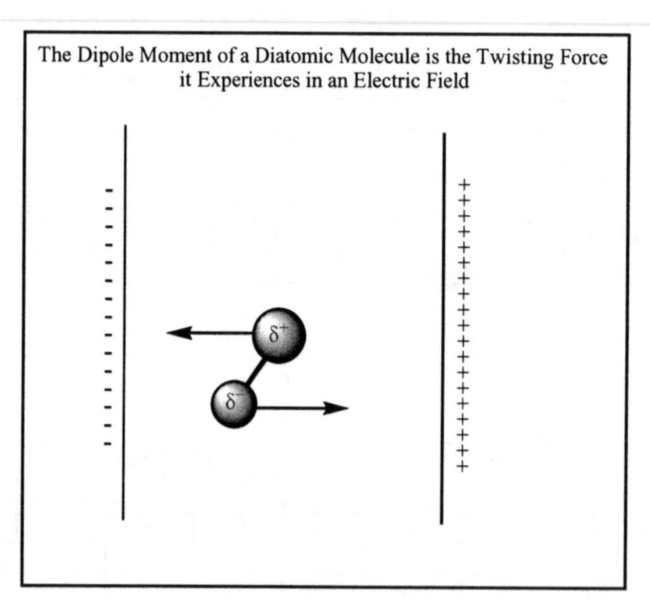


Figure 6.5 A polar diatomic molecule twists in an electric field to align itself with the field.

The measured dipole moments of molecules are stated in debye units (D); one debye is defined as 1 D = 3.33 × $10^{-30}$ coulomb meters (C m). Experimental dipole moments for diatomic molecules can be interpreted using the above equation. In a pure covalent bond q = 0 and there is no dipole moment. In an ionic bond an entire electron is transferred from one atom to the other and the charges are +e and -e; the dipole moment for one electron transfer is:

$$\mu(\text{one electron transfer}) = e\,R$$

A dipole moment calculation is shown in Example 6.3.

**Example 6.3**: Calculate the dipole moment in debye units of a diatomic ionic compound whose atoms have charges of +e and -e corresponding to a one electron transfer, and whose bond length is 150 pm.

Strategy: Use μ(one electron transfer) = e R. The charge on the electron e = $1.602 \times 10^{-19}$ C. Divide by the proper conversion factor to change C m to D.

$$\mu(\text{one electron transfer}) = e\,R$$

$$= 1.602 \times 10^{-19}\,\text{C} \times (150 \times 10^{-12}\,\text{m}) \quad = 2.4 \times 10^{-29}\,\text{C m}$$

$$\text{converting}, = 2.4 \times 10^{-29}\,\text{C m} \div (3.33 \times 10^{-30}\,\text{C m D}^{-1})$$

$$= \underline{7.2\,\text{D}}$$

To obtain the extent of polarity in the bond of a diatomic molecule we do the following: (1) measure the bond length and the dipole moment of the compound, (2) calculate the dipole moment of the molecule assuming a one electron transfer, (3) express the experimental dipole moment as a percentage of the calculated one-electron-transfer dipole moment. Doing this gives the percentage ionic character of the bond.

$$\text{Percentage ionic character} = \frac{\text{Experimental dipole moment}}{\text{Calculated one-electron transfer dipole moment}} \times 100\%$$

In Table 6.6 are shown experimental data for the bond lengths and dipole moments of some diatomic molecules. Also shown are the calculated dipole moments based on the assumption of one-electron transfer, and the percentage of ionic character in the bond. Finally, in the rightmost column, the bond type is classified.

Table 6.6 **Experimental and Calculated Dipole Moments of Some Diatomic Molecules**

| Diatomic Molecule | Experimental Bond Length (pm) | Experimental Dipole moment (D) | Calculated dipole* (D) | Percentage of ionic character | Bond type |
|---|---|---|---|---|---|
| $H_2$ | 74 | 0 | 3.59 | 0% | Nonpolar |
| $N_2$ | 110 | 0 | 5.28 | 0% | Nonpolar |
| $Cl_2$ | 199 | 0 | 9.56 | 0% | Nonpolar |
| HI | 160 | 0.44 | 7.72 | 6% | Slightly polar |
| HBr | 141 | 0.82 | 6.80 | 12% | Moderately polar |
| HCl | 128 | 1.11 | 6.12 | 18% | Moderately polar |
| HF | 92 | 1.82 | 4.41 | 41% | Very polar |
| LiBr | 220 | 7.3 | 10.4 | 70% | Ionic |
| LiCl | 202 | 7.1 | 9.69 | 74% | Ionic |
| NaCl | 236 | 9.0 | 11.3 | 79% | Ionic |
| KCl | 267 | 10.3 | 12.8 | 80% | Ionic |
| LiF | 156 | 6.3 | 7.51 | 84% | Ionic |

* The calculated dipole moment is based on the assumption that one electron charge is separated.

In an ionic bond, one (or more) electrons are transferred from one atom to the other, whereas in covalent bonding the electrons that create the bond remain more or less symmetrically distributed between the two atoms. If we regard pure ionic and pure covalent bonds as special cases at the ends of a **bond-type continuum**, then the majority of bonds fall somewhere between them.

The dipole moments of polyatomic molecules can be measured in exactly the same way as those of diatomic molecules. If the polyatomic molecule has a high symmetry, its dipole moment may be exactly zero. Unsymmetrical polyatomic molecules usually have dipole moments. To a good approximation, the dipole moment of a polyatomic molecule is the **vector sum** of its individual bond dipoles.

The bent water molecule has a dipole moment, but the linear carbon dioxide molecule does not. In water the H-O-H angle is 104.5°, whereas the O=C=O angle in carbon dioxide is exactly 180°. The symmetry in $CO_2$ means that the two bond dipoles =C=O and O=C= cancel.

For some simple molecules, a single measurement of their dipole moment is sufficient to reveal their structure. For instance, as shown in Figure 6.6, the molecule $BF_3$ has a zero dipole moment, despite the fact that a B-F bond has an estimated percentage ionic character of 50%. The only conceivable structure that gives a bond dipole vector sum of zero is planar with 120° bond angles as shown In Figure 6.6.

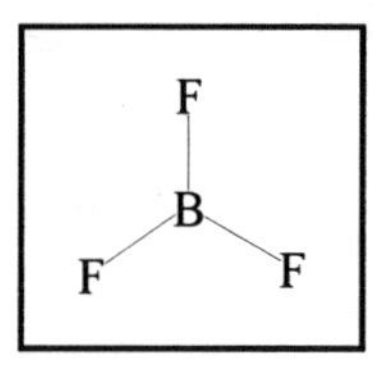


Figure 6.6 Boron trifluoride is a planar molecule shaped like an equilateral triangle with a boron atom in the middle. For simple polyatomic molecules with high symmetry, knowing that the molecule has a zero dipole moment is sufficient evidence to establish the molecular structure of the compound.

## Electronegativity

What is it that controls the percentage of ionic character in a bond? The deep answer is, of course, quantum mechanics. A simpler approach, traditionally used by chemists, makes use of an arbitrary property assigned to each atom of an element called electronegativity. The **electronegativity** of an atom is a unit-less value that measures the relative tendency of that atom to attract electrons. Table 6.7 shows Pauling's 1960 electronegativity values.

Table 6.7 **Pauling's 1960 Electronegativity Values**

| | | | | | | | | | | | | | | | | | |
|---|---|---|---|---|---|---|---|---|---|---|---|---|---|---|---|---|---|
| H<br>2.1 | | | | | | | | | | | | | | | | | He<br>- |
| Li<br>1.0 | Be<br>1.5 | | | | | | | | | | | B<br>2.0 | C<br>2.5 | N<br>3.0 | O<br>3.5 | F<br>4.0 | Ne<br>- |
| Na<br>0.9 | Mg<br>1.2 | | | | | | | | | | | Al<br>1.5 | Si<br>1.8 | P<br>2.1 | S<br>2.5 | Cl<br>3.0 | Ar<br>- |
| K<br>0.8 | Ca<br>1.0 | Sc<br>1.3 | Ti<br>1.5 | V<br>1.6 | Cr<br>1.6 | Mn<br>1.5 | Fe<br>1.8 | Co<br>1.8 | Ni<br>1.8 | Cu<br>1.9 | Zn<br>1.6 | Ga<br>1.6 | Ge<br>1.8 | As<br>2.0 | Se<br>2.4 | Br<br>2.8 | Kr<br>- |
| Rb<br>0.8 | Sr<br>1.0 | Y<br>1.2 | Zr<br>1.4 | Nb<br>1.6 | Mo<br>1.8 | Tc<br>1.9 | Ru<br>2.2 | Rh<br>2.2 | Pd<br>2.2 | Ag<br>1.9 | Cd<br>1.7 | In<br>1.7 | Sn<br>1.8 | Sb<br>1.9 | Te<br>2.1 | I<br>2.5 | Xe<br>- |
| Cs<br>0.7 | Ba<br>0.9 | Lu<br>1.1 | Hf<br>1.3 | Ta<br>1.5 | W<br>1.7 | Re<br>1.9 | Os<br>2.2 | Ir<br>2.2 | Pt<br>2.2 | Au<br>2.4 | Hg<br>1.9 | Tl<br>1.8 | Pb<br>1.8 | Bi<br>1.9 | Po<br>2.0 | At<br>2.2 | Rn<br>- |
| Fr<br>0.7 | Ra<br>0.9 | Lr | Rf | Db | Sg | Bh | Hs | Mt | Uun | Uuu | Uub | | Uuq | | Uuh | | Uuo<br>- |

| | | | | | | | | | | | | | |
|---|---|---|---|---|---|---|---|---|---|---|---|---|---|
| La<br>1.1 | Ce<br>1.1 | Pr<br>1.1 | Nd<br>1.1 | Pm<br>1.1 | Sm<br>1.1 | Eu<br>1.1 | Gd<br>1.1 | Tb<br>1.1 | Dy<br>1.1 | Ho<br>1.1 | Er<br>1.1 | Tm<br>1.1 | Yb<br>1.1 |
| Ac<br>1.1 | Th<br>1.3 | Pa<br>1.5 | U<br>1.7 | Np<br>1.3 | Pu<br>1.3 | Am<br>1.3 | Cm<br>1.3 | Bk<br>1.3 | Cf<br>1.3 | Es<br>1.3 | Fm<br>1.3 | Md<br>1.3 | No<br>1.3 |

At a superficial level, the polarity of a bond (A-B) depends on the electronegativity difference ($\Delta \mathbf{X} = |\mathbf{X}_A - \chi_B|$) of its atoms. As the electronegativity difference increases, electron transfer increases, and the bonding becomes more ionic and less covalent. In a polar bond (one with some ionic character), the more electronegative atom will be negatively charged and the less electronegative one positively charged. Bond polarity is a function of $\Delta \mathbf{X}$. The largest electronegativity differences occur between the alkali and alkaline earth metals (from the left side of the periodic table) and the halogens (from the right side of the periodic table), so their bonds are highly polar, consistent with the ionic nature of the alkali halides. If two elements form bonds with more than 50% ionic character, their compound is a solid with the typical ionic properties. Consequently, as a practical matter any bond with more than 50% ionic character, corresponding to an electronegativity difference of 2, belongs to an ionic substance.

Figure 6.7 shows a plot of percentage ionic character of a bond between two atoms versus the electronegativity difference between those atoms. To estimate the percentage ionic character of any chemical bond you calculate the electronegativity values of the two atoms (from the data of Table 6.7), by subtracting the smaller from the larger and read the approximate percentage ionic character from Figure 6.7.

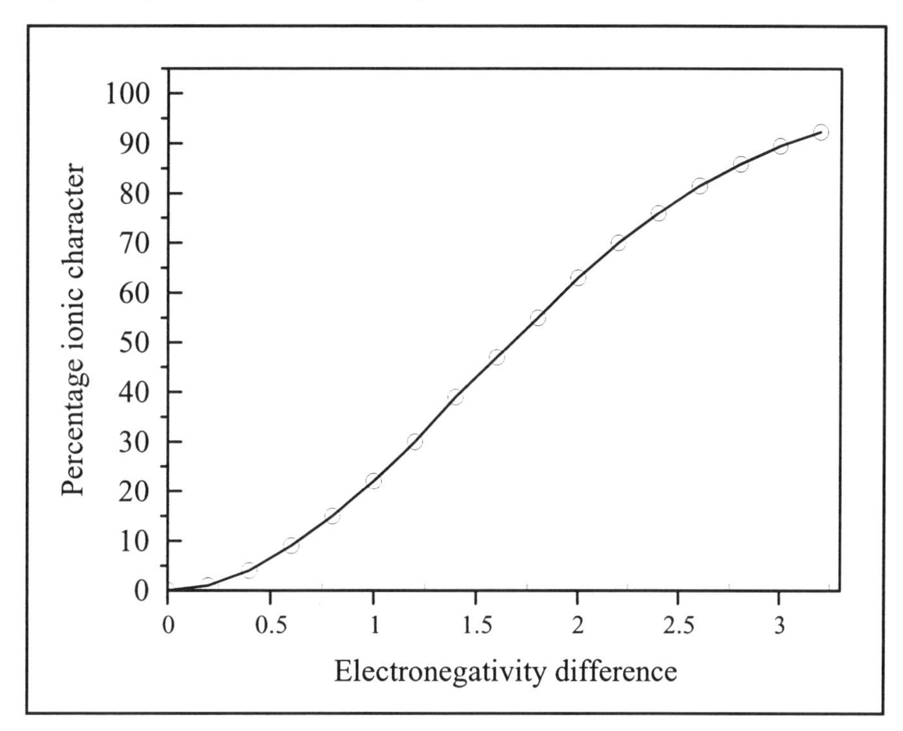


Figure 6.7 The percentage of ionic character of a bond plotted as a function of the electronegativity difference of the two atoms that form the bond. The property of atoms called electronegativity is a constructed rather than real property. However electronegativity differences between the bonded atoms correlate well with the experimental dipole moments of diatomic molecules. The curve shown here is based on an equation originally proposed by Linus Pauling in 1939. It is only approximate. As Pauling wrote in 1960, “We cannot hope to formulate an expression for the partial ionic character of bonds that will be accurate.” Quantum mechanics is not a simple theory.

**Example 6.4**: Use electronegativity differences and Figure 6.7 to fill in the empty column in the table below. Compare the percentages you obtain using electronegativities with the values in the preceding column that were calculated from the formula

$$\text{Percentage ionic character} = \frac{\text{Experimental dipole moment}}{\text{Calculated one-electron transfer dipole moment}} \times 100\%.$$

| Diatomic Molecule | Percentage of ionic character calculated from the formula above. | Percentage of ionic character from calculated electronegativity difference and Figure 6.7 | Bond type |
|---|---|---|---|
| HI | 6% | | Slightly polar |
| HBr | 12% | | Moderately polar |
| HCl | 18% | | Moderately polar |
| HF | 41% | | Very polar |

Strategy: Calculate the electronegativity difference of each bonded pair of atoms using the formula ($\Delta X = |X_A - \chi_B|$). Read the percentage ionic character from the curve in Figure 6.7.

| Diatomic Molecule | $X_A$ | $\chi_B$ | $\|\Delta X\|$ | Percentage of ionic character from Figure 6.7 |
|---|---|---|---|---|
| HI | 2.1 | 2.5 | 0.4 | 4% |
| HBr | 2.1 | 2.8 | 0.7 | 12% |
| HCl | 2.1 | 3.0 | 0.9 | 19% |
| HF | 2.1 | 4.0 | 1.9 | 58% |

Comment: The agreement is reasonable for HI, HBr, and HCl, but HF's value is obviously out of line. Pauling worked for a number of years to find a better fit before concluding that no simple equation would fit all four hydrogen halide molecules.

## Bond Angles

The final experimental property of bonds to consider is bond angles. That means we'll be talking about molecules with at least three atoms—which is most molecules. **Bond angle** is defined in Figure 6.8.

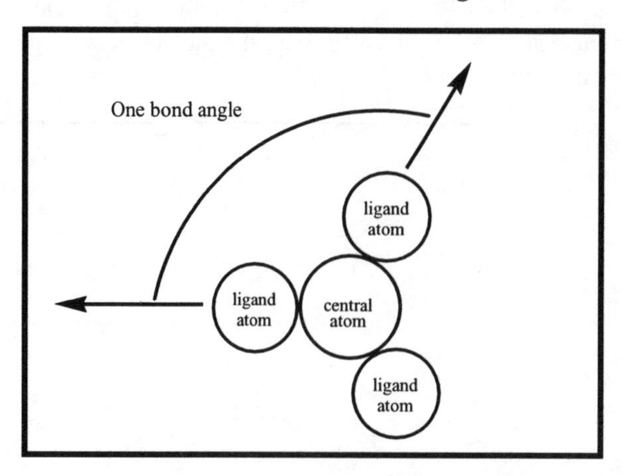


Figure 6.8 Bond angles are easily visualized if you think about atoms as hard spheres.

Like bond lengths, bond angles are most frequently measured by x-ray crystallography. The number of bond angles at a given atom depends on the number of ligand atoms that surround a central atom. The number of ligand atoms that can surround a given central atom depends on two factors: (1) the valency of the central atom, and (2) the size of the central atom compared to the size of the ligand atoms—the radius ratio.

$$\text{Radius ratio} = \frac{\text{radius of the central atom}}{\text{radius of the ligand atoms}}$$

If the radius ratio is large, then many ligands can "fit" around the central atom. If the radius ratio is small, only a few can. The radius ratio principle is illustrated in two dimensions in Figure 6.9.

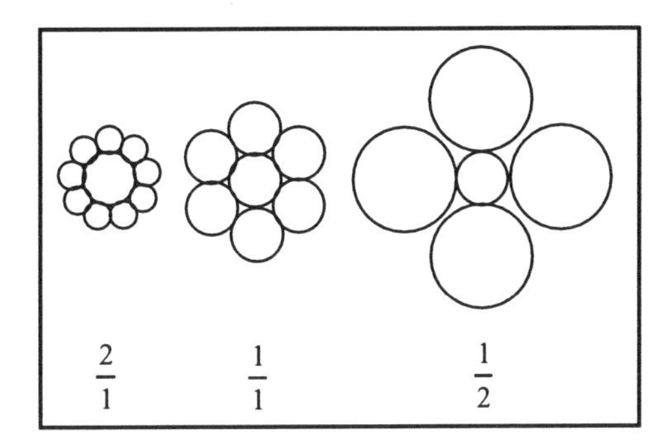


Figure 6.9 Radius ratio illustrated in two dimensions. Despite the optical illusion, the "central atoms" are all the same size. The smaller the radius ratio the fewer circles can surround the central circle.

The largest conceivable bond angle is 180°. Linear polyatomic molecules such as triatomic carbon dioxide, O=C=O, and tetratomic acetylene molecule, H-C≡C-H, have only 180° bond angles. The smallest bond angle is apparently not recorded. The bond angle in the cyclic hydrocarbon cyclopropane in Figure 6.10 is exactly 60°. Some bond angles are even less than 40°. For example, the C-Pt-C bond angle in a compound made from platinum and an acetylene derivative measures only 38°. It's safe to say that almost all bond angles are between 35° and 180°.

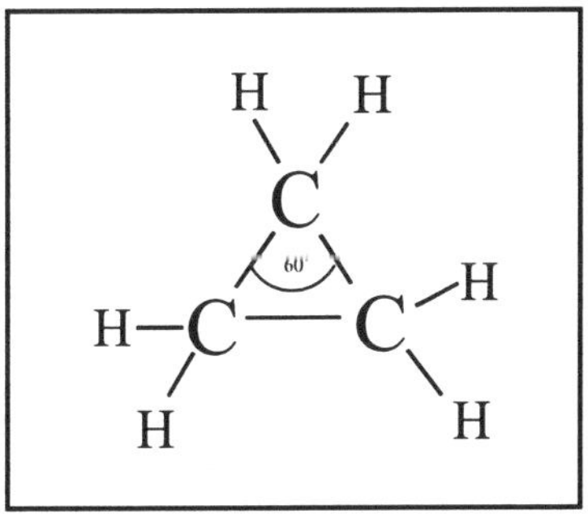

Figure 6.10 Cyclopropane, $C_3H_6$, has an unusually small C-C-C angle of 60°. Compared with other saturated hydrocarbons it is unusually reactive, reflecting the destabilizing effect of the small angle.

All the bond lengths and all the bond angles around all the central atoms in a molecule constitute its **molecular geometry** or **molecular structure**. Typically the structure is measured by x-ray crystallography, which provides a list of the x, y, and z coordinates of the atomic positions—it's then a matter of simple algebra to calculate all the bond lengths and angles from the coordinate list (Appendix G).

Now that we have laid out the key properties of bonds, and introduced the topic of molecular structure, it's time to look at the historical and quantum mechanical theories of bonding.

## Section 6.2 Static Electron Models for Chemical Bonding

A **static electron model** is one that uses stationary electrons or electron pairs to explain the properties of bonds and molecules.

The notion that electrons are responsible for chemical bonding belongs to the first decade of the twentieth century. By 1900-1905, William Ramsey (1852-1916), Gilbert N. Lewis (1875-1946), and others had begun to speculate about how electrons might create chemical bonds. The first attempts to base chemical bonding on electrons involved electrons arranged in definite ways around the atom. The electrons were imagined to take up definite static positions, and both Lewis and Walther Kossel (1888-1956) recognized that a static model of electron bonding using eight electrons would explain the well-known valencies of the elements (Figure 6.11). Lewis used cubes and Kossel used rings to show this "octet rule." The **octet rule** states that "Bonded atoms tend to be surrounded by eight electrons." You've already heard about the octet rule: it's simply the notion that elements try to imitate the electron counts of the noble gases with their $s^2p^6$ electron configurations.

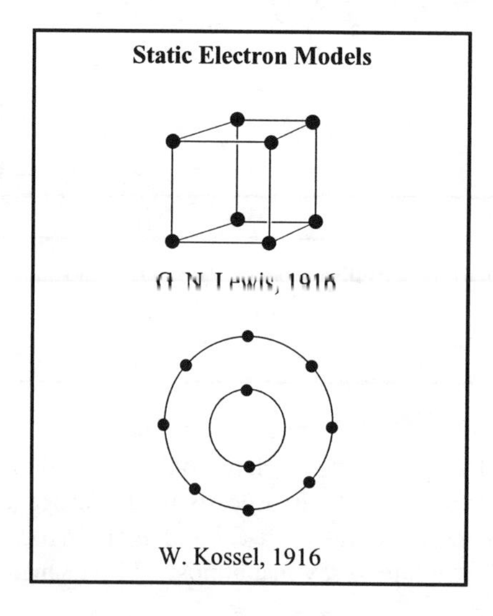


Figure 6.11 Static Electron Models

After Bohr described the quantum atom around 1912, interest in the problem of understanding chemical bonding as an electron phenomenon was greatly stimulated. By 1916 there were independent proposals of electron-based bonding by Lewis, Kossel, and Irving Langmuir (1881-1957). In 1919 Langmuir stated that electron sharing between two atoms is the basis for chemical bonding. He coined the word *covalent* to describe this concept. At about this time there arose the notion that a single chemical bond is equivalent to a shared pair of electrons between two atoms.

### Electron Configurations, Core, and Valence Electrons

The octet rule and the electron pair bond are related to the electron configurations of atoms that we've already studied. In the Lewis view of molecules, each atom achieves a stable, closed shell electron configuration by sharing electrons, and can be regarded as having two electrons of *two types*: valence and core. Broadly speaking, an atom's **valence electrons** are those involved in forming chemical bonds. An atoms **core electrons** are the others, or those not involved in forming chemical bonds. The distinction between valence and core electrons is one of convenience and convention, not a fundamental one. However, it is necessary to make the distinction, because when chemists write static electron or electron-dot diagrams they have to know which electrons to put on and which to leave off. Because different chemists use different definitions of core and valence electrons we'll define them with a couple of working rules here:

- Rule 1 for valence electrons: The valence electrons of an atom of a main group element (excepting He and Ne, which don't form molecules) are its s- and p-electrons with the greatest principal quantum numbers (they are sometimes called the valence shell or outer shell electrons).

- Rule 2 for valence electrons: The valence electrons of an atom of a d-block transition element are its s- and d-electrons with the greatest principal quantum numbers (its outermost s- and d-electrons). This rule applies only to elements in the left hand half of the d-block.

With those rules you can figure out core and valence electrons for many elements from their electron configurations, as illustrated in Example 6.5.

**Example 6.5**: Write the electron configurations of phosphorus, titanium, and selenium. In each, identify the core and valence electrons.

Strategy: From the periodic table, or by looking it up (Appendix E), write the electron configurations. Identify valence electrons by identifying the s and p, or s and d electrons, with the greatest principal quantum number.

P = $1s^22s^22p^63s^23p^3$, it can be written as $[Ne]3s^23p^3$

where [Ne] represents its core of $1s^22s^22p^6$.

A phosphorus atom has 10 core and 5 valence electrons.

Ti = $1s^22s^22p^63s^23p^64s^23d^2$, it can be written as $[Ar]4s^23d^2$

where [Ar] represents its core of $1s^22s^22p^63s^23p^6$

A titanium atom has 18 core and 4 valence electrons.

Se = $1s^22s^22p^63s^23p^64s^23d^{10}4p^4$, it can be written as $[Ar]4s^23d^{10}4p^4$

where [Ar] represents $1s^22s^22p^63s^23p^6$. The core electrons of selenium are $1s^22s^22p^63s^23p^63d^{10}$

A selenium atom has 28 core and 6 valence electrons.

To show the relationship between bonds and valence electrons, Lewis and Kossel originally used individual dots (Figure 6.11). Chemists now pair the dots to show paired electrons while leaving single dots to represent unpaired electrons. Modern Lewis dot representations of atoms are shown in Figure 6.12. Observe that the representations use both paired and unpaired electrons in accord with electron configurations and Hund's rules. Krypton exists in some molecules, is capable of forming bonds, and has valence electrons. Neon and argon fail to form bonds, so they are shown without valence electrons.

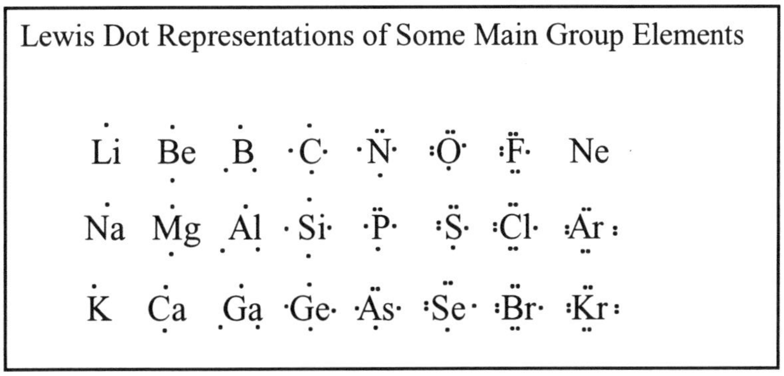

Lewis Dot Representations of Some Main Group Elements

| | | | | | | | |
|---|---|---|---|---|---|---|---|
| Li | Be | B | C | N | O | F | Ne |
| Na | Mg | Al | Si | P | S | Cl | Ar |
| K | Ca | Ga | Ge | As | Se | Br | Kr |

Figure 6.12 The Lewis Dot Representations of some Main Group Elements

Lewis dot diagrams for negative ions are made by adding one dot (electron) for each negative charge, and for positive ions by subtracting one dot (electron) for each positive charge.

## Lewis Dot Diagrams for Molecules

Lewis dot diagrams for molecules are the classic static electron model. Compounds are represented by joining unpaired electrons from two Lewis dot atoms to create an electron pair bond and a molecule. Usually, one electron (dot) comes from each of the atoms that are bonded; less frequently, two electrons from a single atom form the bond. Lewis dot diagrams are the traditional way of representing shared electron pairs in a covalent bond. The hydrogen molecule is the simplest example:

H• + H• → H:H

The bond is represented by the **:** symbol. Representing the electrons as a pair in this manner fits the experimental fact that the hydrogen molecule is diamagnetic, meaning the two electrons in the molecule are indeed paired. By "sharing" an electron, each hydrogen atom in $H_2$ has achieved a noble gas configuration—in this case the $s^2$, or helium, configuration. This single bond is called an **electron pair bond**. In its stable chemical compounds hydrogen "obeys" the doublet (like the octet) rule.

**Diamagnetic** substances interact only weakly with a magnetic field and contain no unpaired electrons; quantum mechanics explains that when two electrons are "paired" in a single orbital their magnetic fields cancel one another. **Paramagnetic** substances contain one or more **unpaired electrons**. Diamagnetic substances are weakly repelled by magnetic fields, whereas paramagnetic substances are attracted by them. When two atoms form an electron pair bond, two paramagnetic atoms create a diamagnetic molecule.

When forming compounds, second row elements tend to acquire the closed shell configuration of $s^2p^6$ with its eight electrons. A **closed shell** is a set of electrons that complete the filling of a principal quantum level. Closing a shell is equivalent to obeying the octet rule. The formation of diatomic fluorine ($F_2$) illustrates this process:

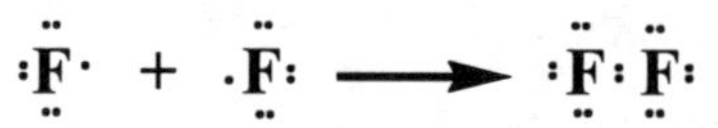

:F̤̈· + ·F̤̈: ⟶ :F̤̈:F̤̈:

In $F_2$, just as in $H_2$, the unpaired electrons from two separate atoms combine to form an electron pair bond. Viewed from either fluorine atom, it seems as if that atom has a "share" in eight valence electrons: seven of its own and an eighth from the bond. Thus, in forming $F_2$, both fluorine atoms achieve an octet. The pair of dots in the bond is said to be a **bonding electron pair**, or a bond pair (BP). Each fluorine atom also has three nonbonding pairs of valence electrons called **lone pairs** (LP). In the fluorine molecule there are a total of six lone pairs and one bond pair.

An oxygen atom has six valence electrons and a hydrogen atom has one. The combination of an oxygen atom with two hydrogen atoms to form water is shown below using Lewis dot diagrams:

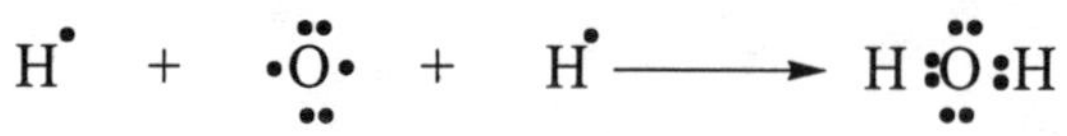

H• + •Ö̤• + H• ⟶ H:Ö̤:H

The Lewis dot diagram for water illustrates that O gets the two additional valence electrons it needs to complete its octet by sharing one electron with each of two separate hydrogen atoms. The O in $H_2O$ has two bonding and two lone pairs of valence electrons (there are also two core electrons on the oxygen atom). The relative positions of the paired and unpaired electrons are not considered important in drawing Lewis diagrams. For example, the lone pairs on an oxygen atom could equally be written at an angle of 90°. But in molecules the lone pairs of electrons play a key role in determining their three dimensional geometry. We'll discuss molecular geometry shortly in connection with bond angles.

Nitrogen and carbon have respectively 5 and 4 valence electrons. The Lewis diagrams of ammonia ($NH_3$), methane ($CH_4$), and ethane ($C_2H_6$) can be written:

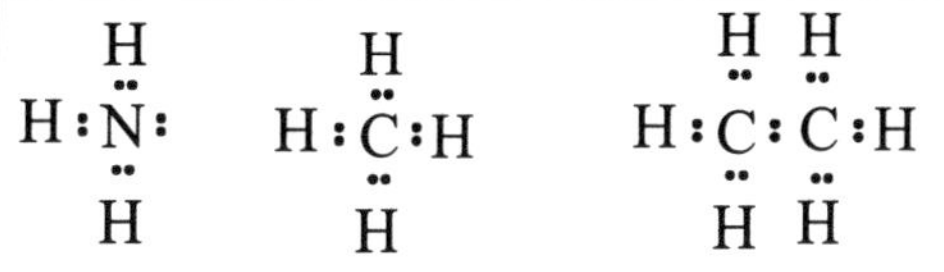

| H<br>H:N̈:<br>H | H<br>H:C̈:H<br>H | H H<br>H:C:C:H<br>H H |
|---|---|---|
| $NH_3$ | $CH_4$ | $C_2H_6$ |

You will probably have noticed by now that the valencies of H, O, C, F, etc., are explained by the Lewis dot diagrams and the octet rule. The valency of these elements is simply the number of electrons each needs to complete its octet. When Lewis dot diagrams show that a molecule has one or more lone pairs we get more insight into chemical bonding than from the simple valence concept. For example, ammonia's Lewis dot diagram shows that $NH_3$ has a lone pair of electrons on its nitrogen atom. This lone pair makes ammonia a potential electron pair donor: it can "give" its lone pair to an atom or ion that needs them or can accept them. Electron pair donation explains the well-known fact that ammonia is a base and reacts with acids. The reaction of $NH_3$ with $H^+$ (the acid) illustrates the formation of a coordinate-covalent bond. A **coordinate-covalent bond** arises when an electron pair from one atom is donated to a second atom. In the example below it's the hydrogen ion that accepts the electron pair. The left pointing arrow (←) in the line below designates the coordinate-covalent bond.

$H^+ + :NH_3 \rightarrow H^+\leftarrow NH_3$ which written more familiarly is $NH_4^+$.

**Example 6.6:** Use the octet rule to draw the Lewis dot diagram of the molecule formed between one atom of sulfur and two atoms of chlorine.

Strategy: A sulfur atom has six valence electrons. A chlorine atom has seven. Combine the three atoms so that the octet rule is obeyed. An equation to represent the process is:

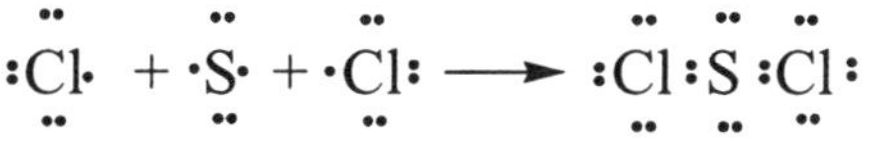

:C̈l· + ·S̈· + ·C̈l: ⟶ :C̈l:S̈:C̈l:

## The Static Electron Model of Double and Triple Bonds

Double and triple bonds can be represented in Lewis diagrams by writing more than one pair of electrons between the boded atoms. Diatomic nitrogen has a triple bond and is quite adequately represented as :N:::N: or :N≡N:

The hydrocarbons ethylene and acetylene have respectively two and three pairs of electrons between their carbon atoms. An organic molecule with at least one double or triple bond is unsaturated. An organic molecule is **unsaturated** if any of its carbon atoms is bonded to fewer than four different atoms. The historical term was "valence unsaturation," signifying that the carbon atom in question in some manner had not used up its valency. As shown below, each carbon atom in the ethylene molecule has three ligand atoms and each carbon atom in the acetylene molecule has two ligand atoms.

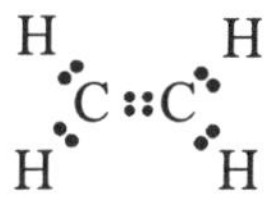

| H H<br>C::C<br>H H | H:C:::C:H |
|---|---|
| Ethylene, $C_2H_4$ | Acetylene, $C_2H_2$ |

A triple bond is stronger and shorter than a double bond, and a double bond is stronger and shorter than a single bond. The Lewis dot diagram of ethylene, $H_2C::H_2$, implies that the two bonds in the double bond are identical. The

Lewis dot structure of acetylene implies that the three bonds in the triple bond of acetylene, H-C:::C-H, are identical. But the facts are different. Abundant chemical and physical evidence shows that the two bonds in ethylene's double bond are not identical. Their strengths are different and their chemical reactivities are different. Likewise for the three bonds in acetylene. This is a situation that static electron models cannot resolve but which can be explained using the methods of quantum mechanics we'll meet later in this chapter.

Incidentally, ethylene is an important chemical compound. Annual per capita US consumption is about 200 pounds a year, most of it winding up eventually in plastics or polymers of some sort. The chemical reactivity of its double bond is what makes ethylene such a useful chemical raw material. Ethylene undergoes many, many addition reactions. Here's one.

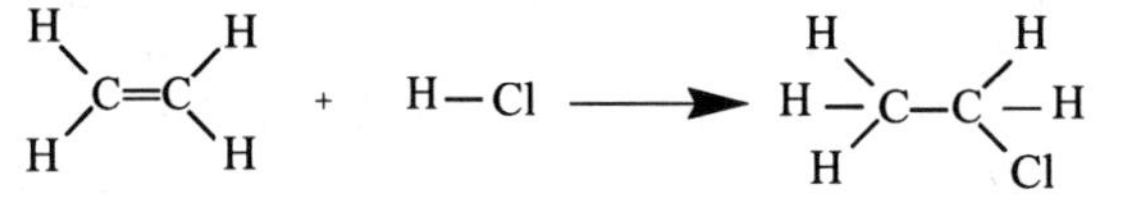

$$H_2C{=}CH_2 + H{-}Cl \longrightarrow H{-}CH_2{-}CH_2{-}Cl$$

ethylene — Valence Unsaturated; ethyl chloride or chloroethane — Valence Saturated

The formation of diatomic oxygen from two oxygen atoms can be represented as

$$:\ddot{O}\cdot \;\; \cdot\ddot{O}: \longrightarrow O::O$$

Counting the shared electrons as contributing to each atom yields an octet of paired electrons around each oxygen atom. However, the experimental fact is that $O_2$ is paramagnetic. Recall, paramagnetic substances must contain at least one unpaired electron. So the Lewis dot diagram of $O_2$ cannot be correct. It turns out that we need quantum mechanical insight to explain oxygen's paramagnetism and its unpaired electrons.

Diatomic oxygen being paramagnetic is another example of the failure of a static electron model. Static electron models are obviously wrong, but they do produce two key ideas: (1) that chemical bonds are electron pairs, and (2) that electron octets are important. Why obviously wrong? Because they are sometimes at odds with the experimental facts, and because both the Bohr model and quantum mechanics deny stationary electrons.

Chemists continue to use the historical ideas of the electron pair bond and the octet rule because they are simple and they do give significant chemical insight. But the ideas have really now been subsumed by the combined power of the Schrödinger equation and fast computers. As noted earlier in this chapter, already by 1930 Linus Pauling and others were beginning to apply quantum mechanical ideas to the problem of chemical bonding. But for the next 30-40 years the mathematics of quantum mechanics could hardly be done. Perhaps the durability of the static electron models owes more to the former lack of computer power than it does to the inherent value of the models.

## Section 6.3 Quantum Mechanics and Molecules

### The Hydrogen Molecule

We begin with the simplest possible diatomic molecule, $H_2$. The electron in a hydrogen atom is a matter wave in a definite atomic orbital, with a definite shape and a definite energy. It is a simple extension to say that the electrons in a hydrogen molecule are matter waves in a definite molecular orbital, with a definite shape and a definite energy. A **molecular orbital** is an electron probability function that extends over two or more nuclei.

Figure 6.13 shows how, in quantum mechanical terms, we might imagine the formation of a H-H bond. As two hydrogen atoms approach one another and begin to form a bond we can visualize that their electron clouds start to overlap. Overlap is a quantum mechanical idea; orbitals from two different atoms that share some region in space are said to **overlap**. That overlap is shown by the lens-shaped region at the bottom of Figure 6.13. The electrons, spinning in opposite directions so that that may form a pair, are shown by arrows.

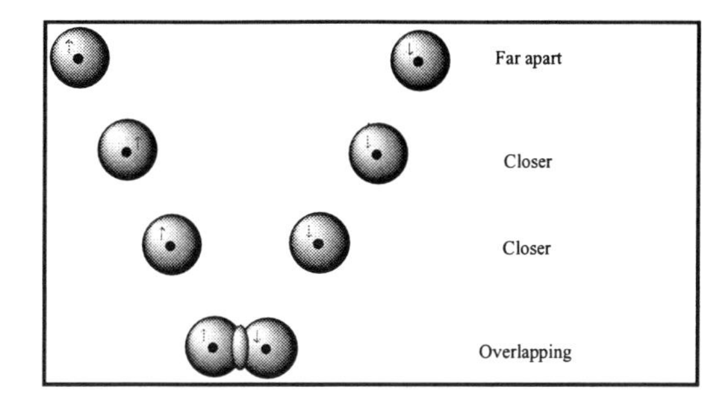


Figure 6.13 Overlap is a quantum mechanical idea. Two initially separated hydrogen atoms at the top of the diagram approach. The arrows show the $1s^1$ electron of each atom, the black central dots show the position of each atom's nucleus, and the 1s orbital on each atom is represented by an orbital picture. In the bottom frame the two orbitals on the separated atoms have moved into a position of overlap.

**Modern Quantum Chemistry**

In 1929, Paul Dirac, one of the founders of quantum theory, wrote: "The fundamental laws necessary for the mathematical treatment of large parts of physics and the whole of chemistry are thus fully known, and the difficulty lies only in the fact that application of these laws leads to equations that are too complex to be solved." That's no longer true (at least for small molecules).

The 1998 Nobel Prize in Chemistry was awarded to Walter Kohn (1923-) and John Pople (1925-) for their work in quantum chemistry and the development of computational procedures for calculating the properties of molecules and chemical reactions. Kohn was cited for his development of *density-functional theory* (which allows quantum mechanical calculations to be based on the average number of electrons present at any point in space, a method far more economical than treating all the electrons individually). Pople published the first version of his computer program GAUSSIAN in 1970. Descendants of that program are now in use at thousands of universities and companies all over the globe. In the early 1990s Pople was able to incorporate Kohn's density-functional theory into his own computational procedures, enabling the treatment of increasingly complex molecules.

Quantum calculations are today widely applied to many chemical problems. The principal objectives of such calculations are to determine molecular structure and map chemical reactions. Quantum calculations afford a deeper understanding of molecular structure and molecular processes than can be obtained solely from experiments. In modern chemistry, theory and experiment work together in the search for chemical truth. Here are three examples of the way modern quantum chemical calculations are used:

- Detailed structural calculations are made on an organic molecule containing carbon, hydrogen, oxygen, and nitrogen. A computer map is made of the surface electrostatic potential of the entire molecule. This information is used by a pharmaceutical company to assess how the organic molecule will interact with a human protein. In this way, molecules with useful pharmaceutical properties can be optimized.
- Radio waves coming from the vast dust clouds of interstellar space reveal information about the rotations of molecules. But these molecules are unlike any on earth—they have formulas and structures that would be instantly changed by the earth's environment. How to identify these molecules? Make quantum calculations based on assumed structures and compare the calculated rotational energies with frequencies seen by the radio telescope, using $E = h\nu$. Together, quantum theory and astronomical experiments reveal the molecular composition of interstellar matter.
- Chlorofluorocarbons released into the atmosphere cause depletion of the ozone layer that protects us from solar ultraviolet radiation. Quantum mechanical computations reveal the details of the chemical reactions involved, opening the possibility that the reactions might be controlled and ozone depletion slowed or reversed.

Because electrons in atoms are matter waves, overlapping means that interference occurs as the two hydrogen atoms merge into a molecule. Interference has the usual two possibilities: it can be constructive or destructive. We can imagine that those two alternative kinds of interference transform the two atomic orbitals into two molecular orbitals. One of the molecular orbitals is lower in energy than either of the atomic orbitals; the other of the molecular orbitals is higher in energy than either of the atomic orbitals. The formation of two molecular orbitals from two atomic orbitals is shown conceptually in Figure 6.14.

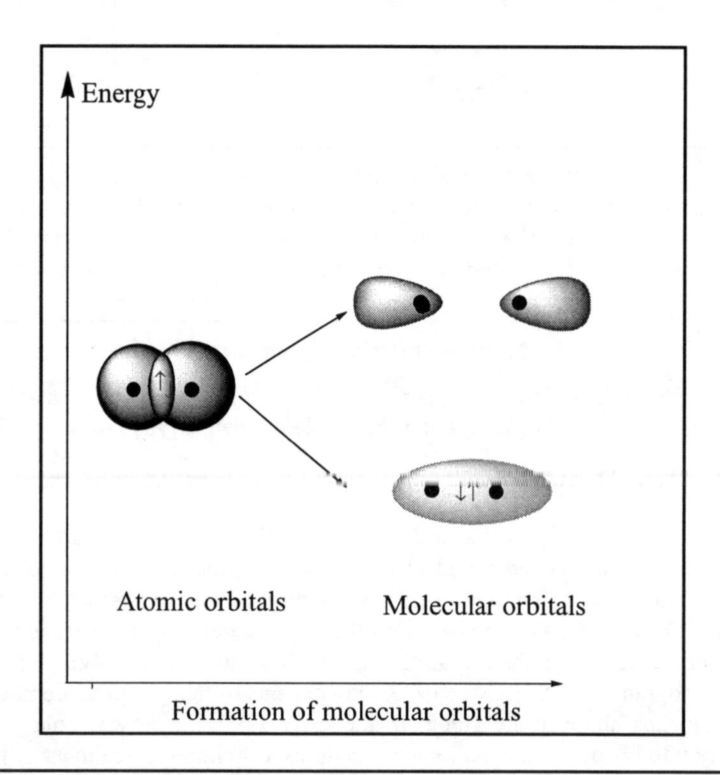


Figure 6.14 The formation of two molecular orbitals from a pair of overlapping atomic orbitals. As two hydrogen atoms approach, we can imagine that their individual atomic orbitals smoothly transform to a pair of bonding and antibonding molecular orbitals. The lower right orbital is called a bonding sigma orbital and designated $\sigma$. The upper right orbital is called an antibonding bonding sigma orbital and designated $\sigma^*$.

Constructive interference, the ellipse-shaped region at the lower right of Figure 6.14, puts the main region of molecular density in the right place—where the bond ought to be—and creates a **bonding molecular orbital**. Destructive interference, both the pear-shaped regions (two lobes) in the upper right, puts the main region of molecular electron density away from the bond axis, in the wrong place, and creates an **antibonding molecular orbital**. In the hydrogen molecule both electrons occupy the bonding molecular orbital (MO). Note that there is "conservation of orbitals;" two atomic orbitals transform to two molecular orbitals.

Despite the fact that the bonding and antibonding molecular orbitals are separated by energy, they share much of the same space. Figure 6.15 repeats the essential information of Figure 6.14 but show the original atomic orbitals and the molecular orbitals in the proper relationship in space.

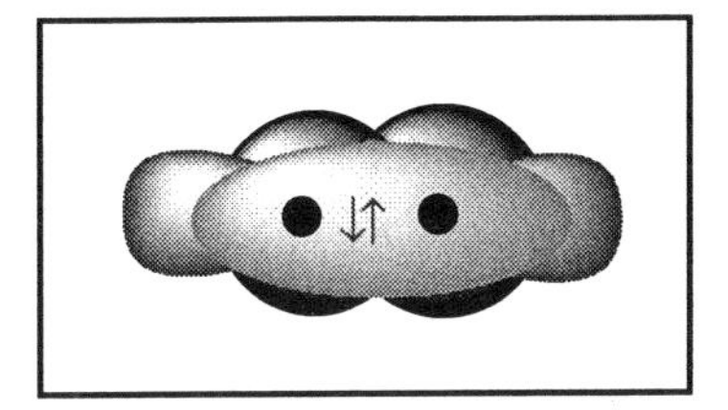

Figure 6.15 The four orbitals involved in the formation of an H-H bond shown in approximately the correct spatial relationship. At the back are the two original 1s atomic orbitals. Extending farthest to the left and right are the two lobes of the single σ* antibonding molecular orbital. In front is the ellipse-shaped region of the bonding σ molecular orbital. In an actual hydrogen molecule the atomic orbitals have vanished. The bonding molecular orbital contains two electrons, and the antibonding molecular orbital is present but empty.

Molecular orbitals can be calculated as approximations to solutions of the Schrödinger equation. Depending on the size of the molecule involved the calculations can be done with more or less accuracy. What's possible and how accurately it can be done changes with each new generation of computers. At the start of the 21st century, the properties of molecules with, say, five atoms and fewer than 50 electrons can be calculated with reasonable accuracy. Properties of molecules with tens of atoms and 100s of electrons can be calculated with moderate success. Calculations involving molecules with thousands of atoms require many calculational shortcuts, and the properties obtained are only approximately correct.

An important point: A bonding molecular orbital is *lower* in energy than the atomic orbitals that it came from, and an antibonding molecular orbital is *higher* in energy than the atomic orbitals that it came from. When a hydrogen molecule forms, both electrons "fall" from a higher energy atomic level to a lower energy bonding molecular orbital. The "falling" liberates energy, that is how quantum mechanics interprets the property of bond energy or bond strength. The antibonding orbitals are higher in entrgy than the original atomic orbitals. Electrons that enter antibonding orbitals need energy, which in effect gets taken from the bond, making the bond weaker. So an antibonding orbital is well named—an electron in it acts to counteract bonding.

A molecular orbital that shows circular symmetry when viewed along a bond axis is designated a sigma (σ) orbital. Because its overlap region lies with circular symmetry along the bond axis and between the two hydrogen nuclei, hydrogen's bonding molecular orbital is designated a bonding sigma (σ) orbital. Because its overlap region lies with circular symmetry along the bond axis but *not* between the two hydrogen nuclei, hydrogen's antibonding molecular orbital designated an antibonding sigma (σ*) orbital. In Chapter 5 we saw how to build up the elements by filling electrons into an atomic energy level diagram. Now we'll see *in an entirely parallel manner* how we can build up diatomic molecules by filling electrons into a molecular energy diagram. Figure 6.16 shows the molecular energy level diagram for a hydrogen molecule.

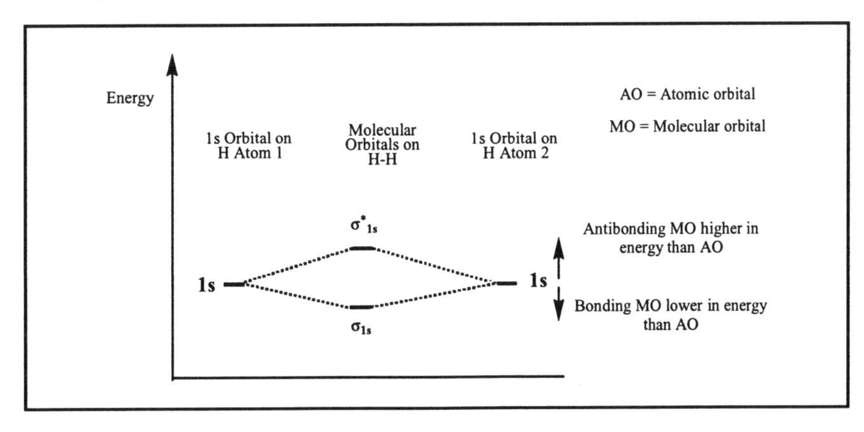


Figure 6.16 The energy level diagram for diatomic molecules and ions formed by hydrogen and helium.

To interpret Figure 6.16 you imagine two hydrogen atoms approaching one another from the far left and the far right. The atomic 1s energy levels are at the far left and right of the diagram. As the hydrogen atoms come together, the atomic energy levels transform into two molecular orbitals—the dotted lines are intended to show the atomic orbitals merging, splitting, and finally transforming into molecular orbitals shown in the central region of the diagram. Once formed, the hydrogen molecule has a bonding 1s σ-orbital and an antibonding 1s σ*-orbital. Filling a molecular orbital energy level diagram parallels filling an atomic energy level diagram (Figures 5.23, 5.24, and 5.27): you do it from the bottom up applying Hund's rules.

## Electron Configurations of $H_2$ and $He_2$

Now that we've got an orbital energy diagram with two energy levels (Figure 6.16) we can handle four electrons and so can write electron configurations for $H_2$ and $He_2$ and their positive ions. (The ions $H_2^+$ and $He_2^+$ are not available in ordinary substances, but they are easily observed and studied spectroscopically in gas discharge tubes.) We'll generate molecular electron configurations by the usual method of filling them into the diagram from the bottom up while paying attention to Hund's rules. The results are shown in Table 6.8, where dia- means the molecule is diamagnetic and lacks unpaired electrons and para- means the molecule is paramagnetic and has at least one unpaired electron. Other than the notation being a little fancier, molecular electron configurations are just like atomic electron configurations.

Table 6.8 **Electron Configurations and Properties of $H_2^+$, $H_2$, $He_2^+$ and $He_2$**

| Entity | Total electrons | Bonding electrons | Antibonding electrons | Ground state MO configuration | Unpaired electrons (theory) | Experimental Magnetic state | Bond order (theory) | Experimental bond energy kJ mol$^{-1}$ |
|---|---|---|---|---|---|---|---|---|
| $He_2$ | 4 | 2 | 2 | $\sigma_{1s}^2 \sigma^*_{1s}{}^2$ | 0 | dia- | 0 | 4 |
| $He_2^+$ | 3 | 2 | 1 | $\sigma_{1s}^2 \sigma^*_{1s}{}^1$ | 1 | para- | 0.5 | 240 |
| $H_2$ | 2 | 2 | 0 | $\sigma_{1s}^2$ | 0 | dia- | 1 | 436 |
| $H_2^+$ | 1 | 1 | 0 | $\sigma_{1s}^1$ | 1 | para- | 0.5 | 270 |

The **bond order** of a diatomic molecule is one half of the difference between the number of its electrons in bonding orbitals and the number of its electrons in antibonding orbitals. Every pair of electrons in a bonding orbital adds one to the bond order. Every pair of electrons in an antibonding orbital reduces the bond order by one. Bond orders of 1, 2, and 3 correspond to the single, double, and triple bonds of the static electron model.

$$\text{Bond order} = \frac{\text{Number of electrons in bonding orbitals} - \text{number of electrons in antibonding orbitals}}{2}$$

What is the significance of Table 6.8? It's the agreement between the theory of molecular orbitals and the experimental properties of the molecules and ions in the table. Note that the predicted number of unpaired electrons matches the experimental dia- or paramagnetism of the entities. Note also in the two rightmost columns that the bond orders derived from the energy level diagram correspond very well with the experimental bond energies. We're going to develop molecular orbital theory just a little further, and we'll see even more impressive agreement between the same pairs of properties for the considerably more complicated diatomic molecules that we tackle next.

Finally, before moving on from Table 6.8, observe that *removing* one electron from $He_2$ to make $He_2^+$ causes the bond to go from zero strength to about half a bond (4 kJ mol$^{-1}$ to 240 kJ mol$^{-1}$). Furthermore *removing* one electron from $He_2^+$ to make $He_2^{2+}$ causes the bond to go from about half a bond to a full bond. Who'd have thought *removing* electrons would increase bond strength? Well, if the electron being removed comes out of an *antibonding* orbital that's exactly what happens. That's *why* it's called an antibonding orbital!

**The Parallel Between Atomic and Molecular Energy Levels**

In Chapter 5 we met the aufbau principle in which the electron configurations of atoms were imagined to be generated by filling electrons into an atomic energy level diagram. In this Chapter we undertake a parallel treatment for molecules and ions in which their electron configurations are imagined to be generated by filling electrons into a molecular energy level diagram. In the section just concluded we've done it for the simplest chemical entities involving diatomic species of hydrogen and helium. In the following section we'll go just a little further—across the second row of the periodic table and the diatomic entities of those elements.

## The Quantum Mechanical View of Second Row Diatomic Molecules

For hydrogen and helium we needed only to consider the atomic 1s orbitals to understand the process of bond formation. But an ethylene molecule contains two carbon atoms, and carbon is in the second row of the periodic table. For the second row elements: Li, Be, B, C, N, O, F, and Ne, it's the atomic 2s and 2p orbitals that hold the valence electrons. So, our quantum mechanical treatment of the bonding between these atoms must begin by asking how do 2s and 2p orbitals overlap to produce bonding and antibonding orbitals?

For the 2s atomic orbitals the overlap picture is easy. They overlap just like the 1s-1s hydrogen to form a 2σ and 2σ* bonding and antibonding pair that are the same shape as the 1σ and 1σ* but are higher in energy and somewhat larger. So to visualize this pair, go back to Figure 6.14 and imagine it slightly larger.

As shown in Figure 6.17, the overlap of p-orbitals is more complicated. In the lower half of Figure 6.17 the $p_x$ orbital from one atom overlaps the $p_x$ orbital from the second atom. The geometry of this overlap is "straight on." Straight-on overlap puts the region of bond electron density centered right on the bond axis. The bonding molecular orbital resulting from this overlap is shown on the lower right. It's called the $\sigma_{2p}$ orbital and is a bonding orbital because its energy level is lower than the the corresponding 2p energy levels on the separated atoms. The corresponding antibonding orbital, the $\sigma_{2p}$*, with two lobes, is shown immediately above it.

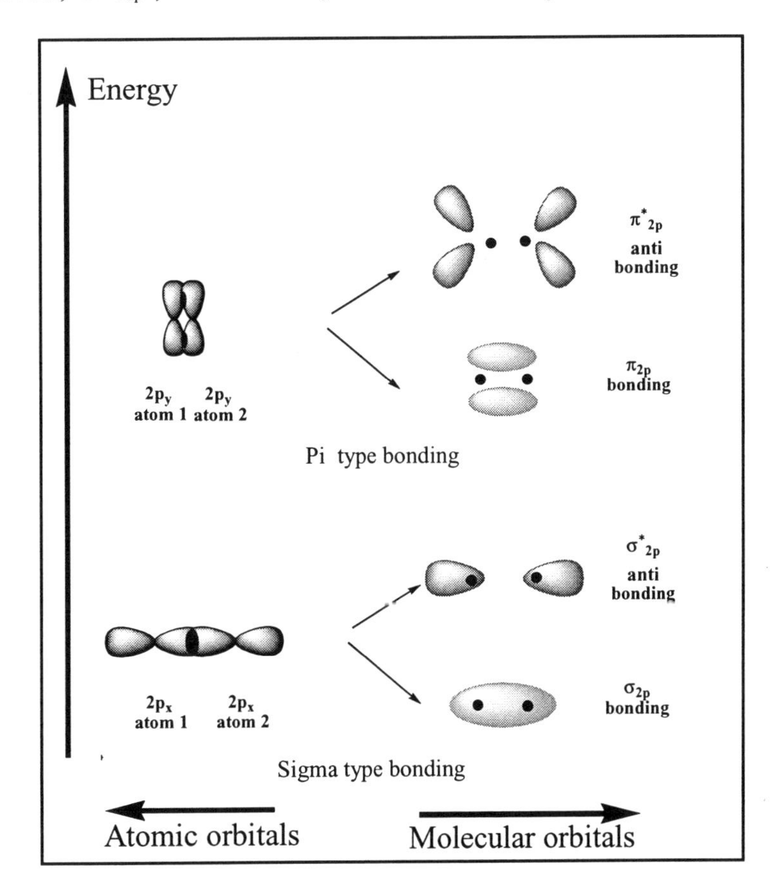


Figure 6.17 Molecular orbitals formed by the overlap of p-type orbitals. Because of their more compliacted shapes, overlap between p-orbitals is somewhat more complex than between pairs of s-orbitals (shown in Figure 6.14). However, the quantum mechanical basis remains. As before, the formation of molecular orbitals results from the interference between pairs of atomic orbital electron matter waves.

In the upper half of Figure 6.17 you see that the overlap of the $p_y$ orbital from one atom and the $p_y$ orbital from the second atom occurs in a "parallel" fashion. The bonding molecular orbital resulting from this overlap is called the $\pi_{2py}$ orbital and is a bonding pi orbital because its energy level is lower than the corresponding $2p_y$ energy levels on the atoms; note that it has two lobes. Above it is the antibonding $\pi^*_{2py}$ orbital; you'll note that even though it's just one orbital it has four lobes. In contrast to sigma orbitals, pi orbitals lack circular symmetry when viewed down the bond axis. The $p_z$ orbitals are omitted from this diagram; other than pointing in a different direction (they would stick in and out of the plane of the paper) they overlap just like the $p_y$ atomic orbital pair.

Figure 6.18 shows how the molecular orbitals might arise from atomic p-orbitals as two separate atoms of second-row elements approach along the x axis. The $p_x$ orbitals are directed straight on, and so will form a $\sigma_{2px}$ bonding molecular orbital and a $\sigma^*_{2px}$ antibonding molecular orbital. The other p-orbitals, the $p_y$ and the $p_z$, approach one another in a parallel fashion. Atomic orbitals $p_y$ and $p_y$, and also atomic orbitals $p_z$ and $p_z$, overlap and produce two pi bonding orbitals, $\pi_{2py}$ and $\pi_{2pz}$, and two pi antibonding orbitals, $\pi^*_{2py}$ and $\pi^*_{2pz}$.

At the end of Section 6.2 it was pointed out that the Lewis dot diagram of ethylene, $H_2C::H_2$, implies that the two bonds in the double bond are identical and that of acetylene, H-C:::C-H, implies that the three bonds in the triple bond are identical. But the experimental facts deny these implications. We are now in a position to see how quantum mechanics explains that the two pairs of electrons (written as two pairs of dots in a static electron model) are different because one pair is in a sigma bonding orbital whereas the other is in a pi bonding orbital. Chemists say the double bond in ethylene has one sigma bond and one pi bond. In a similar way, the triple bond in acetylene is interpreted to have of one sigma bond and two pi bonds. A **sigma bond** is one created by a pair of electrons in a sigma type molecular orbital. A **pi bond** is one created by a pair of electrons in a pi-type molecular orbital.

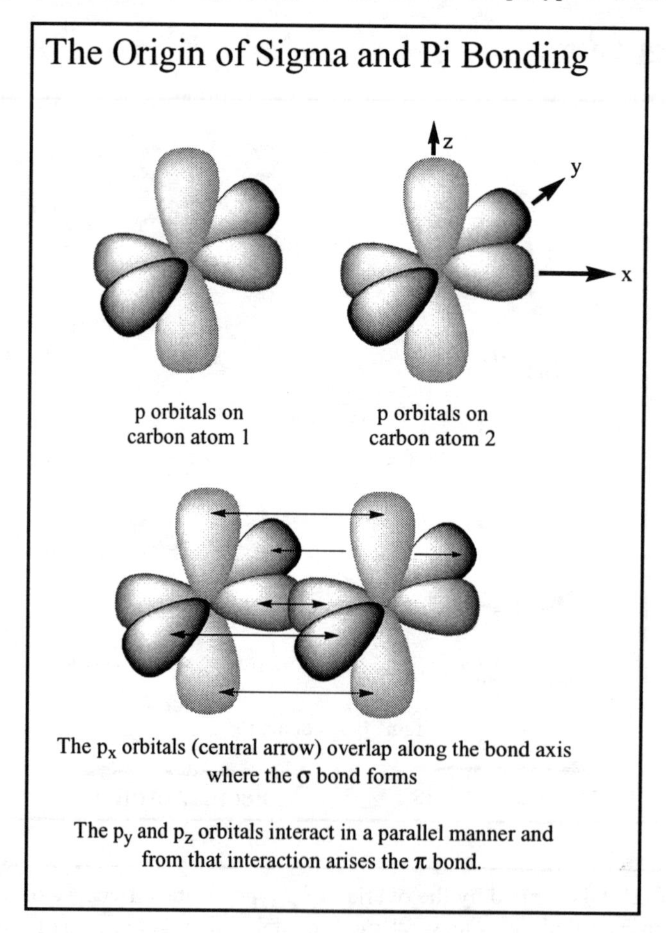


Figure 6.18 Overlap of the p-type orbitals from two atoms. Quantum mechanics explains that when two atoms are joined by more than one bond then at least two types of bonding are involved. This situation has important consequences in organic chemistry, where the chemical reactivity of unsaturated hydrocarbons allows them to be the basis for an entire chemical industry.

## Energy States in Second Row Diatomic Molecules

The concluding portion of this section extends the molecular orbital treatment and presents a generalized energy level diagram for the diatomic molecules formed by the second row elements. Just as the energy level diagram for the entire periodic table, in Figure 5.27, traded generality for detailed accuracy, so the same tradeoff is made in Figure 6.19 below.

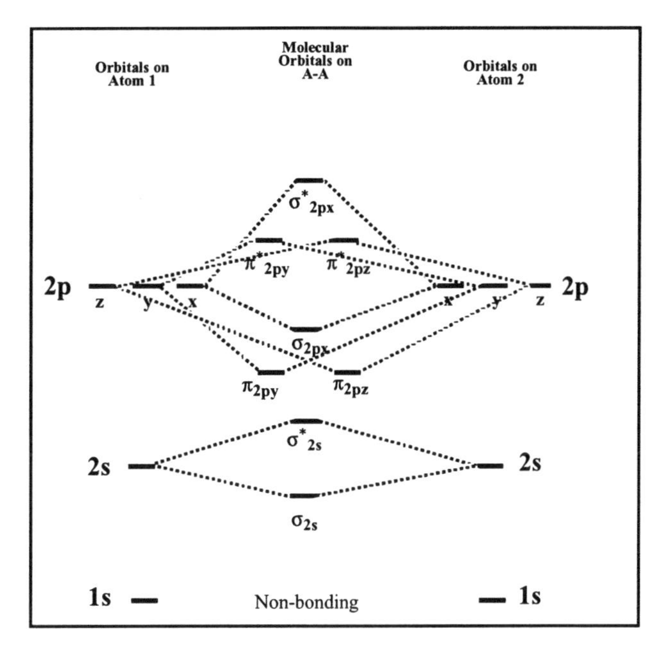


Figure 6.19 The Approximate Quantum Mechanical Energy Level Diagram for the Diatomic Molecules of the Elements Lithium through Neon

Figure 6.19 is of the same type as Figure 6.16. It shows how the 2s and 2p energy levels of two atoms create molecular orbitals as the atoms approach one another and their atomic orbitals overlap. As before, the atomic energy levels of the separated atoms are shown on the left and right. As the atoms come together the atomic orbitals can be imagined to interfere, merge, and form the molecular orbitals with energies and labels as shown in the middle of the diagram.

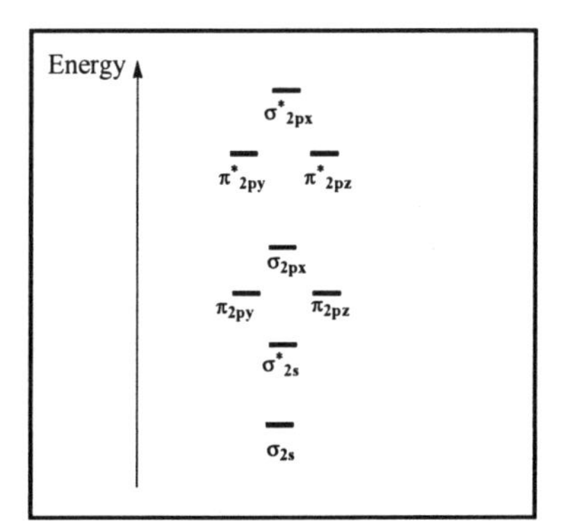


Figure 6.20 Generalized Molecular Orbital Energy Levels for the Diatomic Molecules of the Elements Lithium through Neon. "Generalized" means that all the diatomic molecules and ions of Li to Ne have an energy diagram such as this, but that the exact positions of the energy levels vary from case to case.

Figure 6.20 shows just the molecular energy levels taken from Figure 6.19. It is the filling of electrons into this molecular energy level diagram that will provide a quantum mechanical description of the diatomic molecules formed by the eight elements Li-Ne with 1 through 7 valence electrons respectively, and electron configurations from $1s^22s^1$ to $1s^22s^22p^6$. In all eight of these elements the $1s^2$ are core electrons and are **nonbonding** and neither assist nor diminish bonding; the nonbonding 1s orbitals are therefore not shown on Figure 6.20. To understand a diatomic molecule one does an "aufbau" or filling up of the energy levels shown in Figure 6.20.

## Electron Configurations of $Li_2$ to $Ne_2$

Now that we've got an orbital energy diagram for the diatomic molecules $Li_2$ to $Ne_2$, we get their molecular electron configurations by filling electrons into the diagram from the bottom and using Hund's rules as shown in Example 6.7. The results of that process for the electron configurations of $Li_2$ to $Ne_2$ are shown in Table 6.9.

**Example 6.7**: Use the generalized diatomic molecular energy level diagram (Figure 6.20) to calculate oxygen's bond order, and show how the diagram explains that ordinary oxygen molecules are paramagnetic.

Strategy: Fill electrons into the generalized energy level diagram. First, count the numbers of electrons in bonding and antibonding orbitals. Second, count the number of unpaired electrons in the configuration.

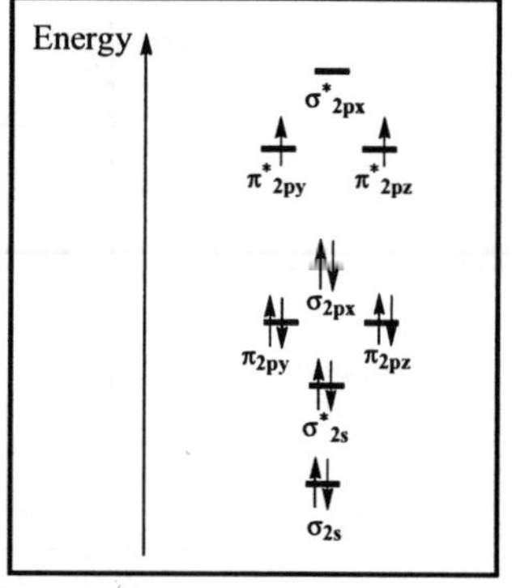


Oxygen has six valence electrons. So, for $O_2$, 12 electrons must be filled into the diagram as shown. The oxygen molecule's valence electron configuration is $\sigma_{2s}^2\,\sigma^{*2}_{2s}\,\pi_{2py}^2\,\pi_{2px}^2\,\sigma_{2px}^2\,\pi^{*1}_{2py}\,\pi^{*1}_{2py}$

There are eight "bonding electrons" and four "antibonding" electrons. So applying the rule gives bond order = (8-4)/2 = 2. Which shows that the oxygen-oxygen bond is more or less a double bond.

The are two unpaired electrons in diatomic oxygen's $\pi^*_{2py}$ and $\pi^*_{2py}$ orbitals. They account for oxygen's paramagnetism.

What is the significance of Table 6.9? It's the agreement between the theory of molecular orbitals and the experimental properties of the molecules and ions in the table. Note that the predicted number of unpaired electrons in every case matches the experimental dia- or paramagnetism of the entities. Note also in the two rightmost columns that the bond orders derived from the energy level diagram correspond well with the experimental bond energies.

Table 6.9 **Configurations and Properties of Second Row Homonuclear Diatomic Molecules and Ions**

| Entity | "tve" | "be" | "abe" | Ground state MO electron configuration | Unpaired electrons (theory) | Experimental magnetic state | Bond order (theory) | Experimental bond energy kJ mol$^{-1}$ |
|---|---|---|---|---|---|---|---|---|
| $Ne_2$ | 16 | 8 | 8 | $\sigma_{2s}^2\,\sigma^{*2}_{2s}\,\pi_{2py}^2\,\pi_{2pz}^2\,\sigma_{2px}^2\,\pi^{*2}_{2py}\,\pi^{*2}_{2pz}\,\sigma^{*2}_{2px}$ | 0 | dia- | 0 | 4 |
| $F_2$ | 14 | 8 | 6 | $\sigma_{2s}^2\,\sigma^{*2}_{2s}\,\pi_{2py}^2\,\pi_{2pz}^2\,\sigma_{2px}^2\,\pi^{*2}_{2py}\,\pi^{*2}_{2pz}$ | 0 | dia- | 1 | 130 |
| $O_2$ | 12 | 8 | 4 | $\sigma_{2s}^2\,\sigma^{*2}_{2s}\,\pi_{2py}^2\,\pi_{2pz}^2\,\sigma_{2px}^2\,\pi^{*1}_{2py}\,\pi^{*1}_{2pz}$ | 2 | para- | 2 | 503 |
| $O_2^+$ | 11 | 8 | 3 | $\sigma_{2s}^2\,\sigma^{*2}_{2s}\,\pi_{2py}^2\,\pi_{2pz}^2\,\sigma_{2px}^2\,\pi^{*1}_{2py}$ | 1 | para- | 2.5 | 650 |
| $N_2$ | 10 | 8 | 2 | $\sigma_{2s}^2\,\sigma^{*2}_{2s}\,\pi_{2py}^2\,\pi_{2pz}^2\,\sigma_{2px}^2$ | 0 | dia- | 3 | 955 |
| $N_2^+$ | 9 | 7 | 2 | $\sigma_{2s}^2\,\sigma^{*2}_{2s}\,\pi_{2py}^2\,\pi_{2pz}^2\,\sigma_{2px}^1$ | 1 | para- | 2.5 | 850 |
| $C_2$ | 8 | 6 | 2 | $\sigma_{2s}^2\,\sigma^{*2}_{2s}\,\pi_{2py}^2\,\pi_{2pz}^2$ | 0 | dia- | 2 | 615 |
| $B_2$ | 6 | 4 | 2 | $\sigma_{2s}^2\,\sigma^{*2}_{2s}\,\pi_{2py}^1\,\pi_{2pz}^1$ | 2 | para- | 1 | 300 |
| $Be_2$ | 4 | 2 | 2 | $\sigma_{2s}^2\,\sigma^{*2}_{2s}$ | 0 | dia- | 0 | 59 |
| $Li_2$ | 2 | 2 | 0 | $\sigma_{2s}^2$ | 0 | dia- | 1 | 107 |

"tve" = total valence electrons
"be" = count of bonding electrons
"abe" = count of antibonding electrons
Bond order = (number of bonding electrons - number of antibonding electrons)/2

Despite the fact that we've used a generalized energy level diagram, we get a sensible fit between most of the predicted and experimental values. Obviously, the 59 kJ $mol^{-1}$ bond energy of beryllium is not zero, and the bond energies of $N_2^+$ and $O_2^+$ seem uncomfortably far apart for diatomic ions both with a bond order of 2.5. However, the overall agreement, even in this rudimentary example, should convince you of the power of quantum mechanics.

If you understand the significance of Table 6.9 you have a genuine insight into the way quantum mechanics is applied to modern chemistry. To solve chemical problems by quantum mechanical methods requires dealing with polyatomic molecules, and sometimes with large polyatomic molecules. Although the details are far beyond our scope here, the essence of what is done by modern quantum chemistry computer programs is to calculate an approximate set of molecular orbitals for any molecule or ion. From these orbitals and their energy states are derived molecular properties such as dia- and paramagnetism, bond energies, bond lengths, and bond angles. [❁kws +"quantum mechanics" +molecule +calculation]. One important objective of modern quantum mechanics is to calculate molecular structures. Depending on the size of the molecule, quantum mechanics gives a more or less accurate calculation of its structure, including all its bond angles. For a small molecule, such as water, the quantum mechanically calculated bond lengths and angles are as accurate as the experimental ones. We'll discuss molecular structure in the concluding sections of this chapter. Finally, the results of quantum mechanics are used for practical purposes, such as directing the design of a drug, or interpreting the mechanism of an important biochemical reaction. We can predict in the 21st century that chemical quantum mechanics will be heavily driven by its biological and medical applications.

In Section 2.4 we learned that the molecule benzene contains delocalized chemical bonds that are intermediate in their properties between single and double bonds. It was not until the successful application of quantum mechanics to chemical bonding that chemists could understand these delocalized bonds. **Delocalized bonds** are ones created by electrons in pi orbitals that extend over more than two atoms. The bonding in benzene, $C_6H_6$, is represented in Figure 6.21. On the left is the traditional depiction showing alternating single and double bonds around the ring of six carbon atoms that constitute the molecule. On the right, the circle is used to depict the continuous pi bond that circles the entire molecule.

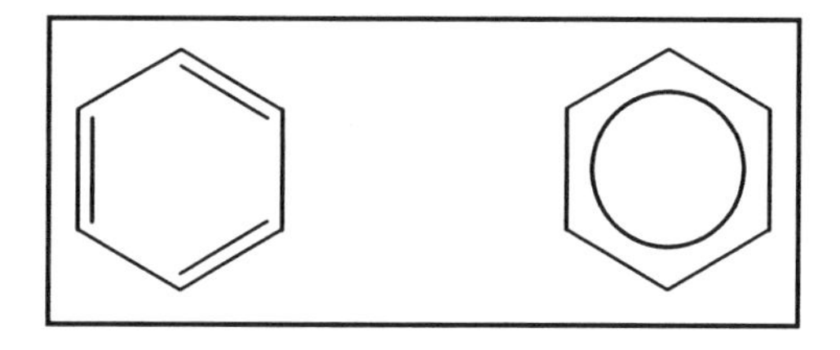

Figure 6.21 On the left is the traditional representation of benzene. The more modern depiction on the right reflects understanding of delocalized molecular orbitals derived from the theory of quantum mechanics. Although these diagrams may not look like much, it took a profound advance in chemistry to move from the formula on the left to the formula on the right.

## Section 6.4 Molecular Geometry and VSEPR

A molecule's bond lengths and bond angles, and the way all its atoms are disposed in space, constitute its **molecular geometry** or **molecular structure**. Historically, chemists developed an understanding of molecular structures primarily through studies of organic chemistry. At the beginning of the chapter we noted that the tetrahedral geometry of many carbon atoms had been established by 1874, before crystallography, before quantum mechanics, and even before the definitive proof of the existence of molecules! So knowledge of chemical structure predated the discovery of the role of electrons in forming chemical bonds, and it was only after considerable time that chemists would develop electron-based theories of structure. In this section we describe a static electron model for chemical structure; more specifically, we see how we can use the forces of repulsion among electrons to account for the bond angles around central atoms.

The aim of both static electron models and quantum mechanics is to minimize the energy of the system of atoms that makes up a molecule. When the system's energy is minimized the atoms are in their most stable average positions, and the bond lengths and angles will have their experimental values. In a static electron model, bond angles are imagined to be controlled by repulsions among the valence shell electron pairs. That's called VSEPR: the **V**alence **S**hell **E**lectron **P**air **R**epulsion "theory." Both bond pairs (BP) and lone pairs (LP) contribute to determining the bond angles.

**Informal Example:**

Imagine a child's globe. That's the "central atom." Now imagine two baseballs that must touch the globe but be as far apart as possible. Do you see that they sit at the opposite ends of a diameter of the globe? Put one on the equator in the middle of the Atlantic Ocean and the other on the equator in the middle of the Pacific ocean.

A third baseball approaches horizontally to touch down of the surface. All three baseballs want to be as far apart as possible. Imagine how the three adjust. They're all on the equator 120° of latitude apart.

How about a fourth baseball? It descends from above, down toward the North Pole. The first three head due south and keep moving until number four lands. After number four has landed all four have perfect symmetry. That's how to make tetrahedral geometry.

Shown in Figure 6.22 are the five basic or ideal geometries that define bond angles. The five cases involve a central atom surrounded by and bonded to 2, 3, 4, 5, or 6 other atoms. An **ideal geometry** is one that has the maximum possible symmetry for the specific number of ligands around the central atom. Molecules with more than six ligands around a cental atom are rare. It's the radius ratio effect—a large central atom and small ligands are necessary if seven or eight ligands are going to squeeze around a central atom.

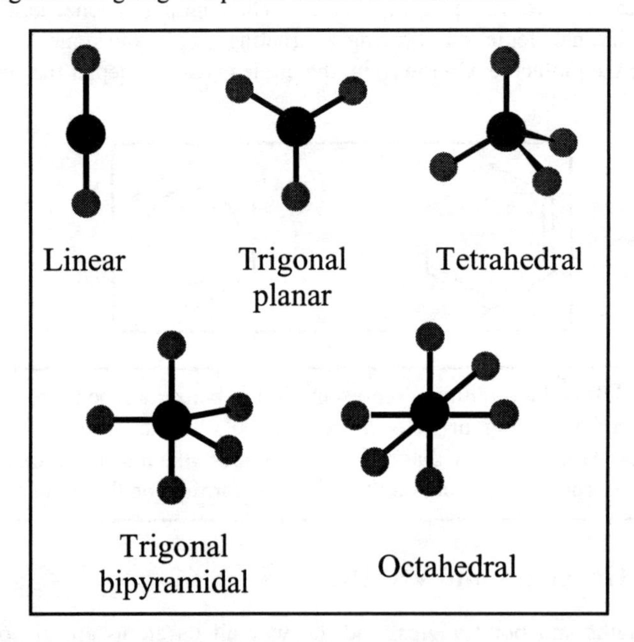


Figure 6.22 The five basic types of ideal geometry shown in pseudo three-dimensional display. The corresponding bond angles are: Linear, 180°; Trigonal, 120°; Tetrahedral, 109° 28′; Trigonal bipyramidal, 90°, 120°, and 180°; Octahedral: 90° and 180°.

The naming of the basic geometries comes from the classic Greek solids, as shown in Figure 6.23. Students often wonder why the octahedral structure has only six ligands while the prefix octa- obviously means eight, The answer is that while an octahedron has eight faces it has only six points. The ligand atoms are imagined to "sit" at the points of a regular octahedron, and that's the reason we call it an octahedral structure.

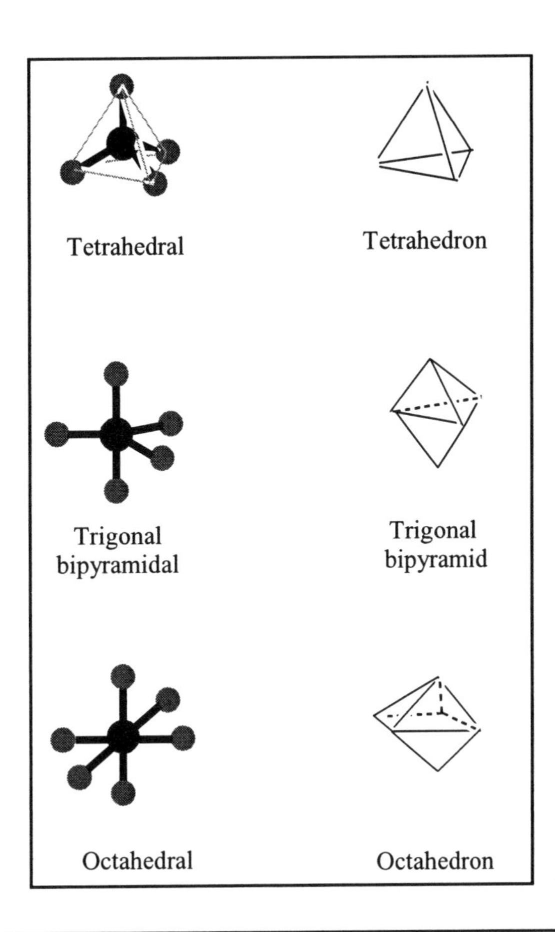


Figure 6.23 The three-dimensional solids associated with three types of ideal geometry.

VSEPR "theory" holds that bond angles around a central atom are controlled by the collective repulsions among both bond pairs and lone pairs in the central atom's valence shell. When all the pairs are collectively as far apart as possible, the energy has reached its minimum possible value. We can tell where the bond pairs are because with x-ray crystallography we can see the atoms at each end of the bond. The lone pairs are harder to see, and they don't themselves create experimentally measurable bond angles, but their effects on the structure are perfectly evident.

## Valence Shell Lone Pairs and Geometry

In VSEPR, a valence shell lone pair of electrons acts like a ligand. Indeed, its effect on the bond angles is typically greater than that of a bond electron pair because, lacking a nucleus on the other end of the bond, a lone pair is likelier to be closer to the central atom than is a bond pair. So the lone pair has a greater repulsive effect on the bond angles.

In a water molecule, there are two bond pairs and two lone pairs. Water's structure is that of a bent molecule based on tetrahedral ideal geometry, as shown in the upper left of Figure 6.24. The experimental value of water's bond angle is 104.5° or significantly less than the tetrahedral angle of 109° 28′.

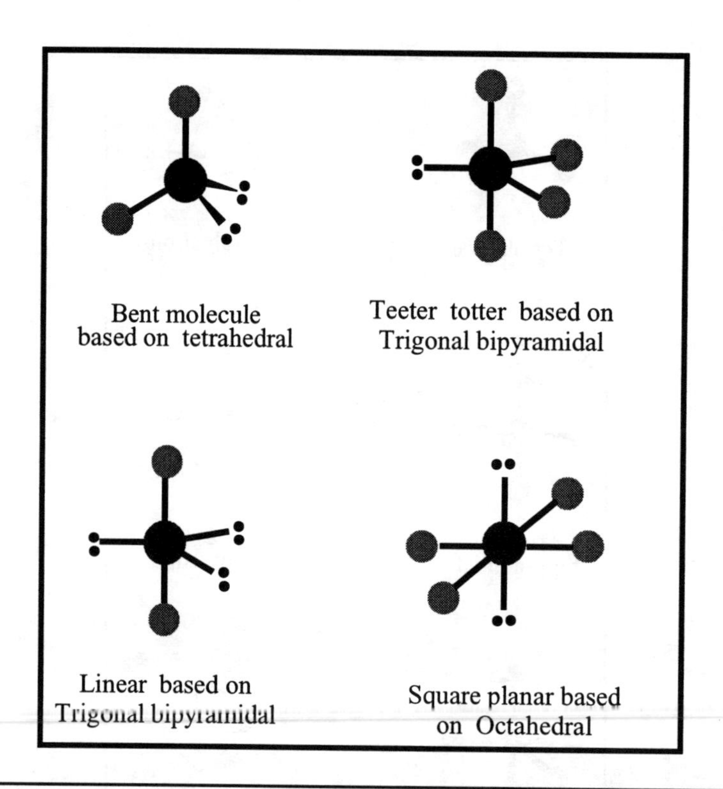


Figure 6.24 The effect on molecular structure of lone pair valence shell electrons (Lewis dots). Although difficult to observe directly, the role of lone pairs in determining molecular structure is unambiguous.

X-ray crystallography reveals that distortions of bond angles of 1-10° from the "ideal" geometry angles are common. These deviations are rationalized by VSEPR and calculated by quantum mechanics. For example, the water molecule geometry with its 104.5° bond angle is considered in VSEPR to be a tetrahedral angle of 109.5° "squeezed down" to 104.5° because the lone pairs are "more repulsive" than the bond pairs. The actual molecular geometry, looking only at bonds, ligands, and bond pairs, is in these cases different from the "electron geometry." The **electron geometry** around a central atom is the disposition of the electron pair probability distributions in space.

**A Personal Note:** In 1962, Neil Bartlett at Princeton announced that he had made a chemical compound of the noble gas xenon. This was electrifying news because I (and all other chemistry students up to that time) had been taught that the noble gases don't form *any* chemical compounds. Soon, the preparation of $XeF_4$ was reported in the chemical literature. Early in 1963 my mentor at Maryland, Professor Sam Grim, told me that $XeF_4$ would have a square planar structure. He was right. How did he do it? He applied VSEPR ideas.

Xenon is a noble gas and so has either zero or eight valence electrons. If it has zero, that means no bonds; if it's going to form any sort of chemical compound, it will have eight. In other words, for a xenon atom to form bonds some of its valence electrons must get unpaired. Thus, the Lewis dot diagram of xenon must "open up" two pairs of electrons to create four unpaired electrons that can combine with the four fluorine ligands. That process is shown in Figure 6.25. Observe that there are twelve valence electrons around the xenon atom. The Xe atom in the $XeF_4$ molecule does **not** obey the octet rule.

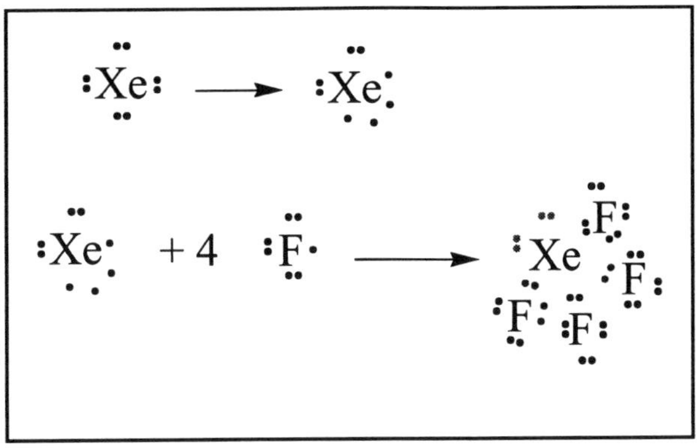

Figure 6.25 The formation of $XeF_4$ depicted by means of Lewis-dot diagrams. The actual square planar structure of $XeF_4$ (depicted on the bottom right of Figure 6.24) arises from the six valence pairs of electrons disposed around the central xenon atom, with the two lone pairs occupying opposite points of the octahedron.

The two lone pairs go above and below the square plane of fluorine atoms to give a square planar geometry based on octahedral electron geometry. There's a picture of that structure in the bottom right hand corner of Figure 6.24. VSEPR gives a nice rationalization of $XeF_4$'s structure and shows what static electron models can do at their best. Of course quantum mechanical calculations reach the same conclusion. But that's brute force, not human insight.

## Section 6.5 Molecular Geometry and Quantum Mechanics

In this section we are going to discuss orbital hybridization. This idea is now close to 70 years old, having been introduced by Pauling around 1930. Nowadays, it's a bit fossilized, and perhaps not entirely necessary. But hybridization is deeply embedded in chemical language and thinking, especially in organic chemistry, so it's probably worthwhile for an engineer to be familiar with it.

Combining orbitals on a single atom is called hybridization. **Hybridization** is a quantum mechanical concept; it's the idea that hydrogen-like orbitals can be combined to yield new orbitals that point in such directions that the shapes of molecules can be rationalized. Hybridization transforms some number of hydrogen orbitals on a single atom into an equal number of hybrid orbitals. The hybrid orbitals are "directed in space;" they point in particular directions. Bond angles in molecules are then considered to derive from the underlying angles among the directed hybrid orbitals. So the procedure of hybridization is a quantum mechanical way of discussing molecular geometry. Hybridization is not something that really happens, rather it a way for humans to think that has a proper quantum mechanical basis but does not involve any serious or complicated quantum mechanical calculations. Hybridization is neither a fact nor a proper theory, it's sort of a help to chemical thinking.

### $sp^3$ Hybridization

We've already seen that quantum mechanics can combine wave functions and generate different wave functions by overlapping orbitals. In this way, original orbitals are transformed into new ones. That was the basic procedure we used either to make sigma orbitals by combining pairs of s or $p_x$ orbitals from two atoms, or to make pi bonds by combining pairs of $p_y$ or $p_z$ orbitals. Hybridization combines orbitals on one atom. The first case we'll consider is that of the combining of one s and three p orbitals to give a set of four hybrid orbitals called the $sp^3$ hybrid orbital set. That's shown in Figure 6.26.

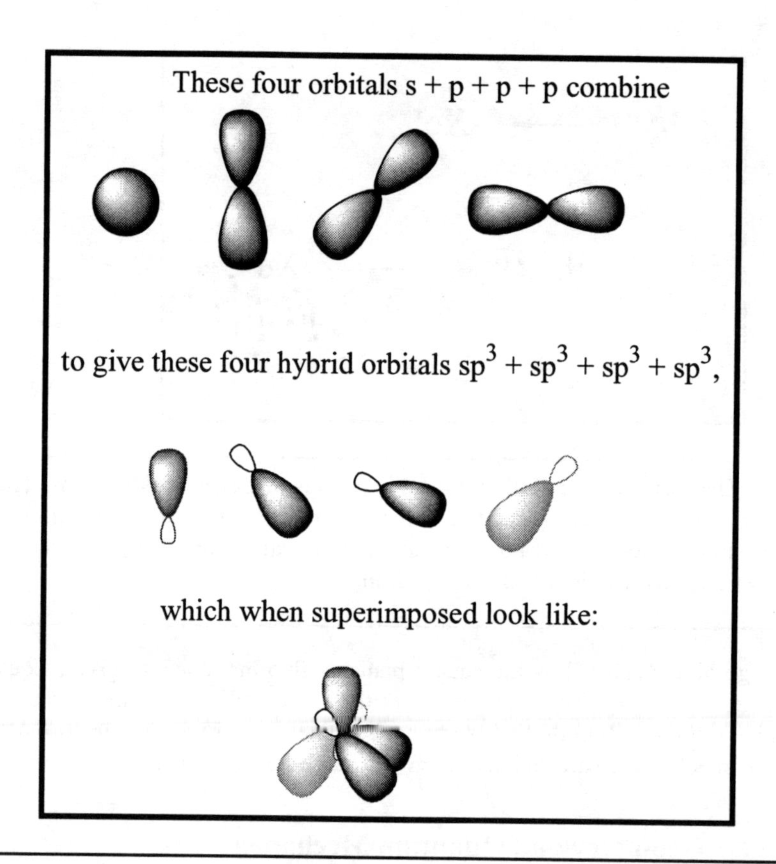


Figure 6.26 How $sp^3$ hybridization accounts for ideal tetrahedral geometry.

Experimentally it's found that most saturated carbon atoms exhibit near-ideal tetrahedral geometry. Quantum mechanical calculation shows that carbon's 2s and three 2p orbitals can be mathematically combined to generate four "hybrid" orbitals directed tetrahedrally as shown in Figure 6.26. Because the calculated hybridization geometry matches the experimental facts, chemists say that saturated carbon atoms are "$sp^3$ hybridized." The simplest alkane, $CH_4$, exists with true ideal geometry. At the top of Figure 6.26 are the 2s- and three 2p-orbitals separated in space so you can see them clearly—they are really superimposed. The wave functions of these four orbitals can be mathematically combined to generate the four new hybrid orbitals shown separated in space in the center of the diagram. Finally, at the bottom you see the hybrid orbitals superimposed as they are imagined to be on a saturated carbon atom. Figure 6.26 applies to any saturated carbon atom in any organic compound. That's a lot of carbon atoms in a whole lot of compounds.

If hybridization accounts for the molecular geometry, how do the bonds themselves arise? The answer is by orbital overlap and the pairing of atomic unpaired electrons to make bonding pairs. Again chemists invoke a quantum mechanical idea but use it in a qualitative way. Figure 6.27 shows the overlap of hydrogen 1s orbitals (gray circles) with the $sp^3$ hybrid orbitals imagined to exist on the carbon atom. Experimentally, all four C-H bonds in methane behave as are single bonds. In this overlap model of the chemical bonding in methane each bond pair is imagined to arise via the combination of a unpaired electron in one of the carbon atom's $sp^3$ hybrid orbitals with the unpaired 1s electron in a hydrogen atom. So every bond is a one electron pair bond, or an expected single bond—agreeing neatly with the experimental facts about methane.

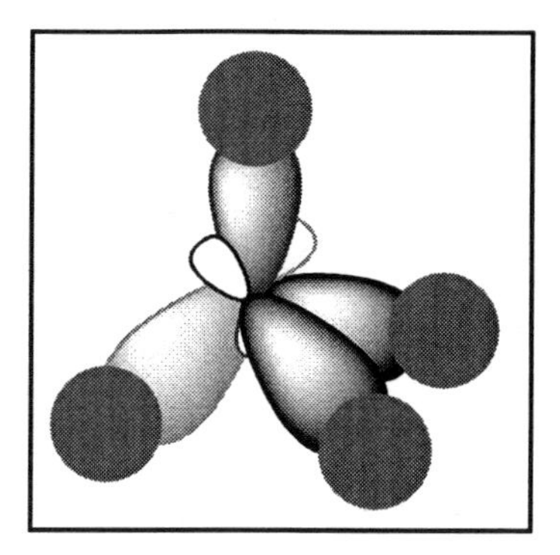

Figure 6.27 $sp^3$ Hybridization and the structure of methane, $CH_4$. Because the geometry of calculated $sp^3$ hybrid orbitals is tetrahedral, the overlap of the 1s hydrogen orbitals (gray circles) of four different hydrogen atoms with those hybrid orbitals neatly accounts for methane's experimental structure.

## $sp^2$ Hybridization and Ethylene

Unsaturated carbon compounds containing a double bond (the alkenes) plays a very important role in organic chemistry, for example, ethylene ($H_2C{=}CH_2$). Experimentally it's found that ethylene's carbon atoms and any other such atoms involved in double bonds show trigonal planar geometry. That's a lot of compounds. Trigonal planar geometry means that the central carbon atom and the four atoms bonded to it lie in a flat surface and that the ideal bond angles are 120°. Quantum mechanical calculation shows that carbon's 2s and two of its three 2p orbitals can be mathematically combined to generate three "hybrid" orbitals directed in a trigonal planar fashion as shown in Figure 6.28.

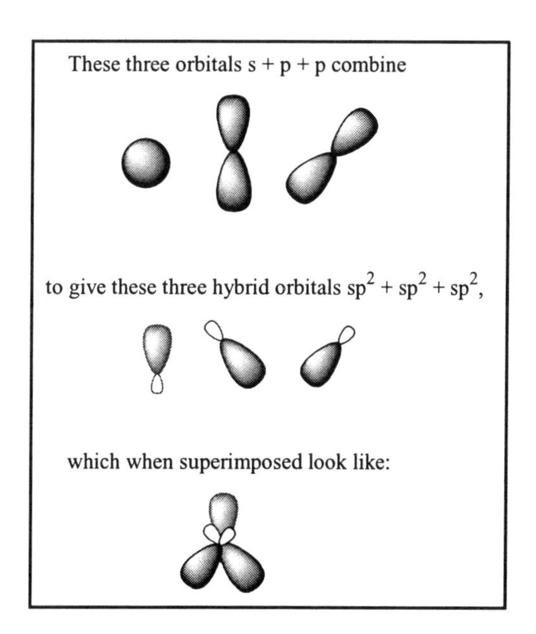


Figure 6.28 How $sp^2$ hybridization accounts for ideal trigonal geometry.

By examining Figure 6.29 you can see how $sp^2$ hybridization accounts for the trigonal geometry observed experimentally for carbon atoms that are forming a double bond. The overlap region in the center of the diagram represents the sigma bond formed by the overlap of two $sp^2$ hybrid orbitals from separate carbon atoms. Each carbon also overlaps with two hydrogen 1s orbitals, shown at the ends of the diagram; that's a total of three sigma bonds for each carbon atom.

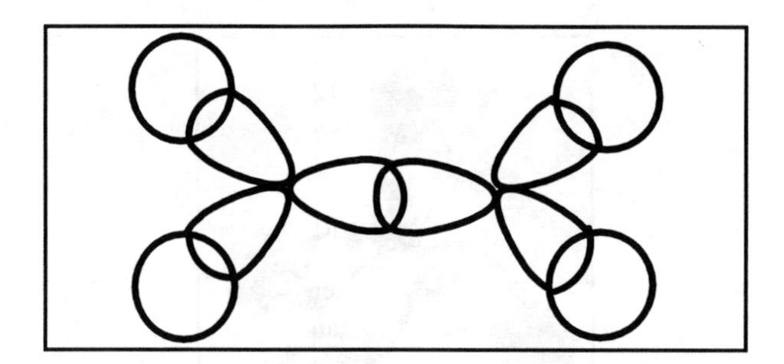

Figure 6.29 $sp^2$ Hybridization and the $H_2C{=}CH_2$ molecule. In this diagram are represented a total of ten orbitals in contour display. The four circles in the corners represent the 1s orbitals of the four hydrogen atoms. The two sets of trigonally disposed egg-shaped contours represent the two sets of $sp^2$ hybrid orbitals on the carbon atoms. The five sigma bonds in the molecule are represented by the five places where orbital overlap occurs.

Figure 6.29 shows the ethylene molecule, $H_2C{=}CH_2$, viewed from the top. Thus, the five overlap regions of ethylene, as pictured, account for the formation of five sigma bonds, and these five bonds are said to constitute the sigma framework of the ethylene molecule. A sixth bond also exists in the ethylene molecule. It is the second bond of the C=C double bond and is a pi bond. The pi bond in ethylene is created by electrons in a pi molecular orbital that lies above and below the molecular plane. The pi orbital is not shown in Figure 6.29. With each carbon atom in ethylene being ascribed $sp^2$ hybridization, two out of its three p orbitals are accounted for, leaving one unhybridized p orbital remaining on each atom. It is these p orbitals that are imagined to combine to create both a bonding pi orbital and an antibonding pi orbital. The unhybridized p orbitals are shown in Figure 6.30.

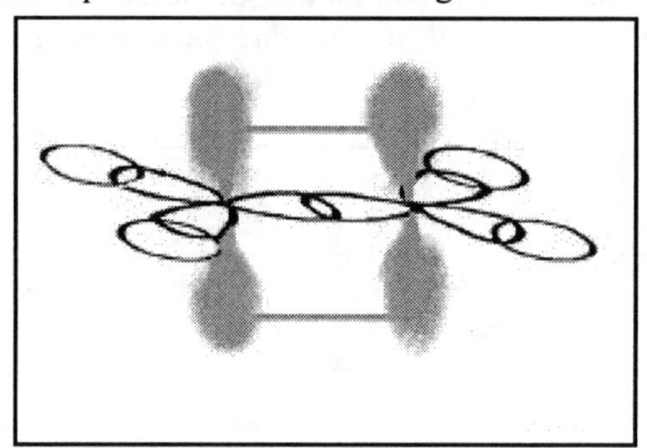

Figure 6.30 Unhybridized p orbitals (shown gray), one on each carbon atom, are imagined to be origin of the pi orbital in ethylene. One electron from each of these orbitals will join to make a bonding pair that will reside in the bonding pi orbital and create the second of the two bonds that make the double bond.

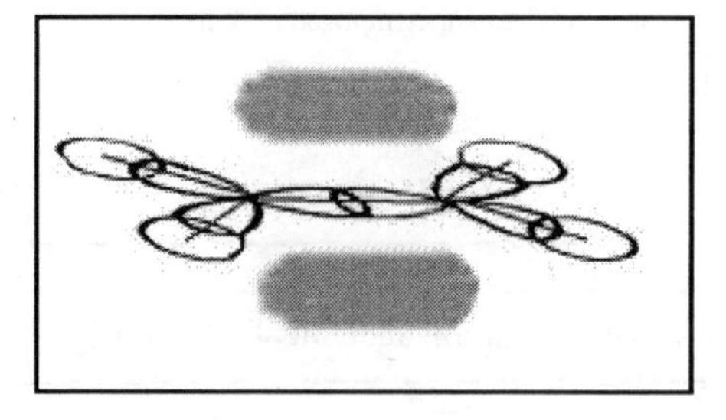

Figure 6.31 Bonding in ethylene. The five sigma bonds in the sigma framework are shown as lines. The bonding pi orbital, with its two lobes, one above and one below the plane of the ethylene molecule, is shown in gray.

Figure 6.31 is not a bad quantum mechanical depiction of ethylene; we've come a long way from croquet balls and wooden sticks. The $sp^2$ hybridization explains ethylene's 120° angles, and by representing the double bond as a combination of sigma bond and a pi bond the hybridization also explains why ethylene undergoes addition reactions and why actual breaking of the C-C bond is rare.

## sp Hybridization and Acetylene

The final case to consider is the triple bond. Unsaturated carbon compounds containing a triple bond (the alkynes) are important in organic chemistry. The simplest organic molecule that contains a triple bond is acetylene. Experimentally it's found that acetylene's carbon atoms and any other such atoms involved in triple bonds show linear geometry. In acetylene itself, $C_2H_2$, H-C≡C-H, all four atoms lie in a line.

For two carbon atoms that are triply bonded the hybridization ascribed to the carbon atoms is sp. The sp hybridization scheme is shown at the top of Figure 6.32. Hybridization of one s and one p orbital produces two sp hybrid orbitals directed 180° apart in space. At the bottom of Figure 6.32 is a respectable quantum mechanical depiction of acetylene. There are three sigma bonds in H-C≡C-H, both of the H-C bonds, and one of the three bonds in the triple bond. Acetylene's sigma bonds are indicated by lines. There are also two pi bonds in acetylene. The two lobes of *one* of the pi bonds are shown as gray ellipses. The second pi bond looks just like the one pictured, except that its lobes lie above and below the plane of the paper. If you imagine grabbing the molecule, holding one hydrogen atom in each hand like a rolling pin, and then rolling through a 90° angle you'll move the $p_y$ pi bond into the position of the $p_z$ pi bond.

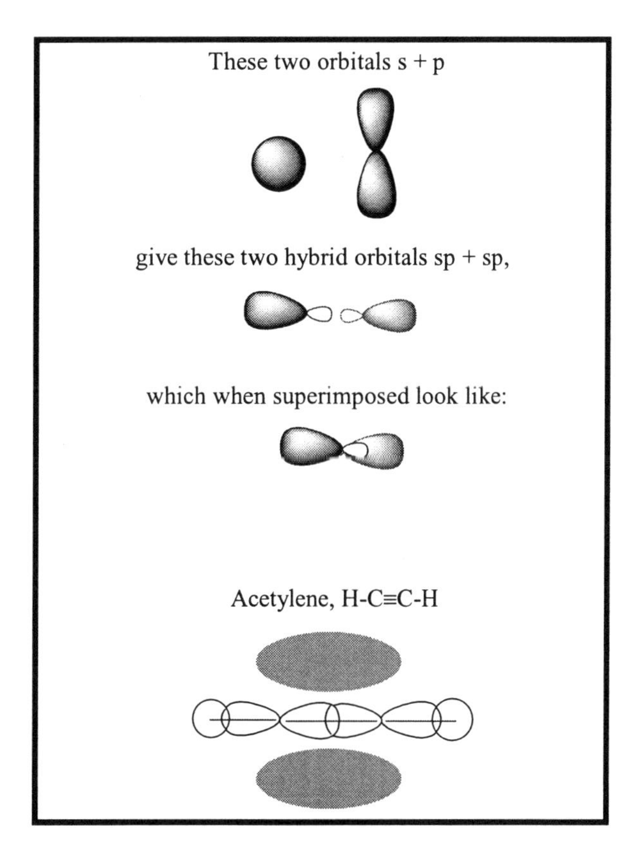


Figure 6.32 sp Hybridization (top) and the bonding in acetylene. Only one of the two pi orbitals is shown; its two lobes are represented by the gray ellipses. The two lobes of the second pi orbital lie above and below the plane of the paper and are not shown in this diagram.

Conclusion: in this chapter we've made a quick survey of chemical bonding—from its beginning less than 150 years ago to modern quantum mechanics and molecular visualization. No matter what materials and substances engineers work with, those materials and substances are held together by chemical bonds. When we talk about liquid and solid materials in later chapters we'll return to many of these ideas.

## Essential Knowledge—Chapter 6

**Glossary words:** Chemical bonds, valency, Chime, equilibrium bond length, bond energy, bond strength, homonuclear diatomic molecule, x-ray crystallography, atomic radius, covalent radius, ionic radius, bond polarity, non polar, pure covalent bond, polar covalent bond, dipole moment, percentage ionic character, bond-type continuum, electronegativity, bond angle, radius ratio, molecular geometry, molecular structure, static electron model, octet rule, valence electrons, core electrons, Lewis dot diagrams, electron pair bond, diamagnetic. paramagnetic, unpaired electrons, closed shell, bonding electron pair, lone pair, coordinate-covalent bond, valence unsaturation, molecular orbital, overlap, bonding molecular orbital, sigma orbital, pi orbital, antibonding molecular orbital, sigma bond, pi bond, nonbonding orbital, delocalized bonds, molecular geometry, molecular structure, VSEPR, ideal geometry, electron geometry, hybridization.

**Key Concepts**: Static electron models for chemical bonding. Electron configurations and their relationship to core and valence electrons. How to draw Lewis dot diagrams. Quantum mechanics and molecular orbitals; how atomic orbitals combine. Energy level diagram for $H_2$ and $He_2$. Interpreting the generalized energy level diagram for second row diatomic molecules and its connection to bond energy and magnetic properties of diatomics. Writing molecular electron configurations for $Li_2$ to $Ne_2$. Molecular geometry and VSEPR. The distinction between molecular geometry and electron geometry. Molecular geometry and hybridization. $sp^3$, $sp^2$, and sp hybridization.

**Key Equations**:

$$\text{Percentage ionic character of a bond} = \frac{\text{Experimental dipole moment}}{\text{Calculated one-electron transfer dipole moment}} \times 100\%$$

$$\text{Radius ratio} = \frac{\text{radius of the central atom}}{\text{radius of the ligand atoms}}$$

$$\text{Bond order} = \frac{\text{Number of electrons in bonding orbitals} - \text{number of electrons in antibonding orbitals}}{2}$$

## Questions and Problems

### Chemical Bonding Bond Strengths or Energies

6.1. The potential energy of two atoms varies depending on the distance between the atoms. Make a sketch showing this variation.

6.2. The diatomic hydrogen molecule has a bond energy of 436 kJ $mol^{-1}$ and a bond length of 74.6 pm. The corresponding values for diatomic nitrogen are 945.3 kJ $mol^{-1}$ and 109.8 pm. Sketch potential energy curves for these two molecules on the same graph.

6.3. Rank the following bonds in order of their strengths from the lowest bond energy to the greatest: H-O, H-Cl, H-N, C-H, H-H, C=C, and H-I.

6.4. Some bond energies are reported as averages over many molecules, others are reported a just one value. Why?

### Bond Lengths

6.5. Estimate the lengths of the following bonds: H-N, O-C, O-O, and H-S.

6.6. Use data from the table of atomic radii to calculate the lengths of the following bonds: Cl-Cl, Br-Br, Br-Cl, I-F, and, I-Cl. Which is the longest and which the shortest?

6.7. State the trend in atomic radius across and down the periodic table.

6.8. Assign the cesium +1 ion a radius of 100 size units. On that scale what are the sizes of the ions $Rb^+$, $K^+$, $Na^+$, and $Li^+$.

6.9. Compare ionic and covalent radii of the element in groups 1 and 17 of the periodic table.

### Bond Polarity

6.10. Which of the following bonds is non polar? H-F, C-F, F-F, Cl-Cl, P-Br. Rank them from least polar to most polar.

6.11. Use electronegativity values to estimate the percentage ionic character of the following bonds: a. H-H, b. C-H, c. C-Br, d. C-Cl, e. C-F, f. F-F, and g. H-F.

6.12. Use the data in the table of electronegativities (Table 6.7) to deduce the percentage of ionic and covalent character in each of the following bonds: a. Cl-Cl, b. H-Br, and c. Li-Br.

6.13. Calculate the dipole moment in debye units of a diatomic molecule with a bond length of 180 pm whose atoms have charges of +e and -e.

6.14. A diatomic molecule has a dipole moment of 5.6 debye. Assuming that its atoms have charges of +e and -e, what is the bond length of the diatomic molecule? Is this a reasonable bond length for a diatomic molecule?

6.15. Calculate the fractional ionic character of a bond with dipole moments and bond lengths of 1.4 D and 127 pm, 1.25 D and 159 pm, and 0.56 D and 211 pm.

### Core and Valence Electrons

6.16. Define or describe in 25 words or less the term valency.

6.17. Distinguish between core and valence electrons.

6.18. Define the term "noble gas core."

6.19. How many valence electrons and core are there in each of the following elements: a. carbon, b. aluminum, c. calcium, d. actinium.

6.20. How many valence electrons are there in each of the following elements: a. sulfur, b. argon, c. vanadium, and d. barium.

6.21. What will be the charges on the ions commonly formed by these elements that are located close to one of the noble gases in the periodic table: a. calcium, b. sulfur, c. rubidium, d. scandium, e. tellurium? How many core and valence electrons are there in each of these atoms?

6.22. Referring back to Chapter 5, define the term isoelectronic. For each of the following pairs of atoms and ions state whether or not the pair is isoelectronic: a. $Na^{+}$ and Ne, b. Li and He, c. $Be^{2+}$ and He, d. $Sc^{3+}$ and Ar, e. $La^{3+}$ and $Xe^{+}$, and f. $Fe^{3+}$ and $Mn^{2+}$.

6.23. How many electrons must be removed from a potassium atom so that the resulting ion is isoelectronic with a neutral sulfur atom?

6.24. Use the periodic table to write electron configurations of the following atoms or ions: a. N, b. $F^{-}$, c. $Al^{3+}$, d. Ar, e. K, f. V.

## Magnetic Properties of Atoms and Molecules

6.25. Sketch the atomic orbital energy diagram for beryllium and boron. How many unpaired electrons are there in each atom?

6.26. How many pairs of electrons are in a potassium atom? How many unpaired electrons?

6.27. Deduce the number of unpaired electrons in each of the following atoms or ions: a. Ca, b. $Sc^{2+}$, c. V, d. Cr, e. Mn, f. Cu, g. $Br^{-}$, h. Ru, i. Pt. State which are diamagnetic and which are paramagnetic.

6.28. Select the element whose atoms are diamagnetic: a. Na, b. Cl, c. Ca, or d. Ga?

## Lewis Dot Diagrams

6.29. Sketch the Lewis dot diagrams of water and hydrogen peroxide. State the number of bond pairs and lone pairs in each of these molecules.

6.30. Sketch the Lewis dot diagrams of $N_2H_4$ (hydrazine), and methyl chloride $CH_3Cl$. State the number of bond pairs and lone pairs in each of these molecules.

6.31. Sketch the Lewis dot diagrams of $F_2C{=}CF_2$, $H_2N{-}CH_3$, $H{-}C{\equiv}C{-}CH_3$. State the number of bond pairs and lone pairs in each of these molecules.

6.32. Draw the Lewis-dot diagram of two of the nitrogen oxides, NO and $NO_2$. Both of these molecules are stable but rather reactive. Suggest a reason for their reactivity.

6.33. $NO_2$ dimerizes to form $N_2O_4$. Write a chemical reaction using Lewis dot diagrams for the reactants and products to show the bonding that occurs on dimerization.

6.34. Sulfur, silicon, and iodine readily form polyatomic ions with "expanded octets." Draw Lewis dot diagrams for: a. $SF_4$, b. $SiF_6^{2-}$, c. $IF_3$.

6.35. Cl and F form three neutral binary compounds: ClF, $ClF_3$, and $ClF_5$. I and F form four binary compounds, three of which are analogs of the chlorine fluorides. Speculate about the formula of the fourth iodine fluoride molecule and draw its Lewis diagram.

6.36. Aluminum bromide in the solid state consists of $Al_2Br_6$ molecules. Two of the six bromine atoms are in bridging positions between the aluminum atoms and two further bromine atoms are covalently bonded to each of the aluminum atoms. Sketch the Lewis diagram for $Al_2Br_6$ and indicate the coordinate covalent bonds.

6.37. Write a reasonable Lewis structure for the addition compound formed when $SO_3$ and $H_2O$ combine via a coordinate covalent bond. This addition compound can rearrange to form sulfuric acid. Draw the Lewis structure of sulfuric acid.

## Molecular Orbitals

6.38. Molecular orbitals arise from the interference of atomic matter waves. Discuss.

6.39. Sketch the two molecular orbitals that form from the 1s orbitals as two hydrogen atoms approach one another. Label them as the bonding and antibonding orbitals.

6.40. What is it about a bonding orbital that makes it bonding?

6.41. What is it about an antibonding orbital that makes it antibonding?

6.42. Make a sketch of the generalized molecular energy diagram for elements Li through Ne. How many total electrons will be need to fill all of the energy levels in this diagram completely ? State the names of two orbitals in this diagram that are degenerate. State another pair of orbitals in this diagram that are degenerate.

6.43. Consider the diatomic molecule $B_2$. How many valence electrons are there in this molecule? How many electrons are in bonding orbitals? How many electrons are in antibonding orbitals? How many are unpaired? What is the bond order of the molecule?

6.44. Describe in your own words a sigma orbital and a pi orbital. How many electrons can enter these orbitals. How many total molecular orbitals are there on any diatomic molecule?

6.45. In any diatomic molecule, how does the number of bonding orbitals compare with the number of antibonding orbitals?

6.46. In any stable, diatomic molecule, how does the number of electrons in bonding orbitals compare with the number of electrons in antibonding orbitals. Discuss with respect to the specific case of the Be-Be diatomic molecule.

## VSEPR and Molecular Geometry

6.47. State the octet rule. State the names of the five basic molecular geometries described in Chapter 6. Only one basic geometry can satisfy the octet rule. Which one and why?

6.48. State all the conceivable bond angles in an ideal trigonal bipyramidal molecule.

6.49. Derive the value of the tetrahedral angle using the 2500-year-old methods of geometry. See also problem 2.67.

6.50. Draw Lewis dot diagrams for the following molecules that violate the octet rule. a. $PCl_5$, b. $BF_3$, c. $SCl_4$, d. $XeF_2$, and e. $SF_6$.

6.51. Use the method of VSEPR to speculate about the bond angles in the molecules: a. $PCl_5$, b. $BF_3$, c. $SCl_4$, d. $XeF_2$, and e. $SF_6$.

6.52. Use the method of VSEPR to speculate about the bond angles in the ions: a. $I_3^-$, b. $NO_2^-$, c. $NH_4^+$, and d. $BF_4^-$.

## Atomic Orbital Hybridization

6.53. Describe the atomic orbitals involved in the formation of $sp^3$ hybrid orbitals. Make a sketch of the hydrogen like atomic orbitals and the resulting hybrid atomic orbitals.

6.54. Describe the atomic orbitals involved in the formation of $sp^2$ hybrid orbitals. Make a sketch of the hydrogen-like atomic orbitals and the resulting hybrid atomic orbitals.

6.55. Describe the atomic orbitals involved in the formation of sp hybrid orbitals. Make a sketch of the hydrogen-like atomic orbitals and the resulting hybrid atomic orbitals.

## Sigma and Pi Bonding

6.56. Sketch the formation of a sigma bond from: two s-type atomic orbitals, two $p_x$-type atomic orbitals, two $p_y$-type atomic orbitals, and two $p_z$-type atomic orbitals.

6.57. Draw the Lewis dot diagram of carbon dioxide and then work out how many sigma and pi bonds there are in this molecule.

6.58. Draw the Lewis dot diagram of formaldehyde ($CH_2O$), which has a double bond between the carbon and oxygen atoms. Work out how many sigma and pi bonds there are in this molecule.

6.59. Sketch the structure of the molecule $H_3C\text{-}C{\equiv}C\text{-}F$. How many sigma and how many pi bonds in this molecule?

6.60. Why is the pi bond between the carbon atoms in ethylene more reactive than the sigma bond between the same carbon atoms?

6.61. a. Does the static electron model of benzene work? b. How many sigma and pi bonds are there in benzene? c. Make two *different* sketches of the benzene molecule using only line bonds. Discuss how together these two sketches might be used to rationalize electron delocalization. d. What does a circle inside a hexagon have to do with benzene?

## Problem Solving

6.62. Use the concepts of radius ratio and Lewis dot structure to select the binary compound that contains two halogen elements and the greatest possible number of total atoms.

6.63. Speculate about the number of valence electrons in atoms of nickel and protactinium.

6.64. Give a complete account of the entire bond structure in acetylene. Discuss the sigma framework of bonds and the pi bonding orbitals. Explain precisely how your account rationalizes all the bond angles and all the bond lengths in the molecule.

6.65. Sulfur trioxide is a planar molecule with three 120° O-S-O angles. Experimentally, the three S-O bonds are found to be equal in length. Sketch the Lewis dot structure for sulfur trioxide. What relative bond lengths does your sketch predict? If your prediction is not correct, how can it be fixed?

6.66. What is the size of a hydride ion, $H^-$? Is it stable? [❁kws +"hydride ion"].

6.67. Fill the molecular energy level diagram for the peroxide ion, $O_2^{2-}$. What is the order of the oxygen-oxygen bond? Is the ion paramagnetic or diamagnetic?

6.68. The table of covalent radii omits helium. Why?

6.69. Assuming three atoms all mutually touching and the hard sphere model, a. what would be the C-Pt-C bond angle b. what would be the H-Ra-H bond angle?

6.70. The dipole moment of water is 1.84 D. From this value and information provided in the chapter make an estimate of the H-O-H angle in water.

6.71. The fluoromethane molecule ($CH_3F$) has an experimental dipole moment of 1.85 D. It is a tetrahedral molecule with three C-H bonds and one C-F bond. The experimental C-F bond length is 133 pm (the value calculated from atomic radius sum is 64 + 77 = 141 pm). Calculate the percentage ionic character of the C-F bond by making a ratio of its experimental dipole moment (converted from D to C m) to the 1 electron transfer dipole moment. Use electronegativity difference values and Figure 6.7 to estimate the percentage ionic character of the C-H bond and C-F bond. Discuss the results.

# Chapter 7

# Chemical Energy and the First Law of Thermodynamics

## Introduction

**Thermodynamics** is the study of heat and work and their relation. Heat and work are different forms of energy. The SI unit of energy, work, or quantity of heat is the derived unit called the joule. Recall, 1 joule = 1 kg $m^2 s^{-2}$.

**Chemical energy** is energy produced as a result of chemical reactions. Chemical energy is obtained when a fuel is burned to produce heat; **heat** is energy that is transferred between two points because of a temperature difference. The study of chemical energy is called **thermochemistry**. Thermochemistry is part of thermodynamics. Fuel burning reactions are important; without them, there would be no modern, industrial societies. The power plant shown in Figure 7.1 provides both steam-heat and electrical power. Oil, coal, gasoline, natural gas, wood, and food are all familiar sources of chemical energy. The United States obtains more than 90% of its energy from chemical fuels. Non-chemical energy sources, such as nuclear power plants, hydroelectric generators, windmills, and solar energy collectors, account for the remainder. Almost all chemical energy derives from the combustion reactions that were introduced in Section 3.1. The products of combustion reactions are primarily carbon dioxide and water, along with smaller amounts of carbon monoxide, and the oxides of sulfur and nitrogen. Solid fuels that contain noncombustible minerals also produce ash.

Figure 7.1 A coal burning power plant. Most of the energy we use to run our industrial economy comes from the combustion of chemical fuels.

**Caution**

The terminology of thermodynamics is very specialized. Thermodynamics uses many ordinary words with precisely defined technical meanings. One example is the word *state*—which has too many technical meanings.

Previously we've met *state* of aggregation of solid, liquid, and gas, and in the preceding chapter we met quantum mechanical energy *states* of atoms and molecules. The thermodynamic *state* of a system is defined in Section 4.6 in connection with the work done by expanding gases. Make sure you have a clear understanding of that section before you go on.

Thermodynamics is fundamental to all science and is specially important in engineering science. About 200 years ago, studies by engineers of the efficiency of steam engines led to fundamentally new conceptions of how the natural world behaves. In this chapter, we'll introduce the first law of thermodynamics and show how it is applied to chemical reactions.

**Energy** is the property that gives a system or object the ability to do work. **Work** is energy used to move an object: work = force × distance. In connection with our studies of gases in Chapter 4 we calculated kinetic energies ($KE = \frac{1}{2}mv^2$) and saw how gases can be modeled using the kinetic-molecular theory. Objects have kinetic energy because of their motion. Objects have potential energy because of the forces that act on them. Atoms and molecules have both kinetic and potential energy—kinetic energy because of their motion, and potential energy because of the forces they exert on one another.

The main experimental technique of thermochemistry is calorimetry. **Calorimetry** is the study of the heat released or absorbed by chemical reactions and other processes. A **calorimeter** is an instrument used to measure heat. For reasons that we'll detail later, the amount of heat released by a chemical reaction may depend on whether the reaction is done at constant pressure or at constant volume. For this reason it is convenient to divide calorimeters into two classes: **constant pressure calorimeters** and **constant volume calorimeters**. A simple constant pressure calorimeter is shown in Figure 7.2. Foamed polystyrene is an excellent thermal insulator, so whatever's in the cup can be treated as a thermodynamic system.

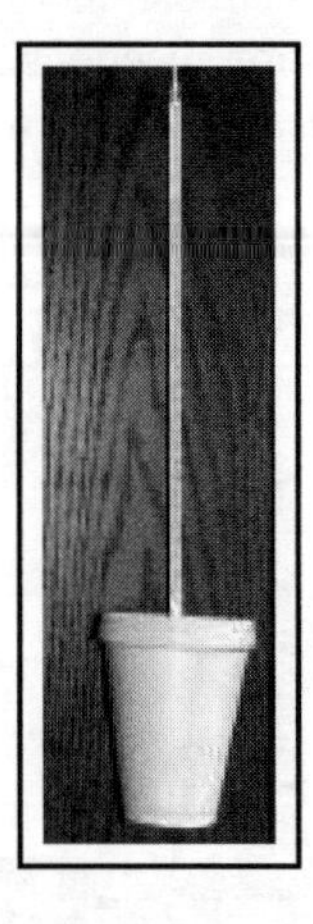

Figure 7.2 A simple, constant pressure calorimeter. It consists of a thermally well-insulated container exposed to the pressure of the atmosphere. "Coffee cup" calorimeters are widely used in student laboratory exercises in thermochemistry. The cup and its contents constitute a thermodynamic system (see Section 4.6).

Figure 7.3 shows a disassembled "bomb" calorimeter. The bomb calorimeter is a constant volume calorimeter. The sample to be tested is placed in a strong stainless steel vessel (the "bomb") pressurized with oxygen at approximately 100 atmospheres pressure and equipped with an igniter—a kind of spark plug. The bomb is submerged in a bucket of water, and the heat released when the fuel is ignited increases the water's temperature. The energy of the combustion can be calculated from this temperature increase.

Figure 7.3. A bomb calorimeter shown disassembled. The bomb itself is at the lower left. The bucket in the upper left will be filled with water to absorb the released heat. The large black object is the well-insulated container in which the submerged bomb is placed before ignition. A schematic diagram of this bomb calorimeter is shown in Figure 7.11.

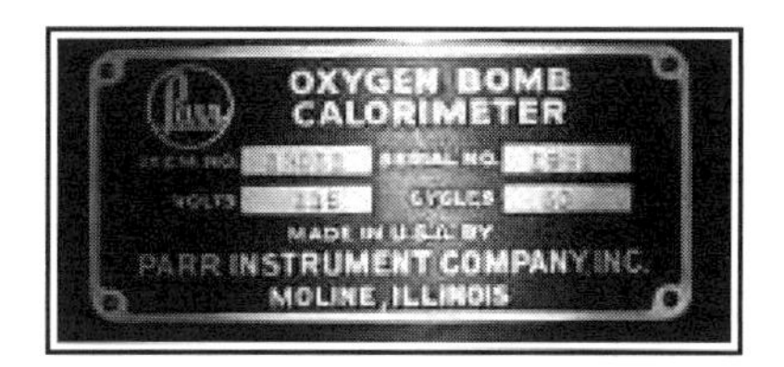


Figure 7.4. Oxygen bomb calorimeters are widely used to measure the chemical energy of fuels and foods.

Results obtained from many calorimetric experiments have been organized into **standard tables** of thermodynamic data. Such tables are an important scientific resource. They can be used to estimate how many tons of coal, cubic feet of natural gas, or gallons of oil must be burned to run a factory, heat a home, or provide electricity for a city. The tables also provide a basis for choosing the most useful rocket fuels and for estimating the energy values of foods. In the next chapter we'll see that the same tables can be used to predict the position of equilibrium of any chemical reaction.

Standards organizations around the world have the responsibility of producing and codifying accurate thermodynamic data. In the US, that's the National Institute of Standards and Technology, NIST, (formerly the National Bureau of Standards), located in Gaithersburg, Maryland. Obtaining accurate, reliable thermodynamic data is an ongoing effort for NIST and other national standards organizations. For example, it has lately been important to measure accurate thermodynamic data for fluorocarbons to help commercial producers develop "ozone friendly" refrigerant blends.

A brief outline of the history of thermodynamics is shown in the box. Thermodynamics is perhaps the one branch of science that owes more to engineers than any other.

**Outline of the History of Thermodynamics**

1665 Perhaps for the first time it is written that heat is a form of energy. "If this motive Energie ... must be called heat. I contend not." Joseph Glanvill.

1760 Joseph Black (1728-1799) studies the temperature changes of substances when they are heated. He discovers the properties of specific heat and latent heat. His ideas greatly influence the engineer James Watt.

1798 Benjamin Thompson (Count Rumford, 1753-1814), born in Massachusetts, leaves the US for Europe in 1783, having fought on the wrong side in the Revolutionary War. Later, in Bavaria, he supervises cannon boring operations and realizes that the frictional heat released depends only on the amount of work done. He is the first to conclude that heat is a form of motion. Still later, he founds the Royal Institution in London and has an unhappy marriage to Antoine Lavoisier's widow, Marie.

1799 Sir Humphry Davy (1778-1829) rubs pieces of ice together and confirms Benjamin Thompson's view that friction creates heat.

1824 Nicholas Sadi Carnot (1796-1832), engineer, physicist, and a lieutenant on the French General Staff, thinks deeply about the efficiency of steam engines, does a *gedanken* (thought) experiment, imagines that heat is the rapid movement of small particles, and discovers the second law of thermodynamics. His work, motivated by a desire to understand and improve the low efficiency of engines, is not appreciated until 25 years later, in part because the first law of thermodynamics is not yet understood.

1842 Julius R. Meyer (1814-1878), a German ship's physician, notices differences between human metabolism in hot and cool climates. He hypothesizes that heat and physical work are equivalent and is the first to state the law of conservation of energy—the first law of thermodynamics.

1843-1847 James Prescott Joule (1818-1889) experimentally measures the mechanical equivalent of heat by stirring water and observing its temperature rise. Joule introduces the Btu. He takes a four-foot thermometer along to the Chamonix Falls on his honeymoon to check the water temperature at top and bottom. Unfortunately, temperature rise due to the release of potential energy by the falling water is obscured by evaporative cooling.

1847 Hermann Ludwig von Helmholtz (1821-1894), independent of Mayer and Joule, states the law of conservation of energy. Three independent discoveries of the law show that there is sometimes in science an idea whose time has come.

1848 William Thomson (1824-1907), professor of Natural Philosophy at Glasgow University, develops Joule's ideas and realizes that there is a natural temperature scale based on thermodynamics. In 1892 he becomes Lord Kelvin. The kelvin unit bears his name.

1850 Rudolf Clausius (1822-1888), German mathematical physicist and professor, states the second law and introduces the entropy function as a state property. In 1865 he immortalizes the first and second laws as: 1. the energy of the universe is constant, and 2. the entropy of the universe tends to maximize.

1875 Josiah Willard Gibbs (1839-1903), American mathematician and physicist, writes "The Equilibrium of Heterogeneous Substances." He thereby extends thermodynamics to chemistry and founds the branch of the subject called physical chemistry.

1906 Walther Nernst (1864-1941) proposes the third law of thermodynamics and shows that it is possible to calculate the equilibrium position of a chemical system from thermodynamic data.

During chemical reactions, some chemical bonds are broken and others are formed. Bond breaking absorbs energy; bond forming releases energy. When more energy is released by bond formation than is absorbed by bond breaking, the "reaction system" produces excess energy that is released to the environment. A chemical reaction that releases energy to its environment is **exothermic**. A chemical reaction that absorbs energy from its environment is **endothermic.**

## Section 7.1 Energy in Fuels and Foods

An important application of thermochemistry is to make precise measurements of the chemical energy in fuels and foods. The heats of combustion of some fuels, foods, and compounds are listed in Table 7.1. Values of substances with variable composition are stated as ranges. The values in the table are derived from experiments using bomb calorimeters.

Table 7.1
**Energy Values of Foods and Fuels**

| Fuel or Food (*shows a food group) | Heat of Combustion kJ $g^{-1}$ |
|---|---|
| Hydrogen | 141.8 |
| Methane | 55.1 |
| n-Octane | 47.7 |
| Diesel fuel/gasoline | 40-45 |
| Animal fats and oils* | 39-40 |
| Coal | 25-35 |
| Charcoal | 34 |
| Ethyl alcohol | 27.8 |
| Wood | 15-18 |
| Cornstarch (carbohydrate)* | 17.1 |
| Gelatin | 15.7 |
| Table sugar | 15.6 |
| Trinitrotoluene (TNT) | 15.1 |
| Protein* | 12-16 |
| Glycine | 13.1 |
| Dynamite | 7 |
| Gunpowder | 3 |

A general, unbalanced, chemical equation for the combustion of a food or fuel can be written as:

$$1 \text{ g of Fuel or food} + \text{Oxygen(g)} \rightarrow CO_2(g) + H_2O(l) + N_2(g) + \text{heat}$$

Both the reactants and the products are at 25°C and 1 bar pressure. One bar is equivalent to 1.01325 atmosphere (exactly). Lots of older data are reported for 25°C and 1 atmosphere pressure. Thus, a food or fuel's **energy value** is the energy (heat) released by the complete combustion of 1 g of the food or fuel according to the above equation (balanced for the specific food or fuel) at 298 K and a pressure of 1 bar.

The three traditional food groups of human nutrition are marked with an asterisk in Table 7.1. Hydrogen provides the greatest energy per gram of any fuel; that's why liquid hydrogen is used along with liquid oxygen as rocket fuel. Animal fats and oils are excellent fuels; they rank between gasoline and coal in energy value. Common sense and your personal experience probably confirm the fact that wood has less energy value than coal or oil; that's demonstrated in Example 7.1. You may be surprised to see that explosives (TNT, dynamite, and gunpowder) have rather low energy values. Explosives must contain internal oxygen, so they can self-react in the absence of air; consequently, explosives contain relatively less carbon and hydrogen than fuels without oxygen. Note that the chemical energy of alcohol is less than that of octane for the same reason—alcohol molecules, unlike octane molecules, contain some oxygen; the relationship between these fuels and their chemical compositions is illustrated in Example 7.2.

**Example 7.1**: A large electric generating plant burns an average of 1,250 tons per hour of a medium grade coal. Estimate the amount of wood that must be burned per hour to maintain the same rate of electric power production.

Strategy: Use the heats of combustion from Table 7.1 to establish the ratio of combustion energy available from the two fuels. Then multiply by that factor. The approximate energy value of coal is 30 kJ $g^{-1}$. The approximate energy value of wood is 16.5 kJ $g^{-1}$

The ratio of these is 30/16.5 = 1.8 times as much energy value in coal as in wood

$$= 1{,}250 \text{ tons coal hour}^{-1} \times 1.8 \frac{\text{tons wood}}{\text{tons coal}}$$

$$= \underline{2300 \text{ tons wood hour}^{-1}}$$

Almost twice as much wood as coal would be required. The answer is reported to two significant figures, but just one significant digit would better reflect the variability of the chemical energy of real fuels.

**Example 7.2**: The energy value of ethane ($C_2H_6$) is 51.8 kJ $g^{-1}$ and of ethylene ($C_2H_4$) 50.2 kJ $g^{-1}$. Table 7.1 shows the energy values of various hydrocarbons and ethyl alcohol. a. Discuss the comparative energy values of hydrocarbons on a per gram basis, and, b. estimate the energy value of ethyl alcohol from the energy value of ethylene.

Strategy a: To compare the energy values of ethane, ethylene, and the various hydrocarbons listed in Table 7.1, examine the energy values of pure hydrogen and pure carbon. Recognize that the energy values of hydrocarbons lie on a scale intermediate between pure hydrogen and pure carbon.

The energy value of hydrogen is 141.8 kJ $g^{-1}$. The energy value of carbon (charcoal) is 34 kJ $g^{-1}$

The energy values of hydrocarbons vary with their H:C ratios. The more hydrogen atoms relative to carbon atoms the more energy per gram. In units of kJ $g^{-1}$ that's:

$$H_2(141.8) > CH_4\,(55.1) > C_2H_6\,(51.8) > C_2H_4\,(50.2) > C_8H_{18}\,(47.7) > C\,(34)$$

Strategy b: To estimate the energy value of ethyl alcohol using ethylene's value, *assume* that ethyl alcohol is composed of ethylene and water:

$$C_2H_4 + H_2O \rightarrow C_2H_6O$$

So with water having zero energy value, 1 mole of ethyl alcohol ($C_2H_6O$, 46.07 g) is the *energy equivalent* of 1 mol of ethylene ($C_2H_4$, 28.02 g). With that in mind, a mole of ethyl alcohol will have the same energy value as a mole of ethylene, but because it weighs more, the alcohol's energy value per gram will be less. The calculated energy value of ethyl alcohol will be the ethylene value multiplied by the ratio of the masses of the two compounds, or 28.02/46.07.

Calculated energy value of ethyl alcohol = $50.2 \times \frac{28.02}{46.07}$ kJ $g^{-1}$ = $\underline{30.5 \text{ kJ g}^{-1}}$

Conclusion: The estimated value of 30.5 kJ $g^{-1}$ agrees reasonably well with the experimental value of 27.8 kJ $g^{-1}$ (Table 7.1). The discrepancy shows that ethylene and ethyl alcohol have somewhat different molar energy values.

This example clearly shows why (when compared on a per gram basis) organic compounds that contain oxygen have a substantially lower energy value than do hydrocarbons.

The chemical energy of a food can be dramatically illustrated by soaking it in liquid oxygen, and lighting it with a match. The photograph in Figure 7.5 shows the combustion of crumbled oatmeal cookies, saturated with liquid oxygen. All sorts of foods can be burnt in this way. Low density, powdery foods, such as puffed cheese flavored morsels, seem to produce the biggest flames, but even something dense, such as a Peanut Butter Cup®, can be ignited if it is thoroughly doused in liquid oxygen.

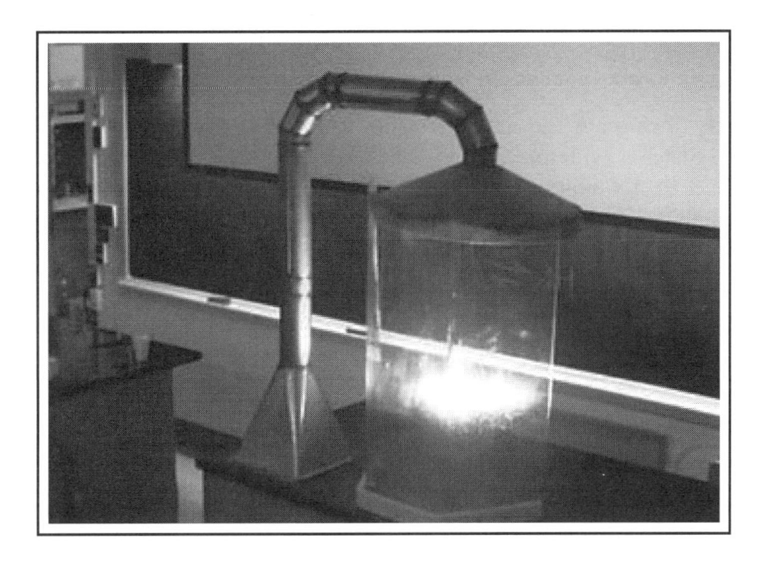

Figure 7.5. Several cookies saturated with liquid oxygen produce a dramatic flare when ignited. Note the safety shield and the venting hood. Worker health and safety are always important issues to consider when using chemical materials and energetic fuels. Compare for example the Materials Safety Data Sheet in Appendix I.

Thermochemistry shows up in everyday life as the calorie counts of foods, as illustrated in Figure 7.6.

**Nutrition Facts**
Serving Size 1 bar (42 g)
Servings Per Container 6

| Amount Per Serving | |
|---|---|
| **Calories** 220 | Calories from Fat 100 |
| | % Daily Value* |
| **Total Fat** 11 g | 17% |
| Saturated Fat 7 g | 35% |
| **Cholesterol** less than 5 mg | 1% |
| **Sodium** 30 mg | 1% |
| **Total Carbohydrate** 27 g | 9% |
| Dietary Fiber less than 1 g | 3% |
| Sugars 21 g | |
| **Protein** 3 g | |

Vitamin A 0% • Vitamin C 0%
Calcium 6% • Iron 2%

*Percent Daily Values are based on a 2,000 calorie diet. Your daily values may be higher or lower depending on your calorie needs:

| | Calories: | 2,000 | 2,500 |
|---|---|---|---|
| Total Fat | Less than | 65g | 80g |
| Sat Fat | Less than | 20g | 25g |
| Cholesterol | Less than | 300mg | 300mg |
| Sodium | Less than | 2,400mg | 2,400mg |
| Total Carbohydrate | | 300g | 375g |
| Dietary Fiber | | 25g | 30g |

INGREDIENTS: SUGAR; FLOUR; COCOA BUTTER; NONFAT MILK; CHOCOLATE; REFINED PALM KERNEL OIL; MILK FAT; LACTOSE

Kit Kat
CRISP WAFER

Figure 7.6 In the United States, food energy values are labeled in calories. In most other countries, kilojoules are the favored energy unit for food labels.

The three food groups are fats, carbohydrates, and proteins. Some foods belong to a single group, others are mixtures of ingredients from two or three food groups. Food labels typically give the mass of each of the three food groups present in a food portion. The energy value of the portion is the sum of the products of the mass of each food group times that group's chemical energy.

Energy value of portion (kJ) = g(fat) × 39.5 kJ $g^{-1}$ + g(carbohydrate) × 17.1 kJ $g^{-1}$ + g(protein) × 14 kJ $g^{-1}$,

where the numbers in the formula above are taken from Table 7.1

To convert the SI units to calories, use the conversion factor: 4.184 joule (exactly) = 1 calorie. An annoying complication in the United States is widespread use of the term "Calories," or "large calories," as an improper and confusing alternative to the correct SI unit of kilocalories; 1 Calorie (capital C) = 1000 calories (lower case c). Using the SI unit kilojoule is unambiguous. Remember to substitute kilocalorie when you read the word Calorie (or often just calorie) on a food label. A food energy value calculation and its units conversion is shown in Example 7.3

**Example 7.3**: The label of a 42 g KitKat® bar states that the bar contains 11g of fat, 27 g of carbohydrate, and 3 g of protein. The label also states that the entire bar has an energy value of 220 Calories. Calculate the energy value of the bar. How well does the calculated value agree with the label value?

Strategy: Substitute into the energy value formula.

Energy value of bar = g(fat) × 39.5 kJ $g^{-1}$ + g(carbohydrate) × 17.1 kJ $g^{-1}$ + g(protein) × 14 kJ $g^{-1}$

= (11 × 39.5 + 27 × 17.1 + 3 × 14) kJ $g^{-1}$

= (430 + 460 + 40) = 930 kJ (two significant figures)

Converting to calories, Energy value = 930 kJ ÷ 4.184 J $cal^{-1}$

= 220 kcal

and that's 220 Calories (meaning SI kilocalories). So to within the appropriate number of significant figures the calculated energy value is identical to the energy value stated on the label.

**Metabolic rate** is how fast a human (or other organism) consumes chemical energy. The metabolic rate of a resting, young adult human is approximately 80 watts, corresponding to an average resting daily energy requirement of about 7,000 kJ $day^{-1}$ (≈1,700 kcal $day^{-1}$). Males tend to have a slightly higher metabolic rate than females. Moderate activity raises human metabolic rates one-and-a-half to two times their resting rates. A moderately active, young adult requires an energy intake of about 2,500 kcal $day^{-1}$ or about 10,000 kJ $day^{-1}$.

Depending on how fast you go, walking increases metabolic rates by a factor of 3-4. Doing manual labor and participating in sports increase metabolic rates by a factor of 5-15. The relationship of physical activity to food consumption is illustrated in the following example:

**Example 7.4**: A young adult's resting metabolic rate is 80 watts. A strenuous game of basketball increases the rate by a factor of 7.5. How many kilocalories does this person use while playing for 30. minutes? How long can this person play using just the energy of one KitKat® bar (220 kcal)?

Strategy: Use the relationship, Kilocalories used = rate of use × time, where the rate of use is the base metabolic rate times the factor 7.5. Convert to calories and compare with the energy value of the bar.

Energy used = 80 J $s^{-1}$ × 7.5 × 30. min × 60 s $min^{-1}$ × (1/4.184) cal $J^{-1}$ × 0.001 kcal $cal^{-1}$

= 260 kcal

So, if 260 kcal gives 30 minutes, the 220 kcal from the bar gives 30 × (220 ÷ 260) or about 25 minutes of strenuous basketball playing.

Thermodynamic properties in standard tables are usually listed on a molar basis, rather than a weight basis. But it is a simple matter to convert from grams to moles using the relationship $n = \frac{m}{MW}$, as shown in Example 7.5.

**Example 7.5**: Glucose, $C_6H_{12}O_6$, is a carbohydrate with a heat of combustion of 2816 kJ mol$^{-1}$. Compare its chemical energy with the value given for sucrose (table sugar) in Table 7.1 of 15.6 kJ g$^{-1}$.

Strategy: Calculate the molar weight of glucose from its chemical formula to be 180.2 g mol$^{-1}$. Divide the molar heat of combustion by the molar weight:

$$\text{Heat of combustion per gram} = 2816 \text{ kJ mol}^{-1} \div 180.2 \text{ g mol}^{-1}$$

$$= \underline{15.63 \text{ kJ g}^{-1}}$$

This value agrees well with the value of 15.6 kJ g$^{-1}$ stated in Table 7.1.

## Section 7.2 Work and Heat

### Work

In Section 4.6, we saw that the work done by the expansion of a gas against a fixed, external pressure is $w = -P_{ext}\Delta V$. In thermodynamics, by convention, work done by a system is defined as negative work and work done on a system is defined as positive work. Because an expanding gas loses energy and does work on the environment, its PV work is $-P\Delta V$. Note that $\Delta V$ ($V_{final} - V_{initial}$) is positive for an expansion.

**Example 7.6**: A gas expands from an initial volume of 12.50 L to a final volume of 21.80 L against a constant pressure of 11.5 atm. Calculate the work involved.

Strategy: Plug the values in the appropriate units into the equation: $w = -P_{ext}\Delta V$

$$w = -11.5 \text{ atm} \times (21.8 \text{ L} - 12.5 \text{ L})$$

$$w = -11.5 \times 9.30 = 107 \text{ L atm}$$

$$w = -107 \text{ L atm} \times 101.3 \text{ J L}^{-1} \text{ atm}^{-1} \times 0.001 \text{ kJ J}^{-1}$$

$$\text{Work} = \underline{-10.8 \text{ kJ}}$$

The work has a negative value because the gas loses energy during its expansion.

### Heat

Fuel-burning chemical reactions always release energy to the environment; this energy is manifested as either heat or as both heat and work. In thermodynamics, the terms heat and work are both used to describe the transfer of energy. Recall that heat is energy that flows from one point to another because of a temperature difference, while work is energy that is used to move objects.

Starting with James Watt in the 1760s, and Sadi Carnot in 1824, engineers have tried to understand engines so as to maximize the work that they can do for a given input of heat. This effort was a major intellectual driving force of the industrial revolution. When a fuel is burned in practical engines, the fraction of the chemical energy converted to work is called the fractional **efficiency**; multiplied by 100 it's the **percentage efficiency**:

$$\text{Percentage efficiency} = \frac{\text{work obtained}}{\text{heat required}} \times 100$$

Modern electric power generating plants convert up to 37% of their fuels' chemical energy into electrical energy. Automobile engines typically convert 10-15% of their fuel's chemical energy into work. Biochemical engines, such as people, are less efficient; only about 5-10% of your food's chemical energy is converted into work.

The motor cycle in Figure 7.7 is water-cooled. Most of the chemical energy from its gasoline is released as waste heat through the radiator you can see on the right of the picture. It has to be thus, because trying to operate a motor at a higher temperature, where its efficiency is better, melts it. The desire to operate heat engines at ever higher temperatures accounts for the use of exotic metal alloys in jet engine turbines. Diesel engines run somewhat hotter than gasoline engines, so diesel engines are somewhat more efficient than gasoline engines.

Figure 7.7. This motorcycle is water-cooled. Engines must be cooled lest the heat they generate destroy them. Because of this need for cooling, a substantial fraction of the heat supplied to the engine must be "wasted" to the coolant. Despite that apparent paradox, engineers have long known that practical engines have to operate in this manner.

When fuels burn, not in an engine but in an unconfined way, such as an open fire, essentially all their chemical energy is converted into heat. The only work done by an open fire is the small amount done by the combustion gases expanding to provide room for themselves under the blanket of the earth's atmosphere. In a bomb calorimeter all the energy of a chemical reaction is obtained as heat. The gases formed inside a bomb calorimeter *cannot* expand and thus cannot do any work at all.

Thermodynamics treats the world as a set of processes that convert systems from one state to a second state. Work and energy are the means by which a system communicates with its environment. We can describe the situation diagrammatically as follows:

**A Chemical Reaction as a Thermodynamic Process:**

**reactants** **yield** **products**

**System in state 1** $\rightarrow$ **System in state 2**

A chemical reaction is a process that transforms a system in one state (reactants) to the same system in a second state (products). The atoms are conserved. To specify the system, you have to state the temperature, the pressure, and the amounts of the various products and reactants. During the process, usually heat and perhaps some work are exchanged with the environment.

The diagram below shows in a parallel fashion that a steam engine is a thermodynamic process. It was thinking along these lines that led Sadi Carnot to the second law of thermodynamics. In a steam engine, such as a power plant (Figure 7.8), the chemical energy from fuel combustion is used to heat water in a boiler and generate the high-pressure, high-temperature condition, that constitutes state 1 of the steam. Expansion with cooling is the process by which the working fluid (steam in a steam engine) goes from state 1 to state 2. State 2 is characterized by being lower pressure and lower temperature than state 1.

**A Steam Engine as a Thermodynamic Process:**

**hot higher pressure steam yields colder lower pressure steam**

**System in state 1 $\rightarrow$ System in state 2**

A steam engine enables a process (expansion) that transforms a system in one state (hot, higher pressure steam) to the same system in a second state (colder, lower pressure steam). To specify the system, you have to state the temperature, the pressure and the amount of steam. During the process, work is done on the environment (the turbine), and heat is released to the environment when the exhausted steam is condensed.

A substance continually recycled in a thermodynamic process is called the **working fluid**. In a steam turbine, water is the working fluid. In an air conditioner, the fluorocarbons constitute the working fluid. During different stages of the thermodynamic cycle the working fluid is alternatively in the gaseous and liquid states.

Figure 7.8. The steam expansion turbine of this power plant is seen here shut down for maintenance. High-pressure, high-temperature steam enters the turbine through the pipe seen on the right. In the summer this plant produces only electric power. In the winter it produces both electric power and steam for local heating. Twice a year, in Fall and Spring, the plant switches from single mode to dual mode operation. Not visible behind the turbine is the electrical generator.

The diagram in Figure 7.9 summarizes the laws of thermodynamics as applied to an engine. Heat from some fuel burning reaction is used to heat steam. The hot steam is used to generate work by expansion in two different engines: one of higher efficiency and one of lower efficiency. Observe that in both engines the total amount of energy is constant: 100 = 63 + 37, and 100 = 85 + 15. However, the more efficient engine converts a greater proportion of the input chemical energy into work.

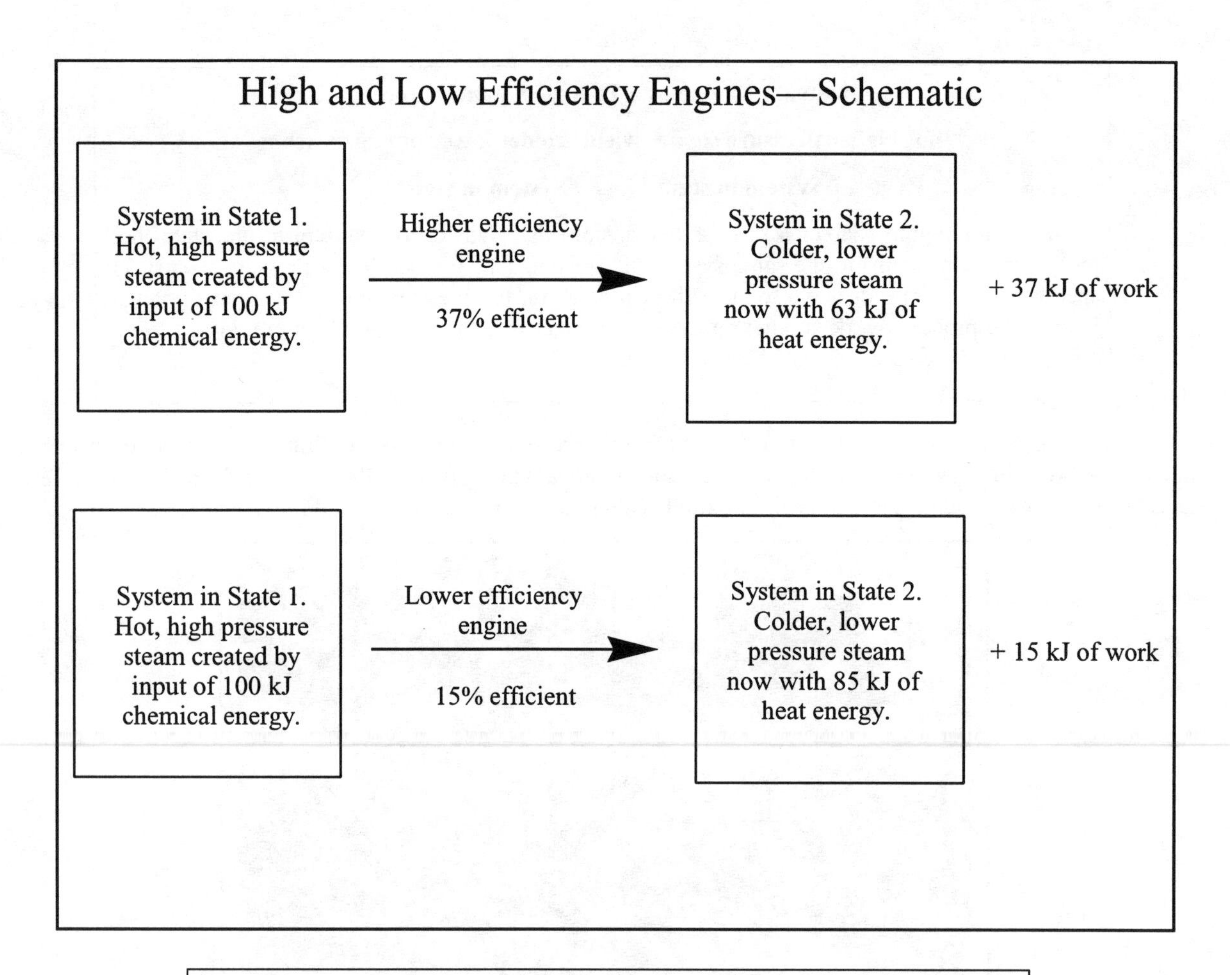


Figure 7.9. A schematic representation of a high versus low efficiency engine.

Reversible Gas Expansion (Work depends on the path)

A reversible thermodynamic process, such as a reversible isothermal (constant temperature) gas expansion, is not something we can achieve in practice. But conceptually, such an expansion is very important because it's the reversible gas expansion that defines exactly what maximum work can be achieved when a gas expands isothermally from any initial state to any final state. In the real world we'll always get less work, never more.

We saw earlier that gas expansion against a constant external pressure is given by the equation Work = $-P_{ext}\Delta V$. A special case of that equation arises if the gas expands into a vacuum. In that case no work is done because $P_{ext} = 0$.

The maximum work of reversible expansion is achieved when expansion takes place against an external opposing pressure that is always just infinitesimally less than the current gas pressure. It's easy to see why we can't do it in practice; among other problems, it would occur infinitely slowly. Calculus, however, gives us a way to handle the situation. We integrate over the volume change with pressure as a variable: $w_{max} = \int_{V_1}^{V_2} P$ dv. If we then make the assumption that we're dealing with an ideal gas, it's possible to integrate as follows:

$$w_{max} = -\int_{V_1}^{V_2} \frac{nRT}{V}\, dV = -nRT \int_{V_1}^{V_2} \frac{dV}{V} = -nRT \ln \frac{V_2}{V_1}$$

Three cases of the work of the isothermal expansion of an ideal gas are shown below in the form of P-V diagrams, where the area under each curve represents the work done and the bold line represents the pressure-volume path of the system.

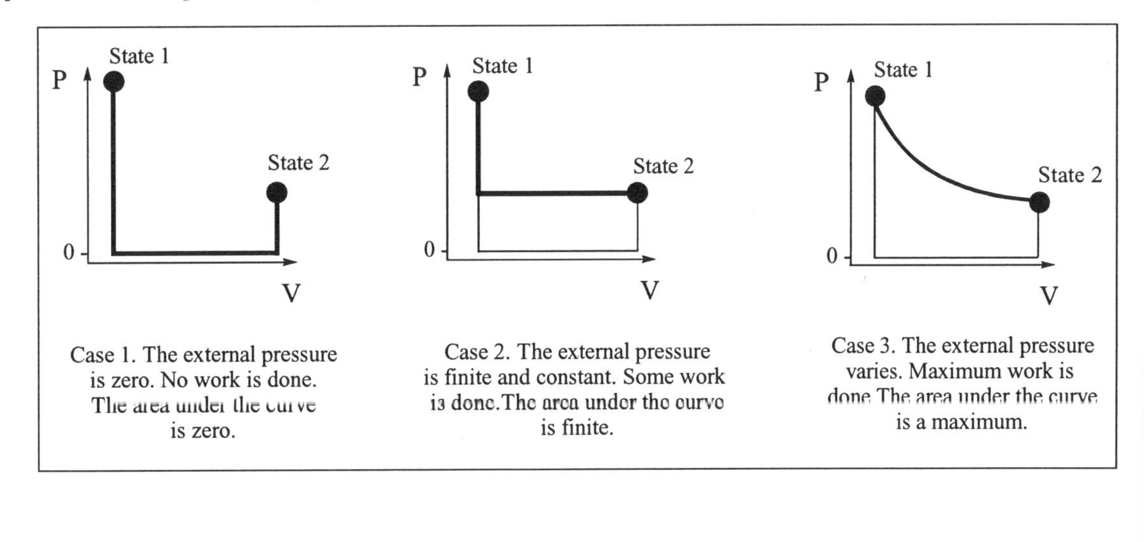


Case 1. The external pressure is zero. No work is done. The area under the curve is zero.

Case 2. The external pressure is finite and constant. Some work is done. The area under the curve is finite.

Case 3. The external pressure varies. Maximum work is done. The area under the curve is a maximum.

## Heat Capacity

In general terms, the heat capacity of a substance is the way its temperature changes when heat is added to it. In this section we see that the absorption of heat is not a straightforward matter and examine what happens to thermodynamic systems when they absorb heat.

Heat always flows from higher temperature regions to lower temperature regions; it always flows from the hotter to the colder of two bodies placed in contact. This obvious fact of human experience is sometimes called the **zeroth law of thermodynamics**. Heating a pure substance raises its temperature, except in the special case where the substance undergoes a phase change, in which case the temperature remains constant. **Phase changes** are changes in the state of aggregation of a substance (such as gas-to-liquid, liquid-to-solid, gas-to-solid, or their reverse). You already know from personal experience that phase changes occur at constant temperature; for example, boiling water stays at 100 °C despite the fact that the water is absorbing heat. Likewise, an iced drink stays cold, despite the fact that the drink is absorbing heat, until the last of the ice has melted. Because chemical reactions can be exothermic or endothermic, heat added to, or taken from, a chemical reaction system may not significantly change the system's temperature.

Figure 7.10 shows a constant heating rate time-temperature graph at constant pressure. It shows the temperature of a pure substance as the substance is heated at a constant rate. At first the solid warms and its temperature rises, not at a constant rate but close to it. Then, melting (fusion) begins and for some time the temperature remains constant. During melting, imagine that the input heat increases the potential energy among the molecules rather than increasing their kinetic energy. After melting has occurred the liquid warms. Eventually the boiling point of the substance is reached, and again for a much longer time the temperature remains constant. The vaporization line is much longer that the fusion line because process of vaporization requires much more energy than the process of melting. Finally, the substance has converted to the gas phase and the gas heats.

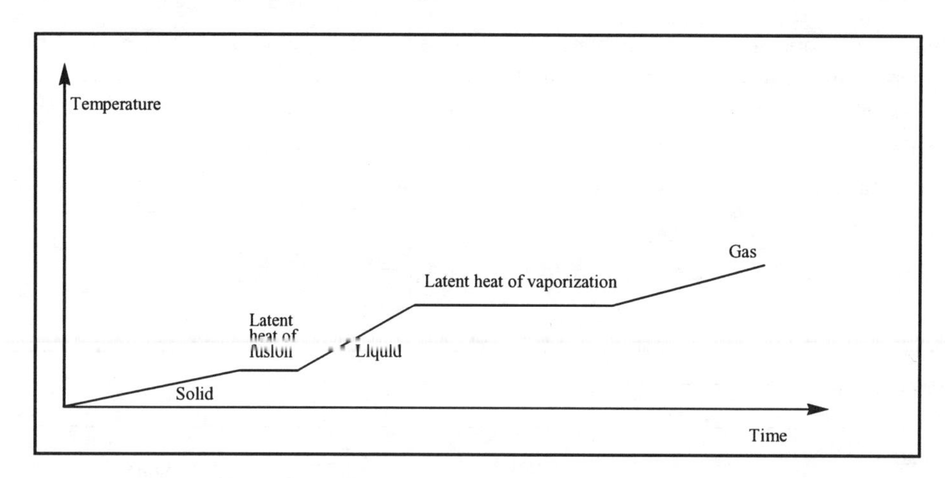


Figure 7.10. A typical solid substance shows quite complicated behavior when heated at a constant rate. At certain times its temperature rises while at other times its temperature remains constant. The diagram here is for the process at constant pressure.

Two *different* constant heating rate graphs can be obtained for every substance: one at constant volume, and one at constant pressure. In the solid and liquid regions the two graphs are essentially indistinguishable. In contrast, heating a gas at constant pressure causes *less* temperature increase than does heating the same gas at the same rate at constant volume. The reason? At constant pressure a gas expands and does some work—so its energy increases less than it would at constant volume. Under constant volume conditions (such as in a bomb calorimeter) despite being heated the gas cannot expand and thus can do no work. Recall that work = $-P\Delta V$, so a gas doing work expends energy and if $\Delta V$ is zero the work must be zero.

In Figure 7.10, the slopes of the lines for heating solid, liquid, and gas depend on their heat capacities. Operating at a constant pressure, the temperature rise of a system (a coffee cup calorimeter, for example) depends on the amount of heat, q, added to it. The system's constant-pressure **heat capacity**, $C_p$, which has units of J $K^{-1}$ or J $°C^{-1}$, determines the size of the temperature rise. For ordinary calorimetry, C is assumed to be temperature independent, when the formula $q = \int CdT$ can be integrated to give the expression: $q = C_p(T_2 - T_1)$ or $q = C_p\Delta T$. But note that this equation is valid *only* if there is neither a phase change nor a chemical reaction during the thermodynamic process being studied.

When we speak about a substance with its mass specified in grams, $C_p$ has the units of J $g^{-1}$ $K^{-1}$ or J $g^{-1}$ $°C^{-1}$ and is called the **specific heat capacity** (often just specific heat) of the substance. The equation used to calculate q, the heat change of a substance, is:

$$q = C_p \times g \times \Delta T$$

$$\text{heat change} = q = \text{specific heat} \times \text{mass in grams} \times \Delta T$$

The constant pressure specific heat capacities of some substances at 25°C are shown in Table 7.2. The **molar heat capacity** with units of J °C$^{-1}$ mol$^{-1}$ of a substance is its specific heat capacity times its molar weight.

$$C_p(\text{molar}) = C_p \times MW$$

If $T_2$ is greater than $T_1$ and $\Delta T$ (= $T_2$ - $T_1$) is positive, then heat is gained by the system. Conversely, if $T_2 < T_1$ and $\Delta T$ is negative, then heat is lost by the system. The temperature *change* may be expressed either in Celsius degrees or in kelvins, because the sizes of those units are identical.

Table 7.2 **Constant-Pressure Specific Heat Capacities of Some Substances in J g$^{-1}$ K$^{-1}$ at 25°C**

| Gases | | Liquids | | Solids | |
|---|---|---|---|---|---|
| Hydrogen | 14.2 | Water | 4.184 | Rubber | 1.90 |
| Helium | 5.2 | Hydrogen peroxide | 2.43 | Wood | 1.80 |
| Oxygen | 0.91 | Gasoline | 2.2 | Iron | 0.44 |
| Air | 0.90 | Mercury | 0.140 | Aluminum | 0.89 |
| Argon | 0.52 | Ethyl alcohol | 0.581 | Sodium chloride | 0.88 |
| Ethane | 1.6 | n-Octane | 0.578 | Granite | 0.80 |
| Sulfur dioxide | 0.63 | Acetic acid | 1.90 | Gold | 0.13 |
| Carbon dioxide | 0.84 | Benzene | 1.7 | Platinum | 0.14 |

Observe the wide range of values in Table 7.2 and note that substances with lighter particles (lower atomic or molecular weight) need more energy to heat them by a given amount. This relationship is clearly revealed by the contrast between the two monatomic noble gases helium (5.2 J g$^{-1}$ K$^{-1}$) and argon (0.52 J g$^{-1}$ K$^{-1}$).

Heat capacities measured at constant volume are labeled $\mathbf{C_v}$; heat capacities measured at constant pressure are labeled $\mathbf{C_p}$. The $C_p$ and $C_v$ values of a gas, are significantly different, but for liquids and solids $C_p$ and $C_v$ are essentially the same.

The amount of heat released in a calorimetry experiment is measured by the temperature rise of the water sink. Using a water sink to measure a quantity of heat is shown in Example 7.7. A **heat sink** is an object or substance used to absorb heat. Computer engineers are much concerned with heat sinks because the heat generated by microelectronic devices must be removed sufficiently fast that the proper operation of the device is not threatened by heat build up.

**Example 7.7**: A 27.6 g sample of water is heated from 22.34°C to 25.57°C by the heat released from a combustion reaction. Assuming the water is well insulated and exchanges no heat with its environment, how much heat energy has it absorbed? Water's heat capacity is 4.184 J g$^{-1}$ K$^{-1}$ at 25°C.

Strategy: Substitute into the formula q = specific heat × mass × $\Delta T$ in the proper units and solve for q.

$$q = 4.184 \text{ J g}^{-1}\ {}^{\circ}\text{C}^{-1} \times 27.6 \text{ g} \times (25.57 - 22.34\ {}^{\circ}\text{C})$$

$$= \underline{372 \text{ J}}$$

Example 7.8 is slightly more complicated. It shows how you can calculate the final equilibrium temperature reached by mixing two different substances, that have different masses, different starting temperatures, and different specific heat capacities. In this example cold water acts as a heat sink and hot metal acts as the heat source. Conservation of energy requires that the heat lost by the hot substance equals the heat gained by the cold substance.

**Example 7.8**: The heat capacity of solid aluminum is 0.890 J $g^{-1}$ °$C^{-1}$. A 99.7 g block of aluminum at 90.0°C is placed in a bath containing 1002 g of water at 15.0°C. What will be the final temperature reached by both the water and the aluminum? Assume that no heat is lost to the environment during the experiment.

Strategy: Recognize that heat is transferred from the hot aluminum to the cold water until the temperatures of the two become identical at x, the final temperature. So the two q's must be equal in magnitude (but opposite in algebraic sign). Use x and the formula q = specific heat × mass × $|\Delta T|$ twice. The aluminum temperature falls from 90.0°C to x°C so its $\Delta T$ is (90.0 - x) °C. The water temperature rises from 15.0°C to x°C so its $\Delta T$ is (x -15.0) °C. Finally, solve for x.

$$\text{heat lost by Al} = 0.890 \text{ J g}^{-1}\,°\text{C}^{-1} \times 99.7 \text{ g} \times (90.0 - x)\ °\text{C}$$

$$\text{heat gained by water} = 4.18 \text{ J g}^{-1}\,°\text{C}^{-1} \times 1002 \text{ g} \times (x - 15.0)\ °\text{C}$$

$$\text{heat lost by Al} = \text{heat gained by } H_2O$$

$$0.890 \text{ J g}^{-1}\,°\text{C}^{-1} \times 99.7 \text{ g} \times (90.0 - x)\ °\text{C} = 4.18 \text{ J g}^{-1}\,°\text{C}^{-1} \times 1002 \text{ g} \times (x - 15.0)\ °\text{C}$$

$$7990 - 88.7\,x = 4190\,x - 62800$$

$$4280\,x = 70800, \text{ and } \underline{x = 16.5°\text{C}}$$

The final temperature is 16.5°C.

Looking ahead: Thermodynamics is a rigorous discipline formulated using natural properties and differential calculus. Because there are two kinds of specific heats of gases ($C_p$ and $C_v$) we can *imagine* that there are two kinds of energy changes in gases: an energy change at constant pressure, $\Delta H$, called **enthalpy change**; and an energy change at constant volume, $\Delta U$, called **internal energy change**. Partial derivatives are the natural way to express thermodynamic relationships. In conventional notation, the relationships between heat capacity and these energies are:

$$C_p = \left(\frac{\partial H}{\partial T}\right)_p \text{ and } C_v = \left(\frac{\partial U}{\partial T}\right)_v$$

Compared to gases, liquids and solids don't expand much when heated, so for them $-P\Delta V$ is small and they can't do much work. The variation of a solid or liquid substance's heat capacity with temperature is often modest, as illustrated by the data for liquid water presented in Table 7.3.

Table 7.3 **The Specific Heat Capacity of Water at Several Temperatures**

| Temperature °C | Specific heat capacity J $g^{-1}$ $K^{-1}$ |
|---|---|
| 0°C | 4.218 |
| 25°C | 4.180 |
| 50°C | 4.181 |
| 75°C | 4.193 |
| 100°C | 4.216 |

The molar heat capacities of gases are conveniently represented by empirical equations in the form of a power series $C_p = a + bT + cT^2 + d\,T^3 + \ldots$ where T is the kelvin temperature and a, b, c, d … are empirical (curve fitted) constants. If you check at the NIST web site you'll find these called Shomate equations. [❁kws +NIST +webbook]. For example the molar heat capacity of hydrogen is well fitted between 300 K and 1500 K by the empirical equation:

$$C_p(\text{hydrogen gas}) = 29.06 - 0.8364 \times 10^{-3}\ T + 2.012 \times 10^{-6}\ T^2\ \text{J mol}^{-1}\ \text{K}^{-1}$$

where T is in kelvins. The variation of hydrogen gas's molar heat capacity is illustrated in Example 7.9.

**Example 7.9**: Calculate the constant pressure heat capacity ($C_p$) of hydrogen gas at 300. K and 1500. K.

Strategy: Substitute into the empirical equation for hydrogen's constant pressure heat capacity ($C_p$) posted at the NIST web site.

at 300 K, $C_p = 29.06 - 0.8364 \times 10^{-3} \times 300. + 2.102 \times 10^{-6} \times 300.^2$

$= 29.06 - 0.25 + 0.19$

$= \underline{29.00\ \text{J mol}^{-1}\ \text{K}^{-1}}$

at 1500 K, $C_p = 29.06 - 0.8364 \times 10^{-3} \times 1500. + 2.102 \times 10^{-6} \times 1500.^2$

$= 29.06 - 1.25 + 4.73$

$= \underline{32.54\ \text{kJ mol}^{-1}\ \text{K}^{-1}}$

You see that the molar heat capacity of hydrogen increases by only about 10% over a 1200 K temperature range.

Explaining in detail the specific heats of substances shown in Table 7.2 involves understanding how energy is taken up by the molecules of substances. For a monatomic gas added energy simply becomes additional kinetic energy. Reconsider the equation:

$$KE = \tfrac{3}{2}RT$$

that was described in Chapter 4. This equation shows that the average molecular energy of monatomic gas is independent of the gas's molecular weight and depends only on the temperature. For the simple case of a monatomic ideal gas, its molecules have only translational energy (translational simply means going from point to point). If the gas is at temperature $T_2$, then we can imagine that it has gained its energy by being heated from absolute zero to $T_2$ under constant volume conditions. Then we can apply: $q = C_v \times \text{mass} \times \Delta T$, where $\Delta T = T_2$ because $T_1 = 0$; doing this also assumes that $C_V$ is constant over the temperature range 0 K to $T_2$ K. If we choose a gas sample with a mass equal to its molar mass, then both q and KE represent the quantity of energy needed to heat one mole of the gas from 0 K to $T_2$. Thus we have:

$$\tfrac{3}{2}RT_2 = C_v \times MW \times T_2$$

The temperature cancels, and rearranging gives the specific heat at constant volume of a monatomic ideal gas as: $C_v = \frac{3R}{2MW}$. Here is a units check: J g$^{-1}$ K$^{-1}$ = J mol$^{-1}$ K$^{-1}$ ÷ g mol$^{-1}$, which is proper.

The experimental constant volume molar heat capacities of the ideal gases are all 12.5 J mol$^{-1}$ K$^{-1}$ over a wide temperature range. Using the relationship we've just derived, $C_v(\text{molar}) = \frac{3R}{2MW} \times MW = 1.5 \times R = 1.5 \times 8.314$ J mol$^{-1}$ K$^{-1}$, or $C_v$(molar) calculated = 12.5 J mol$^{-1}$ K$^{-1}$. The agreement between the experimental and calculated values demonstrates that the kinetic theory of gases allows accurate calculation of the heat capacities of the noble gases.

Rotational and Vibrational Energies of Polyatomic Gas Molecules

A gas molecule with two or more atoms has the possibility of both vibrating and rotating. These modes of motion both serve as energy sinks and are both quantized. In Chapter 6 we pointed out that quantized transitions among the rotational modes of polyatomic molecules allowed for their deep-space identification via radio telescope measurements.

These additional energy sinks make it difficult to give simple rules about the specific heat capacities of molecular substances, no matter whether gases, liquids, or solids. In contrast, metals behave as though they are monatomic substances, and their molar heat capacities are all approximately the same.

What are some practical consequence of a gas's specific heat? One consequence is that light gases (low MW gases) are effective heat sinks. So divers who are using helium-based gas mixtures have to be careful not to lose too much body heat via breathing. Also, light gases make a good heat exchange fluid. For this reason an early nuclear fission reactor in Colorado was helium cooled.

As noted in the preceding box, the molar heat capacities of metals are all approximately the same. Experimental evidence to prove that point is shown in Table 7.4. The rightmost column in Table 7.4 is the molar heat capacity of each of the listed metals.

Table 7.4 **Specific Heat Capacities of Some Metals**

| Metal | Specific heat $J\ g^{-1}\ K^{-1}$, at 298 K | Atomic weight $g\ mol^{-1}$ | Specific heat × Atomic weight $J\ mol^{-1}\ K^{-1}$ |
|---|---|---|---|
| Iron | 0.44 | 55.8 | 25 |
| Aluminum | 0.89 | 27.0 | 24 |
| Gold | 0.13 | 197.0 | 26 |
| Platinum | 0.14 | 195.1 | 27 |
| Mercury | 0.14 | 200.6 | 28 |

Table 7.4 shows that metals can be regarded as monatomic substances; the specific heat capacity of metals is inversely proportional to their atomic weights. Today, we'd say they have a roughly constant molar heat capacity. In the nineteenth century this idea was called the **law of Dulong and Petit** and was stated as "the product of a metal's specific heat (in cal $g^{-1}\ K^{-1}$) and its atomic weight is the constant 6.4." Today we'd convert that to joules as $6.4 \times 4.184 = 27$, and say that the product is about 27. Historically, the Law of Dulong and Petit was important in deciding the atomic masses of metals when the stoichiometry of their compounds was uncertain. Its results helped Mendeleev correctly place several metals in the periodic table.

## Constant Volume Calorimetry (Bomb Calorimeter)

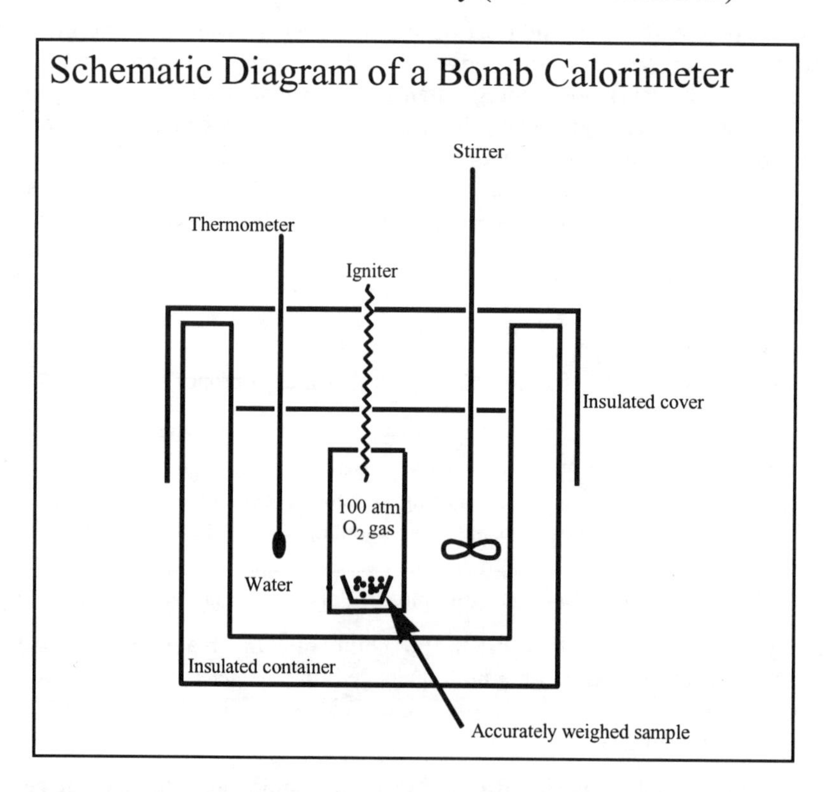


Figure 7.11. Schematic representation of an oxygen bomb calorimeter. Photographs of such a device and its component parts are shown in Figure 7.3.

Accurate thermodynamic measurements can be made with an oxygen bomb calorimeter such as the one pictured in Figure 7.3. A schematic diagram of such a device is shown in Figure 7.11. The bomb itself is a strong, corrosion-free, stainless steel vessel. Bomb calorimeters are especially useful for studying combustion reactions and can be used for many other reactions. To study a combustion reaction, the sample to be burned is sealed inside the bomb and the bomb pressurized with gaseous oxygen to about 100 atmospheres. A large excess of oxygen guarantees that combustion is complete. A heated metal wire acts like a spark plug and ignites the sample when the experimenter is ready to begin. The energy released by the combustion heats the calorimeter and the water it is submerged in.

Accurate measurements require (1) that the calorimeter be well insulated so that little or no heat escapes, (2) that $\Delta T$ be no more than a few degrees, so that the heat capacity of the calorimeter ($C_{calorimeter}$) can be safely assumed to be constant, and (3) a good quality thermometer. The basic formula to obtain q is simple.

$$\text{Heat produced, } q = C_{calorimeter}\,\Delta T$$

Before any measurements can be made with a particular calorimeter, its heat capacity must be measured; in other words, the calorimeter must be calibrated. Calibration can be done by burning an accurately weighed sample of a standard substance. Burning a standard substance inside a calorimeter with the intent to establish its heat capacity is called **calibration** or **standardization**. A **standard substance** is one whose properties are known and certified. NIST sells a wide range of standard substances, including benzoic acid ($C_6H_5COOH$)—a useful combustion standard. (An alternative way to calibrate a calorimeter is by measuring the temperature rise caused by an electrical heater placed inside it.)

Benzoic acid, when combusted at constant volume at 25°C, releases 3226 kJ $mol^{-1}$ via the reaction:

$$C_6H_5COOH(s) + \frac{15}{2}\,O_2(g) \rightarrow 7\,CO_2(g) + 3\,H_2O(l)$$

Example 7.10 shows the calculation of a calorimeter's heat capacity from a calibration experiment using benzoic acid.

**Example 7.10**: To calibrate a calorimeter, 0.8582 g of benzoic acid was burned in it. The temperature of the calorimeter rose by 4.55°C. Calculate the heat capacity of the calorimeter. Standard grade benzoic acid, when combusted at 25°C, releases 3226 kJ $mol^{-1}$. The molar weight of benzoic acid is 122.12 g $mol^{-1}$.

Strategy: Use the molar weight of benzoic acid to figure the number of moles burned and multiply that by the molar heat capacity. The heat capacity of the calorimeter will be that value divided by the temperature change.

Hence the energy generated by the combustion of 0.8582 g is

$$\text{Energy released} = \frac{0.8582\text{ g}}{122.12\text{ g mol}^{-1}} \times 3226\text{ kJ mol}^{-1} = 22.67\text{ kJ}$$

and the heat capacity of the calorimeter, $C_{cal} = q \div \Delta T = 22.67\text{ kJ} \div 4.55°C$

$$C_{cal} = \underline{4.98\text{ kJ °C}^{-1}}$$

After the heat capacity of a calorimeter has been measured, the calorimeter can be used to study other reactions and measure the energy values of unknown foods and fuels. Use of a standardized calorimeter to measure the energy value of a substance is illustrated in Example 7.11.

**Example 7.11**: A bomb calorimeter with a heat capacity of 6.427 kJ $°C^{-1}$ was used to study the combustion of naphthalene ($C_{10}H_8$) at 25°C. A 0.3140 g sample of naphthalene was burned and the calorimeter temperature increased by 1.93°C. Calculate the energy released per gram of naphthalene, and the energy released per mole of naphthalene.

Strategy: Use the temperature rise and the heat capacity of the calorimeter to calculate the heat released by combustion. Divide that result by the sample mass to express the energy on a per gram basis. Finally, multiply by the molar weight of naphthalene (128.17 g $mol^{-1}$) to express the energy release on a mole basis. The reaction involved in this experiment is:

$$C_{10}H_8(s) + 12\ O_2(g) \rightarrow 10\ CO_2(g) + 4\ H_2O(l)$$

Energy released = $C_{cal}\ \Delta T$ = 6.427 kJ $°C^{-1} \times$ 1.93°C

= 12.4 kJ

Energy of combustion per gram = 12.4 kJ ÷ 0.3140 g

= 39.5 kJ $g^{-1}$ (that's in the range of gasoline and diesel fuel)

The energy of combustion released per mole of naphthalene is

Energy of combustion per mole = 39.5 kJ $g^{-1} \times$ 128.17 g $mol^{-1}$ = 5060 kJ $mol^{-1}$

We'll conclude our discussion of heat by calculating the efficiency of a coal burning power plant. First, in Example 7.12, we'll calculate the energy value of the coal that the plant uses.

**Example 7.12**: The coal used at a power plant has an energy content of 13,250 Btu $lb^{-1}$ in the traditional engineer's units. What is the coal's energy value in kJ $g^{-1}$?

Strategy: Solve this problem by units conversion.

x kJ $g^{-1}$ = 13,250 Btu $lb^{-1} \times$ 1055.06 J $Btu^{-1} \times$ (1/453.6) lb $g^{-1} \times$ 0.001 kJ $J^{-1}$

= 30.82 kJ $g^{-1}$

Comparison with Table 7.1 shows this to be a reasonable value. Coal's energy value is in the range 25-35 kJ $g^{-1}$

The power plant described in Example 7.12 consumes an average of 99 tons of coal per day and is rated as a 6.5 megawatt plant. Together with the energy value of the coal burned, these values can be used to calculate the plant's efficiency, as shown in Example 7.13.

**Example 7.13**: Calculate the efficiency (the percentage of chemical energy converted to electrical energy) for a power plant that burns 99 tons of coal per day with an energy value of 30.8 kJ $g^{-1}$ and produces 6.5 megawatts of electrical power. 1 US ton = 2,000 lb.

Strategy: A watt is a joule per second, so use that to calculate the number of daily joules of electrical energy. Do a units conversion using 30.8 kJ $g^{-1}$ to calculate the daily amount of chemical energy production. Finally, express the former as a percentage of the latter.

Daily electric power production

$$= 6.5 \times 10^{6}\ J\ s^{-1} \times 3600\ s\ hour^{-1} \times 24\ hour\ day^{-1}$$

$$= 5.6 \times 10^{11}\ J\ day^{-1}$$

Daily chemical energy production

$$= 30.82\ kJ\ g^{-1} \times 1000\ J\ kJ^{-1} \times 453.6\ g\ lb^{-1} \times 2{,}000\ lb\ ton^{-1} \times 99\ tons\ day^{-1}$$

$$= 2.8 \times 10^{12}\ J\ day^{-1}$$

Percentage efficiency = $(5.6 \times 10^{11}\ J\ day^{-1}) \div (2.8 \times 10^{12}\ J\ day^{-1}) \times 100\% = \underline{20\%}$

Comment: If the plant produced only electricity, 20% would be a rather low efficiency. However this particular plant also delivers steam heat to its local neighborhood. The economics of delivering both electric power and local heat are often attractive.

The engineer Sadi Carnot deduced that the maximum percentage efficiency of a steam engine is $(T_2-T_1)/T_2 \times 100$, where $T_2$ is the temperature of the hot compressed steam in Kelvins and $T_1$ is the temperature of the cooler, expanded steam in Kelvins. In real engines $T_2$ will be limited by the properties of materials, so the efficiency inevitably has a practical limit. An efficiency calculation using the Carnot formula is shown in Example 7.14.

**Example 7.14**: Calculate the maximum percentage efficiency of an electric power plant where $T_2$ is hot, superheated steam at 400.°F and $T_1$ is cold river water at 45°F.

Strategy: Convert the temperature to kelvins and plug into % efficiency = $(T_2-T_1)/T_2 \times 100$

$T_2$ = 400.°F or 204°C or 477 K    $T_1$ = 45°F or 7°C or 280 K

Fractional efficiency = (477 - 280)/477 = 0.41

Percentage efficiency = 0.41 × 100 = $\underline{41\%}$.

Optimizing power plant efficiency remains today an important objective of mechanical engineering. As a traditional aspect of this objective, mechanical engineers have historically made very detailed investigations of the properties of water and steam. [❁kws +"steam tables"]

Comment: If you think about the formula used above, it becomes clear that the larger the value of $T_2$ the greater will be the efficiency of the engine. However, it's obvious there's a practical limit. Eventually, as $T_2$ is pushed higher and higher, something melts.

## Section 7.3 The First Law of Thermodynamics

Having now laid a foundation for thermochemistry, and established that heat and work are closely linked forms of energy, we are now ready to discuss the first law of thermodynamics. The first law is law of conservation of energy. It describes the energy change of a thermodynamic system when the system exchanges energy with its environment in the form of either heat or work, or of both heat and work simultaneously. The mathematical statement of the **first law of thermodynamics** is:

$$\Delta U = q + w$$

where U is called the **internal energy** of a system, q is the heat exchanged between the system and its surroundings, and w is the work exchanged between the system and its surroundings. The first law is a mathematical statement of the principle of energy conservation. ("Surroundings" is the technical word for environment in thermodynamics; patent lawyers would call the word surroundings a "term-of-art.")

Bomb calorimeters convert chemical energy *completely* into heat. No work is done as a consequence of a reaction that takes place inside a constant volume calorimeter because such a calorimeter by definition allows no volume change. Recalling the formula for work, $w = -P\Delta V$, it is obvious that if $\Delta V = 0$ then $-P\Delta V = 0$ and no work is done. For a constant volume process we write q as $q_v$ and state that $\Delta U = q_v$. In contrast, in an engine, chemical energy is converted partly into work and partly into heat. Work is performed when hot gases drive pistons against an external pressure or resistance, or spin a turbine. However, as we have already seen, a greater share of the combustion energy is inevitably wasted as heat to the coolant, exhaust gases, friction, etc., because engines are never very efficient.

When a system undergoes a process, the change of any state property is the difference between the final value and the initial value of that property:

$$\Delta(\text{Property}) = \text{Property}_{\text{final}} - \text{Property}_{\text{initial}}$$

or, for the first law, $\Delta U$, the change in internal energy of the system is $U_{\text{final}} - U_{\text{initial}}$

The changes that occur to any of a system's properties are completely defined by the initial and final states of the system. Such properties are called **state properties** or **state functions** to emphasize the fact that they *are* completely defined by the state of the system and that changes in them are independent of the way in which the process is carried out—they are **path independent**. (The concept of the path of a process was introduced in the earlier box about reversible gas expansion.) In the first law equation, $\Delta U$ is path independent and U is a state property. In contrast, both q and w are path dependent and neither is a state function.

To emphasize: heat and work are not properties of a system; they are not state functions. They represent energy in transit between a system and its surroundings. For a change, $\Delta U$, in a practical engine from the hotter high pressure state to the cooler low pressure state, q and w may vary depending on the path, but q+w is fixed. In an all-heat no-work path, q is 100% of $\Delta U$. In an efficient, practical engine q is perhaps 65% of $\Delta U$ and w is 35% of $\Delta U$.

Algebraic signs are an important consideration when we apply the first law of thermodynamics. Note that $\Delta U$ is treated always from *the point of view of the system*. Table 7.5 shows the algebraic signs of heat and work used in thermodynamics. An **endoergic** process is one in which work is *done on* a thermodynamic system, increasing the system's internal energy. An **exoergic** process is one in which work is *done by* a thermodynamic system, decreasing the system's internal energy. The use of thermodynamic sign conventions is illustrated in Example 7.15.

Table 7.5 **Sign Conventions for Heat and Work for Four Separate Processes**

| | | | |
|---|---|---|---|
| $q > 0$ | heat added to the system | $\Delta U$ is positive | endothermic |
| $q < 0$ | heat removed from the system | $\Delta U$ is negative | exothermic |
| $w > 0$ | work done on the system | $\Delta U$ is positive | endoergic |
| $w < 0$ | work done by the system | $\Delta U$ is negative | exoergic |

**Example 7.15**: A system has 3700 J of work done on it and loses 4.9 kJ of heat. Calculate the change in internal energy of the system.

Strategy: Substitute the values in the proper units, with the proper signs, into the first law of thermodynamics: $\Delta U = q + w$. Work done on the system is positive: the system gains energy. Heat lost by the system is exothermic: the system loses energy.

$$\Delta U = -4.9 \text{ kJ} \times 1000 \text{ J kJ}^{-1} + 3700 \text{ J}$$

$$= -4900 \text{ J} + 3700 \text{ J}$$

$$= \underline{-1200 \text{ J or } -1.2 \text{ kJ}}$$

## Internal Energy

The total energy of a system is called its internal energy, U, and is the totality of the energy in all forms that the system contains. The value of U itself cannot be known because it is a complicated sum of all the varieties of potential and kinetic energy that exist among the atoms, ions, molecules, etc., that make up the system. Fortunately, we're not particularly interested in U; rather, it's change, $\Delta U$, that we want to measure.

There's one simple case where we can make a sensible estimate of U, and that's a monatomic ideal gas. For the sake of argument, we will assume all its internal energy is the kinetic energy of the atoms themselves, so U is just the total kinetic energy. A calculation of U for helium is shown in Example 7.16.

**Example 7.16**: 3.000 mol of helium gas is heated from 0.0°C to 100.0°C at a constant volume. What is the change in the internal energy, U, of the gas? Helium is a monatomic gas; assume that it is ideal. Assume further that the internal energy of helium is just its kinetic energy.

Strategy: Recognize that the only form of internal energy in a monatomic ideal gas is kinetic energy. Recall that the molar energy is $KE = \frac{3}{2}RT$.

$$U_{initial} = 3.000 \text{ mol} \times (3/2) \times 8.314 \text{ J K}^{-1} \text{ mol}^{-1} \times 273.2 \text{ K}$$

$$= 1{,}0220 \text{ J}$$

$$U_{final} = 3.000 \text{ mol} \times (3/2) \times 8.314 \text{ J K}^{-1} \text{ mol}^{-1} \times 373.2 \text{ K}$$

$$= 1{,}3960 \text{ J}$$

$$\Delta U = U_{final} - U_{initial} = 13960 \text{ J} - 10220 \text{ J} = \underline{3740 \text{ J}}$$

$\Delta U$ is positive. Heat energy enters the system when it is warmed. At constant pressure it would take more heat to get the same temperature change. Why? Because the gas will expand and do some work, thereby losing energy to its surroundings.

## Section 7.4 Enthalpy

Many chemical reactions occur under constant pressure conditions with an accompanying volume change. Typical of this situation is the large number of reactions that take place in vessels open to the atmosphere. Examples of reactions that take place at or near a pressure of one atmosphere are: bench-top reactions, many industrial reactions, and the majority of biological chemical reactions. As we saw when we compared $C_p$ and $C_v$, constant pressure processes often do work. So it's helpful to have two energy values: U for constant volume processes and H for constant pressure processes. H is called **enthalpy** and for any thermodynamic system is defined as:

$$H = U + PV$$

Because all quantities on the right side of this equation (U, P, and V) are state properties, the enthalpy is also a state property. Thus, for any process or reaction

$$\Delta H = H_{final} - H_{initial} = \Delta U + \Delta(PV)$$

and the enthalpy change, $\Delta H$, is independent of the reaction's path. To apply this equation first calculate $\Delta(PV)$. For a process that has an ideal gas as its working fluid, we may use the ideal gas equation, $PV = nRT$, in the form $\Delta(PV) = nR(\Delta T)$ or $\Delta(PV) = n\,R\,(T_{final} - T_{initial})$. This equation provides a convenient way to obtain the $\Delta(PV)$ term for the process of heating or cooling an ideal gas.

It is easy to see why enthalpy is useful for a process or reaction that takes place at constant pressure. For a constant pressure process, $P_{initial} = P_{final} = P$, so we have

$$\begin{aligned}\Delta(PV) &= P_{final}V_{final} - P_{initial}V_{initial}\\ &= P\,V_{final} - P\,V_{initial}\\ &= P\,(V_{final} - V_{initial})\\ &= P\Delta V\end{aligned}$$

However, $P\Delta V$ is equal to -w for a constant pressure process. Hence, for a constant pressure process the enthalpy change becomes

$$\Delta H = \Delta U - w$$

and, using the first law definition, $\Delta U = q + w$,

we get $\Delta H = (q + w) - w$

so $\Delta H = q$ for a constant pressure process.

In other words, the heat transferred to or from a system during a constant pressure process is equal to the system's enthalpy change ($\Delta H$). Sometimes we write $\Delta H = q_p$, where $q_p$ is the heat change of a process at constant pressure.

The enthalpy changes of essentially *all* possible chemical reactions can be calculated using the vast, existing data base of standard enthalpy data. **Heats of combustion** at constant pressure are the standard enthalpy changes of combustion reactions. For a chemical reaction that occurs at constant pressure the enthalpy change is

$$\Delta H_{reaction} = H_{products} - H_{reactants}$$

$\Delta H_{reaction}$ is called the **reaction enthalpy change,** or slightly less precisely the **heat of reaction**. Reactions with $\Delta H < 0$ release heat and are exothermic reactions; endothermic reactions have $\Delta H > 0$ and absorb heat.

We have seen that for a constant volume process, $\Delta U = q_v$. Therefore, for a chemical reaction, $\Delta U$ is the heat of that reaction measured at constant volume. Heats of reaction at constant volume are measured in closed calorimeters, such as the bomb calorimeter

Open calorimeters operate under constant pressure conditions and measure reaction enthalpies. Experimentally, the constant pressure of the atmosphere is used simply by equipping the calorimeter with a loosely fitting stopper or lid. The coffee cup calorimeter pictured in Figure 7.2 is often used in student laboratory experiments because it's cheap, easy to use, and able to give reasonably accurate results. Using a Thermos bottle as a constant pressure calorimeter, and a good quality thermometer, gives measurements reliable to 4 or 5 significant figures.

The defining equation for enthalpy, $H = U + PV$, can be applied to any chemical reaction and written for the reaction's enthalpy change as:

$$\Delta H_{reaction} = \Delta U_{reaction} + \Delta(PV)_{reaction}$$

and, at constant pressure, $\Delta H = \Delta U + P\Delta V$. Thus, the difference between the heat of reaction at constant pressure ($\Delta H$) and the heat of reaction at constant volume ($\Delta U$) depends on the sign and magnitude of the amount of work done during the constant pressure process.

Changes in the volumes of solids and liquids can usually be ignored compared with changes in the volumes of gases, so a good approximation of $P\Delta V$ can be obtained by considering only the gaseous reactants and products. If we use the ideal gas law for the gases, this becomes

$$P\Delta V_{reaction} = n_{gas}\text{ (products) } RT - n_{gas}\text{ (reactants) } RT \quad = \quad \Delta n_{gas}RT$$

and so constant pressure work, $P\Delta V = \Delta n_{gas}RT$

where $\Delta n_{gas}$ is the number of moles of gaseous products minus the number of moles of gaseous reactants. If $\Delta n_{gas}$ is zero, then $P\Delta V_{reaction}$ is zero and $\Delta H_{reaction} = \Delta U_{reaction}$. This is the situation with coal burning, where the balanced chemical equation $C(s) + O_2(g) \rightarrow CO_2(g)$ shows that there is no change in the number of moles of gas during the reaction. The calculation of the energy change of a constant pressure process is illustrated in Example 7.17.

**Example 7.17**: We saw in the previous example that when 3.000 mole of helium, treated as a monatomic, ideal gas, is heated from 0.0°C to 100.0°C at constant volume, its internal energy increases by 3740 J. Calculate the enthalpy change for the same process performed at constant pressure.

Strategy: Use the equations $\Delta(PV) = n\,R\,\Delta T$ and $\Delta H = \Delta U + \Delta(PV)$ .

$$\Delta(PV) = 3.000 \text{ mol} \times 8.314 \text{ J K}^{-1} \text{ mol}^{-1} \times 100.0 \text{ K}$$

$$= \underline{2494 \text{ J}}$$

Therefore, using $\Delta H = \Delta U + \Delta(PV) = 3740 \text{ J} + 2494 \text{ J}$

$$= \underline{6{,}230 \text{ J}}$$

Note that 6,230 J is nearly double 3,740 J. So this example shows that it takes considerably more energy to heat a mole of helium gas at constant pressure than it does to heat it at constant volume. The difference arises because the expanding helium does work, and the energy equivalent to that work must be supplied to the gas in the form of heat.

Example 7.18 demonstrates that the expansion work done during combustion is typically small when compared with the energy released by the combustion reaction.

**Example 7.18**: Calculate the expansion work done at 298 K and 1 bar during the combustion of ethanol ($C_2H_5OH$) when the combustion yields carbon dioxide gas and gaseous water as its products. Compare that value with the standard enthalpy of combustion of ethanol, -1,244 kJ $mol^{-1}$.

Strategy. Write the balanced equation and obtain $\Delta n_{gas}$. Substitute that value into the equation $n_{gas}RT$.

$$C_2H_5OH(l) \;+\; 3\,O_2(g) \;\rightarrow\; 2\,CO_2(g) + 3\,H_2O(g)$$

$$\Delta n_{gas} = (2 + 3) - 3 = 2 \text{ moles}$$

$$\text{So, work done} = n_{gas}RT$$

$$= 2 \times 8.314 \text{ J mol}^{-1} \text{ K}^{-1} \times 298 \text{ K} \times (0.001 \text{ kJ J}^{-1}) = \underline{4.96 \text{ kJ}}$$

Compared to the heat release of 1,244 kJ $mol^{-1}$ the expansion work of 4.96 kJ is tiny. Clearly, open air combustion produces negligible work compared to its heat release.

## Hess's Law

During the 19th century the Russian chemist G. H. Hess (1802-1850) discovered that the heat liberated at a fixed pressure by a stepwise sequence of reactions equals the heat liberated by the single reaction that involved all the steps together. **Hess's law** states: the enthalpy change of an overall reaction equals the sum of the enthalpy changes of the steps by which it is carried out. Even though the law's discovery by Hess pre-dated the laws of thermodynamics, it is a consequence of them, not a fundamentally independent law. Hess's law has been confirmed by many thermochemical experiments.

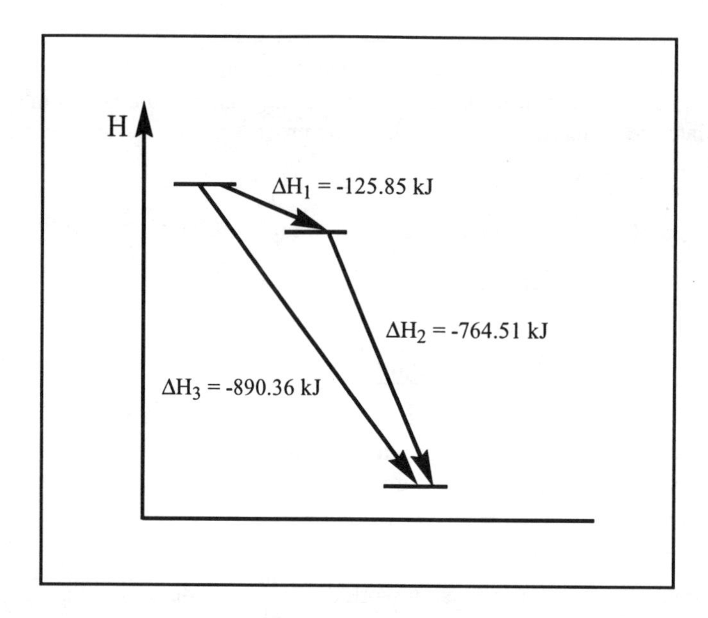


Figure 7.12. Schematic representation of Hess's law. The enthalpy change of a process is identical whether it occurs by a one- or a two-step path. The values shown here are for reactions occurring at 298 K and a pressure of 1 bar.

To illustrate Hess's law we'll use the two step process represented schematically on an energy diagram in Figure 7.12 that occurs via the chemical equations shown below. The reactions take place at 298K and a pressure of 1 bar. Step 1 is the incomplete combustion of methane to make methanol(g). Step 2 is the complete combustion of methanol to yield $CO_2(g)$ and $H_2O(l)$. Steps 1 and 2 together constitute path 1. Path 2 is the complete, single-step combustion of methane to make $CO_2(g)$ and $H_2O(l)$.

| | | |
|---|---|---|
| Step 1 | $CH_4(g) + \frac{1}{2} O_2 \rightarrow CH_3OH(g)$ | $\Delta H_1 = -125.85$ kJ |
| and Step 2 | $\underline{CH_3OH(g) + 1.5\, O_2 \rightarrow CO_2(g) + 2\, H_2O(l)}$ | $\Delta H_2 = -764.51$ kJ |
| Sum for path 1 | $CH_4(g) + 2\, O_2 \rightarrow CO_2(g)\ \ 2\, H_2O(l)\ \ \Delta H = \Delta H_1 + \Delta H_2$ | $= -890.36$ kJ |
| Path 2 (direct) | $CH_4(g) + 2\, O_2 \rightarrow CO_2(g)\ \ 2\, H_2O(l)\ \ \Delta H_3 = -890.36$ kJ | |

Observe that when the equations of steps 1 and 2 are added, the $CH_3OH(g)$ that occurs on both sides of the equation cancels, leaving a sum that is identical to path 2. Observe also, that the enthalpy change of paths 1 and 2 are equal: $\Delta H_1 + \Delta H_2 = \Delta H_3$ ( -125.85 - 764.51 = -890.36 kJ), which is Hess's law. The law is a good example of the principle that the change in a thermodynamic state property during a process is path independent.

Hess's law has the useful practical application of allowing us to find enthalpy changes for reactions that cannot be carried out in the laboratory. For example, it enables us to find the enthalpy of combustion of graphite to carbon monoxide. In practice, you can't burn graphite just to make CO, the reaction always produces some $CO_2$. In other words, the enthalpy of formation of CO cannot be obtained directly from experiments. However, graphite can be completely burned to $CO_2$, and pure CO can be burned to $CO_2$ and the results of those experiments (carried out at 298 K and a pressure of 1 bar) can be combined via Hess' law as shown by the equations:

| | |
|---|---|
| $C(graphite) + \frac{1}{2}O_2(g) \rightarrow CO(g)$ | $\Delta H_1 = ?$ |
| $\underline{CO(g) + \frac{1}{2}O_2(g) \rightarrow CO_2(g),}$ | $\underline{\Delta H_2 = -284 \text{ kJ}}$ |
| $C(graphite) + O_2(g) \rightarrow CO_2(g)$ | $\Delta H_3 = -394$ kJ |

Because $\Delta H_2$ and $\Delta H_3$ can be readily measured, then $\Delta H_1$ (which cannot be measured experimentally) is readily calculated from Hess's law, $\Delta H_1 + \Delta H_2 = \Delta H_3$. So $\Delta H_1 = \Delta H_3 - \Delta H_2$, and $\Delta H_1 = -110$ kJ. The calculation of another experimentally inaccessible enthalpy change via heats of combustion and Hess's law is shown in Example 7.19.

**Example 7.19**: The heats of combustion of graphite and diamond (at 298 K and a pressure of 1 bar) are respectively -393.51 kJ $mol^{-1}$ and -395.41 kJ $mol^{-1}$. Calculate ΔH for the process of transforming graphite to diamond.

Solution: The two combustion reactions are

$$C(graphite) + O_2(g) \rightarrow CO_2(g),\ \Delta H = -393.51 \text{ kJ}$$

$$C(diamond) + O_2(g) \rightarrow CO_2(g),\ \Delta H = -395.41 \text{ kJ}$$

If the second reaction is reversed, the final and initial states are interchanged; doing this reverses the sign of ΔH:

$$CO_2(g) \rightarrow C(diamond) + O_2(g),\ \Delta H = +395.41 \text{ kJ}$$

When this is added to the first reaction, $CO_2$ and $O_2$ cancel. The result, according to Hess's law, is

$$C(graphite) \rightarrow C(diamond),\ \Delta H = -393.51 \text{ kJ} + 395.41 \text{ kJ} = \underline{+1.90 \text{ kJ}}$$

## Section 7.5 Standard Enthalpies of Formation

Hess's law makes it possible to calculate ΔH for essentially all reactions, both real *and* imaginable. To summarize an enormous quantity of thermodynamic data in a convenient form, chemists use the concepts of enthalpy of formation and standard state. The **standard enthalpy of formation**, $\Delta H_f^o$, of a substance is the enthalpy change when the compound in its standard state is formed from its elements in their standard states. The equation for the process is balanced and written so that the only product is one mole of the compound. A general representation of standard enthalpy of formation at 298 K is:

Elements in their standard states at 298 K → Compound in its standard state at 298 K

**Standard state** means at a pressure of 1 bar, with the element or compound in its normal state of aggregation: gas, liquid, or solid, and if solid in a specified solid form. The temperature of standard state is chosen to be 298 K and the tabulated data were obtained at that temperature. The standard physical state for a solid element with allotropes is generally taken as its most stable allotrope. Graphite is the standard state of carbon because, as you see from Table 7.6, it is the thermodynamically most stable form of carbon at standard state. The standard heats of formation of elements are zero. Historically, much enthalpy data was obtained from reactions at a pressure of one atmosphere. The differences between standard enthalpy data gathered at 1 bar and 1 atmosphere are of concern only to specialists.

The Most Thermodynamically Stable State of Carbon

The standard reaction C(graphite) → C(diamond) at 298 K has an enthalpy change of +1.9 kJ $mol^{-1}$. The positive sign of that value tells you that energy is entering the thermodynamic system. Chemically, we'd say a mole of diamond has slightly more energy than a mole of graphite.

Thus, in everyday language we say graphite, because it is at a lower energy state, is a more stable form of carbon than diamond. More precisely, we'd say that graphite is more *thermodynamically* stable than carbon.

Here are example equations showing standard enthalpy of formation reactions:

| | | | |
|---|---|---|---|
| Carbon monoxide: | $C(graphite) + \frac{1}{2} O_2(g) \rightarrow CO(g)$ | $\Delta H_f^o$ of $CO(g)$ | = -110.53 kJ |
| Liquid water: | $H_2(g) + \frac{1}{2} O_2(g) \rightarrow H_2O(l)$ | $\Delta H_f^o$ of $H_2O(l)$ | = -285.83 kJ |
| Liquid ethanol: | $3 H_2(g) + \frac{1}{2} O_2(g) + 2 C(s) \rightarrow C_2H_5OH(l)$ | $\Delta H_f^o$ of $C_2H_5OH(l)$ | = -277.69 kJ |

Table 7.6 shows just a tiny fraction of the gigantic data base that constitutes the standard enthalpies of formation of compounds (see also Appendix C). For now you should understand the columns labeled $\Delta H^0_f$, and $C^0_p$. In Chapter 8, we'll discuss $\Delta G^0_f$ and $S^0$.

Table 7.6 **Selected Standard Thermodynamic Properties of Pure Substances at 298 K**

| Formula (state, name) | $\Delta H^0_f$ kJ mol$^{-1}$ | $\Delta G^0_f$ kJ mol$^{-1}$ | $S^0$ J K$^{-1}$ mol$^{-1}$ | $C^0_p$ J K$^{-1}$ mol$^{-1}$ |
|---|---|---|---|---|
| C(cr,graphite)* | 0 | 0 | 5.74 | 8.527 |
| C(cr,diamond)* | 1.9 | 2.9 | 2.38 | 6.113 |
| C(g) | 716.68 | 671.26 | 158.1 | 20.84 |
| CO(g) | -110.53 | -137.17 | 197.67 | 29.42 |
| $CO_2$(g) | -393.51 | -394.36 | 213.74 | 37.11 |
| $CH_4$(g,methane) | -74.81 | -50.72 | 186.26 | 35.31 |
| $C_2H_2$(g,ethyne) | 226.73 | 209.2 | 200.94 | 43.93 |
| $C_2H_4$(g,ethene) | 52.25 | 68.12 | 219.45 | 43.56 |
| $C_2H_6$(g,ethane) | -84.68 | -32.82 | 229.60 | 52.63 |
| $CH_3OH$(g,methanol | -200.66 | -162.00 | 239.70 | 43.89 |
| $CH_3OH$(l,methanol) | -238.66 | -166.36 | 126.8 | 81.6 |
| $C_2H_5OH$(g,ethanol) | -235.1 | -168.49 | 282.70 | 65.44 |
| $C_2H_5OH$(l,ethanol) | -277.69 | -174.78 | 160.7 | 111.46 |
| $H_2$(g) | 0 | 0 | 130.68 | 28.82 |
| H(g) | 217.97 | 203.25 | 114.71 | 20.78 |
| $H^+$(g) | 1,536.2 | -- | -- | -- |
| $H_2O$(g) | -241.82 | -228.57 | 188.82 | 33.577 |
| $H_2O$(l) | -285.83 | -237.13 | 69.91 | 75.29 |
| $H_2O_2$(g) | -136.31 | -105.57 | 232.7 | 43.1 |
| $H_2O_2$(l) | -187.78 | -120.35 | 109.6 | 89.1 |
| $O_2$(g) | 0 | 0 | 205.14 | 29.36 |
| $O_3$(g,ozone) | 142.7 | 163.2 | 238.93 | 39.2 |
| O(g) | 249.17 | 231.73 | 161.055 | 21.91 |

* cr = crystal, also written as s

To calculate $\Delta H^0$ for any reaction you use the standard thermodynamic data (Table 7.6) for each of its reactants and each of its products and apply Hess's law. The equation is:

$$\Delta H^\circ = \sum_{\text{moles}} \Delta H^\circ_f(\text{products}) - \sum_{\text{moles}} \Delta H^\circ_f(\text{reactants})$$

where the word moles signifies that the number of moles of each reactant and product must be the coefficient of the corresponding substance in the balanced chemical equation for the desired reaction. Figure 7.13 illustrates the relationship. The use of this equation is shown in Example 7.20.

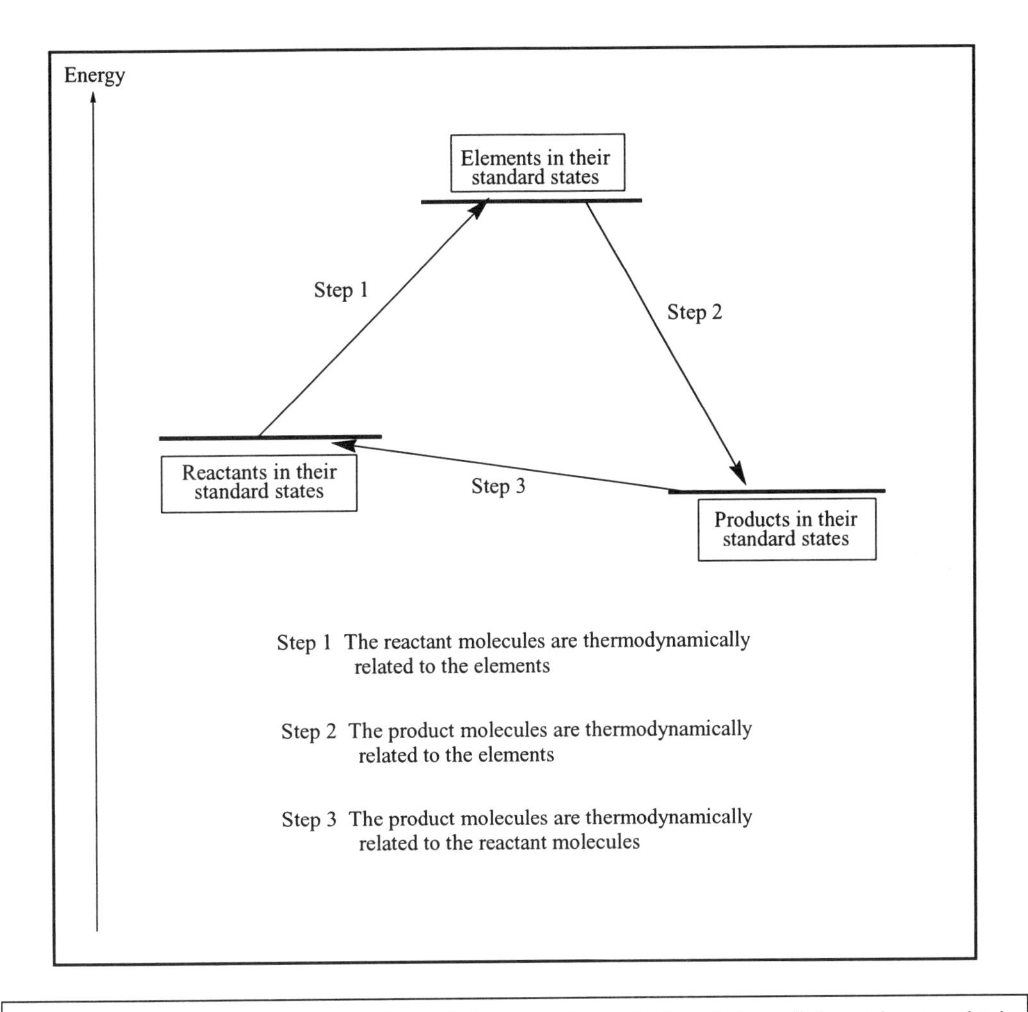


Figure 7.13. Schematic representation of the use of standard enthalpy of formation to obtain standard enthalpy of reaction.

Note that Figure 7.13 shows a cyclic process, and because after step 3 we are back where we started, we can write $\Delta H(\text{step 1}) + \Delta H(\text{step 2}) + \Delta H(\text{step 3}) = 0$. We can rewrite that as $-\Delta H(\text{step 3}) = \Delta H(\text{step 1}) + \Delta H(\text{step 2})$ and relate the diagram to the formula above as follows:

| | |
|---|---|
| $-\Delta H(\text{step 3})$ | $\Delta H^0$ |
| $\Delta H(\text{step 1})$ | $-\sum \Delta H^0_f\,(\text{reactants})$ |
| $\Delta H(\text{step 2})$ | $\sum \Delta H^0_f\,(\text{products})$ |

**Example 7.20**: Use standard data tables of enthalpies of formation to calculate the standard enthalpy of combustion of propane, $C_3H_8$. The standard enthalpy of formation of propane is -103.85 kJ $mol^{-1}$. The molar weight of propane = 44.1 g $mol^{-1}$. Also calculate propane's energy value in kJ $g^{-1}$.

Strategy: Look up the needed data in the table and plug it into the equation:

$$\Delta H^0 = \sum_{\text{moles}} \Delta H^0_f(\text{products}) - \sum_{\text{moles}} \Delta H^0_f(\text{reactants})$$

$$C_3H_8(g) \; + \; \tfrac{10}{2}\, O_2(g) \; \rightarrow \; 3\, CO_2(g) \; + \; 4\, H_2O(l)$$

$$1 \times -103.85 \qquad 5 \times 0 \qquad 3 \times -393.51 \qquad 4 \times -285.83$$

$$\Delta H^0 = [(3 \times -393.51) + (4 \times -285.83)] - [(-103.85)]$$

$$\Delta H^0 = \underline{-2220.00 \text{ kJ mol}^{-1}} \text{ of propane}$$

The standard enthalpy of combustion per mole of propane gas is -2220.00 kJ. Its large and negative combustion enthalpy reveals that propane is a good fuel.

On a gram basis, the chemical energy is 2220.00 kJ $mol^{-1}$ ÷ 44.1 g $mol^{-1}$ = 50.3 kJ $g^{-1}$.

That number fits sensibly in Table 7.1, lying between methane, $CH_4$, at 55.1 kJ $g^{-1}$ and octane, $C_8H_{18}$, at 47.7 kJ $g^{-1}$.

The enthalpy of formation of every element in its standard state is by definition zero. The majority of compounds have a negative $\Delta H^0_f$, which means that the formation of these compounds from their elements is an exothermic reaction; such compounds are said to be **thermodynamically stable** with respect to their decomposition. Some compounds have positive values of their $\Delta H^0_f$. That means that the formation of these compounds from their elements is an endothermic reaction; such compounds are said to be **thermodynamically unstable** with respect to their decomposition. Ordinary explosives have a positive values of their standard enthalpy of formation.

Standard enthalpies of formation of ions in solution are also listed in data tables. We won't discuss these here. But for solutions it is necessary to add one more standard state. That state is a solution concentration of 1 molar. We'll need that supplementary standard state definition later when we talk about electrochemistry in Chapter 14.

To reiterate, standard conditions are defined by specifying T as 298 K and pressure as 1 bar. Standard state is defined by specifying a stable physical state at 1 bar and any specified temperature.

## Section 7.6 Bond Energies

Atoms are held together in molecules and polyatomic ions by bonds. **Bond energy** is the enthalpy required to break a mole of bonds at standard pressure and 298 K. The simplest case is for diatomic molecules whose bond energy is that required to convert one mole of the gaseous molecules into two moles of gaseous atoms at 298 K and one bar. Thus,

$$H_2(g) \rightarrow 2\, H(g) \quad \Delta H = +436 \text{ kJ mol}^{-1} = \text{Bond energy } H_2$$

$$O_2(g) \rightarrow 2\, O(g) \quad \Delta H = +495 \text{ kJ mol}^{-1} = \text{Bond energy } O_2$$

By convention, bond energies are stated with a positive algebraic sign. So immediately be aware that bond energy discussions handle the algebraic sign opposite to the way we've done it for enthalpy changes. The author restored consistency in his own mind by mentally replacing the term "bond energy" by the term "unbond energy."

Converting molecules with three or more atoms requires the breaking of more than one bond. For example, the energy required to convert methane, $CH_4$, into its component atoms requires the breaking of four C-H bonds:

$$CH_4(g) \rightarrow C(g) + 4\, H(g) \qquad \Delta H = 1{,}660 \text{ kJ}$$

The average C-H bond energy in methane is thus defined as one fourth of the energy required to convert a mole $CH_4(g)$ to atoms, or 1,660/4 = 415 kJ $mol^{-1}$. The average energy of C-H bonds in other organic molecules is usually within 5% of this value. The average of a large number of C-H bond energies in many different molecules is 413 kJ $mol^{-1}$, and this is taken as the tabulated **average bond energy** (BE). The average bond energies for many pairs of atoms have been measured (Table 7.7). In Chapter 2 we met bond energy in connection with valency and the classification of the types of chemical bonds. In Chapter 6 we discussed the quantum mechanical basis for chemical bonding. Here we learn that the theoretical basis for the measurement of bond energy lies in thermodynamics ($\Delta H$)

Table 7.7 **Bond Energies (kJ $mol^{-1}$) for Selected Bonds**

| | | | |
|---|---|---|---|
| C-H | 413 | O=O (in $O_2$) | 495 |
| H-H | 436 | N≡N (in $N_2$) | 945 |
| O-H | 463 | C-O single | 360 |
| F-F | 155 | C=O (in $CO_2$) | 799 |
| H-F | 567 | C≡O (in CO) | 1,077 |
| H-Cl | 431 | C-C single | 347 |
| C-F | 485 | C=C double | 615 |
| C-Cl | 328 | C≡C triple | 812 |
| C-S | 301 | C-N single | 305 |
| Cl-Cl | 242 | C=N double | 615 |
| N-H | 389 | C≡N triple | 891 |

There are many ways to measure bond energies experimentally. Using the tables of standard thermodynamic data is a traditional method. Those values are in turn based on calorimetry and other methods. Ultraviolet spectroscopy and mass spectrometry measurements give bond energy information. In modern chemistry, accurate bond energies for small molecules can be obtained from quantum mechanical calculations.

## Estimating Enthalpies of Reactions from Bond Energies

In Section 7.5 we saw how enthalpies of reaction can be calculated from tables of standard enthalpies of formation of products and reactants. Estimates of the gas phase enthalpy change of any reaction can be made using bond energies and the structural formulas of the reactants and products. The equation you need is:

$$\Delta H(\text{estimated}) = \underset{\text{moles}}{\sum \text{BE (reactants)}} - \underset{\text{moles}}{\sum \text{BE (products)}}$$

Note that this equation is in the form reactant minus products. Figure 7.14 shows a schematic representation of the meaning of the formula.

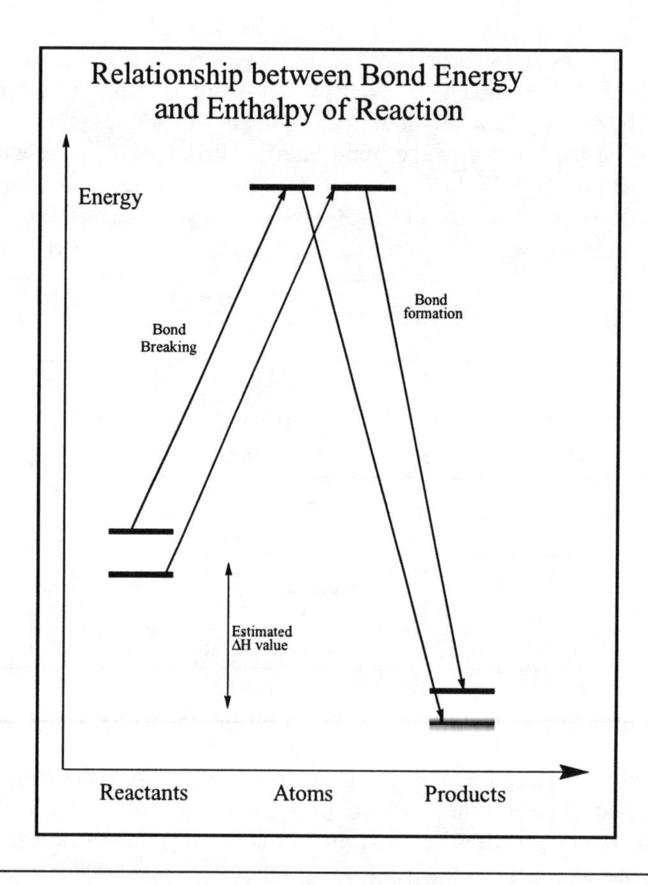


Figure 7.14. Schematic representation of the use of bond energies to estimate the enthalpy change of a gas phase reaction. Bond breaking requires an energy input. Bond formation generates an energy output. If more energy comes out than goes in (as shown here) then the reaction is exothermic. The energy state of the system as products is lower than the energy state of the system as reactants. The system loses energy, and heat is released from the system to its surroundings.

The reaction of combustion of methane to carbon dioxide gas and water vapor is shown below using structural formulas:

```
      H
      |
  H—C—H(g)  + 2 O=O(g)   →   O=C=O(g)   + 2 H—O—H(g)
      |
      H
```

In this reaction, energy must be supplied to break four moles of C-H bonds in $CH_4$ and two moles of the O=O bonds in $O_2$; energy is liberated when two moles of C=O bonds in $CO_2$ and a total of four moles of O-H bonds (two in each mole of $H_2O$) are formed. The steps are:

| Bond breaking | Step 1 | $CH_4(g) \rightarrow C(g) + 4\,H(g)$ | energy in |
|---|---|---|---|
| Bond breaking | Step 2 | $2\,O_2(g) \rightarrow 4\,O(g)$ | energy in |
| Bond making | Step 3 | $C(g) + 2\,O(g) \rightarrow CO_2(g)$ | energy out |
| Bond making | Step 4 | $4\,H(g) + 2\,O(g) \rightarrow 2\,H_2O(g)$ | energy out |

We can estimate the net amount of heat absorbed or liberated by using average values for bond energies given in Table 7.7 by substituting into:

$$\Delta H(\text{estimated}) = \underset{\text{moles}}{\Sigma \text{ BE (reactants)}} - \underset{\text{moles}}{\Sigma \text{ BE (products)}}$$

$$\Delta H(\text{estimated}) = [4 \times BE(C\text{-}H) + 2 \times BE(O{=}O)] - [2 \times BE(C{=}O) + 4 \times BE(O\text{-}H)]$$

$$= [4 \text{ mol} \times 413 \text{ kJ mol}^{-1} + 2 \text{ mol} \times 495 \text{ kJ mol}^{-1}) - [2 \text{ mol} \times 799 \text{ kJ mol}^{-1} + 4 \text{ mol} \times 463 \text{ kJ mol}^{-1}]$$

$$= \underline{-808 \text{ kJ}}$$

The details are omitted here, but if you use the heats of formation of $CH_4(g)$, $CO_2(g)$, and $H_2O(g)$ to calculate $\Delta H$, you obtain a value for $\Delta H$ for the reaction of -803 kJ, quite close to the value obtained from the average bond energies.

If the bonds in the reactants are collectively stronger than those in the products, the reaction is endothermic. Conversely, if the bonds in the products are collectively stronger than those in the reactants, the reaction is exothermic.

The combustion of hydrocarbon fuels involves the breaking of C-H bonds and the formation of O-H bonds and C=O double bonds. It is the relatively high bond energy of O-H and C=O bonds compared with C-H and $O_2$ bonds that makes fuel burning a highly exothermic process. Example 7.21 shows estimating an enthalpy of reaction from bond energies.

**Example 7.21**: Use the bond energies to estimate the gaseous enthalpy change of the following reaction: Compare the heat of the reaction with that calculated from heats of formation, which is -98.32 kJ mol$^{-1}$. Comment on the difference.

```
      H                              Cl
      |                              |
H — C — H(g)  +  Cl — Cl(g)  →  H — C — H(g)  +  H — Cl(g)
      |                              |
      H                              H
```

Strategy: Apply the equation,

$$\Delta H(\text{estimated}) = \underset{\text{moles}}{\Sigma \text{ BE (reactants)}} - \underset{\text{moles}}{\Sigma \text{ BE (products)}}$$

Bonds broken: C-H, and Cl-Cl　　　Bonds formed: C-Cl, and H-Cl

$$\Delta H(\text{estimated}) = (413 \text{ kJ mol}^{-1} + 242 \text{ kJ mol}^{-1}) - (328 \text{ kJ mol}^{-1} + 431 \text{ kJ mol}^{-1})$$

$$\Delta H(\text{estimated}) = \underline{-104 \text{ kJ mol}^{-1}}$$

The calculation of $\Delta H$ using bond energies is an estimate, and we see that it differs from the more exact value determined from heats of formation by about 6% for this reaction.

## Essential Knowledge—Chapter 7

**Glossary words:** Thermodynamics, chemical energy, heat, thermochemistry, work, calorimetry, calorimeters, standard thermodynamic tables, exothermic, endothermic, energy values of foods and fuels, metabolic rate, efficiency of a process, percentage efficiency, working fluid, reversible gas expansion, zeroth law of thermodynamics, phase changes, specific heat capacity, molar heat capacity, law of Dulong and Petit, standardization, standard substance, first law of thermodynamics, internal energy of a system, state properties, state functions, path dependency of processes, endoergic, exoergic, enthalpy, heat of combustion, reaction enthalpy change, heat of reaction, Hess's law, standard enthalpy of formation, standard state, thermodynamically stable, thermodynamically unstable, bond energy, average bond energy.

**Key Concepts:** Nature of energy in fuels and foods. The relationship between thermodynamics and steam engines. Use of thermodynamic language and careful definitions of work and heat, etc. Heat capacity: its measurement, origin, and application. Use and operation of calorimeters. First law of thermodynamics and the distinction between path dependent and path independent properties. Applications and use of internal energy and enthalpy. The equivalence of heat and work. Hess's law and the concept that state properties are unchanged in a circuit around a cyclic process. Use of standard enthalpies of formation to get enthalpy changes of reactions. Use of bond energies to estimate enthalpy changes of reactions.

**Key Equations:**

Energy of food portion (kJ) = g(fat) × 39.5 kJ $g^{-1}$ + g(carbohydrate) × 17.1 kJ $g^{-1}$ + g(protein) × 14 kJ $g^{-1}$

Work of expansion of a gas against constant pressure: $w = -P_{ext}\Delta V$.

Percentage efficiency of a machine or steam engine = $\frac{\text{work obtained}}{\text{heat required}} \times 100$

Maximum isothermal work of ideal gas expansion: $w_{max} = \int_{V_1}^{V_2} P\,dv$.

Relationship between molar and specific heat capacity $C_p(\text{molar}) = C_p \times MW$

Power series fit for molar heat capacity: $C_p = a + bT + cT^2 + d\,T^3 + \ldots$

Heat change of system, constant P, $\Delta T$ small: $q = C_p(T_2 - T_1)$, or $q = C_p\Delta T$ or $q = C_p \times g \times \Delta T$

Specific heat at constant volume of a monatomic ideal gas as: $C_v = \frac{3R}{2MW}$

Heat and temperature change of calorimeter: $q = C_{calorimeter}\,\Delta T$

First law of thermodynamics: $\Delta U = q + w$.

Definition of enthalpy: $H = U + PV$

Enthalpy change of constant pressure process: $\Delta H = \Delta U + P\Delta V$.

Hess's law: $\Delta H° = \sum_{\text{moles}} \Delta H^0_f(\text{products}) - \sum_{\text{moles}} \Delta H^0_f(\text{reactants})$

Enthalpy change from bond energy: $\Delta H(\text{estimated}) = \sum_{\text{moles}} BE\ (\text{reactants}) - \sum_{\text{moles}} BE\ (\text{products})$

## Questions and Problems

### Thermodynamics, Foods and Fuel Energy

7.1. Rank the following chemical fuels in order of decreasing chemical energy in kJ $g^{-1}$: wood, hydrogen, diesel fuel, coal.

7.2. State in a couple of sentences the scientific contributions of each of the following persons as described in Chapter 7: a. Joseph Black, b. Count Rumford, c. Humphrey Davy, d. Sadi Carnot, e. James Joule, f. Julius R. Meyer, g. Hermann Ludwig von Helmholtz, h. William Thompson (Lord Kelvin), i. Rudolf Clausius, and, j. Josiah Willard Gibbs.

7.3. Write balanced chemical equations for the complete combustion of toluene, sucrose, and glycine. If you don't know their chemical formulas try an internet search. [☸kws +toluene +"chemical formula"]

7.4. The label on a box of cereal states that a one ounce serving contains 6 grams of protein, 22 grams of carbohydrate, and no fat. Estimate how many calories there are in a one-ounce serving and compare your estimate with the label's statement of 130 kcal $oz^{-1}$.

7.5. An analysis of pecan nuts is 9.6% protein, 70.5% fat, 15.3% carbohydrate, and 4.6% water. How many calories are in 2 ounces (56.7 g) of pecan nuts?

7.6. Estimate the mass of pecans needed to provide sufficient chemical energy to meet the daily caloric need of a average young person.

7.7. The energy value of hydrogen is 141.8 kJ $g^{-1}$ and the energy value of diamond is 33.9 kJ $g^{-1}$. Compare the energy values of the listed fuels with the energy value of a stoichiometric mixture of its elements. Calculate the percentage discrepancy between them. Discuss your results.

7.8. Calculate the maximum efficiency of a steam engine that operates between 175°C and 18°C. What is the working fluid in such an engine. Explain.

### Thermodynamic Definitions

7.9. Define the terms: a. thermodynamic state, b. a process, c. the temperature of standard state, d. standard pressure, and e. thermodynamic efficiency. Strategy: look for the definitions in either the text book or the glossary.

7.10. What is the relationship between a standard pressure of 1 bar and a pressure of 1 atmosphere?

7.11. What is NIST and what does it have to do with thermodynamics?

7.12. State the zeroth law of thermodynamics. State the Clausius definitions of the first and second laws of thermodynamics.

7.13. What is the Carnot cycle?

7.14. Why does the kelvin scale of temperature have its name?

### Heat Capacity

7.15. Distinguish between specific heat capacity and molar heat capacity. State the SI units of each. Use internet resources to find out what common engineering units are used for these quantities and state the conversion factors between SI and engineering units.

7.16. Rank the specific heat capacities of the following substances at 25°C from smallest to greatest: water, hydrogen, carbon dioxide, ethyl alcohol, rubber, and iron.

7.17. For each of the following substances calculate the temperature rise ($\Delta T$) at constant pressure of 1.00 g of the substance when it absorbs 1,000 J of heat energy: gold, platinum, helium gas, water, and granite. List your answers in decreasing order of $\Delta T$.

7.18. Calculate the final temperatures if 500. J of heat is added to 10.0 g of each of the following substances at constant pressure: a. liquid water, b. solid aluminum, and c. solid iron.

7.19. Use values of specific heat capacities at constant pressure to calculate the molar heat capacities of the following elements: a. hydrogen, b. helium, c. aluminum, d. iron, e. gold, and f. platinum.

7.20. How much heat is required to warm 1.00 kg of liquid water from room temperature (22.5°C) to 67.4°C at constant pressure?

7.21. What will be the final temperature if 1.00 kg of liquid water at 88.3°C is mixed with 100. g of liquid water at 22.1°C in a calorimeter?

7.22. A 5.67 g sample of iron pellets is heated to 99.4°C and then quickly and carefully added to 57.8 g of water at 19.3°C in a thermos bottle. What will be final temperature when the system reaches thermal equilibrium?

7.23. Discuss the historical significance of the Law of Dulong and Petit. With the benefit of 150 years of hindsight explain why the existence of such a law makes sense to you.

7.24. Why are the terms latent heat of fusion and latent heat of vaporization so named? Explain why for a given substance the latter is usually a lot larger that the former.

7.25. Use its Shomate equation to calculate the constant-pressure specific heat of hydrogen gas ($C_p$) at 425 K.

## Calorimetry

7.26. a. Describe a bomb calorimeter and make a list of all of its component parts. b. What is a combustion standard? c. Why does NIST sell combustion standards?

7.27. The combustion of 0.6744 g of benzoic acid raises the temperature of a bomb calorimeter by 2.17°C. Calculate the heat capacity of the bomb calorimeter.

7.28. The same calorimeter as in the previous question is used to combust 0.500 g of chocolate cake. The temperature rises by 1.03°C. How many Calories (capital C) are in a 4 ounce slice of the cake?

7.29. Coal used to fire a power plant has an energy value of 13,250 Btu $lb^{-1}$. A 0.2500 g sample of this coal is combusted in a bomb calorimeter with a heat capacity of 4,589 J $K^{-1}$. Calculate the temperature rise of the calorimeter.

7.30. A 50.0 watt electric heater is used to measure the heat capacity of a calorimeter. After the heater has been running for 100.0 sec, the temperature of the calorimeter rises by 0.703°C. What is the heat capacity of the calorimeter?

7.31. Two fuel samples were combusted using the same calorimeter. When 0.5402 g of the first was burned, the temperature of the calorimeter increased by 1.57°C. When 0.3263 g of the second was burned, the temperature increase was 1.23°C. The chemical energy of the first sample was 23.6 kJ $g^{-1}$. Calculate the chemical energy of the second fuel sample.

## Work

7.32. List three ways that useful energy is obtained from the expansion of hot gases in our society. State three energy sources that do not involve the work of hot gases expanding. (Review Section 4.6 if necessary.)

7.33. How much work is required to raise a 25 kg mass by a height of 25 m? What do engineers use the unit foot-pound force for?

7.34. A 1.00 mol sample of an ideal gas is in a cylinder with a movable piston at a temperature of 25°C. The external pressure on the piston is kept constant at 3.5 atmospheres. The gas is heated until its volume triples. How much work does the gas do? Express your answer in both L atm and kJ.

7.35. How much work does a gas perform if it is allowed to double its volume by expanding into an evacuated vessel?

7.36. A syringe with a movable, frictionless piston contains 3.00 mmol of an ideal gas with a volume of 34 mL at 2.00 atm. The external pressure on the piston is slowly reduced to 0.500 atm, and the gas's temperature is held constant at 298 K. a. Calculate the final volume of the gas. b. Calculate the maximum amount of work that the expansion could do. In other words, calculate the reversible work.

7.37. A solid has a volume of 0.1500 L. Initially at one atmosphere pressure, it is squeezed by pressure of 100 atm decreasing its volume by a factor of 0.9980. Calculate the work done on the solid during compression. Explain thermodynamically why liquids and solids are able to do very little work.

7.38. For solids and liquids, $C_p$ is approximately equal to $C_v$. But for gases $C_p > C_v$. Explain what these relationships have to do with work and with the previous problem.

7.39. When a liquid boils, its vapor does work by expanding against the blanket of the atmosphere. Calculate the work done by 1.00 mole of water as it boils (at 100°C) at a pressure of 1.00 atm? Assume the density of liquid water is 1.0 g $mL^{-1}$ and that steam is an ideal gas. Express your answer in both L atm and kJ.

## First Law of Thermodynamics

7.40. a. How did James Prescott Joule experimentally establish the law of conservation of energy? b. What experiment—with an inconclusive result—did he undertake during his honeymoon?

7.41. Which of the following descriptions adequately defines a system in a definite state? a. Four grams of neon gas at a pressure of 3 atm. b. 54.0 mL of 2.5 M $H_2SO_4$ solution in water at a pressure of one atmosphere and a temperature of 13°C.

7.42. A 1.00 mole sample of an ideal gas is originally at 0°C and 1.00 atm. Its state is changed to T = 500. K and P = 3.00 atmosphere. Calculate $\Delta T$, $\Delta P$, and $\Delta V$ for the gas.

7.43. A sample of an ideal gas contains 4.00 moles and is originally at 300. K and 10.0 atm pressure. It is subjected to a change in state for which $\Delta T$ = - 50.0 K, and $\Delta V$ = 10.0 L. Calculate the final pressure, temperature, and volume of the gas.

7.44. If a 50.0 watt electric heater heats water for 2.50 s, how much change occurs in the internal energy of the water? (1 watt = 1 J $s^{-1}$)

7.45. a. Calculate the change in the internal energy of 1.00 mole of an ideal gas if it is heated from 0.0°C to 150.°C under constant volume conditions. b. Would the value of $\Delta U$ be different if the process were at constant pressure instead of constant volume?

7.46. A gas is compressed (without exchanging heat with its environment) using a constant external pressure of 2500. Torr. The gas's volume decreases by 2.33 L. Calculate the change in the internal energy of the gas in L atm and in J.

7.47. As a consequence of a thermodynamic process 15.0 J of heat is added to a system, and 10.0 J of work is done on the system. How much change ($\Delta U$) occurs in the internal energy of the system during the process?

7.48. As a consequence of a thermodynamic process 355 J of heat is transferred to a gas, while simultaneously the gas does 2.50 L atm of work by expanding against a piston. Calculate the change in the gas's internal energy in joules.

7.49. The first law of thermodynamics is $\Delta U = q + w$. $\Delta U$ is path independent; q and w are path dependent; q + w is path independent. Explain.

## Enthalpy

7.50. Define every term in the equation H = U + PV. State under precisely what conditions $\Delta H = \Delta U$ for a thermodynamic process.

7.51. Write balanced equations for enthalpy of formation reactions for: a. propane, b. ethylene, c. urea, and d. sucrose, $C_{12}H_{22}O_{11}(s)$.

7.52. Describe how Hess's law and a table of standard enthalpies of formation can be used to obtain the standard heats of many reactions.

7.53. Use the table of standard enthalpies of formation to calculate $\Delta H^0$ for each of the following reactions:

7.54. Which of the following has the largest (least negative) ΔH of formation: $CH_3OH(l)$ or $CH_3OH(g)$? Why?

7.55. Is ethyne thermodynamically stable?

7.56. Use $\Delta H^0_f$ values in the table of selected thermodynamic properties to compute the heats of vaporization of water and methanol.

7.57. Use the following two thermochemical equations and their standard enthalpy changes to calculate the standard enthalpy change of the third reaction.

$$H_2O(l) + \tfrac{1}{2} O_2(g) \rightarrow H_2O_2(l) \quad \Delta H^\circ_1 = 94.7 \text{ kJ mol}^{-1}$$
$$H_2O(l) \rightarrow H_2O(g) \quad \Delta H^\circ_2 = 44.0 \text{ kJ mol}^{-1}$$
$$H_2O_2(l) \rightarrow H_2O(g) + \tfrac{1}{2} O_2(g) \quad \Delta H^\circ_3$$

7.58. Use the standard enthalpy of combustion of naphthalene ($C_{10}H_8$, s), $\Delta H^\circ = -5154 \text{ kJ mol}^{-1}$, and the standard heats of formation of the other reactants and products from the table of thermodynamic data, to calculatc thc standard heat of formation of naphthalene. The combustion equation is: $C_{10}H_8(s) + 12\ O_2(g) \rightarrow 10\ CO_2(g) + 4\ H_2O(l)$.

7.59. Use standard heats of formation to calculate the standard heat of combustion of methanol, $CH_3OH(l)$, forming $CO_2(g)$ and $H_2O(l)$.

7.60. Use standard enthalpies of formation to calculate the standard heat of combustion of ethylene gas.

7.61. Calcium metal burns according to the following equation: $Ca(s) + \tfrac{1}{2} O_2(g) \rightarrow CaO(s)$. When 1.00 g of calcium metal is burned in a constant pressure calorimeter with a heat capacity of 15.20 kJ $K^{-1}$, the temperature increases by 1.04°C. Calculate the enthalpy of formation of calcium oxide. Compare your answer with the NIST value of -635.09 kJ $mol^{-1}$.

## Bond Energies

7.62. Calculate ΔH for the reaction: $H(g) + F(g) \rightarrow HF(g)$ from the following thermochemical data:

$$H_2(g) + F_2(g) \rightarrow 2\ HF(g) \quad \Delta H = -518.8 \text{ kJ}$$
$$H_2(g) \rightarrow 2\ H(g) \quad \Delta H = 435.1 \text{ kJ}$$
$$F_2(g) \rightarrow 2\ F(g) \quad \Delta H = 158.2 \text{ kJ}$$

7.63. Use the table of average bond energies (Table 7.7) to estimate the standard enthalpy of the reaction

$$N{\equiv}N(g) + 3\ H\text{-}H\ (g) \rightarrow 2\ \begin{matrix} H \\ | \\ N\text{-}H \\ | \\ H \end{matrix}\ (g)$$

7.64. The heat of atomization of ethylene gas is 2,250 kJ $mol^{-1}$. Assuming the C-H bond energy is 413 kJ $mol^{-1}$ calculate the energy of the carbon-carbon double bond in ethylene.

## Problem Solving

7.65. How much energy is consumed by burning a 100 watt light bulb for 8.0 hours? (1 watt = 1 J $s^{-1}$.) Assuming 30% efficiency for converting chemical energy to electrical energy, how many grams of fuel oil must be burned to supply the energy?

7.66. Discuss the use of significant figures in energy values listed in Table 7.1.

7.67. Use data for the $C_p$ of water to calculate the percentage difference between how much heat is needed to heat water by 1°C at 25°C and to heat water by 1°C at 75°C.

7.68. Total energy consumption in the United States is about $10^6$ kJ per person per day. a. If that energy was the heat of combustion of coal, how many pounds of coal per day per person is that? b. If the energy came solely from the combustion of gasoline, how many gallons would it be? The density of gasoline is about 6.2 lb gallon$^{-1}$.

7.69. Write a balanced equation for the reaction in which the single reactant glucose yields ethyl alcohol and carbon dioxide. The equation you have written represents the fermentation of glucose. Knowing that, can you figure out if fermentation is an exothermic or endothermic process?

7.70. $CFCl_3$ and $CF_2Cl_2$ were once used as refrigerants and propellant gases in aerosol products. In the stratosphere they decompose to produce Cl atoms which react with ozone, $O_3$, depleting the ozone, perhaps with serious environmental consequences. The O-Cl bond energy is estimated to be 210 kJ mol$^{-1}$, and the each of the two delocalized oxygen-oxygen bonds in ozone (O≃O≃O) is estimated to have an energy of 303 kJ mol$^{-1}$. What are the enthalpy changes of the following reactions?

$$Cl(g) + O_3(g) \rightarrow ClO(g) + O_2(g)$$
$$ClO(g) + O_3(g) \rightarrow Cl(g) + 2\,O_2(g)$$

7.71. The US economy consumes an estimated $1 \times 10^{18}$ kJ of energy per year. It is estimated with present methods of gas production that the remaining recoverable US natural gas supply is 500 TCF. (1 TCF is a trillion cubic feet, $10^{12}$ ft$^3$, measured at 14.73 psi and 60°F.) Assuming that natural gas is methane, and that it can be used with 50% efficiency to produce useful heat and work, how long could the US meet its entire energy need just from natural gas.

7.72. A conservative estimate of recoverable US coal reserves places them at 250 billion tons ($2.5 \times 10^{11}$ tons). Using data from the previous problem and assuming that 50% of the energy from coal can be converted to useful heat and work, how long could the US meet its entire energy need just from coal?

7.73. A large coal-burning power plant generates 1000 megawatts of electric power (1 megawatt = $1 \times 10^6$ J s$^{-1}$). To do this it generates 3000 megawatts of heat (this plant is 33% efficient). How many tons of good quality coal (chemical energy = 35 kJ g$^{-1}$) does the plant burn every day. If the coal contains 1.5% sulfur, how many tons of sulfur dioxide ($SO_2$) does the plant produce each day?

7.74. Select the molecule that is likely to yield the least combustion energy per gram and the one likely to yield the most: $C_5H_{12}$, $C_5H_{12}O$, $C_5H_{12}O_2$, and $C_5H_{12}O_3$. (Hint: think how its oxygen content affects the chemical energy of a fuel.)

7.75. Answer the previous question on the basis of comparing these compounds mole for mole.

7.76. Chemical energies of saturated hydrocarbons rank in the sequence: methane > ethane > propane > butane > pentane, etc. Why?

7.77. In Appendix C there is a brief description of the CODATA group. Use internet resources to find out everything you can about CODATA.

7.78. In Section 7.6 an in-text example shows that the enthalpy of combustion of methane according to the reaction equation $CH_4(g) + 2\,O_2(g) \rightarrow CO_2(g) + 2\,H_2O(g)$ is estimated to be -808 kJ mol$^{-1}$. In the chapter it is asserted that the thermochemical value for the identical reaction is -803 kJ mol$^{-1}$. Test this assertion using standard thermochemical data.

7.79. "Double glazing" uses two panes of window glass with a gas trapped between them. Argon-filled window glass is often advertised as having superior heat insulating qualities to air-filled window glass. Use internet resources to find an advertisement for argon-filled double-glazed window glass. Discuss the scientific basis for the claim.

# Chapter 8

# The Second Law of Thermodynamics and Chemical Equilibrium

## Introduction

The term **chemical equilibrium** describes the balance among the amounts of all reactants and products in a chemical reaction system. Chemical equilibrium is a fact of nature: candles burn out, batteries run down, acids get neutralized by bases. We can start such a reaction and see it apparently conclude; yet the reaction *does not stop*. At the conclusion, a state of chemical equilibrium has been reached, and the rate of the forward reaction has merely become equal to the rate of the reverse reaction. These ideas will be explained in this Chapter.

A chemical reaction system is said to be **in equilibrium** when the amounts of its reactants and products are constant. The position of equilibrium of a chemical reaction system is fundamentally related to the reaction's energy change. Because chemical thermodynamics deals with such energy changes, the laws of thermodynamics turn out to be fundamental to understanding chemical equilibrium.

The simplest conceivable equilibrium reaction system is: $H_2 \rightleftarrows 2\ H$. The symbol $\rightleftarrows$ means that the reaction is *simultaneously* going in both the forward and reverse directions. The simplest possible measurement of a chemical equilibrium is to know how much molecular hydrogen ($H_2$) and how much atomic hydrogen (H) are present in a sample of pure hydrogen at any specified temperature and pressure. Experiments with this reaction show that the square of the partial pressure of atomic hydrogen divided by the partial pressure of molecular hydrogen is a constant at any specified temperature. It's called the **equilibrium constant** of the reaction at the specified conditions of T:

$$K_p = \frac{(P_H)^2}{(P_{H_2})} = \text{hydrogen's equilibrium constant using partial pressures (T is specified).}$$

The equilibrium constant can be written equally well, but with alternative units, as

$$K_c = \frac{[H]^2}{[H_2]} = \text{hydrogen's equilibrium constant using molar concentrations (T is specified).}$$

If you think about the diatomic hydrogen molecule, it's easy to see why the position of chemical equilibrium and reaction energy must be related. Hydrogen molecules have kinetic energy because their motion, they have bond energy because of their covalent bonding, and the two energies are linked because the molecules are continually colliding. At any particular temperature, some collisions provide sufficient energy to break the H-H bond. Obviously, the stronger the bond the fewer atoms are formed by collisions; the energy of the bond breaking reaction and the number of hydrogen atoms that exist in the system are necessarily linked.

Chemical thermodynamics is a powerful tool for organizing and manipulating chemical data. In this chapter we will study the use of tables of thermodynamic data to calculate $K_p$ and $K_c$ for *any* reaction at *any* temperature. A small selection of this vast body of available thermodynamic data is presented in Appendix C.

We'll begin by looking at the second law of thermodynamics. The second law involves the subtle concept of entropy (S) that we will examine from both a thermodynamic and statistical point of view. Next, we'll discuss the third law of thermodynamics that allows us conveniently to choose a reference standard state for the entropy of substances. Then, knowing about the second and third laws, we'll study the connection between thermodynamics and chemical equilibrium and meet the important Gibbs equation $\Delta G = \Delta H - T\Delta S$. For reactions that take place at constant pressure, it is the Gibbs equation that enables us to calculate the reaction's $K_p$ and $K_c$ values.

Entropy is a purely bulk (macroscopic) thermodynamic property of systems. However, it can be interpreted statistically as the amount of "randomness" in a system. For a beginner, the statistical interpretation of entropy is usually easier to grasp than the macroscopic interpretation. **Statistical mechanics** is the branch of modern science that explains the bulk properties of materials based on the collective properties of an assembly of the material's component molecules. It provides a fully satisfactory link between thermodynamics, statistics, and molecular behavior, but lies beyond our scope.

The first indication of the existence of what we now call the second law of thermodynamics was in the pioneering work of Sadi Carnot in 1824. Carnot, you'll recall, was a French military engineer. The state property that is entropy is implicit in Carnot's ideas, but the concept of entropy was not clearly and explicitly stated until 1850 by Rudolf Clausius after William Thompson (Lord Kelvin) had shown that Carnot's work demanded the existence of an absolute (Kelvin) temperature scale. In a chemistry book written for engineers it is worthwhile to sketch out Carnot's ideas:

## The Carnot Cycle and Entropy

Carnot thought deeply about the efficiency of steam engines. In his mind, he did imaginary experiments to understand why the conversion of heat into work was an inefficient process. He imagined the working fluid in an engine (steam, for example) to undergo transformations from high temperature and pressure to low temperature and pressure, and then (by heating and compressing it) returning the fluid to its original high-temperature and high-pressure state. The word **cycle** refers to a system passing through several intermediate states before being returned to its original state; for emphasis we could call it a **thermodynamic cycle**. Carnot was guided in his thinking by practical knowledge of what engines could and could not do. Dominant among these practical considerations was the fact that it is impossible to make heat *spontaneously* flow "uphill" from a colder object to a warmer object. (Uphill flow can be done, but, as we know from our summertime electricity bills for air-conditioning, only at great expense of energy.) Carnot realized that the highest possible temperature difference between the heat source and the heat sink yielded the greatest possible conversion of heat to work, and that the chemical composition of the working fluid was fundamentally irrelevant. Carnot's main conclusions are summarized in the following box:

### Carnot's Ideas

1. No cyclic process of an engine can be more efficient than a reversible cycle. A **reversible cycle** is one that does the maximum amount of work; one that has the highest percentage efficiency. This statement is equivalent to the practical idea that there is fundamental upper limit to the efficiency of an engine.

2. All "reversible" cycles have the same maximum efficiency. This statement means no matter how the engine is designed, all designs face the same ultimate limit to their efficiency.

3. The efficiency of an engine doesn't depend on its working fluid. A steam engine is as useful as an ammonia engine or as a carbon dioxide engine, etc.

4. No engine can produce work by removing heat from a colder source and transferring the heat to a hotter source. Note that this idea does *not* violate the law of conservation of energy (the first law of thermodynamics) but it does violate all of human experience. Because it violates human experience it is a statement of the second law of thermodynamics. Imagine a boat that takes in sea water, extracts heat from the sea water, and sails on forever, leaving a trail of cold water in its wake; that's a wonderful—but impossible—idea.

For a reversible process (a process arranged to produce the maximum possible work; a process that operates at the maximum possible efficiency) $q_{rev}$ is the heat involved when the process does maximum work, $w_{max}$. A reversible process is one that is infinitesimally near a state of equilibrium at each stage of the process: that is, one that can be reversed at any point by an infinitesimal change in an independent variable. The concept of $q_{rev}$ or reversible heat leads to one definition of a system's **entropy change** at constant temperature as:

$$\Delta S = \frac{q_{rev}}{T}$$

The units of entropy are J $K^{-1}$, or energy divided by temperature. When we talk about the entropy of chemical compounds we usually use molar entropies.

Entropy changes are related to heat transfer in and out of thermodynamic systems. So it is usually a straightforward calorimeter experiment that provides the measure of entropy changes of specific processes. For a reversible process the first law can be written: $\Delta U = q_{rev} + w_{max}$.

If a system is kept at a constant absolute temperature, T, and an amount of heat, $dq_{rev}$, is transferred to it very slowly, its entropy changes by

$$dS = dq/T$$

This equation is a general definition of the **entropy change** of a system. The entropy of any system increases when heat is added to it. For a given amount of heat, the entropy change is greater the *lower* the absolute temperature of the system: heat added to a cold system is relatively more disruptive than the same heat added to a hot system. For the reversible addition of a finite quantity of heat that is accompanied by an increase in temperature, the entropy change of the system is $\Delta S = \int dq_{rev}/T$.

## Section 8.1 The Second Law of Thermodynamics

It's odd, but there are many statements of the second law of thermodynamics and most of them are not blindingly obvious. Most of the definitions in the box below come from chemistry text books.

**The Second Law of Thermodynamics—Some Statements of It**

*The entropy of the universe tends to a maximum.* (Rudolf Clausius, 1865)

The principle of Thompson: *It is impossible by a cyclic process to take heat from a reservoir and convert it into work without, in the same operation, transferring heat from a hot to a cold reservoir.* (Walter Moore,1957).

The principle of Clausius: *It is impossible to transfer heat from a cold to a hot reservoir without, in the same process, converting a certain amount of work into heat.* (Walter Moore, 1957).

*The energy of the world remains constant; the entropy of the world tends to a maximum value.* Attributed to Clausius. (E. A. Moelwyn-Hughes, 1957).

*It is impossible for energy to move unaided from a given potential to a higher one.* (Charles M. Thatcher, 1962).

*The entropy of the isolated system increases in all real processes (and is conserved in ideal processes).* Said by Planck to be "the most general statement of the second law." (Edward F. Obert and Richard A. Gaggoli, 1964).

*Any process that takes place causes an increase in the entropy of the universe.* (J. Rex Goates and J. Bevan Ott, 1971).

*It is impossible for a system operating in a cycle and connected to a single heat reservoir to produce a positive amount of work in the surroundings.* (Gilbert W. Castellan, 1971)

*It is not possible to convert heat into work by means of an isothermal cyclic process.* (Arthur W. Adamson, 1973).

*All spontaneous or natural processes increase the entropy of the universe.* (John Hill and Ralph Petrucci, 1999)

### Entropy Changes of Some Specific Processes

Odd definitions of the second law notwithstanding, the measurement of entropy changes is an absolutely concrete experimental (usually calorimetric) procedure. For pure substances the equations are: $\Delta S = \int C_p/T\ dT$ for warming at constant pressure, $\Delta S = \Delta H_{fus}/T$ for melting, and $\Delta S = \Delta H_{vap}/T$ for vaporization. Where "fus" refers to fusion or melting, and "vap" refers to vaporization. If $C_p$ is independent of T, then as we saw in the preceding chapter, the enthalpy change for warming a substance is simply its heat capacity times the temperature change, or $\Delta H = C_p(T_2-T_1)$. For a *small* temperature change, ΔH divided by the average temperature approximates the entropy change. Generally correct is $\Delta S = \int C_p/T\ dT$ which, if $C_p$ is independent of temperature, integrates to $\Delta S = C_p \ln (T_2/T_1)$. If $C_p$ is an empirical, power-series equation (Shomate equation) we simply integrate that equation. The entropy change of a chemical reaction is $\Delta S = S_{products} - S_{reactants}$.

The property of entropy is perhaps best approached by some specific concrete examples. We will discuss three specific cases of entropy change at constant pressure: 1. the entropy change of phase changes, 2. the entropy change of a pure substance as it warms from the starting point of absolute zero to any temperature, and 3. the entropy change of *any* standard state chemical reaction. We will outline them now and return to them throughout this chapter.

1. **Entropy of phase change** is the change of entropy during changes in the states of aggregation of substances among solid, liquid, and gas. The changes (s) $\rightleftarrows$ (l), (s) $\rightleftarrows$ (g), and (l) $\rightleftarrows$ (g) are equilibrium processes at constant pressure. Here $\Delta S = q_{rev}/T$, where $\Delta S$ is the entropy change, $q_{rev}$ is the reversible heat of the process (at constant pressure) and T is the kelvin temperature. Because they are equilibrium processes, $q_{rev} = \Delta H$ so we'll use $\Delta S = \Delta H/T$. Incidentally, the energy changes of these processes are the latent heats that were first understood by Joseph Black and James Watt in the 18th century. Later, we'll see that a given substance has the least entropy in the solid state of aggregation, intermediate entropy in the liquid state of aggregation, and the most entropy in the gaseous state of aggregation. This sequence introduces the idea that the entropy of a substance is greater the more dispersed are its molecules.

2. **Temperature dependence of entropy**. The entropy of a substance falls with falling temperature. It falls to zero for perfectly crystalline substances at absolute zero. The kelvin scale of temperature, that starts at 0 K, is called the thermodynamic temperature scale. Entropy may be interpreted as the randomness or disorder of a substance or system.

3. **The entropy change during chemical reactions**. Except for the handful of cases where it is accidentally zero, all chemical reactions involve entropy changes. The entropy change of a chemical reaction can be obtained by manipulating the tabulated standard entropy data.

## The Statistical Interpretation of Entropy

Thermodynamics itself makes no assumptions about the nature of matter. Systems simply contain "stuff" with thermodynamic properties that are "just measured." However, as the talk about "randomness" has been hinting, we humans can get some insight about entropy if we imagine the behavior of the particles in a thermodynamic system. Loosely, thinking about molecules "explains" entropy.

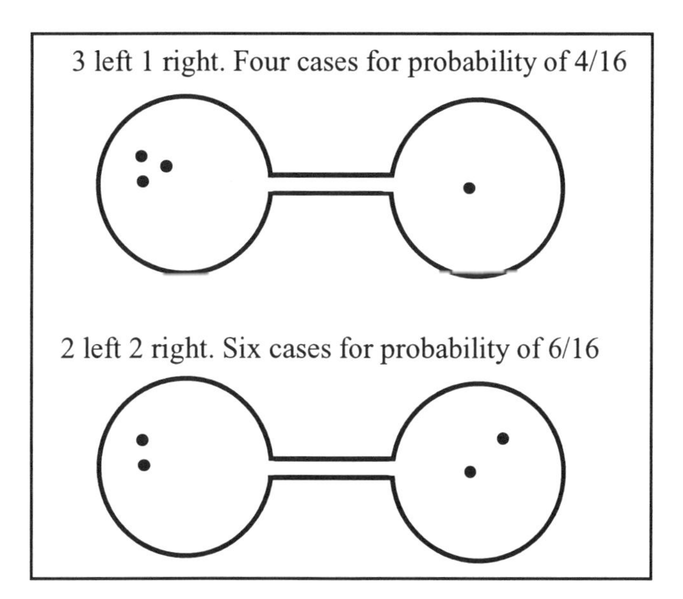


Figure 8.1. A system composed of four particles each of which can be in one or other of two containers. Thinking about the properties of this imaginary system gives great insight into the nature of the property called entropy. The sixteen total possible states for this system are listed in Table 8.1.

Figure 8.1 shows two different situations of four particles or molecules distributed between two flasks. A common, simple illustration of the statistical nature of entropy is to imagine a system composed of a handful of molecules that can move between two flasks. One particular arrangement of the molecules is called a **microstate** of the system. A set of indistinguishable microstates is called a **macrostate**. The entire list of the possible microstates and macrostates of four particles in two flasks is listed in Table 8.1.

Clausius said the entropy of the universe tends to a maximum. The statistical interpretation of entropy is: systems arrange themselves to be in the *most probable* macrostate.

Table 8.1

**Microstates and Macrostates of Four Particles in Two Flasks**

| | Microstate | Macrostate | Number of microstates in macrostate |
|---|---|---|---|
| 1 | LLLL | 4 L 0 R | 1 |
| 2 | LLLR | 3 L 1 R | 4 |
| 3 | LLRL | 3 L 1 R | |
| 4 | LRLL | 3 L 1 R | |
| 5 | RLLL | 3 L 1 R | |
| 6 | LLRR | 2 L 2 R | 6 |
| 7 | LRLR | 2 L 2 R | |
| 8 | LRRL | 2 L 2 R | |
| 9 | RLRL | 2 L 2 R | |
| 10 | RLLR | 2 L 2 R | |
| 11 | RRLL | 2 L 2 R | |
| 12 | RRRL | 1 R 3 L | 4 |
| 13 | RRLR | 1 R 3 L | |
| 14 | RLRR | 1 R 3 L | |
| 15 | LRRR | 1 R 3 L | |
| 16 | RRRR | 0 R 4 L | 1 |

Each one of the sixteen arrangements in Table 8.1 is a possible microstate. Every individual microstate has the same probability of 1/16. Note that the distribution of aggregate microstates is binomial (1:4:6:4:1). The most probable macrostate with 2L and 2R has a probability of 6/16. However, if you allow that 3:1 and 1:3 are the same, then the most probable state is one in which three molecules are in one box and one molecule is in the other box (4/16 + 4/16 = 8/16), or a 50% probability that there are three molecules in one box and one in the other.

The probability that all four molecules will be in the same flask (either left or right) is 2/16, and the probability that all four will be in one specific flask is 1/16. In general, the total number of arrangements possible for n molecules distributed between two flasks is $2^n$. In the case of n molecules, there is only one arrangement in which all the molecules are in the left. The algebraic formula for the number of arrangements of n molecules that have L molecules in the left flask and R molecules in the right flask is the fraction:

$$\text{Number of arrangements} = \frac{n!}{(L!\,R!)}$$

where n is the total number of molecules (n = L + R), and n! is the factorial of n.

The formula above shows that the number of arrangements that lead to equal numbers of molecules in each flask increases dramatically as the number of molecules increase. For ten molecules the number of arrangements with five molecules in each flask is 252, compared to only 10 cases with one of the ten molecules in the left flask. With fifty molecules it is $1.26 \times 10^{14}$ times more likely that the system will exist with 25 molecules in each flask than with all 50 molecules in the left flask.

Practical gas samples contain huge numbers of gas molecules (Avogadro's number *is* a big number). Imagine two flasks, one containing a mole of gas and the other a vacuum. When the valve connecting the flasks is opened, the gas expands so that, within experimental measurement error, there are equal numbers of molecules and equal pressure in each flask.

**Informal Example**: A human trapped in a closed room asphyxiates when all the air molecules move randomly to the side of the room away from the human. Possible? In principle, yes. But if there are $10^4$ moles of air in the room, the probability of it happening is about 1 chance in $e^x$, where $x = 6 \times 10^{27}$. Spontaneous human asphyxiation is indeed a very, very rare phenomenon.

The expansion of an ideal gas to fill a vacuum is driven by its molecules seeking the state of maximum statistical probability. In the statistical context, the **second law of thermodynamics** states that a system's entropy increases as its particles seek and find the system's most probable macrostate.

From your personal experience you already know about the tendency of things to mix spontaneously and stay mixed. For example, dye dropped in a swimming pool eventually colors the pool uniformly, and the pool *stays* uniformly colored. Mixing dye in water provides a simple visualization of an entropy change.

Figure 8.2a A few drops of dye solution dropped in water begins the process of diffusing.

Figure 8.2b After a couple of hours the dye has become uniformly distributed throughout the container. It's now "at equilibrium." Yet diffusion continues now as it did before. It's just that we think nothing is happening because nothing *seems* to change. If we could climb aboard a single dye molecule and travel with it, we'd find that over time we'd visit every region of the solution. Chemical equilibrium is like that, too. Microscopic processes continue when macroscopic ones have ceased.

Figures 8.2a and 8.2b show one drop of a dye solution added to a beaker of ordinary water. In Figure 8.2a, the dye has only been added for a few seconds. You can see the swirls of dye, but the bulk of the water is at this time unaffected by the dye. In Figure 8.2b, after a couple of hours, the dye is uniformly dispersed throughout the water.

This mixing is a spontaneous process (more about that shortly) and proceeds because the later system (state 2) is more probable, and so has greater entropy than the initial system (state 1). You've never seen a pool of dyed water spontaneously clear itself, nor the brown caramel color of a cola drink collect on one side of the glass, leaving the other side colorless. Spontaneous unmixing of homogeneous solutions doesn't happen. It's not forbidden by the first law of thermodynamics—but it is forbidden by the second law.

## Entropy of Vaporization

The entropy of vaporization of a pure substance at a specified temperature, T, is its enthalpy of vaporization divided by T in kelvins, $\Delta S = \frac{\Delta H_{vap}}{T}$. Example 8.1 illustrates that while entropy is a subtle concept, its actual calculation is straightforward.

**Example 8.1**: Calculate the entropy of vaporization of butane at its normal boiling point.

Strategy: Normal boiling point means at one bar pressure. Plug into $\Delta S = \frac{\Delta H_{vap}}{T_b}$, so we'll need $\Delta H_{vap}$ and $T_b$. Looking them up in standard tables or on the internet gives $\Delta H = 22.44\ \text{kJ mol}^{-1}$ and $T_b = -0.4°\text{C}$. Be sure to convert the temperature to kelvins. So, T (K) = 273.1 + (-0.4) = 272.7 K.

$$\Delta S = \frac{\Delta H}{T}$$

$$\Delta S = \frac{22440\ \text{J mol}^{-1}}{272.7\ 7\text{K}}$$

$$\Delta S = \underline{83.29\ \text{J mol}^{-1}\ \text{K}^{-1}}$$

Conclusion: When butane boils its entropy increases.

A single entropy of vaporization calculation, such as the one in Example 8.1, is not valuable in isolation. The broader question is: How do entropies of vaporization change from substance to substance? The answer to that question is shown in Table 8.2, in which the abbreviation BP stands for boiling point.

Table 8.2 **Entropies of Vaporization of Some Organic Compounds at their Normal Boiling Points**

| Compound | Formula | BP (°C) | BP K | $\Delta H_{vap}$ at BP kJ mol$^{-1}$ | $\Delta S_{vap}$ J mol$^{-1}$K$^{-1}$ |
|---|---|---|---|---|---|
| Non-Hydrogen Bonding Liquids | | | | | |
| Methane | $CH_4$ | -161.4 | 111.8 | 8.17 | 73.1 |
| Ethane | $C_2H_6$ | -88.6 | 184.6 | 15.54 | 84.1 |
| Propene | $C_3H_6$ | -32.8 | 240.4 | 20.05 | 83.4 |
| Propane | $C_3H_8$ | -42 | 231.2 | 19.04 | 82.3 |
| Butane | $C_4H_{10}$ | -0.4 | 272.8 | 22.44 | 82.3 |
| Isobutane | $C_4H_{10}$ | -11.9 | 261.3 | 21.3 | 80.8 |
| Benzene | $C_6H_6$ | 80.1 | 353.3 | 30.42 | 86.9 |
| Freon®12 | $CCl_2F_2$ | -29.8 | 243.3 | 19.97 | 82.0 |
| Freon®13 | $CClF_3$ | -81.4 | 191.8 | 15.5 | 80.8 |
| Hydrogen Bonding Liquids | | | | | |
| Methanol | $CH_3OH$ | 64.6 | 337.8 | 35.21 | 104.2 |
| Ethanol | $C_2H_5OH$ | 78.4 | 351.6 | 38.56 | 109.7 |
| 1-Propanol | $C_3H_7OH$ | 97.1 | 370.3 | 41.44 | 111.9 |
| 2-Propanol | $C_3H_7OH$ | 82.3 | 355.5 | 39.85 | 112.0 |
| Water | $H_2O$ | 100.0 | 373.2 | 40.66 | 108.9 |

The first thing to notice in Table 8.2 is that the entropies of vaporization of liquids at their normal boiling points in the rightmost column are approximately the same for all liquids. This was recognized in the middle of the nineteenth century and is called Trouton's Rule, after its discoverer the English physics professor Frederick T. Trouton (1863-1922). Why should this be so? Well, vaporization of one mole of any liquid in Table 8.2 involves a few milliliters of the liquid transforming to many liters of gas.

In other words, vaporization is a process in which 10-100 mL of liquid become 10 - 40 L of gas. So, to a first approximation, the increase in randomness is more or less the same, no matter what liquid is being vaporized. The data in Table 8.2 show that. Of course there are individual variations, and it is interesting to see how they can be interpreted.

If you study the entropy values in Table 8.2 you'll see that close chemical relatives have similar entropies of vaporization: propane and butane, for example. But butane and its isomer isobutane have slightly different $\Delta S_{vap}$ values. Methane's value is low because with a low boiling point it makes gas at lower temperature and with a smaller volume than any of the other compounds. At the bottom of the table, alcohols with their characteristic -OH functional groups (see Table 2.7 for an extensive listing of functional groups) have distinctly higher entropies of vaporization. Table 8.2 is divided into two sections: one section for non-hydrogen bonding liquids and one section for hydrogen bonding liquids. We'll return to hydrogen bonding at the beginning of Chapter 10.

Because the entropy change is greater, we may conclude that vaporization of an alcohol destroys more order than does the boiling of propane or butane. Thus, there must be more order in liquids whose molecules contain -OH groups than in liquids composed just of hydrocarbon molecules. That conclusion makes sense, because R-OH molecules have hydrogen bonds among their molecules, and the hydrogen bonds create some organization in solution. When a substance such as ethanol boils, it gains entropy *both* by the destruction of organization in its liquid phase and by its molecular separation into the gas phase. You see that thermodynamics, which itself says nothing about molecules, can actually be quite helpful in understanding molecular behavior.

## Section 8.2 The Third Law of Thermodynamics

The second case of entropy change that we'll take up involves the gain of entropy that a substance experiences simply by being heated. If a substance gains entropy by being heated it, loses it by being cooled, and that simple proposition leads to the third law.

The **third law of thermodynamics** states: the entropy of a perfectly crystalline substance is zero at absolute zero. Why perfectly crystalline? Well a perfect crystal is a perfectly ordered array of entities. So because entropy is a measure of disorder, it's convenient to take this perfectly ordered state as the starting point for measuring the actual entropy of a real substance. Algebraically we write $S^0_0 = 0$. The third law of thermodynamics entered science as Nernst's heat theorem around the beginning of the 20th century. [☸kws +Nernst" +heat theorem"].

The practical consequence of the third law is that it sets up a standard state for entropy values. The third law allows us to relate the entropies of compounds at 25°C to a zero value at -273.15°C. Defining S in this way means elements in their standard states at 298 K have *non-zero* values of their standard entropies (in contrast to $\Delta H^0_f$ values for elements which we saw in Chapter 7 are defined to be zero). Furthermore, all standard entropies are positive (again in contrast to $\Delta H^0_f$ values, which for compounds, can be either positive or negative depending on the compound).

Table 8.3 repeats Table 7.6 and is a highly abbreviated table of thermodynamic properties that includes standard entropies in column 4 (see also Appendix C).

Table 8.3 **Standard Thermodynamic Properties of Pure Substances at 298.1 K**

| Formula (state, name) | $\Delta H^0_f$ kJ mol$^{-1}$ | $\Delta G^0_f$ kJ mol$^{-1}$ | $S^0$ J K$^{-1}$ mol$^{-1}$ | $C^0_p$ J K$^{-1}$ mol$^{-1}$ |
|---|---|---|---|---|
| C(c,graphite) | 0 | 0 | 5.74 | 8.527 |
| C(c,diamond) | 1.9 | 2.9 | 2.38 | 6.113 |
| C(g) | 716.68 | 671.26 | 158.1 | 20.84 |
| CO(g) | -110.53 | -137.17 | 197.67 | 29.42 |
| $CO_2$(g) | -393.51 | -394.36 | 213.74 | 37.11 |
| $CH_4$(g,methane) | -74.81 | -50.72 | 186.26 | 35.31 |
| $C_2H_2$(g,ethyne) | 226.73 | 209.2 | 200.94 | 43.93 |
| $C_2H_4$(g,ethene) | 52.25 | 68.12 | 219.45 | 43.56 |
| $C_2H_6$(g,ethane) | -84.68 | -32.82 | 229.60 | 52.63 |
| $CH_3OH$(g,methanol | -200.66 | -162.00 | 239.70 | 43.89 |
| $CH_3OH$(l,methanol) | -238.66 | -166.36 | 126.8 | 81.6 |
| $C_2H_5OH$(g,ethanol) | -235.1 | -168.49 | 282.70 | 65.44 |
| $C_2H_5OH$(l,ethanol) | -277.69 | -174.78 | 160.7 | 111.46 |
| $H_2$(g) | 0 | 0 | 130.68 | 28.82 |
| H(g) | 217.97 | 203.25 | 114.71 | 20.78 |
| $H^+$(g) | 1,536.2 | -- | -- | -- |
| $H_2O$(g) | -241.82 | -228.57 | 188.82 | 33.577 |
| $H_2O$(l) | -285.83 | -237.13 | 69.91 | 75.29 |
| $H_2O_2$(g) | -136.31 | -105.57 | 232.7 | 43.1 |
| $H_2O_2$(l) | -187.78 | -120.35 | 109.6 | 89.1 |
| $O_2$(g) | 0 | 0 | 205.14 | 29.36 |
| $O_3$(ozone) | 142.7 | 163.2 | 238.93 | 39.2 |
| O(g) | 249.17 | 231.73 | 161.055 | 21.91 |

Carefully examine the $S^0$ values in Table 8.3 above. Graphite and diamond have low values because they are solids. Three liquid-gas pairs in the table, water, methanol, and ethanol, show that each gas has substantially more entropy than the liquid from which it formed at standard temperature. So even the limited data of Table 8.3 demonstrate that S(g)> S(l) > S(s), as we noted earlier. Entropy of vaporization is the difference between a substance's gas and liquid entropies. For methanol it's 239.70 - 126.8, or 112.9 J mol$^{-1}$ K$^{-1}$. In the Trouton's rule table (Table 8.2) the value for the entropy of vaporization of methanol is 104.2 J mol$^{-1}$ K$^{-1}$. Why the slight difference? The value in the table is at 25°C, and the Trouton's rule value is at 64.6°C.

> Entropy can be thought of as a measure of randomness. At 298 K gases are more random than liquids and liquids more random than solids. Examination of the table of standard entropies of pure substances at 298 K shows:
>
> $$S(g) > S(l) > S(s)$$

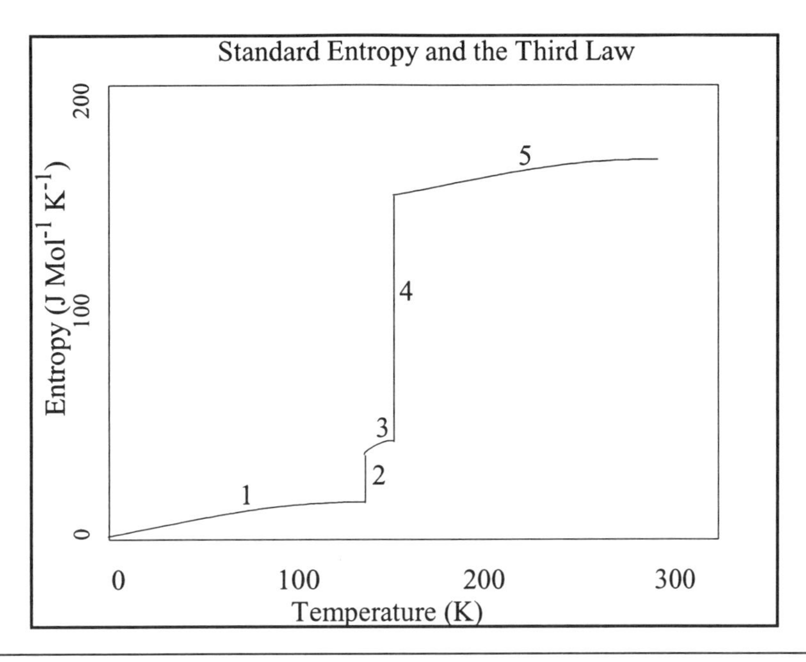


Figure 8.3 The temperature dependence of the entropy of a pure substance as it warms from absolute zero. The entropy gain of a gas can be imagined to have arisen in five distinct processes or steps.

For a substance that is a gas at 25°C, Figure 8.3 to shows how it "accumulates" its entropy as its temperature rises farther and farther above absolute zero. For a substance that is a gas at 298 K and 1 bar, its standard entropy is the sum of the following five steps. The symbol $T_f$ means melting (or freezing) point and the symbol $T_b$ means boiling point.

| | | |
|---|---|---|
| 1. Solid (1 mol) warms from 0 K to the melting point. | $\Delta S = \int C_p/T\ dT$ | $\Delta S_1$ |
| 2. The solid melts at $T_f$. | $\Delta S_{melting} = \Delta H_{melting}/T_{melting}$ | $\Delta S_2$ |
| 3. The liquid warms from $T_f$ to $T_b$. | $\Delta S = \int C_p/T\ dT$ | $\Delta S_3$ |
| 4. The liquid boils at $T_b$. | $\Delta S_{vaporization} = \Delta H_{boiling}/T_{boiling}$ | $\Delta S_4$ |
| 5. The gas warms to 298 K (standard temperature) | $\Delta S = \int C_p/T\ dT$ | $\Delta S_5$ |

For the five steps: $\Delta S = \Delta S_1 + \Delta S_2 + \Delta S_3 + \Delta S_4 + \Delta S_5$. It's easy to see from Figure 8.3 that if a substance is a gas it will have a much larger standard entropy than if it is a solid or liquid.

## The Standard Entropy Change of Reactions

From a table of standard entropies we can calculate the standard entropy change of any process. The **standard entropy change** ($\Delta S^0$) of a reaction is obtained by subtracting the sum of the standard entropies of the reactants from the sum of the standard entropies of the products of a reaction.

$$\Delta S^0 = \sum_{\text{moles}} S^0\,(\text{products}) - \sum_{\text{moles}} S^0\,(\text{reactants})$$

The use of the above equation is illustrated in Example 8.2.

**Example 8.2**: Calculate the standard entropy change, $\Delta S^0$, at 298 K of the reaction: $C_2H_6(g) + O_2(g) \rightarrow CO_2(g) + H_2O(g)$. The equation is unbalanced and the water produced is in the gaseous state of aggregation.

Strategy: Balance the reaction equation and then apply the $\Delta S^0$ formula:

$$\Delta S^0 = \sum_{\text{moles}} S^0\,(\text{products}) - \sum_{\text{moles}} S^0\,(\text{reactants})$$

The balanced equation is:

| | $1\ C_2H_6(g)$ | + | $3.5\ O_2(g)$ | $\rightarrow$ | $2\ CO_2(g)$ | + | $3\ H_2O(g)$ |
|---|---|---|---|---|---|---|---|
| $S^0$ | 229.60 | | $3.5 \times 205.138$ | | $2 \times 213.74$ | | $3 \times 188.825$ |

Below the table are written the standard entropies of the reactants and products taken from Table 8.3. The stoichiometric coefficients of the reaction equation are used as multipliers to convert molar entropy to actual entropy for each reactant or product. Note that the element oxygen has a *non-zero* standard entropy.

Substituting into the $\Delta S^0$ equation gives:

$$\Delta S^0 = (2 \times 213.74 + 3 \times 188.825) - (229.60 + 3.5 \times 205.138)$$

$$\Delta S^0 = 994.00 - 947.58 \quad \text{or} \quad \Delta S^0 = \underline{+46.42\ \text{J mol}^{-1}\ \text{K}^{-1}}$$

A small entropy increase is reasonable for a gas reaction that goes from 4.5 moles of reactants to 5 moles of products.

Because gases have considerably more standard entropy per mole than solids or liquids, the entropy changes of reactions are heavily influenced by gases. Reactions that form gaseous products from solids and liquids have large positive entropy changes; conversely, reactions in which gases form solids or liquids have large negative entropy changes. Reactions that have small volume changes have small entropy changes.

**Example 8.3**: At 25°C the standard molar entropies of $N_2(g)$, $H_2(g)$, and $NH_3(g)$ are respectively 191.5, 130.7, and 192.5 J $K^{-1}$ $mol^{-1}$. Calculate the standard entropy change, $\Delta S^0$, for the reaction: $N_2(g) + 3\ H_2(g) \rightarrow 2\ NH_3(g)$

Strategy. Substitute the given values into the expression

$$\Delta S^0 = \sum_{\text{moles}} S^0\,(\text{products}) - \sum_{\text{moles}} S^0\,(\text{reactants})$$

$$\Delta S^0 = 2 \times (192.5\ \text{J K}^{-1}\ \text{mol}^{-1}) - 1\ \text{mol} \times (191.5\ \text{J K}^{-1}\ \text{mol}^{-1}) - 3\ \text{mol} \times (130.6\ \text{J K}^{-1}\ \text{mol}^{-1})$$

$$= \underline{-198.3\ \text{J K}^{-1}}$$

$\Delta S^0$ for this reaction is negative. The reaction corresponds to a reduction in the number of moles of gas (4 moles of gaseous reactants make 2 moles of gaseous products); the decrease in disorder corresponds to an entropy decrease.

## Section 8.3 Spontaneity and Chemical Reactions

### Position of Equilibrium of a Chemical Reaction

The **position of equilibrium** of a chemical reaction system refers to the balance between the amounts of products and reactants present in the system.

In the simplest conceivable reaction, $H_2 \rightleftarrows 2$ H, the equilibrium position is measured by the balance of $H_2$ and H. At ordinary temperatures (300 K) the reaction system is almost entirely in the form of $H_2$ molecules; the position of equilibrium **lies to the left**. At high temperatures (5000 K) the system is almost entirely in the form of H atoms; the position of equilibrium **lies to the right**. In a chemical reaction system that is **in or at equilibrium** there is no change over time of the pressures or concentrations of the reactants or products. However, the forward and reverse reactions have not stopped. At equilibrium the rates of the forward and reverse reactions have simply become equal.

At intermediate temperatures comparable amounts of $H_2$ and H coexist in the reaction system. This is a **reversible** reaction and its precise balance of concentrations among the reactants and products, i.e., where the system's equilibrium *lies,* is controlled by the temperature, pressure, and the thermodynamic properties of the reactants and products.

Incidentally, the Kinetic-Molecular Theory of gases helps explain why hydrogen's molecules break up into atoms at high temperature: higher temperature increases molecular speed, and so intermolecular collisions grow ever more violent as the temperature increases. Eventually, if the temperature is sufficiently raised, the collisions are so violent that molecules can hardly exist, only atoms. In this situation we say the position of equilibrium lies **far to the right**; we also say that the reaction has essentially **gone to completion**.

**Chemical Equilibrium and Thermodynamics in a Nutshell**

Chemical equilibrium is both a balance of concentrations and a balance between entropy change and energy change: the reaction $H_2 \rightarrow 2H$ is endothermic; energy must be added to the system to break the H-H bonds. But simultaneously the entropy of the system increases: one mole of $H_2$ gas is converting to two moles of H gas. Chemical equilibrium in $H_2 \rightleftarrows 2H$ clearly depends on *both* the energy change and the entropy change of the reaction. The system's temperature is the other important factor. As we will see later, the Gibbs equation ($\Delta G = \Delta H - T\Delta S$) that describes reaction systems in chemical equilibrium at constant pressure involves three factors: energy, entropy, and temperature.

## Spontaneity

The word spontaneity is another of those superficially everyday words that has been appropriated by thermodynamics and given a special meaning. A **spontaneous process** or a **spontaneous reaction** is one that has an inherent, thermodynamic tendency to occur. For the $H_2 \rightleftarrows 2$ H reaction with the partial pressures at 1 bar we can say:

At low temperatures the reaction $2\ H \rightarrow H_2$ is spontaneous
At high temperatures the reaction $H_2 \rightarrow 2\ H$ is spontaneous

Chemical equilibrium is studied by mixing reactants together and measuring what happens to the reactant and product concentrations as time passes. If the reaction system is not at equilibrium, its reactant and product concentrations will change to approach equilibrium. The change that occurs is called a **shift** in equilibrium and is a spontaneous process. A **spontaneous process** is a reaction system's approach to equilibrium.

**The Combustion Reaction of Propane Gas is Spontaneous**

If it's spontaneous, how come the propane doesn't catch fire when we start the gas flow to the grill? Well, because to be thermodynamically spontaneous doesn't mean to be spontaneous in the ordinary sense.

Once the flame's lit you can see that combustion is spontaneous. But to get the flame started you have to supply a tiny, initiation energy. Technically, it's called the activation energy and we'll meet it in Chapter 15.

The approach to equilibrium for the $H_2 \rightleftarrows 2H$ reaction at an intermediate temperature is shown in Figure 8.4. Sloping lines of concentration, changing with time, mean that the reaction is shifting. When the concentration lines flatten out, equilibrium has been reached.

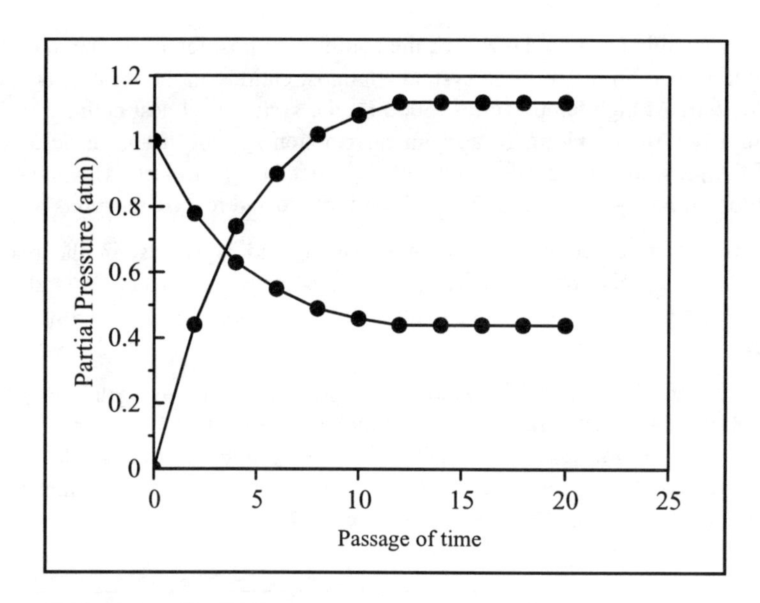


Figure 8.4 The approach to equilibrium for the $H_2 \rightleftarrows 2H$ reaction. The reaction begins with a pressure of $H_2$ of one atmosphere and drops along the path of the falling line. The partial pressure of H follows the rising line. Note, in accord with the reaction's stoichiometry, the slope of the $P_H$ line is opposite in sign and double in value to the slope of the $P_{H_2}$ line.

Take some time to study the diagram in Figure 8.4 and you'll see that where the process begins on the left (at zero time) the $H_2$ pressure is 1 and the H pressure is zero. As the approach to equilibrium spontaneously begins, the $H_2$ pressure drops while simultaneously the H pressure increases at twice the rate. The rate of formation of H is twice the rate of decomposition of $H_2$ because of the 2 in the balanced chemical equation; the H pressure at any time is twice the drop in the $H_2$ pressure. For example, when the $H_2$ pressure has dropped from 1 atm to 0.7 atm the H pressure has risen to $2 \times (1 - .7) = .6$ atm.

The adjective spontaneous applied to a process implies *nothing* about the rate at which the system approaches equilibrium. Some spontaneous processes occur rapidly, others slowly, and some never get started because they need energy to activate them. For example, it is good that the spontaneous reaction of gasoline with oxygen requires a spark to initiate combustion—were this not the case we could not carry fuel around in our gas tanks. Gasoline is spontaneously combustible, but it burns only if given a kick of energy to get combustion started. The topic of how fast chemical reactions take place is called *chemical kinetics* (we'll take up this subject in Chapter 15). We already met the example of a dye spontaneously spreading throughout water; here are two more physical processes that illustrate the nature of spontaneity:

**Situation 1** A gas is exposed to a vacuum. The gas spontaneously expands to fill the vacuum. The reverse process, in which a gas spontaneously contracts to form a vacuum, never occurs.

**Situation 2** Heat flows from hot to colder regions, never the reverse. Transfer of heat from a hot body to a cold one is spontaneous; the reverse is unnatural.

At any time, a chemical reaction system will be engaged in one of three processes: 1. reactant concentrations are decreasing while product concentrations are increasing, 2. product concentrations are decreasing while reactant concentrations are increasing, and 3. the amounts of reactants and products are unchanging with time. What we say about these three situations is summarized in Table 8.4.

| Table 8.4 **Possible Conditions of a Reactive Chemical System** | | | |
|---|---|---|---|
| What is Happening | We say either… | or… | or |
| 1. Reactant concentrations are decreasing while product concentrations are increasing | Reaction is spontaneous in the forward direction | Reaction is nonspontaneous in the reverse direction | The reaction is shifting right. |
| 2. Product concentrations are decreasing while reactant concentrations are increasing | Reaction is spontaneous in the reverse direction | Reaction is nonspontaneous in the forward direction | The reaction is shifting left. |
| 3. The amounts of reactants and products are unchanging with time | Reaction is in equilibrium | Reaction is nonspontaneous in either direction | The reaction is not shifting. |

One footnote to Table 8.4. Suppose we mix propane gas and oxygen. That system is clearly not reacting, yet it has enormous chemical spontaneity. Isn't that an example of an equilibrium system? No, it's not. The reaction is too slow to be detected at ordinary temperature, but it will proceed vigorously if given a tiny initiation. We give this situation a special name: we say that, prior to ignition, the system is in a **metastable state**.

## Entropy Change and Spontaneous Processes

Every spontaneous process is accompanied by an increase in disorder. The combined entropy of a system and its surroundings goes up during spontaneous processes. For example, when a gas expands, the location of any of its molecules becomes more uncertain, so the gas is less ordered. Similarly, when a gas is heated, the motion of its molecules becomes more random and the gas's entropy increases.

Spontaneous processes always bring a system toward equilibrium. So it follows from the second law of thermodynamics that the equilibrium state of any system is the state that produces maximum entropy in the universe (that's how thermodynamicists actually talk). In the equation below the term **surroundings** just means everything that isn't the system. Surroundings is another one of those specialist terms; previously I've often used environment as a substitute. For any change of state in a in a thermodynamic system:

$$\Delta S_{universe} = \Delta S_{system} + \Delta S_{surroundings} \text{ and if,}$$

$\Delta S_{univ}$ is Positive—Process is spontaneous; reverse process is nonspontaneous

$\Delta S_{univ}$ is Negative—Process is nonspontaneous; reverse process is spontaneous

**Example 8.4**: Water spontaneously turns to ice when placed in a freezer. Discuss the entropy changes that occur during the process of freezing.

Strategy: Apply the second law of thermodynamics and common sense.

The water clearly experiences a decrease in its entropy both because its temperature is lowered and because it turns from liquid to solid. However, heat is extracted from the freezer by the refrigerant fluid and is in turn released at the plate coil behind the refrigerator. That heats the air, increasing its entropy. The relatively small decrease in entropy of the system is overbalanced by an entropy increase in the environment.

Whenever a physical process or a chemical reaction occurs spontaneously, you can be sure that the entropy of the universe is increasing. But it is, to say the least, inconvenient to have to measure the entropy change of the universe to discover if the reaction at hand is spontaneous. What we want is a system-based property for spontaneity—not a universe-based property! For this purpose Gibbs introduced the thermodynamic property called free energy.

# Section 8.4 Free Energy and the Spontaneity of Chemical Reactions

The concept of free energy is yet another thermodynamic idea that grows out of Carnot's analysis of the efficiency of engines. Energy in a hot system is useful because it can be partially converted to work of expansion, resulting in an expanded gas at a lower temperature. The considerable energy that remains in the now expanded system is less available to do work because that requires a still lower temperature. Eventually, after some number of steps, the lower temperature becomes the temperature of the environment, and the extraction of work from the system is no longer possible. Each step in the sequence represents a decrease in the quality of the energy. Every natural spontaneous process is accompanied by a similar degradation. We say that at each step the free energy of the system is decreasing.

Reiterating, the spontaneity of chemical reactions at constant temperature and pressure is controlled by two factors: (1) being exothermic tends to make a reaction spontaneous (the reaction's $\Delta H$ is negative), and (2) causing an entropy increase tends to make a reaction spontaneous (the reaction's $\Delta S$ is positive). It is the balance of these two tendencies that controls whether any given reaction is spontaneous, nonspontaneous, or at equilibrium,

$G = H - TS$ is the equation that defines G, the **Gibbs free energy function**. G is a state function because H, T, and S are state functions. G is useful because it provides a criterion for the spontaneity of processes that take place at constant temperature and pressure. Actually, G itself is never measured. Rather, it is the change in G, $\Delta G$, that we use. The Gibbs equation may be written:

$$\Delta G = \Delta H - T\Delta S$$

for processes that occur at constant pressure and temperature, where G, H, and S are properties of the *system*. For any constant T and P process, it can be shown that if $\Delta S_{universe}$ is positive then $\Delta G_{system}$ is negative (see, for example, Problem 8.58). So we can now state the thermodynamic criteria for reaction spontaneity based on a thermodynamic property of our system:

$\Delta G$ is positive, the reaction is **nonspontaneous** (the reverse reaction is spontaneous)
$\Delta G$ is negative, the reaction is **spontaneous** (the reverse reaction is nonspontaneous)
$\Delta G$ is zero, the reaction is in **equilibrium**

When is it a *free* energy function? Think about the fuel burning reaction that provides heat energy to run a power plant. Not all of that energy is "free" or "available" to do work. In a power plant, we can use some (the available energy), but not all (the unavailable energy). The unavailable goes as waste heat into the river or up the cooling towers of the plant. If a plant is 33% efficient it's 67% inefficient—so when you heat your house electrically you heat some river or the atmosphere twice as much! Free energy is available energy.

The origin and definition of $\Delta G$ may still seem a little mysterious to you. But you should be absolutely clear where values of $\Delta G^0$ (at standard states, and a specified temperature—usually 298 K) for any and all reactions come from—they come from NIST (or more locally from Table 8.3 and Appendix C). As we've seen before, while thermodynamic ideas are complex, calculating thermodynamic properties from standard data tables is simple arithmetic; so obtaining $\Delta G^0$ for any standard state reaction is straightforward. Using the standard thermodynamics data sources you follow a procedure identical to the one you've already seen for $\Delta H^0$. The equation is:

$$\Delta G^0 = \sum_{\text{moles}} \Delta G^0_f\,(\text{products}) - \sum_{\text{moles}} \Delta G^0_f\,(\text{reactants})$$

where $\Delta G^0_f$ is any substance's **standard free energy of formation**. $\Delta G^0_f$ is defined in a perfectly familiar manner: the **free energy of formation** of a compound, written as $\Delta G_f^o$, of a substance is the free energy change when the compound in its standard state is formed from its elements in their standard states. The equation for the process is balanced and written so that the only product is one mole of the compound. Here are examples of free energy of formation reactions, with the elements in their standard states on the left and the compound in its standard state on the right:

Standard free energy of formation of $CO(g)$:

$$C(graphite) + \tfrac{1}{2}O_2(g) \rightarrow CO(g) \qquad \Delta G_f^0 = -137.17 \text{ kJ mol}^{-1}$$

Standard free energy of formation of $H_2O(l)$:

$$H_2(g) + \tfrac{1}{2}\, O_2(g) \rightarrow H_2O(l) \qquad \Delta G_f^0 = -237.13 \text{ kJ mol}^{-1}$$

Standard free energy of formation of liquid ethanol:

$$2\, C(s) + 3\, H_2(g) + \tfrac{1}{2}\, O_2(g) \rightarrow C_2H_6O(l) \qquad \Delta G_f^0 = -174.78 \text{ kJ mol}^{-1}$$

Examples 8.5 and 8.6 show how standard free energies from data bases are used to calculate the standard free energy changes of chemical reactions.

**Example 8.5**: At 25°C, the standard free energies of formation of NO(g), $O_2$(g), and $NO_2$(g) are respectively 86.8, 0, and 51.9 kJ $mol^{-1}$. Calculate the standard free energy change for the reaction NO(g) + ½ $O_2$(g) → $NO_2$(g)

Strategy. Substitute the given values, multiplied by the equation's coefficients, into the expression

$$\Delta G^0 = \underset{\text{moles}}{\sum} \Delta G^0_f(\text{products}) - \underset{\text{moles}}{\sum} \Delta G^0_f(\text{reactants})$$

$$\Delta G^0 = 1 \text{ mol} \times (51.9 \text{ kJ mol}^{-1}) - 1 \text{ mol} \times (86.8 \text{ kJ mol}^{-1}) - \tfrac{1}{2} \text{ mol} \times (0 \text{ kJ mol}^{-1})$$

$$= \underline{-34.9 \text{ kJ mol}^{-1}}$$

**Example 8.6**: Calculate the standard free energy of combustion of ethane using standard free energy of formation data. Assume that the products are gaseous water and carbon dioxide.

Strategy: Write the balanced equation, look up the appropriate values in Table 8.3, and substitute into the equation

$$\Delta G^0 = \underset{\text{moles}}{\sum} \Delta G^0_f(\text{products}) - \underset{\text{moles}}{\sum} \Delta G^0_f(\text{reactants})$$

$$C_2H_6(g) + 3.5\, O_2(g) \rightarrow 2\, CO_2(g) + 3\, H_2O(g)$$

$$\Delta G^0_f \text{ (values from table)} \quad -32.82 \quad + \quad 0 \quad \rightarrow \quad -394.359 \quad -228.572$$

$$\text{Thus, } \Delta G^0 = (2 \times -394.359 + 3 \times -228.572) - (-32.82 + 3.5 \times 0)$$

$$\Delta G^0 = (-1474.434) - (-32.82) = \underline{-1441.61 \text{ kJ mol}^{-1}} \text{ (of ethane)}$$

**Example 8.7**: Consider the combustion reaction: $C_2H_6$(g) + 3.5 $O_2$(g) → 2 $CO_2$(g) + 3 $H_2O$(g). In example 8.2 we calculated that $\Delta S^0$ of this reaction is +46.42 J $mol^{-1}$ $K^{-1}$. Calculate $\Delta H^0$ using standard enthalpies of formation and then use the Gibbs equation: $\Delta G^0 = \Delta H^0 - T\Delta S^0$ to calculate $\Delta G^0$. Compare this value with the one calculated from tabulated $\Delta G^0_f$ values in the preceding example (8.6).

Strategy: First find the $\Delta H^0$ of the reaction from $\Delta H^0_f$ values in Table 8.3, using the equation below. Plug that value and the $\Delta S^0$ from Example 8.2 into the Gibbs equation at standard conditions.

$$\Delta H^0 = \underset{\text{moles}}{\sum} \Delta H^0_f(\text{products}) - \underset{\text{moles}}{\sum} \Delta H^0_f(\text{reactants})$$

$$\Delta H^0 = (2 \times -393.509 + 3 \times -241.818) - (-84.68) = 1427.79 \text{ kJ mol}^{-1}$$

calculate $\Delta G^0$ from Gibbs equation: $\Delta G^0 = \Delta H^0 - T\Delta S^0$

$$\Delta G^0 = -1427.79 \text{ kJ mol}^{-1} - 298.15 \text{ K} \times 0.04642 \text{ kJ mol}^{-1} \text{ K}^{-1}$$

$$= -1427.79 \text{ kJ mol}^{-1} - 13.84 \text{ kJ mol}^{-1} = \underline{-1441.63 \text{ kJ mol}^{-1}}$$

The value of -1441.63 kJ $mol^{-1}$ calculated in Example 8.7 agrees well with the value of -1441.61 kJ $mol^{-1}$ calculated from standard free energies of formation in Example 8.6. The good agreement reflects the efforts of the members of the standards community to make the thermodynamic data bases as precise and accurate as possible.

The values of $\Delta H^0$ and $\Delta S^0$ of a chemical reaction do not depend strongly on temperature. Although it's not strictly correct, it is instructive to imagine that $\Delta H^0$ and $\Delta S^0$ are temperature independent and examine to the effect of changing temperature on the reaction's equilibrium position. We can distinguish four cases, as shown in the Table 8.5.

Table 8.5

**The Effect of Temperature on a Reaction's Equilibrium Position—Four Cases**

$$\Delta G^0 = \Delta H^0 - T\Delta S^0$$

If $\Delta H^0$ is negative and $\Delta S^0$ is positive then $\Delta G^0$ is always negative.
Reaction spontaneous to the right under standard conditions at *all* temperatures.

If $\Delta H^0$ is negative and $\Delta S^0$ is negative the sign of $\Delta G^0$ depends on temperature.
Reaction spontaneous to the right under standard conditions at *low* temperatures.

If $\Delta H^0$ is positive and $\Delta S^0$ is positive the sign of $\Delta G^0$ depends on temperature.
Reaction spontaneous to the right under standard conditions at *high* temperatures.

If $\Delta H^0$ is positive and $\Delta S^0$ is negative then $\Delta G^0$ is always positive.
Reaction spontaneous to the left under standard conditions at *all* temperatures,
alternatively, reaction nonspontaneous to the right at all temperatures.

Example 8.8, which concludes this section, shows a rather peculiar application of thermodynamics, the calculation of water's boiling point from standard thermodynamic data by means of the Gibbs equation. However, the example confirms that water's $\Delta H$ and $\Delta S$ of vaporization are not strongly temperature dependent.

**Example 8.8**: 1. Use standard thermodynamic data at 298 K to calculate the enthalpy of vaporization of water. (The standard enthalpy of formation of liquid water is -285.83 kJ mol$^{-1}$; the standard enthalpy of formation of gaseous water is -241.82 kJ mol$^{-1}$.) 2. Use standard thermodynamic data at 298 K to calculate the standard free energy of vaporization of water. (The standard free energy of formation of liquid water is -237.13 kJ mol$^{-1}$; the standard free energy of formation of gaseous water is -228.57 kJ mol$^{-1}$.) 3. Use standard thermodynamic data at 298 K to calculate the standard entropy change of the vaporization of water; (the standard entropy of liquid water is 69.91 J mol$^{-1}$ K$^{-1}$; the standard entropy of gaseous water is 188.82 J mol$^{-1}$ K$^{-1}$.) 4. Estimate the ordinary boiling point of water; i.e. the boiling point at one atmosphere pressure.

Strategy: Substitute the values into the appropriate expressions for $\Delta H^0$, $\Delta G^0$, and $\Delta S^0$and obtain those values. Then, substitute those values into the Gibbs equation and solve for temperature. The boiling point of water is that temperature (at one atmosphere) when liquid and gaseous water are in equilibrium. For the equilibrium $H_2O(l) = H_2O(g)$, $\Delta G^0$ is zero, so we can calculate T from the Gibbs equation.

$$1.\ \Delta H^0 = -241.82 \text{ kJ mol}^{-1} - (-285.83 \text{ kJ mol}^{-1}) = +44.01 \text{ kJ mol}^{-1}$$

$$2.\ \Delta G^0 = -228.57 \text{ kJ mol}^{-1} - (-237.13 \text{ kJ mol}^{-1}) = +8.56 \text{ kJ mol}^{-1}$$

$$3.\ \Delta S^0 = 188.82 \text{ J mol}^{-1} \text{K}^{-1} - 69.91 \text{ J mol}^{-1} \text{K}^{-1} = 118.91 \text{ J mol}^{-1} \text{K}^{-1}, \text{ or } 0.11891 \text{ kJ mol}^{-1} \text{K}^{-1}$$

4. So far all straightforward, but now we come to the interesting part of this example. At water's normal boiling point (1 bar), liquid and gaseous water are in equilibrium. So we know that $\Delta G^0 = 0$. We'll substitute that value and assume that neither $\Delta H^0$ nor $\Delta S^0$ changes with temperature. Doing that gives:

$$\Delta G^0 = \Delta H^0 - T\Delta S^0$$

$$0 = \Delta H^0 - T\Delta S^0, \text{ and } T = \frac{\Delta H^0}{\Delta S^0}$$

$$T = \frac{+44.01 \text{ kJ mol}^{-1}}{0.11891 \text{ kJ mol}^{-1}\text{K}^{-1}} = \underline{370.1 \text{ K or } 97.0°\text{C}}$$

That result, being pretty close to the expected 100°C, confirms our assumption of the lack of temperature dependence of $\Delta H$ and $\Delta S$.

## Section 8.5 Equilibrium Constants

The next step in our development is to relate equilibrium constants to standard thermodynamic data and specifically to $\Delta G^0$. For the hydrogen equilibrium ($H_2 \rightleftarrows 2H$) at a fixed temperature there is an equilibrium constant $K_p$. We'll see how to generalize in just a moment, but here again is the equilibrium constant for hydrogen's dissociation with the reactant and product concentrations measured in pressures:

$$K_p = \frac{(P_H)^2}{(P_{H_2})}$$

Recall that an equilibrium constant is an experimental fact. Example 8.9 illustrates the calculation of the $K_p$ value for the $H_2 \rightleftarrows 2H$, equilibrium depicted in Figure 8.4.

**Example 8.9**: Calculate $K_p$ for the reaction $H_2 \rightleftarrows 2H$, using the equilibrium data shown in Figure 8.4.

Strategy: Recognize that equilibrium has been reached when the lines become flat, showing that the partial pressures of the reactant and product are unchanging. Take the data from the figure (the units are atmospheres) and substitute into $K_p = \frac{(P_H)^2}{(P_{H_2})}$.

Estimating from the figure, $P_{H_2} = 0.45$ and $P_H = 1.1$

$$K_p = \frac{(P_H)^2}{(P_{H_2})} \text{ or } K_p = \frac{1.1^2}{0.45} = \underline{2.7} \text{ atm}$$

The same basic mathematical approach applies to our study of all equilibrium chemical systems. The following box shows that chemical equilibrium concepts have wide applications.

**Some Topics in Chemical Equilibrium**

One of the difficulties in teaching chemical equilibrium is that it is hard to do justice to both the formidable intellectual content of the topic and its incredible breadth of application. It's hard for experienced chemists, let alone beginning engineering students, to see the role of $\Delta G^0$ in our daily lives. Here are a few of them.

**Living Systems**

All life is an exercise in chemical equilibrium. In biology there's even a word for it: *homeostasis*, or the normal response of an organism to maintain its internal stability over a wide range of environmental conditions. For biochemistry, $\Delta G$ is overwhelmingly the most important thermodynamic function; first, because most biochemistry takes place at 1 atmosphere and near 298 K, and second, because biochemical processes occur at or near chemical equilibrium. For example, the acidity of human blood, as measured by its pH, is held in a very narrow range by chemical equilibria at work in the blood. (The equilibrium effect is called buffering.) The equilibrium between the free and oxygenated forms of the oxygen carrier hemoglobin (Hb), $Hb + O_2 \rightleftarrows HbO_2$, serves as the basis of oxygen transport and exchange in animals.

**Chemical Engineering and the Materials Economy**

During the 20th century, humans living in the developed economies have come to control their material environment in a way unimaginable to former generations. At the beginning of the century it was chemical equilibrium principles, applied by engineers to the problems of fertilizer manufacture, that both laid the foundations of the subject and began a mammoth industry. In the second half of the 20th century chemical engineering applied the same principles to develop manufacturing methods for the entire range of synthetic chemical materials and their precursors. Glance at the list of compounds in Table F-1 in Appendix F. Note the enormous quantities of each of them that support our daily lives, and consider that each is made in a plant operated by chemical engineers applying chemical equilibrium principles. For that latter reason, the principles outlined in this chapter feature prominently in any elementary chemical engineering text book.

**The Future of Planet Earth**

Surely, you say, chemical equilibrium can't be *so* important that the very future of the planet depends on our understanding it? Well, maybe. Air pollution modeling, the behavior of the ozone layer, and understanding greenhouse gases and global warming, are all problems in applied gas-phase equilibrium. To the extent that human activities are changing the atmosphere in a manner that threatens the future well-being of the planet, to that extent is it important to understand chemical equilibrium.

In Chapter 9 we'll discuss specific topics of chemical equilibrium and work a number of examples. While those examples will not be quite as overarching as the ones in this box, the principles of chemical equilibrium are the same for all reactions and processes.

## The General Reaction Equilibrium Expression

Chemical equilibrium reactions take place in either the gas or the liquid phase. Both types are enormously important. Gas concentrations are usually expressed as partial pressures; solution concentrations are usually expressed as molarities. Here is a general two-reactant, two-product gas phase reaction:

$$a\ A(g) + b\ B(g) \rightleftarrows c\ C(g) + d\ D(g)$$

The general equilibrium constant expression for such a gas phase reaction in an existing state of equilibrium is

$$K_p = \frac{(P_C)^c(P_D)^d}{(P_A)^a(P_B)^b}$$

The general equilibrium constant expression written in terms of molarity is

$$K_c = \frac{[C]^c[D]^d}{[A]^a[B]^b}$$

For a reaction system composed of ideal gases, $K_p = K_c(RT)^{\Delta n}$, where $\Delta n$ ($\Delta n = c + d - a - b$) is the difference between the total number of moles of product gases and the total number of moles of reactant gases. For an ideal gas, molarity = n/V = P/RT, or P = MRT. Use the relationship $P_i = M_iRT$ to derive for yourself the equation $K_p = K_c(RT)^{\Delta n}$ that relates a gas phase equilibrium reaction constant expressed in partial pressure units to the same constant expressed in molarity units.

For reactions that take place in liquid solution, chemical equilibrium can be handled using the equation for $K_c$ above, with the molarities now being the concentrations of the reactants and products in the liquid phase.

## Relationship Between ΔG and the Equilibrium Constant

In this section we state without proof the fundamental defining equation that links the free energy change and the reactant and product concentrations in a chemically reactive thermodynamic system at constant temperature and constant pressure. This is perhaps the most important single relationship in all of chemical thermodynamics.

Consistent with experimental evidence and the laws of thermodynamics, the fundamental defining equation that relates the free energy changes of a chemical reaction to the concentrations of the components of the reaction system, at a given temperature T, is:

$$\Delta G = \Delta G^0 + RT \ln\frac{Q}{Q^0} \quad \text{where the units are usually kJ mol}^{-1}$$

In words, the equation states that the free energy change of the reaction ($\Delta G$) equals the standard free energy change of the reaction ($\Delta G^0$) plus the quantity $RT \ln\frac{Q}{Q^0}$, where Q is the system's **reaction quotient**. The reaction quotient Q is like K in form and has the same units as K, but Q has a different value than K *unless the system has already arrived at equilibrium*. For a reaction in which the products and reactants are ideal gases, with their concentrations expressed as partial pressures in the units bars, and with $\Delta n = c + d - a - b$, the reaction quotient is labeled $Q_p$ and is written with its units as:

$$Q_P = \frac{(P_C)^c(P_D)^d}{(P_A)^a(P_B)^b} \quad (\text{bar})^{\Delta n}$$

$Q_P{}^0$ is the system's standard reaction quotient, or the system's Q when all the reactants and products are in their standard states of 1 bar pressure. $Q_P{}^0$ has a value of 1 and the same units as K and $Q_P$. The expression for $Q_p{}^0$, the standard reaction quotient of an equilibrium reaction, in pressure units is:

$$Q_P{}^0 = \frac{(1\ \text{bar})^c(1\ \text{bar})^d}{(1\ \text{bar}\ ^a(1\ \text{bar})^b} = (1\ \text{bar})^{c+d-a-b} = (1\ \text{bar})^{\Delta n}$$

The fundamental defining equation that relates the free energy change of a gas phase chemical reaction to the partial pressures of the components of the reaction system, at a specified temperature T, is thus:

$$\Delta G = \Delta G^0 + RT \ln\frac{Q_P}{Q_P^0} \quad \text{kJ mol}^{-1}$$

In practice, chemists usually simplify the above equation to $\Delta G = \Delta G^0 + RT \ln Q_P$, while simultaneously adopting the convention that $Q_P$ has no units. What's really been done, of course, has been to divide $Q_P$ by $Q_P{}^0$, which means divide by one and cancel the units. We'll follow the standard practice from now on and just write the equation as $\Delta G = \Delta G^0 + RT \ln Q$, omitting the units of Q when we substitute values, though both Q and K have units when we write them as freestanding values.

We now state explicitly the thermodynamic equation (first, expressed in pressure units for a gas phase reaction and second in molarity units for a either a gas or liquid phase reaction) that links the concentrations of the reactants and products in chemical systems to the reaction's free energy change. It is

$$\Delta G = \Delta G^0 + RT \ln Q_p \quad \text{or} \quad \Delta G = \Delta G^0 + RT \ln\frac{(P_C)^c(P_D)^d}{(P_A)^a(P_B)^b} \quad \text{for a gas phase reaction, or}$$

$$\Delta G = \Delta G^0 + RT \ln Q_c, \quad \text{or} \quad \Delta G = \Delta G^0 + RT \ln\frac{[C]^c[D]^d}{[A]^a[B]^b} \quad \text{for a gas or liquid phase reaction}$$

In Chapter 9, we will make much use of the above equations and you'll see many examples of how they are applied. But let's immediately examine the equation by testing its properties in two different ways. Test 1 will be to substitute concentrations at standard conditions, where all reactants and product concentrations are set to 1.0 molar; that's the defining condition for $Q_c^0$, of course. Test 2 will be to substitute the concentrations attained when the system has reached chemical equilibrium.

For Test 1, the system is at standard state conditions and all the concentrations are 1 molar, so $Q_c^0$ is 1

$$Q_c^0 = \frac{[C]^c[D]^d}{[A]^a[B]^b} = \frac{1^c \times 1^d}{1^a \times 1^b} = 1$$

Because ln 1 = 0, the equation $\Delta G = \Delta G^0 + RT \ln Q_c$, reduces to $\Delta G = \Delta G^0$. That is, of course, exactly as it should be at standard conditions.

In Test 2, the system has reached equilibrium, ΔG is zero, and $Q_c$ has become $K_c$. So we write the equation $\Delta G = \Delta G^0 + RT \ln Q_c$ as:

$$0 = \Delta G^0 + RT \ln K_c$$

or,

$$-RT \ln K_c = \Delta G^0$$

or,

$$K_c = e^{-\frac{\Delta G^0}{RT}}$$

In the form shown above (and its parallel gas phase form, $K_p = e^{-\frac{\Delta G^0}{RT}}$) we see what is arguably the most important single equation of chemical thermodynamics. The equation allows the direct calculation of the numerical value of the equilibrium constant for *any* chemical reaction from the vast body of known thermodynamic data. To do this, $\Delta G^0$ is obtained from standard thermodynamic data tables and substituted into the equation. The procedure is illustrated in Examples 8.10 and 8.11.

**Example 8.10**: What is the equilibrium constant at 298 K of a gas phase reaction that has a standard free energy change, $\Delta G^0$, of -10 kJ mol$^{-1}$?

Note that because $\Delta G^0$ is negative the standard state reaction is spontaneous, but note also that the size of $\Delta G^0$ is modest (bond energies average about 400 kJ mol$^{-1}$ for example).

Strategy: Substitute into $K_p = e^{-\frac{\Delta G^0}{RT}}$ and solve for $K_p$.

Substituting

$$K_p = e^{+\frac{10\ \text{kJ mol}^{-1}\ 1000\ \text{J kJ}^{-1}}{8.314\ \text{J mol}^{-1}\ \text{K}^{-1} \times 298.15\ \text{K}}}$$

So, $$K_p = e^{\left(+\frac{10000}{2479}\right)}$$

and, $$K_p = e^{+4.034} = \underline{56.48}\ \text{(unspecified units)}$$

The equilibrium ratio [Products]/[Reactants] is 56. That's substantially larger than 1, so we can assert that the position of equilibrium lies to the right. The units of $K_p$ will be $(1\ \text{bar})^{\Delta n}$, as described in the text. But we can't specify them without an actual reaction equation.

**Example 8.11**: For the reaction $NO(g) + \frac{1}{2} O_2(g) \rightarrow NO_2(g)$ at 25°C the standard free energy of reaction is -34.9 kJ mol$^{-1}$. Calculate $K_p$ for the reaction.

Strategy. Substitute into $K_p = e^{-\frac{\Delta G^0}{RT}}$ and solve for $K_p$.

$$K_p = e^{\frac{+34.9\ \text{kJ mol}^{-1}\ 1000\ \text{J kJ}^{-1}}{8.314\ \text{J mol}^{-1}\ \text{K}^{-1} \times 298.15\ \text{K}}}$$

So, $$K_p = e^{\left(+\frac{34900}{2479}\right)}$$

and, $$K_p = e^{+14.0} = \underline{1.31 \times 10^6\ \text{bar}^{-1/2}}$$

If the pressures are measured in bars then the unit of $K_p$ will be bar$^{-1/2}$ The equilibrium lies on the right. Observe how this value fits correctly in Table 8.5.

Table 8.6 summarizes the relationship between equilibrium position (as measured by the equilibrium constant, K) and the free energy change of the reaction as obtained from the standard thermodynamic data tables. Table 8.6 refers to any reaction at 298 K, but such a table can easily be made for any other temperature.

Table 8.6 **$\Delta G^0$ versus K at 298 K**

| $\Delta G^0$ (kJ mol$^{-1}$) | K | Position of Equilibrium |
|---|---|---|
| -200 | $1 \times 10^{35}$ | Complete on right |
| -100 | $3 \times 10^{17}$ | Essentially complete on right |
| -50 | $6 \times 10^{8}$ | Largely complete on right |
| -25 | $2 \times 10^{4}$ | Lies well on right |
| -10 | 56 | Lies on right |
| 0 | 1 | Reactants and products both plentiful |
| 10 | 0.02 | Lies on left |
| 25 | $4 \times 10^{-5}$ | Lies well on left |
| 50 | $2 \times 10^{-9}$ | Largely complete on left |
| 100 | $3 \times 10^{-18}$ | Essentially complete on left |
| 200 | $1 \times 10^{-35}$ | Complete on left |

We can summarize the values in Table 8.6 by the following three statements:

K is big, equilibrium lies on the right
K is small, equilibrium lies on the left
K is near 1, reactants and products both plentiful

## Writing K Expressions for Reactions

The procedure to write an equilibrium constant expression is simply to substitute into $K_p = \frac{(P_c)^c(P_d)^d}{(P_a)^a(P_b)^b}$ or $K_c = \frac{[C]^c[D]^d}{[A]^a[B]^b}$ as appropriate. Here are some example of equilibrium reactions and their corresponding equilibrium constant expressions. In each case, the units of the equilibrium constant derive from the algebra-of-quantities.

Equilibrium 1: Formation of atomic hydrogen from molecular hydrogen (gas phase)

$$H_2(g) \rightleftarrows 2\,H(g) \qquad K_p = \frac{(P_H)^2}{(P_{H_2})}\ \text{bar}$$

Equilibrium 2: Formation of $N_2O_4$ from $NO_2$ in (gas phase)

$$2\,NO(g) \rightleftarrows N_2O_4(g) \qquad K_p = \frac{(P_{N_2O_4})}{(P_{NO_2})^2}\ \text{bar}^{-1}$$

Equilibrium 3: Formation of ozone from oxygen (gas phase)

$$3\,O_2(g) \rightleftarrows 2\,O_3(g) \qquad K_p = \frac{(P_{O_3})^2}{(P_{O_2})^3}\ \text{bar}^{-1}$$

Equilibrium 4: Combustion of methane (gas phase, water produced in gaseous state)

$$CH_4(g) + 2\,O_2(g) \rightleftarrows CO_2(g) + 2\,H_2O(g) \qquad K_p = \frac{(P_{CO_2})(P_{H_2O})^2}{(P_{CH_4})(P_{O_2})^2}\ \text{(no unit)}$$

Equilibrium 5: The autoionization of water (pure liquid phase)

$$H_2O \rightleftarrows H^+ + OH^- \qquad K_c = \frac{[H^+][OH^-]}{[H_2O]}\ \text{mol L}^{-1}\ \text{or M}$$

$$\text{alternatively, } 2\,H_2O \rightleftarrows H_3O^+ + OH^- \qquad K_c = \frac{[H_3O^+][OH^-]}{[H_2O]^2}\ \text{(no unit)}$$

Equilibrium 6: The reaction of ammonia with water (aqueous solution)

$$NH_3(aq) + H_2O \rightleftarrows NH_4^+ + OH^- \qquad K_c = \frac{[NH_4^+][OH^-]}{[NH_3][H_2O]}\ \text{(no unit)}$$

alternatively, $NH_3(aq) + H^+ \rightleftarrows NH_4^+$ $\quad K_c = \frac{[NH_4^+]}{[NH_3][H^+]}$ L mol$^{-1}$ or M$^{-1}$

Equilibrium 7: The ionization of acetic acid (aqueous solution)

$$CH_3COOH \rightleftarrows CH_3COO^- + H^+ \quad K_c = \frac{[CH_3COO^-][H^+]}{[CH_3COOH]} \text{ mol L}^{-1} \text{ or M}$$

## Writing Equilibrium Constants for Stepwise Reactions

Many common reactions occur in a succession of reaction steps. There is an equilibrium constant for each of the individual steps, and also a combined equilibrium constant for the collective series of steps. Diprotic acids represent a common example of this situation. The diprotic oxalic acid is shown in Figure 8.5.

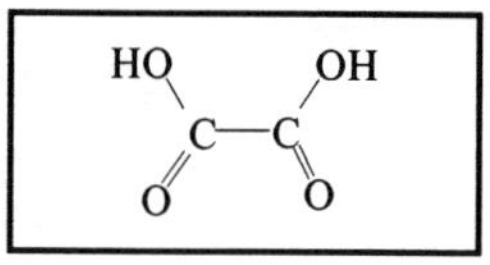

Figure 8.5 Oxalic acid is a diprotic acid. Each of the two hydrogen atoms in the oxalic acid molecule can neutralize a base.

The two-step ionization of oxalic acid can be represented by the following equilibrium constant expressions:

Step 1 $\quad H_2C_2O_4 \rightleftarrows HC_2O_4^- + H^+ \quad K_{a1} = \frac{[H^+][HC_2O_4^-]}{[H_2C_2O_4]} = 5.90 \times 10^{-2}$ mol L$^{-1}$

Step 2 $\quad HC_2O_4^- \rightleftarrows C_2O_4^{2-} + H^+ \quad K_{a2} = \frac{[H^+][C_2O_4^{2-}]}{[HC_2O_4^-]} = 6.40 \times 10^{-5}$ mol L$^{-1}$

Combined Steps $\quad H_2C_2O_4^- \rightleftarrows C_2O_4^{2-} + 2H^+ \quad K_a = \frac{[H^+]^2[C_2O_4^{2-}]}{[H_2C_2O_4]}$

$$= (5.90 \times 10^{-2}) \times (6.40 \times 10^{-5})$$

$$= 3.78 \times 10^{-6} \text{ mol}^2 \text{ L}^{-2}$$

The equilibrium constant for an overall, two-step reaction is the product of the individual equilibrium constants of the two steps. Note also that we use the notation $\mathbf{K_a}$ to represent the ionization of an acid into hydrogen ion and its corresponding anion. $K_a$ is called the **dissociation constant** of an acid. The notations $K_{a1}$ and $K_{a2}$ are use for the individual equilibrium constants of the two dissociation steps. If any equilibrium reaction is written in reverse, then the products and reactants interchange. Thus, the equilibrium constant becomes its reciprocal. In summary:

$$K_a = K_{a1} \times K_{a2}$$

$$K_{forward} = (K_{reverse})^{-1}$$

You'll begin to see the value of equilibrium constants if you compare the two dissociation constants for the ionization of oxalic acid. The first is about 1000 times bigger than the second. This fact means that oxalic acid at its first ionization step generates hydrogen ion much more readily and is thus a much stronger acid than at its second ionization step. Ranking the relative strengths of acids is an important application of chemical equilibrium principles.

Measuring series of K values for different series of related chemical compounds was an important part of chemistry during the 20th century because studies of such relationships yielded detailed information about how chemical reactivity is linked to chemical structure; the procedure is called **structure-property correlation**. We will discuss quantitative structure property relationships (QSPR) in Chapter 9.

## Equilibrium Constants and Heterogeneous Equilibrium

Many chemical reactions are heterogeneous. That is to say, they involve more than one phase. Consider the reaction of limestone when it's heated to make quicklime:

$$CaCO_3(s) \rightarrow CaO(s) + CO_2(g)$$

For this reaction we could write a pressure-based equilibrium constant expression as:

$$K_p = \frac{(P_{CaO})(P_{CO_2})}{(P_{CaCO_3})}$$

But because both $CaCO_3(s)$ and $CaO(s)$ are pure solid phases *and* the solids are present in the reaction system, their vapor pressures $P_{CaCO_3}$ and $P_{CaO}$ are both constant (at any specified temperature). These pressures are generally too small to measure, but because they are constant we can simply incorporate them into the equilibrium constant by writing:

$$P_{CO_2} = \frac{K_p(P_{CaCO_3})}{(P_{CaO})}$$ or, collecting up the terms on the right into a single value, $P_{CO_2} = K_p'$

So the equilibrium constant for the reaction is just the partial pressure of carbon dioxide in the reaction system. Actually, the equilibrium system behaves much like vapor pressure equilibrium, constant at any specific temperature and increasing with increasing temperature.

In general, when we write the equilibrium constant for a reaction involving pure condensed phases, we simply set the partial pressures (or concentrations) of those phases equal to one, which is mathematically the same as incorporating them into the equilibrium constant expression.

Conclusion: In this chapter we've taken a big step. We've moved from calorimetric measurements and thermodynamics to a study of chemical equilibrium, and we've seen how these two superficially distinct approaches to nature are in fact deeply linked. We've also seen that the arcane topic of chemical equilibrium is in fact fundamental to the processes of life, to a manufacturing economy, and even to the future of the planet itself. In the upcoming chapter we'll get down to discussing the details of chemical equilibrium in some specific situations.

## Essential Knowledge—Chapter 8

**Glossary words:** Chemical equilibrium, equilibrium constant, statistical mechanics, Carnot Cycle, thermodynamic cycle, reversible cycle, microstate, entropy of vaporization, standard entropy, third law of thermodynamics, Trouton's rule, standard entropy change of reaction, spontaneity, spontaneous process, spontaneous reaction, position of chemical equilibrium, free energy, standard free energy of formation, Gibbs free energy function, general equilibrium expression, reaction quotient, reaction quotient at standard state, heterogeneous equilibrium, acid dissociation constant.

**Key Concepts:** The nature of chemical equilibrium and the equilibrium constant. Recognition of the statistical interpretation of entropy. The Carnot cycle and the role of cyclic processes in thermodynamics. Understand the second law of thermodynamics. The third law and its role in allowing the definition of standard entropy. The concept of spontaneity and the relationship between the second law and a spontaneous process. Standard entropy change in chemical reactions. The position of equilibrium of chemical reaction systems. The significance of the Gibbs equation. Free energy as a system-based criterion for spontaneity. Writing equilibrium constants in pressure and molarity units. Writing equilibrium constant expressions for heterogeneous reaction systems.

**Key Equations:**

$K_p = \frac{(P_H)^2}{(P_{H_2})}$ = hydrogen's equilibrium constant using partial pressures (T is specified)

$K_c = \frac{[H]^2}{[H_2]}$ = hydrogen's equilibrium constant using molar concentrations (T is specified)

Entropy change at constant temperature: $\Delta S = \frac{q_{rev}}{T}$  Entropy of phase change: $\Delta S = q_{rev}/T$

Entropy change of system that is warmed: $\Delta S = \int dq_{rev}/T$

Number of arrangements of n molecules in two flasks: $= \frac{n!}{(L!\,R!)}$

The entropy of vaporization: $\Delta S = \frac{\Delta H_{vap}}{T}$

Standard entropy change of a reaction: $\Delta S^0 = \sum_{moles} S^0\,(\text{products}) - \sum_{moles} S^0\,(\text{reactants})$

Gibbs equation: $\Delta G = \Delta H - T\Delta S$

Standard Gibbs free energy change of a reaction: $\Delta G^0 = \sum_{moles} \Delta G^0_f\,(\text{products}) - \sum_{moles} \Delta G^0_f\,(\text{reactants})$

General equilibrium constant expressions: $K_p = \frac{(P_C)^c(P_D)^d}{(P_A)^a(P_B)^b}$ and $K_c = \frac{[C]^c[D]^d}{[A]^a[B]^b}$

Free energy change of a reaction at non standard conditions: $\Delta G = \Delta G^0 + RT \ln\frac{Q}{Q_0}$

Definition of standard reaction quotient, gas phase reaction: $Q^0 = \frac{(1\text{ bar})^c(1\text{ bar})^d}{(1\text{ bar}^a(1\text{ bar})^b} = (1\text{ bar})^{c+d-a-b} = (1\text{ bar})^{\Delta n}$

Definition of reaction quotient, gas phase reaction: $Q_p = \frac{(P_C)^c(P_D)^d}{(P_A)^a(P_B)^b}$

Relationship between free energy change and equilibrium constant: $\Delta G = \Delta G^0 + RT \ln\frac{[C]^c[D]^d}{[A]^a[B]^b}$

Relationship between standard free energy change and equilibrium constant: $-RT \ln K_c = \Delta G^0$

## Questions and Problems

### Introduction

8.1. a. What was the first indication in human history of the existence of the second law of thermodynamics? b. What temperature scale did that law imply? c. Under what circumstances did the concept of entropy come into being? d. What is statistical mechanics?

8.2. Define or describe in 25 words or less each of the following: a. a thermodynamic cycle, b. the equilibrium constant of a reaction, c. the Gibbs equation, d. a macroscopic property, and e. a reversible cycle.

8.3. Define each symbol in the formula $\Delta S = \frac{q_{rev}}{T}$. Is it appropriate to use this equation to determine the entropy change of a phase change?

### Second Law of Thermodynamics

8.4. Why are there so many possible statements of the second law?

8.5. Describe the entropy change of a pure substance: when it melts, when it boils, when it freezes, when it condenses, and when it sublimes.

8.6. What would you expect to happen to the entropy of an ideal gas as the temperature is increased at a constant pressure? Give a molecular interpretation to explain your answer.

8.7. How many ways can five molecules be distributed between two flasks of equal volume such that four are in the left flask and one in the right flask?

8.8. How many ways can six molecules be distributed equally between two flasks of equal volume? What is the percentage probability that three will be in each flask? Hint: use the binomial theorem.

8.9. The specific heat capacity at constant pressure of argon is 0.5213 J $g^{-1}$ $K^{-1}$ at a pressure of 1 atmosphere at 300 K. Argon's entropy at 300 K is 3.877 J $g^{-1}$ $K^{-1}$. Use these data to calculate the entropy of argon at 500 K.

8.10. The specific heat capacity at constant pressure of carbon dioxide is 0.8527 J $g^{-1}$ $K^{-1}$ at a pressure of 1 atmosphere at 300 K. Carbon dioxide's entropy at 300 K is 4.858 J $g^{-1}$ $K^{-1}$. Use these data to calculate the entropy of carbon dioxide at 500 K.

8.11. The specific heat capacity of carbon dioxide at constant pressure in the units of J $g^{-1}$ $K^{-1}$ at a pressure of one atmosphere can be obtained from the empirical equation (Shomate type equation) $C_p = 0.590 + 0.989 \times 10^{-3}\ T - 0.337 \times 10^{-6}\ T^2$, where T is the temperature in kelvins. Calculate the specific heat of carbon dioxide using this equation and compare the result with the value stated in the preceding question.

8.12. The enthalpy of vaporization of 1,2-dibromopropane ($C_3H_6Br_2$) at its boiling point of 141.9°C is 35.61 kJ $mol^{-1}$. Calculate its entropy of vaporization and discuss the result in terms of Trouton's rule.

8.13. The enthalpy of vaporization of dimethylamine ($C_2H_7N$) at its boiling point of 6.9°C is 26.40 kJ $mol^{-1}$. Calculate its entropy of vaporization and discuss the result in terms of Trouton's rule.

8.14. The enthalpy of fusion of nitrobenzene ($C_6H_5NO_2$) is 22.50 cal $g^{-1}$. Express this value in the units of kJ $mol^{-1}$ and then calculate the molar entropy of fusion of nitrobenzene at its normal melting point of 5.7°C.

8.15. State an interpretation of Trouton's rule.

8.16. Why do hydrocarbons better obey Trouton's rule than do alcohols?

### Third Law of Thermodynamics

8.17. The third law of thermodynamics provides a reference state for entropy. Discuss.

8.18. Use the data in Table 8.3 to calculate the entropy change for the vaporization of methanol at standard conditions. Compare this value with the one in the Trouton's rule table (Table 8.2). Discuss the result.

8.19. What is entropy of mixing?

8.20. Starting from absolute zero, how many steps of entropy addition are there for a substance that is a gas at standard temperature.

## Standard Entropy Change of Chemical Reactions

8.21. For which of the following reactions do you expect $\Delta S^0$(reaction) to be positive? You are *not* being asked to use data tables here.

a. $O_2(g) + 2\ H_2(g) \rightarrow 2\ H_2O(l)$
b. $H_2(g) + Cl_2(g) \rightarrow 2\ HCl(g)$
c. $CH_4(g) + 2\ O_2(g) \rightarrow CO_2(g) + 2\ H_2O(l)$
d. $CH_4(g) + 2\ O_2(g) \rightarrow CO_2(g) + 2\ H_2O(g)$
e. $CaCO_3(s) \rightarrow CaO(s) + CO_2(g)$
f. $NH_3(g) + HCl(g) \rightarrow NH_4Cl(s)$

8.22. Predict the sign (positive or negative) and relative magnitude (large versus small) of the entropy changes expected for the following reactions:

a. $Zn(s) + 2\ HCl(aq) \rightarrow ZnCl_2(s) + H_2(g)$
b. $Mg(s) + Cl_2(g) \rightarrow MgCl_2(s)$
c. $C(graphite) \rightarrow C(diamond)$
d. $F_2(g) + 2\ NaCl(s) \rightarrow Cl_2(g) + 2\ NaF(s)$
e. $2\ NH_4NO_3(s) \rightarrow 2\ N_2(g) + 4\ H_2O(g) + O_2(g)$
f. $CuO(s) + H_2(g) \rightarrow Cu(s) + H_2O(g)$

8.23. Why do thermodynamic tables tabulate absolute entropies of pure substances, but not absolute enthalpies or free energies?

8.24. Use standard entropy data to calculate the standard entropy changes of the following (possibly unbalanced) reactions

a. $C(graphite) \rightarrow C(diamond)$
b. $CH_4(g) + O_2(g) \rightarrow CO_2(g) + H_2O(g)$
c. $CH_4(g) + O_2(g) \rightarrow CO_2(g) + H_2O(l)$
d. $CH_3OH(l) + O_2(g) \rightarrow CO_2(g) + H_2O(l)$
e. $H_2(g) + O_3(g) \rightarrow H_2O(g)$
f. $H_2(g) \rightarrow 2H(g)$

## Spontaneity and Chemical Reactions

8.25. Explain the terms: a. the position of equilibrium, b. to lie on the right, c. to lie on the left, d. to go to completion, and e. a spontaneous process.

8.26. Give two examples of spontaneous physical processes. How does the second law of thermodynamics explain the spontaneity of processes that are neither endo- nor exothermic?

8.27. Consider the gas reaction: $2\ SO_2(g) + O_2(g) \rightleftarrows 2\ SO_3(g)$. In a closed vessel, a study of this reaction shows that the concentration of $SO_3$ is increasing. a. Is the system in equilibrium? b. If not, is the forward reaction proceeding faster than the reverse reaction?

8.28. A chemical reaction mixture is in a state of equilibrium. Conditions change so that the rate of the forward reaction becomes faster than the rate of the reverse reaction. What happens to the concentrations of the reactants? What happens to the concentrations of the products? How does the position of equilibrium shift?

8.29. Is the freezing of water a spontaneous process? How about its melting? On what do the answers to these two questions depend?

## Gibbs Free Energy

8.30. a. Who was Gibbs? b. What is the purpose of the equation that bears his name?

8.31. State the four cases of the effect of temperature on the balance between $\Delta H$ and $T\Delta S$ in the Gibbs equation.

8.32. Calculate $\Delta G^0$ at 25°C for the reaction: $2\ H_2(g) + O_2(g) \rightarrow 2\ H_2O(g)$ given that $\Delta H^0 = -484$ kJ mol$^{-1}$ and $\Delta S^0 = -88.9$ J mol$^{-1}$ K$^{-1}$. Is this reaction spontaneous under standard state conditions at 25°C?

8.33. Using standard free energies of formation, calculate $\Delta G^0$ at 25°C for each of the following (possibly unbalanced) reactions:

a. $H_2(g) + C_2H_4(g) \rightarrow C_2H_6(g)$
b. $H_2O(g) + C(graphite) \rightarrow H_2(g) + CO(g)$
c. $C_2H_4(g) + H_2O(l) \rightarrow C_2H_5OH(l)$
d. $C_2H_2(g) + CO(g) \rightarrow C(graphite) + H_2O(l)$
e. $CO(g) + \frac{1}{2} O_2(g) \rightarrow CO_2(g)$

Which of these reactions are spontaneous under standard state conditions at 25°C? Which is most favorable and which least?

8.34. Use the data from Table 8.3 to calculate the standard free energy for the complete combustion of acetylene (ethyne) at standard conditions.

8.35. Use the data from Table 8.3 to calculate the standard free energy for the complete combustion of diamond at standard conditions.

8.36. Is the formation of diatomic oxygen from ozone a spontaneous or nonspontaneous process at standard conditions? If it is a spontaneous process does it necessarily happen at a finite rate?

8.37. Rationalize the facts that the standard free energy of formation of C(g) is +671 kJ mol$^{-1}$ (a large positive value) whereas the standard free energy of formation of $CO_2(g)$ is -394 kJ mol$^{-1}$ (a large negative value).

## Equilibrium Constants

8.38. Write the equilibrium constant, $K_p$, expression for each of the following reactions:

a. $H_2(g) + C_2H_4(g) \rightarrow C_2H_6(g)$
b. $H_2O(g) + C(graphite) \rightarrow H_2(g) + CO(g)$
c. $C_2H_4(g) + H_2O(g) \rightarrow C_2H_5OH(g)$
d. $C_2H_2(g) + CO(g) \rightarrow C(graphite) + H_2O(l)$
e. $CO(g) + \frac{1}{2} O_2(g) \rightarrow CO_2(g)$

8.39. Write the equilibrium constant, $K_c$, expression for each of the following reactions:

a. $H_2(g) + C_2H_4(g) \rightarrow C_2H_6(g)$
b. $H_2O(g) + C(graphite) \rightarrow H_2(g) + CO(g)$
c. $C_2H_4(g) + H_2O(g) \rightarrow C_2H_5OH(g)$
d. $C_2H_2(g) + CO(g) \rightarrow C(graphite) + H_2O(l)$
e. $CO(g) + \frac{1}{2} O_2(g) \rightarrow CO_2(g)$

8.40. Write the equilibrium constant expressions for the reactions:

$3\ H_2(g) + N_2(g) \rightleftarrows 2\ NH_3(g)$
$2\ NH_3(g) \rightleftarrows 3\ H_2(g) + N_2(g)$

What is the relationship between the two reactions? What is the relationship between the equilibrium constants of the two reactions?

8.41. Write the $K_p$ expression for each of the following heterogeneous reactions:

a. $2\ AgI(s) \rightleftarrows 2\ Ag(s) + I_2(g)$
b. $2\ HI(g) \rightleftarrows H_2(g) + I_2(g)$
c. $2\ AgI(s) + H_2(g) \rightleftarrows 2\ Ag(s) + 2\ HI(g)$

8.42. State five "real world" situations in which equilibrium constants are important.

## ΔG and Equilibrium Constants

8.43. Carefully state the difference between $Q_p$ and $K_p$.

8.44. Consider the equation $\Delta G = \Delta G^0 + RT \ln \frac{[C]^c[D]^d}{[A]^a[B]^b}$. If the reactant and product concentrations are at their standard states, what is the value of ΔG and why? If the reaction is at equilibrium, what is the value of ΔG?

8.45. In Section 8.5 it's stated that for a reaction system composed of ideal gases, $K_p = K_c(RT)^{\Delta n}$, where Δn is the difference between the total number of moles of product gases and the total number of moles of reactant gases. Derive the equation $K_p = K_c(RT)^{\Delta n}$

8.46. The equilibrium reaction system $A(g) \rightleftarrows 2\ B(g)$ has a $\Delta G^0$ value of +15 kJ mol$^{-1}$. Calculate the value of $K_p$ for the reaction.

8.47. The equilibrium reaction system $A(g) \rightleftarrows 2\ B(g)$ has a $\Delta G^0$ value of -132 kJ mol$^{-1}$. Calculate the value of $K_c$ for the reaction.

8.48. The equilibrium reaction system $HA(aq) \rightleftarrows H^+(aq) + A^-(aq)$ has a $\Delta G^0$ value of +21.3 kJ mol$^{-1}$. Calculate the value of $K_c$ for the reaction.

8.49. A series of weak acids that dissociate according to the general equation $HA(aq) \rightleftarrows H^+(aq) + A^-(aq)$ have $\Delta G^0$ values of a. +21.3 kJ mol$^{-1}$, b. +24.9 kJ mol$^{-1}$, and c. +31.8 kJ mol$^{-1}$. Which is the strongest acid, a, b, or c?

8.50. A chemical reaction has an equilibrium constant, $K_c$, of $3.4 \times 10^{-5}$ at 298.15 K. a. Where does the position of equilibrium lie? b. What is $\Delta G^0$ of the reaction?

## Problem Solving

8.51. The specific heat capacity at constant pressure of carbon dioxide in the units of J g$^{-1}$ K$^{-1}$ at a pressure of one atmosphere can be obtained from the empirical equation (Shomate type equation) $C_p = 0.590 + 0.989 \times 10^{-3}\ T - 0.337 \times 10^{-6}\ T^2$, where T is the temperature in kelvins. At 300 K this empirical equation gives a Cp value of 0.854 J g$^{-1}$ K$^{-1}$. a. Use the equation $\Delta S = C_p \ln (T_2/T_1)$, which assumes that $C_p$ is not dependent on temperature, to calculate the entropy change when carbon dioxide is heated from 300 K to 1000 K. b Integrate the expression $\Delta S = \int C_p/T\ dT$ using the empirical power series equation and calculate the entropy change a second time. Compare the two results.

8.52. A kitchen refrigerator runs on electricity. Imagine perfectly sealing the kitchen with insulating tape. No heat can escape or enter. Now run the refrigerator with the door open. Does the temperature of the kitchen go up, down, or remain constant.

8.53. Discuss the principle of Clausius.

8.54. Use internet resources to find the standard enthalpies of formation and the standard entropies of toluene, naphthalene, and cyclopentadiene.

8.55. Vapor pressure is an equilibrium phenomenon. Discuss.

8.56. At 600 K the equilibrium dissociation constant, $K_p$, of the reaction $H_2 \rightleftarrows 2\ H$ is $3.6 \times 10^{-33}$ atm. 1.00 mol of $H_2$ is injected into a 10.5 L oven at 600 K and allowed to reach equilibrium. Calculate the total number of free hydrogen atoms in the reaction system.

8.57. At 2000 K the equilibrium dissociation constant of the reaction $Cl_2 \rightleftarrows 2\ Cl$ is 0.570 atm. A reaction equilibrium mixture weighs 4.5 milligrams and is contained in a 67.8 mL reaction chamber at 2000 K. Calculate the total pressure inside the reaction chamber.

8.58. Use internet resources to find the term "fugacity." Write its definition and explain what fugacity has to do with chemical equilibrium.

8.59. A glass is a frozen, supercooled liquid. It is in a nonequilibrium state both at room temperature and absolute zero. What is the implication of this situation for the standard entropy of glass?

8.60. In Chapter 8 it was asserted that if $C_p$ is independent of T, then the enthalpy change for warming a substance is simply its heat capacity times the temperature change, $\Delta H = C_p(T_2\text{-}T_1)$ and that for a *small* temperature change, $\Delta H$ divided by the average temperature approximates the entropy change, $\Delta S$. The coorect equation to use is $\Delta S = \int C_p/T\ dT$ which, if $C_p$ is independent of temperature integrates to give $\Delta S = C_p \ln (T_2/T_1)$. Use both of these equations to calculate the constant pressure entropy change when exactly one mole of helium is heated from 273 K respectively to 274, 278, 283, 323, and 373 K. Criticize the assertion. From Table 7.2 you can figure that the molar heat capacity of helium is 20.9 J $mol^{-1}\ K^{-1}$. How small is *small* in this context?

8.61. In Section 8.4 it is stated: "For any constant T and P process, it can be shown that if $\Delta S_{universe}$ is positive then $\Delta G_{system}$ is negative." Show it.

8.62. For non hydrogen bonding organic liquids, Trouton's rule states that their entropies of vaporization are all about 84 J $mol^{-1}K^{-1}$. Use the data in Table 8.2 for these liquids to make a graph of their entropy of vaporization plotted against their boiling points. Is a trend apparent? Discuss.

# Chapter 9

# Gas and Solution Equilibrium

## Introduction

Chemical equilibrium is one of the bedrock topics of any beginning chemistry course. In the preceding chapter we briefly described the importance of chemical equilibrium. The first half of this chapter is largely concerned with gas phase reactions, such as those by which ammonia is manufactured. The second half of this chapter is concerned with equilibria in solution and focusses on acid-base equilibrium and pH.

Figure 9.1 The CF Industries ammonia plant at Donaldsonville, Louisiana is the largest and most modern ammonia production facility in the US. A 1998 plant expansion raised its production capacity to more than 1,000 tons of $NH_3$ per day. Ammonia is an important fertilizer. The foundations of the science of chemical equilibrium were largely laid by chemical engineers as they developed ammonia plants during the first decade of the twentieth century.

In the preceding chapter we mentioned three topics: biochemical reactions in living organisms, industrial chemical engineering reactions and the large scale production of materials, and equilibrium chemical reactions in the atmosphere with implications for the earth's future. More specifically, we saw how thermodynamics databases contain the information needed to calculate the equilibrium properties of any chemical reaction, you've seen how you can get a K from a $\Delta G^0$. In this chapter we'll apply equilibrium principles to reactions involving gases and to reactions in aqueous solution.

Many branches of science make use of applied chemical equilibrium. Biologists make extensive use of it; for example, pH is a dominant factor in controlling the reactions that occur in fluids inside and outside cells. Geochemists use it to understand the formation of mineral deposits. Agronomists must understand how soil equilibria affect the availability of nutrients, so that they may assess and improve the suitability of soils for growing crops. Engineers of many types are concerned with corrosion phenomena—preventing corrosion is controlling the chemical degradation of metals. Understanding pH equilibrium allows chemists to formulate shampoo (the "pH balance" of innumerable TV commercials), and to maintain water quality in rivers, lakes, and swimming pools. The applications of pH are extremely wide ranging.

## Section 9.1 The Approach to Equilibrium

We'll begin with the simplest case: one reactant makes one product A → B, starting with only reactant A in the reaction system, and allowing it to come to equilibrium, A ⇄ B, with both A and B present. The situation is shown graphically in Figure 9.2, where the reactant follows the falling curve and the product the rising one.

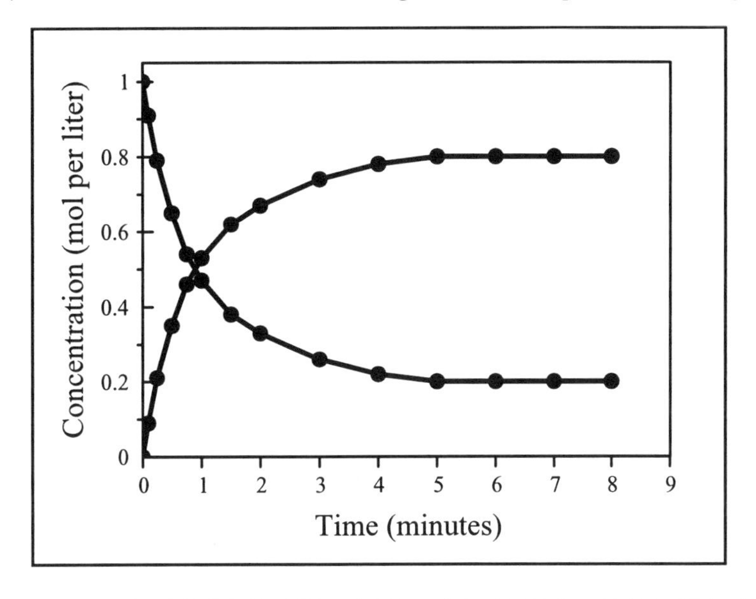


Figure 9.2 The approach to equilibrium for the general chemical reaction A ⇄ B. At the beginning of the reaction, reactant A drops rapidly in concentration while product B increases in concentration in accord with the stoichiometry of the reaction equation. As time passes, the rates of change of the reactant and the product slow down. Eventually, concentrations stop changing, though the forward and backward reactions continue at equal rates.

The calculation of the equilibrium constant from the data in Figure 9.2 is simple:

In this case $K_c = \frac{[C]^c[D]^d}{[A]^a[B]^b}$ simplifies to $K_c = \frac{[B]}{[A]}$ and it's only necessary to substitute the equilibrium molar concentrations from the graph.

$$\text{At equilibrium, } K_c = \frac{[B]}{[A]} = \frac{0.80}{0.20} = 4.0 \text{ (no units)}$$

A chemical reaction has reached equilibrium (is at equilibrium) when the concentrations of *all* the reactants and *all* the products are unchanging. Figure 9.3 represents the approach to equilibrium for the general reaction A ⇄ 2B + C. The equilibrium constant of this reaction has the form:

$$K_c = \frac{[B]^2[C]}{[A]}$$

You can see that the system reaches a state of equilibrium after five minutes. You can also see that the position of equilibrium lies on the right, because when equilibrium has been reached the product concentrations are substantially greater than the residual reactant concentration. During the approach to equilibrium, the concentration of B rises twice as fast as the concentration of C, and that the concentration of C changes in a manner exactly opposite to that of A. These relationships are a direct consequence of the reaction's stoichiometry.

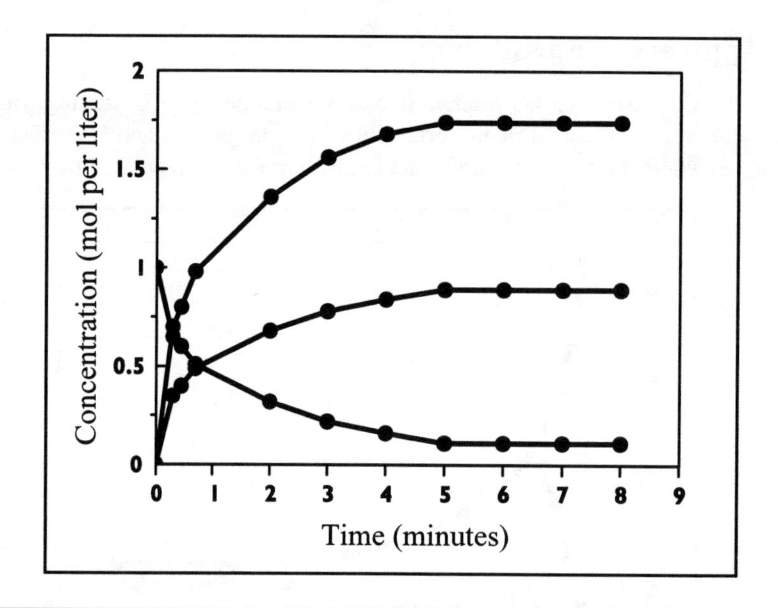


Figure 9.3 The approach to equilibrium for the general chemical reaction A ⇄ 2B + C. The concentration of the reactant A is represented by the falling line. Note that the top line (product B) rises twice as fast as the middle line (product C) in accord with the stoichiometry of the reaction.

If the equilibrium constant of the reaction is known, it is not difficult to calculate the equilibrium concentrations of the reactants and products. The procedure involves stoichiometry and algebra. It is summarized in the following box and illustrated in Example 9.1.

**General Strategy for Solving Equilibrium Calculations**

The steps below are a general strategy for solving problems in which initial concentrations are specified for a reaction system, and the reaction's equilibrium constant is known.

1. Write the balanced chemical equation for the reaction.
2. Write the equilibrium constant expression for the reaction and set it equal to the numerical value of the equilibrium constant (K).
3. Define an unknown, x, for the *change* in concentration or pressure of any single reactant or product. Calculate the change in concentration or pressure of every other reactant or product using that x and the stoichiometric coefficients of the balanced equation. You now have a list of all the equilibrium concentrations in terms of x.
4. Substitute the list of equilibrium concentrations into the K expression.
5. Solve the algebraic equation using any appropriate method.

**Example 9.1**: The equilibrium constant of the reaction A ⇄ 2B + C shown in Figure 9.3 is 25.6 $mol^2\ L^{-2}$ at a constant but unspecified temperature. The reaction begins with reactant A at a concentration of 1.00 M and no products. After five minutes the reaction establishes equilibrium. Calculate the equilibrium concentrations of the reactant and both products.

Strategy: Set up the equilibrium shift in terms of the change in concentration of the reactant, A. Specifically, let x be the decrease in A's concentration when equilibrium has been established. Next, figure the reactant concentrations in terms of x using stoichiometry. Finally, substitute all these algebraic quantities into the K expression and solve for x.

| | A | ⇄ | 2B | + | C |
|---|---|---|---|---|---|
| At start | 1.00 | | 0.00 | | 0.00 |
| At equilibrium | 1.00 - x | | 2 x | | x |

$$K_c = \frac{[B]^2[C]}{[A]}$$

$$25.6\ mol^2\ L^{-2} = \frac{[2x]^2[x]}{[1.000-x]}$$

This is a cubic equation that has to be solved by successive approximation or some such method. Successive approximation gives x = 0.890. So [B] = 1.78 and [C] = 0.890, while [A] itself is 1.000 - 0.890 = 0.110 M. It is easy to demonstrate that the solution is correct by substituting in the equilibrium concentrations and showing that the value they produce matches the stated K value.

$$K_c = \frac{(1.780)^2(0.890)}{(0.110)} = 25.6\ mol^2\ L^{-2}$$

Check: The equilibrium concentration values confirm that Figure 9.3 is correctly drawn.

## Reestablishing Equilibrium After a Concentration Change

We'll now change the reactant concentration, taking as our starting point the equilibrium system reached in Example 9.1 and illustrated in Figure 9.3. Additional reactant A is added to the system. Instantly, the A concentration leaps upward, but the newly boosted A concentration soon drops, while the B and C concentrations rise. After some time the system reestablishes equilibrium. Let's see the details:

After the system of Example 9.1 has reached equilibrium, an additional amount of reactant, equal to a concentration of 0.270 M, is added (the system is "slugged"). The concentration of reactant A is thus 0.110 M + 0.270 M = 0.380 M. These changes are illustrated in Figure 9.4

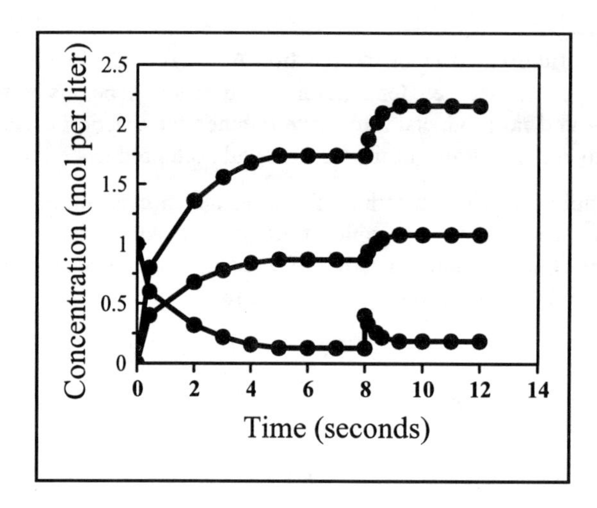


Figure 9.4 Reestablishment of equilibrium for the chemical reaction A ⇄ 2B + C in an existing state of equilibrium when it is "slugged" with an added quantity of reactant. (This figure is an extension of Figure 9.3.) For a while the system is "out of equilibrium." But part of the slug converts to products. A new equilibrium is established with *different* concentrations but *the same* equilibrium constant. The "old" equilibrium was established between 5 and 8 seconds; the "new equilibrium" is established between 9 and 12 seconds. Those are the regions on the diagram where the lines are flat.

Observe that the concentration of reactant A instantly jumps up after we slug the system. Immediately, the reaction system begins to shift to the right: the product concentrations increases, and the jumped-up reactant concentration decreases. In less than 2 minutes equilibrium is reestablished—which you recognize by the concentrations once again becoming unchanging. All three concentrations are *higher* at the new equilibrium than they were at the old equilibrium, but the $K_c$ is, of course, just the same. The new concentrations must plug into the equilibrium constant expression to give (within experimental uncertainty) the old equilibrium constant. The equilibrium concentrations are [A] = 0.191 M, [B] = 2.15 M, and [C] = 1.07 M. Substituting those values gives:

$$K_c = \frac{(2.15^2) \times (1.)07}{(0.191)} = 25.9 \text{ mol}^2 \text{ L}^{-2}$$

The calculation of these new equilibrium concentrations is done in Problem 9.18 at the end of the chapter.

## Section 9.2 Factors that Affect Chemical Equilibrium

### Le Chatelier's Principle

In this section we'll examine the consequences of making various changes to a system that is already in an existing state of chemical equilibrium—doing this is called disturbing the equilibrium. To **disturb an equilibrium** is to change one or other property of the reaction system. Disturbing an equilibrium is also a procedure used to control the extent of chemical reactions. The three changes that we'll examine are:

1. Changing the concentration of a reactant or product in an equilibrium system.
2. Changing the total pressure of an equilibrium system.
3. Changing the temperature of an equilibrium system.

The traditional development of this topic is to use a historical approach following Le Chatelier, whose principle is a useful way of stating the consequences of the law of chemical equilibrium: that $K_c$ depends only on T, and is independent of pressure and concentration. More fundamentally, thermodynamics fully explains all the experimental facts and corrects the minor imprecision of the principle. Interestingly, we now know that there are exceptions to his principle some special cases of equilibrium.

If an equilibrium reaction system is disturbed, changes will occur so that it again approaches equilibrium. The principle that describes how the system reestablishes equilibrium was first stated by the French chemist, Henri Louis Le Chatelier (1850-1936). It states that:

> When an equilibrium system is disturbed, it will react in a way that tends to minimize the disturbance.

Le Chatelier's principle applies to *all* systems that are in an existing state of chemical equilibrium.

When systems are disturbed, the previously unchanging concentrations of reactants and products change. If the concentrations of products increase while the concentrations of reactants decrease, we say the equilibrium **shifts to the right**. If the concentrations of products decrease while the concentrations of reactants increase, we say the equilibrium **shifts to the left**.

Reactants $\rightarrow$ Products A shift to the right
Reactants $\leftarrow$ Products A shift to the left

Reactants $\rightleftarrows$ Products No shift. System in equilibrium.

## The Effect of Changing a Reactant or Product Concentration

For the general two-reactant, two-product equilibrium reaction system, a A + b B $\rightleftarrows$ c C + d D, we can now summarize the consequences of disturbing the equilibrium by changing the concentration of either a reactant or product at constant temperature and volume. Adding a single reactant reduces the concentration of the other reactant and increases the concentrations of both products. Adding a single product reduces the concentration of the other product and increases the concentrations of both reactants. Those effects are summarized in Table 9.1.

Table 9.1 **Summary of Le Chatelier Effects**

| Disturbance | Consequence as Equilibrium is Reestablished | |
|---|---|---|
| add A | [B] decreases, | [C] and [D] increase |
| add B | [A] decreases, | [C] and [D] increase |
| add C | [D] decreases, | [A] and [B] increase |
| add D | [C] decreases, | [A] and [B] Increase |

Table 9.1 shows concentrations in molarity units. However, pressure is also a concentration unit, and the rules in the table apply equally as well for gas phase reactions.

Chemists frequently use Le Chatelier's principle to increase the yield of a desired product. You will recall from Chapter 3 the concept of the limiting reagent in a chemical reaction. The application of Le Chatelier's principle to increase yield can be thought of as deliberately forcing a particular reagent to be limiting. Imagine we want to carry out a homogenous reaction,

$$A + B \rightleftarrows C$$

Suppose that reactant A is either in short supply, or is expensive, but that B is cheap and readily available. To produce the maximum yield of C, the practical solution is to use a large excess of reactant B—perhaps many times more than the stoichiometric amount.

Having now described the effect of molar concentration changes on an existing equilibrium, we'll discuss the related effects of pressure changes on equilibrium reactions involving gases.

## The Effect of Changing Total Pressure

The effects of pressure change on an existing equilibrium are negligible for reactions that only involve solids and liquids. As pointed out earlier, solids and liquids are nearly incompressible, so changes in pressure do not significantly affect the concentrations of solutions.

Changes in pressure also have a negligible effect on an existing equilibrium for gas phase reactions that occur with zero volume change. Zero volume change gas phase reactions are those for which the total number of moles of reactant gases equals the total number of moles of product gases. For example, the reaction A(g) + 2B(g) $\rightleftarrows$ 3C(g) is a gas phase reaction with no volume change because there are three moles of total reactants (1 + 2 = 3) and three moles of product. Thus, the position of equilibrium of this reaction is unaffected by changing the total pressure.

However, for an existing gas equilibrium reaction that occurs *with* a change in the number of moles of gas, there is a significant shift in the position of equilibrium when the pressure is changed. Le Chatelier's principle tells that increase of pressure causes the equilibrium to shift in the direction that reduces the number of moles of gas present.

If you are given a gas phase chemical equilibrium in the form of a balanced equilibrium equation, you can judge the effect of changing the system's pressure by simply counting the moles of gas on each side of the equation. Listed in Table 9.2 are some examples of gas phase reaction shifts. Many industrial gas phase reactions are operated at extremely high pressures precisely because doing that shifts the position of equilibrium to favor the desired products.

Table 9.2 **Effect of Increasing the Total Pressure of Several Equilibrium Gas Reactions**

|  |  |
|---|---|
| $H_2(g) \rightleftarrows 2H(g)$ | Shifts left under increased pressure |
| $2\ NO_2(g) \rightleftarrows N_2O_4(g)$ | Shifts right under increased pressure |
| $2\ H_2(g) + O_2(g) \rightleftarrows H_2O(g)$ | Shifts right under increased pressure |
| $C(s) + H_2O(g) \rightleftarrows CO(g) + H_2(g)$ | Shifts left under increased pressure |

The Haber Process

Gas phase equilibrium reactions are of great importance in industrial chemistry and none more so than the Haber process for making ammonia. Ammonia production is a good practical example of an important equilibrium chemical reaction that is operated at high pressure to shift its equilibrium favorably. Figure 9.1 pictures the largest ammonia plant in the US, [❁kws +"CF Industries" +ammonia]. The Haber process (sometimes called the Haber-Bosch process) for producing ammonia, is based on the equilibrium:

$$N_2(g) + 3\ H_2(g) \rightleftarrows 2\ NH_3\ (g)$$

Because there are fewer moles of gas on the right than on the left (2 versus 4), the products occupy less volume than the reactants under any specified temperature and pressure conditions. Therefore, increasing pressure causes the disappearance of some nitrogen and hydrogen, and the formation of some ammonia. So you see immediately why chemical equilibrium principles urge running the process at the highest possible pressures—several hundred atmospheres.

The Haber Process is named for its developer, Fritz Haber (1868-1934). Nitrogen is an unreactive element, but it is needed in a chemically available form as one of three key fertilizer elements in modern agriculture (potassium and phosphorus are the other two). The conversion of atmospheric nitrogen to a usable chemical form is called **nitrogen fixation**. Ammonia is the fifth most-produced industrial chemical compound, with the average American consuming about 150 pounds each year; many other nitrogen containing compounds are synthesized from ammonia.

Haber was a German, and the stimulus for the full development of his process was World War I, when Germany was cut off from its traditional source of nitrates in the guano deposits of Peru and Chile. Guano deposits are accumulated sea-bird droppings (guanay is Spanish for cormorant); the nitrate, of course, originated by biological nitrogen fixation before moving up the food chain to the birds. In addition to their use for fertilizer, nitrogen compounds are important military explosives. So with Germany cut off from its traditional prewar transatlantic source of chemical nitrogen, Haber's process became vital to the German war effort because ammonia could be further processed, via nitric acid, to provide needed military explosives. In our time, explosives manufacturers make large quantities of "nitro" compounds for use as rocket propellants.

Haber published *Thermodynamics of Technical Gas Reactions* in 1908. His book is a classic text of chemical engineering as applied to gas phase reactions. Any gas phase reaction at any modern refinery is based on the same thermodynamic principles that Haber investigated. [❁kws +"petroleum refining" +"chemical engineering"]

The design and operation of industrial chemical plants is the province of chemical engineering. The design of modern ammonia plants incorporates a long series of incremental technical developments, including: operating with high performance catalysts, with second-by-second computer control, and with large reactors where the economy of scale becomes increasingly significant, etc. The highly competitive nature of the ammonia business has acted as a powerful spur to technical refinement of the basic idea of equilibrium shift.

Decomposition Reaction of Dinitrogen Tetroxide

The gas phase equilibrium reaction between dinitrogen tetroxide and nitrogen dioxide is affected by pressure changes. We'll use this reaction to show examples of equilibrium calculations for a gas phase reaction. The chemical equation and the equilibrium constant expression (in partial pressure units) for this decomposition reaction are:

$$N_2O_4(g) \rightleftarrows 2\ NO_2(g)$$

colorless brown

for which the equilibrium constant expression is $K_p = \frac{P^2_{NO_2}}{P_{N_2O_4}}$

At 25°C the equilibrium constant, $K_p$, for the reaction is 0.146 atm, a value showing that the equilibrium lies somewhat on the left of the reaction ($0.146 \approx \frac{1}{6}$,so the denominator is roughly six times the numerator). Remember, $K_p$ is independent of pressure but does depend on temperature. We can calculate the equilibrium pressures starting from either pure reactant or pure product. To do this, we use the method of stoichiometry and algebra previously described, as illustrated in Example 9.2.

**Example 9.2**: A closed glass vessel is maintained at 25°C. Into it is injected pure $N_2O_4$ so that the initial pressure is 2.57 atm. The pressure immediately begins to rise as $NO_2$ is formed. Calculate the equilibrium partial pressures of the reactant and product, and the total pressure once equilibrium has been established.

Strategy: Write the balanced reaction and its $K_p$ expression. Define x as the pressure drop of the $N_2O_4$. Apply stoichiometry to get the algebraic change in product pressure as +2x. Substitute the values and solve the algebra. The balanced equation is $N_2O_4(g) \rightleftarrows 2\ NO_2(g)$ for which $K_p = \frac{P^2_{NO_2}}{P_{N_2O_4}} = 0.146$ atm

| | $N_2O_4(g)$ | $\rightleftarrows$ $2\ NO_2(g)$ |
|---|---|---|
| At start | 2.57 atm | 0 atm |
| At equilibrium | 2.57 - x | +2x |

$\frac{(2x)^2}{(2.57-x)} = 0.146$ atm. This quadratic solves to give x = 0.289 atm

The equilibrium pressures are $P_{N_2O_4}$= 2.28 atm, $P_{NO_2}$ = 0.578 atm, and $P_{tot}$ = 2.86 atm

The reaction $N_2O_4(g) \rightleftarrows 2\ NO_2(g)$ shifts to the left with increasing total pressure. The calculation of such a shift is illustrated by Example 9.3.

**Example 9.3**: The total pressure on the equilibrium gas reaction system $N_2O_4(g) \rightleftarrows 2\ NO_2(g)$ is doubled from 2.86 atm (as calculated in Example 9.2) to 5.72 atm. After the pressure has been increased, the reaction system shifts left and a new equilibrium is established. Calculate the new equilibrium partial pressures of reactant and product and demonstrate that the stated direction of reaction shift is indeed correct.

Strategy: Recognize that the partial pressures of the reactant and product also instantaneously double to $P_{N_2O_4}$= 4.56 atm and $P_{NO_2}$ = 1.156 atm. Define the subsequent change in the $N_2O_4$ pressure as x. From stoichiometry it follows that the change on the $NO_2$ pressure will be -2x. Substitute the changed values into the equilibrium constant expression and solve the resulting equation for x.

The balanced equation is $N_2O_4(g) \rightleftarrows 2\ NO_2(g)$

$$\text{for which } K_p = \frac{P^2_{NO_2}}{P_{N_2O_4}} = 0.146 \text{ atm}$$

| | $N_2O_4(g)$ $\rightleftarrows$ | $2\ NO_2(g)$ |
|---|---|---|
| At start | 4.56 atm | 1.156 atm |
| At equilibrium | 4.56 + x | 1.156 - 2x |

$\frac{(1.156 - 2x)^2}{(4.56 + x)} = 0.146$ atm. This quadratic solves to give x = 0.163 atm

The equilibrium pressures are $P_{N_2O_4}$= 4.74 atm, $P_{NO_2}$ = 0.796 atm, and $P_{tot}$ = 5.54 atm

Check these values by substituting into the $K_p$ expression

$$= \frac{(1.156 - 2 \times 0.163)^2}{(4.56 + 0.163)} = .146 \text{ atm (thus okay)}$$

Observe, as predicted by the principle of Le Chatelier, that increasing the total pressure of the equilibrium reaction system causes a shift to the left. The fact that x has a positive value shows that indeed it shifts left.

We'll return to the role of oxygen gas pressure and the heterogeneous oxygen-hemoglobin equilibrium at the end of Section 9.3, after we've discussed pH.

## The Effect of Temperature Change on Equilibrium

You already know from our several discussions of the $H_2 \rightleftarrows 2\ H$ equilibrium that temperature changes can have a major effect on an equilibrium chemical reaction system. So temperature changes change equilibrium constants. At sufficiently high temperatures all molecules break down: first to molecular fragments and eventually to atoms. The equilibrium constants of all gas phase reactions approach infinity at sufficiently high temperature.

In Examples 9.2 and 9.3 we used the $N_2O_4(g) \rightleftarrows 2\ NO_2(g)$ equilibrium to show how changes in the total pressure caused shifts in the equilibrium's position. This same reaction is the basis for a well known visible demonstration of the reversible, shifting nature of chemical equilibrium. Just looking at $N_2O_4(g) \rightleftarrows 2\ NO_2(g)$ you can speculate that the single molecule $N_2O_4$ will be favored at lower temperatures, whereas the two molecules 2 $NO_2$ will be favored at higher temperatures.

Raising the temperature does indeed have exactly that effect. When temperature is raised, the pressure and concentration of dinitrogen tetroxide go down while the pressure and concentration of nitrogen dioxide go up. What is more remarkable is that you can see with your own eyes the evidence for the change, because $NO_2$ is a brown gas and $N_2O_4$ is a colorless gas, and the intensity of the color of the reaction system changes as the equilibrium shifts.

Figure 9.5a The brown color shows that the reaction system $N_2O_4(g) \rightleftarrows 2\ NO_2(g)$ contains a substantial proportion of $NO_2(g)$ at room temperature.

Figure 9.5b The reaction $N_2O_4(g) \rightleftarrows 2\ NO_2(g)$ exists primarily as $N_2O_4(g)$ at the temperature of liquid nitrogen, -195.8°C.

The equilibrium reaction system is sealed inside a glass bulb with a volume of about 20 mL. In Figure 9.5a, at room temperature, brown $N_2O_4$ can be seen. After cooling, via a short dip in liquid nitrogen, the colorless dinitrogen tetroxide predominates in Figure 9.5b. Alternating the temperature back and forth causes the change from brown to colorless to reverse endlessly. This particular sealed gas reaction vessel has been used for over 20 years to demonstrate chemical equilibrium.

The way an equilibrium reaction system shifts in response to a temperature change depends on whether it is exothermic or endothermic. You can predict the shift caused by a temperature change if you *imagine* that heat itself behaves as a chemical reactant or a product.

> Heating a system where heat appears on the left hand side (on the reactant side) of the equation causes products to form and reactants to disappear. In other words heating an endothermic equilibrium reaction system shifts it right:
>
> $$\text{Reactants} + \text{heat} \rightarrow \text{Products} \quad \text{(endothermic reaction)}$$
>
> Heating a system where heat appears on the right hand side (on the product side) of the equation causes reactants to form and products to disappear. In other words heating an exothermic equilibrium reaction system shifts it left:
>
> $$\text{Reactants} \leftarrow \text{Products} + \text{heat} \quad \text{(exothermic reaction)}$$

A more formal thermodynamic approach to estimating the temperature dependence of equilibrium constant involves using $K_p = e^{-\frac{\Delta G^o}{RT}}$ to calculate $K_p$ at standard conditions and 298 K and then using the van't Hoff equation to calculate $K_p$ at temperatures other than 298 K. The van't Hoff equation is

$$\ln \frac{K_2}{K_1} = \frac{\Delta H^0(T_2-T_1)}{RT_1T_2}$$

where $K_2$ is the equilibrium constant at temperature $T_2$, $K_1$ is the equilibrium constant at temperature $T_1$, and R is the ideal gas constant. Using this data we can first calculate the equilibrium constant for the reaction at 298 K and then at any other temperature. The procedure described in this paragraph is illustrated in Example 9.4.

**Example 9.4**: Calculate the $K_p$ for the equilibrium $N_2O_4(g) \rightleftarrows 2\ NO_2(g)$ at 25°C and 50°C. The needed thermodynamic data for the $N_2O_4\ (g) \rightleftarrows 2\ NO_2(g)$ reaction at 25°C are:

| | $\Delta H^0_f$ kJ mol$^{-1}$ | $\Delta G^0_f$ kJ mol$^{-1}$ |
|---|---|---|
| $N_2O_4$ (g) | 9.16 | 97.82 |
| $NO_2$(g) | 33.2 | 51.30 |

and the changes in the state properties in the reaction at 25°C are:

$\Delta H^0 = 2 \times 33.2 - 9.16 = 57.2$ kJ mol$^{-1}$

$\Delta G^0 = 2 \times 51.3 - 97.82 = +4.78$ kJ mol$^{-1}$

Strategy: First substitute into $K_p = e^{-\frac{\Delta G^o}{RT}}$ to get the equilibrium constant at 298 K. Second, substitute into the van't Hoff equation using 25°C (298 K) and 50°C (323 K).

So,
$$K_p = e^{-\frac{\Delta G^o}{RT}} = e^{-\frac{4.78\ kJ\ mol^{-1} \times 1000\ J\ kJ^{-1}}{8.314\ J\ mol^{-1}\ K^{-1} \times 298\ K}} = e^{-1.93}$$

$K_p$ = <u>0.146 atm</u> (at 25°C)

That's $K_1$, at 25°C, using it and the van't Hoff equation we can get $K_2$ from the substitution:

$$\ln \frac{K_2}{K_1} = \frac{\Delta H^0(T_2-T_1)}{RT_1T_2}$$

$$\ln \frac{K_2}{K_1} = \frac{57200\ J\ mol^{-1}(323-298)}{8.314\ J\ mol^{-1}\ K^{-1} \times 298 \times 323}$$

So, $\ln \frac{K_2}{0.146} = 1.79$

and, $K_2 = 0.146 \times e^{1.79} =$ <u>0.87 atm</u>

Table 9.3 shows calculated $K_p$ values at various temperatures, using the procedure of Example 9.4. The higher the temperature, the larger the $K_p$ value. Liquid nitrogen's boiling point is -196.8°C and at that temperature the equilibrium lies overwhelmingly on the left. Thus, the equilibrium reaction system contains essentially no brown $NO_2$ when dipped in liquid nitrogen and is colorless. At high temperatures, the decomposition is complete and the equilibrium constant becomes infinitely large.

| Table 9.3 **Variation of $K_p$ with temperature for the reaction $N_2O_4(g) \rightleftarrows 2\ NO_2(g)$** | |
|---|---|
| T (°C) | $K_p$ atm$^{-1}$ |
| very high | $\infty$ |
| 150 | 133 |
| 100 | 15 |
| 75 | 4 |
| 50 | 0.87 |
| 25 | 0.15 |
| 0 | 0.02 |
| -25 | 0.009 |
| ... | ... |
| -196 | $1 \times 10^{-30}$ |

The temperature dependence of the equilibrium constant of any reaction can be calculated using the procedure of Example 9.4. The van't Hoff equation gives only approximations to $K_p$, but they are good enough for our purposes here.

**Chemical Equilibrium and Chemical Analysis**

Chemists apply equilibrium principles to select particular reactions for particular purposes. For example, choosing the proper reaction is important in methods of traditional analytical chemistry that rely on stoichiometry. You've already seen how the reaction in aqueous solution between an acid and a base can be used for analysis of acid or base concentrations (see Section 3.6, Stoichiometry of Reactions in Solution). A reaction used for analysis (such as an acid-base titration) should be quantitative. A **quantitative reaction** is one with a sufficiently large $K_c$ such that at equilibrium the reactant concentrations are so small that they can be neglected. Chemists say that quantitative reactions "go to completion."

## Section 9.3 The Water Autoionization Equilibrium and pH

It is *not* possible to remove the ions from water—no matter how pure it is. The reason is that water is *always* in equilibrium with hydrogen ions and hydroxide ions. The presence of ions in water make any aqueous solution electrically conducting; the purest water has a small but definite electrical conductivity.

One experimental proof of water's autoionization comes from the measurement of the electrical conductivity of a repeatedly distilled sample of water. The experimental situation is shown schematically in the diagram below.

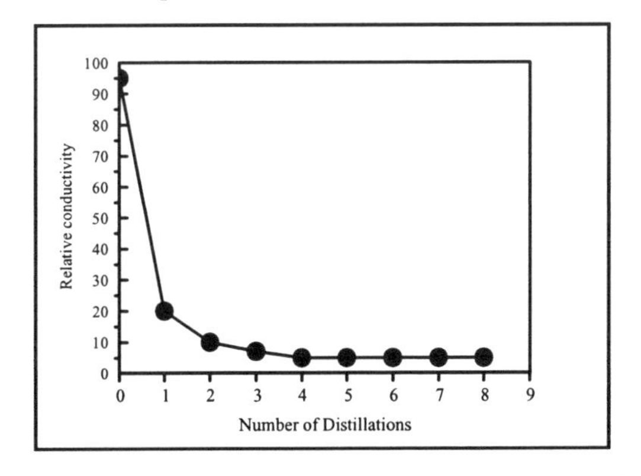


Figure 9.6 The conductivity of water after successive distillations confirms that you cannot *ever* remove all the ions from it. An equilibrium dissociation reaction always makes ions in water.

Observe from Figure 9.6 that even after repeated distillations the conductivity of water does not become zero. Water that contains dissolved ionic compounds has a significant electrical conductivity. Thus, sea water with a high concentration of sodium ions and chloride ions is highly conducting. River water is much less conductive than sea water, but is usually fairly conductive. In contrast, distilled water has a low conductivity—alternatively we can say that distilled water has high electrical resistance. Distillation purifies water; ionic substances are nonvolatile (incapable of turning into a vapor) and they separate from pure water during the distillation procedure—they remain in the distillation vessel while the water distills out and is recovered in purer form in the condenser. But the experimental fact is that repeated distillations do *not* reduce water's conductivity to zero. The residual conductivity of repeatedly distilled water is direct evidence for water's autoionization. The **autoionization of water** is its inevitable, spontaneous formation of hydrogen and hydroxide ions. The autoionization of water is an equilibrium process:

$$H_2O \rightleftarrows H^+ + OH^-$$

for which the equilibrium constant expression is written as:

$$K_c = \frac{[H^+][OH^-]}{[H_2O]}$$

Furthermore, the molarity of water, $[H_2O]$, can be regarded as being essentially constant at 1000 g $L^{-1}$ ÷ 18.0 g $mol^{-1}$, or 55.6 mol $L^{-1}$. Because $[H_2O]$ is constant, water's equilibrium constant can be simplified and re-labeled to give:

$$K_w = [H^+][OH^-] = 1.0069 \times 10^{-14} \text{ mol}^2 \text{ L}^{-2} \text{ at } 25°C$$

where $K_w$ is the **ion-product constant of water**, equal to $1.0069 \times 10^{-14}$ $mol^2$ $L^{-2}$ at 25°C. The equilibrium concentrations of $H^+$ and $OH^-$ ions in pure water can be calculated from the above equation. In pure water, $[H^+] = [OH^-]$ and each is equal to the square root of $K_w$, or $1.00 \times 10^{-7}$ M. Such small ionic concentrations explain why pure water is such a poor conductor of electricity. The ion-product constant of water varies with temperature as shown in Table 9.4.

Table 9.4 **Variation of the Ion-Product Constant of Water with Temperature**

| Temperature °C | Ion-Product Constant |
|---|---|
| 15 | $4.5082 \times 10^{-15}$ |
| 20 | $6.8077 \times 10^{-15}$ |
| 25 | $1.0069 \times 10^{-14}$ |
| 30 | $1.4689 \times 10^{-14}$ |
| 35 | $2.0893 \times 10^{-14}$ |

Water's ionization is an endothermic process, revealed by the data of Table 9.4, which shows that the autoionization reaction shifts to the right (K gets larger) as the temperature is raised. Because the value of $K_w$ at 25°C is $1.00 \times 10^{-14}$ M it is convenient to set up a logarithmic scale of acid-base behavior using the positive value of the exponent of the hydrogen ion molarity at that temperature. This scale is the pH scale.

## The pH Scale

The pH of a solution is a readily measured property. Perhaps you've already used pH paper such as that pictured in Figure 9.7a. Voltage measuring devices that read out directly in pH units are called pH meters. pH meters (Figure 9.7b) were first developed about 70 years ago and are now widespread, cheap, and reliable.

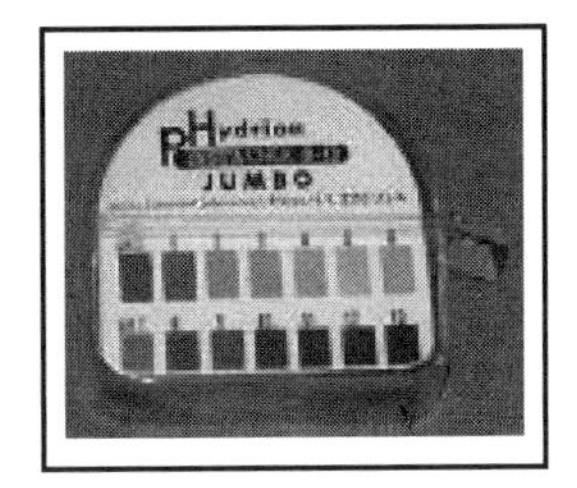

Figure 9.7a pH paper is inexpensive and widely used. Perhaps you've used pH paper to test a swimming pool or fish tank. The color of the paper depends on an equilibrium acid-base reaction in the water being tested.

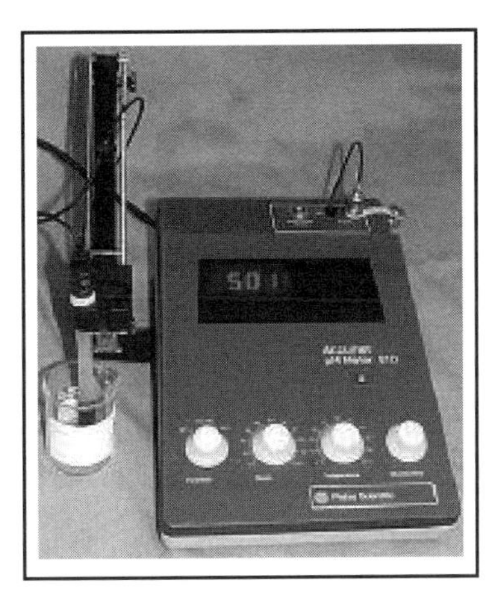

Figure 9.7b A pH Meter. Modern pH meters are cheap, reliable, rugged, and easy to use.

The pH scale was introduced into chemistry in 1909 by the Danish chemist S. P. L. Sørensen, director of the laboratories of Carlsberg, the well-known Copenhagen brewing company. The reason that the letter "p" came to be used is obscure, but a plausible suggestion is that it is the initial letter of "puissance," the French word for strength. Hence, pH loosely means "strength of hydrogen." The hydrogen ion molarity of a solution and its pH are defined by the following two equivalent equations:

$$pH = -\log_{10}[H^+] \qquad [H^+] = 10^{-pH}$$

In an entirely parallel manner, the hydroxide ion concentration of a solution is defined by the two equivalent equations:

$$pOH = -\log_{10}[OH^-] \qquad [OH^-] = 10^{-pOH}$$

The relation between pH and pOH is derived by taking logarithms of both sides of the equation that defines $K_w$, $K_w = [H^+][OH^-]$, and multiplying both sides by -1:

$$-1 \times \log(K_w) = -1 \times \log([H^+][OH^-])$$

$$-1 \times \log(1.00 \times 10^{-14}) = -\log[H^+] - \log[OH^-]$$

$$\text{Thus,}\ pH + pOH = 14 \quad (\text{at } 25°C)$$

In aqueous solutions, hydrogen ion concentrations may vary over many orders of magnitude. A strongly acid solution may have a $[H^+]$ of up to perhaps 10 M; a neutral solution (at 25°C) has $[H^+] = 10^{-7}$ M; and a strongly basic solution may have a hydrogen ion concentration, $[H^+]$, as low as $10^{-15}$ M. In the concentrations above, exponents of ten vary between +1 and -15. It is these exponents that form the pH scale, but because the numbers are mostly negative, their signs are reversed to give a mostly positive scale that ranges from -1 to 15.

### The pH of Solutions of Strong Acids and Bases

For any acid that completely ionizes in solution, the acid's hydrogen ion concentration can be directly calculated from stoichiometry (Section 3.6). The hydroxide ion concentration of the same solution can be obtained from $K_w$. Unless the acid concentration is small (comparable to $10^{-7}$ M), the autoionization of water does not contribute significantly to the hydrogen ion concentration. The pH scale can be summarized as:

| pH | -1 | 0 | 1 | 2 | 3 | 4 | 5 | 6 | 7 | 8 | 9 | 10 | 11 | 12 | 13 | 14 | 15 |
|---|---|---|---|---|---|---|---|---|---|---|---|---|---|---|---|---|---|
| acidic or basic | strongly acidic | | | | weakly acidic | | | | neutral | weakly basic | | | | strongly basic | | | |

For any base that completely ionizes in solution, the hydroxide ion concentration can be directly calculated from stoichiometry. Once the hydroxide ion concentration is known, the hydrogen ion concentration can be obtained from the equation $[H^+] = K_w/[OH^-]$—or via the relationship pH + pOH = 14. Unless the base concentration is small (comparable to $10^{-7}$ M), the autoionization of water does not contribute significantly to the hydroxide ion concentration of a base solution. A strong acid or base is one that is essentially 100% ionized when it is dissolved in water. Common strong acids and bases are listed in Table 9.5.

Table 9.5 **Common Strong Acids and Bases**

| Strong Acids (100% ionized) | |
|---|---|
| HCl (hydrochloric acid) | $HNO_3$ (nitric acid) |
| HBr (hydrobromic acid) | $HClO_4$ (perchloric acid) |
| HI (hydroiodic acid) | $H_2SO_4$ (sulfuric acid) |
| **Strong Bases (100% ionized)** | |
| NaOH, KOH, and other hydroxides of group 1 metals | |
| $Ca(OH)_2$, $Sr(OH)_2$, $Ba(OH)_2$, | |

Bottles of the three most common strong acids are pictured in Figure 9.8. Hydrochloric acid is still often called muriatic acid and as such is readily available in hardware stores for use as a concrete cleaner, etc. The etymology of "muriatic" is interesting. The Latin word *muriaticus* means pickled in brine; brine in Latin is *muria*. Brine is an aqueous solution of sodium chloride. Hence the chlorine-containing acid that derives from pickling salt is muriatic acid.

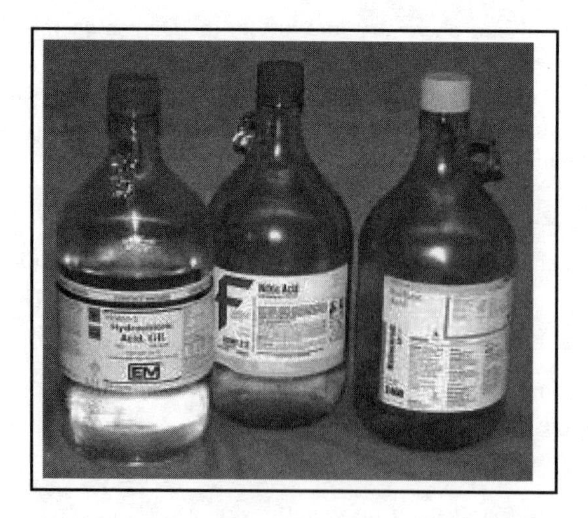


Figure 9.8 Gallon containers of hydrochloric, nitric, and sulfuric acids for laboratory use are reasonably pure. These acids are all large scale industrial compounds. Annual per capita US consumption of HCl, $HNO_3$, and $H_2SO_4$ is respectively about 40, 80, and 380 pounds.

Examples 9.5, 9.6, and 9.7 demonstrate calculations of the pH, pOH, $[H^+]$, and $[OH^-]$ of strong acid and strong base solutions. Note that pH and pOH are inevitably linked because of chemical equilibrium; if one goes up the other goes down, that's pH balance!

**Example 9.5**: Calculate the pH and the pOH of a solution of 0.213 M nitric acid.

Strategy: Recognize that nitric acid is 100% ionized, and thus its hydrogen ion concentration is equal to its molarity. Use the definition, $pH = -\log_{10}[H^+]$, to get the pH and subtract that from 14 to get the pOH.

$$pH = -\log_{10}(0.213) = \underline{0.67}$$

$$pOH = 14.00 - 0.67 = \underline{13.33}$$

**Example 9.6**: Calculate the pH and the pOH of a solution of 0.0745 M potassium hydroxide solution.

Strategy: Recognize that potassium hydroxide is 100% ionized, and thus its hydroxide ion concentration is equal to its molarity. Use the definition, $pOH = -\log_{10}[OH^-]$, to get the pOH and subtract the pOH from 14 to get the pH.

$$pOH = -\log_{10}(0.0745) = \underline{1.13}$$

$$pH = 14.00 - 1.13 = \underline{12.87}$$

**Example 9.7**: Calculate the hydrogen ion concentration and the hydroxide ion concentration of a solution that has a pH of 3.75.

Strategy: Use the definition $[H^+] = 10^{-pH}$ to calculate the hydrogen ion molarity. Next, substitute that pH into $pOH = 14 - pH$ and solve for the pOH. Finally, plug the pOH value into the equation $[OH^-] = 10^{-pOH}$

$$[H^+] = 10^{-pH} = 10^{-3.75} = \underline{1.78 \times 10^{-4}\ M}$$

$$pOH = 14 - 3.75 = 10.25; \quad [OH^-] = 10^{-10.25} = \underline{5.62 \times 10^{-11}\ M}$$

To check the answers, multiply the hydrogen ion and hydroxide ion concentrations:

$$\text{Check:} \quad 1.78 \times 10^{-4}\ M \times 5.62 \times 10^{-11}\ M = 1.00 \times 10^{-14} \ \text{(okay)}$$

The pH scale is logarithmic. Example 9.8 shows you how to determine a hydrogen ion concentration ratio from two specified pH values.

**Example 9.8**: Solution A has a pH of 2.3 and solution B has a pH of 4.6. Which is more acidic, and by how many times, as measured by the ratio of the solutions' hydrogen ion molarities?

Strategy: Recognize first that the solution with the lower pH is more acidic. Next, use the definition $[H^+] = 10^{-pH}$ to calculate the molarities. Finally, calculate the ratio.

Solution A, $[H^+] = 10^{-2.3} = 5.0 \times 10^{-3}$ M    Solution B, $[H^+] = 10^{-4.6} = 2.5 \times 10^{-5}$ M

$$\text{Hydrogen ion molarity ratio} = (5.0 \times 10^{-3}\ M \div 2.5 \times 10^{-5}\ M) = \underline{200}$$

So measured by its hydrogen ion concentration, solution A is two hundred times as strong as solution B.

## Some Applications of pH

The concept of pH is used widely in fields other than chemistry, and many books have been written describing its uses. The pH values of some ordinary substances are shown in Table 9.6.

Table 9.6 **The pH Values of Some Foods**

| | |
|---|---|
| Limes | 1.8 - 2.0 |
| Vinegar | 2.4 - 3.4 |
| Wines | 2.8 - 3.8 |
| Apples | 2.9 - 3.3 |
| Soft drinks | 3.0 - 5.0 |
| Beers | 4.0 - 5.0 |
| Tomatoes | 4.1 - 4.4 |
| Spinach | 5.1 - 5.7 |
| Cow's milk | 6.4 - 6.8 |
| Human milk | 6.6 - 7.6 |

### pH in Everyday Life

The pH of Soils: The pH of soil is a significant factor in determining what plant species will or will not thrive, and soil scientists have made extensive studies of pH and plant growth. For example, azaleas and orchids grow best at pH 4.0 - 5.0, while lettuce and celery prefer the 6.0 - 7.0 range. Probably the greatest general influence of pH on plant growth is through its effects on the availability of nutrients.

Acid Rain: Natural systems, especially forests and lakes, are damaged when environmental conditions become too acidic. Lowered pH caused by acid rain poses a major environmental threat.

Corrosion: Corrosion is unwanted chemical attack on metals or alloys. Many engineers perforce become involved in corrosion control, for example in protecting pipelines, sea going boats, or offshore oil rigs. There are many other engineering aspects of pH. In power plants with steam boilers, the problems of corrosion are potentially severe, and close attention is paid to monitoring the pH of the boiler water.

pH and Biochemistry: In all living systems, the pH of cell and body fluids is vitally important. The blood of mammals is maintained at pH 7.40 by chemical equilibria involving carbon dioxide, carbonic acid, and the bicarbonate ion. Inside living cells, the equilibrium between hydrogen phosphate ($HPO_4^{2-}$) and dihydrogen phosphate ($H_2PO_4^-$) ions maintains constant pH. Maintaining a constant pH is an equilibrium process called buffering, and a solution designed to maintain a constant pH is called a buffer solution. We'll look at buffering shortly, after we've taken up the topic of weak acids. We'll also look at the role of pH in the hemoglobin-oxygen equilibrium.

## Indicators and the Measurement of pH

A **pH indicator** is a compound (a dye) that changes color as the pH of its environment varies. It exists in two differently colored forms in equilibrium with one another: a molecular form and an ionized form. If the molecular form of the indicator is represented by HIn and the ionized form by $In^-$ ($In^-$ is an organic anion), its equilibrium reaction is $HIn \rightleftarrows H^+ + In^-$. An indicator can therefore be regarded as a weak acid. The equilibrium constant, $K_{in}$ is written as:

$$K_{in} = \frac{[H^+][In^-]}{[HIn]}$$

The relative concentrations of HIn and $In^-$, and thus the color of the indicator, are controlled by the hydrogen ion concentration. The Le Chatelier principle states that adding acid ($H^+$) will shift the reaction left. Conversely, adding $OH^-$ reduces $H^+$ and shifts the reaction right. For an indicator that changes color at pH 8, the speciation diagram curve is shown in Figure 9.9. Note that each species has a fraction of ½ at pH 8. A **speciation diagram** (or **species distribution curve**) shows how the concentrations of ions and molecules in an aqueous solution change as the pH is changed.

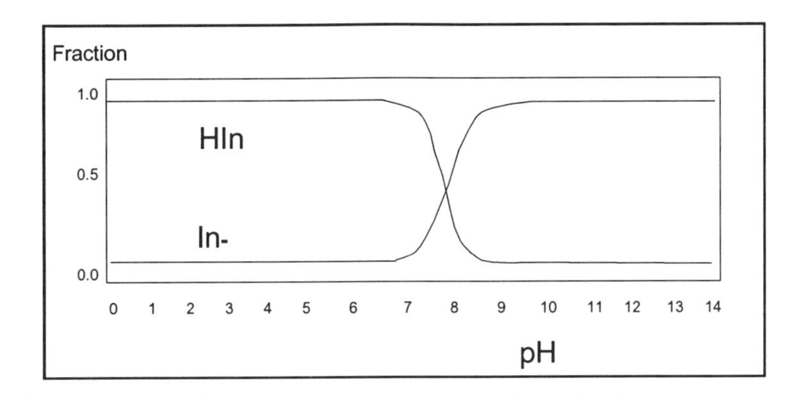


Figure 9.9 A speciation diagram. The concentrations of the species HIn and $In^-$ vary as a function of pH. Because the two species have different colors, the shifting equilibrium between them creates a color change as the pH is altered. This equilibrium shift can then be exploited to make an acid-base indicator.

Indicators are intensely colored dyes, so only small amounts of them are needed. The equilibrium between the two forms of the indicator is determined by the pH of the solution. Phenolphthalein is a common acid-base indicator. The structures of the colorless and pink forms of phenolphthalein are shown in Figure 9.10.

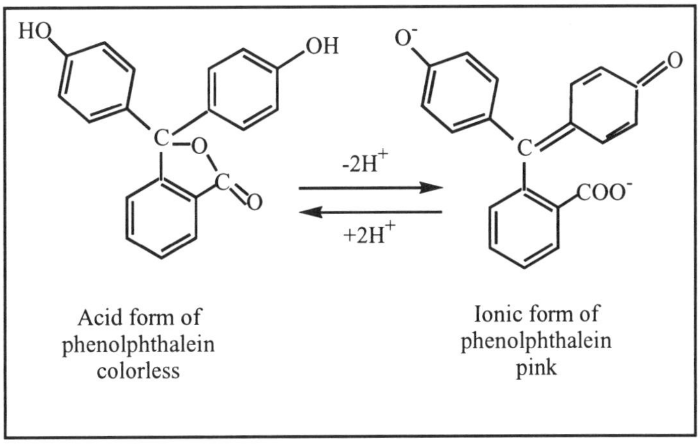

$-2H^+$ / $+2H^+$

Acid form of phenolphthalein colorless

Ionic form of phenolphthalein pink

Figure 9.10 The common indicator phenolphthalein contains three aromatic rings and two hydrogen atoms that make it a diprotic acid. The equilibrium shown above could be represented much more simply as $H_2P \rightleftarrows P^{2-} + 2\,H^+$.

Adding hydrogen ion (acid) to a solution of phenolphthalein shifts the equilibrium to the left and the indicator becomes colorless. Adding hydroxide ion (base) lowers the hydrogen ion concentration, shifts the reaction to the right, and causes the appearance of a pink color. Table 9.7 shows a few of the many available acid-base indicators. Typically, these indicators are organic compounds containing benzene rings and having 20 - 60 carbon atoms. In Table 9.7, equilibrium constants for indicators dyes are shown as $pK_{in}$ values. The subscript "in" refers to indicator. pK is defined using K in a parallel fashion to the definition of pH.

$$pK = -\log_{10} K \qquad \text{or} \qquad K = 10^{-pK}$$

Table 9.7 **Some Common Acid-Base Indicators**

| Indicator | pH range of color change | $pK_{in}$ | Color at lower pH | Color at higher pH |
|---|---|---|---|---|
| Methyl violet | 0 - 2 | 1.0 | Yellow | Violet |
| Metanil yellow | 1.2 - 2.3 | 1.8 | Red | Yellow |
| Methyl orange | 2.9 - 4.0 | 3.5 | Red | Yellow |
| Bromocresol purple | 5.2 - 6.8 | 6.0 | Yellow | Purple |
| Bromthymol blue | 6.0 - 7.6 | 6.8 | Yellow | Blue |
| Phenol red | 6.4 - 8.4 | 7.4 | Yellow | Magenta |
| Thymol blue | 8.0 - 9.6 | 8.8 | Yellow | Blue |
| Phenolphthalein | 8.3 - 10.0 | 9.1 | Colorless | Pink |
| Alizarin yellow G | 10.1 - 12.0 | 11 | Yellow | Red |

An indicator has a distinct color in the lower end of its color change range, and a different distinct color in the upper end of its color change range. The two colors of each indicator are shown in Table 9.7. The color change actually occurs across a range of pH, and within that range significant amounts of HIn and $In^-$ coexist, producing a blend of the two colors. Examples of this color blending are shown (in black-and-white) in Figures 9.11 and 9.12. The $pK_{in}$ value shown in the table is the pH at which equimolar amounts of the two forms of the indicator are present. The $pK_{in}$ value is the midpoint of the range of color change. Incidentally, pH paper is made by impregnating paper with a dye mixture.

Figure 9.11 Bromocresol purple is yellow in acid solution, blue in basic solution, and a green blend at pH 6.

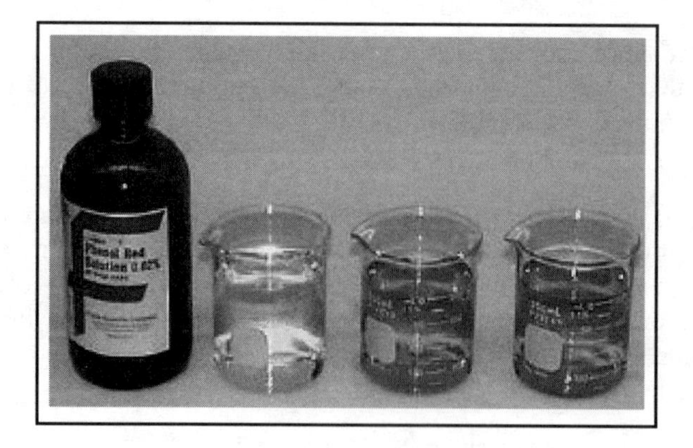

Figure 9.12 Phenol red is yellow in acid conditions and magenta in basic conditions. Phenol red is indeed red around pH 7, but the observed color is actually a blend of the yellow and magenta forms of the dye.

## The Hemoglobin-Oxygen Equilibrium

Many real world chemical processes involve multiple simultaneous equilibria. The hemoglobin-oxygen equilibrium is a heterogeneous reaction involving the uptake of oxygen in the lungs and its subsequent release in various body tissues. For all vertebrates and some other animals this equilibrium and its shifts are literally vital.

Hemoglobin (Hb), is a biochemical entity composed of four protein (polypeptide) chains held by van der Waals bonds in a roughly tetrahedral arrangement (Figure 9.13). The polypeptide chains are folded in such a way as to enable the molecule to carry oxygen reversibly. The hemoglobin entity is roughly spherical with a diameter of about 5500 pm. It has a molar weight of about 64,500 g $mol^{-1}$ and contains roughly 5000 atoms. Each of the four protein molecules has an attached heme group with the molecular formula $C_{34}H_{32}N_4O_4Fe$. Its iron atom is in an octahedral geometric environment with five of the octahedral positions occupied by atoms from the heme group and a sixth position available for reversible oxygen binding. In this way nature has evolved a remarkable oxygen transport system; it is a triumph of natural design and engineering.

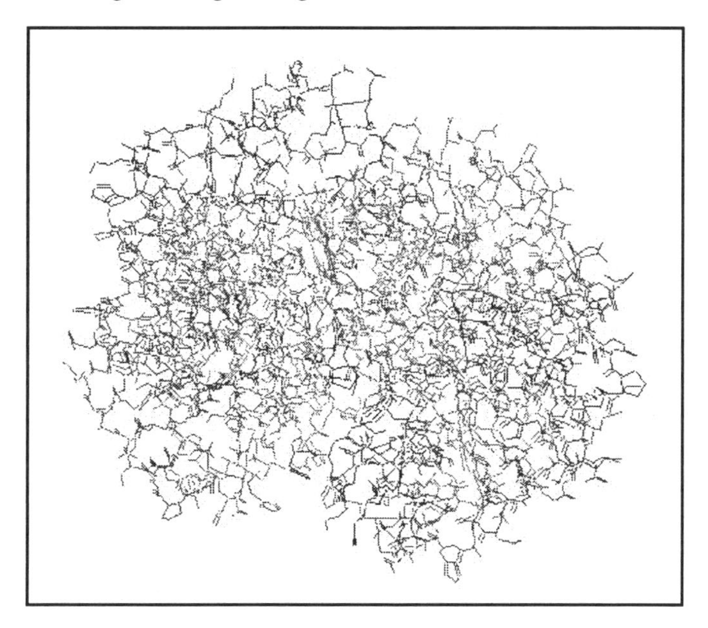

Figure 9.13 Hemoglobin. Obviously, this two-dimensional black-and-white representation is totally inadequate. Today, the way to study protein structure is on the web using a chemical visualization plug-in. [☸kws +"hemoglobin" +structure]

The structure of hemoglobin was first deduced in 1958 by the methods of x-ray crystallography. The crystal structures of several hundred different types of hemoglobin and its mutations are currently known. Figure 9.13 was derived from atomic coordinates from the protein data bank. [☸kws "protein data bank"]. Interested students will find a remarkable world of protein structures awaits them if they search for structures on the web and use one of the visualization plug-ins that have recently become available.

The four steps of the reversible equilibrium reaction in hemoglobin can be written:

$$Hb(aq) + O_2(g) \rightleftarrows Hb.O_2(aq)$$

$$Hb.O_2(aq) + O_2(g) \rightleftarrows Hb.(O_2)_2(aq)$$

$$Hb.(O_2)_2(aq) + O_2(g) \rightleftarrows Hb.(O_2)_3(aq)$$

$$Hb.(O_2)_3(aq) + O_2(g) \rightleftarrows Hb.(O_2)_4(aq)$$

The greater the partial pressure of oxygen the more these equilibrium reactions shift to the right and the greater becomes the saturation fraction of the hemoglobin. When all the hemoglobin exists as $Hb.(O_2)_4(aq)$ the saturation fraction has become 1.0. In the human lungs, hemoglobin absorbs oxygen. Blood flow carries the oxygen-laden hemoglobin to the tissues where the oxygen is released.

The mechanism by which hemoglobin's affinity for oxygen changes is quite complex, but we can give a simplified version here. Hemoglobin responds to allosteric activation. **Allosteric activation** occurs when molecules or ions interact with hemoglobin away from the active site of oxygen in the heme groups and subtly alter the shape of the molecule, changing its oxygen affinity. Some of the factors that affect hemoglobin's oxygen affinity are: 1. other $O_2$ molecules bonded to it (remarkably, each added oxygen molecule increases hemoglobin's affinity for another; this is called cooperative binding by oxygen.), 2. pH, and 3. the concentration of $CO_2$. This action by pH and $CO_2$ is collectively called the Bohr effect (Christian Bohr, 1904). A general equilibrium hemoglobin-oxygen reaction taking the Bohr effect into account may thus be written

$$H^+\text{–}Hb\text{–}CO_2 + O_2(g) \rightleftarrows O_2Hb + H^+ + CO_2$$

The relation between the saturation factor of hemoglobin and the partial pressure of oxygen is shown in Figure 9.14.

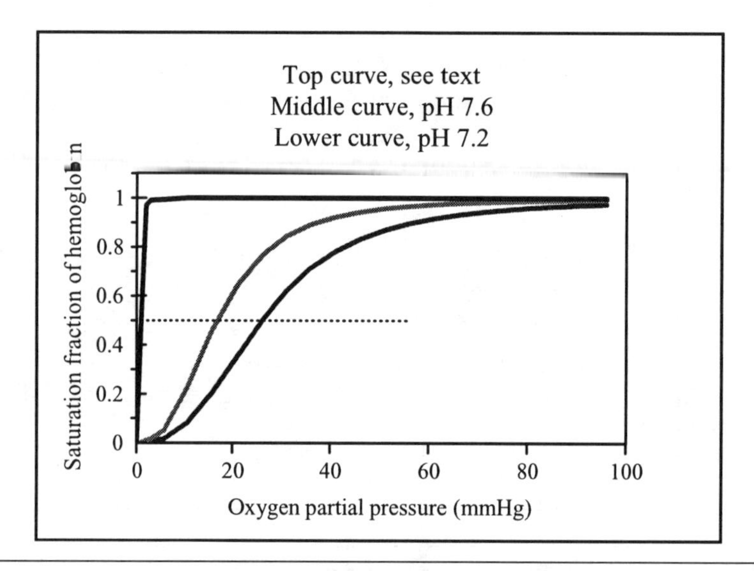


Figure 9.14 Hemoglobin-Oxygen Dissociation Curves. Changing pH shifts the hemoglobin-oxygen equilibrium. In this example, an oxygen partial pressure of 26 Torr produces 50% hemoglobin saturation (dotted line) at pH 7.2, but only 17 Torr is needed to achieve 50% saturation at pH 7.6. The higher pH in vertebrate lungs promotes the pick up of oxygen molecules by hemoglobin; in metabolizing tissue, the pH is lower and the release of oxygen from hemoglobin is stimulated. Shifting pH is only one of a remarkable range of biochemical mechanisms that nature has evolved to regulate chemical equilibria in living systems.

Paradoxically, hemoglobin is *too* good an oxygen binding molecule. Since 1967 it's been known that human blood contains equal molar concentrations of hemoglobin and bisphosphoglycerate ion (Figure 9.15). One BPG ion binds near the center of hemoglobin and alters its overall structure so that the binding of oxygen is diminished. The role of BPG is evidenced by the fact that adding BPG to oxygen-saturated, BPG-free hemoglobin causes the release of most of the oxygen. In other words, the equilibrium below shifts a long way to the right:

$$BPG + O_2Hb \rightleftarrows Hb\text{-}BPG + O_2(g)$$

In Figure 9.14 the top line shows oxygen uptake by hemoglobin at either pH in the absence of BPG; observe that the Bohr effect allowing the reversible equilibrium would not be possible without the allosteric role of BPG. So BPG is critical. Without BPG, hemoglobin would bind oxygen so tightly that it could not be released in the lungs.

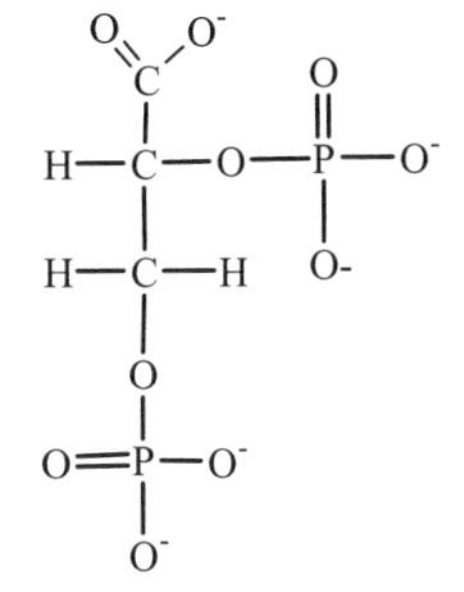

$C_3H_3O_{10}P_2^{5-}$
2,3-Bisphosphoglycerate anion

Figure 9.15 The 2,3-bisphosphoglycerate anion (BPG). One of these anions is located close to the center of every hemoglobin molecule. The role of BPG to distort the hemoglobin structure and prevent it from being *too good* an oxygen binder.

Before we leave hemoglobin we'll note an interesting case of competitive equilibrium. It has long been known that fetal hemoglobin differs from adult hemoglobin in that it binds oxygen better. The reason for this became clear after the discovery of BPG and the observation that fetal hemoglobin is less affected by BPG than is adult hemoglobin; fetal hemoglobin has a higher affinity for oxygen than does maternal hemoglobin. Because of this higher affinity, the growing fetus is able to capture oxygen from the mother's blood in a competitive equilibrium. Nature is ingenious.

## Section 9.4 Weak Acid Equilibrium

Weak acids are those that only partially dissociate in water. They are encountered frequently in our daily lives (Figure 9.16). Common examples include citric acid in fruits and fruit juices, acetic acid in vinegar, and carbonic acid in carbonated drinks. Weak household bases include ammonia solution, baking soda, and antacids,

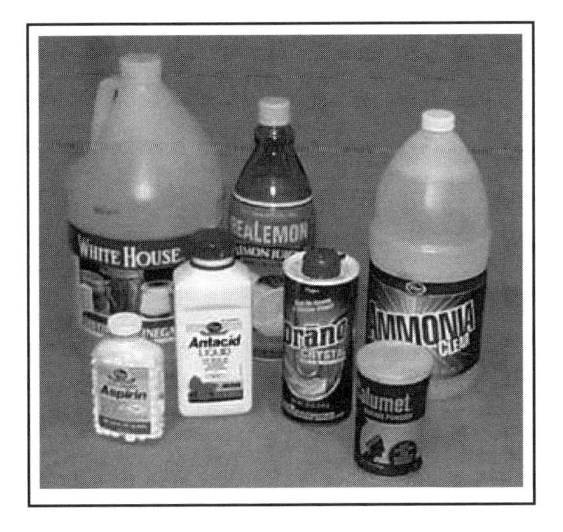


Figure 9.16 Household acids and bases.

If HA is a weak acid that partly dissociates to yield hydrogen ion and the anion $A^-$, then the equilibrium reaction is:

$$HA \rightleftarrows H^+ + A^- \quad \text{where, } K_a = \frac{[H^+][A^-]}{[HA]}$$

$K_a$ is called the dissociation constant for the acid HA. The dissociation constants of many, many weak acids have been measured. Some are shown in Table 9.8. Note that the acids in Table 9.8 are all weak and are ranked from top to bottom from the strongest to the weakest. One important use of equilibrium constants is to rank weak acids (and, of course, other classes of chemical compounds).

The larger the value of $K_a$ the greater the strength of a weak acid. Alternatively, the smaller the value of the acid's $pK_a$ the stronger the acid. The $pK_a$ of a weak acid is defined in the usual manner as $-\log_{10}K_a$; pKa values are shown in the rightmost column of Table 9.8. The pH of a solution of any concentration of any weak acid can be calculated from the acid's $K_a$ value, as shown in Example 9.9. The general procedure is exactly the same as introduced in Section 9.1.

Table 9.8
**Acid Dissociation Constants $K_a$ and $pK_a$ Values for Some Weak Acids in Aqueous Solution at 25°C**

| Acid | Dissociation Reaction | $K_a$ | $pK_a$ |
|---|---|---|---|
| Chloroacetic acid | $HC_2H_2O_2Cl \rightleftarrows H^+ + C_2H_2O_2Cl^-$ | $1.4 \times 10^{-3}$ | 2.85 |
| Hydrofluoric acid | $HF \rightleftarrows H^+ + F^-$ | $6.5 \times 10^{-4}$ | 3.19 |
| Nitrous acid | $HNO_2 \rightleftarrows H^+ + NO_2^-$ | $4.5 \times 10^{-4}$ | 3.35 |
| Benzoic acid | $HC_7H_5O_2 \rightleftarrows H^+ + C_7H_5O_2^-$ | $3.5 \times 10^{-5}$ | 3.45 |
| Lactic acid | $HC_3H_5O_3 \rightleftarrows H^+ + C_3H_5O_3^-$ | $1.4 \times 10^{-4}$ | 3.86 |
| Ascorbic acid (Vitamin C) | $HC_6H_7O_6 \rightleftarrows H^+ + C_6H_7O_6^-$ | $8.0 \times 10^{-5}$ | 4.10 |
| Acetic acid | $HC_2H_3O_2 \rightleftarrows H^+ + C_2H_3O_2^-$ | $1.8 \times 10^{-5}$ | 4.74 |
| Butanoic acid | $HC_4H_7O_2 \rightleftarrows H^+ + C_4H_7O_2^-$ | $1.5 \times 10^{-5}$ | 4.82 |
| Propanoic acid | $HC_3H_5O_2 \rightleftarrows H^+ + C_3H_5O_2^-$ | $1.4 \times 10^{-5}$ | 4.85 |
| Hypochlorous acid | $HOCl \rightleftarrows H^+ + OCl^-$ | $3.1 \times 10^{-8}$ | 7.51 |
| Hydrocyanic acid | $HCN \rightleftarrows H^+ + CN^-$ | $4.9 \times 10^{-10}$ | 9.31 |

**Example 9.9**: Calculate the pH of a 0.367 M solution of hydrofluoric acid (HF).

Strategy: Recognize that hydrofluoric acid is a weak acid and obtain its $K_a$ from Table 9.8. Write the balanced ionization reaction and its $K_a$ expression. Define x as the concentration drop of the acid when it ionizes. Use stoichiometry to get the algebraic change in product concentrations. Finally, substitute into the $K_a$ expression, solve for $[H^+\}$ and take the negative base 10 log of that to get pH.

Hydrofluoric acid equilibrium: $HF = H^+ + F^-$ and $K_a = \frac{[H^+][F^-]}{[HF]}$

| | HF | $\rightleftarrows$ | $H^+$ | + | $F^-$ |
|---|---|---|---|---|---|
| At start | 0.367 | | 0 | | 0 |
| At equilibrium | 0.367 - x | | x | | x |

Substituting: $6.5 \times 10^{-4} = \frac{[H^+][F^-]}{[HF]} = \frac{x^2}{0.367 - x}$

This is a quadratic equation and can be solved by any method you choose. The result is:

$x = 0.0151$, $= [H^+]$. So $pH = -\log_{10}(0.0151) = \underline{1.8}$

## Hydrolysis of the Salts of Weak Acids

When a weak acid reacts with a strong base such as sodium hydroxide or potassium hydroxide, the product is the salt of the weak acid. For example:

$$NaOH + HC_2H_3O_2 \rightleftarrows NaC_2H_3O_2 + H_2O \qquad \text{sodium acetate formed}$$

$$KOH + HF \rightleftarrows KF + H_2O \qquad \text{potassium fluoride formed}$$

Salts are ionic substances that dissolve in water *because* of the ions they form. Thus, solutions of the salts of weak acids are always 100% dissociated in solution. But because the anion comes from a weak acid, it actually reacts with water and the pH of the solution becomes mildly basic. The **hydrolysis** (literally splitting by water) of the salt of a weak acid is the reaction of the salt's anion with water. Here are the hydrolysis reactions of acetate ion and fluoride ion:

$$C_2H_3O_2^- + H_2O \rightleftarrows HC_2H_3O_2 + OH^-$$

$$F^- + H_2O \rightleftarrows HF + OH^-$$

Experimentally it is found that salts such as sodium acetate and sodium fluoride dissolve in water to give solutions with a pH greater than 7.

A weak acid and its related anion are called a **conjugate acid-base pair**. The relationship between them is shown by the following examples that merely reiterate what you know already as acid dissociation.

$$HC_2H_3O_2 \rightleftarrows C_2H_3O_2^- + H^+$$

$$HF \rightleftarrows F^- + H^+$$

$$\text{conjugate acid} = \text{conjugate base} + \text{hydrogen ion}$$

So why do we need new terminology? The answer is that the anions of weak acids really do behave as bases in solution, and the new terminology is simply a recognition of that fact. We'll soon return to conjugate acid-base pairs. Anions from strong acids do not suffer hydrolysis; strong acids are 100% ionized.

We can even treat water itself as a conjugate acid-base pair if we write the autoionization of water as

$$H_2O + H_2O \rightleftarrows H_3O^+ + OH^-$$

Where $H_3O^+$ (**hydronium ion**) is the conjugate acid of water and $OH^-$ is the conjugate base of water. Because water can act as both an acid and a base it is an amphoteric substance. **Amphoterism** is the phenomenon of a single compound showing both acid and base characteristics. Example 9.10 shows a hydrolysis calculation.

**Example 9.10**: Calculate the pH of a solution of 0.200M sodium nitrite.

Strategy: Recognize that this is the salt formed from a strong base (NaOH) and a weak acid ($HNO_2$, nitrous acid). Write the hydrolysis equation for the reaction of the nitrite ion with water. Define x to be the decrease in the nitrite ion concentration caused by hydrolysis; calculate the changes in other concentrations using x and stoichiometry. Finally, substitute the algebraically related quantities into the $K_a$ expression. The hydrogen ion concentration is inserted as $[H^+] = 10^{-14}/[OH^-]$ using the relation $K_w = [H^+][OH^-]$. Finally, use the definition pOH $= -\log_{10}[OH^-]$ to get pOH and the relation pH = 14 - pOH to get pH.

$$NO_2^- + H_2O \rightarrow HNO_2 + OH^-$$

| | $NO_2^-$ | $HNO_2$ | $OH^-$ |
|---|---|---|---|
| At start | 0.200 | 0 | 0 |
| At equilibrium | 0.200 - x | x | x |

The acid dissociation constant for nitrous acid is $K_a = \frac{[H^+][NO_2^-]}{[HNO_2]}$

Water's concentration is constant and is omitted because it is incorporated as part of the $K_a$ value.

Substituting the related hydroxide concentration gives $K_a = \frac{[10^{-14}/OH^-][NO_2^-]}{[HNO_2]}$

Substituting values: $4.5 \times 10^{-4} = \frac{[10^{-14}/x][0.200 - x]}{[x]}$

Again, we get a quadratic equation that can be solved by any method you choose. By the method of successive approximation I got the result $x = 2.1 \times 10^{-6}$, which I checked by back substitution to recalculate K.

$$\text{Check: } K_{calculated} = \{(10^{-14}/(2.1 \times 10^{-6})) \times (0.200 - 2.1\times 10^{-6})\} \div (2.1\times 10^{-6})$$

$$= 4.5 \times 10^{-4} \text{ (okay)}$$

Finally use the definition pOH $= -\log_{10}[OH^-]$ to get pOH and the relation pH = 14 - pOH to get pH

$$pOH = -\log_{10}(2.1 \times 10^{-6}) = \underline{5.7} \quad pH = 14 - 5.7 = \underline{8.3}.$$

Note, as we expected, the salt of a weak acid hydrolyses in water to give a slightly basic solution.

## Changing pH During a Weak Acid-Strong Base Titration

In the titration of a weak acid (in the flask) with a strong base (delivered from the burette) there is a region in which adding base does not produce much change in the pH. This is called the "buffer region." **Buffering** is the phenomenon in which a solution resists change in its pH when *either* acid or base is added. A **buffer solution** is one that resists changes in its pH either by added acid or by added base. That's a functional definition of buffer solution (what it does, as opposed to how it is made).

The buffer region during the titration is shown in Figure 9.17. If you wish to calculate the changing pH during this type of titration you have to apply the principles of weak acid (or weak base) equilibrium.

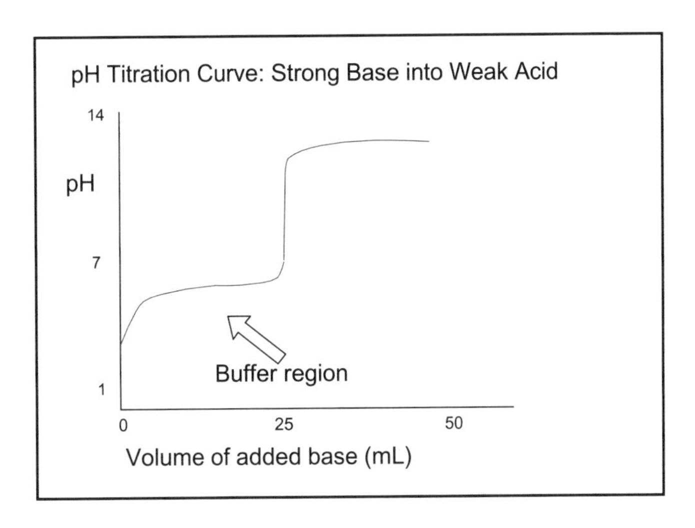


Figure 9.17 The titration curve of the reaction between a weak acid and a strong base shows a buffer region where added base causes relatively small pH change. Here buffer-type chemical equilibrium is at work.

A buffer solution contains *both* partners of a conjugate acid-base pair. That's a compositional definition of a buffer solution (what it's composed of, as opposed to what it does). One partner neutralizes added acid; the other partner neutralizes added base. Thus, a buffer solution is one that can in effect neutralize either acid or base that is added to it. Figure 9.18 shows commercial buffer solutions. The solutions have traces of dyes to code them; these solutions are naturally colorless. Such buffer solutions are used to standardize pH meters.

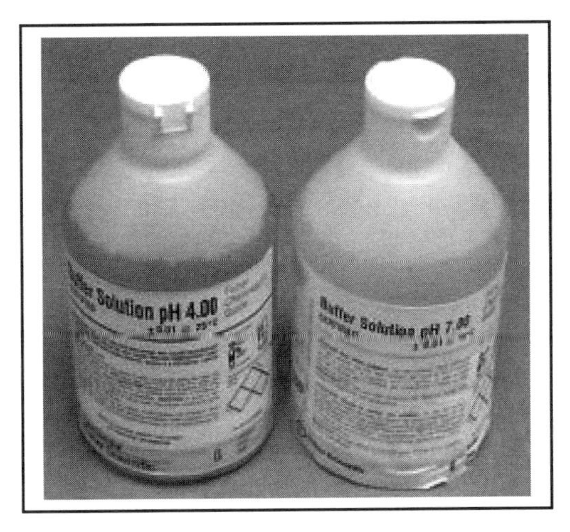


Figure 9.18 Commercial buffer solutions are sold for the purpose of standardizing pH meters.

If you look carefully at the close-up of the pH 4.00 buffer label in Figure 9.19, you'll see that the reliability of this buffer is "traceable to N. I. S. T." That means that sometime during its preparation it was compared with an NIST standard, or more likely to a secondary standard that itself was compared with an NIST standard.

"Formerly N.B.S." on the label refers to the fact that the National Institute of Science and Technology was renamed fifteen years ago from its earlier name of "National Bureau of Standards." What do you think a "mold inhibitor" does?

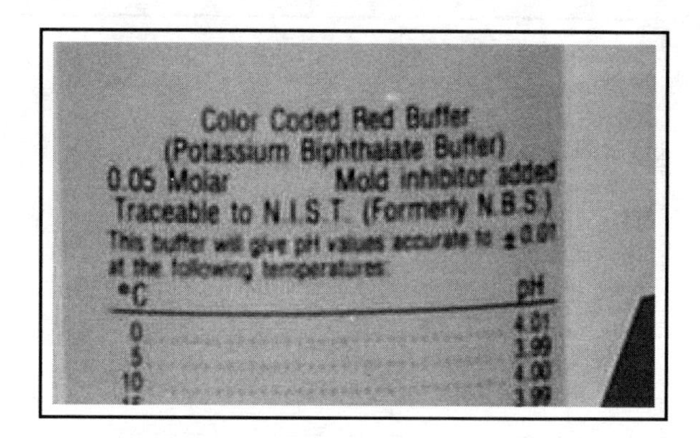


Figure 9.19 This buffer solution claims to be directly linked to a national NIST standard.

You'll also see that the solution has different pH values at different temperatures. This fact should not surprise you, because we've already pointed out that $K_w$, along with all other equilibrium constants, changes with changing temperature. The label has an expiration date; this is a product with a shelf-life. The topic of chemical kinetics that we'll meet in Chapter 15 tackles the issue of shelf-life and the rates of chemical reactions.

## The Procedure to Make a Buffer Solution

Because a buffer solution must be able to neutralize both added base and added acid, it contains two components. One component is a weak acid and the other is its conjugate base. The most reliable buffer action occurs when the two components are present in equimolar amounts.

A specific procedure to prepare an acetate buffer solution is take a liter of a 1.0 molar solution of acetic acid and add 1.0 mol of solid sodium acetate. That produces a solution containing more or less 1.0 molar each of $C_2H_3CO_2H$ and $C_2H_3CO_2^-$. The compositional definition of a standard buffer solution is that it is a mixture (more or less equimolar) of a weak acid and the salt of that weak acid.

The example above gives an acidic buffer, pH > 7. To get a basic buffer solution pH < 7, you would mix a weak base and the salt of that weak base, again in roughly equimolar quantities. A mixture of ammonium hydroxide ($NH_4OH$) and ammonium chloride ($NH_4Cl$) is an example of a basic buffer solution.

With this background you can better understand why a buffer region arises during a weak acid with strong base titration. At the start of the titration there is pure acid in the reaction flask. But at some stage, sufficient base solution has been added to neutralize *one-half* of the original acid. At this point, equimolar amounts of the acid and its salt are present. So the solution in the reaction flask precisely fits the definition of a buffer solution.

## The Henderson-Hasselbalch Equation

Because buffer solutions involve weak acid equilibrium, calculations of buffer pH values can be made using the weak acid equilibrium principles you have recently learned about. It turns out that rewriting the equilibrium constant expression gives an equation of a conveniently simple form. It's called the Henderson-Hasselbalch equation. The Henderson-Hasselbalch equation enables you to calculate the pH of any buffer solution. It also allows you to calculate the pH during an acid-base titration when either the acid or base is weak—though we are not going to do that here. We'll derive the equation using weak acid equilibrium principles:

$$K_a = \frac{[H^+][A^-]}{[HA]}$$

The needed inputs for the Henderson-Hasselbalch equation are the value of $K_a$, and the molar concentrations of the acid HA, and its conjugate base the anion $A^-$. Rewriting gives:

$$K_a = [H^+] \times \frac{[A^-]}{[HA]}$$

Taking base 10 logs of both sides:

$$\log_{10}[K_a] = \log_{10}[H^+] + \log_{10}\frac{[A^-]}{[HA]}$$

and, using the definitions:

$$-pK_a = -pH + \log_{10}\frac{[A^-]}{[HA]}$$

Rearranging gives the conventional form of the Henderson-Hasselbalch equation:

$$pH = pK_a + \log_{10}\frac{[A^-]}{[HA]}$$

**Example 9.11**: A buffer solution is made to be 0.735 M in sodium acetate and 0.422 molar in acetic acid. Calculate the pH of this buffer solution.

Strategy: The value of $pK_a$ for acetic acid is 4.74. Substitute that value and the given concentrations into the Henderson-Hasselbalch equation. Sodium acetate is 100% ionized, so its concentration is the concentration of the acetate ion; acetic acid is only slightly ionized, so its concentration can be taken as the given value.

$$pH = 4.74 + \log_{10}\frac{[0.735]}{[0.422]}$$

$$pH = 4.74 + 0.24 = \underline{4.98}.$$

The Henderson-Hasselbalch equation gives us a simple method to get the pH of an equimolar buffer solution. With the conjugate acid and conjugate base concentrations being equal, their ratio $[A^-]/[HA]$ becomes 1, and since log 1 is zero we get, for an equimolar buffer solution:

$$pH = pK_a.$$

Earlier we saw a similar situation for a acid-base indicator, which you recall is a weak acid. At the mid-point of its color change, during a titration, the HIn and $In^-$ concentrations are equimolar, so at the mid-point of the range of its color change $pH = pK_{in}$, paralleling the equation above.

**Editorial comment by an anonymous reviewer:** "I've never understood how two biochemists could achieve immortality by simply writing someone else's equation in logarithmic form and naming it after themselves."

## Speciation Diagrams

In Figure 9.9 we saw how the molecular and ionic forms of an acid base indicator can be represented on a speciation diagram. A **speciation diagram** is a graph that shows how concentrations in an acid-base equilibrium change as a function of pH. The horizontal axis is pH and the vertical axis shows the fraction of each species that exists at that pH. For example, malonic acid, $H_2C_3H_2O_4$, is a weak diprotic organic acid that exhibits the following two dissociation equilibria:

Step 1 $H_2C_3H_2O_4 \rightleftarrows HC_3H_2O_4^- + H^+$ $\quad K_{a1} = \frac{[H^+][HC_3H_2O_4^-]}{[H_2C_3H_2O_4]} = 1.6 \times 10^{-3}\ \text{mol L}^{-1}$, $pK_a = 2.8$

Step 2 $HC_3H_2O_4^- \rightleftarrows C_3H_2O_4^{2-} + H^+$ $\quad K_{a2} = \frac{[H^+][C_3H_2O_4^{2-}]}{[HC_3H_2O_4^-]} = 2.1 \times 10^{-6}\ \text{mol L}^{-1}$, $pK_a = 5.7$

The speciation diagram of malonic acid is shown in Figure 9.20. Note that the crossover points on the diagram occur at the $pK_a$ values of 2.8 and 5.7. The structural formula of malonic acid is shown in Figure 9.21.

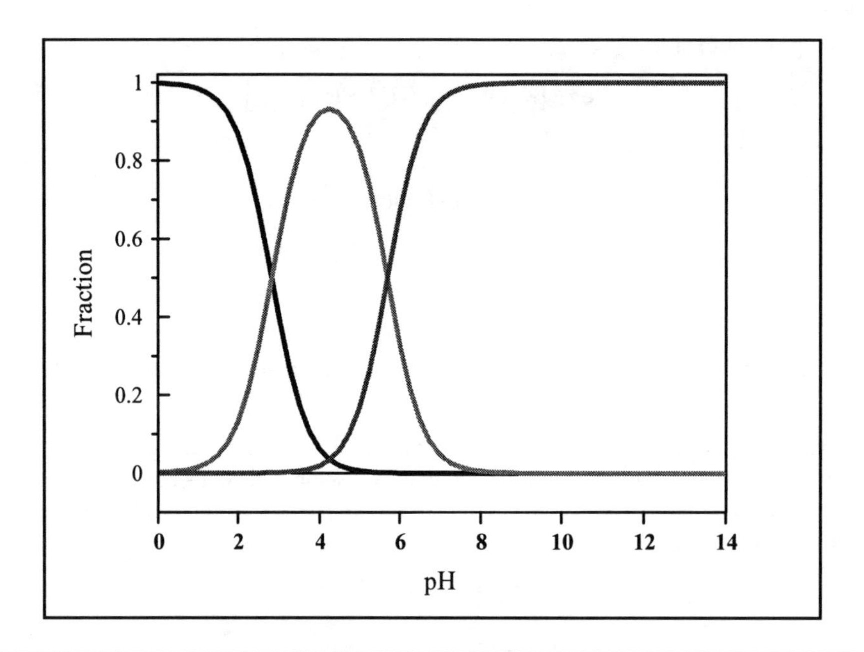


Figure 9.20 The speciation diagram of malonic acid. Undissociated malonic acid, $H_2C_3H_2O_4$, is shown by the line that starts at the upper left. Singly dissociated hydrogen malonate ion, $HC_3H_2O_4^-$, is shown by the line that peaks at pH 4. Doubly dissociated malonate ion, $C_3H_2O_4^{2-}$, is shown by the line that finishes in the upper right.

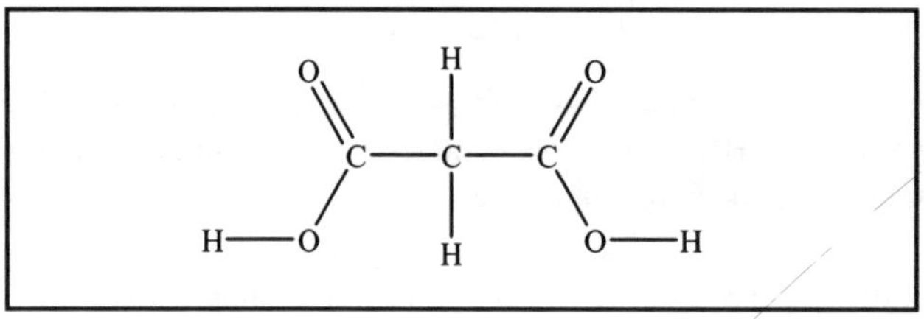

Figure 9.21 The molecular structure of malonic acid $H_2C_3H_20_4$. The two hydrogen atoms attached to oxygen atoms are acidic. The two hydrogen atoms attached to carbon atoms remain attached in acid-base reactions.

The speciation diagram of phosphoric acid is shown in Figure 9.22. Speciation diagrams are valuable to chemists and engineers who are concerned with issues of water quality in natural bodies of water. For example, the quantity and type of phosphate species in the Chesapeake Bay is one of innumerable factors that count in determining the biological health of that body of water.

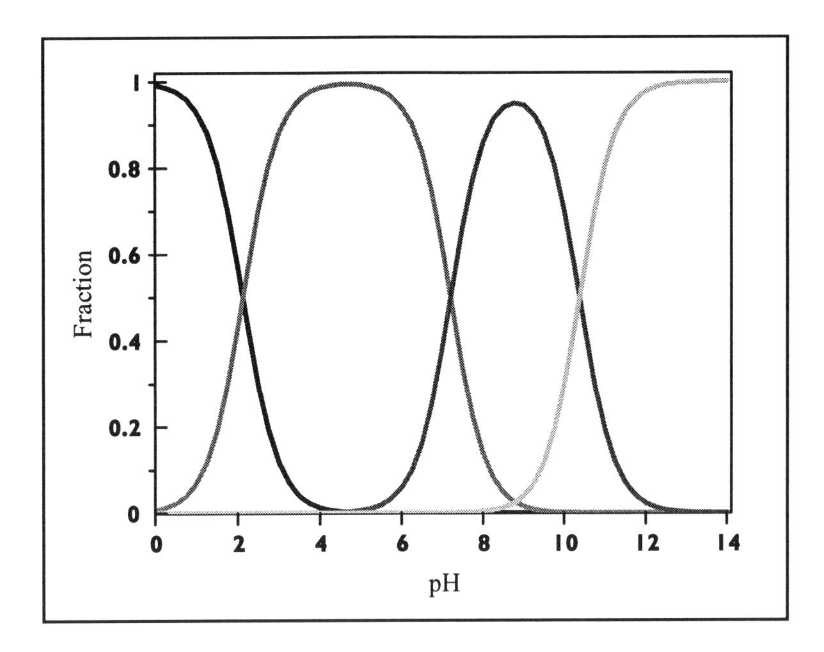


Figure 9.22 The speciation diagram for the triprotic acid, phosphoric. The species are: $H_3PO_4$, the line that begins at the upper left; $H_2PO_4^-$, the broad humped line that peaks around pH 5; $HPO_4^{2-}$, the sharper humped line that peaks around pH 9; and $PO_4^{3-}$, the the line that finishes in the upper right.

## Section 9.5 Solubility and Precipitation Equilibrium

As we saw in Section 3.1, a precipitation reaction occurs when we mix aqueous solutions containing ions that combine to make an insoluble solid. A crude way to deal with precipitation and the solubility of substances in water is simply to tabulate which are soluble and which not.

**Solubility Rules for Common Classes of Inorganic Compounds**

1. Salts containing the alkali metal cations or ammonium are soluble in water
2. Salts containing the following anions arc soluble in water:
   $NO_3^-$, $ClO_3^-$, $ClO_4^-$, $C_2H_3O_2^-$
3. Salts containing the following anions are soluble in water with a small number of exceptions as listed:
   $Cl^-$ except AgCl, $Hg_2Cl_2$, $PbCl_2$, CuCl (that are insoluble)
   $Br^-$ except AgBr, $Hg_2Br_2$, $PbBr_2$, CuBr, $HgBr_2$ (that are insoluble)
   $I^-$ except AgI, $Hg_2I_2$, $PbI_2$. CuI, $HgI_2$ (that are insoluble)
   $SO_4^{2-}$ except $CaSO_4$, $Ag_2SO_4$ (that are moderately soluble)
   and $BaSO_4$, $SrSO_4$, $PbSO_4$, $Hg_2SO_4$ (that are insoluble)
4. Salts containing $OH^-$, $S^{2-}$, $CO_3^{2-}$, $CrO_4^{2-}$, $PO_4^{3-}$, $SO_3^{2-}$ ions are insoluble in water (except in combination with the cations listed in Rule 1 above)

Dissolving a part of a solid in a liquid to form ions creates an equilibrium system. This process is an important example of heterogeneous equilibrium and arises any time an ionic solid comes into contact with a solvent. The most important applications of solubility equilibrium involve substances dissolving in water with the formation of ions. For example, studies of soil and natural waters are in large part the study of solubility equilibria.

The general equation that represents equilibrium between a solid and its solution is: solid ⇄ solution. The forward process is **dissolving** and the reverse is **precipitation**. At equilibrium, the rates of dissolving and precipitation are equal and the solution is **saturated** with solute. If the solute concentration is below its equilibrium value, the solution is **unsaturated**. If the solute concentration is greater than its equilibrium value, the solution is **supersaturated**. A supersaturated solution is not thermodynamically stable, and, the solute will eventually precipitate.

## Solubility Product

A **solubility product** is the equilibrium constant expression for an ionic compound that dissolves to some extent in water. If $M_mX_n$ is an ionic substance slightly soluble in water, it dissolves according to the general equation:

$$M_mX_n(s) \;=\; m\,M^{n+}(aq) + n\,X^{m-}(aq)$$

The general $K_{sp}$ expression is

$$K_{sp} = [M^{n+}]^m[X^{m-}]^n$$

Note that the solid is omitted from the denominator for the reasons described earlier. Typical values calculated from thermodynamic data of $K_{sp}$ are in Table 9.9 at the end of this section. Silver(I), AgCl(s), dissociates in water into one silver(I) ion and one chloride ion. The ionization equation and solubility product expressions are

$$AgCl(s) \rightleftarrows Ag(aq)^+ + Cl^-(aq)$$

and $K_{sp} = [Ag^+]\,[Cl^-]$ (at equilibrium = saturated solution)

Calcium fluoride, $CaF_2(s)$, dissociates in water into one calcium ion and two fluoride ions. The ionization equation and solubility product expressions are

$$CaF_2(s) \rightleftarrows Ca^{2+}(aq) + 2\,F^-(aq)$$

and $K_{sp} = [Ca^{2+}]\,[F^-]^2$ (at equilibrium = saturated solution)

Lithium carbonate, $Li_2CO_3(s)$, dissociates in water into two lithium ions and a carbonate ions. The ionization equation and solubility product expressions are

$$Li_2CO_3(s) \rightleftarrows 2\,Li^+(aq) + CO_3^{2-}(aq)$$

and $K_{sp} = [Li^+]^2\,[CO_3^{2-}]$ (at equilibrium, = saturated solution)

Precipitation reactions are discussed in Section 3.1. Here's a refresher equation in aqueous solution. All the ionic reactants and products are in solution, and the precipitate is a heterogeneous solid.

$$Ag^+ + NO_3^- + Na^+ + Cl^- \rightleftarrows Na^+ + NO_3^- + AgCl\downarrow$$

$Na^+$ and $Cl^-$ are spectator ions and AgCl is a solid, so the K is written *for the reverse reaction* as:

$$K_{sp} = [Ag^+][Cl^-]$$

Table 9.9 shows calculated solubility product constants. Yes, that's *calculated* values. Their root source lies in the experimental calorimetric (or electrochemical) determination of the free energies of formation of the components. But the values in the table are written in the form of traditional equilibrium constants. Table 9.9 is a good example of thermodynamics at work.

**Example 9.12**: Lead(II) chloride has a $K_{sp}$ value of $1.7 \times 10^{-5}$ $M^3$ at 298 K. Calculate the concentration of a saturated solution of $PbCl_2$ in grams per liter. MW of $PbCl_2$ = 277 g $mol^{-1}$.

Strategy: Write the balanced ionization equation and the $K_{sp}$ expression. Define x to be the concentration of the lead ion and 2x to be the concentration of the chloride ion. Substitute these values into the $K_{sp}$ expression and solve for x. Finally, convert molarity to grams per liter.

Lead chloride, $PbCl_2(s)$, dissociates in water into one $Pb^{2+}$ ion and two $Cl^-$ ions. The ionization equation and solubility product expression are

$$PbCl_2(s) \rightleftarrows Pb^{2+}(aq) + 2\ Cl^-(aq)$$

Equilibrium concentrations are x 2x

and, $K_{sp} = [Pb^{2+}]\ [Cl^-]^2$ (at equilibrium = saturated solution)

Plugging x and 2x into the equilibrium constant expression gives:

$$1.7 \times 10^{-5}\ M^3 = (x) \times (2x)^2$$

$$\text{So } 4x^3 = 1.7 \times 10^{-5}\ M^3$$

$$\text{and } x = 1.6 \times 10^{-2}\ mol\ L^{-1}$$

$$\text{Converting to grams} = 1.6 \times 10^{-2}\ mol\ L^{-1} \times 277\ g\ mol^{-1}$$

$$= \underline{4.5\ g\ L^{-1}}$$

Notice that we can make comparisons of the solubilities of compounds by examining their relative $K_{sp}$ values. A solid with a $K_{sp}$ value of $10^{-35}$ is extremely insoluble; a solid with a $K_{sp}$ of $10^{-7}$ is still pretty insoluble, but a lot more soluble that the one with $K_{sp}$ of $10^{-35}$. Example 9.13 shows how we can use solubility equilibrium constants to predict whether or not a precipitate will form when various solutions are mixed.

**Example 9.13**: A solution is prepared by mixing 100 mL of 0.0003 M potassium bromide and 100 mL of 0.00006 M silver(I) nitrate. Does a precipitate form?

Strategy: Consider the possible insoluble compound that might form. Based on the solubility rules, the only possible precipitate would be silver bromide. Silver(I) bromide has a $K_{sp} = 5.35 \times 10^{-13}$ $M^2$. Calculate the product of the ion concentrations after mixing and compare it with the $K_{sp}$ value. A precipitate will form if the product exceeds $K_{sp}$.

$$Q_c = [Ag^+]\ [Br^-] = (\tfrac{1}{2} \times 0.00030\ M) \times (\tfrac{1}{2} \times 0.000060\ M)$$

where the two factors of ½ account for the fact that mixing equal volumes of two solutions mutually halves the concentrations of the solutes.

$$\text{So, } Q_c = 4.5 \times 10^{-9}\ M^2$$

The ratio of $Q_c$ to $K_{sp}$ is $4.5 \times 10^{-9}\ M^2 \div 5.35 \times 10^{-13}\ M^2 = 8400$.

$Q_c$ is much larger than $K_{sp}$, so a precipitate will form and precipitation will continue until the ion product is reduced to being equal to $K_{sp}$.

**Table 9.9 Calculated Solubility Product Constants**

The values listed are *calculated* from the free energies of formation of the compounds and their ions at 298.15 K.

$$M_mX_n(s) = m\,M^{n+}(aq) + n\,X^{m-}(aq)$$

$$\Delta G^o = m\,\Delta G^o_f\,(M^{n+}, aq) + n\,\Delta G^o_f\,(X^{m-}, aq) - \Delta G^o_f\,(M_mX_n, s)$$

$$\ln K_{sp} = -\frac{\Delta G^o}{RT}$$

| Compound | Formula | $K_{sp}$ |
|---|---|---|
| Aluminum phosphate | $AlPO_4$ | $9.83 \times 10^{-21}$ |
| Barium fluoride | $BaF_2$ | $1.84 \times 10^{-7}$ |
| Calcium phosphate | $Ca_3(PO_4)_2$ | $2.53 \times 10^{-33}$ |
| Calcium hydroxide | $Ca(OH)_2$ | $4.68 \times 10^{-6}$ |
| Calcium sulfate | $CaSO_4$ | $7.10 \times 10^{-5}$ |
| Copper(I) sulfide | $Cu_2S$ | $2.26 \times 10^{-48}$ |
| Copper(II) sulfide | $CuS$ | $1.27 \times 10^{-36}$ |
| Iron(II) hydroxide | $Fe(OH)_2$ | $4.87 \times 10^{-17}$ |
| Iron(III) hydroxide | $Fe(OH)_3$ | $2.64 \times 10^{-39}$ |
| Lead(II) chloride | $PbCl_2$ | $1.70 \times 10^{-5}$ |
| Lead(II) fluoride | $PbF_2$ | $7.12 \times 10^{-7}$ |
| Lithium carbonate | $Li_2CO_3$ | $8.15 \times 10^{-4}$ |
| Magnesium hydroxide | $Mg(OH)_2$ | $5.61 \times 10^{-12}$ |
| Mercury(I) iodide | $Hg_2I_2$ | $5.33 \times 10^{-29}$ |
| Mercury(II) iodide | $HgI_2$ | $2.82 \times 10^{-29}$ |
| Nickel(II) hydroxide | $Ni(OH)_2$ | $5.47 \times 10^{-16}$ |
| Silver(I) bromide | $AgBr$ | $5.35 \times 10^{-13}$ |
| Silver(I) chloride | $AgCl$ | $1.77 \times 10^{-10}$ |
| Silver(I) iodide | $AgI$ | $8.51 \times 10^{-17}$ |
| Strontium fluoride | $SrF_2$ | $4.33 \times 10^{-9}$ |
| Zinc sulfide | $ZnS$ | $2.93 \times 10^{-25}$ |

## Essential Knowledge—Chapter 9

**Glossary words:** Approach to equilibrium, slugging a system, disturbing an equilibrium, equilibrium shifts, position of equilibrium, Le Chatelier's principle, homogenous reaction, Haber process, nitrogen fixation, quantitative reaction, autoionization, pH, ion-product constant, pH scale, pH paper, strong acid and bases, van't Hoff equation, pH indicator, speciation diagram, hemoglobin-oxygen equilibrium, bisphosphoglycerate ion, hydrolysis, weak acids and bases, conjugate acid-base pair, hydronium ion, amphoterism, buffering, buffer solution, NIST standard, speciation diagram, solubility product.

**Key Concepts:**
General strategy for solving equilibrium calculations. Equilibrium shifts and reestablishing equilibrium after changes of concentration, pressure, or temperature. Nature of pH and the water equilibrium. Relation of water equilibrium to the strength of acids and bases. Operation of a pH indicator. Principles of buffering. Concept of conjugate acids and bases. Methods for the measurement of pH. Changing pH during titration. Nature of solubility equilibrium and the solubility rules.

**Key Equations:**
General equilibrium constant expression in solution: $K_c = \frac{[C]^c[D]^d}{[A]^a[B]^b}$

van't Hoff equation: $\ln\frac{K_2}{K_1} = \frac{\Delta H^0(T_2-T_1)}{RT_1T_2}$

Water's dissociation: $K_c = \frac{[H^+][OH^-]}{[H_2O]}$ Ion-product of water: $K_w = [H^+][OH^-]$

Definitions of pH: $pH = -\log_{10}[H^+]$ and $[H^+] = 10^{-pH}$

Definitions of pOH: $pOH = -\log_{10}[OH^-]$ and $[OH^-] = 10^{-pOH}$

Relation of pH and pOH: $pH + pOH = 14$ (at 25°C)

The equilibrium constant, $K_{in}$ of an indicator: $K_{in} = \frac{[H^+][In^-]}{[HIn]}$

Definitions of pK: $pK = -\log_{10} K$ and $K = 10^{-pK}$

Weak acid equilibrium: $HA \rightleftarrows H^+ + A^-$ where, $K_a = \frac{[H^+][A^-]}{[HA]}$

The Henderson-Hasselbalch Equation: $pH = pK_a + \log_{10}\frac{[A^-]}{[HA]}$

The general solubility product expression: $K_{sp} = [M^{n+}]^m[X^{m-}]^n$

## Questions and Problems

### Introduction and the Approach to Equilibrium

9.1. Consider the gas phase reaction: 2 $SO_3(g) \rightleftarrows 2\ SO_2(g) + O_2(g)$. In a closed vessel, a study of this reaction shows that the concentration of $SO_2$ is increasing. Is the system in equilibrium? If not, is the forward reaction proceeding faster than the reverse reaction?

9.2. Discuss the practical applications of chemical equilibrium principles.

9.3. When chemical reaction systems reach equilibrium all chemical reaction has stopped. Discuss.

9.4. Under what circumstances was the scientific discipline of chemical equilibrium developed?

9.5. What does it mean to "slug" an equilibrium reaction system?

9.6. List three ways in which a chemist might disturb a chemical equilibrium.

### Le Chatelier's Principle

9.7. State Le Chatelier's principle. Discuss the types of disturbances that can be exerted on a reaction at equilibrium. What is meant by the phrase "reaction equilibrium system"?

9.8. It is well known that ice placed under high pressure will melt. Use the principle of Le Chatelier to deduce what this implies about the volume change during the freezing of ice. Explain what your deduction has to do with icebergs.

9.9. Consider the gas-phase reaction 3 $O_2(g) \rightleftarrows 2\ O_3(g)$. Explain how the equilibrium shifts in this reaction as the total pressure is increased.

9.10. The following equilibrium reaction is thought to exist in a mixture of nitric and sulfuric acids:

$$H_2SO_4(aq) + HNO_3(aq) \rightleftarrows HSO_4^-(aq) + NO_2^+(aq) + H_2O(l)$$

State what happens to the concentrations of each of the other products and reactants when an equilibrium mixture is diluted with a small amount of water.

9.11. Diluting a solution of a monoprotic weak acid give a more dilute solution but one with a greater percentage dissociation. Discuss this statement in the light of Le Chatelier's principle.

### Gas Phase Reactions

9.12. The gas reaction A $\rightleftarrows$ B has a $K_p$ value of 0.370 at 530 K. The equilibrium reaction system has a total pressure of 18.3 atm. Calculate the partial pressures of A and B.

9.13. The gas reaction A $\rightleftarrows$ B has a $K_p$ value of 0.370 at 530 K. Reaction begins with 12.7 moles of pure A (and no B) in a 20.0 L reaction chamber at 530 K. Calculate the partial pressures of A and B when equilibrium has been established.

9.14. The gas reaction A $\rightleftarrows$ 2B has a $K_p$ value of $1.22 \times 10^4$ atm at 435°C. The equilibrium reaction system has a total pressure of 103 atm. Calculate the partial pressures of A and B.

9.15. The gas reaction A + B $\rightleftarrows$ 2C starts with $P_A = 2.0$ atm and $P_B = 1.0$ atm. It shifts right until 80.% of the B has reacted, when equilibrium has been reached. Calculate $P_A$, $P_B$, and $P_C$, and the value of $K_p$ at equilibrium.

9.16. Make a time-concentration graph to illustrate the process described in the preceding question.

9.17. Consider the general reaction 2 A(g) + B(g) $\rightleftarrows$ 3 C(g) + D(g). The concentration of B is increased by injecting some fresh B into a closed system containing an equilibrium mixture. After equilibrium has been allowed to reestablish itself, the concentration of D has increased by 0.002 mol $L^{-1}$. What were the changes in the concentrations of A , and C?

9.18. The equilibrium constant of the reaction A $\rightleftarrows$ 2B + C shown in Figure 9.3 is 25.6 mol$^2$ L$^{-2}$ at a constant but unspecified temperature. The equilibrium concentrations in this reaction system are [A] = 0.110 M, [B] = 1.78 M and [C] = 0.890 M. The reaction system is slugged with an additional amount of reactant, equal to a concentration of 0.270 M, instantaneously raising the [A] concentration to 0.0380 M. The system then undergoes a shift and reestablishes equilibrium. Calculate the new equilibrium concentrations.

9.19. Increasing the external pressure on a gas-phase equilibrium causes an instantaneous increase in the concentrations of all the reactants and products. Explain this statement.

9.20. Consider the gas phase equilibrium 2 A $\rightleftarrows$ B. It has a $K_p$ of 50.0 atm$^{-1}$ and the total pressure of the system is 4.56 atm. Calculate the equilibrium pressures of A and B.

9.21. What is the Haber process? Why is it operated at high pressure?

## The Autoionization of Water

9.22. Describe what is meant by the autoionization of water. Write a chemical equation to represent the process. Explain why the conductivity of pure water increases as the temperature increases.

9.23. At 25°C, an aqueous solution contains a hydrogen ion concentration that is twice the hydroxide ion concentration. Use the ion-product constant of water to calculate the two concentrations.

9.24. Water's autoionization constant, $K_w$, is $1.0069 \times 10^{-14}$ mol$^2$ L$^{-2}$ at 25°C. From that value calculate the pH of pure water at 25°C. Calculate the percentage discrepancy between your calculated value and 7.

9.25. The pH of an aqueous solution of a weak monoprotic acid ($pK_a$ = 9.1) is 5.10 pH. If a sample of the solution is diluted with an equal volume of pure water, what will be the pH of the resulting solution?

9.26. Heavy water ($D_2O$) has deuterium atoms in place of the hydrogen atoms of ordinary water. Write a dissociation equation for heavy water. The ion-product constant for heavy water at 25°C is $1.35 \times 10^{-15}$ mol$^2$ L$^{-2}$. Calculate the deuterium ion ($D^+$) and deuteroxide ion ($OD^-$) concentrations in pure heavy water at 25°C.

9.27. The conductivity of pure water is proportional to the total concentration of ions formed by autoionization. Use the values of $K_w$ to calculate the ratio of water's conductivity at 30°C to that at 15°C.

## pH Calculations for Strong Acids and Bases

9.28. Calculate the pH of solutions with hydrogen ion concentrations of: a. 0.281 M, b. 2.38 M, c. 0.0506 M, d. $1.2 \times 10^{-9}$ M, e. $1.59 \times 10^{-13}$ M.

9.29. Calculate the pOH and [$OH^-$] for each solution in the preceding problem.

9.30. A solution is made by dissolving 50.0 g of NaOH(s) in sufficient water to produce 2.00 L of solution. Assuming 100% dissociation, calculate the pH and pOH of the solution.

9.31. At 25°C, a 4.567 L solution contains 45.6 g of HCl. Calculate the pH and pOH of the solution.

9.32. Calculate the $H^+$ and $OH^-$ ion concentrations at 25°C in the following aqueous solutions: a. 0.100 M NaOH, b. 0.323 M HCl, c. 0.041 M $Ca(OH)_2$, d. 3.05 M HCl.

9.33. What weight of $Ca(OH)_2$ must be present in 500.0 mL of an aqueous solution if it is the only solute and the solution pH is 10.2.

9.34. 1.50 moles of pure NaOH and 2.00 moles of pure HCl are added to 1.000 L of water at 25°C, well-stirred, and allowed to cool to 25°C. Calculate the $H^+$ and $OH^-$ molar concentrations of the final solution.

9.35. 25.0 mL of a 0.234 molar solution of hydrochloric acid is mixed with 33.5 mL of 0.135 molar potassium hydroxide solution. Calculate the pH of the mixture after it reaches equilibrium. (Hint: write the reaction equation and calculate the amount of excess reactant.).

9.36. Calculate the pOH of a 0.200 molar solution of hydrochloric acid.

9.37. Use Table 9.4 that shows the change in $K_w$ with temperature, to calculate the pH of absolutely pure water at 35°C.

9.38. Calculate the ratios of the hydrogen ion molarities (A/B) of two solutions. Solution A has a pH of 3.4 and solution B has a pH of 4.7.

9.39. Calculate the ratios of the hydrogen ion molarities (A/B) of two solutions. Solution A has a pOH of 11.5 and solution B has a pOH of 12.2.

## Weak Acid Equilibrium

9.40. Calculate the hydrogen ion concentration and the pH of a 0.100 M solution of hydrocyanic acid at 25°C.

9.41. A 0.100 M solution of a monoprotic weak acid has a hydrogen ion concentration of 0.00205 M. Calculate $K_a$ and $pK_a$ of the acid.

9.42. Calculate the percent of acetic acid that is ionized in 1.00, 0.100, 0.0100, and 0.00100 M aqueous solutions at 25°C. Explain your results using Le Chatelier's principle.

9.43. The pH of a vinegar sample is 2.85 at 25°C. Calculate the molarity of the vinegar. Assuming that the density of vinegar is 1.0 g $mL^{-1}$, calculate the weight percentage of acetic acid in the vinegar.

9.44. Write the base dissociation reaction and the equilibrium constant expression for the weak base ammonium hydroxide.

## Hydrolysis Reactions

9.45. a. Explain why the aqueous solution of a salt of a strong base and a weak acid, such as sodium cyanide, is basic. b. Write the equation that represents the hydrolysis of the cyanide ion.

9.46. Write balanced chemical equations to show each of the following hydrolysis reactions: a. potassium cyanide dissolving in water to yield a basic solution, b. lithium fluoride dissolving in water to yield a basic solution, and, c. potassium carbonate dissolving in water to yield a basic solution.

9.47. Solid ammonium chloride dissolves in water to yield an acidic solution. This reaction is an example of the hydrolysis of a cation. Write a balanced chemical equation to represent the hydrolysis reaction.

9.48. Write an equation to represent the hydrolysis reaction of sodium nitrite as it dissolves in water. Identify the two conjugate acid-base pairs involved in the hydrolysis reaction.

9.49. Calculate the pH of a 1.0 M aqueous solution of sodium acetate at 25°C.

9.50. Calculate the percentage of the acetate ion that is hydrolyzed in a 0.100 M aqueous solution of sodium acetate at 25°C.

9.51. Calculate the pH of a 0.076 M aqueous solution of sodium cyanide at 25°C.

## Buffer Solutions

9.52. Describe the characteristic chemical property of any buffer solution.

9.53. What is meant by the term buffer action? What is a "buffer" in everyday language?

9.54. Describe the two chemical components that must be present in a solution to make it a buffer solution.

9.55. Potassium hydrogen phthalate (KHP) is the acid salt of the weak diprotic acid, phthalic acid, $H_2C_8H_4O_4$ ($H_2P$). A dilute solution of KHP is a widely used pH = 4.00 buffer solution. Explain why this is a buffer solution. It's the buffer solution pictured in Figure 9.19.

9.56. Consider a buffer solution that is 0.20 M in acetic acid and 0.25 M in potassium acetate. a. What is the pH of the buffer? Assume there is no change in volume of the solution in parts b-e. b. What is the pH if 0.010 mole of NaOH(s) is added to 1.00 L of the buffer? c. What is the pH if 0.010 mole of HCl(g) is added to 1.00 L of the buffer? d. What would be the pH change if 0.010 mole of HCl(g) is added to 1.00 L of pure water? e. What would be the pH change if 0.01 mole of NaOH(s) is added to 1.00 L of pure water?

9.57. Explain how the Henderson-Hasselbalch equation can be used to calculate the shape of the pH titration curve for a weak acid with a strong base.

9.58. Discuss why the equivalence point of a weak acid - strong base titration is never at pH 7.

9.59. Using a strong base titrant, explain how the shape of the titration curve of a weak diprotic acid, such as malonic acid, differs from the titration curve of a weak monoprotic acid, such as acetic acid.

9.60. Write solubility product expressions for the ionization reactions of each of the following slightly water soluble salts: gallium(III) hydroxide, calcium oxalate, cobalt (II) phosphate, bismuth(III) sulfide, and lead(II) sulfate.

9.61. Calculate the concentrations of the barium and fluoride ions in a saturated solution of barium fluoride at 25°C.

9.62. Calculate the concentrations of the lead and carbonate ions in a saturated solution of lead carbonate at 25°C.

9.63. A sodium fluoride solution has a concentration of 0.0500 mol $L^{-1}$. Solid barium fluoride powder is shaken into this solution until an equilibrium has been established and the solution is saturated. Calculate the concentration of barium ions in the saturated solution.

9.64. The solubility product of silver chromate, $Ag_2CrO_4$, at 25°C, is $4.05 \times 10^{-12}$ $mol^3$ $L^{-3}$. Calculate the solubility of silver chromate in: a. pure water, b. 0.0500 M silver nitrate solution, and c. 0.0500 M potassium chromate solution.

9.65. Derive the mathematical relationship between molar solubility in pure water expressed as x mole $L^{-1}$ and $K_{sp}$, for each of the following compounds: strontium carbonate, cerium(III) fluoride, copper(II) phosphate, and silver(I) sulfate.

9.66. Use solubility product data to decide whether or not a precipitate will form when the following solutions are mixed: 500.0 mL of 0.0003 molar silver(I) nitrate and 500.0 mL of 0.00005 molar calcium chloride.

9.67. Use solubility product data to decide whether or not a precipitate will form when the following solutions are mixed: 25.8 mL of 0.0000001 molar lead(II) nitrate and 13.2 mL of 0.0009 molar potassium iodide. $K_{sp}$ of lead(II) iodide is $8.49 \times 10^{-9}$.

9.68. A solution in equilibrium with solid silver(I) sulfide was found to have silver ion and sulfide ion concentrations of $1.26 \times 10^{-17}$ M and $6.3 \times 10^{-18}$ M, respectively. Calculate the $K_{sp}$ of silver(I) sulfide, $Ag_2S$.

## Problem Solving

9.69. Distilled water is thought of as being a very pure form of water. However, the pH of laboratory distilled water at 25°C is usually in the range 5-6. Discuss possible reasons that the pH is not 7, as it should be in truly pure water.

9.70. Natural buffering protects lakes in the southeastern US from the severest effects of acid rain, whereas lakes in the northeast are largely unbuffered. Discuss the soil and geologic factors that cause this difference.

9.71. At a definite but unspecified pressure the equilibrium constant ($K_c$) for the dissociation reaction $I_2(g) \rightleftarrows 2\ I(g)$, is $2.2 \times 10^{-4}$ M. A reaction vessel with a volume of 20.0 L contains 200.0 g of a mixture of $I_2$ and I in a state of chemical equilibrium. Calculate the molar concentrations [$I_2$] and [I]. [Hint: assume you begin with only $I_2$ and allow the equilibrium to be established.]

9.72. If nitric oxide and bromine are mixed, they react according to $2\ NO(g) + Br_2\ (g) \rightleftarrows 2\ NOBr(g)$. An experimental study of this reaction at 500. K showed that an equilibrium mixture contained gases with the following partial pressures: $P_{Br_2}$ = 33.9 Torr, $P_{NO}$ = 96.5 Torr, and $P_{NOBr}$ = 100.8 Torr. Calculate $K_p$ and $K_c$ for this reaction. From Section 8.5 we know for a reaction system composed of ideal gases, $K_c = K_p(RT)^{-\Delta n}$, where $\Delta n$ is the difference between the total number of moles of product gases and the total number of moles of reactant gases; in this case $\Delta n = -1$. Use that expression to answer the second part of the question.

9.73. At 2000 K, the gas phase dissociation reaction $2\ HCl(g) \rightleftarrows H_2(g) + Cl_2(g)$ has an equilibrium constant of $K_p$ = $2.75 \times 10^{-6}$. A sample of HCl(g) is in a closed vessel at 2000 K. Calculate the fraction of it that will dissociate. Explain why it is not necessary to know either the volume of the vessel or the amount of HCl in the system to solve this problem. [Hint: There is redundant information in this problem.]

9.74. $K_p$ for the reaction $H_2(g) + I_2(g) \rightleftarrows 2\ HI(g)$ is 50.2 (no units) at 445°C. $\Delta H^0_f$ of HI gas is 25.94 kJ $mol^{-1}$. Use this information to calculate $\Delta S^0$ of the reaction.

9.75. An equilibrium reaction has $\Delta G^0$ = -20.0 kJ $mol^{-1}$ and a $\Delta H^0$ = -15.0 kJ $mol^{-1}$ Calculate $K_p$ of the reaction at 298 K. Use that value and the van't Hoff equation to estimate $K_p$ at 50°C.

# Chapter 10

# Liquids and their Mixtures

## Introduction

To restate our theme: the bulk properties of materials are determined by the particles that compose the materials and the forces among those particles. This chapter deals with the properties and chemistry of liquids—pure liquids, mixtures of liquids, and mixtures that are liquid. We'll begin with a discussion of pure liquids. At room temperature, almost all pure liquids are molecular substances. A **molecular substance** or a **molecular liquid** is one composed of molecules. The properties of pure molecular liquids are controlled by the intermolecular forces among their molecules. Mercury is a rare exception to the rule that pure liquids at room temperature are molecular in character. Mercury is an atomic liquid metal.

In Section 4.5 we discussed the vapor pressure of liquids. In this chapter we'll see how thermodynamics interprets vapor pressure and relates it to the liquid's enthalpy of vaporization. The relationship reveals much about intermolecular forces in liquids and involves the Clausius-Clapeyron equation. Like Sadi Carnot, Benoit Clapeyron (1799-1864) was a French engineer—but an engineering professor, not a military engineer. So engineers not only find many practical uses for liquids, they have contributed to our understanding of their fundamental nature.

Figure 10.1 The Drake Well—Titusville, Pennsylvania. A replica of Edwin Drake's derrick and well house built at the actual site. The original burned on 7 October 1859—two months after Drake became the first person ever to bring in an oil well. [❁kws +"Edwin Drake"] Gasoline and fuel oil are liquids consumed in large volumes.

The chapter begins by examining some pure liquids and then moves on to discuss mixtures of liquids. The widespread use of gasoline, oil, and other liquid fuels is vital to the functioning of our economy. These liquids are derived from fossil deposits of petroleum, or crude oil (Figure 10.1). **Petroleum** is a complex mixture of organic compounds that occur naturally in geologic formations. The separation and modification of the molecules in crude oil to make fuels involves some of the largest chemical engineering projects in the world. One key process in making useful liquid fuels is distillation. **Distillation** is the process of heating liquid mixtures, recondensing the vapors that form, and repeating the boiling and recondensation steps until components of the mixture separate according to their vapor pressures or boiling points. The process of distillation was known to alchemists as early as the first century AD. However, the making of distilled alcoholic liquor from fermented vegetable substances did not arise until about 1000 years later. A nineteenth century still is pictured in Figure 10.2.

Figure 10.2 A pot still. A fermented mash is place in the pot on the right. Heated to boiling (in this case by coal), vapor enriched in alcohol passes down the slanting tube to a condenser. Cold water from the reservoir at the left condenses the alcohol vapor to a liquid, which is collected in the flask on the floor at the front of the apparatus.

To conclude the chapter we'll survey some complicated mixtures of liquids and complicated liquid mixtures. Engineering applications often depend critically on liquids: lubrication is principally a phenomenon involving liquid films reducing friction between two solid surfaces. The viscosity (or "weight") of a motor oil is a liquid property that you probably know about: motor oils must be carefully selected to meet their desired service applications. Hydraulic fluids are widely used for power transmission in machines, so their properties interest engineers. In chemical engineering, it is vital to understand and to be able to control the flow of liquids in oil refineries and chemical production plants.

Many useful liquid products contain soaps or surfactants. **Surface active agents** or **surfactants** are substances that concentrate at liquid surfaces. The properties of many liquid mixtures can be dramatically altered by adding small amounts of surfactants. Chemists who specialize in surfactant chemistry design and select specific chemical compounds and mixtures to modify the properties of liquid mixtures. An enormous range of useful mixtures ranging from foods to soaps contain surfactants. The most familiar liquid surfactant mixtures are the ones we use for household cleaning, like the dishwashing product in Figure 10.3 that contains both anionic and nonionic surfactants. We'll look at examples of these in the second part of the chapter.

Most industrial liquids are used in various chemical processes. Most of the liquids listed in Table 10.1 are unknown to the average citizen, who consumes a total of about 800 pounds of them each year.

Figure 10.3 Surfactants are so common that we tend to overlook their unusual practical properties.

Table 10.1 **Industrial Liquids**
*Pounds per person per year in 1997 based on US population of 250 million

| Rank | Material | Formula | lbs* | Important use or uses |
|---|---|---|---|---|
| 2 | Sulfuric acid | $H_2SO_4$ | 380 | Fertilizer and chemical manufacture, etc. |
| 9 | Phosphoric acid | $H_3PO_4$ | 100 | Fertilizer, metal cleaning, food products |
| 13 | Nitric acid | $HNO_3$ | 70 | Fertilizer and explosives manufacture |
| 19 | Ethyl benzene | $C_6H_5–C_2H_5$ | 51 | Chemical intermediate for styrene manufacture |
| 20 | Styrene | $C_6H_5–CH=CH_2$ | 46 | Rubber and polystyrene manufacture |
| 22 | Hydrochloric acid | $HCl$ | 34 | Chemical synthesis, steel pickling, oil well acidizing |
| 24 | p-Xylene | $H_3C–C_6H_4–CH_3$ | 31 | Chemical synthesis, for terephthalic acid |
| 25 | Cumene | $C_6H_5–CH(CH_3)_2$ | 25 | Phenol and acetone manufacture |
| 32 | Acrylonitrile | $CH_2=CH-CN$ | 14 | Resin and fiber manufacture |
| 33 | Ethylene dichloride | $CH_2Cl–CH_2Cl$ | 10 | Intermediate for PVC production |
| 37 | Benzene | $C_6H_6$ | 9 | Chemical synthesis |
| 41 | Isopropyl alcohol | $C_3H_7OH$ | 6 | Solvent, intermediate for acetone |
| 42 | Aniline | $C_6H_5NH_2$ | 5 | Intermediate for polyurethane manufacture |
| 46 | o-Xylene | $H_3C–C_6H_4–CH_3$ | 4 | Chemical synthesis, for phthalic anhydride |
| 47 | 2-Ethylhexanol | $C_8H_{17}OH$ | 3 | Plastics additives, detergents |
| 49 | Bromine | $Br_2$ | 2 | Sanitizing compounds |
| | Total | | 790 | |

For comparison, the per capita US consumption of liquid fuels, primarily gasoline, aviation, diesel, and fuel oil, was approximately 3000 pounds in 1997.

## Section 10.1 The Properties of Pure Liquids

Because all of the molecules in a pure liquid are alike, the intermolecular forces are well defined. In Section 4.8 we discussed van der Waals forces in connection with the nonideal behavior of gases caused by intermolecular forces. These same van der Waals forces are at work in liquids. The special interaction called *hydrogen bonding* is also important in some liquids—especially water. However, because the behavior of liquids is more complicated than the behavior of gases, there is no simple "ideal liquid equation" comparable to the ideal gas equation.

Some of the properties of pure liquids that we can relate to their intermolecular forces are:

- boiling and melting points
- miscibility (or immiscibility) with other liquids
- surface tension
- viscosity
- rates of evaporation,
- flash points—if combustible
- vapor pressure and enthalpy of vaporization
- solvent power

The **solvent power** of a liquid (what it will and will not dissolve) is strongly related to its intermolecular forces. The rate of evaporation of a liquid is closely linked to its vapor pressure, which, in turn, is controlled by the magnitude and extent of the intermolecular forces. The enthalpy of vaporization of a liquid is a pretty good thermodynamic measure of the forces among its molecules. Broadly speaking, the more energy it takes to separate a liquid's molecules into the gas phase the stronger the intermolecular forces in that liquid.

Two other properties of liquids which are quite useful to understanding their intermolecular forces are the critical temperature and the critical pressure of the gas that the liquid forms. We'll defer our discussion of those two properties to Chapter 12, where we examine the relationships among the gas, liquid, and solid phases of substances.

## Intermolecular Forces in Liquids

We can place the intermolecular forces among the molecules in a liquid into three categories: 1. dispersion forces, 2. dipole forces, and 3. hydrogen bonding.

1. Dispersion forces exist between *any* two molecules; such forces even occur among the atoms of the noble gas helium. The fact that helium gas eventually condenses to a liquid when cooled to near absolute zero (4 K) is sufficient evidence that dispersion forces are real. What is the origin of such forces? Imagine that the electrons in an atom or molecule can "drift" and create a temporary electrical imbalance. A **temporary dipole** is a short-lived imbalance of electrical charge in an atom or molecule. Inducing such an unbalanced charge in a molecule is to **polarize** it. **Dispersion forces** are intermolecular forces caused by temporary dipole–temporary dipole interactions between pairs of molecules. The electrostatic origin of dipole forces is straightforward: the oppositely charged ends of two dipoles from different molecules (whether temporary or permanent) tend to attract. In general, the larger the molecules the greater the dispersion forces among them. Intermolecular forces are greater among the molecules of easily polarized substances than among molecules of substances difficult to polarize. An easily polarized molecule has more electron "drift."

Because of dispersion forces the boiling points of a related series of liquids, such as the alkanes, steadily increase with each added carbon atom. Figure 10.4 shows the boiling points of the normal, or extended chain, alkanes $C_nH_{2n+2}$ for n = 1 to 19. Along the horizontal axis, each succeeding alkane molecule has one more methylene group, $-CH_2-$, than the preceding one.

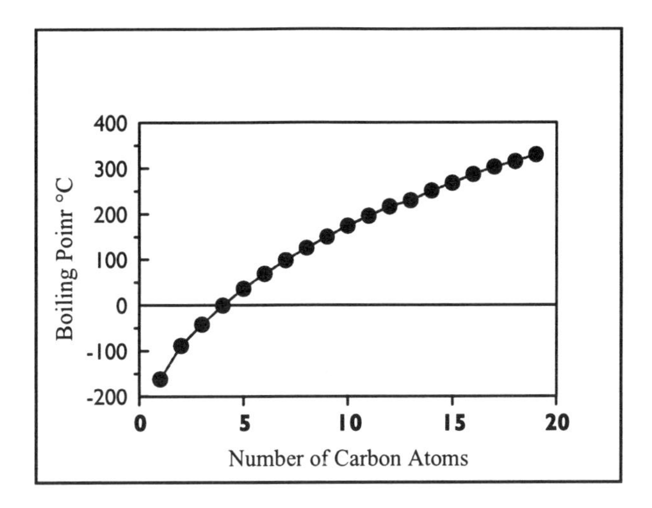


Figure 10.4 The boiling points of the normal alkanes increase with their size. $C_nH_{2n+2}$ for n = 1 to 19

2. Dipole forces. All molecules have temporary dipoles; some have permanent dipoles (review Table 6.6). **Dipole forces** operate among molecules that have permanent electric dipoles. Liquids composed of molecules with permanent dipoles are called **polar liquids**. Because water itself is a polar liquid, other polar liquids are often miscible with water. Acetone is an example of a polar liquid that mixes with water in any proportion. The two exemplify a liquid pair said to be **infinitely miscible**, i.e., any volume of water will mix with any volume of acetone to form a liquid-liquid solution.

3. Hydrogen bonding forces. **Hydrogen bonding** occurs among molecules that have O-H bonds or N-H bonds, and among the molecules in the compound H-F. Hydrogen bonding is a specific kind of intermolecular dipole attraction. Water and liquid ammonia molecules experience extensive hydrogen bonding. It is the extensive hydrogen bonding among water molecules that makes water such an unusual substance. (We hardly think of water as unusual because it is so common in our everyday life. But indeed it is a highly unusual liquid.) With a boiling point of 100°C and a mole weight of only 18.0 g $mol^{-1}$, water must have especially strong intermolecular forces. By comparison, methane ($CH_4$) with a mole weight of 16.0 g $mol^{-1}$, and ethane ($C_2H_6$) with a mole weight of 30 g $mol^{-1}$, have boiling points respectively of -183°C and -172°C—nearly 300° *lower* than water (they are both non polar, with only dispersion forces among their molecules). Table 8.2 presented some thermodynamic evidence for hydrogen bonding in certain liquids. Hydrogen bonding has special implications for biochemistry: the two strands of the double helix of DNA are joined by hydrogen bonds.

## Properties of Selected Pure Liquids

The properties of pure liquids can be easily measured and catalogued. You can find them in standard compilations and databases and on the web. Table 10.2 summarizes the properties of some common pure liquids: normal melting point, normal boiling point, surface tension (in millinewtons per meter at 25°C), viscosity (in Pascal seconds at 20°C), and standard enthalpy of vaporization at the liquid's normal boiling point. Many thousands of liquids might have been tabulated, but we'll use this handful to make some general points. In the table the liquids are listed in order of their boiling points, with diethyl ether, the most volatile, at the foot of the table.

Table 10.2 **Properties of Some Selected Pure Liquids**

| Liquid | MP °C | BP °C | Surface tension mN $m^{-1}$ (298 K) | Viscosity Pa s (298 K) | $\Delta H_{vap}$ kJ $mol^{-1}$ (at normal BP) |
|---|---|---|---|---|---|
| Glycerol | 20.0 | 290.0 | 63.4 | 1490 | 105.8 |
| Ethylene glycol | -11.5 | 198.0 | 47.7 | 19.90 | 67.8 |
| Acetic acid | 16.6 | 117.9 | 27.8 | 1.18 | 23.7 |
| Toluene | -95.0 | 110.6 | 28.5 | 0.59 | 30.0 |
| Water | 0.0 | 100.0 | 72.0 | 1.00 | 40.7 |
| Cyclohexane | 6.5 | 80.7 | 25.5 | 1.02 | 30.0 |
| Ethanol | -117.3 | 78.5 | 22.8 | 0.80 | 38.6 |
| Ethyl acetate | -83.6 | 77.1 | 23.9 | 0.46 | 31.9 |
| Carbon tetrachloride | -23.0 | 76.5 | 27.0 | 0.97 | 29.8 |
| Acetone | -95.4 | 56.2 | 23.7 | 0.33 | 29.1 |
| n-Propyl amine | -83.0 | 47.8 | 22.4 | 0.60 | 29.6 |
| Diethyl ether | -116.0 | 35.0 | 17.0 | 0.36 | 26.5 |

There is no single, simple theory to explain the precise values of the data in Table 10.2. However, for small molecules, various computer modeling programs can provide quite accurate simulations of liquid behavior. But merely studying the limited data in the table shows you rather well the role of intermolecular forces and the related molecular ordering on the bulk properties of liquids. With some exceptions, the properties increase from the liquid with the weakest intermolecular forces (diethyl ether at the bottom) to the liquid with the strongest intermolecular forces (glycerol at the top). Incidentally, you'll recall from the discussion in Section 8.1 that the entropies of vaporization of substances and Trouton's rule show that the hydrogen bonding alcohols have more internal order than liquids which have only dispersion forces among their molecules.

A liquid substance whose molecules associate to create long range order has a high viscosity. **Long range order** is created when 1000's of molecules aggregate in an orderly way inside a liquid. Glycerol and ethylene glycol are examples of liquids with long range order. Sugar is a molecule with multiple -OH groups, and sugar molecules can form multiple hydrogen bonds. This molecular behavior of sugar explains why concentrated sugar solutions (syrups) are viscous. **Viscosity** is the property that measures the resistance to flow of a liquid; it's SI unit is the pascal second. Some practical aspects of viscosity modification are described in the box titled "Viscosity Modification" to be found at the end of the chapter. Hydrogen bonding is often the dominant factor controlling the properties of pure liquids. Water's spectacularly high surface tension shows that it must have very strong intermolecular forces, but its low viscosity shows that water molecules are by themselves unable to organize to form much long range structure. The extensive hydrogen bonding in ethylene glycol and glycerol is confirmed by their high boiling points. Ethylene glycol is a diol—a molecule with two -OH groups, $CH_2OH\text{-}CH_2OH$. Glycerol is a triol—a molecule with three -OH groups, $CH_2OH\text{-}CHOH\text{-}CH_2\text{-}OH$.

In the surface of a liquid, the intermolecular forces manifest themselves as surface tension. **Surface tension** is the bulk force in the surface of a liquid that derives from intermolecular forces that hold the liquid together; it is force that makes water, mercury, and other liquids form balls or droplets. Pictures from space of astronauts stroking big, floating, wobbly weightless balls of liquids are excellent graphic demonstrations of surface tension. The listed surface tension values are in the units of millinewtons per meter, or force per length. If you imagine a line 1 meter long in the surface of a liquid, then the liquid's surface tension corresponds to the force needed to separate the liquid surface horizontally along that line.

## Computer Estimation of Liquid Properties

In Chapter 6 we discussed the role that computers have played in chemical quantum mechanics. However, computers have also been widely used for the prediction of chemical properties by classical empirical methods. For example, if you look at the boiling point data of Figure 10.4, you'll see it's not difficult to estimate what the boiling point of $C_{25}H_{52}$ will be. Extend that concept to all liquids, with all types of intermolecular bonding forces, and you'll see the possibilities.

Because liquids and liquid mixtures are useful as solvents for commercial chemical reactions, considerable effort has been devoted to building databases of liquids which combine experimental properties with empirically calculated properties. For example, each molecule is assigned a "dispersion parameter," a "polarity parameter," and a "hydrogen bonding parameter." These are empirical values designed to give rough measure of each of the types of intermolecular forces among the liquid's molecules. In recent years, many elaborate property simulation programs have been developed.

Computer simulation of this type goes by the name **quantitative structure property relationships (QSPR)**. The objective of QSPR is to take a written chemical structure and to estimate the molecular properties of that substance from it. There is no single rule or procedure for doing this. Rather, each programmer designs a calculation based on a substantial existing body of literature, using a mixture of experimental data, empirical parameters, and statistical analysis. All of this may seem a little mysterious, but the results for pure liquids can be quite good. Figure 10.5 compares the experimental boiling points with the QSPR calculated boiling points of 300 solvents. If the agreement were perfect, the result would simply be a straight line with a slope of 1. The information in Figure 10.5 is not particularly valuable to a working chemist; the real value of QSPR programs is that they can just as easily calculate the properties of liquid *mixtures*. For example, being able to predict the proper liquid mixture to use as a reaction solvent could be enormously helpful to an organic chemist trying to improve the yield of a critical intermediate step in the synthesis of a valuable, patented drug.

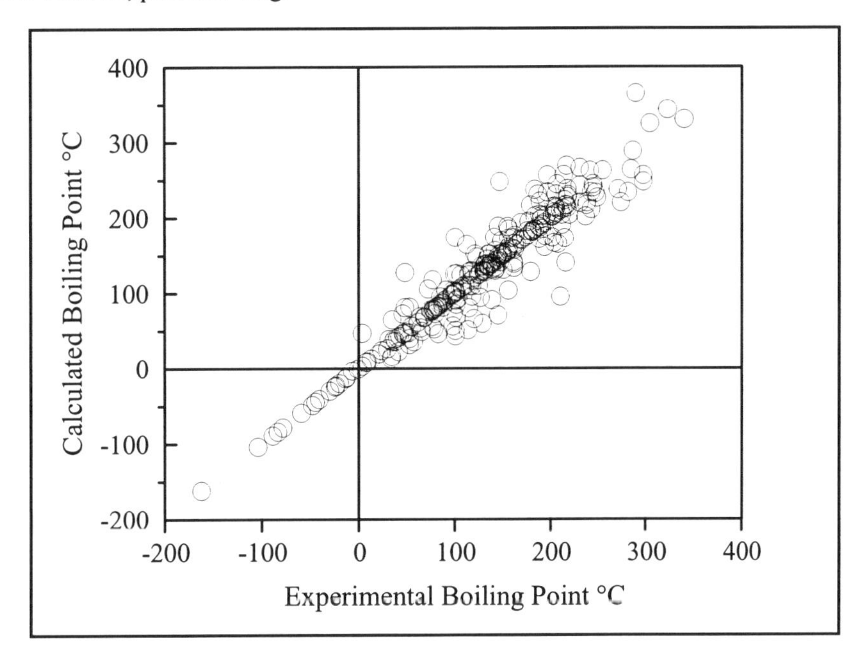


Figure 10.5 QSPR-calculated and experimental boiling points for 300 solvent liquids. The calculation of molecular properties from chemical structure by empirical methods can be quite effective. Graph by the author using data generated by the program mappro.exe.

It's obvious that what can be done for physical properties such as boiling point can also be done for chemical properties. That field is called **quantitative structure activity relationships (QSAR)** and aims to predict the chemical properties of compounds based their chemical structures. Techniques of QSAR have been, and continue to be, of great interest to chemical companies whose growth and success depends on continually finding new molecules with commercial applications, for example new drugs and biocides. You saw a tiny glimpse of what QSAR involves when we used the Clausius-Clapeyron equation to calculate enthalpies of vaporization from vapor pressure data. [❁kws +QSAR]. There are even computer models that combine QSAR methods and quantum mechanical models. [❁kws +QSAR +"quantum mechanics"].

## Miscibility

**Miscibility** is a property of two liquids that mix; if two liquids mix we say they are **miscible**. Many pairs of liquids, such as oil and water or alkanes and water, don't mix; they are **immiscible** liquids. We first met the terms solvent and solute in Section 3.6 when we calculated the molarity of aqueous solutions. When two liquids are mixed and mutually dissolve, it is conventional to call one the **solvent** and the other the **solute**. The designations are arbitrary, but the one present in larger proportion is often called the solvent.

The two main factors that control mutual solubility between a pair of liquids are: 1. the specific chemical nature of the solvent and solute, and 2. the temperature. Changing pressure usually has little effect on liquid-liquid solubility, though it has a substantial effect on the solubility of a gas in a liquid.

At root, it is the nature of the forces within the solvent and within the solute that control how the pair interact when they are mixed. Solvent-solute pairs can be insoluble, slightly soluble, or miscible over the entire concentration range (infinitely miscible). If A and B are two molecular liquids, their mutual solubility is generally controlled by how the A⇔A and B⇔B forces compare with the A⇔B forces. This situation is illustrated for liquid-liquid pairs in Table 10.3.

Table 10.3 **Properties of Liquid-Liquid Pairs**

| | |
|---|---|
| Octane ($C_8H_{18}$) and nonane ($C_9H_{20}$) | infinitely miscible |
| Water and ethanol | infinitely miscible |
| Water and ether ($C_2H_5$-O-$C_2H_5$) | slight mutual solubility |
| Water and octane ($C_8H_{18}$) | immiscible (two layers) |
| Water and oil | immiscible (two layers) |

Handbooks of chemistry often contain tables showing the extent of miscibility of many liquid pairs, and solubility is one of the standard outputs of QSPR programs. The well-worn, practical rule of experience, learned by generations of chemists, is that "Like dissolves like."

## Section 10.2 Liquid-Vapor Equilibrium

In Section 4.5 we described water's equilibrium vapor pressure over the temperature range 0°C to 100°C. Every liquid exerts a vapor pressure. At room temperature, volatile liquids have large vapor pressures—they evaporate. Nonvolatile liquids have low vapor pressures and evaporate slowly if at all—oil spilled on a garage floor never seems to evaporate.

**Liquid-Vapor Equilibrium** (Figure 4.13) is a dynamic process in which the rates of evaporation and recondensation are equal. Liquid-vapor equilibrium is controlled by intermolecular forces. The vapor pressure of a pure liquid does not depend on the volume of the liquid or the volume of its container, but it does depend on temperature. Vapor pressure increases with increasing temperature, as shown by the liquid's vapor pressure curve. Every liquid has a characteristic vapor pressure curve, as illustrated in Figure 10.6 The vapor pressure curve of a pure liquid can be explained by thermodynamics and be used to calculate the liquid's enthalpy of vaporization ($\Delta H_{vap}$).

Liquids boil when their vapor pressure reaches and just exceeds atmospheric pressure. 2-Methylpentane ($CH_3$–$CH(CH_3)$–$C_3H_7$) is a volatile substance: a **volatile substance** is one that evaporates quickly. 2-Methylpentane's low intermolecular forces are evidenced by its low boiling point of 60.3°C. Ethyl alcohol (ethanol) has intermolecular forces of intermediate strength between those of 2-methylpentane and those of water. Acetic acid molecules actually dimerize via hydrogen bonding, so its boiling point is the highest of the four liquids plotted in Figure 10.6

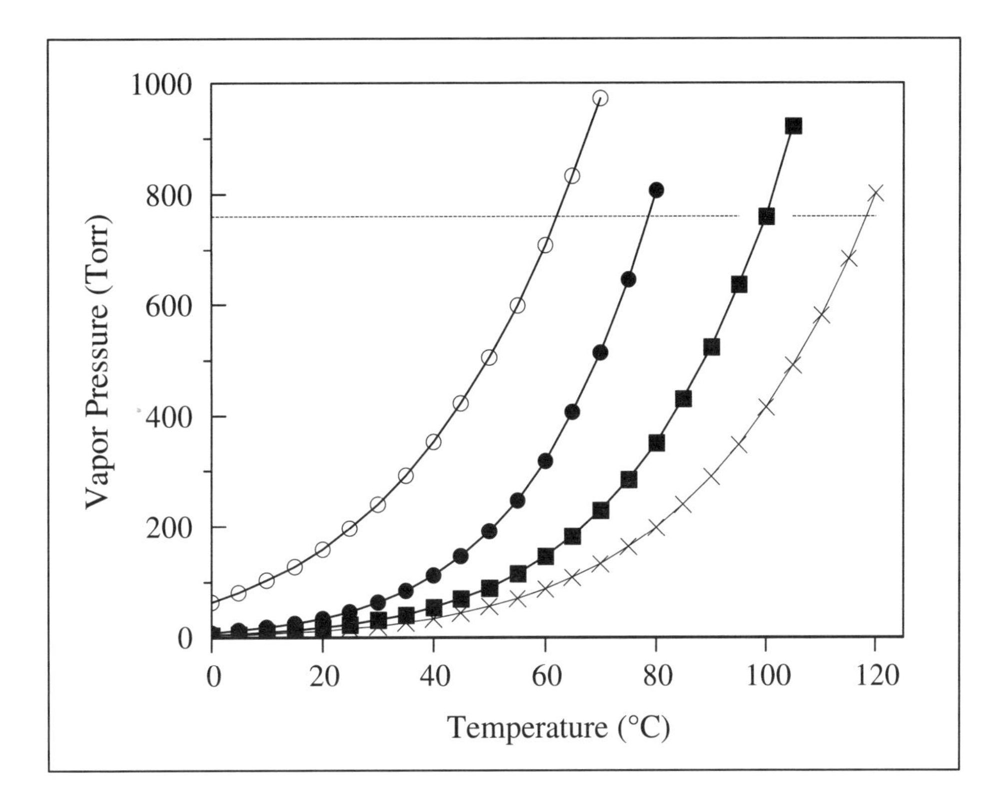


Figure 10.6 Vapor pressure curves. The vapor pressures of liquids rise exponentially with increasing temperature. The normal boiling point of a liquid is the temperature at which its vapor pressure exceeds 760 Torr (the horizontal dotted line).

—O—O—O— 2-methylpentane, —●—●—●— ethyl alcohol, —■—■—■— water, —x—x—x— acetic acid

The data points represented by the curves in Figure 10.6 can each be made to fit a straight line if ln P is plotted against 1/T. The general expression is:

$$\ln P - \frac{B}{T} + A$$

where B and A are empirical constants, each specific to a particular pure liquid. Remember: an empirical equation with empirical constants is one that fits data. It works. In the next section we see how thermodynamics interprets the empirical constant, B.

**Example 10.1**: A useful empirical equation for the vapor pressure of water is: ln P = 20.5 - 5170/T, with P in mmHg and T in kelvins. The equation works reasonably well from 0°C to 100°C. Use this equation to estimate the vapor pressure of water at 122°F (= 50°C).

Strategy: Plug 'n' chug. Watch the temperature unit: must be kelvins. Compare Table 4.3.

$$\ln P = 20.5 - 5170/T = 20.5 - 5170/323.2$$

$$\ln P = 20.5 - 16.0 = 4.5$$

$$P = e^{4.5}$$

$$P = \underline{90.4 \text{ mmHg (Torr)}}$$

The tabulated value is 92.5 mmHg. So the estimated value above looks pretty reasonable.

## The Clausius-Clapeyron Equation

The Clausius-Clapeyron equation is $\ln P = A - \Delta H_{vap}/RT$. The equation is derived from thermodynamic arguments which we will not go through here. "A" is a unitless empirical constant that we will not discuss. What the Calusius-Clapeyron equation represents is an empirical equation *with its slope interpreted*. Comparing the thermodynamic equation $\ln P = A - \Delta H_{vap}/RT$ with the empirical equation $\ln P = 20.5 - 5170/T$, you can see that for water:

$$-\Delta H_{vap}/R = -5170$$

$$\text{Thus,} \quad \Delta H_{vap} = 5170 \times R$$

$$\Delta H_{vap} = 5170 \text{ K} \times 8.314 \text{ J mol}^{-1} \text{ K}^{-1} \times 0.001 \text{ kJ J}^{-1}$$

$$\Delta H_{vap} = \underline{43.0 \text{ kJ mol}^{-1}}$$

The value 43.0 kJ $mol^{-1}$ represents an average of water's enthalpy of vaporization across the temperature range to which it applies (0°C - 100°C). To get water's $\Delta H_{vap}$ values at specific temperatures you could use the steam tables (beloved of mechanical engineers, because knowing accurate properties of water is crucial to designing and building steam engines) and convert from °F and Btu per pound mass to SI metric units. Converted results from the steam tables are shown in Table 10.4.

Table 10.4 **The Enthalpy of Vaporization of Water at Several Temperatures**

| Temperature | $\Delta H_{vap}$ |
|---|---|
| 0°C | 44.9 kJ $mol^{-1}$ |
| 25°C | 44.0 kJ $mol^{-1}$ |
| 50°C | 42.9 kJ $mol^{-1}$ |
| 70°C | 41.9 kJ $mol^{-1}$ |
| 100°C | 40.6 kJ $mol^{-1}$ |

As you can see, there is reasonable agreement between the value of 43.0 kJ $mol^{-1}$ estimated across the vapor pressure range and the experimental values measured at individual temperatures.

Because a plot of ln P versus 1/T is a straight line, it is possible to obtain $\Delta H_{vap}$ of any liquid by measuring its vapor pressures, $P_1$ and $P_2$, at two temperatures, $T_1$, and $T_2$. The data are manipulated by writing the Clausius-Clapeyron twice and solving the resulting pair of expressions simultaneously:

$$\ln P_2 = A - \frac{-\Delta H_{vap}}{RT_2}$$

$$\ln P_1 = A - \frac{-\Delta H_{vap}}{RT_1}$$

Subtracting the lower equation from the upper gives:

$$\ln P_2 - \ln P_1 = \ln (P_2/P_1) = \frac{-\Delta H_{vap}}{R}\left(\frac{1}{T_2} - \frac{1}{T_1}\right)$$

which is the two-point fit version of the Clausius-Clapeyron equation. This equation allows us to estimate the enthalpy of vaporization for any pure liquid from just two vapor pressure measurements. You mean we measure pressures and get an energy change? Yes. That's one power of thermodynamics: it can relate properties that don't seem as though they are related.

Acetone is a common industrial and laboratory solvent. It is also used around the home and can be purchased at any hardware store. A can of acetone is pictured in Figure 10.7. Acetone can also be purchased at drug stores for use as a solvent to remove fingernail polish. Perhaps you have used acetone yourself. If you have, you might have noticed that it often feels "cold." Because it is volatile, acetone rapidly evaporates, drawing the heat needed for its $\Delta H_{vap}$ from your skin, producing a cooling effect. For large industrial users, acetone is shipped via rail tankers, each of which might contain 40,000 gallons of the solvent. Safety engineers need to be knowledgeable about the potential hazards of shipping volatile, flammable compounds.

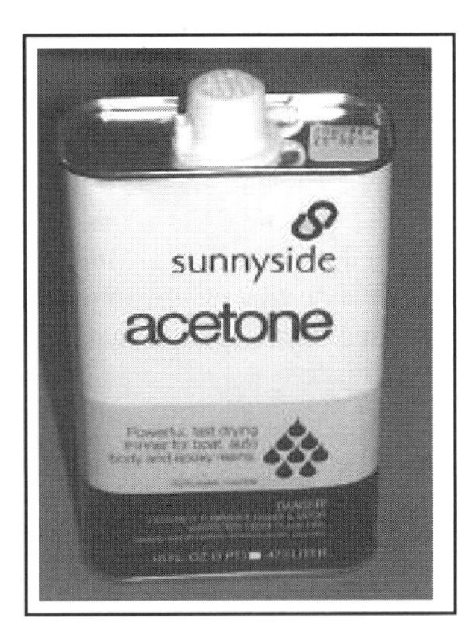


Figure 10.7 Acetone, $CH_3–CO–CH_3$, is a useful solvent. It is widely used industrially as well around the home. The description on this can says "Powerful, fast drying thinner for boat, auto body and epoxy resins."

Example 10.2 shows the calculation of a $\Delta H_{vap}$ from vapor pressure data:

**Example 10.2**: The vapor pressure of acetone is 400. mmHg at 39.5°C and 760. mmHg at 56.5°. Use these data to estimate the enthalpy of vaporization of acetone, and compare it with the experimental value at 56.5°C.

Strategy: Use the two-point fit to the Clausius Clapeyron equation and solve for $\Delta H_{vap}$. Substitute the data values after converting the Celsius temperatures to kelvins.

$$\ln (P_2/P_1) = \frac{-\Delta H_{vap}}{R}\left(\frac{1}{T_2} - \frac{1}{T_1}\right)$$

$T_1 = 39.5 + 273.2 = 312.7$ K, $T_2 = 56.5 + 273.2 = 329.7$ K

$P_1 = 400.$ mmHg $P_2 = 760.$ mmHg

So now plug 'n' chug: $$\ln (760./400.) = \frac{-\Delta H_{vap}}{R}\left(\frac{1}{329.7} - \frac{1}{312.7}\right)$$

$$0.642 = \frac{-\Delta H_{vap}}{R} \times (-1.65 \times 10^{-4})$$

$$\frac{-\Delta H_{vap}}{R} = 0.642 \div (-1.65 \times 10^{-4})\ K^{-1} = -3890\ K$$

$$\text{So } \Delta H_{vap} = 3890\ K \times 8.314\ J\ mol^{-1}\ K^{-1} \times 0.001\ kJ\ J^{-1}$$

$$\text{and } \Delta H_{vap} = \underline{32.3\ kJ\ mol^{-1}}$$

The thermochemical value of acetone's $\Delta H_{vap}$ is 30.99 kJ $mol^{-1}$ at 25°C and 29.10 kJ $mol^{-1}$ at 56.5°C.

This is pleasantly good agreement for a value derived from an empirical pressure equation compared with a calorimetrically measured number. Similar good agreement is found for most pure liquids comparing their $\Delta H_{vap}$ values measured either by vapor pressure studies or by calorimetry.

Industrial solvents are liquids with many practical applications, as described in the following box:

### Industrial Solvents

The term **industrial solvents** is usually applied to organic liquids used industrially to dissolve materials, clean materials, suspend solid particles in a liquid medium, or change the physical properties of a liquid. Many organic liquids are used as industrial solvents. Most industrial solvents have boiling points between 70°C and 250°C. Specific uses of industrial solvents include: metal cleaning (vapor degreasing, for example); making paints, lacquers, and varnishes; making inks; and treating and staining wood, especially in the furniture manufacturing industry.

The label of the pictured lacquer thinner states that it is a mixture of toluene, acetone, ethyl acetate, isopropyl alcohol, and petroleum distillates. Toluene is methylbenzene, $C_6H_5CH_3$, acetone was described earlier, ethyl acetate is the ester formed by reaction of ethyl alcohol and acetic acid, isopropyl alcohol can be purchased in any drug store as "rubbing alcohol." The term "petroleum distillate" means a hydrocarbon mixture derived from distilling petroleum. Kerosene is probably the best known petroleum distillate solvent.

Users of industrial solvents must be aware of personnel exposure issues, potential fire hazards, air pollution regulations, disposal regulations, etc. This is especially applicable if heated solvents are being used. Engineers in many manufacturing industries use solvents. [❁kws +"metal degreasing" +vapor"].

## Empirical Vapor Pressure Equations for Various Liquids

The two-point fit to the Clausius-Clapeyron equation, shown in Example 10.1, can be applied to any pure liquid. To show you more examples the author took vapor pressure data and wrote a short computer program to do the needed two-point fit to $\ln (P_2/P_1) = \frac{-\Delta H_{vap}}{R}\left(\frac{1}{T_2} - \frac{1}{T_1}\right)$ using the two temperatures at which $P_1 = 100$ mmHg and $P_2 = 760$ mmHg. $T_2$ is thus the normal boiling point of the liquid and $T_1$ is lower. The results are shown in Table 10.5. Note that the empirical constants A and B are slightly different for water when it's done as a two-point fit, compared with fitting at many points across the vapor pressure curve, as in Example 10.1.

Table 10.5 **Two-Point Fit Parameters to the Clausius-Clapeyron Equation for Various Liquids**

| Liquid | T °C at which VP is 100 mmHg | T °C at which VP is 760 mmHg (= BP) | $-\Delta H_{vap}/R$ (calculated) kJ mol$^{-1}$ | A (no units) | $\Delta H_{vap}$ calculated kJ mol$^{-1}$ | $\Delta H_{vap}$ (thermo-chemical) at normal BP |
|---|---|---|---|---|---|---|
| Water | 51.9 | 100 | 5120 | 20.3 | 42.5 | 40.7 |
| Ethanol | 34.9 | 78.4 | 5476 | 22.2 | 45.5 | 40.5 |
| Cyclohexane | 25.5 | 80.7 | 3880 | 17.6 | 32.3 | 32.8 |
| Ethyl acetate | 27.0 | 77.1 | 4260 | 18.8 | 35.4 | 34.7 |
| n-Propyl amine | 0.5 | 48.5 | 3720 | 18.2 | 30.9 | 31.0 |
| Acetic acid | 63 | 118.1 | 4841 | 19.0 | 40.3 | 41.7 |
| 2-Methylpentane | 8.1 | 60.3 | 3643 | 17.6 | 30.3 | 32.1 |

These empirical equations can be used to estimate the vapor pressure of any of the liquids at any desired temperature across that liquid's range of vapor pressure. That's shown in the following example.

**Example 10.3**: Estimate the vapor pressure of ethyl acetate at 59.3°C.

Strategy: Write an empirical equation based on data in Table 10.5 and then plug 'n' chug.

For ethyl acetate: $\ln P = 18.8 - 4260/T$

$\ln P = 18.8 - (4260/332.5) = 18.8 - 12.8$

$P = e^{6.0}$ $\underline{P = 391\ \text{mmHg}}$.

The Handbook value is 400. mmHg. So in this case the Clausius Clapeyron works quite well.

## Section 10.3 Distillation, Ideal Liquid Mixtures, and Raoult's Law

**Distillation** is a method of separating a mixture of two liquids, or of separating a complex mixture of liquids into fractions. It involves heating the liquid and subsequently condensing its vapors and works because the liquids in a mixture have different vapor pressures and boiling points, so the vapor that comes from the liquid is richer in the more volatile components. A **fraction** or **distillate fraction** is a portion of a distilled liquid that boils over a narrow range of temperature. When a liquid mixture is heated, the more volatile components (the ones with the greater vapor pressure) concentrate in the gas phase. The vapors are condensed and then re-boiled in a **fractionation column**. The condensed liquid, like the vapor from which it formed, is enriched in the more volatile components. Successive steps of boiling and re-condensation eventually yield either pure compounds or fractions containing compounds with a narrow boiling point range.

Distillation is a very important process in chemistry. Figure 1.6 shows a picture of a still from approximately 800 AD. Stills were probably invented in Alexandria in 200-600 AD. A pot still is shown in Figure 10.2. Figure 10.8 shows a simple laboratory distillation setup, and distillation on a modern industrial scale is shown in Figure 10.9.

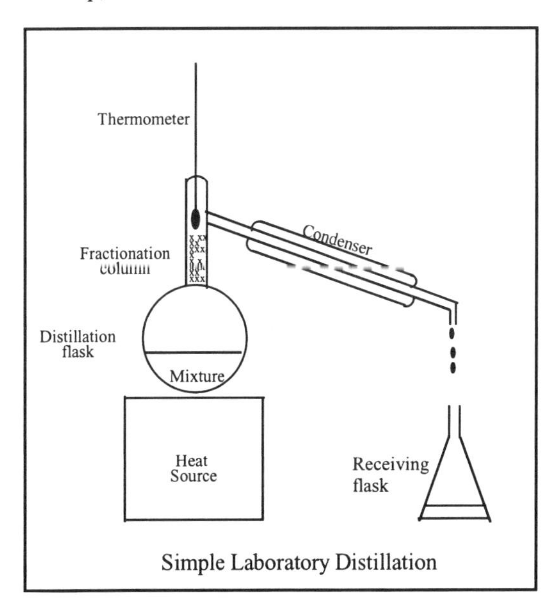


Figure 10.8 Simple Laboratory Distillation.

The liquid mixture in the distillation flask in Figure 10.8 is heated and its vapors enter the fractionation column. Multiple condensations and reboilings separate the vapor mixture so that the lowest boiling compounds first emerge from the flask. The progress of the distillation is monitored by observing the temperature. After passing the thermometer, the vapors enter the water-cooled condenser and run down into the receiving vessel. The author has a clear recollection of the first distillation (chloroform) he ever performed—in high school some 40 years ago.

Figure 10.9 Distillation on an industrial scale. Distillation towers at the Motiva refinery in Lousiana. Summer 1999.

Fractional distillation (Figure 10.9) is an enormous scale operation in the petrochemical industry. It is the primary method for separating the complex mixture of organic molecules that compose crude oil. Table 10.6 shows major petroleum fractions and the names, boiling ranges, and approximate number of carbon atoms in the molecules that make up the fraction. These molecules are primarily saturated hydrocarbons, so you can figure out the number of hydrogen atoms using the formula $C_nH_{2n+2}$. Compare these boiling points with those shown for pure alkanes in Figure 10.4.

Table 10.6 **Petroleum Fractions or Distillates**

| Name | Boiling Range | Carbon number |
|---|---|---|
| Gas and liquefied gas | Below 25°C | $C_1$-$C_4$ |
| Gasoline | 20-200°C | $C_4$-$C_{13}$ |
| Kerosene | 175-275°C | $C_9$-$C_{16}$ |
| Fuel & diesel oil | 200-400° | $C_{15}$-$C_{25}$ |
| Heavy fuel oil | 350-450°C | $C_{20}$ and up |
| Lubricating oil | 350-450°C | $C_{20}$-$C_{70}$ |
| Asphalt | >450°C | large |

## Raoult's Law

Raoult's law is the fundamental law that explains distillation: a more volatile component is enriched in the vapor above a liquid mixture. **Raoult's law** states that when a solution of two chemically similar liquids is made, the vapor pressure of the mixture is the sum of the individual vapor pressures taken in proportion to their mole fractions.

$$P_{tot} = P_A^0 X_A + P_B^0 X_B$$

Where $P_{tot}$ is the vapor pressure of the mixture, $P_A^0$ and $P_B^0$ are the vapor pressures of pure A and pure B, and $X_A$ and $X_B$ are the mole fractions of components A and B in the liquid phase.

Only pairs of chemically similar liquids come close to obeying the law precisely—those liquid pairs that form an **ideal solution**. In an ideal solution A⇔A, B⇔B, and A⇔B forces are much alike. For many pairs of liquids, the total vapor pressure is either much greater or much smaller than the value calculated from the equation. The three cases are sketched in Figure 10.10.

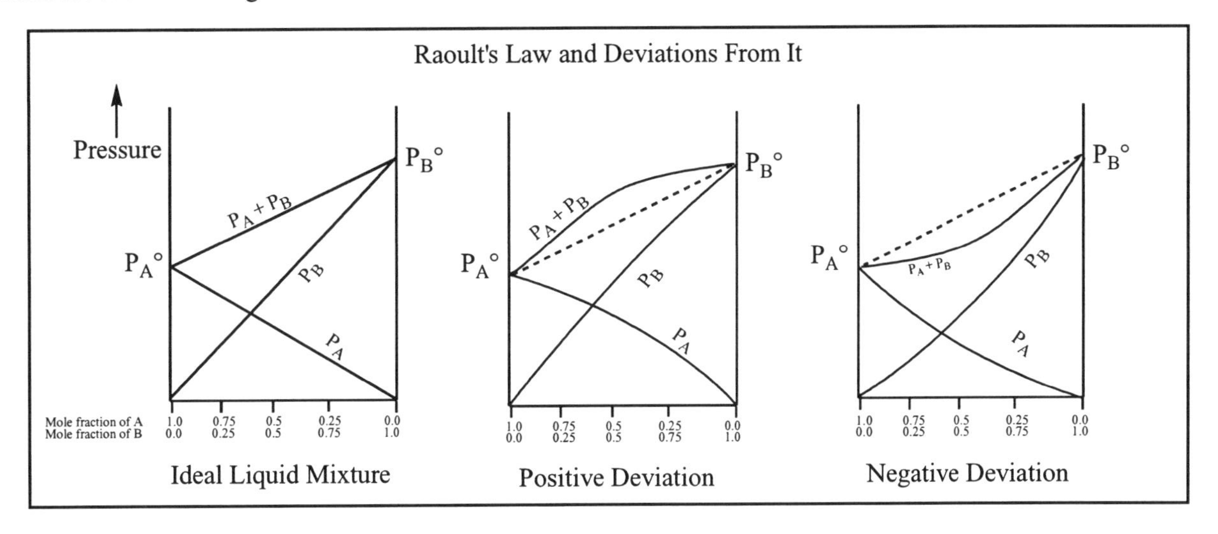


Figure 10.10 Graphical representation of Raoult's law and the two ways in which real liquid mixtures differ from the law. Positive deviation is so called because the vapor pressure of the mixture is greater than the value calculated from the law. Negative deviation is the converse situation.

Chemists say that non-ideal liquid pairs either show positive or negative deviations from Raoult's law. If the A and B liquid pair has a positive Raoult's law deviation, then the A⇔B forces are probably weaker that the A⇔A and B⇔B forces. If the A and B liquid pair has a negative Raoult's law deviation, then the A⇔B forces are probably stronger that the A⇔A and B⇔B forces. In other words, deviations from Raoult's law provide insight into the inter-molecular forces between different compounds. Example 10.4 shows a Raoult's law calculation.

**Example 10.4**: Calculate the vapor pressure of a 50.0 weight percent solution of benzene and toluene at 25°C. Assume that the mixture is ideal (a good assumption because benzene and toluene are chemically very similar). At 25°C the vapor pressure of benzene is 96.4 mmHg and the vapor pressure of toluene is 35.2 mmHg. The mole fraction of A is defined as $n_A/(n_A + n_B)$. MW benzene ($C_6H_6$) is 78.11 g mol$^{-1}$; MW toluene $C_7H_8$ is 92.13 g mol$^{-1}$.

Strategy: Assume a 100 gram sample and calculate the mole fraction of each component. Then plug into the Raoult's law expression.

moles benzene = 50.0/78.11 = 50.0 ÷ 78.11 g mol$^{-1}$ = 0.640 mol

moles toluene = 50.0/92.13 = 50.0 ÷ 92.13 g mol$^{-1}$ = 0.543 mol

mole fraction benzene = 0.640/(0.640+0.543) = 0.640 ÷ 1.183 = 0.540

By difference the mole fraction of toluene is 1 - 0.540 = 0.460.

Substituting these values gives:

$$P_{tot} = 96.4 \times 0.540 + 35.2 \times 0.460$$

$$P_{tot} = 61.7 + 16.2 = \underline{77.9 \text{ mmHg}}.$$

## Section 10.4 Practical Liquid Mixtures

We've seen a good deal about liquid mixtures in the first part of this chapter, and some of those mixtures, such as lacquer thinner, have been fairly complicated. With the use of surfactants it is possible to expand the types of liquid mixtures that can be made.

## Surfactants

A **surfactant** is chemical material whose molecules tend to concentrate at a liquid-liquid interface or a liquid-gas interface. There are perhaps 10,000 commercially available surfactant materials; many are themselves complex mixtures. Substances that dissolve in water are **hydrophilic**; substances that dissolve in kerosene or oil are **lipophilic**. Substances that "dislike" being in water are **hydrophobic**; substances that "dislike" being in oil are **lipophobic**. In terms of the basic classification of intermolecular forces, the features of a molecule that make it hydrophilic are: having the ability to hydrogen bond, having a high polarity, and having a charged functional group. The features of a molecule that make it hydrophobic are regions that interact primarily by dispersion forces. In the great majority of cases, hydrophobic regions of surfactants are hydrocarbon-like.

The molecules of a surface active agent are **amphipathic**: they have both hydrophilic and hydrophobic characteristics. Surfactant molecules have both hydrophilic regions (hydrogen bonding, polar, or ionic portions) and hydrophobic regions (hydrocarbon chains). The balance between the type and size of these two regions defines the surfactant's properties. Surfactant chemists use a measure called **HLB** (hydrophilic-lipophilic balance) to catalog surfactants. HLBs range from 0-40, with 40 being the most hydrophilic. Ordinary soap has an HLB value of about 12, and the calculation of a molecule's HLB is a standard feature of the quantitative structure activity (QSAR) programs that were mentioned earlier.

Ordinary soap is a surfactant that has been known for several thousand years. It is made from animal fat and an alkali such as fire ashes. Chemically, soap is the sodium salts of a mixture of carboxylic acids; it's usually a mixture of $C_{12}$-$C_{16}$ molecules. An example equation for the formation of soap from a fatty acid [❁kws +"fatty acid"] is:

$$C_{11}H_{23}COOH \quad + \quad NaOH \quad \rightarrow \quad C_{11}H_{23}COO^- \quad + \quad Na^+ \quad + \quad H_2O$$

dodecanoic acid + sodium hydroxide → sodium dodecanoate (soap) + water

In this case the hydrophobic portion of the soap molecule is the dodecanoate group, $C_{11}H_{23}$–, and the hydrophilic portion is the carboxylate anion $-COO^-$. For surfactants in general, the hydrophilic portion of the molecule can contain a negative charge (**anionic surfactants**), a positive charge (**cationic surfactants**), or merely polar or hydrogen bonding groups (**a nonionic surfactant**).

Figure 10.11 shows the chemical structure of a widely used nonionic surfactant called nonylphenoxypolyethoxyethanol—7 mole EO. EO is a common abbreviation for ethylene oxide, the molecule $C_2H_4O$, that is used to manufacture many nonionic surfactants, and "7 mole EO" means that the surfactant molecule incorporated seven moles of ethylene oxide during its manufacture. Commercial surfactant makers often use the EO naming method; organic chemists have other, more systematic, methods.

| Figure 10.11 **Nonylphenoxypolyethoxyethanol—7 mole EO (ethylene oxide)** | | | |
|---|---|---|---|
| $C_9H_{19}–C_6H_4–O–(CH_2–CH_2–O)_7–H$ | | | |
| Hydrophobic region of surfactant molecule, $C_9H_{19}–C_6H_4–$ ← | | Hydrophilic region of the surfactant molecule, $–O–(CH_2–CH_2–O)_7–H$ → | |
| $C_9H_{19}–$ | $–C_6H_4–O–$ | $–(CH_2–CH_2–O)_7–$ | $–H$ |
| nonyl group | group derived from phenol | repeat of unit derived from seven moles of ethylene oxide (EO or $C_2H_4O$) | terminal hydrogen atom |

Figure 10.12 shows the anionic surfactant sodium dioctylsulfosuccinate. The $Na^+$ cation is at the top, the anionic functional group of the molecule $-SO_3^-$ is in the center, and two hydrophobic octyl groups, $-C_8H_{17}$, are on the left and right sides. So this particular surfactant molecule has one hydrophilic region linked to two separate hydrophobic regions. It is a good wetting agent—as we'll describe shortly.

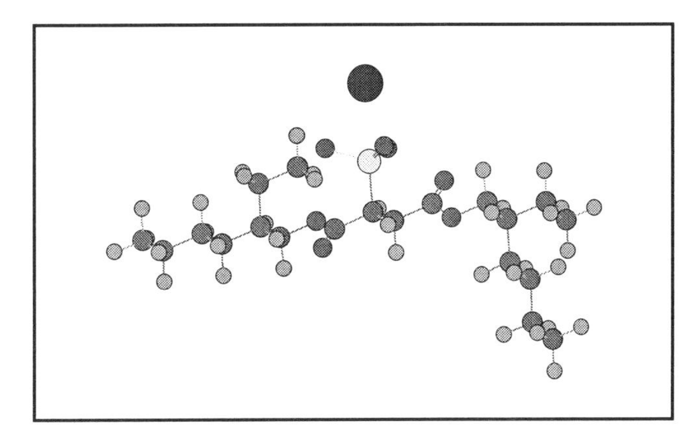

Figure 10.12 The anionic surfactant sodium dioctylsulfosuccinate. It is an inexpensive wetting agent. It and its close relatives have been around for nearly 80 years. It was originally introduced commercially into the textile industry, where the rapid wetting of cotton fiber improves production efficiency in dye baths and other operations.

**Reducing Water's Surface Tension**

One of the valuable practical properties of soaps or synthetic detergents is their ability to lower the surface tension of water. It is this ability that in part accounts for the cleaning power of soapy water. At the molecular level, surface tension reduction is interpreted as an accumulation of surfactant molecules in the water's surface. Because the intermolecular forces among the soap molecules are less than those among water molecules, the surface tension drops. The manipulation of surface forces using surfactants has many practical and technological applications.

Mixed with water, surfactant molecules rapidly concentrate in its surface. General purpose surfactants can reduce the surface tension of water from 72 mN $m^{-1}$ to about 30 mN $m^{-1}$. It is this lowering that prevents water beading on a surface. Advertisers promoting dish washing products often refer to the phenomenon as "sheeting action."

Surface tension reduction of water means that laundry is cleaned better as a result of improved oil removal and better liquid penetration and that the dying of cotton and synthetic fibers can be done rapidly and with a uniform dye level across the fibers. Fire fighters use small amounts of synthetic surfactants to make their water penetrate better into materials that are fueling fires, etc.

An extreme surface tension lowering of water is valuable in a few situations, of which the best known occurs when foam is to be used to combat oil, kerosene, or gasoline fires. Foamed water is an excellent fire fighting medium, but for it to be effective it must spread across the burning fuel's surface. Fluorinated surfactants are used to reduce water's surface tension below 20 mN $m^{-1}$, and the foam produced spreads because this is lower than the surface tension of the fuel (octane, for example, has a surface tension of 21.8 mN $m^{-1}$). The technique was pioneered in the 1960's by US Navy researchers to fight fires aboard aircraft carriers. Incidentally, such foams are stabilized and thickened by partially hydrolyzed protein, so they exemplify a useful, water-based product in which both the viscosity and surface tension of water are much modified. In the jargon, modern products developed in the 1980s are called FFFP's, or Film-Forming FluoroProteins.

[❉kws +"fire fighting" +fluorocarbon]

The most important cationic surfactants are those with four hydrocarbon groups linked to a positively charged nitrogen atom. They are widely used as germicides, disinfectants and sanitizers. Figure 10.13 shows N-benzyl-N-dodecyl dimethyl ammonium chloride. It's a quarternary benzyl ammonium halide. Quarternary means the nitrogen atom is forming four separate covalent bonds. The wiggly line on the right is the hydrophobic dodecyl chain $-C_{12}H_{25}$.

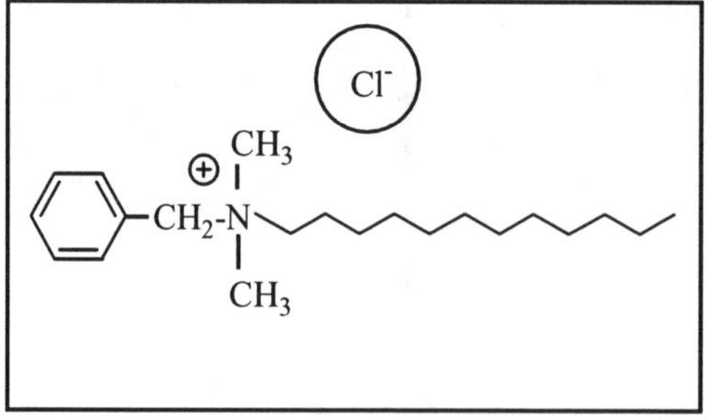

Figure 10.13 Benzyl dodecyl dimethyl ammonium chloride is a cationic surfactant or "quat," as this class of surfactants is popularly known. The methyl groups ($-CH_3$) are above and below in the diagram. The benzyl group ($C_6H_5-CH_2-$) is to the left, and the dodecyl group ($C_{12}H_{25}-$) to the right. The fact that it's a cationic surfactant is indicated by its positive charge. The negative chloride counterion floats free in solution.

The fact that cationic surfactants have excellent germicidal properties can be directly attributed to their chemical structure. Although the detailed mechanism not well understood, the cationic surfactants disrupt the cell envelope of bacteria. The consequence is to stop the normal functioning of bacteria and to destroy them. So here is a biological example of the chemical structure of a molecule creating its bulk properties: germicidal activity in this case.

Modern soap powders contain synthctic anionic surfactants, although the surfactant usually makes up only a few percent by weight of the finished commercial product. Other ingredients include compounds to buffer the pH, antibacterial agents, UV fluorescent dyes to cause washed fabrics to glow and look "whiter-than-white." A dozen years ago, detergent powders contained large amounts of pentasodium tripolyphosphate ($Na_5P_3O_{10}$), but the effect of phosphate as a nutrient (fertilizer) in bodies of water, such as the Chesapeake Bay, led to legislation banning the use of phosphate-containing formulations. Today, soap powders mostly contain organic replacements for the traditional phosphates. Modern soap powders also use biodegradable surfactants—ones that serve as nutrients for microorganisms during biological waste water treatment and so get eaten. In the 1950s non-biodegradable surfactant molecules created huge amounts of foam at water treatment plants, resulting in a severe public health hazard when the wind blew the pathogen-laden foam into nearby communities.

Figure 10.14 Surfactant-based products are prominently on display in supermarkets.

## Liquid-Liquid Emulsions

Perhaps the clearest example of the hydrophile-lipophile nature of surfactants is their ability to create water-oil emulsions. Pine oil is a naturally derived hydrocarbon oil obtained by processing trees. It is an excellent disinfectant (the pine trees produce the oil as a form of chemical defense). To make the pine oil mix with water, it is pre-blended with a surfactant. When the surfactant-oil blend is added to water by the consumer, a cloudy emulsion forms.

Figure 10.15 It's easy to observe the formation of a stable oil-in-water emulsion. Just add a little pine oil disinfectant to water and watch what happens.

The emulsion consists of small oil droplets surrounded by surfactant molecules that stabilize the droplets in their aqueous environment. Figure 10.16 shows the hydrophobic portions of the surfactant molecules in the oil droplet and the hydrophilic portions in the aqueous phase.

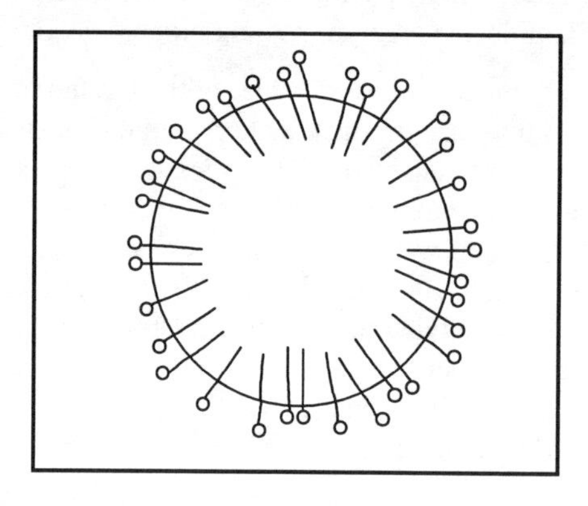

Figure 10.16 A surfactant-stabilized oil droplet in water. The hydrophobic "tails" of the surfactant molecules penetrate the oil. The polar end groups of the surfactant are in the aqueous phase.

The oil droplets are small; they approach the size of the wavelength of visible light (550 nm), scattering the light, and creating the milky appearance of the emulsion. Nonionic surfactants are suitable for this application. The 7 mole EO molecule shown in Figure 10.11 would work fairly well. If a very stable emulsion were required, an anionic surfactant would be blended into the emulsion. If you think about it you can see why: emulsions are susceptible to "breaking" when their oil droplets eventually coalesce and form a separate liquid phase. The polar groups of an anionic surfactant would coat the oil droplets with a negative charge, and this charge will make it more difficult for the drops to approach one another sufficiently closely for coalescence. Laundry soap removes oily stains by the same sort of mechanism.

Incidentally, milk is milky and clouds are cloudy precisely because they both contain droplets of the proper size to scatter light. The adjective *milky* refers to milk's whiteness, which is in turn caused by light-scattering fat droplets stabilized in an aqueous medium by natural surfactants in the milk. Skim milk contains less fat than regular milk and is less milky. Of course in clouds it's water droplets rather than fat droplets that create their whiteness.

Two examples of oil-water emulsions of particular interest to engineers are those used as cutting fluids and those used as hydraulic fluids. **Cutting fluids** are the milky liquids sprayed onto the work piece during the machining of metal parts. The water cools the work piece while the emulsified oil provides lubrication between the part and the cutting tool. Emulsified **hydraulic fluids** are used applications in such as the roof support pistons in underground coal mining. Here, a water-based fluid is desired because of the potential fire hazards of a purely organic-based material. [❁+"cutting fluid"]

When kerosene and water are blended with the proper mix of an anionic and a nonionic surfactant, the result is a very stable, viscous emulsion, often sold as a waterless hand cleaner. Gojo® is a popular brand. The term "waterless" refers to the cleaner's use by applying it directly, massaging the grease, and wiping with a cloth or paper towel; it can also be washed off with water. However, there is water in waterless hand cleaner—it's one of the listed label ingredients. The cleaner works because the kerosene in it helps to dissolve the grease, while the surfactants allow the formation of an emulsion that can be washed away.

The Gojo® ingredient list shows: water, isoparaffin, soap, mineral oil, surfactant, amorphous mineral particles, cetyl alcohol, propylene glycol, aloe extract, fragrance, lanolin, preservative, pumice, tocopheryl acetate, FD&C [FD&C = food dyer's and colorer's classification] yellow #5 and yellow #6. This product is basically a thick, surfactant-stabilized emulsion of hydrocarbon oil and water with some mild abrasive added (amorphous mineral particles and pumice). Minor ingredients include skin emollients (aloe and lanolin), modified vitamin E (tocopheryl acetate)—as a preservative, and two yellow dyes.

Figure 10.17 Thickened oil-water emulsions are useful for cleaning the grimy hands of auto mechanics.

Surfactants that create oil-water emulsions are often called **emulsifiers**. Many food products consist of oil-water emulsions, such as almost any "creamy" food. Salad dressings, mayonnaise, and ice cream are three examples of surfactant-stabilized commercial food products. The ingredient list for the blue cheese salad dressing shown in Figure 10.18 is long and formidable and, in addition to blue cheese milk culture, includes soybean oil, xanthan gum, cellulose gel, calcium disodium EDTA, and lactic acid. The major emulsifying ingredients are propyleneglycolalginate a semi-synthetic surfactant, made by the chemical modification of a seaweed-derived substance, and Polysorbate-60® another semi-synthetic surfactant made by the ethoxylation of esters derived from sorbitol (a sugar derivative) and the unsaturated carboxylic acid, oleic acid.

Figure 10.18 Creamy food products, such as these salad dressings, contain emulsifiers—surfactants designed to produce stable oil-water emulsions.

Accurate food ingredient labeling has been required in the US for over 50 years. During the past 20 or so years, various states have passed chemical "right-to-know" laws. To comply with these laws, many products, in addition to foods, are required to carry label statements and warnings about particular ingredients. There are thousands of examples. Figure 10.19 shows the label from a concrete patching formulation as an example. Here's its label statement:

Ingredients per New Jersey Right to Know Act Crystalline Silica 14808-60-7, Calcium Carbonate 1317-65-3, Nonylphenoxyployethoxyethanol 9016-45-9 and TSRN 618608-5024P

Figure 10.19 Sand, chalk and the nonionic surfactant nonylphenoxypolyethoxyethanol are the named ingredients on this label of a concrete patching formulation.

Nonylphenolpolyethoxylates (such as those disclosed on the label in Figure 10.19) are a widely used class of surfactants; they are in the class of the nonionic surfactant pictured in Figure 10.11. Applications that rely on nonionic surfactant ingredients include detergents, wetting agents, emulsifiers and stabilizers (for emulsions, etc.), dispersants (to stabilize solid suspensions), etc. The surfactant in the patching compound probably serves both to stabilize the chalk suspension, of which the compound is largely composed, and to enable the product to make close contact with the surface to be patched. Incidentally, under the generic pharmaceutical name of "Nonoxynol" the 9 mole and 11 mole ethoxylates find use as a spermaticide in various commercial contraceptive products.

## Wetting Agents

You are probably well aware of the tendency of water droplets to bead up on the surface of a newly waxed automobile or on waxed paper. One purpose of "waxing" a car is to make its surface hydrophobic—wax is a hydrophobic, hydrocarbon based material. When water droplets are on a hydrophobic surface there is a force balance between the gravitational pull on the liquid and the intermolecular forces inside the water that tend to draw it up into a ball. The result is that the angle θ is established, as shown in Figure 10.20.

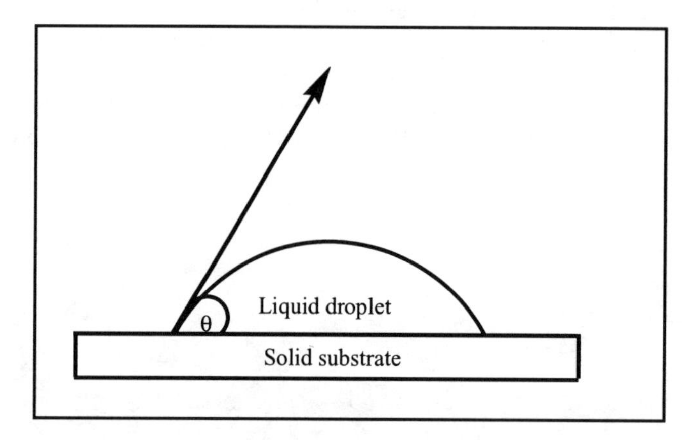


Figure 10.20 The formation of a bead of water on a hydrophobic surface. The angle, θ, is called the contact angle.

The **contact angle** between any liquid and any solid surface is the angle where the liquid's surface meets the solid's surface. For a pure liquid on a pure solid, the contact angle is a direct measure of the intermolecular forces between the two phases. However, adding surfactants can drastically alter the interaction because the surfactant molecules themselves concentrate in the liquid's surface and in the liquid-solid interfacial region. A surfactant that promotes the spread of a liquid across a solid surface is called **wetting agent**. Complete spreading occurs when the contact angle is reduced to 0°.

With contact angle in mind, you can understand how intermolecular forces control the shape of the meniscus formed by a liquid in a tube. If the liquid wets the tube, then it will spread upward forming a concave surface. Conversely, if the liquid does not wet the tube, its own internal forces will tend to draw it into a ball, forming a convex surface.

Water wets glass (especially if it is clean) and forms a concave meniscus; water-water interactions and water-glass interactions are comparable in strength. Both water in the common plastic polyethylene and mercury in glass form a convex meniscus. In these two cases the cohesive forces within the liquid exceed the forces between the liquid's atoms or molecules and the solid, inside surface of the tube. Figure 10.21 shows a schematic diagram for three cases. In the case of wetting, on the left, we call it capillary rise. **Capillary rise** is the phenomenon of water spontaneously rising up a small diameter tube (a capillary tube) until the mass of water raised by liquid-solid forces exactly balances gravitational attraction. The narrower the tube, the greater the extent of capillary rise.

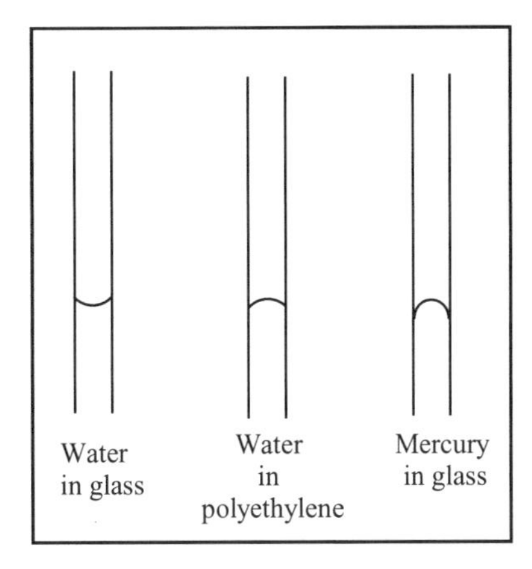


Figure 10.21 The curved surface formed by a liquid in a tube is the meniscus. The shape of a meniscus is controlled by the balance of forces between the solid and the liquid.

In one of the great mistakes of modern chemistry, around 1965, it was thought that an entirely new form of water—polywater—could be made in narrow capillaries. Not so. [☸kws +polywater]

Figure 10.22 Many leafy surfaces are hydrophobic, as you can tell if you've seen water form beads on them. Wetting agents are used in agricultural sprays to ensure good leaf surface coverage.

Leafy surfaces such as the one shown in Figure 10.22 are often waxy and hydrophobic. It is desirable that sprayed, agriculture pesticide formulations spread across leaf surfaces. To ensure good spreading, the pesticides are formulated into products that contain wetting agents.

**A Home Experiment:**

An interesting home experiment is "float" a needle on water by gingerly laying it horizontally on a the surface of a glass of water. Best is a steady hand and a pair of tweezers. After you've watched the needle float for a while, barely touch the water's surface with toothpick that has been dipped in a liquid detergent. The surface tension abruptly falls, the contact angle drops to zero, and the needle gets pulled through the surface by gravity.

## Dust Wetting

Another use of wetting agents is to control the amount of airborne dust in underground coal mines. Modern coal mining technology involves the use of high powered, rotating toothed drums that tear the coal from the coal face. This operation produces a great deal of dust, and conventional coal mining equipment uses many water jets directed at the cutting areas to minimize dust clouds. Exposure to coal dust and mineral particles is the cause of the miner's disease pneumonoconiosis, known colloquially as "black-lung."

Coal dust is hydrophobic and a pile of it placed on water will float, as shown in Figure 10.23. However, the addition of a wetting agent such as sodium dioctyl sulfosuccinate (the anionic surfactant pictured in Figure 10.12) provides for better contact between the dust particles and the water droplets in the control spray.

Figure 10.23 Coal dust, and many other types of powder, will float on water because the forces between water molecules and the surface of the dust particles are weak. The coal shown here has been sized to pass through a 325-mesh screen. That's about a 40 μm screen opening size.

Figure 10.24 shows that coal dust particles stream through the water surface after a surfactant has been added. The rate at which a precisely defined (specified mass and specified particle size range) coal sample "sinks" below the surface is a useful comparative test for the wetting rates of various surfactants, although it is not altogether clear that such simple laboratory tests are directly relevant to the practical coal mining situation.

Figure 10.24 Sunken coal dust. A successful commercial product sinks the dust in 30 seconds.

Some of the earliest synthetic wetting agents first came on the market in the 1930s. A remarkable experiment to dramatize the properties of these, then, brand new materials, involved adding sodium dioctyl sulfosuccinate to a basin of water in which a duck was swimming. The duck sank. (Nowadays, we are more sensitive about the way we treat creatures.) Duck flotation is made possible by air bubbles trapped in the hydrophobic regions of its feathers. wetting agents break the bubbles: so this became the sunken duck experiment.

Finally, on the related subject of *non-wetting*. Lt. Commander Gerald Pawle tells that during World War II the British initiated experiments with coal dust floated on water to camouflage rivers at night and so deny their use as navigational aids to enemy aircraft. Attempts to camouflage the river Thames, observed by no less a personage than Winston Churchill, were defeated by tidal currents that piled up the coal dust in shoreline films. However, success was achieved on a quiet canal in Coventry, when an old gentleman and his dog walked into a canal—under the impression that it was a newly paved road.

## Foams and Gels

One of the most familiar properties of soap and surfactants is that they make water foamy. A **liquid foam** consists of gas bubbles stabilized in a liquid matrix by surfactant molecules. In soapy water, the bubbles are filled with air. In commercial aerosol products, the bubbles are filled with the propellant gases in the product formulation. The propellants are pressurized liquids that expand, making bubbles when the consumer puts the product use by depressing the valve. For some foam applications it is important to have a specific type of surfactant, such as the fluorocarbon surfactants used in fire fighting formulations that were described earlier.

Figure 10.25 Foams and gels are so commonplace that we take their remarkable properties for granted.

A **gel** is jelly-like substance often produced by adding structure-forming compounds to liquid water. **Structure-forming** occurs when the molecules of a liquid become somewhat organized over a long range. Cornstarch, used for example in gravy making, forms a gel by putting long, structured, carbohydrate chains in water and thereby increasing water's viscosity. If you've seen gravy that has been stored overnight in a refrigerator you'll know about its gel properties. Gel formation is an important aspect of food chemistry.

The most spectacular example of gelation comes with the traditional protein-derived product called gelatin (Figure 10.26). Gelatin, which is obtained by boiling collagen (from animal skin, tendons, ligaments, bones, etc.), dissolves in hot water to make a low viscosity solution. However, if that solution is left undisturbed and allowed to cool, then the long chain protein molecules form three dimensional threadlike structures that hydrogen bond large numbers of water molecules. The result is the formation of the familiar jellies that we eat. It is remarkable that only a few grams of gelatin are required to gel several hundred milliliters of water.

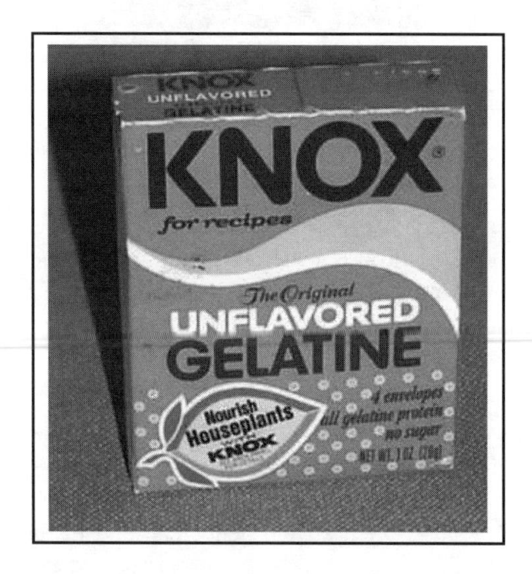


Figure 10.26 A few grams of gelatin is sufficient to gel several hundred milliliters of water. The water molecules in an aqueous gel are constrained from moving by hydrogen bonds to the protein chains in the gelatin.

## Polyelectrolytes

As mentioned in Section 2.6, a **polymer** is by definition a substance composed of very large molecules. Our major discussion of polymers, which we encounter mostly as solids, will come in Chapter 11. But polyelectrolytes are a special class of polymers that often occur as aqueous solutions and so belong with our chapter on liquids. **Polyelectrolytes** are water soluble polymers with charged substituent groups along the polymer chain; both anionic and cationic exist. A schematic representation of an anionic polyelectrolyte is shown in Figure 10.27. Traditionally, polymers made from acrylamide ($CH_2$=CH–CO–$NH_2$) are also included in the family of polyelectrolytes though they are uncharged; polyacrylamides are called nonionic polyelectrolytes.

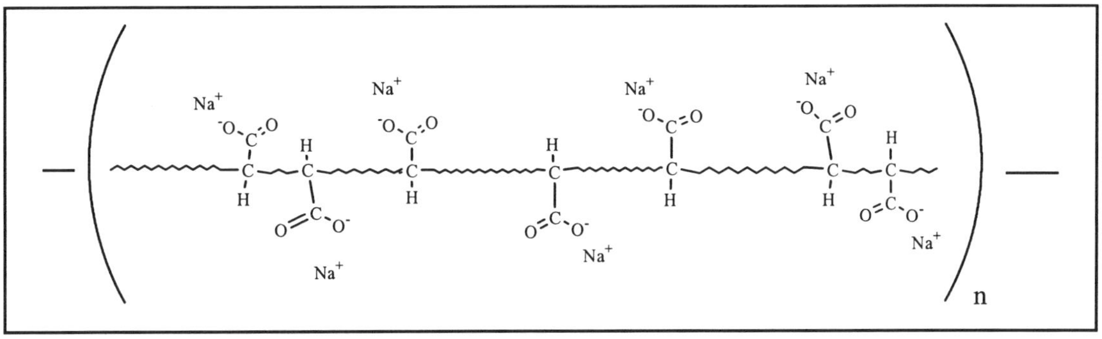

Figure 10.27 An anionic polyelectrolyte represented schematically.

Polyelectrolytes are find extensive application in drinking water treatment, industrial raw and process water treatment, municipal sewage treatment, mineral processing and metallurgy, oil drilling and recovery, color removal, paper and board production, etc. Polyelectrolytes are both flocculants as well as deflocculants depending upon their molar weight. A **flocculant** is a solid liquid separating agent while a **deflocculant** is a dispersing agent. Major applications of flocculants derive from their inherent solid-liquid separating efficiency. In all their applications their solid-liquid separation ability is commercially exploited. Many commercial polyelectrolytes are available, varying over a wide range of chemical composition, charge density (the proportion of charged substituents along the chain) and molar weight.

How do they work? We have all seen turbid or muddy water, and you know that water can remain indefinitely turbid from suspended mineral particles unless treated. So treating water treatment for human consumption generally requires the treatment engineer to make the suspended particles agglomerate so that they can fall under the influence of gravity and thus be separated. Stokes's law predicts that spherical particles suspended in a fluid medium settle at a rate proportional to the fourth power of their radius. Thus large particles will settle much faster than smaller ones.

Most suspended particles in water have a net negative surface charge. So Coloumbic repulsions among the particles act to prevent them coagulating. The first step in the process of flocculation is the neutralization of the net charge carried by each particle, which is accomplished by means of a cationic polyelectrolyte with a high charge density. Once charge neutralization has taken place, suspended particles can undergo initial coagulation. The second step is to add much longer chain polyelectrolytes with a low charge density to "bridge" the "primary flocs." Flocculation occurs when the destabilized particles are induced to collect into larger aggregates. The aggregation is followed by rapid settling in accord with Stokes's law.

Even from this brief discussion it will be clear that skill at using polyelectrolytes is valuable for many civil engineers, biological systems engineers, process engineers and others. [❁kws +polyelectrolytes +applications].

### Viscosity Modification

Controlling or changing the viscosity of fluids has many practical applications. One of the best known, and most important, involves engine oils. The viscosity of unmodified motor oil decreases as the motor heats up to operating temperature. To counter this phenomenon, additives are used that help the oil retain the viscosity needed for satisfactory performance. Such "**viscosity index improvers**" are usually synthetic organic polymers. Their precise mode of action is not known. It is speculated that at low temperatures the polymer chains form tightly-coiled, individual particles that have little or no effect on the fluid properties, while at high temperatures the polymer molecules uncoil, and the tangling of their chains causes the viscosity to increase. In any event, as temperature increases, the increase in viscosity of the additive offsets the decrease in viscosity of the oil. Consequently, the viscosity of the mixture remains fairly constant over a temperature range.

By good fortune, the same additives also help prevent motor oil from becoming excessively viscous during cold winter conditions. Any oil has a **pour point**, or a temperature at which it ceases to flow from a container. One way to lower an oil's pour point is to refine the oil to lower its content of dissolved solids (wax). However, such refining is expensive, and the use of a pour point depressant additive is a good alternative. It is speculated that the mode of action of the additive is to modify the crystal structure of any solid wax as it begins to crystallize. Thus, a single additive both increases oil's high temperature viscosity and lowers oil's low temperature viscosity.

Most paints and spray painting formulations contain thickeners, for the purpose of making the paints easier to apply by preventing "runs." A special case is thixotropic thickening, which is done by means of certain natural clays as well as with synthetic materials. A thixotropic fluid will flow under stress, but will rebuild viscosity when the stress is removed—an obvious advantage for brush painting. The molecular interpretation of thixotropy is that applied energy temporarily destroys molecular order, which "heals" after a time interval.

Other common examples of viscosity modification are found in the kitchen. Aerosol spray oven cleaners are formulated from a moderately concentrated sodium hydroxide solution with detergents and a thickening agent. The purpose of the thickening agent is to increase the viscosity of the spray so that it will cling to the inside surfaces of the oven. The use of flour or starch for thickening brown meat juices to make gravy is an old technique. Modern food processing employs a wide range of agents to modify the texture of prepared foods. Starch, or modified starch, is the most used food additive. It thickens sauces and pie fillings, gels jelly beans, stabilizes salad dressing, and helps frosting retain its moisture. For more specialized purposes, gums derived from plants or seaweed are useful in providing viscosity control in foods. For example, algin derived from kelp gives a smoothness to ice cream.

The interaction of polymer molecules with each other and with their suspending fluid provides a number of interesting examples of viscosity modification. Clearly, such interactions have widespread and important technological applications. While the principles of chemical bonding and intermolecular forces must generally cover these situations, much of the technology still falls in the area of "art" rather than theory.

In this chapter we've seen how the bulk properties of liquids and liquid mixtures are controlled by the intermolecular forces among their molecules and we've illustrated that with a blend of theory and practical examples. In the following chapter we'll see how these same forces, and others, are at work in solids. The forces in solids are generally much stronger than those in liquids, and this increased strength accounts for the mechanical properties of solid materials. Much of what we'll take up next can be called **materials science**.

## Essential Knowledge—Chapter 10

**Glossary words:** Molecular substance, molecular liquid, petroleum, temporary dipole, dispersion forces, dipole forces, polar liquids, infinitely miscible, hydrogen bonding, surface tension, viscosity, long range order, quantitative structure property relationships (QSPR), quantitative structure activity relationships (QSAR), miscibility, immiscible liquids, solvent and solute, liquid-vapor equilibrium, volatile substance, industrial solvents, distillation, distillate fraction, fractionation column, ideal solution, surfactant, hydrophilic and lipophilic. hydrophobic and lipophobic, amphipathic, hydrophilic-lipophilic balance (HLB), anionic surfactants, cationic surfactants, nonionic surfactant, liquid-liquid emulsions, cutting fluids, hydraulic fluids, emulsifiers, wetting agent, contact angle, polywater, liquid foam, gel, structure-forming, polyelectrolytes, flocculant, viscosity index improver, pour point.

**Key Concepts:** That the bulk properties of pure liquids are controlled by intermolecular forces. The vapor pressure of a liquid is related to its enthalpy of vaporization through the Clausius-Clapeyron equation. Distillation involves liquid-vapor equilibrium and is important for the production of fuels and the purification of liquids. The miscibility of liquids depends on intermolecular forces. Some chemically similar liquids form ideal mixtures and follow Raoult's Law. There are many important practical liquid mixtures. Surfactants reduce water's surface tension, allow it to form emulsions, and can be used to produce foam. Liquids with modified viscosities have useful applications. Polyelectrolytes have important uses in water treatment and related areas.

**Key Equations:**

The empirical equation for a liquid's vapor pressure: $\ln P = \frac{B}{T} + A$

The Clausius-Clapeyron Equation: $\ln P = A - \Delta H_{vap}/RT$.

Relationship between a liquid's enthalpy of vaporization and the empirical B constant: $\Delta H_{vap}/R = B$

Two-point fit for the Clausius-Clapeyron Equation: $\ln P_2 - \ln P_1 = \ln (P_2/P_1) = \frac{-\Delta H_{vap}}{R}\left(\frac{1}{T_2} - \frac{1}{T_1}\right)$

Raoult's law: $P_{tot} = P_A^\circ X_A + P_B^\circ X_B$

## Questions and Problems

### Introduction

10.1. Define a molecular liquid and explain why at room temperature, and a pressure of one bar, pure liquids are almost always molecular substances.

10.2. The chapter concludes with a box describing viscosity modification. State five reasons why an engineer might need to understand and apply the property of viscosity.

10.3. When was distillation first practiced? How is distillation used to make motor fuel?

10.4. What is the relationship between the boiling point of a liquid and the intermolecular forces among its molecules?

10.5. State the three major types of forces among the molecules of a liquid.

10.6. State the nature of the intermolecular forces in the following liquids: a. water, b. benzene, c. acetone, d. kerosene, e. ethyl alcohol, and f. n-propyl amine The three types are dispersion, dipole-dipole, and hydrogen bonding forces. Some of the liquids have more than one type of force.

10.7. Carbon monoxide (CO) has a higher melting point that diatomic nitrogen ($N_2$). Each of these molecules has an identical number of protons and electrons. The molecules are isoelectronic. Explain why they have different melting points.

10.8. The molar enthalpies of vaporization of water, methyl alcohol, and ethyl alcohol are all about 40 kJ $mol^{-1}$. The molar enthalpy of vaporization of methane is only about 9 kJ $mol^{-1}$. Discuss the intermolecular forces responsible for these differences.

10.9. Describe the nature and origin of dispersion forces, dipole-dipole forces, and hydrogen bonding forces.

### Intermolecular Forces in Liquids

10.10. State the name and chemical formula of each of the following: a. a common liquid with a very high surface tension, b. a common liquid with a very low surface tension, c. a common liquid with a very high viscosity, d. a common liquid with a very high molar enthalpy of vaporization.

10.11. What is meant by the term "pour point" of a liquid? Why is modification of a liquid's pour point important to automotive engineers?

10.12. Describe the factors that make a given pair of liquids either miscible or immiscible.

10.13. Use the general principles of intermolecular forces to speculate whether or not the following pairs of liquids are miscible or immiscible: a. water and glycerol, b. water and hexane, c. hexane and benzene, ethanol and ethylene glycol, e. benzene and toluene.

10.14. Warming a liquid usually results in its flowing more readily. Suggest a molecular interpretation for this.

10.15. Raw syrup is usually heated in large open pans to concentrate the sugar content of the syrup by evaporating some of the water. How does this process affect the syrup's viscosity?

10.16. Rank the following liquids in order of increasing surface tension: water, cyclohexane, carbon tetrachloride, glycerol, ethylene glycol.

### Liquid-Vapor Equilibrium

10.17. Ethyl acetate has a vapor pressure of 41.9. mmHg at 10.0°C and a vapor pressure of 595 mmHg at 70.0°C. Use these data to estimate the enthalpy of vaporization of ethyl acetate. Compare your answer with the value shown in Table 10.5.

10.18. Use the empirical equation for ethyl acetate's vapor pressure to estimate its vapor pressure at 62°C.

10.19. The empirical equation for water's vapor pressure is ln P = 20.5 - 5170/T, where P is in mmHg and T is in kelvins. One morning, at a camp high in the mountains, water boiled at 192°F. What was the atmospheric pressure that morning?

10.20. A second empirical equation ln P = 20.3 - 5120/T, where P is in mmHg and T is in kelvins, is stated for water's vapor pressure. Why is this equation slightly different from the one used in the previous problem?

10.21. When the external pressure on an ordinary liquid is sufficiently reduced below atmospheric pressure the liquid will boil. How low must the pressure be to make ethyl alcohol boil at 20°C?

10.22. An organic liquid has a vapor pressure of 45.7 mmHg at 25.0°C and a vapor pressure of 90.4 mmHg at 38.3°C. Use the Clausius-Clapeyron equation to estimate its molar enthalpy of vaporization.

10.23. Isoflurane, $CHF_2$-O-$CHClCF_3$, introduced as an anesthetic in 1971, is a stable gas. It has a normal boiling point of 48.5°C and at 20.0°C its vapor pressure is 250 mmHg. Estimate its vapor pressure at 65.0°C. The structural formula of isofluorane is:

$CHF_2$-O-$CHClCF_3$

```
   H     H   F
   |     |   |
F—C—O—C——C—F
   |     |   |
   F    Cl   F
```

10.24. List the following substances in order of increasing molar enthalpies of vaporization: methyl alcohol, water, methane, argon, helium. Rationalize your ranking using arguments based on comparative intermolecular forces.

## Distillation and Raoult's Law

10.25. Define or describe in 25 words or less each of the following: a. distillation flask, b. condenser, c. distillate fractions, and d. petroleum distillates.

10.26. List four petroleum distillates, their boiling range, their uses, and the carbon count of the molecules they contain.

10.27. State Raoult's law as it applies to a mixture of two miscible, volatile liquids.

10.28. A solution of two liquids that obeys Raoult's law is called an ideal liquid mixture. Such mixtures are rare. Why? What are positive and negative deviations from Raoult's law? What molecular explanation do chemists give for positive and negative deviations from Raoult's law?

10.29. Carbon tetrachloride ($CCl_4$) and chloroform ($CHCl_3$) are a pair of liquids of similar chemical composition whose mixtures obey Raoult's law rather closely. At 50°C the vapor pressure of pure carbon tetrachloride is 317 mmHg and the vapor pressure of pure chloroform is 526 mmHg. Calculate the vapor pressure of a mixture of 20.0 g of carbon tetrachloride and 50.0 g of chloroform.

## Surfactants

10.30. Define or describe in 25 words or less each of the following: a. surfactant, b. hydrophobicity, c. hydrophilicity, d. lipophobicity, e. lipiphilicity, and f. amphipathic.

10.31. Distinguish among nonionic, anionic, and cationic surfactants. What is the characteristic polar group of a cationic surfactant? Name two different functional groups that are found in anionic surfactants.

10.32. What is ethylene oxide? How is it used in the manufacture of nonionic surfactants? Polyoxyethylene oxides are polyethers. Explain.

10.33. Describe the formation of an emulsion when a commercial pine oil disinfectant is added to water. Make a sketch of an oil droplet suspended in water and stabilized by a surfactant. Explain why such oil droplets do not coalesce to form an immiscible liquid layer.

10.34. Describe the composition and uses of cutting fluids and nonflammable hydraulic fluids.

10.35. What is a contact angle? On a plastic surface, how do the contact angles of ethyl alcohol and water compare?

10.36. Name two uses of wetting agents. State the name of a wetting agent.

10.37. Describe the phenomenon of polyelectrolyte-induced flocculation.

10.38. Polyelectrolyte additives are useful in improving the properties of paper products. Use internet resources and find out why.

10.39. Give examples *not* already stated in Chapter 10 of three practical applications of foams.

10.40. Give examples *not* already stated in Chapter 10 of two practical applications of gels.

## Problem Solving

10.41. A butane cigarette lighter contains about 25 $cm^3$ of liquefied butane with a density of 0.60 g $cm^{-3}$. What volume would the lighter require if it held butane gas at 1.00 atm and 25°C? Assume that butane gas behaves ideally. The mole weight of butane is 58.12 g $mol^{-1}$.

10.42. Describe how the enthalpy of vaporization of water changes with changing temperature. Use internet resources to find the earliest possible historical reference to water's enthalpy of vaporization.

10.43. Define the word polarizability. Which noble gas has the most easily polarized atoms? Which noble gas is least easily polarized?

10.44. Which hydrogen halide molecule interacts most strongly by dipole-dipole interactions? [Hint: review Chapter 6 if necessary.]

10.45. A typical intermolecular hydrogen bond has an interaction energy of 15 kJ $mol^{-1}$. What percentage is this of the H-H covalent bond energy?

10.46. Use internet resources to distinguish between dynamic viscosity and kinematic viscosity. Make a sketch of an Ostwald viscosimeter. Define the words: Brookfield, centipoise, and centistokes.

10.47. A closed container is equipped with a pressure gauge and a pressure relief valve. The container has in it a pure substance. The valve is briefly opened and some substance escapes. What happens to the pressure inside the container?

10.48. Select any commercial nonfood product with a labeled ingredient list showing at least four ingredients. Describe the purpose of each ingredient in the product.

10.49. Use internet resources to discover commercial suppliers and producers of QSAR software.

10.50. Do an internet search to discover if there are any ionic liquids at standard pressure and 25°C.

# Chapter 11

# Solids

## Introduction

At the end of the preceding chapter we discussed polyelectrolytes and gels. These substances occupy a middle ground between true liquids and true solids. In this chapter, we'll discuss well defined solids, but we'll also discuss glass, leather, rubber, and plastic films—materials that are equally as difficult to classify, in a simple scheme that labels substances as either liquid or solid, as polyelectrolytes and gels. Indeed, one could argue that rubber isn't very solid at all. However, we mustn't be too concerned about the difficulty of cataloging materials. Rather, keep in mind that the theme plays on: the bulk properties of materials are determined by the particles that compose them and the forces among those particles.

Solids can be *arbitrarily* placed into five broad categories: ionic solids, covalent solids, molecular solids, metallic solids, and polymeric solids. In this chapter we'll survey these classes of solids and look more particularly at solids with engineering applications. Molecular solids are similar to molecular liquids and we'll look at these only briefly. Likewise, we'll only glance at ionic solids and at metals—though metals, especially, are an important class of solids with widespread engineering applications. Much of the chapter concerns materials science. Broadly speaking, **materials science** is the scientific study of solid materials. Materials science is a wide ranging branch of chemistry and physics, a dominant aspect of modern technology, and a well defined area of engineering science. Materials science can be broadly subdivided into three areas: inorganic materials, organic materials, and biological, materials. However, sometimes, two or more materials are be combined for practical applications in what are termed composite materials or composites. A **composite** is any material made from at least two discrete substances, such as concrete or fiberglass-reinforced plastics, but the term is usually used to describe modern, industrially-manufactured composites, such as carbon fiber-reinforced plastics, which combine the high strength and stiffness of the fiber with the low weight and fracture resistance of the polymer matrix. Abalone shells, which have remarkable strength, are a natural composite. The possibilities for new materials are enormous; here we can just sketch out an overview.

The **inorganic materials** include rocks and minerals, glasses, and cements. Chemically, almost all the important inorganic materials contain –O–Si–O– atomic linkages. Inorganic materials are typically covalent solids (or network covalent solids) and they often have valuable and unique properties. The dominant group of network covalent solids are the **natural silicate materials** and **silicate minerals**—the rocks that compose the earth's crust. Most glass and ceramic materials derive from natural silicate minerals. Calcium silicates are the main ingredient of Portland cement—the most used human made substance on earth. Silicate materials can be regarded as inorganic polymers.

The **organic materials** are primarily synthetic organic polymers. We first met polymers briefly in Section 2.4; now we'll further examine this important class of materials. A **polymer** is a substance composed of very large molecules; it's a material that has long chains of atoms making up its structure. A good interpretation of the word "poly-mer" is "many-units;" it's a substance composed of many units or made by joining them together. The units are small molecules; a single chain forming unit is called a **monomer**. The word polymer was coined in 1830 by the Swedish chemist Jons Jakob Berzelius (1779-1848). Organic polymeric solids divide into two major groups: **natural polymers** and **synthetic polymers**.

The **biological materials** are natural polymers, sometimes called biopolymers. **Biopolymers** are materials derived from the three chemical groups: polysaccharides, proteins, and nucleic acids.

Our studies in this chapter will involve a good deal of organic chemistry. Synthetic polymers are made by taking advantage of the functional groups of monomer molecules in monomer linking reactions. Recall, Table 2.7 described functional groups in organic chemical compounds. In this chapter we'll see how chemical reactions involving many of these functional groups are exploited for the practical purpose of manufacturing the synthetic materials we use to enrich our daily lives.

## Section 11.1 Types of Solids

We may classify solids into five major types, based on the dominant forces that hold them together. The classification is not completely satisfactory, but it's useful. The different types of forces cause each class of solids to have characteristic properties. Table 11.1 shows the five classes.

Table 11.1 **Chemical Classification of Solids**

| Class | Molecular solids | Ionic solids | Metallic Solids | Covalent solids | Polymeric solids |
|---|---|---|---|---|---|
| Composed of | Molecules | Ions | Atoms | Atoms | Long chains of atoms |
| Main forces involved | Covalent bonds within the molecules. Intermolecular forces among them. | Coulombic forces. | Orbitals that extend throughout the solid | Covalent bonds extending throughout the solid. | Covalent bonds form the chains. Intermolecular forces among the chains. Chain entanglement not a force but a factor. |
| Melting Point | Low | High | High (not Hg) | High | Thermoplastics moderate. Thermosets have no melting point, they decompose when heated. |
| Mechanical strength | Weak | Strong | Variable. Most are malleable and ductile. | Very strong | Widely variable. Controlled by blending and additives. Elastomers (rubbers) are a special case. |
| Electrical properties | Non conducting | Conducting if melted. | Highly conducting | Insulators. * | Insulators.* |
| Water solubility | Yes if highly polar or hydrogen bonding. | Many are soluble. | No. | No. | Depends on type. Most are insoluble. |
| Examples | Wax, sugar, dry ice, | Salt. Salts in general. | Lead, gold, nickel. | Silica, diamond. | Polystyrene, polyethylene, starch, proteins, etc. |

* Both covalent solids and polymeric solids can be "doped" or chemically treated to produce semiconductors or even conductors. Such doped materials are the basis for semiconductor devices, computer chips, etc.

An alternative classification of solids is based on four major classes of engineering materials. An **engineering material** is any kind of stuff that engineer's make things with. Table 11.2 the four major classes of engineering materials, some examples, and the approximate dates of introduction of the industrial age synthetic materials.

Table 11.2 **Four Major Types of Engineering Materials**

| Class | Historic natural solids | Manufactured industrial solids |
|---|---|---|
| Ceramics and Glasses | Stone, flint, obsidian. | Portland cement (1830)<br>Engineering ceramics (1960-). |
| Composites | Straw-brick, wood. | Graphite and fiber reinforced plastics (1960-).<br>Fiberglass and Fiberglas® (1931-). |
| Polymers | Wood, leather, fibers, membranes. | Bakelite (1907). Polyvinyl chloride (1925).<br>Nylon (1935). (Polyethylene 1938). Etc. |
| Metals | Gold, copper, bronze, iron. | Ti and Zr alloys (1950).<br>Modern super alloys (1980-). |

## Molecular Solids

In **molecular solids** the building block is the molecule itself; individual molecules are held together by relatively weak van der Waals forces, and sometimes these are strengthened by hydrogen bonds. Recall from the preceding chapter that we can conveniently classify the forces among molecules into three categories: 1. Dispersion Forces, 2. Dipole Forces, and 3. Hydrogen Bonding forces.

Because their intermolecular binding forces are weak, molecular crystals have low melting points and are soft. They are generally poor conductors of heat and electricity. Examples of molecular solids are wax, aspirin, sugar, fats, steroids, most drugs, and any solid formed by freezing a molecular liquid. Ordinary molecular solids don't find engineering applications that require mechanical strength, though they can have special purpose applications, as, for example, in the lost wax method of making metal castings. [❁kws "lost wax method"]. In the model of carbon dioxide in Figure 11.1 you can see individual $CO_2$, O=C=O, molecules each modeled as a black ball flanked by two lighter colored balls.

Figure 11.1 A model of solid carbon dioxide ("dry ice"). Solid carbon dioxide consists of individual $CO_2$ molecules held in an array by van der Waals forces. These forces are weak, as demonstrated by the fact that under ordinary conditions dry ice sublimes unless stored at a temperature below -78°C.

Many molecular solids are crystalline. In a **crystalline material** the constituent entities (molecules in a molecular solid) arrange themselves in a characteristic repeating array with extensive long range order. In some crystals the repeating structure extends across the entire macroscopic size of the crystal. Recall, we discussed the closely related notion of a perfect crystal in connection with our study of the Third Law of Thermodynamics.

Some molecular solids are glassy or amorphous. An **amorphous** solid is one that lacks long range order. The ordinary definition of amorphous is simply something lacking a crystalline form; the technical definition given here is narrower. We can imagine that an amorphous solid is formed when the molecules of a liquid being cooled are not able to organize themselves sufficiently before the temperature gets so low that they get stuck in place; this happens especially when a cooled liquid experiences a sharp viscosity increase. For example, if you cool a mixture of water and ethylene glycol (antifreeze) in dry ice, a sticky, syrup like amorphous substance forms that eventually cools further to become a *glassy* solid. Glass is the classic amorphous solid, hence the adjective glassy and the use of the word "glass" in a general way for any amorphous solid. One way to think of amorphous molecular solids is as "freeze-framed liquids."

What really *is* ordinary glass?

Glass is regarded as the classic amorphous solid, or as a supercooled liquid, and is still believed by some to flow and to crystallize if left long enough. All these characterizations are wrong. One observer has recently pointed out: "Glass is a rigid solid with a lower degree of molecular order (higher entropy) than a crystal but with greater molecular order (lower entropy) than a liquid."

Molecular solids have variable solubility characteristics. Those held together primarily by dispersion forces tend to be soluble in alkane (or other hydrocarbon) solvents that themselves are composed of molecules that interact primarily via dispersion forces. A common household example of a solid dissolved in a hydrocarbon solvent is rubber cement—which consist of rubber dissolved in hexane ($C_6H_{14}$). Hexane is a volatile alkane solvent that evaporates quickly and does not "wrinkle" paper because its own intermolecular interactions via dispersion forces are different from the primarily hydrogen bonding interactions in paper. In contrast, water wrinkles paper because it's a hydrogen bonding liquid and cellulose's chains carry many –OH groups to which water molecules can hydrogen bond. So pouring water on paper disrupts the hydrogen bonding among the cellulose molecules which change shape and don't return to the same geometry when they dry out.

Molecular solids that can form hydrogen bonds are often water soluble. Sugar is a molecular solid that is highly soluble in water because of hydrogen bonding. A solution of sugar, or any other molecular substance, in water does not conduct electricity. A solid that dissolves in water to give a non-conducting solution is classified as a **non-electrolyte.**

## Ionic Solids

**Ionic solids** are held together by strong Coulombic forces among all their anions and cations. They have high melting points and are typically hard and brittle. Figure 11.2 shows lighter-colored metal ions in a cubic array with darker-colored halide anions. Wooden bonds are needed to hold the model together, but there are no such localized bonds in the real crystal—just generalized Coulombic attractions and repulsions.

Figure 11.2 A model of an ionic crystal. Wooden sticks are needed to hold this model of an ionic compound together. In reality, the bonding forces in ionic crystals are generalized. Every cation attracts every anion, no matter where each is in the crystal lattice. Similarly, each ion repels every other ion of the same charge.

Figure 11.3 is a picture of a physical model of the "space-filling" variety. You will be familiar with space-filling representations if you've used one of the molecular visualization programs such as Chime. Amusingly, the actual model is made from painted, foamed polyurethane balls. So we here have an example of an inorganic material, such as sodium chloride, represented by a synthetic organic material.

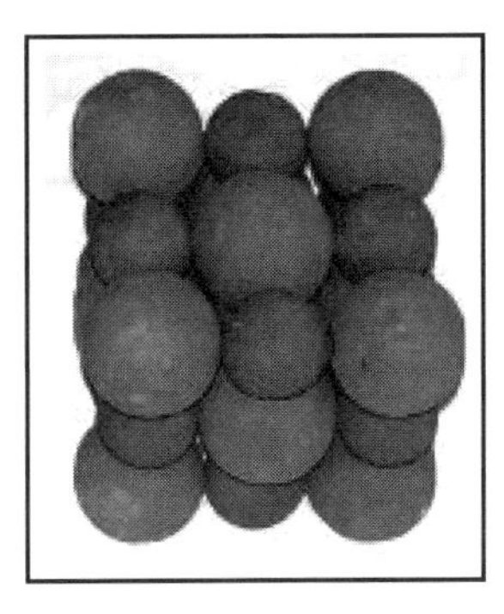

Figure 11.3 A space-filling model of an ionic crystal. The foamed plastic balls of different sizes represent oppositely charged ions as touching "hard" spheres.

Knowing their general structure, it is easy to understand in a general way the reasons for both the strength of ionic crystals and why they shatter. Their fracture requires that layers of adjacent ions with opposite electrical charges be stripped from one another. But if there is just a *one layer displacement* in an ionic crystal, which happens when it is put under stress, then ions with the same charge suddenly face one another, the strong attractive forces turn to equally strong repulsive ones and the ionic crystal shatters.

Many ionic compounds are water soluble—although others—with extremely strong interionic forces in their crystal lattice are not. Thus, ionic compounds with the $A^+B^-$ stoichiometry are more likely to be water soluble than those with the $A^{2+}B^{2-}$ stoichiometry because the electrostatic forces between doubly charged ions are four times as great as those among singly charged ions. Simple ionic solids are not used as structural engineering materials; however, many ceramics involve negatively charged chains held together by positive ions, and many silicate network covalent solids have cations among their anionic chains. Simple ionic compounds may have good strength, but as described above, they are brittle. Alumina (aluminum oxide. $Al_2O_3$), widely used as an abrasive, is nominally an ionic compound. However in this application alumina is probably better regarded as a network covalent solid. A similar distinction can be made for many ceramics. You see again the difficulty of making a simple classification of solids: lots of real materials are not so simple.

## Metallic Solids

**Metallic solids** are held together by metallic bonding. Metals can be imagined as an array of cations embedded in a cloud of free electrons. The cloud of electrons is a gigantic "molecular orbital" extending across the full dimensions of the metal's crystals with the cations are studded throughout the cloud. The electrons are mobile and that mobility creates the characteristic metallic properties of high thermal and electrical conductivity. The strength of these metallic bonds varies widely, as demonstrated by the range of melting points of metals with mercury being a liquid at room temperature to tungsten with a melting point of over 3400°C.

Metals and alloys (mixtures of metals) are an important class of engineering materials. In a very real sense, modern civilization could only emerge after the bronze and iron ages. In 16th and 17th centuries, in the county of Shropshire, in the west of England, declining forests curtailed the availability of charcoal and stimulated the use of coal for smelting iron. This development ushered in the industrial revolution. So it would be hard to overestimate the role of metals and their alloys in modern economies. **Metallurgy** is the science of metals. It ranges across topics such as the manufacture of metal and metal alloy products; the crystallographic and atomic structures of metals; strengths of metals; grains and grain boundaries in metals; nucleation phenomena; etc. Metallurgy is a branch of engineering science; we'll say more about it in Chapter 13

From the point of view of engineering design there are three key properties of metals, one or more of which is wanted in any useful metal alloy. 1. Formability: they can be manufactured in the desired shape. 2. Strength without brittleness: glass is many times stronger than aluminum but we don't build glass airplanes. 3. Having required electrical and magnetic properties: which are properties found in few materials other than metals.

To most people it comes as a surprise to learn that most metals have a definite crystalline structure, a property that makes them quite different from other ordinary materials such as glass, wood, plastics, and paper. The atoms in metals are arranged in a periodically repeating, geometric array. There is a lot of long range order in metals. The hard-sphere model is quite useful for describing the crystal structure of metals. Metals form some of the simplest crystals, because they are composed of identical atoms. Most metals are **close-packed**. A close packed structure is one in which the atoms occupy the minimum volume possible. “Loose” and close packing is illustrated schematically in two-dimensions in Figure 11.4.

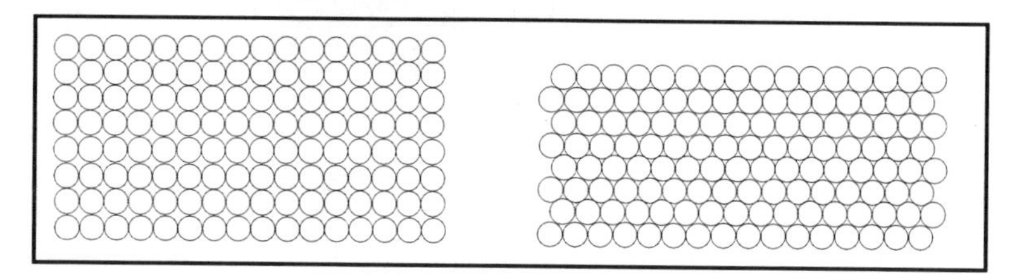

Figure 11.4 Loose and close packing illustrated schematically in two dimensions. Each array contains 128 atoms, but the close packed structure on the right has a smaller footprint. Close packing brings the greatest density and least internal void space for a given metal.

A three dimensional array of points showing the locations of a crystal’s atoms is called a **crystal lattice**. A crystal lattice can be represented by its unit cell. A **unit cell** is the simplest unit that produces the entire lattice when continuously repeated in three dimensions. Fourteen basic kinds of unit cells, called the Bravais lattices, are possible in nature [✿kws +“Bravais lattice”]. Other imaginable arrangements can be reduced to one of the basic fourteen by changing coordinate systems. Figure 11.5 shows three common atomic arrangements in metals.

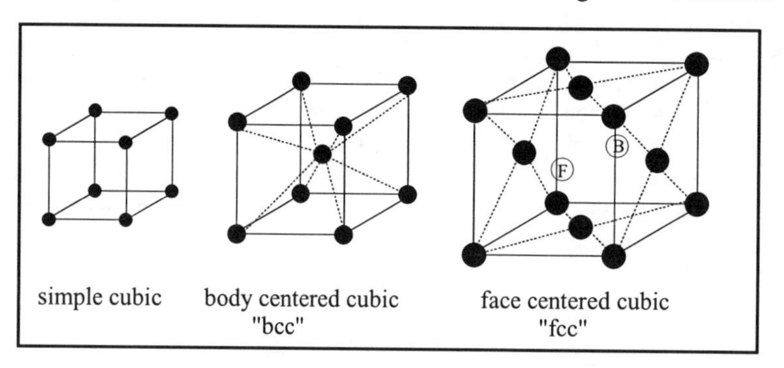


Figure 11.5 Three possible cubic unit cells for metals. The lines define the unit cells. The black circles represent the atoms but are not drawn to scale. The transparent atoms labeled F and B in the fcc structure are respectively at the front and back of the unit cell. There are 11 other basic unit cells found in nature.

The detailed structures of crystals are found using the important technique of x-ray crystallography that is described in Appendix G.

The metal with the simplest crystal structure is polonium in its room temperature form (polonium, and many other metals, exist in more than one form, called **allotropes**). In a simple cubic crystal, as shown on the left hand side of Figure 11.5, the atoms are stacked one on top of the other in all three perpendicular directions with each individual atom being surrounded by six identical neighbors. Its number of nearest neighbors in the crystal is called the **coordination number** of an atom. Polonium has a coordination number of six. Each atom is on a corner where eight unit cells meet, and thus one-eighth of the atom is in each cell, and there is one atom per unit cell overall.

We can imagine that the atoms or ions in a crystal are hard, non-penetrating spheres (the hard-sphere model). When spheres are packed in a lattice, gaps or voids are left between them. If the spheres are packed in a simple cubic lattice, such as that in polonium, the spheres take up 52.36% of the total space in the crystal and 47.64% is void space. Polonium is unusual, all other metals have their atoms packed so that a larger fraction of the total space is occupied and there is less void space.

Tighter packing is achieved by each atom being surrounded by a larger number of nearest neighbors; that is, the coordination number of the metal is larger. Sixteen metals, including all the group 1 metals, barium from group 2 and a few of the d-block and f-block metals crystallize with their atoms having a coordination number of eight. An eight-coordinated atom can be seen at the center of the body centered cube in the middle of Figure 11.5). In a body centered cubic lattice the spheres take up 68.02% of the total space in the crystal and 31.98% is void space.

Geometry dictates that the largest possible coordination number for equal sized spheres is twelve and arrangements with 12-coordination are close-packed. An example of 12-coordination packing is shown on the right of Figure 11.5. This unit cell is called face centered cubic (fcc). An alternative name for the face centered cubic structure is cubic close packed (ccp). It (ccp), and its close relative the hexagonal close packed (hcp) structure, are the main structural types found for metals at ordinary pressures and temperatures. About 40 metals exist with close packed atomic structures; for example, cobalt, copper and platinum are cubic close packed whereas vanadium, barium, and tungsten are hexagonal close packed. In a close packed lattice, such as face centered cubic, the spheres take up 74.04% of the total space in the crystal and 25.96% is void space.

These structural characteristics of metals are of paramount importance to metallurgists and mechanical engineers. Not only because its structure and packing affect the density and other physical properties of the metal, but because it is defects and dislocations in a metal's structure that opens it up to cracks and metal fatigue, and also allows corrosion to begin. With metals, there's a very clear relationship between the metal's structure and how the metal is likely to perform in a demanding service application. However, it is not well understood what underlying factors make a metal favor one crystal structure over another, and many metals can exist in more than one structural form depending on the temperature, pressure, and their method of preparation.

## Covalent Solids

Covalent solids are essentially giant molecules in which all the atoms are covalently bonded to their neighbors, which in turn are connected to their neighbors, and eventually throughout the entire structure. In a sense, a diamond or a sand grain is just a single molecule. Covalent solids are extremely hard and have high melting points. Diamond, in Figure 11.6, composed of only carbon atoms, and silica ($SiO_2$, sea sand), in Figure 11.7, are the classic examples of covalent solids. In Figure 11.7 the silicon atoms are gray-colored and the oxygen atoms are white. The thermosetting resins that we'll meet shortly are also covalent solids.

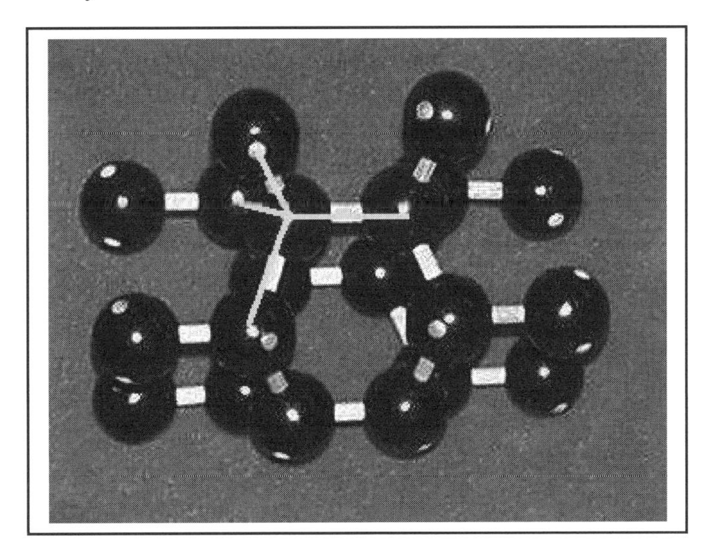

Figure 11.6 A ball and stick model of the structure of diamond. Every carbon atom is surrounded by four other carbon atoms with tetrahedral geometry. The tetrahedral structure of one carbon atom is outlined in the picture.

Figure 11.7 A ball and stick model of the structure silica, $SiO_2$. The silicon atoms are the smaller gray spheres. The oxygen atoms are the larger, lighter spheres.

Engineering abrasives are covalent solids for the obvious reason that they have the desired properties of mechanical strength, hardness, and durability. Some examples of abrasive materials are: quartz (naturally occurring $SiO_2$); garnet ($Ca_3Al_2(SiO_4)_3$, and similar structures); fused zirconia ($ZrO_2$); corundum ($Al_2O_3$); emery (a mixture of corundum and magnetite ($Al_2O_3$ and $Fe_3O_4$); silicon carbide (SiC); diamond (C); and, boron carbide ($B_{13}C_2$) and boron nitride (BN).

Purely covalent inorganic solids are actually fairly uncommon. Diamond's allotrope, graphite, shown in Figure 11.8, is a good example of a solid material whose unique properties derive directly from its structure. In graphite, there are strong, covalent bonds *within* each sheet, but relatively weak forces *between* the sheets. Ordinary graphite absorbs oxygen molecules between its sheets and the sheets slide under a shear force—which is why ordinary graphite is a good lubricant. Many naturally occurring minerals contain significant portions of covalent structure, but also have ionic regions. We'll return to such structures in Section 11.4.

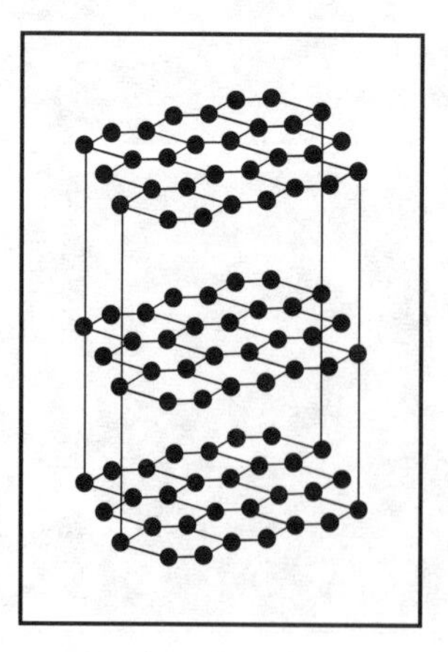

Figure 11.8 Graphite is typical of many natural solids in that its structure and properties derive from more than just a single type of bonding. Graphite consists of sheets of hexagonally-arranged, pi-bonded carbon atoms. Within the sheets, the pi electrons delocalize over long distances and thus, like a metal, graphite is a good conductor of electricity. Having good electrical conductivity is a rare property for a nonmetal, but you can see how graphite's structure with its globally delocalized electrons allows that to happen.

## Section 11.2 Natural, Semi-Synthetic, and Synthetic Polymers

In the preceding section we discussed the four basic classes of solids. For the remainder of this chapter we will be concerned with the large and structurally complex class of materials called polymeric solids. In terms of their chemical composition, we can divide them into three groups: inorganic polymer solids, organic polymer solids, and biopolymers. It is also convenient to divide them into the two categories: natural and synthetic, which we will do in this section. Humans have been practicing chemical modification of biopolymers for many years, creating an intermediate class called semi-synthetic materials.

The group of inorganic polymer solids is dominated by the silicate materials. In the form of rocks and minerals these constitute by far the largest mass of polymers on our planet. Many silicate materials, such as sand and crushed rock for roadbeds, are used in their natural form, but others are processed to make a wide variety of glass and ceramic materials. The relationship between natural and synthetic silicates can be approximately represented by the "equation:"

$$\text{natural silicate minerals} \xrightarrow{\text{heat and/or chemical processing}} \text{synthetic glass and ceramics}$$

Because they are produced by living organisms, biopolymers are by definition natural. They include the fibrous parts of plants and vegetables such as wood (a natural composite of lignin and cellulose), skin and muscle tissue (proteins), natural rubber (polyisoprene), and the nucleic acids DNA and RNA (polynucleotides).

We don't known when humans first learned how to make useful semi-synthetic polymers. Perhaps leather was the first—animal skins, smoked or salted. Tawing, the treatment of skins with natural alum $[(NH_4)_2SO_4 \cdot Al_2(SO_4)_3 \cdot 12H_2O]$, was widely used in ancient Egypt and continued into the middle ages. Likewise, glass dates from at least 4000 BCE and pottery from not much later. The chemical processing of natural fibers to make paper and textiles has almost as long a history.

On June 15, 1844 Charles Goodyear (1800-1860) was awarded a US patent that described a process to modify natural rubber chemically. The process involved heating natural rubber with sulfur and basic lead carbonate ($Pb(OH)_2 \cdot PbCO_3$). His invention solved the two problems then plaguing the practical use of natural rubber: it gets too brittle and inflexible in winter and becomes too soft and sticky in the summer. Goodyear called his process *vulcanization*, and vulcanized rubber became the first semi-synthetic polymer. A **semi-synthetic** polymer is one derived from nature but chemically modified. Vulcan was the Roman god of fire, later associated with volcanoes; so Vulcan's name's appropriate for a process that involves heat and sulfur.

A number of commercially important semi-synthetic polymers are manufactured from cellulose. Cellulose is a natural polymer that composes the fibrous part of wood. Ordinary paper is composed mainly of cellulose fibers. The monomer units in wood's cellulose are familiar to you as sugars. **Cellulose** is a natural polymer composed of long chains of sugar monomers. A simple sugar, such as glucose shown in Figure 11.9, has the molecular formula $C_6H_{12}O_6$.

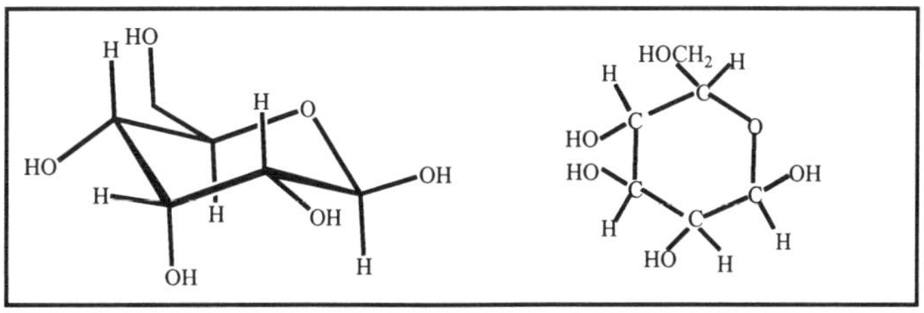

Figure 11.9 Glucose. Two depictions of a single glucose unit. On the left is a conformational structure which, by means of perspective, attempts to show the atoms in more or less the positions they take in three dimensions. The depiction on the right is less faithful, but it shows all the individual atoms and their connections. As with most organic molecules, glucose's structure can be rationalized with the simple valencies C = 4, O = 2, and H = 1 as you can check for yourself.

The linking of sugar units to make polymeric cellulose can be represented by the chemical equation:

$$n\ C_6H_{12}O_6 \rightarrow (C_6H_{10}O_5)_n\ +\ n\ H_2O$$

A water molecule is lost when two sugar units combine—when water is split out during linking, the reaction is called a **condensation reaction**. The "n" in the equation is the number of sugar units; in cellulose "n" ranges widely with 2000 being a typical value. The structure of a tiny portion of a cellulose chain is shown in Figure 11.10.

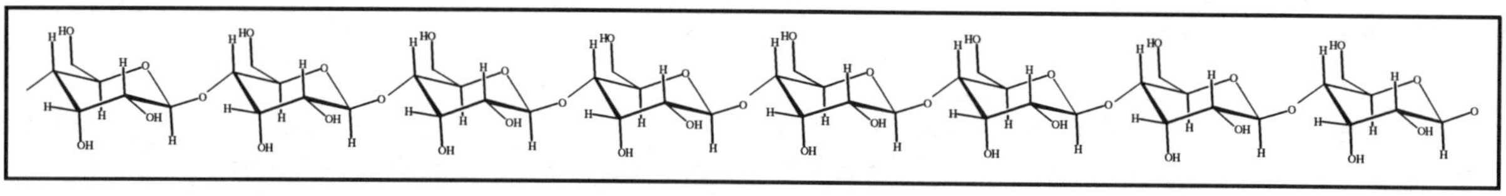

Figure 11.10 A portion of a cellulose chain. Each sugar unit has three –OH functional groups available for reaction. Cellulose, with the empirical formula $C \cdot H_2O$, is a carbohydrate. It is also a polysaccharide, a polymer of sugar units. Starch is a closely related carbohydrate. Unlike cellulose, starch can be disassembled by enzymes in your saliva, and you can taste the sugar if you suck for a while on a soda cracker.

The manufacture of semi-synthetic cellulose derivatives involves the reaction of the –OH alcohol groups along the cellulose chain with acids to convert the –OH groups to ester groups –OR. For example the esterification reaction (see below) of wood fibers with acetic acid (or its more active relative acetic anhydride, $(CH_2CO)_2O$) forms the triacetate ester of each sugar unit. **Cellulose acetate** is a semi-synthetic polymer made by the reaction of acetic acid with cellulose. It is used to manufacture the filter tips of cigarettes. The reaction of cellulose with nitric acid yields the trinitrate ester of cellulose, called nitrocellulose or guncotton. **Guncotton** is the trinitrate ester of cellulose made from cotton or a similar natural polysaccharide; it is a useful explosive and propellant.

An **esterification reaction** occurs when an alcohol functional group and a carboxylic acid functional group react to yield an ester and water. Esterification is a condensation reaction because it splits out water. The general reaction for the formation of an ester and the specific reaction for the formation of the ester ethyl acetate from ethyl alcohol and acetic acid are:

General esterification reaction

$$R{-}O{-}H \quad + \quad R'{-}\underset{\underset{O}{\|}}{C}{-}O{-}H \longrightarrow R{-}O{-}\underset{\underset{O}{\|}}{C}{-}R' \quad + \quad H_2O$$

general alcohol + general acid → general ester + water

Esterification reaction to make ethyl acetate

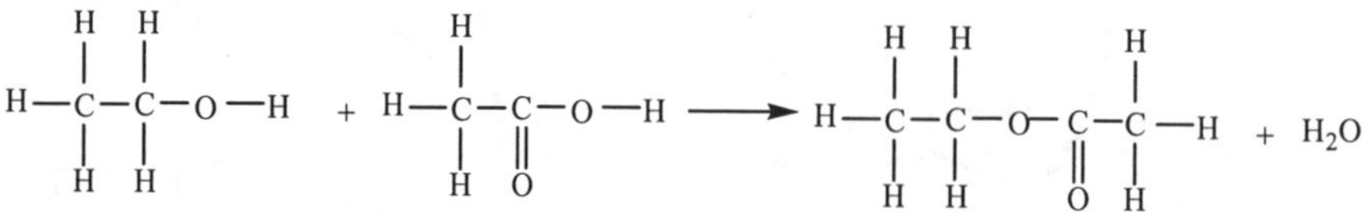

$$H{-}\overset{H}{\underset{H}{C}}{-}\overset{H}{\underset{H}{C}}{-}O{-}H \; + \; H{-}\overset{H}{\underset{H}{C}}{-}\underset{\underset{O}{\|}}{C}{-}O{-}H \longrightarrow H{-}\overset{H}{\underset{H}{C}}{-}\overset{H}{\underset{H}{C}}{-}O{-}\underset{\underset{O}{\|}}{C}{-}\overset{H}{\underset{H}{C}}{-}H \; + \; H_2O$$

ethyl alcohol + acetic acid → ethyl acetate + water

The word ester was coined by German speaking chemists. It's a contraction of the two German words *Essig* + *Aether*, which loosely translates as vinegar-ether.

**Synthetic polymers** are the deliberately produced substances that today compose most of the materials that pervade our daily lives. Almost all synthetic polymers begin as small petrochemically derived molecules. A **petrochemical** is one obtained from petroleum, usually in an oil refinery. The list of useful stuff made from synthetic polymers is seemingly endless: plastic films, synthetic fibers, plastics for molding, plastics for automobile parts, plastics to make computer keyboards, polymers for paints and coatings, plastic wrapping and packaging, plastic foams for cushioning and thermal insulation, plastics compatible with human blood for use in artificial hearts, etc. Synthetic polymer materials are often known by their trademarked commercial names. Some that you probably know are: Spandex®, Teflon®, Dacron®, Lexan®, Styrofoam®, Plexiglass®, Kevlar®, Nomex®, and on and on.

The manufacture of synthetic polymers grew during the second half of the twentieth century into an enormous, global industry. The author recollects that in his childhood the term "plastic" was pejorative. Anything made of plastic was cheap, but of poor quality and durability. The efforts of polymer chemists and engineers have radically changed that perception. Historically, applied chemistry produced the world's first science-based industry. An industry in the US that has grown over the past 150 years from an infant to become America's largest manufacturing based industry and a leader in setting worldwide manufacturing standards. The US chemical industry has been highly successful for many reasons: a strong US technology base, a large US market size, excellent US resource endowments, a ready supply of entrepreneurial capital, sensible government regulation (on the whole), and the independent spirit of many individual Americans. Now, at the beginning of the 21st century, the production of synthetic polymers has become the largest single enterprise of the US chemical industry.

Many examples of the available range of synthetic polymer materials can be seen in any supermarket. Garbage bags and wrapping materials come in a bewildering range of types. Not *one* of the commercial polymer products that you see on the store shelves in Figure 11.11 was available during the author's childhood. The revolution in everyday materials has been quiet and undramatic—but it has been a revolution nonetheless. Advances in our understanding of cosmology and the discovery of planets elsewhere in the universe may satisfy a deep human need to inquire. But it's synthetic polymers that make our everyday lives rich with so much stuff and create the problem of how to dispose of it all.

Figure 11.11 Plastic consumer wrapping materials were unknown in the 1950's.

To conclude this section we'll look at the polymer story we can uncover in the evening meal pictured in Figure 11.12 and listed in Table 11.3. You'll see from this table that the polymer chemistry of everyday living is not so simple.

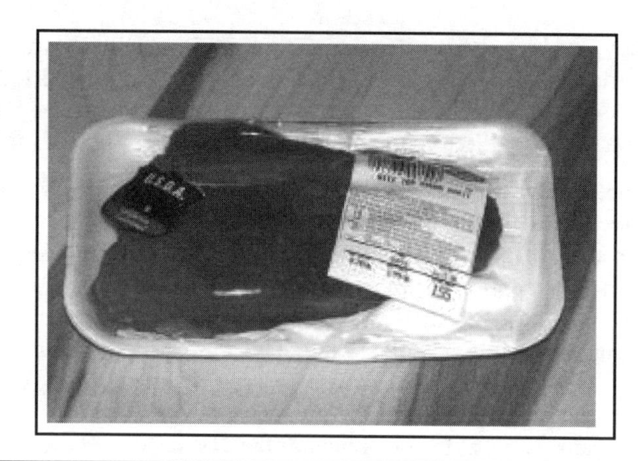

Figure 11.12. Lots of polymers here. Some natural, some not. See Table 11.3 for details. Muscle tissue is composed of about 70 - 80% water, 20% protein, 5% lipid (fat), 1% carbohydrate, 0.3% nucleic acids and 1% inorganic material.

Table 11.3 **Polymeric Materials in a Package of Meat**

| | Item | Polymer composition | Classification |
|---|---|---|---|
| 1 | the meat (20%) | protein (polymino acid) | natural polymer |
| 2 | the meat (1%) | carbohydrate | natural polymer |
| 3 | the meat (0.3%) | nucleic acids (DNA and RNA) | natural polymer |
| 4 | the label | paper (polysaccharide) | modified natural polymer |
| 5 | the tray | foamed polystyrene | synthetic polymer |
| 6 | the film wrap | polyvinyl chloride with plasticizers | modified synthetic polymer |
| 7 | the tray pad | polyethylene outer layer | synthetic polymer |
| 8 | the tray pad | modified starch absorbent filler | semi-synthetic polymer |

## Section11.3 Silicates

Silicates constitute the dominant category of inorganic polymeric materials. **Silicates** are solids (occasionally liquid solutions) that contain linked silicon-oxygen tetrahedra as their basic structural unit. The silicate anion itself, $SiO_4^{4-}$, is shown in conventional perspective view in Figure 11.13. It may be considered as the basic structural unit of silicate chemistry. More complex silicates are regarded as built up from silicon-oxygen tetrahedra linked in chains, sheets, or a three-dimensional network.

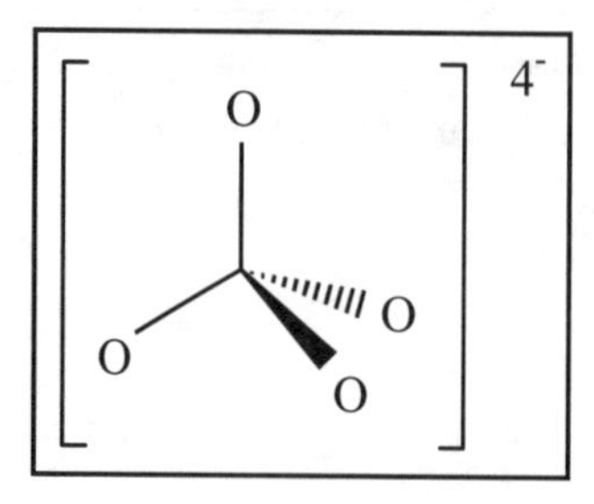


Figure 11.13 The tetrahedral silicate anion, $SiO_4^{4-}$, is the basic structural unit of silicates.

Figure 11.14 shows the oxygen atoms as open circles and the silicon atom as a solid circle—the Si atom is shown as a black circle and in space is below the open-circle oxygen atom above it.

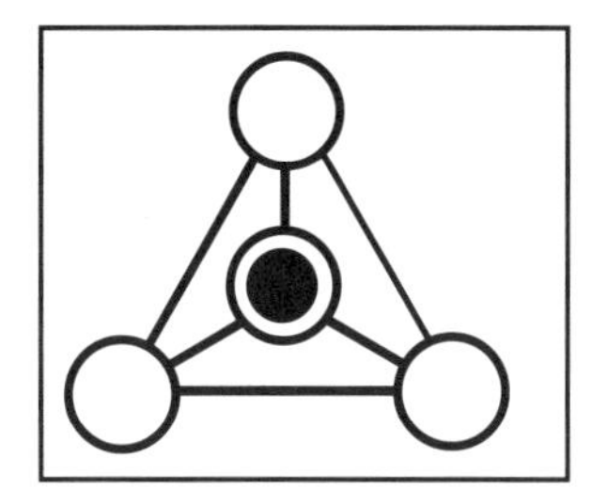

Figure 11.14 The silicon-oxygen tetrahedron is represented as if viewed from above. The black circle represents the silicon atom which is in reality below the oxygen atom

Silicates are the only class of inorganic compounds that approach organic compounds in their variety. Silicates have been studied as minerals for many years, and there are thousands of individually recognized and named silicate minerals. A visit to a mineral or geological museum is invariably both an aesthetic and educational experience. Someone once said that the aesthetic attraction of crystals is the planeness of their faces. The natural, cubic shaped, pyrite (FeS) crystals shown in Figure 11.15 were uncovered during construction at the Dallas airport in 1972. [❁kws +"mineral gallery"]

Figure 11.15 The mineral pyrite is iron(II) sulfide, FeS. Because of its color, it's often called fool's gold. The cubic-shaped crystals of this mineral, which you can see, accurately reflect that the ions within the crystal are arranged in a cubic array.

Silicon-containing minerals and glasses are silicate compounds that involve one or more types of silicon-oxygen polymeric ion. Cements typically contain O-Si-O linkages as do many ceramics. **Ceramics** are a wide range of inorganic materials, many are silicate based, that are stable over a wide range of environmental conditions; most ceramics are composed of crystals with varying compositions. Ceramics are usually prepared by melting together the appropriate recipe of ingredients, mixing, and then allowing the liquid melt to cool. Pictured in Figure 11.7, silica (quartz, silicon dioxide, $SiO_2$) is widely distributed throughout the world as beach sand—it is the least weatherable constituent of many rocks. Table 11.4 shows some types of solid silicates.

Table 11.4 **Types of Silicate Structures**

| Silicate Structure | Example Minerals |
|---|---|
| single chains | pyroxenes, |
| double chains | amphiboles, |
| sheets | biotite or mica |
| three-dimensional | quartz, feldspars |

The bulk properties of natural silicate minerals provide an excellent example of the fact that the underlying chemical bonding controls the bulk properties of materials. For example, **asbestos** is a chain silicate that reveals its internal structure by forming fibers. "Rock-wool" thermal insulation is made by abrading fibers from chain silicate minerals such as asbestos. However, because of its implication as a agent causing lung cancer, most forms of asbes-tos are today subject to stringent regulation and have been replaced in many of their traditional uses by synthetic materials such as aramids (to be discussed later). A portion of a silicate chain is shown in Figure 11.16. Note that every time two silicon-oxygen tetrahedra join an oxygen atom is lost.

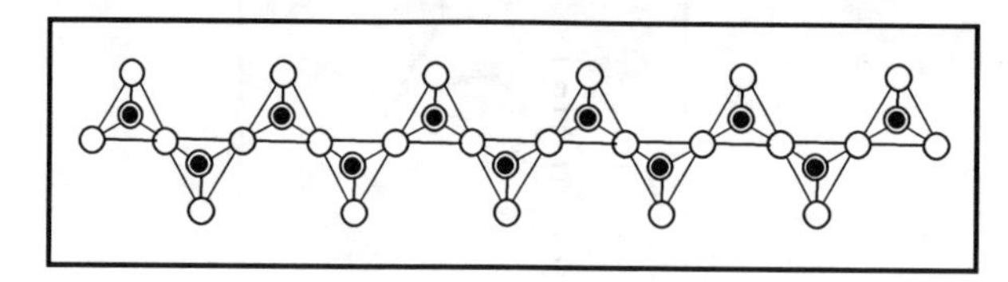

Figure 11.16 Chains of single strands of silicon-oxygen tetrahedra are the structural characteristic of the class minerals called pyroxenes.

**Mica** is a sheet silicate whose sheets are held together (like those in graphite) by fairly weak bonds. The ready fracture of mica into flakes clearly reveals this internal bonding. Six silicate anions can connect in a ring to make the anion $Si_6O_{18}^{12-}$ shown in Figure 11.17. One can imagine it happening via the balanced equation:

$$6\ SiO_4^{4-} + 12\ H^+ \rightarrow Si_6O_{18}^{12-} + 6\ H_2O$$

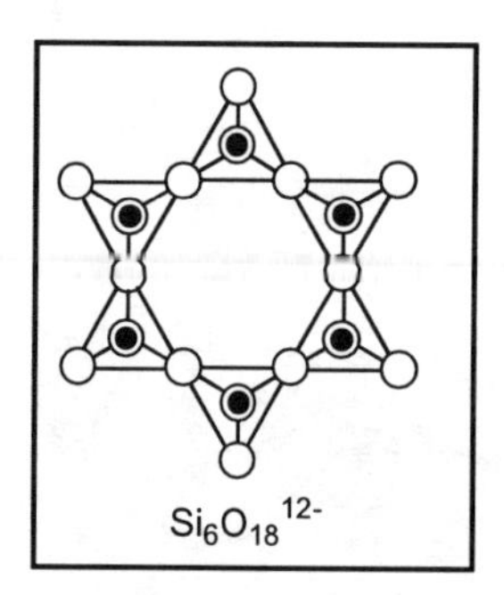


Figure 11.17 The anion $Si_6O_{18}^{12-}$. You can imagine that sheet silicates are made up many such rings linked in a plane as shown in Figure 11.18.

Linking of the $Si_6O_{18}^{12-}$ units eventually leads to the formation of two-dimensional sheet as shown in Figure 11.18.

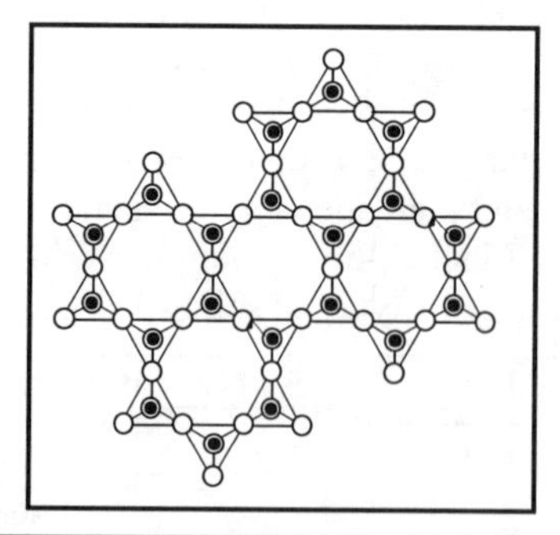

Figure 11.18 Sheet silicates are made up of $Si_6O_{18}^{12-}$ anions linked in a sheet.

Text books of mineralogy catalog 1000's of named minerals, usually along with their chemical formulas, and their crystal properties. A handful of these are shown in Table 11.5.

Table 11.5 **Some Important Silicate Minerals**

| Class | Structural Unit | Examples |
|---|---|---|
| Quartz | $SiO_2$ covalent network polymer | Quartz, quartzite, sandstone. |
| Feldspars | Three dimensional framework structure | Orthoclase ($KAlSi_3O_8$), Albite ($NaAlSi_3O_8$) |
| Pyroxenes | Chains of $SiO_4^{4-}$ units sharing two oxygen atoms linked by metal ions | Augite [$Ca(Mg,Fe)Si_2O_6$] |
| Amphiboles | Double chains of $SiO_4^{4-}$ units linked by metal ions | Hornblende [$Ca(Mg,Fe)_5(OH)_2Si_8O_{22}$] Chrysotile asbestos [$Mg_6Si_4O_{10}(OH)_8$] |
| Phyllosilicates | Sheets of $SiO_4^{4-}$ units linked by metal ions. | Various clay minerals and micas. Biotite or mica [$K_2Mg_6(Mg,Fe,Al)_6(Si,Al)_8O_{20}$] Kaolinite clay [$Al_4(OH)_8Si_4O_{10}$] |
| Orthosilicates | Isolated $SiO_4^{4-}$ units linked by metal ions. | Olivine [$(Mg,Fe)_2SiO_4$], Kyanite ($Al_2SiO_5$), Zircon ($ZrSiO_4$), Garnets[$Ma_3Mb_2(SiO_4)_3$], where Ma = $Ca^{2+}$, $Mg^{2+}$, or $Fe^{3+}$, and Mb = $Al^{3+}$, $Cr^{3+}$, or $Fe^{3+}$. |

Many natural silicate materials have valuable and practical industrial uses. Hard, stone age natural materials such as flint (a dark-colored form of $SiO_2$) were probably among the first engineering materials used by early humans as axes and choppers [❁kws +"Grimes Graves"]. Because of their durability, much of what we know about stone age human cultures comes from studies of these archeological artifacts. In the industrial age, a wide range of silicate minerals is used for a variety of purposes. A few of those are listed Table 11.6.

Table 11.6 **Some Commercially Important Natural Silicate Materials**

| Name | Chemical type | Uses |
|---|---|---|
| Quartz | $SiO_2$ | Crystal oscillators, glass making, abrasives, construction. |
| Clays | Complex alumino silicates | Paper, paint, pigments, ceramics, drilling muds, viscosity modification. |
| Cements | Fired clay minerals | Construction. |
| Asbestos | Complex magnesium silicate | Brake linings (formerly), heat resistant materials, flooring. |
| Feldspar | Na, K, alumino silicates. | Glass making. |
| Garnet | Iron alumino silicate. | Abrasives, sandblasting, water filtration. |
| Kyanite | $Al_2SiO_5$ | Spark plugs and other ceramics. |
| Mica | Complex sheet silicate | Electronics industries for spacers, capacitors etc. |
| Perlite | Complex silicate | Abrasives, filter aids, acoustical tile, etc. |
| Pumice | Complex silicate | Abrasive, light weight building material. |
| Talc | $Al_2O_3{\cdot}4SiO_2{\cdot}H_2O$ | Ceramics, plastics filler, cosmetics, rubber lubricant. |

## Portland Cement

Cement, or **Portland cement**, is a mixture of various calcium silicates and calcium sulfate that undergoes hydrolysis and hydration reactions at low temperature to form a rocklike solid used in construction. It is named for its similarity to a rock that was once mined at Portland on the south coast of England. Typical raw materials for Portland cement include limestones, clays, shales, sand, iron ore, and gypsum. However, the composition of Portland cement varies over wide limits, depending on what raw material is available near the production site. The composition of cement is traditionally specified in terms of oxides. A typical Portland cement is mainly a calcium aluminum silicate and might have the composition shown in Table 11.7.

Table 11.7 **Typical Composition of Portland Cement**

| | |
|---|---|
| $CaO$ | 63.9% |
| $SiO_2$ | 21.1% |
| $Al_2O_3$ | 5.8% |
| $Fe_2O_3$ | 2.9% |
| $MgO$ | 2.5% |
| $SO_3$ | 1.7% |
| $Na_2O$ | 1.4% |
| Other | 0.7% |

The balanced chemical equation for the cementing reaction may be written as:

$$2\ Ca_3SiO_5(s) + 3\ H_2O(l) \rightarrow Ca_6Si_2O_{10}.3H_2O(s)$$

However, in addition to this reaction, many related reactions actually occur, and to write a single equation is to oversimplify a complex process.

**Concrete** is a stone-like material made by mixing cement powder with aggregates such as sand or gravel, adding water, and waiting for the mass to harden. If you like, concrete is the classic composite material. The chemistry of inorganic cements is an important specialized area of calcium chemistry. To build the highways and cities of modern industrial states requires the annual production of millions of tons of cement and concrete.

## Synthetic Inorganic Silicates

The best known synthetic inorganic silicates are the various grades of glass. Glass is a hard, amorphous, usually transparent, material made by heating together various metal oxides and natural silicates. As expected, the basic structural unit of glass is the $SiO_4^{4-}$ tetrahedron, with various cations being added during manufacture. Heating silica sand ($SiO_2$) with a mixture of sodium carbonate and lime produces soda-lime glass. About 90% of all glass manufactured is of the soda-lime type; window glass is a common example. Window glass has the approximate composition: $SiO_2$, 72%; $Na_2O$, 14%; $CaO$, 13%; and $Al_2O_3$, 1%. An enormous range of specialty glasses are available. Examples of these include colored glass, crystal (glass with a high lead content), toughened glass, etc.

Various commercial grades of sodium silicate solutions are made by heating silica with sodium hydroxide solution:

$$SiO_2(s) + 2\ NaOH(aq) \rightarrow Na_2SiO_3(aq) + H_2O(l)$$

The composition of liquid sodium silicate can vary over a wide range, and some of these compositions are water soluble. The water soluble sodium silicates have the general formula $Na_2O.mSiO_2.nH_2O$. Where the value of m ranges from about 0.5 to 4 and n is 1-5. Liquid silicate solutions find many applications in industry where they are used as adhesives for paper, binders in acid-proof cement, to make protective coatings on cement and concrete, and as ingredients in detergent compounds. The label of a brand of machine dishwasher detergent (Figure 11.19) states that the formulation contains sodium silicate among many other ingredients.

Figure 11.19 Silicates are an ingredients of many cleaning products such as the dishwashing powder pictured here.

Open-lattice silicates (which are members of the zeolite mineral family) have relatively accessible channels between the silicon-oxygen chains of their structures. Zeolites are important cracking catalysts at oil refineries. **Cracking** is the chemical reaction of breaking down of large hydrocarbon molecules into smaller ones; the conversion of crude oil to gasoline involves cracking. Organic molecules of the proper size can penetrate into the zeolite structure where they are catalytically cracked. The chemical-engineered reactor where this is done is called the **catcracker**. A more mundane use of zeolites is as absorbents for liquids and odors in such household products as kitty litter. Silicones (an entirely different class of silicon-containing compounds) are discussed toward the end of this chapter.

The classification of solids into five broad groups (Table 11.1) is convenient, but as already pointed out, not all solids fit neatly into just one class. Many common solids involve more that one type of bonding. For example, the simple inorganic salts, such as the carbonates, sulfates, phosphates, etc., have covalent bonding within the polyatomic ion, while the ion *as a unit* is held in the solid by coulombic forces. The model in Figure 11.20 of calcite (one of many natural mineral forms of calcium carbonate, $CaCO_3$) shows the calcium ions as white balls, while the planar carbonate anions are shown as black carbon atoms covalently bonded to three gray oxygen atoms. Recall that the fibrous silicate mineral, asbestos, is composed of extended chains of covalently bonded silicon and oxygen atoms, with the chains being held together electrostatically by ions.

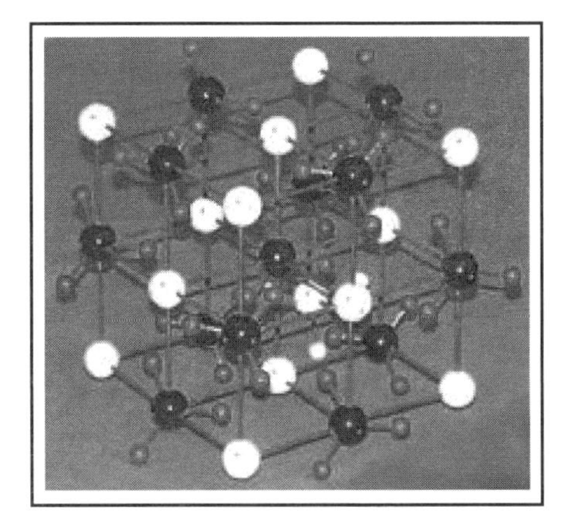

Figure 11.20 A ball-and-stick model; of calcium carbonate. The carbonate ions are covalently bonded internally, but held to the calcium ions by Coulombic forces.

## Section 11.4 Synthetic Organic Polymers

A polymer is composed of molecules with long chains of atoms. As we noted at the beginning of the chapter, the introduction of synthetic organic polymers has changed daily life enormously over the past 50 – 60 years. As I write, I am wearing a polyester shirt and slacks, vinyl shoes, typing on plastic keys that send electrical signals through a plastic encased cable, to a computer with a plastic case, with polyethylene insulation ... well, you get the idea.

Synthetic polymer materials are generically called **plastics**. The term **resin** is often used interchangeably with the term plastic. A plastic material that melts when heated is called a **thermoplastic** resin or polymer. A plastic material that cannot melt is a **thermosetting** resin or polymer.

A **polymerization** reaction is one that forms large molecules by joining many small molecules. Large molecules are called **macromolecules** and the study of polymers is sometimes called **macromolecular science**. Recall from Chapter 3, and from our discussion of esterification earlier in this chapter, that an **addition** reaction has occurred when two smaller molecules combine to make a single, larger molecule. A polymerization reaction involves repeated addition reactions onto a growing polymer molecule; if the polymer is a chain, then each addition reaction is said to cause **chain extension**. Organic molecules that contain at least one double bond can undergo addition polymerization, for example all the valence unsaturated hydrocarbons in the alkene, $C_nH_{2n}$ class. **Addition polymerization** is the process of unsaturated organic molecules joining to form chains. A **monomer** is a substance whose individual molecules can be reacted to form a polymer.

Polymer chemistry is a 20th century phenomenon. But most of the development of the science occurred during the second half of the century. In the "history box" for this chapter (page 396) it's noted that polyethylene and polystyrene were developed in the 1930s, while the condensation polymer nylon was developed during the 1940s.

The 1950s brought the development of polyurethanes and polycarbonates and the introduction of the Zeigler-Natta type catalysts that brought polymer synthesis to the level of a first rank scientific enterprise; also in the 1950s came block copolymers: polymers begun with one monomer and then further polymerized with a second monomer. A **block copolymer** has a single backbone chain, but different regions of that chain are formed from different monomers.

The 1960s was a time when many new, superior performance-polymers were developed. Among those were: aramids, polyphenylene oxide, polysulfones, and polyimides. This decade also brought the first polyethylene made by the Ziegler-Natta method.

The 1970s and 1980s were a period of consolidation, and the rate of introduction of entirely new polymer materials slowed. Developments in this period were more aimed at optimizing what could be accomplished with the existing palette of materials. Much work was done in developing polymer blends and finding ways to combine known materials to tailor properties and performance to specific targets.

The 1990s have brought new metallocene type catalysts and ever better physical methods to investigate and characterize polymers. Some polymer chemists think we "ain't see 'nuffin, yet." And given the spectacular developments of the past fifty years, who knows? They may very well be right. One thing is certain, the US consumes more and more plastic each year. Current plastics' consumption is about 350 pounds per capita and growing at about 5% per year. Figure 11.21 shows the consumption curve for the last 30 years.

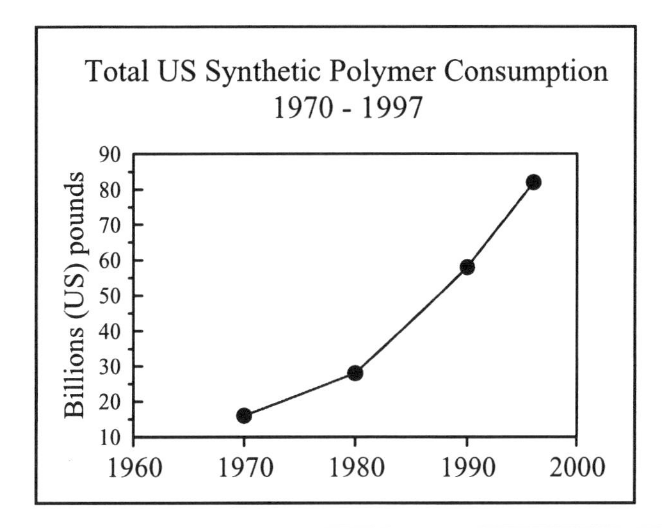


Figure 11.21 The growth of synthetic polymer consumption in the US.

The currently available range of synthetic polymers is breathtakingly wide. Here we can only glance at the incredible profusion of synthetic materials. Obviously, modern engineering practice requires considerable chemical sophistication for the careful selection of proper selections of the materials to be used for specific purposes. One way to tell which polymers are used for what purposes is to check the recycling icon stamped or molded on many plastic items. The recycling codes are shown in Table 11.8 along with the chemical names of the polymers, a description of the characteristics properties of the class, and some representative uses.

Table 11.8 **Society of the Plastics Industries Plastic Recycling Codes**

| Code | Composition | Characteristic properties | Typical uses |
|---|---|---|---|
| 1 PETE | Code 1 PETE Polyethylene Terephthalate | Tough and shatterproof. Transparent. | Soft drink containers. Food and medicine containers. Distilled alcoholic beverage containers. |
| 2 HDPE | Code 2 HDPE High Density Polyethylene | Flexible, translucent. Not expensive. | Milk, water, and juice containers. Grocery bags. Toys. Liquid-detergent bottles. Motor oil and antifreeze containers. Boil-in-bag pouches. |
| 3 V | Code 3 V Poly Vinyl Chloride | Durability and oil resistance. Flexibility and ability to stretch. Widely variable. PVC is capable of wide modification. | Piping and plumbing accessories. Home siding. Meat wrap. Vegetable oil and salad dressing bottles. Clear food packaging, Mouthwash and shampoo bottles. |
| 4 LDPE | Code 4 LDPE Low Density Polyethylene | Inert. Moisture proof. Inexpensive. | Bread, grocery, garbage, and dry-cleaning bags. Coating for paper. |
| 5 PP | Code 5 PP Polypropylene | Stiffness. Heat and chemical resistance compared to polyethylene. | Ketchup bottles, syrup bottles, yogurt containers, margarine and deli tubs, flexible caps. Furniture. Carpet fiber. |
| 6 PS | Code 6 PS Polystyrene | Crystal clear. Brittle. Insulation properties when foamed. | Audio and videocassettes, compact disk jackets. Cafeteria trays. Disposable foam dishes and cups. Meat and produce trays. |
| 7 OTHER | Code 7 OTHER Other or mixed | Various. Films with metallic layers offer good resistance to oxygen penetration. | Usually multilayer packaging. Most chip and snack food containers. |

**A Brief History of Synthetic Polymers**

3000 BCE. Sumerians and other Mesopotamian cultures use the natural polymers asphalt and pitch for floor covering. Asphalt is used to the present day for floor tiles and we all travel across miles of asphalt bonded stone aggregate on the nation's highways—a use for asphalt seemingly first proposed by the poet John Milton in 1667.

2000 BCE. Various natural polymers and gums are used by the Egyptians as varnishes.

1000 BCE? Cultivation of the silkworm begins in China. Silk is a polymeric secretion. Later, European demand for silk has important social consequences.

1600 Secretions from the lac insect become the basis for shellac coatings.

1790 Silk coated with rubber is used for hydrogen balloons such as the one Jacques Charles (of Charles' law) ascended in.

1823 Charles Macintosh patents a process to sandwich a rubber sheet between two cloth sheets and thereby make a waterproof fabric. He is remembered today whenever we refer to a raincoat as a "macintosh or mac."

1830 Polymers are named by the Swedish chemist Jons Jakob Berzelius (1779-1848).

1844 Charles Goodyear wins a US patent for the vulcanization of rubber. His process involves cross-linking by sulfur atoms—a chemical modification that substantially improves the qualities of natural rubber. Vulcanized rubber is the first semi synthetic polymer.

1861 Thomas Graham, a Scottish chemist, states that polymers are composed of molecules with long chains of atoms. It will be well into the 20th century before this correct view is finally adopted.

1869 John Hyatt reacts wood fibers with nitric acid to form the semi-synthetic nitrocellulose polymer. It proves to be a useful explosive.

1907 First commercial synthetic polymers introduced. Leo Baekeland makes hard, thermosetting plastics from formaldehyde and phenol.

1920 Hermann Staudinger, a German chemist, becomes convinced that polymers are macromolecules and embarks on an energetic research program to prove it. In 1953 he is awarded to Nobel prize for his ground breaking work on polymer chemistry.

1927 Accidental discovery of "plexiglass" by Otto Röhm in Germany.

1930 Hermann Staudinger makes polystyrene.

1933 In March of this year M. W. Perrine working at ICI in England produces polyethylene in a high pressure reaction designed to react ethylene with benzaldehyde. By 1939 a patent has issued and the production of polyethylene is reaching 100 tons per year. Its first use was for radar cable insulation.

1935 W. H. Carothers of DuPont discovers nylon. Patents on nylon issued in 1938 when the *New York Times* ran an article titled: "New Silk Made on Chemical Base Rivals Quality of Natural Product."

1942 Cut off by the war with Japan from its traditional rubber sources the US petroleum industry launches an ambitious drive to produce synthetic rubber. Emulsion polymerization of 75% butadiene and 25% styrene mixtures leads to the production of the fully synthetic elastomers known as SBR.

1945-Present The production, design, and application of synthetic polymers grows to become the dominant aspect of applied chemistry. Some of these changes are described in this chapter. Collectively, chemists and materials engineers turn in a virtuoso performance. All sorts of stuff keeps getting better and cheaper. By the end of this period polymer chemists outnumber all other kinds. Everyday life is changed remarkably.

The term **engineering polymers** is a collective noun for synthetic polymers used to make objects, parts, or components. The single biggest outlet for engineering polymers is in manufacturing automobile components and body parts, with an average of about 200 pounds of plastic materials being found in a new automobile, but engineering polymers are used throughout the economy. Many working engineers have to write specifications for the materials they need. Being able to do that requires the engineer to understand and interpret the technical data specifications of the candidate materials. Below is an example product "promo" for the engineering plastic Crastin®. [❁kws +Dupont +crastin]. The purpose of including it here is show why many engineers need to understand some chemical terms. You'll find a lot of background information with [❁kws "engineering polymers"].

**A Crastin® "Promo"**

"Crastin® PBT thermoplastic polyester resins are high performance plastic molding resins which are based on polybutylene terephthalate (PBT) polymers. By adding reinforcing agents and tougheners, the inherent low creep, dimensional stability and good electrical properties of PBT can be augmented to provide a family of molding resins with exceptional strength, stiffness and toughness. Crastin® PBT resins have mechanical properties and physical characteristics which make them well suited for plastic components in the Automotive, Electrical and Electronics, and Appliance Industries. Parts are fabricated using conventional injection molding techniques. Crastin® PBT can increase molding productivity with fast cycle times and consistent lot-to-lot melt viscosity."

## Polyethylene

The empirical formula for polyethylene is $(CH_2)_x$ where x can be several hundred thousand. It is the long polymer chains that give polyethylene its characteristic properties of toughness and flexibility. You'll appreciate these properties if you've set out garbage in strong yet remarkably light weight and inexpensive polyethylene bags. In contrast, petroleum wax (Figure 11.22) with a formula of approximately $C_{60}H_{122}$, and almost the same empirical formula as polyethylene, is a brittle material quite unsuited for making films. Long chains of carbon atoms are the key to polyethylene's properties.

Figure 11.22 Wax is a saturated hydrocarbon with a chemical composition not much different from polyethylene. But is short chain length compared to polymers gives it very different physical properties.

The simplest alkene is ethylene or ethene, $H_2C{=}CH_2$. The formation of the **addition polymer** polyethylene from ethylene can be written as a stepwise process and represented by the following equations:

Dimerization: $H_2C{=}CH_2 + H_2C{=}CH_2 \rightarrow H_2C{=}CH\text{-}CH_2\text{-}CH_3$

two ethylene monomers ($C_2H_4$) → a dimer ($C_4H_8$)

Chain extension: $H_2C{=}CH_2 + H_2C{=}CH\text{-}CH_2\text{-}CH_3 \rightarrow H_2C{=}CHCH_2\text{-}CH_2\text{-}CH_2\text{-}CH_3$

monomer + dimer → a trimer ($C_6H_{12}$)

More Chain extension:

$H_2C{=}CH_2 + H_2C{=}CHCH_2\text{-}CH_2\text{-}CH_2\text{-}CH_3 \rightarrow H_2C{=}CHCH_2\text{-}CH_2\text{-}CH_2\text{-}CH_2\text{-}CH_2\text{-}CH_3$

monomer + trimer → a tetramer ($C_8H_{16}$)

Yet more chain extension occurs as successive ethylene molecules attach to the growing polymer chain. When a few monomer units have joined, say 5 – 50, the resulting product is called an oligomer. An oligomer is a molecule formed by polymerization that links only a few monomer units. (Related informal example: an oligarchy is rule by the *few*.) Eventually a long chain polymer can be formed. Chain lengths of 5000 – 100,000 monomer units are common.

The overall addition polymerization reaction for ethylene can be written:

$$x\ H_2C{=}CH_2 \quad \rightarrow \quad (\ CH_2)_{x/2}–$$

x ethylene monomers → polyethylene with a chain length of x/2 carbons

In a **linear** polyethylene molecule the carbon atoms form extended chains in which every carbon atom is connected to two other carbon atoms and two hydrogen atoms. The single chain in a linear polymer is called the **backbone** chain. At either end of the chain is an **end group**. The polymerization of ethylene yields polyethylene. Linear polyethylene is composed of long chains of methylene, -$CH_2$-, units as pictured in Figure 11.23. With hydrogen end groups, a "formula" for polyethylene might be $H–(–CH_2–)_{7000}–H$ but the chain length is variable.

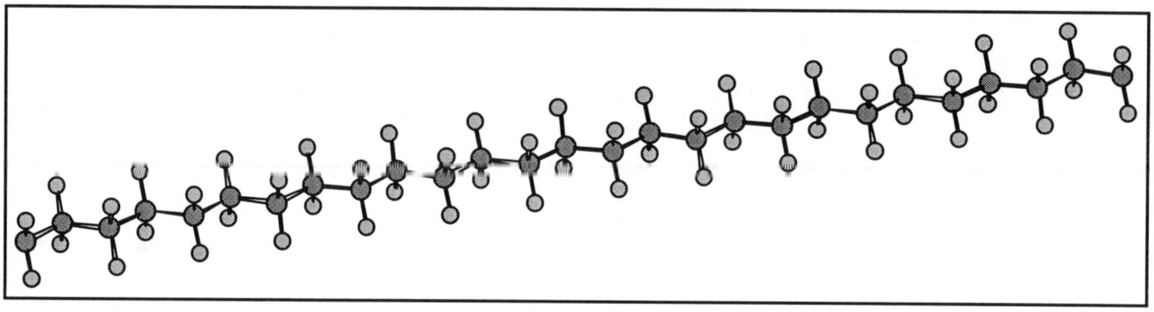

Figure 11.23 Linear polyethylene is composed of long chains of methylene, $–CH_2–$ units.

One important commercial form of polyethylene is **linear low density polyethylene** (LLDPE). The molecules in LLPDE consist of long sequences of linked methylene groups with occasional short side chains or branches.

A **branched** polymer molecule is one with chain branching. In a branched polyethylene, most carbon atoms are connected to two other carbon atoms and two hydrogen atoms, but some carbon atoms are connected to three other carbon atoms resulting in the molecule having branches. A branched polyethylene molecule is shown in Figure 11.24 where there are two side chains attached to the backbone chain. Every point of attachment of a side group marks a chain branch.

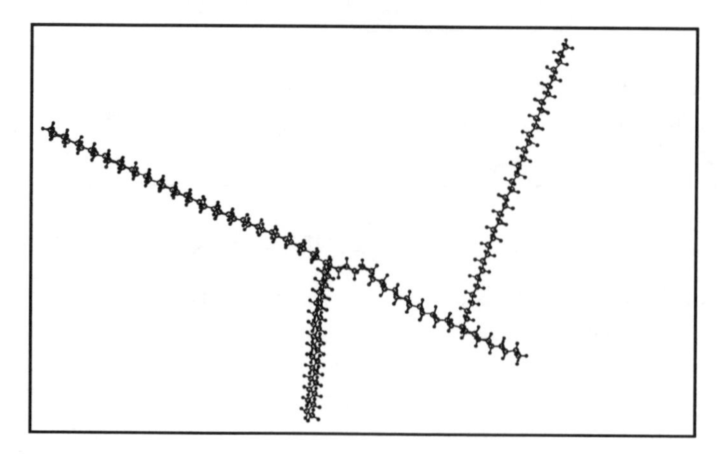

Figure 11.24 A fragment of branched polyethylene.

## Polyethylene Copolymers

Currently, many commercial addition polymers are copolymers. A **copolymer** is a polymer made from a mixture of monomers. Modern LLDPE production technology gives polymer manufacturers the ability to introduce a controlled degree of branching into the polyethylene backbone. The practical consequence of this control is that desired "engineering properties" can be designed into the finished product at the production stage.

Typically, modern commercial polyethylene is manufactured by copolymerizing mixtures of ethylene and other alkenes. For co-polymerization with ethylene the two most widely used co-monomers are 1-butene, $H_2C=CH-CH_2-CH_3$ and 1-octene, $H_2C=CH-C_6H_{13}$. The 1-octene molecule consists of a hexyl chain ($-C_6H_{13}$) attached to a pair of doubly bonded carbon atoms. Thus, when 1-octene is a co-monomer it creates six-carbon branching side chains in the polymer. Commercial practice is to use from less than 1% to up to 15% of 1-butene or 1-octene in the copolymerization of polyethylene. The differences between polymers with 0.2% and 15% 1-octene content are summarized in Table 11.9. The table demonstrates our thesis again: The useful bulk properties of polymer materials reflect the nature of the intermolecular forces among the long chain molecules that compose them.

Table 11.9 **A Comparison of the Physical Properties of Two Copolymers**

| Monomer Mixture | Polymer Properties | Polymer Use |
|---|---|---|
| 99.8% ethylene and 0.2% 1-octene | Tough, limited flexibility. Relatively more crystalline. | Plastic supermarket shopping bags. |
| 85% ethylene and 15% 1-octene | Tough and flexible. Relatively less crystalline. | As a foamed insert for athletic shoes. |

Because they "hang off" the backbone chain the side chain groups are **pendant groups**. The extent of branching in polyethylene has a substantial effect on the plastic's bulk properties. The two polymers described in Table 11.9 are drawn schematically in Figure 11.25.

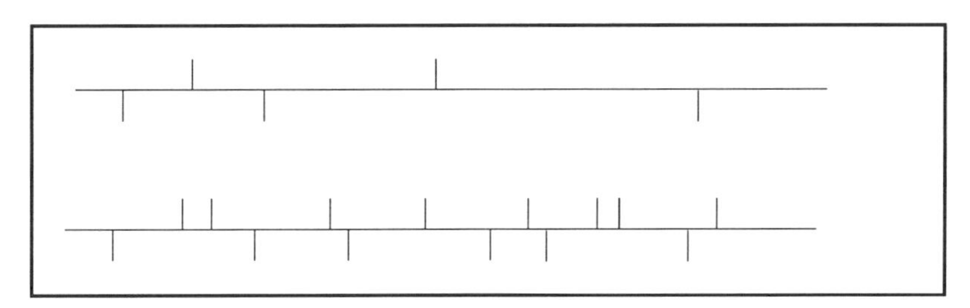

Figure 11.25. A polyethylene chain with fewer pendant groups (top), and with more pendant groups (bottom). The polymers have quite different physical properties.

The presence of side chains in LLDPE interferes with the crystallization of the polymer backbone chains. So the chains do not "pack" efficiently in the solid polymer and its bulk density is relatively low. Low density polyethylene (LDPE, it has highly branched chains) is somewhat arbitrarily designated a density of 0.94 g $mL^{-1}$ or less. The effect of the side chains on crystallization is also shown by the two ethylene-octene copolymers shown in Figure 11.25. The upper polymer in the diagram, the one with fewer side chains, is more crystalline, which manifests itself as a bulk property by the polymer being stiffer.

A different commercial form of polyethylene is high density polyethylene (HDPE). Its molecules are not branched and its bulk density ranges from 0.941 – 0.965 g $mL^{-1}$. HDPE when cooled from a softened state, forms highly crystalline structures that make the plastic tough and give it shatter-resistant mechanical properties, good chemical resistance, and resistance to most solvents. HDPE is used for wire and cable insulation, milk and detergent containers, housewares, plastic films, etc. LLDPE is used for inexpensive films and other less demanding applications.

## Other Addition Polymers

Polyethylene is by far the largest volume synthetic plastic, accounting for about one-third of total annual polymer production. However, taken together polypropylene, polyvinyl chloride, and polystyrene exceed polyethylene in volume. The structures of the monomers of these addition polymers are shown in Table 11.10. Recent production figures for addition polymers are shown in Table F-7 in Appendix F. Summary consumption information for the four major addition polymers is shown in Table 11.11. On average, every US citizen uses a total of about 250 pounds of these four addition polymers each year.

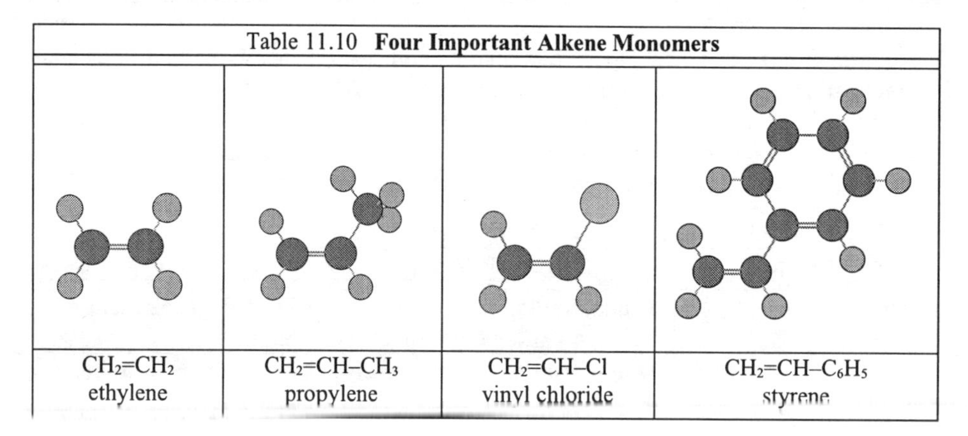

Table 11.10 **Four Important Alkene Monomers**

| $CH_2{=}CH_2$<br>ethylene | $CH_2{=}CH{-}CH_3$<br>propylene | $CH_2{=}CH{-}Cl$<br>vinyl chloride | $CH_2{=}CH{-}C_6H_5$<br>styrene |
|---|---|---|---|

Table 11.11 **1998 US Consumption of Major Addition Polymers**

| | Percent of all plastics by weight | Annual US consumption (pounds) | US per capita (pounds) |
|---|---|---|---|
| Polyethylene (all types) | 35% | $27{,}800 \times 10^6$ | 110 |
| Polypropylene | 17% | $14{,}000 \times 10^6$ | 52 |
| Polyvinyl chloride & copolymers | 19% | $15{,}000 \times 10^6$ | 55 |
| Polystyrene and ABS | 12% | $9{,}300 \times 10^6$ | 38 |
| Totals | 83% | $66{,}000 \times 10^6$ | 245 |

Interestingly, the ethylene, propylene, vinyl chloride, and styrene monomers are all *toxic*, especially vinyl chloride, which polymerizes to give polyvinyl chloride or PVC. It's important that polymerization is NOT a reversible reaction, otherwise how could we use synthetic polymers for food packages, water piping, and children's toys? Because compounds with a double bond were once called olefins (literally, oil-formers; they form "oils" when they react with bromine), their polymers are often termed **polyolefins**.

Although all but unknown except to the specialists, many of the greatest achievements of modern chemistry have been made in finding cheap, reliable ways to polymerize ethylene, propylene, vinyl chloride, and styrene. Most of the research and development work was done by large petrochemical companies, driven by the high profitability of successful polymer products. Because of the industrial nature of this work, polymerization catalysts are the subject of a vast patent literature. Not unexpectedly, in a competitive global market place, with millions of dollars at stake, patent disputes and lawsuits over polymerization catalysts have been common

Commercial polymerization methods rely on highly specialized catalysts and a deep understanding of the detailed mechanisms by which the polymers form. In the 1950s polymerization catalysts based on titanium, vanadium, and other metal compounds [❁kws +"Ziegler-Natta" +catalyst] enabled polymerization to be run at moderate temperatures and pressures. This class of catalysts has been augmented in the past five years by a new class of catalysts called metallocenes. Metallocene catalysts yield high quality polymer products that extend the range of polyolefin applications to the point that they are beginning to compete directly with polyester and other traditionally more expensive plastics.

Although the actual mechanism may be different, addition polymerization reactions of a substituted ethylene, $H_2C{=}CHR$ (R = $-CH_3$, -Cl, or $-C_6H_5$)can be schematically represented by the equations:

Dimerization: $H_2C{=}CHR + H_2C{=}CHR \rightarrow H_2C{=}CR\text{-}CH_2\text{-}CH_2R$
two monomers → a dimer

Chain extension: $H_2C{=}CHR + H_2C{=}CR\text{–}CH_2\text{–}CH_2R \rightarrow H_2C{=}CRCH_2\text{–}CHR\text{–}CH_2\text{–}CH_2R$
monomer + dimer → a trimer

More Chain extension:

$H_2C{=}CHR + H_2C{=}CRCH_2\text{–}CHR\text{–}CH_2\text{–}CH_2R \rightarrow H_2C{=}CRCH_2\text{–}CHR\text{–}CH_2\text{–}CHR\text{–}CH_2\text{–}CH_2R$
monomer + trimer → a tetramer

The polymerization of each of these monomers can yield considerable structural variation. For example if we call one end of the monomer the head and the other the tail, then do the monomers join head-to-head or head-to-tail? Or do they join both head-to-head and tail-to-tail, and, if both, is the joining pattern random or systematic? How about the stereochemistry—at what angle do the pendant groups point in space? Possibilities such as these make polymer science an interesting subject. [❁kws +atactic +syndiotactic]

At the other end of the volume scale of commercial addition polymers lies Teflon®. Compared with the "big four" addition polymer described above, Teflon® is produced only in minuscule amounts. Teflon® is an addition polymer formed from the molecule $F_2C{=}CF_2$ called tetrafluoroethylene (TFE). Its formation equation is:

$$F_2C{=}CF_2 \rightarrow -(CF_2)-$$

The resulting polymer is composed of linked $-CF_2-$ units. It is the many C–F bonds that give it its characteristic low surface energy and its ability to resist anything sticking to it.

By varying the degree of hydrogen and fluorine on an olefin monomer it's possible to create polymers that balance excellent properties with reasonable cost. A typical use of such specialty polymers to make elastomers with extreme chemical resistance for severe-duty flexible seals. Such polymers are produced in small production volume, yet because of their premium price they return a high profit.

## Condensation Polymers

If 83% of the polymer market is occupied by addition polymers, the other 17% is the condensation polymers. Information about the US consumption of condensation polymers is shown in Table 11.12.

Table 11.12 **1998 US Consumption of Major Condensation Polymers**

| | Percent of all plastics by weight | Annual US consumption (pounds) | US per capita (pounds) |
|---|---|---|---|
| Polyester (thermoplastic) | 5% | $4{,}300 \times 10^6$ | 17 |
| Phenolic resin | 4% | $3{,}900 \times 10^6$ | 14 |
| Urea resin | 3% | $2{,}600 \times 10^6$ | 10 |
| Polyester resin | 2% | $1{,}700 \times 10^6$ | 6 |
| Polyamide (nylons) | 2% | $1{,}300 \times 10^6$ | 5 |
| Epoxy resins | 1% | $600 \times 10^6$ | 3 |
| Melamine resin | 0.4% | $300 \times 10^6$ | 1 |
| Totals | 17% | $14{,}000 \times 10^6$ | 56 |

Polyesters are produced in both thermoplastic and thermosetting forms. Thermoplastic polyester is used largely for making fibers, and along with nylon dominates the US fiber market (see Table F-9 in Appendix F). Thermosetting polyester is used largely to make bottles. Together the two types of polyester account for about 7% of US plastics production.

When a water molecule (or some other small molecule such as $NH_3$ or HCl) is split out when two molecules join, the linking is a **condensation** reaction. Condensation reactions are used to make a variety of commercial, synthetic polymers. Two important classes of condensation polymers are polyesters and polyamides. Dacron® is a polyester; nylon is a polyamide. Nylon was once a trademark, but it slipped into common usage and was lost to its originator (DuPont).

The condensation reaction between an alcohol and a carboxylic acid forms an ester, as described earlier in this chapter. The condensation reaction between an amine and a carboxylic acid results in the formation of an amide. Note that the name amide is a wordplay on **amine** + ac**id**.

General amide forming reaction

$$R{-}NH_2 + R'{-}C(=O){-}O{-}H \longrightarrow R{-}NH{-}C(=O){-}R' + H_2O$$

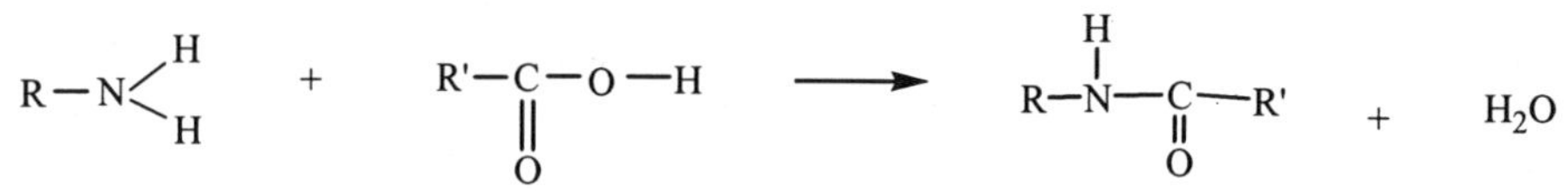

general amine + general acid → general amide + water

Amide forming reaction to make N- methyl acetamide

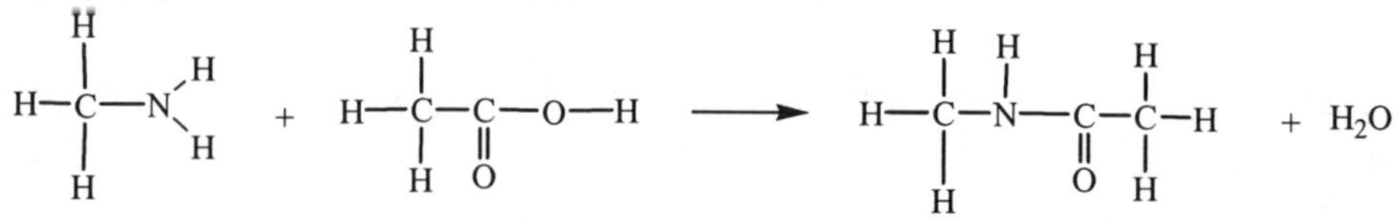

$$CH_3{-}NH_2 + CH_3{-}C(=O){-}O{-}H \longrightarrow CH_3{-}NH{-}C(=O){-}CH_3 + H_2O$$

methyl amine + acetic acid → N- methyl acetamide + water

In biochemistry the amide functionality , R'-CONH-R is called the **peptide** linkage. It is responsible for the condensation reactions among amino acids that form polypeptides or proteins. The formation of esters and amides serve as the basis for condensation polymerization reaction provided that the monomers are difunctional. A **difunctional** molecule is one that contains two functional groups. Reacting difunctional molecules enables the formation of long chains linked by ester groups (**polyesters**) or long chains linked by amide groups (**polyamides**). Figure 11.26 shows some difunctional molecules used commercially to make polyesters and polyamides.

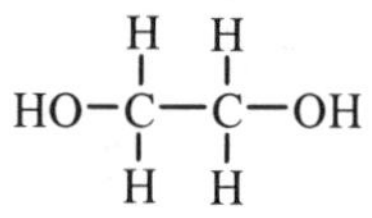

HO-CH₂-CH₂-OH — Ethylene glycol, a difunctional alcohol

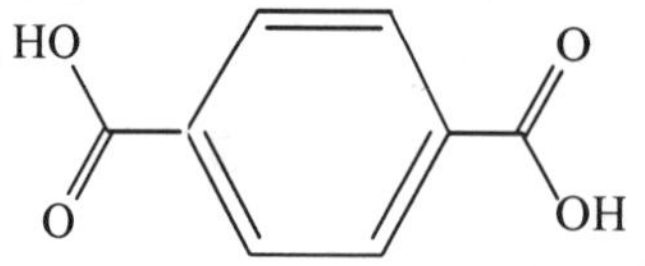

HOOC-C₆H₄-COOH — Terephthalic acid, a difunctional acid

$HOOCCH_2$-$CH_2$-$CH_2$-$CH_2$-COOH — Adipic acid, a difunctional acid

$H_2NCH_2$-$CH_2$-$CH_2$-$CH_2$-$CH_2$-$CH_2$-$NH_2$ — Hexamethylene diamine, a difunctional amine

$H_2N$-$CH_2$-$CH_2$-$CH_2$-$CH_2$-$CH_2$-COOH — 6-amino hexanoic acid, a difunctional amino acid

Figure 11.26 Difunctional molecules used commercially to manufacture condensation polymers

The polyester made from ethylene glycol and terephthalic acid is particularly useful. It is called polyethylene(terephthalate) or PETE. Clear soft drink containers and synthetic fibers for clothing are the large markets for PETE. The manner in which the difunctional molecules in Figure 11.26 combine to make condensation polymers is described in the following section.

### The Jigsaw Piece Analogy for Condensation Polymerization

The process of condensation polymerization can be illustrated schematically using objects shaped like jigsaw puzzle pieces as illustrated in Figure 11.27. When you imagine condensation polymerization in this manner it is obvious why difunctional molecules are necessary to create the needed long chains.

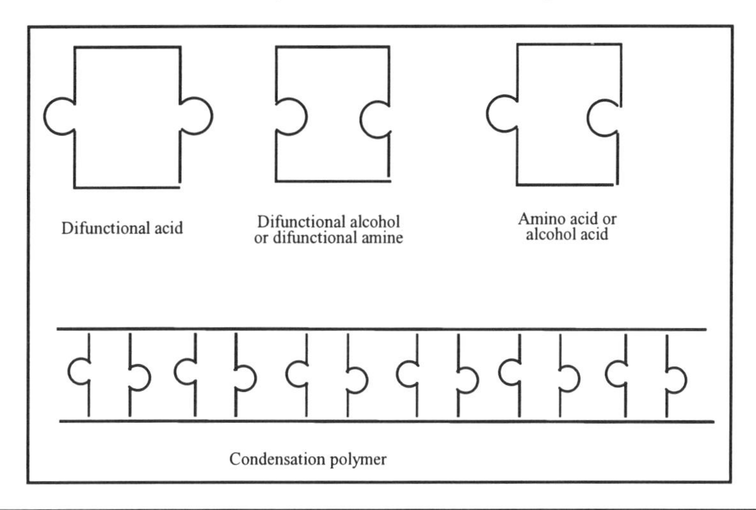


Figure 11.27 The jigsaw piece analogy for condensation polymerization. Condensation requires difunctionality because all the monomer units have to be incorporated into a long chain.

An important polyester is the one produced by condensation of ethylene glycol and terephthalic acid. An important polyamide is the one produced by the condensation reaction of hexamethylene diamine and adipic acid; it's called Nylon-66®. The designation 66 means a nylon made from a 6-carbon diamine and a 6-carbon diacid.

### The Nylon Rope Trick

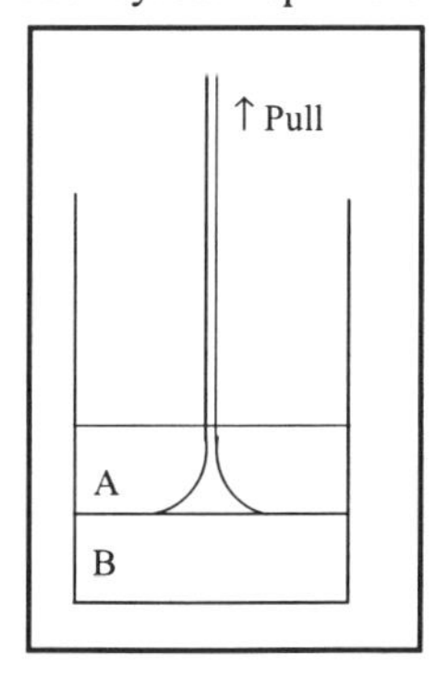


Figure 11.28 Nylon rope can be pulled from the interface where two immiscible liquids, each containing a dissolved reactive monomer, meet.

An ingenious and interesting example of polymerization is the lecture demonstration popularly known as the "nylon rope trick," and shown schematically in Figure 11.28 It involves the condensation reaction between hexamethylene diamine and sebacoyl chloride. Sebacoyl chloride is a derivative of the 8-carbon diacid $HOOC–(CH_2)_6–COOH$ called sebacoic acid. The derived chloride has the formula $ClOC–(CH_2)_6–COCl$ and is used in the demonstration, in preference to the acid itself, simply because the chloride derivative reacts much faster at room temperature with the diamine than does its parent acid. The reaction occurs at the interface formed by solutions of the two reactant in an immiscible pair of solvents. Sebacoyl chloride is dissolved in hexane and forms the light upper layer show as layer A in Figure 11.28. Reactant hexamethylene diamine is dissolved in water and forms the lower layer B in the diagram. The polymerization reaction is:

$$x\ ClOC\text{–}(CH_2)_6\text{–}COCl + x\ H_2N(CH_2)_6NH_2 \rightarrow (\text{–}CO\text{–}(CH_2)_6\text{–}CONH(CH_2)_6NH\text{–})_x + 2x\ HCl$$

The polymer can be pulled out of the reaction vessel using a pair of tweezers. As the polymer film is pulled from the interface through the upper layer, more reactant molecules diffuse to the liquid interface making still more polymer. The consequence is that substantial lengths of "rope" can be pulled from the reaction mixture.

## Additives for Plastics

Like many commercial polymers, polypropylene, polyvinyl chloride, polystyrene, and polytetrafluoroethylene (PTFE) are modified by incorporating co-monomers; so the finished materials are actually copolymers. The term plastic additives refers to substances incorporated into plastics to improve them in some significant way. A substantial technology exists to make modified plastics and it has considerable commercial significance. There are many reasons why plastics additives are used in commercial plastics. Here's a partial listing.

- pigments give them color
- antioxidant compounds slow their decomposition
- plasticizers increase their flexibility [❁kws +"dioctyl phthalate"]
- UV absorbing molecules act like sun block to protect them from sunlight [❁kws +"UV] stabilizer"]
- flame retardants and flame suppressants make them safer in fires [❁kws +"flame retardant"]
- process aids improve things such as their moldabilty and the ease with which molded articles come out of the mold. [❁kws +"release agent"]
- rubber intended for automobile tires contains a good deal of carbon black to provide improved wear resistance and block light that can cause degradation

## Cross Linking and Rubber

The US annual per capita consumption of synthetic rubber is about 20 pounds. Synthetic rubbers were investigated during the 1930s but serious, industrial scale production of synthetic rubber did not begin in the US until after December 7th, 1941 when the US economy was cut off from its traditional sources of tree-grown rubber in southeast Asia. Today, synthetic rubber dominates the market place. Chemically, cross linking is the key to rubber chemistry.

Substantial modifications can be made to polymer materials if chains from separate molecules are cross-linked. To **cross link** is to carry out chemical reactions that form covalent bonds from one polymer chain to another. Cross linking is shown schematically in Figure 11.29. The shorter chains link the longer ones together. Of course, any real cross linked polymer will be three-dimensional, not flat as shown in this sketch. When Goodyear vulcanized rubber with sulfur he was in fact cross linking molecules of natural rubber by means of C–S–C (carbon-sulfur-carbon) bonds.

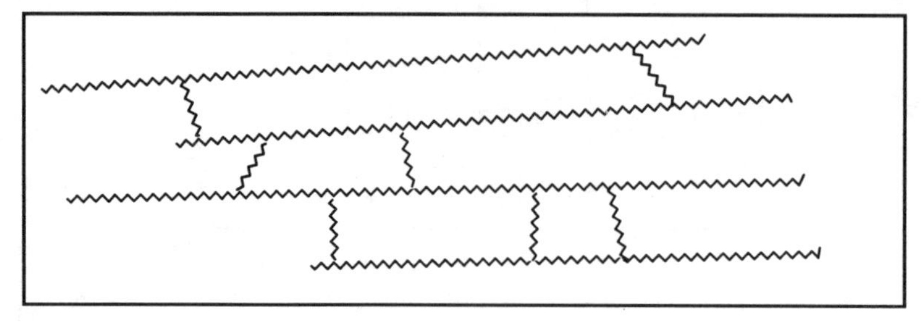

Figure 11.29 Vulcanization of natural rubber involves producing cross-links among natural polyisoprene chains.

**Rubber** is a synthetic or semi-synthetic polymer composed of flexible, long chain, cross-linked molecules. Synthetic polymers with rubbery properties are **elastomers**. Most elastomers contain double bonds in their structure.

The most important property of rubber is obviously its elasticity. Two key factors in giving an elastomer its properties are cross linking and having a heat distortion temperature be well below the temperature at which the rubber will be used. **Natural rubber** is cis-polyisoprene—a natural product of rubber trees that can be chemically regarded as the addition polymer produced from the monomer isoprene. The prefix *cis* refers to the specific geometric isomer that constitutes natural rubber. The natural polymer is:

$$(-CH_2-\underset{\substack{|\\ CH_3}}{C}H=CH-CH_2-)_n$$

A polymer's **heat distortion temperature** is measured experimentally by an ASTM (American Society for Testing and Materials) protocol that loads a ½" × ½" × 8" bar of the polymer on a bridge supporting a blade. As the temperature rises, the sample softens and when a 1 mm deflection of blade occurs, that's the distortion temperature.

What makes a material rubbery? Let's begin by thinking about a gas. The property of elasticity is characteristic of gases: if you squeeze a gas it pushes back; gas filled shock absorbers work on exactly that principle. Compressing a gas lowers its entropy and when the pressure is released the gas re-expands because expansion is a spontaneous, entropy increasing process. Stretched rubber allowed to contract undergoes a spontaneous endothermic process. You can test that for yourself by stretching a band, allowing a minute or two to let the now warm band return to room temperature, and then checking its temperature immediately after you allow it to contract. It's cooled. Stretching an elastomer decreases its entropy, as illustrated schematically in Figure 11.30.

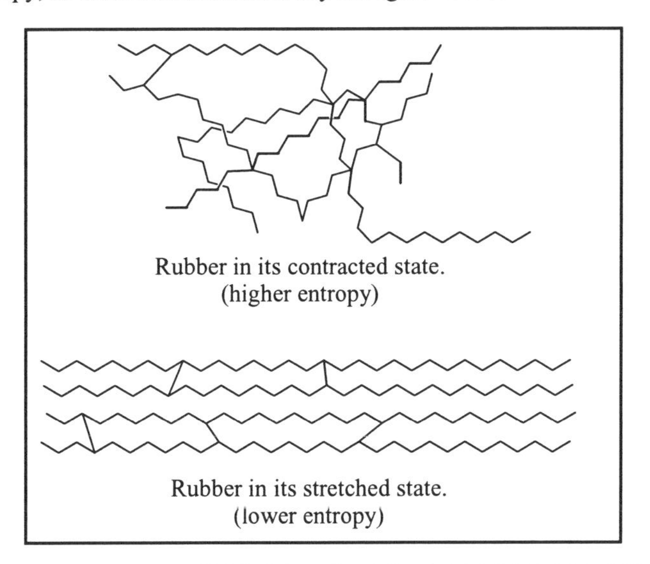


Figure 11.30 The spontaneous contraction of rubber is an entropy driven process.

The parallel chains are much more ordered than the entangled ones. Stretching *decreases* the entropy of the elastomer. When the force is relieved there's an entropy driven contraction. There's actually a nice home experiment you can do that teaches something both about rubber and thermodynamics. Put a rubber band on a hook and hang a weight from it so the band is well stretched. Then heat the band with a hair dryer. It contracts sharply. Remember $\Delta G = \Delta H - T\Delta S$? With $\Delta S$ being positive during contraction (Figure 11.30), increasing T makes contraction more spontaneous (larger negative $\Delta G$ value) than at room temperature. That's the thermodynamic explanation of why heating a stretched rubber band causes it to contract. Perhaps now you see why it takes having both a low distortion temperature and cross links make a rubber. The polymer chains must be able to move, but they mustn't move too far.

A different experiment with rubber, and one that chemistry professors like to demonstrate, is to observe the effect on it of extreme cold. Dumping racquet balls in liquid nitrogen for a few minutes cools them well below their heat distortion temperature, so that when thrown against a hard surface they shatter like glass. They shatter like glass because they are now glassy! However rubber turning to glass on cooling is reversible, and the re-warmed shards of rubber show normal elasticity.

There are many commercial varieties of rubber. Rubber vehicle tires are often a blend of natural and synthetic rubbers. The largest production volume commercial rubber is produced by the emulsion copolymerization of 1,3-butadiene and styrene mixtures. The product is a fully synthetic elastomer known as SBR or styrene-butadiene-rubber. [❁kws "styrene-butadiene rubber"]. It is an alternating copolymer. The reaction for its production can be represented by the equation:

$$n\ C_6H_5\text{–}CH{=}CH_2 + n\ H_2C{=}CH\text{–}CH{=}CH_2 \rightarrow (\text{–}H_2C\text{–}C(C_6H_5)\text{–}CH\text{–}CH{=}CH\text{–}CH_2\text{–})_n$$

## Thermosetting Plastics

Recall that a plastic material that cannot melt is a **thermosetting** resin or polymer. Thermosets are a useful group of plastics. The major categories of thermosets and their consumption data is shown in Table 11.13. Thermoset plastic materials become hard or "set" either during synthesis or when exposed to heat (or sometimes light). Once "cured," thermoset resins never again become plastic or moldable. Thermoset plastics are used principally for bathroom fixtures, electronic circuit boards, heat and electrical insulation, as adhesives and glues, and for super-tough two part epoxy paints. "Super glue" is an acrylic monomer which when exposed to moisture polymerizes and cross links instantly. Despite their advantages, the market for thermoset plastics has been eroded by thermoplastics, which are cheaper to produce and can be recycled. The oldest, and still one of the most useful, types of thermoset plastics are the phenolic resins of which bakelite was the first to be discovered.

**Bakelite—The First Completely Synthetic Plastic**

The world's first completely synthetic plastic material was discovered in 1907. That year, in Yonkers, New York, the Belgian born chemist Leo Baekeland mixed phenol ($C_6H_5OH$) and formaldehyde ($H_2C{=}O$) creating a tough, polymeric mass that he named Bakelite after himself. Working cleverly, he figured out how to control heat and pressure so as to make a liquid form of the polymer that would subsequently irreversibly harden while at the same time adopting the exact shape of its container. It could be molded to any desired shape. Today chemists call this class of materials **phenolic resins**, and by weight phenolics remain today the sixth ranked category of commercial polymer materials. Phenolics are **thermosets**: plastics that retain their shape when heated.

Many bakelite products were made during the 1920s and 1930s. Today, there is a thriving antique trade in collecting old, Bakelite items, and some are quite valuable. [❁kws bakelite]

Silly example questions: Can you melt wood with a blow torch? Can you imagine what happens when you heat a Formica® counter top (chemically a melamine resin) with a blow torch? Why doesn't it melt? Wood and melamine resin char when heated. Melamine is a highly cross linked thermoset resin. As such, it is a network covalent solid and would have a very high melting point, except that its melting point cannot be reached because when heated it chars (oxidatively decomposes) long before it melts.

Table 11.13 **1998 US Consumption of Thermosetting Reins**

| | Percent of all plastics by weight | Annual US consumption (pounds) | US per capita (pounds) |
|---|---|---|---|
| Phenolic resin | 4% | $3{,}900 \times 10^6$ | 14 |
| Urea resin | 3% | $2{,}600 \times 10^6$ | 10 |
| Polyester resin | 2% | $1{,}700 \times 10^6$ | 6 |
| Epoxy resins | 1% | $600 \times 10^6$ | 3 |
| Melamine resin | 0.4% | $300 \times 10^6$ | 1 |
| Totals | 11% | $9{,}100 \times 10^6$ | 34 |

Urea resins (urea-formaldehye resins) are often used to bond plastics to wood. If there's a plastic counter top in your kitchen it was probably formed from a thin sheet of melamine resin glued to a plywood or particle board core. Sometimes the plastic-to-wood bond is made simply by applying pressure at ambient temperature. Stronger bonding can be achieved at high temperature—up to 85°C and with the use of surfactant that enhances the wetting of the hot resin on the surfaces being joined.

The jigsaw puzzle piece analogy for the formation of a thermoset polyester is shown in Figure 11.31.

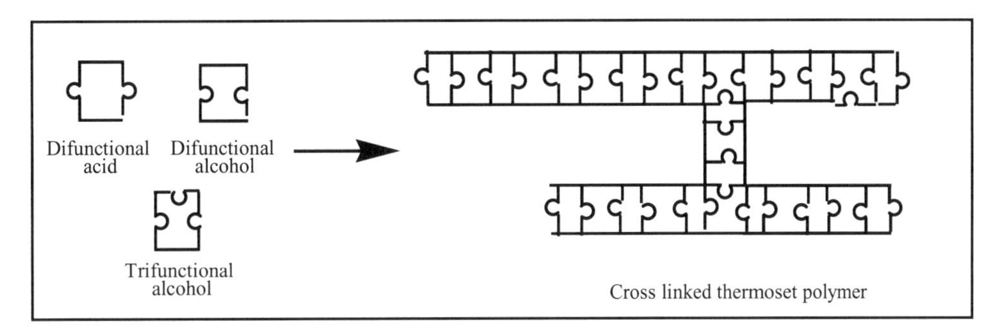


Figure 11.31 Having some jigsaw pieces with three tabs allows polymer branches and provides a means of cross linking. Experimentally, this can be as simple as mixing in some glycerol (a trifunctional alcohol) with the ethylene glycol (a difunctional alcohol) during the manufacture of a polyester resin.

Epoxies are pairs of monomers or oligomers that react to give thermosetting plastics. If the monomers are well chosen then the resulting thermoset epoxy plastic will have high performance characteristics. For household use, epoxies are typically sold as two component systems, as illustrated in Figure 11.32. In effect, when using this product, the consumer is mixing the two reactants that cause polymer formation. They get their name because one of the reactants in epoxy resin formation contains the epoxide functional group.

Figure 11.32 One of the two components of this bonding material is an epoxide. The epoxide functional group contains two carbon atoms and one oxygen atom in a three-membered ring. Ethylene oxide ($CH_2CH_2O$) is the simplest known epoxide.

Like the acid chloride group in sebacoyl chloride (for making nylon rope), the epoxide group is chemically reactive. You can understand that fact if you realize that the internal bond angles in the epoxide ring are near 60° and are thus, being greatly less that the preferred tetrahedral angle of about 109°, much strained. The high reactivity of epoxide rings means that the polymerization reaction occurs at room temperature once the reactants are blended.

## Transparency and Optically Transparent Plastics

For certain applications it is desirable to have a glasslike transparent polymer. The plastic champagne glasses used at weddings are a good example of why this is so. However, for products such as milk containers and garbage bags, transparency is not so important.

Whether or not a plastic looks clear or cloudy depends on the size of the particles that form from a heated, liquid polymer melt as it cools and solidifies. As you know, if those particles are sufficiently large they cause ordinary light to scatter—this scattering is exactly the same phenomenon that makes clouds cloudy, and both milk and a pine oil emulsions milky.

Until recently, the cheapest polymers such as polyethylene and polypropylene could be made only in cloudy forms. However, during the past ten years it has become possible to induce these materials to crystallize with sufficiently small crystals that the final polymer is transparent; doing this has the obvious commercial advantage: saving money. The method by which transparency is achieved is interesting: small amounts of hydrogen bonding molecular substances are added to the liquid polymer—sorbital derivatives that originate as sugars are suitable substances. As the polymer melt slowly cools, the sorbital derived molecules form a long network of chain like structures; the process is reminiscent of how protein chains cause ordinary Jell-O® to congeal. These threadlike structures allow many crystals to simultaneously begin to form, in a process known as *nucleation*. With many crystals forming and growing simultaneously, the melt is depleted of liquid before any large crystals can form, and so the final solid, being composed of only relatively small crystals, is reasonably transparent.

For making eyeglasses it is necessary to use a hard, scratch resistant plastic. The needed material in this application is a **polycarbonate**. The synthesis steps for the manufacture of a polycarbonate resin are shown in Figure 11.33.

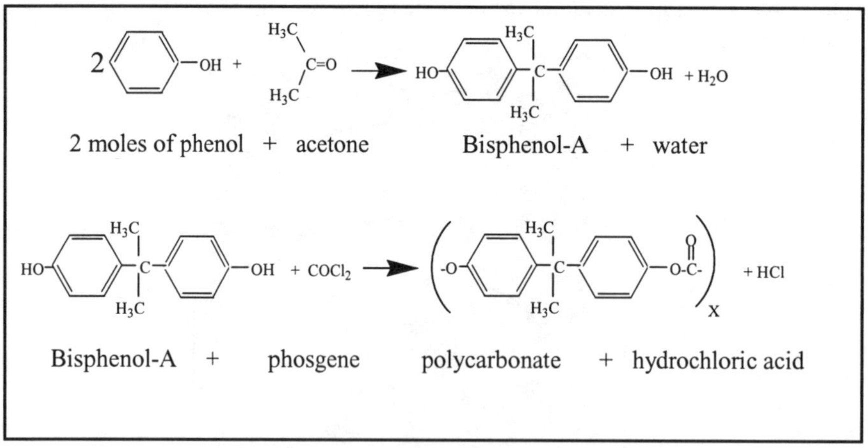

2 moles of phenol + acetone → Bisphenol-A + water

Bisphenol-A + phosgene ($COCl_2$) → polycarbonate + hydrochloric acid ($HCl$)

Figure 11.33 The chemical route to polycarbonate manufacture. Multistep synthesis to make desired products from available raw materials is a crowning achievement of organic chemistry

Making the monomer that is historically called bisphenol-A begins with the two petrochemicals acetone and phenol; they react in the presence of concentrated hydrochloric acid. The product bisphenol-A is then combined with phosgene in a condensation polymerization process. Other hard plastics with good scratch resistance are Lucite® and Plexiglass®. The latter is polymethylmethacrylate or PMMA.

The object pictured in Figure 11.34 is used by the author as a paperweight. It has a curious history. It is polymethylmethacrylate. Originally, PMMA was developed during World War II as a hard, transparent plastic for making canopies for fighter planes and machine gun enclosures on bomber planes. Many years ago I purchased a bottle of the methylmethacrylate monomer, for reasons long since forgotten. Eventually, the unused portion spontaneously polymerized in the brown glass bottle in which the sample had arrived. From time-to-time I would show the bottle to people, although it was difficult to see inside. Finally, one day by accident, I dropped the bottle and shattered the glass. The result you see in Figure 11.34.

Figure 11.34 An accidentally polymerized sample of the transparent plastic polymethylmethacrylate. Acrylic acid is $CH_2$=$CH_2$-COOH and polymers derived from it and its relatives are collectively called "acrylics." Acrylics come in a variety of guises. Inexpensive latex paint and white glue are often formulated from acrylic latex polymer emulsions.

## Polyurethanes

Polyurethanes are a class of synthetic organic polymers that contain the urethane (or carbamate) functional group as the chemical unit that links together their monomer units. With R– and R'– being alkyl or aromatic groups, the structure of the urethane group is:

```
        OR'
        |
  R–N–C =O
    |
    H
```

Commercial polyurethanes are produced via the polymerization of diisocyanates ONC–R–NCO with diols. The key reaction is an addition:

$$\text{R–N=C=O} + \text{HO–R'} \rightarrow \text{R–NH–C(=O)–O–R'}$$

In addition to having value as coating materials, polyurethanes enjoy a large market for the manufacture of flexible foams used in the furniture and bedding industries. In fact, most cushioning material now available consists of flexible polyurethane foams. Polyurethanes are also produced as rigid foams, and these are used primarily for purposes of thermal insulation. Other applications of polyurethanes are as architectural finishes, for shock absorbing in automobile bumpers, and for leather-like fabrics. Foamed polyurethanes require fire retardant additives if they are to be used in items such as chairs and sofas, because unless properly treated, polyurethanes release toxic gases when heated.

## Aramids

Aramids are a family of nylons and are chemically classified as polyamides. Aramids are so-named because they are **ar**omatic **amid**es. Two well known aramids are Kevlar® and Nomex®. Kevlar® is poly(p-phenyleneterephthalamide) while Nomex®is poly(m-phenyleneterеisophthalamide). Their closely related structures are shown in Figure 11.35. In Kevlar® the two nitrogen atoms are separated on the benzene ring by two carbon atoms (they are in para, or ring positions 1 and 4 with respect to one another). In Nomex® the two nitrogen atoms are separated on the benzene ring by one carbon atom (they are in meta or ring positions 1 and 3 with respect to one another).

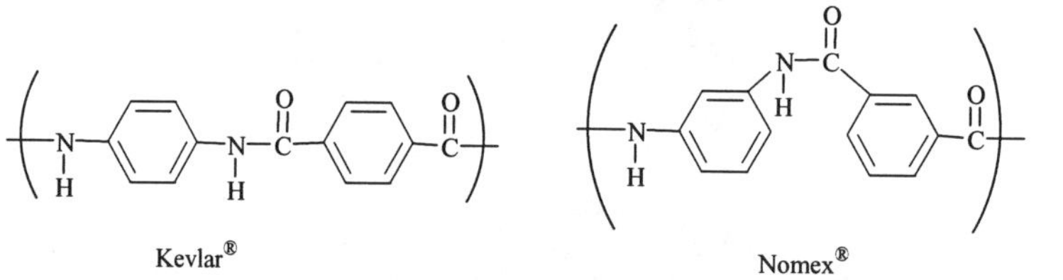

Kevlar®   Nomex®

Figure 11.35 Aramid polymers have outstanding strength and heat resistant characteristics because their chains pack exceptionally well, yielding highly crystalline structures.

The polymer chains of aramids extend and pack exceptionally well into crystals. This gives them remarkably good mechanical properties and extremely high melting points. In addition, all of the C–H bonds in these polymers are between hydrogen and a carbon atom in an aromatic ring; these aromatic C–H bonds are far less reactive with oxygen than are C–H bonds in a non aromatic polymer, so aramids can be used at temperatures where ordinary nylon would have long since burned. Kevlar® is used to make bullet proof vests, puncture resistant bicycle tires, trampolines, brake and clutch friction pads (replacing asbestos in that application), composites, fiber optic cable coatings, heavy duty belts and hoses. Nomex® combines excellent heat and flame resistance with good textile qualities; it's used to make protective clothing, hot-gas filters, and as electrical insulation for demanding applications.

Incidentally, making these polymers is quite a trick because you can't melt them and they don't dissolve in conventional solvents. The manufacturing procedure involves using amine solvents (such as N-methyl pyrrolinidone) modified with inorganic salts such as lithium and calcium chloride. To spin fibers, the polymer is converted to a form of liquid crystals in concentrated sulfuric acid for processing!

## Silicones

**Silicones** are synthetic polymeric silicon compounds. The general formula of silicones is $R_nSiO_{(4-n)/2m}$. Where R is a hydrocarbon group, often methyl, n = 1 - 3 and m is two or greater. Methyl silicone polymers can be created with oil like properties, and are valued for their resistance to thermal degradation in automobile engines. The repeating unit in linear polydimethyl siloxane fluids is:

$$
\begin{array}{ccccccc}
 & & CH_3 & & CH_3 & \\
 & & | & & | & \\
- & O - & Si & - O - & Si & - \\
 & & | & & | & \\
 & & CH_3 & & CH_3 &
\end{array}
$$

Silicones serve as oils, greases, and lubricants, for making specialty rubbers, and a host of other uses. The ever popular child's toy "silly putty" is a mixture of silicone oil and boric acid. It was developed around 1950 by James Wright, a GE Company engineer. Silly putty acts as a rubber over short time frames, and a ball of it will bounce 20% higher than a comparable rubber ball. However, over longer time frames, it flows and will not hold a definite shape against the force of gravity.

## Section 11.5 Biopolymers

Molecular biology is the study of life at its most fundamental level and is the branch of chemistry called biochemistry. From a small number of chemical elements, nature has constructed an incredible universe of living plants and animals. Living organisms have highly organized internal structures made from complex molecules. The multitude and complexity of natural biological processes are awe-inspiring. No other area of chemical research has such potential to affect the human future as does biochemistry.

**Biopolymers** are long chain molecules involved in the processes of life and created by living organisms; biopolymers are also called natural polymers. In the introductory section of this chapter we discussed cellulose, which is a member of the broader class of biopolymers called polysaccharides which are composed of repeating sugar (or saccharide) units. In this section we will briefly describe two other classes of biological materials: proteins and nucleic acids. **Proteins** are natural polymers derived from alpha amino acids; we mentioned leather (which is a protein) in connection with our discussion of semi-synthetic polymers, and the protein muscle tissue of a bovine is pictured in Figure 11.12. **Nucleic acids** are natural polymers composed of **nucleotides**, which in turn are polymerizable entities each derived from three subunits: a sugar residue, an organic base, and a phosphate group.

**Anonymous Editorial Comment**

A colleague of the author recently wrote to him: "It would be nice to see some biochemistry in the short course for engineers. Many of these folks will be critical members of biotech procedures and enterprise over the next few decades. A feeling for molecular events in cells and bodies will stand them in good stead. Also, the whole array of life's subtle actions is likely to catch and hold their interest. It's all more charming than plastic, don't you think?"

### Proteins

Proteins are amino acid polymers that with very few exceptions are constructed from twenty amino acids. The structures, names, and conventional three-letter abbreviations of the twenty common amino acids are shown in Figure 11.36.

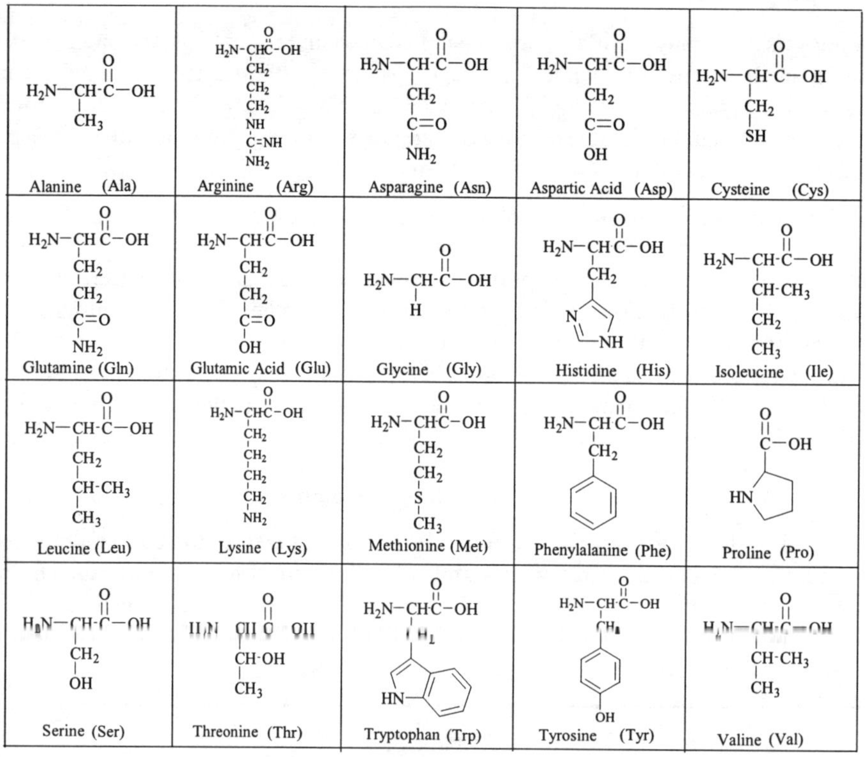

| | | | | |
|---|---|---|---|---|
| Alanine (Ala) | Arginine (Arg) | Asparagine (Asn) | Aspartic Acid (Asp) | Cysteine (Cys) |
| Glutamine (Gln) | Glutamic Acid (Glu) | Glycine (Gly) | Histidine (His) | Isoleucine (Ile) |
| Leucine (Leu) | Lysine (Lys) | Methionine (Met) | Phenylalanine (Phe) | Proline (Pro) |
| Serine (Ser) | Threonine (Thr) | Tryptophan (Trp) | Tyrosine (Tyr) | Valine (Val) |

Figure 11.36 The twenty key amino acids of living organisms. Almost all natural proteins consist of polymers of these twenty molecules.

The general formula of the amino acids in Figure 11.36 is $H_2N–CHR–COOH$, and they can be placed in four chemical categories. 1. Those with nonpolar, hydrophobic R groups: gly, ala, val, leu, pro, ile, phe, trp, and met. 2. Those with polar but uncharged R groups: ser, thr, cys, asn, gln, and tyr. 3. Those with a second acid functional group in R: asp and glu. 4. Those with a second basic functional group in R: lys, arg, and his. Nature exploits these various R-group properties to produce an enormously wide range of protein types. The formation of a **polypeptide** (a polyamino acid) is shown in Figure 11.37. The structure of the protein hemoglobin is shown in Figure 9.13.

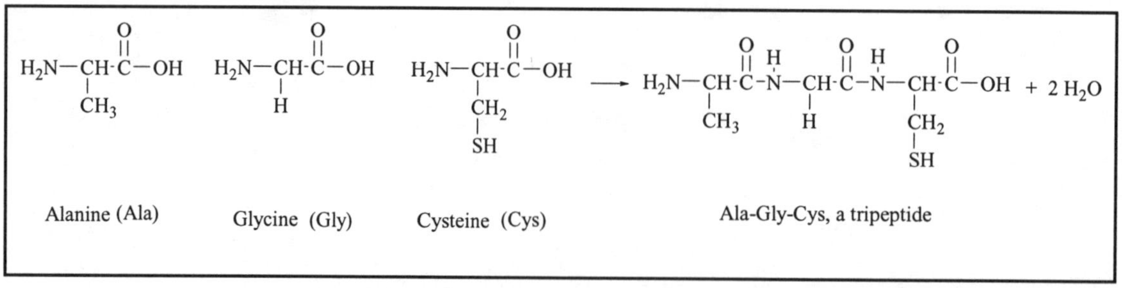

$$H_2N-CH(CH_3)-C(=O)-OH + H_2N-CH(H)-C(=O)-OH + H_2N-CH(CH_2SH)-C(=O)-OH \rightarrow H_2N-CH(CH_3)-C(=O)-N(H)-CH(H)-C(=O)-N(H)-CH(CH_2SH)-C(=O)-OH + 2\,H_2O$$

Alanine (Ala) Glycine (Gly) Cysteine (Cys) Ala-Gly-Cys, a tripeptide

Figure 11.37 Proteins arise from the condensation of amino acids. The product Ala-Gly-Cys is called a polypeptide, specifically in this case a tripeptide because it's composed of three amino acid residues.

There are perhaps 100,000,000,000 ($10^{11}$) different proteins in all living organisms. They play many biological roles. Enzymes are the largest class of proteins; they catalyze biochemical reactions. Transport proteins, such as hemoglobin, bind specific molecules and carry them where needed. Some hormones, biochemical messengers, are proteins; in mammals, the protein insulin controls the transport of glucose into cells. Structural proteins are widely distributed in nature: collagen is a the major component of connective tissue and bone; α-keratin is the major component of skin, hair, and feathers. Proteins (actin and myosin) are what cause muscles to contract. The self protecting immune system response of higher organisms involves the formation of protective proteins (antibodies) which chemically combine with harmful agents. The way to examine proteins is by manipulating them on-screen with a chemical visualization program. [❁kws "protein data bank"].

## Nucleic Acids

Deoxyribonucleic acid (DNA) is an informational polymer. It is the substance that contains the biochemical information needed to construct an organism—to build a human being. In 1953, James Watson (1928-) and Francis Crick (1916-) determined its now well-known double helix structure based partly on x-ray diffraction data obtained by Rosalind Franklin (1920-1958) and Maurice Wilkins (1916-). One form of ribonucleic acid (RNA) transcribes the information from DNA in the cell nucleus and is therefore a polymer copy of DNA. RNA transports the molecular information permanently stored in the DNA to the ribosomes of cells where proteins are made following the instructions of the genetic code. The way to view the DNA molecule is by manipulating it on-screen with a chemical visualization program [❁kws +DNA +structure].

A **nucleotide** is a chemical entity formed by the condensation reaction of a phosphoric acid molecule, a cyclic sugar molecule, and an organic base. The formation of a nucleotide is shown in Figure 11.38. The sugar in RNA is ribose; the sugar in DNA is deoxyribose. There are five different bases that occur in the nucleotide units of DNA and RNA. The four that occur in DNA are adenine (A), guanine (G), thymine (T), and cytosine (C). The four that occur in RNA are adenine (A), guanine (G), and cytosine (C), and uracil (U). The component residues of DNA and RNA nucleotides are summarized in Table 11.14. A **residue** is any portion of a larger molecule recognized as being derived from a smaller molecule that went into making the larger molecule.

Table 11.14 **The Nucleotides of DNA and RNA and their Component Residues**

| Biopolymer | Acid | Sugar | Base | Symbol |
|---|---|---|---|---|
| DNA | Phosphoric | Deoxy ribose | Adenine | A |
| DNA | Phosphoric | Deoxy ribose | Guanine | G |
| DNA | Phosphoric | Deoxy ribose | Cytosine | C |
| DNA | Phosphoric | Deoxy ribose | Thymine | T |
| | | | | |
| RNA | Phosphoric | Ribose | Adenine | A |
| RNA | Phosphoric | Ribose | Guanine | G |
| RNA | Phosphoric | Ribose | Cytosine | C |
| RNA | Phosphoric | Ribose | Uracil | U |

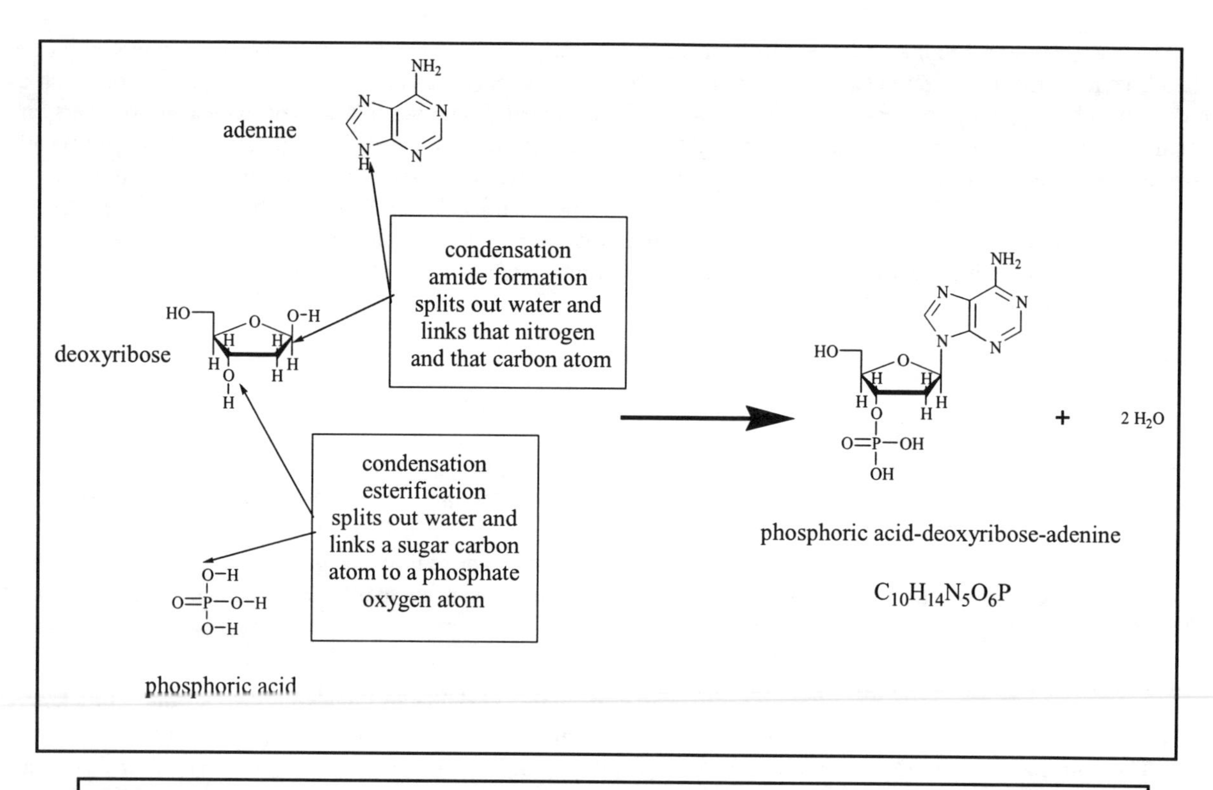


Figure 11.38 The formation of a nucleotide. This one is a DNA nucleotide. A nucleotide is a single entity formed from three residues: 1. A phosphoric acid molecule or phosphate ion, 2. A cyclic 5-carbon sugar (deoxyribose in DNA, shown here, or ribose in RNA), and 3, an organic base.

The structures of the four DNA nucleotides are shown in Figure 11.39. The structures of the RNA nucleotide are shown in Figure 11.40.

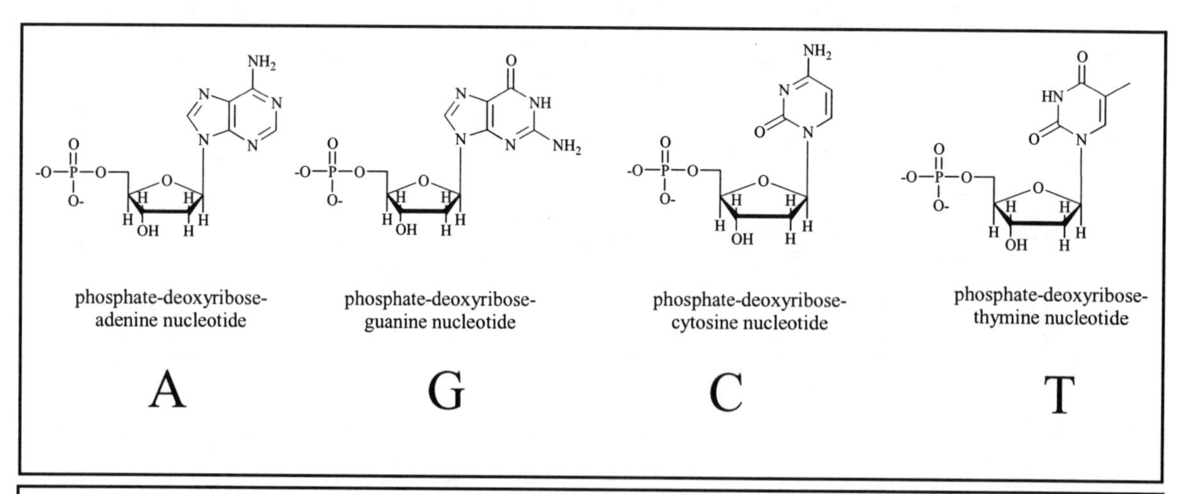


Figure 11.39 The four nucleotides in DNA. The organic base residue in each case is the cyclic moiety containing nitrogen atoms at the upper right of each nucleotide.

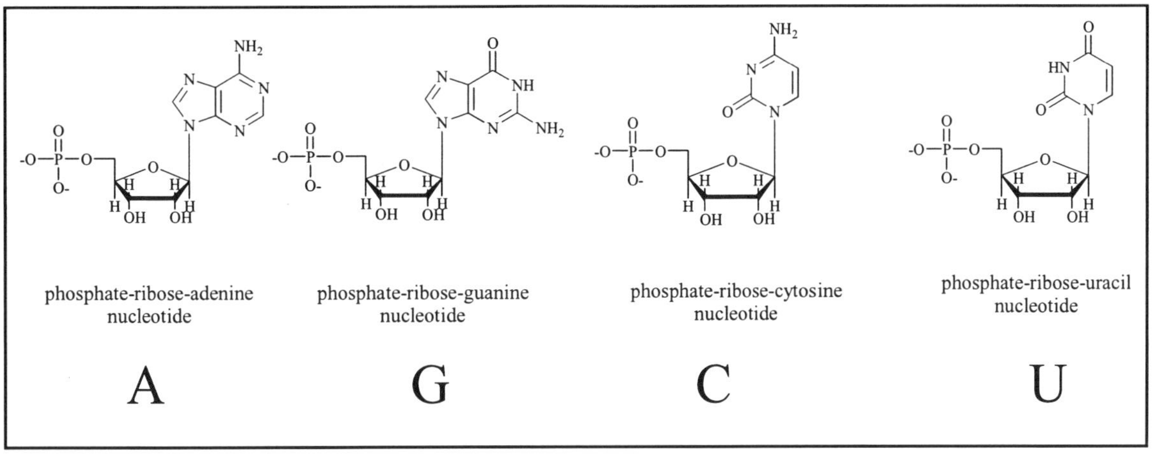

| phosphate-ribose-adenine nucleotide | phosphate-ribose-guanine nucleotide | phosphate-ribose-cytosine nucleotide | phosphate-ribose-uracil nucleotide |
|---|---|---|---|
| A | G | C | U |

Figure 11.40 The four nucleotides in RNA. The organic base residue in each case is the cyclic moiety containing nitrogen atoms at the upper right of each nucleotide. Compare this figure with Figure 11.39 and note that here there is an extra oxygen atom on the front right of the ribose sugar ring. The D in DNA stands for deoxyribose, and deoxy refers to the absent oxygen atom.

DNA's structure consists of two helical polynucleotide chains coiled around a common axis. The two separate strands of the DNA molecule are linked by base pairing as shown in Figure 11.41. A **base pair** consists of two organic bases with complementary structures capable of being linked by hydrogen bonds. The mechanism of base pairing involves hydrogen bonding between the two members of the pair by means of cleverly matched geometry. Thus, in double stranded DNA adenine matches with thymine (A—T) and guanine matches with cytosine (G—C). The geometry of hydrogen bonding is shown in Figure 11.42.

DNA is a strong acid. Each nucleotide shown in Figure 11.41 has a hydrogen ion associated with it. The DNA obtained from the different cells of any organism is identical. Human DNA has total of about 3 billion base pairs. The base pairs have an average mole weight of 600 g $mol^{-1}$. In the coiled double stranded DNA, each base pair takes on average a linear distance of 0.33 nm. So the entire human DNA would stretch out a distance of about one meter if the chain were fully extended.

In higher organisms, in our cells for example, DNA exists in the chromosomes. Packaged in the chromosomes DNA comes in multiply coiled and wrapped structures (nucleosomes), DNA's negative charge being neutralized by positively charged proteins (histones) loaded with lysine and arginine residues.

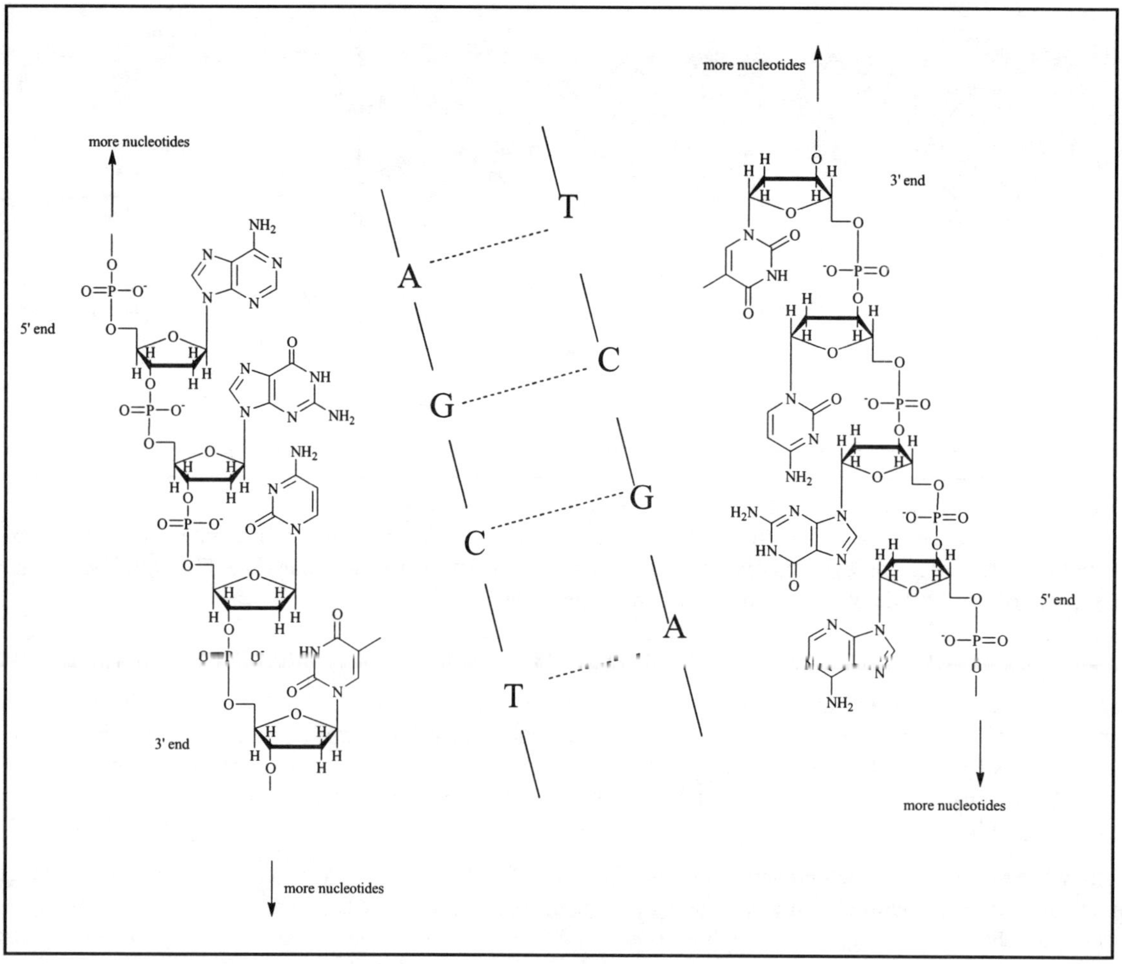

Figure 11.41. Double stranded DNA. Each chain consists of a series of nucleotides linked by sugar phosphate condensation bonds. The particular sequence of nucleotides stores the genetic information of an individual organism. Such sequencing is probably the ultimate example of atomic structure controlling bulk properties. In this case atomic structure, through the base sequence, controls life itself.

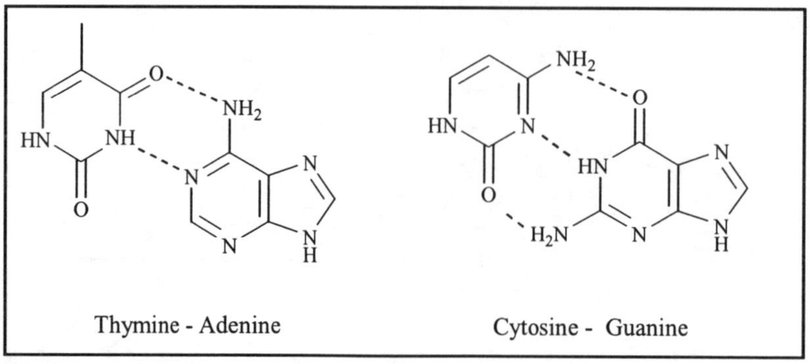

Thymine - Adenine

Cytosine - Guanine

Figure 11.42 Base pairing. The pairs of DNA bases T—A and C—G have geometries such that each forms hydrogen bonds with its partner in a complementary manner.

With the exception of identical twins, each human being has his or her own characteristic DNA sequence. This sequence, which can be obtained from small samples of biological fluids or fragments, provides a unique indicator of that individual. During the past ten years, such DNA sequencing has had important consequences for forensic science, and for understanding human history and the peopling of our planet.

## The Central Theory

Now that we know something about the nature of proteins and nucleic acids we can briefly discuss the fundamental chemical mechanism of life. This idea has come to be called the central theory of molecular biology. It consists of three parts: 1. **Transcription**, in which the information stored in DNA is transferred to RNA and thus made available for protein synthesis. 2. **RNA processing**, in which the transcribed RNA is modified in preparation for protein synthesis. And, 3. **Translation**, the process in which the information in mRNA is biochemically interpreted to make the appropriate polypeptides and eventually proteins. The central theory is represented schematically in Figure 11.42.

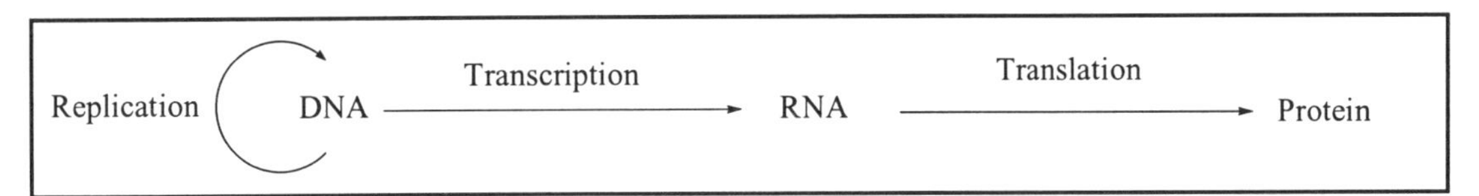


Figure 11.43 The central theory of molecular biology. Biological information flows from DNA to RNA to protein. Genetic information is passed by replication of the DNA, which divides during cell division with each separate strand serving as a template for a complementary DNA strand. In this way one DNA molecule becomes two.

The hereditary information carried by DNA is organized into chromosomes. Bacteria contain a single chromosome, consisting of a single DNA double helix. The fruit fly has four chromosomes composed of 170 million base pairs. Yeast has sixteen chromosomes and 14 million base pairs. The 3.1 billion base pairs of humans are divided among 24 different chromosomes.

The process of translation defined by the central theory is carried out by means of the genetic code. The **genetic code** is the method by which base sequences in an organism's DNA becomes expressed in the amino acid sequences of its proteins. The genetic code uses three-base sequences or triplets, called **codons**, to code for specific amino acids. Messenger RNA (mRNA) transcribes the DNA information by base pairing along a specific gene. At a molecular level, a **gene** is any DNA sequence that is transcribed in an organism. Each of the three RNA bases can be U, C, A, or G, so a codon can have one of 4 x 4 x 4 = 64 different "values." These are designated by three-letter symbols UUU, UAG, ACG, etc., to describe the three bases they contain. The relationship between the codons and the amino acids is shown in Table 11.15. Because there are 64 codons but only 20 amino acids there are duplications. Certain codons code for "stop," which is the molecular way of telling when the end of a particular protein has been reached, and AUG codes for "start," as well as the amino acid methionine. A second class of RNA molecules, transfer RNA (tRNA), recognizes the mRNA codons and carries the appropriate amino acid to its place in the growing polypeptide (chain extension). That assembling process takes place on ribosomes which contain the third class of RNA, ribosomal RNA (rRNA).

Recombinant DNA technology alters the genes of organisms so they will produce proteins that were not in their original repertoire. For example, the gene for human insulin has been inserted into the DNA of a bacterium (*E. coli*) and today most insulin needed by diabetics is manufactured by bacteria. Similarly, the use of genetically modified (GM) crops offers advantages of disease resistance and superior quality. However, the use of such technology has been controversial. Our ability to manipulate human genetic material raises profound questions and we may reflect with wonder on the DNA molecule, its role in evolution, and how our control of it will fundamentally change the future of human beings.

Table 11.15 **The Genetic Code**

| UUU | UUC | UUA | UUG | UCU | UCC | UCA | UCG | UAU | UAC | UAA | UAG | UGU | UGC | UGA | UGG |
|---|---|---|---|---|---|---|---|---|---|---|---|---|---|---|---|
| Phe | Phe | Leu | Leu | Ser | Ser | Ser | Ser | Tyr | Tyr | Stop | Stop | Cys | Cys | Stop | Trp |
| | | | | | | | | | | | | | | | |
| CUU | CUC | CUA | CUG | CCU | CCC | CCA | CCG | CAU | CAC | CAA | CAG | CGU | CGC | CGA | CGG |
| Leu | Leu | Leu | Leu | Pro | Pro | Pro | Pro | His | His | Gln | Gln | Arg | Arg | Arg | Arg |
| | | | | | | | | | | | | | | | |
| AUU | AUC | AUA | AUG | ACU | ACC | ACA | ACG | AAU | AAC | AAA | AAG | AGU | AGC | AGA | AGG |
| Ile | Ile | Ile | Met | Thr | Thr | Thr | Thr | Ans | Asn | Lys | Lys | Ser | Ser | Arg | Arg |
| | | | | | | | | | | | | | | | |
| GUU | GUC | GUA | GUG | GCU | GCC | GCA | GCG | GAU | GAC | GAA | GAG | GGU | GGC | GGA | GGG |
| Val | Val | Val | Val | Ala | Ala | Ala | Ala | Asp | Asp | Glu | Glu | Gly | Gly | Gly | Gly |

The triplets of capital letters correspond to three-base sequences in mRNA. Below them are the conventional abbreviations for the amino acids. Thus, for example the codon UAC codes for the amino acid tyrosine.

**The Human Genome Project**

At the present time we are close to the end of an enormous scientific project to know the entire base pair sequence (the genome) of the 24 different chromosomes of individual humans. A key part of that project is to map human genes on the entire human genetic material. Many genes are known to be associated with specific human diseases. The actual number of genes is hotly debated and subject to an internet contest called genesweep. [❁kws gene-sweep]. Guesses range from 27,000 to 200,000 genes on the human chromosome, with the average guess being about 62,500 genes. The diverse guesses remind us that even when we know the entire base sequence of the human genome there will still remain the problem of interpreting it and figuring out how it all works.

Sequencing the human genome has been greatly aided by high powered computer technology and the future for computer-biochemical joint ventures looks to be quite interesting. Over the next decade, the computer manufacturer IBM intends to develop a a compute dubbed "Blue Gene" to try to model the way that proteins fold into the unique and special shapes that give them their properties. Blue Gene is reportedly 1000 time more powerful than its predecessor Deep Blue, which beat the human chess champion Gary Kasparov in a match in February, 1996. Blue Gene will make a strong claim to be the ultimate QSPR machine.

To conclude this chapter we can remark that it has mainly focused on the chemical structure of materials and how the inherent chemical properties of atoms, molecules, and ions lead to the bulk properties of materials that give them useful engineering applications. In closing, we observe that while engineers are people who make things with stuff, biopolymers are things to make stuff to make engineers.

## Essential Knowledge—Chapter 11

**Glossary words:** Materials science, composite, polymer, monomer, natural polymer, synthetic polymer, biopolymers, molecular solids, ceramics, glasses, crystalline, amorphous, ionic solid, metallic solid, metallurgy, close-packed, allotropes, coordination number, covalent solid, semi-synthetic and synthetic polymers, vulcanization, cellulose, esterification reaction, condensation reaction, cellulose acetate, guncotton, petrochemical, silicate mineral, quartz, Portland cement, concrete, cracking, polymeric solid, plastics, resins, thermoplastic resin, polymerization, macromolecular, addition reaction chain extension, addition polymerization, block copolymer, engineering polymers, addition polymer, oligomer, linear polyethylene, backbone chain, end group, low density polyethylene, linear low density polyethylene, high density polyethylene, branched polymer, copolymers, pendant groups, polypropylene, polyvinyl chloride (PVC), polystyrene, ABS polymer, polyolefin, polytetrafluoroethylene, condensation polymer, thermoplastic polyester, phenolic resin, urea resin, polyester resin, polyamide (nylon), epoxy resins, natural rubber, elastomers, cis-polyisoprene, heat distortion temperature, thermosetting resin, Bakelite, polycarbonate, polymethylmethacrylate, polyurethanes, aramids, silicones, biopolymers, amino acid, polypeptides, proteins, nucleotides, nucleic acid, deoxyribonucleic acid (DNA), ribonucleic acid (RNA), informational polymer, mRNA, tRNA, residue, base pairing, central theory, transcription, RNA processing, translation, genetic code, codons, gene.

**Key Concepts:** The nature of basic forces inside solids and the classification of solids. The ongoing theme that the bulk properties of solids reflect the forces within them. The nature of polymers both natural and synthetic and their importance in forming the materials of the world. The range of inorganic silicates and their geological and geochemical significance. The production, manufacture, and uses of synthetic addition and condensation polymers. Recognition of classes of thermoplastic and thermosetting resins and their commercial applications. Having an appreciation of the commercial significance of plastics additives. Understanding the phenomenon of cross linking and its role in the formation and production of elastomers. Understanding the range of biopolymers as represented by polysaccharides, polypeptides, and nucleic acids. Understanding in a broad context the central theory of genetic transmission. Being able to demonstrate a clear understanding of the genome project and the scientific basis for recombinant DNA technology.

**Key Equations:**

The length of the face-diagonal, x, of a cubic unit cell with edge d: $x = d\sqrt{2}$

The length of the body-diagonal, y, of a cubic unit cell with edge d: $y = d\sqrt{3}$

## Questions and Problems

### Introduction

11.1. Define or describe in 25 words or less each of the following terms: a. materials science, b. polymer, c. inorganic polymer, d. synthetic polymer, e. semi-synthetic polymer.

11.2. Define a composite and give examples.

11.3. Classifying solids is not easy. Discuss.

### Classes of Solids

11.4. State the class of solids that is electrically conducting. What nonmetallic element is significantly electrically conducting? Explain why this nonmetal is able to conduct electricity. Why is this same nonmetal a useful solid lubricant?

11.5. Assign each of the following substances to one of the five classes of solids: a. calcium nitrite, b. carbon tetrachloride, c. rubber, d. magnesium sulfate, e. an amalgam (look up the word in a dictionary or on the internet), f. firebrick, g. solid nitrogen, and h. the plastic often used to top kitchen counters (state a brand name for such a plastic).

11.6. State the name and chemical formula of two ionic compounds that are soluble in water. State the names and formulas of two ionic compounds that are insoluble in water. To name insoluble compounds you will find helpful the solubility rules presented in Section 9.6.

11.7. List the names of five compounds that find use as abrasives. What are the main chemical characteristics that are needed for a given compound to be a useful abrasive.

11.8. State the three types of unit cells found in metals. Make a table that shows for each the number of atoms in the cubic unit cells, the atomic coordination number, and the hard-sphere percentage void volume.

11.9. A cubic unit cell has an edge length of d. Calculate the lengths of a. the diagonal across any face of the cube, and b. the diagonal across the body of the cube.

11.10. A simple cubic unit cell (Figure 11.15) contains one hard-sphere atom in each unit cell. Calculate the percentage void volume in a metal that adopts the simple cubic atomic structure.

### Natural, Semi-Synthetic, and Synthetic Polymers

11.11. What is a carbohydrate? What function does the carbohydrate cellulose serve in living organisms?

11.12. State the relationship between a monosaccharide, disaccharide, and polysaccharide.

11.13. Write a balanced chemical equation to represent the reaction of a sugar molecule with three molecules of acetic acid to produce a triacetate ester of the sugar.

11.14. State the name of a class of natural polymers that contain nitrogen.

11.15. Glycine is the amino acid $H_2N\text{-}CH_2\text{-}COOH$. Make a sketch of polyglycine. What is the empirical formula of polyglycine?

11.16. What is nitrocellulose? Why is it explosive? What must be the algebraic sign of its standard enthalpy of formation? Hint: review section 7.4.

11.17. Propanol and propanoic acid (older name: propionic acid) react to form the ester propyl propanoate. Both the acid and the alcohol here contain three linked carbon atoms. Write a balanced equation for this reaction using structural formulas that show the bonds in the reactant and product molecules.

11.18. What is the origin of the word ester? Is cellulose a polyester, a polyether or neither? State the general formula of the ether functional group.

11.19. According to the table of organic functional groups (Table 2.7), and other information this chapter, which of the following classes of organic chemical compounds contain nitrogen: amines, amides, ketones, carboxylic acids, nitro compounds, and urethanes.

## Silicates

11.20. According to the chapter, all of the following materials are composed of silicon-oxygen tetrahedral structural units, joined in various ways, and containing the appropriate cations. State the manner in which the silicon-oxygen tetrahedra are joined in: a. pyroxenes, b. amphiboles, c. mica, and d. feldspar.

11.21. State the chemical formulas of orthoclase and chrysotile asbestos.

11.22. Garnets are silicate minerals with a common chemical structure but a variable chemical composition. Explain this situation and give examples.

11.23. Silicate minerals are valuable commercial materials. State the name of the natural silicate that you would select for each of the following applications: a. glass making, b. to make cement, c. to make automobile spark plugs, d. to make brake linings, e. to make dielectric media for capacitors, and f. for flooring tiles.

11.24. Use the formula stated in the chapter to calculate the percentage of sodium, calcium, aluminum, silicon, and oxygen, in a typical window glass.

11.25. Define or describe in 25 words or less each of the following: a. amorphous, b. hydration, c. concrete, d. crystal glass, and e. liquid silicate.

## Synthetic Organic Polymers

11.26. Define or describe in 25 words or less each of the following: a. polymerization, b. addition polymerization, c. condensation polymerization, d. monomer, e. chain extension.

11.27. Write the names of 10 compounds identified as petrochemicals. Read Appendix F and identify the names of five companies that make petrochemicals.

11.28. Commercial plastics such as Teflon® have an ® following their name. What does ® mean?

11.29. Sketch the chemical structures of the following oligomers: the trimer formed by ethylene, the pentamer formed by propylene, and the tetramer formed by styrene. State the chemical formula $C_xH_y$ of each of the preceding oligomers.

11.30. Identify the contribution to polymer science of each of the following: a.Charles Goodyear, b. Jons Jakob Berzelius, c. Thomas Graham, d. Leo Baekeland, e. Hermann Staudinger, f. M. W. Perrine, and g. Wallace H. Carothers.

11.31. Define or describe in 25 words or less each of the following terms: a. linear polymer, b. branched polymer, c. backbone chain, d. pendant group, e. side chain, f. cross linking, g. copolymer, and h. copolymerization.

11.32. Distinguish between a thermosetting plastic and a thermoplastic. Give three examples of each class that you use in you daily life. To do this consider the desirable chemical features of each class and how you use the objects in you daily life. You are also invited to answer this question experimentally with a fabric or soldering iron. However, if you do such experiments don't damage anything of value.

11.33. State the names of the four largest commercial volume addition polymers.

11.34. Explain why there are two different classes of polyester resins listed in Table 11.12.

11.35. State the empirical formulas of polypropylene, polyvinyl chloride, and polystyrene.

11.36. What are "end-groups?"

11.37. Discuss what makes rubber rubbery. Provide both a molecular and a thermodynamic explanation.

11.38. Define the term "heat distortion temperature" of a polymer.

11.39. Name three classes of additives that are incorporated into commercial plastic materials.

11.40. Explain the physics that lies behind any particular plastic material being either opaque or transparent. How is nucleation used to control or alter the optical transparency of a plastic?

11.41. The word "carbonate" is used in two different sections of Chapter 11. The structure of the mineral calcite, which is a natural form of calcium carbonate, is pictured and the manufacture of a polycarbonate plastic is described in detail. Explain how two such different compounds can share the name.

11.42. State the molecular formulas of: phenol, acetone, bisphenol-A.

11.43. Write a balanced chemical equation to represent the formation of an amide from ethylamine and formic acid (HCOOH).

11.44. Write a balanced chemical equation to show the esterification reaction that occurs between one mole of ethylene glycol and two moles of acetic acid. Can these two reactants be chemically combined to make a polyester? Why? Why not?

11.45. Describe in your own words the "nylon rope trick."

11.46. What is a difunctional molecule and why are such molecules needed if one wishes to manufacture a condensation polymer?

11.47. Write the molecular formula for terephthalic acid.

11.48. What is the linking group of atoms used by chemists to synthesize polyurethanes? Are polyurethanes substances with elastomeric properties? State the major use of polyurethane polymers.

11.49. Ethylene oxide ($C_2H_4O$) is the smallest molecule that contains an epoxide functional group. Draw the structure of an ethylene oxide molecule and use VSEPR principles combined with principles of Euclidean geometry to speculate about all of the bond angles in the molecule. What do you think they are?

11.50. By analogy with the previous question, state the molecular formula of propylene oxide.

11.51. Why are molecules that contain epoxide functional groups particularly useful in making thermosetting resins around the household. The "2-ton epoxy" pictured in the Chapter is an example of a product that contains molecules with epoxy functional groups.

11.52. Describe the circumstances under which synthetic rubber was developed.

11.53. Sketch the structures of cis-polyisoprene and SBR.

11.54. Many automobile enthusiasts like to use synthetic engine oils in lieu of hydrocarbon based engine oils. What is the empirical formula of a likely synthetic motor oil?

## Biopolymers

11.55. Use three-letter amino acid abbreviations to show all different possible tripeptides derived from one molecule each of phenylalanine, tyrosine, glycine.

11.56. Write an equation to show alanine forming a polypeptide.

11.57. Which of the twenty common alpha amino acids contain a benzene ring?

11.58. Draw the structures of the ring forms of ribose and deoxyribose as found in RNA and DNA. Calculate their molecular formulas and molar weight.

11.59. What are the key differences between the biochemical building blocks of DNA and RNA?

11.60. What is meant by the terms 5' (five-prime) and 3' (three-prime) ends of a DNA chain?

11.61. What forces are involved in holding together the two strands of the double helix together in DNA?

11.62. What would be the base sequence on a DNA strand that would be generated on a template strand with the following base composition: TCG GAA TCA AAT CGG.

11.63. What is a codon and what is its importance in protein synthesis?

11.64. Briefly describe the role which DNA, mRNA, tRNA and ribosomes play in protein synthesis.

11.65. Proteins are polypeptides but polypeptides are not proteins. Discuss.

11.66. Write the amino acid residue sequence that would be formed in a protein from the following sets of codons on an mRNA strand assuming that the sequence giving is in the 5'-3' orientation: UUU CAU GGC AGA CGA GGG UAG.

11.67. What is recombinant DNA?

## Problem Solving

11.68. Use internet resources to locate the nearest cement plant to your home.

11.69. State the chemical formulas of the following gemstones: a. beryl, b. emerald, c. topaz, d. opal, and e. turquoise.

11.70. Use internet resources to download any patent dealing with a polymerization catalyst.

11.71. If nothing sticks to the Teflon®, how come the Teflon® sticks to the pan?

11.72. Related to the previous question, do an internet search and write a biographical sketch of Roy Plunkett.

11.73. What are cermets?

11.74. Describe the resin identification coding system of icons.

11.75. Use internet resources to identify the uses and composition of a. Riston®, b. UHMWPE, c. styrene-acrylonitrile copolymer (SAN), d. and UF (urea-formaldehyde).

11.76. Discuss the effect of branching on the extent of crystallinity of a polymer. Describe how the bulk properties of a polymeric material relate to its crystallinity.

11.77. Write an essay entitled "ABS Materials for Engineering Applications".

11.78. Review the topic of crystallography in Appendix G. How might a polymer's crystallinity be measured by means of x-ray crystallography?

11.79. What is PETE's empirical formula?

11.80. Calculate the percentage by mass of chlorine in polyvinyl chlorine.

11.81. A body centered cubic unit cell (Figure 11.15) contains two hard-sphere atoms in each unit cell, one in the body and one-eighth of an atom at each corner. The body diagonal of the cube has a length 4r, where r is the radius of the atoms. Calculate the percentage void volume in a metal that adopts the body centered cubic atomic structure.

11.82. A face centered cubic unit cell (Figure 11.15) contains four hard-sphere atoms in each unit cell, half of an atom in each face and one-eighth of an atom at each corner. The face diagonal of the cube has a length 4r, where r is the radius of the atoms. Calculate the percentage void volume in a metal that adopts the face centered cubic atomic structure.

# Chapter 12

# Phase Diagrams and Solutions

## Introduction

In Chapter 4 we studied gases and in the past two chapters we've looked at liquids and solids. The relationships among these three states of aggregation of matter (gas, liquid, and solid) can be brought together on a phase diagram. The **phases** of a **pure substance** are simply its states of aggregation. Every pure substance has its own, unique phase diagram. The **phase diagram** of a pure substance is a pressure-volume-temperature map of a substance's solid, liquid, and gas phases. In this chapter we will study mainly pressure-temperature maps. In the jargon, a pure compound is called a one-component system. Mixtures of two compounds are called two-component systems. The phase diagram of a two-component system is a three-dimensional pressure-temperature-composition map; typically phase diagrams on paper are two-dimensional slices through the three-dimensional representation space. Multi-component systems are mixtures of three or more substances. Every pure substance *and* every mixture has its own phase diagram. That's a lot of phase diagrams.

Multi-component phase diagrams are especially important in understanding the properties and behavior of metal alloys. Thus, mechanical engineers are trained to interpret and to apply phase diagrams. The properties of an alloy are strongly influenced by the microstructure that arises as a solid cools and crystallizes from a melt. Perhaps you know that many metals are "tempered" by quenching them in cold water to give them a superior microstructure. **Physical metallurgy** is the study of the atomic arrangement and microstructure on a metal's physical properties.

The properties of a pure liquid solvent are changed when a solute is dissolved in it to make a solution. In Chapter 10 we saw that Raoult's law can be applied to mixtures of liquids to account for their distillation behavior and to give insight about the nature of intermolecular forces. In this chapter we'll look at how nonvolatile solutes change the properties of the solvent. A **nonvolatile solute** is one that does not vaporize when its solution is boiled. We'll consider four related **colligative properties**: boiling point elevation, vapor pressure lowering, freezing point depression, and the formation of osmotic pressure. They are called the colligative (literally, collected together) properties because any solute added to a solvent changes the four properties in a similar manner. (Alternatively, they are properties of a *collection* of particles, not a single particle. One water molecule doesn't have a boiling point.) The key underlying concept is that when *any* solute is dissolved in a solvent the concentration of the *solvent* decreases. So, adding any solute to any solvent changes the phase diagram of the pure solvent in a definite way.

To conclude the chapter we'll see that the colligative properties of nonelectrolytes and electrolytes are different. A **nonelectrolyte** is a substance that dissolves in water without the formation of ions. An **electrolyte** is a substance that dissolves in water with the formation of ions. A **strong electrolyte** ionizes completely when it dissolves in water. A **weak electrolyte** ionizes partially when it dissolves in water. Historically, the nature of ionization in aqueous solution was first revealed by studies of colligative properties near the end of the nineteenth century.

## Section 12.1 Characteristics of Phases and Limiting Laws

### Phases

The chemical properties of a single molecule or ion are the same whether it's in the solid, liquid, or gas phase. However, the *bulk* properties of a compound are quite different in the three phases. The characteristics that distinguish the three phases are: 1. Solids have fixed volume and shape; liquids have fixed volume but assume the shape of the container that they occupy; gases have neither fixed volume nor shape; they completely and uniformly fill any container they occupy. 2. Solids and liquids are difficult to compress; gases are easily compressed. Solids and liquids are called **condensed phases**. 3. Solids do not flow; both liquids and gases readily flow under stress. Gases and liquids are called **fluid phases**.

The properties of phases, and the conditions that cause a compound to change from one phase to another, can be explained in two different but complementary ways: 1. a molecular interpretation, and 2. a thermodynamic interpretation. The molecular interpretation (Figure 12.1) holds that the properties of phases are controlled by the balance between intermolecular forces (that tend to hold molecules together) and kinetic energy (that tends to move the molecules apart). At sufficiently low temperatures and sufficiently high pressures all substances are solids. In a solid, kinetic energies are small compared with intermolecular attractions, so the molecules remain rigidly in fixed positions. In solids, the distances between molecules are small—the atoms or molecules are more or less touching. The thermodynamic interpretation holds that changes among phases are controlled by free energy changes that can be calculated (at any given temperature) from the Gibbs equation.

At intermediate pressures and temperatures, molecular kinetic energies are comparable to the intermolecular attractions, and liquids exist. Liquids have densities similar to the solids from which they come. But, in the liquid phase the molecules are not fixed at specific positions. Rather, random molecular motion (Brownian motion) occurs in liquids. Each molecule moves only a short distance before colliding with a neighbor.

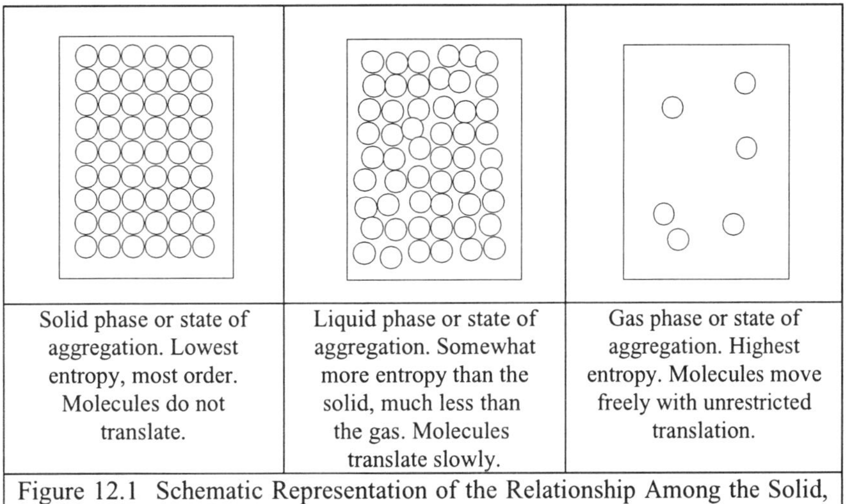


Figure 12.1 Schematic Representation of the Relationship Among the Solid, Liquid, and Gas Phases of a Pure Molecular Substance at the Triple Point Temperature.

At sufficiently low pressures and sufficiently high temperatures all substances are gases. Having a low density distinguishes gases from the condensed phases. In a gas, molecules travel considerable distances before colliding with another molecule. Because the molecules of a gas are far apart, each molecule behaves more or less independently of the other molecules (except during collisions). As the pressure of a gas is lowered the distances among its molecules increase. In the limit, if we lower the pressure to zero, the gas molecules become infinitely far apart and utterly independent. Thus, all gases become ideal gases as they approach zero pressure.

The ideal gas equation is a limiting law. Recall, a **limiting law** is one that is only approximately correct for real substances, but that becomes exactly true if a property of the substance is taken to an extreme limit. All of the colligative properties of a solution obey limiting laws. But for the laws of colligative properties the limit is the limit of infinite dilution. That is to say, as we measure any colligative property in successively more and more dilute solution, the graph we make gets straighter and straighter until the law finally becomes exact.

## A Description of Phase Diagrams

Phase diagrams are maps of the phase properties of materials. Specifically, they are graphs labeled to show what phases exist at conditions specified along the graph's axes. Thus, for a pure compound (where the composition is always 100%) the phase diagram shows the three phase regions on the pressure-temperature map designated G (gas), L (liquid), and S (solid). The phase diagram of a multi-component system is often presented with one of the axes being composition.

Phase diagrams are used to understand and to control the properties of specific substances. Perhaps their principal application is to understand the properties and behavior of metal alloys and ceramic materials for engineering applications. They also have important applications in chemistry and geology. For example, geologists use phase diagrams to study how solid rocks form from molten rock.

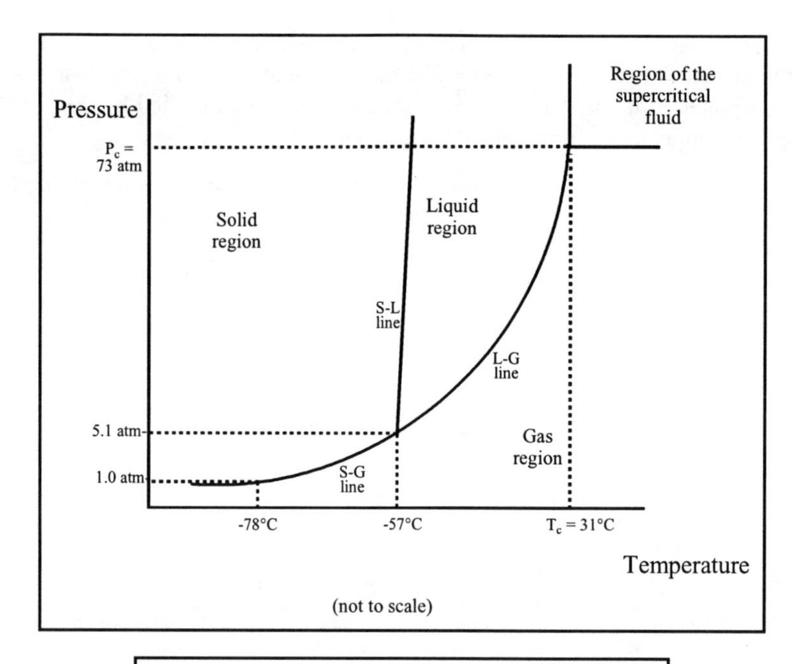


Figure 12.2 The phase diagram of carbon dioxide.

Figure 12.2 shows the phase diagram of carbon dioxide. It is not sketched to scale. Pressure is plotted on the vertical axis; temperature is plotted on the horizontal axis. Observe that there are three labeled **phase regions** (solid, liquid, and gas) separated linea. The lines are named for the regions they separate; along those lines the two separated phases are in equilibrium. Thus, the S-L line separates the solid and liquid regions of the phase diagram. The point at which the S-G, S-L, and L-G lines meet is the triple point. The **triple point** on a phase diagram is the unique temperature and pressure at which all three phases are in mutual equilibrium. For carbon dioxide, the triple point is at -57°C and 5.1 atm.

Observe that at a pressure of 1 atmosphere the S-G line has a temperature of -78°C. This point on the S-G line represents the normal sublimation of solid carbon dioxide (dry ice); carbon dioxide ordinarily sublimes because its vapor pressure in the air is well bellow the vapor pressure of the solid. **Sublimation** is the process in which a solid turns directly to a gas. Carbon dioxide sublimes at low temperatures. However at ambient temperature, in a carbon dioxide fire extinguisher the $CO_2$ is liquid. For example, by locating the point -10°C and 60 atm on Figure 12.2, you can see that carbon dioxide is in its liquid region. That point is located about where the "n" in "liquid region" is.

In the upper right of the diagram is the region of the **supercritical fluid** (SCF). The bottom left corner of this region is defined by the critical constants of carbon dioxide and its ΔH of vaporization is zero here. In the SCF region there is no distinction between liquid and gas; it is a single phase. The **critical temperature** of a substance is the temperature below which a gas can be condensed to a liquid merely by compression; gases above their critical temperature cannot be pressure condensed. The **critical pressure** of a substance is pressure at which the liquid and gas states of a substance are in equilibrium at the critical temperature of the substance. The critical temperature ($T_c$) of carbon dioxide is 31°C and the critical pressure ($P_c$) is 73 atmospheres.

The values of a substance's critical temperature and pressure can be directly related to the intermolecular forces within it. The critical temperatures of the gaseous elements can be used to illustrate this point, as shown in Table 12.1.

| Table 12.1 **Critical Temperatures of Gaseous Elements** | | | |
|---|---|---|---|
| Element | Boiling Point (K) | Critical T (K) | Number of electrons |
| He | 4.23 | 5.26 | 2 |
| Ne | 27.1 | 44.5 | 10 |
| Ar | 87.3 | 150.9 | 18 |
| Kr | 119.8 | 209.4 | 36 |
| Xe | 165 | 289.8 | 54 |
| Rn | 211 | 378 | 86 |
| $N_2$ | 77.36 | 126.6 | 14 |
| $O_2$ | 90.18 | 154.8 | 16 |
| $F_2$ | 53.5 | 85 | 18 |
| $Cl_2$ | 172 | 239 | 34 |
| $Br_2$ | 266 | 332 | 70 |
| $I_2$ | 387 | 458 | 106 |

As you can see from Table 12.1, the critical temperatures of the gaseous elements are invariably somewhat higher than their normal boiling point. Because monatomic and diatomic elements are composed of non-polar atoms or molecules, both their boiling point and critical temperature are controlled by dispersion (temporary dipole) forces. Note from the table that the greater the number of electrons in the atom or molecule the higher its critical temperature. The interpretation of this behavior is that those molecules with more electrons have bigger temporary dipole forces, they are more polarizable, and so it takes a higher temperature to prevent their molecules from "sticking together" and condensing into the liquid state.

The data of Table 12.1 could well have been included in Table 10.2, along with the other properties that were listed in Table 10.2 to illustrate the nature of the intermolecular forces in liquids. However, when we discussed Table 10.2 we had not yet encountered the concept of critical properties.

**The Use of Supercritical Fluids in Industry**

A supercritical fluid (SCF) has properties that are intermediate between those of the liquid and vapor states of the substance. Near its critical point the properties of a substance are particularly dependent on pressure and temperature. Thus, if the pressure and temperature are varied, the properties of the supercritical fluid can be changed continuously over a wide range between those typical of the liquid and those typical of the gas. Analytical chemists and chemical engineers are exploiting supercritical fluids to perform useful separations.

Carbon dioxide forms a supercritical fluid that is widely used in the food industry. The critical temperature (31°C) and critical pressure (73 atm) of carbon dioxide can be easily achieved in industrial processing equipment. Above the critical point, the solubilities of organic compounds change rapidly as a function of pressure and temperature. As the pressure is raised, carbon dioxide becomes more liquid-like, and its solvent properties improve. Liquid $CO_2$ and supercritical $CO_2$ are called dense-phase $CO_2$. Using dense-phase $CO_2$ it is possible to extract heat-sensitive compounds at relatively low temperatures with minimal degradation. The dissolved substances can be easily recovered by allowing the supercritical fluid to vaporize. Carbon dioxide extraction is favored in the food industry because it leaves no residues in the product, and doesn't create air or water pollution. Any residual carbon dioxide escapes as a gas.

The best known commercial use of supercritical carbon dioxide in the food industry is for the removal of caffeine from coffee beans, where it has replaced the undesirable chlorinated solvent, dichloromethane ($CH_2Cl_2$). The beans are submerged in supercritical carbon dioxide and the caffeine dissolves in the fluid. Later the caffeine is recovered. Supercritical carbon dioxide is also used to extract flavor agents from hops for use in beer production, to extract coloring and flavoring agents from spices, and to extract oils and fats from seeds.

In analytical chemistry, supercritical fluids are being widely used to extract compounds selectively from complex matrices. Examples include pesticide residues in soils, and additives in such products as tobacco, chewing gum, and candy bars. These methods are well developed and have a high level of acceptance in the community of analytical chemists.

Proponents of SCF extraction predict a bright future for supercritical fluids. "The future of supercritical carbon dioxide is just around the corner," says one enthusiast. He predicts that it will be used for dry cleaning, for making high strength concrete, for extracting cancer drugs from natural substances, and for preserving old books by flushing them out.

Commercial applications of dense-phase carbon dioxide began in the early 1980's. Dry-cleaning and paint spraying are two processes that are already commercialized.

## The Phase Diagram of Water

Because water is a ubiquitous substance its phase diagram is especially important. The conventional scale distorted water phase diagram is shown in Figure 12.3. By definition, the temperature at water's triple point, where all three phases meet at the lower left, is 273.16 K or 0.01°C (exactly). The difference, -273.15°C, is defined as the absolute zero of temperature. These definitions resulted from the work of the General Conference on Weights and Measures (CGPM), and were adopted in 1967. The pressure at water's triple point is 4.585 mmHg (Torr). Applying the CGPM definition makes water's normal boiling point an *experimental* quantity, and at 99.975°C it's very close to 100°C at a pressure of 1 bar. In the same way, water's freezing point is also an experimental value. The measured freezing point of water is $0.0000 \pm 0.0001$°C at 1 bar pressure.

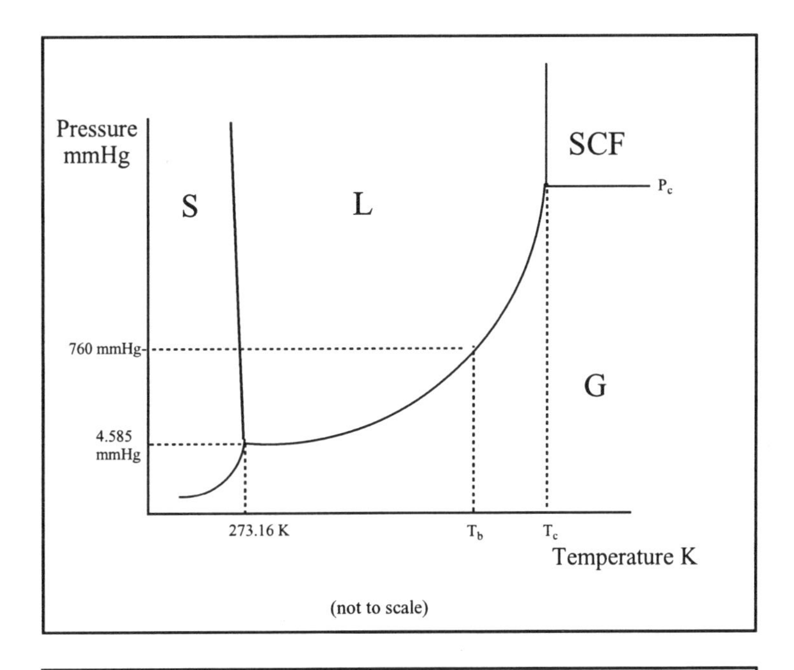


Figure 12.3 The phase diagram of water. $T_b$ is water's boiling point of 99.975°C and $T_c$ is water's critical temperature of 374°C. $P_c$ is water's critical pressure of 218 bar.

Water's critical temperature is 374°C and its critical pressure is 218 bar. Recall that the critical temperature of a gas is the highest temperature at which it can be condensed to a liquid by compression.

## Changes of Phase on Water's Phase Diagram

Figure 12.4 shows water's phase diagram with many of the phase changes annotated by arrows. You are already familiar with most of these changes because your everyday life brings you into frequent contact with water.

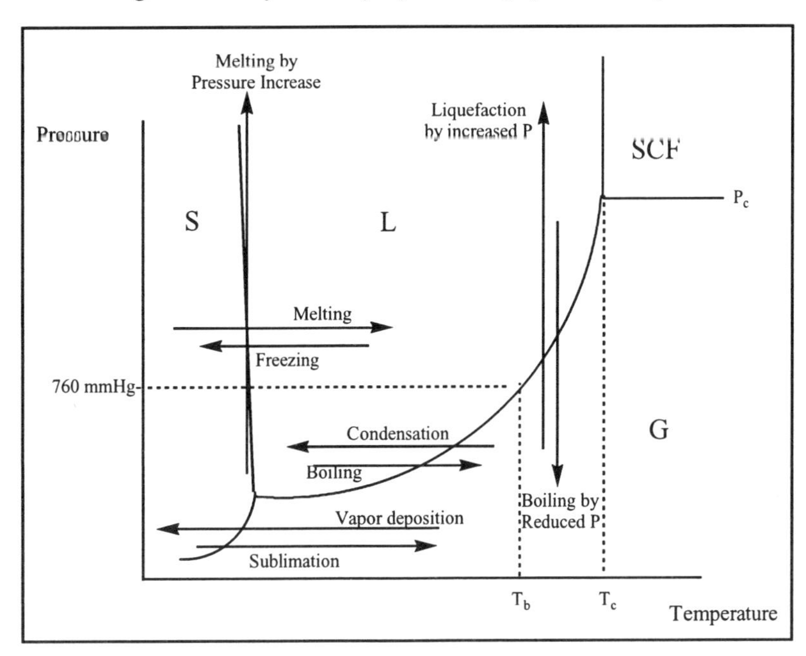


Figure 12.4 Changes of phase illustrated on the phase diagram of water.

The arrows show there are two ways to vaporize water. The black arrow labeled "vaporization" represents the normal boiling of water by raising its temperature—moving horizontally from left to right on the diagram corresponds to increasing the temperature at a fixed pressure. The gray arrow pointing downwards at the right of the diagram shows that water can also be vaporized by lowering its pressure at a fixed temperature. Vaporization at reduced pressure happens whenever water or a solution is discharged into a vacuum—such as when liquid waste is released from a space vehicle. On earth, we see the results of the same effect if we click a butane lighter or press the valve on a household aerosol product. Moving vertically from bottom to top on the diagram corresponds to increasing the pressure at a fixed temperature. A pressure cooker works by raising both the pressure *and* the temperature at which food is prepared; it is the increased temperature that shortens cooking times compared with conventional cooking.

At pressures below water's triple point pressure of 4.585 Torr, ice and gaseous water are in equilibrium without the involvement of the liquid phase. Thus, at pressures below 4.585 Torr, ice sublimes and water vapor undergoes solid deposition. Such changes are frequently seen during winter. Beautifully formed, leaf-like ice crystals on a car's windows, seen on a winter morning, form by vapor deposition. **Vapor deposition** is the process whereby a gas turns directly to a solid. Similarly, during the daylight hours, the same ice crystals may disappear without ever melting—the ice crystals will begin to sublime as the temperature rises during the day because at those warmer temperatures the vapor pressure of ice will exceed the partial pressure of water vapor in the atmosphere.

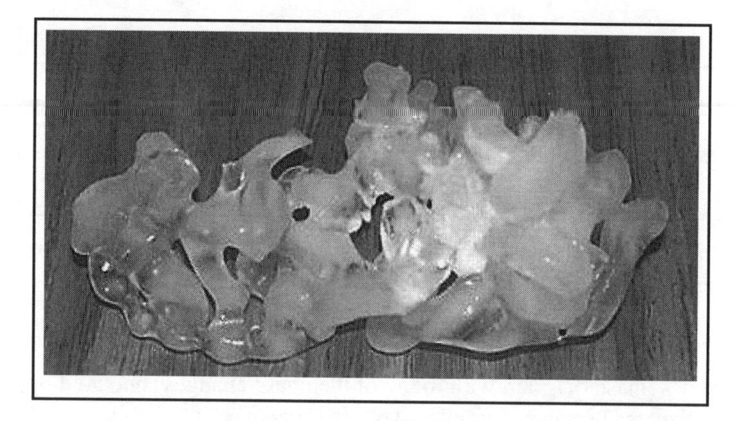

Figure 12.5 "Sculpted ice." Ice cubes stored too long in a freezer become congealed and sublime away leaving rounded surfaces.

The phenomenon of ice sublimation is familiar to cooks who store frozen foods too long and find that the foods suffer "freezer burn"; you also see it in the smooth, sculpted ice cubes (Figure 12.5) that have been stored too long in your freezer. **Freezer burn** is the desiccation of foods caused by ice's sublimation. When done carefully and deliberately, low temperature food desiccation is called **freeze drying**.

Water's phase diagram is unusual in having an S-L line with a negative slope. Thus, at sufficiently high pressures ice will melt. It makes a good story to suggest that pressure-induced melting of ice is the basis for skating, but it takes over 100 atm to lower the melting point of water to - 1°C, so skating is probably more a consequence of ice's surface molecular properties and the effect of friction. However, it does seem to be true that skating on ice depends on the formation of a liquid or liquid-like layer to lubricate the blade as it passes across the ice's surface. A modern view regards this surface layer as being composed of only about half as many water molecules as in a layer in the bulk ice, with those molecules vibrating up and down but not side to side. If water is an odd liquid, then ice is also an odd solid. [❁kws +"Ice skating" +physics]. On the other hand, if a strong metal wire is weighted, and laid across a large block of ice, it cuts its way through the block leaving the block intact. [❁kws +regelation]

**Example 12.1**: Use the phase diagrams of carbon dioxide and water in Figures 12.2 and 12.3 to determine their stable phases at 1 atm pressure and 50°C, at 5 atm and -10°C, and at 30 atm and -25°C.

Strategy: Locate each point on the appropriate phase diagram and see in which phase that point occurs. The results are shown in the table below:

| Conditions | Stable Phase | |
|---|---|---|
| | Carbon dioxide | Water |
| 1 atm and 50°C | gas | liquid |
| 5 atm and -10°C | gas | solid |
| 30 atm and -25°C | liquid | solid |

## The Phase Diagram of an Aqueous Solution

When a nonvolatile solute is dissolved water, the properties of the solution are different from the properties of pure water. These differences are summarized by Figure 12.6. The L-G line drops lower than the same line of the phase diagram of pure water, while the S-L line moves to the left. These shifts in the phase diagram summarize and explain the difference between the properties of pure water and those of an aqueous solution. It is traditional to describe the changes by stating the properties of the solution compared with the properties of pure water. Figure 12.6 forms the basis for our investigation of the four colligative properties.

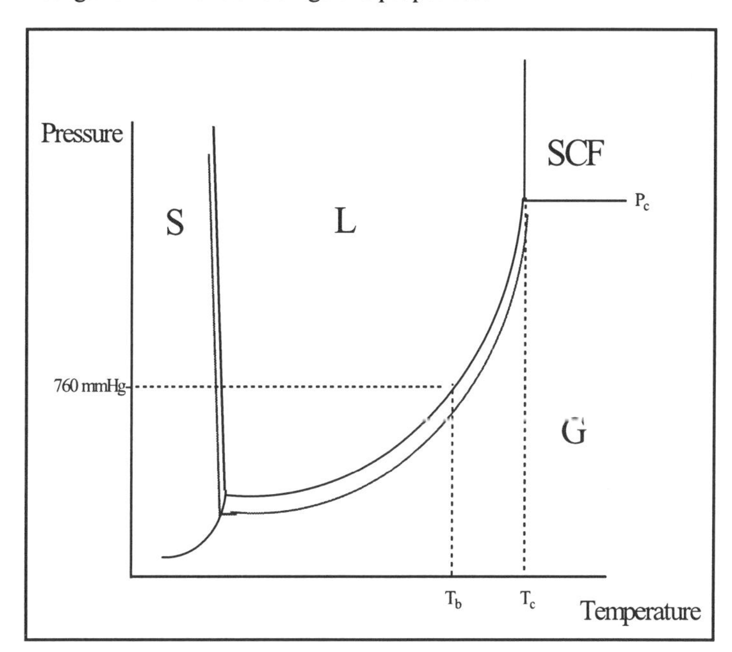


Figure 12.6 The phase diagram of an aqueous solution of a nonvolatile substance in water. The nonvolatile solute here is immiscible with ice, and freezing the solution causes pure ice to form.

The extent (effect) of a colligative property is controlled (at a given T) primarily by two factors: 1. the concentration of the solute, and 2. the amount of ionization (if any) of the solute. Historically, chemists have taken advantage of colligative property phenomena to measure molar weights.

A key to thinking about colligative property phenomena is to realize that a solute dissolved in a solvent lowers the concentration of the *solvent* itself. For example, the vapor pressure of salt water is lower than the vapor pressure of pure water because pure water is more concentrated (in terms of water) than salt water. We will return to this point shortly.

## Phase Diagrams and Metallurgy

Metals exhibit a broad range of properties and are used in many different ways. (Much of what is said in this section also applies to ceramic materials.) Metals' single most important use is as high strength materials for construction, industrial, and manufacturing applications.

Metals form an interesting and important group of engineering materials because they have valuable mechanical and physical properties, can be easily fabricated into various shapes, and are readily available. Metals are recyclable and can be remelted and refashioned into new products after their previous useful life-cycle is complete. Students in materials science who concentrate on metals typically study extractive metallurgy, crystallography, phase diagrams, corrosion technology, and other aspects of metals technology. Phase diagrams are basic tools for the metallurgist. The design of new alloys requires knowledge of unexplored areas of multi-component phase diagrams. The experimental determination of these diagrams is time-consuming and difficult. Moreover, the number of potentially interesting phase diagrams is tremendous. Hence, metallurgy today relies heavily on computer simulations to do the thermodynamic analysis needed to relate bulk metal properties to variables such as pressure, temperature, stress, and composition. But making a desirable alloy is not just an exercise in thermodynamic equilibrium. In real life, materials are seldom in their equilibrium state. As liquid metals cool and solidify, they tend to grow microcrystals at different rates. It is for this reason that a metal made by rapid quenching may have dramatically different properties from the same metal made by slow cooling.

Because of its widespread use and historical role, iron is the most important single metal. It combines with carbon to make carbon steel, and steel is perhaps the most important metal. It would be difficult to overestimate the historical and practical importance of studies of iron-carbon alloys; learning how to control and use the properties of steels has been one of the major technological triumphs of the past 300 years. Photomicrographs of polished carbon steel surfaces are shown in Figure 12.7. Metallurgists study such photomicrographs to learn about the microstructure of metals and alloys. You can see for yourself that steel has a complicated microstructure. A rather over simplified portion of the temperature-composition diagram of carbon steel is shown in Figure 12.8. Note that there are various solid phases such as the $\alpha$, P, $\gamma$, and eutectic phases.

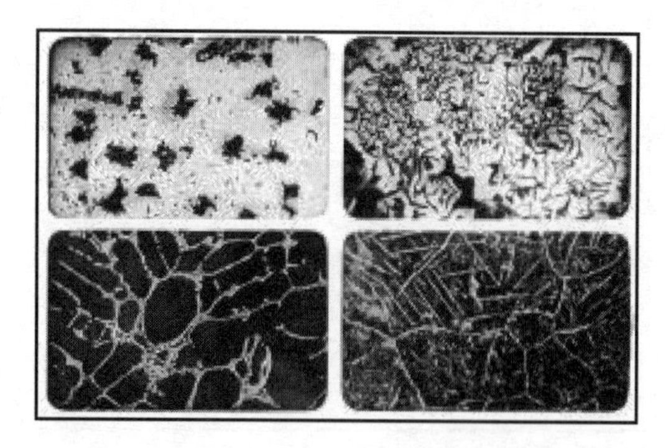

Figure 12.7 Grain structures in metals. Examining polished metal surfaces under a microscope is one of the time honored procedures of metallurgy.

A eutectic phase is a solid formed with the same composition as the liquid melt from which it solidifies. The curious name comes from the Greek *eutektos* meaning easily melted. For mixtures of particular metals, the eutectic phase is the solid mixture with the lowest melting point. Ordinary solder is a solid alloy with a low melting point.

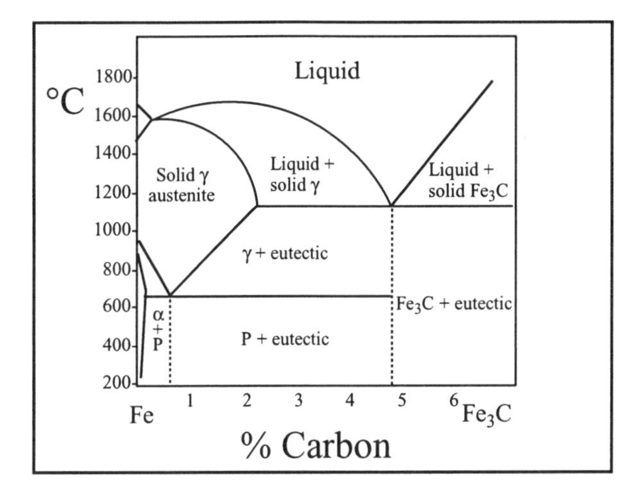


Figure 12.8 The temperature-composition iron-carbon phase diagram. This diagram is the basis for understanding the properties of steel.

## Section 12.2 Solutions Revisited

Before we take up a detailed discussion of colligative properties we'll review what we know about solution concentrations. Solution concentrations were introduced in Chapter 3 when we met the units mass percentage and molarity. In Chapter 10 we discussed the distillation of liquid mixtures and saw that Raoult's law is written in terms of mole fractions. In addition to these three units chemists also use the unit of molality. **Molality** (m) is solution concentration expressed as moles of solute divided by the mass of the solvent in kilograms.

$$\text{molality (m)} = (\text{moles solute}) \div \text{kg of solvent}$$

In water, which has a density of close to 1.0 g $mL^{-1}$, molality and molarity have almost the same value if the solute is dilute. Dilute is not precisely defined; it means a few grams of solute in a liter or kilogram of water. In concentrated solutions molality is greater than molarity. In our discussion here of colligative properties we'll restrict ourselves to aqueous solutions, though there are, of course, many non-aqueous solvents and solutions. Example 12.2 shows molarity and molality calculations.

**Example 12.2**: A 24.0% by weight solution of perchloric acid ($HClO_4$) has a density of 1.16 g $mL^{-1}$. Calculate the molarity and molality of the solution.

Strategy: For molarity, assume an exactly one liter sample of the solution. Use the density to find that sample's mass, and the given percentage to find the mass of solute; finally use the molar weight (100.5 g $mol^{-1}$) to calculate molarity. For molality use 100.0% - 24% as the percentage of water solvent and thus obtain the mass of solvent. Express that in kg and plug into the molality formula.

For molarity:

$$1.000 \text{ L of the solution weighs } 1.000 \text{ L} \times 1.16 \text{ g mL}^{-1} \times 1000 \text{ mL L}^{-1} = 1160 \text{ g}$$

$$\text{of this, 24.0\% is solute, so the mass of HClO}_4 = 1160 \text{ g} \times (24.0/100) = 278 \text{ g}$$

$$\text{Molarity} = (\text{moles solute}) \div \text{Volume of solution in liters}$$

$$\text{Molarity} = (278 \text{ g} \div 100.5 \text{ g mol}^{-1}) \div 1.000 = 2.77 \text{ molar}$$

For molality:

$$\text{In our 1.00 liter sample, the mass of water is } (76.0/100) \times 1160 \text{ g} = 882 \text{ g or } 0.882 \text{ kg}$$

$$\text{Molality} = (\text{moles solute}) \div \text{mass of solvent in kilograms}$$

$$\text{Molality} = (278 \text{ g} \div 100.5 \text{ g mol}^{-1}) \div 0.882 \text{ kg} = \underline{3.14 \text{ molal}}$$

Many substances are soluble in water, and you already know that it is intermolecular forces between the solvent and solute that control whether or not a substance will or will not dissolve in water. Substances that dissolve in water fall into one of three categories: 1. Polar substances whose molecules interact with the dipole forces of water; acetone is an example of this type. 2. Hydrogen bonding substances whose molecules can form hydrogen bonds with water molecules; alcohol and sugar are examples of this type. 3. Ionic substances that dissolve in water with the formation of both positive and negative ions; salt (NaCl), and magnesium sulfate ($MgSO_4$) are examples of this type. The properties of aqueous solutions are extremely important. For example, we need to study aqueous solutions to understand our own bodies, rivers, lakes, the oceans, etc. But before going on to those we'll briefly consider gas-in-liquid solutions.

## Gas-in-Liquid Solutions

That gases dissolve in water has important consequences. For example, fish rely on dissolved oxygen gas. Gases such as oxygen become more soluble in water at low temperatures and high pressures. An everyday example of the temperature effect on air's water solubility occurs when you take a glass of cold tap water to bed and finish only part of it before falling asleep. In the morning, the presence of tiny air bubbles in the glass reveals that as the water warmed up to room temperature the dissolved gas it contained passed out of solution. The solution became over saturated as it warmed. In a related phenomenon, fisherman are aware that in summertime fish favor the cooler, deeper regions of lakes where there is more dissolved oxygen.

The effect of pressure on a gas's water solubility is well illustrated by carbonated drinks. The slight hiss you hear on opening a can of soda shows that the contents were under pressure, and the vigorous bubbling confirms that lowering the pressure significantly decreases the solubility of the carbon dioxide in the colored, flavored water that is the pop. Of course, vigorously shaking a carbonated drink prior to opening it shows that many carbonated beverages are supersaturated. The merry antics that go on with sparkling wines in the locker rooms of celebrating sports teams show that effect rather convincingly. In general, the greater the pressure the more soluble is a gas in a liquid. At low pressure the extent of solubility is directly proportional to the pressure—that's Henry's law. The law states

$$X_g = kP$$

where $X_g$ is the mole fraction of the gas in the liquid, P is the pressure, and k is a constant called the Henry's law constant that depends on the temperature and the liquid-gas pair being studied. The English chemist William Henry (1775-1836) measured the solubilities of several gases in water in the early 1800's. Some measured values of the Henry's law constant are given in Table 12.2 and 12.3.

| Table 12.2 **Henry's Law Constant for Several Gases in Water at 25°C** | |
|---|---|
| Gas | k<br>mole fraction $Torr^{-1}$ |
| Nitrogen | $1.54 \times 10^{-8}$ |
| Oxygen | $3.03 \times 10^{-8}$ |
| Hydrogen | $1.87 \times 10^{-8}$ |
| Carbon dioxide | $8.00 \times 10^{-7}$ |

| Table 12.3 **Henry's Law Constant for Oxygen in Water at Various Temperatures** | |
|---|---|
| Temperature<br>°C | k<br>mole fraction $Torr^{-1}$ |
| 0 | $5.18 \times 10^{-8}$ |
| 5 | $4.54 \times 10^{-8}$ |
| 10 | $4.20 \times 10^{-8}$ |
| 15 | $3.62 \times 10^{-8}$ |
| 20 | $3.28 \times 10^{-8}$ |
| 25 | $3.03 \times 10^{-8}$ |
| 30 | $2.76 \times 10^{-8}$ |

Henry's law summarizes experimental observations that can be understood in terms of a process in which gas phase molecules are in dynamic equilibrium with dissolved molecules in a saturated solution. Increasing the partial pressure of the gas increases the rate at which its molecules strike the surface of the liquid and hence the rate at which the gas dissolves. Equilibrium is established when dissolving rate equals the rate at which molecules leave the solution. Thinking about this situation you'll see that the vaporization rate increases in direct proportion to the concentration of dissolved molecules—which is the conceptual basis for Henry's law.

**Example 12.3**: Calculate the concentration of a saturated solution of oxygen dissolved in water at 10°C. Assume that the pressure is 1.0 atm and that air contains 20.0 mole% oxygen. State the answer in parts per million.

Strategy: Use Table 12.3 to find the appropriate value of Henry's law constant. It's $4.20 \times 10^{-8}$ mole fraction $Torr^{-1}$. Take the pressure as 0.20 atm, that's the partial pressure of oxygen in the air. Substitute these two values into the Henry's law expression and solve for the mole fraction of oxygen. Finally, do a units conversion to change from mole fraction to ppm.

$$X_g = kP$$

$$X_g = 4.20 \times 10^{-8} \text{ mole fraction Torr}^{-1} \times 0.20 \times 760 \text{ Torr atm}^{-1} \quad = \quad 6.4 \times 10^{-6} \text{ mole fraction}$$

If you take the reciprocal of the mole fraction you get the number of moles of water in which one mole of oxygen is dissolved. Thus, there is one mole of dissolved oxygen in 160,000 moles of water. Strictly it's in 160,000 + 1, but obviously we ignore that 1.

To convert that to ppm I'll use the relationship $\text{ppm} = \frac{\text{mass of solute}}{\text{mass of solution}} \times 10^6$

Where the masses derive from the relationship mass = moles × molar weight

Doing that, $\text{ppm } O_2 = \frac{1 \text{ mol} \times 32 \text{ g mol}^{-1}}{160000 \text{ mol} \times 18 \text{ g mol}^{-1}} \times 10^6 \quad = \underline{11 \text{ ppm}}.$

## Henry's Law and Killer Lakes

The idea of a killer lake at first sounds like something out of Monty Python. But there are three such crater lakes, and they are a serious natural hazard for the people who live downslope from them. Two of these lakes, Nyos and Monoun, are in the west African country of Cameroon. On August 21st, 1986 Lake Nyos exploded, killing 1,700 people and 3,000 cattle. The cause of death: asphyxiation, when a dense, rolling blanket of carbon dioxide erupted from the lake and swept down pushing the air away, leaving people and creatures without oxygen.

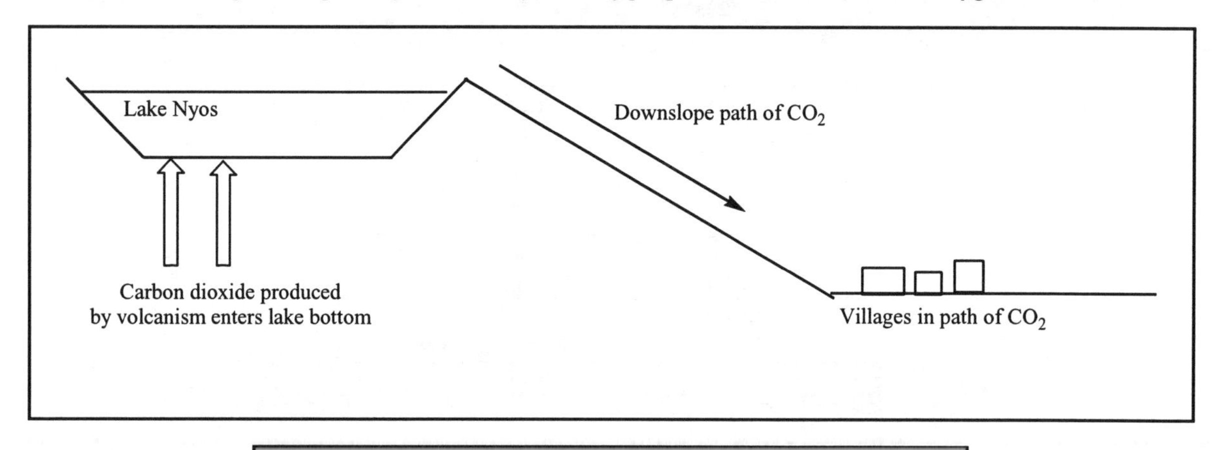


Figure 12.9 Schematic diagram of the 1986 Lake Nyos eruption.

Lake Nyos is fed with carbon dioxide and other gases from geothermal springs at its bottom, 208 meters from its surface. Warm water normally rises, but in Lake Nyos it remains trapped in the deep because of its heavy mineral content. In the deep regions the carbon dioxide is held in solution by the water pressure of about 20 atm. Every thirty years (it's estimated) sufficient carbon dioxide has entered the lake to create a condition of saturation. After saturation, some triggering event such as a rock slide or heavy rain, allows the deepest layers to rise, and out gassing begins. Once started it's an explosive, runaway process, like opening a shaken soda can but on a monumental scale.

The carbon dioxide exists in a natural acid base equilibrium with water according to the equation $H_2O + CO_2 \rightleftarrows H^+ + HCO_3^-$. That equilibrium reaction makes the lake acidic, and more acidic at depth where the carbon dioxide concentration is highest. Water chemistry data from different depths of Lake Nyos taken during May 1998 are shown in Table 12.4.

Table 12.4 **Lake Nyos Conditions, May 1998**

| Depth (meters) | Estimated Pressure (bar) | pH | Oxygen (mg $L^{-1}$) | $CO_2 + HCO_3^-$ as mg $L^{-1}$ $HCO_3^-$ |
|---|---|---|---|---|
| 0 | 1.0 | 8.7 | 8.1 | 39 |
| 15 | 1.5 | 6.9 | 6.1 | 40 |
| 30 | 3.0 | 6.3 | 2.3 | 42 |
| 50 | 4.8 | 5.5 | 0 | 370 |
| 100 | 9.6 | 5.3 | 0 | 550 |
| 150 | 15 | 5.2 | 0 | 690 |
| 200 | 19 | 5.0 | 0 | 1,100 |
| 208 | 20 | 5.0 | 0 | 1,200 |

The data in Table 12.4 fully support our discussion of the water conditions in Lake Nyos. The pH falls steadily in accordance with the acid base equilibrium, which shifts further to the right with increasing carbon dioxide concentration. The carbon dioxide concentration steadily increases, roughly in accord with Henry's law equation, and the oxygen concentration drops to zero about 40 meters below the surface. You are invited to graph these data in one of the end of chapter problems. The data represent a reasonable experimental proof of Henry's law.

## Section 12.3 Colligative Properties

Recall that there are four related colligative properties: boiling point elevation, vapor pressure lowering, freezing point depression, and the formation of osmotic pressure. We'll take them up one by one. Traditionally, they are handled by comparing the solution property to the same property of the pure solvent.

In dilute solutions (say, less than 1% solute by mass) the change of the property from pure solvent to solution follows fairly reliable, simple laws. As the solutions become more and more dilute the behavior follows the laws closer and closer. In the limit of "infinitely dilute solution," the solutions become "ideal"—they follow the laws precisely. In contrast, at higher concentrations there is a breakdown of the simple linear relationship between solute concentration and the change in the solvent's properties

If you recall our study of the ideal gas equation in Chapter 4, you will remember that the pressure of a gas depends directly on the number of molecules it contains. The equation $PV = nRT$ shows that, *ceteris paribus* (Latin for "other things being equal"), halving the number of gas molecules halves the pressure. The colligative properties behave in a parallel manner; the extent of the change in any colligative property depends on the number of particles of solute and *not* on their mass.

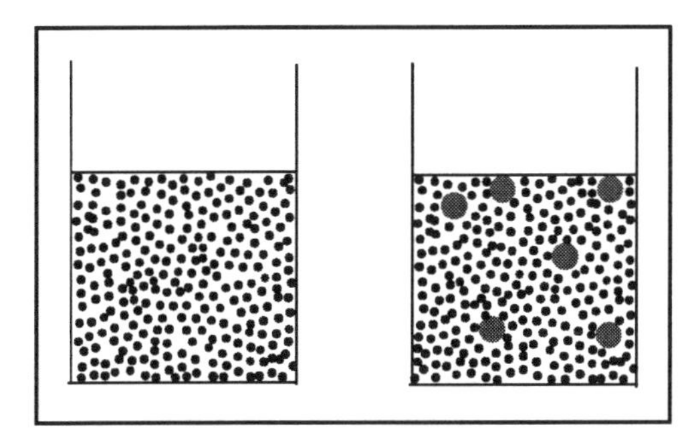

Figure 12.10 Adding any solute to any solvent *lowers* the concentration of the *solvent*. There are fewer solvent molecules in the same volume on the right than on the left.

The dependence of colligative phenomena on the number of dissolved particles is illustrated by Figure 12.10. A pure solvent has a definite concentration that depends on its density and molar weight. It is 55.4 molar in the case of pure water at room temperature ($998\ g\ L^{-1} \div 18.0\ g\ mol^{-1} = 55.4\ M$). *Any* added solute will lower water's molarity below 55.4 M, and this reduced solvent concentration will help us understand colligative properties. The colligative properties do not depend on the size of the dissolved particles—as we'll soon see, it is how many there are that counts.

### Vapor Pressure Lowering

We saw in Chapter 10 that ideal liquid mixtures, with vapors that obey the ideal gas law, follow Raoult's law. For a two-component liquid mixture Raoult's law is written $P_{tot} = P_A^0 X_A + P_B^0 X_B$, where $X_A$ and $X_B$ are the respective mole fractions of A and B in the liquid phase and $P_A^0$ and $P_B^0$ are the vapor pressures of pure A and B at the specified temperature.

When a solution is prepared using a solid solute or a nonvolatile liquid solute, the vapor pressure of the solvent is reduced in proportion to the amount of dissolved solute. Raoult's law in this case contains only one term and may be written:

$$P = P^0 X$$

where P is the actual vapor pressure, $P^0$ is the vapor pressure of the pure solvent at the specified temperature, and X is the mole fraction of the *solvent*. Experimental results of vapor pressure lowering as a function of solution concentration in molarity units are graphed in Figure 12.11.

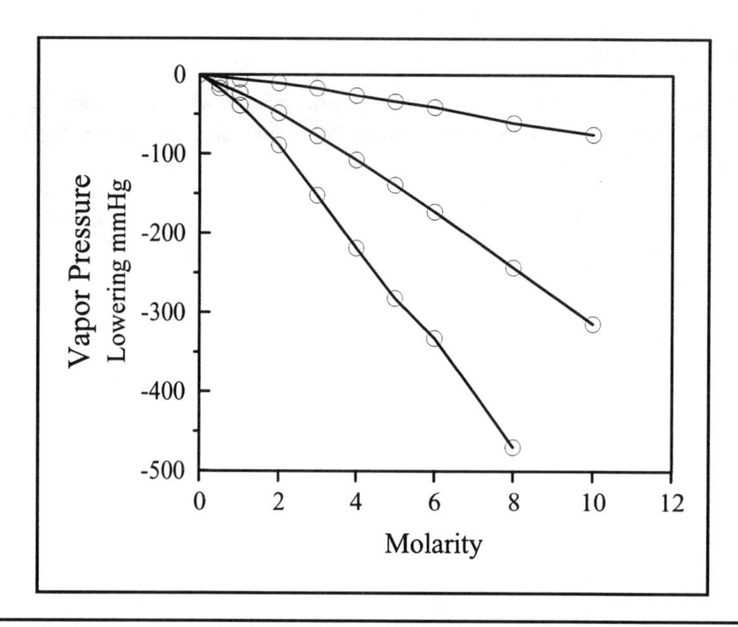


Figure 12.11 Vapor pressure lowering in water by ethylene glycol (top), sodium hydroxide, NaOH (middle), and cobalt(II) nitrate, $Co(NO_3)_2$ (bottom).

Note that the three different solutes have different sized effects. At any particular solute molarity ethylene glycol, a nonelectrolyte, has the least effect; sodium hydroxide, a 1:1 electrolyte that ionizes to give two moles or particles per mole of solute ($NaOH \rightarrow Na^+ + OH^-$), has roughly double the effect of ethylene glycol; and, cobalt(II) nitrate, which ionizes to give three moles of ions per mole of solute ($Co(NO_3)_2 \rightarrow Co^{2+} + 2\ NO_3^-$), has roughly three times the effect of ethylene glycol.

There are two main conclusions we can draw from the data in Figure 12.11: 1. vapor pressure lowering depends on the number of particles that a solute creates when it dissolves in water, and 2. there is a clear relationship between increasing solute concentration and decreasing vapor pressure—though it is not a simple, straight line relationship, except when approaching infinite dilution toward the left of the diagram. The use of Raoult's law to calculate vapor pressure lowering is shown in Examples 12.4 and 12.5.

**Example 12.4**: Calculate the vapor pressure at 35°C of a solution that is 12.0% by weight of glycerol ($C_3H_8O_3$, MW = 92.0 g mol$^{-1}$) in water (MW = 18.0 g mol$^{-1}$). The vapor pressure of water at 35°C is 42.4 mmHg.

Strategy: Recognize that glycerol is a nonvolatile liquid so that a one term Raoult's law calculation is needed. For convenience, assume a 100 g sample to calculate the mole fraction of water (the solvent).

Mole fraction = (88.0/18) ÷ (88.0/18 + 12.0/92.0) = 0.974 mole fraction

$P = P^o X$ = 42.4 mmHg × 0.974 = <u>41.3 mmHg</u>

**Example 12.5**: Urea, $CO(NH_2)_2$, (MW = 60.0 g mol$^{-1}$) is the chief excretory product of humans and other mammals. It is also an important industrial fertilizer. The vapor pressure of a urea solution can be estimated using Raoult's law. At 25°C the vapor pressure of pure water is 23.8 mmHg. Estimate the vapor pressure of a 45.0% solution by weight of urea in water at 25°C. Urea is a solid, nonvolatile material.

Strategy: Recognize that urea is a nonvolatile liquid so that a one term Raoult's law calculation is needed. Assume a 100 g sample to calculate the mole fraction of water (the solvent).

$$X_1 = (n_1)/(n_1+n_2) \quad P_1° = 23.8 \text{ mmHg}$$

$$\text{Mole fraction} = (55.0/18.0)/(55.0/18.0 + 45.0/60.0) = 0.804 \text{ mole fraction}$$

Substitute into $P = P°X$ where $P° = 23.8$ mmHg and $X = 0.804$

Vapor pressure = 23.8 mmHg × 0.804 = 19.1 mmHg

## Boiling Point Elevation and Freezing Point Depression

Whenever a nonvolatile solute is dissolved in a solvent, the boiling point of the solution is higher than the boiling point of the pure solvent; the boiling point is said to be elevated. Similarly, the freezing point of the solution is lower than the freezing point of the pure solvent (for the situations we'll consider here). The freezing point is said to be depressed. Boiling point elevation and freezing point depression measurements are traditionally related to the solute's molality. For dilute solutions, the equations are:

$\Delta T_b = k_b m$, the increase in boiling temperature is proportional to the solute's molality

$\Delta T_f = k_f m$, the decrease in freezing temperature is proportional to the solute's molality

where $T_b$ and $T_f$ are the liquids normal boiling and freezing points, $k_b$ is the **molal boiling point elevation constant** of any solute in a specified solvent, and $k_f$ is the **molal freezing point depression constant** of any solute in a specified solvent (values of $k_f$ are negative). $\Delta T_b$ is the increase in the solvent's boiling point caused by the solute. $\Delta T_f$ is the decrease in the solvent's freezing point caused by the solute. Once again for a colligative property we are dealing with a limiting law, and the straight line implied by these relationships occurs only in dilute solutions. As determined by experiment, each pure liquid solvent has its own characteristic values of $k_a$ and $k_b$; some of these values are shown in Table 12.5.

Table 12.5 **Molal Freezing Point and Boiling Point Constants**

| Liquid | Freezing Point °C | $k_f$ °C m$^{-1}$ | Boiling point °C | $k_b$ °C m$^{-1}$ |
|---|---|---|---|---|
| Water | 0 | -1.86 | 100 | 0.52 |
| Benzene | 5.5 | -4.90 | 80.1 | 2.53 |
| Phenol | 41 | -7.40 | 182 | 3.56 |
| Naphthalene | 80.3 | -6.94 | 218 | 5.80 |

What causes boiling point elevation? As we saw in the previous section, when a nonvolatile solute dissolves in a pure liquid, the vapor pressure is lowered. It follows, that to boil the solution it must be heated to a higher temperature than is needed to boil the pure liquid. This increase of temperature is called **boiling point elevation**; it's $\Delta T_b$, the boiling point of the solution minus the boiling point of the pure solvent. This elevated boiling temperature is reached when the vapor pressure of the *solution* becomes equal to the external pressure. Thus, vapor pressure lowering and boiling point elevation are different manifestations of the same underlying phenomenon.

Boiling point elevation is a useful practical phenomenon for automobile radiators. The same dissolved ethylene glycol that provides wintertime antifreezing also serves to raise the boiling point of radiator fluid. Example 12.6 shows a boiling point elevation calculation related to the vapor pressure change of a solution.

**Example 12.6**: Estimate the boiling point elevation of a 12.0 weight percent solution of glycerol in water at a pressure of 1 bar (760 mmHg). a. Use Raoult's law and the vapor pressure equation for water. b. Use water's molal elevation boiling point constant of 0.52 °C $m^{-1}$. Discuss the results.

Strategy both parts: Recognize that glycerol, a highly hydrogen bonded molecule, is a nonvolatile solute.

Strategy part a: The presence of glycerol will lower water's vapor pressure according to the equation $P = P^0X$. For the solution to boil, P must be 760 mmHg. So $P^0$ must be greater than 760 mmHg and we can figure that value if we find the mole fraction X and substitute into the equation. In other words, we use Raoult's law to figure what vapor pressure of pure water will give 760 mmHg when reduced by the mole fraction. (In Example 12.4 we already calculated for the same solution that X = 0.974). Finally, for part a, substitute the value of P into the empirical equation for water's vapor pressure near its boiling point ln P = 19.71 + 4881/T where P is in mmHg and T is in K. Doing that gives the temperature at which the vapor pressure of the solution is 760 mmHg, or its boiling point.

Substitute into $P = P^0 X$ where P = 760 mmHg and X = 0.974

$P^0 = (P/X) = 760. \div 0.974$ so, $P^0 = 780.$ mmHg

and plugging that into ln P = 19.71 - 4881/T gives:

$\ln 780. = 19.71 - 4881/T$

or $T = -4481 \div \{(\ln 780.) - 19.71\} = 374.0$ K

so the boiling point is elevated to 374.0 - 273.15°C = 100.9°C

The boiling point elevation is 100.9°C - 100°C = 0.9°C

Note that the lack of sufficient significant figures hurts in this case. The calculated boiling point elevation of 0.9°C is reported to just one significant digit.

Strategy part b: Calculate the molality of the glycerol in the solution and then substitute into the boiling point elevation expression, $\Delta T_b = k_b m$. Assume a kilogram of solution: it contains 880.0 g of water and 120.0 g of glycerol ($C_3H_8O_3$, MW = 92.0 g $mol^{-1}$).

molality (m) = (moles solute) ÷ kg of solvent

molality = (120.0/92.0) moles ÷ 0.880 kg = 1.48 molal

$\Delta T_b = k_b m = +0.52$ °C $molal^{-1} \times 1.48$ molal

= 0.77°C

The agreement between the two different calculations is satisfactory, recognizing that we are dealing with approximate equations throughout.

You've seen the treatment of the phenomenon of boiling point elevation A parallel treatment is used to deal with the phenomenon of freezing point depression. Freezing point depression is also an important practical phenomenon. We use it to protect our water cooled engines from wintertime damage. Any solute we'll consider here lowers the freezing point of a solvent. This lowering of temperature, $\Delta T_f$, is called **freezing point** depression; it's the freezing point of the pure solvent minus the freezing point of the solution.

What causes freezing point depression? The presence of the solute dilutes the solvent. So when you cool water with antifreeze in it, there are fewer water molecules in a given volume than in pure water; the water molecules can't so easily get assembled into ice. Example 12.7 shows a typical freezing point depression calculation.

**Example 12.7**: Calculate the freezing point of an aqueous solution that contains 35.0% of ethylene glycol.

Strategy: Calculate the molality of the solution and substitute that value into the expression $\Delta T_f = k_f\ m$. Ethylene glycol has a molar weight of 62.0 g $mol^{-1}$. Water's molal freezing point depression constant is -1.86°C $m^{-1}$. Assume a kilogram of solution: it contains 650.0 g of water and 350.0 g of ethylene glycol.

$$\text{molality (m)} = (\text{moles solute}) \div \text{kg of solvent}$$

$$\text{molality} = (350.0/62.0)\ \text{moles} \div 0.650\ \text{kg} = 8.68\ \text{molal}$$

$$\Delta T_f = k_f\, m = -1.86\ °\text{C molal}^{-1} \times 8.68\ \text{molal}$$

$$= \underline{-16.2°\text{C}}$$

Note: Because the melting point of water is 0°C then $\Delta T_f = T_f$

The experimental data for the freezing points of ethylene glycol-water solutions are shown in Figure 12.12. The concentration is expressed in two different units: In the upper half of the diagram the weight percentage of ethylene glycol is shown by the straight diagonal line; the corresponding molarity is shown on the horizontal axis. In the lower half of the diagram the curved line represents the actual freezing point of the solution (it's also the freezing point lowering because water's freezing point is 0°C).

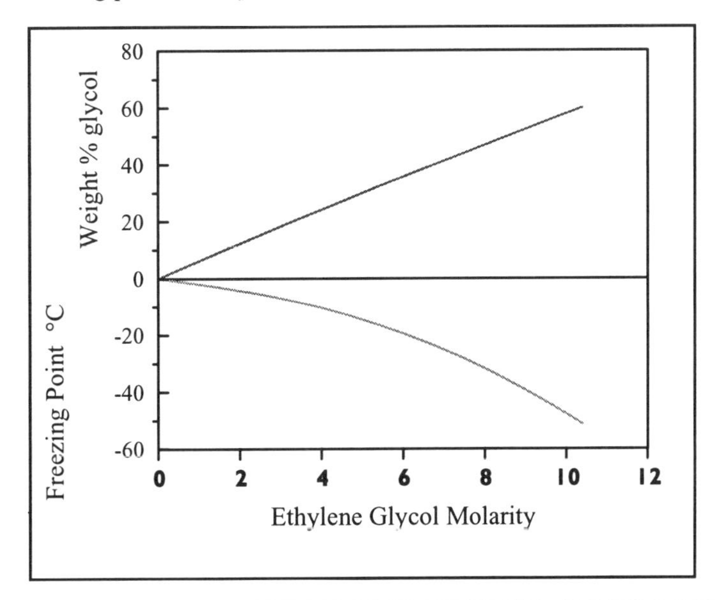


Figure 12.12 The freezing points of ethylene glycol-water solutions.

To work out how much ethylene glycol you'll need to protect a vehicle down to any desired temperature, you could use Figure 12.12 (though it's much easier to use a prepared table). For example, if you need protection down to -16°C you start at -16° on the y-axis and move horizontally to the right reaching about 6 molal. Then you move vertically upward to the diagonal line, and when your reach it, left again to the y-axis at about 35%. Doing this you can judge that the experimental freezing point of a 35% aqueous ethylene glycol solution is about -16°C, in good agreement with the value calculated in Example 12.7 (-16.2 °C) for a 35% solution.

We're all familiar with traditional ethylene glycol ($HO-CH_2-CH_2-OH$) antifreeze shown on the left in the picture in Figure 12.12. The white jugs on the right of the picture contain a low-toxicity antifreeze based on propylene glycol ($CH_3-CHOH-CH_2OH$). Ethylene glycol is a poisonous compound; propylene glycol has relatively low toxicity. The appearance on the market of a propylene glycol-based antifreeze is one of two relatively recent developments in the commercial antifreeze market. The other has been a movement to recycle ethylene glycol from radiator fluid.

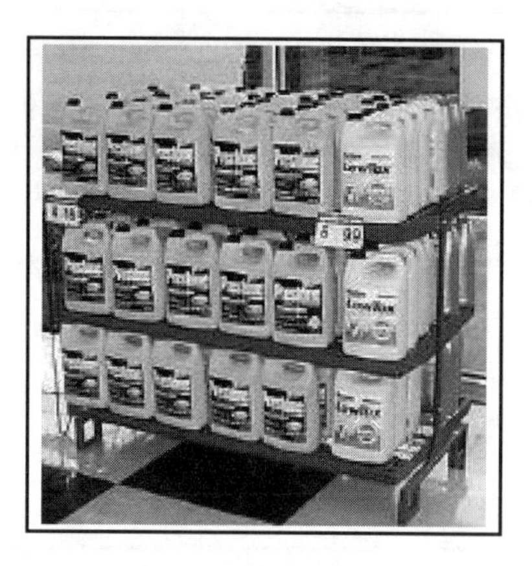

Figure 12.13 This store display offers ethylene glycol in the gray containers on the left and propylene glycol in the white containers on the right. Compared with ethylene glycol, propylene glycol is more expensive, much less toxic, and less effective as an antifreeze agent.

Windshield washer fluid is winterized using methanol as an antifreeze. Methanol is volatile but, that's not too important in winter. Also, methanol is a fair solvent and helps to clean the windshield.

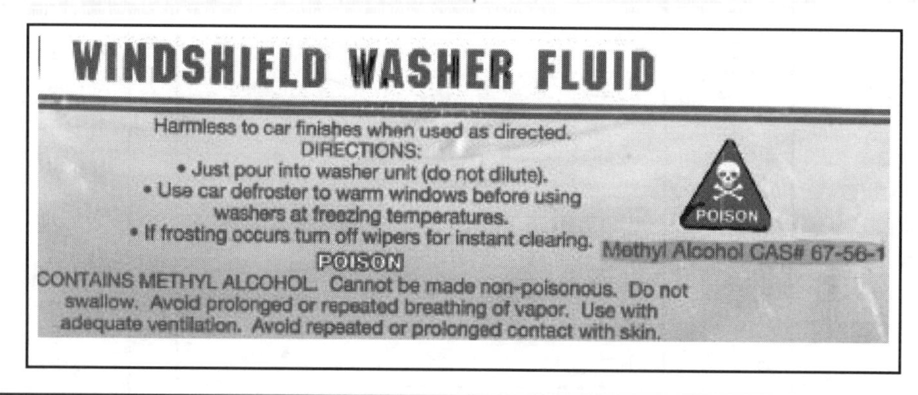


Figure 12.14 Methyl alcohol is used as both an antifreeze and cleaning agent in windshield washer fluid. Because methyl alcohol's molar weight is half that of ethylene glycol, methyl alcohol is twice as good an antifreeze agent (on a pound-for-pound basis). So why not use in radiators? It's too volatile. It boils to easily and generates too high a vapor pressure.

Figure 12.14 reproduces part of the label of a commercial windshield washer fluid formulation. Note that the methanol it contains is poisonous. The notation CAS# on the label above stands for "Chemical Abstract Service Number." In the wake of environmental concerns in the period immediately after Earth Day in 1970, there developed a strong movement to identify chemical compounds. CAS numbers were one response to that concern. Chemical Abstracts itself is a large chemical information organization (part of the American Chemical Society) that is now about 100 years old. The antifreeze effect of methanol is shown by calculation in Example 12.8.

**Example 12.8**: In summer, windshield washer fluid is usually just a dilute solution of soapy water with a trace of blue dye. In winter, some methanol (MW = 32.0 g $mol^{-1}$) is added to provide for antifreeze action. Calculate the freezing point of a 35.0% by weight solution of methanol in water, assuming that the solution obeys the law $\Delta T = m\ k_f$

Strategy: Use the data to calculate the molality of methanol in the solution. It is safe to assume that together water and methanol compose essentially 100% of the fluid. Substitute the calculated molality into the freezing point depression equation using water's $k_b$ value (-1.86 °C $m^{-1}$).

The solution is (100.0% - 35.0%) = 65.0% water. So, 1000 g of solution contains 350 g ÷ 32.0 g $mol^{-1}$ = 10.9 mol methanol and 650 g, or 0.650 kg water. The molality of the solution is thus

$$m = 10.9 \text{ mol methanol} \div 0.650 \text{ kg } H_2O = 16.8 \text{ m}$$

Plugging into $\Delta T = m\ k_f$, gives $\Delta T = -1.86\ °C\ m^{-1} \times 16.8\ m = \underline{-31.2°C}$

## The van't Hoff i-Factor

Around 1885, the Dutch chemist Jacobus Henricus van't Hoff (1852-1911) deduced why different solutes have different sized effects on colligative properties. Many observations had been made during the previous two centuries, and all the colligative properties were well known in 1885, but until van't Hoff's work there was no comprehensive theory to explain them. van't Hoff initially saw how to do it for osmotic pressure, (that's the topic of our next section). A correspondence with Arrhenius quickly led him to the recognition that the same explanation could be extended to cover all the colligative properties. Their key insight was to recognize that salts dissolved in water form solutions that contain ions.

At that time, the spontaneous formation of ions by a dissolving salt was a controversial idea not well accepted by the chemical establishment. Once the van't Hoff i-factor was widely understood, the role of ionization became much better, though still not universally, accepted. (Sometimes it happens that older scientists remain opposed to a new idea. Eventually, they pass from the scene and opposition to the idea dies with them. This happened with the concept of ionization in solution.) The van't Hoff i-factor is the ratio of a measured colligative property of a substance that ionizes (an electrolyte) compared with the same colligative property of a substance that doesn't ionize (a nonelectrolyte). To account for this phenomenon, van't Hoff rewrote the boiling point elevation and freezing point lowering solutions as:

$\Delta T_b = i\ k_b\ m$, the increase in boiling temperature is proportional to the solute's molality times a factor

$\Delta T_f = i\ k_f\ m$, the decrease in freezing temperature is proportional to the solute's molality times a factor

Where the $\Delta T$'s are experimental measurements and $k_f$ m and $k_b$ m are calculated. So i is a "fudge-factor." It's chosen to *force* experiment and calculation to agree (the laws don't give perfect straight lines). Actually, it's a rather wonderful fudge factor because it reveals the truth about ionization in water. Typical experimental results for the freezing points of solutions are shown in Figure 12.14.

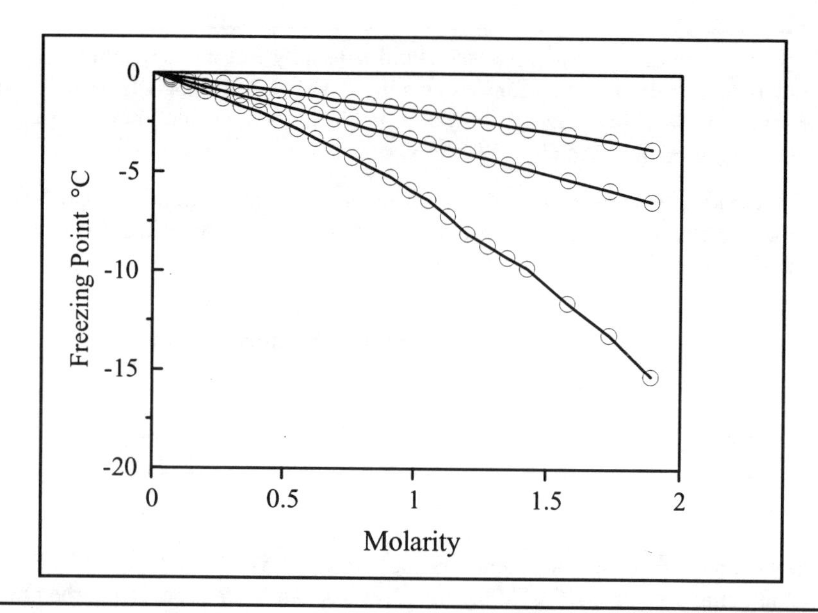


Figure 12.15 Experimental freezing points of three aqueous solutions plotted versus molarity. Methanol ($CH_3OH$, a nonelectrolyte, top), potassium chloride (KCl, a 1:1 electrolyte, middle), calcium chloride ($CaCl_2$, a 1:2 electrolyte, bottom).

The experimental freezing point data in Figure 12.15 reinforce the earlier data for vapor pressure lowering (Figure 12.10) and show that exactly the same phenomenon is at work. Methanol is a nonelectrolyte and so a 0.1 molar methanol solution is indeed 0.1 molar, and for it, i = 1. In contrast, the number of ions that an ionic compound (electrolyte) creates in solution acts as a multiplier for any colligative property. Observe that $KCl \rightarrow K^+ + Cl^-$ has roughly double the effect of methanol, while $CaCl_2 \rightarrow Ca^{2+} + 2\ Cl^-$ has roughly triple the effect of methanol. The van't Hoff i-factor is a the number of moles of ions that form from one mole of an ionizable solute (approximately).

Figure 12.15 also well illustrates again the concept of a limiting law, with the lines becoming increasingly straight at ever lower concentrations. At concentrations above about 0.1 M the lines become markedly curved, with their curvature increasing with increasing concentration. Note also that the simple 1:2:3 ratio is not maintained at the higher concentrations. To explain aqueous solutions at concentrations greater than 0.1 molar is a complicated problem; it can be done fairly well using theories of ionic behavior and computer modeling.

## Osmotic Pressure

Osmosis is a phenomenon in which a pressure is generated across a semi-permeable membrane when a solution and a pure solvent, or two solutions of different concentration, are on opposite sides of the membrane. A **semi-permeable membrane** is a structure or part of a device that allows water molecules to flow through it while being impenetrable to other molecules and ions. Osmosis occurs because solvent molecules preferentially pass through a semi-permeable membrane between solutions of different concentrations.

The phenomenon of osmosis was noted as long ago as 1748 by the Abbé Jean Antoine Nollet (1700-1770) who covered a bottle of alcohol with a pig's bladder (a natural, biological semi-permeable membrane) and placed the whole thing under water. Water flowed into the alcohol-filled bladder, its volume swelled, and the bladder eventually burst. Bladders, and many other biological structures, are semi-permeable membranes. **Osmotic pressure** is the pressure that must be exerted on the solution with the lower solvent concentration just to prevent the process of osmosis from occurring. Osmosis is shown schematically in Figure 12.16.

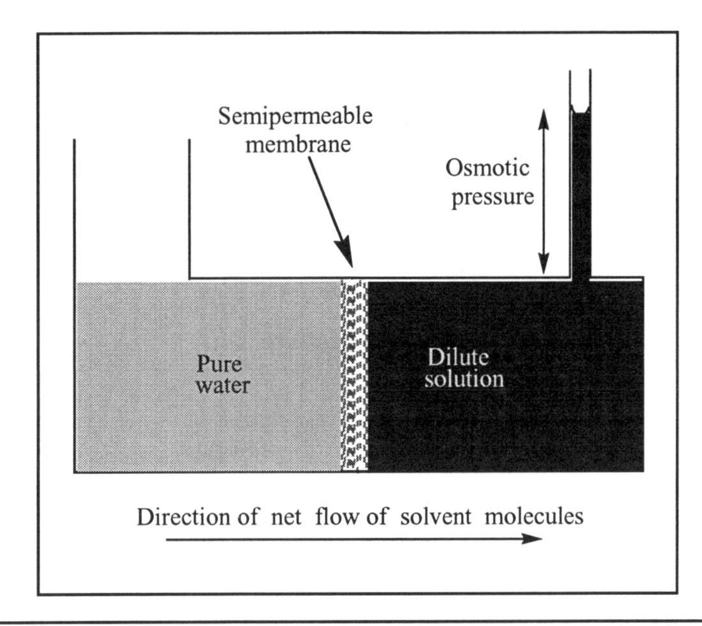


Figure 12.16 A schematic representation of the phenomenon of osmosis.

A dilute solution is separated from pure water by a semi permeable membrane. The liquid level on the right is higher than that on the left. The level rises in the right compartment because water molecules pass preferentially into it. As the experiment continues, the level on the right rises until finally its height generates sufficient pressure to stop further movement across the membrane. That pressure is the osmotic pressure of the solution. At the root of the phenomenon of osmosis lies the fact that the solvent in a solution is *less concentrated* than it is in the pure solvent itself. So water will continue to move from the more concentrated to the less concentrated environment until opposed by a sufficient pressure. Of course, we'll remind ourselves that because osmosis is an equilibrium process the movement of water molecules never stops; it's just that once osmotic equilibrium has been established, water molecules simply go back and forth through the membrane at equal rates.

Despite their formidable name, semi-permeable membranes are common. The membranes of plant and animal cells are semi-permeable, and the process of osmosis is vital to life. Prunes are dried plums; they are a concentrated solution of sugar and pulp in a skin. The skins serve as a semi-permeable membrane. In Figure 12.17 you see the results of an overnight kitchen experiment to soak prunes in water. The upper two prunes were soaked in tap water, the lower two came directly out of the bag. Even more convincing than the swollen size of the soaked prunes is their degraded taste. Fresh from the bag the prunes taste sweet and succulent; the swollen prunes taste watery and not very appetizing. A related food phenomenon occurs when a cucumber placed in a salt solution exudes water and wrinkles up as a consequence the cucumber's water passing out into the pickling solution.

The twin facts that pickles shrink in brine (an electrolyte solution) to get wrinkles and prunes swell in plain water to lose their wrinkles are thus both attributable to the phenomenon of osmosis.

Figure 12.17 Prunes undergo osmosis if placed in plain water. They swell up and lose their wrinkles. The two prunes at the top, placed in water over night, are considerably larger than the two below, fresh from the bag.

Many synthetic semi-permeable membranes are commercially available; some are made from cellulosic (plant- or wood-derived) fibers, others are synthesized from organic chemicals. The way that molecules or ions either pass through or are blocked by semi-permeable membranes is not fully understood. At least part of the explanation, however, lies in the size of openings in the membrane compared with that of the molecules or ions in the solution. Many semi-permeable membranes appear to have channels of molecular dimensions. Molecules small enough to pass through these channels find the membrane permeable while larger molecules find the membrane impermeable.

When sufficient pressure is applied to the sea water across a semi-permeable membrane, it forces water molecules into the fresh water side, leaving the salt behind. That's called reverse osmosis. Reverse osmosis plants are costly but are used in dry climates—such as in the Persian Gulf—and on ships that spend long periods at sea. The past ten years has seen the development of small, portable electric or hand-operated reverse osmosis systems. Following shipwrecks there have been several instances of human survival that are attributed to hand cranked reverse osmosis devices. [❁kws +"reverse osmosis" +shipwreck]

The mathematical law describing osmotic pressure has the form: $\Pi = MRT$, where $\Pi$ is the osmotic pressure generated against a pure solvent across a semi-permeable membrane by a solution of molarity M. R is the gas constant and T is the kelvin temperature. Osmotic pressures are usually surprisingly large, as shown in Example 12.9.

**Example 12.9**: Calculate the osmotic pressure generated against pure water by a 2.15 molar sugar solution across a suitable semi-permeable membrane at 25°C.

Strategy: Substitute values into the osmotic pressure expression:

$$\Pi = MRT$$

$$\Pi = 2.15 \text{ moles L}^{-1} \times .0821 \text{ liter atm mol}^{-1} \text{ K}^{-1} \times 298 \text{ K}$$

$$= \underline{52.6 \text{ atmospheres}}$$

The molar weight of large molecules such as proteins can be measured by osmotic pressure because it is such a leveraged phenomenon. Even an extremely dilute aqueous solution can generate an easily measured osmotic pressure. Example 12.10 shows a typical molar weight calculation.

**Example 12.10**: A protein is prepared as a 0.91 g $L^{-1}$ solution. At 25°C this solution has an osmotic pressure of 3.8 mm of water. Recall from Chapter 4 that 34 feet of water is equivalent to one atmosphere (34 $ftH_2O$ = 1 atm). Calculate the molar weight of the protein.

Strategy: Use the osmotic pressure equation to calculate the molarity of the protein solution. Then, use that and the g $L^{-1}$ concentration to obtain the MW.

$$M = \Pi \div (RT) = \frac{(3.8\ mmH_2O) \div (34\ ftH_2O\ atm^{-1} \times 12\ inch\ ft^{-1} \times 25.4\ mm\ inch^{-1})}{0.0821\ L\ atm\ mol^{-1}\ K^{-1} \times 298\ K}$$

$$M = 1.5 \times 10^{-5}\ mol\ L^{-1}$$

$$\text{Molar weight} = g\ L^{-1} \div mol\ L^{-1} = 0.91\ g\ L^{-1} \div 1.5 \times 10^{-5}\ mol\ L^{-1} = \underline{61{,}000\ g\ mol^{-1}}$$

**Example 12.11**: The osmotic pressure of sea water is about 25 bar. Estimate the mass percentage of salt in the water. Sea water has a density of about 1.02 g $mL^{-1}$. Assume that sodium chloride is the only solid dissolved in sea water.

Strategy: Use the osmotic pressure equation to calculate the total ion molarity of the salt. Halve that value because the van't Hoff i-factor for $NaCl \rightarrow Na^+ + Cl^-$ is approximately 2. Convert molarity to mass percentage. One atmosphere ≈ one bar.

$$M = \Pi \div (RT) = \frac{25\ atm}{0.0821\ L\ atm\ mol^{-1}\ K^{-1} \times 298\ K}$$

$$= 1.02\ M \quad \text{of total ions}$$

Dividing by two to get the NaCl molarity 1.02 ÷ 2 = 0.51 M

For conversion assume 1 L of seawater. It has a mass of 1000 mL × 1.02 g $mL^{-1}$ = 1020 g

The mass of 0.51 mols of NaCl is 0.51 mol × 58.5 g $mol^{-1}$ = 29.8 g

$$\text{Mass percentage} = \frac{29.8}{1020} \times 100 = \underline{2.9\%} \text{ (two sig figs)}$$

Sea water around the world averages about 3.5% dissolved solids, of which about 80% is sodium chloride.

## Supersaturation

Sometimes it is possible to make solutions in which the solute concentration is greater than its solubility. Such a solution is *not* an equilibrium system: it is **supersaturated**. One common way of making a supersaturated solution is by changing the temperature. The solubilities of most solids in liquids increase as the temperature rises. When a warm saturated solution of a solid in a liquid solvent is cooled it sometimes happens that solid fails to separate and the solution becomes supersaturated. It may then be necessary to add a "seed" crystal to the solution to induce crystallization. Honey is a concentrated and often supersaturated solution of sugar in water. Perhaps you have seen a sample of honey that "crystallized" during storage. Figure 12.18 shows a crystal that grew "accidentally" from a sample of maple syrup that was stored for a long time.

Figure 12.18 This long stored sample of maple syrup eventually produced sugar crystals several centimeters in length. Ordinary syrup is a metastable thermodynamic system.

A supersaturated solution of sodium acetate is put to work as a convenient portable heat source (Figure 12.19). The solution is sealed in a strong plastic container. The container includes a small metal disk which when manipulated initiates crystallization. Because crystallization of sodium acetate is an exothermic process the container gets hot. Such devices serve as hand warmers for people who have to be out in cold weather.

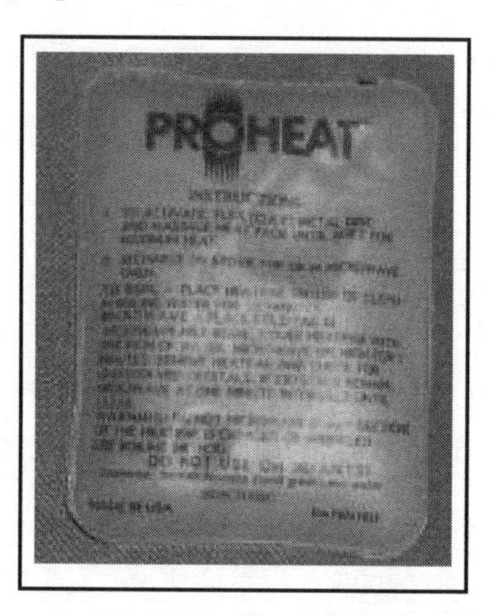


Figure 12.19 Sealed containers of saturated sodium acetate solution provide a portable form of "instant heat." Crystallization is initiated by the user manipulating a small metal disk sealed in the pouch, which causes the solution to crystallize, warming as it does so. To regenerate the device it is submerged for a few minutes in boiling water.

## Essential Knowledge—Chapter 12

**Glossary words:** Phase diagram, condensed phases, fluid phases, physical metallurgy, nonvolatile solute, colligative properties, nonelectrolyte, electrolyte, strong and weak electrolyte, limiting laws, phase regions, triple point, sublimation, supercritical fluid, critical temperature, critical pressure, vapor deposition, freezer burn, freeze drying, eutectic phase, Henry's law, vapor pressure lowering, boiling point elevation, freezing point depression, molal boiling point elevation constant, molal freezing point depression constant, Chemical Abstracts Number, van't Hoff i-factor, osmotic pressure, semi-permeable membrane, supersaturation.

**Key Concepts:** The nature of phase diagrams and the relationships among phases. Nature of supercritical fluids. How phase diagrams of mixtures differ from the phase diagrams of pure substances. The connection between the phase diagram of an aqueous solution and the four classic colligative properties. Using the colligative properties for practical purposes. The nature of ionization in solution as demonstrated by the various classes of electrolytes. Historic role of electrolytes in solution theory and the i-factor. The phenomenon of supersaturation.

**Key Equations:**

molality (m) = (moles solute) ÷ kg of solvent

Henry's law: $X_g = kP$

Formula for vapor pressure lowering: $P = P^o X$

Boiling point elevation, nonvolatile, nonelectrolyte: $\Delta T_b = k_b m$

Freezing point depression, nonvolatile, nonelectrolyte: $\Delta T_f = k_f m$

Boiling point elevation, electrolyte: $\Delta T_b = i k_b m$

Freezing point depression, electrolyte: $\Delta T_f = i k_f m$

Osmotic pressure: $\Pi = MRT$

## Questions and Problems

### Phases

12.1. Which states of aggregation are referred to as the fluid phases and which as the condensed phases? Why?

12.2. The colligative properties follow limiting laws. What is a limiting law?

12.3. The density of liquid water at 0°C is 1.000 g $cm^{-3}$, and the density of ice at 0°C is 0.917 g $cm^{-3}$. Calculate the change in volume when a mole of water freezes to ice at 0°C. Discuss the effect of applying pressure to ice in light of Le Chatelier's principle.

12.4. When ice melts, what is the volume of the water formed, expressed as a percentage of the volume of the ice?

12.5. The L-G line on a phase diagram is a vapor pressure curve. Discuss.

12.6. Describe how the phase diagram of a dilute solution of a nonvolatile solute in water differs from the phase diagram of pure water.

12.7. The triple point of water is the basis of the thermodynamic scale of temperature. It was so decided in 1967. Discuss.

12.8. If the Celsius scale of temperature goes from 0°C to 100°C for water's freezing and boiling points, how can it be that the modern value of water's boiling point is not exactly 100°C? Explain.

12.9. What are the critical temperatures of helium and neon? Define critical temperature. Explain why the critical temperature of these elements is related to their atomic numbers. Is it also related to interatomic dispersion forces?

### Phase Equilibrium and Phase Changes

12.10. Describe or define in twenty-five words or less each of the following: a. phases, b. states of aggregation, c. heat of fusion, d. heat of vaporization, e. phase diagram, f. triple point, g. sublimation, h. supercritical fluid.

12.11. a. Describe a temperature-composition phase diagram. b. Is the phase diagram of carbon steel of this type? c. Why is the carbon-iron phase diagram important?

12.12. For each of the following water-soluble substances state whether it behaves as a strong electrolyte, a weak electrolyte, or a nonelectrolyte: a. sodium nitrate, b. ammonium sulfate, c. urea, d. table sugar, e. acetic acid, and f. ammonia.

12.13. a. What are the two methods by which water can be boiled? b. What is vapor deposition?

12.14. Discuss the phenomenon of ice skating from a scientific point of view. Do an internet search to find out why ice is slippery.

12.15. Describe the commercial uses of supercritical carbon dioxide.

12.16. a. Define a eutectic phase. b. US nuclear submarine reactors are cooled with a metal alloy that is a mixture of sodium metal and potassium metal. Why is the coolant alloy a eutectic mixture?

12.17. What is austenite?

### Solution Concentrations

12.18. For review, write the definitions of a. molarity, b. mole fraction, and c. ppm.

12.19. An aqueous solution of lead(II) nitrate, $Pb(NO_3)_2$, contains 4.00% by weight of the solute. The solution has a density of 1.0352 g $mL^{-1}$. Calculate a. the molarity, and b. the molality of the solution.

12.20. A solution of sodium bromide, NaBr, is 9.50% by mass in water. The solution has a density of 1.0779 g $mL^{-1}$. Calculate the molarity and the molality of the solution.

12.21. A 0.50% by mass solution of acetone in water has a density of 0.993 g $mL^{-1}$. Calculate the concentration of acetone in a. g $L^{-1}$, and b. ppm.

12.22. A 1.277 molar solution of potassium nitrate, $KNO_3$, has a density of 1.0871 g $mL^{-1}$. a. Calculate the weight percentage of the solute. b. Calculate the molality of the solution. ($KNO_3$ = 101.1 g $mol^{-1}$)

## Henry's Law

12.23. State Henry's law and the units of the Henry's law constant. Use the data of Table 12.2 to decide which of the gases in that table is most soluble in water at 25°C.

12.24. The solubility of gases in liquids changes significantly with changing pressure. However, the solubilities of solids in liquids is scarcely affected by changing pressure. Suggest a molecular explanation for this difference.

12.25. Calculate the concentration of nitrogen in water at 25°C. Use its partial pressure in air and the appropriate Henry's law constant.

12.26. When cold water is slightly warmed, it often becomes supersaturated in dissolved oxygen, leading to the formation of bubbles. Explain.

12.27. The Henry's law constant for carbon dioxide dissolving in water at 25°C is $8.00 \times 10^{-7}$ mole fraction per Torr. Calculate the mole fraction of dissolved carbon dioxide in water at 500 and 1000 Torr at 25°C.

12.28. The pressure of a diver's breathing gas increases by approximately 760 Torr for every 34 feet of depth. Calculate the concentration of nitrogen in 25°C water at the surface (where nitrogen's partial pressure is about 600 Torr) and at depth of 600 feet. Use your answer to explain why divers must undergo a slow decompression when they return to the surface from deep dives.

12.29. Calculate the mole fractions of dissolved $CO_2$ in a soft drink container at 25°C: a. when the bottle is capped under a partial pressure of $CO_2$ of 4 atm, and b. after an uncapped bottle has come to equilibrium with the open air where the partial pressure of $CO_2$ is $4 \times 10^{-4}$ atm.

12.30. Oxygen dissolved in water is important for oxygen-breathing aquatic life. Oxygen's concentration is usually expressed in parts per million (ppm), or grams of oxygen per million grams of water. At 25°C the Henry's law constant for oxygen in water is $3.03 \times 10^{-8}$ mole fraction per Torr. Calculate the concentration of oxygen in ppm in water at 25°C when the partial pressure of oxygen in the air is 200 Torr.

## Colligative Properties

12.31. Briefly describe the meaning of the word colligative. Name the colligative properties. Why are the colligative properties the subject of limiting laws.

12.32. Explain in molecular terms why the presence of a solute generally lowers the freezing point and increases the boiling point of the solution compared with that of the pure solvent.

12.33. What scientist was the first to provide a thermodynamic explanation of colligative properties? When did he explain them?

12.34. Explain how the measurement of colligative properties can be used as the basis of molar weight determinations.

## Vapor Pressure Lowering

12.35. The vapor pressure of water at 100°C is 760 mmHg. Use Raoult's law to estimate the vapor pressures of aqueous solutions at 100°C that contain the following mole fractions of a nonvolatile, non electrolyte solute: a. 0.01, b. 0.03, c. 0.07, and d. 0.1.

12.36. A 0.0163 mole fraction solution of NaCl solute in water has a boiling point of 101.0°C. Calculate the molality of the NaCl. The solution has a density of 0.997 g $mL^{-1}$.

12.37. Calculate the vapor pressure of a 22.0% weight percent solution of ethylene glycol at 25°C.

12.38. Calculate the vapor pressure of a 1.0 molal solution of cobalt(II) nitrate using Raoult's law. Compare your result with the value read Figure 12.10.

## Boiling Point Elevation and Freezing Point Depression

12.39. A solution is made by dissolving 50.4 g of urea in 741 g of water. Calculate the freezing point of the solution assuming that it is ideal. Calculate the boiling point of the solution.

12.40. Camphor ($C_{10}H_{16}O$) was historically used as a solvent for molar weight determinations because it has a large freezing point depression constant of -40.0 °C $m^{-1}$. Camphor's melting point is 178.8°C. A mixture of 10.0 mg of a substance and 1.00 g of camphor melts at 176.6°C. Calculate the molar weight of the substance.

12.41. The molar weight of caffeine was measured for its freezing point depression in camphor. A caffeine sample weighing 18.4 g was mixed with 250.0 g of camphor, and the mixture heated until it melted. On cooling, the solution was observed to begin to freeze at 163.6°C. Calculate the molar weight of caffeine. (See the previous problem for needed data.)

12.42. Caffeine contains 49.5% C, 28.9% N, 16.5% O, and 5.1% H. A solution of 1.000 grams of caffeine dissolved in 100.0 grams of water has a freezing point of -0.09°C. What are the molar weight and the molecular formula of caffeine?

## The van't Hoff i-factor

12.43. Define the van't Hoff i-factor.

12.44. What will be the i-factors for aqueous solutions of: a. ethylene glycol, b. sodium nitrate, and c. calcium chloride?

12.45. Which solution has the lower freezing point: 1.50 g of NaCl in 1000 g of water or 1.00 g of $CaCl_2$ in 1000 g of water? Why?

12.46. List the following aqueous solutions in order of increasing boiling point: 1 M NaCl, 1 M $Na_2SO_4$, and 1 M $C_6H_{12}O_6$.

## Osmosis

12.47. Briefly describe an osmosis experiment.

12.48. Calculate the osmotic pressure generated by a 1.0 M solution of a mixture of non-electrolytes that are separated from pure water by a semi-permeable membrane, at 298 K. What is the height of a column of water that will generate the same pressure?

12.49. A solution was prepared by dissolving 1.00 mg of an enzyme in a total volume of 100.0 mL of water. The solution had an osmotic pressure of 3.10 Torr at 29°C. Calculate the molar weight of the enzyme.

## Problem Solving

12.50. Calculate the freezing point depression of a solution which is 14.0% by weight of methanol in water.

12.51. What is the freezing point of an aqueous solution that has a boiling point of 101.76°C?

12.52. What weight percent mixture of ethylene glycol ($C_2H_6O_2$) and water should be used to fill a radiator, which has an 8 liter capacity, to protect it from freezing down to a temperature of -30°C.

12.53. The molar weight of acetic acid determined from its colligative properties in benzene solution is about 120 g $mol^{-1}$—a result that is interpreted to arise because of dimerization. Similar measurements using acetone ($C_3H_6O$) as the solvent give a molar weight for acetic acid of about 60 g $mol^{-1}$. Use your knowledge of inter-molecular forces to explain why acetic acid exists in dimeric form in benzene but monomeric form in acetone.

12.54. Describe the process of reverse osmosis. Use internet resources to discover the meaning of the acronym BWRO.

12.55. At 25°C the vapor pressure of pure water is 23.756 Torr. If 3.5 g of a compound whose molar weight is 198 g $mol^{-1}$ is dissolved in 100.0 g of water, calculate the expected vapor pressure lowering in Torr. Also estimate the osmotic pressure of the solution. Which colligative property serves as the basis for the preferred method for determination of molar weights? Why?

12.56. An unknown organic compound is a nonelectrolyte known to be composed of C, N, O, and H. Combustion analysis showed that it contained 67.3% C, 4.62% N and 6.93% H. A solution made by dissolving 1.20 g of the unknown compound in 25.0 g of benzene had a freezing point of 4.7°C Determine a. the empirical formula, b. the molar weight, and, c. the molecular formula of the unknown compound.

12.57. Deep sea divers who descend to a depth below about 15 meters breathe a gas mixture made from helium and oxygen rather than air (a nitrogen-oxygen mixture). Doing this prevents a condition called "rapture of the deep." What is this condition and what does it have to do with Henry's law?

12.58. It is well known that a bottle of soda water shaken before it is opened foams vigorously. Suggest a molecular interpretation for the phenomenon.

12.59. A saturated solution of oxygen in water at 0°C is warmed to 30°C. What percentage of the dissolved oxygen is released from the solution? The Henry's law constants for solutions of oxygen in water at different temperatures are given in Table 12.3. Assume that the partial pressure of oxygen above the solution remains constant at 0.205 atm.

12.60. List the following aqueous solutions in order of increasing boiling point: 1 M NaCl, 1 M $Na_2SO_4$, 1 M $H_2SO_4$, 1 M $C_6H_{12}O_6$, and 1 M cobalt(II) nitrate solution.

12.61. What is a eutectic phase?

12.62. Describe the use of phase diagrams in metallurgy.

12.63. Use their relative molar weights to compare the antifreeze effectiveness of ethylene glycol (density 1.116 g $mL^{-1}$) and propylene glycol (density = 1.040 g $mL^{-1}$) on a pound-for-pound basis.

12.64. Make a graph of the data presented in Table 12.4 of the conditions in Lake Nyos, Cameroon in May 1998.

# Chapter 13

# The Periodic Table and Redox

## Introduction

In this chapter we'll make a brief survey of the periodic table. The chapter's contents are mainly descriptive inorganic chemistry. **Descriptive inorganic chemistry** catalogs the full range of properties and uses of the elements and their compounds: their origin and abundance; their sources, production, and uses; their reactions and particularly their redox reactions; their biological/biochemical, ecological, environmental, and geologic aspects. That's a daunting list.

Over the past 300 years, chemists have amassed an enormous store of descriptive information. To reduce that store to a single chapter is an impossible task. But descriptive chemistry forms a vital part of chemistry. The study of the comparative properties of the elements has had a profound impact on the thinking of chemists and on the development of modern chemistry. Mendeleev's periodic table was described in Chapter 1 and later explained by quantum mechanics as we saw in Chapter 6. Mendeleev organized the periodic table almost entirely on the basis of descriptive and comparative chemistry. Viewed from our perspective, it is easy to believe that Bohr's atomic model and quantum mechanics solved the problem of electron configuration. However, when the author reads chemistry books published around 1880-1910 he's invariably impressed that chemists well informed about descriptive chemistry had anticipated many (but not all) of Bohr's chemical predictions.

Three main themes tie together the disparate material of the Chapter: periodicity, electron configurations, and redox. 1. **Periodicity** is defined as trends and relationships among the physical and chemical properties of the elements: the properties of an element in relation to its position in the periodic table. 2. **Electron configurations of the elements**. These were discussed in Section 5.6, and the experimental configurations are displayed on a periodic table in Appendix H. 3. **The redox chemistry of the elements**. Recall from Chapter 3 that an atom, molecule, or ion that loses electrons is said to have been **oxidized**. An atom, molecule, or ion that gains electrons is said to have been **reduced**.

We'll begin by discussing the metals. Next, we will discuss redox reactions and show how to balance redox equations and we'll conclude by discussing the nonmetals.

## Section 13.1 Metals

### Overview of the Periodic Table by Broad Classification

One helpful way to describe relationships among the elements is to color, shade, or label the periodic table. The Table 13.1 shows the major classifications of the metals. A quick glance at the table reveals that, despite its position above the alkali metals, hydrogen is a nonmetal. The rule of periodicity is simple: elements with the same classification have similar properties. It's a gross oversimplification, but it's not a bad start. Here's a description of the main regions occupied by metals on the periodic table.

Recall from Section 1.3 that the elements are divided into **main group** elements and the **transition** elements. Table 13.1 shows the elements divided into nine categories based on their positions in the periodic table. Each category is briefly described following Table 13.1.

| Table 13.1 **Classification of Elements on the Periodic Table** | |
|---|---|
| M | Main group metals. |
| N | Nonmetals. |
| S | Semi-metals or metalloids. |
| T1 | First row d-block transition metals. |
| T2 | Second row d-block transition metals. |
| T3 | Third row d-block transition metal. |
| T4 | Fourth row transition metals. (Radioactive, not much chemistry here). |
| L | The lanthanides. |
| A | The actinides. (Radioactive, some chemistry here). |

| 1 | 2 | 3 | 4 | 5 | 6 | 7 | 8 | 9 | 10 | 11 | 12 | 13 | 14 | 15 | 16 | 17 | 18 |
|---|---|---|---|---|---|---|---|---|---|---|---|---|---|---|---|---|---|
| N | | | | | | | | | | | | | | | | | N |
| M | M | | | | | | | | | | | S | N | N | N | N | N |
| M | M | | | | | | | | | | | M | S | N | N | N | N |
| M | M | T1 | T1 | T1 | T1 | T1 | T1 | T1 | T1 | T1 | T1 | M | S | S | N | N | N |
| M | M | T2 | T2 | T2 | T2 | T2 | T2 | T2 | T2 | T2 | T2 | M | M | S | S | N | N |
| M | M | T3 | T3 | T3 | T3 | T3 | T3 | T3 | T3 | T3 | T3 | M | M | M | S | S | N |
| M | M | T4 | T4 | T4 | T4 | T4 | T4 | T4 | T4 | T4 | T4 | | M | | S? | | N? |
| | | L | L | L | L | L | L | L | L | L | L | L | L | L | L | | |
| | | A | A | A | A | A | A | A | A | A | A | A | A | A | A | | |

## Major Regions of Metals in the Periodic Table

1. The **main group** metals of groups 1 and 2 are on the left of the periodic table. The electron configurations of the elements are $ns^1$ and $ns^2$, which account for the group oxidation states of +1 and +2. They are very reactive and have generally simple chemical behavior. The alkali metals of group 1 exhibit strong vertical periodicity—by which we mean that the elements are like those above and below them. In group 2, beryllium is rather distinctive, magnesium is less distinctive, and Ca-Sr-Ba form a **triad.** A triad is three adjacent elements with the middle one having properties that are near the average of the outer pair; triads provide the best example of vertical periodicity. Four of the metals: Na, K, Mg, and Ca are abundant, widespread and important. Potassium—along with the nonmetals nitrogen and phosphorus—is one of the three fertilizer elements.

2. The **main group** metals of groups 13-15 are the seven elements Al, Ga, In, Tl, Sn, Pb, and Bi. They exhibit reasonably simple chemistry. Their group electron configurations are $s^2p^1$ $s^2p^2$ and $s^2p^3$. Aluminum fits least well with the others; it actually shows a fairly strong resemblance to magnesium. Other than aluminum these metals each form two series of compounds in which the metal differs by two oxidation states. For example gallium and indium form 1+ and 3+ states, while tin and lead form 2+ and 4+ states. Aluminum, tin and lead among these metals have important economic value. Annual demand and use patterns of the main group metals and metalloids are shown in Table 13.2.

3. The three rows of **d-block transition metals: first row** Sc-Zn, **second row** Y-Cd, and **third row** Hf-Hg. The reactivity of these metals varies widely. The chemistry of some elements (for example Sc and Zn) is fairly simple. But the chemistry of many is complex: 1. they show multiple oxidation states, 2. they tend to form many complex ions, and 3. they form compounds with a wide variety of colors and with variable magnetic properties. **Complex ions** are metal ions bonded to one or more molecules or ions (collectively called ligands). Many of our most valuable industrial **catalysts** are d-block transition metals, their compounds, or their complex ions. Some transition metals are noble (chemically unreactive); examples are gold, silver, and the platinum metals (Ru-Rh-Pd-Os-Ir-Pt). But the noble

metals do form thousands of compounds—many more than the noble gases. Annual demand and use patterns of the d-block transition metals are shown in Table 13.5, which follows our discussion of the main group metals.

4. The f-block transition metals. The **lanthanides** (La-Lu) exhibit relatively simple chemistry. The chemistry of the actinides (Ac-Element 112) is complex and further complicated by the fact that they are all radioactive and at higher atomic numbers (>95) impossible to obtain in significant quantity.

Table 13.2 **Estimated Annual US Consumption and Use Pattern of Main Group Metals and Metalloids**

| Element | Demand (tons) | Summary Consumption Pattern (Percentages are approximate) |
|---|---|---|
| Li | 3,000 | Al production 40%. Ceramics 20%. Grease 15%. Rubber production 15% |
| Na | $10 \times 10^6$ | As NaCl. Chlor-alkali 55%. Deicing 30%. Others 20%. |
| K | $10 \times 10^6$ | Fertilizer 95%. Glass 2%. Soaps 2%. Other 1% |
| Rb | 2 | Glass 20%. Chemicals 20%. Other 60%. |
| Cs | 4 | Electrical 40%. Chemical 40%. Other 20%. |
| Be | 500 | Electrical 50%. Electronics 40%. Aerospace 10%. |
| Mg | 170,000 | Al alloys 50%. Castings 25%. Steel desulfurization 15%. Antirust 5%. |
| Ca | $3 \times 10^6$ | Chemical and industrial 80%. Construction 15%. Agricultural 5% |
| Sr | 17,000 | Glass 70% Pyrotechnics 20%. Other 0%. |
| Ba | 3,000,000 | As $BaSO_4$. Drilling 95%. Glass 5%. |
| Ra | 0 | Formerly used for cancer radiation treatment. |
| B | 600,000 | Glass 60%. Fiberglass 30%. Minor uses 10%. |
| Al | 4,000,000 | Cans 40%. Construction 20%. Transport 30%. Aerospace 10% |
| Ga | 10 | Opto-electronics 60%. Integrated circuits 40%. |
| In | 50 | Coatings 45%. Alloys/solders 35%. Semiconductors 15%. Other 5%. |
| Tl | 1 | Alloys 100%. |
| Si | 900,000 | Ferrosilicon 60%. 30% steel production. 10% Other 10% |
| Ge | 15 | Fiber optics 40%. Polymer catalysts 20%. Solar 20%. Other 20%. |
| Sn | 50,000 | Cans 35%. Electrical 25%. Transportation 15%. Construction 15%. |
| Pb | 70,000 | Batteries 60%. Alloys/ammunition 30%. Other 10% |
| As | 25,000 | Wood preserving 95%. Herbicides 5%. |
| Sb | 13,000 | Fire retardants 60%. Batteries 40%. |
| Bi | 1,500 | Pharmaceuticals 50%. Alloys 50%. |
| Te | 60 | Iron/steel 50%. Catalysts 25%. 10% alloys. 10% Photoreceptors. |

## Periodicity and Descriptive Chemistry of the Alkali Metals

The metals lithium, sodium, potassium, rubidium, cesium, and francium constitute the first column (Group 1) of the periodic table. These metals are called **alkali metals** because their oxides react with water to form basic solutions (For example, $Na_2O + H_2O \rightarrow 2\ NaOH$). Group 1 is one of the most regular families in the periodic table. By regular we mean that the trends in the elements' physical and chemical properties are smooth. The group name of the alkali metals derives from the Arabic term, *al qali*, for ashes. The ashes in a pot are "potash," ash from wood burning, with the approximate chemical formula $K_2CO_3$. Potash is a rich source of the alkali metal potassium (not "potashium," which suggests inebriation).

Periodicity can be studied by taking a group of elements, listing many of their properties, and studying or graphing the trends. Table 13.3 shows summary properties of the alkali metals.

Table 13.3 **Properties of the Alkali Metals**

| | Lithium | Sodium | Potassium | Rubidium | Cesium |
|---|---|---|---|---|---|
| Electron configuration | $[He]2s^1$ | $[Ne]3s^1$ | $[Ar]4s^1$ | $[Kr]5s^1$ | $[Xe]6s^1$ |
| Atomic number | 3 | 11 | 19 | 37 | 55 |
| Atomic weight | 6.94 | 23.0 | 39.1 | 85.5 | 132.9 |
| Ionization energy kJ $mol^{-1}$ | 520.1 | 495.7 | 418.6 | 402.9 | 375.6 |
| Reduction Potential $M^+(aq) + e^- \rightarrow M(s)$ $E^\circ$ (volts) | -3.04 | -2.71 | -2.93 | -2.98 | -2.92 |
| Density (g $mL^{-1}$) at 20°C | 0.53 | 0.97 | 0.86 | 1.53 | 1.9 |
| Melting/Boiling points (°C) | 181/1350 | 98/881 | 63/766 | 39/688 | 29/705 |
| Atomic radius of M (pm) | 135 | 154 | 196 | 211 | 225 |
| Ionic radius of $M^+$ (pm) | 68 | 97 | 133 | 147 | 167 |
| Flame test color | red | yellow | violet | red | blue |

The pure alkali metals are soft and silvery white and are good conductors of electricity. Many of their properties exhibit a regular trend from lithium to cesium and available evidence suggests that francium falls into line. But, francium's isotopes are all radioactive. Its longest lived isotope, $^{223}Fr$, has a half-life of only 22 minutes, so its chemistry is difficult to study and is not well known.

You'll see from Table 13.3 that many properties of rubidium are close to the average of that property for potassium and cesium. K-Rb-Cs form a **triad**. In the first half of the 18th century the existence of triads such as Ca-Sr-Ba and Cl-Br-I was important in guiding the development of the periodic table.

The alkali metals are so active that they react immediately with water. In Figure 13.1 you see a 500 mg sample of potassium metal dropped into a few milliliters of water in a dish. The potassium has melted and vaporized because of the heat of its reaction with water. The flame is potassium vapor burning in air. Obviously, potassium metal can never be found in nature. It exists only in compounds.

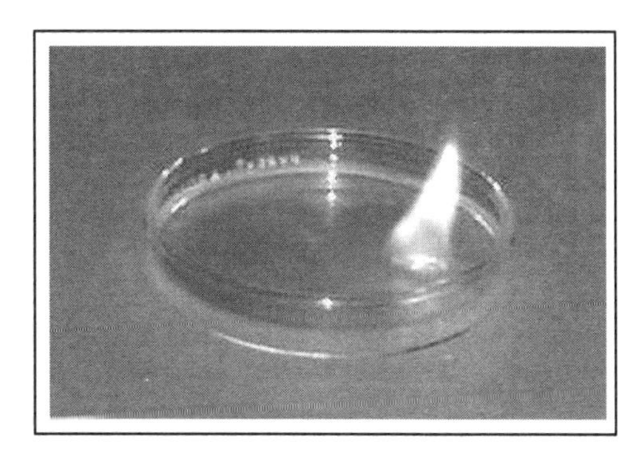

Figure 13.1 As seen here, a small piece of potassium reacts vigorously when dropped into water.

Every alkali metal atom has one electron beyond a noble gas core configuration, $[core]ns^1$. This configuration causes Group 1 metals to have relatively small first ionization energies and low electronegativities; it also accounts for the emission spectra of the elements. Alkali metal compounds impart a characteristic color to a flame or firework; recall from Section 1.2 that rubidium and cesium were discovered by Bunsen in this way. Because of their ready tendency to give up their valence electron to any other element with even modest electronegativity, i.e., to almost any nonmetal, the Group 1 metals are highly reactive and are never found uncombined in nature. They exist as singly charged positive ions in their compounds. Sodium and potassium are widespread and important elements. Lithium is less important, and rubidium and cesium are minor elements with few significant uses.

## The Alkaline Earth Metals

Like the alkali metals, the elements Be, Mg, Ca, Sr, Ba, and Ra, in the second column (Group 2) of the periodic table, form a coherent group. These elements are called the **alkaline earth metals** because their oxides are basic and because they occur widely in mineral deposits throughout the earth's crust. Calcium (3.4% crustal abundance) and magnesium (1.9% crustal abundance) are the two most widespread metals in this group; the other elements in the group are rare. Alkaline earth metals react to form compounds that contain 2+ ions, for example:

$$Mg(s) + Cl_2(g) \rightarrow MgCl_2(s)$$

$$2\ Ba(s) + O_2(g) \rightarrow 2\ BaO(s)$$

Some properties of the alkaline earth elements are summarized in Table 13.4

Table 13.4 Properties of the Alkaline Earth Metals

| | Beryllium | Magnesium | Calcium | Strontium | Barium | Radium |
|---|---|---|---|---|---|---|
| Electron configuration | $[He]2s^2$ | $[Ne]3s^2$ | $[Ar]4s^2$ | $[Kr]5s^2$ | $[Xe]6s^2$ | $[Rn]7s^2$ |
| Atomic number | 4 | 12 | 29 | 38 | 56 | 88 |
| Atomic weight | 9.01 | 24.3 | 40.1 | 87.6 | 137.3 | -226 |
| First ionization energies kJ $mol^{-1}$ | 899 | 738 | 590 | 549 | 503 | 509 |
| Second ionization energies kJ $mol^{-1}$ | 1,757 | 1,451 | 1,145 | 1,064 | 965 | 975 |
| Reduction Potential $M^{2+}(aq) + 2e^- \rightarrow M(s)$ E° (volts) | -1.85 | -2.37 | -2.87 | -2.89 | -2.91 | -2.92 |
| Density g $mL^{-1}$ (20°C) | 1.85 | 1.74 | 1.55 | 2.63 | 3.62 | 5.5 |
| Melting point °C | 500 | 649 | 839 | 768 | 727 | 700 |
| Boiling point °C | 1,287 | 1,105 | 1,494 | 1,381 | 1,850 | 1,700 |
| Atomic Radius of M (pm) | 90 | 130 | 174 | 192 | 198 | 230 |
| Ionic radius of $M^{2+}$ (pm) | 35 | 82 | 99 | 112 | 134 | 148 |

The triad, Ca-Sr-Ba, provides an excellent example of vertical periodicity. Many of the physical and chemical properties of strontium are close to the average of those of calcium and barium: see, for example, density, ionic radius, and atomic radius in the preceding table. Average properties are an indication of periodicity—not a precise measure. Beryllium, the topmost element, is distinct from the other members of the group. Beryllium shows similarities to the metalloid boron (on its right in the periodic table after the transition metals) and to aluminum (diagonally below it to the right after the transition metals). Magnesium is also somewhat distinct from the lower three elements. The chemical and physical properties of magnesium and its compounds are frequently similar to those of lithium, another example of a diagonal periodicity. The magnesium 2+ ion and the lithium 1+ ion are similar in size and tend to mimic one another in many chemical reactions. Magnesium also shows some similarities to zinc. The lowest element in the alkaline earth group is radium. It is better known for its radioactivity than for its chemical properties.

**Example 13.1**: Use several properties of the calcium-strontium-barium triad of elements to illustrate strong vertical periodicity.

Strategy. To illustrate vertical periodicity, calculate the average properties of calcium and barium, and compare the average with the value for strontium.

a. Melting point: Average for Ca and Ba is (839°C + 727°C) ÷ 2 = 783°C

The melting point of Sr is 768°C

b. Ionic ($2^+$) radius: Average for $Ca^{2+}$ and $Ba^{2+}$ is (99 pm + 134 pm) ÷ 2 = 117 pm

The ionic radius of $Sr^{2+}$ is 112 pm

c. Density of solid metal: Average for Ca and Ba is

$(1.55\ g\ cm^{-3} + 3.62\ g\ cm^{-3}) \div 2 = 2.59\ g\ cm^{-3}$

The density of Sr is $2.63\ g\ cm^{-3}$

In each case, the value for Sr is close to the average of the properties of Ca and Ba, so strong vertical periodicity is exhibited by these three elements.

Many mineral forms of calcium and magnesium carbonates are known. Magnesium and calcium dipositive ions are also present in many silicate minerals. Like the active metals of group 1, the alkaline earth metals were not extracted from their compounds until after the development of electrolysis for reduction of active metals (see Table 1.4).

**Calcium's** most important compound, lime, is a large volume chemical product with a per capita US annual consumption of about 170 pounds. Its importance in the chemical industry is due to its alkalinity and its ready availability. It is manufactured by calcining crushed limestone, chalk, or oyster shells (all of which are mainly calcium carbonate). **Calcining** means heating in a kiln to drive off carbon dioxide. Many types of kilns are in use; the chemical reaction is:

$$CaCO_3(s) \xrightarrow{1200°C} CaO(s) + CO_2(g)$$

Lime in the form of calcium oxide (CaO) quickly absorbs moisture to form hydrated lime, ($Ca(OH)_2$) or calcium hydroxide.

$$CaO(s) + H_2O(l) \rightarrow Ca(OH)_2(s)$$

Both forms are called lime. The unhydrated variety is sometimes called quicklime and the hydrated variety is sometimes called slaked lime. The biggest consumers of lime are the iron and steel industry and the chemical industry. In purifying iron from its ores, lime is used in-slag forming reactions to remove acidic oxides by reactions such as

$$CaO(l) + SiO_2(l) \rightarrow CaSiO_3(l) \text{ (calcium silicate slag)}$$

$$CaO(l) + P_2O_5(l) \rightarrow Ca_3(PO_4)_2(l) \text{ (calcium phosphate slag)}$$

Lime also is used for water treatment, as an alkali, as an ingredient in cements and mortar, and in agriculture to raise the pH of soils. Gypsum, a hydrated form of calcium sulfate, $CaSO_4{\cdot}2H_2O$, is a mineral found throughout the world. It loses water when gently heated to form a "half-hydrate":

$$2\ CaSO_4.2H_2O(s) \rightarrow 2\ CaSO_4{\cdot}\tfrac{1}{2}H_2O(s) + 3\ H_2O(g)$$

The half-hydrate of calcium sulfate is called **plaster of Paris**. A water paste of the half-hydrate hardens to form the familiar "plaster cast." Hardening occurs because of re-hydration to the dihydrate; during plaster cast formation the equation above reverses. Commercial plaster for home and industrial use usually incorporates other ingredients (such as glue, hair, or mineral wool) to make it more workable and allow it to dry more uniformly.

Calcium carbonate in purified form is used as an antacid (Tums®), and is an ingredient in many over-the-counter stomach remedies. Calcium carbonate is also a source of dietary calcium. The main human nutritional need for calcium is in bone formation. A pregnant woman must consume an adequate quantity of calcium compounds to supply her growing fetus. Calcium carbonate also occurs as many minerals such as marble.

**Magnesium** is a lightweight, high-strength metal. Magnesium was named for Magnesia, a location in Central Greece where magnesium minerals are found. Lavoisier included the magnesium compound Epsom salt ($MgSO_4$) in his 1789 table of elements. In the US, magnesium is obtained both from sea water and mineral sources. Magnesium metal is chemically reactive but can be used in contact with oxygen and water, despite its reactivity, because it becomes covered with a tough, thin, adherent oxide film that protects it from further chemical attack (this is a characteristic that magnesium shares with aluminum—a horizontal relationship). Magnesium metal and its alloys are used for aircraft, missiles, and automobile parts ("mag wheels" are made from magnesium alloy). Magnesium metal is also used to provide cathodic protection for aluminum, for flares (magnesium strips burn with a brilliant white flame), and for chemical reduction. Metals manufactured by reduction of one of their compounds by magnesium metal include beryllium, titanium, hafnium, zirconium, uranium, and plutonium.

The largest use of magnesium compounds is as magnesium oxide (MgO, which is called magnesia) in refractory (heat resistant) linings for high temperature crucibles used in the production of iron, steel, and other metals. Other uses for magnesium oxide and carbonate (usually in powdered forms) are for rubber, inks, ceramics and fertilizers. Aqueous solutions of Epsom salt ($MgSO_4 \cdot 7H_2O$) have been used by humans for many years to relieve sore muscles. Milk of Magnesia®, an aqueous suspension of powdered magnesium hydroxide, $Mg(OH)_2$, is used as an antacid and a laxative. The antacid product Maalox® is so named because it is a mixture of magnesium and aluminum oxides.

The $Mg^{2+}$ ion is an essential nutrient for life. In plants, it is a component of chlorophyll, a key molecule in photosynthesis. In animals, magnesium ions are vital in the control of metabolism and muscle function. It is present in intracellular and extracellular fluids in the human body and is essential for the proper functioning of several enzymes.

Beryllium was first isolated independently by Wohler and Bussy in 1828, by reduction of beryllium chloride with potassium metal. There are several dozen natural beryllium minerals, of which beryl, familiar as the gemstones aquamarine and emerald, is the best known. Beryl, which gives beryllium its name, has the formula $3BeO.Al_2O_3.6SiO_2$ or $Be_3Al_2Si_6O_{18}$. Beryllium metal first became available in sizable quantities around 1957; it is prepared commercially by magnesium reduction:

$$BeF_2 + Mg \rightarrow MgF_2 + Be$$

Beryllium is alloyed with copper to make non-sparking tools and electrical switches for use in environments where explosive gas mixtures may be present. Beryllium metal is used in nuclear weapons as a neutron reflector and moderator. It's strength and light weight cause it to be used for structural components of the space shuttle, where its high cost is overcome by the importance of the application. Beryllium is relatively transparent to x-rays and windows made from it are used in x-ray generating tubes such as that shown in Appendix G.

**Strontium** is a minor element with few distinctive uses. It is found in nature in the rare minerals celestite ($SrSO_4$), and strontianite ($SrCO_3$). Strontium is produced commercially from celestite, more than 90% of which comes from four countries: Iran, Mexico, Spain, and Turkey. Metallic strontium can be prepared by various reduction procedures. There are few commercial applications of metallic strontium, although it has been used as a getter. A **getter** is a reactive metal that is used to remove the last traces of oxygen gas from a vacuum tube (it **"gets"** the oxygen). The principal use of strontium is as strontium carbonate glass for the front of television picture tube screens; here, it functions as a radiation shield for the x-rays produced when the electron beam impacts on the phosphors of the tube. Barium carbonate competes to some extent with strontium in this use, but strontium is both lighter in weight and more efficient. Another use of strontium is in pyrotechnics. It gives red flares their characteristic color, and strontium nitrate is used in red fireworks. Other minor uses of strontium compounds are in greases, ceramics, and paints.

Strontium has four isotopes. Its best known isotope is the radioactive $^{90}Sr$ with a half-life of 28.1 years. This nuclide was released to the global environment in significant amounts during the time of atmospheric nuclear weapons testing in the 1950s. One of the most dangerous aspects that release was that $^{90}Sr$ entered the food chain. Because of its periodic relationship to calcium, strontium is a "bone-seeker" and accumulates in human skeletons where its decay may cause bone cancer and leukemia.

**Barium**: The mineral, barite ($BaSO_4$), has been known for many centuries because it occurs in many mineral deposits. The first American production of barite was in 1845 in Fauquier County, Virginia. Barite's earliest use was as a pigment for white paint. In the 1980s, Nevada became the largest barium producing state in the US. The dominant use of barite is for oil well **drilling fluids**. In this application, powdered barite is mixed with water and other agents to produce a fluid called a "mud." Drilling mud has a high density. It is circulated during well drilling to lubricate the drilling bit, to remove the drilling debris, and to prevent the escape of gas under pressure from the well. The name barium comes from the Greek word for heavy (*barys*) and refers to the high density (4.50 g $mL^{-1}$) mineral barite. Lavoisier included barite in his 1789 table of elements.

About 97% of US barium consumption is as barite and only about 3% is converted to other chemical uses. Barium hydroxide is a useful base. Barium sulfate, $BaSO_4$, is used as a shadowing agent for x-ray examination of soft tissue, such as the stomach; $BaSO_4$ is the so-called "barium diet." Soluble barium salts are very toxic, but barium sulfate *can* be used in this application because it is extremely insoluble. Barium compounds are also used as paint pigments and to control the passage of x-rays through television screen glass. Barium metal is manufactured by the high temperature reduction of barium oxide with aluminum:

$$3\ BaO(s) + 2\ Al(l) \rightarrow 3\ Ba(s) + Al_2O_3(s)$$

There are some similarities between the properties of barium $[Xe]6s^2$ and those of lead $[Xe]6s^25d^{10}4f^{14}6p^2$. For example, soluble compounds of both metals are extremely poisonous.

All the isotopes of **radium** are radioactive. Radium was first recovered from pitchblende, a uranium ore, and the story of its discovery in 1898 by Marie and Pierre Curie is well known. The Curies were able to recover approximately 1 g of radium from 7 tons of ore. Radium is present in all uranium minerals as a **natural decay product**, and could be recovered from this source if needed. The significance of radium is now largely historical. Although formerly used for radiation treatment of cancer, it has been supplanted by various synthetic nuclides. The longest lived isotope of radium ($^{226}Ra$) has a half life of approximately 1600 years.

## Metallurgy

Historically, the ability to work and use metals is associated with the beginnings of human settlement and culture. Bronze is an alloy primarily of copper and tin. The bronze age lasted from roughly 3500-1000 BC. Around 1000 BC the iron age began. For the reasons described in Chapter 12, metals and alloys are valuable materials. Copper (a fairly unreactive metal) can be obtained merely by "roasting" its sulfide ore in air. A few unreactive metals occur "native." A **native element** is one found in nature in chemically uncombined state. Gold, silver, mercury, and even copper are native metals.

The most chemical aspect of metallurgy is **smelting** or reducing metal ores to metals. The manner in which a metal is obtained from its ore depends on the activity of the metal. Metals in chemical compounds have positive oxidation states. Metals as metals have zero oxidation states. Thus, the reduction of a metal involves chemical processing to lower its positive oxidation state to zero.

Highly active metals (K, Na, Al, etc.) are obtained by electrolytic methods to be described in Chapter 14. Slightly less active metals can be obtained by reduction of one of their compounds by a more active metal. As we've already seen, many "exotic metals" (Be, Pu, etc.) are obtained in such a manner.

Iron (a moderately active metal) has been manufactured for hundreds of years using carbon (charcoal or coke) as the reducing agent. Figure 13.2 shows German iron workers circa 1550. It comes from Georgius Agricola's famous book *De Re Metallica*, a compendium of useful and practical information about metals and their properties. With the decline of the English forests, the iron masters of the eighteenth century turned from charcoal to the use of coke for iron making. Coke is a vesicular (like a solid foam) form of carbon made by heating coal; its use ushered in the industrial revolution. Figure 13.3 shows one of Abraham Darby's iron furnaces. This particular furnace, now preserved, is believed to have been built on foundations that go back to the late seventeenth century. [❁kws "worldofcoke"]

Figure 13.2 Sixteenth century German iron workers. Illustrated in the famous work *De Re Metallica*. The book was once translated by the US President, and engineer, Herbert Hoover.

Figure 13.3 An Abraham Darby iron furnace. Circa 1750, Shropshire, England. You can judge the scale from the two people standing just to the right of the modern roof. Furnaces such as this one ushered in the industrial revolution.

In 1779 Darby, in the county of Shropshire, England, designed, cast, and erected an iron bridge over the river Severn (Figure 13.4). He was the third generation of owners of the famous Coalbrookedale iron works. The iron bridge represented the first use of cast iron for an industrial scale project. It is still open today for foot traffic. Construction of the bridge had a considerable impact. In America, Tom Paine proposed building an American iron bridge in Pennsylvania with thirteen ribs—one rib for each of the original States of the union. The redox reaction $2\ Fe_2O_3 + 3\ C \rightarrow 4\ Fe + 3\ CO_2$ was used to reduce the iron ore to make iron for the bridge.

Figure 13.4 *The* iron bridge. First monument to the power of chemistry and engineering to transform the world.

## Uses of the d-block Transition Metals

Obviously this chapter would become impossibly long if we discussed the properties of the remaining metals in the same detail that we've discussed groups I and II. So in Table 13.5 we'll just summarize some information about the uses and the consumption patterns of the transition metals. The data in Table 13.5 are *rough* estimates derived from various sources.

Table 13.5 **Estimated Annual US Consumption and Use Pattern of the d-block Transition Metals**

| Metal | Demand (tons) | Summary Consumption Pattern (Percentages are approximate) |
|---|---|---|
| | | First Row d-block Transition Metals |
| Sc | 0.5 | Al alloys for baseball/lacrosse bats. Mercury lamps. $^{46}Sc$ as tracer. |
| Ti | 800,000 | Dioxide pigment for paints, paper, and plastics 95%. Metal 5%. |
| V | 13,000 | Hardening element for steel 90%. Catalysts 10%. |
| Cr | 900,000 | Stainless steel 65%. Chemicals, refractories, plating 35%. |
| Mn | $2 \times 10^6$ | Steels 95%. Chemicals and batteries 5%. |
| Fe | $175 \times 10^6$ | Steel and other ferrous alloys. Essentially 100%. |
| Co | 12,000 | High T alloys 35%. Cutting tools 20%. Magnets 22%. Chemicals 20%. |
| Ni | 500,000 | Corrosion resistant alloys 90%. Catalysts and chemicals 10%. |
| Cu | $4 \times 10^6$ | Electrical 60%. Construction 15%. Alloys 30%. Chemicals 1%. |
| Zn | $1.5 \times 10^6$ | Galvanizing 35%. Brasses, die casting alloys 35%. Chemicals 20%. |
| | | Second Row d-block Transition Metals |
| Y | 25,000 | With lanthanides: alloys, ceramics, superalloys, lasers, catalysts etc. |
| Zr | 130,000 | 95% as zircon for refractories & foundry molds. Zr cladding. |
| Nb | 10,000 | >90% for high temperature, high strength steels. |
| Mo | 60,000 | Steels 90%. Chemicals 10%. |
| Tc | 0.001 | Medical radioisotope. |
| Ru | 5 | Electronics 60%. Oxide coating for Ti electrodes for Cl 30%. |
| Rh | 4 | Platinum alloys. Catalysts. Plating. |
| Pd | 40 | Catalyst: auto 25%, chemical 25%. Dental 25%. Electrical 20%. |
| Ag | 10,000 | Photography 50%. Contacts and conductors 25%. Plating 10%. |
| Cd | 60,000 | Plating 30%. Batteries 20%. Alloys 10%. Pigments 10%. |
| | | Third Row d-block Transition Metals |
| La* | 25,000 | * All lanthanides. Alloys. Ce for steel. Laser and electronic uses. |
| Hf | 50 | Nuclear control rods 50%. Alloys 50%. |
| Ta | 1,500 | Ta oxide capacitors 60%. Tool alloys 20%. Chemical equipment 10%. |
| W | 20,000 | W Carbide 65%. Alloys 20%. Lamps 5%. Chemicals 2%. |
| Re | 2 | Pt/Re catalysts for petrochemicals 90%. Alloys 10%. |
| Os | 0.2 | Minor chemical uses. |
| Ir | 2 | Crucible manufacture. Electrical contacts. Catalysts. |
| Pt | 45 | Catalysts. Crucible manufacture. Electrical alloys. Jewelry. |
| Au | 200 | Jewelry and artistic uses 50%. Electrical 40%. Dental 10%. |
| Hg | 1,300 | Electrical 40%. Chlor-alkali 35%. Instruments 20%. Dental 10%. |

Looking through Table 13.5 will give you some sense of how important transition metals are in a modern industrial economy. It's interesting that each element has some unique property that makes it the favorite for some specialized application. Even the synthetic radioactive element technetium finds medical applications. The US is a disproportionately large consumer of the world's metallic resources. During the 1990s the US consumption of most of the metals in the table was 10-40% of global consumption.

## Oxidation States of Metals

We'll study detailed rules for finding oxidation numbers and oxidation states in the following section. All metals in their compounds have positive oxidation states. Nonmetals (except fluorine) exhibit both negative and positive oxidation states. Table 13.6 shows the oxidation states of the main group metals. The maximum oxidation number of the main group metals is equal to their number of valence electrons (Section 6.2).

Aluminum, like the metals of groups 1 and 2, has only one oxidation state and that is 3+. All of the other elements on the right hand side of Table 13.6 show two oxidation states, as indicated at the top of the table.

| Table 13.6 **Oxidation States of Main Group Metals** | | | | | |
|---|---|---|---|---|---|
| +1 | +2 | | +3<br>+1 | +4<br>+2 | +5<br>+3 |
| Li | Be | | | | |
| Na | Mg | . . . | Al | | |
| K | Ca | | Ga | | |
| Rb | Sr | | In | Sn | Sb |
| Cs | Ba | | Tl | Pb | Bi |
| Fr | Ra | | | | |

There are too many oxidation states of the transition metals to fit them conveniently into a periodic table. Table 13.7 shows the oxidation states of the d-block transition metals in a different format—a format with a distinguished tradition. Dmitri Mendeleev himself was the first to use it as he worked out the periodic table. More important and frequently encountered oxidation states are shown in bold.

| Table 13.7 **Oxidation States of d-Block Transition Elements in Mendeleev Format** | | | | | |
|---|---|---|---|---|---|
| First Row | | Second Row | | Third Row | |
| Sc | 0 **3** | Y | 0 **3** | La | 0 **3** |
| Ti | 0 2 **3** **4** | Zr | 0 1 2 3 **4** | Hf | 0 3 **4** |
| V | 0 1 2 3 **4** 5 | Nb | 0 1 2 4 **5** | Ta | 0 1 2 4 **5** |
| Cr | 0 1 2 **3** 4 5 **6** | Mo | 0 1 2 3 4 5 **6** | W | 0 1 2 3 4 5 **6** |
| Mn | 0 1 **2** 3 **4** 5 6 7 | Tc | 0 1 2 3 4 5 **6** 7 | Re | 0 1 2 3 4 5 6 **7** |
| Fe | 0 1 **2** **3** 4 6 | Ru | 0 1 2 **3** 4 5 6 7 8 | Os | 0 1 2 3 **4** 5 6 7 8 |
| Co | 0 1 **2** **3** 4 | Rh | 0 1 2 **3** 4 6 | Ir | 0 1 3 **4** 5 6 |
| Ni | 0 1 **2** 3 4 | Pd | 0 **2** 4 | Pt | 0 2 **4** 5 6 |
| Cu | 0 **1** **2** 3 | Ag | 0 **1** 2 3 | Au | 0 1 2 **3** |
| Zn | 0 **2** | Cd | 0 **2** | Hg | 0 **1** **2** |

The arrowhead shapes in Table 13.7 are in themselves good evidence for periodicity. Oxidation states maximize around 8, halfway across the transition metal series. Note also that the very highest oxidation states occur in the second and third row metals. In the left hand half of the d-block (periodic table groups 3-7) the maximum oxidation state of a transition metal is equal to its number of valence electrons (Section 6.2). A transition metal in a high oxidation state is likely to be a good oxidizing agent; the reason, in part, is that it has a stable lower oxidation state—reached when it gains electrons via reduction. We'll see later that reactions such as manganese +7 → +2 and chromium +6 → +3 are common.

## Section 13.2 Oxidation and Reduction

### Oxidation Numbers and Oxidation States

An **oxidation number** is a value assigned to an atom in a molecule or ion according to conventional rules. Oxidation number is closely related to the experimental oxidation states of the elements in their compounds that we discussed earlier. Most oxidation numbers are integers—but some are fractional.

An **oxidation state** is a value assigned to an atom in a molecule based on its chemical properties and reactivity. Often, the terms oxidation number and oxidation state used with equivalent meaning. We will use them interchangeably here.

Are oxidation numbers are "real." The answer is "sort of." When we say that metals in high oxidation states are good oxidants it's a perfectly useful chemical generalization. However, if we are using oxidation states to balance some complicated, obscure redox reaction it's perhaps sufficient to regard them as a chemical bookkeeping device.

Here are five uses for oxidation numbers and oxidation states:

1. To name inorganic compounds—the Stock nomenclature. In his system, the oxidation state of the metal is placed in Roman numerals: titanium(IV) sulfate, iron(II) oxide, etc.
2. To systematize the study of the chemistry of the elements. Compounds that contain the same element with the same oxidation number frequently undergo similar chemical reactions and have similar properties.
3. To correlate the reactivity of an element with its different oxidation states.
4. To study, and particularly to balance, redox reactions.
5. To represent electrochemical reactions (Chapter 14).

An atom, molecule, or ion that loses electrons is said to have been **oxidized**. An atom, molecule, or ion that gains electrons is said to have been **reduced**. A **reduction-oxidation reaction** (or just **redox reaction**) is one in which one reactant is oxidized while another is *simultaneously* reduced. In Section 3.1 we introduced the general form of a redox reaction as:

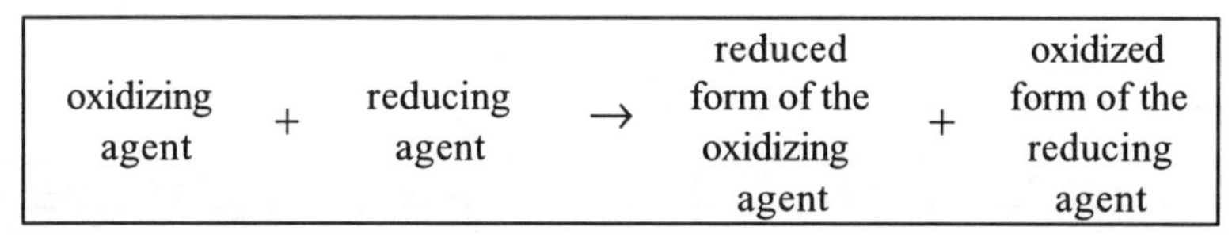


or, even simpler:

| oxidant | + | reductant | → | reduced oxidant | + | oxidized reductant |
|---|---|---|---|---|---|---|

**Example 13.2**: Show some simple redox reactions.

Strategy: Pick some elements and monatomic ions that gain or lose electrons.

$Cl_2 + 2e^- \rightarrow 2Cl^-$ A chlorine molecule is reduced to two chloride ions

$Sn^{4+} + 2e^- \rightarrow Sn^{2+}$ A tin(IV) ion is reduced to a tin(II) ion

$Zn \rightarrow Zn^{2+} + 2e^-$ A zinc atom is oxidized to a zinc ion

$Fe^{2+} \rightarrow Fe^{3+} + e^-$ An iron(II) ion is oxidized to an iron(III) ion

Redox reactions showing electron gain or loss are given in Example 13.2. Redox reactions written to show electrons are called **half-reactions** with the symbol $e^-$ being used.. In chemical equations where electrons are shown, if the electrons are on the left of the equation the reactant is getting reduced, if the electrons are on the right the reactant is getting oxidized. Two half-reactions with equal numbers of electrons on opposite sides can be combined to give a net reaction. The procedure of combining pairs of redox half-reactions to make a net reaction is shown in Example 13.3.

**Useful mnemonic**: As a memory aid remember OIL-RIG.

OIL = oxidation is loss of electrons

RIG = reduction is gain of electrons

Because redox reactions involve the transfer of electrons, we can study such reactions via electrical measurements and can harness the electrons for use in ordinary electric batteries. That's called electrochemistry and it's the subject of Chapter 14.

**Example 13.3**: Use the half-reactions shown in Example 13.2 to illustrate how pairs of half-reactions can be combined into net redox reactions.

Strategy: Combine the first and third examples and the second and the fourth examples. In the second case the oxidation half reaction must be doubled so that the electrons cancel.

$Cl_2 + 2e^- \rightarrow 2Cl^-$ A chlorine molecule is reduced to two chloride ions
$\underline{Zn \rightarrow Zn^{2+} + 2e^-}$ A zinc atom is oxidized to a zinc ion

$Cl_2 + Zn \rightarrow Zn^{2+} + 2Cl^-$ Net reaction. Both of the above stated processes occur simultaneously.

$Sn^{4+} + 2e^- \rightarrow Sn^{2+}$ A tin(IV) ion is reduced to a tin(II) ion
$\underline{2\,Fe^{2+} \rightarrow 2\,Fe^{3+} + 2e^-}$ An iron(II) ion is oxidized to an iron(III) ion (doubled half reaction)

$Sn^{4+} + 2\,Fe^{2+} \rightarrow Sn^{2+} + 2\,Fe^{3+}$ Net reaction. Both of the above stated processes occur simultaneously.

Comment: When combining redox half-reactions, multiply one or both half-reactions by an integer so that the electrons cancel when the half-reactions are added.

The term oxidation obviously originated because reaction with oxygen is the classic oxidation reaction. Burning charcoal with air, $C + O_2 \rightarrow CO_2$, must be one of the oldest deliberately performed chemical reactions. In it, the oxidizing agent oxygen oxidizes the reducing agent carbon. Carbon is oxidized and oxygen is reduced.

In metallurgy, a very important reaction is the reduction of iron oxide to iron using carbon. What is it about iron that is reduced during reduction? It is iron's oxidation number, which is reduced from +3 to 0 in the reaction immediately below. To be reduced is to have one's oxidation number lowered. During the iron age, carbon would have been used in the form of charcoal. In the modern world, coke is used in the reduction (smelting) of iron ore.

$Fe_2O_3 + 3\,C \rightarrow 3\,CO + 2\,Fe$ charcoal smelts (reduces) iron oxide to iron metal.

During the manufacture of iron, iron(III) oxide is reduced, while charcoal acts as a reducing agent. Equally well one could say that carbon is oxidized, while iron(III) oxide acts as the oxidizing agent.

The oxidation of hydrogen by oxygen to produce water is a popular lecture demonstration. A hydrogen-filled balloon is approached at arm's length with a lighted candle (Figure 13.5a). The redox reaction is accompanied by a loud bang and a flash of yellow flame (Figure 13.5b). The ignition of a 2:1 stoichiometric hydrogen:oxygen mixture gives an especially violent explosion. Depending on your point of view, water is either oxidized hydrogen or reduced oxygen or both.

Figure 13.5a A hydrogen-filled balloon and a lighted candle securely taped to the end of a meter stick.

Figure 13.5b The reaction between oxygen and hydrogen gases is fast, noisy, and brilliant.

Many chemical reactions are redox reactions. *All* combustion and explosive reactions are redox reactions. An explosive is a compound or mixture that undergoes rapid self-redox to form lots of hot gas. A wide range of explosives can be purchased at any gun shop in the US.

The descriptive chemistry of the transition metals (and also of the nonmetals, to be described shortly) is largely a study in redox chemistry. Oxidation states are useful to catalog and rank elements and chemical compounds in terms of their ability to undergo oxidation and reduction.

The concept of oxidation-reduction is a valuable generalization in chemistry. Chemists extend the concept of redox widely by using assigned oxidation numbers. Complete rules for assigning oxidation numbers are lengthy; an abbreviated set of hierarchical rules follows. Hierarchical means that if two rules apply, the higher ranking rule takes precedence. For example, the oxidation number of fluorine in $F_2$ is assigned as zero (rule 1) not as -1 (rule 3), because rule 1 takes precedence over rule 3.

## Abbreviated Rules for Assigning Oxidation Numbers

**Rule 1.** Elements have an oxidation number of zero. The oxidation number of the element in a monatomic ion is the charge on the ion.

Examples:

O in $O_2$ and $O_3$ oxygen has an oxidation number of zero
Na, Ca, Cd, Hg and all metals (as metals) have an oxidation number of zero.
The diatomic halogens $F_2$, $Cl_2$, etc. have an oxidation number of zero.
The oxidation state of sodium as $Na^+$ is +1—likewise $Li^+$, $K^+$, $Rb^+$, and $Cs^+$, etc.
The oxidation state of barium as $Ba^{2+}$ is +2—likewise $Be^{2+}$, $Mg^{2+}$, and $Ca^{2+}$, etc.
The oxidation state of fluorine as $F^-$, is -1—likewise $Cl^-$, $Br^-$, and $I^-$, etc
The oxidation state of oxygen as $O^{2-}$, is -2—likewise $S^{2-}$, and $Se^{2-}$, etc.

**Rule 2**. The algebraic sum of the oxidation numbers of the atoms in a molecule is zero, and in a polyatomic ion is the net charge of the ion.

Examples:

In $H_2O$ $(2 \times +1) + -2 = 0$ in agreement with rules 3 and 4
S in the sulfate ion $SO_4^{2-}$ is +6, as proved by $(1 \times +6) + (4 \times -2) = -2$, which is the -2 net charge on the ion
Cl in the perchlorate ion $ClO_4^-$ is +7, as proved by $(1 \times +7) + (4 \times -2) = -1$, which is the -1 net charge on the ion
The oxidation state of carbon in formaldehyde $CH_2O$ is zero, as proved by $(1 \times 0) + (2 \times +1) + (1 \times -2) = 0$

**Rule 3**. Main group metals of Groups 1 and 2 have oxidation states of +1 and +2 respectively (in their compounds). Fluorine always has an oxidation state of -1. Aluminum always has an oxidation state of +3.

Examples:

In KCN the potassium is +1. That balances the -1 charge of the cyanide ion $CN^-$.
In $NaN_3$, the sodium is +1. The N must be therefore - $\frac{1}{3}$
In $Mg_3N_2$ the magnesium is +2. The nitrogen must therefore be -3
In $Na_3AlF_6$ The sodium is +1 and the fluorine -1. The aluminum is +3 as proved by the $(3 \times +1) + (1 \times +3) + (6 \times -1) = 0$

**Rule 4**. Hydrogen has an oxidation number of +1 (in its compounds). Rare exceptions: in its binary compounds with metals, hydrogen has an oxidation number of -1.

Examples:

In $HClO_4$ the chlorine must be +7 because $+1 + 7 + (4 \times -2) = 0$
In $NH_4^+$ the nitrogen must be -3 because $-3 + (4 \times +1) = +1$ (the net charge on the ion)

**Rule 5**. Oxygen has an oxidation number of -2 (in its compounds). Common exception (covered by rule 4): oxygen has an oxidation number of -1 in hydrogen peroxide, $H_2O_2$. Rare exceptions: other peroxides (-1) and some binary compounds with metals (-1 and other values).

Examples:

In CO the carbon must be +2
In $CO_2$ the carbon must be +4
In $NO_3^-$ the nitrogen must be +5 because $+5 + (3 \times -2) = -1$ (the net charge on the ion)
In $ZrO_2$, the Zr must be +4 because $+4 + (2 \times -2) = 0$
In $Cr_2O_7^{2-}$ the Cr must be +6 because $(2 \times +6) + (7 \times -2) = -2$

**Example 13.4**: Write oxidation numbers above the atoms of each reactant and product in the balanced equation $HNO_3 + 4\ H_2S \rightarrow 4\ S + NH_3 + 3\ H_2O$. Identify the oxidizing and reducing agents. State what gets oxidized and what reduced.

Strategy: Apply the rules and write the assigned oxidation numbers of each element above the compound's formula. Examine the oxidation numbers to deduce what is going on.

$$\overset{+1\ +5\ -2}{HNO_3} + \overset{+1\ -2}{4\ H_2S} \rightarrow \overset{0}{4\ S} + \overset{-3\ +1}{1\ NH_3} + \overset{+1\ -2}{3\ H_2O}$$

The atoms that change oxidation number are nitrogen and sulfur:

Nitrogen changes from +5 to -3. Each atom of N has its oxidation number reduced by eight. Nitrogen gets reduced and is the oxidizing agent.

Sulfur changes from -2 to 0. Each atom of sulfur has its oxidation number increased by two. Sulfur gets oxidized and is the reducing agent.

## The Procedure to Balance a Redox Reaction

One application of oxidation numbers is to balance redox reactions. Balancing can be accomplished by following the prescribed set of steps as laid out in this section. The reason that redox equations can be balanced using oxidation numbers is that electrons lost by the oxidant must exactly balance electrons gained by the reductant. We have already established that principle for half-cell reactions, see Example 13.3.

Step 1: **Obtain the skeleton reaction**. Usually, one is given the skeleton equation, or told what the reactants and products are. The skeleton reaction will contain most of the reactants and products—maybe all of them. For a redox reaction that takes place in water, $H_2O$ can be added as needed on either side of the equation. If the reaction occurs in acid solution then both or either $H_2O$ and $H^+$ ion can be added to either side as needed; if the reaction occurs in basic solution, then both or either $H_2O$ and $OH^-$ can be added to either side as needed.

Step 2: **Assign oxidation numbers according to the rules**. Follow the procedure shown in Example 13.4.

Step 3A: **Identify the atom whose oxidation number goes up during the reaction**. This atom is being oxidized and releases a number of electrons equal to its initial oxidation state minus its final oxidation state.

Step3B: **Identify the atom whose oxidation number goes down during the reaction**. This atom is being reduced and gains a number of electrons equal to its initial oxidation state minus its final oxidation state.

Step 4: **Add coefficients to the equation to balance the gain and loss of electrons** as they were counted in steps 3A and 3B. The redox part is now complete; the next step involves balancing the remaining reactants and products by inspection.

Step 5: **Complete the balancing using algebra**. Add $H_2O$, $H^+$, or $OH^-$, as needed in aqueous solution to balance O, H, and electrical charge.

Step 6: **Check that the final equation is balanced**.

Redox often occurs in aqueous solution. Depending on the way we adjust the pH, the reaction might occur either in acidic or in basic solution. It is always easiest to balance redox reactions in acid solution—even though a specific reaction actually occurs in basic solution. In the latter case, balance it as though it occurred in acid solution and add $OH^-$ to both sides at the end.

## Examples of Balancing Redox Equations

In this section we apply the rules for oxidation states and the rules for balancing redox equations. Doing this is a chemical tradition extending over many years. The algebraic method described in Section 3.2 can easily be applied to these equations—particularly with the help of computer-run algebra solving routines. However, the procedures described here do give some actual chemical insight. Which is best? Take your choice. Incidentally, there is yet another popular method called the ion-electron method that we won't discuss.

Modern US pennies are zinc filled and copper clad. This fact can be demonstrated by using concentrated nitric acid to strip their copper cladding.

Figure 13.6a Nitric acid and copper pennies before reaction begins.

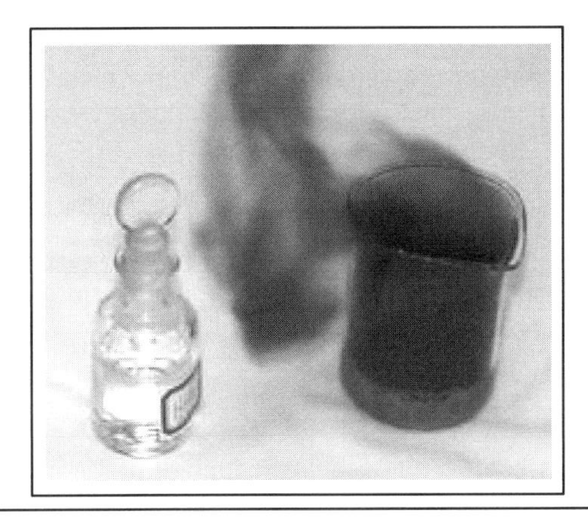

Figure 13.6b. The reaction release a large volume of $NO_2$ gas. It is dark brown in color.

The reactants are shown in Figure 13.6a. Figure 13.6b shows the reaction a few seconds after the nitric acid solution has been poured on the copper. The reaction proceeds vigorously with the liquid bubbling and turning greenish-blue. Abundant brown fumes of dense nitrogen dioxide gas are produced. Incidentally, you can dissolve the zinc out of the inside of a penny using hydrochloric acid if you first file some the copper off the edge of the coin. Hydrochloric acid dissolves zinc readily, but not copper. How come nitric acid will attack copper but hydrochloric won't? Nitric is an oxidizing acid, hydrochloric isn't. Best of all is to mix the two acids to make *aqua regia* ("royal water"). It was known to the alchemists from about 1200 because of its power to dissolve gold. Incidentally, the idiom "to pass the acid test" refers to gold's ability to resist acids (though not *aqua regia*). If a sample of purported gold dissolves, it has literally failed the "acid" test of being treated with an acid solution. How to balance the equation for the reaction pictured in Figure 13.6a is shown in Example 13.5. Balancing the slightly more complicated redox reaction for the oxidation of ethylene glycol by potassium permanganate is shown in Example 13.6. Although a coefficient of 1 is normally omitted from equations, in some that follow we deliberately write it, to emphasize that it's actually a known value at the stage of balancing we've reached.

**Example 13.5**: Write a balanced redox equation to show copper metal dissolving in a solution of nitric acid with the formation of copper(II) ion, $Cu^{2+}$, nitrogen dioxide, $NO_2$, and water. Write nitric acid as $H^+$ and $NO_3^-$.

Strategy: Write the skeleton equation and then follow the rules for balancing redox equations.

Oxidation states: 0 +1 +5 -2 +2 +4 -2 +1 -2

$$Cu + H^+ + NO_3^- \rightarrow Cu^{2+} + NO_2 + H_2O$$

So nitrogen the oxidant goes from +5 to +4 and gains one electron

Copper the reductant goes from 0 to +2 and loses 2 electrons

Next, we use the ratio 1:2 as coefficients for the reductant and oxidant. Doing this balances the 2 electrons lost with the 1 electron gained. Rewriting the skeleton equation with the 1:2 ratio and adding appropriate coefficients to the products $Cu^{2+}$ and $NO_2$ gives:

$$1\,Cu + H^+ + 2\,NO_3^- \rightarrow 1\,Cu^{2+} + 2\,NO_2 + H_2O$$

The next step is to balance the element oxygen by changing the coefficient of the product water. There are six oxygen atoms on the reactant side, so add a coefficient of 2 in front of water to make the total on the right add to six.

$$1\,Cu + H^+ + 2\,NO_3^- \rightarrow 1\,Cu^{2+} + 2\,NO_2 + 2\,H_2O$$

The final step is to add a coefficient of 4 in front of the reactant hydrogen ion. Doing this balances the four hydrogen atoms on the right.

$$1\,Cu + 4\,H^+ + 2\,NO_3^- \rightarrow 1\,Cu^{2+} + 2\,NO_2 + 2\,H_2O$$

Checking the charge shows that on the left it is +2 (4 × +1 plus 2 × −1) and on the right +2 as well. So the final equation is balanced both in net charge and net atoms.

**Example 13.6**: Write a balanced redox equation to show the oxidation of ethylene glycol ($C_2H_6O_2$) to carbon dioxide and water by permanganate ion in acid solution. The manganese-containing product of the reaction is $Mn^{2+}$. The products of oxidation are carbon dioxide and water.

Strategy: Write the skeleton equation and then follow the rules for balancing redox equations.

Oxidation states -1 +1 -2 +7 -2 +2 +4 -2 +1 -2

$$C_2H_6O_2 + MnO_4^- \rightarrow Mn^{2+} + CO_2 + H_2O$$

So manganese the oxidant goes from +7 to + 2; that's a gain of five electrons

Carbon the reductant goes from -1 to +4 for a loss of five electrons, but we have to double that to 10 because the reactant has a carbon subscript of two: $2\,C^- \rightarrow 2\,C^{4+}$ represents a total loss of 2 × 5 or 10 electrons.

Next, we use the ratio 1:2 as coefficients for the reductant and oxidant. Doing this balances the 10 electrons lost with the 5 electrons gained. Rewriting the skeleton equation with the 2:1 ratio and adding appropriate coefficients to the products $Mn^{2+}$ and $CO_2$ gives:

$$1\,C_2H_6O_2 + 2\,KMnO_4 \rightarrow 2\,Mn^{2+} + 2\,CO_2 + H_2O$$

Now note that there is a net charge of positive 6 on the right (2 × +1 plus 2 × +2). To balance this charge add 6 hydrogen ions (it's an acid solution) on the left and complete by inspection to get

$$6\,H^+ + 1\,C_2H_6O_2 + 2\,MnO_4^- \rightarrow 2\,Mn^{2+} + 2\,CO_2 + 6\,H_2O$$

Which is balanced both in net charge and net atoms.

Nitric acid and nitrate salts are good oxidizing agents because the nitrogen in them is in the highest possible oxidation state, +5, for nitrogen. Reduction of $N^{5+}$ often produces $N_2$ gas. This ability makes nitrates suitable for explosives ,as described in the following box.

## Gunpowder: The Classic Redox Reaction

Redox reactions have had a profound effect on human history. The introduction of gunpowder changed the course of warfare, and, until the advent of atomic weapons in the 20th century, the history of explosives was the history of ever more efficient redox reactions.

Natural flammable materials were used for war in Europe from as early as 500 BC and in China from as early as the 10th Century AD and probably much earlier. The fabled "Greek fire" was important in defense of Byzantium in the 7th century AD. Greek fire was probably composed mainly of naphtha, a chemical mixture obtained from seeps of natural crude oil. In China, it was known by the 11th century that nitre (potassium nitrate, $KNO_3$) boosted naphtha's incendiary properties. The mode of transmission of this information to the West is unclear, but by about 1300 AD, mixtures of charcoal and sulfur (as reductants) and nitre (as the oxidant) were being made in Europe for use in small arms.

Pure sulfur was long known in Europe—for example it occurs native in the volcanic regions of the island of Sicily, and charcoal is ancient. But while the reductants needed for gunpowder could be readily obtained in the Middle Ages, obtaining $KNO_3$ presented much greater difficulties. Its commonest source was from the earth underneath pigsties and stables, where urea from animal excreta is oxidized to nitrate by bacterial action. $KNO_3$–containing earth was extracted with water to which lime or potash had been added. The extracted liquor was concentrated by boiling until sodium chloride (the main impurity) crystallized, and hot $KNO_3$-liquor was obtained by filtering or decanting. On cooling, reasonably pure nitre crystals were obtained.

As with all explosives manufacture, mixing the ingredients is dangerous. Because the discharge of black powder involves a reaction among solids, the ingredients must be finely ground. Since about 1500 AD, the preferred formulation for fast-burning black powder has been 75% $KNO_3$ , 15% C, and 10% S. In the twentieth century, black powder has become obsolete as a blasting agent, but it remains the best fuse and igniter composition ever developed.

An approximate redox equation to represent the reaction of black powder is:

$$10\,KNO_3\,(s) + 12\,C(s) + 4\,S(s) \rightarrow 5\,K_2CO_3\,(s) + 4\,SO_2\,(g) + 5\,N_2\,(g) + 7\,CO(g)$$

The equation illustrates the main requirements of an explosive: it is a solid or liquid substance that converts quickly and substantially to gases, with the liberation of a lot of energy. Such reactions are invariably redox reactions.

## Useful Oxidizing Agents

The compounds of transition metals in high oxidation states have long been favored as laboratory oxidizing agents—used mainly for the purpose of volumetric analysis. Prominent among these are potassium permanganate, $KMnO_4$, where the manganese is in the +7 oxidation state, and potassium dichromate, $K_2Cr_2O_7$, where the two chromium atoms are in the +6 state of oxidation (Figures 13.7 and 13.8). (Sutton's 1871 Handbook of Volumetric Analysis devotes most of an entire chapter to these two oxidants and their uses.)

To use either $KMnO_4$ or $K_2Cr_2O_7$ as an oxidant for analysis, it's dissolved in water to make a standard solution with which to titrate any substance that can be stoichiometrically oxidized. For example, in this manner potassium dichromate can be used to test for alcohol in human breath or blood, and potassium permanganate can be used to test for ethylene glycol that reaches the environment in connection with aircraft deicing. These applications are illustrated in Example 13.7, which balances the stoichiometric reaction used in the former, and in Example 13.8 in the following section, which illustrates an actual analysis calculation.

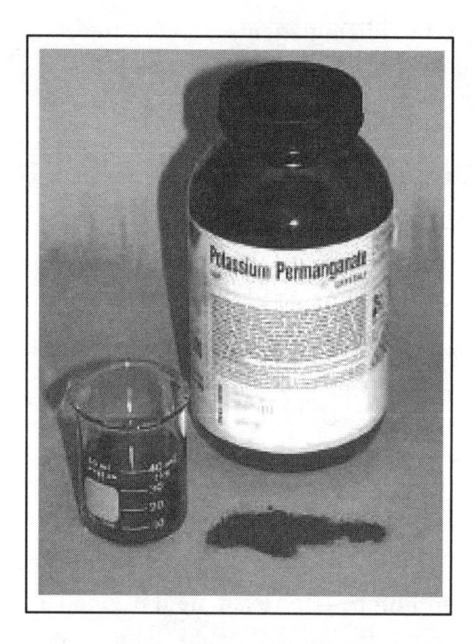


Figure 13.7 Potassium permanganate is an especially convenient oxidant to use because it has an intense dark purple color. The color disappears when $MnO_4^-$ is reduced. At the endpoint of the redox titration a pale pink/purple color remains, showing that the reducing agent has just been consumed by the oxidant in the previous drop of titrating solution. Thus, conveniently, potassium permanganate acts as its own indicator.

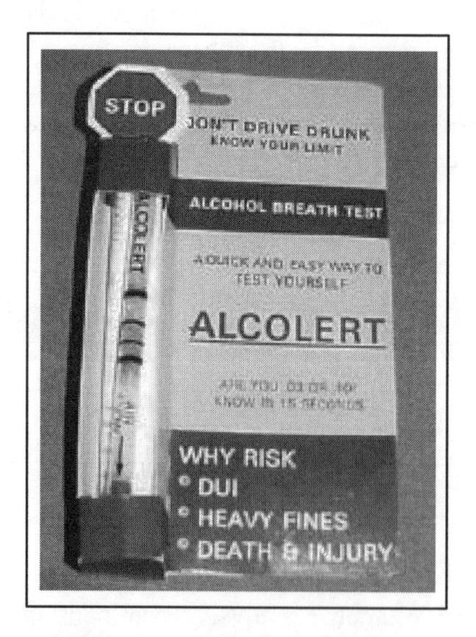


Figure 13.8 The redox chemistry of dichromate ion is put to work in devices that measure a person's breath alcohol level. The person getting tested blows over dichromate-impregnated granules in a tube. The farther down the tube the orange-to-green color change progresses, the more alcohol in the breath.

**Example 13.7**: Write a balanced redox equation to show the oxidation of ethyl alcohol ($C_2H_6O$) to carbon dioxide and water by potassium dichromate in acid solution. The chromium-containing product of the reaction is $Cr^{3+}$. This reaction can be used as a method of volumetric analysis to establish the blood alcohol level of a person suspected of drunk driving.

Strategy: Write the skeleton equation and then follow the rules for balancing redox equations.

Oxidation states: +1 +6 -2   -2 +1 -2   +1   +3   +4 -2   +1 -2

$$K_2Cr_2O_7 + C_2H_6O \rightarrow K^+ + Cr^{3+} + CO_2 + H_2O$$

For the oxidation part: $2\ C^{2-} \rightarrow 2\ C^{4+}$ that's a loss of 2 × 6 electrons, or a 12 electron loss (loss of 12 $e^-$).

For the reduction part $2\ Cr^{6+} \rightarrow Cr^{3+}$ that's a gain of 3 electrons × 2, or a 6 electron gain (gain of 6 $e^-$).

Next, we use the ratio 2:1 as coefficients for the oxidant and reductant. Doing this balances the 6 electrons gained with the 12 electrons lost. Rewriting the skeleton equation with the 2:1 ratio and adding coefficients to the products gives:

$$2\ K_2Cr_2O_7 + 1\ C_2H_6O \rightarrow 4\ K^+ + 4\ Cr^{3+} + 2\ CO_2 + H_2O$$

Now note that there is a net charge of positive 16 on the right (4 × +1 plus 4 × +3). To balance this charge add 16 hydrogen ions (it's an acid solution) on the left.

$$16\ H^+ + 2\ K_2Cr_2O_7 + 1\ C_2H_6O \rightarrow 4\ K^+ + 4\ Cr^{3+} + 2\ CO_2 + H_2O$$

Completing redox-balanced equation by inspection we get

$$16\ H^+ + 2\ K_2Cr_2O_7 + 1\ C_2H_6O \rightarrow 4\ K^+ + 4\ Cr^{3+} + 2\ CO_2 + 11\ H_2O$$

Which is balanced both in net charge and net atoms.

Ammonium dichromate is an explosive, but one that decomposes fairly slowly. Ammonium ion serves as the reductant and dichromate as the oxidant. Figure 13.9 shows the well known decomposition reaction of $(NH_4)_2Cr_2O_7$ forming green $Cr_2O_3$ as its solid product; $N_2$ and $H_2O(g)$ are leaving as gases.

Figure 13.9 The rather charming redox reaction: $(NH_4)_2Cr_2O_7 \rightarrow Cr_2O_3 + N_2(g) + 4\ H_2O(g)$

Hydrogen peroxide is a common oxidant used around the home; most people are familiar with the dark bottles found in many bathroom cabinets. Drugstore hydrogen peroxide is typically a 3% by weight solution of $H_2O_2$ in water. Chemists often use 30% solutions. Solutions of greater than 90% hydrogen peroxide concentration are exciting. They decompose according to the equation $2\ H_2O_2(l) \rightarrow 2\ H_2O(g) + O_2(g)$ and were used to power the attitude control jets of the Mercury spacecraft. Squirting a liquid 90% hydrogen peroxide solution onto a metallic catalyst produces hot puffs of gas that turn the vehicle. The author once saw 99% hydrogen peroxide poured on the ground with spectacular results, lots of hissing and vast clouds of steam.

## Redox Stoichiometry and Volumetric Analysis

Using redox for chemical analysis is based on the principle of reaction stoichiometry via titrations. Redox titrations are an important example of applied volumetric analysis—perhaps used somewhat less now than in the past because instrumental methods are so widespread—but still useful because they are often portable and inexpensive. Volumetric analysis is based on solution stoichiometry . Example 13.8 concludes our discussion of solution stoichiometry which began with the final section of Chapter 3.

**Example 13.8**: Ethylene glycol ($C_2H_6O_2$, MW = 62.0 g $mol^{-1}$) is used as a deicing agent for airplanes. Around 1990, an inadequately deiced plane crashed as it took off from Washington, DC. However, deicing operations have the potential to cause pollution problems, for example, when oversprayed ethylene glycol runs off into the Potomac River. Redox titration can be used measure the extent of run-off.

A 100.0 mL sample of Potomac River water was titrated with 0.0684 M potassium permanganate solution. It took 13.1 mL to reach the endpoint of the titration. Assume that the only oxidizable substance in the sample is ethylene glycol and calculate the concentration of ethylene glycol in mg $L^{-1}$.

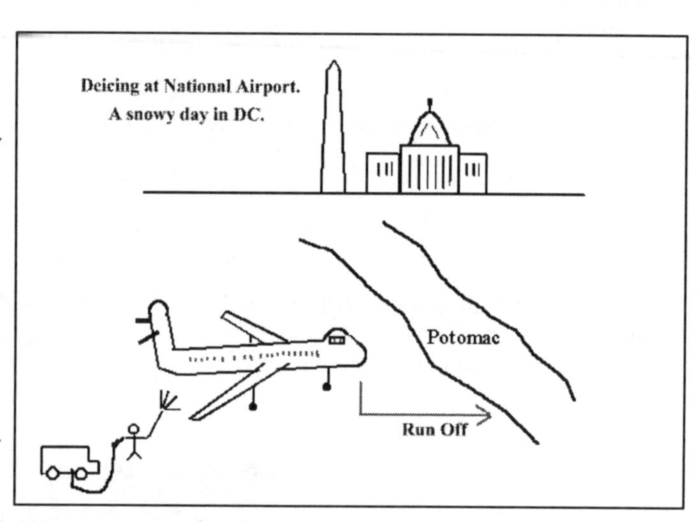


Strategy: Write the balanced redox equation and apply stoichiometry principles. The reaction was described and balanced earlier. The equation shows that 2 moles of potassium permanganate are needed for each mole of ethylene glycol:

$$6\ H^+ + 2\ KMnO_4 + 1\ C_2H_6O_2 \rightarrow 2\ K^+ + 2\ Mn^2 + 2\ CO_2 + 6\ H_2O$$

Moles of potassium permanganate = 13.1 mL × 0.001 L $mL^{-1}$ × 0.0684 mol $L^{-1}$ = $8.96 \times 10^{-4}$ mol

Moles of ethylene glycol = one half of this (stoichiometry) = $4.48 \times 10^{-4}$ mol

Mass of ethylene glycol = mols × molar weight = $4.48 \times 10^{-4}$ mol × 62.0 g $mol^{-1}$ × 1000 mg $g^{-1}$ = 27.8 mg in sample

Finally, divide the mass just obtained by the sample volume to get the desired unit

Ethylene glycol concentration = 27.8 mg ÷ (100.0 mL ÷ 1000 mL $L^{-1}$) = <u>278 mg $L^{-1}$</u>

That's much too high a concentration for comfort.

## Section 13.3 The Descriptive and Redox Chemistry of the Nonmetals

### Overview of the Nonmetals

This section begins with an overview of nonmetals (Tables 13.8), followed by a discussion of each group of nonmetals from right to left across the periodic table in the order: Group 18 —The Noble Gases; Group 17—The Halogens; Group 16—Oxygen and Sulfur; and Group 17—Nitrogen and Phosphorus. We will not discuss carbon, because most of its compounds are organic, and anyway we met them in Chapter 2 and during our discussion of polymeric solids in Chapter 11. Neither will we discuss here silicon, some of whose many mineral compounds we described in Chapter 11.

Including element 118, there are 17 nonmetals. Excepting C and Si leaves 15. Of those, seven are noble gases with very little chemistry.

Table 13.8 **An Overview of the Nonmetals**

| Group Number | | | | | Group Description |
|---|---|---|---|---|---|
| 14 | 15 | 16 | 17 | 18 | Group 18—Noble gases. Chemically unreactive. A few commercial applications. |
| | | | | 2<br>He | Group 17—Halogens. Very reactive. Many important applications. |
| 6<br>C | 7<br>N | 8<br>O | 9<br>F | 10<br>Ne | Group 16—Chalcogens. Fairly reactive. Oxygen and sulfur widespread and important. |
| | 14<br>P | 16<br>S | 17<br>Cl | 18<br>Ar | Group 15—Nitrogen and phosphorus. Very important applications. |
| | | 34<br>Se | 35<br>Br | 36<br>Kr | Group 14—Carbon. Very important. Discussed earlier. |
| | | | 53<br>I | 54<br>Xe | |
| | | | | 86<br>Rn | |
| | | | | 118<br>Uuo | |

Table 13.9 shows the consumption and use patterns of the nonmetals. Note that five nonmetals (O, P, Cl, N, and S) are used in large quantity. You should not take either the demand or the use pattern too seriously. Industrial statistics are reported on a wide range of different bases and a lot of judgment (guesswork) went into reducing the available data to just one line.

Table 13.9 **Estimated Annual US Consumption and Use Pattern of Nonmetals**

| Element | Demand (tons) | Summary Consumption Pattern (Percentages are approximate) |
|---|---|---|
| He | 15,000 | Cryogenic 25%. Purging 20% Welding cover 20%. Other 15%. |
| Ne | small | Neon tubes 75%. Lasers and other 25%. |
| Ar | 100,000? | Steelmaking and welding cover gas. 95%. Light bulbs and other 5%. |
| Kr | small | Fluorescent lamps and high speed flash lamps. 95%. |
| Xe | small | Lighting systems. 50%. Bubble chambers. 20%. Nuclear reactors 15%. |
| F | 500,000 | For Al production 80%. Flux 10%. Fluoro-organics 5%. Other 5%. |
| Cl | $42 \times 10^6$ | As NaCl. 45% electrolysis. Deicing 30%. Industrial 10%. Food 3%. etc. |
| Br | 250,000 | Fire retardants 25%. Well drilling 20% Sanitation 20%. Gasoline 15%. |
| I | 1,400 | Sanitation 40%. Pharmaceutical 25%. Rosin 13%. Catalyst 9%. Feed 7%. |
| O | $70 \times 10^6$ | Synthesis gas 60%. Steel 25%. Chemicals 10%. Other 5%. |
| S | $12 \times 10^6$ | Superphosphate 75%. Sulfuric acid 20%. Copper ore treatment 5%. |
| Se | 400 | Glass 35%. Metallurgy 30%. Pigments 20%. Electronics 10%. Other 5%. |
| N | $17 \times 10^6$ | As $NH_3$. Fertilizer 90%. Chemical, blasting, and other 10%. |
| P | $56 \times 10^6$ | As $Ca_3(PO_4)_2$ rock. Fertilizer 93%. Metal treatment 3%. Food etc. 3%. |
| C | large | Fuels 95%. Organic chemical compounds 5%. |

## The Noble Gases

The elements of the noble gas family (He, Ne, Ar, Kr, Xe) are distinguished by their lack of chemical reactivity. Their shared lack of reactivity is a prime example of periodicity. The term noble gas is quite old, being used as at least as early as W. A. Noyes 1914 General Chemistry textbook, which incidentally uses the name *niton* for xenon. Some of the properties of the noble gases, and the approximate number of known stable chemical compounds of each noble gas, are summarized in Table 13.10.

Table 13.10 **The Noble Gases—Some Properties**

| | He | Ne | Ar | Kr | Xe | Rn |
|---|---|---|---|---|---|---|
| Atomic Number | 2 | 10 | 18 | 36 | 54 | 86 |
| Outer Electron Configuration | $1s^2$ | $2s^22p^6$ | $3s^23p^6$ | $4s^24p^6$ | $5s^25p^6$ | $6s^26p^6$ |
| First Ionization Energy (kJ mol$^{-1}$) | 2,372 | 2,080 | 1,520 | 1,351 | 1,170 | 1,037 |
| Boiling Point (°C) | -268.9 | -246.1 | -185.9 | -153.5 | -108.1 | -62.2 |
| Approximate number of stable compounds known. | none | none | 1 | ca. 10 | <100 | * |

*Radon is radioactive and its compounds difficult to study.

In the 1870s and 1880s analysis of sunlight revealed a series of lines in its spectrum that did not occur in the spectrum of any element then known on earth. In this way, helium was discovered in the Sun—its name is taken from the Greek word for the sun. It had been known as early as the end of the 18th century that ordinary air contains traces of gases that resisted chemical combination by means of any reactions known at the time. Intensive studies of air around 1900 led to the isolation and separation (by means of fractional distillation) of the entire noble gas family. They were added *en masse* to the periodic table—the only group of elements in the periodic table to have been added in this manner. To accommodate them, a whole new group was added to the original Mendeleev version of the table.

The noble gases provide a good example of vertical periodicity. Their ionization energies decrease in a regular manner, while their boiling points increase likewise. The most important feature of the chemistry of the noble gases is that there is *so little of it.* This situation is attributed to the closed shell structure of the atoms of the noble gases. Their lack of chemical reactivity might be regarded as *experimental proof of the correctness of quantum theory*; it certainly is powerful experimental evidence.

No stable compounds of helium, neon, or argon are known. Krypton forms only a handful of stable compounds, and xenon a modest number. The first known compound containing a xenon-sulfur bond was reported in the Spring of 1998. All known radon isotopes are radioactive, so its compounds can be studied only with considerable difficulty. In agreement with the decreasing ionization energies listed in the table, the group trend is for noble gas compounds to become more stable going down the group. Therefore, if stable isotopes of radon or element 118 existed, they would have the most extensive chemistry of any of the noble gases. The recently discovered element 118 has a half life so short that it will be difficult to measure its properties and impossible to prepare any compounds of it—even though we would safely predict that many such compounds would exist if element 118's nucleus was not so unstable.

## Uses of the Noble Gas Elements

Helium is the second most abundant element in the universe. However, there is only a trace of helium in the earth's atmosphere. Helium atoms are sufficiently light that many reach escape velocity (compare Example 4.18), and the earth steadily loses helium to space. That loss, however, is counterbalanced by the decay of radioactive minerals via alpha particle emission (you recall that the alpha particle is a helium nucleus) a process which serves as a continuous source of helium in underground rock formations. All gases emanating from the earth contain helium; some natural gas wells yield as much as 7% helium. The helium is extracted from natural gas in plants operated by the US Government. Liquid helium is valuable as a **cryogen** (a cold source) for cooling superconducting magnets, for example (Figures 13.10 and 13.11). Helium gas is used in lifting balloons.

Figure 13.10 Filling a superconducting magnet (in the rear) with liquid helium from the super well-insulated traveling container in the foreground.

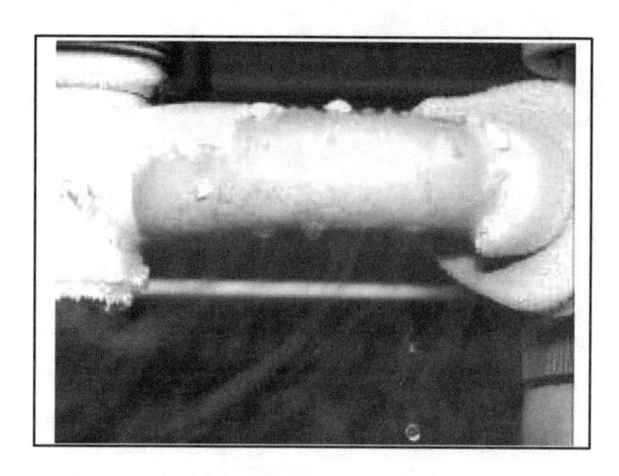

Figure 13.11 Liquid helium (boiling point 4 K) is transferred through a pipe with a vacuum jacket. However, even the vacuum jacket is not sufficient to prevent condensation of air (boiling point of nitrogen 77 K; boiling point of oxygen 90 K). In this picture you see droplets of liquid air condensing on the transfer pipe and dripping from it. Two falling drops are caught in mid air by the photographer.

Argon is by far the most abundant of the noble gases in the earth's atmosphere—dry air contains 0.93% argon by volume. Argon (BP = -185.9°C) boils between nitrogen (BP = -195.8°C) and oxygen (BP = -183.0°C). Argon is thus widely available as a by-product of the fractional distillation of liquid air; several hundred air separating plants are operated in the US. Thus, large quantities of liquid and gaseous argon are available, typically in 99.996% purity. There are three main uses of argon: 1. as the inert blanketing gas for arc welding of many metals; 2. as an inert purging gas in the manufacture of stainless steels; and 3. as an inert atmosphere for incandescent light bulbs.

The noble gases other than argon have minor uses, mainly deriving from their optical characteristics. You have seen red-colored neon discharge tubes used for advertising signs, perhaps the red light of a helium-neon laser at a supermarket check out, or the bright blinking of a xenon flash lamp at a light show. Xenon-filled, plastic lamps have recently been developed for the automobile headlamp market. Not only are they lighter weight than glass lamps, they produce high light flux and have a long service life. Krypton also finds use in high intensity lighting.

## Compounds of the Noble Gases

In its stable compounds, xenon frequently occurs in combination with the electronegative elements fluorine and oxygen. In Section 6.4 we discussed the valence shell electron pair repulsion theory of molecular structure. VSEPR works well for the compounds of the noble gases; in the case of xenon, with a number of stable compounds whose structures have been determined experimentally, we see that there is good agreement between VSEPR predictions and the facts. Table 13.11 shows some xenon compounds and their structures.

Table 13.11 **Some Compounds of Xenon**

| Oxidation number of Xenon | Compound | Melting Point °C | Experimental Structure | Predicted by VSEPR Theory | Bonding pairs (BP) and Lone pairs (LP) |
|---|---|---|---|---|---|
| +2 | $XeF_2$ | 129 | Linear | Linear | 2BP, 3LP |
| +4 | $XeF_4$ | 117 | Square planar | Square planar | 4BP, 2LP |
| +6 | $XeF_6$ | 50 | Distorted octahedron | Distorted octahedron | 6BP, 1LP |
| +8 | $Ba_2XeO_6$ | >300 | $XeO_6^{4-}$ octahedron | Octahedron | 6BP, 0 LP |

Xenon occurs with oxidation numbers of +2, +4, +6, and +8 (which is its group number, 18, minus 10). Noble gas compounds are all highly reactive. To say this is the converse of saying that noble gas compounds are difficult to prepare.

## The Halogens

The group name halogen was a learned borrowing from the Greek that means "salt-former." The group 17 halogen elements (F, Cl, Br, I) are characterized by having [core]$ns^2np^5$ outer shell electron configurations, and by the chemical tendency to act as powerful oxidants, gaining an electron to become halide ions. The symbol X is commonly used to represent any halogen element, and when they act as oxidants the equation for their reduction is written:

$$X_2 + 2e^- \rightarrow 2X^-$$

Some properties of the halogens are shown in Table 13.12. You can see for yourself that Cl-Br-I form a triad. Chlorine, bromine, and iodine all have many uses and form many useful compounds. Chlorine is produced in large quantities (currently about 100 lbs per person annually in the US), and is used for purposes such as water treatment, plastics manufacture, paper and cotton bleaching, and pesticide and solvent manufacture. Bromine (currently produced at about 150,000 tons annually in the US) is used to manufacture gasoline additives, sanitizing compounds, and fire extinguisher fluids. Iodine (currently consumed at about 3000 tons annually in the US, some of which is imported) is used in photographic chemicals, pharmaceuticals, and disinfectants The electron configuration of the halogen elements gives them a high electron affinity. Consequently, each of the halogens exists as a negative ion in naturally occurring compounds or solutions. A solution of a metal halide salt is often called a **brine**. Sea water contains chloride, bromide, and iodide ions, along with positively charged ions—mostly sodium ions.

Table 13.12 **Properties of the Halogens**

| | Fluorine | Chlorine | Bromine | Iodine |
|---|---|---|---|---|
| Electron Configuration | $[He]2s^22p^5$ | $[Ne]3s^23p^5$ | $[Ar]3d^{10}4s^24p^5$ | $[Kr]4d^{10}5s^25p^5$ |
| Form | Colorless Gas | Green Gas | Brown Liquid | Purple Solid |
| Boiling or melting point | BP = -188.1°C | BP = -34°C | BP = 59.5°C | MP = 113.6°C |
| Oxidation states | 0,-1 | 0,-1,+1,+3,+5,+7 | 0,-1,+1,+3, +5,+7 | 0,-1,+1,+3, +5,+7 |
| Bond energy (kJ $mol^{-1}$) | 158.8 | 242.6 | 192.8 | 151.1 |
| Ionization energy (kJ $mol^{-1}$) | 1,681 | 1,256 | 1,143 | 1,009 |
| Electron Affinity (kJ $mol^{-1}$) | 333 | 349 | 325 | 296 |
| $X^-$ radius (pm) | 133 | 184 | 196 | 220 |

### Periodicity and the Sources of the Halogen Elements

The similarity of the commercial sources of chlorine, bromine, and iodine—as water soluble sodium salts—is an illustration of how periodicity can be responsible for the way elements are deposited in the earth's crust.

Chlorine is mainly produced from rock salt (mined, mineral sodium chloride). Sea water contains sodium chloride, but large scale commercial production from this source is not competitive with production from rock salt solutions (synthetic brines). Bromine, however, is sufficiently abundant, and is sufficiently valuable that some is commercially recovered from sea water. Iodine and bromine are also produced from natural brines that are impure solutions of sodium bromide and sodium iodide in water. Natural brines are commonly associated with oil fields, and also occur in commercially significant quantities beneath the state of Michigan. Dow Chemical has its headquarters here because Herbert Dow founded it where the raw materials were. The halogens illustrate the general rule that main group elements in a vertical group have strong chemical similarities.

Why is fluorine not geochemically associated with the other halogens? The commercial sources of fluorine, the halogen at the top of periodic group 17, are minerals composed of calcium fluoride ($CaF_2$). Calcium fluoride is insoluble in water (unlike the other calcium halides). Because calcium is widespread, fluoride ions that reach natural waters are almost all precipitated. Over geologic time, the precipitated calcium fluoride has formed the ores that are now mined. This distinction of fluorine from the other halogens reminds us that periodicity refers only to the general similarities among elements in a group, and does not mean that they have identical properties. The combination of similarities and differences is really what makes chemistry a rich discipline in which to study and do research.

The halogens illustrate another general rule about the periodic table: the topmost element in a group has chemical properties that are different from those of the lower elements in the group. In the alkali metal group, lithium differs significantly from the others and beryllium differs from the other alkali earth metals. The solubilities of lithium and beryllium salts are usually quite different from the other salts of their respective group elements.

(Incidentally, that's also true for the d-block elements, where the chemistry of first row element is always somewhat different from the second and third row elements—which often have similar chemistries.)

## Some Reactions of the Halogens

We can summarize the chemical similarities among the halogens by citing four classes of reactions in which they all participate to give comparable products:

1. Reaction with metals to form salts (X = F, Cl, Br, I)

$$2\,Na + X_2 \rightarrow 2\,NaX$$
$$2\,Fe + 3\,X_2 \rightarrow 2\,FeX_3$$
$$Ti + 2\,X_2 \rightarrow TiX_4$$

2. Reaction with hydrogen to form hydrohalic acids

$$H_2 + X_2 \rightarrow 2HX$$

3. Reaction with water to form simultaneously hydrohalic and hypohalic acids

$$X_2 + H_2O \rightarrow HOX + HX$$

4. Reaction with hydrocarbons to form halogenated hydrocarbons of varying degrees of halogenation, e.g. with methane:

$$CH_4 + X_2 \rightarrow CH_3X + HX$$
$$CH_3X + X_2 \rightarrow CH_2X_2 + HX$$
$$CH_2X_2 + X_2 \rightarrow CHX_3 + HX$$
$$CHX_3 + X_2 \rightarrow CX_4 + HX$$

Many other general reactions of the halogens are known; the four types cited above illustrate how periodicity allows chemists to summarize the properties of groups of element in generalized equations. When writing general reactions, chemists frequently employ **collective abbreviations** to stand for entire classes of elements. We have already encountered HA for a general weak acid: M is often used for metals and X for the halogens.

## Oxidation States of Nonmetals

The oxidation states of the nonmetals are shown collectively in Table 13.13. We've already seen that chlorine, bromine, and iodine in Group 17 exhibit oxidation numbers of -1, 0, +1, +3, +5, and +7. The positive states are high oxidation states, so compounds that contain halogens in these states are strong oxidizing agents. We'll see shortly that sulfur in Group 16 shows the range -2, 0, +2, +4, and +6, and that nitrogen in Group 15 shows states that range from -3 to +5. The lower the electronegativity of a nonmetal the more likely it is to form compounds in which it adopts a high oxidation state. The fact that nonmetals can adopt negative oxidation states distinguishes them from the metals. By combining in a single compound a nonmetal in both a high and low oxidation state, something is produced that can undergo self-redox. Most commercial explosives are of this type.

Table 13.13 **Main Oxidation States of the Nonmetals**

| Group 14 | Group 15 | Group 16 | Group 17 | Group 18 |
|---|---|---|---|---|
| | | | | He<br>0 |
| C<br>-4, -2, 0, +2, +4 | N<br>-3, -1, 0, +1, +3, +5 | O<br>-2, 0, +2 | F<br>-1, 0 | Ne<br>0 |
| | P<br>-3, -1, 0, +1, +3, +5 | S<br>-2, 0, +2, +4, +6 | Cl<br>-1, 0, +1, +3, +5, +7 | Ar<br>0 |
| | | | Br<br>-1, 0, +1, +3,+5, +7 | Kr<br>0, +2 |
| | | | I<br>-1, 0, +1,+3, +5,+7 | Xe<br>0,+2, +4, +6, +8 |
| | | | 85 At<br>? | Rn<br>? |

## Oxygen and Sulfur. The Group 16 Elements

The elements of Group 16 of the periodic table (O, S, Se, Te, Po) are sometimes called the **chalcogens**, from the Greek meaning those elements found occurring with copper. The smooth trends of nonmetallic properties found in the halogen and noble gas groups are not present in Group 16. Polonium, at the bottom of the group, has distinctly metallic characteristics such as high electrical conductivity. Tellurium and selenium are semiconductors while sulfur and oxygen are nonmetallic nonconductors.

**Oxygen** is the third most abundant element in the solar photosphere and the most abundant element in the earth's crust. It is present in chemically combined form in almost all rocks and minerals and, of course, in water. The earth's atmosphere began to accumulate oxygen about 2500 million years ago, with the evolution of photosynthesis. The present atmospheric concentration of oxygen is thought to have been quite stable for millions of years, maintained by the global ecological balance between photosynthesis and organic decay—processes that respectively produce and consume oxygen. This oxygen cycle is one of the major ecological redox-reaction processes of the biosphere. Lavoisier gave oxygen its name, taken from the Greek for "acid-producer," under the mistaken notion that all acids contain oxygen.

Oxygen is obtained commercially by the fractional distillation of liquid air. Oxygen is a reactive substance. It is paramagnetic due to the presence of two unpaired electrons in its molecule. It is available in pressurized gas cylinders and in liquid form. The largest consumer of oxygen is the steel industry, which uses it to feed blast furnaces, and in the steel making process itself, where oxygen's main role is to burn carbon from the crude, unrefined iron that comes from the blast furnace. Steel plants usually have oxygen plants adjacent to them. The gas is simply supplied via a pipe. Other uses of oxygen are: as a liquid oxidant for rocket fuels; in making synthesis gas (a mixture of CO and $H_2$) for methanol production; for the synthesis of ethylene oxide, hydrogen peroxide, nitric acid, and other chemicals; as an oxidant in fuel cells and wastewater treatment; and in medical applications.

Oxygen occurs as two **allotropes**: the familiar diatomic oxygen (dry sea-level air is 20.95% by volume oxygen), and the much less abundant triatomic allotrope ozone ($O_3$). Oxygen is the most important of all oxidizing agents because it is abundant and available at low cost. Ozone, a pale blue gas that condenses to a dark blue liquid at -111.5°C, is a much more powerful oxidizing agent than $O_2$. Next to fluorine, ozone is the strongest of the oxidizing agents listed in the table of $E^0$ values (Table 14.2 in the upcoming chapter).

Ozone is routinely generated in the laboratory by passing $O_2(g)$ between metal plates, where electrical discharges provide the energy for the reaction. Photocopy machines sometimes generate small amounts of ozone; perhaps you have noticed its characteristic odor. Ozone is sometimes used as a substitute for oxidizing agents such as $Cl_2$, for example, in the treatment of municipal water supplies. It has the advantage that the products of ozone oxidation are rarely harmful, unlike those of chlorine oxidation (chlorination). A disadvantage of ozone for water treatment is its higher cost. Many years ago the author participated in an experiment to make liquid ozone. It is a beautiful, deep-blue liquid. Another allotrope, $O_4$, was reported to have been made during 1999.

## Sulfur

Sulfur is one of the half-dozen elements known to the ancients. It occurs in elemental form in nature as a characteristic, low melting yellow "rock" composed of rhombic crystals (MP = 114°C). Solid sulfur can be set on fire, and its ancient name, **brimstone**, literally means "the stone that burns." Sulfur is found in many parts of the world, usually associated with oil-bearing deposits. It is formed in nature as a result of both microbial and chemical action. Relatively pure sulfur can almost always be found near volcanoes. Sulfur is a relatively unreactive solid, and, contrary to popular misconception, pure sulfur is essentially odorless. However, in air it burns with a pale blue flame to yield the gas sulfur dioxide, $SO_2$, that produces a choking sensation when inhaled. A wide variety of sulfate-containing minerals are known, examples being gypsum ($CaSO_4$) and barite ($BaSO_4$).

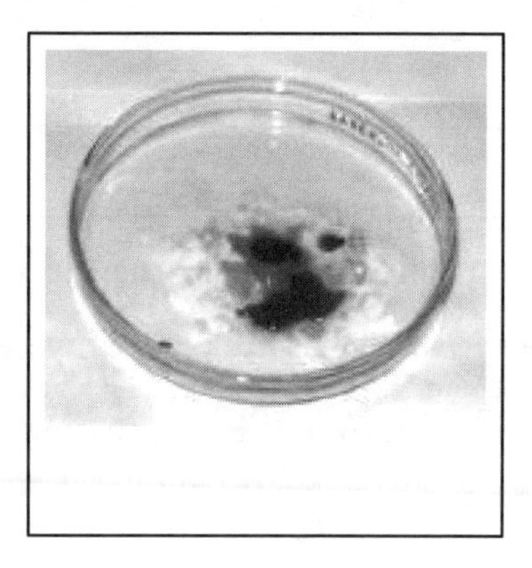

Figure 13.12. When sulfur burns, as shown here, some of the sulfur melts into a dark colored liquid, while at the same time producing clouds of choking sulfur dioxide gas.

Sulfur exists in a number of allotropic solid forms. The most stable form, rhombic sulfur, is composed of molecules containing cyclic $S_8$ molecules. When heated to 95.6°C, rhombic sulfur undergoes a slow transition to monoclinic sulfur, the crystals of which are also composed of $S_8$ molecules but packed differently from the rhombic variety. If cooled quickly, the solid sulfur initially remains in the monoclinic form but slowly reverts to the more stable rhombic form over a long period of time. Extensive studies of the phase diagram of sulfur have been made, and over 30 allotropic forms are known.

During the 20th century the main source of sulfur in the US has been the sulfur deposits near the salt domes of the Louisiana-Texas Gulf Coast. This sulfur is steam-mined using the Frasch process. Steam mining involves drilling down into sulfur formations and injecting steam at a sufficiently high pressure that its boiling point is above 114°C, the melting point of sulfur. Since 1950 the same process has been used in Mexico and Poland. About 60% of the US sulfur requirement is met by Frasch extraction; the other 40% is recovered from "sour" gases arising during petroleum processing, or from sulfur oxides recovered as a byproduct of copper smelting.

Most specifically, sour gas is natural gas that contains hydrogen sulfide. When sour gas is burned, the hydrogen sulfide reacts to give sulfur dioxide, an acidic (hence, sour) product. More generally, the term refers to any hydrocarbon gas contaminated by sulfur. The annual per capita US consumption of sulfur is about 120 pounds.

Sulfur is found in a wide range of compounds of which sulfuric acid is the most important. It forms binary compounds with many elements. One common oxidation state is -2. With metals it forms sulfides of a predominantly ionic character and with nonmetals sulfides of predominantly covalent character. Sulfur frequently exhibits positive oxidation states of +2, +4, and +6. Compounds that contain sulfur in positive oxidation states are oxidants. Some compounds and ions that contain sulfur in its common oxidation states are shown in Table 13.14.

| Table 13.14 **Compounds of Sulfur** | |
|---|---|
| | |
| Oxidation Number of Sulfur | Typical Compounds |
| 6 | $SO_3$, $H_2SO_4$, $SO_4^{2-}$, $SF_6$ |
| 4 | $SO_2$, $H_2SO_3$, $SO_3^{2-}$, $SF_4$ |
| 2 | $SCl_2$ |
| 0 | S, $S_8$ |
| -2 | $H_2S$, $S^{2-}$, $CS_2$ |

## Sulfuric Acid

The main use of sulfur is as a raw material for sulfuric acid, which is used for a wider variety of industrial applications than any other single synthetic chemical compound. Current US consumption of sulfuric acid is about 50 million tons per year, roughly equivalent to about one pound per day for every US citizen. Because its industrial uses are so widespread, the volume of sulfuric acid produced in a national economy has long been regarded as a country's economic barometer. The production of sulfuric acid from elemental sulfur is accomplished through the Contact Process, in which sulfur is converted to sulfur trioxide in two steps using air as the oxidant:

$$S(s) + O_2(g) \rightarrow SO_2(g) \quad \text{Step 1}$$

the preceding step is carried out in a sulfur burner, and is followed by

$$2\ SO_2(g) + O_2(g) \rightarrow 2\ SO_3(g) \quad \text{Step 2}$$

which uses a vanadium oxide/molten potassium salt catalyst and high temperature (about 450°C). The second step of the reaction is usually achieved in several passes over catalyst beds. Then the sulfur trioxide is converted to pyrosulfuric acid ($H_2S_2O_7$) in absorbing towers. The absorbing liquid is 99% sulfuric acid.

$$SO_3(g) + H_2SO_4(aq) \rightarrow H_2S_2O_7(aq)$$

Finally, the pyrosulfuric acid (also called oleum) is diluted to generate sulfuric acid.

$$H_2S_2O_7(aq) + H_2O(l) \rightarrow 2H_2SO_4(aq)$$

Modern plants typically convert greater than 99.8% of the sulfur to sulfuric acid. This high efficiency is necessary for economic reasons and to minimize loss of sulfur gases to the environment. About 700 sulfuric acid plants were operating in the US in the 1990. Most plants use the acid for fertilizer production captively—that is, the sulfuric acid produced is reacted on site, sold. The reaction between phosphate rock and sulfuric acid to make the so-called superphosphate fertilizer, $CaH_4(PO_4)_2.H_2O$, can be represented approximately by the equation:

$$\underset{\text{phosphate rock}}{CaF_2.3Ca_3(PO_4)_2(s)} + 7\ H_2SO_4(aq) + 3\ H_2O(l) \rightarrow \underset{\text{superphosphate fertilizer}}{2\ CaH_4(PO_4)_2.H_2O(s)} + 2\ HF(aq) + 7\ CaSO_4(s)$$

The calcium sulfate ($CaSO_4$) produced as a by-product of this reaction is called anhydrite. Some of this anhydrite is used in gypsum board for construction, but much of it is built up into man made white "mountains" that are a feature of the northwest Florida landscape. Also, from the reaction above some of the fluorine, in the form of fluosilicic acid ($H_2SiF_6$), is recovered for use in making fluorine compounds—about 5% of the US's fluorine demand is met this way. In addition to fertilizer manufacture, other high volume uses for sulfuric acid are phosphoric acid manufacture, making explosives, for pickling metals to clean them, and paper making.

## Nitrogen and Phosphorus. The Group 16 Elements

The elements of Group 15 show a wide range of chemical properties. Nitrogen and phosphorus at the top of the group are clearly nonmetallic in their behavior and are widespread and important elements. Arsenic and antimony are relatively minor elements best classified as metalloids. Bismuth is best classified as a metal.

## Nitrogen

Nitrogen's discovery is generally credited to Daniel Rutherford (1772), though the gas was found independently at about the same time by several others. As **azote**, it was included by Lavoisier in his 1789 table of the chemical elements. In the universe, nitrogen is a relatively abundant element (.01% by weight of the solar photosphere, for example). Nitrogen also accounts for about 0.01% by weight of the earth's crust, where it occurs principally in the form of soluble nitrate salts. Prior to the development of the Haber process these salts were mined as fertilizers. Except for the noble gases, diatomic nitrogen is the most inert chemical element.

Nitrogen is obtained commercially by fractional distillation of liquid air. The inertness of diatomic nitrogen is put to practical application when $N_2(g)$ is used as a blanket to exclude air and moisture in many chemical processes and storage facilities. Liquid nitrogen is widely used in the food processing industries for quick freezing, for portable refrigeration, and to produce an oxygen free atmosphere to retard spoilage. Small bottles of nitrogen gas are sold to wine connoisseurs so they can puff the gas into an opened bottle to preserve the contents.

Conversion of the chemically inert $N_2(g)$ into more reactive forms is called **nitrogen fixation**. Chemically, fixation generally means the reduction of $N^0$ to $N^{3-}$. With the help of microorganisms, nitrogen fixation occurs naturally in association with the root systems of many plants. The commercial chemical fixation of nitrogen by the Haber process is one of the bedrock chemical processes of modern technological societies and the key to intensive food production. The reaction of the Haber process is:

$$N_2(g) + 3H_2(g) \rightarrow 2NH_3(g)$$

The ammonia is either consumed directly as fertilizer or converted to nitric acid and then to nitrate salts for fertilizer or other use. In 1997 there were 26 US ammonia producers with 42 plant locations. The largest (in Donald sonville LA, shown in Figure 9.1) produces about 1000 tons of ammonia daily. Total annual US ammonia production amounts to roughly 150 pounds of ammonia each year for every US citizen. The cost of ammonia has ranged 0.1 - 0.2 \$ $kg^{-1}$ for the past several years.

## Compounds of Nitrogen

Nitrogen forms a wide range of compounds, including binary compounds with most of the other elements. It occurs with oxidation numbers from -3 (in ammonia, for example) up to +5 (in nitric acid and the nitrate ion). Nitrogen forms either three or four covalent bonds in the majority of its compounds. In inorganic chemistry, nitrate and nitrite salts of most metals have been prepared. Several nitrogen oxides are well known, and some of them are important in the production of nitric acid. In organic and biochemistry, nitrogen is present in amines and the amino acids; the main use to which green plants put nitrogen fertilizer is the production of amino acids for proteins. The range of nitrogen's oxides is shown in Table 13.15.

Table 13.15 **The Oxides of Nitrogen**

| Nitrogen Oxidation Number | Formula of Oxide | Name |
|---|---|---|
| 5 | $N_2O_5$ | dinitrogen pentoxide |
| 4 | $N_2O_4$ | dinitrogen tetroxide |
| 4 | $NO_2$ | nitrogen dioxide |
| 3 | $N_2O_3$ | dinitrogen trioxide |
| 2 | $N_2O_2$ | dinitrogen dioxide |
| 2 | NO | nitrogen monoxide or nitric oxide |
| 1 | $N_2O$ | dinitrogen monoxide or nitrous oxide |

Ammonia is converted to nitric acid via oxidation with air. The process is complicated (lot's of high-powered chemical engineering) and involves internal recycling and transition metal catalysts such as Pt or $CoO_3$ to get the needed equilibrium shifts and needed reaction rates. Despite the practical complexities, the net equation for the conversion is simple:

$$NH_3(g) + 2\,O_2(g) \rightarrow HNO_3(aq) + H_2O(l) \qquad \Delta H = -437\ kJ\ mol^{-1}$$

Much nitric acid is reacted with ammonia to produce ammonium nitrate.

$$HNO_3(aq) + NH_3(g) \rightarrow NH_4NO_3(aq)$$

The solution is evaporated to give $NH_4NO_3(s)$, a crystalline solid that is a more convenient form of fixed nitrogen than gaseous ammonia. Ammonium nitrate is highly soluble in water and is used in aqueous solution as a convenient liquid fertilizer. You can understand why solid ammonium nitrate is such a good explosive if you understand that it contains nitrogen both in its lowest oxidation state of -3 (in the ammonium ion) and its highest oxidation state of +5. The explosive reaction is complex. A simple approximate equation is:

$$NH_4NO_3(s) \rightarrow 2\ H_2O(g) + N_2O(g)$$

The notorious Oklahoma City and World Trade Center bombings of the 1990s were ammonium nitrate explosions. On its own, ammonium nitrate is a pretty safe compound. But when mixed with a little oil (ANFO or ammonium-nitrate-fuel-oil) and detonated it is deadly.

## Phosphorus

Phosphorus was the first of the chemical elements to be discovered by a deliberate chemical reaction. In 1669, the merchant Hennig Brand of Hamburg evaporated human urine to a syrupy, phosphate-rich liquid, heated this liquid to red heat with a mixture of sand and coal, and distilled phosphorus from the reaction vessel. What possessed him to undertake this remarkable experiment is lost to history.

Like sulfur, phosphorus has several solid allotropes. The common industrial form of phosphorus is **white** (also called "**yellow**") **phosphorus** that exists as discrete tetramers, $P_4(s)$, with a tetrahedral structure. White phosphorus is a waxy solid, soluble in organic solvents. It is highly flammable and spontaneously ignites if exposed to air. A phosphorus solution allowed to evaporate on a sheet of paper eventually bursts into flame—as demonstrated in Figure 13.13. Small amounts are kept stored under water. A second allotrope is **red phosphorus,** obtained by heating the white allotrope to about 250°C in the absence of air, or by exposing white phosphorus to ultraviolet light. Red phosphorus has properties quite different from white. The red form is much higher melting, insoluble in common organic solvents, and much less reactive. These properties are largely accounted for by its structure; red phosphorus is an amorphous solid that consists of $P_4$ tetrahedra joined at their corners.

Figure 13.13 Spontaneous combustion in action. Professor Luther Brice demonstrates the incendiary nature of phosphorus allowed to dry on paper from a solution in carbon disulfide.

Slow oxidation of phosphorus, via the reaction $P_4 + 5\ O_2 \rightarrow P_4O_{10}$, emits visible light. This light emission is called phosphorescence, hence the element's name.

## Compounds and Uses of Phosphorus

The abundance of phosphorus in the universe is relatively low. However, in the earth's crust, it is the twelfth most abundant element at 0.12% by mass. Its commercially exploitable deposits invariably are in the form of calcium phosphate minerals—usually of biogenic origin (formed as a result of life). In the US two main areas of mining are the regions around Pocatello, Idaho and the counties to the east of Tampa, Florida. Both these deposits are geologic "boneyards." The Florida one is of relatively recent origin and contains many phosphorus-rich fish bones—fossilized shark's teeth are abundant in this deposit.

Elemental phosphorus is manufactured by heating phosphate minerals, coke, and sand ($SiO_2$) in an electric furnace. The phosphorus in the mineral is reduced to elemental form that is distilled off. An equation for the process is

$$2\,Ca_3(PO_4)_2(s) + 6\,SiO_2(s) + 10\,C(s) \rightarrow 6\,CaSiO_3(l) + 10\,CO(g) + P_4(g)$$

Thus, the modern industrial process is reminiscent of Brand's original phosphorus synthesis. In the US, elemental phosphorus is produced in Idaho, from where it is shipped in hot, liquid form in insulated tank cars. In its end uses phosphorus is almost always combusted to give phosphorus(V) oxide in a quite pure form.

$$P_4 + 5\,O_2 \rightarrow 2\,P_2O_5$$

Shipping the phosphorus as the element still makes reasonable economic sense; air and water are available cheaply at the destination. The reaction of phosphorus(V) oxide with water produces phosphoric acid:

$$P_4O_{10}(g) + 6H_2O(l) \rightarrow 4\,H_3PO_4(aq)$$

However, during the past decade, the "wet-process" has been developed to produce phosphoric acid directly from the ore; doing this is cheaper than going through the element as described by the two equations above. The most important use of phosphorus compounds is as fertilizers. Much of the Florida phosphate-rock production is either shipped directly for this purpose or, as we discussed earlier, converted to more soluble forms via processing with sulfuric acid to make a more effective fertilizer.

When phosphorus burns in air, the oxide forms a cloud of white particles (as seen in Figure 13.13) and the resulting dense smoke is used for military pyrotechnics and smokescreens. Recent demand for elemental phosphorus has been growing because it is used in the manufacture of the popular glyphosate herbicides (Round-Up®). As described below, elemental phosphorus is useful in the manufacture of safety matches.

## The Chemistry of Matches

The common safety match is a familiar example of applied chemistry. Matches are an ingenious use of a pyrochemical reaction between two dry solids—the match head and the striking strip. The (usually) brown, striking strip of match books and boxes is impregnated with red phosphorus. The match head (which may be pigmented many different colors) contains a combustible mixture of oxidizing agents and fuels (a formulation is shown in Table 13.16). When the match is struck, friction releases a small amount of phosphorus on to the match head, the phosphorus ignites and fires the match head.

Table 13.16 **A Typical Match Head Composition**

| Ingredient | Amount | Purpose |
|---|---|---|
| | | |
| animal glue | 10% | binder/fuel |
| starch | 2% | serves as thickener during manufacture |
| potassium chlorate | 50% | oxidant |
| calcium carbonate | 3% | neutralizer—helps reduce gases |
| sand or silica | 30% | abrasive and filler |
| sulfur | 5% | fuel/reductant |
| potassium dichromate | trace | burning rate catalyst |

In the laboratory, or as a lecture demonstration, a carbon disulfide ($CS_2$) solution of white phosphorus placed on a small pile of potassium chlorate and allowed to evaporate produces a detonator—that is a chemical mixture that will explode because of the energy of mechanical shock when the mixture is struck. Such a detonation is captured in the photograph shown in Figure 13.14.

Figure 13.14 Pressure shock-sensitive detonation. A small pile of potassium chlorate is impregnated with a phosphorus solution and allowed to dry. When struck with a wooden meter stick it explodes with a loud crack, often blowing the end off the meter stick. An approximate, balanced equation for the reaction is: $3\ P_4 + 10\ KClO_3 \rightarrow 3\ P_4O_{10} + 10\ KCl$

## Some Important Phosphorus Compounds

Phosphorus forms phosphate and polyphosphate salts, and binary compounds with most of the elements. Phosphorus halides and oxohalides are useful chemical intermediates in the synthesis of phosphorus oxoacids. There is an extensive organic chemistry of phosphorus. As an alloying ingredient, phosphorus is important in such materials as phosphor-bronze. Chlorinated alkyl phosphate esters are useful fire retardants for textiles. Organophosphorus compounds are an important class of insecticides and find use as nerve gases.

Phosphorus is essential for all living organisms. Reactions involving adenosine triphosphate (ATP) and adenosine diphosphate (ADP) supply chemical energy for living organisms, and the nucleic acids are esters of phosphoric acid. Paradoxically, phosphorus is highly toxic in many chemical forms—such as in its white allotropic state or, as the phosphine ($PH_3$) that is used as grain fumigant and as an environmentally benign replacement for the greenhouse gas methyl bromide. One reaction for the formation of phosphine is:

$$Mg_3P_2 + 6\ H_2O \rightarrow 3\ Mg(OH)_2 + 2\ PH_3$$

The MSDS (material safety data sheet) for magnesium phosphide is in Appendix I.

The inadvertent fertilizing effect of phosphate on streams and lakes has sometimes caused undesired growth of algae and green plants. The problem has been well publicized in such bodies of water as Lake Erie and the Chesapeake Bay. For this reason, some States in the US have banned the use of phosphate salts in detergent compositions. In Virginia, the use of phosphates in home laundry detergents was banned in 1992.

## Prices of the Elements

Although we cannot consider chemical economics in any detail as part of our study of chemistry, we should point out that they are important in controlling the materials that can be used for specific purposes. The price of any element at a given time broadly reflects three factors: (1) the inherent difficulty of obtaining its ore and extracting the metal; (2) the commercial demand for the element; and (3) the price and availability of substitutes. The US government maintains a strategic stockpile of several vital elements, and it is anticipated that releases from this stockpile would serve to maintain orderly markets during periods of economic disruption. The replacement of metals (such as zinc, lead, tin, etc.) by plastics in many traditional applications has been a trend for many years, and we may expect this trend to continue. The metalloid selenium once enjoyed significant use in photocopiers as the photoelectric drum material. In 1999, substitute materials have displaced selenium, and there is a significant amount of selenium recycling from outmoded copiers.

Table 13.17 shows estimated prices of the transition metals—they vary widely. These figures are only *approximations*. Many of the metals are available in various grades and specifications. In general, higher purity commands a higher price. Furthermore, market factors may change prices daily. For example, the prices of copper, zinc, gold, silver, platinum, and several other metals are quoted on the world's commodity markets.

Table 13.17 **Estimated 1999 Prices of the d-block Transition Metals ($ kg$^{-1}$)**

| Sc | Ti | V | Cr | Mn | Fe | Co | Ni | Cu | Zn |
|---|---|---|---|---|---|---|---|---|---|
| 5,000 | 10 | 1,200 | 1 | 0.5 | 0.1 | 50 | 7 | 2.5 | 1 |
| | | | | | | | | | |
| Y | Zr | Nb | Mo | Tc | Ru | Rh | Pd | Ag | Cd |
| 2,000 | 300 | 40 | 10 | High | 600 | 9,000 | 2,500 | 100 | 5 |
| | | | | | | | | | |
| La | Hf | Ta | W | Re | Os | Ir | Pt | Au | Hg |
| 615 | 1,000 | 70 | 10 | 1,000 | 10,000? | 3,000 | 6,000 | 8,000 | 5 |

Other elements range from being free (atmospheric oxygen) to having incredibly high prices. Oak Ridge National Laboratory has a price list on line. [❁kws +"Oak Ridge" +americium]. The radioactive isotope $^{243}Am$ is $185 million kg$^{-1}$, $^{249}Bk$ is $185 thousand million kg$^{-1}$, and $^{240}Pu$, at only $56,000 kg$^{-1}$ is quite a bargain—provided you have the proper license to handle this special nuclear material with bomb-making potential.

## Essential Knowledge—Chapter 13

**Glossary words:** Descriptive inorganic chemistry, periodicity, electron configurations of the elements, redox chemistry, redox reactions, main group elements, transition elements, complex ions, catalysts, lanthanides, alkali metals, triad, alkaline earth metals, calcining, quicklime, slaked lime, plaster of Paris, getter, drilling fluids, native element, smelting, oxidation numbers and oxidation states, oxidizing agent (oxidant), reducing agent (reductant), half-reactions, volumetric analysis, cryogen, brine, chalcogens, allotropes, brimstone, nitrogen fixation, yellow phosphorus, red phosphorus.

**Key Concepts:** Have an overview of the periodic table by broad classification and its major regions of elements. Understand the significance of periodicity in its broad sense. Appreciate the economic roles of metals and their range of uses and applications. Oxidation and reduction and the use of oxidation numbers. Applications of oxidation numbers as a means of classifying chemical compounds and for balancing redox equations. The use of redox stoichiometry for volumetric analysis. Appreciate the economic roles of nonmetals and their range of uses and applications. The industrial and economic role of sulfuric acid, ammonia, phosphate fertilizers and other major compounds of the nonmetallic elements. Have a sense of the prices of the chemical elements.

**Key Equations:**

None

## Questions and Problems

### Metals

13.1. Sodium and potassium metals were first prepared within a few days of one another. Can you think of any good reasons for this?

13.2. Draw graphs showing the variations of the alkali metals' melting points, boiling points, atomic radii, and ionic radii with atomic number. Show that each of these properties has a periodic relationship as described in the chapter.

13.3. Commercial fertilizers are labeled as weight percent N:K:P. Thus a 20:10:5 fertilizer contains 20% nitrogen, 10% potassium, and 5% phosphorus. The salt dipotassium ammonium phosphate ($K_2NH_4PO_4$) contains all three of the major fertilizing elements. Calculate the N:K:P rating for dipotassium ammonium phosphate.

13.4. What triad of alkaline earth elements exhibit strong vertical periodicity? Cite four items each of physical and chemical data that confirm your answer.

13.5. Write balanced chemical equations to show the manufacture of lime and slaked lime from limestone. What are three important uses of lime?

13.6. Write the chemical formula for beryl. If beryllium can be recovered from beryl with a 55% overall yield, how many tons of beryl must be processed in order to obtain 100. lbs of beryllium?

13.7. Describe four commercial uses of beryllium as the metal, an alloy, or in the form of one of its compounds.

13.8. Beryllium resembles aluminum, diagonally below it to the right after the transition metals in the periodic table. Give experimental facts to support this statement and more generally to illustrate the nature of diagonal periodicity.

13.9. What is meant by a "getter?" Write the chemical reaction or reactions involved when barium acts as a getter.

13.10. Define or describe in 25 words or less each of the following terms: a. metalloid, b. transition metal, c. lanthanide, d. actinide, e. vertical periodic relationship, and f. diagonal periodic relationship.

13.11. Use internet resources to figure out what the element boron has to do with twenty mule teams.

13.12. The chemical reaction that takes place when a mixture of lye and aluminum shot is used to unclog a blocked drain produces $Al(OH)_4^-$ and hydrogen gas. Write an unbalanced equation for the reaction.

13.13. Aluminum is a relatively active metal as shown by its ability to reduce many active metal halides. However, enormous quantities of aluminum are used in construction, where the element is exposed to the environment. Explain why aluminum can be used so widely.

13.14. Speculate as to how early humans might have developed an industry to make bronze alloys several thousand years ago. Can you think of a sound chemical reason why the bronze age preceded the iron age?

13.15. State one important use for each of three of the second row transition metals.

13.16. Why do most transition metals exhibit a variety of positive oxidation states while main group metals most often exhibit only one non zero oxidation state?

13.17. What oxidation state will the metal have for the following configurations and metals (assuming that there are no valence shell s electrons in any case)? a. Molybdenum in a $d^1$ configuration, b. titanium in a $d^2$ configuration, c. manganese in a $d^6$ configuration, d. copper in a $d^9$ configuration, and e. platinum in a $d^4$ configuration.

13.18. Deduce the maximum oxidation states of the following transition metals from the left hand side of the transition metal region of the periodic table: Ti, Y, Nb, Tc, and Hf.

13.19. Zirconium is neutron transparent. Use internet resources to find out what this means. What is zircalloy? How is it used?

13.20. Identify the transition metal described by each of the following statements: a. a first row transition metal with a maximum oxidation state of +5 that exhibits all lower positive states; b. a second row transition metal that can exist in nine different oxidation states; c. a third row transition metal whose most stable oxidation state is +6; d. a transition metal that is stable only in the oxidation states 0, +2 , and +4.

13.21. What distinguishes a transition metal from a main group metal?

13.22. Metals of all types are formed by reduction of their ores. Why is electrochemical reduction not used to recover any of the commercially significant transition metals?

## Redox and Redox Equations

13.23. Carbon in its compounds adopts many oxidation states. Rank the following chemical compounds from the one with the lowest oxidation state of carbon to the one with the greatest: $CO_2$, $CH_4$, $CH_2O$, $C_2H_5OH$, $C_2H_6O_2$, and $C_3H_8O_3$.

13.24. Tin(II) is a useful reducing agent. Write a balanced chemical equation showing the reduction of iron(III) sulfate to iron(II) sulfate in acid solution using tin(II) chloride.

13.25. Write a balanced chemical equation to show the reduction of iron ore by carbon monoxide to yield iron.

13.26. Powdered iron is a reactive substance. The sparks that shower when vehicle loses a wheel and iron scrapes across a concrete surface are iron particles burning and forming iron(III) oxide. Write a balanced equation for this reaction.

13.27. Discuss what is meant by the term self-redox.

13.28. Write a balanced chemical equation showing the combustion of propane ($C_3H_8$). Identify the oxidant and the reductant in this reaction.

13.29. Assume that black powder is a mixture of potassium nitrate ($KNO_3$), charcoal, and sulfur in exactly the stoichiometric amounts needed for the reaction $10\ KNO_3 + 12\ C + 4\ S \rightarrow 5\ K_2CO_3 + 4\ SO_2 + 5\ N_2 + 7\ CO$ Calculate the weight percentages of $KNO_3$ , C, and S that should be used in the mixture.

13.30. Use the rules to assign oxidation numbers to each of the following compounds. State the rule or rules that have been used. a. $K_3N$, b. $BaBr_2$ , c. $MnF_2$ , d. $PH_3$ , and e. $PF_5$

13.31. Use the rules to assign oxidation numbers to the following polyatomic ions. State the rule or rules that have been used.a. $Eu^{2+}$ , b. $NH_2^-$ , c. $S_2O_8^{2-}$ , d. $BF_4^-$ , and e. $SO_3^{2-}$

13.32. Use the rules to assign oxidation numbers to the following organic compounds. State the rule or rules that have been used: a. $C_2H_6$, b. $C_3H_4$, c. $C_4H_{10}O_4$, d. $CH_4O$, and e. $H_2C_2O_4$

13.33. Assign the oxidation numbers of each atom in the following. State in each case the rules that have been used: a. $POF_3$, b. $SF_2$ , c.$TiO_2$, d. $Al(NO_3)_3$, e. $Cu_3(PO_4)_2$, and f. $CHO_2^-$. For compounds that contain polyatomic ions use the charge on the polyatomic ion to deduce the charge of the cation they are in combination with.

13.34. Oxidation numbers are periodic. Discuss.

13.35. Write the following reaction as a net ionic equation: $KMnO_4 + C_7H_6O_4 + H_2SO_4 \rightarrow K_2SO_4 + MnSO_4 + CO_2 + H_2O$. Balance the net ionic reaction.

13.36. Write the formulas of a. sodium sulfate, b. sodium sulfite, and c. sodium thiosulfate. Assign oxidation numbers to the sulfur in each compound. Which of the is the best oxidant? Which is the best reductant?

13.37. Based on a comparison of the oxidation numbers of the chlorine atoms, rank the following compounds in order from the most reactive oxidant to the least reactive: $LiClO$, $KClO_4$, $NH_4ClO_3$, $BaCl_2$, and $NaClO_2$.

13.38. The explosive decomposition of ammonium nitrate can be represented by the reaction equation: $NH_4NO_3(s) \rightarrow 2\ H_2O(g) + N_2O(g)$, Ammonium nitrate is a widely used, inexpensive explosive. Identify the oxidizing agent and the reducing agent. Also identify the product or products of oxidation and the product or products of reduction.

13.39. Balance the following redox reactions.

a. $KIO_3 + KI + H^+ \rightarrow I_2 + K^+ + H_2O$
b. $MnO_2 + Cl^- + H_2SO_4 \rightarrow MnSO_4 + Cl_2 + H_2O + SO_4^{2-}$
c. $KMnO_4 + C_6H_6O_2 + H_2SO_4 \rightarrow K_2SO_4 + MnSO_4 + H_2O + CO_2$
d. $C_{12}H_{22}O_{11} + HNO_3 \rightarrow H_2C_2O_4 + NO_2 + H_2O$
e. $KMnO_4 + H_2O_2 + H_2SO_4 \rightarrow K_2SO_4 + MnSO_4 + H_2O + O_2$
f. $K_2Cr_2O_7 + C_3H_7OH + H^+ \rightarrow Cr^{3+} + CO_2 + K^+ + H_2O$

13.40. Balance the following three redox reactions.

a. $Cu + H^+ + NO_3^- \rightarrow Cu^{2+} + NO_2 + H_2O$
b. $Cu + H^+ + NO_3^- \rightarrow Cu^{2+} + NO + H_2O$
c. $Cu + H^+ + NO_3^- \rightarrow Cu^{2+} + NH_3 + H_2O$

13.41. 75.4 mL of a 0.0462 molar $K_2Cr_2O_7$ solution is required for the titration of 48.50 mL of an acidic solution of $H_2O_2$ to the end point. $O_2$ and $H_2O$ are the products produced by the reaction of the $H_2O_2$. What is the molarity of the $H_2O_2$ solution?

## Nonmetals

13.42. Why are the Noble Gases sometimes called the Inert Gases? Write their electronic configurations and discuss what feature of their electronic structure is used to rationalize the fact that they form only a small number of compounds?

13.43. What are the primary uses of each of the Noble Gases?

13.44. State the oxidation number of Xe in each of these compounds: a. $XeO_3$, b. $XeO_2F_2$, c. $XeOF_4$, d. $(XeOF_3)^+$.

13.45. The Xe-F bond energy is estimated to be 126 kJ $mol^{-1}$ and that of the Xe-O bond about 84 kJ $mol^{-1}$. Use bond energy data to estimate $\Delta H$ for the following gaseous reactions,
a. $XeO_3(g) \rightarrow Xe(g) + \frac{3}{2}O_2(g)$
b. $XeF_2(g) \rightarrow Xe(g) + F_2(g)$
c. $XeF_4(g) \rightarrow Xe(g) + 2\,F_2(g)$
d. $XeF_6(g) \rightarrow Xe(g) + 3\,F_2(g)$

13.46. Using the results of the previous question suggest a reason that $XeO_3$ is explosively unstable while the xenon fluorides are relatively stable.

13.47. $XeO_3$ cannot be prepared by direct reaction of Xe with $O_2$. It can, however, be prepared by reacting $XeF_6$ with $H_2O$. Write a balanced reaction for the preparation of $XeO_3$.

13.48. What are the most important commercial uses of each of the halogens?

13.49. Write balanced equations for the reaction of a halogen with: a. A group 1 metal. b. A group 2 metal. c. A transition metal. d. $H_2$. e. $H_2O$.

13.50. What is the composition of "chlorine-bleach?" How is it prepared?

13.51. a. Explain the trend in boiling points of the diatomic halogens. b. Explain the trend in first ionization energies of the halogen atoms in their gaseous state.

13.52. Write examples of compounds which illustrate the range of oxidation numbers shown by O. How does oxygen attain a positive oxidation number?

13.53. Write examples of compounds which show the full range of oxidation states shown by S.

13.54. Write chemical equations which show the production of sulfuric acid from elemental sulfur.

13.55. How is $N_2$ obtained for commercial use? Describe three commercial uses of $N_2$.

13.56. Write examples of compounds which show the full range of oxidation states shown by nitrogen.

13.57. Write chemical equations which show the production of nitric acid from ammonia.

Strategy: Read the information in the chapter.

The net equation is: $NH_3(g) + 2\,O_2(g) \rightarrow HNO_3(aq) + H_2O(l)$

13.58. Classify each of the elements in Groups 13 through 16 as nonmetal, metalloid, or metal.

## Problem Solving

13.59. The US annual demand for helium gas is about 85 million cubic meters. Assuming that the helium is measured at 1 atmosphere pressure and 298 K, how many tons of helium is this?

13.60. Suppose platinum costs $400 per troy ounce and that rhodium costs $330 per avoirdupois ounce. Which is more expensive on a mass basis? On a mole basis?

13.61. Calculate the percentage of phosphorus in $CaF_2 \cdot 3Ca_3(PO_4)_2$ rock. How many tons of pure phosphoric acid can be made from 100 tons of this rock, assuming that the percentage yield of the process is 94%. The molar weight of the rock is 1010 g $mol^{-1}$.

13.62. Although the topic was not developed in Chapter 13, the colors of transition metal compounds arise because of the selective absorption of light. What color light is a red compound absorbing? How about a green compound?

13.63. Draw electron dot (Lewis) structures for each of the entities: $XeF_2$, $XeF_4$, $XeF_6$. and $Ba_2XeO_6$. Compare your answers with the experimental structures described in this chapter.

13.64. Use internet resources to find the link list called "Metallic Minerals Resources." Discuss the possible uses of this site by a working engineer.

13.65. During the 20th century, we have seen a steady replacement of one material with another. Pick a piece of sports or recreation equipment such as a golf club, bicycle frame, baseball bat, or hockey stick. Use internet resources to track that piece of equipment over time. What material was it traditionally made from? What replaced that? What is now used?

13.66. Borosilicate glass has "stealth" applications. Explain.

13.67. The density of monatomic liquid xenon is about 3.6 g $mL^{-1}$. Use this information, xenon's atomic weight, Avogadro's number and conversion factors to estimate the radius of a xenon atom in pm.

13.68. Use internet resources to locate the USGS minerals yearbook. What is USGS? Discuss the consumption pattern of tellurium and selenium.

13.69. In 1998, 426 billion cubic feet (bcf) of hydrogen gas was produced in the US. Assume that the volume of this gas was measured at 25°C (298 K) and 1 atmosphere pressure. What was the mass of this hydrogen in metric tons?

13.70. According to the Clorox Company a Clorox® bleach solution contains 5.25% of sodium hypochlorite in water. Calculate the solution's concentration of hypochlorite ion in mol $L^{-1}$. According to an MSDS sheet (Mallinckrodt Baker Company) the density of the commercial product is 1.11 g $mL^{-1}$. The molar weight of NaClO is 74.4 4 g $mol^{-1}$.

13.71. Chlorine bleach is a dilute solution of sodium hypochlorite and ordinary salt in water. It's made by dissolving chlorine gas in a dilute solution of sodium hydroxide according to the reaction equation $2\,NaOH + Cl_2 \rightarrow NaClO + NaCl + H_2O$. The commercial solution contains 5.25% by weight of NaClO. a. What is the weight percent of NaCl in the solution? b. What is the weight percentage of the total dissolved solutes?

A chlorine bleach solution has a density of 1.11 g $mL^{-1}$. The solution contains 5.25% by weight of NaClO and 4.11% by weight of NaCl. According to an MSDS sheet (Mallinckrodt Baker Company) the freezing point of this solution is -6.1°C. Calculate the freezing point of the solution and the percentage discrepancy between the calculated value and -6.1°C. NaCl = 58.5 g $mol^{-1}$ and NaClO = 74.4 g $mol^{-1}$.

# Chapter 14

# Electrochemistry

## Introduction

**Electrochemistry** is the science of electricity and chemistry. That's an extremely wide definition—as we've seen, all chemical interactions are fundamentally electrical. In a more restricted sense, electrochemistry has come to mean the study of electrolyte solutions and processes at electrodes. Those are the subjects of this chapter.

There are two broad aspects of electrochemistry: *electrolysis reactions* and *battery reactions*. An **electrolysis reaction** uses an electric current to bring about a chemical reaction. The commercial manufacture of aluminum metal from aluminum ore is one of many examples of electrolysis. A **battery reaction** uses a redox chemical process to generate an electric current. Automobile batteries and the range of batteries available at any drug store are examples of devices that perform battery reactions.

Electrolysis reactions and battery reactions are **electrochemical reactions**—reactions either caused by the flow of electric current or creating an electric current. The device or apparatus in which an electrochemical reaction takes place is called an **electrochemical cell**—or just a **cell**: a device that enables the exchange of electrical and chemical energy. Electrochemical reactions are *always* redox reactions. Inside the cell, current is carried by the movement of ions; in the external circuit electrons flow through an electrical conductor (typically a metal wire). The discharging and recharging of an automobile battery are examples of electrochemical reactions. An electrochemical reaction is a redox reaction arranged so that the electron flow from the reductant to the oxidant passes through an external circuit. We may call this "*redox by wire*."

We saw in the preceding chapter that the chemically active elements cannot be prepared by chemical means. It is obvious that there is no chemical agent that can either oxidize the most powerful oxidants or reduce the most powerful reductants; electrical methods must be used instead. Thus, historically, the preparation of the active metals Na (in 1807), K (in 1807), Mg (in 1808), and Ca (in 1808), had to await the development of a suitable electrical power source. The extremely reactive nonmetal fluorine was finally isolated electrochemically in 1886 after a tragic 120-year record of failed attempts in which many chemists were injured and at least two were killed.

Today, there are a number of commercially important electrolysis reactions. Collectively they are called the **electrochemical industries**. The manufacture of aluminum, the co-production of chlorine and sodium hydroxide, and the purification of copper are important electrochemical industries. Electrochemical reactions follow the principles of stoichiometry we studied in Chapter 3. The extent of an electrolysis reaction depends on the number of electrons that flow into and out of the electrochemical cell: the bigger the current and the longer the time, the greater the extent of electrolysis. In this chapter we'll see examples of how to relate the time and amperage of current flow to the masses of reactants and products of electrochemical reactions.

An important property of an electrochemical cell is the **cell voltage**—the voltage at which it operates. For an electrolysis reaction, there is always some minimum voltage need to carry out the reaction. A battery produces a particular voltage depending on what it is made of—depending on what oxidant and reductant are used, and their concentrations.

Cells are conveniently imagined as a **reduction half-cell** joined to an **oxidation half-cell**. The voltage of a half-cell is called its **half-cell potential**. Values of the standard voltages (or potentials) of half-cells can be combined pair-wise to give the voltages of any conceivable cell. One reason half-cell potentials are useful to chemists is because they measure chemical reactivity, more reactivity means bigger potentials. Half-cell potentials are measured by comparison with the standard hydrogen electrode (referred to that electrode) at standard thermodynamic conditions and a specified temperature. The standard hydrogen electrode reaction is $H_2(g) = 2\ H^+(aq) + 2\ e^-$ under standard state conditions ($P_{H_2} = 1$ bar, $[H^+] = 1$ molar) with a defined voltage of 0.000... volts at 298 K.

First we'll establish the basic processes in electrochemistry and define some terms. We've already begun to do that in the foregoing discussion. Next we'll look at some of the electrochemical industries and the chemical reactions involved there. Doing that will lead into a discussion of the stoichiometry of electrochemistry—a topic that entered chemistry in the form of Faraday's laws of electrolysis. From there we'll take up the subject of standard half-cell potentials and relate electrochemical measurements to chemical reactivity. We will also see how a changing solution concentration creates a changing electrical potential that enables the change in concentration to be measured. The governing equation comes from thermodynamics and is called the Nernst equation. The Nernst equation is perhaps most widely applied to measure changing acidity, as in a pH meter. To conclude the chapter we will discuss corrosion, an electrochemical phenomenon of considerable importance to many engineers. [❁kws "NACE" or +"National Association of Corrosion Engineers"]

## Section 14.1 The History of Electrochemistry

Figure 14.1 shows an in-class demonstration of the electrolysis of a solution of sodium sulfate to produce hydrogen gas and oxygen gas. The net electrolysis reaction is $H_2O \rightarrow H_2 + \frac{1}{2} O_2$. This reaction was first carried out in London in 1800 by a team of experimenters composed of a physician and a *civil engineer*. The two half-cell reactions taking place are:

$$2\ Na^+ + 2\ H_2O + 2\ e^- \rightarrow NaOH\ + \tfrac{1}{2}\ H_2(g)$$

$$SO_4^{2-} + H_2O \rightarrow H_2SO_4\ +\ \tfrac{1}{2}\ O_2 + 2\ e^-$$

Electrolysis reactions of this type require direct current (dc). Alternating current (ac), which we use at home, reverses direction 60 times a second and only moves ions in the solution back and forth—not to one of the electrodes.

Figure 14.1 The electrolysis of sodium sulfate.

**A Short History of Electrochemistry**

1580-1590 William Gilbert (1540-1603), physician to Queen Elizabeth I, of England, studies static electricity and magnetism. It was already known that amber when rubbed acquires the ability to pick up small pieces of paper, etc. The Latin word for amber is *electrum*. Thus amber gives electricity its name, which in turn gave the electron its name.

1660 Otto von Guericke (1602-1686) invents a frictional machine to generate a continual supply of electricity. It is soon shown that charged bodies might either repel or attract one another.

1729 Stephen Gray (1696-1736) shows that some substances do conduct electricity and some don't. He is thus the discoverer of the conduction of electricity.

1735 Charles Francis De Cisternay Du Fay (1698-1739), Director of the French Dyeing Industry, notes that there are two kinds of electrical charge. We call these positive and negative.

1749 Benjamin Franklin (1706-1790) describes the "wonderful effect of pointed bodies both in *drawing off* and *throwing off* the electrical fire." He thereby identifies lightning as an electrical discharge and invents the lightning conductor. He also establishes the "one fluid" theory of electricity. In modern science, this theory becomes the notion that an excess of electrons creates a negative charge and a deficiency of electrons creates a positive charge.

1791 Luigi Galvani (1737-1798) examines the twitching of the legs of dead frogs caused by a static electricity machine. The phenomenon has been known for at least 100 years, but Galvani finds that twitching could also be induced when the frog is placed on a conducting plate of one metal and touched with a second metal. In so doing he discovers electric current and establishes that different metals have different electrical activities.

1800 Alessandro Volta (1745-1827) writes to Sir Joseph Banks, President of the Royal Society of London, describing the "voltaic pile," a stack of silver plates alternating with copper or zinc plates separated by cloth soaked in brine. The pile is a major advance because it provides a simple, convenient source of direct electrical current.

1800 Only a month after Volta's letter reaches England, the London surgeon Anthony Carlisle (1768-1840) and the London civil engineer William Nicholson (1753-1815) perform the first electrolysis of water to yield hydrogen and oxygen gases.

1807-1808 Sir Humphry Davy (1778-1829) isolates the elements sodium, potassium, magnesium, and calcium using the voltaic pile as his electrical energy source.

1830-1832 Michael Faraday (1791-1867) discovers the "laws of electrolysis" and, with the classical scholar William Whewell (1794-1866), devises the language of electrochemistry: anode, cathode, electrolyte, etc.

1836 Development of the cell that bears his name by John Frederic Daniell (1790-1845). This cell becomes important for the new telegraphy industry. Later, in 1853, J. F. Fuller modifies and greatly improves the cell by substituting zinc sulfate solution for sulfuric acid as the electrolyte.

1844 The "interpoint" (as opposed to the internet) comes into existence when Samuel F. Morse (1791-1872) sends the first ever telegraphic message "What hath God wrought!" between Baltimore and Washington, DC.

1868 Georges Leclanché (1839-1882) introduces the zinc/manganese dioxide battery. With subsequent modifications this becomes today's ordinary "dry cell" battery.

1878 Gaston Planté (1834-1889) develops the lead-acid battery still used today in automobiles. By now, the design and manufacture of commercial batteries, primarily for the telegraph, has become an important industry.

1886 American Charles Martin Hall (1863-1914) and Frenchman Paul L. Héroult independently develop the process to make aluminum metal that now bears their names.

1888 The use of nickel plating becomes widespread.

1890-1894 The American chemist H. Y. Castner (1858-1899) and the Austrian chemist Carl Kellner (1850-1905) independently patent the chlor-alkali process to manufacture chlorine and sodium hydroxide via the electrolysis of salt water. Manufacturing plants are built where supplies of electricity and salt are cheap and plentiful. By 1894 a small chlor-alkali plant is operating in Saltville, Virginia. The Saltville plant serves as the model for much larger plant that is built in 1896 at Niagara Falls.

## Description of Electrochemical Cells

Electrochemical cells come in an array of types, all the way from tiny ones that fit inside hearing aids to multi-ton monsters that are used in industry. Just from observing store displays of batteries, you know that even household electrochemical cells come in a wide range of sizes and styles. Because of the demand for portable electronic devices, many new types of batteries are currently being introduced to the market place. We know the history of the first published diagram of an electrochemical cell; it's shown in Figure 14.2 and comes from Alessandro Volta's letter to Sir Joseph Banks, dated March 20th, 1800 and sent from Como, now in Italy. It pictures what Volta called "the crown of cups."

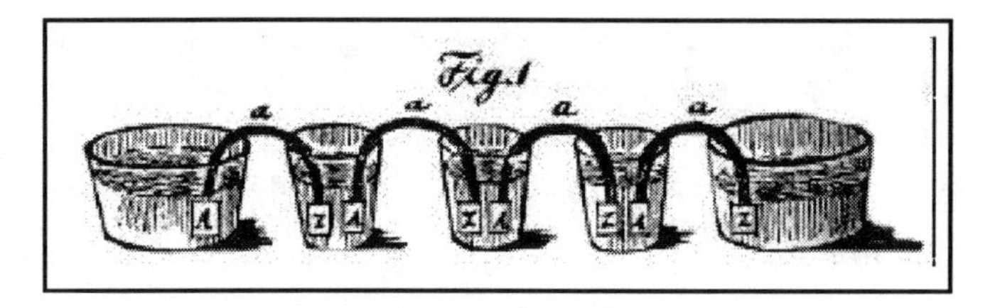


Figure 14.2 Volta's "crown of cups." The first published diagram of an electrochemical cell.

In Figure 14.2, *A* is a metal such as copper, *Z* is a metal such as zinc. Pairs of copper and zinc strips are soldered at the positions labeled "a" and immersed in salt solutions. A weak, continuous electric current can be drawn by connecting the end cups to an electrical circuit. The battery reaction is in effect the reaction of zinc and water: $Zn + 2 H_2O \rightarrow Zn(OH)_2 + H_2$

Figure 14.3 shows a generic electrochemical cell. It has two compartments. Oxidation occurs in one compartment and reduction in the other. The compartments are the cell's half-cells. Externally, the half-cells are connected through an electrical circuit consisting of wires, a meter, and a load—the load is the light bulb, radio, or whatever the cell is powering. Direct electric current—a stream of electrons—flows through the external circuit. If electrolysis is being performed, a dc power supply drives the current. If the cell is functioning as a battery then, *it* generates the current.

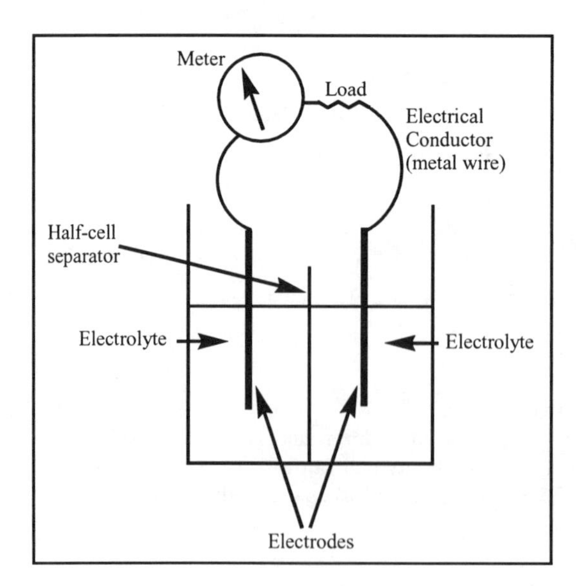


Figure 14.3 A generic electrochemical cell. Despite their vastly different outward appearances, all electrochemical cells operate on the same principle.

Inside the cell, the two half-cell compartments are connected just by the electrolyte solution itself or by an ion conduit. An **ion-conduit** is a device that allows the passage of ions, while not allowing the half-cell electrolytes to mix; it's not needed if the two half-cells contain the same electrolyte, because then mixing doesn't matter. Examples of ion conduits are: paper, cloth, a porous solid such as unglazed pottery, and a gelled aqueous solution of salt (called a salt bridge). Electrons enter and leave the cell via two electrically conducting **electrodes** usually made of metal or graphite.

## Section 14.2 Electrolysis Reactions

**Electrorefining** is the process of purifying a metal via electrolysis. Figure 14.4 shows the arrangement of a cell used for electrorefining copper. All cells contain two electrodes: an **anode** that has a positive charge and attracts anions and a **cathode** that has a negative charge and attracts cations. An anode gathers electrons from the electrolyte and feeds them into the external circuit; an **anode half-cell reaction** is an oxidation. A cathode feeds electrons from the external circuit into the cell; a **cathode half-cell** reaction is a reduction.

In electrorefining, the anode is made of impure copper—typically produced from a copper ore by a roasting process. The cathode is made of pure copper. At the beginning of purification the anode is large and the cathode is small. As purification progresses, the anode shrinks and the cathode grows in size because of electroplating. **Electroplating** is the deposition of a metal on an electrode; in addition to its use in electrorefining, plating is also used to provide decorative and/or protective coatings for metal objects.

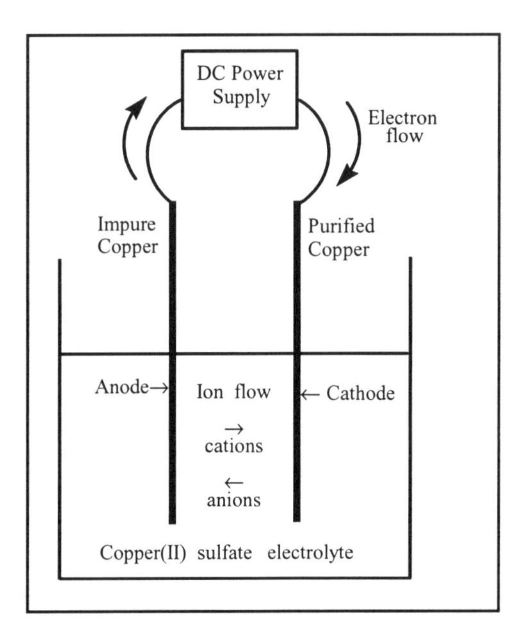


Figure 14.4. A simple electrolysis cell for purifying copper. There is no net chemical reaction. The cell removes copper atoms from the anode, transfers them through the electrolyte as copper ions and deposits them as copper atoms on the cathode.

The electrochemical half-cell reactions for the electrorefining of copper can be written:

| | | |
|---|---|---|
| At the cathode: | $Cu^{2+} + 2e^{-} \rightarrow Cu^{0}\downarrow$ | A reduction reaction |
| At the anode: | $SO_4^{2-} + Cu^{0} \rightarrow SO_4^{2-} + Cu^{2+} + 2e^{-}$ | An oxidation reaction |
| Net reaction: | Nothing! | |

There is no *net* chemical reaction: the cell is just an interesting way of moving copper from one place to another. So why do it at all? Purification. In the oxidation reaction, the sulfate appears on both sides—it could have been omitted, but writing it explicitly reminds us that anions go to the anode and in this case are necessary to mediate the electrochemical process.

Two elements, selenium and tellurium, are recovered as by-products of the electrorefining of copper. As the anode is slowly oxidized away, tiny particles that contain these elements fall by gravity to the bottom of the electrolyte as a goopy mass. It's called "anode slime." Figure 14.5 shows schematically the situation when electrorefining has been completed.

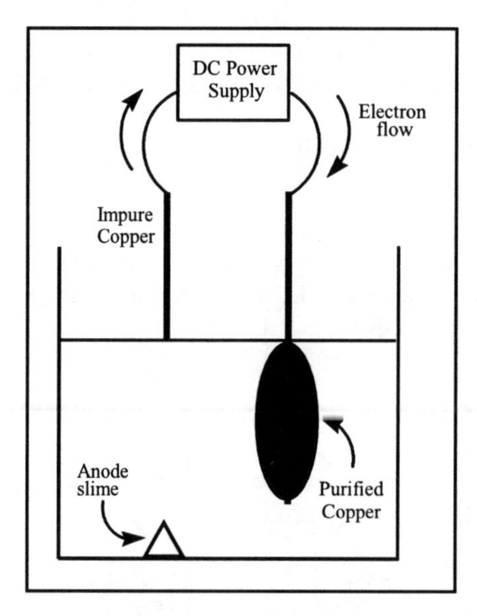


Figure 14.5 At the end of electrorefining the cathode has grown in size while the anode has disappeared leaving behind impurities called "anode slime." Estimated annual US consumption of selenium and of tellurium are respectively 400 tons and 60 tons. Much of this demand is met by recovering these elements from anode slime.

## Electrochemical Industries

A number of important electrochemical industries are briefly described in Table 14.1. The manufacture of aluminum metal and the chlor-alkali process are the two dominant electrochemical industries of our time. The use of electroplating is also an important electrochemical industry in its own right.

Table 14.1 **Electrochemical Industries**

| Process | Method | Equation | Scale |
|---|---|---|---|
| Manufacture of Aluminum Metal | Electrolytic reduction | $Al^{3+} + 3e^- \rightarrow Al$ | Extremely large |
| Manufacture of Chlorine | Electrolytic oxidation | $2\ Cl^- \rightarrow Cl_2$ | Extremely large |
| Manufacture of sodium hydroxide | Electrolytic reduction and reaction of product with water | $Na^+ + e- \rightarrow Na$ <br> $Na + H_2O \rightarrow NaOH$ | Extremely large |
| Manufacture of magnesium metal | Electrolytic reduction. | $Mg^{2+} + 2e^- \rightarrow Mg$ | Very large |
| Purification of copper. (electrorefining) | Combination of electrolytic oxidation and electrolytic reduction | Cu (impure) → Cu (pure) | Very large |
| Manufacture of sodium hypochlorite | Electrolysis of NaCl solution with mixing | $NaOH + Cl_2 \rightarrow NaCl + NaClO$ | Medium |
| Manufacture of Fluorine | Electrolytic oxidation. | $KHF_2 \rightarrow K + ½H_2 + F_2$ | Small |
| Manufacture of sodium chlorate | Electrolysis of NaCl solution with mixing | $NaOH + Cl_2 \rightarrow NaClO_3$ | Small |
| Manufacture of potassium hydroxide | Electrolytic reduction and reaction of product with water | $K^+ + e^- \rightarrow K$ <br> $K + H_2O \rightarrow KOH$ | Small |
| Production of hydrogen and oxygen | Electrolysis of an aqueous electrolyte | $H_2O \rightarrow H_2 + ½O_2$ | Small |
| Purification of gold (electrorefining) | Combination of electrolytic oxidation and reduction | Au (impure) → Au (pure) | Very small |

## The Chlor-Alkali Process

The chlor-alkali process is the manufacture of chlorine and sodium hydroxide from salt water (brine). The importance of the chlor-alkali industry is evident from the annual average US per capita consumption of about 100 pounds of chlorine and 90 pounds of sodium hydroxide.

The process reactions are:

At the cathode: $2\ e^- + 2\ Na^+ + 2\ H_2O(l) \rightarrow 2\ NaOH + H_2(g)$ Reduction
At the anode: $2\ Cl^- \rightarrow Cl_2(g) + 2\ e^-$ Oxidation
Net reaction: $2\ NaCl(aq) + 2\ H_2O(l) \rightarrow 2\ NaOH + H_2(g) + Cl_2(g)$

Electric current passing through a brine (sodium chloride solution) generates sodium hydroxide and chlorine gas. In modern commercial plants the cathode is made of graphite and the anode of a titanium alloy. The two compartments are kept physically (but not electrically) separated by means of a porous medium. It is necessary to keep them separated, or else the product chlorine reacts with the product sodium hydroxide. (However, mixing is required if the desired product is sodium hypochlorite.) A filter made up of mineral particles is one commercial separation medium. Figure 14.6 shows the banks of cells of a chlor-alkali plant operated at Edgewood Arsenal, Maryland, by the US Army during World War I, and Figure 14.7 shows a modern plant.

Figure 14.6 Banks of electrolytic cells used to manufacture chlorine and sodium hydroxide.

Figure 14.7 The Pioneer Chlor-Alkali plant, St. Gabriel, Louisiana. This modern plant is located on a 300-acre site near the Mississippi river. It uses a mercury-cell electrolysis process to produce low-salt grade sodium hydroxide. In addition to sodium hydroxide, it makes the other two classic products of brine electrolysis: chlorine and hydrogen gases. The silo-shaped structure is used to store rock salt (NaCl). The main plant with its banks of electrolysis cells is in the large dark building at the rear.

## Section 14.3 Stoichiometry and Electrochemistry—Faraday's Laws

The quantitative relationships between the flow of electricity and chemical charge were worked out by Michael Faraday (1791-1867) in the 1830s. We can restate his laws in modern terms in the following way: The mass of a product of an electrochemical reaction is proportional to:

1. The current, i, in amps.
2. The time, t, for which the current flows.
3. The molar weight of that product.
4. The number of electrons required to oxidize or reduce one reacting ion.

Reactions and equations involving electrons follow the ordinary principles of chemical stoichiometry. To reduce one mole of silver(I) ions requires one mole of electrons: $Ag^+ + e^- \rightarrow Ag(s)$. To reduce one mole of copper(II) ions requires two moles of electrons: $Cu^{2+} + 2\ e^- \rightarrow Cu(s)$. To reduce one mole of aluminum ions to aluminum metal requires three moles of electrons: $Al^{3+} + 3\ e^- \rightarrow Al(s)$. In general, to reduce a mole of any metal ion will require one mole of electrons per positive charge. The amount of a product can be equally well expressed as moles or grams.

The algebraic product of a current in amps and a time in seconds is an electrical charge (q) measured in the units of Coulombs, q = it. The Coulomb (C) is the SI metric unit of electrical charge. It is the charge carried by a current of exactly one amp flowing for exactly one second (the amp and second are base SI units).

$$\text{amps} \times \text{seconds} = \text{Coulombs}$$

$$1\ \text{amp-sec} = 1\ \text{C} \quad (\text{or } 1\ \text{A s})$$

Quantitatively, chemical stoichiometry is related to electric flow by the Faraday constant (F). The **Faraday constant** is the charge on a mole of electrons and its 1997 best value is $9.648\ 530\ 9\ (29) \times 10^4\ \text{C mol}^{-1}$ (an experimental value, not an exact number). To five significant figures, sufficient for our purposes, F is 96485 coulombs $\text{mol}^{-1}$. The charge on a single electron is simply the Faraday constant divided by Avogadro's number, as shown in Example 14.1. For an electric current of i amps, flowing for t seconds, the number of moles of electrons that passes down the wire is $n = \frac{it}{F}$ or $n = \frac{q}{F}$.

**Example 14.1**: What's the charge on a single electron?

Strategy: Recognize from the units that it's the Faraday constant divided by the Avogadro number.

Charge on one electron = $96485\ \text{C mol}^{-1} \div 6.022 \times 10^{23}\ \text{mol}^{-1}$

$= \underline{1.602 \times 10^{-19}\ \text{C}}$

Typical calculations involving Faraday's laws of electrolysis are shown in Examples 14.2 and 14.3. Example 14.2 shows how to calculate the mass of copper produced by a specified current over a specified period of time; such calculations can be used to predict how much of a metal will be electroplated during an electrolysis process. Example 14.2 illustrates that the mass of metal electroplated depends on its atomic weight and its electrical charge.

**Example 14.2**: Calculate the mass of copper(II) that can be electrorefined by a current of 1.35 A in a 24.0 hour time period.

Strategy: Calculate the number of coulombs of charge using the relationship $C = i \times t$. Use the Faraday constant to convert the total charge to moles of electrons. Divide that value by two to obtain the number of moles of copper produced, because the stoichiometric equation, $Cu^{2+} + 2\ e^- \rightarrow Cu(s)$, shows that it takes two moles of electrons to produce one mole of copper. Finally, use copper's atomic weight (63.5 g $mol^{-1}$) to calculate the mass of copper.

$$\text{Number of coulombs used} = 1.35\ \text{A} \times 24.0\ \text{hours} \times 3600\ \text{s hour}^{-1} = 1.17 \times 10^5\ \text{C}$$

$$\text{Number of moles of electrons} = \text{number of coulombs} \div \text{F (coulombs mol}^{-1})$$

$$\text{Moles of electrons} = 1.17 \times 10^5\ \text{C} \div 96485\ (\text{coulombs mol}^{-1}) = 1.21\ \text{mol e}^-$$

$$\text{Moles of copper produced} = \tfrac{1}{2}\ \text{mol Cu per mole of e}^- \times 1.21\ \text{mol e}^- = 0.604\ \text{mol}$$

$$\text{Mass of copper produced} = 0.604\ \text{mol} \times 63.5\ \text{g mol}^{-1}$$

$$= \underline{38.4\ \text{grams of copper}}$$

**Example 14.3**: An electrochemical reduction produced 223 mg of copper metal from copper(II) ions. In a separate experiment, what mass of silver metal would have been produced from Ag(I) ions by the identical current passing for an identical time period?

Strategy: Use ratio and proportion recognizing two differences: 1. The mass of a silver atom is 107.9/63.5 times heavier than a copper atom; the fraction is their atomic mass ratio. 2. Twice as many silver atoms will be plated because silver(I) ions need only one electron for reduction whereas copper(II) ions need two electrons.

$$\text{Mass of silver} = \text{mass of copper}\ \frac{\text{atomic weight of silver}}{\text{atomic wright of copper}} \times \frac{\text{1 silver atom per electron}}{\text{0.5 copper atom per electron}}$$

$$\text{Mass of silver} = 223\ \text{mg} \times (107.9/63.5) \times 2/1 = \underline{758\ \text{mg}}$$

Figure 14.8 Chromium plating is highly valued because it produces an attractive and long lasting coating. Electroplating has many engineering applications.

Electroplating is used to protect and decorate metals. Chromium plating is widely used both for its protective effect and its decorative value. The faucet shown in Figure 14.8 has been electroplated with chromium to give it a brilliant appearance and a lifetime of service use. Incidentally, the sink in Figure 14.8 is made of stainless steel, iron alloyed with about 15% chromium. Gold plating was commercialized in England in the 1840's and originally applied to make gold plated tableware. Perhaps as a computer user you are familiar with the modern use of gold plating to ensure good electrical contact between circuit boards and connectors. Nickel plating iron to make it corrosion resistant was developed at the end of the nineteenth century. Brass is an alloy of zinc and copper, and under the right conditions, the electrolysis of a mixture of salts of zinc and copper produces a golden colored electroplate. Brass bedsteads, birdcages, and similar items are manufactured in this way. [❁kws +electroplating]. Example 14.4 shows how the stoichiometry of an electroplating reaction can be used to calculate the actual thickness of the metal plate.

**Example 14.4**: A current of 10.0 milliamp is passed through a nickel(II) sulfate solution for 30.0 minutes. a. Calculate the mass of nickel produced, and b. if this nickel forms a plate on a surface of total area of 1.00 square meters, calculate the plate's thickness in picometers (pm). The density of the nickel plate is 8.90 g $cm^{-3}$. Estimate the thickness of the film in atomic diameters. The diameter of a nickel atom is about 250 pm.

Strategy: Write the redox equation, $Ni^{2+} + 2\ e^- \rightarrow\ Ni(s)$ and use that equation and Faraday's constant to calculate the moles and then the mass of nickel plated. Use nickel's density to convert that mass to a volume and divide by the area of the plate to convert the volume to a thickness. Finally, divide the thickness by the diameter of a nickel atom to estimate the thickness of the plate in atomic diameters.

a. The number of moles of electrons that pass is:

$$(10.0 \times 10^{-3}\ \text{amp} \times 30.0\ \text{min} \times 60\ \text{sec min}^{-1}) \div 96485\ \text{amp sec mol}^{-1}$$

$$= 1.87 \times 10^{-4}\ \text{mol e}^-$$

From stoichiometry, 1 mol $e^-$ produces ½ mol nickel. The atomic weight of nickel is 58.7 g $mol^{-1}$, so the mass of the nickel plate is

$$1.87 \times 10^{-4}\ \text{mol e}^- \times \tfrac{1}{2}\ \text{mol nickel (mol e}^-)^{-1} \times 58.7\ \text{g mol}^{-1}$$

$$= \underline{5.48 \times 10^{-3}\ \text{g Ni}}$$

b. The area of the nickel plate times its thickness equals its volume, and the volume of the nickel plate is its mass divided by its density.

$$\text{Volume} = \text{mass} \div \text{density}$$

$$= 5.48 \times 10^{-3}\ \text{g} \div 8.90\ \text{g cm}^{-3} = 6.15 \times 10^{-4}\ \text{cm}^3$$

But: thickness of plate × area of plate = volume of plate

So, $\text{thickness} \times 1.00\ \text{m}^2 \times 100^2\ \text{cm}^2\ \text{m}^{-2} = 6.15 \times 10^{-4}\ \text{cm}^3$

$$\text{thickness} = \text{volume} \div \text{area} = 6.15 \times 10^{-4}\ \text{cm}^3 \div 10^4\ \text{cm}^3$$

$$= 6.15 \times 10^{-8}\ \text{cm} \times 10^{10}\ \text{pm cm}^{-1}$$

$$= \underline{615\ \text{pm}}$$

Because 615 pm ÷ 250 pm is 2.5, the plate consists of only two or three layers of atoms.

Electroplating is an old but specialized technology. As with any other important commercial technology, its use raises issues of proprietary information, as described in the following box.

**Proprietary Information**

It is difficult to find out about modern commercial plating practices, because electroplaters tend not talk much about the methods they use. In a competitive economy, the company that produces the best electroplate will probably make the most profit. Modern battery manufacture, described later in this chapter, is another example of an industry based on a great deal of proprietary information. **Proprietary information** is knowledge held within a company of how to run a process or how to build a device. When that sort of information is held by an individual it's one type of **intellectual property**. Commercial organizations often take substantial measures to protect their proprietary information.

A different way of protecting a commercial interest is to use the patent system. A **patent** is a license agreement between an inventor and a government organization. In the US, patents are granted by the US Patent and Trademark office (uspto.gov). The US government is interested in promoting public knowledge of inventions to improve the efficiency of the economy. In a patent an inventor discloses his invention for all to read—and in return the government grants a 17 year exclusive license for the patent holder to profit from the invention.

While there are many individual inventions each year, most inventors work for large companies and, as employees, assign "patent rights" to their employer. When you accept your first job as a working engineer you'll likely be asked to sign legal papers dealing with proprietary information and patent assignment.

## Aluminum Production

The commercial production of aluminum in the US has been estimated to consume no less than 5% of the the country's electrical power production. This is an enormous figure and immediately tells you why aluminum recycling is such an economically attractive proposition. Used aluminum can be re-melted for new uses without requiring additional electrical power for its reduction.

Virgin aluminum is a made from bauxite ore in two steps: 1. ore purification, and 2. electrolytic reduction. Bauxite ore, which occurs in Australia, Brazil, Jamaica, Guinea, etc., is typically 40% - 60% aluminum oxide, $Al_2O_3$, derived from the natural weathering of clay minerals. After grinding, the ore is dissolved in strong alkali (sodium hydroxide) solution and filtered free of its major impurities—usually iron oxides that are disposed of as "red mud." The now soluble aluminum is then reprecipitated as aluminum hydroxide by adding acid. Finally, the aluminum hydroxide is converted to pure aluminum oxide by heating it at around 1500 K. The equations for the steps are:

Dissolving $$Al_2O_3{\cdot}nH_2O + 2\,NaOH \rightarrow 2\,NaAlO_2 + (n+1)\,H_2O$$

Reprecipitation $$NaAlO_2 + H_2O + H^+ \rightarrow Al(OH)_3 + Na^+$$

Heating at 1500 K $$2\,Al(OH)_3 \rightarrow Al_2O_3 + 3\,H_2O(g)$$

The electrolytic decomposition of purified aluminum oxide to produce aluminum metal (shown in Figure 14.9) is done in a molten cryolite ($Na_3AlF_6$) bath at a temperature of about 1300 K. A carbon-lined steel tank serves as both the container and the cathode. Carbon also serves as the anode material. Approximately correct half-cell reactions can be represented by the equations:

At the cathode: $$12\,Na^+ + 4\,Na_3AlF_6 + 12\,e^- \rightarrow 4\,Al + 24\,NaF$$

At the anode: $$2\,Al_2O_3 + 24\,NaF + 3\,C \rightarrow 12\,Na^+ + 4\,Na_3AlF_6 + 3\,CO_2 + 12\,e^-$$

Net: $$2\,Al_2O_3 + 3\,C \rightarrow 4\,Al + 3\,CO_2$$

The cells themselves are made from large steel boxes with dimensions up to 9 m long, 3 m wide and 1 m deep. They are lined with carbon made from pitch or anthracite coal and baked into place. Carbon anodes are made in a similar manner. Cell linings are designed to last about three years. There is an enormous amount of experience and specialized knowledge involved in the large scale manufacture of aluminum.

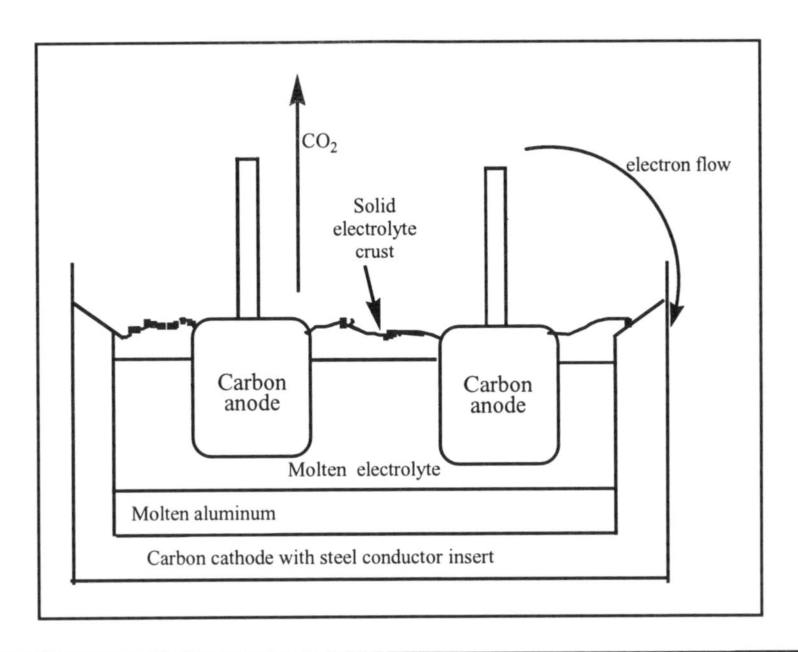


Figure 14.9 Schematic diagram of an electrolysis cell for aluminum production. Some commercial cells are as large as a house.

**Example 14.5**: A typical US home might consume 1000 kilowatt hours (kWh) of electrical energy per month at a cost of $100, although individual figures vary widely. Calculate the mass of aluminum that could be prepared electrolytically by 1000 kilowatt hours of electricity.

Strategy: First carry out a units conversion to change kWh to joules; the needed relation is 1 watt = 1 J $s^{-1}$. Then convert from joules to coulombs using the relationship joules = coulombs × volts, or 1 C = 1 J $V^{-1}$; household electricity is delivered at 110 volts. Next use the Faraday constant to calculate the number of moles of electrons involved. Then use the stoichiometry of the reduction reaction, $Al^{3+} + 3\ e^- \rightarrow Al(s)$, and divide the moles of electrons by three to find the moles of aluminum. Finally use aluminum's atomic weight, 27.0 g $mol^{-1}$ to convert moles to mass.

kWh → joules $1000\ kW\ h \times 1000\ W\ kW^{-1} \times 3600\ s\ h^{-1} = 3.6 \times 10^9$ watt seconds

$3.6 \times 10^9$ watt seconds $\times 1\ J\ s^{-1}\ watt^{-1} = 3.6 \times 10^9\ J$

joules → coulombs $3.6 \times 10^9\ J \div (1\ J\ V^{-1}\ C^{-1} \times 110\ V) = 3.3 \times 10^7\ C$

coulombs → moles $3.3 \times 10^7\ C \div 96486\ C\ mol^{-1} = 339$ moles $e^-$

Divide by three 339 moles $e^-$ ÷ 3 mol $e^-$ per mole Al = 113 mols Al

Convert to mass 113 mols Al × 27.0 g $mol^{-1}$ = 3050 g Al

That is about seven pounds of aluminum. Considering the amount of aluminum used in our economy, you can see it's a plausible estimate that 5% of US electric power generation goes into aluminum production.

## Section 14.4 Cell Potential

In this section we describe how cell potentials are measured, how cell potentials are standardized, and how relative values are obtained for the potentials (voltages) of half-reactions.

A **cell potential** (E) is the voltage generated by a single electrochemical cell. A **half-cell potential** is the voltage of either the oxidation or reduction half-cell reaction. **Standard** cell and half-cell potentials are measured at standard thermodynamic conditions: pressure = 1 bar, solution concentrations = 1 molar, pure substances in their standard states, and so on.

The voltage generated by an electrochemical cell depends on the *relative* chemical reactivity of the redox pair, i.e., on which specific oxidant is coupled with which specific reductant to make the cell. The operating voltage of an electrochemical cell is the sum of two half-cell voltages: the anode or oxidation half-cell ($E_{ox}$) plus the cathode or reduction half-cell ($E_{red}$). The relationship is simply:

$$E_{cell} = E_{ox} + E_{red}$$

If the cell is operating under standard conditions we show that fact by using the usual superscripts, and the equation becomes

$$E^0_{cell} = E^0_{ox} + E^0_{red}$$

It is not possible to measure a single half-cell potential in isolation; any real, operating cell must have both an oxidant and reductant half-cell. But half-cell voltages can be assigned separately, by joining each half-cell in turn to an arbitrarily defined standard reference electrode.

The reference standard for electrochemical voltage is the **standard hydrogen electrode** (SHE), Figure 14.10. By definition, it has a half-cell voltage of 0.00 V at 298 K. It consists of an electrode partly submerged in 1 M hydrogen ion solution over which is bubbled hydrogen gas at one atmosphere pressure. The electrode itself is activated platinum—platinum prepared so that it has a catalytic surface to facilitate the half-reaction.

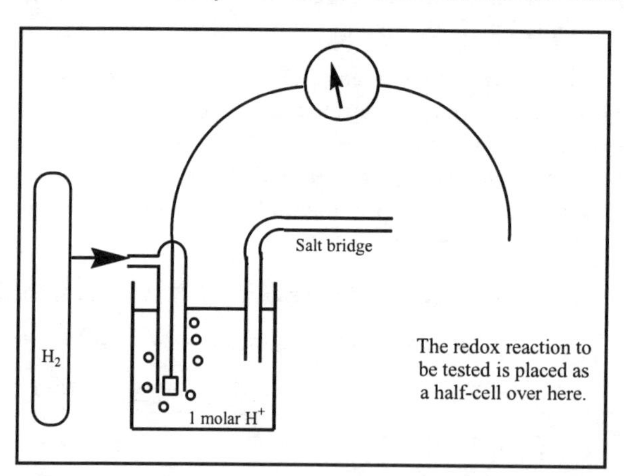


Figure 14.10 Schematic diagram of the standard hydrogen electrode. At 298 K it's defined to have a voltage of zero.

To use a standard hydrogen electrode, it's joined as one half-cell to a second half-cell consisting of the redox reaction to be tested. Because the voltage of the standard hydrogen electrode is zero, the voltage of the joined cell is simply the standard voltage of the second half-cell cell being tested (with the proper algebraic sign). .

The **standard reduction potentials** of many reactions have been measured. Table 14.2 shows them for acidic solutions with $[H^+]$ = 1.0 molar. The values are called reduction potentials because they are for electrochemical reactions with the reactant being reduced—the electrons are on the left. Reversing any half-cell reaction changes the sign of its voltage and changes the reduction potential to an oxidation potential: $E_{ox} = - E_{red}$.

| Table 14.2 **Standard Reduction Potentials in Acid Solution at 25°C** | |
|---|---|
| Oxidizing Agent ⇄ Reducing Agent | $E^0_{red}$ Volts |
| $Li^+(aq) + e^- \rightleftarrows Li(s)$ | -3.04 |
| $K^+(aq) + e^- \rightleftarrows K(s)$ | -2.94 |
| $Ba^{2+}(aq) + 2e^- \rightleftarrows Ba(s)$ | -2.91 |
| $Ca^{2+}(aq) + 2e^- \rightleftarrows Ca(s)$ | -2.87 |
| $Na^+(aq) + e^- \rightleftarrows Na(s)$ | -2.71 |
| $Mg^{2+}(aq) + 2e^- \rightleftarrows Mg(s)$ | -2.36 |
| $Al^{3+}(aq) + 3e^- \rightleftarrows Al(s)$ | -1.68 |
| $Mn^{2+}(aq) + 2e^- \rightleftarrows Mn(s)$ | -1.18 |
| $Zn^{2+}(aq) + 2e^- \rightleftarrows Zn(s)$ | -0.76 |
| $Cr^{3+}(aq) + 3e^- \rightleftarrows Cr(s)$ | -0.74 |
| $Fe^{2+}(aq) + 2e^- \rightleftarrows Fe(s)$ | -0.41 |
| $Cr^{3+}(aq) + e^- \rightleftarrows Cr^{2+}(aq)$ | -0.41 |
| $Cd^{2+}(aq) + 2e^- \rightleftarrows Cd(s)$ | -0.40 |
| $PbSO_4(s) + 2e^- \rightleftarrows Pb(s) + SO_4^{2-}(aq)$ | -0.36 |
| $Tl^+(aq) + e^- \rightleftarrows Tl(s)$ | -0.34 |
| $Co^{2+}(aq) + 2e^- \rightleftarrows Co(s)$ | -0.28 |
| $Ni^{2+}(aq) + 2e^- \rightleftarrows Ni(s)$ | -0.24 |
| $AgI(s) + e^- \rightleftarrows Ag(s) + I^-(aq)$ | -0.15 |
| $Sn^{2+}(aq) + 2e^- \rightleftarrows Sn(s)$ | -0.14 |
| $Pb^{2+}(aq) + 2e^- \rightleftarrows Pb(s)$ | -0.13 |
| $2H^+(aq) + 2e^- \rightleftarrows H_2(g)$ | 0 |
| $AgBr(s) + e^- \rightleftarrows Ag(s) + Br^-(aq)$ | 0.07 |
| $S(s) + 2H^+(aq) + 2e^- \rightleftarrows H_2S(aq)$ | 0.14 |
| $Sn^{4+}(aq) + 2e^- \rightleftarrows Sn^{2+}(aq)$ | 0.15 |
| $SO_4^{2-}(aq) + 4H^+(aq) + 2e^- \rightleftarrows SO_2(g) + 2H_2O$ | 0.16 |
| $Cu^{2+}(aq) + e^- \rightleftarrows Cu^+(aq)$ | 0.16 |
| $Cu^{2+}(aq) + 2e^- \rightleftarrows Cu(s)$ | 0.34 |
| $Cu^+(aq) + e^- \rightleftarrows Cu(s)$ | 0.52 |
| $I_2(s) + 2e- \rightleftarrows 2I^-(aq)$ | 0.53 |
| $Fe^{3+}(aq) + e^- \rightleftarrows Fe^{2+}(aq)$ | 0.77 |
| $Hg_2^{2+}(aq) + 2e^- \rightleftarrows 2Hg(l)$ | 0.80 |
| $Ag^+(aq) + e^- \rightleftarrows Ag(s)$ | 0.80 |
| $2Hg^{2+}(aq) + 2e^- \rightleftarrows Hg_2^{2+}(aq)$ | 0.91 |
| $NO_3^-(aq) + 4H^+(aq) + 3e^- \rightleftarrows NO(g) + 2H_2O$ | 0.96 |
| $AuCl_4^-(aq) + 3e^+ \rightleftarrows Au(s) + 4Cl^-(aq)$ | 1.00 |
| $Br_2(l) + 2e^- \rightleftarrows 2Br^-(aq)$ | 1.08 |
| $O_2(g) + 4H^+(aq) + 4e^- \rightleftarrows 2H_2O$ | 1.23 |
| $MnO_2(s) + 4H^+(aq) + 2e^- \rightleftarrows Mn^{2+}(aq) + 2H_2O$ | 1.23 |
| $Cr_2O_7^{2-}(aq) + 14H^+(aq) + 6e^- \rightleftarrows 2Cr^{3+}(aq) + 7H_2O$ | 1.33 |
| $Cl_2(g) + 2e^- \rightleftarrows 2Cl^-(aq)$ | 1.36 |
| $ClO_3^-(aq) + 6H^+(aq) + 5e^- \rightleftarrows ½\ Cl_2(g) + 3H_2O$ | 1.47 |
| $Au^{3+}(aq) + 3e^- \rightleftarrows Au(s)$ | 1.50 |
| $MnO_4^-(aq) + 8H^+(aq) + 5e^- \rightleftarrows Mn^{2+}(aq) + 4H_2O$ | 1.51 |
| $PbO_2(s) + SO_4^{2-}(aq) + 4H^+(aq) + 2e^- \rightleftarrows PbSO_4(s) + 2H_2O$ | 1.69 |
| $H_2O_2(aq) + 2H^+(aq) + 2e^- \rightleftarrows 2H_2O$ | 1.76 |
| $Co^{3+}(aq) + e^- \rightleftarrows Co^{2+}(aq)$ | 1.95 |
| $F_2(g) + 2e^- \rightleftarrows 2F^-(aq)$ | 2.89 |

The size of the voltages in Table 14.2 tells about the chemical activity of the reactants and products in the listed redox reactions (at 25°C and standard conditions in acid solution). Whether or not a chemical entity is a good oxidizing or reducing agent broadly correlates with its position in the table of standard redox potentials as shown in Table 14.3. The algebraic sign of the voltage tells about the likely spontaneous direction of the redox reaction. That information is summarized in Table 14.4.

Table 14.3 **Locations of Chemical Entities in the Table of Reduction Potentials**

| Region of $E^0$ table | Inhabited by the entity |
|---|---|
| At the top of the table on the left of the redox equation | Cations very hard to reduce |
| At the top of the table on the right of the redox equation | The strong reductants |
| At the bottom of the table on the right of the redox equation | Ions/molecules very hard to oxidize |
| At the bottom of the table on the left of the redox equation | The strong oxidants |

Table 14.4 **Chemical Activity and Standard Reduction Potentials**

| Value of $E^0$ | Comment about the equation: reactant + $e^-$ ⇄ product |
|---|---|
| -1.5 to -3.0 volts | Reaction strongly spontaneous in reverse direction. Reactant very difficult to reduce. Product is a powerful reducing agent |
| -0.5 to -1.5 volts | Reaction moderately spontaneous in reverse direction. Reactant somewhat difficult to reduce. Product is a fairly good reducing agent. |
| +0.5 to 0.0 volts | Reaction weakly spontaneous in reverse direction. Reactant easy to reduce. Product is a weak reducing agent. |
| 0.0 to -0.5 volts | Reaction weakly spontaneous in forward direction. Reactant is a weak oxidizing agent. Product is easy to oxidize. |
| +0.5 to +1.5 volts | Reaction moderately spontaneous in forward direction. Reactant is a fairly good oxidizing agent. Product is somewhat difficult to oxidize. |
| +1.5 to +3.0 volts | Reaction strongly spontaneous in the forward direction. Reactant is a powerful oxidizing agent. Product is very difficult to oxidize. |

## Combining Half-Cell Potentials

If you reverse any reduction potential equation, such as one of those listed in Table 14.2, you simultaneously change the sign of the reaction's voltage, and the reduction potential is converted to an oxidation potential, as shown by the examples below:

| This reduction potential equation, | when reversed, becomes | this oxidation potential equation. |
|---|---|---|
| $Li^+(aq) + e^- \rightleftarrows Li(s)$, $E^0 = -3.04$ V | | $Li(s) \rightleftarrows Li^+(aq) + e^-$, $E^0 = +3.04$ V |
| $Ni^{2+}(aq) + 2e^- \rightleftarrows Ni(s)$, $E^0 = -0.24$ V | | $Ni(s) \rightleftarrows Ni^{2+}(aq) + 2e$, $E^0 = +0.24$ V |
| $Cl_2(g) + 2e- \rightleftarrows 2Cl^-(aq)$ $E^0 = +1.36$ V | | $2Cl^-(aq) \rightleftarrows Cl_2(g) + 2e-$, $E^0 = -1.36$ V |

Real cells consist of two half-cells, one operating to perform reduction and the other operating to perform oxidation. The voltage of a real cell operating at 25°C at standard conditions is found by combining the equations and the voltages of its two half-cells in accord with stoichiometry. You can combine any two reduction half-cells in three steps:

Step 1: Reverse either one of them to make it an oxidation reaction.

Step 2: Use multiplying coefficients to balance the electrons, such that the number of electrons in the reduction equation equals the number of electrons in the oxidation equation.

Step 3: Add the two reaction equations *and* add their voltages.

When you combine two half-cell reactions it doesn't matter which one you reverse. There are two possible ways, but the two equations you get are chemically reversed and have voltages with opposite algebraic signs. So combining half-cell reactions gives only one answer, though it can be written two ways. Note, voltage is an intensive property; so when you multiply equations by factors to balance their electrons you do NOT change the reactions' voltages.

After combining half-cell reactions, a positive voltage means that the reaction as written is thermodynamically spontaneous. A negative voltage means that the reaction as written is thermodynamically nonspontaneous, or alternatively that it will be thermodynamically spontaneous in the reverse direction. A **thermodynamically spontaneous electrochemical reaction** has a positive cell voltage. A **thermodynamically nonspontaneous electrochemical reaction** has a negative cell voltage. Incidentally, battery reactions are thermodynamically spontaneous. Electrolysis reactions are thermodynamically nonspontaneous.

Example 14.6 illustrates the use of the table of standard reduction potentials. Combining half-cells is illustrated in Examples 14.7 to 14.10.

**Example 14.6**: Use Table 14.2 and state the fifth best reductant and the third best oxidant.

Strategy: Understand where the best oxidants and reductants are to be found on the table, and count.

$H_2O_2$ with a standard voltage of +1.76 V is the third best oxidant listed.

Sodium metal, Na(s), with a standard voltage of +2.71 V (when reversed from the table) is the fifth best reductant listed.

**Example 14.7**: What's the voltage of the best standard cell reaction that we can imagine? "Best" means the coupled pair of redox reactions that gives the largest possible positive voltage.

Strategy: Combine the best reductant in Table 14.2 with the best oxidant. The best reductant is lithium metal at the top of the table on the right hand side. Diatomic fluorine at the bottom of the table on the left hand side is the best oxidant.

Step 1: Reverse the lithium half cell reaction to get the largest possible reduction potential; change the sign of its voltage. Step 2: Double that equation to make the electrons balance 2:2. Step 3: Add the two half reactions:

$$2\,Li(s) \rightleftarrows 2\,Li^+ + 2\,e^- \qquad E^\circ = +3.04 \text{ volts}$$

$$F_2 + 2\,e^- \rightleftarrows 2F^- \qquad E^\circ = +2.89 \text{ volts}$$

$$2\,Li(s) + F_2 \rightleftarrows 2\,Li^+ + 2F^- \qquad E^\circ = +5.93 \text{ volts}$$

Thus, the best voltage imaginable for a single cell is about 6 volts. However, a practical version of such a cell probably cannot be built because of the extreme chemical reactivity of the redox pair.

How can there be a 12 volt battery? It has more than one cell.

**Example 14.8**: When chlorine gas is bubbled into a solution of sodium iodide (NaI) the solution immediately darkens, revealing that iodide ion ($I^-$) has been oxidized to iodine, $I_2$, which in this case dissolves in the solution. Chlorine is the oxidizing agent and iodide ion the reducing agent. Select half reactions for the redox from Table 14.2 and combine them to get the voltage of the net reaction at standard conditions at 298 K.

Strategy: Select the appropriate equations from Table 14.2 and combine them according to the steps outlined in the text. The equations are: $Cl_2(g) + 2e^- \rightleftarrows 2Cl^-(aq)$, $E^0 = +1.36$ V, and $I_2(s) + 2e- \rightleftarrows 2I^-(aq)$, $E^0 = +0.53$ V. The description in the question tells us to reverse the iodine reaction.

Oxidation reaction: $Cl_2(g) + 2e^- \rightleftarrows 2Cl^-(aq) \qquad E^0 = +1.36$ V

Reduction reaction: $2I^-(aq) \rightleftarrows I_2(s) + 2e- \qquad E^0 = -0.53$ V

Net reaction: $Cl_2(g) + 2I^-(aq) \rightleftarrows 2Cl^-(aq) + I_2(s) \qquad E^0 = +0.83$ V

The positive sign of the voltage confirms that the reaction is indeed spontaneous at standard conditions and 298 K.

**Example 14.9**: Does zinc dissolve in acid solution at standard conditions ($[H^+]$ = 1 M, $[Zn^{2+}]$ = 1 M, pressure = 1 bar) at 298 K?

Strategy. To answer this question you need to complete the skeleton equation Zn(s) + $H^+ \rightarrow$ products using the reactions in Table 14.2. Find the needed half reactions, combine them properly, and see if the voltage is positive or negative. A positive voltage means the reaction is spontaneous and that zinc does dissolve.

Here are the equations from the table. The Zn must be reversed. The electrons are already in balance.

$$2\,H^+(aq) + 2e^- \rightleftarrows H_2(g) \quad 0.00V \quad \text{and} \quad Zn^{2+}(aq) + 2e^- \rightleftarrows Zn(s)\ -0.76V$$

Combining: $2\,H^+(aq) + 2e^- \rightleftarrows H_2(g) \quad 0.00V$

$\underline{Zn(s) \rightleftarrows Zn^{2+}(aq) + 2e^-\ +0.76V}$

Net: $2\,H^+(aq) + Zn(s) \rightleftarrows H_2(g) + Zn^{2+}(aq) \quad E_{cell} = \underline{+0.76V}$

Positive voltage. Answer, yes.

**Example 14.10**: Calculate the Daniell cell's standard voltage using the table of standard half-cell potentials.

Strategy: Read to learn what a Daniell cell is. The Daniell cell was an important power source during the early days of telegraphy at the middle of the 19th century. It's a battery made by connecting a $Cu/Cu^{2+}$ half-cell to a $Zn/Zn^{2+}$ half-cell. As the cell operates, copper plates out on the copper electrode; in the other half-cell, zinc metal dissolves with the formation of zinc ions. We'll assume standard state conditions and 298 K to answer the question. Write the two half-cell reactions and combine them appropriately—that means reverse the $Zn/Zn^{2+}$ pair.

At the cathode $Cu^{2+}(1.0\ M) + 2\,e^- \rightleftarrows Cu(s) \quad E^0 = +0.34\ V$

At the anode $Zn(s) \rightleftarrows Zn^{2+}(1.0\ M) + 2\,e^- \quad E^0 = +0.76\ V$

Net: $Cu^{2+}(1.0\ M) + Zn(s) \rightleftarrows Cu(s) + Zn^{2+}(1.0\ M) \quad E^0 = +1.10\ V$

The Daniell cell generates a potential of 1.10 volts under standard conditions at 298 K.

## The Activity Series of the Metals

The table of standard reduction potentials enables us to predict and correlate chemical reactivity. The more difficult it is to obtain a metal from its ion (as judged by it half-cell potential), the better a reducing agent is the metal. Zinc, at -0.76 V, is harder to obtain from zinc ion than is copper, at +0.34 V, from copper ion. It follows that zinc metal is a better reducing agent than copper metal.

In general, the more negative the standard reduction potential of a metal, the better a reducing agent and the more chemically reactive it is. The relative reactivities of metals can thus be placed on a quantitative basis by ranking them according to their standard electrochemical half-cell voltages:

$$Li > K > Ba > Ca > Mg > Be > Al > Zn > Fe > Cu > Ag > Au$$

The sequence above is traditionally called the **activity series** of the metals. Every metal will reduce the ions of any metal lying to its right. Reduction reactions with active metals at the left hand end of this series are routinely used to prepare "exotic" metals. Two commercially used reactions are:

$$BeF_2 + Mg = Be + MgF_2$$

$$2\,PuCl_3 + 3\,Ca = 3\,CaCl_2 + 2\,Pu$$

## Section 14.5 Thermodynamics and Electrochemistry

To begin, a quick refresher about units:

**A Note About Units**

Coulomb (C) is the SI unit of electrical charge. It is the charge carried by a current of exactly one amp flowing for exactly one second (where the amp and second are defined base units of SI). 1C = 1 amp sec.

Chemical stoichiometry is linked to flow of electric current through the Faraday constant (F), 9.648 530 9 (29) × $10^4$ C $mol^{-1}$

1. amps × volts = watts (power)
2. 1 watt = 1 J $s^{-1}$ (a watt is a joule per second: definition)
3. 1 C = amp × sec (SI definition of electric charge.)

So, rewrite equation 1 volts $\frac{\text{watts}}{\text{amps}}$

From 2, watts = J $s^{-1}$

From 3, amps = C $s^{-1}$

And, plugging the latter two into the rewritten 1

$$\text{Volts} = \frac{\text{watts}}{\text{amps}} = \frac{\text{Js}^{-1}}{\text{Cs}^{-1}}$$

Thus, 1 volt = 1 J $C^{-1}$ or volts × coulombs = joules

In the preceding section we described how to calculate the voltages of any cell operating at standard state conditions using a table of $E^0$ values. An operating cell is a thermodynamic system, so there is a direct connection between the cell's voltage and the thermodynamic properties of the system.

Electrical work is the quantity of charge that flows multiplied by the voltage at which the cell operates (q × E). The quantity of charge that flows during an electrochemical reaction is the total charge of all the electrons or the number of moles of electrons (n) times the Faraday constant, F (Example 14.1). Setting the electrical work equal to the free energy change gives the mathematical relationship. It's as useful as it is simple:

$$\Delta G^0 = - n E^0 F$$

where, $\Delta G^0$ is the standard Gibbs free energy change of the cell, n is the number of moles of electrons involved in the balanced electrochemical equation (the electron coefficient), and F is the Faraday constant. The units work out because the voltage $E^0$ can be expressed as J $C^{-1}$ and the Faraday can be expressed at C $mol^{-1}$. The more positive the cell voltage the more negative the Gibbs free energy change. We've already pointed out that electrochemical reactions with large positive voltages are spontaneous.

Thus (at a specified T) Gibbs free energy change ($\Delta G^0$) and cell voltage ($E^0$) are equivalent ways to state the thermodynamic spontaneity of an electrochemical reaction system. They have opposite signs as a consequence of the way they are defined. Relating $E^0$ and $\Delta G^0$ is shown in Example 14.11.

**Example 14.11**: The Daniell cell operates with a standard potential of +1.10 volts and has an electron coefficient of 2 moles of $e^-$ per mole of reaction. Calculate the standard Gibbs free energy change of the electrochemical process in kJ $mol^{-1}$.

Strategy: Plug the appropriate quantities and units into $\Delta G° = - n E0F$. The unit volt is replaced by its equivalent J $C^{-1}$.

$$\Delta G^0 = (-2 \text{ mol mol}^{-1}) \times (1.10 \text{ J C}^{-1}) \times 96485 \text{ C mol}^{-1}$$

$$\Delta G^0 = \ -2 \times 1.10 \text{ J C}^{-1} \times 96485 \text{ C mol}^{-1} \times 10^{-3} \text{ kJ J}^{-1} \ = \ \underline{-212 \text{ kJ mol}^{-1}}$$

Recall from thermodynamics that there is also a direct relationship between standard Gibbs free energy change of a reaction and the reaction's equilibrium constant, $K_c$. The mathematical form of the relationship is $\Delta G^0 = -RT\ln K_c$. Table 8.6 shows how $K_c$ varies with varying $\Delta G^0$ values. Table 14.5 amplifies the earlier table by adding another column—the $E^0$ values; it is calculated on the basis that the electron coefficient of the electrochemical reaction is 2 (n=2).

Table 14.5 **Relation Between $E^0$, $\Delta G^0$, and K, (n = 2), T = 298 K**

| $E^0$ (V) | $\Delta G^0$ (kJ) | K |
|---|---|---|
| | | |
| 2.00 | -386 | $4.6 \times 10^{67}$ |
| 1.00 | -193 | $6.8 \times 10^{33}$ |
| 0.50 | -96 | $8.2 \times 10^{16}$ |
| 0.00 | 0 | 1.0 |
| -0.50 | 96 | $1.2 \times 10^{-17}$ |
| -1.00 | 193 | $1.5 \times 10^{-34}$ |
| -2.00 | 386 | $2.2 \times 10^{-68}$ |

## The Nernst Equation

We have now established that the voltage of an electrochemical cell is related to the Gibbs free energy change of the reaction in the cell. Our next step is to see how the cell voltage changes as the concentrations of the reacting chemicals change. You already *know* that the voltage does change, because you've used batteries and know that they "run out." Running out is the informal way of say that the battery voltage has changed—it's fallen to zero!

In Chapter 8 we saw how the concentrations of gas phase and liquid phase reactions affect the free energy change of a reaction. You'll recall that the equation is $\Delta G = \Delta G^0 + RT\ln Q_c$. In this section we've seen that free energy is linked to the voltage of a cell by the equation $\Delta G = - nFE$. So it's combining these that leads to the Nernst equation. Substituting $\Delta G = - nFE$ and $\Delta G^0 = - nFE^0$ into $\Delta G = \Delta G^0 + RT\ln Q_c$ gives:

$$- nFE = -nFE^0 + RT\ln Q_c$$

Which is the Nernst equation: $E = E^0 - \frac{RT}{nF} \ln Q_c$

The equation (named after the German chemist Walter Nernst, 1864-1941) relates the actual operating voltage (E) of a cell to its standard voltage ($E^0$) and the concentrations of the reacting chemicals (in the term $Q_c$) in the cell at temperature, T; R is the gas constant, F the Faraday constant, and n the electron coefficient of the reaction.

The Nernst equation enables us to calculate the half-cell voltages and cell voltages under nonstandard conditions and can account for changing temperature, changing pressure for gaseous reactants and products, and changing solution concentrations of dissolved reactants and products. To make the connection explicit we'll substitute the equilibrium constant expression from Section 8.5 for the general two-reactant, two-product reaction:

$$a\text{A} + b\text{B} \rightleftarrows c\text{C} + d\text{D} \text{ is}$$

$$Q_c = \frac{[C]^c[D]^d}{[A]^a[B]^b} \text{ for system not in equilibrium}$$

$$\text{and } K_c = \frac{[C]^c[D]^d}{[A]^a[B]^b} \text{ for system in equilibrium}$$

So the complete form of the Nernst equation, explicitly showing the concentration terms, is:

$$E = E^0 - \frac{RT}{nF} \ln \frac{[C]^c[D]^d}{[A]^a[B]^b}$$

where E is the actual cell voltage, $E^0$ is the standard cell voltage, R is the constant 8.314 J $mol^{-1}$ $K^{-1}$, standard T is 298.15 K, n is the number of moles of electrons involved in the balanced electrochemical equation, F is 96485 J $C^{-1}$, and the terms in square brackets are the molar concentrations of the reactants and products in the electrochemical equation (or the pressure if a gas is involved, as in the case of the SHE). The ratio of the concentration terms is $Q_c$ if the reaction is not at equilibrium and $K_c$ if the reaction is in equilibrium.

We can make the same tests of the Nernst equation that we made of $\Delta G = \Delta G^0 + RT\ln Q_c$ in Section 8.5. Test 1was to substitute concentrations at standard conditions, where all reactants and product concentrations are set to 1.0 molar. Test 2 was to substitute the concentrations attained when the system has reached chemical equilibrium, which is when its voltage has fallen to zero.

1. At standard state conditions all the concentrations are 1.0 molar, so the reaction quotient $Q_c = 1$ and, because ln 1 is zero, the whole right hand term becomes zero. Thus at standard state conditions E becomes $E^0$ just as it should.

2. At equilibrium the voltage is zero (the battery has gone dead) and $Q_c$ is $K_c$. So, when the discharge condition has arrived, $Q_c$ has become $K_c$ and the two terms on the right hand side of the Nernst equation are equal:

$$E^0 = \frac{RT}{nF} \ln\left(\frac{[C]^c[D]^d}{[A]^a[B]^b}\right)_{eq}$$

One valuable use of the Nernst equation in to measure the changing voltage of an electrochemical cell and use that to measure a concentration change inside the cell. In practice, the Nernst equation enables electrochemical measurement over a huge range of solution concentrations. Such measurements are the basic procedure used by a pH meter and by the many other types of meters that provide electrochemical measurements of concentration. For example, electrochemical oxygen sensors—which are calibrated using air—are capable of measuring parts per billion (ppb) of oxygen.

When using the Nernst equation for concentration measurement near standard conditions and 298 K, it is traditional and convenient to collect the constants in the Nernst equation at standard T and write the equation in an abbreviated form:

Substituting, R = 8.314 J $mol^{-1}$ $K^{-1}$, T = 298.15 K, F = 96485 J $V^{-1}$ $mol^{-1}$, and $\ln Q = 2.303 \log_{10} Q$

So, $$E = E^0 - \frac{8.314\ \text{J mol}^{-1}\ \text{K}^{-1} \times 298.15\ \text{K}}{96485\ \text{J V}^{-1}\ \text{mol}^{-1}} \times 2.303 \times \log Q_c$$

Doing the arithmetic, gives $$E = E^0 - \frac{0.0592}{n} \log Q_c$$

So we write, $$E = E^0 - \frac{0.0592}{n} \log_{10} \frac{[C]^c[D]^d}{[A]^a[B]^b}$$

Note that the concentrations of the products C and D are in the numerator of the log term, while the concentrations of the reactants A and B are in the denominator. Thus, as a cell reaction proceeds, and products accumulate while reactants are consumed, the log term gets larger. Thus, the whole right hand term represents a larger subtraction from $E^0$ as the reaction proceeds. The voltage of any cell drops as its reactants convert increasingly to products, as the battery runs down. When the battery is completely exhausted E has fallen to zero.

Typical results of calculations using the above equation are shown in Example 14.12.

**Example 14.12**: Use the Nernst equation to calculate the electrochemical potential (voltage) for the reaction of zinc metal with acid $[H^+]$ at various concentrations.

Strategy: Select a range of values and substitute them into the equation

$$E = E^0 - \frac{0.0592}{n} \log \frac{[C]^c[D]^d}{[A]^a[B]^b}$$

The reaction is: $Zn(s) + 2\,H^+ \rightleftarrows Zn^{2+} + H_2$, for which $Q = \frac{[Zn^{2+}]P_{H_2}}{[H^+]^2}$, where the Zn(s) is omitted (for the usual reason: this is a heterogeneous equilibrium), the concentrations of the $Zn^{2+}$ and the $H^+$ are in molarity, and the pressure of the gaseous hydrogen is in atmospheres. The value n = 2 because 2 moles of electrons are needed to oxidize one mole of zinc.

In this case the Nernst equation becomes:

$$E = E^0 - \frac{0.0592}{2} \log \frac{[Zn^{2+}]P_{H_2}}{[H^+]^2}$$

Combining the standard half-cell reactions gives the standard cell voltage:

$$2H^+(aq) + 2e^- \rightleftarrows H_2(g) \quad 0.00\ V$$

$$Zn(s) \rightleftarrows Zn^{2+}(aq) + 2e^- \quad +0.76\ V$$

So the standard potential is +0.76 V. Below is a table of E values at selected concentrations chosen to illustrate how the equation works.

| | $[Zn^{2+}]$ | $P_{H2}$ | $[H^+]$ | pH | E |
|---|---|---|---|---|---|
| Set 1 | 1.0 | 1.0 | 1.0 | pH = 0 | 0.760 (=$E^0$) |
| | 2.0 | 1.0 | 1.0 | pH = 0 | 0.739 |
| | 3.0 | 1.0 | 1.0 | pH = 0 | 0.727 |
| | 4.0 | 1.0 | 1.0 | pH = 0 | 0.719 |
| | 5.0 | 1.0 | 1.0 | pH = 0 | 0.712 |
| Set 2 | 1.0 | 1.0 | 1.0 | pH = 0 | 0.760 (=$E^0$) |
| | 1.0 | 2.0 | 1.0 | pH = 0 | 0.739 |
| | 1.0 | 3.0 | 1.0 | pH = 0 | 0.727 |
| | 1.0 | 4.0 | 1.0 | pH = 0 | 0.719 |
| | 1.0 | 5.0 | 1.0 | pH = 0 | 0.712 |
| Set 3 | 1.0 | 1.0 | 0.1 | pH = 1 | 0.692 |
| | 1.0 | 1.0 | 0.01 | pH = 2 | 0.624 |
| | 1.0 | 1.0 | 0.001 | pH = 3 | 0.556 |
| | 1.0 | 1.0 | $1.0 \times 10^{-6}$ | pH = 6 | 0.351 |
| | 1.0 | 1.0 | $1.0 \times 10^{-7}$ | pH = 7 | 0.283 |

The data of set 1 could be applied to provide an electrical sensor for zinc ion concentration in solution. The data of set 2 could be applied as a hydrogen pressure transducer. The data of set 3 could be applied to measure the hydrogen ion concentration, or pH, of a solution.

## Section 14.6 Devices to Produce Electric Current

### Traditional Batteries

Several of the most common cells and batteries were developed during the 19th century to supply electric current for the expanding telegraph industry. Today, with electric power available at any wall plug, we forget that battery current was the only convenient source of electricity until well into the 20th century.

Curiously, the search for efficient batteries to power the telegraph in an earlier age is paralleled in our own time because of the explosion in the use of battery-powered portable devices. In the year 2000 it's estimated that there are 65 million cell-phones in use in the US, and worldwide there will be more than 50 million battery powered laptop computers. It's predicted that the market for batteries will continue to grow vigorously in the 21st century. Many different reductant-oxidant pairs have been incorporated into commercial batteries. Deciding what battery design is best for a particular service job is an interesting example of an engineering decision made on the basis of chemical properties.

When several cells are connected **in series** ("in series" means that the anode of one is connected to the cathode of the next, and so on) the voltage of the series is the sum of the voltages of the individual cells. In the early days of electrochemistry, batteries of cells were used to give both increased voltage and higher capacity. The term battery derives from a root meaning to beat or to batter. A battery of guns consists of several artillery pieces working together for better effect. It was Benjamin Franklin, in a letter written in 1748, who first applied the term battery to a series of electrical cells working together. Over time, the meaning of the word **battery** has changed to refer to any device with *one* or *more* cells used to produce electric current.

Here's list of some battery requirements; no single battery can meet all of them.

**Questions about the Desired Characteristics of a Battery**

What is its energy density per mass?
Is its voltage sufficiently stable as the battery discharges?
Is it rechargeable? If so, does it have a sufficiently rapid recharge rate?
Does it lose charge on storage?
Can it be fully recharged to its original capacity?
Does it have a satisfactory recharge rate?
How many recharges is the battery capable of?
What are the materials costs?
Does it show satisfactory performance at high (and/or low) temperatures?
Can it be made thin and of a convenient shape?
Does it pass safety tests: heat to 150°C, nail penetration, crushing, etc?

The **energy density** of a battery per mass is a measure of its energy to weight ratio; it is usually expressed in the unit of kilowatt-hours per kilogram. In the material that follows we'll discuss some of the more important commercial batteries.

### The Lead-Acid Battery

A lead-acid battery gets its name from its two lead-containing electrodes and its electrolyte of sulfuric acid (about 6 molar). A 12 volt automobile battery consists of six cells in series each generating about 2 volts. Each cell contains two lead alloy electrodes—one is impregnated with finely divided lead metal (Pb) and the other with lead(IV) oxide ($PbO_2$). The discharge half-cell reactions consist of the oxidation of lead(0) to lead(II) at the anode, and the simultaneous reduction of lead(IV) to lead(II) at the cathode. When the battery is being discharged, the following reactions occur:

anode $Pb(s) + HSO_4^- \rightarrow PbSO_4(s) + H^+ + 2e^-$ $E^0 = 0.4\ V$

cathode $PbO_2(s) + 3H^+ + HSO_4^- + 2e^- \rightarrow PbSO_4(s) + 2H_2O$ $E^0 = 1.7\ V$

Overall $Pb(s) + PbO_2(s) + 2H_2SO_4 \rightarrow 2PbSO_4(s) + 2H_2O$ $E^0 = 2.1\ V$

The oxidant ($PbO_2$), reductant (Pb), and the products of the redox reaction ($PbSO_4$) are all solids. Because they are solids, they cannot move into direct contact. Thus, the cell needs neither a salt bridge nor a porous divider to operate successfully. The hydrogen ion concentration is not 1 M—so this is not a standard state conditions cell.

Known since 1878, the lead storage battery has been in use in cars since about 1915, when the first self-starters were introduced. Lead-acid batteries (Figure 14.11) are rugged and inexpensive, and can be recharged many times. Lead has good recycle value as evidenced by its thriving scrap market. But they are heavy and lead-acid battery-powered vehicles, such as golf carts, require a large fraction of their mass to be devoted to their batteries.

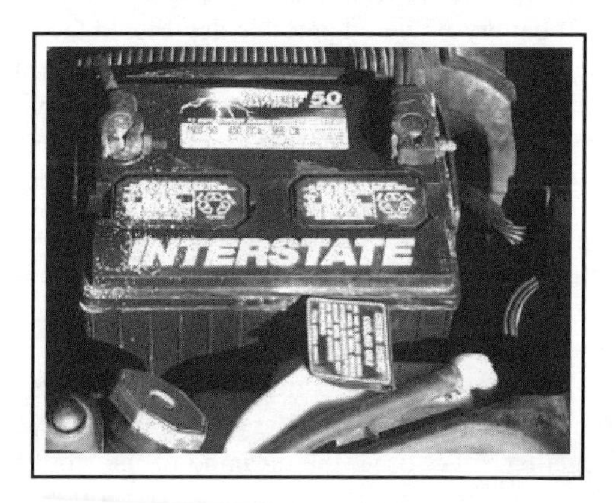


Figure 14.11 Developed around 1878, the lead-acid battery remains the universal choice for automobiles. Its drawbacks of weight and toxic ingredients, are more than compensated for by its reliability, ease of recharge, and modest cost.

As a lead-acid battery is discharged in use, the water formed (see the equation above) dilutes the sulfuric acid solution and lowers the solution's density. Thus, a hydrometer can be used to measure the state of each cell of the battery—to test if the battery is "charged" or "discharged." The acid in a fully charged battery has a density of about 1.28 g $mL^{-1}$. When the density falls below about 1.20 g $mL^{-1}$, the battery is in a discharged state.

Although the technology of the lead-acid battery is old, there have been some relatively recent wrinkles such as using gelling agents to thicken the electrolyte, and improving electrode design to allow for fully sealed batteries. The connecting posts on a car battery are made of lead. With the demise of lead pipes for plumbing and the removal of lead compounds from gasoline, battery connectors are one of the few remaining places where we are aware of lead in our daily lives.

The lead-acid battery is sometimes called a secondary cell. A **secondary cell** is one that can be recharged. After around 1870, steam-powered generators became, available and a rechargeable cell became known as a secondary power supply.

### Common Dry Cell Battery (Leclanché cell)

The common dry-cell flashlight battery is a modern embodiment of the Leclanché cell which dates from 1867. The oxidant is $MnO_2$, the reductant Zn, and the electrolyte is ammonium chloride, $NH_4Cl$. A cardboard separator keeps the two half-cell reactions apart. Typically the metal container for the device is the zinc anode itself, so the battery slowly dissolves itself from the inside out as it discharges. To prevent leaks, the zinc container is usually protected on its outer surface by a plastic wrap. Because it is used once, and then disposed of, it is called a **primary cell**.

Inside the battery is a moist paste consisting of $MnO_2$, $NH_4Cl$, carbon, thickening agents, and water, surrounding an inert graphite cathode. The cathode half-reaction is complex and depends on the amount of current being drained from the battery. The cathode compartment is separated from the anode housing by a cylinder of cardboard that serves as an inexpensive ion conduit. Approximate equations for dry-cell battery operation are:

| | | |
|---|---|---|
| anode | $Zn \rightarrow Zn^{2+} + 2\ e^-$ | $E^0 = 0.76$ volts |
| cathode | $\underline{2\ MnO_2 + 2NH_4^+ + 2\ e^- \rightarrow Mn_2O_3 + 2NH_3 + H_2O}$ | $E^0 = 0.74$ volts |
| Net: | $Zn + 2\ MnO_2 + 2NH_4^+ \rightarrow Zn^{2+} + Mn_2O_3 + 2NH_3 + H_2O$ | $E^0 = 1.5$ volts |

The dry-cell battery is so called because its electrolyte is a viscous paste, rather than an a free-flowing liquid. In contrast to earlier commercial cells, the dry-cell is much less likely to spill or accidentally discharge liquids, and it can be manufactured as a sealed unit. The dry-cell battery is not rechargeable. Applying a recharging voltage does not cause a reversal of the half-reactions; in fact it may produce gases and cause the cell to rupture, possibly dangerously. Although they cannot be recharged, dry-cells are used widely because of their low cost. About 10% of batteries sold in the US are of the dry-cell type. Figure 14.12 shows a dry-cell battery with a lot of cells.

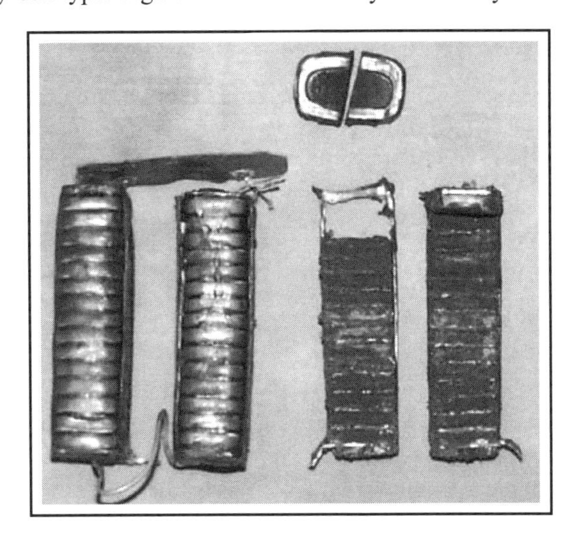

Figure 14.12 A disassembled dry-cell battery. Its 45 cells were arranged in three banks of 15 cells each, one bank of which has been cut in half. At the top you can see one individual cell—it's about the size of a silver dollar. The ring is the zinc metal and the dark material in the center is manganese dioxide ($MnO_2$) oxidant and the carbon anode. When operating, the cell produced 45 × 1.5, or 67.5 V

## Alkaline Zinc Manganese Dioxide Cells

The alkaline zinc manganese dioxide reaction is by far the leading US battery technology in terms of the sheer numbers of batteries sold. Commonly, cells based on this redox reaction are labeled as "alkaline." Alkaline cells offer more capacity for a given cell size than do the Leclanché types, although they cost 50% more and weigh 25% more. Approximate half-cell equations are:

| | |
|---|---|
| anode | $Zn + 2\ OH^- \rightarrow ZnO + 2\ e^- + H_2O$ |
| cathode | $\underline{H_2O + 2\ MnO_2 + 2\ e^- \rightarrow Mn_2O_3 + 2\ OH^-}$ |
| net | $Zn + 2\ MnO_2 \rightarrow Mn_2O_3 + ZnO$ |

The electrolyte in this cell is a viscous aqueous liquid that contains potassium hydroxide and dissolved zinc oxide, $Zn(OH)_4^{2-}$. There have been improvements in alkaline cells over the past 20 years, but these improvements have been entirely because of changes in packaging and manufacturing techniques. Today's cells are lighter and more durable, but their underlying redox chemistry remains the same.

Unfortunately, batteries of this type require a trace of mercury on their zinc electrode to serve as an "anti-passivation" additive. Passivation occurs when an electrode corrodes in such a way that it becomes an insulator. If the mercury is absent, the cell's capacity is drastically reduced. Legislation is likely to limit the mercury content of these cells to 250 ppm. However, finding an environmentally suitable replacement for the mercury would be a better solution.

## Newer Batteries

We already met the lead-acid battery. Other rechargeable batteries find use in phones, cameras, and computers.

### The Nicad (Ni-Cd) Battery

Obviously, nicad batteries are named because of the nickel and cadmium used in their manufacture. During the past 30 years their use has grown sharply, for applications such as toys and electric razors. Nicad cells contain an alkaline electrolyte. Their approximate electrode reactions can be represented by the following equations:

Cathode $2\ NiOOH + 2\ H_2O + 2\ e^- \rightarrow 2\ Ni(OH)_2 + 2\ OH^-$

Anode $\underline{Cd + 2\ OH^- \rightarrow Cd(OH)_2 + 2e^-}$

Net: $2\ NiOOH + Cd + 2\ H_2O \rightarrow 2\ Ni(OH)_2 + Cd(OH)_2$

The net reaction shown is the discharge reaction. During charging the net reaction is reversed and the original oxidant and reductant are restored by an electrolysis reaction. The products of discharge, the cadmium hydroxide and nickel hydroxide, adhere well to the respective electrodes, and the battery can be recharged many times before it becomes inoperable. Approximately 1500 million nicad batteries were manufactured in the US in 1997.

The demand for high battery output, longer service life before recharge, and greater power density has led in the past ten years to the development of nickel-metal hydride (NiMH) batteries (approximately 500 million manufactured in the US in 1997) and lithium ion (Li-ion) batteries (approximately 200 million manufactured in the US in 1997).

### Nickel Metal Hydride Batteries

NiMH batteries have the advantages of high energy density and greater run times between recharges. Their disadvantages are that they can discharge up to 30% per month in storage and are expensive. These cells contain a complex hydrogen storage alloy anode. The anode alloy comes in two types: the $AB_2$ type that is typically a $TiZr_2$ alloy with a small amount of chromium or other metal, and a type designated $AB_5$ based on $LaNi_5$. Additives are important in controlling the properties of the metal hydride electrode—but much of what is known about these cells is proprietary. The electrode reactions are approximately represented by the following equations:

Cathode: $2\ NiOOH + 2\ H_2O + 2\ e^- \rightarrow 2\ Ni(OH)_2 + 2\ OH^-$

Anode: $\underline{TiZr_2H_2 + 2\ OH^- \rightarrow TiZr_2 + 2\ H_2O + 2\ e^-}$

Net: $2\ NiOOH + TiZr_2H_2 \rightarrow TiZr_2 + 2\ Ni(OH)_2$

The net reaction shown is during battery discharge. The reverse reaction occurs during battery charging.

### Lithium-Ion Battery

The lithium-ion battery is a relatively new technology that offers excellent power-to-weight ratio and a high voltage. The anode is based on graphite that can absorb lithium ions between the hexagonal planes of carbon atoms. The cell is shown schematically in Figure 14.13. Though the actual electrode reactions in this battery are complex, they can be approximately represented by the following half reaction equations:

Cathode: $Li^+ + LiCoO_2 + e^- \rightarrow Li_2CoO_2$

Anode: $\underline{LiC_6 \rightarrow C_6 + Li^+ + e^-}$

Net: $LiCoO_2 + LiC_6 \rightarrow C_6 + Li_2CoO_2$

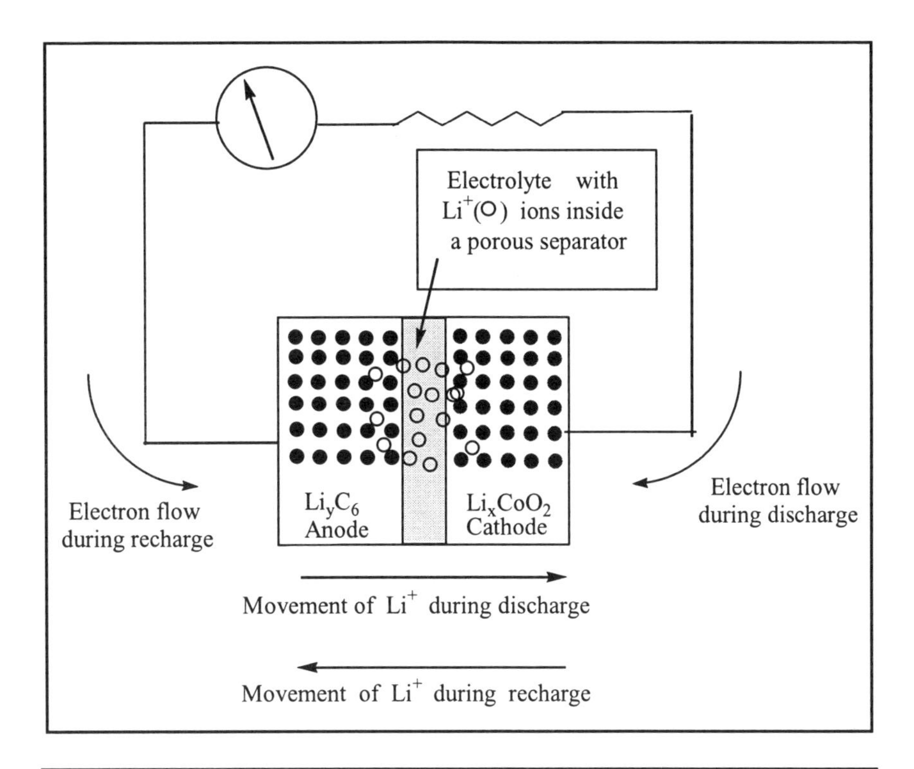


Figure 14.13 The lithium ion battery depends for its operation on the flows of lithium ions through a specially designed porous separator.

Miscellaneous Types of Batteries

There are hundreds of battery types. We'll mention a few others in this section. **Primary lithium batteries** use lithium metal as the reductant and combine powerful reduction potential (recall that Li is at the top of the table of reduction potentials) with a low atomic weight. So the energy density of primary lithium cells is high.

Lithium-thionyl chloride ($SOCl_2$) cells have the highest energy density of any primary lithium battery and are sold in hermetically sealed cases. In low-current applications these cells have a service life of 15-20 years making them valuable for medical applications such a powering pacemakers implanted inside humans—where changing the battery requires surgery.

The zinc/mercury(II) oxide cell, with the net reaction Zn(s) + HgO(s) = Hg(l) + ZnO(s), generates a voltage of 1.344 volts that remains highly stable during discharge. This property makes the mercury cell useful in instruments that rely on constant cell potential for their operation. But it is both an expensive and environmentally unsafe battery. These cells have been largely replaced by zinc-air batteries (to be discussed in the next section) since about 1990.

## Fuel Cells

Ordinary batteries quit when their reactants are depleted and, if they cannot be recharged, are useless. In contrast, fuel cells operate indefinitely—provided fuel is supplied to them.

A **fuel cell** is a device that generates electrical energy from a redox reaction in which one or both of the reactants is continually supplied to the cell from the outside. Fuel cells that use air or oxygen as the oxidant have high energy densities because the air (oxygen) does not have to be packaged along with the battery. A conventional battery parallels an explosive in that a battery is packaged with both the oxidant and reductant inside. A an air-breathing fuel cell parallels a gasoline engine in that it will operate in any environment where air is available.

Zinc-air batteries have grown to become an important class of commercial batteries since their introduction in the mid-1980s. Their net reaction is Zn(s) + $O_2$(g) = 2 ZnO(s). They are energized only when a sealing tab is removed and oxygen from the air passes through a gas-permeable, liquid-tight membrane into the cell. The zinc-air cell generates full power after only a few seconds. These characteristics the batteries of choice for pagers and cardiac monitors. They are, however, susceptible to the weather and tend to lose efficiency when the humidity is either high or low.

Perhaps the best potential application for fuel cell technology is to perform electrochemical combustion of liquid hydrocarbon fuels. The largest fuel cell ever built generated almost 5 megawatts of power (almost as much as the power plant pictured in Figure 7.1). It used phosphoric acid as its electrolyte, methane-ethane-propane gas mixtures as its fuel, and oxygen from air as the oxidant. Such fuel cells have excellent potential to operate with low pollution emission and high efficiency, but until recently they have remained uneconomic compared with conventional power plants.

The best-known fuel cell is the $H_2/O_2$ type that was used as the primary power source in the Apollo manned lunar space flights and in subsequent shuttle missions. A schematic drawing of the cell is shown in Figure 14.14. Its reactions are:

$$\text{Anode:} \quad 2\,H_2 + 4\,OH^- \rightarrow 4\,H_2O + 4\,e^- \qquad E^0 = 0.83\text{ V}$$

$$\text{Cathode:} \quad \underline{O_2 + 2\,H_2O + 4\,e^- \rightarrow 4\,OH^-} \qquad E^0 = 0.41\text{ V}$$

$$\text{Net:} \quad 2\,H_2 + O_2 \rightarrow 2\,H_2O \qquad E^0 = 1.24\text{ V}$$

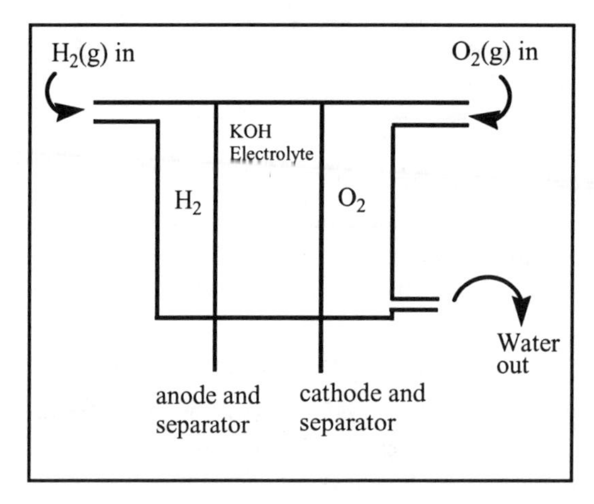


Figure 14.14 A schematic representation of a fuel cell.

This particular fuel cell requires an electroactive catalyst. An **electroactive catalyst** is a material that provides an electrically conducting surface and facilitates the breakdown of diatomic hydrogen and oxygen into reactive atoms. Example of such catalysts are the metals platinum, palladium, and nickel. The most practical electrodes are porous graphite rods impregnated with one or more of the electroactive catalysts. The use of purified fuel gases is important, because the catalysts are easily poisoned by small amounts of impurities, especially by carbon monoxide and sulfur-containing compounds.

The $H_2/O_2$ fuel cell offers a tremendous advantage on space flights where payload size and weight are critical. Recall from the Table 7.1 (energy values of food and fuels) that hydrogen has by far the greatest energy per gram of any chemical fuel. The fuel cells used on the Apollo flights weighed about 500 lb and operated continuously for 440 hours. They produced a large percentage of the power needed by the spacecraft and 100 liters of byproduct water for the astronauts. In the more than 30 years since the first Apollo flight, large fuel cells have failed to find commercial applications. For example, using fuel cells to power automobiles has been long contemplated, but there have been persistent problems with system weight, volume, cost, and lifetime. Finally, at the opening of the 21st century this situation may be changing.

What are the reasons that fuel cell technology may be at the point of takeoff? Better materials and better understanding of materials is the short answer. One promising technology is based on sulfonate (containing $-SO_3H$ groups) **polymer electrolyte membranes** (PEMs). A PEM is a thin ($\approx 150\ \mu m$) film of ion-conductive polymer coated on both sides with a thin film ($\approx 50\ \mu m$) of 10 nanometer sized carbon particles impregnated with platinum atoms. The

PEM is sandwiched between sheets of porous carbon paper, and that in turn is sandwiched by a graphite conducting layer. This membrane-electrode assembly is finally bringing fuel cells to the point of practical applications. No new electrochemical principles are involved in PEMs, their success derives from their potential to overcome the problems of cell weight, cost, etc.

The US Army has for several years been testing fuel cells for mobile combat needs, and during the past couple of years fuel cell-powered buses have been put into service. Most encouraging has been increased activity at the automobile companies. One major car manufacturer has announced plans to put a fuel cell-powered vehicle on the market by 2004. [❁kws +"fuel cell" +automobile]

## Section 14.5 Corrosion—An Electrochemical Phenomenon

One of the most readily observable examples of an electrochemical reaction is the corrosion of metals, particularly the rusting of iron. **Corrosion** is the reaction of a metal with substances in its environment or the atmosphere. Corrosion damages metal and eventually destroys its structural integrity. Oxygen in combination with water is the most common cause of corrosion.

**Rusting** is a chemical reaction involving iron metal, oxygen, and water. The water must be electrically conducting, salt water, for example. Dissolved salts in water serve to complete the electrochemical circuit and thereby speed the corrosion process. Rusting is a serious economic problem. It has been estimated that the replacement of corroded metals costs the US almost $100,000,000 per year. Among others, rusting can be represented by the following four chemical equations:

$$4\ Fe(s) + 4\ H_2O(l) + 2\ O_2(g) \longrightarrow 4\ Fe(OH)_2(s)$$

$$4\ Fe(OH)_2(s) + 2\ H_2O(l) + O_2(g) \longrightarrow 4\ Fe(OH)_3(s)$$

$$4\ Fe(s) + 6\ H_2O(l) + 3\ O_2(g) \longrightarrow 4\ Fe(OH)_3(s)$$

$$Fe(OH)_3(s) \longrightarrow FeO(OH)(s) + H_2O(l)$$

The rates of these reactions are increased when the $O_2$ is dissolved in water. Iron does not rust appreciably if it's kept dry. You probably know that automobiles operated in cold climates, where salts (usually NaCl or $CaCl_2$) are used to de-ice the roads in winter, tend to have more rust than cars from warm dry climates. Iron(II), $Fe^{2+}$, is formed on iron at specific localized regions of the surface of the metal, called anodic regions because electrons flow from these regions into the iron. Anodic regions are most often found where the metal is under strain. For example, metallic automobile parts often begin rusting in areas where they have been damaged by collision, bent, and lost their paint coating. The mechanical strain increases the energy of the metal, and therefore the tendency of the oxidation reaction to occur. The $Fe^{2+}$ undergoes further oxidation to form hydrated iron(III) oxide, which is rust. Corrosion of iron under mechanical strain is an important type; it is called stress corrosion. Corrosion is also accelerated by higher temperatures; that's why tail pipes and mufflers are particularly prone to corrosion. The overall chemical equation for the rusting of iron is:

$$4\ Fe + 3\ O_2(g) + 2x\ H_2O \rightarrow \underset{\text{rust}}{2Fe_2O_3 \cdot xH_2O(s)}$$

Rust does not have a definite chemical formula. It is hydrated iron(III) oxide with x in the equation above being typically 2-3. Typical rust is porous, so molecular oxygen can penetrate through it to the iron surface below and continue the corrosion process. Furthermore, rust does not adhere strongly to the surface of the iron where it is formed. It scales off easily, leaving the newly exposed metal surface subject to further attack.

### Controlling Corrosion

Corrosion is discussed here and in Appendix I. While corrosion cannot be completely prevented, it is possible to delay the process. Slowing down corrosion is accomplished in a variety of ways. One way is to coat the metal with a material that is itself less susceptible to oxidation. So long as the coating remains intact, it provides protection for the metal underneath. The most common coating is paint. But unless it is regularly replaced, all paint eventually develops cracks or peels off, allowing corrosion to occur. Another way to protect a metal surface is to coat it with a second metal that undergoes oxidation less easily than the base metal. One example is the "tin can," familiar from its use as a food container. The metal of a tin can consists of an iron core with a thin layer of tin electrodeposited on its surface. The standard oxidation potentials of the two metals are:

$$Fe \rightarrow Fe^{2+} + 2\,e^- \quad E^0 = 0.44\ V$$

$$Sn \rightarrow Sn^{2+} + 2\,e^- \quad E^0 = 0.14\ V$$

Tin is oxidized less easily than iron. Protection of the iron surface is provided as long as the tin layer remains intact. Once holes appear in the tin coating, however, the exposed iron becomes the target for attack because it oxidizes more easily than tin. It is not uncommon to find old, discarded cans that have been exposed to moisture with holes "eaten" through them.

Another method of protection, that is superior to coating by a less active metal, is to use a more active metal as the coating material. The most common example of this type is a galvanized container that has a coating of zinc on an iron core. Zinc coatings can be put on steel by dipping the steel in large baths of molten zinc. The standard oxidation potentials of these two metals are:

$$Fe \rightarrow Fe^{2+} + 2\,e^- \quad E^0 = 0.44\ V$$

$$Zn \rightarrow Zn^{2+} + 2\,e^- \quad E^0 = 0.76\ V$$

In this case, zinc is oxidized more easily than iron. Even when holes appear in the coating, the zinc (which is more easily oxidized) reacts in preference to the iron. The iron core does not corrode until all of the zinc surface coating has disappeared. The guard rails along thousands of miles of US highways are made of zinc coated steels. If zinc is more active than iron, why doesn't zinc corrode? Well, it does—*slightly*. Zinc electrodes can become passivated (recall that small amounts of mercury are needed in zinc batteries to prevent passivation). Passivation occurs when zinc grows a corrosion-protective surface film.

The use of a more active metal to protect a less active one is known as **sacrificial cathodic protection**. The more active metal is also said to act as a **sacrificial anode**; it is sacrificed to provide protection for the object to which it is attached and electrically connected. The metal used has an oxidation potential greater than the metal being protected, which ensures that the removed electrons come from the more active of the two metals. To protect the Alaskan pipeline from corrosion, pieces of magnesium are connected to the pipe at regular intervals. The magnesium itself is oxidized and is eventually consumed. It must be replaced periodically, but doing this is much cheaper and easier than replacing the pipe itself. A similar procedure is used for protection of ships and crabpots, using zinc metal electrodes. In the case of crabpots, a relatively pure piece of zinc is attached to a crab pot made of impure zinc (actually galvanized steel). The purer form of zinc oxidizes more readily than the impure, thereby protecting the main body of the pot until the sacrificial anode is ultimately consumed (Figure 14.15). The protection of a steel vessel by a sacrificial zinc electrode is shown schematically in Figure 14.16.

Figure 14.15 Sacrificial electrodes—before and after. The electrode on the left is new; the one on the right saw two years service in Virginia coastal waters. (Zincs courtesy of Mr. Frank Wagner, Newport News.)

The basic principle of sacrificial protection is shown in Figure 14.16. When oxidative attack occurs at some iron atom anywhere on the vessel, electrons from the zinc are “sent” through the conducting hull of the boat and simultaneously a a zinc ion leaves. The net result is that iron the does not corrode, while the zinc slowly does.

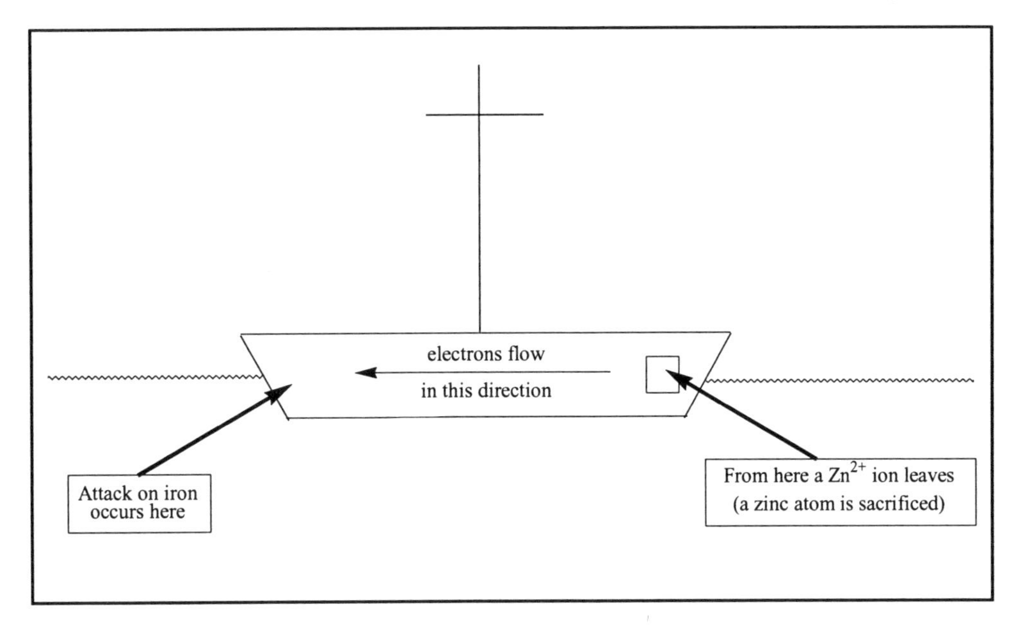


Figure 14.16 The operation of a sacrificial electrode.

## Essential Knowledge—Chapter 14

**Glossary words:** Electrochemistry, electrolysis reaction, battery reaction, electrochemical reaction, electrochemical cell, electrochemical industries, cell voltage, reduction half-cell, oxidation half-cell, half-cell potential, ion-conduit, electrorefining, anode, cathode, half-cell reaction, electroplating, chlor-alkali process, Faraday constant, proprietary information, intellectual property, patent, bauxite, standard cell potential, standard hydrogen electrode, standard reduction potentials, activity series of the metals, Daniell cell, Nernst equation, batteries, connected in series, energy density, battery discharge, battery recharge, secondary cell, primary cell, dry cell, nicad, fuel cell, electroactive catalyst, polymer electrolyte membrane, corrosion, rusting, sacrificial cathodic protection, sacrificial anode.

**Key Concepts:** Construction and principles of operation of electrochemical cells. Terminology of electrochemistry. Typical industrial electrochemical reactions. Relationship between electron stoichiometry and chemical stoichiometry as expressed through Faraday's laws. Necessary units conversion relationships to relate chemical and electrical phenomenon. Nature of proprietary information and of a patent. Experimental basis and practical use of the table of standard half-cell potentials. Derivation and practical applications of the Nernst equation. Knowledge of important types of batteries and the factors affecting battery choice for specific applications. Fuel cell technology. Electrochemical foundations of corrosion and corrosion control.

**Key Equations:**

Moles of electrons for given current and time: mols = $\frac{it}{F}$

Reversal of half-cell potential: $E_{ox} = - E_{red}$.

Relationship of voltage to energy: 1 volt = 1 J $C^{-1}$ or volts × coulombs = joules

Definition of the watt: 1 W = 1 J $s^{-1}$ (a watt is a joule per second: definition)

Relationship between standard free energy change and standard reduction potential: $\Delta G° = - n E^0F$

Complete form of the Nernst equation: $E = E° - \frac{RT}{nF} \ln \frac{[C]^c[D]^d}{[A]^a[B]^b}$

Relationship between standard cell potential and equilibrium constant: $E^0 = \frac{RT}{nF} \ln\left(\frac{[C]^c[D]^d}{[A]^a[B]^b}\right)_{eq}$

Abbreviated form of the Nernst equation: $E = E° - \frac{0.0592}{n} \log_{10} \frac{[C]^c[D]^d}{[A]^a[B]^b}$

## Questions and Problems

### Introduction

14.1. Define or describe in twenty-five words or less each of the following: a. electrochemical reaction, b. electrolysis, c. battery reaction, d. cell potential, and e. electrochemical cell.

14.2. In the history of the discovery of the elements (Table 1.4), what elements were discovered during the foundation period of electrochemistry? Why were they first isolated at that time?

14.3. Write the equation for the decomposition of water by electrolysis as first performed in London in 1800. What were the professions of the two men who did this reaction?

14.4. State the nature and objective of five electrochemical industries. Name three items that you purchase in a supermarket that are connected with an electrochemical process.

14.5. Distinguish between: a. an oxidation half reaction and a reduction half reaction, and b. anions and cations.

### Electrochemical Cells

14.6. Describe or define in twenty-five words or less each of the following: a. electrode, b. anode, c. cathode, d. ion flow, e. ion conduit, f. salt bridge, g. external circuit, h. electric charge, i. coulomb, and j. volt.

14.7. Write half-reactions and the net reaction to represent: a. the electrorefining of copper, b. the production of aluminum, and c. the production of chlorine and sodium hydroxide.

14.8. Make a sketch of a Daniell cell and carefully label each of its components.

14.9. Sketch a cell which makes use of the following spontaneous reaction: $2\ MnO_4^- + 16\ H^+ + 10\ Cl^- \rightarrow 2\ Mn^{2+} + 5\ Cl_2 + 8\ H_2O$. Write the anode and cathode reactions and state the direction of electron and ion flow in the cell. Why would platinum electrodes be preferred to copper electrodes in this cell?

14.10. Describe how a salt bridge operates. Why can't a salt bridge be replaced by a metallic conductor?

14.11. An electrolysis reaction is carried out using direct current. Explain why alternating current is not suitable. What device is used to convert ac to dc?

### Faraday's Laws and Units in Electrochemistry

14.12. Calculate the mass of copper than can be electrorefined by a current of 2.80 amps in a 48.0 hour time period.

14.13. An electrochemical reduction produces 1.45 grams of Zn(s) from $Zn^{2+}$. If the identical electric charge were used to produce Ag(s) from $Ag^+$ ion, what mass of silver would be plated?

14.14. A certain bauxite ore contains 55% aluminum oxide. If the overall process is 91% efficient, what mass of aluminum metal will be obtained from 1000 kg of this ore?

14.15. a. State the definition of the watt. b. Why is it so named? c. State the definition of the watt in terms of electric current and voltage. d. What is the relationship between watts and joules? e. Explain how cell voltage can be converted to give the energy of an electrochemical reaction.

14.16. A current of 25.0 mA is passed through a nickel sulfate solution for 124 minutes. Calculate the mass of nickel that plates on the cathode. Calculate the thickness of the nickel plate. The cathode has a surface area of 54.3 $cm^2$ and the density of the nickel plate is 8.90 g $cm^{-3}$.

14.17. The electrolysis of a sulfuric acid solution produces hydrogen and oxygen. If a current of 349 mA is passed through sulfuric acid for 231 minutes, what volume of each of the product gases will be obtained? Assume one atmosphere pressure and a temperature of 22.3°C.

### Half-cell Potentials. Cell voltages.

14.18. Describe the construction of a standard hydrogen electrode. Why does this half-cell have a voltage of zero?

14.19. What is it about the table of standard reduction potentials that makes them reduction potentials? What would a table of oxidation potentials look like?

14.20. Name the types of chemical entities located in the upper and lower, left and right regions of the table of standard reduction potentials.

14.21. Describe the procedure to combine any two half-cell reactions to make a net cell reaction. How do the voltages from the table combine to give the net cell voltage?

14.22. Write the activity series of the metals as derived from their standard reduction potentials.

14.23. What is the meaning of a negative $E^0$ for a half reaction in which a metal ion is reduced to the metal?

14.24. List the following in terms of decreasing strength as oxidizing agents (assume all ions are present at 1 molar concentration and get reduced to the corresponding elements): $Cu^{2+}$(aq), $Ca^{2+}$(aq), $Ag^{+}$(aq), $H^{+}$, and $Al^{3+}$(aq).

14.25. List the following in terms of decreasing strength as oxidizing agents (assume the ions are present at 1 M concentration in acid solution): $MnO_2$(s), $ClO_3^-$(aq), $PbO_2$(s), and $MnO_4^-$(aq).

14.26. List the following metals in terms of decreasing strength as reducing agents: Pb, Ag, K, Mg, Ni, and Ba.

14.27. Calculate the standard cell potential for each of the following electrochemical reactions: a. the reduction of Fe(II) to Fe by Zn(s), b. the oxidation of Fe(II) to Fe(III) by permanganate ion in acid solution, and c. the reduction of $Cd^{2+}$ to $Cd^0$ by barium metal.

14.28. Use the table of standard reduction potentials to select three reagents capable of reducing $Br_2$ to $Br^-$.

14.29. Use the table of standard reduction potentials to predict what reaction or reactions will occur when $Br_2$(l) is added to a solution which is 1 M NaCl and 1 M NaI.

14.30. Copper metal does not react with 1 M HCl but does react with 1 M $HNO_3$. Explain.

## Cell Potentials and Chemical Thermodynamics

14.31. Calculate the $\Delta G^0$ value for the electrochemical reaction in cells that produce +1.5 volts and -1.5 volts. Assume that the number of moles of electrons involved is 2 for each reaction. Calculate the K's for these same reactions. Assume a temperature of 298 K.

14.32. Write the Nernst equation and identify each of the terms in it.

14.33. Using the standard reduction potentials calculate the value of the equilibrium constant for each of the following reactions:

a. $Cu + Zn^{2+} \rightarrow Zn + Cu^{2+}$
b. $Pb + 2\,H^+ \rightarrow Pb^{2+} + H_2$
c. $MnO_4^- + 5\,Fe^{2+} + 8\,H^+ \rightarrow Mn^{2+} + 5\,Fe^{3+} + 4\,H_2O$

14.34. Determine the potentials at 25°C for the following half reactions at the specified concentrations.

a. $Cu^{2+}$(0.20 molar) + 2 $e^-$ $\rightarrow$ Cu
b. $Co^{2+}$(1.10 molar) $\rightarrow$ $Co^{3+}$(0.80 molar) + 1 $e^-$
c. $MnO_4^-$(2.00 molar) + 8 $H^+$(4.00 molar) + 5 $e^-$ $\rightarrow$ $Mn^{2+}$(0.50 molar) + 4 $H_2O$

14.35. Determine the $[Zn^{2+}]/[Cu^{2+}]$ ratio for the Daniell cell which will produce a voltage of 1.40 V.

14.36. Describe how pH can be measured via an electrochemical reaction.

## Batteries

14.37. Distinguish between a primary and secondary battery. Is this distinction fundamental or historical?

14.38. What reaction takes place during the discharge of a lead storage battery? What reaction occurs during the recharging of the battery.

14.39. Write down the anode reaction and cathode reaction that occur in the Leclanché dry cell. Another identified cathode reaction has $MnO_2$ and $NH_4^+$ as the reactants and the product $Mn_2O_4^{2-}$ ion. Complete and balance this cathode reaction.

14.40. What combination of the half-cell reactions listed in Table 14.2 would produce a cell with the largest voltage? Compute the standard voltage that would be produced. Why is this combination of half reactions not used in a commercial battery?

14.41. Write the net cell reactions for a nickel-metal hydride battery and for a lithium ion battery. What is the reason for the rapid introduction of these batteries at the present time?

14.42. What advantages do zinc-air batteries offer compared to batteries in current use? What disadvantages? What earlier battery did they replace.

14.43. The electrodes in the $H_2/O_2$ fuel cell are catalytic electrodes. They provide a metal contact for conduction of electrons and also enhance the reaction rate by catalytically cleaving the bonds in $H_2$ and $O_2$. Write the half-cell reactions and the overall reaction which occur in an alkaline medium in a fuel cell.

## Corrosion as an Electrochemical Process

14.44. What is a sacrificial anode. Explain how such an anode prevents corrosion of the metal to which it is attached.

14.45. Steel guard rails are galvanized by coating them with zinc. Explain how doing this prevents corrosion of the iron.

14.46. Describe how zinc is used by Virginia crabbers to extend the life of their crab pots.

14.47. A basic solution of $Au(CN)_2^-$ is used in electroplating gold onto a base metal for protection. $O_2(g)$ is produced at one electrode and gold metal at the other. Deduce the half reactions and the overall reaction which occurs in this process.

## Problem Solving

14.48. Assuming standard state conditions, select a reagent that will:

a. oxidize $I^-$ to $I_2$ but will not oxidize $Cl^-$ to $Cl_2$
b. reduce $Na^+$ to Na(s).
c. reduce $Ni^{2+}$ to Ni but will not reduce $Zn^{2+}$ to Zn.

14.49. If a Leclanché dry cell is connected to a device that draws power quickly, the cell may go dead. However, after a period of time it may revive and produce additional power. Speculate on the reason or reasons that it behaves in this way.

14.50. How can baking soda help a motorist stranded on a very cold day with a dead battery?

14.51. If electricity costs $1.07 per kilowatt hour, what is the electrical cost to produce 1 kg of aluminum by the Hall-Heroult process operated at a potential of 4.5 V and 92% current efficiency?

14.52. Use internet resources to seek recent information about PEMs.

14.53. Use internet resources to compile an absolutely up-to-date list of the types of experimental vehicles actually now running under fuel cell power.

# Chapter 15

# Rate Processes

## Introduction

In Section 2.2 we discussed the rate process of radioactive decay. In this concluding chapter we take up chemical rate processes. In a **rate process** a concentration (or other property) of a reacting system changes with time. How fast a reaction is proceeding is its **reaction rate**. For chemical reactions it's usually a concentration change with time that is measured. Typical rate processes that engineers need to characterize are: the corrosion rates of metals, the service lifetimes of elastomeric gaskets, and the breakdown rates of plastics exposed to UV light.

Any commercial product with a "best if used by" date stamp is undergoing a rate process. Foods, photographic film, and medications are examples of everyday items that undergo slow deterioration. There are many other examples of materials with a limited shelf life. In selecting materials for a particular task, the design engineer has to consider both shelf life and service life.

Chemical reaction rates are important in industrial chemistry, biology, and our daily lives. Some chemical reactions occur extremely slowly, others occur extremely rapidly. Iron's rusting is a slow reaction and explosions are rapid reactions. Reaction rates depend on the nature of the reactants, the interactions of reactants and products with solvents, the reactants' (and sometimes the products') concentrations, the temperature, and whether or not a catalyst is present. Careful studies of reaction rates allow all these factors to be understood. A **catalyst** is a substance that affects the rate of a reaction without itself being consumed. We use automobiles equipped with **catalytic converters** to speed the final combustion of the fuel and reduce air pollution. We use antimony compounds as flame retardants in plastics to slow the rate at which the plastic will burn.

Figure 15.1 represents the progress over time of the reaction $A \rightarrow 2B + C$. It is a **time-concentration graph**. The rate of change of the reactant concentration is negative—the reactant is being used up. The rates of change of the product concentrations are positive—products are being formed.

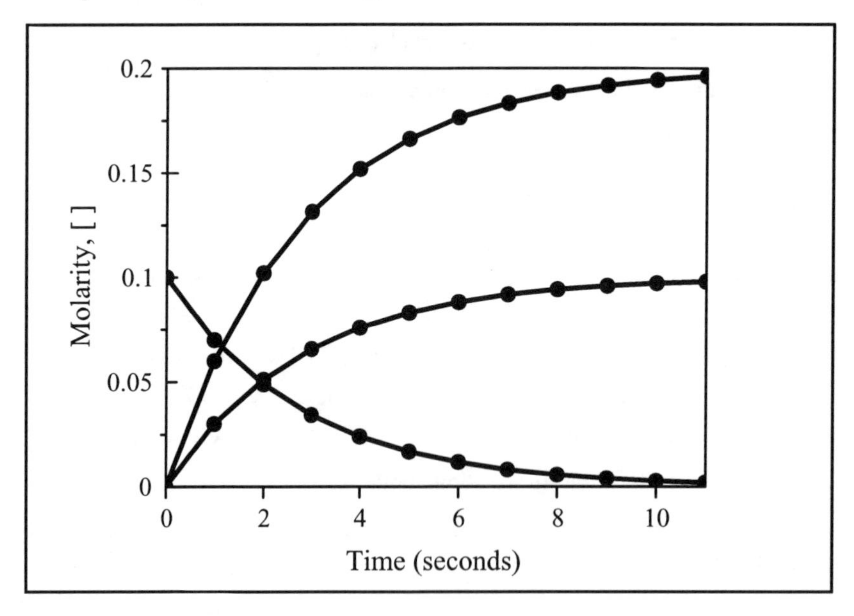


Figure 15.1 The change in reactant and product concentrations over time for the reaction: $A \rightarrow 2B + C$. The reactant concentration follows the rate law: rate = $-k[A]$. [A] is the falling line; [B] is the upper rising line; [C] is the lower rising line. The time unit is chosen arbitrarily.

The stoichiometry of the reaction shown in Figure 15.1 allows us to make the following statements: When one mole of reactant A has reacted, then one mole of product C will have been formed. When one mole of reactant A has reacted, then two moles of product B will have been formed.

$$\text{Rate of formation of C} = 1 \times \text{Rate of disappearance of A}$$
$$\text{Rate of formation of B} = 2 \times \text{Rate of disappearance of A}$$

The drop in [A] is mirrored by a rise in [C]. [B] rises twice as fast as [A] drops, because the A:B stoichiometric ratio is 1:2.

The rates of chemical reactions follow mathematical laws called **rate laws**. Often, rate laws have an exponential-type form. Careful analysis of time-concentration data allows the precise shape and the precise mathematical form of the rate law to be determined. An analysis of the rate law can then yield an understanding of the mechanism of the reaction. A **reaction mechanism** is the step-by-step sequences of events by which the reactants become transformed into the products. Here's the rationale and procedure for studies of chemical kinetics:

1. Get the concentration-time data
2. Figure out what mathematical expression the data fit
3. Deduce the rate law
4. Work out a mechanism consistent with the rate law
5. Do something useful: speed up, slow down, or control the reaction

In this chapter we'll consider the important subject of chemical kinetics with an emphasis on the iodine-clock reaction.

**Examples of Reactions We Would Like to Speed Up**

1. Drug delivery
2. Cooking (e.g. pressure cooking)
3. Paint drying
4. Destruction of air pollutants in automobile exhaust
5. Most industrial chemical processes
6. Breakdown of synthetic plastics in land fills, etc.

**Examples of Reactions We Would Like to Slow Down**

1. Food decay (that's a big reason for food additives)
2. Rubber decay (why are rubber bands "pale" and auto tires black?)
3. Human aging (who has heard of antioxidants?)
4. Fading of clothes (detergent additives help to prevent this)
5. Fires (fire retardants are incorporated in many consumer products)
6. Destruction of the ozone layer (this might be a big problem)
7. Rusting of metals objects composed of iron, etc.

## Section 15.1 Kinetics and Chain Reactions

In Section 4.7 we saw that one of the postulates of the kinetic molecular theory of gases is that gas molecules are in constant motion. The word kinetics comes ultimately from the Greek root *kinein*, to move. **Chemical kinetics** is the study of the rate processes of chemical reactions.

Fundamentally, chemical rate processes are controlled by the rates at which molecules collide. At the molecular level, a reaction is a collision of two molecules or molecular fragments. If the collision occurs with sufficient kinetic energy and with the particles in a suitable orientation, then existing chemical bonds break and new ones form. For the gas phase reaction of a hydrogen atom ($H^{\bullet}$) with a chlorine molecule we would write:

$$H^{\bullet} + Cl\text{-}Cl \rightarrow H\text{-}Cl + Cl^{\bullet}$$

where the collision breaks a Cl-Cl bond and more or less simultaneously an H-Cl bond is formed. A **reaction mechanism** is a detailed account of the microscopic process by which reactants get converted into products. Most reactions have multi-step mechanisms. For example, combustion of an octane molecule can be represented by the equation:

$$2\ C_8H_{18}(g) + 25\ O_2(g) \rightarrow 16\ CO_2(g) + 18\ H_2O(g)$$

But logic dictates that the reaction must occur by many steps. The chance that two octane molecules and twenty-five oxygen molecules react in a single, concerted step is incredibly small.

> **Informal Example:** Road accidents are common. Chain reactions occur on foggy highways when, after an initial wreck, succeeding vehicles pile on as they arrive at the accident scene. What is the chance that two eighteen-wheelers and twenty five cars would *simultaneously* collide? It's about as probable as spontaneous human asphyxiation. So too with the chance of 2 $C_8H_{18}(g)$ + 25 $O_2(g)$ happening simultaneously.

## Chain Reactions

In Section 2.1 we briefly discussed the principle of a nuclear chain reaction. Chemical chain reactions are like nuclear ones, except that by comparison they occur at a snail's pace.

The first step of a chemical chain reaction begins with an **initiation** event. That event produces a reaction intermediate that brings about the next step of the reaction, called **propagation**. A **reaction intermediate** is a chemical species that is temporarily formed during a chemical reaction but is not itself either a reactant or a product of the reaction. Propagation regenerates another reaction intermediate, and there can be many consecutive propagation reactions until the process is finally brought to a halt by a **termination** reaction. Consider the reaction A + B → C + D for which we can write the following four reaction steps:

| | | |
|---|---|---|
| A → Intermediate 1 | Chain initiation step | |
| Intermediate 1 + B → Intermediate 2 + Product D | First step of propagation | } one link |
| Intermediate 2 + A → Intermediate 1 + Product C | Second step of propagation | |
| Intermediate 1 + Intermediate 2 → loss of intermediates | Chain termination step | |

The reaction sequence is initiation » propagation 1 » propagation 2 » termination. But the two propagation steps constitute a link that can repeat many times to create a long chain. If we designate the two propagation steps as one link in the chain reaction, as shown above. Here in Table 15.1 are the details of the chain reaction:

Table 15.1 **The Steps of a Chain Reaction**

| | | |
|---|---|---|
| A → Intermediate 1 | Initiation step | |
| Intermediate 1 + B → Intermediate 2 + Product D | First step of propagation | link 1 |
| Intermediate 2 + A → Intermediate 1 + Product C | Second step of propagation | link 1 |
| Intermediate 1 + B → Intermediate 2 + Product D | First step of propagation | link 2 |
| Intermediate 2 + A → Intermediate 1 + Product C | Second step of propagation | link 2 |
| Intermediate 1 + B → Intermediate 2 + Product D | First step of propagation | link 3 |
| Intermediate 2 + A → Intermediate 1 + Product C | Second step of propagation | link 3 |
| Intermediate 1 + B → Intermediate 2 + Product D | First step of propagation | link 4 |
| Intermediate 2 + A → Intermediate 1 + Product C | Second step of propagation | link 4 |
| ...(possibly many more links) | | |
| Intermediate 1 + Intermediate 2 → loss of intermediates | Termination step | |

With an understanding of this mechanism you can see how the rate of the overall reaction is controlled by the rates of the various steps:

- The rate of the overall reaction will depend on the rate of initiation. The more frequent are initiation events the faster will be the reaction.
- The rate of the overall reaction will depend on the rate of termination. The more frequent are termination events the slower will be the reaction.

- The rate of the overall reaction will depend on the relationship between the propagation steps and the termination step. If the propagation reactions are fast compared with termination, the reaction chains will be long and the overall rate fast. If the propagation reactions are slow compared with termination, the reaction chains will be short and the overall rate slow.

## The Chlorination of Methane

The discussion in the previous section is abstract. In this section we'll examine a concrete example of a chain reaction. Chlorinated hydrocarbons have the general formula $C_xH_yCl_z$. [❁kws "chlorinated hydrocarbons"]. They are widely used for metal cleaning, as solvents in the production of electronic devices, and as reaction intermediates. The prototype chlorination reaction occurs between methane and chlorine, for which the overall chemical equation is:

| $CH_4$ | + | $Cl_2$ | → | $CH_3Cl$ | + | HCl |
|---|---|---|---|---|---|---|
| methane | + | chlorine | | chloromethane | + | hydrogen chloride |

Both heat and light, especially ultraviolet light, increase the rate of this reaction. The initiation step is the formation of a pair of chlorine atoms as the first reactive intermediate ($Cl^•$ ):

| $Cl_2$ | → | $2\ Cl^•$ | The initiation step |
|---|---|---|---|
| chlorine molecule | | chlorine atoms | |

The first propagation step is a collision in which a chlorine atom abstracts a hydrogen atom from a methane molecule, forming a second reaction intermediate—a methyl radical—and a molecule of hydrogen chloride as shown below:

| $CH_4$ | + | $Cl^•$ | → | $CH_3^•$ | + | HCl | First propagation step |
|---|---|---|---|---|---|---|---|
| methane | | chlorine atom | | methyl radical | | hydrogen chloride | |

A methyl radical is a reactive chemical entity. It is an intermediate and has a transient existence during chemical reactions. In a **methyl radical**, the carbon atom is surrounded by three bond pairs of electrons and one unpaired electron; in other words, the carbon atom in a methyl radical has only seven valence electrons—which accounts for its extreme reactivity. Incidentally, a chlorine atom has only seven valence electrons, so $Cl^•$ is a free radical.

The second propagation step occurs when a methyl radical reacts with a chlorine molecule, yielding a methyl chloride molecule and a new chlorine atom:

| $CH_3^•$ | + | $Cl_2$ | → | $CH_3Cl$ | + | $Cl^•$ | Second propagation step |
|---|---|---|---|---|---|---|---|
| methyl radical | | chlorine molecule | | chloromethane | | chlorine atom | |

Note immediately that if you add the two propagation steps you get the correct equation for the overall reaction.

| $CH_4$ | + | $Cl^•$ | → | $CH_3^•$ | + | HCl | First propagation step |
|---|---|---|---|---|---|---|---|
| $CH_3^•$ | + | $Cl_2$ | → | $CH_3Cl$ | + | $Cl^•$ | Second propagation step |
| $CH_4$ | + | $Cl_2$ | → | $CH_3Cl$ | + | HCl | Sum of first and second steps |

Finally, one possible termination reaction is:

| $CH_3^•$ | + | $Cl^•$ | → | $CH_3Cl$ | Possible termination step |
|---|---|---|---|---|---|
| methyl radical | | chlorine atom | | chloromethane | |

A mechanism that involves free radical propagation steps is called a **free radical chain mechanism** The chlorination of methane is thus a **free radical chain reaction**. A **free radical** is a reactive fragment of a molecule containing at least one unpaired electron and usually appearing as a reaction intermediate. In its historic origin, what we now call an organic functional group was once called a radical, so when a "radical" in that sense is not attached to a stable molecule, it's a "free radical," free from the molecule it would otherwise be part of.

In the scheme above, the termination step is shown as the combination of two reactive intermediates. However, in practice, a chain terminates far more often by a reactive intermediate combining with a contaminant or impurity in the reaction system. So, free radical chain reactions can be speeded by removing contaminants and slowed by deliberately adding contaminants; both practices are widely applied. Some practical aspects of controlling free radical chain reactions are described in the following section.

## Changing the Rates of Free Radical Chain Reactions

Many important reaction mechanisms involve highly reactive free radicals. Free radical chains are involved in combustion reactions, in human aging, in the synthesis of many commercially useful organic compounds, and in the degradation reactions that cause products of all kinds to have limited shelf lives.

High reactant purity is an important factor in many emulsion polymerization reactions, where the needed length of the polymer chain (and hence the bulk properties of the polymer) requires uncontaminated monomers. In such a case, to increase the rate of a free radical chain reaction, chemists provide a means for the easy formation of radicals and remove from the reaction mixture traces of contaminants that are themselves highly reactive with free radicals. A **free radical initiator** is a compound that serves as a source of free radicals. A common example is dibenzoyl peroxide, which contains a relatively weak -O-O- single bond in the molecule $C_6H_5\text{-CO-O-O-CO-}C_6H_5$. When heated as a pure compound, dibenzoyl peroxide explodes. However, under its conditions of use as an initiator, the molecule decomposes to form two free radicals

$$C_6H_5\text{-CO-O-O-CO-}C_6H_5 \rightarrow 2\ C_6H_5\text{-CO-O}^{\bullet}$$

followed by $C_6H_5\text{-CO-O}^{\bullet} \rightarrow CO_2 + C_6H_5^{\bullet}$ (which is very reactive)

If the reactants are well purified, the propagation chains will be long and the reaction will be fast. These principles are frequently applied to polymerization reactions. So, many commercial applications use peroxide-type initiators and high purity feed stock. (By the way, you might have heard already of dibenzoyl peroxide. It's an ingredient in popular teenage facial cleansing preparations.)

To slow the rate of a free radical chain reaction, chemists take action both to prevent the formation of chain initiating radicals and to trap them if they do form.

The chief strategy to prevent initiation of a free radical chain reaction is to control the effect of light on a reaction system or on a material. Sunlight, and particularly the ultraviolet (UV) part of sunlight, breaks bonds in molecules with the formation of atoms or free radicals:

$$Cl_2 \quad + \quad h\nu \rightarrow 2\ Cl^{\bullet}$$

where hν represents a single photon of light. You are probably familiar with UV absorbing molecules as an ingredient of sun screens—lotions applied prior to sunbathing that mitigate the effects of UV light on human skin. Molecules that absorb UV can thus greatly extend both the shelf life and the service life of chemical or polymeric materials. An ideal UV absorber is a molecule that absorb all the radiation between 290 and 400 nm while transmitting all the visible light. For polymers, the UV absorber must have a low volatility, be capable of being well dispersed into the finished item, and be inexpensive and nontoxic. 2-(2'hydroxyphenyl)-benzotriazole is a good UV absorber. It's molecular structure and that of a common sun-blocking molecule are shown in Figure 15.2.

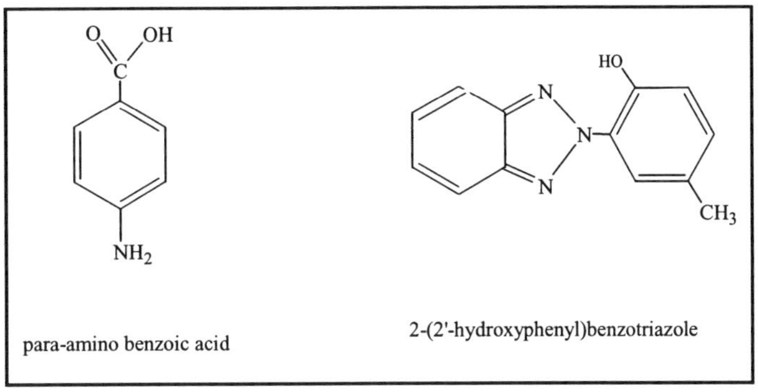

para-amino benzoic acid

2-(2'-hydroxyphenyl)benzotriazole

Figure 15.2 The sun blocking molecule para-amino benzoic acid is a simple aromatic amino acid. The benzotriazole derivative on the right is an additive used to extend the service life of a plastic.

A second strategy to control the rate of free radical chain reaction is to provide an ingredient, a **free radical trap**, that reacts vigorously with free radicals and so interrupts the ongoing free radical chain. Free radical chain mechanisms initiated by atmospheric oxygen or ozone are responsible for a variety of decomposition reactions, so radical traps are often called **anti-oxidants** or **anti-ozonants**. Materials subject to this kind of decomposition include rubber, plastics, adhesives, gasoline, and foods. Organic compounds are added to all of these materials to slow down the rate at which they are degraded. Like UV absorbers, typical antioxidants are substituted aromatic organic molecules. Two examples are shown in Figure 15.3.

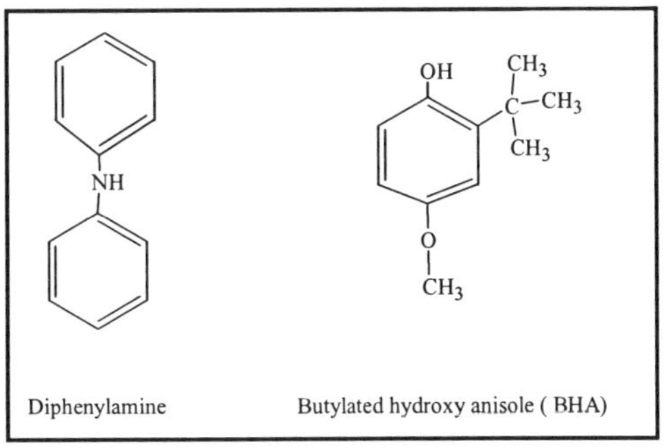

Diphenylamine

Butylated hydroxy anisole ( BHA)

Figure 15.3 Diphenylamine is an antioxidant additive used in rubber tires. BHA is a food additive used to extend the food's shelf life.

Diphenylamine and its derivatives with related molecular structures are valuable rubber stabilizers. Diphenylamine has also been used as a stabilizer for high explosives. It is likely that the tragic explosion in the gun turret of the USS Iowa in 1989 was precipitated by the breakdown of the diphenylamine stabilizer over the more than 50-year storage life of the guncotton (nitrocellulose) propellant.

Figure 15.4 Firing the big guns of a battleship is a very fast chemical decomposition reaction. Antioxidants added during propellant manufacture help stabilize the propellant against degradation during storage.

Frequently you can find the initials BHA (butylated hydroxy anisole) on food labels. BHA in a food product prevents oxygen-initiated free radical chain decomposition reactions. Radical traps intended for food use have the special requirement that they must be safe for human consumption. Butylated hydroxy anisole was first disclosed in patents assigned to Universal Oil products in 1949—so there is over half a century of experience with this molecule, and it is regarded as being completely safe. [❁kws "generally recognized as safe" or GRAS]. Sometimes, mixtures of radical traps work better than individual compounds. For example, the American Meat Institute Foundation recommends an antioxidant solution made by dissolving 20% of butylated hydroxy anisole, 6% of propyl gallate, and 4% of citric acid in 70% of propylene glycol solvent.

**A Personal Note:** The author's first patent (assigned to FMC Corporation) described free radical trapping compounds to stabilize hydrogen peroxide against decomposition in aqueous solutions containing copper ions. The initiation reaction is $H_2O_2 \rightarrow 2\ {}^{\bullet}OH$. The decomposition reaction is $2H_2O_2 \rightarrow 2H_2O + O_2$. The value of stabilization is that it makes possible the cleaning of copper items at a reasonable cost. In the mid-1970s, a mixture of hydrogen peroxide and sulfuric acid replaced the previously used and potentially environmentally damaging mixture of sodium dichromate and sulfuric acid. Incidentally, I discovered that this was a free radical chain reaction when I noticed that the reaction rate increased with the lights on and decreased with the lights off.

## Curing Reactions

If excluding UV light slows some reactions, it makes sense that we can deliberately use UV light to accelerate other reactions. In industry, **curing** generally means increasing the rate at which a product can be formed; more specifically it means cross linking a polymer. It was discovered as long ago as 1914 that dibenzoyl peroxide speeds the curing of natural rubber and so is a useful vulcanizing agent. Today, there are probably hundreds of different compounds that can be used as rubber vulcanization accelerators. [❁kws "rubber accelerator"].

Other UV-cured systems include pre-polymers that can be polymerized *in situ* by exposing them to UV light. Various printing processes take advantage of light or UV polymerization to prepare printing plates [❁kws photolithography]. In the electronics industry, materials used this way are called **photoresists**. Perhaps you've wondered why a printed circuit is called "printed." The answer is that it made using a process that involves selective UV polymerization which "prints" the negative image of the needed pattern on a copper laminated circuit board. The unpolymerized material is cleaned away, and subsequent etching of the exposed copper leaves the polymer-protected copper to form the desired circuit. [❁kws +Riston +DuPont].

Curable materials are used in dentistry. Maybe you've had dental work where the final step was having a small UV lamp placed inside your mouth to irradiate a repaired tooth and cause the repair material to harden. With this procedure, the dentist can apply the repair material while it is workable, and, once the material is properly applied, the repair is made permanent by UV curing.

## Section 15.2 Rate Laws

A **rate law** is a mathematical fit for time-concentration data. For many reactions involving one or two reactants, A + B → Products (P), the rate expression for the disappearance of reactant A can be written in the generalized form:

$$Rate_A = -k[A]^m[B]^n$$

where $[A]^m$ is the molar concentration of reactant A raised to the mth power and $[B]^n$ is the molar concentration of reactant B raised to the nth power. The SI unit of rate is mol $L^{-1}$ $s^{-1}$. The **specific rate constant** of the reaction, k, is a constant for any specific reaction at any specified temperature. The quantities m and n are determined for each reaction by experiment. The units of the specific rate constant depend on the exponents m and n. The exponent m is called the **order** with respect to reactant A. The exponent n is called the **order** with respect to reactant B. Together, m + n is the **overall order** of the reaction. The rate at which reactant B disappears, and the rate at which product P is formed, are related to the rate at which A disappears through the reaction's stoichiometry as described at the beginning of the chapter in the discussion about Figure 15.1.

The first step of a kinetics experiment is to gather time-concentration data. To do that you need 1. a clock, 2. a thermostat, and 3. a method of analysis. Almost *any* property of a reaction system that changes with time can be made the basis for an analytical method. Some examples of analytical methods are: pressure change (for gas reactions); color change; sample and titrate, etc. To deduce the rate law of any chemical reaction requires a careful analysis of rate-concentration data, as illustrated by Figure 15.5.

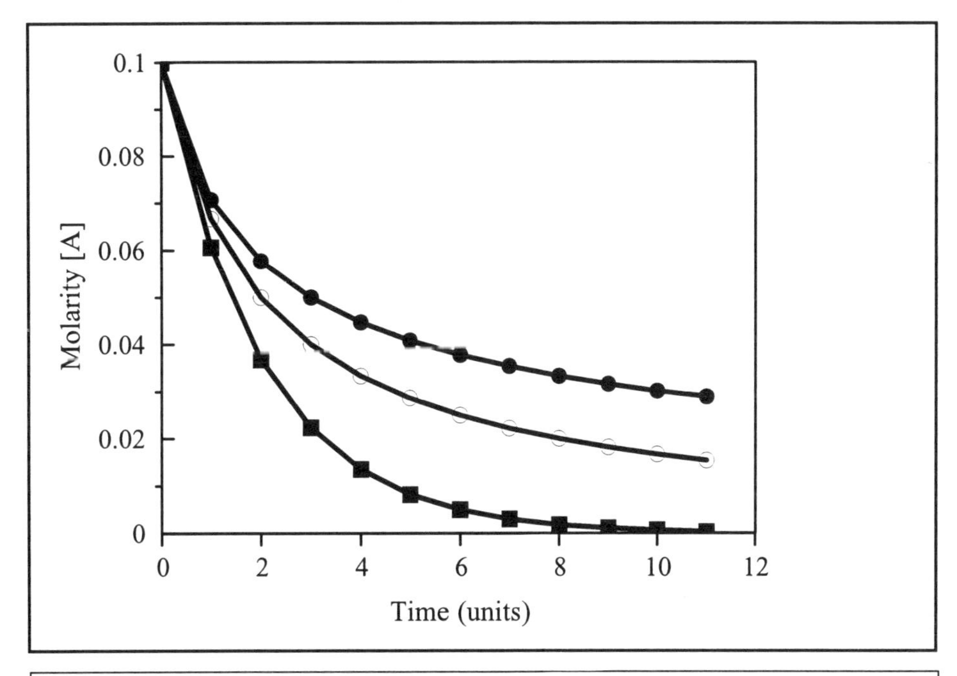


Figure 15.5 Calculated concentration-time curves for three rate law expressions. The shape of the curve changes only subtly with changing reaction order.

—●—●—●— $Rate_A = -50. \times [A]^3$
—○—○—○— $Rate_A = -5.0 \times [A]^2$
—■—■—■— $Rate_A = -0.5 \times [A]^1$

In every case in Figure 15.5 the initial molarity of the reactant was chosen to be 0.10 molar and the decreasing concentrations are calculated for a 11-unit time period according to the first order ($Rate_A = -0.5 \times [A]^1$), second order ($Rate_A = -5.0 \times [A]^2$), and third order ($Rate_A = -50. \times [A]^3$) rate laws. The second order reaction is much slower than the first order reaction; the third order reaction is much slower than the second order reaction. The values of k were selected solely to show the data conveniently on a single diagram. In summary:

| | | | |
|---|---|---|---|
| Third order reaction | top line in Figure 15.5 | $k = 50.\ L^2\ mol^{-2}\ s^{-1}$ | $rate = 50.\ [A]^3$ |
| Second order reaction | middle line in Figure 15.5 | $k = 5.0\ L\ mol^{-1}\ s^{-1}$ | $rate = 5.0\ [A]^2$ |
| First order reaction | bottom line in Figure 15.5 | $k = 0.5\ s^{-1}$ | $rate = 0.5\ [A]$ |

The purpose of this diagram is to show you that the shape of the time-concentration curve varies only slightly from order to order. Historic methods of deriving reaction orders involved various graphical procedures. As with many other situations, modern computer programs can derive reaction orders more or less automatically from experimental data.

Table 15.2 shows some characteristics of rate law expressions: their orders, their specific rate constants, the units of those rate constants, etc. Deriving the units of a particular rate constant is show in Example 15.1.

Table 15.2 **Characteristics of Rate Laws**

| Individual Orders m n | Overall Order | Rate Law | Units of k | Description |
|---|---|---|---|---|
| 0 0 | 0 | Rate = k | $mol\ L^{-1}\ s^{-1}$ | zero order in A and B |
| 1 0 | 1 | Rate= k[A] | $s^{-1}$ | first order in A, zero order in B |
| 1 1 | 2 | Rate= k[A][B] | $L\ mol^{-1}\ s^{-1}$ | first order in both A and B |
| 2 0 | 2 | Rate= $k[A]^2$ | $L\ mol^{-1}\ s^{-1}$ | second order in A, zero order in B |
| 2 1 | 3 | Rate= $k[A]^2[B]$ | $L^2\ mol^{-2}\ s^{-1}$ | second order in A, first order in B |
| 1 2 | 3 | Rate= $k[A][B]^2$ | $L^2\ mol^{-2}\ s^{-1}$ | first order in A, second order in B |
| 2 2 | 4 | Rate= $k[A]^2[B]^2$ | $L^3\ mol^{-3}\ s^{-1}$ | second order in both A and B |

**Example 15.1**: What are the units of the rate constant of a third order reaction?

Strategy: The SI unit of reaction rate is $mol\ L^{-1}\ s^{-1}$. A third order reaction has the form Rate= $k[A]^3$. Use this equation and the algebra-of-quantities to get the desired unit.

The units-only equation is

$$mol\ L^{-1}\ s^{-1} = \text{unknown units} \times (mol\ L^{-1})^3$$

$$mol\ L^{-1}\ s^{-1} = \text{unknown units} \times mol^3\ L^{-3}$$

$$\text{unknown units} = (mol\ L^{-1}\ s^{-1}) \div mol^3\ L^{-3}$$

$$= \underline{L^2\ mol^{-2}\ s^{-1}}$$

The rate of a **zero order reaction** is independent of the concentrations of any reactants. The rate law is simply Rate = k. In other words, changing the concentrations of any reactant has *no effect* on the reaction's rate. How can this be? Well, a readily understood zero order reaction is a gas reaction catalyzed by and occurring at a solid surface. In this situation, no matter how many molecules are in the gas phase, they can only react as fast as they arrive (and the products leave) the surface. Hence, it is the area of the surface that ultimately controls the reaction rate (provided, of course, that the pressure is sufficiently high that all the active sites on the catalyst's surface are occupied by reacting molecules). In practical applications catalysts that are surface active, are used with their surface area maximized. You recall that for this reason, the platinum catalysts used in current-generation fuel cells consisted of platinum impregnated on 10 μm diameter carbon particles.

## First Order Rate Processes

A first order process follows an exponentially decreasing path and is conveniently described by half-life: the length of time it takes for one half of any original sample to react, or for a reactant concentration to fall to one half of the original concentration. Many chemical reactions and all radioactive decays are simple first order rate processes. In such a process, the rate is proportional to the concentration of a single reactant raised to the first power. Mathematically (for either a chemical or radioactive process) the rate law for that reactant is:

$$\text{Rate} = -k\,[\ ] \qquad \text{or} \qquad \text{rate} = -k\,c$$

where k is the specific rate constant and is negative because reactants are disappearing, [ ] means the molar concentration of a compound, and c is amount or concentration. Every simple first order chemical reaction taking place at a specific temperature has its own k value.

In Section 2.2 we illustrated the use of the integrated form of the first order rate law to calculate the extent of decay of radionuclides. We use exactly the same rate law for first order chemical rate processes and will show here, in a simple application of calculus, how the integrated form is derived:

Rate is change of concentration with time: $\frac{dc}{dt} = -kc$

Separating the variables gives: $\frac{dc}{c} = -kdt$

and integrating gives: $\int \frac{dc}{c} = -k \int dt$

$$\therefore \ln c - \ln c_0 = -k\,(t - t_0)$$

But, $t_0 = 0$ and $\ln c - \ln c_0 = \ln (c/c_o)$

so, $c = c_0\, e^{-kt}$ (as we saw in Section 2.2), where

$c$ is the concentration remaining after time t
$c_0$ is the original amount or concentration of the sample
k is the specific rate constant.
$(c/c_0) \times 100$ is the percentage remaining.

Chemical reactions that follow the above exponential law are **first order reactions**.

The time needed for the ratio $c/c_0$ to become one-half is called the half-life, $t_{½}$. Substituting those quantities into the integrated form of the rate law gives:

$$½ = e^{kt_{\frac{1}{2}}}$$

Which rearranges to show that the specific rate constant k and the half life are related by the expression $k = (\ln 2)/t_{½}$. We used that expression in Example 2.2, and now see here how it is derived. The SI unit of a first order specific rate constant is $s^{-1}$. Figure 15.6 shows the shape of a first order decay process as the percentage of reactant remaining as time passes. Solving the exponential decay equation is a straightforward plug ‘n’ chug, as demonstrated in Example 15.2.

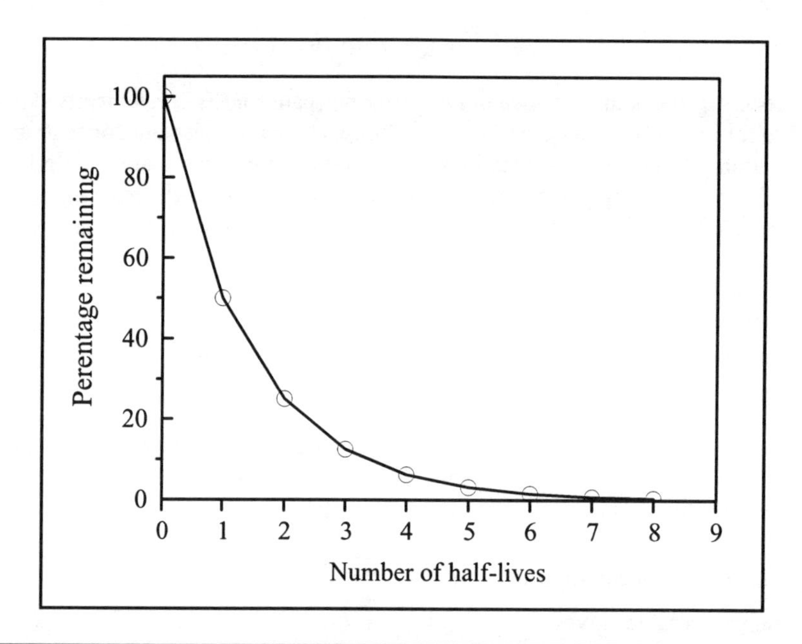


Figure 15.6 The change of concentration of a reactant during any first order rate process. The amount or concentration of the reactant falls to one half of its previous value in every time interval. This diagram essentially repeats Figure 2.7.

**Example 15.2**: In an aqueous solution, in the presence of $FeCl_3$ as a catalyst, hydrogen peroxide decays, $H_2O_2 \rightarrow H_2O + \frac{1}{2} O_2$, via a first order reaction with a half life of 89.5 minutes. What will be the concentration of a 0.0828 molar solution after 30.0 minutes under these conditions.?

Strategy: First use the stated half-life to calculate the value of k, the specific reaction constant. Second, substitute into the integrated form of the first order rate law expression.

To solve the equation first calculate $k = (\ln 2)/t_{1/2}$

$$k = 0.693/t_{1/2} = 0.693/89.5 \text{ min} = 0.00774 \text{ min}^{-1}$$

Substituting into $c = c_0e^{-kt}$

$$c = 0.0828 \text{ mol L}^{-1} \times e^{(-0.00774 \text{ min}^{-1} \times 30.0 \text{ min})}$$

$$c = \underline{0.0793 \text{ molar}}$$

## Finding the Order of a Reaction from Initial Reaction Rates

Chemists often deduce reaction orders by measuring the *initial* rates of reactions. An **initial reaction rate** is one measured soon after the reaction begins, before significant quantities of products have accumulated.

Figuring out a reaction's order from raw time-concentration data can be tricky. However, even a beginner can do it when initial reaction rate data are available for a series of separate kinetic experiments. To do it you need to understand the concept of **isolating a reactant**. Isolating a reactant means repeating experiments in which all possible reactant concentrations except one are held constant.

With one reactant initial concentration isolated, a change in the rate of the reaction must have been caused by the variation in that one initial concentration *not* held constant. The precise value of the change depends on that reactant concentration's exponent in the rate law. For a reaction with several reactants, we do many experiments isolating each reactant in turn. The procedure is illustrated in Example 15.3.

**Example 15.3**: The table below shows four separate measurements (I, II, III, and IV) of the initial rate of formation of product P from reactants A and B via the reaction $A + B \rightarrow P$. All the data were obtained at the same, fixed temperature.

| Experiment | Initial [A] | Initial [B] | Initial Rate (mol $L^{-1}$ $s^{-1}$) |
|---|---|---|---|
| I | 0.150 | 0.150 | 0.107 |
| II | 0.300 | 0.150 | 0.427 |
| III | 0.300 | 0.300 | 0.855 |
| IV | 0.600 | 0.030 | 0.342 |

Deduce the rate law from the experimental data using the method of initial rates. The rate law for the reaction is assumed to have the form given by equation. $Rate_A = -k[A]^m[B]^n$

Strategy: Take the data in a pair-wise fashion so that, in turn, reactant A and reactant B are isolated.

First compare the data of experiments I and II. Here the value of [A] has doubled because 0.300 ÷ 0.0150 = 2.00. Simultaneously, the initial rate of the reaction has gone up four fold because 0.427 ÷ 0.107 = 3.99. With initial [B] fixed, the rate law is Rate = $-k'[A]^m$, where k' is $k[B]^n$. Perhaps you can already see that the doubling has to be squared if the rate is to go up fourfold. You can also solve the problem algebraically:

rate of experiment 1 = 0.107 mol $L^{-1}$ $s^{-1}$ = $k'[A]^m$ equation (i)

and the rate of experiment 2 is 0.427 mol $L^{-1}$ $s^{-1}$ = $k'[2A]^m$ equation (ii)

Dividing equation (ii) by equation (i) gives

$$\frac{.427}{.107} = \frac{[A]^m}{[2A]^m} \quad \text{or} \quad 3.99 = \frac{A^m \times 2^m}{A^m}$$

So $2^m = 3.99$ and m, the order with respect to [A], is 2.

Second, compare the data of experiments II and III to isolate reactant B. Here the concentration of B has doubled (0.300 ÷ 0.150 = 2.00) while the rate has also doubled (0.855 ÷ 0.427 = 2.00). The only way that can happen is if n = 1. So the reaction is first order in B. The rate equation for the reaction is

$$Rate_A = -k[A]^2[B]$$

It is a third order reaction overall.

But wait a minute, you say. We haven't used the data from experiment IV. Well, we didn't need to, because in this case three separate experiments were sufficient to get the individual reactant orders and the overall order of the reaction. However, we can put the data of experiment IV to work if we use any of the data pairs to calculate k, followed by using that k and the concentration data of experiment IV to calculate the rate. Those two further steps are shown in Examples 15.4 and 15.5.

**Example 15.4**: Use the data in Example 15.3 to calculate k for the reaction $A + B \rightarrow P$ described in that example.

Strategy. Take the rate law equation and substitute the concentration data for experiment I. Solve for k. The rate law is $Rate_A = -k[A]^2[B]$.

Substituting: 0.107 mol $L^{-1}$ $s^{-1}$ = k × (0.150 mol $L^{-1})^2$ × 0.150 mol $L^{-1}$

rearranging: k = 0.107 mol $L^{-1}$ $s^{-1}$ ÷ {(0.150 mol $L^{-1})^2$ × 0.150 mol $L^{-1}$}

k = 31.7 mol$^{-2}$ $L^2$ $s^{-1}$

**Example 15.5**: Use the data in Example 15.3 and the calculated value of k from Example 15.4 to predict the rate of reaction in experiment IV in Example 15.3. Compare that value with the experimental result.

Strategy. Once again write down the rate law equation. Substitute in the values of the reactant concentrations and the value of k. The rate law is Rate = $k[A]^2[B]$.

For experiment IV Rate = $31.7 \text{ mol}^{-1} \text{ L s}^{-1} \times (0.600 \text{ mol L}^{-1})^2 \times (0.030 \text{ mol L}^{-1})$

Calculated rate = $0.342 \text{ mol L}^{-2} \text{ s}^{-2}$

The calculated value is the same as the experimental value. We have confirmed that the proper rate law has been derived for the reaction.

## Section 15.3 The Iodine Clock Reaction

The term clock reaction is used to describe any chemical reaction which undergoes a sudden change. Figure 15.7 pictures the before-and-after of the iodine clock reaction.

Figure 15.7 The iodine clock reaction is always a crowd pleaser in lecture demonstrations. A clear solution, like the one on the left, instantly turns deep blue, like the one on the right.

The iodine clock reaction takes place between iodate ion ($IO_3^-$) (in excess) and bisulfite ion ($HSO_3^-$) (the limiting reactant) in the presence of starch. The moment when all the bisulfite ion is consumed is signaled by a color change. There are three reactions to be considered.

Main reaction: $IO_3^- + 3\,HSO_3^- \rightarrow I^- + 3\,SO_4^{2-} + 3\,H^+$

Reaction that starts when $HSO_3^-$ runs out: $6\,H^+ + IO_3^- + 5\,I^- \rightarrow 3\,I_2 + 3\,H_2O$

Indicator reaction: $I_2$ + starch $\rightarrow$ Blue-black colored “complex”

The chemical reactions begin when aqueous solutions of the reactants are mixed. The moment when the bisulfite ion concentration reaches zero is dramatically visible as the formation of a dark blue color in a water-white solution. A schematic representation of the concentration changes during an iodine clock reaction is shown in Figure 15.8.

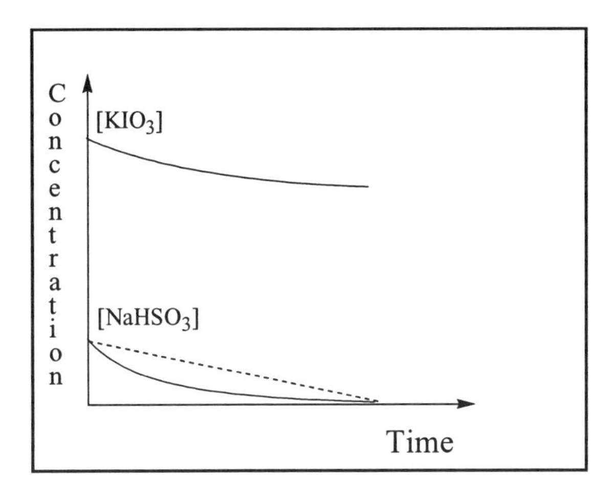


Figure 15.8 A schematic representation of the kinetics of the iodine clock reaction. In this particular experiment, the $[KIO_3]$ begins in substantially greater concentration than $[HSO_3^-]$. However, the reaction stoichiometry requires that $[HSO_3^-]$ drops three times faster than $[KIO_3]$. The color changes when the lower curved line hits the x-axis. The slope of the dotted line represents the average rate of disappearance of $[KIO_3]$ over the duration of the experiment.

Figure 15.8 shows that the concentration of bisulfite drops until it reaches zero, when the indicator reaction kicks in. The average rate of the reaction is measured by the slope of the dotted line in Figure 15.8:

$$\text{Average rate} = \Delta\text{Concentration} \div \Delta\text{time}.$$

Table 15.3 reports a series of measurements of the iodine clock reaction made by the author at room temperature. The experiments were conducted so as to keep the initial $HSO_3^-$ concentration fixed, while varying the initial $IO_3^-$ concentration over a wide range. The experimental procedure was to use 10.0 mL of $HSO_3^-$ solution (column 3) with a fixed initial concentration of $0.960 \times 10^{-3}$ mol $L^{-1}$ (column 5) and mix it with 30.0 mL of $KIO_3$ solution (the sum of columns 1 and 2) that had been successively diluted with make-up water (column 1) to keep the total volume constant (column 1 plus column 2), while providing a range of initial $KIO_3$ concentrations (column 4).

Table 15.3 **Studies of the Iodine Clock Reaction**

Average Rates of $HSO_3^-$ Disappearance at a Fixed Initial $HSO_3^-$ Concentration and a Variable Initial $KIO_3$ Concentration

| $H_2O$ mL | $KIO_3$ mL | $HSO_3^-$ mL | Initial $[KIO_3]$ | Initial $[HSO_3^-]$ | Time (sec) Raw data | Average rate mol $L^{-1}$ $s^{-1}$ |
|---|---|---|---|---|---|---|
| 0.0 | 30.0 | 10.0 | $7.01 \times 10^{-3}$ | $0.960 \times 10^{-3}$ | 18.4, 18.0 | $52.7 \times 10^{-6}$ |
| 5.0 | 25.0 | 10.0 | $5.84 \times 10^{-3}$ | $0.960 \times 10^{-3}$ | 21.1, 20.8 | $45.7 \times 10^{-6}$ |
| 10.0 | 20.0 | 10.0 | $4.68 \times 10^{-3}$ | $0.960 \times 10^{-3}$ | 26.5, 26.0 | $36.6 \times 10^{-6}$ |
| 15.0 | 15.0 | 10.0 | $3.51 \times 10^{-3}$ | $0.960 \times 10^{-3}$ | 34.9, 35.2 | $27.4 \times 10^{-6}$ |
| 20.0 | 10.0 | 10.0 | $2.34 \times 10^{-3}$ | $0.960 \times 10^{-3}$ | 55.5, 53.6, 55.2 | $17.5 \times 10^{-6}$ |
| 25.0 | 5.0 | 10.0 | $1.17 \times 10^{-3}$ | $0.960 \times 10^{-3}$ | 126.0, 128.1 | $7.56 \times 10^{-6}$ |
| 30.0 | 0.0 | 10.0 | 0 | $0.960 \times 10^{-3}$ | $\infty$ | 0 |

The average rate of reaction (column 7), measured from the moment of mixing the two solutions to the moment at which the blue color appeared (column 6) is plotted in Figure 15.8 as a function of the potassium iodate concentration.

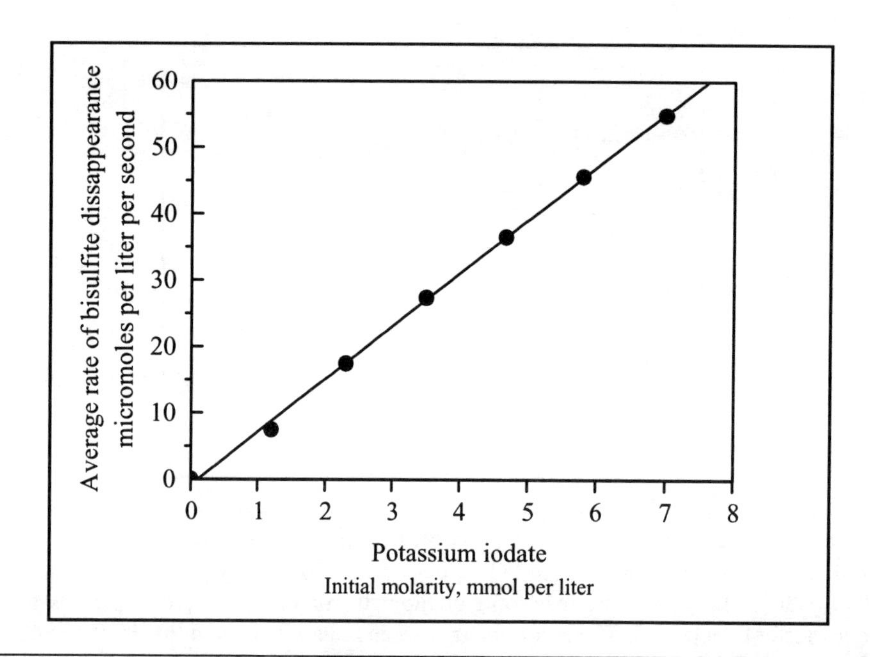


Figure 15.9 A study of the iodine clock reaction shows that the average rate of disappearance of $HSO_3^-$ is directly proportional to the initial concentration of potassium iodate in the reaction mixture.

The straight line in Figure 15.9 shows quite clearly that there is a direct relationship between the rate of the reaction and the initial concentration of the iodate ion. The more iodate ion in the initial reaction solution the faster a fixed amount of bisulfite ion is consumed. The iodine clock reaction is first order with respect to iodate ion concentration.

Unfortunately, this reaction is not simple, and we will not pursue the questions of its overall rate law and its order with respect to the bisulfite ion concentration. However, this is an easy an interesting kinetics experiment, and its results are easily interpreted for one reactant.

## Section 15.4 The Arrhenius Equation

### Temperature Effects on Reaction Rates

Experimentally it is found that the rates of almost all chemical reactions increase smoothly with increasing temperature. As a rule of thumb, reaction rates double for every ten degree increase in the Celsius temperature. Reaction rates increase because their rate constants (k) increase. In Table 15.4 you can see for the first order reaction $2\,N_2O_5(g) \rightarrow 4\,NO_2(g) + O_2(g)$ how k increases with temperature.

Table 15.4 **Change of a Reaction Rate with Temperature**

| Temperature (°C) | Rate constant, k, ($s^{-1}$) |
|---|---|
| 0 | $0.0787 \times 10^{-5}$ |
| 25 | $3.46 \times 10^{-5}$ |
| 35 | $13.5 \times 10^{-5}$ |
| 45 | $49.8 \times 10^{-5}$ |
| 55 | $150 \times 10^{-5}$ |
| 65 | $487 \times 10^{-5}$ |

The effect of changing temperature on the rate of this chemical reaction is substantial. Across the range of a 65°C temperature rise the rate constant of this particular reaction increases by a factor of over 6000 ($487 \times 10^{-5} \div 0.0787 \times 10^{-5}$). The strong effect of temperature on reaction rates can be explained by molecular collisions.

Violent molecular collisions cause bond breaking and lie at the heart of every chemical reaction. From the kinetic-molecular theory of gases we know that as temperature increases the average collision becomes more violent, while at the same time the frequency of collisions grows. So faster reaction rates at higher temperatures derive from the more violent and more frequent collisions found there.

The temperature dependence of the specific rate constant of a chemical reaction is described by the **Arrhenius equation**. The equation is an experimental fact. Svante Arrhenius introduced the equation in the 1880s to describe the effect of temperature on rate constants:

$$k = Ae^{-E_a/RT}$$

where R is the gas constant
T is the absolute temperature
The pre-exponential factor, A, is called the frequency factor
and $E_a$ is called the activation energy for the reaction

Observe that T occurs in the denominator of a negative exponent. Thus, as T increases, the exponent becomes less negative and the value of k becomes larger (for given values of A and $E_a$), in accord with the experimental fact that reaction rates get faster at higher temperatures.

The activation energy and the frequency factor for a reaction are determined experimentally from studies of the temperature dependence of the reaction's rate constant. If we take the natural logarithm of each side of this equation, we get

$$\ln k = \ln A - E_a/(RT)$$

A plot of ln k versus 1/T gives a straight line whose slope is equal to $(-E_a/R)$, and whose intercept is ln A. The usual procedure to determine the activation energy for a reaction is to measure values of the specific rate constant, k, at several values of T, plot ln k versus 1/T, and determine $E_a$ from the slope of the best straight line through the data points. However, it is only necessary to make two measurements because of the mathematical situation that it requires only two points to define a straight line. The derivation below parallels the two-point fit to the Clausius-Clapeyron equation that we developed in Section 10.2.

If the rate constant is measured at two temperatures, $E_a$ and A can be calculated. Suppose k is equal to $k_1$ and $k_2$ at temperatures $T_1$ and $T_2$, respectively. Then:

$$\ln k_1 = \ln A - E_a/(RT_1)$$

$$\ln k_2 = \ln A - E_a/(RT_2)$$

These two equations can be solved simultaneously to find values for A and $E_a$. Subtracting of the top equation from the bottom yields a two-point fit form of the Arrhenius equation:

$$\ln \frac{k_2}{k_1} = -\frac{E_a}{R}\left(\frac{1}{T_2} - \frac{1}{T_1}\right)$$

The application of the two-point fit form of the Arrhenius equation is straightforward, as shown in example 15.6.

**Example 15.6**: The rate of a reaction doubles when its temperature is increased from 298 to 306 K. Calculate the activation energy of the reaction. $R = 8.314\ J\ mol^{-1}\ K^{-1}$.

Strategy: Recognize that if the rate doubles ,then $k_2$ is twice $k_1$ and the ratio $k_2/k_1$ is 2. Substitute the given data into the Arrhenius equation and solve.

$$\ln \frac{2}{1} = -\frac{E_a}{R}\left(\frac{1}{307} - \frac{1}{298}\right)$$

$$-\frac{E_a}{R} = -0.693 \div (3.257 \times 10^{-3} - 3.356 \times 10^{-3}) = 7040\ K$$

So $E_a = 8.314\ J\ mol^{-1}\ K^{-1} \times 7040\ K \times 0.001\ kJ\ J^{-1}$ = 58.1 kJ mol

## The Effect of Molecular Orientation on Reaction Rate

If chemical reactions arise from molecular collisions, a moment's thought will reveal that exactly how the molecules collide will be a critical factor. Imagine that the formation of a bond between two colliding entities is like two spacecraft docking. Successful docking requires precise maneuvering, just as does successful bond formation. It is usually the role of the catalyst to bring the reacting molecules into the proper orientation. It is the precision with which catalysts bring reactant molecules in to the proper "docking" orientation that accounts for the spectacular ability of catalysts to speed chemical reactions. Incidentally, in the Arrhenius equation, orientation is inherent in the pre-exponential factor, in the A term.

In biochemistry it is the role of enzymes to provide the orientating templates to catalyze reactions. In industrial chemistry it is metal surfaces, molecular-sized channels in complex solids, or metal atoms caged in specific molecular geometries, that provide the orienting role.

We'll come back to industrial catalysts in the concluding section of the book after examining activation energy in more detail in the following section.

## Section 15.5 Activation Energy

Figure 15.10 shows along its horizontal axis how potential energy varies as reactants are converted to products during a chemical reaction—that's labeled "the progress of the reaction." You can imagine the diagram showing two molecules 1. moving towards one another, 2. colliding, 3. undergoing bond breaking and making, and 4. moving apart as *different* molecules. At the moment of greatest impact, at the top of the peak, we say that the reacting molecules have formed an **activated complex** or that they exist in a **transition state**

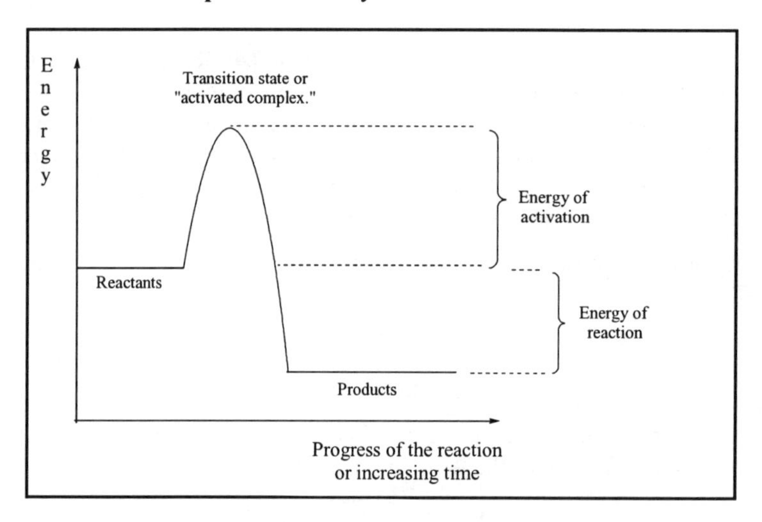


Figure 15.10 Activation energy and the progress of a reaction.

The horizontal line on the left of Figure 15.0 represents the energy state of the reactants. The horizontal line on the right represents the energy state of the products. Because, in this case, the energy of the products is lower than that of the reactants, this example is an exothermic reaction; energy $E_r$ is liberated from the system to its environment during the reaction. The reverse reaction (Products → Reactants) is endothermic. The height of peak that separates the reactants and products is the **activation energy** of the reaction. Figures 15.11 and 15.12 show two further activation energy diagrams with their descriptions in the figure legends.

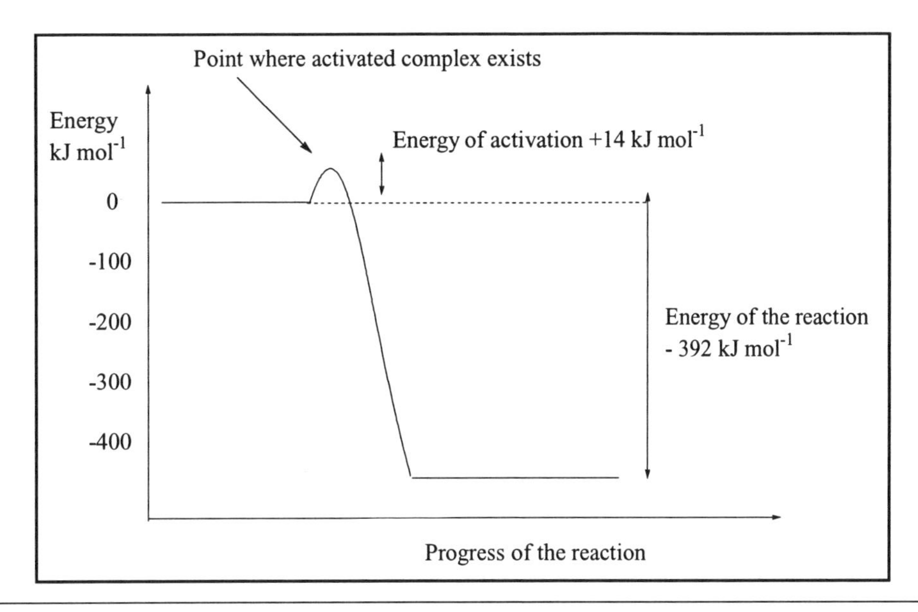


Figure 15.11 Activation energy and the progress of a highly exothermic, easily activated reaction. The reactants in this hypothetical example would have to be stored at a very low temperature because they have such a small activation energy.

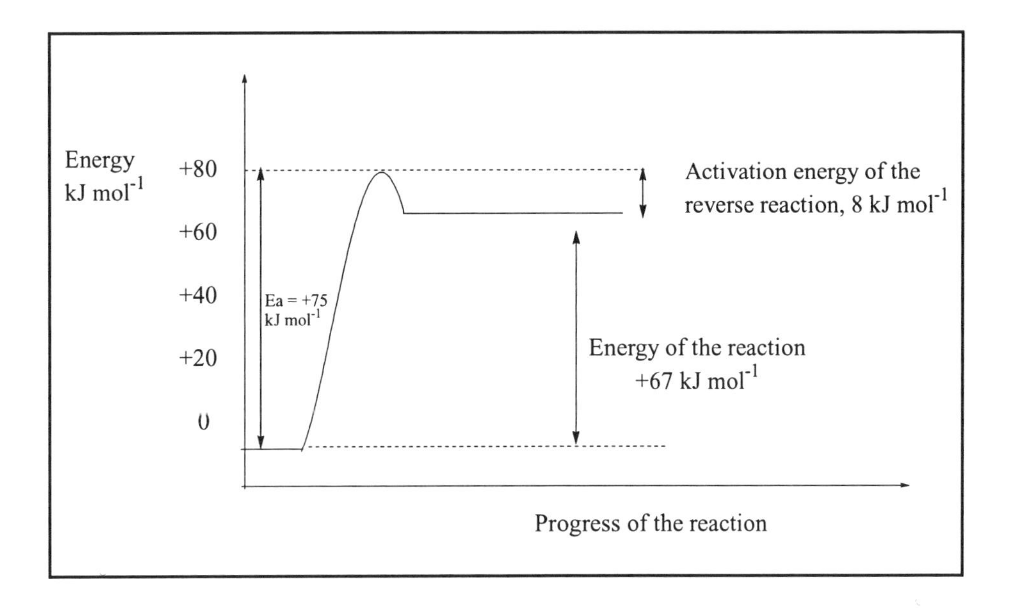


Figure 15.12 Activation energy and the progress of an endothermic reaction with a moderate activation energy. It would be difficult to capture the products of this hypothetical reaction because with the activation energy of the reverse reaction being so small they quickly revert to the form of the original reactants.

It is a straightforward experiment to measure the activation energy of a chemical reaction by the methods of chemical kinetics. You've already seen how to do it. All that's needed is a two-point fit to the Arrhenius equation; just measure the reaction's rate constant at two temperatures. It seems remarkable, doesn't it? Just two rate measurements at different temperatures are sufficient to tell you how much energy is needed for two properly oriented molecules to react.

In Section 8.4 we discussed chemical reactions, such as the combustion of methane, that are thermodynamically spontaneous but need an ignition source to get them started. The concept of activation energy explains why a reaction that's thermodynamically spontaneous is not just plain old spontaneous. At room temperature, collisions between $O_2$ and $CH_4$ molecules are not sufficiently energetic to breach the activation energy barrier. During collisions the reactants climb part way up the slope on the left of Figure 15.10, only to roll back down after failing to reach the top. However, once you ignite (activate) an $O_2/CH_4$ mixture, there is an explosion because the energy of reaction is sufficient to activate all the other molecules.

Implicit in the preceding discussion, and unsurprisingly, activation energies for reactions are comparable to chemical bond energies (Section 7.6). The activation energy is needed to break bonds (Table 15.5). In an obvious case, the activation energy of a diatomic molecule equals its bond energy.

Table 15.5 **Activation Energies of Some Reactions**

| Reaction | $E_a$ (kJ mol$^{-1}$) |
|---|---|
| $I_2(g) \rightarrow 2\ I(g)$ | 151 |
| $CH_3\text{-}CH_3(g) \rightarrow 2\ CH_3^{\cdot}(g)$ | 368 |
| $N_2(g) \rightarrow 2\ N(g)$ | 946 |
| $O_3(g) + NO(g) \rightarrow O_2(g) + NO_2(g)$ | 9.6 |
| $Br^{\cdot}(g) + CH_4(g) \rightarrow HBr(g) + CH_3^{\cdot}(g)$ | 75 |
| $NO_2(g) + CO(g) \rightarrow NO(g) + CO_2(g)$ | 132 |
| $N_2(g) + 3\ H_2(g) \rightarrow 2\ NH_3(g)$ | 427 |
| $2\ CH_3^{\cdot}(g) \rightarrow CH_3\text{-}CH_3(g)$ | 0 |

The largest value in Table 15.5, 946 kJ mol$^{-1}$, corresponds to breaking the triple bond in diatomic nitrogen. As we've noted before, molecular nitrogen is unreactive, and it takes a lot of activation energy to get it to react. The second equation in the table at 368 kJ mol$^{-1}$ represents the breaking up of an ethane molecule into two methyl radicals. We saw earlier in this chapter, in connection with free radical chain reactions, that methyl radicals are reactive, and that fact is confirmed by the activation energy need to make them. But the combination of two methyl radicals (perhaps in a chain termination event) has zero activation energy.

## Section 15.6 Polymerization Catalysts

As described in Chapter 11, synthetic organic polymers have come to hold a dominant position as the materials of everyday life. The principles of chemical kinetics, and especially the use of catalysts, have been central to the rise of the dynamic and burgeoning polymer business. In this concluding section we give a very brief description of industrial chemical catalysis.

### Ziegler-Natta Catalysts

Until 1950, polyethylene could be produced only by processes involving high pressure reactions of the type represented by the equation:

$$\tfrac{n}{2}H_2C{=}CH_2(g) \rightarrow -(CH_2)-_n(s)$$

Between 1950 and 1954 Karl Ziegler (1878-1973) and Guilio Natta (1903-1979) (they shared the 1963 Nobel prize) independently developed a related family of catalysts capable of carrying out a range of olefin polymerization reactions at low temperatures and pressures. Furthermore, the stereochemistry (arrangement in space) of the polymer chain can be controlled by members of this catalyst family, yielding desirable qualities in the finished polymers. The development such catalysts was a key step in making polyethylene a cheap and readily available plastic material. Conceptually the process is simple:

In step one, a single ethylene molecule joins the catalyst

$$\text{catalyst–H} + H_2C{=}CH_2 \rightarrow \text{catalyst–}CH_2\text{–}CH_3,$$

followed by a second ethylene molecule inserting itself between the catalyst and the first ethylene molecule

$$\text{catalyst–}CH_2\text{–}CH_3 + H_2C{=}CH_2 \rightarrow \text{catalyst–}CH_2\text{–}CH_2\text{–}CH_2\text{–}CH_3.$$

Eventually after n+2 total steps, the long polymer chain shown below has been produced by the insertion of n ethylene molecules between the catalyst and the first two ethylene molecules.

$$\text{catalyst} -CH_2-CH_2-CH_2-CH_3 + n\ H_2C{=}CH_2 \rightarrow \text{catalyst}-(CH_2-CH_2-)_nCH_2-CH_2-CH_2-CH_3$$

One type of **Ziegler-Natta catalysts** are mixtures of titanium tetrachloride ($TiCl_4$) and trialkyaluminum compounds, such as triethylaluminun $Al(C_2H_5)_3$. For almost 40 years catalysts of the Ziegler-Natta type provided the chief means to make polyolefins commercially.

## Metallocene Catalysts

Around 1950 the molecule cyclopentadiene ($C_5H_6$), shown in Figure 15.13, was being distilled in an iron still at the research center of the British Oxygen Company in South London. Eventually when the still was dismantled, it was found to contain a large amount of an extremely stable orange, crystalline material which sublimed when heated.

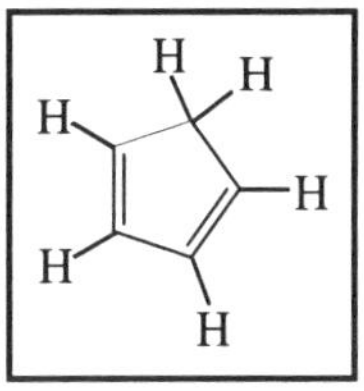

Figure 15.13 Cyclopentadiene. Distillation of this compound in an iron still around 1950 lead to the accidental formation and discovery of ferrocene. The history of industrial chemistry is a story of accidents aggressively pursued, leading in many cases eventually to great economic rewards for the pursuing companies.

The orange substance, called ferrocene, turned out to be the first member of a remarkable new class of compounds, the metallocenes. If cyclopentadiene reacts with a strong base it yields a proton and forms a cyclopentadienyl anion:

$$C_5H_6 \rightarrow C_5H_5^- + H^+$$

Two of these anions attach to a dipositive iron atom, $Fe^{2+}$, producing ferrocene, a stable **organometallic sandwich compound** with the structure shown in Figure 15.14.

Figure 15.14 Ferrocene. The first discovered sandwich compound and founder of a line of important industrial catalysts called the metallocenes.

Metallocenes are called π-complexes because they are characterized by a unique type of metal-to-carbon bond that involves the overlap of σ-type orbitals on the metal with π-type orbitals on the cyclopentadienyl ligand.

During the period 1960-1985 there were many studies of metallocenes. Most of the transition metals form them, and many different organic rings and organic fragments can be attached to various transition metals. A new, rich, and varied chemistry came about, with some of its developments producing commercial applications: for example, tricarbonylmethylcyclopentadienyl manganese, $(C_5H_4\text{–}CH_3)(CO)_3Mn$, is an antiknock fuel additive.

Around 1985 metallocenes were discovered that serve as remarkably good polymerization catalysts. By attaching the right functional groups in the proper stereochemical alignment, a metallocene catalyst can be made that polymerizes a specific olefin. Modern crystallography makes it easy for catalyst chemists to become intimate with the geometry of the catalytic site on the catalyst molecule. The consequence of this knowledge is to allow for the of design catalysts that add monomers to the growing polymer chain in a highly controlled manner, thereby producing a polymer with desirable bulk qualities.

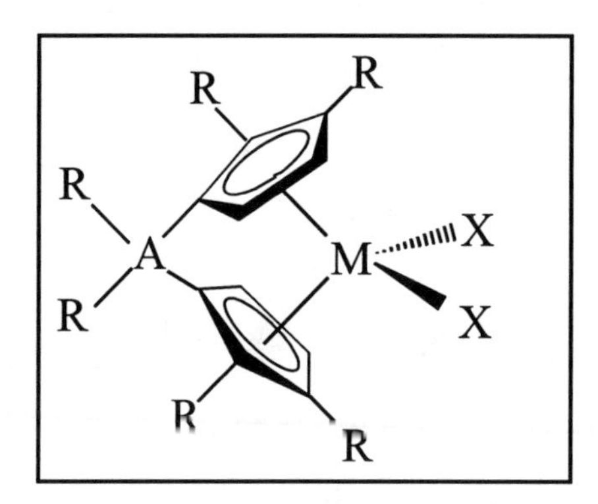


Figure 15.15 A generic metallocene catalyst. By suitable selection of X and A, chemists skilled at catalysis can design molecules well-suited to the task of producing commercially desirable polymers from olefin monomers.

In Figure 15.15 the two cyclopentadienyl rings are clearly visible. Commercial catalysts may use other ring systems. R is an alkyl group, often just methyl ($-CH_3$), the simplest alkyl group. A is an optional bridging atom that may or may not connect the two cyclopentadienyl rings. If it's not present, the rings are not bridged. If present, bridging atoms are typically carbon or silicon. M is the metal atom at the heart of the catalyst. It is here that the growing polymer chain is attached, with the cyclopentadienyl ring system serving as a sort of conveyer belt bringing each new monomer molecule to the active site in exactly the proper orientation. M is usually an atom of one of the Group 4 transition metals: titanium, zirconium, or hafnium. X is an anion, perhaps a chloride ion.

The first new polymers produced using metallocene catalysts reached the market place in 1991. Applications for metallocene-based synthetic polymers range from durable polypropylene auto parts and molded containers to films used in shopping bags, garbage bags, heavy duty shipping sacks, flexible food packaging, and diapers. At the present time they are the fastest growing class of commodity polymers.

In Chapter 11 it is mentioned that the production of synthetic polymers has involved big patent suits. That's been true since the earliest days of the Ziegler-Natta catalysts. In 1983 two Phillips Petroleum chemists were finally awarded the patent for the production of crystalline polypropylene after a more than 20-year court battle.

ExxonMobil manufactures metallocene catalysts at its plant in Mont Belvieu, Texas. It manufactures polyolefins and resins at its plant in Baton Rouge, Louisiana (Figure 15.16).

Figure 15.16 The ExxonMobil polyolefins plant in Baton Rouge, Louisiana.

## Essential Knowledge—Chapter 15

**Glossary words**: Rate process, reaction rate, time-concentration graph, rate law, reaction mechanism, chain reaction, initiation step, propagation step, reaction intermediate, termination reaction, chlorination, free radical, methyl radical, free radical chain mechanism free radical chain reaction, free radical, free radical initiator, free radical trap, anti-oxidants, anti-ozonants, curing reactions, photoresists, photolithography, specific rate constant, reaction order, order with respect to a reactant, overall reaction order, zero order reaction, first order rate processes, radiodecay, initial reaction rate, isolating a variable, clock reaction, iodine clock, Arrhenius equation, pre-exponential or frequency factor, activated complex, transition state, polymerization catalyst, Ziegler-Natta catalyst, metallocene catalyst, organometallic sandwich compound.

**Key Concepts**: General nature of chemical kinetics and its relation to the concepts of shelf-life and service life. Chain reactions and the ways in which they can be both speeded and slowed. Reaction mechanism and bond breaking and making as the fundamental basis for chemical rate processes. How time-concentration data are obtained and analyzed as a way to discovering reaction mechanisms. The mathematical basis of rate laws. The method of isolating a variable and its application to initial reaction rate data. The operation of a clock reaction. Basis of the Arrhenius equation and its use to measure activation energies of reactions. The effects of temperature and molecular orientation on reaction rate. Chemical kinetics and its application to polymerization catalysts. The role of catalysis in the production of polyolefins and ultimately in making engineering materials.

**Key Equations**:

Rate law expression for many reactions $A + B \rightarrow P$: Rate = $k[A]^m[B]^n$

First order rate law: rate = $-k\,c$. Integrated form of the law $c = c_0\, e^{-kt}$

Relationship between the half-life of a first order reaction and its specific rate constant: $k = (\ln 2)/t_{1/2}$

Average rate of reaction: Average rate = $\Delta$Concentration $\div$ $\Delta$time.

Arrhenius equation: $k = Ae^{-E_a/RT}$

Two-point fit form of the Arrhenius equation: $\ln \frac{k_2}{k_1} = -\frac{E_a}{R}\left(\frac{1}{T_2} - \frac{1}{T_1}\right)$

## Questions and Problems

### Kinetics and Chain Reactions

15.1. Each of the following situations gives a clue to the rate of a chemical reaction. In each case, discuss what kind of a reaction might be involved: a. a shiny, new copper roof turns green, b. blacktop asphalt paving in a parking lot turns gray, c. a steel tool left outside overnight is turned bright orange by the dew, d. the outside page of a newspaper left lying in a driveway becomes yellowed, and e. a person's brown hair becomes blonde during a summer spent at the beach.

15.2. Define or describe in 25 words or less each of the following: a. a rate process, b. a reaction rate, c. a time-concentration graph, d. a rate law.

15.3. State three chemical processes that we would like to speed up. Why?

15.4. State three chemical processes that we would like to slow down. Why?

15.5. How is the rate of disappearance of each reactant related to the rate of appearance of each product in the following reactions?

a. $H_2(g) + Cl_2(g) \rightarrow 2\ HCl(g)$
b. $2\ NO(g) + O_2(g) \rightarrow 2\ NO_2(g)$
c. $2\ N_2O_5(g) \rightarrow 4\ NO_2(g) + O_2(g)$
d. $2\ O_3(g) \rightarrow 3\ O_2(g)$

15.6. A chain reaction involves initiation, propagation, and termination steps. Carefully explain what each step involves and give examples of each type.

15.7. The chlorination of methyl chloride, $CH_3Cl$, yields dichloromethane, $CH_2Cl_2$. Write a reaction mechanism for this reaction which involves a free radical chain.

15.8. State definitions of: a. a free radical, b. a free radical chain reaction, and c. a radical trap.

15.9. Describe a UV absorber: how it is used and why it works. Use internet resources to learn why UV light is dangerous for humans. What specific effects can overexposure UV light have?

15.10. Use internet resources and write a 100-word discussion of photoresists.

15.11. Describe the uses of butylated hydroxy anisole and dibenzoyl peroxide.

### Rate Laws

15.12. The rate of a chemical reaction is $4.0 \times 10^{-5}$ mol $L^{-1}$ $sec^{-1}$. Express this rate in molecules $m^{-3}$ $hour^{-1}$.

15.13. List three physical properties that could be used to monitor the concentration change of a reactant or a product continuously during a reaction. What instruments or equipment would be needed in each case?

15.14. Describe the difference between an instantaneous reaction rate and an average reaction rate over a specified interval of time.

15.15. Explain why a zero order reaction has a rate independent of any reactant concentration.

15.16. The reaction $2\ H_2O_2(l) \rightarrow 2\ H_2O(l) + O_2(g)$ is first order in $H_2O_2$. Write the rate expression for the disappearance of hydrogen peroxide in this reaction.

15.17. If the initial concentration of the reactant is tripled in the reaction $A \rightarrow P$, what effect will this have on the initial rate disappearance of A if the order with respect to reactant A is: zero, one, two, and three?

15.18. When the initial concentration of a reactant is doubled (and all other concentrations are held constant), the initial reaction rate increases by a factor of four. What will be the effect on the initial rate if the concentration of that reactant is increased by a factor of 3.5?

15.19. For the reaction: $H_2 + I_2 \rightarrow 2\ HI$, determine the rate law from the following table of initial concentrations and rates.

| | $[H_2]$ mol $L^{-1}$ | $[I_2]$ mol $L^{-1}$ | Initial Rate mol $L^{-1}$ $s^{-1}$ |
|---|---|---|---|
| Experiment 1 | 0.100 | 0.100 | 0.012 |
| Experiment 2 | 0.200 | 0.100 | 0.024 |
| Experiment 3 | 0.200 | 0.200 | 0.048 |

Determine the value of the rate constant, k.

15.20. Determine the value of the rate constant, k, for the reaction, $2\ A + B \rightarrow$ Products, from the following table of initial concentrations and rates.

| | [A] mol $L^{-1}$ | [B] mol $L^{-1}$ | Initial Rate mol $L^{-2}$ $s^{-1}$ |
|---|---|---|---|
| Experiment 1 | 0.100 | 0.100 | 0.009 |
| Experiment 2 | 0.200 | 0.100 | 0.036 |
| Experiment 3 | 0.200 | 0.200 | 0.072 |

Calculate the value of the rate constant, k. What will be the initial rate when [A] and [B] are both equal to 0.30 M?

15.21. The reaction of t-butyl bromide, $(CH_3)_3CBr$, with KOH is first order in t-butyl bromide. The value of the rate constant, k, is 0.010 $s^{-1}$ at 55°C in a solvent which is 80% ethanol and 20% water. Compute the rate of the reaction when the concentration of $(CH_3)_3CBr$ is 0.500 mol $L^{-1}$.

15.22. State the units of the specific rate constant, k, for a zero, first, second, and third order reaction.

15.23. The decomposition of hydrogen peroxide is catalyzed by $Fe^{3+}$ ions and has the reaction equation: $2\ H_2O_2 \rightarrow 2\ H_2O + O_2(g)$. The following data were obtained at 25°C by titrating samples of a solution to determine $[H_2O_2]$ at five different times:

| $[H_2O_2]$ (mol $L^{-1}$) | time (minutes) |
|---|---|
| 0.200 | 0.0 |
| 0.126 | 20.0 |
| 0.080 | 40.0 |
| 0.050 | 60.0 |
| 0.032 | 80.0 |

Use a graphical procedure to demonstrate that this reaction is first order in $H_2O_2$. Obtain the value of the rate constant from the graph.

15.24. The experimentally determined rate law for the disappearance of either reactant in the reaction: $CHCl_3(g) + Cl_2(g) \rightarrow CCl_4(g) + HCl(g)$ is rate = $-k\ [CHCl_3][Cl_2]^{1/2}$. What are the units of k?

15.25. The reaction of methyl bromide with $OH^-$ ($CH_3Br + OH^- \rightarrow CH_3OH + Br^-$) is first order in $CH_3Br$ and first order in $OH^-$. The value of the overall second order rate constant is 0.0214 L $mol^{-1}$ $s^{-1}$ at 55°C in a mixture of 80% ethanol and 20% water. Calculate the rate of the reaction when the concentration of $CH_3Br$ is 1.0 mol $L^{-1}$ and that of $OH^-$ is 2.0 mol $L^{-1}$. (Compare this with 15.21 above.)

15.26. The reaction, $A + B \rightarrow$ Products, is a second order reaction that is first order with respect to A and first order with respect to B. The reaction takes place in aqueous solution at 25°C. When [A] = 0.563 mol $L^{-1}$ and [B] = 0.740 mol $L^{-1}$, the rates of disappearance of A and B are 0.00245 mol $L^{-1}$ $min^{-1}$. What will be the rate of disappearance after the following changes at 25°C: to 100.0 mL of the reaction mixture is added 100.0 mL of pure water; and, to 50.0 mL of the reaction mixture is added 100.0 mL of pure water?

## Iodine Clock Reaction

15.27. Write chemical equations for the main, secondary, and indicator reactions of the iodine clock reaction.

15.28. Explain, in the case of the iodine clock reaction, how by using water for successive dilution one can perform a series of experiments in which one reactant is kept at a fixed concentration while another reactant is varied.

## Temperature Effects on Reaction Rates

15.29. A rule of thumb is that reaction rate will double for each 10 K rise in temperature. What is the energy of activation of a reaction that exactly doubles from 298 K to 308 K?

15.30. The rate constant of a reaction at 40°C is 3 times the rate constant of the same reaction at 10°C. What is the activation energy of the reaction?

15.31. By what factor will the rate constant for a reaction change when the temperature is increased from 290 K to 310 K, if the reaction has an activation energy of 10 kJ $mol^{-1}$, 50 kJ $mol^{-1}$, 100 kJ $mol^{-1}$, and 200 kJ $mol^{-1}$

## Activation Energy

15.32. The reaction: $O_3(g) + O(g) \rightarrow 2\ O_2(g)$ has a $\Delta H_{Reaction}$ of -392 kJ $mol^{-1}$ and an activation energy of 14.0 kJ $mol^{-1}$. Sketch a potential energy plot for this reaction and indicate on the plot where the transition state occurs.

15.33. The gas phase reaction: $CH_4 + Br \rightarrow CH_3 + HBr$ has an activation energy of 75 kJ $mol^{-1}$ and a $\Delta H_{Reaction}$ of 67 kJ $mol^{-1}$. Sketch an energy profile for the reaction and determine the energy of activation for the reverse reaction.

## Problem Solving

15.34. Reactions in which a chemical is catalyzed by a solid catalyst are frequently zero order in the material undergoing reaction. Explain why this might be expected.

# Appendix A—Units

## Units in Chemistry

A unit is what we "measure a property in." Learning how to use and manipulate units is one of those fundamental, toolbox skills that an engineer will use for a lifetime—so you are exhorted to learn the skill now and learn it right. Every physical quantity or measurement is taken to be the algebraic product of a numerical value and a unit.

A physical quantity (measurement) = its value × its unit

For example, the highway speed limit is 65 (the numerical value) × mile per hour (the unit)

IUPAC (the International Union of Pure and Applied Chemistry) recommends this method of treating a physical quantity—as the product of its value and its unit (Green Book, 2nd Edition, page 107, 1995), and suggests that an appropriate name for the method is "**algebra-of-quantities**." That's a splendid suggestion.

To learn more about IUPAC do ❁kws [+IUPAC +homepage]. The book has a green cover—which gives it its informal name.

Consider, for example, the length 38 mm. According to the algebra-of-quantities, this quantity is interpreted as 38 (a number) ´ 1 mm (a unit).

| 38 | × | mm |
|---|---|---|
| the numerical value | × | the unit |

So 38 mm is the product of the number 38 and the unit mm. Chemistry teachers have long described the method using names such as dimensional analysis or the factor-label method. You might remember one of these names from high school chemistry. Calling the method the algebra-of-quantities is straightforward, easily understood, and consistent with international practice.

Two Rules:

1. Addition and subtraction can be performed *only* with quantities that have *identical* units. (You can't add apples to oranges.)
2. Multiplication and division of quantities produces answers with units formed by algebraically combining the original units.

Every working chemist and engineer knows that if you get proper units in your answer there's a good chance you did the problem correctly. If you don't get them, you certainly did the problem incorrectly.

## Origin of Conversion Factors

A **conversion factor** is a quantity (with a value and a unit) that is used to change the units of a measurement. Let's look at a sample problem using the algebra-of-quantities method. One common unit of length in the United States is the inch (abbreviated as in). One inch is defined (by the US Congress) to be exactly equal to 2.54 cm. As shown below, this definition yields two fractions—both equal to one.

Definition: 1 in = 2.54 cm (exactly) which yields either $\frac{2.54 \text{ cm}}{1 \text{ in}} = 1$ or $\frac{1 \text{ in}}{2.54 \text{ cm}} = 1$

Thus, from the definition we get two conversion factors (the ratios above). To convert inches to cm you multiple by 2.54 cm $inch^{-1}$. To convert cm to inches multiply by $\frac{1}{2.54}$ in $cm^{-1}$. Note the use of the exponent -1 to represent what is in the denominator of the fraction. If you think about the two alternative conversion factors above you will discover an obvious rule.

> Rule: The conversion factor that converts A to B is the reciprocal of the conversion factor that converts B to A.

The rule makes excellent sense because multiplying by a conversion factor *must* at heart be multiplying by one: A measured value *cannot* be fundamentally changed merely by changing its units.

The units of conversion factors can be written in different ways; for example, cm/inch can be written as cm $inch^{-1}$ or cm per inch or $\frac{cm}{inch}$, etc. In this book we'll choose to use the exponent form, cm $inch^{-1}$. Using exponents is an unambiguous method, whereas some other methods can lead you astray in complicated expressions, leaving you to decide for yourself which units are in the denominator and which in the numerator.

Here's a table of common conversion factors.

| Some Common Conversion Factors | | |
|---|---|---|
| | | |
| 1 inch | = | 2.54 cm (exactly) |
| 1 liter | = | 1000 $cm^3$ (exactly) or 1000 mL |
| 1 pound | = | 0.45359237 kg (exactly) |
| 1 pound | = | 16 ounces (exactly) |
| 1 mile | = | 5280 feet (exactly) |
| 1 joule | = | 4.184 calories (thermochemical, exactly) |
| 1 gallon (US) | = | 3.78541 L (exactly) |
| 1 standard atmosphere | = | 101325 Pa (exactly) |
| 1 standard atmosphere | ≈ | 760.000 mmHg |
| 1 g | ≈ | $6.0221367 \times 10^{23}$ u |
| 1 light-year (Ly) | ≈ | $9.460528 \times 10^{15}$ m |

As used for conversion factors, = is understood to mean "is equivalent to." According to the algebra-of-quantities, 1000 mL cannot literally *equal* 1000 $cm^3$ (because for two quantities to be equal both their values *and* their units must be identical) but in everyday language we all slip into the ellipsis "equals" for "is equivalent to." When the = sign is used the conversion is exact; the value on the right has an infinite number of significant figures (by definition). A wiggly equals sign, ≈, shows that the value is limited to the number of figures shown; relationships shown above with a ≈ have a finite number of significant figures because they involve experimental measurements.

There is a detailed discussion of significant figures in Appendix B. But even at this point you should understand how you can tell the number of significant figures of each quantity in the table above. Here's an example problem:

**Appendix Example 1**: Use the preceding table and state the number of significant digits in the numbers 0.45359237 kg $lb^{-1}$ and 760.000 mmHg (standard atm)$^{-1}$

Strategy: Learn and understand the difference between an exact, or defined, value and a measured, or experimental, value.

0.45359237 is defined as an exact number. It has an infinite number of significant figures. It could be equally well be written 0.4535923700000000000000...

Because of the wiggly equals sign in the relationship 1 standard atmosphere ≈ 760.000 mmHg, the relationship is not exact. So 760.000 has six significant figures.

Whenever you work a problem involving quantities with units, always ask yourself what the units of the answer should be? Asking that question will help you avoid mistakes and will improve your test scores. Throughout this book we'll try to use proper numbers of significant figures, and to use the units throughout example problems—so you can see for yourself that we are manipulating the data and units correctly. Incidentally, there are three significant figures in the answer to Example 3 below, because 76.4 has three significant figures. The other four values (1, 2.54, 12, and 3) are all exact numbers. An **exact number** (or exact value) has zero uncertainty. A **measured number** (or measured value) is the result of some experiment and has a limited number of significant figures and a *non zero* uncertainty.

**Appendix Example 2**: Convert 3.67 inches (in) to a length in centimeters (cm).

Strategy: Write an equation with the unit of the quantity you want to calculate (cm) on the left and the unit of the quantity you know (inch, or in) on the right:

$$? \text{ cm} = 3.67 \text{ in}$$

To make the conversion, the quantity on the right side must be multiplied by a conversion factor that has units of inch in its denominator and cm in the numerator

$$? \text{ cm} = 3.67 \text{ in} \times \text{(value of conversion factor)} \times \frac{\text{cm}}{\text{in}}$$

Obviously, the unit inch will cancel (that's implied in the term **algebra-of-quantities**), leaving the desired unit, cm. Thus the conversion factor must have units of cm $\text{inch}^{-1}$. The conversion factor (the value in the equation above) that will convert inches to cm is 2.54, where the 2.54 is an exact number (by definition). Thus,

$$\begin{aligned} \text{length in cm} &= 3.67 \text{ in} \times 2.54 \text{ cm in}^{-1} \qquad \text{(remember } y \times y^{-1} = 1) \\ &= 9.32 \text{ cm (Rounded to three significant figures)} \end{aligned}$$

The same concepts apply in more complicated cases. If you understand this one example you understand them all. Recall the rule that when manipulating quantities, all of the units must cancel except for those appropriate for the final result.

**Appendix Example 3**: Convert 76.4 cm to a length in yards (yd).

Strategy: Use the three conversion factors: 2.54 cm $\text{in}^{-1}$ (exactly), 12 inch $\text{foot}^{-1}$ (exactly), and 3 feet $\text{yd}^{-1}$ (exactly).

$$? \text{ yd} = 76.4 \text{ cm} \times \frac{1 \text{ in}}{2.54 \text{ cm}} \times \frac{1 \text{ foot}}{12 \text{ in}} \times \frac{1 \text{ yd}}{3 \text{ ft}} = \underline{0.836 \text{ yd}}$$

All of the units cancel except for the desired unit of yards. There are three significant digits in the given value so there are three significant digits in the reported answer.

The following example shows that a conversion factor and its reciprocal are sometimes written with different numbers of digits:

**Appendix Example 4**: How many significant figures has the conversion factor that converts centimeters to inches?

Strategy: Take the reciprocal of the conversion factor 2.54 cm $inch^{-1}$ (exactly).

The factor is $\frac{1}{2.54}$ = 0.393700787... inch $cm^{-1}$. It has an infinite number of significant figures.

Note: The conversion factor 2.5400000000... also has an infinite number of significant figures, but we usually omit those trailing zeroes.

**Appendix Example 5**: If the conversion factor from mm to meters is 0.001 m $mm^{-1}$ what is the conversion factor from meters to mm?

Strategy: Use the rule that the conversion factor that converts A to B is the reciprocal of the conversion factor that converts B to A.

The conversion factor for meters to mm is simply the reciprocal of 0.001 m $mm^{-1}$.

Reciprocal of 0.001 m $mm^{-1}$ = 1000 mm $m^{-1}$

Note in this context, both 0.001 and 1000 are exact numbers (both have an infinite, ∞, number of significant figures).

## Units and Measurements

Units arise because when you measure a value or quantity you measure it *in* something. What you measure it *in* is its **unit**. You can study and characterize materials by doing experiments on them and by making measurements of their properties. Weighing them and measuring their volumes are two types of measurements commonly made to determine **amounts of substances**.

The ratio of a substance's property of mass divided by its property of volume is a third property of the substance called the substance's density.

**Appendix Example 6**: The mass of a cube of metal is 113 grams (g). The cube has an edge length of 2.34 cm. Calculate the density of the metal.

Strategy: First calculate the volume of the cube by cubing its edge length. Second, calculate density by dividing mass by volume: d = m ÷ V.

edge length = 2.34 cm, and cubing both sides gives Volume = $2.34^3$ $cm^3$

Using d = m ÷ V, the density = 113 g ÷ $2.34^3$ $cm^3$, so d = 8.82 g $cm^{-3}$

Note there are three significant figures in the answer because both given values have three and we used division as our calculation method.

For a pure, ideal gas, measuring and then stating its temperature, volume, and pressure specifies the amount of substance of the gas present. Recall, the number of moles of an ideal gas is $n = \frac{PV}{RT}$. The mole is a "counting" unit that specifies the total number of particles present in the gas sample. The unit mole is used to specify the amount of substance in a sample; it is a base unit of the SI metric system. Any measurement requires the existence of a standard. A standard is anything that is commonly accepted to provide a quantitative comparison for what you want to measure. Thus, a standard often defines the unit of a measurement. An ordinary ruler provides a standard of length. A kitchen measuring cup provides a standard of volume. Because measurements have been made in many cultures throughout history, there are many redundant units, and the conversion of one system of measuring units to another is a common need. For example, distance or length can be measured in units of miles, nautical miles, feet, kilometers, light-years, or many others.

## Fundamental Physical Constants

A **fundamental physical constant** is some property of nature with a constant value. For example, the speed of light, or the mass or charge of a proton.

A table of some of the fundamental physical constants of nature is shown below. Many of these constants are measured values, so their uncertainty is limited to the number of significant figures shown. For example, the charge of the electron (elementary charge) is known to 9 significant figures. Some of the values below are defined to be exact. For example, the speed of light, standard atmospheric pressure, and standard gravitational attraction are all defined (and hence, exact) values.

| **Some Physical Constants (IUPAC, 1993 Values)** | | |
|---|---|---|
| Avogadro's number | ≈ | $6.0221367 \times 10^{23}$ entities $mol^{-1}$ |
| Planck constant | ≈ | $6.6260755 \times 10^{-34}$ J s |
| Elementary charge | ≈ | $1.60217733 \times 10^{-19}$ C |
| Electron rest mass | ≈ | $9.1093897 \times 10^{-31}$ kg |
| Proton rest mass | ≈ | $1.6726231 \times 10^{-27}$ kg |
| Neutron rest mass | ≈ | $1.6749286 \times 10^{-27}$ kg |
| Pi (π) | = | 3.14159265359... (exactly) |
| Natural log of 10 (ln) | = | 2.30258509299... (exactly) |
| Unified atomic mass unit (u) | ≈ | $1.6605402 \times 10^{-27}$ kg |
| Faraday constant | ≈ | 96485.309 C $mol^{-1}$ |
| Standard atmosphere | = | 101325 Pa (exactly) |
| Speed of light | = | 299792458 m $s^{-1}$ (exactly) in vacuum |
| Gas constant | ≈ | 8.314510 J $K^{-1}$ $mol^{-1}$ |
| Gas constant | ≈ | 0.08205783 L atm $K^{-1}$ $mol^{-1}$ |
| Rydberg constant | ≈ | 109737.1534 $cm^{-1}$ |
| Standard acceleration | = | 9.80665 m $s^{-2}$ (exactly) |
| Absolute zero | = | -273.15°C (exactly) |

As described earlier, an ordinary equals sign (=) means that the value of the property is exact and any number of zeroes can be added as appropriate. A wiggly equals sign (≈) means that the property is measured and known only as accurately as possible; these values are the **correct values** of the properties in the table.

## Box of Conversion Factors

Chemists and engineers routinely need to use conversion factors and fundamental constants of nature. The following box is the one that the author places on tests for student use during examinations. The box includes a mixture of needed conversion factors and fundamental constants. Your instructor will advise you about his or her methods.

One US gallon ≈ 3785 mL  1 Liter = 1000 $cm^3$ = $10^{-3}$ $m^3$  1 pound ≈ 453.6 g
1 mile = 5280 feet (exactly)  1 foot = 12 inches (exactly)  1 inch = 2.54 cm (exactly)
1 L atm ≈ 101.3 J  Gas constant (R) ≈ 0.08206 L atm $K^{-1}$ $mol^{-1}$ or ≈ 8.3145 J $K^{-1}$ $mol^{-1}$
Absolute zero = -273.15 °C (exactly)  Planck's constant ≈ 6.626 × $10^{-34}$ J s
Speed of light = 2.99792458 × $10^8$ m $s^{-1}$ (exactly)
Avogadro's number ≈ 6.0221 × $10^{23}$ $mol^{-1}$ (entities $mol^{-1}$)
760 Torr = 760 mmHg = 1 atm ≈ 1.013 × $10^5$ N $m^{-2}$ ≈ 14.70 psi
Standard atmosphere = 101325 Pa (exactly) ≈ 1.013 × $10^5$ N $m^{-2}$
1 calorie = 4.184 J (exactly)  1 Btu = 1055.06 J (exactly)
1 faraday ≈ 96485 C $mol^{-1}$ = charge in coulombs of 1 mol of electrons or protons

## The SI Metric System

### Official Definitions of the Seven SI Base Units

Notes:

1. CGPM = General Conference on Weights and Measures.
2. From the IUPAC "Green Book," second edition, 1993, page 70.
3. Caesium is usually spelled cesium in the US.
4. Metre is usually spelled meter in the US.

**Metre**: The metre is the length of path traveled by light in vacuum during a time interval of 1/299 792 458 of a second (17th CGPM, 1983).

**Kilogram**: The kilogram is the unit of mass; it is equal to the mass of the international prototype of the kilogram (3rd CGPM, 1901).

**Second**: The second is the duration of 9 192 631 770 periods of the radiation corresponding to the transition between the two hyperfine levels of the ground state of the caesium-133 atom (13th CGPM, 1967).

**Ampere**: The ampere is that constant current which, if maintained in two straight parallel conductors of infinite length, of negligible circular cross-section, and placed 1 metre apart in vacuum, would produce between these conductors a force equal to 2 × 10-7 newton per metre of length (9th CGPM, 1948).

**Kelvin**: The kelvin, unit of thermodynamic temperature, is the fraction 1/273.16 of the thermodynamic temperature of the triple point of water (13th CGPM, 1967).

**Mole**: The mole is the amount of substance of a system which contains as many elementary entities as there are atoms in 0.012 kilogram of carbon-12. When the mole is used, the elementary entities must be specified and may be atoms, molecules, ions, electrons, other particles, or specified groups of such particles (14th CGPM, 1971).

**Candela:** The candela is the luminous intensity, in a given direction, of a source that emits monochromatic radiation of frequency 540 × $10^{12}$ hertz and that has a radiant intensity in that direction of (1/683) watt per steradian (16th CGPM, 1979).

As noted previously, the history of measurements has been long and colorful. For example, the inch is so-called because it is ultimately derived from the Latin *uncia*, meaning a twelfth, and reached us via Old English as *ynce*. Even today, the US still differs from most of the rest of the world in using English units for distance and mass. But manufacturing industry now operates globally, so many US industries work in metric sizes.

Commercial activity continues to be a driving force for improved standards. Thus, most countries have established national laboratories or national organizations with the responsibility of maintaining standards. In the United States, that organization is the National Institute of Standards and Technology (NIST). It will come as no surprise to you that NIST is part of the US Department of Commerce. You may even be aware of the International Standards Organization (ISO) if you have seen product notices that say something like "ISO 9000 compliant."

The international scientific community as a whole has made considerable progress in achieving a single set of standards for worldwide use, and the modern world is gradually progressing toward uniform adoption of SI metric units (from the French, Systeme International d'Unites). Chemists generally also use SI units. The International Union of Pure and Applied Chemistry (IUPAC) is the umbrella organization for terminology and systematization in chemistry.

In the US, NIST is the coordinating agency for the International metric system. NIST publishes the *Guide for the Use of the International System of Units (SI)*. You can get this 74 page document as a downloadable pdf file from http://physics.nist.gov/Pubs/SP811/sp811.html, or search for ❁kws[+nist +sp811].

The SI system is based on seven well-defined base units. Each base unit is defined in terms of a carefully considered, but ultimately arbitrary, standard. The first six of these base units are important in chemistry. The "official" definitions of the seven base units of the SI metric system are in the nearby box.

During your chemistry course you'll see lots of examples using meters, kilograms, and seconds. The mole, of course, is *the* chemist's unit and you'll use it repeatedly—especially in connection with stoichiometry. For temperatures you'll use Celsius degrees and kelvins—the base unit of thermodynamic temperature. The author can remember (even now with embarrassment) over 40 years ago in school screwing up a gas law problem by forgetting to convert °C to K. You'll meet amperes in electrochemistry. The only base unit you won't use during your chemistry course is the candela.

The names and symbols of the seven base units of the quantitiies in the SI metric system are shown in the following table.

**Names and Symbols for the SI Base Units**

| Quantity | Unit | Symbol |
|---|---|---|
| length | meter | m |
| mass | kilogram | kg |
| time | second | s |
| electric current | ampere | A |
| thermodynamic temperature | kelvin | K |
| amount of substance | mole | mol |
| luminous intensity | candela | cd |

## SI Metric Prefixes

Prefixes are used to form decimal multiples and submultiples of the SI base units. Note the exception to this rule for the base unit kilogram, in which case the gram is used. In recent years there has been considerable discussion about replacing the physical kilogram as the international standard of mass. The kilogram is today the only base unit defined in terms of a specific physical object. For many years, chemists have used an alternative mass scale based on the atomic mass unit (u) which is defined as one-twelfth the mass of an atom of C-12. Probably in the 21st century we'll see a new basis for a mass standard. In your chemistry course you'll learn about mass scales. The conversion factor between the atomic mass unit (u) and the gram (g) is $6.0221367 \times 10^{23}$ u $g^{-1}$. It is Avogadro's number.

**Prefixes in SI**

| Submultiples of base unit | Multiples of base unit |
|---|---|
| $10^{-24}$ yocto (y) | $10^{24}$ yotta (Y) |
| $10^{-21}$ zepto (z) | $10^{21}$ zetta (Z) |
| $10^{-18}$ atto (a) | $10^{18}$ exa (E) |
| $10^{-15}$ femto (f) | $10^{15}$ peta (P) |
| $10^{-12}$ pico (p) | $10^{12}$ tera (T) |
| $10^{-9}$ nano (n) | $10^{9}$ giga (G) |
| $10^{-6}$ micro (μ) | $10^{6}$ mega (M) |
| $10^{-3}$ milli (m) | $10^{3}$ kilo (k) |
| $10^{-2}$ centi (c) | $10^{2}$ hecto (h) |

## Derived SI Units

Many other useful measuring units are formed from the SI base units; these are called derived SI units. A **derived SI unit** is any unit made from an algebraic combination of SI base units. For example the derived SI unit of area is the $m^2$, the derived unit of volume is the $m^3$, and the derived unit of speed is meter per second m $s^{-1}$. Appendix Example 6 showed how to obtain the derived unit of density expressed in the units g $cm^{-3}$. Some derived SI units are given special names, usually the name of a scientist associated with the development of the unit or the property the unit measures. SI and non-SI units are in widespread use in the United States and it is necessary for engineers to be familiar with both. Different units for the same quantity or property are related by conversion factors.

**Some Useful Derived SI Units**

| Property | Unit name | Symbol | Base units |
|---|---|---|---|
| Frequency | hertz | Hz | $s^{-1}$ |
| Volume | liter | L | $10^{-3}$ $m^3$ |
| Concentration | molarity | M | $10^{-3}$ mol $m^{-3}$ (= mol $L^{-1}$) |
| Energy | joule | J | kg $m^2$ $s^{-2}$ |
| Force | newton | N | kg m $s^{-2}$ |
| Pressure | pascal | Pa | kg $m^{-1}$ $s^{-2}$ |
| Power | watt | W | J $s^{-1}$ |
| Electric charge | coulomb | C | A s |
| Voltage | volt | V | J $C^{-1}$ |
| Entropy | e.u. | S | J $K^{-1}$ $mol^{-1}$ |
| Electrical dipole | no name | p,μ | C m |

## Background Reading

You can obtain a free copy of the NIST special publication about metric units (SP811) by downloading it from the NIST site described on page A-8. That's perhaps a useful reference to keep on your hard disk.

Engineers have often been part of the history of measurement. An entertaining book on the subject is *The Science of Measurement*, by H. A. Klein. Dover Books, New York 1974. The book is a good general survey, written with a light touch. One of the many uses of internet searches is to locate old books. Try ☸kws [+"used books"]. About a dozen copies of Klein's book were on offer when the author looked in the spring of 2000.

Footnote:

Units conversion is a subject that hardly ever reaches the level of public debate. However, it did just that in September, 1999, when NASA's Mars Climate Orbiter crashed into the Martian surface at an estimated 50 mph and was destroyed.

Improper units were the cause of the crash. Engineers at Lockheed Martin in Denver failed to convert critical spacecraft control values from English customary units to SI units and NASA personnel at the Jet Propulsion Laboratory in Pasadena failed to check. So in the event, critical engine firing instructions transmitted to the spacecraft were wrong.

# Appendix B—Uncertainty

## Experimental Uncertainty, or

## "What do we Know and How Well do we Know It?"

Anyone who measures anything needs to be aware of the quality of his or her measurements, and to know how to deal with quality using appropriate methods. The key questions are "What do we Know?" and "How Well do we Know It?" These are questions that you will ask throughout you professional career. When you report the results of trials or experiments your readers or listeners (or boss) will want to know "How good are your data?" They will be interested in your values (in your numbers), but they will also want to know how reliable those values are.

The amount of confidence that we have in the value of a measurement is its experimental uncertainty; the quality of our measurements is described by their experimental uncertainty. Experimental uncertainty is the inevitable limit to our knowledge of the value of measured quantities.

Part of your task as an undergraduate is to sort out what information you'll need for your lifetime. Knowing about experimental uncertainty is information for your lifetime. A working engineer needs to be able to assess the quality of his or her own data and the quality other people's data and data from standard sources. To apply other people's data properly you must be able to assess how reliable it is. If you learn the methods of handling uncertainty now, you will find that they remain useful throughout your professional career.

The question that heads this section seeks two separate pieces of information: (1) What is the best estimate of the quantity we are trying to measure, and (2) How good do we think is our measurement of that quantity? Good measurements and results have a low experimental uncertainty; bad ones have a high experimental uncertainty. In the scientific community, anyone who reports a measured value has the responsibility to specify the uncertainty of that value and thereby advertise its quality.

In many situations, the best estimate of a measured value is simply the average of many, repeated, independent measurements of the value. The word mean is a widely used alternative to the word average. Two methods are widely used to specify the experimental uncertainty of a measured value:

- Method 1: Use the appropriate number of significant figures (a crude method).
- Method 2: Use statistics-based methods deriving from the normal curve of error. This is the best method.

Understanding experimental uncertainty, and being able to use it, is a skill needed in *all* scientific and technical fields. The statistics-based method of assessing the quality of measurements goes by many colloquial names: uncertainty, experimental uncertainty, error analysis, errors, data error, accuracy, precision, and standard deviation, etc. Indeed, the wide range of names for the method reveals that professionals in many different scientific and engineering disciplines deal with experimental uncertainty.

**The Historical Beginning of Experimental Uncertainty**

The mathematician Karl Friedrich Gauss (1777-1855) was perhaps the first to apply statistics to scientific measurements. The astronomer Zach wanted to be able to find the asteroid Ceres again when it reemerged after passing behind the sun, so he asked Gauss to predict the position of Ceres and where he should point his telescope. In consequence, Gauss developed the least-squares method of data analysis and made an excellent prediction. Gauss's method remains today the basis for the statistical interpretation of experimental uncertainty.

In practice, assuming a competent experimenter, the main uncertainty of an experimental result or a single measurement is the inherent limit of the measuring device (or devices) used. Regardless of its design and construction, every measuring instrument gives uncertain results. Generally speaking, good instruments give results with low experimental uncertainty, poor instruments give results with high experimental uncertainty.

Although every *measured* number has uncertainty, counted numbers and integers have none: they are exact numbers, not measured ones. An **exact number** is one with *no* uncertainty. Examples of exact numbers are: the 8 musicians in an octet; metric system prefixes such as 1000 things in a kilothing; and many conversion factors such as 12 inches per foot and 60 seconds per minute. How can you tell if a written number is or is not exact? Often you can't. If the writer does not specify which it is, the best you can do is make a guess based on the context.

## Accuracy and Precision

Experimental uncertainty is usually discussed in terms of two related concepts: accuracy and precision. Accuracy and precision have closely related but different meanings.

**Accuracy** is how close a measured quantity is to the correct value. The **correct value** for a property is the best value humans can determine. In the United States, the National Institute of Standards and Technology (NIST) maintains actual standards (such as a physical kilogram) and a large number of **reference materials** (called **primary reference standards**) used to establish accurate values by direct comparison. Usually, these reference materials are employed to calibrate instruments. An instrument is calibrated by adjusting it so that it produces the correct reading for a reference sample. Somewhat less reliable standards for many materials are available for less cost from commercial supply companies. These are called **secondary standards**.

**Precision** refers to the reproducibility of a measurement; that is, it describes how closely repeated measurements agree with one another. Precision is determined by the instruments that are available, how well they are calibrated, and how carefully they are used. The following "target" diagram illustrates the difference between accuracy and precision. The statistical quantity called standard deviation expresses the same idea as you see in the diagram. We will discuss standard deviation later in this appendix.

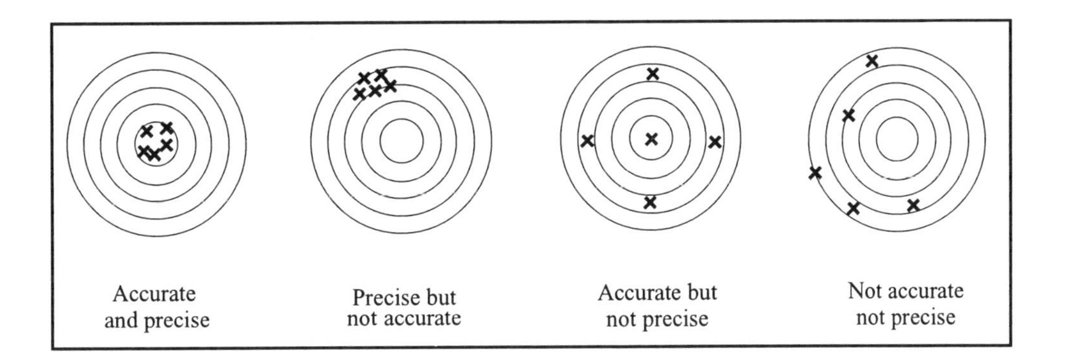


The unit used for atomic weight is the **unified atomic mass unit** given the symbol u. The "official" 1998 IUPAC value of chlorine's atomic weight is 35.4527 ± 0.0009 u. (The ±0.0009 is a standard deviation—to be discussed in the following section.) Chlorine's atomic weight is the average weight (mean weight) of the two stable isotopes of chlorine, Cl-35 and Cl-37, taken in proportion to their abundance in nature. Because of the variation in the natural abundance of chlorine's two isotopes, human's can only know chlorine's atomic weight within an experimental uncertainty that is expressed by six significant figures.

The accuracy/precision idea is shown in the following table. You should compare the numbers in that table with the same ideas presented visually in the diagram above. You'll know when you understand the ideas—when it all fits together and makes sense in your mind.

| Atomic weight | Standard deviation | Comment | Percentage discrepancy between this value and the correct value |
|---|---|---|---|
| | | | |
| 35.4527 | ± 0.0009 | Correct value | 0% |
| 35.45 | ±.02 | Accurate and precise | 0.008% |
| 36.77 | ±.02 | Not accurate but precise | 4% |
| 35.5 | ±2 | Accurate but not precise | 0.1% |
| 42 | ±5 | Not accurate not precise | 20% |

At the end of this appendix is the reference to *Guidelines for Evaluating and Expressing the Uncertainty of NIST Measurement Results*, by Barry N. Taylor and Chris E. Kuyatt (1993). You'll see from that document that professional **metrologists** (people who study measurements as a career) are very detail conscious. You'll also realize that the treatment in this appendix barely skims the surface of a complicated subject.

## Factors that Control the Precision of a Value

- Analog Scale uncertainty: A scale such as the one on an ordinary ruler is an analog device. When such a scale is read, precision is limited by the smallest division of the scale. The precision is usually estimated to be one-tenth of the smallest division on the scale. For example, if a thermometer is calibrated in 0.1°C increments, a person can usually estimate to the nearest 0.01°C if he or she takes the average of several measurements. Naturally, different people may estimate the hundredths place differently. Lacking other information, a useful rule of thumb is to take the experimental uncertainty of a *single* measurement as ±½ the smallest scale division of the measuring device.
- Digital scale uncertainty: Many instruments have digital displays. Digital displays can be read "exactly" but *not* without uncertainty. Their readings are typically limited by electronic "noise" in the instrument's circuits. The precision of a digital device can be estimated by making repeated measurements on a single sample, or by observing variations in the display over an extended period of time. A reasonable estimate of the uncertainty of a digital device is ±1 in the last digit of the read-out. Digital scale error is sometimes called digitization error.
- Sample variability: Precision may also be limited by natural variations in the property being measured. For example, the uncertainty in the atomic weights of the elements arises because their isotopic composition varies from sample to sample.

Note: If you read an instrument incorrectly it is not an error, it is a *mistake*. I prefer to avoid using the word error because many people use the phrase "experimental error" (confusingly for beginning students) to refer to experimental uncertainty. However, the usage is complicated by the fact that nobody ever calls the "normal curve of error" the "normal curve of uncertainty," (though it would make life easier for students if they did). We'll meet this curve shortly.

Even if an instrument *is* read correctly, the accuracy of a measurement is affected by both systematic uncertainties and random uncertainties. **Systematic uncertainties** are those incorporated uniformly into every measurement. They usually can be traced to a faulty instrument or improper calibration of the instrument, and are always present if an unreliable standard has been used. **Random uncertainties** arise from uncontrollable fluctuations at the limit of measurement.

In Appendix Example 7 you see a picture of a typical laboratory balance such as the one you will use in the laboratory, The example shows how a crude estimate can be made of experimental uncertainty by making repeated measurements of a value: in this case the mass of a single object (a coin).

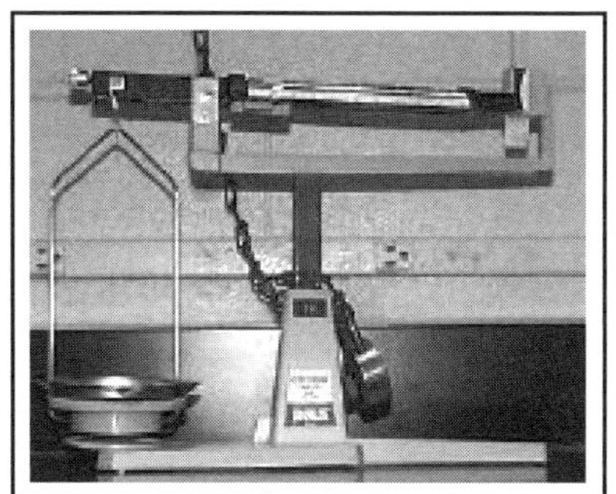

**Appendix Example 7**: A student measured the mass of a coin three times using a single-pan balance. The balance scale is marked in 0.1 g increments and the student estimates the readings to nearest hundredth of a gram. The three readings were 4.91 g, 4.93 g, and 4.92 g. Estimate the balance's precision.

Strategy: Use the rule that the estimated precision of a data set is one half of the range of the data. Range is the difference between the largest and smallest values in the data set.

The value that would normally be reported for the mass as a result of these three observations is the average, 4.92 g. The complete range covered by the three separate readings is ±0.02 g 4.93 g - 4.91 g). One half the range of the data is an estimate of the data's precision, which is an estimate of the precision of the measurement.

## Expressing Uncertainty in Measurements

Two ways to specify experimental uncertainty in a calculated result are: (1) the method of significant figures and, (2) the method of standard deviations. Both methods are widely used. The method of significant figures is somewhat simpler and more widely applicable than the method of standard deviations. However, in either case, the method of standard deviations is the fundamental method; the method of standard deviations is one aspect of applied statistics.

**Significant figures** show uncertainty according to conventional rules (i.e., people or professional societies simply agree how to do it). Generally speaking, digits that have been measured are **significant digits**, zeros that have *not* been measured are not significant. Non-significant zeroes that locate a decimal point are called **place holders**. **Standard deviation** is a statistically based estimate of uncertainty that can be interpreted in terms of a statistical frequency (chance) that the best possible measured value will lie between specified limits.

To reiterate: when we report an experimental value, we are obligated to convey two separate pieces of information about the value: the actual magnitude of the value and an estimate of the uncertainty of the value. We are also required to report the correct units. The uncertainty is specified using *either* significant figures or standard deviations. Significant figures are assigned according to widely accepted, but not unique, rules. In the absence of other information, the unccrtainty of a reported value that has been repeatedly measured may be estimated as ±1 in the rightmost significant digit. It is better actually to calculate the standard deviation, as you are shown later in this appendix.

The three things needed for a proper measurement are:

1. The best estimate of its value
2. The best estimate of the uncertainty of the value
3. The units of the measurement

Uncertainty arises both from the natural limitations of our instruments and from nature itself. The values of the atomic weights of the elements provide a good example of uncertainty that arises from nature. Modern mass spectrometers routinely measure the masses of atoms to nine or ten significant figures, but the atomic weights on the periodic table mostly have far fewer significant figures. This situation occurs because an element's isotopic mixture varies slightly depending on where we obtained the element. This natural variation limits the number of significant figures in its atomic weight. On the Periodic Tablc, in other appendixes of this book, and indeed anywhere that measured values are reported, you will find extensive use of both significant figures and standard deviations.

## The Method of Significant Figures

In general, the **least significant digit** of a reported value its the rightmost significant digit. The least significant digit in the value of a *single measurement* will normally be the digit obtained by judging the measurement to one-tenth of the smallest scale division of the measuring device. The actual number of significant figures to be reported will thus depend on the value of the measurement and the measurement scale. For example, if a thermometer is marked in degrees Celsius, a temperature measured with it would normally be reported to the nearest one-tenth degree. On the Celsius scale, 25.6°C has three significant figures, 9.8°C has two, and 0.3°C has one. Converted to the kelvin scale (K = °C + 273.15), each of these temperature values would be reported to four significant figures: 298.8 K, 283.0 K, and 273.5 K.

As noted earlier, some numbers are exact. Recall that an exact number has an infinite number of significant figures and zero uncertainty. There are 12 inches in a foot and 2.54 cm in an inch; because they are defined to be exact numbers, 12 and 2.54 *in this context* have an infinite number of significant figures.

Note: (It's a small point, but an irritating one if you ever run into it.) Different authorities use different conventions for expressing significant figures. For example, the number 12,700 may be interpreted to have three, four, five, or even an indeterminate number of significant figures. Different professions and different individuals have their own wrinkles. According to the rules stated below, 12,700 has three significant figures. Note, however, that any set of rules for using and combining significant figures must ultimately derive from the statistical concept of experimental uncertainty and should so far as possible be consistent with statistics.

Many technical professionals rely on the American Society for Testing and Materials (ASTM) as their authority. For example, ASTM publishes a "Standard Practice" that tells a manufacturer how to use significant digits when specifying a property of a commercial product. The Standard Practice uses 0.6% moisture in sodium bicarbonate as an example of a specification value with one significant digit. [❁kws "ASTM Standards"].

## Rules for Significant Figures

1. Whole number conversion factors, and certain defined quantities, are exact and have an infinite number of significant figures. The atomic number of boron is exactly 5; there are exactly 1000 milligrams in a gram, and exactly 4.184 J per calorie. In each of these quantities as many trailing zeros as desired may be added following the decimal point.

2. Any non zero digit is significant. For example 1543 has four significant digits; 137.33 has five significant digits.

3. Zeros that lie between non-zero digits are significant. Thus in 8507 J, since the 7 is significant (rule 2) the zero must also be significant; the value has 4 significant figures. Likewise, all zeroes in each of the following numbers are significant: .707, 909.08, and 40.7.

4. Place holding zeros on the left of the first non-zero digit are not significant, neither is the leading zero in a number less than one. Thus, 0.222 has three significant digits and 0.004678 has four.

5. Trailing zeros following a decimal point are significant. The number 58.70 u has four significant digits; 0.30000 L has five significant digits—the 3 and the four trailing zeros. In 0.0001900 g there are four significant digits.

6. Trailing zeros in a number without a decimal point are presumed to be place holders and not significant. Thus 2380 has three significant digits, 1500 has two significant digits, and 800 has only one significant digit. To specify 800 to two significant figures requires that it be written in exponential notation as $8.0 \times 10^2$. To specify 800 to three significant figures one may write it as $8.00 \times 10^2$, or write it as 800., using the decimal point to show that the zeros are significant digits rather than place holders.

## Combining Uncertainties Using Significant Figures

When values from different measurements are combined in arithmetic operations, the uncertainty of the result depends both on the uncertainty of each individual measurement and on the way in which the measurements are combined. (Note: If all the measured values have estimated standard deviations, you can use well-founded statistical procedures described in the following section. However, often we must rely solely on significant figures to estimate uncertainties in values that come from calculations.) Specific rules must be followed to obtain the proper number of significant figures in the result. The superficial logic of the rules for handling significant figures is that that no arithmetic operation can conceivably improve (reduce) an experimental uncertainty. The fundamental logic of the rules derives from statistical treatment of uncertainty. Be aware, however, that even if you correctly apply the rules, significant figures sometimes only give a crude estimate of experimental uncertainty.

Uncertainty estimates of the results of addition and subtraction using of significant figures can be made using the following rule:

> Rule: The uncertainty of the result of an addition or subtraction is estimated by its number of significant figures. The result of an addition or subtraction can have no significant digits farther to the right than in the quantity used that has the leftmost least significant digit.

To apply this rule, replace every needed non-significant digit by a question mark. Adjust the numbers with their decimal points in the same column and carry down the question mark in any column in which it appears. In the following example, it's the trailing zero in the number 2340 that shows that the 4 in 2340 is the leftmost least significant digit.

| Adding Numbers with Significant Figures | | |
|---|---|---|
| 2340 | (three significant figs) | 234?.?? |
| 24.12 | (four significant figs) | 24.12 |
| 11111.0 | (six significant figs) | 11111.0? |
| 13475.12 | | 1348?.?? |

On the left, the result of the addition has 7 *apparent* significant figures. However, the question marks shown on the right reveal that only four of those are significant. Adding a measured value to a question mark gives a question mark. To get the final, reported value you must round to 13480. In rounding, 0,1,2,3,4 round down and 5,6,7,8,9 round up. In the example above, the figure 8 in the result comes from rounding up a non-significant 5.

Uncertainty estimates of the results of multiplication and division using significant figures can be made using the following rule:

> Rule: The uncertainty of the result of multiplication or division is estimated by its number of significant figures; the number of significant digits in the result is the same as that in the multiplier or divisor with the fewest number of significant digits.

This notion is close to the metaphor that any chain is only as strong as its weakest link. The reliability of a calculated value will be only as good as the *least* unreliable value that went into the calculation. For example:

| 2.8770 | × | 3.4 | × | 0.01224 | = | 0.12 |
|---|---|---|---|---|---|---|
| (5 sig figs) | | (2 sig figs) | | (4 sig figs) | | (2 sig figs) |

## The Method of Standard Deviation

### Estimating a Standard Deviation from a Series of Repeated Measurements

Recall that **standard deviation** is a statistically based estimate of uncertainty that can be interpreted in terms of a statistical frequency (chance) that the best possible measured value will lie between specified limits; it's a measure of the spread of uncertainty associated with a measurement. If you take a basic course in statistics (strongly recommended) you'll get a proper treatment of the statistics of experimental uncertainty. The author regards this appendix as not much more than a crude, crash course to get you started with this important topic.

It is a simple matter to get a good estimate of standard deviation. You keep making measurements of the same property until you've got a list of measurements and then you plug-and-chug through the standard deviation estimating formula. Probably you've got a routine for doing such a calculation on your handheld calculator.

The **estimated standard deviation** of a measured value is an approximation to its true, statistical standard deviation. Estimated standard deviations can be made for any data set; true standard deviations can only be approached with very large data sets. It is best to assume that all practical standard deviations are merely estimates.

For a series of n repeated measurements $x_1, x_2, x_3... x_n$ the average value is $x_{av}$.

$$x_{av} = \frac{x_1 + x_2 + x_3 + x_4 + ... + x_n}{n}$$

The deviation of a single measurement ($x_i$) is the difference between it and the average value of all the measurements ($x_i - x_{av}$). Note: the symbol $\bar{x}$ is widely used to represent the mean value of x.

The estimated standard deviation of a series of repeated measurements (that is, many experimental observations) represents the precision of the average value of the measurements. Statisticians call standard deviation σ (sigma). To accurately measure standard deviation requires much experimental data. Most of the time in chemistry we obtain only a limited set of experimental data; calculating from a limited set of data we obtain only an estimate of σ. An estimated standard deviation is given the symbol s.

- σ is the proper statistical measure of the precision with which we know or report an experimental value. Chemists rarely (if ever) measure σ; to do so needs far more experimental data than are typically available. Chemists get what statisticians call sample data.
- s is an estimate of standard deviation obtained from a small data set—from sample data. The estimated standard deviation (s) of the average value of a sample set of measurements is given by the formula:

$$\text{s (estimated standard deviation)} = \sqrt{\frac{\sum_{i=1}^{i=n}(x_i - x_{av})^2}{n-1}}$$

In the formula above, n is the number of independent measurements of our quantity or value. The formula for the statistical standard deviation (σ) is

$$\sigma \text{ (standard deviation)} = \sqrt{\frac{\sum_{i=1}^{i=n}(x_i - x_{av})^2}{n}}$$

Observe that the only difference between the two above expressions is that s uses n-1 as the denominator whereas σ uses n as the denominator. With an increasingly large number of measurements n and n-1 become closer and closer. The value 1 is significantly different from the value 2, whereas 998 and 999 differ only slightly. Hence, as the number of measurements is increased, s slowly becomes identical to σ.

Using n-1 instead of n in the denominator makes s > σ. Thus, we get a conservative estimate. To be able to make realistic but conservative estimates is an important talent that engineers need to develop for their professional careers.

**Appendix Example 8**: Four (n = 4) repeated measurements of a mass are 1.34, 1.23, 1.28, and 1.31 g. Calculate the average mass to be reported and its estimated standard deviation (s).

Average ($x_{av}$) = (1.34 + 1.23 + 1.28 + 1.31) ÷ 4 = 5.16 ÷ 4 = 1.29 g

| Measurement number (i) | Value ($x_i$) | Deviation ($x_i$ - $x_{av}$) | Square deviation |
|---|---|---|---|
| | | | |
| 1 | 1.34 | 0.05 | 0.0025 |
| 2 | 1.23 | -0.06 | 0.0036 |
| 3 | 1.28 | -0.01 | 0.0001 |
| 4 | 1.31 | 0.02 | 0.0004 |
| | | | 0.0066 = sum |

s = square root [(0.0066) ÷ (4 - 1)] = 0.047

Reported value = 1.29 ± 0.05 g

**Suggested Problem**: Use a calculator to obtain the standard deviations of each of the five six-point data sets in the table below:

| Point # → | 1 | 2 | 4 | 3 | 5 | 6 |
|---|---|---|---|---|---|---|
| | | | | | | |
| Set 1 | 45.3 | 44.8 | 50.5 | 51.8 | 44.9 | 48.2 |
| Set 2 | 46.4 | 43.8 | 41.4 | 42.2 | 46.7 | 44.8 |
| Set 3 | 39.4 | 35.5 | 36 | 37.2 | 41.2 | 38.9 |
| Set 4 | 36.9 | 38.7 | 32.8 | 33.4 | 36.6 | 36.9 |
| Set 5 | 25.9 | 27.8 | 27 | 29.3 | 29.2 | 27.7 |

## Using Standard Deviations During Adding and Subtracting

The standard deviation of the result of an addition or subtraction is given by the following rule: The standard deviation of the result ($s_R$) of addition or subtraction is the square root of the sum of the squares of the standard deviations of the components ($s_A$ and $s_B$). Recall, we use the symbol s here (not $\sigma$) because chemists (and engineers) almost invariably deal with *estimated* standard deviations. The standard deviation of the result (R ± $s_R$) of an addition or subtraction of A ± $s_A$ and B ± $s_B$ is obtained using the formula

$$s_R = \sqrt{s_A^2 + s_B^2}$$

Note that the same formula is used both for addition and subtraction. Technically, the above equation is correct provided that $s_A$ and $s_B$ are statistically independent.

**Appendix Example 9**: Calculate the result and estimated standard deviation of the addition (2.3 ± 0.2 + 5.7 ± 0.3).

Strategy: The result of the addition (R) is the sum of the two values. R = A + B, or 2.3 + 5.7 = 8.0. To obtain the estimated standard deviation of R recognize that $s_A = 0.2$ and $s_B = 0.3$. Substitute these values into the estimated standard deviation formula.

$$s_R = \sqrt{(0.2^2 + 0.3^{2)}}$$

$$R = 8.0 \pm \sqrt{(0.2^2 + 0.3^{2)}}$$

So the result of the addition is reported as 8.0 ± 0.4

## Using Standard Deviations During Multiplying and Dividing

The standard deviation of the result ($R \pm s_R$) of a multiplication or division of $A \pm s_A$ by $B \pm s_B$ is obtained using the formula

$$\left(\frac{s_R}{R}\right)^2 = \left(\frac{s_A}{A}\right)^2 + \left(\frac{s_B}{B}\right)^2$$

Technically, the above equation is correct provided that $s_A$ and $s_B$ are statistically independent. The use of the above relationship is shown in Appendix Example 10.

**Appendix Example 10**: Calculate the result and estimated standard deviation of the multiplication 3.67 ± 0.02 × 4.50 ± 0.04.

Strategy: Do the multiplication to get the result (R = 3.67 × 4.50 = 16.5). Use the formula for estimated standard deviation to obtain an estimate of the standard deviation ($s_R$) of the result.

$$\left(\frac{s_R}{R}\right)^2 = \left(\frac{s_A}{A}\right)^2 + \left(\frac{s_B}{B}\right)^2$$

$$\left(\frac{s_R}{16.5}\right)^2 = \left(\frac{0.02}{3.67}\right)^2 + \left(\frac{0.04}{4.50}\right)^2$$

Doing the arithmetic on this expression gives $s_R = 0.17$. Therefore the reported result would be = 16.5 ± 0.2

Note: The standard deviation of a result obtained by ordinary scientists and engineers is normally expressed and reported with just 1 significant figure. So the number of significant figures in your reported result will be decided by the uncertainty estimate. The rule is: there can be no digits in the value farther to the right of the place in which the uncertainty lies. To this extent, the method of standard deviation and the method of significant figures should agree.

An exception to the one-significant-figure-in-uncertainty rule is that uncertainties in important physical constants (but not atomic weights) are often reported by IUPAC to two significant figures. In the event of a "conflict" between uncertainty estimates using the method of significant figures and the method of standard deviations, the latter is by far the better.

**Appendix Example 11**: The value 22.305 ± 0.1 has been incorrectly reported. Correct it.

Strategy: Understand that the standard deviation controls the rightmost significant digit in a reported value. There can be no digits in the value farther to the right of the place in which the uncertainty lies.

The standard deviation of ± 0.1 lies in the tenth's place. The reported value must be rounded to the tenth's place.

The correctly reported value is 22.3

## The Experimental Uncertainty of a Single Measurement

Sometimes, practical realities dictate that a chemist or engineer can only get a single measurement of an important value. A single measurement cannot be treated statistically, yet we need to know how much confidence we can place in it. Lacking other information, a useful rule of thumb is to take the experimental uncertainty of a single measurement as ±½ the smallest scale division of the measuring device. By this rule, a thermometer calibrated to read in whole Celsius degrees has an estimated experimental uncertainty of ±0.5°C. Note: this assumption is useful if one knows something about the measuring instrument. If one simply reads a value in a book, knowing nothing else about it one assumes its uncertainty to be ±1 in the rightmost significant digit.

## Significant Figures, Standard Deviations, and Atomic Weights

Atomic weight values provide an excellent example of the use of significant figures and standard deviations. Twenty selected elements are shown in the following table. In that table, the notation (7) means that the estimated standard deviation is 7 in the rightmost digit of the measured atomic weight. Thus, hydrogen's atomic weight written as $R \pm s_R$ is 1.00794 ± 0.00007. Take some time to study this table and ponder how both significant digits and standard deviation show the real experimental uncertainties of these important values.

Observe that there is a broad relationship between the number of natural isotopes an element has and the experimental uncertainty of its atomic weight. If an element is naturally **mononuclidic** (has only one natural isotope) then its atomic weight has a small experimental uncertainty—these elements have atomic weights reported to 8 or 9 significant digits. On the other hand, if an element has two or more isotopes, then variations in their natural abundance limit our knowledge of the element's atomic weight to 5 or 6 significant digits. In the worst cases, the atomic weights of lithium and lcad can be reported to only 4 significant figures. As described in Chapter 2 and Appendix D, atomic weights with few significant figures can be understood by considering nuclear aspects of the elements in question.

In the following table "notes" refer to the footnotes to the periodic table. Briefly, g means geologically variable, m means modified by humans, hb means modified for hydrogen bomb manufacture, and LL means a long lived radionuclide.

| Atomic Number | Symbol | Element | Atomic weight (u) | Number of significant figures in atomic weight | Standard Deviation | Notes | Number of stable isotopes |
|---|---|---|---|---|---|---|---|
| | | | | | | | |
| 1 | H | Hydrogen | 1.00794(7) | 6 | ±0.00007 | g | 2 |
| 2 | He | Helium | 4.002602(2) | 7 | ±0.000002 | g | 2 |
| 3 | Li | Lithium | 6.941(2) | 4 | ±0.002 | g, m, hb | 2 |
| 9 | F | Fluorine | 18.9984032(5) | 9 | ±0.0000005 | - | 1 |
| 10 | Ne | Neon | 20.1797(6) | 6 | ±0.0006 | g, m | 3 |
| 11 | Na | Sodium | 22.989770(2) | 8 | ±0.000002 | - | 1 |
| 13 | Al | Aluminum | 26.981538(2) | 8 | ±0.000002 | - | 1 |
| 15 | P | Phosphorus | 30.973761(2) | 8 | ±0.000002 | - | 1 |
| 17 | Cl | Chlorine | 35.4527 | 6 | ±0.0009 | m | 2 |
| 21 | Sc | Scandium | 44.955910(8) | 8 | ±0.000008 | - | 1 |
| 25 | Mn | Manganese | 54.938049(9) | 8 | ±0.000009 | - | 1 |
| 26 | Fe | Iron | 55.845(2) | 5 | ±0.002 | - | 4 |
| 27 | Co | Cobalt | 58.933200(9) | 8 | ±0.000009 | - | 1 |
| 34 | Se | Selenium | 78.96(3) | 4 | ±0.03 | - | 6 |
| 36 | Kr | Krypton | 83.80(1) | 4 | ±0.01 | g, m | 6 |
| 42 | Mo | Molybdenum | 95.94(1) | 4 | ±0.01 | g | 7 |
| 79 | Au | Gold | 196.96655(2) | 8 | ±0.00002 | - | 1 |
| 82 | Pb | Lead | 207.2(1) | 4 | ±0.1 | g | 4 |
| 83 | Bi | Bismuth | 208.98038(2) | 8 | ±0.00002 | - | 1 |
| 91 | Pa | Protactinium | 231.03588(2) | 8 | ±0.00002 | LL | 0 |

## The Theory of Experimental Uncertainty

## (Often, and misleadingly, called: The Theory of Experimental Error)

So far in this appendix I have focused on the practical problem of estimating, reporting, and interpreting experimental uncertainty. In this section I will say a little about the background of the statistical method for expressing experimental uncertainty. The **statistical interpretation** of standard deviation is: For a large set of repeated measurements of a single value, about 68% of those measurements will lie within ± 1 standard deviation of the mean, about 95% will lie within ± 2 standard deviations, and about 99.7% will lie within ± 3 standard deviations.

As we've said previously, reporting experimental uncertainty using standard deviations is an application of statistics. Actually, it is an application of something like statistics—working scientists need to communicate in a practical way and they rarely worry about the details that concern the professional statistician. Most professional statisticians would regard the following treatment as naive and oversimplified. But small sample sets and fuzzy data are a daily fact of life for many chemists and engineers. Depicted below is the **normal curve** (the normal curve of uncertainty) for the value 10.0 ± 1 (I chose these values just to make the graph simple). For the obvious reason, the depiction is often called the **bell-shaped curve**.

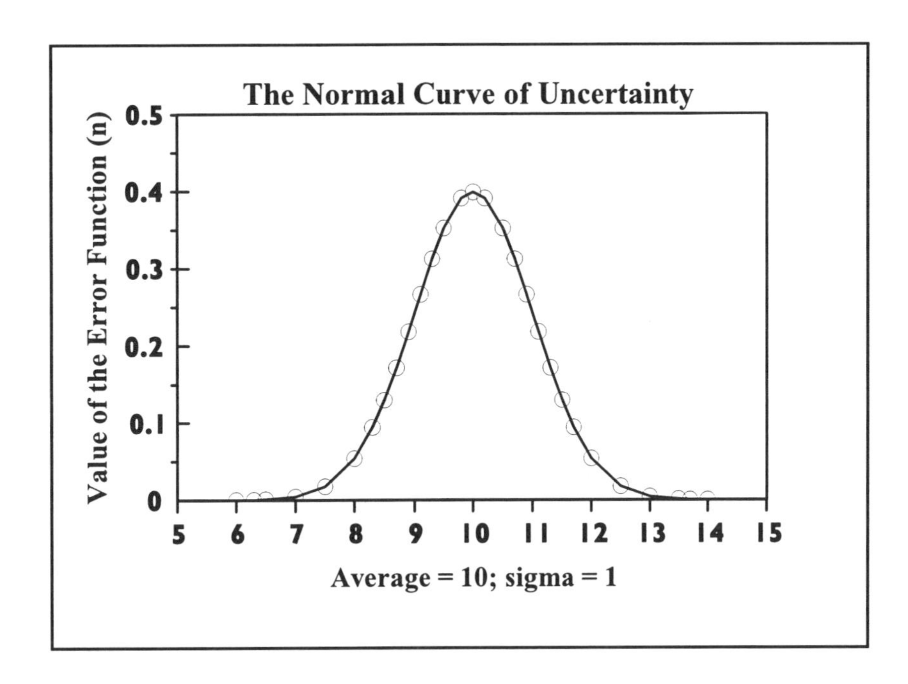


The normal curve of uncertainty is a graph of the function:

$$n(\mu, \sigma) = \frac{1}{\sigma\sqrt{2\Pi}} e^{[-(x-\mu)^2]/2\sigma^2}$$

The quantity n(μ,σ) is the value of the error function for any specific value of the mean or average (μ) and any specific value of the standard deviation (σ). In order to make the above equation concrete, we'll examine the case where the mean value is 10 and the standard deviation is 1. In general, variations in σ compared to the mean alter the shape of the curve. If σ is small compared with the mean, the curve is like a spike; if σ is large compared with the mean, the curve is a hog's back. Changing the mean simply moves the curve left or right along the horizontal axis. The mathematician Pierre Simon, Marquis de Laplace (1749-1827) was the first to explicitly state the correct form of the normal curve formula, $y = e^{-bx^2}$, in 1778.

To help you understand the normal curve a detailed table of values follows. Study this table of values in conjunction with the preceding bell-shaped curve. Looking at the curve you see that the farther you go from the mean, the lower the value of the error function. Columns 1, 2, and 3 are calculated from the formula above. Column 4 shows the range around the mean ± the standard deviation. Column 5 shows the actual area under the curve bounded by the range expressed as a percentage of the total curve area.

| Mean ± the standard deviation Mean ± σ | Actual Range on x-axis | Value of error function y-axis | Range of σ (or, ±σ) | Percentage of area in this ±σ range |
|---|---|---|---|---|
| 10 ± 4.0 σ | 6.00 - 14.00 | 0.0001 | ±4.0 σ | 100.00 |
| 10 ± 3.0 σ | 7.00 - 13.00 | 0.0044 | ±3.0 σ | 99.73 |
| 10 ± 2.5 σ | 7.50 - 12.50 | 0.0175 | ±2.5 σ | 98.76 |
| 10 ± 2.0 σ | 8.00 - 12.00 | 0.0540 | ±2.0 σ | 95.44 |
| 10 ± 1.5 σ | 8.50 - 11.50 | 0.1295 | ±1.5 σ | 86.61 |
| 10 ± 1.0 σ | 9.00 - 11.00 | 0.2420 | ±1.0 σ | 68.23 |
| 10 ± 0.5 σ | 9.50 - 10.50 | 0.3521 | ±0.5 σ | 38.27 |
| 10 ± 0 σ | 10.00 - 10.00 | 0.4000 | ±0.0 σ | 0.00 |

The interpretation of the values in the table above is shown in the diagram below for the $\pm 1\sigma$ range and the $\pm 2\sigma$ range. From the table you can see that if you took a sufficiently large number of repeated measurements, 68.23% would lie within the $\pm 1\sigma$ range, 95.44% within $\pm 2\sigma$, and 99.73% within $\pm 3\sigma$.

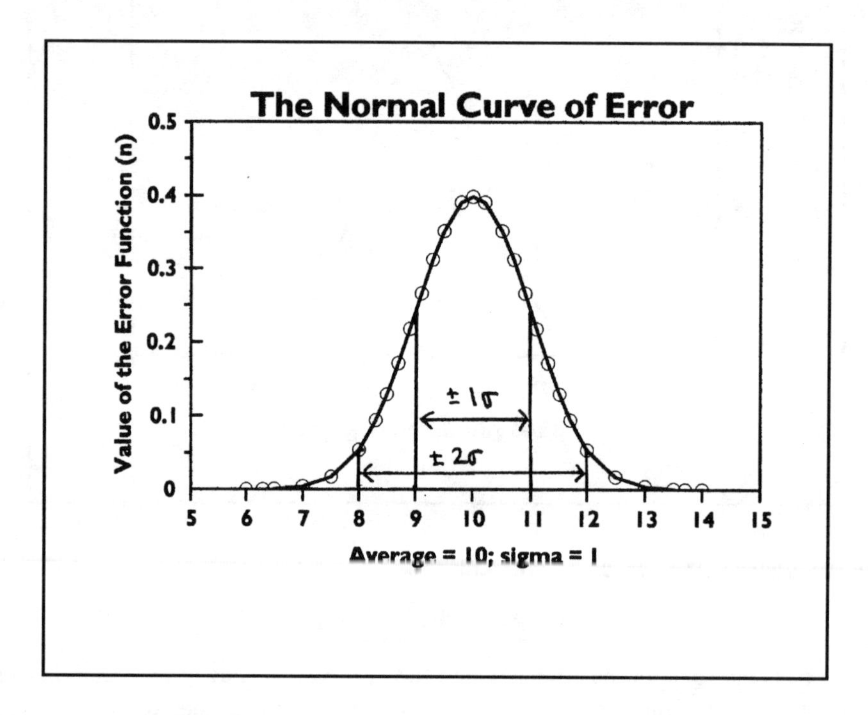


We can say that there is a 68.23% probability that the true value lies within $\pm 1\sigma$ of the mean, there is a 95.44% probability that the true value lies within $\pm 2\sigma$ of the mean, and a 99.73% probability that the true value lies within $\pm 3\sigma$ of the mean.

We never have sufficient data to be 100% certain of our conclusions. As a practical matter, 95% probability ($\pm 2\sigma$) is often taken as the level of "truth." If we are 95% confident of our data then our data are "true." There's nothing special about $\pm 2\sigma$, it's just that we humans like a cutoff point. Labeling food with an expiration date of 22 August makes sense. Labeling it as "Fresh from 12 August for 10 ±3 days" is probably a better way to label it, but the method would be incomprehensible to most people.

## Grading in Big Classes and Experimental Uncertainty

The only time your author ever gets really large sets of data is when he tests students in large classes. The scores of the students represent individual, repeated measurements of the "value" of the test. Over my years of teaching, I've found that for students, classroom scores make the best example of the curve of normal error. Every class becomes a sort of experimental proof of the normal curve. Everyone is interested in scores and grades. So here are data from one of my classes. Incidentally, students sometimes ask "Are you going to curve this course?" To which the correct answer is "No, but you are."

## Histogram of Scores of 147 Examinees

Here are the actual scores on an exam as the author received them from the optical scanning and data analysis program.

Distribution of Scores on an Exam

Average = 64.4, estimated standard deviation (s) = 17.5, sample size 147 students

```
SCORE PERCT  NO.

 25   100     1*
 24    96     5*****
 23    92     2**
 22    88     6******
 21    84    12************
 20    80     8********
 19    76    15***************
 18    72    14**************
 17    68     9*********
 16    64    12************
 15    60    12************
 14    56    10**********
 13    52    11***********
 12    48     7*******
 11    44     5*****
 10    40     5*****
  9    36     4****
  8    32     2**
  7    28     6******
  6    24     1*
______________________________________________
             NO. CASES       10        20
```

**Appendix Example 12**: Use the data from the class scores shown earlier to make an experimental test of the "curve of normal error."

Strategy: Pick a range and count the number of student scores in that range. Express that count as a percentage of the whole class. Compare the percentage obtained with the theoretical value associated with the normal curve.

The class average is 64.4 ± 17.5. Take a estimated standard deviation range about the mean value.

The mean plus one s is 64.4 + 17.5 = 82
The mean minus one s is 64.4 - 17.5 = 47

In the range 47-82 are 7 + 11+ 10 + 12 + 12 + 9 + 14 + 15 + 8 = 98 students

The percentage of students in the $\pm 1\sigma$ range is $(98 \div 147) \times 100\% = 67\%$

The value 67% agrees well with the theoretical 68.23%. We can conclude that this is a "normal" class.

According to the normal curve of error, the $\pm 1\sigma$ range encompasses 68.23% of the area under the curve. In the preceding example, the fact that 67% of the students in the class turned out to be in the $\pm 1\sigma$ range is evidence that the distribution of class scores can be approximated by the normal curve. I did an experiment on this class (I made them take a test) and the results of that experiment confirm the validity of the normal curve of uncertainty error.

If we rotate the preceding histogram counterclockwise 90° you can see that it has a clear resemblance to the graph of the normal curve of error.

Student Test Scores Represented as a Normal Curve

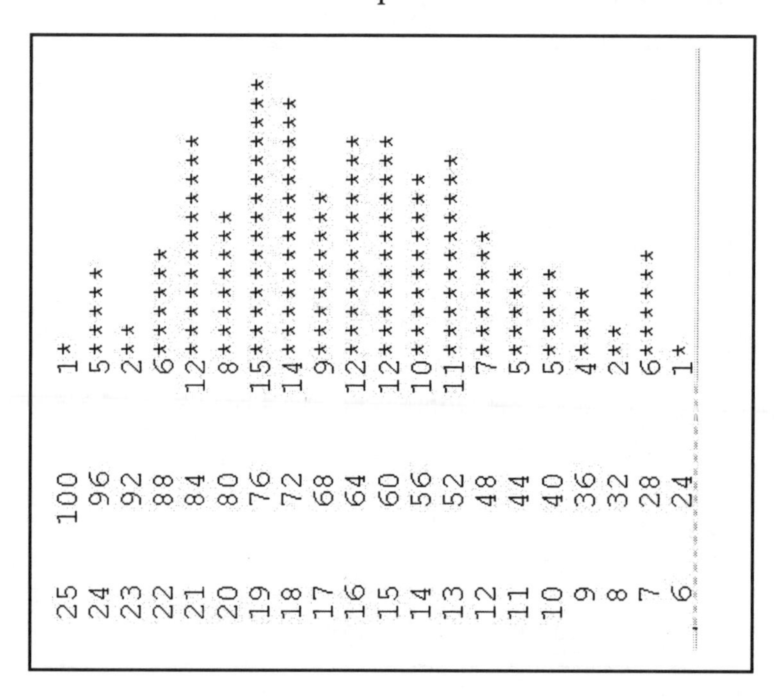


## Statistics and Quality Control in Engineering

In engineering there are many applications of the ideas contained in this appendix. Two important applications are the fields of quality control and tolerancing. **Tolerancing** refers to the need for parts made in one location to "fit" with parts made in another location. The parts must thus be manufactured to "within tolerance." Tolerances are also critical for repair work. A former student wrote to me: "In my line of work, when repairing (e.g. welding cracks) in combustion turbine blades, if we are off by more than 0.003 inches it will cause uneven temperature spread throughout the engine and cause it to shut down. This situation costs big dollars for our customers (and we usually have to absorb a lot of that cost)."

A good internet reference source for statistics for beginning engineering students is at http://qreview.cera.net/chapters.htm, where you'll find the *Quality Engineering Review Course* written by Carl D. Nocera P.E., CQE. CQE is **certified quality engineer**. He has written *QReview,* a set of two quality engineering review books that covers the basic concepts of probability, statistics, and statistical quality control.

### Statement of a Certified Quality Engineer

Date: Wed, 1 Sep 1999 08:52:52 EDT
From: "Carl Nocera" <cnocera@cera.net> Reply-To: <cnocera@cera.net>
To: jglanvil
Subject: The value of engineering statistics

Professor Glanville,

It is really refreshing to hear that you stress the importance and cover the fundamental concepts of statistics and statistical analysis in your freshman chemistry class. To beginning engineering students, statistics may seem like a mundane topic with little practical value. Let me emphasis that the contrary is true. Those students who acquire an understanding of probability and statistics will be a cut above the rest. Engineers in all disciplines, with a knowledge of statistics, are very much in demand. After all, virtually all events whether personal or at work are probabilistic. Statistical methods allow us to make sound decisions while reducing risks.

Companies make multi-million dollar decisions based on statistical analyses. Engineers with outstanding technical capabilities, leadership skills and knowledge of statistical methods are held in high esteem and are pursued by many companies. The pay is not bad either.

Of course it is possible to get an engineering degree and pursue a career without knowing much about statistics. But to really make a difference, a knowledge of statistics, along with the engineer's other skills, is invaluable.

I commend you in helping your students prepare for the future.

Carl D. Nocera P.E., CQE

(Used with permission and grateful thanks.)

Nocera is an independent engineer who teaches about quality control professionally and prepares engineers to take the quality control certification exam. On line, you can find the table of contents of his *QReview*—which is a 12 chapter publication almost entirely concerned with the practical application of statistical methods to engineering problems. Best of all, he's posted the first two chapters to his website.

At Nocera's website I learned several interesting facts, for example, that the Egyptian royal cubit was standardized in 1550 BC at about 20.63 inches ± 0.02 inches and that tolerancing principles were developed in the early 19th century by engineers at US armories in the process of mass producing firearms with interchangeable parts.

## Reference

You can easily obtain the .pdf file titled: *Guidelines for Evaluating and Expressing the Uncertainty of NIST Measurement Results*, by Barry N. Taylor and Chris E. Kuyatt (1993). This document is NIST Technical Note 1297. The .pdf file is downloadable from: http://physics.nist.gov/Document/tn1297.pdf. This 26 page document will tell you a great deal about how America's premiere standards institution handles experimental uncertainty. It is too detailed and technical for us to use here, but it is a good reference source for you.

# Appendix C—Selected Thermodynamic Properties

## Part 1: Selected CODATA Key Values for Thermodynamics

The table below shows the standard enthalpy of formation at 298.15 K and the entropy at 298.15 K, of seven entities at the standard state pressure of 100,000 Pa (1 bar). The data are presented here as an illustration of the use of standard deviations in representing experimental uncertainties in important measurements. The Committee on Data for Science and Technology (CODATA) values given in this table represent the consensus judgment of an international group of experts who encourage the use of these recommended, internally consistent values in the analysis of thermodynamic measurements and the preparation of other thermodynamic tables.

**Selected CODATA Key Values for Thermodynamics at 298 K**

| Formula (state) | Name | $\Delta H^0_f$ kJ mol$^{-1}$ | $S^0$ J K$^{-1}$ mol$^{-1}$ |
|---|---|---|---|
| C(s) | Graphite | 0 | 5.74 ± 0.10 |
| CO(g) | Carbon monoxide | -110.53 ± 0.17 | 197.660 ± 0.004 |
| $CO_2$(g) | Carbon dioxide | -393.51 ± 0.13 | 213.785 ± 0.010 |
| $H_2$(g) | Hydrogen | 0 | 130.680 ± 0.003 |
| $H_2O$(l) | Water | -285.830 ± 0.040 | 69.95 ± 0.03 |
| $H_2O$(g) | Water | -241.826 ± 0.040 | 188.835 ± 0.010 |
| O(g) | Oxygen | 249.18 ± 0.10 | 161.059 ± 0.003 |

## Part 2: Selected Standard Thermodynamic Properties of Pure Substances at 298 K

The table below shows approximate values of the standard enthalpy of formation ($\Delta H^0_f$) at 298.15 K, the standard Gibbs free energy of formation ($\Delta G^0_f$) at 298.15 K, and the standard entropy (S°) at 298.15 K, of some pure substances at a pressure of approximately 1 bar. Data taken from various sources. Notation in the table: (g) = gas, (l) = liquid, and (s) = solid.

**Selected Standard Thermodynamic Properties of Pure Substances at 298 K**

| Formula (state) | Name | $\Delta H^0_f$ kJ mol$^{-1}$ | $\Delta G^0_f$ kJ mol$^{-1}$ | $S^0$ J K$^{-1}$ mol$^{-1}$ |
|---|---|---|---|---|
| C(s) | Graphite | 0.0 | 0.0 | 5.7 |
| C(s) | Diamond | 1.9 | 2.9 | 2.4 |
| C(g) | Carbon | 716.7 | 671.3 | 158.1 |
| CO(g) | Carbon monoxide | -110.5 | -137.2 | 197.7 |
| $CO_2$(g) | Carbon dioxide | -393.5 | -394.4 | 213.8 |
| $CH_4$(g) | Methane | -74.8 | -50.8 | 186.2 |
| $CH_2O$(g) | Formaldehyde | -117.0 | -110.0 | 218.7 |
| $COCl_2$(g) | Phosgene | -218.8 | -204.6 | 283.5 |
| $CH_3Cl$(g) | Chloromethane | -80.8 | -57.4 | 234.6 |
| $CHCl_3$(g) | Trichloromethane | -103.1 | -70.3 | 295.7 |
| $CCl_4$(l) | Tetrachloromethane | -135.4 | -65.3 | 216.4 |

| Selected Standard Thermodynamic Properties of Pure Substances at 298 K | | | | |
|---|---|---|---|---|
| **Formula (state)** | **Name** | **$\Delta H^0_f$ kJ mol$^{-1}$** | **$\Delta G^0_f$ kJ mol$^{-1}$** | **$S^0$ J K$^{-1}$ mol$^{-1}$** |
| $CH_3OH$(g) | Methanol | -200.7 | -162.0 | 239.7 |
| $CH_3OH$(l) | Methanol | -238.7 | -166.4 | 126.8 |
| $CH_3NH_2$(g) | Methylamine | -23.0 | 32.3 | 242.6 |
| $CH_3CHO$(g) | Acetaldehyde | -166.1 | -133.4 | 246.4 |
| $CH_3CHO$(l) | Acetaldehyde | -191.8 | -128.3 | 160.4 |
| $C_2H_2$(g) | Ethyne | 226.7 | 209.2 | 200.9 |
| $C_2H_4$(g) | Ethene | 52.3 | 68.1 | 219.5 |
| $C_2H_6$(g) | Ethane | -84.7 | -32.9 | 229.5 |
| $CH_3CH_2OH$(g) | Ethanol | -234.4 | -167.9 | 282.6 |
| $CH_3CH_2OH$(l) | Ethanol | -277.7 | -174.9 | 160.7 |
| $CH_3OCH_3$(g) | Dimethyl ether | -184.5 | -112.6 | 266.4 |
| $(CH_3)_2CO$(g) | Acetone | -216.6 | -153.1 | 294.9 |
| $(CH_3)_2CO$(l) | Acetone | -247.6 | -155.7 | 200.4 |
| $CH_3COOH$(g) | Acetic acid | -432.3 | -374.0 | 282.5 |
| $CH_3COOH$(l) | Acetic acid | -484.1 | -389.9 | 159.8 |
| $C_3H_8$(g) | Propane | -104.1 | -23.5 | 270.1 |
| $C_3H_6$(g,) | Propene | 20.2 | 62.7 | 266.9 |
| $C_4H_{10}$(g) | Butane | -126.0 | -17.1 | 310.0 |
| $C_4H_{10}$(g) | Methylpropane | -134.6 | -20.9 | 294.6 |
| $C_5H_{12}$(l) | n-Pentane | -146.4 | -8.5 | 349.4 |
| $C_6H_{14}$ (l) | n-Hexane | -167.0 | 0.1 | 388.7 |
| $C_7H_{16}$ (l) | n-Heptane | -187.6 | 8.1 | 427.6 |
| $C_8H_{18}$(l) | n-Octane | -208.2 | 16.4 | 467.2 |
| $C_6H_6$(g) | Benzene | 82.9 | 129.7 | 269.2 |
| $C_6H_6$(l) | Benzene | 49.0 | 124.4 | 173.3 |
| $C_6H_{12}$(g) | Cyclohexane | -123.1 | 31.8 | 298.2 |
| $C_6H_{12}$(l) | Cyclohexane | -156.2 | 26.7 | 204.3 |
| $C_6H_5OH$(s) | Phenol | -165.0 | -50.4 | 144.0 |
| $C_6H_5COOH$(s) | Benzoic acid | -385.1 | -245.3 | 167.6 |
| $C_6H_5NH_2$(g) | Aniline | 86.9 | 166.7 | 319.2 |
| $C_6H_5NH_2$(l) | Aniline | 31.6 | 149.1 | 191.3 |
| $C_6H_{12}0_6$(s) | Glucose | -1273.3 | -910.4 | 212.1 |
| $C_{10}H_8$(g) | Naphthalene | 149.0 | 223.6 | 335.6 |
| $C_{10}H_8$(s) | Naphthalene | 75.3 | 201.0 | 166.9 |
| $Cl_2$(g) | Chlorine | 0.0 | 0.0 | 223.1 |
| Cl(g) | Chlorine | 121.7 | 105.7 | 165.2 |
| $ClO_2$(g) | Chlorine dioxide | 102.5 | 120.5 | 256.8 |

| Selected Standard Thermodynamic Properties of Pure Substances at 298 K | | | | |
|---|---|---|---|---|
| **Formula (state)** | **Name** | **$\Delta H^0_f$ kJ mol$^{-1}$** | **$\Delta G^0_f$ kJ mol$^{-1}$** | **$S^0$ J K$^{-1}$ mol$^{-1}$** |
| $H_2$(g) | Hydrogen | 0.0 | 0.0 | 130.7 |
| H(g) | Hydrogen | 218.0 | 203.3 | 114.7 |
| $H_2O$(g) | Water | -241.8 | -228.6 | 188.8 |
| $H_2O$(l) | Water | -285.8 | -237.1 | 69.9 |
| $H_2O_2$(g) | Hydrogen peroxide | -136.3 | -105.6 | 232.7 |
| $H_2O_2$(l) | Hydrogen peroxide | -187.8 | -1204.0 | 109.6 |
| $H_2S$(g) | Hydrogen sulfide | -20.6 | -33.6 | 205.7 |
| $H_2SO_4$(l) | Sulfuric acid | -812.0 | -690.0 | 156.9 |
| HCl(g) | Hydrogen chloride | -92.3 | -95.3 | 186.9 |
| $N_2$(g) | Nitrogen | 0.0 | 0.0 | 191.6 |
| N(g) | Nitrogen | 472.7 | 455.6 | 153.3 |
| $NH_3$(g) | Ammonia | -46.1 | -16.4 | 192.4 |
| NO(g) | Nitrogen monoxide | 90.3 | 86.5 | 210.8 |
| $NO_2$(g) | Nitrogen dioxide | 33.2 | 51.3 | 240.1 |
| $N_2O$(g) | Nitrous oxide | 82.0 | 104.2 | 219.8 |
| $N_2O_4$(g) | Dinitrogen tetroxide | 9.2 | 97.9 | 304.3 |
| $N_2O_4$(l) | Dinitrogen tetroxide | -19.5 | 97.5 | 209.2 |
| $O_2$(g) | Oxygen | 0.0 | 0.0 | 205.1 |
| $O_3$(g) | Ozone | 142.7 | 163.2 | 238.9 |
| O(g) | Oxygen | 249.2 | 231.7 | 161.1 |
| P(s) | Phosphorus (white) | 0.0 | 0.0 | 41.1 |
| P(g) | Phosphorus | 314.6 | 278.3 | 163.2 |
| $PH_3$(g) | Phosphine | 5.4 | 13.4 | 210.2 |
| $PCl_3$(g) | Phosphorus trichloride | -287.0 | -267.8 | 311.8 |
| $PCl_5$(g) | Phosphorus pentachloride | -374.9 | -305.0 | 364.6 |
| S(s) | Sulfur (rhombic) | 0.0 | 0.0 | 31.8 |
| S(g) | Sulfur | 278.8 | 238.3 | 167.8 |
| $SO_2$(g) | Sulfur dioxide | -296.8 | -300.2 | 248.2 |
| $SO_3$(g) | Sulfur trioxide | -395.7 | -371.1 | 256.8 |
| $SO_3$(l) | Sulfur trioxide | -441.0 | -373.8 | 113.8 |
| $SO_2Cl_2$(g) | Sulfuryl chloride | -364.0 | -320.0 | 311.9 |

# Appendix D—Table of Selected Nuclides/Isotopes

Listed in this table are:

1. Two hundred and sixty-four **stable nuclides**.

2. Twenty-five naturally occurring **pseudostable nuclides** with such long half lives that they are treated as being stable: $^{82}$Se, $^{40}$K, $^{50}$V, $^{87}$Rb, $^{92}$Nb, $^{113}$Cd, $^{115}$In, $^{128}$Te, $^{130}$Te, $^{138}$La, $^{144}$Nd, $^{146}$Sm, $^{147}$Sm, $^{148}$Sm, $^{152}$Gd, $^{176}$Lu, $^{174}$Hf, $^{180}$Ta, $^{187}$Re, $^{186}$Os, $^{190}$Pt, $^{232}$Th, $^{235}$U, $^{238}$U, and $^{244}$Pu.

3. Ten naturally occurring **radionuclides**, continuously created in nature, that exist in significant amounts: $^{14}$C, $^{3}$H, $^{210}$Po, $^{222}$Rn, $^{223}$Fr, $^{226}$Ra, $^{227}$Ac, $^{231}$Pa, $^{233}$U, and $^{234}$U. The first two of these are produced in the atmosphere, the others are obtained from uranium and thorium ores as daughter products of **natural decay chains**.

4. The **longest lived nuclide** of three elements with lower atomic numbers than uranium ($^{98}$Tc, $^{61}$Pm, $^{210}$At). These three elements were each first prepared by laboratory synthesis and subsequently discovered to exist in trace amounts in the earth's crust.

5. Two radionuclides first discovered through laboratory synthesis and subsequently found to be continuously produced in thorium and uranium ores in trace amounts: $^{237}$Np, $^{239}$Pu.

Not included in this list are isotopes of the 21 elements that have been synthesized but not demonstrated to exist in nature. Undoubtedly, a few atoms of elements just beyond plutonium, such as Am and Cm, must exist in samples of uranium and thorium ores—having been formed there by natural nuclear processes. However, given what must be their very low concentrations, the chance of finding experimental proof of their natural existence is small.

## Explanation of Table Columns

Column 1, name of element or the name of a hydrogen isotope.
Column 2, nuclide symbol.
Column 3, number of protons in the nuclide.
Column 4, number of neutrons in the nuclide.
Column 5, the sum of column 3 and column 4 = the nucleon number.
Column 6, the ratio of neutrons to protons in the nuclide, (n/p).
Column 7, the mass of the nuclide in atomic mass units (u).
Column 8, the percent of the element's natural atoms of that type (natural abundance).
Column 9, stability. If nuclide is unstable its half-life is stated.
Column 10, p and n counts; either even number (E) or odd number (O)

Note that many synthetic radionuclides exist; they are not listed here. It is impossible to state an exact number of synthetic radionuclides. The shorter the half-life you are willing to accept the more there are. There are approximately 3000 radionuclides with half lives greater than 0.1 second.

A summary follows the table.

| 1 | 2 | 3 | 4 | 5 | 6 | 7 | 8 | 9 | 10 |
|---|---|---|---|---|---|---|---|---|---|
| Element | Nuclide Symbol | p | n | p+n | n/p | Nuclidic Mass (u) | Nuclidic Abundance number % | Half life (time) | p n even (E) or odd (O) |
| Hydrogen | (two stable isotopes, one continuously formed radioisotope) | | | | | | | | |
| Protium | H-1 | 1 | 0 | 1 | --- | 1.007825035 | 99.985 | Stable | OE |
| Deuterium | H-2 | 1 | 1 | 2 | 1.00 | 2.014101779 | 0.015 | Stable | OO |
| Tritium | H-3 | 1 | 2 | 3 | 2.00 | 3.01604927 | 0.000 | 12.33 y | OE |
| Helium | (two stable isotopes) | | | | | | | | |
| | He-3 | 2 | 1 | 3 | 0.50 | 3.01602931 | 0.000137 | Stable | EO |
| | He-4 | 2 | 2 | 4 | 1.00 | 4.00260324 | 99.999863 | Stable | EE |
| Lithium | (two stable isotopes) | | | | | | | | |
| | Li-6 | 3 | 3 | 6 | 1.00 | 6.0151214 | 7.5 | Stable | OO |
| | Li-7 | 3 | 4 | 7 | 1.33 | 7.0160030 | 92.5 | Stable | OE |
| Beryllium | (one stable isotope) | | | | | | | | |
| | Be-9 | 4 | 5 | 9 | 1.25 | 9.0121822 | 100 | Stable | OE |
| Boron | (two stable isotopes) | | | | | | | | |
| | B-10 | 5 | 5 | 10 | 1.00 | 10.0129369 | 19.9 | Stable | OO |
| | B-11 | 5 | 6 | 11 | 1.20 | 11.003054 | 80.1 | Stable | OE |
| Carbon | (two stable isotopes, one continuously formed radioisotope) | | | | | | | | |
| | C-12 | 6 | 6 | 12 | 1.00 | 12 (exact) | 98.90 | Stable | EE |
| | C-13 | 6 | 7 | 13 | 1.17 | 13.003354826 | 1.10 | Stable | EO |
| | C-14 | 6 | 8 | 14 | 1.33 | 14.003241982 | 0.0 | 5715 y | EE |
| Nitrogen | (two stable isotopes) | | | | | | | | |
| | N-14 | 7 | 7 | 14 | 1.00 | 14.003074002 | 99.634 | Stable | OO |
| | N-15 | 7 | 8 | 15 | 1.14 | 15.00010897 | 0.366 | Stable | OE |
| Oxygen | (three stable isotopes) | | | | | | | | |
| | O-16 | 8 | 8 | 16 | 1.00 | 15.99491463 | 99.762 | Stable | EE |
| | O-17 | 8 | 9 | 17 | 1.13 | 16.9991312 | 0.038 | Stable | EO |
| | O-18 | 8 | 10 | 18 | 1.25 | 17.9991603 | 0.200 | Stable | EE |
| Fluorine | (one stable isotope) | | | | | | | | |
| | F-19 | 9 | 10 | 19 | 1.11 | 18.99840322 | 100 | Stable | OE |
| Neon | (three stable isotopes) | | | | | | | | |
| | Ne-20 | 10 | 10 | 20 | 1.00 | 19.9924356 | 90.48 | Stable | EE |
| | Ne-21 | 10 | 11 | 21 | 1.10 | 20.9938428 | 0.27 | Stable | EO |
| | Ne-22 | 10 | 12 | 22 | 1.20 | 21.9913831 | 9.25 | Stable | EE |
| Sodium | (one stable isotope) | | | | | | | | |
| | Na-23 | 11 | 12 | 23 | 1.18 | 22.9897677 | 100 | Stable | OE |
| Magnesium | (three stable isotopes) | | | | | | | | |
| | Mg-24 | 12 | 12 | 24 | 1.00 | 23.9850423 | 78.99 | Stable | EE |
| | Mg-25 | 12 | 13 | 25 | 1.08 | 24.9858374 | 10.00 | Stable | EO |
| | Mg-26 | 12 | 14 | 26 | 1.14 | 25.982594 | 11.01 | Stable | EE |
| Aluminum | (one stable isotope) | | | | | | | | |
| | Al-27 | 13 | 14 | 27 | 1.08 | 26.9815386 | 100 | Stable | OE |
| Silicon | (three stable isotopes) | | | | | | | | |
| | Si-28 | 14 | 14 | 28 | 1.00 | 27.9769271 | 92.23 | Stable | EE |
| | Si-29 | 14 | 15 | 29 | 1.07 | 28.9764949 | 4.67 | Stable | EO |
| | Si-30 | 14 | 16 | 30 | 1.14 | 29.9737707 | 3.10 | Stable | EE |
| Phosphorus | (one stable isotope) | | | | | | | | |
| | P-31 | 15 | 16 | 31 | 1.07 | 30.9737620 | 100 | Stable | OE |
| Sulfur | (four stable isotopes) | | | | | | | | |
| | S-32 | 16 | 16 | 32 | 1.00 | 31.97207070 | 95.02 | Stable | EE |
| | S-33 | 16 | 17 | 33 | 1.06 | 32.97145843 | 0.75 | Stable | EO |
| | S-34 | 16 | 18 | 34 | 1.13 | 33.9678665 | 4.21 | Stable | EE |
| | S-35 | 16 | 19 | 35 | 1.19 | 35.96708062 | 0.02 | Stable | EO |

| 1 | 2 | 3 | 4 | 5 | 6 | 7 | 8 | 9 | 10 |
|---|---|---|---|---|---|---|---|---|---|
| Element | Nuclide Symbol | p | n | p+n | n/p | Nuclidic Mass (u) | Nuclidic Abundance number % | Half life (time) | p n even (E) or odd (O) |
| Chlorine | (two stable isotopes) | | | | | | | | |
| | Cl-35 | 17 | 18 | 35 | 1.06 | 34.968852721 | 75.77 | Stable | OE |
| | Cl-37 | 17 | 20 | 37 | 1.18 | 36.96590262 | 24.23 | Stable | OE |
| Argon | (three stable isotopes) | | | | | | | | |
| | Ar-36 | 18 | 18 | 36 | 1.00 | 35.96754552 | 0.337 | Stable | EE |
| | Ar-38 | 18 | 20 | 38 | 1.11 | 37.9627325 | 0.063 | Stable | EE |
| | Ar-40 | 18 | 22 | 40 | 1.22 | 39.9623837 | 99.600 | Stable | EE |
| Potassium | (two stable isotopes and one pseudostable isotope) | | | | | | | | |
| | K-39 | 19 | 20 | 39 | 1.05 | 38.9637074 | 93.2581 | Stable | OE |
| | K-40 | 19 | 21 | 40 | 1.11 | 39.9639992 | 0.0117 | $1.28 \times 10^9$ y | OE |
| | K-41 | 19 | 22 | 41 | 1.16 | 40.9618254 | 6.7302 | Stable | OE |
| Calcium | (six stable isotopes) | | | | | | | | |
| | Ca-40 | 20 | 20 | 40 | 1.00 | 39.9625906 | 96.941 | Stable | EE |
| | Ca-42 | 20 | 22 | 42 | 1.10 | 41.9586176 | 0.647 | Stable | EE |
| | Ca-43 | 20 | 23 | 43 | 1.15 | 42.9587662 | 0.135 | Stable | EO |
| | Ca-44 | 20 | 24 | 44 | 1.20 | 43.9554806 | 2.086 | Stable | EE |
| | Ca-46 | 20 | 26 | 46 | 1.30 | 45.953689 | 0.004 | Stable | EO |
| | Ca-48 | 20 | 28 | 48 | 1.40 | 47.952533 | 0.187 | Stable | EE |
| Scandium | (one stable isotope) | | | | | | | | |
| | Sc-45 | 21 | 24 | 45 | 1.14 | 44.9559100 | 100 | Stable | OE |
| Titanium | (five stable isotopes) | | | | | | | | |
| | Ti-46 | 22 | 24 | 46 | 1.09 | 45.9526294 | 8.0 | Stable | EE |
| | Ti-47 | 22 | 25 | 47 | 1.14 | 46.9517640 | 7.3 | Stable | EO |
| | Ti-48 | 22 | 26 | 48 | 1.18 | 47.9479473 | 73.8 | Stable | EE |
| | Ti-49 | 22 | 27 | 49 | 1.23 | 48.9478711 | 5.5 | Stable | EO |
| | Ti-50 | 22 | 28 | 50 | 1.27 | 49.9447921 | 5.4 | Stable | EE |
| Vanadium | (one stable and one pseudostable isotopes) | | | | | | | | |
| | V-50 | 23 | 27 | 50 | 1.17 | 49.9471609 | 0.250 | $>3.9 \times 10^{17}$ y | OE |
| | V-51 | 23 | 28 | 51 | 1.22 | 50.9439617 | 99.75 | Stable | OE |
| Chromium | (four stable isotopes) | | | | | | | | |
| | Cr-50 | 24 | 26 | 50 | 1.08 | 49.9460464 | 4.345 | Stable | EE |
| | Cr-52 | 24 | 28 | 52 | 1.17 | 51.9405098 | 83.789 | Stable | EE |
| | Cr-53 | 24 | 29 | 53 | 1.21 | 52.9406513 | 9.501 | Stable | EO |
| | Cr-54 | 24 | 30 | 54 | 1.25 | 53.9388825 | 2.365 | Stable | EE |
| Manganese | (one stable isotope) | | | | | | | | |
| | Mn-55 | 25 | 30 | 55 | 1.20 | 54.9380471 | 100 | Stable | OE |
| Iron | (four stable isotopes) | | | | | | | | |
| | Fe-54 | 26 | 28 | 54 | 1.08 | 53.9396127 | 5.8 | Stable | EE |
| | Fe-56 | 26 | 30 | 56 | 1.15 | 55.9349393 | 91.72 | Stable | EE |
| | Fe-57 | 26 | 31 | 57 | 1.19 | 56.9353958 | 2.2 | Stable | EO |
| | Fe-58 | 26 | 32 | 58 | 1.23 | 57.9332773 | 0.28 | Stable | EE |
| Cobalt | (one stable isotope) | | | | | | | | |
| | Co-59 | 27 | 32 | 59 | 1.19 | 58.9331976 | 100 | Stable | OE |
| Nickel | (five stable isotopes) | | | | | | | | |
| | Ni-58 | 28 | 30 | 58 | 1.07 | 57.9353462 | 68.077 | Stable | EE |
| | Ni-60 | 28 | 32 | 60 | 1.14 | 59.9307884 | 26.223 | Stable | EE |
| | Ni-61 | 28 | 33 | 61 | 1.18 | 60.9310579 | 1.140 | Stable | EO |
| | Ni-62 | 28 | 34 | 62 | 1.21 | 61.9283461 | 3.634 | Stable | EE |
| | Ni-64 | 28 | 36 | 64 | 1.29 | 63.9279679 | 0.926 | Stable | EE |
| Copper | (two stable isotopes) | | | | | | | | |
| | Cu-63 | 29 | 34 | 63 | 1.17 | 62.9295989 | 69.17 | Stable | OE |
| | Cu-65 | 29 | 36 | 65 | 1.24 | 64.9277929 | 30.83 | Stable | OE |

| 1 | 2 | 3 | 4 | 5 | 6 | 7 | 8 | 9 | 10 |
|---|---|---|---|---|---|---|---|---|---|
| Element | Nuclide Symbol | p | n | p+n | n/p | Nuclidic Mass (u) | Nuclidic Abundance number % | Half life (time) | p n even (E) or odd (O) |
| Zinc | (five stable isotopes) | | | | | | | | |
| | Zn-64 | 30 | 34 | 64 | 1.13 | 63.9291448 | 48.6 | Stable | EE |
| | Zn-66 | 30 | 36 | 66 | 1.20 | 65.9260347 | 27.9 | Stable | EO |
| | Zn-67 | 30 | 37 | 67 | 1.23 | 66.9271291 | 4.1 | Stable | EO |
| | Zn-68 | 30 | 38 | 68 | 1.27 | 67.9248459 | 18.8 | Stable | EE |
| | Zn-70 | 30 | 40 | 70 | 1.33 | 69.925325 | 0.6 | Stable | EE |
| Gallium | (two stable isotopes) | | | | | | | | |
| | Ga-69 | 31 | 38 | 69 | 1.23 | 68.925580 | 60.108 | Stable | OE |
| | Ga-71 | 31 | 40 | 71 | 1.29 | 70.9247005 | 39.892 | Stable | OE |
| Germanium | (five stable isotopes) | | | | | | | | |
| | Ge-70 | 32 | 38 | 70 | 1.19 | 69.9242497 | 21.23 | Stable | EE |
| | Ge-72 | 32 | 40 | 72 | 1.25 | 71.9220789 | 27.66 | Stable | EE |
| | Ge-73 | 32 | 41 | 73 | 1.28 | 72.9234626 | 7.73 | Stable | EO |
| | Ge-74 | 32 | 42 | 74 | 1.31 | 73.9211774 | 35.94 | Stable | EE |
| | Ge-76 | 32 | 44 | 76 | 1.38 | 75.9214016 | 7.44 | Stable | EE |
| Arsenic | (one stable isotope) | | | | | | | | |
| | As-75 | 33 | 42 | 75 | 1.27 | 74.9215942 | 100 | Stable | OE |
| Selenium | (five stable and one pseudostable isotopes) | | | | | | | | |
| | Se-74 | 34 | 40 | 74 | 1.18 | 73.9224746 | 0.89 | Stable | EE |
| | Se-76 | 34 | 42 | 76 | 1.23 | 75.9192120 | 9.36 | Stable | EE |
| | Se-77 | 34 | 43 | 77 | 1.26 | 76.9199125 | 7.63 | Stable | EO |
| | Se-78 | 34 | 44 | 78 | 1.29 | 77.9173076 | 23.78 | Stable | EE |
| | Se-80 | 34 | 46 | 80 | 1.35 | 79.9165196 | 49.61 | Stable | EO |
| | Se-82 | 34 | 48 | 82 | 1.41 | 81.9166978 | 8.73 | $1.4 \times 10^{20}$ y | EE |
| Bromine | (two stable isotopes) | | | | | | | | |
| | Br-79 | 35 | 44 | 79 | 1.26 | 78.9183361 | 50.69 | Stable | OE |
| | Br-81 | 35 | 46 | 81 | 1.31 | 80.916289 | 49.31 | Stable | OE |
| Krypton | (six stable isotopes) | | | | | | | | |
| | Kr-78 | 36 | 42 | 78 | 1.17 | 77.920396 | 0.36 | Stable | EE |
| | Kr-80 | 36 | 44 | 80 | 1.22 | 79.916380 | 2.25 | Stable | EE |
| | Kr-82 | 36 | 46 | 82 | 1.28 | 81.913482 | 11.6 | Stable | EO |
| | Kr-83 | 36 | 47 | 83 | 1.31 | 82.914135 | 11.5 | Stable | EO |
| | Kr-84 | 36 | 48 | 84 | 1.33 | 83.911507 | 57.0 | Stable | EE |
| | Kr-86 | 36 | 50 | 86 | 1.39 | 85.910616 | 17.3 | Stable | EE |
| Rubidium | (one stable isotope and one pseudostable isotope) | | | | | | | | |
| | Rb-85 | 37 | 48 | 85 | 1.30 | 84.911794 | 72.165 | Stable | OE |
| | Rb-87 | 37 | 50 | 87 | 1.35 | 86.909187 | 27.835 | $4.8 \times 10^{10}$ y | OE |
| Strontium | (four stable isotopes) | | | | | | | | |
| | Sr-84 | 38 | 46 | 84 | 1.21 | 83.913430 | 0.56 | Stable | EO |
| | Sr-86 | 38 | 48 | 86 | 1.26 | 85.9092672 | 9.86 | Stable | EE |
| | Sr-87 | 38 | 49 | 87 | 1.29 | 86.9088841 | 7.00 | Stable | EO |
| | Sr-88 | 38 | 50 | 88 | 1.32 | 87.9056188 | 82.58 | Stable | EE |
| Yttrium | (one stable isotope) | | | | | | | | |
| | Y-89 | 39 | 50 | 89 | 1.28 | 88.905849 | 100 | Stable | OE |
| Zirconium | (five stable isotopes) | | | | | | | | |
| | Zr-90 | 40 | 50 | 90 | 1.25 | 89.9047026 | 51.45 | Stable | EE |
| | Zr-91 | 40 | 51 | 91 | 1.28 | 90.9056439 | 11.22 | Stable | EO |
| | Zr-92 | 40 | 52 | 92 | 1.30 | 91.9050386 | 17.15 | Stable | EE |
| | Zr-94 | 40 | 54 | 94 | 1.35 | 93.9063148 | 17.38 | Stable | EE |
| | Zr-96 | 40 | 56 | 96 | 1.40 | 95.908275 | 2.80 | Stable | EE |
| Niobium | (one stable isotope and one pseudostable isotope) | | | | | | | | |
| | Nb-92 | 41 | 51 | 92 | 1.24 | ≈91.90 | 0.000 | $3.2 \times 10^{7}$ y | OO |
| | Nb-93 | 41 | 52 | 93 | 1.27 | 92.90637772 | 100 | Stable | OE |

| 1 | 2 | 3 | 4 | 5 | 6 | 7 | 8 | 9 | 10 |
|---|---|---|---|---|---|---|---|---|---|
| Element | Nuclide Symbol | p | n | p+n | n/p | Nuclidic Mass (u) | Nuclidic Abundance number % | Half life (time) | p n even (E) or odd (O) |
| Molybdenum | (seven stable isotopes) | | | | | | | | |
| | Mo-92 | 42 | 50 | 92 | 1.19 | 91.906809 | 14.84 | Stable | EE |
| | Mo-94 | 42 | 52 | 94 | 1.24 | 93.9050853 | 9.25 | Stable | EE |
| | Mo-95 | 42 | 53 | 95 | 1.26 | 94.9058411 | 15.92 | Stable | EO |
| | Mo-96 | 42 | 54 | 96 | 1.29 | 95.9046785 | 16.68 | Stable | EE |
| | Mo-97 | 42 | 55 | 97 | 1.31 | 96.9060205 | 9.55 | Stable | EO |
| | Mo-98 | 42 | 56 | 98 | 1.33 | 97.9054073 | 24.13 | Stable | EE |
| | Mo-100 | 42 | 58 | 100 | 1.38 | 99.907477 | 9.63 | Stable | EE |
| Technetium | (no stable isotopes, element made synthetically) | | | | | | | | |
| | (trace amounts of this element arise naturally via spontaneous fission) | | | | | | | | |
| | Tc-98 | 43 | 55 | 98 | 1.28 | 97.907215 | | $4.2 \times 10^6$ y | OO |
| Ruthenium | (seven stable isotopes) | | | | | | | | |
| | Ru-96 | 44 | 52 | 96 | 1.18 | 95.907599 | 5.52 | Stable | EE |
| | Ru-98 | 44 | 54 | 98 | 1.23 | 97.905287 | 1.88 | Stable | EE |
| | Ru-99 | 44 | 55 | 99 | 1.25 | 98.9059389 | 12.7 | Stable | EO |
| | Ru-100 | 44 | 56 | 100 | 1.27 | 99.9042192 | 12.6 | Stable | EE |
| | Ru-101 | 44 | 57 | 101 | 1.30 | 100.9055819 | 17.0 | Stable | EO |
| | Ru-102 | 44 | 58 | 102 | 1.32 | 101.9043485 | 31.6 | Stable | EE |
| | Ru-104 | 44 | 60 | 104 | 1.36 | 103.905424 | 18.7 | Stable | EE |
| Rhodium | (one stable isotope) | | | | | | | | |
| | Rh-103 | 45 | 58 | 103 | 1.29 | 102.905500 | 100 | Stable | OE |
| Palladium | (six stable isotopes) | | | | | | | | |
| | Pd-102 | 46 | 56 | 102 | 1.22 | 101.905634 | 1.02 | Stable | EE |
| | Pd-104 | 46 | 58 | 104 | 1.26 | 103.904029 | 11.14 | Stable | EE |
| | Pd-105 | 46 | 59 | 105 | 1.28 | 104.905079 | 22.33 | Stable | EO |
| | Pd-106 | 46 | 60 | 106 | 1.30 | 105.903478 | 27.33 | Stable | EE |
| | Pd-108 | 46 | 62 | 108 | 1.35 | 107.903895 | 26.46 | Stable | EE |
| | Pd-110 | 46 | 64 | 110 | 1.39 | 109.905167 | 11.72 | Stable | EE |
| Silver | (two stable isotopes) | | | | | | | | |
| | Ag-107 | 47 | 60 | 107 | 1.28 | 106.905092 | 51.839 | Stable | OE |
| | Ag-109 | 47 | 62 | 109 | 1.32 | 108.904756 | 48.161 | Stable | OE |
| Cadmium | (seven stable isotopes and one pseudostable isotope) | | | | | | | | |
| | Cd-106 | 48 | 58 | 106 | 1.21 | 105.906461 | 1.25 | Stable | EE |
| | Cd-108 | 48 | 60 | 108 | 1.25 | 107.904176 | 0.89 | Stable | EE |
| | Cd-110 | 48 | 62 | 110 | 1.29 | 109.903005 | 12.49 | Stable | EE |
| | Cd-111 | 48 | 63 | 111 | 1.31 | 110.904182 | 12.80 | Stable | EO |
| | Cd-112 | 48 | 64 | 112 | 1.33 | 111.902757 | 24.13 | Stable | EE |
| | Cd-113 | 48 | 65 | 113 | 1.35 | 112.904400 | 12.22 | $9 \times 10^{15}$ y | EO |
| | Cd-114 | 48 | 66 | 114 | 1.38 | 113.903357 | 28.73 | Stable | EE |
| | Cd-116 | 48 | 68 | 116 | 1.42 | 115.904755 | 7.49 | Stable | EE |
| Indium | (one stable and one pseudostable isotope) | | | | | | | | |
| | In-113 | 49 | 64 | 113 | 1.31 | 112.904061 | 4.3 | Stable | OE |
| | In-115 | 49 | 66 | 115 | 1.35 | 114.903882 | 95.7 | $5.1 \times 10^{14}$ y | OE |
| Tin | (ten stable isotopes) | | | | | | | | |
| | Sn-112 | 50 | 62 | 112 | 1.24 | 111.904826 | 0.97 | Stable | EE |
| | Sn-114 | 50 | 64 | 114 | 1.28 | 113.902784 | 0.65 | Stable | EE |
| | Sn-115 | 50 | 65 | 115 | 1.30 | 114.903348 | 0.34 | Stable | EO |
| | Sn-116 | 50 | 66 | 116 | 1.32 | 115.901747 | 14.53 | Stable | EE |
| | Sn-117 | 50 | 67 | 117 | 1.34 | 116.902956 | 7.68 | Stable | EO |
| | Sn-118 | 50 | 68 | 118 | 1.36 | 117.901609 | 24.23 | Stable | EE |
| | Sn-119 | 50 | 69 | 119 | 1.38 | 118.903311 | 8.59 | Stable | EO |
| | Sn-120 | 50 | 70 | 120 | 1.40 | 119.9021991 | 32.59 | Stable | EE |
| | Sn-122 | 50 | 72 | 122 | 1.44 | 121.9034404 | 4.63 | Stable | EE |
| | Sn-124 | 50 | 74 | 124 | 1.48 | 123.9052743 | 5.79 | Stable | EE |
| Antimony | (two stable isotopes) | | | | | | | | |
| | Sb-121 | 51 | 70 | 121 | 1.37 | 120.9038212 | 57.36 | Stable | OE |
| | Sb-123 | 51 | 72 | 123 | 1.41 | 122.9042160 | 42.64 | Stable | OE |

| 1 | 2 | 3 | 4 | 5 | 6 | 7 | 8 | 9 | 10 |
|---|---|---|---|---|---|---|---|---|---|
| Element | Nuclide Symbol | p | n | p+n | n/p | Nuclidic Mass (u) | Nuclidic Abundance number % | Half life (time) | p n even (E) or odd (O) |
| Tellurium | (six stable isotopes and two pseudostable isotopes) | | | | | | | | |
| | Te-120 | 52 | 68 | 120 | 1.31 | 119.904048 | 0.096 | Stable | EE |
| | Te-122 | 52 | 70 | 122 | 1.35 | 121.903050 | 2.603 | Stable | EE |
| | Te-123 | 52 | 71 | 123 | 1.37 | 122.9042710 | 0.908 | Stable | EO |
| | Te-124 | 52 | 72 | 124 | 1.38 | 123.9028180 | 4.816 | Stable | EE |
| | Te-125 | 52 | 73 | 125 | 1.40 | 124.9044285 | 7.139 | Stable | EO |
| | Te-126 | 52 | 74 | 126 | 1.42 | 125.9033095 | 18.95 | Stable | EE |
| | Te-128 | 52 | 76 | 128 | 1.46 | 127.904463 | 31.69 | $2 \times 10^{21}$ y | EE |
| | Te-130 | 52 | 78 | 130 | 1.50 | 129.906229 | 33.80 | $1.5 \times 10^{24}$ y | EE |
| Iodine | (one stable isotope) | | | | | | | | |
| | I-127 | 53 | 74 | 127 | 1.40 | 126.904473 | 100 | Stable | OE |
| Xenon | (nine stable isotopes) | | | | | | | | |
| | Xe-124 | 54 | 70 | 124 | 1.30 | 123.9058942 | 0.10 | Stable | EE |
| | Xe-126 | 54 | 72 | 126 | 1.33 | 125.904281 | 0.09 | Stable | EE |
| | Xe-128 | 54 | 74 | 128 | 1.37 | 127.9035312 | 1.91 | Stable | EE |
| | Xe-129 | 54 | 75 | 129 | 1.39 | 128.9047801 | 26.4 | Stable | EO |
| | Xe-130 | 54 | 76 | 130 | 1.41 | 129.9035094 | 4.1 | Stable | EE |
| | Xe-131 | 54 | 77 | 131 | 1.43 | 130.905072 | 21.2 | Stable | EO |
| | Xe-132 | 54 | 78 | 132 | 1.44 | 131.904144 | 26.9 | Stable | EE |
| | Xe-134 | 54 | 80 | 134 | 1.48 | 133.905395 | 10.4 | Stable | EE |
| | Xe-136 | 54 | 82 | 136 | 1.52 | 135.907214 | 8.9 | Stable | EE |
| Cesium | (one stable isotope) | | | | | | | | |
| | Cs-133 | 55 | 78 | 133 | 1.42 | 132.905429 | 100 | Stable | OE |
| Barium | (seven stable isotopes} | | | | | | | | |
| | Ba-130 | 56 | 74 | 130 | 1.32 | 129.906282 | 0.106 | Stable | EE |
| | Ba-132 | 56 | 76 | 132 | 1.36 | 131.905042 | 0.101 | Stable | EE |
| | Ba-134 | 56 | 78 | 134 | 1.39 | 133.904486 | 2.417 | Stable | EE |
| | Ba-135 | 56 | 79 | 135 | 1.41 | 134.905665 | 6.592 | Stable | EO |
| | Ba-136 | 56 | 80 | 136 | 1.43 | 135.904553 | 7.854 | Stable | EE |
| | Ba-137 | 56 | 81 | 137 | 1.45 | 136.905812 | 11.23 | Stable | EO |
| | Ba-138 | 56 | 82 | 138 | 1.46 | 137.905232 | 71.70 | Stable | EE |
| Lanthanum | (one stable isotope and one pseudostable isotope) | | | | | | | | |
| | La-138 | 57 | 81 | 138 | 1.42 | 137.907105 | 0.0902 | $1.1 \times 10^{11}$ y | OO |
| | La-139 | 57 | 82 | 139 | 1.44 | 138.906347 | 99.9098 | Stable | OE |
| Cerium | (four stable isotopes) | | | | | | | | |
| | Ce-136 | 58 | 78 | 136 | 1.34 | 135.907140 | 0.19 | Stable | EE |
| | Ce-138 | 58 | 80 | 138 | 1.38 | 137.905985 | 0.25 | Stable | EE |
| | Ce-140 | 58 | 82 | 140 | 1.41 | 139.905433 | 88.48 | Stable | EE |
| | Ce-142 | 58 | 84 | 142 | 1.45 | 141.909241 | 11.08 | Stable | EE |
| Praseodymium | (one stable isotope) | | | | | | | | |
| | Pr-141 | 59 | 82 | 141 | 1.39 | 140.907647 | 100 | Stable | OE |
| Neodymium | (six stable isotopes and one pseudostable isotope) | | | | | | | | |
| | Nd-142 | 60 | 82 | 142 | 1.37 | 141.907719 | 27.13 | Stable | EE |
| | Nd-143 | 60 | 83 | 143 | 1.38 | 142.909810 | 12.18 | Stable | EO |
| | Nd-144 | 60 | 84 | 144 | 1.40 | 143.910083 | 23.80 | $2.1 \times 10^{15}$ y | EE |
| | Nd-145 | 60 | 85 | 145 | 1.42 | 144.912570 | 8.30 | Stable | EO |
| | Nd-146 | 60 | 86 | 146 | 1.43 | 145.913113 | 17.19 | Stable | EE |
| | Nd-148 | 60 | 88 | 148 | 1.47 | 147.916889 | 5.76 | Stable | EE |
| | Nd-150 | 60 | 90 | 150 | 1.50 | 149.920887 | 5.64 | Stable | EE |
| Promethium | (no stable isotopes, element made synthetically) | | | | | | | | |
| | (trace amounts of this element arise naturally via spontaneous fission) | | | | | | | | |
| | Pm-145 | 61 | 84 | 145 | 1.38 | 144.912743 | | 17.7 y | OE |

| 1 | 2 | 3 | 4 | 5 | 6 | 7 | 8 | 9 | 10 |
|---|---|---|---|---|---|---|---|---|---|
| Element | Nuclide Symbol | p | n | p+n | n/p | Nuclidic Mass (u) | Nuclidic Abundance number % | Half life (time) | p n even (E) or odd (O) |
| Samarium | (five stable isotopes and three pseudostable isotopes) | | | | | | | | |
| | Sm-144 | 62 | 82 | 144 | 1.32 | 143.911998 | 3.1 | Stable | EE |
| | Sm-146 | 62 | 84 | 146 | 1.35 | ≈145.9 | 0.000 | $1.03 \times 10^{8}$ y | EE |
| | Sm-147 | 62 | 85 | 147 | 1.37 | 146.914894 | 15.0 | $1.06 \times 10^{11}$ y | EO |
| | Sm-148 | 62 | 86 | 148 | 1.39 | 147.914819 | 11.3 | $8 \times 10^{15}$ y | EE |
| | Sm-149 | 62 | 87 | 149 | 1.40 | 148.917180 | 13.8 | Stable | EO |
| | Sm-150 | 62 | 88 | 150 | 1.42 | 149.917273 | 7.4 | Stable | EE |
| | Sm-152 | 62 | 90 | 152 | 1.45 | 151.919728 | 26.7 | Stable | EE |
| | Sm-154 | 62 | 92 | 154 | 1.48 | 153.922205 | 22.7 | Stable | EE |
| Europium | (two stable isotopes) | | | | | | | | |
| | Eu-151 | 63 | 88 | 151 | 1.40 | 150.919702 | 47.8 | Stable | OE |
| | Eu-153 | 63 | 90 | 153 | 1.43 | 152.921225 | 52.2 | Stable | OE |
| Gadolinium | (six stable isotopes and one pseudostable isotope) | | | | | | | | |
| | Gd-152 | 64 | 88 | 152 | 1.38 | 151.919786 | 0.20 | $1.1 \times 10^{14}$ y | EE |
| | Gd-154 | 64 | 90 | 154 | 1.41 | 153.920861 | 2.18 | Stable | EE |
| | Gd-155 | 64 | 91 | 155 | 1.42 | 154.922618 | 14.80 | Stable | EO |
| | Gd-156 | 64 | 92 | 156 | 1.44 | 155.922118 | 20.47 | Stable | EE |
| | Gd-157 | 64 | 93 | 157 | 1.45 | 156.923956 | 15.65 | Stable | EO |
| | Gd-158 | 64 | 94 | 158 | 1.47 | 157.924099 | 24.84 | Stable | EE |
| | Gd-160 | 64 | 96 | 160 | 1.50 | 159.927049 | 21.86 | Stable | EE |
| Terbium | (one stable isotope) | | | | | | | | |
| | Tb-159 | 65 | 94 | 159 | 1.45 | 158.925342 | 100 | Stable | OE |
| Dysprosium | (seven stable isotopes) | | | | | | | | |
| | Dy-156 | 66 | 90 | 156 | 1.36 | 155.924277 | 0.06 | Stable | EE |
| | Dy-158 | 66 | 92 | 158 | 1.39 | 157.924403 | 0.10 | Stable | EE |
| | Dy-160 | 66 | 94 | 160 | 1.42 | 159.925193 | 2.34 | Stable | EE |
| | Dy-161 | 66 | 95 | 161 | 1.44 | 160.926930 | 18.9 | Stable | EO |
| | Dy-162 | 66 | 96 | 162 | 1.45 | 161.926795 | 25.5 | Stable | EE |
| | Dy-163 | 66 | 97 | 163 | 1.47 | 162.928728 | 24.9 | Stable | EO |
| | Dy-164 | 66 | 98 | 164 | 1.48 | 163.929171 | 28.2 | Stable | EE |
| Holmium | (one stable isotope) | | | | | | | | |
| | Ho-165 | 67 | 98 | 165 | 1.42 | 164.930319 | 100 | Stable | OE |
| Erbium | (six stable isotopes) | | | | | | | | |
| | Er-162 | 68 | 94 | 162 | 1.38 | 161.928775 | 0.14 | Stable | EE |
| | Er-164 | 68 | 96 | 164 | 1.41 | 163.929198 | 1.61 | Stable | EO |
| | Er-166 | 68 | 98 | 166 | 1.44 | 165.930290 | 33.6 | Stable | EE |
| | Er-167 | 68 | 99 | 167 | 1.46 | 166.932046 | 22.95 | Stable | EO |
| | Er-168 | 68 | 100 | 168 | 1.47 | 167.932368 | 26.8 | Stable | EE |
| | Er-170 | 68 | 102 | 170 | 1.50 | 169.935461 | 14.9 | Stable | EE |
| Thulium | (one stable isotope) | | | | | | | | |
| | Tm-169 | 69 | 100 | 169 | 1.45 | 168.934212 | 100 | Stable | OE |
| Ytterbium | (seven stable isotopes) | | | | | | | | |
| | Yb-168 | 70 | 98 | 168 | 1.40 | 167.933894 | 0.13 | Stable | EE |
| | Yb-170 | 70 | 100 | 170 | 1.43 | 169.934759 | 3.05 | Stable | EE |
| | Yb-171 | 70 | 101 | 171 | 1.44 | 170.936323 | 14.3 | Stable | EO |
| | Yb-172 | 70 | 102 | 172 | 1.46 | 171.936378 | 21.9 | Stable | EE |
| | Yb-173 | 70 | 103 | 173 | 1.47 | 172.938208 | 16.12 | Stable | EO |
| | Yb-174 | 70 | 104 | 174 | 1.49 | 173.938859 | 31.8 | Stable | EE |
| | Yb-176 | 70 | 106 | 176 | 1.51 | 175.942564 | 12.7 | Stable | EE |
| Lutetium | (one stable isotope and one long lived isotope) | | | | | | | | |
| | Lu-175 | 71 | 104 | 175 | 1.46 | 174.940770 | 97.41 | Stable | OE |
| | Lu-176 | 71 | 105 | 176 | 1.48 | 175.942679 | 2.59 | $3.6 \times 10^{10}$ y | OO |

| 1 | 2 | 3 | 4 | 5 | 6 | 7 | 8 | 9 | 10 |
|---|---|---|---|---|---|---|---|---|---|
| Element | Nuclide Symbol | p | n | p+n | n/p | Nuclidic Mass (u) | Nuclidic Abundance number % | Half life (time) | p n even (E) or odd (O) |
| Hafnium | (five stable isotopes and one pseudostable isotope) | | | | | | | | |
| | Hf-174 | 72 | 102 | 174 | 1.42 | 173.940044 | 0.162 | $2.0 \times 10^{15}$ y | EE |
| | Hf-176 | 72 | 104 | 176 | 1.44 | 175.941406 | 5.206 | Stable | EE |
| | Hf-177 | 72 | 105 | 177 | 1.46 | 176.943217 | 18.606 | Stable | EO |
| | Hf-178 | 72 | 106 | 178 | 1.47 | 177.943696 | 27.297 | Stable | EE |
| | Hf-179 | 72 | 107 | 179 | 1.49 | 178.9458122 | 13.629 | Stable | EO |
| | Hf-180 | 72 | 108 | 180 | 1.50 | 179.9465457 | 35.100 | Stable | EE |
| Tantalum | (two stable isotopes) | | | | | | | | |
| | Ta-180 | 73 | 107 | 180 | 1.47 | 179.947462 | 0.012 | $>1 \times 10^{13}$ y | OO |
| | Ta-181 | 73 | 108 | 181 | 1.48 | 180.947992 | 99.988 | Stable | OE |
| Tungsten | (five stable isotopes) | | | | | | | | |
| | W-180 | 74 | 106 | 180 | 1.43 | 179.946701 | 0.013 | Stable | EE |
| | W-182 | 74 | 108 | 182 | 1.46 | 181.948202 | 26.3 | Stable | EE |
| | W-183 | 74 | 109 | 183 | 1.47 | 182.950220 | 14.3 | Stable | EO |
| | W-184 | 74 | 110 | 184 | 1.49 | 183.950928 | 30.67 | Stable | EE |
| | W-186 | 74 | 112 | 186 | 1.51 | 185.954357 | 28.6 | Stable | EE |
| Rhenium | (one stable and one pseudostable isotope) | | | | | | | | |
| | Re-185 | 75 | 110 | 185 | 1.47 | 184.952951 | 37.40 | Stable | OE |
| | Re-187 | 75 | 112 | 187 | 1.49 | 186.955744 | 62.60 | $4 \times 10^{10}$ y | OE |
| Osmium | (six stable isotopes and one pseudostable isotopes) | | | | | | | | |
| | Os-184 | 76 | 108 | 184 | 1.42 | 183.952488 | 0.02 | Stable | EE |
| | Os-186 | 76 | 110 | 186 | 1.45 | 185.953830 | 1.58 | $2 \times 10^{15}$ y | EE |
| | Os-187 | 76 | 111 | 187 | 1.46 | 186.955741 | 1.6 | Stable | EO |
| | Os-188 | 76 | 112 | 188 | 1.47 | 187.955830 | 13.3 | Stable | EE |
| | Os-189 | 76 | 113 | 189 | 1.49 | 188.958137 | 16.1 | Stable | EO |
| | Os-190 | 76 | 114 | 190 | 1.50 | 189.958436 | 26.4 | Stable | EE |
| | Os-192 | 76 | 116 | 192 | 1.53 | 191.961467 | 41.0 | Stable | EE |
| Iridium | (two stable isotopes) | | | | | | | | |
| | Ir-191 | 77 | 114 | 191 | 1.48 | 190.960584 | 37.3 | Stable | OE |
| | Ir-193 | 77 | 116 | 193 | 1.51 | 192.962917 | 62.7 | Stable | OE |
| Platinum | (five stable isotopes and one pseudostable isotope) | | | | | | | | |
| | Pt-190 | 78 | 112 | 190 | 1.44 | 189.959917 | 0.01 | $6 \times 10^{11}$ y | EE |
| | Pt-192 | 78 | 114 | 192 | 1.47 | 191.961019 | 0.79 | Stable | EE |
| | Pt-194 | 78 | 116 | 194 | 1.49 | 193.962655 | 32.9 | Stable | EE |
| | Pt-195 | 78 | 117 | 195 | 1.50 | 194.964766 | 33.8 | Stable | EO |
| | Pt-196 | 78 | 118 | 196 | 1.51 | 195.964926 | 25.3 | Stable | EE |
| | Pt-198 | 78 | 120 | 198 | 1.54 | 197.967869 | 7.2 | Stable | EE |
| Gold | (one stable isotope) | | | | | | | | |
| | Au-197 | 79 | 118 | 197 | 1.49 | 196.966543 | 100 | Stable | OE |
| Mercury | (seven stable isotopes) | | | | | | | | |
| | Hg-196 | 80 | 116 | 196 | 1.45 | 195.965807 | 0.15 | Stable | EE |
| | Hg-198 | 80 | 118 | 198 | 1.48 | 197.966743 | 9.97 | Stable | EE |
| | Hg-199 | 80 | 119 | 199 | 1.49 | 198.968254 | 16.87 | Stable | EO |
| | Hg-200 | 80 | 120 | 200 | 1.50 | 199.968300 | 23.10 | Stable | EE |
| | Hg-201 | 80 | 121 | 201 | 1.51 | 200.970277 | 13.18 | Stable | EO |
| | Hg-202 | 80 | 122 | 202 | 1.53 | 201.970617 | 29.86 | Stable | EE |
| | Hg-204 | 80 | 124 | 204 | 1.55 | 203.973467 | 6.87 | Stable | EE |
| Thallium | (two stable isotopes) | | | | | | | | |
| | Tl-203 | 81 | 122 | 203 | 1.51 | 202.972320 | 29.524 | Stable | OE |
| | Tl-205 | 81 | 124 | 205 | 1.53 | 204.974401 | 70.476 | Stable | OE |
| Lead | (four stable isotopes) | | | | | | | | |
| | Pb-204 | 82 | 122 | 204 | 1.49 | 204.973020 | 1.4 | Stable | EE |
| | Pb-206 | 82 | 124 | 206 | 1.51 | 205.974440 | 24.1 | Stable | EE |
| | Pb-207 | 82 | 125 | 207 | 1.52 | 206.975872 | 22.1 | Stable | OE |
| | Pb-208 | 82 | 126 | 208 | 1.54 | 207.976627 | 52.4 | Stable | EE |

| 1 | 2 | 3 | 4 | 5 | 6 | 7 | 8 | 9 | 10 |
|---|---|---|---|---|---|---|---|---|---|
| Element | Nuclide Symbol | p | n | p+n | n/p | Nuclidic Mass (u) | Nuclidic Abundance number % | Half life (time) | p n even (E) or odd (O) |
| Bismuth | (one stable isotope) | | | | | | | | |
| | Bi-209 | 83 | 126 | 209 | 1.52 | 208.980374 | 100 | Stable | OE |
| Polonium | (no stable isotopes, element continuously formed in nature) | | | | | | | | |
| | Po-210 | 84 | 126 | 210 | 1.50 | 208.982404 | | 138.38 days | EE |
| Astatine | (no stable isotopes, element made synthetically) (trace amounts continuously formed in nature) | | | | | | | | |
| | At-210 | 85 | 125 | 210 | 1.47 | 209.987126 | | 8.3 hours | OO |
| Radon | (no stable isotopes, element continuously formed in nature) | | | | | | | | |
| | Rn-222 | 86 | 136 | 222 | 1.58 | 222.017571 | | 3.8325 days | EE |
| Francium | (no stable isotopes, element continuously formed in nature) | | | | | | | | |
| | Fr-223 | 87 | 136 | 223 | 1.56 | 223.019733 | | 21.8 min | EO |
| Radium | (no stable isotopes, element continuously formed in nature) | | | | | | | | |
| | Ra-226 | 88 | 138 | 226 | 1.57 | 226.025403 | | $1.60 \times 10^{3}$ y | EE |
| Actinium | (no stable isotopes, element continuously formed in nature) | | | | | | | | |
| | Ac-227 | 89 | 138 | 227 | 1.55 | 227.027750 | | 21.773 y | OE |
| Thorium | (one pseudostable isotope) | | | | | | | | |
| | Th-232 | 90 | 142 | 232 | 1.58 | 232.0380508 | 100 | $1.41 \times 10^{10}$ y | EE |
| Protactinium | (no stable isotopes, element continuously formed in nature) | | | | | | | | |
| | Pa-231 | 91 | 140 | 231 | 1.54 | 231.035880 | | $3.28 \times 10^{4}$ y | OE |
| Uranium | (two pseudostable isotopes and two continuously produced isotopes) | | | | | | | | |
| | U-233 | 92 | 141 | 233 | 1.53 | 233.039628 | 0.0000 | $1.592 \times 10^{5}$ y | EO |
| | U-234 | 92 | 142 | 234 | 1.54 | 234.0409468 | 0.0055 | $2.45 \times 10^{5}$ y | EE |
| | U-235 | 92 | 143 | 235 | 1.55 | 235.0439242 | 0.7200 | $7.038 \times 10^{8}$ y | EO |
| | U-238 | 92 | 146 | 238 | 1.59 | 238.0507847 | 99.2745 | $4.468 \times 10^{9}$ y | EE |
| Neptunium | (no stable isotopes, element made synthetically) (trace amounts of this element arise naturally via neutron capture) | | | | | | | | |
| | Np-237 | 93 | 144 | 237 | 1.55 | 237.0482678 | | $2.14 \times 10^{6}$ y | OE |
| Plutonium | (one pseudostable isotope exists in trace amounts) (trace amounts of this element arise naturally via neutron capture) | | | | | | | | |
| | Pu-239 | 94 | 145 | 239 | 1.54 | | | $2.411 \times 10^{4}$ y | OE |
| | Pu-244 | 94 | 150 | 244 | 1.60 | 244.064199 | | $8.08 \times 10^{7}$ y | EE |

## Summary of Table of Nuclides Entries

| Stable nuclides | 264 nuclides |
|---|---|
| Twenty-five long lived (or pseudostable) primordial nuclides. | 25 nuclides<br>$^{82}Se$, $^{40}K$, $^{50}V$, $^{87}Rb$, $^{92}Nb$, $^{113}Cd$, $^{115}In$, $^{128}Te$, $^{130}Te$, $^{138}La$, $^{144}Nd$, $^{146}Sm$, $^{147}Sm$, $^{148}Sm$, $^{152}Gd$, $^{176}Lu$, $^{174}Hf$, $^{180}Ta$, $^{187}Re$, $^{186}Os$, $^{190}Pt$, $^{232}Th$, $^{235}U$, $^{238}U$, $^{244}Pu$. |
| Two radionuclides continuously formed by cosmic ray bombardment of the atmosphere. | 2 nuclides<br>$^{3}H$ and $^{14}C$ |
| Six elements lacking both a stable nuclide and a pseudo-stable nuclide that are formed continuously in nature in decay chains from heavier nuclides. | 6 nuclides<br>$^{210}Po$, $^{222}Rn$, $^{223}Fr$, $^{226}Ra$, $^{227}Ac$, $^{231}Pa$ |
| Two isotopes of uranium that are formed continuously in nature. | 2 nuclides<br>$^{233}U$, $^{234}U$ |
| Three, originally synthetic elements, lacking both a stable nuclide and a pseudostable nuclide. These elements were subsequently shown to occur naturally. | 3 nuclides<br>$^{99}Tc$, $^{145}Pm$, $^{210}At$ |
| Two originally synthetic elements now known to be continuously created in trace amounts via natural successive neutron capture processes. | 2 nuclides<br>$^{237}Np$ and $^{239}Pu$ |
| Elements continuously formed naturally in trace amounts via the spontaneous fission of pseudostable nuclides | Tc, Pm, and isotopes of many other elements |

## Count of Stable Nuclides Related to the Oddness or Evenness of their Nucleon Count

| p-n Type | Number |
|---|---|
| Even-Even (EE) | 149 |
| Even-Odd (EO) | 59 |
| Odd-Even (OE) | 52 |
| Odd-Odd (OO) | 4 |
| Total | 264 |

## Five Synthetic Elements Subsequently Found in Nature

**Technetium**. In 1937, technetium became the first element to be synthesized. Chemists had searched from many years for the element, and many erroneous reports of its discovery had been made. However, the radioactivity of all of its isotopes, and their relatively short half-lives, prevented significant amounts of the element from existing in the earth's crust. It was prepared by deuteron bombardment of a molybdenum plate from which was isolated approximately $10^{-10}$ g of $^{99}$Tc. One possible formation reaction is

$$^{97}_{42}\text{Mo} + {}^{2}_{1}\text{H} \rightarrow {}^{99}_{45}\text{Tc}$$

In the early 1940s, technetium was found among the products of uranium fission. Today, kilogram quantities of the element are available from this source. In 1961, trace amounts of $^{99}$Tc were identified in a uranium ore of African origin, where they arose as a consequence of the spontaneous fission of uranium atoms. According to one estimate, there is about 1500 g of technetium in the earth's crust. In 1951, evidence from the solar spectrum revealed the presence of technetium in the sun, and it is now known that technetium exists in many stars.

**Promethium**. As with technetium, many erroneous claims were made for the existence of promethium before its probable first manufacture in 1938 via the bombardment of a neodymium target with deuterons in the nuclear reaction:

$$^{143}_{60}\text{Nd} + {}^{2}_{1}\text{H} \rightarrow {}^{144}_{61}\text{Pm} + {}^{1}_{0}\text{n}$$

Chemical proof of promethium's synthesis came in 1945 from studies of uranium fission products and samples of neodymium subjected to neutron irradiation. Today, over 25 promethium isotopes are known. The longest lived of these, $^{145}$Pm has a half-life of 17.7 years. Despite its only having short lived isotopes, in 1968 natural promethium was shown to be present in uranium ore. Like technetium, promethium forms in such ore because of the spontaneous fission of $^{238}$U atoms. According to one estimate, there is about 780 g of promethium in the earth's crust. Quantities of promethium in the range of grams have been extracted from fission product sources.

**Astatine**. Even in the nineteenth century chemists sought the element astatine in natural minerals and brines containing iodine. This search was unsuccessful, as was the search for astatine during 1900-1939 when other radioactive elements near astatine in the periodic table were found. The reason, of course, is that all isotopes of astatine are very short lived, with the two longest lived being $^{210}$At at 8.3 hours and $^{211}$At at 7.2 hours. Astatine was synthesized in 1940 at the University of California by bombarding bismuth with alpha particles:

$$^{209}_{83}\text{Bi} + {}^{4}_{2}\text{He} \rightarrow {}^{211}_{85}\text{At} + 2\,{}^{1}_{0}\text{n}$$

Very shortly after its synthesis, minute quantities of natural astatine were discovered in uranium minerals, where it is continuously formed. This latter discovery took place in Vienna, Austria in 1943. However, the total amount of astatine present in the earth's crust at any given time is estimated to be only about 30 g.

**Neptunium**. The existence of an element beyond uranium in the periodic table was not suspected until the 1930s. Enrico Fermi actually produced neptunium in 1934, but in a famous scientific missed opportunity, failed to recognize he'd done it. In 1940, $^{239}$Np was synthesized in California by bombarding uranium with neutrons. Neptunium's longest lived isotope is $^{237}$Np with a half-life of $2.14 \times 10^6$ years. Trace quantities of neptunium are formed in nature by reaction sequences such as:

$$^{235}_{92}\text{U} + {}^{1}_{0}\text{n} \rightarrow {}^{236}_{92}\text{U} \qquad ({}^{236}_{92}\text{U half-life} = 2.34 \times 10^7 \text{ years})$$

$$^{236}_{92}\text{U} + {}^{1}_{0}\text{n} \rightarrow {}^{237}_{92}\text{U} \qquad ({}^{237}_{92}\text{U half-life} = 6.75 \text{ days})$$

$$^{237}_{92}\text{U} \rightarrow {}^{237}_{93}\text{Np} + {}^{0}_{-1}\beta \qquad ({}^{237}_{93}\text{Np half-life} = 2.14 \times 10^6 \text{ years})$$

In 1952, the isotope $^{237}$Np was found to occur naturally. Furthermore, the existence of natural $^{239}$Pu implies that at least an extremely small quantity of a second neptunium isotope, $^{239}$Np, must exist naturally. Today, kilogram quantities of neptunium recovered from nuclear reactors are available.

**Plutonium**. The nuclide $^{238}Pu$ was synthesized in California in 1940 by deuteron bombardment of uranium. Very soon after, in 1942, $^{239}Pu$ was discovered in extremely small concentrations in a Canadian uranium ore. The formation process must have been:

$$^{238}_{92}U + ^{1}_{0}n \rightarrow ^{239}_{92}U \qquad (^{239}_{92}U \text{ half-life} = 23.5 \text{ minutes})$$
$$^{239}_{92}U \rightarrow ^{239}_{93}Np + ^{0}_{-1}\beta \qquad (^{239}_{93}Np \text{ half-life} = 2.355 \text{ days})$$
$$^{239}_{93}Np \rightarrow ^{239}_{94}Pu + ^{0}_{-1}\beta \qquad (^{239}_{94}Pu \text{ half-life} = 24{,}110 \text{ years})$$

So plutonium was known to be a natural element almost from the very beginning of plutonium chemistry. Incidentally, the reactions above demonstrate that $^{239}Np$ must exist naturally. Many years later, in 1971, it was reported that $^{244}Pu$ with a half-life of $8.00 \times 10^7$ years is a **primordial nuclide**, i.e., one sufficiently long lived to have endured since the time of the earth's formation. Glenn Seaborg (after whom $^{106}Sg$, seaborgium, is named) once estimated that the amount of primordial $^{244}Pu$ on earth is less than 10 grams. Commercially, $^{239}Pu$ is by far the most import plutonium isotope. Tons of $^{239}Pu$ are produced in nuclear reactors for use as fuel for nuclear reactors and weapons.

**Appendix Example 12**. The radioisotope $^{244}Pu$, with a half-life of $8.3 \times 10^7$ years, was first identified in the debris of the 1952 Mike nuclear test. Several years later a team of chemists at the Los Alamos National Laboratory was able to extract 20 million atoms of $^{224}Pu$ from a sample of commercial thorium ore. The concentration of the $^{244}Pu$ in the ore was estimated to be 1 part in $10^{18}$. Deduce the mass of the ore sample.

Strategy: Estimate the mass of the 20 million atoms in kg and multiply it by $10^{18}$.

Mass of $^{244}Pu = 244 \text{ u atom}^{-1} \times 2 \times 10^7 \text{ atom} = 4.9 \times 10^9 \text{ u}$

Mass of $^{244}Pu$ in kg $= 4.9 \times 10^9 \text{ u} \div 6.02 \times 10^{26} \text{ u kg}^{-1} = 8.1 \times 10^{-18}$ kg

Mass of ore $= 8.1 \times 10^{-18} \text{ kg} \times 10^{18} = \underline{8 \text{ kg}}$ (rounded to one significant figure)

**Appendix Example 13**. The earth's crust is estimated to weigh $2.4 \times 10^{25}$ g and the concentration of $^{235}U$ in the crust is estimated to be 19 μg $kg^{-1}$. The half-life at which $^{235}U$ undergoes spontaneous fission is $9.80 \times 10^{18}$ years. In a $^{235}U$ fission reactor, about 26 mg of $^{99}Tc$ is produced per gram of $^{235}U$ fissioned; assume this same mass fraction of $^{99}Tc$ production for the natural spontaneous fission process. The half-life of $^{99}Tc$ is $2.12 \times 10^5$ years. Use the foregoing data to estimate the mass of natural $^{99}Tc$ in the earth's crust at any moment.

Strategy: Estimate the rate of spontaneous fissioning of $^{235}U$ in the earth's crust. Use that and the other given data to calculate the rate at which $^{99}Tc$ is being formed. The rate at which the $^{99}Tc$ decays is proportional to its mass and is given by c in the first order rate equation $\frac{dc}{dt} = -kc$, where c is the actual mass of $^{99}Tc$ in the crust at any moment. The mass of $^{99}Tc$ can be obtained by setting the rate of its decomposition equal to its rate of formation.

Estimated mass of $^{235}U$ in the crust $= 2.4 \times 10^{25}$ g $\times$ 19 μg $kg^{-1} \times 10^{-9}$ kg $μg^{-1}$
$= 4.6 \times 10^{17}$ g

The half life of $^{235}U$ in minutes
$t_{½} = 9.80 \times 10^{18}$ years $\times$ 365.25 days $year^{-1} \times$ 1440 min $day^{-1}$
$t_{½} = 5.2 \times 10^{24}$ min

k for $^{235}U$ decomposition $= (\ln 2)/t_{½} = 0.693 \div 5.2 \times 10^{24}$ min $= 1.3 \times 10^{-25}$ $min^{-1}$

The rate of fissioning of $^{235}U$ in the crust is $-kc = -4.6 \times 10^{17}$ g $\times 1.3 \times 10^{-25}$ $min^{-1}$
$= -6.0 \times 10^{-5}$ g $min^{-1}$

From the reactor data, the mass fraction of $^{99}Tc$ formed is $\frac{26 \text{ mg of } ^{99}Tc}{1000 \text{ mg of uranium that reacts}}$

Applying that mass fraction, rate of production of $^{99}Tc = 6.0 \times 10^{-5}$ g $min^{-1} \times \frac{26}{1000}$

Estimated rate of production of $^{99}Tc$ from $^{235}U$ in crust $= 1.6 \times 10^{-9}$ g $min^{-1}$

Calculate the half life of $^{99}Tc$ in minutes:
$t_{½} = 2.12 \times 10^{5}$ years $\times$ 365.25 days $year^{-1} \times$ 1440 min $day^{-1}$
$t_{½} = 1.1 \times 10^{11}$ min

k for the $^{99}Tc$ decomposition $= (\ln 2)/t_{½} = 0.693 \div 1.1 \times 10^{11}$ min $= 6.2 \times 10^{-12}$ $min^{-1}$

Rate of decomposition $\frac{dc}{dt} = -kc = -c \times 6.2 \times 10^{-12}$ $min^{-1}$

Setting equal the absolute values of production and decomposition gives:

$$1.6 \times 10^{-9} \text{ g min}^{-1} = c \times 6.2 \times 10^{-12} \text{ min}^{-1}$$

The estimated mass of $^{99}Tc$ in the earth's crust is 250 g.

# Appendix E -- Electron Configurations of the Elements

* Shows a difference between the experimental configuration and that obtained by successive filling of the hydrogen-like energy level sequence 1s<2s<2p<3s<3p<4s<3d<4p<5s<4d<5p<6s <4f<5d<6p<7s<5f<6d<7p. [Noble gas] designates the **noble gas core**. Configurations with a ? are conjectural.

| | | |
|---|---|---|
| 40 | Zr | $[Kr]5s^{2}4d^{2}$ |
| 39 | Y | $[Kr]5s^{2}4d^{1}$ |
| 38 | Sr | $[Kr]5s^{2}$ |
| 37 | Rb | $[Kr]5s^{1}$ |
| 36 | Kr | $[Ar]4s^{2}3d^{10}4p^{6}$ |
| 35 | Br | $[Ar]4s^{2}3d^{10}4p^{5}$ |
| 34 | Se | $[Ar]4s^{2}3d^{10}4p^{4}$ |
| 33 | As | $[Ar]4s^{2}3d^{10}4p^{3}$ |
| 32 | Ge | $[Ar]4s^{2}3d^{10}4p^{2}$ |
| 31 | Ga | $[Ar]4s^{2}3d^{10}4p^{1}$ |
| 30 | Zn | $[Ar]4s^{2}3d^{10}$ |
| 29 | Cu | $[Ar]4s^{1}3d^{10}$* |
| 28 | Ni | $[Ar]4s^{2}3d^{8}$ |
| 27 | Co | $[Ar]4s^{2}3d^{7}$ |
| 26 | Fe | $[Ar]4s^{2}3d^{6}$ |
| 25 | Mn | $[Ar]4s^{2}3d^{5}$ |
| 24 | Cr | $[Ar]4s^{1}3d^{5}$* |
| 23 | V | $[Ar]4s^{2}3d^{3}$ |
| 22 | Ti | $[Ar]4s^{2}3d^{2}$ |
| 21 | Sc | $[Ar]4s^{2}3d^{1}$ |
| 20 | Ca | $[Ar]4s^{2}$ |
| 19 | K | $[Ar]4s^{1}$ |
| 18 | Ar | $[Ne]3s^{2}3p^{6}$ |
| 17 | Cl | $[Ne]3s^{2}3p^{5}$ |
| 16 | S | $[Ne]3s^{2}3p^{4}$ |
| 15 | P | $[Ne]3s^{2}3p^{3}$ |
| 14 | Si | $[Ne]3s^{2}3p^{2}$ |
| 13 | Al | $[Ne]3s^{2}3p^{1}$ |
| 12 | Mg | $[Ne]3s^{2}$ |
| 11 | Na | $[Ne]3s^{1}$ |
| 10 | Ne | $[He]2s^{2}2p^{6}$ |
| 9 | F | $[He]2s^{2}2p^{5}$ |
| 8 | O | $[He]2s^{2}2p^{4}$ |
| 7 | N | $[He]2s^{2}2p^{3}$ |
| 6 | C | $[He]2s^{2}2p^{2}$ |
| 5 | B | $[He]2s^{2}2p^{1}$ |
| 4 | Be | $[He]2s^{2}$ |
| 3 | Li | $[He]2s^{1}$ |
| 2 | He | $1s^{2}$ |
| 1 | H | $1s^{1}$ |

| | | |
|---|---|---|
| 80 | Hg | $[Xe]6s^{2}4f^{14}5d^{10}$ |
| 79 | Au | $[Xe]6s^{2}4f^{14}5d^{9}$ |
| 78 | Pt | $[Xe]6s^{2}4f^{14}5d^{8}$ |
| 77 | Ir | $[Xe]6s^{2}4f^{14}5d^{7}$ |
| 76 | Os | $[Xe]6s^{2}4f^{14}5d^{6}$ |
| 75 | Re | $[Xe]6s^{2}4f^{14}5d^{5}$ |
| 74 | W | $[Xe]6s^{2}4f^{14}5d^{4}$ |
| 73 | Ta | $[Xe]6s^{2}4f^{14}5d^{3}$ |
| 72 | Hf | $[Xe]6s^{2}4f^{14}5d^{2}$ |
| 71 | Lu | $[Xe]6s^{2}4f^{14}5d^{1}$ |
| 70 | Yb | $[Xe]6s^{2}4f^{14}$ |
| 69 | Tm | $[Xe]6s^{2}4f^{13}$ |
| 68 | Er | $[Xe]6s^{2}4f^{12}$ |
| 67 | Ho | $[Xe]6s^{2}4f^{11}$ |
| 66 | Dy | $[Xe]6s^{2}4f^{10}$ |
| 65 | Tb | $[Xe]6s^{2}4f^{9}$ |
| 64 | Gd | $[Xe]6s^{2}4f^{7}5d^{1}$* |
| 63 | Eu | $[Xe]6s^{2}4f^{7}$ |
| 62 | Sm | $[Xe]6s^{2}4f^{6}$ |
| 61 | Pm | $[Xe]6s^{2}4f^{5}$ |
| 60 | Nd | $[Xe]6s^{2}4f^{4}$ |
| 59 | Pr | $[Xe]6s^{2}4f^{3}$ |
| 58 | Ce | $[Xe]6s^{2}4f^{2}$ |
| 57 | La | $[Xe]6s^{2}5d^{1}$* |
| 56 | Ba | $[Xe]6s^{2}$ |
| 55 | Cs | $[Xe]6s^{1}$ |
| 54 | Xe | $[Kr]5s^{2}4d^{10}5p^{6}$ |
| 53 | I | $[Kr]5s^{2}4d^{10}5p^{5}$ |
| 52 | Te | $[Kr]5s^{2}4d^{10}5p^{4}$ |
| 51 | Sb | $[Kr]5s^{2}4d^{10}5p^{3}$ |
| 50 | Sn | $[Kr]5s^{2}4d^{10}5p^{2}$ |
| 49 | In | $[Kr]5s^{2}4d^{10}5p^{1}$ |
| 48 | Cd | $[Kr]5s^{2}4d^{10}$ |
| 47 | Ag | $[Kr]5s^{2}4d^{9}$ |
| 46 | Pd | $[Kr]4d^{10}$* |
| 45 | Rh | $[Kr]5s^{2}4d^{7}$ |
| 44 | Ru | $[Kr]5s^{2}4d^{6}$ |
| 43 | Tc | $[Kr]5s^{2}4d^{5}$ |
| 42 | Mo | $[Kr]5s^{1}4d^{5}$* |
| 41 | Nb | $[Kr]5s^{1}4d^{4}$* |

| | | |
|---|---|---|
| 118 | Uuo | $[Rn]7s^{2}5f^{14}6d^{10}7p^{2}$? |
| 117 | --- | |
| 116 | Uuh | $[Rn]7s^{2}5f^{14}6d^{10}7p^{4}$? |
| 115 | --- | |
| 114 | Uuq | $[Rn]7s^{2}5f^{14}6d^{10}7p^{2}$? |
| 113 | --- | |
| 112 | Uub | $[Rn]7s^{2}5f^{14}6d^{10}$? |
| 111 | Uuu | $[Rn]7s^{2}5f^{14}6d^{9}$? |
| 110 | Uun | $[Rn]7s^{2}5f^{14}6d^{8}$? |
| 109 | Mt | $[Rn]7s^{2}5f^{14}6d^{7}$? |
| 108 | Hs | $[Rn]7s^{2}5f^{14}6d^{6}$? |
| 107 | Bh | $[Rn]7s^{2}5f^{14}6d^{5}$? |
| 106 | Sg | $[Rn]7s^{2}5f^{14}6d^{3}$? |
| 105 | Db | $[Rn]7s^{2}5f^{14}6d^{3}$? |
| 104 | Rf | $[Rn]7s^{2}5f^{14}6d^{2}$? |
| 103 | Lr | $[Rn]7s^{2}5f^{14}6d^{1}$ |
| 102 | No | $[Rn]7s^{2}5f^{14}$ |
| 101 | Md | $[Rn]7s^{2}5f^{13}$ |
| 100 | Fm | $[Rn]7s^{2}5f^{12}$ |
| 99 | Es | $[Rn]7s^{2}5f^{11}$ |
| 98 | Cf | $[Rn]7s^{2}5f^{10}$ |
| 97 | Bk | $[Rn]7s^{2}5f^{9}$ |
| 96 | Cm | $[Rn]7s^{2}5f^{7}6d^{1}$* |
| 95 | Am | $[Rn]7s^{2}5f^{7}$ |
| 94 | Pu | $[Rn]7s^{2}5f^{6}$ |
| 93 | Np | $[Rn]7s^{2}5f^{4}6d^{1}$* |
| 92 | U | $[Rn]7s^{2}5f^{3}6d^{1}$* |
| 91 | Pa | $[Rn]7s^{2}5f^{2}6d^{1}$* |
| 90 | Th | $[Rn]7s^{2}6d^{2}$* |
| 89 | Ac | $[Rn]7s^{2}6d^{1}$* |
| 88 | Ra | $[Rn]7s^{2}$ |
| 87 | Fr | $[Rn]7s^{1}$ |
| 86 | Rn | $[Xe]6s^{2}4f^{14}5d^{10}6p^{6}$ |
| 85 | At | $[Xe]6s^{2}4f^{14}5d^{10}6p^{5}$ |
| 84 | Po | $[Xe]6s^{2}4f^{14}5d^{10}6p^{4}$ |
| 83 | Bi | $[Xe]6s^{2}4f^{14}5d^{10}6p^{3}$ |
| 82 | Pb | $[Xe]6s^{2}4f^{14}5d^{10}6p^{2}$ |
| 81 | Tl | $[Xe]6s^{2}4f^{14}5d^{10}6p^{1}$ |

# Appendix F -- Chemical Business

## The Chemical Industry

Skill at practical chemistry goes back 1000s of years and **applied chemistry** and **industrial chemistry** have a long history. The modern chemical industry plays a dominant role in the global industrial economy.

By the late Middle Ages a substantial demand had arisen for chemical compounds such as soda (sodium carbonate) to treat cloth and nitre (potassium nitrate) for gunpowder. However, the growth of the modern chemical industry, and the modern science of chemistry, occurred largely in tandem with the industrial revolution beginning around 1750. Sulfuric acid is the chemical compound manufactured globally in the greatest quantity. Sulfuric acid has been known since at least the 16th century, but for its large scale commercial production acid resistant equipment is crucial; historically, using the metal lead to create a corrosion-proof reaction chamber was the answer. The first lead-chamber process plant in Europe dates to about 1766. In America, the first lead-chamber process plant started in Philadelphia in 1793. The process of Nicholas LeBlanc to make synthetic sodium carbonate became operational in 1787. The reaction of lime with chlorine to make calcium hypochlorite, $CaO + Cl_2 \rightarrow CaOCl_2$, for cloth bleaching was commercialized in 1799. The manufacture and use of fuel gas for lighting began in the first decade of the 19th century.

The organic chemical industries began in 1845 when August Wilhelm von Hoffman, professor at the Royal College of Science in London, began research on the possible uses of the coal tars formed as an inevitable by-product of gas-making. A decade later, following Hoffman's lead, William Perkin got lucky and produced the world's first synthetic dye. At about the same time, "Colonel" Edwin Drake was drilling his western Pennsylvania oil well in a bold effort to find a source of cheap mineral oil with which to take the big market for lighting held by animal fats and oils. Synthetic dyes became the anvil on which was hammered out the basic understanding of organic chemistry, while Drake's well planted the seed that eventually grew to become the giant, global oil industry. Mastery of organic chemistry, applied to abundant petroleum raw materials in the hands of skillful chemical engineers, created the petrochemical industry and the materials and products that we so easily take for granted in our daily life in the early 21st century.

There's a one-page summary of the modern chemical industry in Chapter 1. Top billing, of course, goes to the petrochemical industry. About 97% of petroleum products are consumed as fuel, the other 3% become chemical compounds and after more chemical reactions eventually create the synthetic organic and polymer materials that we demand in such abundance.

Table F-1 lists the top fifty industrial chemicals as measured by their 1997 US production. To make this information personal it is listed on a per capita US citizen basis; that is, the total production figure in pounds has been divided by the approximate US population of $2.5 \times 10^8$ people. That value constitutes "your share." Note that based on the top 50 chemicals, you consumed not much less than 3000 pounds of chemical production in 1997.

## Table F-1—1997 US Production of some Industrial Materials (pounds person$^{-1}$ year$^{-1}$)

The following table shows the approximate number of pounds per capita of industrial materials consumed by Americans in 1997. If you understand this table, you'll have a good insight into the role of chemical industry in the US economy. The columns of the table are:

Column 1 The rank of the material in this table
Column 2 The common name or the chemical name of the material
Column 3 The chemical formula of the material
Column 4 The number of pounds of the material you "consumed" in the US in 1997
Column 5 The traditional classification of the material
Column 6 An important use or uses of the material

Table F-1 **1997 US Production of some Industrial Materials**
*Pounds per person per year based on US population of 250 million

| Rank | Material | Formula | lbs* | Classification | Important use or uses |
|---|---|---|---|---|---|
| 1 | Phosphate rock | $CaF_2.3Ca_3(PO_4)_2$ | 410 | Mineral | Fertilizer and chemical manufacture |
| 2 | Sulfuric acid | $H_2SO_4$ | 380 | Inorganic chemical | Fertilizer and chemical manufacture, etc. |
| 3 | Ethylene | $C_2H_4$ | 200 | Organic chemical | Polyethylene manufacture, chemical synthesis |
| 4 | Lime (calcium oxide) | $CaO$ | 170 | Mineral | Iron and chemical manufacture, water treatment |
| 5 | Ammonia | $NH_3$ | 145 | Inorganic chemical | Fertilizer |
| 6 | Sulfur | $S$ | 120 | Mineral | Sulfuric acid manufacture |
| 7 | Propylene | $CH_3–CH=CH_2$ | 110 | Organic chemical | Polypropylene manufacture, chemical synthesis |
| 8 | Polyethylene | $–(CH_2)_n–$ | 110 | Thermoplastic polymer | Plastic bags, containers, and bottles |
| 9 | Phosphoric acid | $H_3PO_4$ | 100 | Inorganic chemical | Fertilizer, metal cleaning, food products |
| 10 | Chlorine | $Cl_2$ | 100 | Inorganic chemical | Sterilization, PVC manufacture, bleaching |
| 11 | Sodium carbonate | $Na_2CO_3$ | 90 | Mineral | Glass making, detergents, paper manufacture |
| 12 | Sodium hydroxide | $NaOH$ | 90 | Inorganic chemical | Chemical manufacturing, acid neutralization |
| 13 | Nitric acid | $HNO_3$ | 70 | Inorganic chemical | Fertilizer and explosives manufacture |
| 14 | Ammonium nitrate | $NH_4NO_3$ | 65 | Inorganic chemical | Fertilizer and explosive |
| 15 | Urea | $CO(NH_2)_2$ | 60 | Organic chemical | Fertilizer, livestock feed, resin manufacture |
| 16 | Polyvinyl chloride (PVC) | $–(C_2H_3Cl)_n–$ | 55 | Thermoplastic polymer | Vinyl plastics, piping and siding materials. |
| 17 | Polypropylene | $–(CH(CH_3)–CH)_n–$ | 52 | Thermoplastic polymer | Plastic containers, synthetic fibers |
| 18 | Potash | $K_2O$ | 51 | Mineral | Fertilizer |
| 19 | Ethyl benzene | $C_6H_5–C_2H_5$ | 51 | Organic chemical | Chemical intermediate for styrene manufacture |
| 20 | Styrene | $C_6H_5–CH=CH_2$ | 46 | Organic chemical | Rubber and polystyrene manufacture |
| 21 | Polystyrene | $–(CH(C_6H_5)–CH)_n–$ | 38 | Thermoplastic polymer | Insulation, packaging, CD containers, compounding |
| 22 | Hydrochloric acid | $HCl$ | 34 | Inorganic chemical | Chemical synthesis, steel pickling, oil well acidizing |
| 23 | Ethylene oxide | $C_2H_4O$ | 33 | Organic chemical | Surfactant manufacture |
| 24 | p-Xylene | $H_3C–C_6H_4–CH_3$ | 31 | Organic chemical | Chemical synthesis, for terephthalic acid |
| 25 | Cumene | $C_6H_5–CH(CH_3)_2$ | 25 | Organic chemical | Phenol and acetone manufacture |
| 26 | Ammonium sulfate | $(NH_4)_2SO_4$ | 22 | Inorganic chemical | Fertilizer |
| 27 | Nitrogen | $N_2$ | 22 | Inorganic chemical | Refrigerant, inert gas |
| 28 | 1,3-Butadiene | $H_2C=CH–CH=CH_2$ | 17 | Organic chemical | Rubber manufacture, intermediate for nylon (HMDA) |
| 29 | Oxygen | $O_2$ | 17 | Inorganic chemical | Steel making |
| 30 | Polyester (thermoplastic) | C,H,O, (varies) | 17 | Thermoplastic polymer | Synthetic fibers |
| 31 | Phenolic resin | C,H,O, (varies) | 14 | Thermosetting polymer | Hard plastics |
| 32 | Acrylonitrile | $CH_2=CH-CN$ | 14 | Organic chemical | Resin and fiber manufacture |
| 33 | Ethylene dichloride | $CH_2Cl–CH_2Cl$ | 10 | Organic chemical | Intermediate for PVC production |
| 34 | Titanium dioxide | $TiO_2$ | 10 | Inorganic chemical | Paint pigment |
| 35 | Urea resin | C,H,O,N (varies) | 10 | Thermosetting polymer | Hard plastics |
| 36 | Aluminum sulfate | $Al_2(SO_4)_3$ | 10 | Inorganic chemical | Water treatment |
| 37 | Benzene | $C_6H_6$ | 9 | Organic chemical | Chemical synthesis |
| 38 | Sodium silicate | $Na_2SiO_3$ | 9 | Inorganic chemical | Detergents, catalysts, paper making |
| 39 | Hydrogen | $H_2$ | 7 | Inorganic chemical | Making ammonia, hydrogenation |
| 40 | Polyester resin | C,H,O, (varies) | 6 | Thermosetting polymer | Hard plastics |
| 41 | Isopropyl alcohol | $C_3H_7OH$ | 6 | Organic chemical | Solvent, intermediate for acetone |
| 42 | Aniline | $C_6H_5NH_2$ | 5 | Organic chemical | Intermediate for polyurethane manufacture |
| 43 | Polyamide (nylons) | C,H,O,N (varies) | 5 | Thermoplastic polymer | Fibers, plastics |
| 44 | Sodium chlorate | $NaClO_3$ | 5 | Inorganic chemical | Bleaching wood pulp for paper making |
| 45 | Sodium sulfate | $Na_2SO_4$ | 5 | Inorganic chemical | Detergents, paper making |
| 46 | o-Xylene | $H_3C–C_6H_4–CH_3$ | 4 | Organic chemical | Chemical synthesis, for phthalic anhydride |
| 47 | 2-Ethylhexanol | $C_8H_{17}OH$ | 3 | Organic chemical | Plastics additives, detergents |
| 48 | Epoxy resins | C,H,O, (varies) | 3 | Thermosetting polymer | Adhesives, sealants |
| 49 | Bromine | $Br_2$ | 2 | Mineral | Sanitizing compounds |
| 50 | Melamine resin | C,H,O,N (varies) | 1 | Thermosetting polymer | Hard plastics |
| | Total | | 2,869 | | |

Detailed use patterns can be found by searching for the chemical profile of the individual chemicals. For example [❁kws +"chemical profile" +ethylbenzene].

Notes to Table F-1

Some information in Table F-1 is derived from data in *Chemical and Engineering News*, June 29, 1998, pages 43-45. Data in Table F-1 are probably not reliable to better than 25% uncertainty. Some may be much worse. Study the table not for the values themselves, but to learn about the pattern and vast scale of chemical manufacture that are inherent to a large, modern, industrial economy. Some materials are missing that undoubtedly belong: salt (sodium chloride) for example.

The classifications used in column 5 are the traditional ones deriving from the US Department of Commerce—the US Government agency that was the original source of the data in the table. These classifications are often not the ones that a chemist would select.

**Minerals** are materials directly obtained from natural sources. Often minerals are simply mined. **Organic chemical compounds** contain carbon. **Inorganic chemical compounds** are those that lack carbon content (a handful of carbon containing compounds, such as sodium carbonate, are inorganic rather than organic).

A synthetic **polymer** is a chemical material (made by chemists or engineers) that contains long chains of atoms. Polymers are made by chemically linking small molecules (**monomers**) to make long chain molecules (**polymers**). Synthetic polymer materials contain mainly carbon and hydrogen atoms. The chief source of monomers is the petrochemical industry. Ethylene, propylene, and styrene are examples of monomers; they yield, respectively, the polymers polyethylene, polypropylene, and polystyrene.

Synthetic polymer materials are generically called **plastics**. The term **resin** is often used interchangeably with the term plastic. A plastic material which melts when heated is called a **thermoplastic** resin or polymer. A plastic material that cannot melt is called a **thermosetting** resin or polymer.

Most of the materials listed in the preceding table never reach the consumer. Rather, they are used in chemical processes that eventually yield components or ingredients for consumer products. Thus, don't be surprised to find out that your share of US ethylene in 1997 was 200 pounds, while you probably never saw a pound of ethylene in your life. Polymeric materials derived from ethylene are used in automobile bodies, computer cases, clothing, packaging, etc. So you were probably blissfully unaware that you are a substantial consumer of ethylene.

## The World Top Fifty Chemical Producers

Table F-2 on the following page shows the actual dollar sales of large chemical companies in 1997. As with Table F-1, there are some omissions. For example, Russian and eastern bloc production is absent. Many of the listed companies have other business as well as chemical business; column 5 shows the percentage of their total sales that are chemical. Note that giant oil companies are substantial chemical producers, even though income from their fuel sales far exceeds income from their chemical sales: for example, Shell 92% from fuel versus only 8%, Exxon 89% from fuel versus 11% from chemicals.

In Table F-2 column 4 labeled "your annual share" is entered the sales of US companies divided by 250 million. So if you were an average US consumer, buying only from US companies, this is an estimate of what you are paying each company each year. As with Table F-1, the purpose is not to give exact numbers, but to suggest to you the vast economic scale of the collective chemical enterprise.

Following that, Table F-3 shows the top 75 US chemical producers in 1998.

| Table F-2 The 1997 World Top Fifty Chemical Producers | | | | |
|---|---|---|---|---|
| Rank | Company | Total Sales millions of $ | Your share in 1998 | Percentage of chemical business |
| 1 | BASF (Germany) | $27,046.9 | | 84.1% |
| 2 | DuPont (US) | 21,295.0 | $84 | 47.2% |
| 3 | Bayer (Germany) | 19,178.2 | | 60.5% |
| 4 | Dow Chemical (US) | 19,056.0 | $76 | 95.3% |
| 5 | Hoechst (Germany) | 16,293.8 | | 54.2% |
| 6 | Shell (UK/Netherlands) | 14,251.6 | | 8.3% |
| 7 | ICI (UK) | 13,349.2 | | 73.7% |
| 8 | Exxon (US) | 14,024.0 | $56 | 11.5% |
| 9 | Akzo Nobel (Netherlands) | 9,997.9 | | 81.1% |
| 10 | Elf Aquitaine (France) | 9,954.3 | | 22.8% |
| 11 | Rhone-Poulenc (France) | 9,868.6 | | 64.0% |
| 12 | Sumitomo Chemical (Japan) | 8,196.9 | | 97.2% |
| 13 | Dainippon Ink & Chemicals (Japan) | 7,420.7 | | 87.9% |
| 14 | Clariant (Switzerland) | 7,017.2 | | 100.0% |
| 15 | Norsk Hydro (Norway) | 6,874.1 | | 50.6% |
| 16 | Huls (Germany) | 6,751.2 | | 100.0% |
| 17 | Mitsubishi Chemical (Japan) | 6,740.0 | | 47.0% |
| 18 | General Electric (US) | 6,695.0 | $28 | 7.4% |
| 19 | Union Carbide (US) | 6,502.0 | $28 | 100.0% |
| 20 | SABIC (Saudi Arabia) | 6,414.7 | | 100.0% |
| 21 | Toray Industries (Japan) | 6,342.6 | | 70.6% |
| 22 | Henkel (Germany) | 6,186.9 | | 53.5% |
| 23 | DSM (Netherlands) | 6,122.1 | | 96.3% |
| 24 | BOC (UK) | 6,017.5 | | 92.7% |
| 25 | Amoco (US) | 5,941.0 | $24 | 18.1% |
| 26 | Solvay (Belgium) | 5,906.5 | | 67.9% |
| 27 | Novartis (Switzerland) | 5,737.6 | | 26.7% |
| 28 | ENI (Italy) | 5,668.2 | | 15.9% |
| 29 | Air Liquide (France) | 5,458.1 | | 83.0% |
| 30 | Ciba Specialties (Switzerland) | 5,389.7 | | 100.0% |
| 31 | British Petroleum (UK) | 5,114.6 | | 7.2% |
| 32 | Huntsman Corp. (US) | 5,000.0 | $20 | 100.0% |
| 33 | Total (France) | 4,889.2 | | 14.9% |
| 34 | Praxair (US) | 4,735.0 | $19 | 100.0% |
| 35 | Eastman Chemical (US) | 4,678.0 | $19 | 100.0% |
| 36 | Asahi Chemical (Japan) | 4,590.6 | | 43.3% |
| 37 | Occidental Petroleum (US) | 4,349.0 | $17 | 54.3% |
| 38 | Degussa (Germany) | 4,274.7 | | 48.3% |
| 39 | Allied Signal (US) | 4,254.0 | $16 | 29.4% |
| 40 | Air Products (US) | 4,122.0 | $16 | 88.9% |
| 41 | Zeneca (UK) | 4,120.5 | | 48.4% |
| 42 | Formosa Plastics (Taiwan) | 4,049.7 | | 64.9% |
| 43 | Ashland (US) | 4,047.0 | $16 | 28.5% |
| 44 | Rohm and Haas (US) | 3,999.0 | $16 | 100.0% |
| 45 | Arco Chemical (US) | 3,995.0 | $16 | 100.0% |
| 46 | Showa Denko (Japan) | 3,979.2 | | 60.3% |
| 47 | Roche (Switzerland) | 3,948.9 | | 30.5% |
| 48 | Sekisui Chemical (Japan) | 3,862.6 | | 51.7% |
| 49 | Reliance Industries (India) | 3,691.2 | | 100.0% |
| 50 | Chevron (US) | 3,632.0 | $1 | 8.9% |

Data in Table F-2 from *Chemical and Engineering News*, 20 July 1998.

Table F-3 **The 1998 Top 75 US Chemical Companies**

| Rank 1998 | Company | Chemical sales 1998 ($ millions) | Your share in 1998 | Chemical sales as % of total sales | Industry classification |
|---|---|---|---|---|---|
| 1 | DuPont | 26,202.0 | $105 | 94.4% | Diversified |
| 2 | Dow Chemical | 17,710.0 | $71 | 96.0 | Basic chemicals |
| 3 | Exxon | 10,504.0 | $42 | 9.1 | Petroleum |
| 4 | General Electric | 6,633.0 | $27 | 6.6 | Diversified |
| 5 | Union Carbide | 5,659.0 | $23 | 100.0 | Basic chemicals |
| 6 | Huntsman Chemical | 5,200.0 | $21 | 100.0 | Basic chemicals |
| 7 | ICI Americas | 4,900.0 | $20 | 100.0 | Specialty chemicals |
| 8 | Praxair | 4,833.0 | $19 | 100.0 | Basic chemicals |
| 9 | BASF | 4,800.0 | $19 | 66.7 | Basic chemicals |
| 10 | Eastman Chemical | 4,481.0 | $18 | 100.0 | Basic chemicals |
| 11 | BP Amoco | 4,470.0 | $18 | 13.5 | Petroleum |
| 12 | Air Products | 4,446.7 | $18 | 90.4 | Basic chemicals |
| 13 | Shell Oil | 4,191.0 | $17 | 20.4 | Petroleum |
| 14 | Allied Signal | 4,169.0 | $17 | 27.6 | Diversified |
| 15 | Ashland Oil | 4,087.0 | $16 | 62.5 | Petroleum |
| 16 | Monsanto | 4,032.0 | $16 | 46.6 | Life sciences |
| 17 | Celanese | 3,932.0 | $16 | 100.0 | Basic chemicals |
| 18 | Rohm and Haas | 3,720.0 | $15 | 100.0 | Basic chemicals |
| 19 | Chevron | 3,054.0 | $12 | 10.2 | Petroleum |
| 20 | Occidental Petroleum | 2,975.0 | $12 | 45.1 | Petroleum |
| 21 | Akzo Nobel | 2,901.0 | $12 | 100.0 | Basic chemicals |
| 22 | Solutia | 2,840.0 | $11 | 100.0 | Basic chemicals |
| 23 | IMC Global | 2,696.2 | $11 | 100.0 | Agrichemicals |
| 24 | Dow Corning | 2,568.0 | $10 | 100.0 | Specialty chemicals |
| 25 | Phillips Petroleum | 2,493.0 | $10 | 21.6 | Petroleum |
| 26 | Mobil | 2,428.0 | $10 | 4.5 | Petroleum |
| 27 | Potash Corp. of Sask. | 2,307.8 | $9 | 100.0 | Fertilizers |
| 28 | FMC | 2,220.4 | $9 | 50.7 | Machinery |
| 29 | Elf Atochem | 2,200.0 | $9 | 100.0 | Basic chemicals |
| 30 | American Home Products | 2,194.1 | $9 | 16.3 | Specialty chemicals |
| 31 | Hercules | 2,145.0 | $9 | 100.0 | Specialty chemicals |
| 32 | Witco | 1,941.5 | $8 | 100.0 | Specialty chemicals |
| 33 | Morton International | 1,737.4 | $7 | 68.7 | Specialty chemicals |
| 34 | Engelhard | 1,680.2 | $7 | 40.2 | Diversified |
| 35 | Lubrizol | 1,614.6 | $7 | 100.0 | Specialty chemicals |
| 36 | Millennium | 1,597.0 | $6 | 100.0 | Basic chemicals |
| 37 | Nalco Chemical | 1,573.5 | $6 | 100.0 | Specialty chemicals |
| 38 | Henkel | 1,526.0 | $6 | 100.0 | Specialty chemicals |
| 39 | PPG Industries | 1,524.0 | $6 | 20.3 | Glass products |
| 40 | Ciba Specialty Chemicals | 1,516.0 | $6 | 100.0 | Specialty chemicals |
| 41 | Rhodia | 1,500.0 | $6 | 100.0 | Specialty chemicals |
| 42 | W. R. Grace | 1,463.4 | $6 | 100.0 | Specialty chemicals |
| 43 | Crompton & Knowles | 1,451.6 | $6 | 80.8 | Specialty chemicals |
| 44 | Lyondell Chemical | 1,447.0 | $6 | 100.0 | Petroleum products |
| 45 | Cytec Industries | 1,444.3 | $6 | 100.0 | Basic chemicals |
| 46 | AstraZeneca | 1,437.0 | $6 | 35.9 | Specialty chemicals |
| 46 | Cabot | 1,437.0 | $6 | 87.2 | Basic chemicals |
| 48 | Great Lakes Chemical | 1,394.3 | $6 | 100.0 | Specialty chemicals |
| 49 | H. B. Fuller | 1,347.2 | $5 | 100.0 | Specialty chemicals |
| 50 | CF Industries | 1,200.3 | $5 | 100.0 | Agrichemicals |
| 51 | Clariant | 1,200.0 | $5 | 100.0 | Specialty chemicals |
| 52 | BF Goodrich | 1,195.8 | $5 | 30.3 | Basic chemicals |
| 53 | Farmland Industries | 1,157.8 | $5 | 13.2 | Agricultural supplies |
| 54 | Ethyl | 974.2 | $4 | 100.0 | Basic chemicals |
| 55 | Wellman | 968.0 | $4 | 100.0 | Basic chemicals |
| 55 | DSM | 968.0 | $4 | 100.0 | Basic chemicals |
| 57 | Kerr-McGee | 933.0 | $4 | 66.8 | Petroleum |
| 58 | Degussa | 926.0 | $4 | 43.0 | Specialty chemicals |
| 59 | NL Industries | 894.7 | $4 | 100.0 | Basic chemicals |
| 60 | Georgia Gulf | 875.0 | $4 | 100.0 | Basic chemicals |

| Rank 1998 | Company | Chemical sales 1998 ($ millions) | Your share in 1998 | Chemical sales as % of total sales | Industry classification |
|---|---|---|---|---|---|
| 61 | Arch Chemicals | 862.8 | $3 | 100.0 | Basic chemicals |
| 62 | Terra Industries | 848.1 | $3 | 33.2 | Agrichemicals |
| 63 | Aristech | 830.8 | $3 | 100.0 | Basic chemicals |
| 64 | Intl. Specialty Prod. | 823.9 | $3 | 100.0 | Specialty chemicals |
| 65 | Sterling Chemicals | 822.6 | $3 | 100.0 | Basic chemicals |
| 66 | Albemarle | 820.9 | $3 | 100.0 | Specialty chemicals |
| 67 | Union Camp | 750.0 | $3 | 16.7 | Paper products |
| 68 | Gen Corp | 689.0 | $3 | 39.7 | Rubber products |
| 69 | Alcoa | 670.0 | $3 | 4.4 | Nonferrous metals |
| 70 | Vulcan Materials | 617.8 | $2 | 34.8 | Nonmetallic minerals |
| 71 | Stepan | 610.5 | $2 | 100.0 | Specialty chemicals |
| 72 | General Chemical | 577.3 | $2 | 81.8 | Basic chemicals |
| 73 | Creanova | 550.0 | $2 | 100.0 | Specialty chemicals |
| 74 | Ferro | 544.0 | $2 | 39.9 | Specialty materials |
| 75 | Mississippi Chemical | 519.9 | $2 | 100.0 | Agrichemicals |

Table F-3 **The 1998 Top 75 US Chemical Companies (continued)**

Data in Table F-3 from *Chemical and Engineering News*, 3 May 1999

Data in the following tables (F-4 to F-10) are from *Chemical and Engineering News*, 28 June 1999

Table F-4 **1998 US Inorganic Chemicals Consumption (1000s of tons)**

| Inorganic Chemicals | Formula | 1000s tons | your share pounds |
|---|---|---|---|
| Sulfuric acid, pure | $H_2SO_4$ | 47,576 | 381 |
| Nitrogen (excluding ammonia production) | $N_2$ | 27,600 | 221 |
| Oxygen (excluding ammonia production) | $O_2$ | 26,000 | 208 |
| Phosphoric acid (pure $H_3PO_4$) | $H_3PO_4$ | 19,900 | 159 |
| Ammonia (anhydrous, synthetic, pure) | $NH_3$ | 19,753 | 158 |
| Chlorine | $Cl_2$ | 12,844 | 103 |
| Sodium hydroxide, pure | NaOH | 11,496 | 92 |
| Nitric acid (pure) | $HNO_3$ | 9,374 | 75 |
| Ammonium nitrate, (anhydrous, pure) | $NH_4NO_3$ | 8,620 | 69 |
| Hydrochloric acid | HCl | 4,298 | 34 |
| Ammonium sulfate, (synthetic, pure, non-coke) | $(NH_4)_2SO_4$ | 2,757 | 22 |
| Titanium dioxide, pure and compounds | $TiO_2$ | 1,463 | 12 |
| Aluminum sulfate (commercial) | $Al_2(SO_4)_3$ | 1,247 | 10 |
| Sodium silicate (estimated) | $Na_2SiO_3$ | 1,000 | 8 |
| Hydrogen (nonvented, noncaptive) | $H_2$ | 996 | 8 |
| Sodium sulfate, pure | $Na_2SO_4$ | 592 | 5 |
| Sodium chlorate, pure | $NaClO_3$ | 574 | 5 |
| Hydrogen peroxide | $H_2O_2$ | 512 | 4 |

In Table F-4 "pure" means calculated as 100% of the listed compound. In many cases actual production is in the form of aqueous solutions containing substantially less than 100% of the compound.

| Table F-5 **1998 US Minerals Consumption (1000s of tons)** | | | |
|---|---|---|---|
| Mineral | Formula | 1000s tons | your share pounds |
| Phosphate rock | $CaF_2.3Ca_3(PO_4)_2$ | 49,200 | 394 |
| Lime, sold or used | $CaO$ | 22,491 | 180 |
| Sulfur | $S$ | 15,325 | 123 |
| Sodium carbonate, natural | $Na_2CO_3$ | 11,356 | 91 |
| Potash (as $K_2O$ equivalent) | $K_2O$ | 5843 | 47 |
| Calcium chloride, pure (estimated) | $CaCl_2$ | 700 | 6 |
| Bromine, sold or used | $Br_2$ | 258 | 2 |
| Lithium, (estimated as Li) | $Li$ | 3.2 | 0.02 |

In Table F-5 "pure" means calculated as 100% of the listed compound.

| Table F-6 **1998 US Organic Chemicals Consumption (Millions of lbs)** | | | |
|---|---|---|---|
| Organic Chemical | Formula | $10^6$ lbs | Your share lbs in 1998 |
| Ethylene | $C_2H_4$ | 52,061 | 208 |
| Propylene | $CH_3\text{-}CH{=}CH_2$ | 28,613 | 114 |
| Ethylene dichloride | $CH_2Cl\text{–}CH_2Cl$ | 24,560 | 98 |
| Urea | $CO(NH_2)_2$ | 17,602 | 70 |
| Benzene | $C_6H_6$ | 16,346 | 65 |
| Ethylbenzene (excluding from coke) | $C_6H_5\text{–}CH_3$ | 12,661 | 51 |
| Styrene | $C_6H_5\text{-}CH{=}CH_2$ | 11,390 | 46 |
| Ethylene oxide | $C_2H_4O$ | 8,140 | 33 |
| p-Xylene | $H_3C\text{-}C_6H_4\text{-}CH_3$ | 7,695 | 31 |
| Cumene | $C_6H_5\text{–}CH(CH_3)_2$ | 6,713 | 27 |
| 1,3-Butadiene (rubber grade) | $H_2C{=}CH\text{-}CH{=}CH_2$ | 4,066 | 16 |
| Acrylonitrile | $CH_2{=}CH\text{-}CN$ | 3,120 | 12 |
| Aniline | $C_6H_5NH_2$ | 1,545 | 6 |
| Isopropyl alcohol | $C_3H_7OH$ | 1,449 | 6 |
| o-Xylene | $H_3C\text{-}C_6H_4\text{-}CH_3$ | 1,013 | 4 |
| 2-Ethylhexanol | $C_8H_{17}OH$ | 811 | 3 |

| Table F-7 **1998 US Plastics Consumption (Millions of lbs)** | | | |
|---|---|---|---|
| Thermosetting Resins | $10^6$ lbs | Thermoplastic Resins | $10^6$ lbs |
| Epoxy | 600 | Polyethylene Low density | 7,600 |
| Melamine | 300 | Polyethylene Linear low density | 7,200 |
| Phenolic | 3,900 | Polyethylene High density | 13,000 |
| Polyester | 1,700 | Polypropylene | 14,000 |
| Urea | 2,600 | Polystyrene | 6,200 |
| Total thermosets | 9,100 | Styrene-acrylonitrile | 120 |
| | | Acrylonitrile-butadiene-styrene and other styrene polymers | 3,100 |
| | | Polyamide, nylon type | 1,300 |
| | | Polyvinyl chloride & copolymers | 15,000 |
| | | Thermoplastic polyester | 4,400 |
| | | Total thermoplastics | 71,920 |
| Grand Total, all resins 80,400 million pounds (your share = 320 pounds) | | | |

| Table F-8 **1998 US Synthetic Rubber Production (1000s of metric tons)** | | |
|---|---|---|
| Rubber type | 1000s of metric tons | Your share pounds |
| Styrene-butadiene rubber | 960 | 8 |
| Polybutadiene | 580 | 5 |
| Ethylene-propylene | 321 | 3 |
| Nitrile, solid | 89 | 0.6 |
| Polychloroprene | 72 | 0.6 |
| Other | 454 | 4 |
| Total | 2,476 | 20 |

| Table F-9 **1998 US Synthetic Fiber Production (Millions of pounds)** | | |
|---|---|---|
| Fiber type | Millions of lbs | your share pounds |
| Polyester | 3,911 | 16 |
| Nylon | 2,847 | 11 |
| Olefin | 2,800 | 11 |
| Acetate & rayon | 365 | 1.5 |
| Acrylic | 346 | 1.4 |
| Total | 10,269 | 41 |

# Appendix G—Crystallography and Chemical Structure

## Introduction

Perhaps the single most important task of chemistry is to deduce the atomic structure of *everything*. To measure all the bond lengths and all the bond angles in a substance. The bulk properties of materials derive directly from how atoms and ions are arranged and the forces among them, so chemical structure is the basis for materials and engineering science. To understand chemical structure is to be able to design and modify materials for specific needs. In our everyday lives we use specifically designed materials for thousands of needs. Even more broadly, as the recently completed first draft of the entire human genome emphasizes, it is chemical structure, as embodied in the DNA molecule, that controls the very functioning of human life.

Many solids are crystalline, but not all. Glass is a very familiar example of a non crystalline, or amorphous, solid. If a solid is crystalline, its structure can be deduced by the methods of x-ray crystallography. **X-ray crystallography** is the study of the internal structure of crystals using the phenomenon of x-ray diffraction. Diffraction is a widespread phenomenon of physics that occurs whenever electromagnetic radiation falls on objects having a size approximately the same as the wavelength of the radiation.

## History of the Method

X-ray crystallography came to life in 1912 when Max von Laue (1879-1960), at a conference in Munich, suggested that the then newly discovered x-rays would create diffraction patterns if they were directed at crystalline substances. He had correctly speculated that the sizes of the atomic spacings in crystals were comparable the wavelengths of x-rays. Experimental confirmation came within a few months.

The theory behind the x-ray diffraction experiment was worked out in the 1920s by the father and son team of William Henry Bragg (1862-1942) and William Lawrence Bragg (1890-1971). They showed that studies of the angles of diffraction of the x-rays would reveal the basic repeating unit of the crystal—it's unit cell, and realized that studying the intensity of the various diffracted x-ray beams gave information from which the exact positions of the atoms in the unit cell could be derived. During this period the close packed structures of simple compounds and metals were discovered, and the fundamental value of crystallography for materials science was established. It also became clear that interpreting x-ray diffraction data requires lots of computing—something not available in the 1920s.

In 1928, a Scottish dye company accidentally produced copper phthalocyanine, a commercially valuable blue dye. Its molecule consists of 58 atoms in a planar arrangement, and its crystal structure was solved in 1935, taking advantage of the fact that the dye is essentially a two-dimensional molecule, and was therefore just capable at that time of being handled mathematically. Copper phthalocyanine was to remain for almost 25 years the largest molecule whose detailed structure was known.

Low resolution x-ray methods were exploited in the 1950s to determine the overall, if not detailed structures, of important biological molecules. Low resolution methods show the general arrangements of the atoms in a crystal but not their exact positions. The general structures of various proteins were worked out by Linus Pauling (1901-1994) and others. In 1953, the double helical structure of DNA was deduced based partly on x-ray diffraction data obtained by Rosalind Franklin (1920-1958) and Maurice Wilkins (1916-). The first detailed protein structure, hemoglobin, was determined by Max Perutz (1914-) in a series of increasingly refined measurements culminating around 1963.

Since 1963, crystallography's come a long way, but the developments have been entirely in matters of experimental technique and more and more computational power. Forty years ago, solving just one crystal structure was a project worth a Ph.D. thesis. Today, it is not uncommon for a thesis to present 10-20 structures, and the work will be aimed not so much at solving the structures as at interpreting the role of structure in some chemical phenomenon. Modem x-ray crystallography is an efficient, automatic procedure that in many cases can be conducted by relatively inexperienced persons.

## A Modern Diffractometer

To obtain a diffraction pattern from a crystal, x-rays with a single wavelength (monochromatic radiation) are directed at the crystal, as shown in close-up in Figure G1 and in an overall view in Figure G2. A close up of the detector is shown in Figure G3.

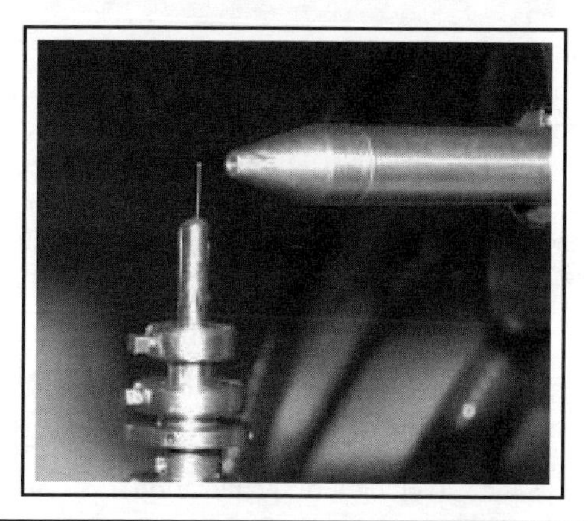

Figure G1 A close up view of a mounted crystal ready for study by x-ray diffraction. The x-ray beam comes through the beam guide on the right. The crystal's dimensions are about 1 mm. You can see that it does not take a very big crystal to do this work; a crystal with a volume of 0.5 $mm^3$ is usually adequate. The crystal is glued onto an amorphous fiber, which being itself non crystalline does not form a diffraction pattern

Figure G2 A modern x-ray diffractometer. The x-ray source is on the far right. The crystal is mounted on a fiber and centered in a gimbal-like device, with a diameter of about 15 inches, that allows the crystal to be oriented at any angle to the incoming x-ray beam. The detector is on the left.

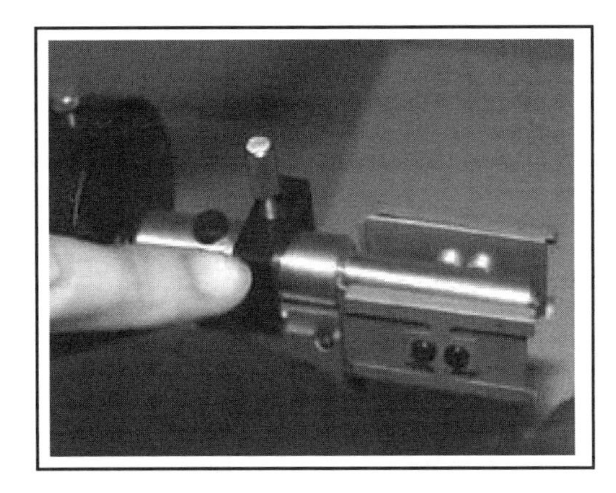

Figure G3 The x-ray detector is a sodium iodide scintillation device. When x-rays strike a properly prepared sodium iodide crystal they cause flashes of light (scintillations). These flashes of light are counted by a sensitive photocell. Modern, high-capacity scintillation counters can achieve 10 million counts per second. This high efficiency means that x-ray crystallographic data can be collected quickly.

X-rays have wave properties, and when a beam of x-rays strikes a crystal, the result is either constructive interference to give bright regions or destructive interference to give dark regions. Whether the interference is constructive or destructive depends on the angle at which the x-rays enter the crystal and on the arrangement of the atoms in the crystal. Thus, a beam of x-rays fired at a crystal will give a diffraction pattern composed of dark regions (reinforced directions where destructive interference has occurred), and light regions (where constructive interference has occurred). As shown in Figure G4, it is possible to visualize this pattern of dark and light regions by exposing photographic film to the diffracted radiation.

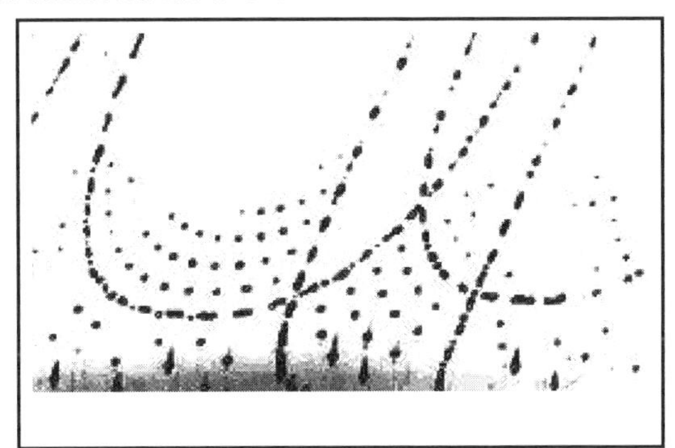

Figure G4 A Weissenberg x-ray diffraction photograph. The pattern consists of spots of exposure where the x-rays constructively reinforced, surrounded by unexposed regions where destructive interference occurred. Modern methods use direct counting, rather than photographic techniques, to study the diffraction patterns.

The determination of crystal structure from diffractometer data is a big calculation. But today such calculations are entirely routine. The specific result of a crystallographic calculation is a list of x, y, z coordinates for the positions of the atoms in the unit cell. From these coordinates, all the bond lengths and angles can be calculated by means of straightforward three-dimensional Cartesian geometry. These same atomic coordinates can then be used as input into a visualization program such as Chime, so that the molecule can be observed and manipulated.

[❁kws "x-ray crystallography", "x-ray diffraction", "protein data bank"]

# Periodic Table (Taken from IUPAC, 1995 -- Updated through June 1999)

http://www.chem.qmw.ac.uk/iupac2/atwt.html

| 1 | 2 | 3 | 4 | 5 | 6 | 7 | 8 | 9 | 10 | 11 | 12 | 13 | 14 | 15 | 16 | 17 | 18 |
|---|---|---|---|---|---|---|---|---|---|---|---|---|---|---|---|---|---|
| $s^1$ | $s^2$ | $s^2d^1$ | $s^2d^2$ | $s^2d^3$ | $s^2d^4$ | $s^2d^5$ | $s^2d^6$ | $s^2d^7$ | $s^2d^8$ | $s^2d^9$ | $s^2d^{10}$ | $s^2p^1$ | $s^2p^2$ | $s^2p^3$ | $s^2p^4$ | $s^2p^5$ | $s^2$ or $s^2p^6$ |
| 1<br>**H**<br>Hydrogen<br>gm 1.00794 | | | | | | | | | | | | | | | | | 2<br>**He**<br>Helium<br>g 4.002602 |
| 3<br>**Li**<br>Lithium<br>gm 6.941 | 4<br>**Be**<br>Beryllium<br>9.012182 | | | | | | | | | | | 5<br>**B**<br>Boron<br>gm 10.811 | 6<br>**C**<br>Carbon<br>g 12.0107 | 7<br>**N**<br>Nitrogen<br>g 14.00674 | 8<br>**O**<br>Oxygen<br>g 15.9994 | 9<br>**F**<br>Fluorine<br>18.9984032 | 10<br>**Ne**<br>Neon<br>gm 20.1797 |
| 11<br>**Na**<br>Sodium<br>22.989770 | 12<br>**Mg**<br>Magnesium<br>24.3050 | | | | | | | | | | | 13<br>**Al**<br>Aluminum<br>26.981538 | 14<br>**Si**<br>Silicon<br>28.0855 | 15<br>**P**<br>Phosphorus<br>30.973761 | 16<br>**S**<br>Sulfur<br>g 32.066 | 17<br>**Cl**<br>Chlorine<br>m 35.4527 | 18<br>**Ar**<br>Argon<br>g 39.948 |
| 19<br>**K**<br>Potassium<br>g 39.0983 | 20<br>**Ca**<br>Calcium<br>g 40.078 | 21<br>**Sc**<br>Scandium<br>44.955910 | 22<br>**Ti**<br>Titanium<br>47.867 | 23<br>**V**<br>Vanadium<br>50.9415 | 24<br>**Cr**<br>Chromium<br>51.9961 | 25<br>**Mn**<br>Manganese<br>54.938049 | 26<br>**Fe**<br>Iron<br>55.845 | 27<br>**Co**<br>Cobalt<br>58.933200 | 28<br>**Ni**<br>Nickel<br>58.6934 | 29<br>**Cu**<br>Copper<br>63.546 | 30<br>**Zn**<br>Zinc<br>65.39 | 31<br>**Ga**<br>Gallium<br>69.723 | 32<br>**Ge**<br>Germanium<br>72.61 | 33<br>**As**<br>Arsenic<br>74.92160 | 34<br>**Se**<br>Selenium<br>78.96 | 35<br>**Br**<br>Bromine<br>79.904 | 36<br>**Kr**<br>Krypton<br>gm 83.80 |
| 37<br>**Rb**<br>Rubidium<br>g 85.4678 | 38<br>**Sr**<br>Strontium<br>g 87.62 | 39<br>**Y**<br>Yttrium<br>88.90585 | 40<br>**Zr**<br>Zirconium<br>g 91.224 | 41<br>**Nb**<br>Niobium<br>92.90638 | 42<br>**Mo**<br>Molybdenum<br>95.94 | 43<br>**Tc**<br>Technetium<br>[98] | 44<br>**Ru**<br>Ruthenium<br>g 101.07 | 45<br>**Rh**<br>Rhodium<br>102.90550 | 46<br>**Pd**<br>Palladium<br>g 106.42 | 47<br>**Ag**<br>Silver<br>g 107.868[illegible] | 48<br>**Cd**<br>Cadmium<br>g 112.411 | 49<br>**In**<br>Indium<br>114.818 | 50<br>**Sn**<br>Tin<br>g 118.710 | 51<br>**Sb**<br>Antimony<br>g 121.760 | 52<br>**Te**<br>Tellurium<br>g 127.60 | 53<br>**I**<br>Iodine<br>126.90447 | 54<br>**Xe**<br>Xenon<br>gm 131.29 |
| 55<br>**Cs**<br>Cesium<br>132.90545 | 56<br>**Ba**<br>Barium<br>137.327 | 57-71<br>* | 72<br>**Hf**<br>Hafnium<br>178.49 | 73<br>**Ta**<br>Tantalum<br>180.9479 | 74<br>**W**<br>Tungsten<br>183.84 | 75<br>**Re**<br>Rhenium<br>186.207 | 76<br>**Os**<br>Osmium<br>g 190.23 | 77<br>**Ir**<br>Iridium<br>192.217 | 78<br>**Pt**<br>Platinum<br>195.078 | 79<br>**Au**<br>Gold<br>196.9665[illegible] | 80<br>**Hg**<br>Mercury<br>200.59 | 81<br>**Tl**<br>Thallium<br>204.3833 | 82<br>**Pb**<br>Lead<br>g 207.2 | 83<br>**Bi**<br>Bismuth<br>208.98038 | 84<br>**Po**<br>Polonium<br>[209] | 85<br>**At**<br>Astatine<br>[210] | 86<br>**Rn**<br>Radon<br>[222] |
| 87<br>**Fr**<br>Francium<br>[223] | 88<br>**Ra**<br>Radium<br>[226] | 89-103<br>† | 104<br>**Rf**<br>Rutherfordium<br>[261] | 105<br>**Db**<br>Dubnium<br>[262] | 106<br>**Sg**<br>Seaborgium<br>[263] | 107<br>**Bh**<br>Bohrium<br>[262] | 108<br>**Hs**<br>Hassium<br>[265] | 109<br>**Mt**<br>Meitnerium<br>[266[ | 110<br>**Uun**<br>Not named<br>[269] | 111<br>**Uuu**<br>Not named<br>[272] | 112<br>**Uub**<br>Not named<br>[277] | | 114?<br>**Uuq**<br>Not named<br>[289] | | 116<br>**Uuh**<br>Not named<br>[289] | | 118<br>**Uuo**<br>Not named<br>[293] |

| | | | | | | | | | | | | | | | |
|---|---|---|---|---|---|---|---|---|---|---|---|---|---|---|---|
| * | 57<br>**La**<br>Lanthanum<br>g 138.9055 | 58<br>**Ce**<br>Cerium<br>g 140.116 | 59<br>**Pr**<br>Praseodymium<br>140.90765 | 60<br>**Nd**<br>Neodymium<br>g 144.24 | 61<br>**Pm**<br>Promethium<br>[145] | 62<br>**Sm**<br>Samarium<br>g 150.36 | 63<br>**Eu**<br>Europium<br>g 151.964 | 64<br>**Gd**<br>Gadolinium<br>g 157.25 | 65<br>**Tb**<br>Terbium<br>158.925[illegible] | 66<br>**Dy**<br>Dysprosium<br>g 162.50 | 67<br>**Ho**<br>Holmium<br>164.93032 | 68<br>**Er**<br>Erbium<br>g 167.26 | 69<br>**Tm**<br>Thulium<br>168.93421 | 70<br>Yb<br>Ytterbium<br>173.04 | 71<br>**Lu**<br>Lutetium<br>174.967 |
| † | 89<br>**Ac**<br>Actinium<br>[227] | 90<br>**Th**<br>Thorium<br>g 232.0381 | 91<br>**Pa**<br>Protactinium<br>231.03588 | 92<br>**U**<br>Uranium<br>gm 238.0289 | 93<br>**Np**<br>Neptunium<br>[237] | 94<br>**Pu**<br>Plutonium<br>[244] | 95<br>**Am**<br>Americium<br>[243] | 96<br>**Cm**<br>Curium<br>[247] | 97<br>**Bk**<br>Berkelium<br>[247] | 98<br>**Cf**<br>Californium<br>[251] | 99<br>**Es**<br>Einsteinium<br>[252] | 100<br>**Fm**<br>Fermium<br>[257] | 101<br>**Md**<br>Mendelevium<br>[258] | 102<br>**No**<br>Nobelium<br>[259] | 103<br>**Lr**<br>Lawrencium<br>[262] |

The typical uncertainty in the listed atomic weights is 1-3 in the least significant digit. See the IUPAC table for the actual, experimental uncertainties.

g Geologically exceptional specimens are known in which the atomic weight is outside the IUPAC specified uncertainty range.

m Modified atomic weight outside the IUPAC specified uncertainty range may be found in commercial samples of this element that have undergone isotope separation or extraction.

[ ] Brackets denote a radioactive element that lacks a characteristic natural mixture of isotopes. The value stated is the nucleon number of the element's most stable nuclide. Th, Pa, and U have no stable isotopes, but their isotopes are so long lived that we can treat their atomic weights as meaningful.

# Periodic Table -- Experimental Electron Configurations of Neutral Atoms in their Ground States

| 1<br>**H**<br>$1s^1$ | s↓ | d↓ | d↓ | d↓ | d↓ | d↓ | d↓ | d↓ | d↓ | d↓ | d↓ | p↓ | p↓ | p↓ | p↓ | p↓ | 2<br>**He**<br>$1s^2$ |
|---|---|---|---|---|---|---|---|---|---|---|---|---|---|---|---|---|---|
| 3<br>**Li**<br>$2s^1$ | 4<br>**Be**<br>$2s^2$ | | | | | | | | | | | 5<br>**B**<br>$2s^22p^1$ | 6<br>**C**<br>$2s^22p^2$ | 7<br>**N**<br>$2s^22p^3$ | 8<br>**O**<br>$2s^22p^4$ | 9<br>**F**<br>$2s^22p^5$ | 10<br>**Ne**<br>$2s^22p^6$ |
| 11<br>**Na**<br>$3s^1$ | 12<br>**Mg**<br>$3s^2$ | | | | | | | | | | | 13<br>**Al**<br>$3s^23p^1$ | 14<br>**Si**<br>$3s^23p^2$ | 15<br>**P**<br>$3s^23p^3$ | 16<br>**S**<br>$3s^23p^4$ | 17<br>**Cl**<br>$3s^23p^5$ | 18<br>**Ar**<br>$3s^23p^6$ |
| 19<br>**K**<br>$4s^1$ | 20<br>**Ca**<br>$4s^2$ | 21<br>**Sc**<br>$4s^23d^1$ | 22<br>**Ti**<br>$4s^23d^2$ | 23<br>**V**<br>$4s^23d^3$ | 24<br>**Cr**<br>‡<br>$4s^13d^5$ | 25<br>**Mn**<br>$4s^23d^5$ | 26<br>**Fe**<br>$4s^23d^6$ | 27<br>**Co**<br>$4s^23d^7$ | 28<br>**Ni**<br>$4s^23d^8$ | 29<br>**Cu**<br>‡<br>$4s^13d^{10}$ | 30<br>**Zn**<br>$4s^23d^{10}$ | 31<br>**Ga**<br>$4s^23d^{10}4p^1$ | 32<br>**Ge**<br>$4s^23d^{10}4p^2$ | 33<br>**As**<br>$4s^23d^{10}4p^3$ | 34<br>**Se**<br>$4s^23d^{10}4p^4$ | 35<br>**Br**<br>$4s^23d^{10}4p^5$ | 36<br>**Kr**<br>$4s^23d^{10}4p^6$ |
| 37<br>**Rb**<br>$5s^1$ | 38<br>**Sr**<br>$5s^2$ | 39<br>**Y**<br>$5s^24d^1$ | 40<br>**Zr**<br>$5s^24d^2$ | 41<br>**Nb**<br>‡<br>$5s^14d^4$ | 42<br>**Mo**<br>‡<br>$5s^14d^5$ | 43<br>**Tc**<br>$5s^24d^5$ | 44<br>**Ru**<br>$5s^24d^6$ | 45<br>**Rh**<br>$5s^24d^7$ | 46<br>**Pd**<br>‡<br>$4d^{10}$ | 47<br>**Ag**<br>$5s^24d^9$ | 48<br>**Cd**<br>$5s^24d^{10}$ | 49<br>**In**<br>$5s^24d^{10}5p^1$ | 50<br>**Sn**<br>$5s^24d^{10}5p^2$ | 51<br>**Sb**<br>$5s^24d^{10}5p^3$ | 52<br>**Te**<br>$5s^24d^{10}5p^4$ | 53<br>**I**<br>$5s^24d^{10}5p^5$ | 54<br>**Xe**<br>$5s^24d^{10}5p^6$ |
| 55<br>**Cs**<br>$6s^1$ | 56<br>**Ba**<br>$6s^2$ | 57-71<br>* | 72<br>**Hf**<br>$6s^24f^{14}5d^2$ | 73<br>**Ta**<br>$6s^24f^{14}5d^3$ | 74<br>**W**<br>$6s^24f^{14}5d^4$ | 75<br>**Re**<br>$6s^24f^{14}5d^5$ | 76<br>**Os**<br>$6s^24f^{14}5d^6$ | 77<br>**Ir**<br>$6s^24f^{14}5d^7$ | 78<br>**Pt**<br>$6s^24f^{14}5d^8$ | 79<br>**Au**<br>$6s^24f^{14}5d^9$ | 80<br>**Hg**<br>$6s^2f^{14}5d^{10}$ | 81<br>**Tl**<br>$6s^24f^{14}5d^{10}6p^1$ | 82<br>**Pb**<br>$6s^24f^{14}5d^{10}6p^2$ | 83<br>**Bi**<br>$6s^24f^{14}5d^{10}6p^3$ | 84<br>**Po**<br>$6s^24f^{14}5d^{10}6p^4$ | 85<br>**At**<br>$6s^24f^{14}5d^{10}6p^5$ | 86<br>**Rn**<br>$6s^24f^{14}5d^{10}6p^6$ |
| 87<br>**Fr**<br>$7s^1$ | 88<br>**Ra**<br>$7s^2$ | 89-103<br>† | 104<br>**Rf**<br>?<br>$7s^25f^{14}6d^2$ | 105<br>**Db**<br>?<br>$7s^25f^{14}6d^3$ | 106<br>**Sg**<br>?<br>$7s^25f^{14}6d^4$ | 107<br>**Bh**<br>?<br>$7s^25f^{14}6d^5$ | 108<br>**Hs**<br>?<br>$7s^25f^{14}6d^6$ | 109<br>**Mt**<br>?<br>$7s^25f^{14}6d^7$ | 110<br>**Uun**<br>?<br>$7s^25f^{14}6d^8$ | 111<br>**Uuu**<br>?<br>$7s^25f^{14}6d^9$ | 112<br>**Uub**<br>?<br>$7s^25f^{14}6d^{10}$ | | 114<br>**Uuq**<br>?<br>$7s^25f^{14}6d^{10}7p^2$ | | | | |
| s↑ | s↑ | d↑ f↓ | d↑ f↓ | d↑ f↓ | d↑ f↓ | d↑ f↓ | d↑ f↓ | d↑ f↓ | d↑ f↓ | d↑ f↓ | d↑ f↓ | p↑ f↓ | p↑ f↓ | p↑ f↓ | p↑ f↓ | p↑ f↓ | s↑ or p↑ |

| | | | | | | | | | | | | | | | |
|---|---|---|---|---|---|---|---|---|---|---|---|---|---|---|---|
| * | 57<br>**La**<br>‡<br>$6s^25d^1$ | 58<br>**Ce**<br>$6s^24f^2$ | 59<br>**Pr**<br>$6s^24f^3$ | 60<br>**Nd**<br>$6s^24f^4$ | 61<br>**Pm**<br>$6s^24f^5$ | 62<br>**Sm**<br>$6s^24f^6$ | 63<br>**Eu**<br>$6s^24f^7$ | 64<br>**Gd**<br>‡<br>$6s^24f^75d^1$ | 65<br>**Tb**<br>$6s^24f^9$ | 66<br>**Dy**<br>$6s^24f^{10}$ | 67<br>**Ho**<br>$6s^24f^{11}$ | 68<br>**Er**<br>$6s^24f^{12}$ | 69<br>**Tm**<br>$6s^24f^{13}$ | 70<br>**Yb**<br>$6s^24f^{14}$ | 71<br>**Lu**<br>$6s^24f^{14}5d^1$ |
| † | 89<br>**Ac**<br>‡<br>$7s^26d^1$ | 90<br>**Th**<br>‡<br>$7s^26d^2$ | 91<br>**Pa**<br>‡<br>$7s^25f^26d^1$ | 92<br>**U**<br>‡<br>$7s^25f^36d^1$ | 93<br>**Np**<br>$7s^25f^5$ | 94<br>**Pu**<br>$7s^25f^6$ | 95<br>**Am**<br>$7s^25f^7$ | 96<br>**Cm**<br>‡<br>$7s^25f^76d^1$ | 97<br>**Bk**<br>$7s^25f^9$ | 98<br>**Cf**<br>$7s^25f^{10}$ | 99<br>**Es**<br>$7s^25f^{11}$ | 100<br>**Fm**<br>$7s^25f^{12}$ | 101<br>**Md**<br>$7s^25f^{13}$ | 102<br>**No**<br>$7s^25f^{14}$ | 103<br>**Lr**<br>$7s^25f^{14}6d^1$ |

To fill an empty configuration periodic table proceed through it adding one electron for each element. Filling proceeds from left to right across each row of the table in the sequence s - f - d - p. The correct configurations of 91/103 elements will be produced. Eleven elements (designated with a double dagger ‡) have slightly different experimental electron configurations than the the automatic filling procedure. The configurations of elements 104-114 are hypothetical. The notation d↑ f↓, etc., indicates that the elements above are filling a d-level and the elements below are filling an f-level.

# The Very Long form of the Periodic Table

## (Werner periodic table)

| 1 | 2 | 3 | | | | | | | | | | | | | | | 4 | 5 | 6 | 7 | 8 | 9 | 10 | 11 | 12 | 13 | 14 | 15 | 16 | 17 | 18 |
|---|---|---|---|---|---|---|---|---|---|---|---|---|---|---|---|---|---|---|---|---|---|---|---|---|---|---|---|---|---|---|---|
| 1<br>H<br>Hydrogen<br>gm 1.00794 | | | | | | | | | | | | | | | | | | | | | | | | | | | | | | | 2<br>He<br>Helium<br>g 4.002602 |
| 3<br>Li<br>Lithium<br>gm 6.941 | 4<br>Be<br>Beryllium<br>9.012182 | | | | | | | | | | | | | | | | | | | | | | | | | 5<br>B<br>Boron<br>gm 10.811 | 6<br>C<br>Carbon<br>g 12.0107 | 7<br>N<br>Nitrogen<br>g 14.00674 | 8<br>O<br>Oxygen<br>g 15.9994 | 9<br>F<br>Fluorine<br>18.9984032 | 10<br>Ne<br>Neon<br>gm 20.1797 |
| 11<br>Na<br>Sodium<br>22.989770 | 12<br>Mg<br>Magnesium<br>24.3050 | | | | | | | | | | | | | | | | | | | | | | | | | 13<br>Al<br>Aluminum<br>26.981538 | 14<br>Si<br>Silicon<br>28.0855 | 15<br>P<br>Phosphorus<br>30.973761 | 16<br>S<br>Sulfur<br>g 32.066 | 17<br>Cl<br>Chlorine<br>m 35.4527 | 18<br>Ar<br>Argon<br>g 39.948 |
| 19<br>K<br>Potassium<br>g 39.0983 | 20<br>Ca<br>Calcium<br>g 40.078 | 21<br>Sc<br>Scandium<br>44.955910 | | | | | | | | | | | | | | | 22<br>Ti<br>Titanium<br>47.867 | [illegible]<br>V[illegible]adium<br>[illegible]415 | 24<br>Cr<br>Chromium<br>51.9961 | 25<br>Mn<br>Manganese<br>54.938049 | 26<br>Fe<br>Iron<br>55.845 | 27<br>Co<br>Cobalt<br>58.933200 | 28<br>Ni<br>Nickel<br>58.6934 | 29<br>Cu<br>Copper<br>63.546 | 30<br>Zn<br>Zinc<br>65.39 | 31<br>Ga<br>Gallium<br>69.723 | 32<br>Ge<br>Germanium<br>72.61 | 33<br>As<br>Arsenic<br>74.92160 | 34<br>Se<br>Selenium<br>78.96 | 35<br>Br<br>Bromine<br>79.904 | 36<br>Kr<br>Krypton<br>gm 83.80 |
| 37<br>Rb<br>Rubidium<br>g 85.4678 | 38<br>Sr<br>Strontium<br>g 87.62 | 39<br>Y<br>Yttrium<br>88.90585 | | | | | | | | | | | | | | | 40<br>Zr<br>Zirconium<br>g 91.224 | [illegible]<br>[illegible]b<br>[illegible]ium<br>[illegible]638 | 42<br>Mo<br>Molybdenum<br>95.94 | 43<br>Tc<br>Technetium<br>[98] | 44<br>Ru<br>Ruthenium<br>g 101.07 | 45<br>Rh<br>Rhodium<br>102.90550 | 46<br>Pd<br>Palladium<br>g 106.42 | 47<br>Ag<br>Silver<br>g 107.8682 | 48<br>Cd<br>Cadmium<br>g 112.411 | 49<br>In<br>Indium<br>114.818 | 50<br>Sn<br>Tin<br>g 118.710 | 51<br>Sb<br>Antimony<br>g 121.760 | 52<br>Te<br>Tellurium<br>g 127.60 | 53<br>I<br>Iodine<br>126.90447 | 54<br>Xe<br>Xenon<br>gm 131.29 |
| 55<br>Cs<br>Cesium<br>132.90545 | 56<br>Ba<br>Barium<br>137.327 | 57<br>La<br>Lanthanum<br>g 138.9055 | 58<br>Ce<br>Cerium<br>g 140.116 | 59<br>Pr<br>Praeseodymium<br>140.90765 | 60<br>Nd<br>Neodymium<br>g 144.24 | 61<br>Pm<br>Promethium<br>[145] | 62<br>Sm<br>Samarium<br>g 150.36 | 63<br>Eu<br>Europium<br>g 151.964 | 64<br>Gd<br>Gadolinium<br>g 157.25 | 65<br>Tb<br>Terbium<br>158.92534 | 66<br>Dy<br>Dysprosium<br>g 162.50 | 67<br>Ho<br>Holmium<br>164.93032 | 68<br>Er<br>Erbium<br>g 167.26 | 69<br>Tm<br>Thulium<br>168.93421 | 70<br>Yb<br>Ytterbium<br>173.04 | 71<br>Lu<br>Lutetium<br>174.967 | 72<br>Hf<br>Hafnium<br>178.49 | [illegible]3<br>[illegible]a<br>[illegible]antalum<br>[illegible]9479 | 74<br>W<br>Tungsten<br>183.84 | 75<br>Re<br>Rhenium<br>186.207 | 76<br>Os<br>Osmium<br>g 190.23 | 77<br>Ir<br>Iridium<br>192.217 | 78<br>Pt<br>Platinum<br>195.078 | 79<br>Au<br>Gold<br>196.96655 | 80<br>Hg<br>Mercury<br>200.59 | 81<br>Tl<br>Thallium<br>204.3833 | 82<br>Pb<br>Lead<br>g 207.2 | 83<br>Bi<br>Bismuth<br>208.98038 | 84<br>Po<br>Polonium<br>[209] | 85<br>At<br>Astatine<br>[210] | 86<br>Rn<br>Radon<br>[222] |
| 87<br>Fr<br>Francium<br>[223] | 88<br>Ra<br>Radium<br>[226] | 89<br>Ac<br>Actinium<br>[227] | 90<br>Th<br>Thorium<br>g 232.0381 | 91<br>Pa<br>Protactinium<br>231.03588 | 92<br>U<br>Uranium<br>gm 238.0289 | 93<br>Np<br>Neptunium<br>[237] | 94<br>Pu<br>Plutonium<br>[244] | 95<br>Am<br>Americium<br>[243] | 96<br>Cm<br>Curium<br>[247] | 97<br>Bk<br>Berkelium<br>[247] | 98<br>Cf<br>Californium<br>[251] | 99<br>Es<br>Einsteinium<br>[252] | 100<br>Fm<br>Fermium<br>[257] | 101<br>Md<br>Mendelevium<br>[258] | 102<br>No<br>Nobelium<br>[259] | 103<br>Lr<br>Lawrencium<br>[262] | 104<br>Rf<br>Rutherfordium<br>[261] | [illegible]05<br>[illegible]b<br>[illegible]nium<br>[illegible]62] | 106<br>Sg<br>Seaborgium<br>[263] | 107<br>Bh<br>Bohrium<br>[262] | 108<br>Hs<br>Hassium<br>[265] | 109<br>Mt<br>Meitnerium<br>[266] | 110<br>Uun<br>Not named<br>[269] | 111<br>Uuu<br>Not named<br>[272] | 112<br>Uub<br>Not named<br>[277] | | | | | | |
| $s^1$ | $s^2$ | $s^2d^1$ | $s^2f^2$ | $s^2f^3$ | $s^2f^4$ | $s^2f^5$ | $s^2f^6$ | $s^2f^7$ | $s^2f^8$ | $s^2f^9$ | $s^2f^{10}$ | $s^2f^{11}$ | $s^2f^{12}$ | $s^2f^{13}$ | $s^2f^{14}$ | $s^2f^{14}d^1$ | $s^2d^2$ | $s^2d^3$ | $s^2d^4$ | $s^2d^5$ | $s^2d^6$ | $s^2d^7$ | $s^2d^8$ | $s^2d^9$ | $s^2d^{10}$ | $s^2p^1$ | $s^2p^2$ | $s^2p^3$ | $s^2p^4$ | $s^2p^5$ | $s^2$ or $s^2p^6$ |

The very long form of the periodic table shows the lanthanides and actinides inside the body of the periodic table. The numbers above the table are the conventional group designations. The notations along the bottom of the table are nominal abbreviated electron configurations. For specific elements, the experimental configurations may differ slightly from the generic ones above. See the electron configuration periodic table for the actual ground state configurations.

To show this table on a normal sheet of paper requires that the type font be reduced to 3-point. So you can see why chemists prefer the long form of the periodic table with the lanthanides and actinides placed below.

Mazurs (Edward G. Mazurs, Graphic Representations of the Periodic System During One Hundred Years, The University of Alabama Press, 2nd. Edition, 1974) attributes the first use of this type of table to Alfred Werner in 1905.

# Mendeleev's Periodic Table circa 1909
# A Typical Text Book Periodic Table

| | Group 0<br>R | Group I<br>RH<br>$R_2O$ | Group II<br>$RH_2$<br>RO | Group III<br>$RH_3$<br>$R_2O_3$ | Group IV<br>$RH_4$<br>$RO_2$ | Group V<br>$RH_3$<br>$R_2O_5$ | Group VI<br>$RH_2$<br>$RO_3$ | Group VII<br>RH<br>$R_2O_7$ | Group VIII<br>--<br>$RO_4$ |
|---|---|---|---|---|---|---|---|---|---|
| 1 | | H=1 | | | | | | | |
| 2 | He=4 | Li=7 | Gl=9.1 | B=11 | C=12 | N=14 | O=16 | F=19 | |
| 3 | Ne=20 | Na=23 | Mg=24.4 | Al=27 | Si=28 | P=31 | S=32 | Cl=35.5 | Fe=56, Ni=58.5 |
| 4 | A=39.9 | K=39.1 | Ca=40 | Sc=44 | Ti=48.1 | V=51.2 | Cr=52.8 | Mn=55 | Co=59.1, Cu=63.3 |
| 5 | -- | (Cu) | Zn=65.4 | Ga=69.9 | Ge=72 | As=75 | Se=79 | Br=80 | Rh=103, Ru=108.8 |
| 6 | Kr=81.8 | Rb=85.4 | Sr=87.5 | Y=89 | Zr=90.7 | Cb=94.2 | Mo=95.9 | --=100 | Pd=108, Ag=107.9 |
| 7 | -- | (Ag) | Cd=112 | In=118.7 | Sn=118 | Sb=120.3 | Te=125.2 | I=126.9 | --- --- |
| 8 | Xe=128 | Cs=132.9 | Ba=137 | La=138.5 | Ce=141.5 | --- | --- | --- | |
| 9 | -- | --- | --- | --- | --- | --- | --- | --- | Ir=191.3, Pt=198.4 |
| 10 | -- | --- | --- | Yb=173.2 | --- | Ta=182.8 | W=184 | --- | Os=200, Au=197.6 |
| 11 | -- | (Au)=196.7 | Hg=200.4 | Tl=204.1 | Pb=206.9 | Bi=208 | --- | --- | --- --- |
| 12 | -- | --- | Ra=225 | --- | Th=233.4 | -- | U=239 | --- | |

Notes:

1. This was the standard form of the periodic table from about 1875 to 1910. Variants of this type of table were in almost every elementary text book during the second half of that time period. Even today, Russian textbooks tend to favor this form of the table and over lunch during the summer of 1999 a Czech chemist in his 80s told me this was the table he grew up with and that it remained his favorite representation. The table depicted here is largely based on that of Ira Remsen in *An Introduction to the Study of Chemistry*, 8th Revised and Enlarged Edition, Henry Holt and Company, New York, 1909, page 263.

2. There are 71 elements shown on the table. Three elements are duplicated: Cu, Ag, Au. The division between what we now call the main group elements and the transition elements was imperfectly understood in 1909. Thus, it was not clear where the "coinage" metals belonged. I think that most modern chemists regard this table as confusing.

3. Observe that the names and symbols of some elements have changed. Cb (columbium) has become Nb (niobium); Gl (glucinium) has become Be (beryllium); and, A has become Ar for argon.

4. In 1909, the noble gases had been known for about 15 years -- so they are in the table. Marie Curie and her colleagues had been isolating radioactive elements, such as radium, In the period 1899-1909. But these radio elements had not yet reached Remsen's edition of 1909.

5. The columns are headed by the formulas of generalized chemical compounds of the elements with hydrogen and oxygen; R = an element in the group. The combining capacity of the elements (valency) was a key property used by Mendeleev to place the elements in the table.

6. Note that the table makes a prediction in group VII where --=100 denotes a missing element with an atomic weight of 100. This element (technetium) remained missing for 30 more years; it has no stable isotopes. Technetium's longest lived isotope has an atomic weight of 97.8 u.

Facsimile of a Periodic Table Handwritten and signed by Mendeleev, circa 1870.

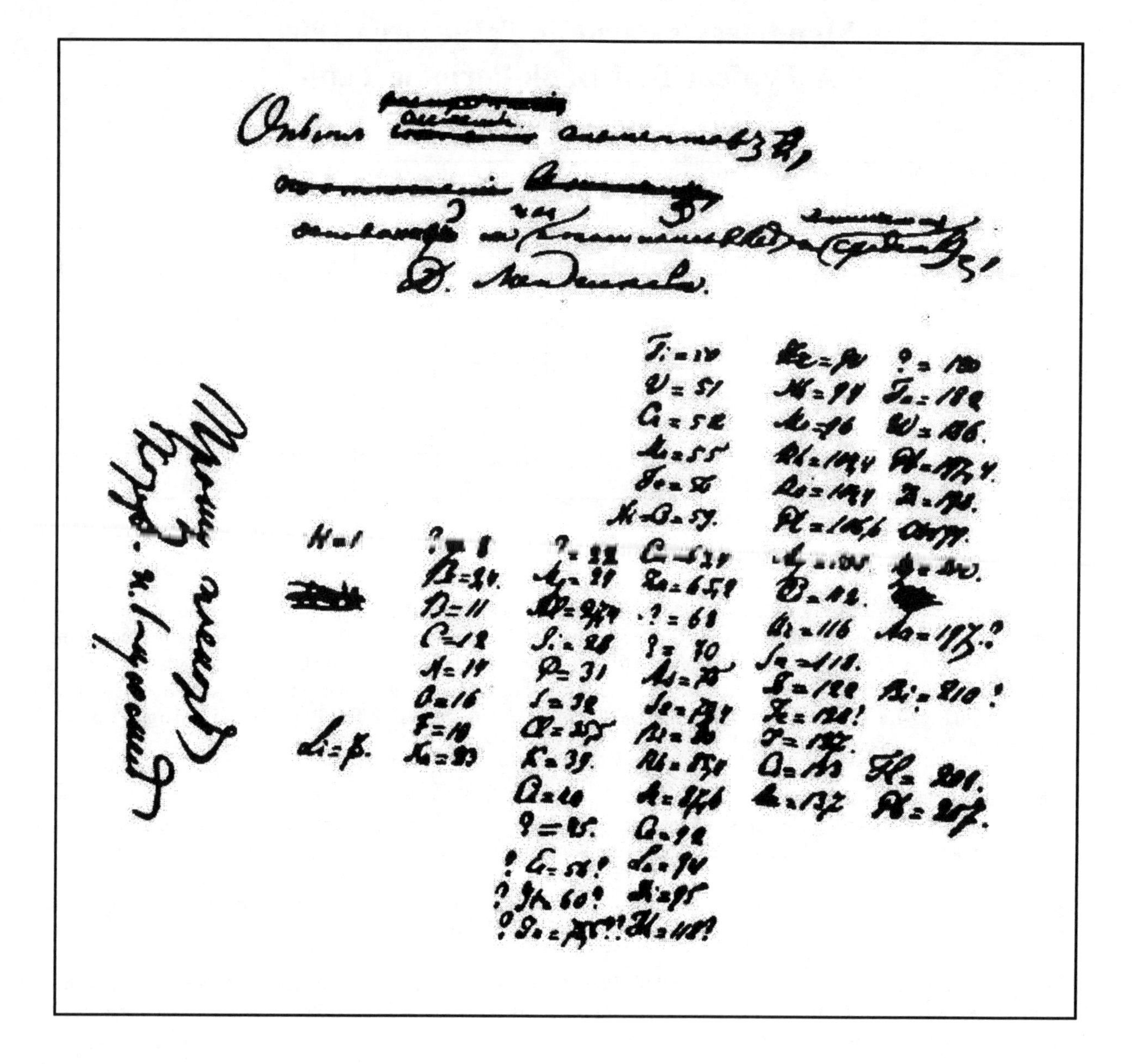

# PtNNTree Type Periodic Table

| 8 | | | | | | | 119<br>? | 120<br>? | | | | | | | 8s |
|---|---|---|---|---|---|---|---|---|---|---|---|---|---|---|---|
| | | | | | 113<br>? | 114<br>nn | 115<br>? | 116<br>nn | 117<br>? | 118<br>nn | | | | | 7p |
| | | | 103<br>Lr | 104<br>Rf | 105<br>Db | 106<br>Sg | 107<br>Bh | 108<br>Hs | 109<br>Mt | 110<br>nn | 111<br>nn | 112<br>nn | | | 6d |
| | 89<br>Ac | 90<br>Th | 91<br>Pa | 92<br>U | 93<br>Np | 94<br>Pu | 95<br>Am | 96<br>Cm | 97<br>Bk | 98<br>Cf | 99<br>Es | 100<br>Fm | 101<br>Md | 102<br>No | 5f |
| 7 | | | | | | | 87<br>Fr | 88<br>Ra | | | | | | | 7s |
| | | | | | 81<br>Tl | 82<br>Pb | 83<br>Bi | 84<br>Po | 85<br>At | 86<br>Rn | | | | | 6p |
| | | | 71<br>Lu | 72<br>Hf | 73<br>Ta | 74<br>W | 75<br>Re | 76<br>Os | 77<br>Ir | 78<br>Pt | 79<br>Au | 80<br>Hg | | | 5d |
| | 57<br>La | 58<br>Ce | 59<br>Pr | 60<br>Nd | 61<br>Pm | 62<br>Sm | 75<br>Re | 64<br>Gd | 65<br>Tb | 66<br>Dy | 67<br>Ho | 68<br>Er | 69<br>Tm | 70<br>Yb | 4f |
| 6 | | | | | | | 55<br>Cs | 56<br>Ba | | | | | | | 6s |
| | | | | | 49<br>In | 50<br>Sn | 51<br>Sb | 52<br>Te | 53<br>I | 54<br>Xe | | | | | 5p |
| | | | 39<br>Y | 40<br>Zr | 41<br>Nb | 42<br>Mo | 43<br>Tc | 44<br>Ru | 45<br>Rh | 46<br>Pd | 47<br>Ag | 48<br>Cd | | | 4d |
| 5 | | | | | | | 37<br>Rb | 38<br>Sr | | | | | | | 5s |
| | | | | | 31<br>Ga | 32<br>Ge | 33<br>As | 34<br>Se | 35<br>Br | 36<br>Kr | | | | | 4p |
| | | | 21<br>Sc | 22<br>Ti | 23<br>V | 24<br>Cr | 25<br>Mn | 26<br>Fe | 27<br>Co | 28<br>Ni | 29<br>Cu | 30<br>Zn | | | 3d |
| 4 | | | | | | | 19<br>K | 20<br>Ca | | | | | | | 4s |
| | | | | | 13<br>Al | 14<br>Si | 15<br>P | 16<br>S | 17<br>Cl | 18<br>Ar | | | | | 3p |
| 3 | | | | | | | 11<br>Na | 12<br>Mg | | | | | | | 3s |
| | | | | | 5<br>B | 6<br>C | 7<br>N | 8<br>O | 9<br>F | 10<br>Ne | | | | | 2p |
| 2 | | | | | | | 3<br>Li | 4<br>Be | | | | | | | 2s |
| 1 | | | | | | | 1<br>H | 2<br>He | | | | | | | 1s |

nn = not named

# Appendix I—A Material Safety Data Sheet (MSDS)

Chemical safety is an issue that occupies the time of many working engineers. Every chemical material is covered by an official document called its Material Safety Data Sheet (MSDS). A sample MSDS for the compound magnesium phosphide, packaged in a variety of forms, is included in this Appendix.

Historically methyl bromide ($CH_3Br$) was the most commonly used as a fumigant. Unfortunately, methyl bromide (like the chlorofluorohydrocarbons discussed at the end of Chapter 4) causes depletion of stratospheric ozone—and does it very efficiently.

Phosphine ($PH_3$) gas is an effective alternative, however, cylinders of 100% phosphine are hazardous as the material is toxic and explosive.

A solution is to use magnesium phosphide ($Mg_3P_2$), because this compound reacts with water to produce phosphine gas as needed according to the reaction equation.

$$Mg_3P_2 + 6H_2O \rightarrow 3Mg(OH)_2 + 2PH_3$$

# MATERIAL SAFETY DATA SHEET: MAGNESIUM PHOSPHIDE- MAGTOXIN® SPOT FUMIGANT, FUMI-CEL®, FUMI-STRIP®

**PROPER DOT SHIPPING NAME: MAGNESIUM PHOSPHIDE, CL 4.3, UN2011, PG I, DANGEROUS WHEN WET, POISON LABELS APPLY**

## SECTION I - PRODUCT INFORMATION

**Manufacturer:**

DEGESCH America, Inc.
275 Triangle Dr.
P. O. Box 116
Weyers Cave, VA 24486 USA

Telephone: (540) 234-9281
Telefax: (540) 234-8225
Internet address: http//www.degeschamerica.com

## EMERGENCY TELEPHONE NOS.:

Emergency - Chemtrec (800) 424-9300
Emergency and Information - DEGESCH America, Inc. (540) 234-9281

**Packaging:** Magtoxin is available in porous blister packs called the Magtoxin Prepac Spot Fumigant. Fumi-Cel is produced by impregnating magnesium phosphide into polyethylene in the form of a 117g plate, each plate liberating 33g of hydrogen phosphide. The Fumi-Strip is formed by attaching together, end-to-end, 20 of the Fumi-Cel plates. Fumi-Cel and Fumi-Strip do not liberate carbon dioxide and ammonia. All metal phosphide products are packed in gas-tight containers.

**Date of Revision:** January 1997

## SECTION II - HAZARDOUS INGREDIENTS INFORMATION

**Identity:**

Magtoxin Spot Fumigant, Fumi-Cel, Fumi-Strip, Magnesium Phosphide, $Mg_3P_2$ - Reacts with water to produce phosphine, hydrogen phosphide, $PH_3$ as shown in Equation 1. Magtoxin is formulated with 66% magnesium phosphide and also contains ammonium carbamate and inert ingredients. Ammonium carbamate releases ammonia and carbon dioxide as shown in Equation 2.

1) $Mg_3P_2 + 6H_2O \rightarrow 3Mg(OH)_2 + 2PH_3$

2) $NH_2COONH_4 \rightarrow 2NH_3 + CO_2$

$Mg_3P_2$ CAS No. 12057-74-8

$PH_3$ CAS No. 7803-51-2

$NH_2COONH_4$ CAS No. 1111-78-0

$NH_3$ CAS No. 7664-41-7

$CO_2$ CAS No. 124-38-9

**NFPA Chemical Hazard Ratings:**

Flammability Hazard 4
Health Hazard 4
Reactivity Hazard 2
Special Hazard W̶

**SARA Physical and Health Hazards:**

Fire
Reactivity
Immediate (Acute)

Magtoxin is available in porous blister packs called the Magtoxin Prepac Spot Fumigant. Fumi-Cel is produced by impregnating magnesium phosphide into polyethylene in the form of a 117g plate, each plate liberating 33h of hydrogen phosphide. The Fumi-Strip is formed by attaching together, end-to-end, 20 of the Fumi-Cel plates. Fumi-Cel and Fumi-Strip do not liberate carbon dioxide and ammonia. All metal phosphide products are packed in gas-tight containers.

**Inhalation Exposure Limits:**

| Component | OSHA PEL TWA (ppm) | ACGIH TLV TWA (ppm) | ACGIH TLV STE L (ppm) | NIOSH IDLH (ppm) |
|---|---|---|---|---|
| Hydrogen Phosphide* | 0.3 | 0.3 | 1.0 | 50 |
| Ammonia | 50 | 25 | 35 | 300 |
| Carbon Dioxide | 5,000 | 5,000 | 30,000 | 40,000 |

*EPA limits are 0.3 ppm TWA during fumigation and 0.3 ppm ceiling at all other times.

## SECTION III - PHYSICAL CHARACTERISTICS

**Boiling Point:**

$Mg_3P_2$ >1000°C
$PH_3$ -87.7°C

**Specific Gravity of Vapors (Air = 1):**

$Mg_3P_2$ N/A
$PH_3$ 1.17

**Vapor Pressure:**

$Mg_3P_2$ 0mm Hg
$PH_3$ 40mm Hg @ -129.4°C

**Solubility in Water:**

$Mg_3P_2$ Insoluble, reacts
$PH_3$ 26 cc in 100 ml water at 17°C

**Appearance and Odor:**

Magtoxin and magnesium phosphide are a dark charcoal gray. The paper covering the polyethylene matrix of the Fumi-Cel and Fumi-Strip is yellow-orange in color. The hydrogen phosphide (phosphine, $PH_3$) gas produced by these products has an odor described as similar to garlic, carbide or decaying fish.

**Specific Gravity:**

$Mg_3P_2$ 2.06

**Melting Point:**

AlP >1000°C

$PH_3$ -133.5°C

## SECTION IV - FIRE AND EXPLOSION HAZARD DATA

**Flash Point:**

Magnesium phosphide and Magtoxin are not themselves flammable. However, they react readily with water to produce hydrogen phosphide (phosphine, $PH_3$) gas which may ignite spontaneously in air at concentrations above its LEL of 1.8% v/v. UEL of hydrogen phosphide is not known. The paper covering and polyethylene matrix of the Fumi-Cel and Fumi-Strip are flammable.

**Extinguishing Media:**

Suffocate flames with sand, carbon dioxide or dry extinguishing chemicals.

**Special Fire Fighting Procedures:**

Do not use water on metal phosphide fires.

**Respiratory Protection:**

Wear NIOSH/MSHA approved SCBA or equivalent respiratory protection.

**Protective Clothing:**

Wear gloves when handling Magtoxin.

**Unusual Fire and Explosion Hazards:**

Hydrogen phosphide-air mixtures at concentrations above the lower flammable limit of 1.8% v/v, $PH_3$ may ignite spontaneously. Ignition of high concentrations of hydrogen phosphide can produce a very energetic reaction. Explosions can occur under these conditions and may cause severe personal injury. Never allow the buildup of hydrogen phosphide to exceed explosive concentrations. Open containers of metal phosphides in open air only and never in a flammable atmosphere. Do not confine spent or partially spent dust from metal phosphide fumigants as the slow release of hydrogen phosphide from these materials may result in the formation of an explosive atmosphere. Spontaneous ignition may occur if large quantities of magnesium phosphide or aluminum phosphide are piled in contact with liquid water. This is particularly true if quantities of these materials are placed in moist or spoiled grain which can provide partial confinement of the hydrogen phosphide gas liberated by hydrolysis.

Fires containing hydrogen phosphide or metal phosphides will produce phosphoric acid by the following reaction:

$$2PH_3 + 4O_2 \rightarrow 3H_2O + P_2O_5 \rightarrow 2H_3PO_4$$

## SECTION V - REACTIVITY DATA

**Stability:**

Magnesium phosphide is stable to most chemical reactions, except for hydrolysis. Magtoxin, Fumi-Cel and Fumi-Strip will react with moist air, liquid water, acids and some other liquids to produce toxic and flammable hydrogen phosphide gas. Magnesium phosphide is more reactive than aluminum phosphide and will liberate hydrogen phosphide more rapidly and more completely at lower temperatures and humidities.

**Incompatibility:**

Avoid contact with water and oxidizing agents.

**Corrosion:**

Hydrogen phosphide gas may react with certain metals and cause corrosion, especially at higher temperatures and relative humidities. Metals such as copper, brass and other copper alloys, and precious metals such as gold and silver are susceptible to corrosion by phosphine. Small electric motors, smoke detectors, brass sprinkler heads, batteries and battery chargers, fork lifts, temperature monitoring systems, switching gears, communication devices, computers, calculators and other electrical equipment may be damaged by this gas. Hydrogen phosphide will also react with certain metallic salts and, therefore, sensitive items such as photographic film, some inorganic pigments, etc., should not be exposed.

**Hazardous Polymerization:**

Will not occur.

## SECTION VI - HEALTH HAZARD INFORMATION

**Routes of Entry:**

Magnesium phosphide and hydrogen phosphide gas from these fumigants are not absorbed dermally. Primary routes of exposure are inhalation and ingestion.

**Acute and Chronic Health Hazards:**

Magnesium phosphide is a highly acute toxic substance. Hydrogen phosphide gas LC50 is about 190 ppm for a one-hour inhalation exposure. The acute oral toxicity of the Magtoxin formulation was found to be 9.1 mg/kg of body weight. Magnesium phosphide is not known to cause chronic poisoning.

**Carcinogenicity:**

Magnesium phosphide is not known to be carcinogenic and is not listed as such by NTP, IARC or OSHA.

**Signs and Symptoms of Exposure:**

Magnesium phosphide fumigant products react with moisture from the air, acids and many other liquids to release hydrogen phosphide (phosphine, $PH_3$) gas. Mild exposure by inhalation causes malaise (indefinite feeling of sickness), ringing in the ears, fatigue, nausea and pressure in the chest which is relieved by removal to fresh air. Moderate poisoning causes weakness, vomiting, pain just about the stomach, chest pain, diarrhea and dyspnea (difficulty in breathing). Symptoms of severe poisoning may occur within a few hours to several days resulting in pulmonary edema (fluid in lungs) and may lead to dizziness, cyanosis (blue or purple skin color), unconsciousness, and death.

### Emergency and First Aid Procedures:

Symptoms of overexposure are headache, dizziness, nausea, difficult breathing, vomiting, and diarrhea. In all cases of overexposure get medical attention immediately. Take victim to a doctor or emergency treatment facility.

If the gas or dust from magnesium phosphide is inhaled:. Get exposed person to fresh air. Keep warm and make sure person can breathe freely. If breathing has stopped, give artificial respiration by mouth-to-mouth or other means of resuscitation. Do not give anything by mouth to an unconscious person.

If magnesium phosphide pellets or powder are swallowed: Drink or administer one or two glasses of water and induce vomiting by touching back of throat with finger, or if available, syrup of ipecac. Do not give anything by mouth if victim is unconscious or not alert.

If powder or granules of magnesium phosphide get on skin or clothing: Brush or shake material off clothes in a well ventilated area. Allow clothes to aerate in a ventilated area prior to laundering. Do not leave contaminated clothing in occupied and/or confined areas such as automobiles, vans, motel rooms, etc. Wash contaminated skin thoroughly with soap and water.

If dust from pellets or tablets gets in eyes: Flush with plenty of water. Get medical attention.

## SECTION VII - SPILL OR LEAK PROCEDURES

### Spill Cleanup Procedures:

If possible, dispose of spilled Magtoxin, Fumi-Cel or Fumi-Strip by use according to label instructions. Freshly spilled material which has not been contaminated by water or foreign matter may be replaced into original containers. Punctured flasks, pouches or containers may be temporarily repaired using aluminum tape. If the age of the spill is unknown or if the product has been contaminated with soil, debris, water, etc., gather up the spillage in small open buckets having a capacity no larger than about 1 gallon. Do not add more than about 0.5kg (l lb.) to a bucket. If on-site wet deactivation is not feasible, transport the uncovered buckets in open vehicles to a suitable area. Wear gloves when handling Magtoxin, Fumi-Cel or Fumi-Strip.

Respiratory protection will most likely be required during cleanup of spilled magnesium phosphide fumigants. If the concentration of hydrogen phosphide is unknown, NIOSH/MSHA approved SCBA or its equivalent must be worn.

Small amounts of spillage, from about 2 to 4 kg (4 to 9 lbs.) may be spread out over the ground in an open area to be deactivated by atmospheric moisture. Alternatively, spilled magnesium phosphide fumigants may be deactivated by the wet method as described in the following.

### Wet Deactivation:

1. Spilled magnesium phosphide fumigants, Magtoxin, Fumi-Cel and Fumi-Strip, may be deactivated with water. Do not use detergent for the deactivation of these products. Fill the container in which the deactivation is to be performed with water to within a few inches of the top.

2. The spilled material is added slowly to the water. Magtoxin Prepacs, Fumi-Cel or Fumi-Strips may ignite during wet deactivation if they are allowed to float to the surface. Add weights or otherwise ensure that they stay submerged until deactivation is complete. At no time should the deactivation container be covered.

3. Due to the reactivity of magnesium phosphide, additions of spilled product to the water should be made slowly and carefully. This should be done in open air and respiratory protection will probably be required.

4. Allow the mixture to stand, with occasional stirring, for about 24 hours. Do not cover the container. The mixture will then be safe for disposal.

5. Dispose of the deactivated material, with or without preliminary decanting, at a sanitary landfill or other suitable site approved by local authorities. Where permissible, the deactivation water containing spent dust may be poured into a storm sewer or out onto the ground.

For Assistance: Contact - DEGESCH America, Inc.
Telephone: (540) 234-9281
Telefax: (540) 234-8225
or
Chemtrec: (800) 424-9300

### Disposal of Spent Magtoxin, Prepacs, Fumi-Cel and Fumi-Strip:

When being disposed of, spilled or partially reacted magnesium phosphide fumigants are considered hazardous wastes under existing Federal Regulations. If properly exposed, the grayish-white residual dust from Magtoxin and spent Fumi-Cel or Fumi-Strip will not be a hazardous waste and normally contains only a very small amount of unreacted magnesium phosphide. This waste will be safe for disposal. However, the residuals from incompletely exposed magnesium phosphide fumigants may require special care.

Triple rinse pouches, tins and pails. Tins and pails may then be offered for recycling or reconditioning, or punctured and disposed of in a sanitary landfill, or by other procedures approved by state and local authorities. Rinsate may be disposed of in a storm sewer, sanitary landfill or by other approved procedures. Or, it is permissible to remove lids and expose empty pails, tins or pouches to atmospheric conditions until the residue is reacted. Then puncture and dispose of in a sanitary landfill or other approved site, or by other procedures approved by state and local authorities.

Some local and state waste disposal regulations may vary from the following recommendations. Disposal procedures should be reviewed with appropriate authorities to ensure compliance with local regulations. Contact your State Pesticide or Environmental Control Agency or Hazardous Waste Specialist at the nearest EPA Regional Office for guidance.

1. Confinement of partially spent fumigant or residual dust, as in a closed container, or collection and storage of large quantities of fumigant may result in a fire or explosion hazard. Small amounts of hydrogen phosphide may be given off from unreacted magnesium phosphide, and confinement of the gas may result in a flash.

2. In open areas, small amounts of spent residual dust may be disposed of on site by burial or by spreading over the land surface away from inhabited buildings.

3. Residuals from magnesium phosphide fumigants may also be collected and disposed of at a sanitary landfill, incinerator or other approved sites or by other procedures approved by Federal, State or Local authorities.

4. From 1 to 2 kg (2 to 4 lbs.) of spent fumigant may be collected for disposal in an open 1-gallon bucket. Caution: Do not collect dust in large drum, dumpsters, plastic bags or other containers where confinement may occur. Transport the buckets in an open vehicle for disposal or deactivation.

**Deactivation of Partially Spent Magtoxin, Fumi-Cel and Fumi-Strip:**

Magtoxin Prepacs, Fumi-Cels or Fumi-Strips which are only partially spent may be rendered inactive by either a "dry" or "wet" deactivation method. The "dry" method entails holding the Prepacs, Cels or Strips out of doors in locked, 30-gallon wire baskets which are available from DEGESCH America, Inc., or your supplier. Protect the partially spent magnesium phosphide fumigants from rain. The deactivated products may then be taken to an approved site for incineration or burial at periodic intervals or whenever the wire container is full. Caution: Storage of partially spent magnesium phosphide in closed containers may result in a fire hazard. Alternatively, partially spent Prepacs, Fumi-Cels and Fumi-Strips may be treated by the "wet" deactivation method as follows:

1. Fill the container in which the deactivation is to be performed with water to within a few inches of the top. Detergent is not necessary for the deactivation of spent magnesium phosphide fumigants.

2. The spent material is added slowly to the water. Magtoxin Prepacs, Fumi-Cels or Fumi-Strips may ignite during wet deactivation if they are allowed to float to the surface. Add weights or otherwise ensure that they stay submerged until deactivation is complete.

3. Partially spent Magtoxin Prepacs, Fumi-Cels or Fumi-Strips may react quite vigorously during wet deactivation if they were exposed under cold and/or dry conditions or if the fumigation period was shortened. It is suggested that a small portion of the product be tested prior to immersing large amounts of material in water if it is suspected that the product contains considerable unreacted magnesium phosphide.

4. Due to the reactivity of magnesium phosphide, additions to the water should be made slowly and carefully. Deactivation should be carried out in open air and respiratory protection may be required.

5. Allow the mixture to stand with occasional stirring. Do not cover the container.

6. Dispose of the deactivated material, with or without preliminary decanting, at a sanitary landfill or other suitable site approved by local authorities. Where permissible, deactivation water containing spent dust may be poured into a storm sewer or out onto the ground.

**Precautions to be Taken in Handling and Storage:**

Store Magtoxin, Fumi-Cel and Fumi-Strip products in a locked, dry, well-ventilated area away from heat. Post as a pesticide storage area. Do not store in buildings inhabited by humans or domestic animals.

**Other Precautions:**

1. Do not allow water or other liquids to contact magnesium phosphide fumigants.
2. Do not pile up large quantities of magnesium phosphide products during fumigation or disposal.
3. Once exposed, do not confine the fumigant or otherwise allow hydrogen phosphide concentration to exceed the LEL.
4. Open containers of Magtoxin, Fumi-Cel or Fumi-Strip only in open air. Do not open in a flammable atmosphere. Hydrogen phosphide in the head space of containers may flash upon exposure to atmospheric oxygen.
5. See EPA approved labeling for additional precautions and directions for use.
6. Magtoxin, Fumi-Cel and Fumi-Strip are restricted use pesticides due to acute inhalation toxicity of highly toxic hydrogen phosphide (phosphine, $PH_3$) gas. For retail sale to and use only by certified applicators or persons under their direct supervision and only for those uses covered by the certified Applicator's Certification.

## SECTION VIII - CONTROL MEASURES

**Respiratory Protection:**

NIOSH/MSHA approved full-face mask with approved canister for phosphine (hydrogen phosphide, $PH_3$) may be worn at concentrations up to 15 ppm. At levels above this or when the hydrogen phosphide concentration is unknown, NIOSH/MSHA approved SCBA or equivalent must be worn.

**Protective Clothing:**

Wear gloves when contact with magnesium phosphide is likely to occur.

**Eye Protection:**

None required.

**Ventilation:**

Local ventilation is generally adequate to reduce hydrogen phosphide levels in fumigated areas to below the TLV/TWA. Exhaust fans may be used to speed the aeration of silos, warehouses, ship holds, containers, etc.

**We believe the statements, technical information and recommendations contained herein are reliable, but they are given without warranty or guarantee of any kind, expressed or implied, and we assume no responsibility for any loss, damage, or expense, direct or consequential, arising out of their use.**

**FORM 19355 (R/1/97)**

# Glossary

**ABET**. Accreditation Board for Engineering and Technology. A private corporation owned by 28 engineering societies. Many States require graduation from an ABET certified institution as part of their engineer licensing process.

**Absolute pressure**. The true pressure of a gas sample. To be distinguished from **gauge pressure**.

**Absolute zero**. The lowest possible temperature. 0 K or -273.15°C.

**Absorption**. Process in which an atom gains energy by capturing a photon of light or electromagnetic energy. The atom is said to absorb the photon and undergoes a **transition**. Compare **emission**.

**Absorption spectrum.** A spectrum that results when a portion of the radiation incident on a sample is absorbed by the sample. Compare **dark line spectrum** and **infrared spectrum**.

**Abundance**. For an isotope it's the fraction or percentage of the atoms of that isotope in a specified sample of an element. Often refers to how much of an isotope is present in a natural sample of an element. See **natural abundance**.

**Accepted value**. Of a property, it's the same as the **correct value**.

**Accuracy**. An important concept in measurement theory. It compares a measured quantity to its **correct value**. The smaller the difference the more accurate the measurement. The most accurate value of a property that humans can measure becomes the correct value.

**ACGIH**. American Council of Governmental Hygienists. Professional organization that, among other activities, estimates the toxicity of airborne chemical compounds.

**Acid**. A corrosive compound with a sharp taste that neutralizes bases. Any compound that produces hydrogen ion ($H^+$) or protons when dissolved in water. See **strong** and **weak acids**. Compare **Arrenhius acid**, **Brønsted-Lowry acid**, and **Lewis acid**.

**Acid dissociation constant**. An experimental value derived by applying equilibrium principles to the liberation of hydrogen ion by an acidic substance. For a monoprotic acid it's the product of the concentrations of the hydrogen ion and the anion divided by the concentration of the **unionized** acid. For a monoprotic acid $HA \leftrightarrows H^+ + A$ the equilibrium constant expression is $K_a = \frac{[H^+][A^-]}{[HA]}$.

**Acid-base indicator**. An organic compound that exists in two different colored forms depending on the pH of its environment. Hence, its change of color can be used to indicate changes in pH. Compare **pH indicator.**

**Acid rain**. Rain with a pH roughly in the range 3-5 pH units. It occurs when human activities or volcanoes release compounds such as $SO_2$ or $NO_2$ to the atmosphere where they eventually combine with moisture to create acids.

**Actinides**. Actinium and the elements thorium (Th) through lawrencium (Lr) that follow actinium (Ac) in the periodic table.

**Activated complex**. In chemical kinetics it refers to a collision-in-progress. It's formed by two entities as they merge and pass over the activation energy barrier. Also called the **transition state** of a chemical reaction.

**Activation energy**. In chemical kinetics, the energy needed in a collision to allow bonds to break and make. The energy needed to get a reaction started.

**Active metals**. In the periodic table the metals of Groups 1 and 2. Any metal with a high chemical reactivity.

**Active site**. In chemical kinetics, the specific location in a catalyst when the catalyzed reaction takes places.

**Activity series of the metals**. A reactivity ranking of metals. A traditional way chemists list metals: most active on the left, least active on the right. In the list, any metal to the left will reduce the ions of a metal to its right.

**Actual yield**. The actual amount of a particular product obtained from a chemical reaction. The actual yield is usually less than the **theoretical yield**.

**Addition polymerization**. Polymerization reaction in which alkene monomerss are linked into a long chain.

**Addition reaction**. One in which two smaller molecules combine to make a single, larger molecule. Widely used to make addition polymers from olefins in which case it is called **addition polymerization**.

**Adduct**. In general, an entity formed when two other entities combine. Often it said of reactions that create a compound containing a **coordinate-covalent** bond.

**Adenine**. One of the bases of **nucleic acids**, $C_5H_5N_5$. Based on its structure it is a purine-type base.

**Adenosine triphosphate**.A biological molecule of great importance. It has the chemical formula $C_{10}H_{16}N_5O_{13}P_3$. It is critical for the release of energy in human muscles, etc.

**Aha! Phenomenon**. When the light goes on and you suddenly make sense of some aspect of nature. The phrase originated in a book by the science writer Martin Gardner. A major reason for studying chemistry. From a student, the most heartwarming response to a chemistry lecture. Successful thinking *"outside-the-box"* can produce the Aha! reaction.

**Agrichemicals**. Fertilizers, biocides, and other chemical compounds or mixtures used to improve agricultural production.

**Air conditioning working fluid**. Typically a mixture of hydrofluorocarbon fluids (**HFCs**). A fluid that is alternatively compressed and evaporated so as to remove heat from a system. See also **cyclic process**.

**Alchemy**. A forerunner of modern chemistry. Hard to define simply. It was curious mixture of practical science, occultism, the search for the philosopher's stone (a magical substance that supposedly would convert base metal to pure gold), and, all too frequently, outright fraud.

**Alcohol**. An organic compound containing the -OH functional group and having the collective formula ROH.

**Aldehyde**. An organic compound containing the -CHO functional group and having the collective formula R-CHO.

**Algebra-of-quantities**. A system of algebra applied to physical quantities. Each physical quantity is regarded as being the product of a numerical value and a unit. Strongly recommended by **IUPAC** for general use in science. It has particular advantage in converting among different units.

**Algebraic method of balancing equations**. Fundamental procedure which enables *any* chemical equation to be balanced. The only requirement is that every element on the reactant side of the equation must be also present on the product side of the equation. A sound, systematic approach to balancing chemical equations.

**Aliphatic compound**. A type of organic compound. A saturated hydrocarbon. A hydrocarbon lacking both double and triple bonds.

**Alkali**. Any substance that behaves as a very strong base is an alkali. Common alkalies are sodium hydroxide (NaOH) and potassium hydroxide (KOH).

**Alkali metals**. The metallic elements Li, Na, K, Rb, Cs, and Fr.

**Alkaline cell**. In electrochemistry, any battery that uses an alkaline electrolyte. A nicad battery, with its potassium hydroxide electrolyte, is an example of an alkaline cell.

**Alkaline earth metals**. The metallic elements Be, Mg, Ca, Sr, Ba, Ra

**Alkanes**. Class of **saturated** hydrocarbons. Those that don't contain any rings of carbon atoms have the general chemical formula $C_nH_{2n+2}$.

**Alkenes**. Class of **unsaturated** hydrocarbons. An alkene molecule contains contain at least one double bond.

**Alkyl group**. A portion of a molecule. An alkane molecule less one hydrogen atom. Alkyl groups that don't contain any rings of carbon atoms have the general chemical formula $-C_nH_{2n+1}$.

**Alkynes**. Class of **unsaturated** hydrocarbons. An alkyne molecule contains contain at least one triple bond.

**Alloy**. A solid mixture containing one or metals and sometimes containing a nonmetal. Often made by mixing the ingredients in the liquid state and cooling the mixture.

**Allotropes**. Different physical and structural forms of a single element. Allotropes differ from one another in the manner in which the atoms of the element are joined.

**Alpha decay**. A radioactive decay mode in which the nucleus of the **parent nuclide** ejects an alpha particle.

**Alpha particle**. A particle composed of two protons and two neutrons (a helium nucleus). The nucleus of a $^4_2He$ atom; it has a charge of 2+ and a nucleon number of 4.

**Altitude**. Height above the earth's surface, measured from average sea level.

**Amalgam**. An **alloy** of mercury and one or more other metals. One of the most familiar amalgams is the dark gray filling often used in dental restoration. In the sense of a joining together the word amalgam has the same root as the word amalgamate.

**American Chemical Society**. The major US professional association for chemists. It plays an influential role maintaining data bases of chemical information and publishes **Chemical Abstracts** and many professional journals.

**Amide**. An organic compound containing the $-CONH_2$ functional group and having the collective formula $R-CONH_2$.

**Amine**. An organic compound containing the $-NH_2$ functional group and having the collective formula $R-NH_2$.

**Amino acid**. A bifunctional organic molecule that contains both the -COOH group and the $-NH_2$ group. Polymerized, amino acids form **polypeptides**.

**Amorphous solid**. A solid that lacks **long range order**. Typically, amorphous solids are glassy in nature. Amorphous literally means "without form."

**Amount of substance**. A fundamental concept of nature. How much of something you have is its amount of substance. Chemists measure amounts of substance in either moles or grams. **Mass** and **count** are two alternative ways to state an amount of substance. See also **chemical amount.**

**Amp**. Widely used contraction of **ampere**.

**Ampere**. "The ampere is that constant current which, if maintained in two straight parallel conductors of infinite length, of negligible circular cross-section, and placed 1 metre apart in vacuum, would produce between these conductors a force equal to $2 \times 10^{-7}$ newton per metre of length (9th **CGPM**, 1948)."

**Amphipathic**. Having two contradictory characteristics. Adjective used to describe the molecules of a surface active agent. The molecules of a surface active agent are simultaneously **hydrophilic** and **hydrophobic**.

**Amphiprotic**. Said of a molecule that can both accept protons (behave as a base) and donate protons (behave as an acid). See also the following glossary entry.

**Amphoteric substance**. One capable of acting either as an acid or as a base. The phenomenon is called amphoterism. See also the preceding glossary entry.

**Amu**. **Atomic mass unit**. In modern chemistry, the **unified atomic mass unit**.

**Analysis**. In chemistry, to carry out an analysis is to determine something about the composition of a substance. See **qualitative analysis** and **quantitative analysis**.

**Analyte**. That entity whose concentration is sought via an analytical procedure. The target of a chemical analysis. What is to be analyzed for.

**Analytical chemistry**. The branch of chemistry devoted to analysis. One of the four traditional branches of chemistry. The others are **organic**, **inorganic** and **physical chemistry**.

**ANFO**. Ammonium nitrate fuel oil. An inexpensive explosive composition. Used, for example, to remove overburden during strip mining or in large scale earth removal operations.

**Angstrom, Å**. A nonstandard unit of length recognized but not encouraged by IUPAC. 1 Å = $10^{-10}$ m.

**Anhydrous**. Literally, lacking water. Said especially of chemical compounds that normally occur in combination with water from which the water has been removed.

**Anion**. An entity with a negative charge. During electrolysis it is attracted to the **anode**.

**Anionic surfactant**. A **surface active agent** whose principal activity comes from a long-chain organic entity to which is attached a functional group with a negative charge.

**Anode**. An electrical conductor used in constructing an **electrochemical cell**. It's the **electrode** that has a positive charge and attracts anions. At the anode, electrons from the electrolyte solution flow into the anode and thence to the external circuit.

**Antibonding molecular orbital**. Type of molecular orbital from which removal of a electron strengthens bonding in the molecule. Also called **antibonding orbital**.

**Antifreeze**. Substance that lowers the freezing point of water. Commercially, ethylene glycol or, occasionally, propylene glycol.

**Anti-knock agent**. Compound added to motor fuel to improve the fuel's combustion characteristics. See **MMT**.

**Anti-oxidants** and **anti-ozonants**. Inhibiting substances added in small amounts to various commercial products to improve the products' resistance to oxidation and improve their shelf or service life. They work by acting as **free radical traps**.

**Applied chemistry**. The use of chemical principles and methods for practical purposes.

**Approach to equilibrium**. Description of the changing concentrations as a reacting system shifts. It refers especially to their slowing rate of change on the way to equilibrium.

**Aqua regia**. Mixture of nitric and hydrochloric acids. Capable of dissolving gold.

**Aqueous solution**. A solution made using water as the **solvent**. Water containing one or more dissolved substances.

**Arene**. An **aromatic** organic **hydrocarbon**.

**Aromatic compound.** Organic compound that contains at least one **benzene ring**. Also called an **arene**. Anthracene, for example

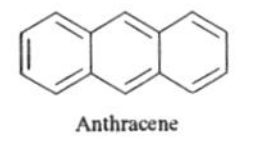
Anthracene

**Aromaticity**. The properties of **benzene ring**-containing hydrocarbons that derive from the presence of the ring.

**Arrenhius acid**. Substance that dissolves in water and ionizes with the production of hydrogen ions.

**Arrenhius base**. Substance that dissolves in water and ionizes with the production of hydroxide ions.

**Arrhenius equation**. In chemical kinetics, the equation that describes the effect of temperature on **rate constants**.

**Aryl group**. In organic chemistry it is any functional group that contains one or more benzene rings. Compare **arene**.

**ASEE**. American Society for Engineering Education. A professional membership society composed largely of engineering professors.

**ASTM**. American Society for Testing and Materials. A US voluntary standards organization that maintains a broad range of test protocols primarily used in industry. Many engineering standards fall under the jurisdiction of ASTM.

**Astrochemistry**. A branch of chemistry devoted to the study of chemical substances and processes beyond the earth.

**Atmosphere**. Pressure unit defined as 101 325 pascal (Pa). Alternatively, 1 atmosphere = 1.01325 bar. Because they are only 1.3% different, bars and atmospheres are sometimes used interchangeably.

**Atmospheric pressure**. The pressure exerted by the molecules of the air on any surface that is exposed to the air.

**Atom**. An atom is the basic unit of ordinary matter. An atom is composed of **protons** and **neutrons** in small, central nucleus and a surrounding cloud of electrons. The nature of any particular atom is determined by its proton count and neutron count. The number of different atoms corresponds to the total number of different **nuclides**.

**Atomic absorption spectroscopy**. Technique that applies atomic **absorption** to measure **analytes** at extremely low concentrations. Pushed to its limit, the technique can find one mercury atom in a sample containing perhaps $10^{13}$ total atoms.

**Atomic energy**. Energy derived from processes of nuclear transmutations. Energy from fission and fusion reactions.

**Atomic mass unit (u or Da)**. Given the symbol $m_u$ by IUPAC and 1 u = one twelfth the mass of a single atom of carbon-12. Sometimes stated as **amu**. The term amu is not recognized or sanctioned by IUPAC. The unit Da is recognized but not sanctioned by IUPAC. Compare **unified atomic mass unit**.

**Atomic mass**. 1. The average mass of the atoms present in a naturally occurring sample of the element, or 2. The mass of a single atom of an element. 3. The average mass of the atoms in any, specified sample. Elements have multiple atomic mass values because all have more than one isotope (even the mononuclidic elements have radioisotopes). As agreed by the IUPAC commission on atomic weights and isotopic abundances in 1979, the relative atomic mass of any specified sample can be defined as the average mass of the atoms present in the sample expressed in the unit u. Compare **atomic weight**. The term **atomic weight** is understood to refer to definition 1 at the beginning of this entry,

**Atomic number**. The number of protons in the nuclei of an element's atoms. Also the number of electrons in an element's *neutral* atoms.

**Atomic orbitals**. The wave functions of the hydrogen atom extended to apply to atoms of all the elements. The electron probability distributions in an atom.

**Atomic radius (plural radii)**. The average radius assigned to an atom on the basis of experimental bond lengths in which that atom participates. The sum of the radii of any pair of bonded atoms is close to the experimental bond length. An atom that forms **single, double**, and **triple** bonds has different radii for each type.

**Atomic spectrum** (plural **spectra)**. Pattern of light, usually visualized as a series of colored lines, created when a sample of atoms is excited (energized).

**Atomic structure**. The concept that atoms are composed of subatomic particles in a definite arrangement. Often used synonymously with **electron configuration**.

**Atomic weight**. The average mass of the atoms in a sample of an element in which the various isotopes are present in their **natural abundances**. This is definition 1 of **atomic mass**. The average mass of the atoms of a natural sample of an element. May be stated in the unit u (average atom)$^{-1}$.

**Aufbau procedure**. A mental process to build up the periodic table based on quantum mechanical principles. Aufbau is a German word that means something like "building up." Aufbau involves constructing an electronic configuration of an entity by placing electrons pair wise into the orbitals of lowest possible energy.

**Austenite**. A type of steel. An important iron-carbon alloy. Specifically, a solid solution of carbon or iron carbide ($Fe_3C$) in iron. Named after an English metallurgist.

**Autoionization**. For water, the equilibrium ionization reaction $H_2O \rightleftarrows H^+ + OH^-$

**Average bond energy**. The enthalpy required to dissociate a given pair of atoms, averaged over many different molecules. The energy needed to *unbond* two bonded atoms.

**Average molar mass**. The **weighted average** of the molar masses of the compounds present in a sample.

**Average molecular weight**. The average molecular weight of a mixture, the individual compounds being weighted according to their mole fractions. $M_{ave} = \Sigma\ x_iM_i$, where $x_i$ is the mole fraction of species i with molecular weight $M_i$.

**Avogadro's law**. Fundamental statement about the nature of gases: If they are measured under the same conditions of pressure and temperature, equal volumes of gases contain equal numbers of molecules.

**Avogadro's number**. The Avogadro constant. Approximately $6.022 \times 10^{23}$ mol$^{-1}$; given the symbol $N_A$. The number of entities in one **mole**. Its primary definition is the number of $^{12}C$ atoms that constitute exactly 12 grams. Also the conversion factor between u and g.

**Backbone chain**. The chain in a linear polymer. **Pendant groups** are attached to the backbone chain.

**Bakelite**. A synthetic **phenolic resin**. Developed in 1907 by Leo Baekeland.

**Balance**. 1. A device used to weigh things. 2. To balance is to make the products of a chemical reaction consistent with the reactants. See **balanced chemical equation**.

**Balanced chemical equation**. An equation in which the number of atoms of each element on the reactant side equals the number of atoms of each element on the product side.

**Balancing by inspection**. Process used to balance simple equations merely by looking at them and applying rules of logic.

**Ball-and-stick-models**. Type of molecular models. Historically made from wooden or plastics balls with holes. Wooden sticks or plastic rods connect the balls.

Also, a computer graphic representation of such a model.

**Ball-and-stick view**. In a molecular visualization, a display that represents individual atoms smaller than their natural size and uses cylinders to show the chemical bonds. **Compare space-filling view.**

**Balmer-Rydberg equation**. $\nu = R_H \left(\frac{1}{n_2^2} - \frac{1}{n_1^2}\right)$. Empirical equation that correlates the lines in the hydrogen spectrum.

**Band or belt of stability**. On a nuclide chart, it's the band- or belt-shaped region in which the stable nuclides are located.

**Band gap**. In a solid, its the energy difference between the solid's **valence band** and **conduction band**. Electrical insulators have very large band gaps; electrical conductors have very small band gaps.

**Bar**. SI named unit of pressure. One bar is defined to be $10^5$ pascal (Pa).

**Barometer**. A device used to measure atmospheric pressure.

**Base**. 1. A corrosive compound with a bitter taste that neutralizes acids. Any compound that produces hydroxide ion ($OH^-$) when dissolved in water. See **strong** and **weak bases**. Compare **Arrhenius base, Brønsted-Lowry base**, and **Lewis base**. 2. In biochemistry a base is one of the five cyclic, nitrogen molecules in the structure of DNA or RNA. The biochemical bases are **adenine, cytosine, guanine, uracil** and **thymine**.

**Base pairing**. Hydrogen bonding between pairs of bases. The specific mechanism by which two DNA molecules join to form a **double helix**.

Thymine - Adenine

Cytosine - Guanine

**Base unit.** In **SI** the base units are the meter, kilogram, second, ampere, kelvin, mole, and candela.

**Battery**. Electrochemical device that serves as a source of electric current. Strictly, two or more electrochemical cells joined in series but long used in the singular sense.

**Battery acid**. Sulfuric acid. The acid used as the electrolyte in conventional automobile batteries. See **lead-acid battery**.

**Battery reaction**. A reaction that can produce electric current. A redox reaction so arranged that the electrons transferred between the oxidant and reductant flow through an external circuit.

**Benzene ring**. In the molecule benzene, $C_6H_6$, The ring of six carbons linked by delocalized bonds of bond order 1.5. Such a ring in any other compound.

**Beta decay**. A radioactive decay mode in which the nucleus of the parent nuclide ejects an electron. The nuclide retains its mass number but increases its positive charge by 1.

**Beta emitter**. Any **radionuclide** that undergoes **beta decay**.

**Beta particle**. An electron created during a radioactive decay process.

**Belt of stability**. See **band of stability**.

**Bifunctional molecule**. Organic molecule that contains two functional groups.

**BHA**. Butylated hydroxy anisole. A widely used **GRAS** food additive

**BHT**. Butylated hydroxy toluene. A widely used **GRAS** food additive

**Binary acids**. Any acid composed of only two elements. For example HCl and HBr.

**Binary compounds**. Those that contain exactly two different elements.

**Binding energy of a nuclide**. The energy equivalent to the mass defect of that nuclide. Actually, it would be better called the *unbinding energy:* The energy needed to decompose a nuclide into its component protons, neutrons, and electrons.

**Binding energy per nucleon**. Concept used so that the energy of all nuclides can be compared. It is the nuclide's binding energy divided by its nucleon number (n+p). When graphed, such values provide insight into the origin and approximate size of nuclear energy.

**Binomial distribution.** In statistics, it's the specification of the results of a series of trials. The coefficients of the expansion of a binomial. In thermodynamics, the binomial distribution applies to the number of possible **microstates** of a **system**.

**Biochemistry**. A vast, important, and rapidly growing field of chemistry. Biochemistry seeks to understand and control the chemical process of living organisms.

**Biopolymers**. Natural polymers. Materials derived from the three chemical groups: **polysaccharides**, **proteins**, and **nucleic acids**.

**Bleach**. In chemistry, a compound that will lighten the color of cotton, paper, hair, etc. Most bleaches are powerful oxidizing agents. Around the household, sodium hypochlorite solution and hydrogen peroxide solution are common bleaches.

**Block copolymer**. One that has a single backbone chain, but different regions of that chain were formed from different monomers.

**Bohr's atomic model**. The first modern theory of how electrons behave in atoms. Developed around 1910.

**Boiling point elevation.** Phenomenon in which the boiling point of a solvent is raised in proportion to the amount of dissolved nonvolatile solute dissolved in that solvent.

**Bomb calorimeter**. A stainless steel vessel used to measure the heat released by chemical reactions.

**Bond**. See **chemical bond**.

**Bond angle**. Angle between any two atoms bonded to a third atom.

**Bond dipole**. The **dipole moment** associated with a particular bond in a molecule.

**Bond energy**. The energy needed to break a mole of bonds of a specified type. The enthalpy required to dissociate a bond. Usually expressed in the units of kJ $mol^{-1}$. See also **average bond energy**.

**Bond moment**. The **dipole moment** associated with a particular bond in a molecule.

**Bond length**. The distance between a pair of bonded atoms. It is measured as the distance between the two nuclei.

**Bond order**. Whether a bond is single, double, triple, or something in between. For a diatomic molecule, excluding nonbonding electrons, it's one half of the difference between the number of electrons in bonding orbitals and the number of electrons in antibonding orbitals.

**Bond pair**. Alternative name for **bonding electron pair**.

**Bond polarity**. Charge separation in a bond. See **polarity** and **percentage ionic character**.

**Bond strength**. The energy liberated when a chemical bond when the bond forms between gaseous atoms. Equally, the energy necessary to break a chemical with the formation of gaseous atoms. Also called **bond energy**. Usually stated in kJ $mol^{-1}$.

**Bond-type continuum**. The notion that every chemical bond can be imagined to exist along a continuous range from purely covalent to purely ionic.

**Bonding electron pair (BP)**. In a **Lewis dot diagram**, it is any pair of dots that represents a chemical bond. Also, any pair of bonds occupying a **bonding molecular orbital**.

**Bonding molecular orbital**. Type of molecular orbital from which removal of a electron weakens bonding in the molecule. Also called a **bonding orbital**.

**Borax**. Hydrated sodium borate, $Na_2(B_4O_7)\cdot 10H_2O$. Used in glass making, cleaning formulations, etc. Obtained from dry lake beds such as Searles Lake in southern California.

**Boyle's law**. Experimental statement about the behavior of gases: As long as the temperature remains constant, the volume of a body of gas varies inversely with the pressure upon it.

**Branched polymer**. A polymer whose molecules exhibit chain branching. See also **side chain**.

**Bright line spectra**. In the visible region of the electromagnetic spectrum these appear as a series of brightly colored lines against a dark background. Typically seen in a spectroscope when observing **excited** atoms.

**Brimstone**. Sulfur. Literally "the stone that burns."

**Brine**. A solution of a metal halide salt. Sea water is "briny" because it contains sodium chloride.

**Brønsted-Lowry acid**. The acid of any **conjugate acid-base pair**.

**Brønsted-Lowry base**. The base of any **conjugate acid-base pair**.

**Brookfield viscosimeter**. Popular type of commercial device for measuring viscosity.

**Brownian motion**. The ceaseless, random motion of molecule or small particle in a liquid or a gas.

**Btu**. Engineering energy unit defined in SI system of units to be exactly 1055.06 joules.

**Buffer action**. The act of resisting pH changes. In chemistry, to buffer means to absorb a **slug**. In everyday language, a buffer state is one that separates two powerful neighbors. Shock absorbing pistons, such as on a locomotive, or automobile shock absorbers, are sometimes called buffers.

**Buffer solution**. One that resists changes in its pH either by added acid or by added base.

**Buffering**. The phenomenon in which a buffer solution resists the change in its pH when *either* acid or base is added to the buffer solution.

**Bulk properties**. The properties of a substance manifested by the collective properties of an assembly of its molecules. Melting point is a bulk property because many molecules together determine its value. Synonym for **macroscopic properties**.

**Butyl group**. A portion of a molecule. The butane molecule less one hydrogen atom, $-C_4H_9$.

**Butylated hydroxy toluene**. A common **free radical trap**.

**Brookfield viscosimeter**. One of a family of commercial devices used to measure the viscosity of liquids.

**Calibration**. Process carried out to adjust an instrument so the instrument produces the correct reading for a reference sample. Once so corrected the instrument is said to be *calibrated*.

**Calorie**. Unit of heat energy. Now obsolete in many parts of the world. Originally, the heat energy needed to raise the temperature of a gram of water from 14.5°C to 15.5°C. Now defined as exactly 4.184 J.

**Calorimeter**. An instrument used to measure heat.

**Calorimetry**. The science of measuring heat and energy changes.

**Candela**. "The candela is the luminous intensity, in a given direction, of a source that emits monochromatic radiation of frequency 540 × $10^{12}$ hertz and that has a radiant intensity in that direction of (1/683) watt per steradian (16th **CGPM**, 1979)."

**Capillary rise**. Wicking. An everyday phenomenon in which a fluid spontaneously rises up a narrow tube or porous substance. Water spontaneously rises up a vertical glass capillary tube placed just below the water's surface. The rise shows that water molecules are even more strongly attracted to glass than they are to other water molecules.

**Carbohydrates**. A class of natural polymers. They have the general **empirical formula** $(CH_2O)_n$—hydrates of carbon—which gives them their name. Sugars, starch, and **cellulose** are common examples.

**Carbon black**. A form of soot. Used commercially in inks and vehicle tires.

**Carbon dating**. Laboratory method used to measure the age of organic materials up to about 50,000 years old. Based on assuming the historic ratio of the nuclides $^{14}C/^{12}C$, combined with a modern measurement of the actual ratio in the sample, and the known half-life of $^{14}C$.

**Carboxylic acid.** An organic compound containing the -COOH functional group and having the collective formula RCOOH.

**Carnot cycle**. Initially, a thought experiment done by Sadi Carnot. The notion that, provided with a fixed amount of heat, there is a maximum limit to the work available from an engine that operates on the cycle: liquid → gas → liquid → gas → liquid…

**CAS**. Chemical Abstracts Service. operated by the **American Chemical Society**.

**CAS Number**. Unique chemical designator assigned to new compounds by the Chemical Abstracts Service. CAS numbers are widely used to track safety, shipping, and hazard criteria for compounds. Compare **MSDS**.

**Catalyst**. A substance that affects the rate of a reaction without itself being consumed.

**Catalytic converters**. Devices used to prevent automobiles from exhausting unburned fuel. They use noble metal catalysts to provide efficient post-combustion oxidation.

**Cathode**. An electrical conductor used in constructing an **electrochemical cell**. It's the **electrode** that has a negative charge and attracts cations. At the cathode, electrons flow to the electrolyte from the external circuit.

**Cathode rays**. Beams of electrons that flow from the negative end (the cathode) to the positive end (the anode) of a vacuum tube when a large voltage is applied to the tube. They provide the basis for the operation of any TV tube or video monitor, hence the abbreviation **CRT** (cathode ray tube).

**Cation**. An entity with a positive charge.

**Cationic surfactant**. A **surface active agent** whose principle activity comes from a long-chain organic entity to which is attached a functional group with a positive charge.

**cc**. Volume unit. Abbreviation for cubic centimeter. 1 cc equals one milliliter and 1000 cc equals one liter.

**Cell potential**. The voltage generated by an electrochemical cell under specified conditions.

**Cell voltage**. The voltage at which an electrochemical cell operates.

**Cellulose**. A natural organic polymer composed of long chains of sugar monomers. Wood contains cellulose.

**Cement**. A powdered substance made by heating limestone, silicates, and/or other natural minerals. Mixed with water it reacts chemically, heats up and eventually forms a solid mass also called cement. Widely used since its introduction by Roman engineers in the first century BC. Also called mortar.

**Centipoise**. An older unit of **dynamic viscosity**. One poise is defined as $10^{-1}$ Pa s, so one centipoise is $10^{-3}$ Pa s. In SI there is no special name designated for dynamic viscosity—this older term and its unit is still encountered in the literature.

**Centistokes**. An older unit of **kinematic viscosity**. One stokes is $10^{-4}$ $m^2$ $s^{-1}$ so one centistokes is $10^{-7}$ $m^2$ $s^{-1}$. In SI there is no special name designated for kinematic viscosity.

**Central atom**. In any entity any atom that is bonded to at least two other atoms can be regarded as a central atom.

**Ceramics**. A wide range of inorganic materials, many silicate based, that are stable over a wide range of environmental conditions. Most ceramics are composed of crystals with varying compositions.

**Cermets**. A class of materials with properties that combine those of ceramics and metals. Useful engineering materials. Four major groups are carbon-containing, oxide-based, boride-based, and carbide-based cermets. Typical cermet applications are as coatings, for aerospace components, measuring tools, etc.

**Certified Quality Engineer (CQE).** Engineer with special training who specializes in applied statistics and its applications to engineering problems and systems.

**CGPM**. General Conference on Weights and Measures (in French). Refers to a particular meeting of the organization that maintains metric system standards.

**cgs system**. The centimeter gram second system of measurements. A forerunner of the modern SI metric system. The cgs system of units is now largely superseded.

**Chain extension.** Making a polymer chain longer. In a polymerization reaction involving repeated additions to a growing polymer molecule, each addition reaction is said to cause chain extension. Often called a **chain extension reaction**.

**Chain reaction**. 1. In chemical kinetics, a reaction mechanism that involves many repeating steps. 2. In nuclear physics, the neutron-mediated mechanism for the release of fission energy.

**Chalcogens**. The elements of Group 16 of the periodic table (O, S, Se, Te, Po). From a Greek root meaning those elements found in nature combined with copper.

**Charge**. 1. Fundamental property of nature. Can be positive or negative. Represented by the SI derived unit **Coulomb**. 2. For an ion, its charge is its number of excess protons or electrons. 3. To charge is to reverse the reaction of an electrochemical cell and restore its ability to support a **load**.

**Charles's law**. Experimental statement about the behavior of gases. At constant pressure, the volume of a fixed mass of gas is directly proportional to the absolute temperature.

**Chemical Abstracts**. Summaries of current chemical literature. Published by the **American Chemical Society**.

**Chemical amount**. Term used synonymously with **amount of substance**. IUPAC deprecates the expression "number of moles" and encourages the use of "chemical amount." I think IUPAC is wrong on this one.

**Chemical analysis**. The laboratory process used to determine what is in a substance (**qualitative analysis**) and in what proportion (**quantitative analysis**).

**Chemical bond**. It said to exist when two atoms join to make a molecule. In the process X + X → X-X the bond is represented by the dash. In more complex molecules, all the pairs of atoms joined in this manner are said to have chemical bonds between them.

**Chemical change**. The consequence of any chemical reaction. After a chemical change some entities (molecules or ions) have disappeared while new ones have formed. However, all the atoms are conserved during a chemical change. See also **chemical reaction**.

**Chemical compound**. A stable chemical entity, with a definite composition, structure, and properties. A pure substance composed of two or more elements with a definite composition and a definite set of physical and chemical properties. Chemical compounds are composed of molecules that are *chemically* alike. Compare **compound**, **molecular substance.**

**Chemical element**. See **element**.

**Chemical energy**. Energy that is evolved or consumed during chemical reactions.

**Chemical equation**. A symbolic, written representation of a chemical reaction. Its general form is reactants → products.

**Chemical engineer**. Person trained in engineering principles applied to chemical science. Professionals who play an important role in the chemical and materials industries, etc.

**Chemical engineering**. Broadly, the branch of engineering that deals with the construction and operation of chemical plants and the commercial development of chemical processes.

**Chemical equilibrium**. Branch of chemistry that studies the balance of reactant and products in chemically reacting systems. A chemical reaction has reached equilibrium (is at equilibrium) when the concentrations of *all* the reactants and *all* the products are unchanging. See also **chemical thermodynamics**.

**Chemical formula**. Shorthand description of a compound showing its elements by their symbols and the numbers of atoms of each by subscripts.

**Chemical instrumental methods**. Wide range of techniques used by chemists to determine chemical structure, identify compounds, perform analytical chemistry, etc., etc. See, for example, **combustion analysis**, **infrared spectroscopy (IR)**, **UV spectroscopy (UV)**, **mass spectrometry (MS)**, **nuclear magnetic resonance spectroscopy (nmr)**, **x-ray crystallography**, **gas chromatography (GC)** and **liquid chromatography (LC)**.

**Chemical kinetics**. The study of the rates of chemical reactions.

**Chemical law**. A summary statement about the chemical behavior of nature. Chemical laws typically derive from a theoretical interpretation of chemical experiments and observations.

**Chemical nomenclature**. The subject of the naming of chemical compounds and entities. A big subject that most engineers should only worry about on a need-to-know basis.

**Chemical property**. Property of a substance revealed by a chemical change. It's reactivity is probably the most important single chemical property of an element or compound.

**Chemical reaction**. The changing of one chemical compound into another, or the combining of elements into a compound. A process that leaves atoms unchanged but transforms one or more substances into other substances by altering how their atoms are combined. See also **chemical change**.

**Chemical reaction system**. A definite chemical reaction treated as a thermodynamic system. Experimentally, it is often an equilibrium reaction carried out in a closed container. Also called a **reaction system.**

**Chemical structure**. Very broad term referring to the atomic or molecular arrangement of the atoms or molecules in a substance. See also **chemical structure** and **molecular structure**.

**Chemical symbols**. The one- or two-letter abbreviations for the elements used in chemical formulas and on the periodic table.

**Chemical synthesis**. Making desired chemical products from available raw materials. Small scale chemical synthesis is done in laboratories. Large scale chemical synthesis is done in chemical plants.

**Chemical thermodynamics**. Branch of chemistry. It deals especially with the energies of chemical reactions and the equilibrium positions of chemical reactions.

**Chemical visualization**. See **molecular visualization**.

**Chime**. A chemical visualization program. A downloadable plug-in that enables molecules to be rotated and studied in pseudo three-dimensional display on a monitor screen.

**Chiral carbon atom**. One bonded to four different substituent atoms or groups and thereby forming a pair of **enantiomers**.

**Chiral purity**. A sample of a chiral compound is 100% chirally pure if it contains only one of two possible **enantiomers**. Chiral purity is very important in drug chemistry. The disastrous consequences of **thalidomide®** were caused by its being chirally impure.

**Chirality**. Natural phenomenon in which pairs of molecules exist as mirror images of one another in different stereoisomeric forms. Loosely, "handedness" in the sense that our left and right hands are related. See also **enantiomers**.

**Chlor-alkali process**. Important industrial electrochemical process. It involves the electrolysis of a sodium chloride solution to produce sodium hydroxide and chlorine gas.

**Chromatography**. The process of separating the components of a mixture by their traveling at different rates. An important instrumental method for chemical separation and analysis. Involves selective absorption of compounds in a **mobile phase** flowing across a **stationary phase**. For example, **thin-layer chromatography**, **gas chromatography**, and **liquid chromatography**.

**Circumstellar molecule**. A molecule or **free radical** known to exist on a comet that circles our sun. About 50 such molecules were known in 2000.

**Cis-trans** isomers. **Stereoisomers** in which pairs of substituent atoms can occupy different relative positions. For example, 3-chloropropene exists as the following pair of cis-trans isomers:

H, $CH_3$, C, ||, C, H, Cl    H, $CH_3$, C, ||, C, Cl, H

**Clausius equation**. Clausius' modification of the ideal gas equation to describe the behavior of real gases. $P(V-nb) = nRT$, where P is the pressure, V is the volume, n is the number of moles of gas, R is the gas constant, T is the temperature, and b is a different constant for each real gas.

**Clausius-Clapeyron equation**. An empirical equation that predicts the variation of a liquid's vapor pressure as a function of temperature. $\ln P = -\Delta H_{vap}/RT + A$

**Close packing**. Geometric idea involving the way in which hard spheres pack in a container. Used in chemistry in understanding the structures of metals, ionic crystals, and the coordination of central atoms by ligands. See **radius ratio**, **cubic close packed**, and **hexagonal close packed**.

**Closed shell**. Term used to describe a set of electrons that complete the filling of a principal quantum level.

**Closed shell configuration**. Said to be adopted by an atom or ion whose outer electron configuration is $ns^2 np^6$.

**Closed system**. A system that contains a fixed amount of material.

**Coefficients**. The numbers used to balance equations in accord with the principle of conservation of atoms.

**Collective abbreviations**. Chemists use many collective abbreviations. For example, X is used to represent any halogen element, and R is used to represent a designated portion of an organic molecule.

**Colligative properties**. Boiling point elevation, vapor pressure lowering, freezing point depression, and the formation of osmotic pressure.

**Colloids**. A wide range of mixtures that lie at the boundary between **homogeneous** and **heterogeneous** substances. Milk and ice cream are examples of colloidal substances.

**Combined gas law**. $\frac{P_1V_1}{T_1} = \frac{P_2V_2}{T_2}$

**Combustion analysis**. Classical technique for the analysis of **organic compounds** that remains important in the 21st century. Bedrock technique of organic analytical chemistry; the most unambiguous method for determining the composition and purity of organic compounds. Uses principles of stoichiometry. It involves burning a weighed amount of a substance under controlled conditions with trapping and weighing of the carbon dioxide and water produced.

**Combustion reaction** The reaction (burning) of a substance (a fuel) with oxygen. Combustion reactions liberate heat and are always exothermic.

**Complete combustion** The situation that occurs when the oxygen available is in excess of what the combustion stoichiometry demands. Under conditions of complete combustion of a hydrocarbon the only reaction products are water and carbon dioxide. Compare **incomplete combustion**.

**Complete reaction**. A reaction in which, as a practical matter, all of the reactants are consumed. See also **go to completion.**

**Complex compounds**. Compounds composed of metal atoms bonded to one or more molecules or ions (collectively called **ligands**).

**Complex ions**. Ions composed of metal ions bonded to one or more molecules or ions (collectively called **ligands**).

**Composite**. A solid material produced by embedding fibers of one substance into a matrix of another substance. Fiberglass is a common composite material

**Composition**. Stating its composition describes what kinds of matter a material contains. Often a list of the percentage by mass of each element in a sample or substance.

**Compound**. A pure substance with a definite chemical formula and structure. Disregarding isotopes, the molecules of a **molecular compound** are identical. See also **chemical compound**. The term compound was introduced by John Dalton when he spoke of "compound atoms."

**Compressibility factor**. A measure of how an actual gas compares to a hypothetical real gas. The compressibility factor, z is given by the formula $z = PV/nRT$. The closer is z to one the "more ideal" the behavior of the gas.

The compressibility factor for any gas depends on its temperature and pressure.

**Compression**. The act of decreasing the volume of a system by applying pressure to it.

**Concentrated** or **strong** solution. One with a large amount of solute compared to the saturation concentration of that solute in that solvent. Compare **dilute**.

**Concentration**. A qualitative term used to describe the amount of a solute present in a solution.

**Concrete**. A stone-like material made by mixing cement powder with aggregates such as sand or gravel, adding water and allowing the mass to harden.

**Condensation reaction**. One in which two molecules combine and simultaneously split out a small molecule (often water). The condensation of bifunctional organic molecules is one important method of producing **synthetic polymers**. Reaction used to make a **condensation polymer.**

**Condensed phases**. Solids and liquids.

**Condenser**. In distillation, it is the part of the apparatus in which hot vapors are converted back to the liquid state. In the laboratory it is typically a double-walled glass tube. Steam power plants have gigantic condensers.

**Conjugate acid-base pair.** Together, a weak acid and the anion it forms when the acid loses a hydrogen ion.

**Conduction band**. In a solid, if the **valence band** is only partially filled with electrons the solid is an electrical conductor. A partly filled valence band is a conduction band. Conduction bands exist in metallic conductors. See also **metallic solids.**

**Contact angle**. The angle formed by a drop of liquid on a solid surface.

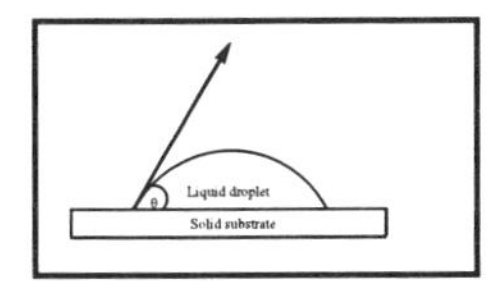


**Continuous spectrum**. One in which radiation of all wavelengths is present. In the visible region of the electromagnetic spectrum, a rainbow is a continuous spectrum.

**Continuum**. A region of a spectrum where individual lines merge. Explained by considering the energy states of an entity, In the hydrogen atom, as energies of the states increase there comes a point where an enormous number of energy states are squashed together.

**Conversion factor**. A quantity with both a value and units. Used to change values from one system of measurement to another. The conversion factor that converts A to B is the reciprocal of the conversion factor that converts B to A. See **algebra of quantities**.

**Coordinate-covalent bond**. Type of covalent bond in which the bonding pair of electrons can be regarded as having come from one of the two bonded atoms (as opposed to a "normal" covalent bond in which each atom contributes one electron). Compare **adduct**.

**Copolymer**. The product of a polymerization of two monomers. A polymer made from more than one monomer.

**Core electrons**. Electrons in an atom that do not participate in chemical bonding. The inner-shell electrons that have only a minor effect on an atom's chemistry. For an atom in the first three rows of the periodic table, the number of core electrons is the atomic number of the last noble gas that precedes it.

**Correct value**. The correct value of a property is the best value of the property that humans can determine.

**Corrosion**. Degradation of metals in the environment. Usually, a chemical reaction involving oxygen, water, and a metal.

**Cosmic radiation**. Cosmic rays from deep space that continually bombard the earth. Cosmic rays include high velocity nuclei, gamma rays, fast electrons, etc.

**Cosmological elements**. Those that arose during the big bang; hydrogen and helium.

**Coulomb**. SI Derived unit of electrical charge. The amount of charge produced by a current of one **amp** flowing for one second.

**Coulombic forces**. The forces among charged particles in accord with **Coulomb's law**.

**Coulomb's Law**. $F \alpha \frac{q_1 q_2}{r^2}$. The force between ions is directly proportional to the charges ($q_1$ and $q_2$) on the ions and inversely proportional to the square of the distance (r) between them.

**Count**. In chemistry, count often refers to specifying an amount of substance by either using moles or stating the number of entities of the specified substance.

**Covalent bond**. One imagined as being created by two atoms sharing electrons.

**Covalent compound**. A compound that exists in the form of molecules as opposed to ions. Ordinary pure liquids are good examples of covalent compounds. Contrast **ionic compound**.

**Covalent radius**. Lengths used to describe the size of bonded atoms. If we imagine that atoms are spherical, the length of the bond between them is the sum of the radii of the two atoms. Studying many bond lengths yields a table of covalent radii. Nonbonded atoms are assigned **van der Waals radii**.

**Covalent solids**. These are essentially giant molecules in which all the atoms are covalently bonded to their neighbors, that in turn are connected throughout the entire structure. Such solids are very strong. Diamond is the classic example of a covalent solid.

**Crastin®**. A family of engineering plastic materials produced by the DuPont Company and finding applications such as for automobile body parts.

**Critical mass**. For a fissile nuclide, it's the smallest mass of the nuclide that undergoes a spontaneous neutron chain reaction. Nuclear weapons work by the rapid compression of subcritical masses to produce a critical mass. Compare **fissile**.

**Critical pressure**. The pressure at which the liquid and gas states of a substance are in equilibrium at the critical temperature of the substance.

**Critical temperature**. Gases above their critical temperature cannot be pressure condensed. Hence, it is the temperature below which a gas can be condensed to a liquid merely by compression.

**Cross linking**. A process by which various polymer chains in a substance are chemically connected. Altering the extent of cross linking in a polymer typically dramatically changes the polymer's bulk properties. The first practical application of cross-linking occurred when Charles Goodyear vulcanized rubber using sulfur to join rubber's natural polymer chains.

**CRT**. A **cathode ray** tube.

**Crude oil**. A complex mixture of hydrocarbon and other organic compounds that occurs naturally in geologic formations. Major source of fuel. Usually obtained from drilled wells.

**Crystal**. See **crystalline material**.

**Crystal glass** is typically glass with a high lead (PbO) content. Such glass has a high index of refraction and looks particularly attractive.

**Crystalline material**. Substance whose entities arrange themselves in a characteristic repeating array with extensive long range order. A substance with a highly ordered internal structure. I forget who it was who said: "the beauty of crystals lies in the planeness of their faces."

**Crystallography**. In chemistry, the study of the structure of materials by means of **x-ray diffraction** experiments. A principal method by which **chemical structure** is determined.

**Cubic close packed (ccp)**. One of two situations in whic a collection of spheres or atoms occupies the smallest possible total volume. The other situation is **hexagonal close packed (hcp)**. See **close packing**.

**Cutting fluids**. Milky liquids sprayed onto the work piece during the machining of metal parts. Cutting fluids are **emulsions**.

**Cyclic molecule**. One that contains some of its atoms joined in a ring.

**Cyclic process**. A process that eventually gets back to where its started. For example, in an air conditioner, when the **working fluid** returns to its starting state.

**Cycloalkane**. A saturated hydrocarbon molecule that contains at least one ring of atoms

**Cytosine**. One of the bases of **nucleic acids**, $C_5H_5N_3O$. Based on its structure it is a pyrimidine-type base.

**Dacron®**. A commercial brand of polyester. Polyethyleneterephthalate.

**Dalton (Da)**. Alternative name for the **unified atomic mass unit**. Widely used in biochemistry. The Dalton is recognized but not sanctioned by IUPAC.

**Dalton's law of partial pressures**. Law of gas behavior noted by John Dalton: The total pressure exerted by a mixture of gases is the sum of the partial pressures of the individual gases.

**Daniell cell**. Historically, one of the most important batteries. Used for telegraphy during the second half of the 19th century. It relies on Zn metal as the reductant and $Cu^{2+}$ ions as the oxidant.

**Dark line spectrum**. A continuous spectrum of visible light interrupted by a pattern of narrow, dark lines. A frequent experimental presentation of an atomic absorption spectrum.

**Daughter nuclide**. In a radiodecay process, the **parent nuclide** disappears with the simultaneous formation of a new, daughter nuclide.

**d-block transition metal**. In the periodic table the elements belonging to the three rows of metals scandium (Sc) to zinc (Zn), the first row; yttrium (Y) through cadmium (Cd), the second row) and, hafnium (Hf) through mercury (Hg), the third row.

**Debye**. A unit of **dipole moment**. The measured dipole moments of molecules are stated in debye units. One debye (D) is $3.33 \times 10^{-30}$ coulomb meters (C m). The use of the debye unit is deprecated by IUPAC but recognized to be in continued wide use because debye values are nicely sized for humans. Named after a Dutch (born American) physicist.

**Decay chain**. See **natural decay chain**.

**Decomposition reaction**. A chemical reaction in which a single compound breaks down or decomposes into two or more fragments.

**Deflocculant**. A dispersing agent. An agent used to suspend particles in a medium and prevent their flocculation.

**Degenerate**. Said of two different electronic states with the same energy.

**Delocalized**. In chemistry, it usually means that an electron in an atom or molecule is not considered to be located in a single place. **Delocalized bond**. A bond formed in part by delocalized electrons. A bond intermediate in its properties between a single and double bond. When a pair of electrons contributes to more than one bond such a pair is said to be **delocalized**. Delocalized bonds are characteristic of the **benzene ring** of atoms.

**Deoxy ribose**. A sugar molecule with the formula $C_5H_{10}O_4$. An important biochemical molecule critical to building the structure of deoxyribonucleic acid. See **DNA** and **nucleic acids**. The structure of deoxy ribose is shown below.

**Depleted uranium**. Uranium from which the isotope $^{235}U$ has been substantially removed by effusion for use as a nuclear fuel. Mostly composed of the isotope $^{238}U$. See **gaseous diffusion**.

**Derived unit**. In the **SI system** of units, derived units are made by combining **base units**.

**Descriptive inorganic chemistry**. Chemistry that involves: the origin and abundances of the elements; the properties of the elements and their compounds; the sources, production, and uses of the elements and their compounds, etc.

**Deuterium**. The isotope $^2H$. Sometimes called **heavy hydrogen**.

**Dew point**. The temperature to which an air sample must be cooled to cause liquid water to condense (form a dew). The temperature at which cooling air becomes saturated with water vapor.

**Diagonal periodicity**. Situation in which two diagonally related elements in the periodic table have similar chemical and physical properties. Examples are Li and Be, B and Si, etc.

**Diamagnetic**. An atom, molecule, or ion is diamagnetic if it has no unpaired electrons. Diamagnetic materials are weakly repelled by magnetic fields.

**Diamagnetic substance**. A substance weakly repelled by magnetic fields.

**Diatomic**. Composed of two atoms. A molecule that contains two atoms is a diatomic molecule.

**Diatomic molecule**. One composed of two atoms such as $H_2$, $O_2$, or $I_2$.

**Dibenzoyl peroxide**. Is the chemical compound $C_6H_5CO\text{-}O\text{-}O\text{-}OCC_6H_5$. Used as a **free radical initiator** and also as a skin cleansing agent.

**Diester**. An organic molecule that contains two ester functional groups.

**Diffraction**. In physics, a phenomenon in which waves interfere. In chemistry, often used in connexion with x-rays in the branch called **crystallography**.

**Diffraction pattern**. The behavior of electromagnetic radiation consequent to a diffraction experiment. In **crystallography** it's the set of secondary x-ray beams that are scattered by the crystal under study.

**Diffraction grating**. Device used to create a colored spectrum from white light. Often a material (such as a piece of glass, or a plastic film) covered with many closely ruled lines. For X-rays, crystals constitute natural diffraction gratings.

**Diffusion**. The process of one fluid mingling with another. An **entropy driven process**.

**Dilute**. To dilute is to add solvent to lower the concentration of a solute.

**Dilute** (or **weak**) **solution**. One with a low solute concentration compared to the saturation concentration of that solute in that solvent. Compare **concentrated**.

**Dilution calculation**. Computing the change in concentration (mol $L^{-1}$) of a solute in a solvent when more solute is added. The mathematical relationship used is: $C_1V_1 = C_2V_2$.

**Dimer**. A molecule formed when two identical molecules join.

**Dimerize**. To undergo a reaction that brings about the formation of **dimers**. Examples of dimerization reactions are: $A + A \rightarrow 2A$ and $2\ C_2H_4 \rightarrow C_4H_8$

**Dipole forces**. Interactions between polar entities. The forces between two molecules each of which has a **dipole moment**.

**Dipole moment**. Characteristic property of a polar entity. For a diatomic molecule it is the product of the charge separation of the bond and the bond's length: $\mu = q\ R$. The bigger a molecule's dipole moment the greater its interaction with an electric field. Compare **polarizability**.

**Diprotic**. Said of an acid that produces two moles of hydrogen ion per mole of the acid. For example: $H_2SO_4 \rightarrow 2\ H^+ + SO_4^{2-}$.

**Discharge**. What happens to a battery when it is used to drive a **load**. To consume the reactants of an **electrochemical cell**.

**Discrepancy**. The difference between two values. The difference between a **correct value** and a **measured value**.

**Dispersion forces**. Intermolecular forces caused by **temporary dipole - temporary dipole** interactions between pairs of molecules. Present among all pairs of molecules, but especially important among the molecules of **nonpolar** liquids. Also called London forces.

**Disproportionation**. In chemistry, a technical term meaning **self-redox**.

**Dissociation**. The process of coming apart. 1. The breaking of one or more chemical bonds in a molecule; fragmentation of a molecule into atoms or groups of atoms. 2. The separation of an ionizable compound into its ions when the compound is put into water.

**Dissolution**. The process by which a solvent dissolves a substance to form a solution.

**Dissolve**. To dissolve a substance is to mix it with a solute to form a solution.

**Distillation**. A process to purify and separate compounds. The process of heating liquid mixtures and recondensing the vapors that form to produce a new liquid with an enriched composition. Historically, the earliest use of distillation was probably to make alcohol. A critical process in the refining of crude oil. Compare **fractional distillation**.

**Distillation flask**. In the laboratory, a device used to distill liquids. A simple distillation flask is a round glass vessel with a neck.

**Distillation fraction**. Liquid mixtures whose various constituents boil over a narrow range of temperature. The product of a **fractional distillation**.

**Disturbing an equilibrium**. Term used to describe the situation when some property of an equilibrium chemical reaction is changed.

**DNA**. Deoxyribonucleic acid. See **nucleic acids**.

**d-orbitals**. Hydrogen-like wave functions in which the secondary quantum number is one. Any $\psi_{(n,\ell,m)}$ with quantum number $\ell=2$.

**Double bond**. A bond between a pair of atoms whose bond energy is approximately twice that of a single bond between a given pair of atoms. A covalent bond in which the atoms share two pairs of electrons.

**Double helix**. The two complementary strands of a DNA molecule are called the double helix.

**Dynamic viscosity**. In SI it's viscosity with the units of Pa s. It has no special name.

**Dyne**. Unit of force equal to $10^{-5}$ newtons. Used in the **cgs system** of measurements. Now outdated.

**Efficiency** In thermodynamics, the efficiency of an engine is the fraction of the heat supplied to the engine that is converted to useable work. Multiplied by 100 it's the **percentage efficiency**:

**Effusion**. A process in which a gas escapes through a pinhole leak into a vacuum. Also called gaseous effusion. Often **gaseous diffusion** is stated when gaseous effusion is meant, as in uranium diffusion.

**Elastomers**. Synthetic polymers with rubbery properties.

**Electric charge**. A property of fundamental particles. The charge on one electron is approximately $1.602 \times 10^{19}$ C and is given a negative sign.

**Electric potential**. The potential energy created by the effect of two charged entities acting on one another. The **voltage** a system generates.

**Electrochemical cell**. A device that enables the exchange of electrical and chemical energy.

**Electrochemical reaction**. One that takes place inside an **electrochemical cell** or during processes such as **corrosion**.

**Electrochemistry**. The science of electricity and chemistry.

**Electrode**. An electrical conductor dipped in an electrochemical cell. There are two electrodes in most electrochemical cells. They provide electrical connection between the electrolyte and the external circuit. Compare **anode** and **cathode**.

**Electrolysis reaction**. One that uses an applied electric current to bring about a chemical reaction.

**Electrolyte** is a substance that dissolves in water with the formation of ions. Also a solution of such a substance, especially the fluids of a battery or the human body.

**Electromagnetic radiation**. The propagation of energy through space by oscillating waves of electric and magnetic fields.

**Electromagnetic spectrum**. The entire range of electromagnetic radiation that stretches from radio waves to gamma radiation.

**Electron affinity**. The amount of energy *released* when an atom captures an electron. For an atom, X, the electron affinity is the energy change, $\Delta E$, for the process $X(g) + e^- \rightarrow X^-(g)$ expressed with a positive algebraic sign.

**Electron capture (EC)**. A radioactive decay mode in which parent nucleus decays by capturing one of the atom's electrons.

**Electron configuration**. A shorthand notation that designates an atom's electrons placed into hydrogen-like atomic orbitals. The specification of the orbitals occupied by the electrons in an atom or ion. For example, the electron configuration of a **ground state** fluorine atom is $1s^2 2s^2 2p^5$.

**Electron density**. Quantum mechanical notion that the square of the wave function at any point in space is a measure of the probability of finding the electron at that point. Also called probability density.

**Electron density map**. A pictorial representation of **electron density**.

**Electron geometry**. The disposition of the electron pair probability distributions in space around a central atom.

**Electron pair**. Two electrons, in different spin states, occupying the same orbital.

**Electron pair bond**. A covalent bond imagined to arise when two bonded atoms share a pair of electrons.

**Electron transition**. It is a change in the energy state of an entity. It is said to occur when an electron changes energy levels (up or down) in an atom.

**Electronegativity**. Property of the atoms of an element. A unit-less value that measures the relative tendency of a bonded atom to attract electrons. Given the symbol $\chi$ (chi). Linus Pauling was the first person to construct an electronegativity scale. On the Pauling scale, electronegativities range from 0.8 - 1 for active metals to 4.0 for the most electronegative element, fluorine.

**Electronic states**. The various energy states of an entity that arise as its electrons undergo **transitions**. Such states are shown on an **energy level diagram**.

**Electroplating**. Industrial process to coat an object with a thin film of inactive metal. Used to produce decorative and protective coatings. Chrome plate is a familiar product of electroplating.

**Electrorefining**. The process of purifying a metal by electrolysis

**Element**. A substance incapable of being chemically split into simpler substances. For the purpose of this definition, protons, neutrons, and electrons are not substances.

**Elemental analysis**. Of a compound it's an experimental method that leads to the specification by weight percentage of the elements in the compound. The measurement of relative amounts of the elements in a compound. See **chemical analysis** and **combustion analysis**.

**Emission**. Process in which an atom loses energy by releasing a photon of light or electromagnetic energy. The atom is said to emit the photon. Compare **absorption**.

**Emission spectra**. Spectra produced by the emission of energy from excited entities. Often they are **bright line spectra**.

**Empirical formula** (of a chemical compound). The simplest ratio of a compound's numbers of atoms obtained by factoring its molecular formula. The simplest whole number ratio of the atoms in a molecule. The molecular formula divided by the highest common factor.

**Empirical equation**. An equation developed to fit experimental data.

**Emulsifiers**. **Surfactants** used to create oil-water emulsions.

**Emulsion**. A mixture that often consists of small oil droplets surrounded by surfactant molecules that stabilize the droplets in water. Most creamy food products are emulsions.

**Enantiomers**. **Stereoisomers** that exist as mirror image pairs. Enantiomers exemplify the property of **chirality**. Often associated with a **chiral** carbon atom. For example the pair:

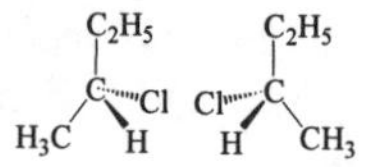

See also **racemic mixture**.

**End groups**. The functional groups or atoms at each end of a long molecule. Used especially in discussions of synthetic polymers.

**Endothermic**. A process that is enthalpy-consuming. Energy flows from the environment to the system in a thermodynamic process.

**Endpoint**. In an indicator moderated acid-base reaction it's the point at which the indicator dye changes color. In general, the point at which a reaction becomes **complete**.

**Energy**. The property of a system or entity that gives a system or object the ability to do work. Energy's base SI unit is the joule. see, for example, **atomic energy**, **kinetic energy**, and **potential energy**.

**Energy density**. 1. A quantity used to measure the ability of fuels to release energy upon combustion; energy released per gram of fuel combusted. 2. The ratio of the power available from a **battery** divided by its mass.

**Energy gap.** See **band gap**.

**Energy level diagram** One dimensional diagram that shows the electronic energy states of an entity. The simplest such diagram is for the hydrogen atom. Compare **hydrogen spectrum**.

**Energy state**. In general terms, the amount of energy in an entity. The **ground state** has the lowest possible energy; **excited states** have more.

**Energy value**. A term used to measure the energy content of foods; energy released per gram of food burned.

**Engineering material**. Loosely, any kind of stuff that engineer's make things with.

**Engineering polymers**. Synthetic polymers used to make objects, parts, or components.

**Enthalpy**. A thermodynamic state property given the symbol H, and defined by the equation: H = E + PV, where E is internal energy, P is pressure, and V is volume.

**Enthalpy of reaction**. The heat change for a reaction that occurs at constant pressure. $\Delta H = q_p$.

**Entity.** A term used in conjunction with the SI definition of the mole. An entity is whatever atom, molecule, ion, etc., is being counted. A mole of any entity contains the Avogadro number of those entities. See also **mole**. Try substituting the phrase "atom, molecule, or ion" whenever you encounter the word entity.

**Entropy**. A thermodynamic state property of systems. Entropy is a purely bulk (macroscopic) thermodynamic property of systems. However, it can be interpreted statistically as the amount of "randomness" in a system. Standard entropies are stated with reference to absolute zero. See **standard entropies**.

**Entropy change**. The change in a system's entropy when the system undergoes a process. The enthalpy change of a **reversible process** divided by the kelvin temperature at which the process takes place, $\Delta S = \frac{q_{rev}}{T}$

**Entropy driven process**. A spontaneous thermodynamic process resulting from an increase of randomness. Usually said of a spontaneous process that occurs without an energy change.

**Entropy of mixing**. The entropy change that occurs when two or more substances are mixed. Compare the **third law of thermodynamics** which demands perfectly crystalline substances to avoid the issue of entropy of mixing.

**Entropy of vaporization**. The entropy change that occurs when a liquid is converted to a gas.

**EPA**. Environmental Protection Agency. US Federal agency that regulates a wide range of environmental issues such as air and water pollution.

**Ephemeral radionuclide**. One whose natural existence can be inferred but which is difficult or actually impossible to prove directly to exist.

**Epoxide**. An organic compound containing the functional group

```
    O
   / \
R-C—C-R
```

and having the collective formula $R_2C_2H_2O$. Epoxides are reactive compounds and act as difunctional molecules. Their reactivity derives primarily because the angles in the ring differ greatly from those of tetrahedral **ideal geometry**. The phenomenon is called **ring strain**.

**Epoxies**. Useful glues. A class of polymers made by reacting pairs of monomers or oligomers that produce thermosetting plastics.

**Equilibrium bond length**. The distance between a pair of bonded atoms. The adjective equilibrium denotes that real bonds are vibrating, making **bond length** a time averaged property.

**Equilibrium constant**. In **chemical equilibrium**. Once a chemical reaction has reached equilibrium the concentrations of *all* the reactants and *all* the products are unchanging. At any specified temperature, the ratios of these constants raised to appropriate powers is the reaction's equilibrium constant. The general equilibrium constant expression for a gas phase reaction a A(g) + b B(g) ⇄ c C(g) + d D(g), in an existing state of equilibrium, is written $K_p = \frac{(P_C)^c(P_D)^d}{(P_A)^a(P_B)^b}$.

**Equilibrium system**. A thermodynamic system whose properties remain constant no matter how long the system is observed.

**Ester**. An organic compound containing the -COO- functional group and having the collective formula RCOOR or RCOOR'.

**Esterification**. A chemical reaction that produces an ester. The organic product of the reaction between a carboxylic acid and an alcohol.

**Esterification reaction**. One in which a molecule having an alcohol functional group reacts with a molecule having a carboxylic acid functional group to yield an ester and water. An example of a **condensation reaction**.

**Estimated standard deviation (ESD)**. The ESD of a measured value is an approximation to its true, statistical standard deviation. Ordinary scientific works rests on estimated, not actual, standard deviations.

**Ether.** 1. An organic compound containing the -O- functional group and having the collective formula ROR or ROR'. 2. Specifically, diethyl ether $H_5C_2$-O-$C_2H_5$.

**Ethyl group**. A portion of a molecule. The ethane molecule less one hydrogen atom, -$C_2H_5$.

**Ethylene oxide**. A highly reactive gas, widely used in chemical synthesis. $C_2H_4O$, the simplest **epoxide**. Its two carbon atoms and the oxygen atom are positioned at the corners of an equilateral triangle. It suffers **ring strain**.

**Eutectic phase**. One type of solid phase that can form when a liquid is cooled. Specifically, a solid formed with the same composition as the liquid melt from which it solidifies.

**Evaporation**. Process during which a substance transforms spontaneously from a liquid to a vapor while simultaneously absorbing environmental heat.

**Evaporites**. Natural minerals. Water soluble compounds left behind when a landlocked lake evaporates. The Great Salt Lake, for example.

**Exact neutralization**. In an acid base reaction, it is said to have occurred if neither an excess of the acid nor an excess of the base remains; exact neutralization requires that chemically equivalent (or **stoichiometrically** equivalent) amounts of an acid and a base were mixed.

**Exact number**. One without uncertainty. Hence, a number not derived from experiment. Contrast **measured number**.

**Excess and limiting reactants**. Chemical reactions can never be set up so that their reactants are present in *exactly* stoichiometric amounts (Avogadro's number is very large). Inevitably, one reactant is present in **excess** and the other is **limiting**.

**Excess reactant**. A reactant present in a reaction mixture in sufficient quantity that it can never be completely consumed. Compare **limiting reactant**.

**Excite**. In chemistry, a verb meaning to add energy. To excite a system or entity is to supply energy to the system or entity.

**Excited state**. The condition of a system or entity to which energy has been added, or which has an excess of energy. A system in a state above its **ground state**.

**Exothermic**. A process that is heat-liberating. Energy flows from the system to the environment in a **thermodynamic process**.

**Experimental science**. Chemistry, for example. One that progresses by means of the interpretation of experiments and observations.

**Experimental uncertainty**. The subject of how reliable are our measurements. In the real world, scientists and engineers can only ever make a limited number of experiments to get a result. Estimating the experimental uncertainty allows them to estimate how far that result can be trusted.

**Explosives**. Chemical compounds that are **thermodynamically unstable**. They are activated by a detonator.

**Extensive properties**. A property that changes when the **amount of substance** of a sample is changed is an extensive property. For example, doubling the amount of water doubles its mass, its number of molecules, and its internal energy. So those three are extensive properties. Compare **intensive property**.

**External circuit**. In electrochemistry, an operating battery supports the external circuit. The wires and electrical devices (the **load**) connected to an electrochemical cell.

**Faraday constant**. The charge on a mole of electrons 9.648 530 9 (29) × $10^4$ C $mol^{-1}$

**Fats**. 1. A food group. 2. Class of organic compounds widely distributed in the biosphere and serving for chemical energy storage. The triglycerol esters of long chain carboxylic acids. Also called triglycerides.

**FDA**. Food and Drug Administration. US federal regulatory agency.

**FERH**. Fundamentals of Engineering Reference Handbook. The data compilation engineering students are allowed to use when they take certification examinations.

**Ferrocene**. The first-discovered **sandwich compound**, $C_5H_5$-Fe-$C_5H_5$. An orange compound with a characteristic odor that is easily sublimed.

**First law of thermodynamics**. Fundamental natural law. Not provable, but not in doubt. It states that $\Delta U = q + w$, where q is the heat absorbed by a system, w is the work done on a system, and $\Delta U$ is the change in the system's internal energy.

**First order reaction.** In **rate process**, exponential decay over time. A reaction whose rate depends on the concentration of a single reactant.

**First transition series**. In the periodic table the elements from scandium (Sc) to zinc (Zn). The topmost of three rows of **d-block** transition metals in the periodic table.

**Fissile**. Capable of being fissioned. A nuclide that breaks apart in a **neutron chain reaction**. The fact that fissile nuclides exist is a necessary condition for the production of fission **atomic energy**. Three fissile nuclides are $^{235}U$, $^{239}Pu$, and $^{232}Th$.

**Fission**. The process of a nuclide breaking down into two smaller fragments and perhaps simultaneously releasing a small number of neutrons. Fission processes are usually accompanied by the release of large amounts of fission energy. See **atomic energy**.

**Fission products**. Daughter radionuclides formed from the fission of a a larger nuclide. On average, the mass of the fission products is about one-half the mass of the parent nuclide. Collectively, these fission products constitute radioactive waste.

**Flame retardants**. Chemical compounds added to commercial products to slow their rate of burning.

**Flash Point**. Characteristic of a flammable liquid. The temperature at which the liquid is sufficiently volatile for its vapor to be ignited by a spark or flame.

**Flocculant**. A solid-liquid separating agent. An agent used to cause matter suspended in water to fall under the force of gravity.

**Fluid phases** Gases and liquids. So-called because they flow.

**f-orbitals**. Hydrogen-like wave functions in which the secondary quantum number is three. Any $\psi_{(n,\ell,m)}$ with quantum number $\ell=3$.

**Formula unit**. The hypothetical "molecular formula" of a substance not composed of molecules. **IUPAC** says not a unit but "an entity specified as a group of atoms by the way the chemical formula is written." For example, we speak of NaCl as salt's "formula unit" because no physical entity corresponding to NaCl actually exists in a lattice composed on $Na^+$ ions and $Cl^-$ ions.

**Fossil fuels**. Coal, oil, and natural gas.

**Fraction. See distillation fraction.**

**Fractional Distillation** is the process of heating liquid mixtures, recondensing the vapors that form, and repeating the boiling and recondensation steps until components of the mixture separate according to their vapor pressures or boiling points. Distillation carried out in such a way that the liquids produced consist of groups of molecules with their boiling points in a narrow range. Such a group of similarly boiling liquids is called a **fraction**.

**Fragmentation pattern**. In **mass spectrometry** it's the set of ionic fragments that form as a molecule is ionized and breaks up. An important technique used for the identification of organic compounds.

**Free energy**. An important thermodynamic state function. For processes that occur at constant temperature it's given by the Gibbs equation $\Delta G = \Delta H - T\Delta S$.

**Free energy of formation**. The free energy change of the reaction that forms a compound in its standard state from the compound's constituent elements in their standard states. A standard thermodynamic property of compounds tabulated by **standards organizations**.

**Free radical**. A reactive fragment of a molecule containing at least one unpaired electron and usually appearing (on earth) as a reaction intermediate. In space, free radicals are commonplace.

**Free radical chain mechanism** A **reaction mechanism** that involves free radical propagation steps.

**Free radical initiator**. A compound that serves as a source of free radicals and initiates chain reactions.

**Free radical propagation**. See **propagation** reaction.

**Free radical trap**. A preservative, often a food preservative. A chemical additive that reacts rapidly with free radicals and prevents long free radical chain events from occurring.

**Freeze drying**. Vacuum process to remove moisture from a material. In everyday life, many food products are preserved by freeze drying.

**Freezer burn**. The desiccation (loss of moisture) of foods caused by ice's **sublimation**. When done carefully and deliberately, low temperature food desiccation is called **freeze drying**.

**Freezing point depression.** Phenomenon in which the freezing point of a solvent is lowered in proportion to the amount of a solute dissolved in that solvent.

**Frequency**. For **electromagnetic radiation** it is the number of times per second that the field oscillates as the wave passes any point.

**Fuel**. A substance capable of reacting with oxygen and liberating heat.

**Fuel cell**. Device that carries out a combustion reaction in such a manner as to generate an electric current directly.

**Functional group**. A group of atoms bonded in a particular way in a molecule and conferring specific properties (functions) on that molecule. Compare **table of functional groups** and **R-Groups**.

**Fundamental physical constant**. A natural constant with the same value anywhere in the cosmos. For example: the speed of light, the mass and charge of an electron, etc.

**Fusion**. 1. Melting. 2. Nuclear process in which smaller nuclides combine to make a larger nuclide accompanied by the release of large amounts of fusion energy. The energy source of our sun and all stars. See **atomic energy**.

**Galvanize**. In applied chemistry, to galvanize is to treat a base metal, such as iron, with zinc for protection.

**Gamma decay**. See **Gamma ray emission**

**Gamma radiation**. High energy, short wavelength electromagnetic radiation usually studied in the context of nuclear decay processes.

**Gamma ray emission**. A process that often accompanies radioactive decay. For example, it occurs when a parent nuclide with a large excess of energy decays to give a daughter nuclide, simultaneously liberating the excess energy.

**Gas chromatography (GC** or **GLC)**. An important technique of chemical analysis. The process of separating the components of a mixture by their traveling at different rates in a flowing gas stream. Because of selective absorption of compounds in a **stationary phase**.

**Gas constant**. A **fundamental physical constant**, $R = 8.314 \text{ J mol}^{-1} \text{ K}^{-1} = 0.08206 \text{ L atm mol}^{-1} \text{ K}^{-1}$.

**Gas discharge tube**. A tube, usually made of glass, equipped with electrodes, containing low pressure gas. Electric current passed through the tube excites the gas and causes its atoms or molecules to discharge light. Neon signs and fluorescent lights are everyday examples of gas discharge tubes.

**Gaseous diffusion**. A process that occurs when a volatile substance placed in a gas spreads throughout the gas. Odors move through still air by means of diffusion. The separation of $^{235}UF_6$ from $^{238}UF_6$ by gaseous diffusion (it's always called that, but the process used was really **effusion**) is an important step in the production of **atomic energy**. See also **Graham's law**.

**Gaseous effusion**. See **effusion**.

**Gaseous state of aggregation.** Exhibited by gases. The least dense form in which ordinary matter can exist.

**Gauge pressure**. Gas pressure measured by a device with a gauge. Conventionally, it is **absolute pressure** plus atmospheric pressure.

**Gay-Lussac's law**. Measured under the same conditions of pressure and temperature, the volumes of gases that are consumed or produced in a chemical reaction occur in ratios of small whole numbers.

**Gels**. Large class of substances that have jelly-like characteristics. Many, such as Vaseline® and inks have useful technological applications

**General formula**. In organic chemistry, a way of designating a related family of chemical compounds—often a family whose members differ by one carbon atom. For example the non-cyclic saturated hydrocarbons have the general formula $C_nH_{2n+2}$, where n is any integer.

**Geochemistry**. Branch of chemistry that overlaps geology. It deals with the distribution of the elements in the earth's crust, the chemistry, chemical structure, and mineralogy of rocks and minerals. The chemistry of rocks and minerals is geochemistry.

**Geometric isomers**. See **cis-trans isomers**.

**Getter**. A reactive metal used to remove the last traces of oxygen gas from a vacuum tube

**Gibbs free energy**. See **free energy**.

**Gibbs equation.** $\Delta G = \Delta H - T\Delta S$. An important thermodynamic relation that defines the thermodynamic criterion of equilibrium for a process that takes place at constant temperature and pressure.

**Good manufacturing practice, (GMP)**. Industrial quality control. In drug manufacturing it is a requirement by the **FDA** (US Food and Drug Administration) that methods of drug production, analysis, and certification must meet state-of-the art technology and be exhaustively validated and documented.

**Go to completion**. A phrase used to describe any reaction in which the reactants turn more or less entirely to products. One with a large equilibrium constant so that at equilibrium the reactant concentrations are essentially zero. Compare **quantitative reaction**.

**Graham's law of diffusion/effusion**. States: "The diffusion/effusion rate of a gas is inversely proportional to its density."

**GRAS**. "Generally recognized as safe." Said of a food additive that was grandfathered into acceptance. One that has been used for a long time and is presumed to be safe for human consumption.

**Gratuitous zero**. Concept used when writing significant digits. Any number less than 1 can be written as a decimal value. A zero placed in front of the decimal point neither changes the value nor the number of significant figures. Hence, it is gratuitous. For example, in 0.825 the zero is gratuitous.

**Ground state**. The lowest possible energy state of an entity or system. In a hydrogen atom, it's the state with the electron occupying the first principal energy shell. Entities or systems with more energy than their ground state are in an **excited state**.

**Group**. 1. In organic chemistry a **functional group**. 2. A vertical family of related elements in the periodic table.

**Guanine**. One of the bases of **nucleic acids**, $C_5H_5N_5O$. Based on its structure it is a purine-type base.

**Gun cotton**. See **nitrocellulose**.

**Haber process**. Important industrial process for the production of ammonia. $N_2(g) + 3\ H_2(g) \rightleftarrows 2\ NH_3\ (g)$. Historically an important example of the application equilibrium and kinetic principles.

**Half-cell potential**. The voltage of a particular half-cell. Standard values are listed in comparison to the **standard hydrogen electrode**.

**Half-cell**. Either the region of reduction or the region of oxidation of an **electrochemical cell**. Physically, a half-cell is where a half-reaction is occurring.

**Half reaction**. A concept used to understand chemical reactions. 1. The chemical reactions that take place separately at the anode and cathode of an electrochemical cell. 2. Any redox reaction can be interpreted as a reduction half reaction combined with an oxidation half reaction.

**Half-life ($t_{½}$)**. For a **radionuclide** or a **first order reaction** it is the length of time needed for one half of the sample to decay—for 100% of an original sample to decay to 50%.

**Halide**. Any of the monatomic ions of the **halogen** elements, $F^-$, $Cl^-$, $Br^-$, $I^-$, and $At^-$.

**Halogens**. The nonmetallic elements F, Cl, Br, I, and At.

**Heat**. Energy that is transferred between two points because of a temperature difference.

**Heat capacity**. The heat required to change the temperature of an object or system. It may be expressed as **molar heat capacity** (J $mol^{-1}$ $K^{-1}$) or **specific heat capacity** (J $g^{-1}$ $K^{-1}$).

**Heat capacity at constant pressure ($C_p$)**. **Heat capacity** measured at constant pressure. The energy input goes both to warm the substance and also to compensate for the expansion work done by the system

**Heat capacity at constant volume ($C_v$)**. **Heat capacity** measured at constant volume. The energy input goes only to warm the substance. No expansion work is done by the system

**Heat distortion temperature**. The temperature at which a slowly heated polymer begins to lose its shape. Measured experimentally by loading a polymer on a blade, slowly heating, and observing.

**Heat of reaction**. The heat consumed or evolved during a chemical reaction. Compare **enthalpy of reaction**, which is usually what is meant.

**Heats of combustion**. Loose description of the enthalpies of combustion reactions.

**Heavy hydrogen**. **Deuterium** or $^2H$.

**Heisenberg Uncertainty Principle**. Fundamental tenet that places natural limits on the amount of knowledge that humans can have about any particle. Its simplest and best-known statement is: "The exact position and exact momentum of a particle cannot be known simultaneously." (The momentum of a particle is the product of its mass and its velocity.)

**Helium-like core**. The two most tightly bound electrons that occupy the 1s orbital ($1s^2$) of an atom or ion.

**Henderson-Hasselbalch Equation**. Is the equation $pH = pK_a + \log_{10}\frac{[A^-]}{[HA]}$

**Henry's law**. $X_g = kP$. Experimental law that the concentration of a gas dissolved in a liquid at any specified temperature is proportional to the partial pressure of the gas.

**Hess's law**. If a chemical reaction can be written as the sum of several separate steps, its overall $\Delta H$ is the sum of the $\Delta H$ values for the individual steps.

**Heterogeneous catalyst**. A catalyst that functions in a different phase or state of aggregation from the reaction it is catalyzing. For example, tail pipe **catalytic converters** in which a solid phase catalyst promotes a gas phase reaction.

**Heterogeneous substance**. Any substance that has internal physical boundaries. A **mechanical mixture** of two or more compounds or substances.

**Hexagonal close packed (hcp)**. One of two situations in which a collection of spheres or atoms occupies the smallest possible total volume. The other situation is **cubic close packed (ccp)**. See **close packing**.

**HFC**. Hydrofluorocarbon. Class of organic substances widely used as **air conditioning working fluids**.

**HLB (hydrophilic lipophilic balance)**. An index used to classify **surfactants**. HLB's range from 0 - 40, with 40 being the most hydrophilic. Ordinary soap has an HLB value of about 12.

**Home experiment**. A safe, innocuous chemistry experiment that a student can do for himself or herself.

**Homogeneous catalyst**. A catalyst the functions in the same phase or state of aggregation as the reaction it is catalyzing.

**Homogeneous substance**. One with a uniform composition. A mixture having the maximum possible extent of mixing. Homogeneous substances divide into two groups: **solutions** and **pure substances**.

**Homonuclear diatomic molecule**. A molecule formed from two atoms of the same element, (A-A).

**Horsepower**. Engineering power unit. Defined in SI metric to be exactly 745.7 watts.

**Hund's rule**. Electrons in a subshell of degenerate orbitals will, so far as possible, occupy the available orbitals singly, with their spins in the same direction.

**Hybridization**. A quantum mechanical concept. The idea that hydrogen-like orbitals can be combined to yield new orbitals that point in such directions that the shapes of molecules can be rationalized.

**Hydrates**. Substances that contain loosely bound water molecules. For many hydrates, gentle heating is sufficient to drive off or vaporize their water of hydration.

**Hydration reaction**. In general, any reaction involving water. More specifically, the setting up of plaster or cement, after mixing the powder with water.

**Hydraulic fluids**. Liquids used in power transmission systems. There are many commercially formulated hydraulic fluids.

**Hydrocarbon**. Any compound with the **general formula** $C_nH_m$.

**Hydrogen bonding**. Strong intermolecular forces that occur among molecules that have O-H bonds or N-H bonds, and among the molecules in the compound H-F. H-bonding is responsible for the unique and unusual properties of liquid water. Important for the double helix structure of **DNA.**

**Hydrogen spectrum**. The series of colored lines observed when light from excited hydrogen is studied by means of a prism or **diffraction grating**. The lines are images of the slit through which the light passes.

**Hydrolysis**. Literally "splitting by water." In general, any chemical reaction brought about by water.

**Hydrophilic**. Literally water-loving. Description for substances that dissolve readily in water.

**Hydrophobic**. Literally water-hating. Description for substances that form two layers or regions when mixed with water. Substances that "dislike" being in water.

**IARC**. International Agency for Research in Cancer. An acronym sometimes encountered on **MSDS** sheets.

**Ideal gas.** A theoretical substance. A gas that obeys the ideal gas law. Real gases approach ideal gas behavior at low pressures and high temperatures.

**Ideal gas equation**. Experimental law of gas behavior: $PV = nRT$, where P is pressure, V is volume, n is number of moles of gas, R is the gas constant, and T is the absolute temperature. There is no such thing as an ideal gas but real gases approach ideal behavior at low pressures and high temperatures. Also called the ideal gas law. The ideal gas law is a **limiting law**.

**Ideal geometry**. Said of any central atom around which the ligands are disposed with the maximum possible symmetry. Four ideal geometries are **trigonal planar**, **tetrahedral**, **trigonal bipyramidal**, and **octahedral**.

**Ideal solution**. A mixture of two chemically similar liquids that closely follows **Raoult's Law**.

**Ideality**. The hypothesis that there is some ideal or perfect behavior that natural systems should adopt.

**IDLH**. Immediately Dangerous to Life and Health: Hazard situation defined by **NIOSH** and **OSHA** for the purpose of respirator selection. It represents the maximum concentration from which – in the event of a respirator failure – one could escape within 30 minutes without experiencing escape-impairing or irreversible health effects.

**i-Factor**. See **van't Hoff i-factor.**

**Immiscible liquids**. A pair of liquids that forms two layers when mixed is said to be an immiscible pair.

**Incomplete combustion**. A fuel burning reaction that occurs when the supply oxygen is limited (and the reaction is fuel rich). Under this condition, combustion produces significant amounts of carbon monoxide, CO, and less heat than the corresponding **complete combustion**.

**Induced dipole**. The **temporary dipole** moment created in a nonpolar molecule when the molecule is placed in an electric field. The same effect occurring among neighboring atoms or molecules. Compare **dispersion forces**.

**Induced fission**. A human caused splitting of a nuclide as in a power plant or nuclear weapon. Fission occurs when neutron bombardment of nuclei splits them.

**Induced radionuclide**. One formed by a neutron capture reaction. For example, in nature, the formation of $^{239}Np$ in uranium ores when neutrons from spontaneous fission events react with $^{238}U$ nuclei. Radionuclides formed when samples are exposed to the neutron flux of a nuclear reactor.

**Industrial chemistry**. Using the raw materials of the planet to produce useful commercial materials is the central and unique role of the chemical industry. Applying chemical principles for commercially and socially useful purposes.

**Industrial solvent**. An organic liquid or liquid mixture used industrially to dissolve, degrease, or clean materials, suspend solid particles in a liquid medium, etc.

**Inert gases**. Older name for the noble gases. He, Ne, Ar, Kr, Xe, Rn. The adjective was changed around 1960 because though they are generally unreactive, the heavier ones do form some not very stable chemical compounds and are thus not strictly inert.

**Inert solvent**. A solvent that serves as a medium for but does not participate in chemical reactions.

**Infrared spectroscopy**. Important **instrumental method** in chemistry. Very useful for identifying or "fingerprinting" organic compounds. Studying the absorption by molecules of IR radiation in the wavelength region 2 μm - 20 μm measures the energies at which the molecules rotate and vibrate. From this information chemists can deduce chemical structure, etc. see **infrared spectrum**.

**Infrared spectrum**. A graph or display of the infrared red radiation absorbed by a compound or substance.

**Initial rate**. In chemical kinetics, it's the rate of the reaction at its very beginning. Studies of initial rates allow chemists to deduce the order of chemical reactions.

**Inorganic compound**. It's any compound that does not contain carbon, along with a few simple carbon containing compounds such as carbon dioxide and carbonate compounds.

**Inorganic chemistry**. The branch of chemistry devoted primarily to the study of the elements and their compounds excluding the element carbon. One of the four traditional branches of chemistry. The others are **analytical, organic**, and **physical chemistry**.

**Inorganic polymer**: A polymer whose backbone chain is not composed of carbon atoms. For example, many silicate materials are inorganic polymers.

**Infrared radiation (IR)**. Electromagnetic radiation that lies beyond the red end of the visible spectrum (>700 nm).

**Intellectual property**. Knowledge held by an individual about how to run a process or build a device, etc. The creative output of a person is that person's intellectual property.

**Intensive properties**. A property that does not change when the amount of substance of a sample is changed is an intensive property. For example, doubling the amount of water does not double its density, color, or temperature. So those three are intensive properties. Compare **extensive property**.

**Instruments**. See **Chemical instrumental methods**.

**Interference**. In physics, the notion that depending on the geometry of the situation two waves can either mutually reinforce or destruct.

**Intermediate**. See **reaction intermediate.**

**Intermolecular force**. The attraction or repulsion of one molecule for another. Intermolecular forces exist between any two molecules in a gas.

**Internal energy**. The total energy of a thermodynamic system, rarely, if ever, measured. Chemists are usually interested in the *changes* in the internal energy of systems.

**Interstellar molecule**. A molecule or **free radical** known from its microwave spectrum to exist in deep space. About 70 such molecules were known in 2000.

**Intramolecular forces**. Literally, forces within molecules. Forces that hold individual molecules together. The **covalent bonds** among the atoms in a molecule.

**Iodine clock**. Name for a family of reactions used to demonstrate the principles of chemical kinetics. So-called because the **endpoint** is observed by the instantaneous formation of a dark blue-black color between iodine and starch.

**Ion**. An entity with an electric charge.

**Ion bombardment reactions**. Nuclear processes in which highly accelerated, light nuclei ions are caused to strike atomic targets. A procedure for bringing about nuclear changes and causing **transmutation**. The standard method, used over the past 50 years, for preparing new elements of ever higher atomic number.

**Ion conduit**. Any natural material or device that allows for the passage of ions through it.

**Ionic bonding**. 1. The force of attraction that holds together two ions of opposite charge. 2. The aggregate of such forces in an **ionic solid**.

**Ionic compound**. One composed of anions and cations. Contrast **covalent compound**.

**Ionic radius, ionic radii**. The typical size of an ion of specified charge. Lengths used to describe the size of bonded ions. If we imagine that ions are spherical, the length of the bond between them is the sum of the radii of the two ions. Studying the crystal structures of many ionic solids yields a table of **ionic radii**. The average radius of an ion of specified charge derived from a comprehensive study of its interionic distances in its solid ionic compounds. The **ionic radius** of an element depends sharply on the charge of the ion.

**Ionosphere**. The region of the earth's atmosphere beginning at an altitude of roughly 25 miles and extending approximately 200 miles into space. It consists of a complex series of ever-changing layers composed of ions and free electrons. Solar energy falling on the upper atmosphere creates the ionosphere.

**Ionic solids**. Those solids held together by strong Coulombic forces among all their anions and cations.

**Ionization reaction**. 1. One in which a substance dissolves in water to form ions. 2. When the excitation of a gaseous atom results in the formation of a gaseous ion. 3. In general any process that produces ions.

**Ionizing radiation**. Any radiation that causes air molecules to ionize when the radiation passes through the air. **α-particles, β-particles**, and **γ–rays** are examples of ionizing radiation.

**Irreproducible results**. Not being able to get the same result twice. The bane of the life of an experimental scientist. There is a tongue-in-cheek publication, in which scientists spoof themselves, called *The Journal of Irreproducible Results*.

**Irreversible process**. In thermodynamics, a process carried out in such a way as to obtain from it less than the theoretically maximum amount of work.

**ISO**. International Standards Organization. The body to which the various national standards organization belong.

**Isoelectronic**. Having the same number of electrons.

**Isolating a reactant**. Experimental procedure. Repeating kinetics experiments in which all the reactant concentrations except one are held constant. The procedure allows study of a single reactant. The changed reactant is thus said to be isolated.

**Isopropyl**. The radical $(CH_3)_2CH$–.

**Isomers**. Molecules composed of identical atoms but having different molecular structure. See, for example, **cis-trans isomers, enantiomers,** and **stereoisomers.**

**Isothermal**. A condition of constant temperature. A thermodynamic process that takes place at constant temperature is an **isothermal process.**

**Isothermal process**. One that takes place at constant temperature

**Isotope enrichment**. Processing an element so that one of its isotopes is concentrated above that isotope's **natural abundance.**

**Isotope separation**. The process of separating the natural isotopes of an element from one another. The normal mode of operation of a **mass spectrometer**. The separation of $^{235}U$ from $^{238}U$ by **gaseous diffusion** is an important step in the production of atomic energy.

**Isotopes**. Atoms of an element that differ in their number of neutrons. Two atoms with an identical number of protons but a different number of neutrons.

**IUPAC**. International Union of Pure and Applied Chemistry. International body with responsibility for naming conventions, symbols, units, etc., in chemistry.

**IUPAC name**. In organic chemistry, the approved name given to classes of compounds with specific functional groups.

**IUPAC symbol**. The letter designation of a property by IUPAC is its IUPAC symbol. As far as possible these are consistent with SI metric practice. For example W for watt and U for internal energy.

**Joule**. The SI unit of energy. $1\ J = 1\ kg\ m^2\ s^{-2}$.

**J-tube**. A tube shaped like the letter J that is convenient for studying the pressure-volume behavior of gases.

**Kelvin**: "The kelvin, unit of thermodynamic temperature, is the fraction 1/273.16 of the thermodynamic temperature of the triple point of water (13th **CGPM**, 1967)."

**Ketone**. An organic compound containing the C=O functional group and having the collective formula $R_2CO$.

**Kilogram**. "The kilogram is the unit of mass; it is equal to the mass of the international prototype of the kilogram (3rd **CGPM**, 1901)."

**Kinetic energy**. Energy due the motion of an object or entity. Given by the formula KE = ½mv$^2$.

**Kinetic molecular theory** (KMT). The theory in which the properties of gases are described in terms of random, perpetual motions of their individual molecules.

**Kinematic viscosity**. In SI it is viscosity with the units of m$^2$ s$^{-1}$. It has no special name.

**Lanthanides**. Lanthanum and the elements cerium (Ce) through lutetium (Lu) that follow lanthanum in the periodic table.

**Lanthanide contraction**. The shrinkage in atomic and ionic radius of elements near hafnium compared to elements that are directly above them in the periodic table.

**Latent heat**. Heat used to melt a solid or boil a liquid. So-called because such heat does not cause a temperature change and is thus hidden or latent.

**Law of combining volumes**. If they are measured under the same conditions of pressure and temperature, the volumes of gases that are consumed or produced in a chemical reaction occur in ratios of small whole numbers. Stated originally by Joseph Gay-Lussac

**Law of conservation of energy**. Energy is a conserved quantity. If energy in one form is consumed, it reappears in exactly the same amount in some other form.

**Law of conservation of mass**. In chemistry its usual statement is that in any chemical reaction the sum of the masses of the products must equal the sum of the masses of the reactants.

**Law of constant composition. Law of definite composition**. Stated by Dalton: "The elemental composition of any pure compound is fixed, independent of the compound's origin." Alternative names for the **law of definite proportion**.

**Law of definite proportion**. In any compound, the relative masses of the elements are constant. This is an alternative name for the **law of constant composition**.

**Law of Dulong and Petit**. The product of a metal's specific heat and its atomic weight is the constant 6.4. Important in its day, mainly of historical interest. Today we would state it as: the product of a metal's specific heat and its atomic weight is the constant 27 J mol$^{-1}$ K$^{-1}$.

**Law of Multiple Proportions**. The experimental situation that when pairs of elements form two or more distinct compounds, then ratios of the fraction of any element's mass fraction in one compound to its mass fraction in the other compound is a simple, whole number ratio. Historically important to the development of Dalton's atomic theory. If two elements form more than one compound, the weight ratios of the elements in the compounds are small whole-number multiples of one another.

**LC$_{50}$**. "Lethal concentration 50%." In toxicology, it refers to the concentration of an agent that under specified test conditions kills 50% of test organisms. Used as a measure of the safety of industrial chemical compounds.

**Le Chatelier's principle**. When an equilibrium system is disturbed, it will react in a way that tends to minimize the disturbance.

**Lead-acid battery**. An ordinary automobile battery. It has lead-containing electrodes and sulfuric acid in its electrolyte.

**Lean**. A fuel condition that exists when there is relative excess of oxygen compared to the available fuel.

**Least significant digit**. For a reported value it is the rightmost significant digit.

**LEL**. Lower explosive limit. The concentration of a fuel in air that is too lean for ignition to occur in the presence of a spark or flame. (**UEL** is the reverse.)

**Lewis acid**. A chemical entity capable of accepting a pair of electrons.

**Lewis base**. A chemical entity capable of donating a pair of electrons.

**Lewis dot diagrams**. Traditional method used by chemists to represent entities and their bonds. In them, atoms are represented by their atomic symbol and valence electrons are represented by dots and pairs of dots.

**Ligand atom**. One of the various atoms bonded to a **central atom**.

**Ligands**. 1. All the entities (atoms, molecules, or ions) bound to a central atom. The term ligand comes from same root word found in ligature or ligation and means binding. 2. In a **complex ions** or **complex compound** they are molecules or ions bonded to the central metal ion or atom.

**Lime**. Calcium oxide (CaO). Important industrial inorganic compound produced by heating limestone. Sometimes called quicklime. See also **slaked lime.**

**Limestone**. Any of a wide range of common natural minerals with the approximate chemical formula $CaCO_3$. Often composed of the shells of long vanished marine organisms. Limestone exposed to geological heat becomes marble.

**Limiting law**. A law which becomes better obeyed in some limiting circumstance. The ideal gas law is a limiting law: at very low pressures and very high temperatures all gases behave ideally.

**Limiting reactant**. A reactant present in a reaction mixture in insufficient quantity. It will be the first of the reactants to be consumed and that will limit the overall extent of the reaction. Compare **excess reactant**.

**Limiting reagent**. An alternative name for limiting reactant.

**Linear polymer**. One in which the carbon atoms and other linking atoms extend in a single, continuous chain. A polymer that is not **branched**.

**Linear low density polyethylene** (LLDPE). A widely used commercial plastic.

**Lipophilic**. Oil-loving. Usually used as a synonym for the **hydrophobic** behavior of a substance.

**Lipophobic**. Oil-hating. Usually used as a synonym for the **hydrophilic** behavior of a substance.

**Liquid chromatography (LC)**. An important technique of chemical analysis. The process of separating the components of a mixture by their traveling at different rates in a flowing liquid stream. Because of selective absorption of compounds in a **stationary phase**.

**Liquid foam**. Mixture that typically consists of gas bubbles stabilized in a liquid matrix by surfactant molecules.

**Liquid solution**. A liquid, homogeneous mixture containing two or more substances.

**Liter**. Unit of volume equal to 1000 cm$^3$. Strictly not a part of the SI system of units but it is both recognized and widely used, especially for liquid volumes. One milliliter is equivalent to one cubic centimeter, **cc**.

**Load**. In electrochemistry, whatever a battery is powering is the battery's load.

**Lobe** A designated sub-region of an **atomic orbital**. So named because of its shape—like an ear lobe.

**London forces**. See **dispersion forces**.

**Lone pairs (LP)**. In a **Lewis dot diagram**, it is any pair of dots that represents a pair of electrons in the valence shell of an atom but not localized in a bond.

**Long range order**. Characteristic of crystals. The situation in a solid when its basic structure is replicated across 1000s of entities. Long range order may also exist to some extent in liquids. Liquids with long range order tend to have a high viscosity

**Longest lived nuclide**. For an element that lacks any stable nuclide the mass of the isotope with the longest half-life is sometimes used as that element's **atomic mass**.

**LOX**. An abbreviation for liquid oxygen. Often used by rocket engineers.

**Lye**. A strongly alkaline material. Historically, lye was obtained from wood ashes. More specifically, sodium hydroxide, $NaOH$. See also **alkali**.

**Macromolecular science**. The study of macromolecules and polymers.

**Macromolecules**. Large molecules. Plastics and synthetic polymers are macromolecules.

**Macroscopic properties**. Those properties of a substance manifested by the collective properties of an assembly of its molecules. Boiling point is a macroscopic property because many molecules together determine its value. Synonym for **bulk properties**

**Macrostate**. In the statistical interpretation of entropy, any set of indistinguishable **microstates**.

**Magnetic quantum number**. It is designated m and follows the rule m = $-\ell$ $-\ell+1$, $-\ell+2$, ..., 0, -1, -2,..., $+\ell$. Where $\ell$ is the secondary quantum number.

**Main group elements**. In the periodic table any of the groups numbered 1, 2, 13, 14, 15, 16, 17, or 18. Elements at the ends of the conventional form of the periodic table.

**Main group metals**. Those **main group elements** that are metals.

**Manometer**. A U-shaped tube filled with a liquid used, for example, to compare pressures of two gas samples, or compare the pressures of a gas sample and atmospheric pressure.

**Many-electron atoms and ions**. Atoms or ions that contain more than one electron.

**Mass**. A property of matter occupying space.

**Mass defect**. For a nuclide it is the sum of the masses of its neutrons, protons, and electrons minus its actual, experimental mass. The mass defect or the **missing mass** can be regarded as the origin of the energy needed to hold the nucleus together as determined via $E = mc^2$.

**Mass number**. An obsolete term (not sanctioned by **IUPAC**) for **nucleon number**—the sum of the protons and neutrons in a single atom or nuclide.

**Mass spectrometer**. An instrument or apparatus for separating ions according to their masses. Historically used to separate isotopes. Widely used today to detect and measure **analytes** and identify organic compounds. Mass spectrometry of a molecular substance gives a characteristic **fragmentation pattern**. See **mass spectrum**.

**Mass spectrometry**. An experiment used to generate a **mass spectrum**.

**Mass spectrum**. A graph or display of the different masses of ions formed from the particles in a sample.

**Material**. A material is any form of matter.

**Materials balance**. A chemical bookkeeping in which all the reactants and products of a chemical reaction, or an industrial process, are accounted for.

**Materials science**. Broadly speaking the scientific study of solid materials. A wide ranging branch of science that has many applications in chemistry, physics, and engineering.

**Matter**. The stuff that makes up our universe; it is anything that has mass and takes up space.

**Matter-waves**. Concept that particles have waves associated with them. In the 1920s a revolutionary notion, now a fundamental aspect of human understanding and manipulation of nature.

**Maxwell-Boltzmann distribution.** A mathematical function that describes the equilibrium distribution of molecular speeds in a gas. A graph of the relative numbers of molecules of a gas that travel at different speeds.

**Mean square speed**. The mean (average) of the squares of the speeds of the molecules of a gas.

**Measured number or value**. One that is obtained in the laboratory or from an actual experiment. Sometimes called an **observed value**. Contrast **exact number**.

**Mechanical mixture**. A substance that is formed by the simple mixing of two materials. Often said of a mixture of two powdered materials such as salt and sand. A mixture made by stirring together two or more powdered substances.

**Melting point analysis**. **Organic chemists** routinely study the melting points of solid organic compounds to test their purity and identify them.

**Metabolic rate**. The rate at which a human (or other organism) consumes chemical energy.

**Metallic solids**. Those solids held together by metallic bonding. Metals can be regarded as regular array of cations floating in a sea of free electrons.

**Metalloids**. Eight elements, B, Si, Ge, As, Sb, Te, Po, and At, that show both metallic and nonmetallic properties and are chemically intermediate between the metals and the nonmetals.

**Metallurgy**. The branch of engineering science that deals with metals, metal alloys, and their properties.

**Metathesis reaction**. One in which groups of atoms or ions are exchanged among the reactants.

**Meter**. "The metre is the length of path traveled by light in vacuum during a time interval of 1/299 792 458 of a second (17th **CGPM**, 1983)."

**Method of successive approximations**. A useful procedure for solving **quadratic equations**--especially those equations that arise during chemical equilibrium calculations. The method involves using physical judgment to simplify the equation to get a first approximation to the solution, followed by refinement of the solution via a recycling procedure.

**Methyl group**. A portion of a molecule. The methane molecule less one hydrogen atom, $-CH_3$.

**Methyl radical**. $CH_3^{\bullet}$, a methane molecule from which a hydrogen atom has been abstracted leaving behind a reactive entity with an **unpaired electron** shown by a dot. See also **free radical**.

**Metric prefix**. Any of the prefixes used to scale units in the International System of units. For example, milli-, kilo-, etc.

**Metric system**. A coherent system of units built from seven SI **base units**.

**Metrologists**. Scientists whose professional work involves them in measurement science. Metrologists are employed at such places as **NIST**.

**Mica**. A sheet silicate **mineral**. Its sheets are held together by fairly weak bonds. The bonding description explains why mica is an easily flaked mineral.

**Microstate**. In entropy, the concept that it is possible to enumerate all the possible ways in which a systems particles can be arranged. A microstate is one such state. Compare **macrostate**.

**Minerals**. Rocks and their constituents. A large class of naturally occurring substances obtained from the earth. Often crystalline and frequently having a definite chemical formula.

**Miscibility**. The property of two liquids that are able to mix—as opposed to forming layers.

**Missing mass**. A alternative name for **mass defect**.

**Mixture**. Broadly, any substance produced by mixing two or more other substances. Compare **mechanical mixture**.

**MMT**. Methylcyclopentadienyl manganese tricarbonyl. An important commercial **anti-knock agent**.

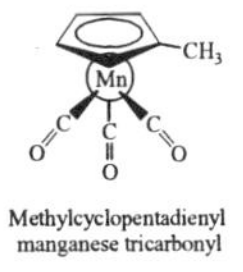


Methylcyclopentadienyl manganese tricarbonyl

**Mobile phase**. In **gas** or **liquid chromatography** it is the gas or liquid. The phase that carries the sample being analysed. In **gas chromatography** it is sometimes called the carrier gas.

**Molality**. Solution concentration expressed as moles of solute divided by the mass of the solvent in kilograms.

**Molar atomic mass**. Alternate term for **atomic weight**.

**Molar heat capacity**. The heat capacity of a mole of substance. Expressed in the units J $mol^{-1}$ $K^{-1}$.

**Molar mass**. The mass of a mole of any specified entity. The **specified entity** in chemistry is usually atoms, molecules, or ions. Molar masses can be specified in the units g $mol^{-1}$ or u (average $entity)^{-1}$.

**Molar weight (MW)**. The mass in grams of a mole of a specified entity with the special restriction that the elements present are in their **natural abundances**. Also, the sum of the **atomic weights** of the elements in an entity expressed in the unit g $mol^{-1}$. Molar weight can also be expressed in the unit g (average $molecule)^{-1}$. A compound has only one molar weight but many molar masses.

**Molarity**. A concentration unit of liquid solutions with units mol $L^{-1}$. The molarity of a **solute** is the number of moles of that solute contained in one liter of the solution.

**Mole**. "The mole is the amount of substance of a system which contains as many elementary entities as there are atoms in 0.012 kilogram of carbon-12. When the mole is used, the elementary entities must be specified and may be atoms, molecules, ions, electrons, other particles, or specified groups of such particles (14th **CGPM**, 1971)." One mole of any entity contains Avogadro's number of those entities.

**Mole fraction**. A ratio used to express the compositions of mixtures. The mole fraction of species i is the number of moles of that species divided by the total number of moles present: $x_i = n_i/n_{tot}$.

**Mole percentage**. Mole fraction multipled by 100 and so expressed as a percentage.

**Mole weight**. Synonym for **molar weight**.

**Molecular biology**. Broad term for the entire field that deals with the molecular processes of living systems.

**Molecular compound**. A **compound** composed of molecules. Disregarding isotopes, the molecules of a **molecular compound** have the same **chemical formula**.

**Molecular display modes**. Alternative ways of displaying molecules on-screen with molecular visualization software. Compare **space-filling** models and **ball-and-stick** models. See also **Chime** and **Rasmol**.

**Molecular formula**. The **chemical formula** of a molecule.

**Molecular geometry**. The bond lengths and bond angles in a molecule constitute its molecular geometry.

**Molecular liquid**. A liquid composed of molecules. For a pure substance to be a liquid, the forces among its molecules must be neither too weak nor too strong. If the forces are too weak, the liquid vaporizes and becomes a gas. If the forces are too strong the substance solidifies.

**Molecular mass**. The mass of a single molecule. Because of isotopic variations, the molecules of most pure compounds show a range of individual masses.

**Molecular weight**. The sum of the **atomic weights** of the atoms that compose a molecule.

**Molecular orbital**. An energy state of a molecule. An electron probability function that extends over two or more nuclei. The wave function of an electron in a molecule or a graph of that function or the square of that function.

**Molecular solid**. A solid in which the building block is a molecule. In such a solid, the individual molecules are held together by relatively weak **van der Waals forces**, sometimes strengthened by **hydrogen bonds**.

**Molecular spectroscopy**. Family of instrumental methods used to study the absorption by molecules in various regions of the electromagnetic spectrum. An important set of extremely powerful and sensitive experimental techniques that make identification of compounds almost routine.

**Molecular structure**. The three-dimensional architecture of a molecule. Collectively, all the bond lengths and all the bond angles around all the central atoms in a molecule. Also called **molecular geometry**.

**Molecular substance**. One composed of molecules with a single chemical formula. If a liquid, it is also a **molecular liquid**

**Molecular visualization**. Procedures using various methods and formats to display molecules and other entities of chemical interest. Modern chemistry enables interactive visualization on a monitor screen. Two **chemical visualization** plug-ins are **Chime** and **Rasmol**.

**Molecular weight**. The sum of the **atomic weights** of the atoms that compose a molecule.

**Molecule.** A stable, uncharged chemical entity composed of atoms held together by chemical bonds. Compare **entity**.

**Monatomic**. Composed of a single atom. Usually used to designate a charged atom as in a *monatomic ion*.

**Monochromatic**. Adjective used to describe electromagnetic radiation that contains only a single wavelength. Literally, single-colored.

**Monomer**. A small molecule capable of linking with itself or other molecules to make a long chain **polymer** compound.

**Mononuclidic**. Composed of a single type of **nuclide**. Adjective used to describe an element that exists naturally in the form of just one stable nuclide. A natural element composed of a single isotope. Fluorine and phosphorus are examples of mononuclidic elements. The atomic weights of mononuclidic elements are expressed with 8 or 9 number of significant figures.

**Monoprotic**. Said of an acid that produces one mole of hydrogen ion per mole of the acid.

**Monoprotic acid**. An acid that produces one mole of hydrogen ion per mole of acid. In general, an acid that ionizes according to $HA \leftrightarrows H^+ + A^-$.

**Most probable speed**. In a gas, the speed at which the greatest number of molecules travel; the speed at which the maximum occurs in the **Maxwell-Boltzmann distribution function**.

**MSDS**. Materials Safety Data Sheet. A summary statement of the toxicity, use parameters, health hazards, disposal requirements, etc., of a chemical compound or chemical mixture. Required by US federal law.

**MSHA**. Mining Safety and Health Administration. US Federal organization that regulates safety and health issues in mining.

**MTBE**. Methyl tertiary butyl ether. An oxygenated motor fuel additive, $H_3C\text{-}O\text{-}C(\text{-}CH_3)_3$, introduced around 1990 in the US to combat air pollution. Subsequently discovered to be a ground water hazard.

**Muriatic acid**. Old name for hydrochloric acid, HCl.

**$N_2O$ injection**. $N_2O$ is nitrous oxide. Used in power boosting combustion engines it is a better fuel oxidant than even oxygen itself.

**n-alkane**. A "normal" alkane. A saturated hydrocarbon with the general formula $C_nH_{2n+2}$ in which all the carbons atoms are bonded in a single extended chain lacking branches.

**Nanotechnology**. The application and control of materials at the nanometer scale. A developing field at the junction of physics, chemistry, and materials engineering.

**Native element**. One found in nature in chemically uncombined state. For example, gold, silver, and mercury are native metals.

**Natural abundance**. Term that refers to the number fraction (or percentage) of each isotope found in natural sample of an element.

**Natural decay chain**. Three sequences of natural radioactive decay processes that begin with the nuclides $^{232}Th$, $^{235}U$, and $^{238}U$ and all eventually conclude with lead isotopes. On the way these chains produce isotopes of the elements Po, At, Rn, Fr, Ra, Ac, and Pa.

**Natural polymers**. Polymeric materials that occur in nature. **Carbohydrates** and **proteins**, for example.

**Natural rubber**. Rubber obtained from rubber trees. As contrasted to synthetic rubber, manufactured in chemical plants.

**NCEES**. National Council of Examiners of Engineers and Surveyors. The national organization that tests engineers as part of their certification process.

**Nernst equation**. Equation used in electrochemistry to calculate the half-cell voltages and cell voltages under nonstandard conditions.

**Net reaction**. For a stepwise chemical reaction, the net reaction is the sum of the steps. Also see **overall reaction**.

**Net neutralization reaction**. A net neutralization reaction equation written omitting the **spectator ions**. It is a summary equation that shows the addition reaction between hydrogen ion and hydroxide ion.

**Neutral solution**. One that is neither acidic nor basic. It has a pH of 7.

**Neutralization**. The reaction that occurs when the characteristic properties of the acid are destroyed by a base and vice versa.

**Neutron chain reaction**. A rapid, branching reaction process. The mechanism by which a fission reactor or nuclear weapon works. Fission produces neutrons that cause more fissions in a branching chain.

**Neutron number**. The number of neutrons in a nuclide. Given the symbol N.

**Neutron rich nuclides**. Those with a neutron/proton ratio larger than required by nature to make the nuclide stable. Radionuclides that lie below and to the right of the **belt of stability**.

**Newton**. The SI derived unit of force equal to 1 kg m $s^{-2}$.

**NFPA**. National Fire Protection Association. An abbreviation encountered on **MSDS** sheets. Organization that among other activities develops shipping codes for hazardous materials. NFPA uses a scale of 0 to 4 to rate various safety concerns. 0 represents the lowest level of concern and 4 represents the highest level of concern. The scale is applied to such properties as flammability, reactivity, health hazards, etc.

**NIOSH**. National Institute of Safety and Health. An acronym encountered on **MSDS** sheets. US federal organization active in promoting human health safety standards.

**NIST**. The US National Institute of Standards and Technology. An important standards organization. Located in Gaithersburg, Maryland. Formerly known as the National Bureau of Standards.

**Nitro compound.** An organic compound containing the $-NO_2$ functional group and having the collective formula $R\text{-}NO_2$.

**Nitrocellulose**. A **Semi-synthetic polymer** made by modifying cellulose with nitric acid to make cellulose trinitrate (**gun cotton**) a useful explosive and propellant.

**Nitrogen fixation**. The conversion of atmospheric nitrogen to a usable chemical form.

**Noble gas**. A chemically unreactive gas. Usually the elements of group 18: He,Ne, Ar, Kr, Xe, and Rn.

**Noble gas core**. Any subset of electrons in an atom or monatomic ion that is isoelectronic with a noble gas atom.

**Noble metal**. A chemically unreactive metal. For example, the platinum metals Ru, Rh, Pd, Os, Ir, and Pt, and gold.

**Nonbonded atoms**. Usually a description of two nearby atoms in a molecule without a chemical bond between them. For example, in $H_2O$ the hydrogen atoms are both bonded to the oxygen but the hydrogen atoms themselves are nonbonded—not bonded to each other.

**Nonbonding electrons**. When electrons in nonbonding orbitals of an entity are removed the bonding in the entity is unaffected.

**Nonbonding orbitals**. Type of molecular orbital from which removal of a electron does not change the bonding in the molecule.

**Nonelectrolyte**. 1. A solid that dissolves in water to give a non-conducting solution. 2. Such a solution.

**Nonionic surfactant**. A **surface active agent** whose principle activity comes from a long-chain organic entity to which is attached an uncharged but polar functional group. For example, in the molecule $C_6H_{13}\text{–}(OCH_2)_8\text{–}OH$ the polar **ether groups** produce the needed water solubility for the surfactant to work.

**Nonpolar**. Property of an entity or substance that lacks internal charge separation. An entity whose centers of positive and negative charge are at the same place.

**Nonspontaneous**. In thermodynamics, a process that has an inherent tendency *not* to occur. The reverse process is **spontaneous**.

**Nonspontaneous reaction**. In thermodynamics, a reaction at constant temperature and pressure with a positive $\Delta G$.

**Nonvolatile solute**. One that does not vaporize when its solution is boiled.

**Normal alkane**. An **n-alkane**. An alkane without chain branches.

**$NO_x$**. An abbreviation usually spoken as "oxides of nitrogen." It refers to the compounds NO and $NO_2$ especially in the context of these compounds as air pollutants.

**NTP or STP**. "Normal (or standard) temperature and pressure." Historic abbreviations usually meaning one atmosphere pressure and 0°C. Obsolete, but still encountered on **MSDS** sheets.

**Nuclear decay modes**. The various ways in which a radioactive nuclide may decay. **Alpha decay**, **beta decay**, etc.

**Nuclear equation**. A written representation of a nuclear process. It usually represents a bombardment process or the radiodecay of an unstable nuclide.

**Nuclear magnetic resonance spectroscopy (nmr)**. Important **instrumental method** used to study the structure of molecules, etc. Actually, a whole family of related techniques. It involves placing samples in powerful magnetic field to create **degenerate** nuclear **spin** states that can be studied in the radiofrequency region of the electromagnetic spectrum. Chemists can interpret nmr spectra to reveal detailed information about the chemical environments of individual atoms in a molecule.

**Nuclear radiation**. See **ionizing radiation**.

**Nuclear stability**. Refers to whether or not a specified nuclide is stable. Stable nuclides do not decay. Unstable nuclides decay by one of a number of **radiodecay modes**.

**Nucleon**. In chemistry, it refers to either a proton or a neutron.

**Nucleic acids**. Natural biological polymers of enormous importance to living organisms. Chemically, they may be regarded as polymerized **nucleotides**.

**Nucleon number**. The sum of the protons and neutrons in a single atom or nuclide. Given the symbol A.

**Nucleosynthesis**. The process of element formation that occurred immediately after the big bang and also occurs in stars (**stellar nucleosynthesis**). Nucleosynthesis of naturally radioactive elements occurs largely during **supernova** events. See **stellar**.

**Nucleotide.** A biochemical molecule derived from one molecule of phosphoric acid, a five-carbon sugar molecule, and an organic, nitrogen-containing compound (a **base**) such as **adenine**, **cytosine**, **guanine**, **uracil** or **thymine**. Nucleotides may be considered as the monomer units of **DNA** and **RNA**.

**Nuclides**. Collectively, all the isotopes of all the elements.

**Nuclide chart**. A useful representation of nuclides and their stability. A proton-count versus neutron-count map. Often, its individual entries are coded by color to show their nuclear stability.

**Nuclidic mass**. The mass of a given nuclide. If a sample consists of a single nuclide then for the atoms *of that sample* their nuclidic mass is identical to their **atomic mass**.

**Number style**. A choice about how to write a number. For example, SI defines the value of the speed of light as 299 792 458 m $s^{-1}$. It could be written as 299792458 m $s^{-1}$, or 299,792,458 m $s^{-1}$, or in **scientific notation** as $2.99792458 \times 10^8$ m $s^{-1}$. Because there is no absolute authority about how numbers should be written, it's best to know the different possibilities.

**Nylon rope trick**. A chemistry lecture demonstration that involves preparing a crude form of nylon at the interface of two immiscible reaction solutions. The rope continuously forms as the nylon is lifted from the reaction environment with a pair of tweezers.

**Observed value**. A **measured value**.

**Octahedral structure**. An ideal geometry based on six ligands symmetrically disposed around a central atom.

**Octane rating**. A test method for motor fuel. The higher a fuel's rating the better its resistance to pre-ignition.

**Octet rule**. The historic idea that molecules formed with the central atoms C, N, O, etc., are surrounded by eight electrons. Later, the concept was refined to four valence shell electron pairs.

**OIL-RIG**. The mnemonic "oxidation is loss - reduction is gain." Useful memory aid for students coming to grips with redox chemistry.

**Olefin**. Older name for an **alkene**. Obsolete, but still widely used. **Alkenes** (organic compounds containing a double bond) were once called olefins (literally, oil-formers; they form "oils" when they react with bromine via an **addition reaction**).

**Oligomer**. A molecule formed by polymerization that links only a few monomer units.

**One-electron transfer dipole moment**. For a diatomic molecule with a bond length of R it is μ(one electron transfer) = e R, when e is the charge on an electron and R is the bond length of the diatomic molecule. Compare **polar covalent bond**.

**Optical rotation**. The phenomenon in which one **enantiomer** of chiral compound rotates the plane of polarized light passed through a solution of the compound. Phenomenon closely elated to the functioning of Polaroid® sun glasses. Technique demanded by the **FDA** for validation of the **chiral purity** of many drugs.

**Orbital**. Formally, a one-electron wave function. Often used loosely to refer to the distribution of a particular electron in an entity.

**Orbital energy**. In an atom or monatomic ion, the energy value of a single-electron. The negative of the orbital energy is approximately equal to the energy required to remove the electron in the given orbital from the entity.

**Orbital energy level diagram**. A diagram that shows the relative orbital energies in a particular atom or ion.

**Orbital occupation diagram**. A diagram that shows how many electrons occupy each of an atom's or ion's orbitals.

**Orbital shape**. A quantum mechanical idea difficult to define but easy to see. The general appearance of an electron density map. A loosely defined term that refers to the way a particular wave function is represented in space.

**Order of a chemical reaction**. In chemical kinetics, it is the mathematical way in which reaction rate depends on concentrations. Specifically, a reaction's order is the sum of the exponents of the concentration terms in the reaction's rate law.

**Ore**. A natural mineral with an economic value. Ores are the source of most elements

**Organic chemistry**. The chemistry of carbon compounds. The largest of the four traditional branches of chemistry. The others are **analytical, inorganic** and **physical chemistry**.

**Organic compound**. Excluding a handful of simple, inorganic compounds it's any compound containing carbon. Contrast **inorganic compound**.

**Organic polymeric solid**. A carbon containing material characterized by having long chains of atoms in its molecules. They divide into two major groups: **natural polymers** and **synthetic polymers**.

**Organic synthesis**. The chemical manufacture of desired **organic compounds** from available **raw materials**. See **petrochemical industry**.

**OSHA**. Occupational Safety and Health Administration. A US federal government agency.

**Osmosis**. A phenomenon in which pressure is generated across a semi-permeable membrane when a solution and a pure solvent, or two solutions of different concentration, are on opposite sides of the membrane.

**Osmotic pressure**. The pressure difference across a semi-permeable membrane when a solution and a pure solvent, or two solutions of different concentration, are on opposite sides of the membrane.

**Ostwald viscosimeter**: A glass device that allows a liquid sample to flow through a narrow capillary. The time taken for a given volume of liquid to flow measures the liquid's viscosity.

**Overall reaction**. If a reaction occurs in a series of steps, the overall reaction is the sum of those steps. In a redox or electrochemical reaction the overall reaction is the sum of the two **half reactions**. Also called the **net reaction**.

**Oxidation**. The process of an entity becoming **oxidized**. Said to have happened to an entity when it loses one or more electrons. The OIL in **OIL-RIG**. More generally, reaction with oxygen.

**Oxidation half-cell**. The region of an electrochemical cell where oxidation occurs.

**Oxidation number**. A value assigned to an atom in a molecule or ion according to conventional rules. Oxidation number is closely related to the experimental oxidation states of the elements in their compounds. Compare **oxidation state**. See also **Stock system.**

**Oxidation state**. A value assigned to an atom in an entity based on its chemical properties and reactivity. Good oxidants tend to have high oxidation states. Often, the terms oxidation number and oxidation state are used with equivalent meaning. Compare **oxidation number**. See also **Stock system.**

**Oxidizing agent (oxidant)**. Entity capable of causing the **oxidation** of a second entity.

**Oxoanion**. A negatively charged, polyatomic ion that contains oxygen.

**Ozone layer**. A region of the atmosphere rich in ozone, $O_3$. High above the earth's surface.

**Pa**. Pressure unit. The **pascal**.

**Paramagnetic**. Property of an entity or substance whose atoms, molecules, or ions contain one or more **unpaired electrons**. Paramagnetic substances are attracted by magnetic fields.

**Parent nuclide**. In a radiodecay process, the parent nuclide disappears with the simultaneous formation of a new, **daughter nuclide**.

**Partial pressure**. In a mixture of gases, the partial pressure of any component is the pressure that gas would exert if it were by itself at the same temperature and in the same volume as the mixture.

**Pascal.** The SI unit of pressure. One pascal is exactly 1 kg $m^{-1}$ $s^{-2}$. Compare **bar**.

**Patent**. A license agreement between an inventor and a government organization. In the US, patents are granted by the US Patent and Trademark office (www.uspto.gov).

**Path**. In thermodynamics, the specific process that a system undergoes as it changes from its initial state to its final state. Changes in **state properties** of the **system** are independent of the path.

**Pauli exclusion principle**. In any many-electron atom or ion, no more than two electrons can occupy any particular orbital. Alternatively, no two electrons in at atom can have identical values of all four of their quantum numbers (n, ℓ, m, and $m_s$).

**pdb**. Initials meaning "protein data base." A standard computer file format for the atomic positions in amino acids, **polypeptides**, and **proteins**. A format recognized by many **molecular visualization** programs.

**PEL**. Permissible Exposure Limits. Set for the workplace by OSHA and **ACGIH**.

**PEM.** A polymer electrolyte membrane. A modern type of **ion conduit**.

**Pendant groups**. In a linear polymer. the groups attached to the **backbone chain.**

**Percentage difference**. The difference between any pair of values expressed as a percentage. $(\frac{x_2 - x_1}{x_2}) \times 100\%$. The denominator may be the smaller of the two values or the average of the two values.

**Percentage discrepancy**. The difference between a **correct value** and a **measured value** expressed as a percentage of the correct value.

**Percentage efficiency**. Of an engine, it's the percentage of the chemical energy converted to work.

**Percentage ionic character**. A value used to describe the extent of polarity of a chemical bond. Bonds that are 0% ionic are 100% **covalent**. Bonds that are 100% **ionic** are zero percent covalent. For diatomic molecules it's their actual dipole moment expressed as a percentage of their maximum calculated dipole moment (their **one-electron transfer dipole moment**). Compare **Bond-type continuum**.

**Percentage yield**. The actual mass of a product formed in a reaction expressed as a percentage of the **theoretical yield** of that product by the reaction.

**Periodic law**. 1. The properties of the elements are periodic functions of their atomic weights. (Mendeleev's original statement). 2. The properties of the elements are periodic functions of their atomic numbers. (The modern reformulation of periodic law).

**Periodic table**. A representation of the chemical elements in the form of a map of their symbols and key properties.

**Periodicity**. Trends and relationships among the physical and chemical properties of the elements. The idea that periodic relationships exist among the properties of the elements when they are arranged in order of increasing atomic number. This study of relationships among the elements is a central organizing principle of chemistry, and a powerful scientific generalization.

**Petrochemical**. A compound obtained or derived from petroleum.

**Petrochemical industry**. The largest global industry. The industry that produces, refines, and distributes petroleum fuels and petrochemicals. The primary source of the raw materials needed for **organic synthesis**.

**Petroleum distillates**. Hydrocarbon liquid mixtures prepared by distillation. Widely used as fuels and solvents.

**Petroleum refining**. Very large scale industrial process. Process that provides the major source of fuels and organic compounds.

**pH indicator**. A dye that changes color as the pH of its environment varies. Compare **acid-base indicator.**

**pH meter**. An electrochemical device that displays the pH of a solution. It is **calibrated** by means of a **buffer solution**.

**pH scale**. Traditional method for expressing the acidity of a solution. pH = $-\log_{10}[H^+]$. At 25°C, pH + pOH ≈ 14.

**Phase**. The state of aggregation of a substance. Solid, liquid, and gas phases are the ordinary phases.

**Phase region**. On a **phase diagram** it is any region or area in which a single phase exists. The possible phases of a pure substance are solid, liquid, gas and supercritical fluid phases. Compare **state of aggregation**.

**Phase diagram**. For a pure compound, it is a pressure-temperature map of a substance's solid, liquid, and gas phases. For mixtures and alloys it is often convenient to use a temperature composition map. Phase diagrams are important in **metallurgy**.

**Phosphate diester linkage**. The functional group $-O-PO_2-O-$ that links **nucleotides** together to make **DNA** and **RNA** chains.

**Phosphor**. A material that emits light when energized. For example, when struck by x-rays or a beam of electrons.

**Photon**. A single quantum of energy in the visible region of the **electromagnetic spectrum**. A "particle of light."

**Photoresists**. UV light sensitive coatings used, for example, to prepare a copper surface for selective etching in the production of circuit boards.

**Physical change**. Any change that occurs to a system but does not alter the chemical entities that compose it.

**Physical chemistry**. The branch of chemistry devoted primarily to measuring and explaining the laws of chemistry. The most theoretical branch of chemistry, it includes chemical thermodynamics, quantum mechanics, etc. One of the four traditional branches of chemistry. The others are **analytical, inorganic**, and **organic chemistry**.

**Physical property**. Property of a substance revealed by a physical change.

**Pi antibonding orbital**. An antibonding molecular orbital in which the electron density lies away from the axis of the two atoms and outside the region between them.

**Place holders**. Non-significant zeroes that locate a decimal point.

**Planck's law**. $E = h\nu$.

**Plasma**. In physics, an electrically neutral gaseous mixture of electrons and ionized atoms and perhaps completely ionized nuclei. Sufficiently heated, any gas becomes a plasma. Sometimes considered a **state of aggregation** of matter.

**Plaster of Paris**. Chemically the half-hydrate of calcium sulfate, $CaSO_4 \cdot \frac{1}{2}H_2O$. A shaped water paste of this compound hardens to form a "plaster cast."

**Plasticizer**. A class of additives used in commercial plastics. Mainly used to improve the flexibility of the finished plastic product.

**Plastics**. Broad group of synthetic polymeric materials with widespread everyday applications.

**Pneumatic trough**. Late 18th century device that revolutionized the study of chemistry and the experimental handling of gases. It makes use of liquids and upturned flasks to keep gases trapped so they can be studied.

**pOH scale**. Traditional method for expressing the basicity of a solution. pOH = - $\log_{10}[OH^-]$. At 25°C, pOH ≈ 14 - pH.

**Polar covalent bond**. A bond intermediate between a pure covalent bond and an ionic bond. A diatomic molecule (AB) that is polar covalent can be represented as $A^{\delta+}$-$B^{\delta-}$ where $\delta^+$ signifies a partial positive charge and $\delta^-$ signifies a partial negative charge. See also **bond-type continuum**.

**Polarimeter**. Instrument used to measure optical rotation of solutions and, hence, compounds. Developed by Jean Baptiste Biot (1774-1862) around 1840. Used for analysis of sugars and drugs via their **optical rotation**. See **chiral purity**.

**Polarity**. Property of an entity or substance that has charge separation. Polar entities tend to be disturbed by electric fields. 1. Separation of electric charge. A **polar covalent bond** has a partial charge between its atoms. 2. A property of a polar molecule or substance. Compare **nonpolar**.

**Polarizability**. The possibility that an entity or object can be polarized or made to undergo **polarization**. In chemistry, the more polarizable a molecule the greater the **temporary dipole** that can be induced in it. Larger intermolecular forces are typically encountered for highly polarizable molecules. See also **dipole moment**.

**Polarization**. 1. What happens to an atom or molecule when it is placed in an electric field. For example, the production of a charge separation in a molecule by an external electric field. 2. Condition of a battery in which too rapid withdrawal of current has caused the battery to stop working; the battery will recover and again produce power if time is allowed for the chemical reaction to catch up with the electrical reaction.

**Poles**. In electrochemistry, the poles of a battery or cell are the electrical contacts to which the **external circuit** is attached.

**Polyamides**. Class of synthetic organic polymers made by reacting bifunctional molecules to form long chains linked by amide groups, -(CONH)-. Nylon® is a well known polyamide.

**Polyatomic ions**. Charged entities that contain two or more atoms.

**Polycarbonate**. Member of class of synthetic polymers whose backbone chain is linked by the carbonate functional group. A transparent, plastic, scratch-resistant material, suitable for making eyeglasses, etc.

**Polyelectrolytes**. Water soluble polymers with charged substituent groups along the polymer chain; both anionic and cationic types are known. Used in water treatment as **flocculants** and **deflocculants**.

**Polyesters**. Class of synthetic organic polymers made by reacting bifunctional molecules to form long chains linked by ester groups, -(COO)-. Dacron® is a well known polyester.

**Polyethers**. Class of synthetic organic polymers made by reacting **epoxides**. They get their name because the polymer has repeating units based on the ether linkage R-O-R. Nonionic surfactants are often polyethers.

**Polymer**. A substance composed of very large molecules. See **natural polymer** and **synthetic polymer**.

**Polymerization**. Any reaction carried out to prepare polymers.

**Polymerization reaction**. One that forms large molecules by joining many small molecules.

**Polymethylmethacrylate**. Polymethylmethacrylate. A strong, transparent, commercial plastic. Plexiglass® is an example.

**Polyolefins**. Polymers made by the addition reactions of alkenes. Compare **olefins**.

**Polypropylene**. Polymer formed by the addition reaction of propylene molecules.

**Polypeptide**. A naturally occurring polyamino acid. The raw material from which proteins are built.

**Polysaccharide**. A natural polymer composed of nine or more simple sugar residues. Starch and cellulose are examples.

**Polystyrene**. Polymer formed by the addition reaction of styrene molecules.

**Polyurethanes**. A class of synthetic organic polymers that contain the urethane (or carbamate) functional group linking together their monomer units. With R- and R'- being alkyl or aromatic groups, the structure of the urethane group is:

```
     OR'
     |
R-N-C=O
  |
  H
```

**Polyvinyl chloride (PVC)**. Widely used synthetic organic polymer. Found in old-style "vinyl" phonograph records, modern plumbing, etc.

**p-orbitals**. Hydrogen-like wave functions in which the secondary quantum number is one. Any $\psi_{(n,\ell,m)}$ with quantum number $\ell$=1.

$2p_x$, $2p_y$, and $2p_z$ orbitals

$3p_x$, $3p_y$, and $3p_z$ orbitals

**Position of equilibrium**. For a chemical reaction system the phrase describes whether either reactants or products are in substantial excess, or if they are present in roughly equal amounts. Compare **shifts to the left** and **shifts to the right**.

**Positron**. The electron's antiparticle. A fundamental particle with the same mass but opposite charge to an electron

**Positron emission**. A radioactive decay mode in which the nucleus of the parent nuclide ejects a positron.

**Potential, electrical**. See **electric potential**.

**Potential energy**. Energy due to position. For example the energy of the water in a lake by reason of its being higher than the river below the dam. $E_P$ = mGh, where m is mass, G is standard gravitational acceleration, and h is the height through which the object of substance will fall.

**Pour point**. The temperature at which an oil (or other liquid substance) ceases to flow from a container when it is cooled. The existence of a pour point is controlled by the change in **viscosity** of a liquid with temperature. An important characteristic of motor oils intended for cold weather service.

**Power**. The rate of energy production or consumption. Its SI base unit is J $s^{-1}$ or Watt. Engineers often use horsepower a a unit for power.

**ppm**. Part per million. A concentration unit convenient for substances present in low concentration in a bulk sample. For dilute aqueous solutions, a concentration in ppm has more or less the same value as it does in the unit mg $L^{-1}$.

**Precipitate**. A solid that forms when certain pairs of aqueous solutions are mixed. Solids that so form in a liquid solution are typically denser than the solution, hence they precipitate or fall to the bottom of their container.

**Precipitation**. In chemistry, the process in which a solid settles from a liquid solution.

**Precipitation reaction**. One that produces an insoluble precipitate. Precipitation reactions are a subclass of **metathesis reactions**.

**Prepolymer**. A formulated composition designed to form a true polymer when it is exposed to UV light. Polymer-based tooth fillings and **photoresists** are two applications of prepolymers.

**Precision**. Refers to the reproducibility of a measurement; that is, it describes how closely repeated measurements agree with one another. Precise measurements are not necessarily **accurate** or **correct**.

**Pressure**. Force per unit area. Pressure in the SI system is called the Pascal and has the base units N $m^{-2}$ (newtons per square meter).

**Pressure condensable**. Said of a gas that can be turned into a liquid simply by the application of pressure.

**Pressure of standard state**. See **standard state pressure**.

**Pressurized water reactor (PWR)**. The main type of commercial nuclear power plant. It boils water to make steam using energy from uranium fission.

**Primary cell**. In electrochemistry, a **battery** not designed to be, or capable of being, recharged.

**Primary radionuclide**. A radioactive nuclide that exists in nature because it has sufficiently long half-life compared to the age of the earth. Natural uranium thorium isotopes and $^{244}$Pu are primary radionuclides.

**Primary reference standard**. A reference material used to establish accurate values. The procedure of **standardization** against the primary involves a direct comparison between a test sample and the primary reference standard. Also called just primary standard.

**Primary standard**. See **primary reference standard**.

**Primordial radionuclide**. See **primary radionuclide**.

**Principal quantum number**. The value of n in the equation $E_n = R_H \frac{h}{n^2}$. The first discovered quantum number. The quantum number of the Bohr atomic model. It follows the rule n = 1, 2, 3, 4....

**Probability density**. Alternative name for **electron density**.

**Process**. 1. In thermodynamics, anything that happens that changes the state of a system. 2. An industrial process used in **industrial chemistry**.

**Products**. Entities that created by a chemical reaction. They are written on the right side of a conventional reaction equation.

**Propagation**. In chemical kinetics, a reaction step that continues a chain reaction is said to propagate the reaction.

**Properties**. The attributes of a material: its mass, color, strength, density, chemical reactivity, etc. Different kinds of material have different properties, so properties distinguish one material from another.

**Proprietary information**. Knowledge held within a company or other organization about how to run a process or build a device, etc. Also called **trade secrets**.

**Propyl group**. A portion of a molecule or radical. The propane molecule less one terminal hydrogen atom, $-CH_2-CH_2-CH_3$.

**Proteins**. A class of natural substances composed mainly of polyamino acid polymers. 1. Any polyaminoacid. 2. A food group. Common examples of proteins are: hair, skin, and enzymes. See **amino acid**.

**Proton**. A positively charged nucleon with a mass of approximately 1.007276470 u.

**Proton number**. The number of protons in a nuclide. Alternative name for **atomic number**. Given the symbol Z.

**Proton rich nuclides**. Those with a neutron/proton ratio greater than required by nature to make the nuclide stable. Radionuclides that lie above and to the left of the **belt of stability** on a nuclide chart are proton rich.

**Pseudostable nuclides**. Nuclides with half-lives in the range of millions of years and longer.

**psi**. Pounds per square inch. Common engineering unit for pressure. In SI metric the psi is defined as $\approx 6.894757 \times 10^3$ Pa. see also **psia** and **psig**.

**psia**. Pounds per square inch absolute. An engineering pressure unit that refers to the actual pressure in a system. Contrast **psig**.

**psig**. Pounds per square inch gauge. An engineering pressure unit that is the actual pressure in the system minus atmospheric pressure. Tire pressure is traditionally reported in psig. Contrast **psia**.

**Pure chemistry**. Chemical research performed for the purpose of better understanding nature.

**Pure covalent bond**. A bond that lacks any ionic character. All **homonuclear diatomic molecules**, $H_2$, $N_2$, $O_2$, $Cl_2$, etc., are nonpolar with pure covalent bonds.

**Pure substance**. Other than elements, pure substances contain at least two different kinds of elements in chemical combination and are called **chemical compounds**.

**Purine**. An organic base with the molecular formula $C_5H_4N_4$. Its structure is important as the prototype for the nucleic acid bases guanine and adenine. Purine's structure is:

**Pyrimidine**. An organic base with the molecular formula $C_4H_4N_2$. Its structure is important as the prototype for the nucleic acid bases thymine, cytosine, and uracil. Pyrimidine's structure is:

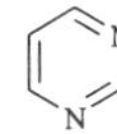

**Pyrochemical reaction**. One used to produce light and heat. A firework reaction.

**Quadratic equations**. Equations having the algebraic form: $ax^2 + bx + c = 0$ that can be solved using the formula $x = \frac{-b \pm \sqrt{b^2 - 4ac}}{2a}$

**Qualitative analysis**. Laboratory process to determine what is in a substance.

**Quantitative analysis.** Laboratory process to determine what proportion of an element or other component is in a substance.

**Quantitative reaction**. One with a large equilibrium constant so that at equilibrium the reactant concentrations are essentially zero. Chemists say that quantitative reactions "**go to completion**."

**Quantitative structure property relationships (QSPR)**. Variety of computer based procedures used to estimate the physical properties of substances from their chemical structure.

**Quantitative structure activity relationships (QSAR)** Variety of computer based procedures used to estimate the chemical properties of substances from their chemical structure.

**Quantization**. The concept that many properties, but most particularly energy, can be expressed in terms of measurable increments. Thus, we say that energy is **quantized**. The smallest energy increment is a **quantum** (plural, quanta).

**Quantum mechanics**. Bedrock theory of nature. Fundamental theory that explains how electrons behave in chemical entities. Mathematically arcane and requiring big computer power to apply, quantum mechanical interpretation underpins much of modern chemistry.

**Quantum numbers**. Integer values originally discovered through studies of atomic spectra. Now regarded as arising from quantum mechanics and, more specifically, as being values that characterize wave functions. In an atom, chemists designate any electron by the set of four quantum numbers, n, l, m, and s.

**Quantum**. A single unit or "particle" of energy.

**Racemic mixture**. A sample of an enantiomeric compound in which both **enantiomers** are present in 50:50 proportion. The original racemic mixture was sodium ammonium tartrate which produces crystals of two chiral forms. In 1848, Louis Pasteur separated such crystals by hand picking, dissolved them in water, and showed with a **polarimeter** that the solutions had optical activities with opposite rotations. This was the moment at which the concept of **chirality** entered chemistry.

**Radical**. 1. In organic chemistry, sometimes used to refer to part of a molecule in the sense of a **group** of atoms. For example the methyl radical $-CH_3$, or the butyl radical $-C_4H_9$. 2. A reactive entity with an unpaired electron, $X^{\cdot}$, for which see **free radical**.

**Radical trap**. See **free radical trap**.

**Radioactive waste**. A complex radioactive mixture composed mainly of short-lived radioactive **beta emitters**. The product of nuclear fuel that has been used for power generation. Radioactive waste is a mixture of **fission products**.

**Radioactivity**. The spontaneous decomposition or decay of an atom's nucleus.

**Radiocarbon (carbon) dating**. Laboratory method used to measure the age of organic materials up to about 50,000 years old. Based on assuming the historic ratio of the nuclides $^{14}C/^{12}C$, combined with a modern measurement of the actual ratio in the sample, and the known half-life of $^{14}C$.

**Radiodecay**. The process of a radioactive nuclide decomposing to produce a daughter nuclide.

**Radionuclides**. Unstable nuclides that are radioactive and spontaneously decay to produce **daughter nuclides**.

**Radius ratio**. An important concept in understanding how a central sphere (the **central atom**) can be surrounded by other spheres (the **ligand atoms**). If the radius of the central atom is large and that of the ligands is small then many ligands can fit around the central atom. If the radius of the central atom is small and that of the ligand atoms is large then few ligands can fit around the central atom. Also useful in explaining the structure and packing in metals and ionic solids.

**Random uncertainties**. Those that arise from uncontrollable fluctuations at the limit of measurement. Uncertainties that lack a **systematic** explanation.

**Raoult's Law**. Law that states when a solution of two chemically similar liquids is made, the vapor pressure of the mixture is the sum of the individual vapor pressures taken in proportion to their mole fractions in the liquid phase, $P_{tot} = P_A^{\circ} X_A + P_B^{\circ} X_B$.

**Rasmol**. A popular **molecular visualization** tool.

**Rate constant**. An indicator of the speed of a chemical reaction. In chemical kinetics, it's the constant, k, in the **rate law**. It depends on the temperature at which the reaction is carried out.

**Rate laws**. Describe reactions that occur over time. Mathematical relationships between reactant concentrations and time.

**Rate process**. One that is said to be occurring if any property of a substance or material changes with time.

**Raw materials**. The raw materials of chemistry are the resources of our planet: metals, minerals, coal, petroleum, sea water, the gases of the atmosphere, and materials obtained from plants and animals.

**Reactants**. Entities that get consumed by a chemical reaction. They are written on the left side of a conventional reaction equation.

**Reaction enthalpy**. $\Delta H$ of a chemical reaction. A reaction's energy change under constant pressure conditions.

**Reaction equilibrium system**. A **chemical reaction system** that has reached equilibrium.

**Reaction intermediate**. A chemical species temporarily formed during a chemical reaction, but not itself either a reactant or a product of the reaction.

**Reaction mechanism**. The step-by-step sequence of events by which the reactants become transformed into the products.

**Reaction order**. See **order of a chemical reaction**.

**Reaction quotient**. For a gas phase reaction a A(g) + b B(g) ⇄ c C(g) + d D(g), it's written $Q_p = \frac{(P_C)^c(P_D)^d}{(P_A)^a(P_B)^b}$. $Q_p$ is the reaction quotient. At equilibrium, $Q_p$ = $K_p$ the equilibrium constant.

**Reaction rate**. The speed at which a reaction is proceeding is its reaction rate.

**Reaction stoichiometry**. The study of the mass and number relationships among reactants and products in a chemical reaction.

**Reaction system**. A definite chemical reaction treated as a **thermodynamic system**. Experimentally, it is often an equilibrium reaction carried out in a closed container. Also called a **chemical reaction system.**

**Reaction yield**. See **yield**.

**Reagent**. A compound or mixture used to bring about a reaction. Solutions prepared in advance for ready use in a laboratory are the laboratory's reagents.

**Real gas**. One that does not obey the ideal gas law.

**Recharge**. To apply direct current to a discharged battery, reverse the electrochemical reaction, and so restore the battery's capacity to produce electric power.

**Redox**. Process or reaction during which one entity gains one or more electrons while *simultaneously* a second entity loses one or more electrons.

**Redox reaction**. One in which one reactant is oxidized while a second reactant is *simultaneously* reduced. It is interpreted as a process in which there is a flow of electrons from one reactant to another.

**Redox rules**. Set of statements telling how oxidation numbers are to be assigned to the atoms of a chemical entity. See also **oxidation number** and **oxidation state**.

**Reduced**. Said to have happened to an entity when it gains one or more electrons. Metals are reduced in order to extract them from their ores. The RIG in **OIL-RIG**.

**Reducing agent (reductant)**. Entity capable of causing the **reduction** of a second entity.

**Reduction**. The process of an entity becoming **reduced**.

**Reduction half-cell**. The region of an electrochemical cell where reduction occurs.

**Reduction-oxidation reaction (redox reaction)**. See **redox reaction**.

**Reference materials** (called **primary reference standards**) used to establish accurate values by direct comparison. Usually, these reference materials are employed to **calibrate** instruments. An instrument is calibrated by adjusting it so that it produces the correct reading for a reference sample.

**Registered trade mark**. It is a word or phrase properly filed and recognized at the US Patent and Trademark office. The symbol ® denotes a registered trade mark. The purpose of the trade mark system is to allow manufacturers and others to protect certain marks and reserve them for their own, exclusive use. See www.uspto.gov.

**Relative humidity**. The partial pressure of water vapor in the air divided by the saturated vapor pressure at the same temperature and expressed as a percentage. Widely used in meteorology.

**Repeatability**. In measurement theory, the ability of an experimenter to get the same result from repeated measurements of one property. Contrast **reproducibility** and **precision**.

**Reproducibility**. Getting the same result more than once is good reproducibility. Achieving reproducible results is an important goal of science. Contrast **irreproducible results**.

**Residue**. Any portion of a larger molecule recognized as being derived from a smaller molecule that went into making the larger molecule. Term widely used in **biochemistry** and **molecular biology**.

**Resins**. Plastics. In chemistry, the term resin generally means a solid, synthetic, organic material.

**Reversible process**. One in which the system is infinitesimally near a state of equilibrium at each step of the process. A **reversible cycle** is one operated in such a manner that each of its steps is a reversible process.

**Reversible reaction**. One in which the products and reactants can exchange if conditions, such as the temperature, are altered. The simplest reversible reaction is written: $H_2 \rightleftarrows 2\,H$.

**Ribose**. A sugar molecule with the formula $C_5H_{10}O_5$. An important biochemical molecule critical to building the structure of ribonucleic acid. See **RNA** and **nucleic acids**. The structure of ribose is shown below.

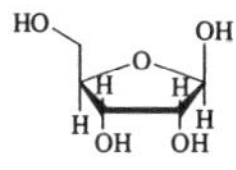


**Rich.** A fuel condition that exists when there is relative excess of fuel compared to the available oxygen.

**Ring strain**. Term used to explain the unexpected chemical reactivity of **cyclic molecules** with internal ring angles that differ substantially from ideal geometry angles. See **epoxides**.

**R-Groups**. In **organic chemistry** the symbol "R" stands for the "rest of the molecule." The actual origin of "R" is probably the word **radical**. For example RH is the **general formula** for the normal alkanes. In this series $-CH_3$ the methyl group gives $HCH_3$ or $CH_4$, which is methane; $-C_2H_5$ the ethyl group gives $HC_2H_5$ or $C_2H_6$, which is ethane; and $-C_3H_7$ the propyl group gives $HC_3H_7$ or $C_3H_8$, which is propane, etc. The designation R' means either a different group than R or possibly the same group. See **alkyl group**.

**RNA**. Ribonucleic acid. See **nucleic acids**.

**Room temperature**. A temperature in the range 15°-30°C.

**Root mean square speed**. The square root of the mean (average) of the squares of the speeds of the molecules. of a gas.

**Rotational motion**. The tumbling motion of a molecule about its center of gravity.

**Rounding**. Important concept in measurement theory. Rounding is reducing the number of figures obtained from a calculation so that the reported answer contains the proper number of significant figures authorized by the given data. Example: 11.2 × 13.787 = 154.4144. But the value 154.4144 must be rounded to 154 to be consistent with the three significant figure reliability of 11.2.

**Rules for oxidation numbers**. A procedure for assigning number values to elements in chemical entities. See also **oxidation number** and **oxidation state**.

**Sacrificial electrode**. An active metal, clamped to a steel object, that prevents corrosion of the steel by itself being oxidized more readily than the steel

**Sandwich compound**. A compound belonging the general class in which a metal atom is sandwiched between two parallel organic molecules each containing unsaturated rings of carbon atoms.

**Salt bridge**. A device that acts as an **ion conduit**. It is a water-based jelly containing a substantial concentration of an electrolyte such as sodium chloride. When electrical charge builds up in an electrochemical cell, ions drift through the bridge.

**SARA**. Federal legislation controlling the clean up of toxic industrial pollution. The acronym stands for Superfund Amendments Reauthorization Act (of 1986).

**Saturated**. 1. Said of an organic compound that lacks **double bonds**, **triple bonds**, and **aromaticity**. 2. When the maximum amount of a solute that can dissolve in a solvent at a given temperature has in fact dissolved, then the solution is said to be saturated by the solute.

**Saturated solution**. A solution that contains the maximum amount of a solute than it is capable of containing. one that is in equilibrium with undissolved solute.

**SCBA**. Self contained breathing apparatus. A valuable device to have in the event of an unplanned chemical release.

**Schrödinger equation**. Fundamental equation of nature. Also called the **wave equation**.

**Scientific law**. A summary statement about the behavior of nature. Scientific laws typically derive from a theoretical interpretation of experiments and observations.

**Scientific notation**. method to represent values using exponential powers of ten. For example, because of its huge size, Avogadro's number. $6.022 \times 10^{23}$ is almost always represented in scientific notation. Compare **number style**.

**Screening.** The effect an electron has on the nuclear charge experienced by another electron in an atom. When an electron is between the nucleus and an outer electron, it screens part of the nuclear charge so the outer electron experiences an effective nuclear charge, $Z_{eff}$, which is less than its full value.

**Second**. "The second is the duration of 9 192 631 770 periods of the radiation corresponding to the transition between the two hyperfine levels of the ground state of the caesium-133 atom (13th **CGPM**, 1967)."

**Second law of thermodynamics**. A fundamental law of nature. Not provable but not in doubt. Capable of being stated in a remarkably large number of ways. An early statement of the law was: *The entropy of the universe tends to a maximum.* (Rudolph Clausius, 1865)

**Second order**. In chemical kinetics, one whose rate law depends on molarity squared. A reaction whose **order** is two.

**Second transition series**. In the periodic table the elements yttrium (Y) through cadmium (Cd). The middle of three rows of **d-block** transition metals in the periodic table.

**Secondary cell**. In electrochemistry, any cell that can be conveniently recharged.

**Secondary radionuclide**. One produced in a **natural decay chain**.

**Secondary quantum number**. It is designated $\ell$ ("ell") and follows the rule $\ell = 0, 1, 2, 3, \ldots (n - 1)$.

**Secondary standards**. Somewhat less reliable standards than **primary reference standards**. Secondary standards for many materials are available for less cost from commercial supply companies.

**Secular equilibrium**. In a chemical reaction or radioactive **decay chain**, a steady state situation when the rate of formation of a substance balances its rate of disappearance. Secular equilibrium may last a long time, but it cannot endure indefinitely because whatever is forming the substance becomes depleted.

**Self-redox**. Reaction in which a chemical entity reacts with itself to form two products, one in a higher and one in a lower oxidation state that the starting entity. For example $2V^{4+} \rightarrow V^{3+} + V^{5+}$.

**Semiconductor**. Solid in which a **conduction band** exists at an energy not much higher than a nonconducting **valence band**. Because of the situation, the electrical properties of the solid can be exploited for practical purposes.

**Semi-permeable membrane**. A structure or part of a device that allows water molecules to flow through it while being impenetrable to other molecules and ions. A membrane necessary for **osmosis**.

**Semi-synthetic polymers**: A material that is in part derived from nature and in part by chemical modification. A class of polymer substances made by modifying a natural biological polymer. For example, cellulose acetate and cellulose nitrate (gun cotton) are examples of chemically modified cellulose.

**Service life**. The anticipated time a device, etc., can be expected to work. Engineers control service lifetimes via chemical specifications for the materials used to construct various devices.

**Shapes of orbitals**. See **orbital shape**.

**Shell**. In quantum mechanics, all the orbitals with the same principal quantum number. In n =1 it's the first principal shell, or the K-shell, If n=2 it's the second principal shell, or L-shell.

**Shielding**. An alternative name for screening

**Shift**. What happens to an equilibrium chemical reaction when it is disturbed. See **disturbing an equilibrium**, **Le Chatelier's principle**, and **shift to the left or right**.

**Shifts to the left**. Said of an equilibrium reaction system if, in response to a disturbance, the concentrations of products decreases while the concentrations of reactants increase.

**Shifts to the right**. Said of an equilibrium reaction system if, in response to a disturbance, the concentrations of products increase while the concentrations of reactants decrease.

**SI metric system**. See **metric system**.

**SI.** The international system of units. See **metric system**.

**Side chain**. In a polymer molecule it's a chain of atoms that branches from the main chain. see also **branched polymer**.

**Sigma orbital**. In a diatomic molecule its a type of molecular orbital having circular symmetry along the bond axis. A **bonding sigma orbital** is located between the atoms it bonds. An **antibonding sigma orbital** is located behind the atoms as so does not contribute to their bonding.

**Significant digits**. Alternative name for **Significant figures.**

**Significant figures**. A conventional way to show experimental uncertainty. Generally speaking, digits or figures that have been measured are **significant digits** or **significant figures.**

**Silicates**. Class of solids (occasionally liquid solutions) that contain linked silicon-oxygen tetrahedra as their basic structural unit. Many natural rocks and minerals are silicates.

**Silicate minerals**. The rocks that compose the earth's crust. The largest class of network covalent solids.

**Silicones**. Synthetic polymeric silicon compounds. Silly Putty® is a well known silicone material.

**Skeleton equation**. Usually, an unbalanced equation showing all or most of the reactants and products. Typically, students are expected to be able to balance skeleton equations provided by their professor.

**Slaked lime**. $Ca(OH)_2$. The product of the reaction of **lime** and water.

**Slug**. 1. "To slug" an equilibrium chemical **reaction system** is to add an additional amount of a reactant or product. Doing this **disturbs the equilibrium**, which after some time returns to a new **position of equilibrium**. 2. In engineering, a unit of force.

**Smelting**. In metallurgy, the process of extracting a metal from its ore by a chemical **reduction** reaction.

**Solar energy**. The radiant energy emitted by the sun.

**Solid foam**. Mixture that consists of gas bubbles trapped in a solid matrix. Synthetic polymer foams are used for such practical applications as thermal insulation, packaging, and upholstery.

**Solid solution**. Any homogeneous solid mixture.

**Solubility product constant**. The equilibrium constant expression for an ionic compound that dissolves to some extent in water. If $M_mX_n$ is an ionic substance slightly soluble in water it dissolves according to the general equation: $M_mX_n(s) = m\,M^{n+}(aq) + n\,X^{m-}(aq)$ then the general $K_{sp}$ expression is $K_{sp} = [M^{n+}]^m[X^{m-}]^n$

**Solubility rules**. A crude way to deal with precipitation and the solubility of substances in water. Often a table specifying which classes of compounds are soluble in water and which are not. A better method is to use **solubility product constants**.

**Solutes**. Substances that dissolve in a **solvent** are called solutes.

**Solutions** are homogeneous mixtures of variable composition. **Liquid solutions** are composed of a solute dissolved in a liquid solvent. Mixtures of gases are the simplest solutions. Occasionally, solids mix homogeneously to make **solid solutions**.

**Solvent**. Any liquid that dissolves another substance to form a solution. A liquid capable of dissolving things, such as a paint solvent. A pure liquid that does the dissolving in the process of forming a solution. See also **industrial solvent** and **inert solvent**.

**s-orbitals**. Hydrogen-like wave functions in which the secondary quantum number is one. Any $\psi_{(n,\ell,m)}$ with quantum number $\ell = 0$.

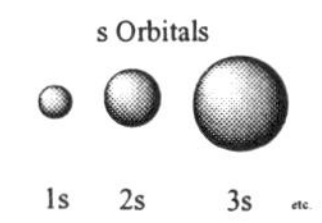


**Space**. What is occupied by matter that has mass.

**Space-filling view**. In a molecular visualization it is a display that shows the relative sizes of the depicted atoms in a natural way. In a space filling view, chemical bonds may be omitted or shown in a rudimentary manner.

**Speciation diagram**. A graph that shows how concentrations change as a function of some variable. Often a graph of changing ion concentrations as a function of pH.

**Species distribution diagram**. Synonym for **speciation diagram**.

**Specific heat capacity**. The heat capacity of one gram of a substance. Expressed usually in the units J $g^{-1}$ $K^{-1}$.

**Specified entity**. A concept implicit in the definition of the **mole** base unit in the SI metric system. Whenever the unit mole is used it must refer to a particular, chemical species called "the entity." The specified entity in chemistry is usually atoms, molecules, or ions.

**Specific rate constant**. See **rate constant**.

**Spectator ions**. Ions present in a reaction mixture but not participating in the reaction. Spectator ions are omitted from the **net** or **overall** reaction. They are onlooker ions, hence they "spectate."

**Spectrometer**. An instrument used to measure the wavelengths of electromagnetic radiation or the masses of ions. See, for example, **infrared spectrum**, **mass spectrum**.

**Spectroscope**. A device for separating electromagnetic radiation according to its frequency or wavelength.

**Spectroscopy**. The study of the absorption and emission of electromagnetic radiation by matter. Spectroscopy produces spectra.

**Spin**. An intrinsic property of electrons (and other particles), which is the basis of the Pauli exclusion principle. Each electron behaves as if it were spinning on an axis. Each electron has two spin states, which are designated by a quantum number, $m_s = \pm\frac{1}{2}$.

**Spin quantum number**. It is designated s and follows the rule $s = +\frac{1}{2}$ or $-\frac{1}{2}$.

**Spontaneity**. In thermodynamics, the notion that processes are inherently favored. Thermodynamic spontaneity is controlled by a balance between energy and entropy factors.

**Spontaneous fission**. A rare radioactive decay mode in which a heavy nucleus spontaneously breaks up into two roughly equal fragments and several free neutrons. It occurs only among the **actinide** and **trans-uranic** elements.

**Spontaneous reaction**. In thermodynamics, at constant temperature and pressure it's a reaction for which $\Delta G$ is negative. Thermodynamically spontaneous processes and reactions are *not* necessarily spontaneous in the ordinary sense of the word. For example, the combustion of gasoline is thermodynamically spontaneous but actual gasoline combustion needs an ignition source—gasoline doesn't catch fire spontaneously.

**Stable isotope**. An isotope that does not decay. One that is not radioactive.

**Stable nuclide**. A nuclide that does not decay. One that is not radioactive.

**Standard**. In general, a standard provides a benchmark for making measurements. Standards can be physical objects, chemical substances, solutions, etc. See also **reference material**.

**Standard deviation**. In measurement theory and statistics, it is an estimate of experimental uncertainty interpreted in terms of a statistical frequency (chance) that the best possible **measured value** will lie between specified limits.

**Standard enthalpy of formation**. Written as $\Delta H_f^o$, for any a substance. It's the enthalpy change when the compound in its standard state (normal form at a pressure of one bar) is formed from its elements in their standard states. ΔH of the reaction in which one mole of a specified compound in its standard state is formed from the needed elements in their standard states. Usually tabulated at 298.15 K.

**Standard entropies** Thermodynamic tables of entropy data. They compare the entropy of substances at 298.15 K with those of the same substance at 0 K. See **third law of thermodynamics**.

**Standard heat of formation**. A somewhat imprecise term used to mean **standard enthalpy of formation**.

**Standard heat of reaction**. ΔH of a reaction when all substances are in their standard states. The term **standard enthalpy of reaction** is more precise.

**Standard hydrogen electrode**. Device used as a standard in measuring standard half-cell voltages.

**Standard pressure**. In modern science this term usually refers to one bar. However the standard atmosphere (which is only slightly different) is still widely used. Standard pressure is inherent in the definition of a substance's standard state. In contrast, the **temperature of standard state** has to be defined separately.

**Standard reduction potentials**. Table of half-cell voltages for many electrochemical reactions measured at standard conditions.

**Standard state pressure**. One bar or $10^5$ pascals. The pressure at which modern tables of thermodynamic properties are reported.

**Standard state**. The **state of aggregation** and specific **chemical structure** of an element or compound that is most stable at a pressure of one bar and at a specified temperature. Tabulated values of standard thermodynamic properties are usually reported at 298.15 K.

**Standard substance**. One whose properties are known and certified. A substance typically used for the calibration of instruments. In the US, **NIST** sells a wide range of standard substances.

**Standard tables**. Physical lists or electronic data bases of measured properties. For example, the results obtained from many calorimetric experiments have been organized into tables of thermodynamic data. Such tables are an important scientific resource. Standard tables are usually managed and maintained by **standards organizations**.

**Standard temperature**. The phrase usually means 25°C or 298.15 K. Occasionally, 0°C or 273.15 K is intended. See **temperature of standard state**.

**Standardization**. Process of testing a standard substance as a means of calibrating the device and method used in the test. For example, burning a standard substance in a calorimeter with the intent to establish the calorimeter's heat capacity. Compare **calibration**.

**Standards organization**. A body with responsibility for overseeing and implementing standards. **NIST** in the US.

**Starch indicator**. The dark blue colored entity formed between iodine and starch. Used to indicate the **endpoint** of the **iodine clock** reaction and certain redox reactions.

**State**. Word used by chemists with too many meanings. **State of aggregation** refers to a substance being a solid, liquid, and gas. **Energy states** exist for entities and are explained by quantum mechanics. The **thermodynamic state** of a system is a specification the properties needed to characterize it uniquely. Compare **state property, state function, state of a system, excited state, ground state, and oxidation state**.

**State functions**. An alternative name for **state properties**.

**State of a system**. The condition of a system. A listing of the values of a sufficient number of its properties such that the system is uniquely defined.

**State of aggregation**. The most obvious bulk property of a substance. The three states of aggregation are solid, liquid and gas.

**State properties.** For any thermodynamic system they are properties that depend only on the **thermodynamic state** of the system. When the state of a thermodynamic system is changed the changes in its state properties are constant—*regardless of the path by which the change is carried out.* Also called **state functions**.

**Static electron model**. One that uses stationary electrons or electron pairs to explain the properties of bonds and molecules.

**Stationary phase**. In **gas** or **liquid chromatography** the stationary phase provides for selective absorption of components from the mobile phase leading to their separation.

**Statistical mechanics**. The branch of modern science that explains the bulk properties of materials based on the collective properties of an assembly of the material's component molecules. It provides a fully satisfactory link between thermodynamics, statistics, and molecular behavior.

**STEL**. Short Term Exposure Limit. The maximum concentration of a particular chemical compound to which workers can be exposed for periods of up to 15 minutes.

**Stellar**. Pertaining to a star. Most of the chemical elements are the product of **stellar nucleosynthesis** or the formation of elements inside stars.

**Stereoisomers**. **Isomers** in which the same atoms are connected by the same bond but where those bonds point in different directions in space. See **cis-trans isomers** and **enantiomers**.

**Stoichiometric amounts**. The relative amounts of reactants for a chemical reaction that correspond exactly to the stoichiometric coefficients in the balanced equation for the reaction.

**Stoichiometric coefficients**. The coefficients in *balanced* chemical equation. They can all be multiplied or divided by any constant and still remain stoichiometric coefficients.

**Stoichiometry**. The study of mass and number relationships in chemistry. It's the boot camp of a freshman chemistry course.

**STP or NTP**. "Standard (or normal) temperature and pressure." Historic abbreviations usually meaning one atmosphere pressure and 0°C. Obsolete, but still encountered on **MSDS** sheets, the older literature, and other places

**Stock system**. Method used to designate the **oxidation number** of an element in a compound; a useful method to name compounds. Uses Roman numerals enclosed in parentheses. Named for Alfred Stock (1876-1946). For example, MnO is manganese(II) oxide and $MnO_2$ is manganese(IV) oxide.

**Strong acids**. Substances composed of hydrogen ions combined with negative ions that completely ionize in water.

**Strong bases**. Substances composed of hydroxide ions combined with positive ions that completely ionize in water. Strong bases are also called **alkalis**.

**Strong electrolyte**. A substance that ionizes completely when it dissolves in water.

**Structural isomers**. Molecules with the same chemical formulas but different chemical structures. They arise if the atoms in a given molecule can be rearranged to give a different stable molecule with a different architecture from the original molecule. Such molecules are said to be isomers of one another. Especially in **organic chemistry,** a large number of possible isomers may exist for a given molecular composition.

**Structure**. Broadly, how atoms are arranged in space. In chemistry, structure refers to the manner in which atoms and molecules bond and the geometry of bonding. See also, **chemical structure**, **molecular structure**, and **atomic structure.**

**Structure-property correlation**. Is the application of chemistry to understand how the chemical or biological reactivity of a compound is linked to the compound's chemical structure. For example it's an important concept in developing new drugs.

**Stuff**. Slightly frivolous name for a **material**. "Engineers are people who make things with stuff."

**Subshell**. A family of **atomic orbitals** each of which has the same principal quantum number and the same secondary quantum number.

**Styrene-butadiene-rubber (SBR)**. A commercially important synthetic rubber. Typically made by the polymerization of 75% butadiene and 25% styrene mixture.

**Sublimation**. A process of phase change from a solid to a gas. Dry ice is so-called because under everyday conditions it sublimes and thus never melts. It is literally ice that stays dry. Contrast **vapor deposition** which is the reverse process.

**Substance**. A material with a particular composition. It might be a pure element, a pure compound, or any mixture of elements and compounds. Substances may be **homogeneous** or **heterogeneous**.

**Successive approximations**. See the **method of successive approximations**.

**Successive neutron captures**. When a nuclide undergoes a process involving a sequence of neutron captures punctuated by $\beta^-$ emissions. A process that accounts for the formation of **trans-uranic** elements in stars, nuclear reactors, and nuclear explosions.

**Suction pump**. A pump that raises a liquid against gravity by removing the gas from the liquid's surface.

**Supercooled**. Said of a liquid sample below its crystallization temperature. A thermodynamically unstable state that may be triggered to undergo a phase change when "seeded." Honey and syrup are often supercooled liquid solutions—sometimes these solutions crystallize on prolonged storage.

**Supercritical fluid (SCF)**. A state of matter intermediate between the liquid and vapor states of a substance. Conventionally, the upper right-hand region of a temperature-pressure **phase diagram.**

**Supernova**. Astronomical phenomenon. A gigantic explosion that occurs when a hot, massive star exhausts its nuclear fuel and recycles itself into the universe. Many elements are synthesized during supernovas.

**Supersaturated solution**. A solution that contains a higher concentration of a solute than a saturated solution at the same temperature. Supersaturated solutions are **thermodynamically unstable**.

**Supersaturation**. Thermodynamically unstable condition in which the concentration of a solute in a solvent exceeds the concentration of a saturated solution at the same temperature.

**Surface active agents** or **surfactants**. Substances that concentrate at liquid surfaces. Soaps and other cleaning products contain surfactants.

**Surface tension**. The bulk force in the surface of a liquid. It derives from the intermolecular forces that hold the liquid together. The force that causes liquids to from drops.

**Surfactants**. Surface active agents.

**Synthesis**. See **chemical synthesis**.

**Synthetic**. In chemistry, synthetic means made by chemists in contrast to natural or obtained from nature. For example, synthetic vanilla is the product of chemical synthesis in a laboratory. Natural vanilla is extracted from beans. Synthetic and natural vanilla are identical compounds.

**Synthetic elements**. So-called because they are created by humans. Elements in the periodic table beyond plutonium (element 94). See also **trans-uranic elements**.

**Synthetic polymers**. Synthetic organic substances. Large class of commercially prepared substances that constitute much of the **stuff** consumed in modern economies. Sometimes called **plastics** or **resins**. The deliberately produced substances that today compose most of the materials that pervade our daily lives.

**System**. In thermodynamics, whatever is the subject of study. Most laboratory experiments require the assembling of a system. See **thermodynamic system.**

**Systematic uncertainties** are those incorporated uniformly into every measurement. They usually can be traced to a faulty instrument or improper calibration of the instrument, and are always present if an unreliable **standard** has been used.

**Table of functional groups**. Listing of common groups of atoms for in the range of organic molecules. Organic molecules can be classified and cataloged using a table of their attached functional groups.

**Temperature of standard state**. 298.15 K (25.0°C). The temperature at which modern tables of thermodynamic properties are reported.

**Temperature-composition phase diagram**. A **phase diagram** in which the axes are temperature and composition.

**Temporary dipole**. A short-lived imbalance of electrical charge in an otherwise **nonpolar** atom or molecule. The dipole induced in a nonpolar molecule when it is placed in an electric field.

**Term-of-art**. Legal phrase meaning that a word or phrase is being used with a specific legal definition. Many terms-of-art are used in chemistry. Thermodynamics is made very difficult for beginners because it uses many ordinary words with very specialized meanings.

**Tetrahedral**. Having the shape of a regular tetrahedron. One of the ideal geometries of molecules. Very common because saturated carbon atoms generally have a tetrahedral structure.

**Tetrahedral angle**. 109° 28'. The angle subtended at the center of a regular tetrahedron by any pair of corners. All the bond angles in methane, $CH_4$, are tetrahedral.

**Thalidomide®**. Drug used in the 1960s with disastrous consequences for the children of pregnant women. Responsible for severe physical birth defects. Lack of **chiral purity** was later found to be the root of the problem.

**Theoretical yield (of a reaction)**. The maximum amount of product that can be obtained presuming that all of the limiting reactant is consumed. If the reactants and products are present in **stoichiometric amounts** then the percentage yield of every product will be 100%.

**Thermal energy**. Energy that a system contains because of the motion and potential energies of its atoms and molecules.

**Thermochemistry**. The study of chemical energy. It is an aspect of thermodynamics.

**Thermodynamics**. Literally the science of heat + work, (thermos + dynamos). Fundamental branch of science widely applied in chemistry, physics, and engineering. The science of steam engines The study of the inter conversion of different forms of energy. The study of heat and work and their relation.

**Thermodynamic cycle**. A a process that eventually gets back to where it started. For example, in an air conditioner, when the working fluid returns to its starting state. A thermodynamic cycle first changes one or more state properties and then returns the state properties back to their original values. Compare **Carnot cycle**, **cyclic process**, and **working fluid**.

**Thermodynamic process**. A process that takes a system from one **thermodynamic state** to a second **thermodynamic state**.

**Thermodynamic properties**. Those used to characterize the state of a thermodynamic system.

**Thermodynamic state**. The specification of the properties of a **thermodynamic system**. For a gas it's a list of sufficient properties that uniquely characterize the gas—the usual four properties are P, V, T, and n or g, but other combinations are possible.

**Thermodynamic system**. Any precisely defined experimental setup we choose to study. See also **system.**

**Thermodynamically stable**. Said of a compound that has a negative value of its standard enthalpy of formation, $\Delta H°_f$. In their standards states, such compounds form from their elements via an exothermic reaction. Thermodynamically stable compounds are stable with respect to their decomposition to their constituent elements.

**Thermodynamically unstable**. Said of a compound that has a positive value of its standard enthalpy of formation, $\Delta H°_f$. In their standards states, such compounds form from their elements via an endothermic reaction. Thermodynamically unstable compounds are unstable with respect to their decomposition to their constituent elements. Ordinary explosives are thermodynamically unstable compounds.

**Thermoplastic resins**. Plastics whose shape distorts when they are heated.

**Thermosets**. Plastics that retain their shape when heated. They are extensively **cross linked** and char before they yield to applied force.

**Thermosetting resin**. A plastic material that cannot melt. **A thermoset**.

**Thymine**. One of the bases of **nucleic acids**, $C_5H_6N_2O_2$. Based on its structure it is a pyrimidine-type base.

**Third law of thermodynamics**. The entropy of a perfectly crystalline substance is zero at absolute zero. Using it enables the construction of tables of standard entropies. It is more a law of convenience than a fundamental law of nature.

**Third order reaction**. In chemical kinetics, a reaction whose rate law depends on molarity cubed. One whose order is three.

**Third transition series**. In the periodic table the elements hafnium (Hf) through mercury (Hg). The lowest of three rows of **d-block** transition metals in the periodic table.

**Thixotropic fluid**. A thixotropic fluid will flow under stress, but will rebuild viscosity when the stress is removed—an obvious advantage for brush painting.

**Time-concentration data**. The raw data of chemical kinetics. A table of changing concentrations measured at definite times. Used to make a **time-concentration graph**.

**Time-concentration graph**. A depiction of the change of concentration of an entity as a function of time. A display of the raw data of chemical kinetics.

**TLV**. Threshold limit value. Used in industrial hygiene to establish a worker exposure limit.

**Tolerancing**. A form of practical engineering statistics. The word refers to the need for parts made in one location to "fit" or be "tolerant with" parts made in another location. Tolerancing principles were developed in the early 19th century by engineers at US armories in the process of mass producing firearms with interchangeable parts.

**Torr**. A pressure unit approximately equal to the force exerted by a one mm column of mercury under standard gravitational force.

**Trade secrets**. Knowledge held within a company or other organization about how to run a process or build a device, etc. Also called **proprietary information**.

**Trade mark**. See **registered trade mark**.

**Transition**. A change in the energy state of an entity. Is said to occur when an electron changes levels (up or down) in an atom.

**Transition state**. In chemical kinetics, the moment that the reactants transform into the products is called the transition state.

**Transition elements**. The elements of periodic groups 3 - 12. All the elements in the periodic table not in the **main groups.**

**Transition metals**. Alternative name for the transition elements—which are all metals.

**Transition state**. See **activated complex**.

**Transmutation**. In alchemy, the notion that one element can be converted into another, especially the notion that a base metal, such as lead, can be converted into gold. Never achieved by any alchemist but well-known in modern chemistry and nuclear physics. Modern transmutation is a far too expensive process to have any commercial prospect for making gold.

**Trans-uranic elements**. Elements in the periodic table beyond uranium (element 92).

**Triad**. Three elements in a single vertical group of the periodic table such that the middle element has physical and chemical properties near the average of the top and bottom elements. Na-K-Rb, Cl-Br-I, etc. Important for the development of the periodic table in the 19th. century.

**Triatomic molecule**. One composed of three atoms such as $H_2O$ or $O_3$.

**Trigonal bipyramidal. structure**. An ideal geometry based on five ligands symmetrically disposed around a central atom.

**Trigonal planar**. An ideal molecular geometry. Geometry of a central atom surrounded by three ligand atoms in a planar arrangement with all the bond angles being 120°.

**Triple bond**. A bond whose bond energy is almost three times that of a single bond between a given pair of atoms. A covalent bond that involves two atoms sharing three pairs of electrons.

**Triple point**. On a phase diagram, the unique temperature and pressure at which all three phases are in mutual equilibrium.

**Triprotic**. Said of an acid that produces three moles of hydrogen ion per mole of the acid. Phosphoric acid for example: $H_3PO_4 \leftrightarrows 3\ H^+ + PO_4^{3-}$

**Tritium**. A **radionuclide** and isotope of hydrogen. $^3H$, the radioactive isotope that contains 1 proton and two neutrons.

**Trouton's rule**. The entropies of vaporization are approximately the same (85 J $mol^{-1}$ $K^{-1}$) for all liquids at their normal boiling points. Recognized around the middle of the nineteenth century by Frederick T. Trouton.

**TWA**. Time weighted average. The average value, over time, of a property that varies with time. Often used in connection with a worker potentially exposed to an environmental health hazard.

**UEL**. Upper explosive limit. The concentration of a fuel in air that is too rich for ignition to occur in the presence of a spark or flame. (**LEL** is the reverse.)

**Ultraviolet light**. Synonymous with **ultraviolet radiation**.

**Ultraviolet radiation** (UV). Electromagnetic radiation that lies beyond the violet end of the visible spectrum Electromagnetic radiation with a wavelength <400 nm.

**Uncertainty**. In statistics, the notion that no matter how carefully a value is measured its **correct** value always lies beyond our grasp. See also **experimental uncertainty.**

**Unified atomic mass unit (u or Da)**. Precise name for what was formerly called the **atomic mass unit**. The adjective unified refers to the fact that there were once two separate atomic mass systems. 1 u equals exactly one-twelfth the mass of a single atom of carbon-12.

**Unionized**. Said of a substance that does not form ions. Used especially to describe a solute that dissolves in water yielding a solution that is not electrically conducting.

**Unit**. What a property is measured "in." Every physical quantity or measurement is the algebraic product of a numerical value and a unit. See **algebra-of-quantities.**

**Unit cell**. In **crystallography**, the basic unit that repeats throughout the entire structure of a crystal. See **long range order**.

**Unpaired electron**. A single electron occupying an orbital. Entities such as $X^{\cdot}$ containing unpaired electrons are **paramagnetic**.

**Unsaturated solution**. A solution that contains less solute than the solution might contain.

**Unsaturated**. 1. An organic molecule is said to be unsaturated if any of its carbon atoms is bonded to fewer than four different atoms. In general, any atom forming fewer bonds than it is capable of forming is said to show unsaturation. More precisely called **valence unsaturation**. An organic molecule is unsaturated if it contains a double bond, a triple bond, or an aromatic ring. 2. A solution that contains less of a solute than it is capable of containing.

**Uracil**. One of the bases of **nucleic acids**, $C_4H_4N_2O_2$. Based on its structure it is a pyrimidine-type base.

**UV absorbers**. Family of compounds capable of absorbing ultraviolet photons. Used as preservatives and sun screens.

**UV spectroscopy**. An important **chemical instrumental method**. Study of molecules using ultraviolet light in the range of 100 nm - 1000 nm. Important instrumental method for chemical analysis, to measure **molecular orbital** energies level, and to elucidate some aspects of chemical structure.

**UV spectrum**. A graphical display of the UV absorption of a sample as a function of UV wavelength. The result of an experiment in **UV spectroscopy**.

**Valence**. A historically important term, originally used to describe the chemical combining power of an element. Used in modern chemistry in the phrases **valence electron** and **valence shell**. Also, often used as a synonym for **valency**.

**Valence electrons**. In an atom, those electrons important in forming chemical bonds. Electrons responsible for chemical bonds. The electrons of an atom that are primarily responsible for its chemical behavior.

**Valence Shell**. For any central atom it's its own **valence electrons** plus electrons shared or captured from atoms (**ligands**) bonded to the central atom.

**Valence Shell Electron Pair Repulsion (VSEPR)**. A simple theory that proposes bond angles in molecules and polyatomic ions are controlled by repulsions among the valence shell electron pairs of the central atoms. Both **bond pairs (BP)** and **lone pairs (LP)** contribute to determining the bond angles.

**Valence unsaturation**. See **unsaturated**.

**Valence band**. In a solid, the valence band is a set of more or less degenerate molecular orbitals (formed from the valence orbitals of the atoms in the solid) that extend throughout the solid. See **conduction band**.

**Valency**. The state of combination of an atom. Often used in the sense of the number of chemical bonds that atom is forming. For example, a pentavalent atom is one forming five bonds, as for example the P atom in $PF_5$. see also **valence**.

**van't Hoff i-factor**. The experimental ratio of a measured **colligative property** and the same colligative property calculated for a nonelectrolyte.

**van der Waals equation**. Equation of state of a gas that improves on the ideal gas equation. $(P + n^2a/V^2) \times (V - nb) = nRT$. Now largely of historical interest, it played a key role in convincing scientists of the reality of molecules around the beginning of the 20th century.

**van der Waals "a" constants**. Empirical constants unique for each gas that occur in the van der Waals' equation. The value of "a" has been interpreted to account for the intermolecular forces among the gas's molecules.

**van der Waals "b" constants**. Empirical constants, unique for each gas, that occur in the van der Waals equation. The value of "b" has been interpreted to account for the size of the gas's molecules.

**van der Waals radius**. A rough measure of the size of an atom. It measures how close two *nonbonded* atoms can approach. The sum of the van der Waals radii of two atoms is their distance of closest approach. Clearly distinguish this radius from those involving bonded atoms. Contrast **covalent radius**.

**Vapor**. A vapor is simply a gas. However, the word vapor is often used to refer specifically to gas created by the evaporation of a liquid. Some chemists define a vapor as a gaseous sample of a pure substance below its critical temperature.

**Vapor deposition**. The process of forming a solid from a gas. The opposite process to **sublimation**.

**Vapor pressure**. The pressure of the vapor in equilibrium with a liquid. Varies with temperature. Volatile liquids have high vapor pressures. Nonvolatile liquids have low vapor pressures.

**Vapor pressure curve**. A graphical representation of the way the vapor pressure of a liquid varies as a function of temperature.

**Vapor pressure lowering**. Phenomenon in which the vapor pressure of a solvent is reduced in proportion to the amount of dissolved nonvolatile solute dissolved in that solvent.

**Vibrational motion**. A periodic relative motion of the atoms within a molecule.

**Viscosity**. Physical property that relates to liquid flow. The resistance of a fluid to flow is its viscosity.

**Viscosimeter**: An instrument used to measure viscosity. **Compare Ostwald** and **Brookfield viscosimeter.**

**Volatile substance**. One that evaporates quickly.

**Volt**. SI derived unit of electrical potential or electromotive force. It has the units of $J\ C^{-1}$.

**Voltage**. In electrochemistry, the **electrical potential** generated by a battery is the battery's voltage.

**Volumetric analysis**. An experimental procedure in which a measured volume of a reactant solution of known concentration provides a quantitative analysis of a target substance (the **analyte**)

**VSEPR**. See **valence shell electron pair repulsion**.

**Watt**. SI derived unit of power. It has the unit $J\ s^{-1}$.

**Wave equation**. Fundamental equation of nature. Also called the **Schrödinger equation**. For the hydrogen atom it's:
$$\frac{\partial^2\psi}{\partial^2 x} + \frac{\partial^2\psi}{\partial^2 y} + \frac{\partial^2\psi}{\partial^2 z} + \frac{8\pi^2 m}{h}\left(E + \frac{e^2}{r}\right)\psi = 0$$

**Wave functions**. In general, wave functions are solutions to the **Schrödinger equation**. For the electron in a hydrogen atom, its wave functions are solutions to the Schrödinger equation. Denoted by the symbol, $\psi$, and having a particular energy value.

**Wavelength**. For **electromagnetic radiation** it is the distance between two successive peaks of the wave.

**Wave-particle duality**. A concept of 20th century physics. Said to be illustrated by any natural phenomenon that exhibits both wave and particle aspects.

**Weak acids**. Substances composed of hydrogen ions combined with negative ions that do not completely ionize in water.

**Weak bases**. Substances composed of hydroxide ions combined with positive ions that do not completely ionize in water.

**Weak electrolyte**. A substance that ionizes partially when it dissolves in water.

**Weight percentage**. The percentage contribution that individual elements make to a compound's mass.

**Weighted average**. The average of a property of a sample such that each component of the sample contributes according to its abundance. See **atomic weight**.

**Wetting**. Process in which a liquid spontaneously spreads across a liquid or solid surface.

**Wetting agent**. A surfactant that promotes the spread of a liquid across a surface.

**Work of expanding gases**. For a constant pressure process it's $w = P_{ext}\ \Delta V$. It's the work that powers most engines.

**Work**. Energy used to move an object: work = force × distance. Work's base SI unit is the joule.

**Working fluid**. Any substance continually cycled in a thermodynamic process. For example, steam and water in a power turbine or the refrigerant fluid in an air conditioner.

**X-ray crystallography**. Important experimental technique used to determine **chemical structure**. It relies on **diffraction** of x-rays by single crystals.

**X-ray diffraction**. Diffraction is the basic physical principle that underlies the technique of **x-ray crystallography**. It involves the interference of electromagnetic waves.

**Yield**. The product or products obtained from a chemical reactions. Reactants yield products. See **actual yield** and **theoretical yield**.

**Zero order reaction**. One whose rate is independent of the concentrations of any of the reactants.

**Zeroth law of thermodynamics**. Fundamental law of nature. The notion that heat always flows from higher temperature regions to lower temperature regions. Not provable, but not in doubt. An obvious truth of human experience.

**Ziegler-Natta catalysts**. Catalysts used commercially to prepare **polyolefins**. For example, mixtures of titanium tetrachloride ($TiCl_4$) and trialkyaluminum compounds, such as triethylaluminun $Al(C_2H_5)_3$.

**Zone of stability**. See **band of stability**.

# Photo/Diagram Credits

*All photographs/images except those listed below by the author. All diagrams prepared by the author.*

## Chapter 1

| | |
|---|---|
| Figure 1.1. | Joseph Black. From *The Lives of the Engineers*, The Steam-Engine, Boulton and Watt, Volume 4. John Murray, Albemarle Street, London 1874. Courtesy Newman Library, Virginia Tech. Public domain |
| Figure 1.2. | James Watt. From the title page of *The Lives of the Engineers*, The Steam-Engine, Boulton and Watt, Volume 4. John Murray, Albemarle Street, London 1874. Courtesy Newman Library, Virginia Tech. Public domain. |
| Figure 1.4. | The Lavoisiers study respiration. From *Famous Chemists -- The Men and Their Work*. William A. Tilden, E. P. Dutton & Company. London, 1921. In author's collection. Public domain. |
| Figure 1.5. | Egyptian metalworker. From *A History of Chemistry -- from the Earliest Times to the Present Day*. James Campbell Brown. Published by J. & A. Churchill, 7 Marlborough Street, London, 1913. Courtesy Newman Library, Virginia Tech. Public domain. |
| Figure 1.6. | A retort of the Arabic period of chemistry. From *A History of Chemistry -- from the Earliest Times to the Present Day*. James Campbell Brown. Published by J. & A. Churchill, 7 Marlborough Street, London, 1913. Courtesy Newman Library, Virginia Tech. Public domain. |
| Figure 1.7. | Paracelsus. From page 140 of *History of Medicine*, Fielding H. Garrison, MD, W. B. Saunders and Company, London and Philadelphia, 1913. Courtesy Newman Library, Virginia Tech. Public domain. |

## Chapter 2

| | |
|---|---|
| Figure 2.2. | Curies' radiograph of key and coin in coin purse. From Figure 27 of *The New Knowledge*, Robert Kennedy Duncan, The A. S. Barnes Company, New York and Chicago, 1919. In author's collection. Public domain. |
| Figure 2.6. | Ivy Mike shot. U.S. Department of Energy photograph. |

## Chapter 3

| | |
|---|---|
| Unnumbered. | Facsimile of the preface from P. J. Macquer (1718-1784), *Elements of the Theory of Chemistry*, published in translation in London, 25 March 1758. Original in author's possession. Public domain. |
| Unnumbered. | John Dalton. From *Famous Chemists -- The Men and Their Work*. William A. Tilden, E. P. Dutton & Company. London, 1921. Original in author's collection. Public domain. |
| Figure 3.7. | Scan of volumetric flask from *A Systematic Handbook of Volumetric Analysis,* by Francis Sutton, J. & A. Churchill, New Burlington Street, London, 1871. Original in author's collection. Public domain. |

## Chapter 4

| | |
|---|---|
| Figure 4.7. | Pneumatic trough. Figure scanned from page 27 of *Fourteen Weeks in Chemistry*, J. Dorman Steele, A. S. Barnes & Company, New York and Chicago, 1873. Public domain. |
| Figure 4.8. | The Magdeburg experiment. Public domain. |

| | |
|---|---|
| Figure 4.11. | Robert Boyle. From *A History of Chemistry -- from the Earliest Times to the Present Day*. James Campbell Brown. Published by J. & A. Churchill, 7 Marlborough Street, London, 1913. Courtesy Newman Library, Virginia Tech. Public domain. |
| Figure 4.15. | A sling psychrometer. Public domain. |

## Chapter 5

| | |
|---|---|
| Figure 5.12. | Atomic Spectra of various types from 1873. From frontispiece of *Fourteen Weeks in Chemistry*, J. Dorman Steele, A. S. Barnes & Company, New York and Chicago, 1873. Public domain. |

## Chapter 10

| | |
|---|---|
| Figure10.2. | Still. From page 196 of *Fourteen Weeks in Chemistry*, J. Dorman Steele, A. S. Barnes & Company, New York and Chicago, 1873. Public domain. |

## Chapter 11

| | |
|---|---|
| Table 11.8. | Plastic recycling icons. Public domain. |

## Chapter 12

| | |
|---|---|
| Figure 12.7. | Etched metals. Diagram facing page 272 from *Creative Chemistry -- Descriptive of Recent Achievements in the Chemical industries*, Edwin D. Slosson, Published by the Century Company, New York, 1919. Public domain. |

## Chapter 13

| | |
|---|---|
| Figure 13.2. | Early iron workers. From de re Metallica. Public domain. |

## Chapter 14

| | |
|---|---|
| Figure 14.2. | Crown of cups. Public domain.. |
| Figure 14.6. | Cells in a chlor-alkali plant. Diagram facing page 217 from *Creative Chemistry -- Descriptive of Recent Achievements in the Chemical industries*, Edwin D. Slosson, Published by the Century Company, New York, 1919. Public domain. |

## Chapter 15

| | |
|---|---|
| Figure 15.4 | Battleship guns. Diagram facing page 32 from *Creative Chemistry -- Descriptive of Recent Achievements in the Chemical industries*, Edwin D. Slosson, Published by the Century Company, New York, 1919. Public domain. |

## Appendix H

| | |
|---|---|
| Unnumbered. | Mendeleev signed periodic table. Public domain. |

# Index



## A



## B

## C

# D

# E

# F



# G

# H



# I

## J



## K



## L



## M

# N



# O

## P

## Q



## R

## S

## T



## U

## V



## W



## X



## Y



## Z

# Periodic Table -- Experimental Electron Configurations of Neutral Atoms in their Ground States

| 1<br>**H**<br>$1s^1$ | s↓ | d↓ | d↓ | d↓ | d↓ | d↓ | d↓ | d↓ | d↓ | d↓ | d↓ | p↓ | p↓ | p↓ | p↓ | p↓ | 2<br>**He**<br>$1s^2$ |
|---|---|---|---|---|---|---|---|---|---|---|---|---|---|---|---|---|---|
| 3<br>**Li**<br>$2s^1$ | 4<br>**Be**<br>$2s^2$ | | | | | | | | | | | 5<br>**B**<br>$2s^22p^1$ | 6<br>**C**<br>$2s^22p^2$ | 7<br>**N**<br>$2s^22p^3$ | 8<br>**O**<br>$2s^22p^4$ | 9<br>**F**<br>$2s^22p^5$ | 10<br>**Ne**<br>$2s^22p^6$ |
| 11<br>**Na**<br>$3s^1$ | 12<br>**Mg**<br>$3s^2$ | | | | | | | | | | | 13<br>**Al**<br>$3s^23p^1$ | 14<br>**Si**<br>$3s^23p^2$ | 15<br>**P**<br>$3s^23p^3$ | 16<br>**S**<br>$3s^23p^4$ | 17<br>**Cl**<br>$3s^23p^5$ | 18<br>**Ar**<br>$3s^23p^6$ |
| 19<br>**K**<br>$4s^1$ | 20<br>**Ca**<br>$4s^2$ | 21<br>**Sc**<br>$4s^23d^1$ | 22<br>**Ti**<br>$4s^23d^2$ | 23<br>**V**<br>$4s^23d^3$ | 24<br>**Cr**<br>‡<br>$4s^13d^5$ | 25<br>**Mn**<br>$4s^23d^5$ | 26<br>**Fe**<br>$4s^23d^6$ | 27<br>**Co**<br>$4s^23d^7$ | 28<br>**Ni**<br>$4s^23d^8$ | 29<br>**Cu**<br>‡<br>$4s^13d^{10}$ | 30<br>**Zn**<br>$4s^23d^{10}$ | 31<br>**Ga**<br>$4s^23d^{10}4p^1$ | 32<br>**Ge**<br>$4s^23d^{10}4p^2$ | 33<br>**As**<br>$4s^23d^{10}4p^3$ | 34<br>**Se**<br>$4s^23d^{10}4p^4$ | 35<br>**Br**<br>$4s^23d^{10}4p^5$ | 36<br>**Kr**<br>$4s^23d^{10}4p^6$ |
| 37<br>**Rb**<br>$5s^1$ | 38<br>**Sr**<br>$5s^2$ | 39<br>**Y**<br>$5s^24d^1$ | 40<br>**Zr**<br>$5s^24d^2$ | 41<br>**Nb**<br>‡<br>$5s^14d^4$ | 42<br>**Mo**<br>‡<br>$5s^14d^5$ | 43<br>**Tc**<br>$5s^24d^5$ | 44<br>**Ru**<br>$5s^24d^6$ | 45<br>**Rh**<br>$5s^24d^7$ | 46<br>**Pd**<br>‡<br>$4d^{10}$ | 47<br>**Ag**<br>$5s^24d^9$ | 48<br>**Cd**<br>$5s^24d^{10}$ | 49<br>**In**<br>$5s^24d^{10}5p^1$ | 50<br>**Sn**<br>$5s^24d^{10}5p^2$ | 51<br>**Sb**<br>$5s^24d^{10}5p^3$ | 52<br>**Te**<br>$5s^24d^{10}5p^4$ | 53<br>**I**<br>$5s^24d^{10}5p^5$ | 54<br>**Xe**<br>$5s^24d^{10}5p^6$ |
| 55<br>**Cs**<br>$6s^1$ | 56<br>**Ba**<br>$6s^2$ | 57-71<br>* | 72<br>**Hf**<br>$6s^24f^{14}5d^2$ | 73<br>**Ta**<br>$6s^24f^{14}5d^3$ | 74<br>**W**<br>$6s^24f^{14}5d^4$ | 75<br>**Re**<br>$6s^24f^{14}5d^5$ | 76<br>**Os**<br>$6s^24f^{14}5d^6$ | 77<br>**Ir**<br>$6s^24f^{14}5d^7$ | 78<br>**Pt**<br>$6s^24f^{14}5d^8$ | 79<br>**Au**<br>$6s^24f^{14}5d^9$ | 80<br>**Hg**<br>$6s^2f^{14}5d^{10}$ | 81<br>**Tl**<br>$6s^24f^{14}5d^{10}6p^1$ | 82<br>**Pb**<br>$6s^24f^{14}5d^{10}6p^2$ | 83<br>**Bi**<br>$6s^24f^{14}5d^{10}6p^3$ | 84<br>**Po**<br>$6s^24f^{14}5d^{10}6p^4$ | 85<br>**At**<br>$6s^24f^{14}5d^{10}6p^5$ | 86<br>**Rn**<br>$6s^24f^{14}5d^{10}6p^6$ |
| 87<br>**Fr**<br>$7s^1$ | 88<br>**Ra**<br>$7s^2$ | 89-103<br>† | 104<br>**Rf**<br>?<br>$7s^25f^{14}6d^2$ | 105<br>**Db**<br>?<br>$7s^25f^{14}6d^3$ | 106<br>**Sg**<br>?<br>$7s^25f^{14}6d^4$ | 107<br>**Bh**<br>?<br>$7s^25f^{14}6d^5$ | 108<br>**Hs**<br>?<br>$7s^25f^{14}6d^6$ | 109<br>**Mt**<br>?<br>$7s^25f^{14}6d^7$ | 110<br>**Uun**<br>?<br>$7s^25f^{14}6d^8$ | 111<br>**Uuu**<br>?<br>$7s^25f^{14}6d^9$ | 112<br>**Uub**<br>?<br>$7s^25f^{14}6d^{10}$ | | 114<br>**Uuq**<br>?<br>$7s^25f^{14}6d^{10}7p^2$ | | | | |

| s↑ | s↑ | d↑ f↓ | d↑ f↓ | d↑ f↓ | d↑ f↓ | d↑ f↓ | d↑ f↓ | d↑ f↓ | d↑ f↓ | d↑ f↓ | d↑ f↓ | p↑ f↓ | p↑ f↓ | p↑ f↓ | p↑ f↓ | p↑ f↓ | s↑ or p↑ |
|---|---|---|---|---|---|---|---|---|---|---|---|---|---|---|---|---|---|
| | * | 57<br>**La**<br>‡<br>$6s^25d^1$ | 58<br>**Ce**<br>$6s^24f^2$ | 59<br>**Pr**<br>$6s^24f^3$ | 60<br>**Nd**<br>$6s^24f^4$ | 61<br>**Pm**<br>$6s^24f^5$ | 62<br>**Sm**<br>$6s^24f^6$ | 63<br>**Eu**<br>$6s^24f^7$ | 64<br>**Gd**<br>‡<br>$6s^24f^75d^1$ | 65<br>**Tb**<br>$6s^24f^9$ | 66<br>**Dy**<br>$6s^24f^{10}$ | 67<br>**Ho**<br>$6s^24f^{11}$ | 68<br>**Er**<br>$6s^24f^{12}$ | 69<br>**Tm**<br>$6s^24f^{13}$ | 70<br>**Yb**<br>$6s^24f^{14}$ | 71<br>**Lu**<br>$6s^24f^{14}5d^1$ | |
| | † | 89<br>**Ac**<br>‡<br>$7s^26d^1$ | 90<br>**Th**<br>‡<br>$7s^26d^2$ | 91<br>**Pa**<br>‡<br>$7s^25f^26d^1$ | 92<br>**U**<br>‡<br>$7s^25f^36d^1$ | 93<br>**Np**<br>$7s^25f^5$ | 94<br>**Pu**<br>$7s^25f^6$ | 95<br>**Am**<br>$7s^25f^7$ | 96<br>**Cm**<br>‡<br>$7s^25f^76d^1$ | 97<br>**Bk**<br>$7s^25f^9$ | 98<br>**Cf**<br>$7s^25f^{10}$ | 99<br>**Es**<br>$7s^25f^{11}$ | 100<br>**Fm**<br>$7s^25f^{12}$ | 101<br>**Md**<br>$7s^25f^{13}$ | 102<br>**No**<br>$7s^25f^{14}$ | 103<br>**Lr**<br>$7s^25f^{14}6d^1$ | |

To fill an empty configuration periodic table proceed through it adding one electron for each element. Filling proceeds from left to right across each row of the table in the sequence s - f - d - p. The correct configurations of 91/103 elements will be produced. Eleven elements (designated with a double dagger ‡) have slightly different experimental electron configurations than the the automatic filling procedure. The configurations of elements 104-114 are hypothetical. The notation d↑ f↓, etc., indicates that the elements above are filling a d-level and the elements below are filling an f-level.